V. Vorsa

MATHEMATICAL METHODS FOR PHYSICISTS
Fourth Edition

MATHEMATICAL METHODS FOR PHYSICISTS

Fourth Edition

George B. Arfken

Miami University
Oxford, Ohio

Hans J. Weber

University of Virginia
Charlottesville, Virginia

Academic Press

San Diego   New York   Boston   London   Sydney   Tokyo   Toronto

Copyright © 1995, 1985, 1970, 1966 by ACADEMIC PRESS

Academic Press
A Division of Harcourt Brace & Company
525 B Street, Suite 1900, San Diego, California 92101-4495

United Kingdom Edition published by
Academic Press Limited
24-28 Oval Road, London NW1 7DX

Library of Congress Cataloging-in-Publication Data

Arfken, George B. (George Brown), date.
   Mathematical methods for physicists / by George B. Arfken,
   Hans-Jurgen Weber. -- 4th ed.
      p.   cm.
   Includes bibliographical references (p.   ) and index.
   ISBN 0-12-059815-9
   ISBN 0-12-059816-7 (International paper edition)
   1. Mathematics.   2. Mathematical physics.   I. Weber, Hans-Jurgen.
II. Title.
QA37.2.A74   1995
515'.1--dc20   94-24911
                                                                 CIP

PRINTED IN THE UNITED STATES OF AMERICA
          97 98 99 00 DO 9 8 7 6 5 4 3 2

To Carolyn and Edith Enzian

CONTENTS

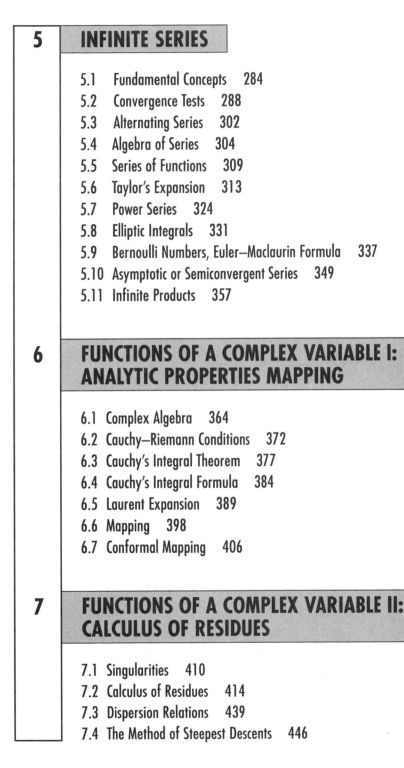

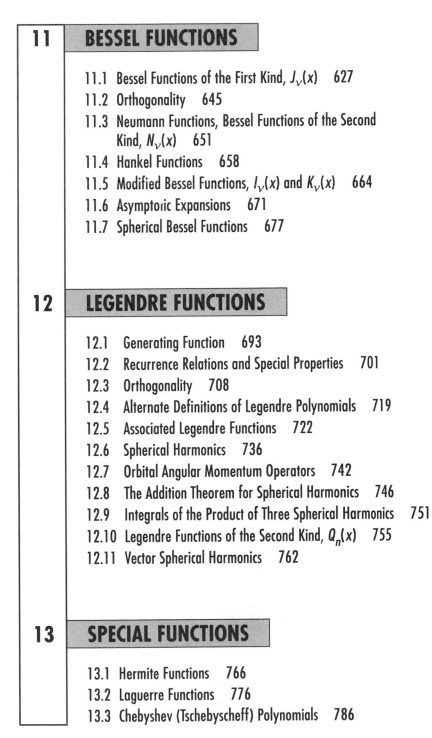

17 | CALCULUS OF VARIATIONS

18 | NONLINEAR METHODS AND CHAOS

PREFACE

There are many additions, revisions and some deletions in this fourth edition of *Mathematical Methods for Physicists*. In detail, the linear properties of scalar and vector products are emphasized to motivate their definitions in Chapter 1. A new Section 1.15 on Dirac's delta function collects portions that were scattered over several chapters.

Chapter 2 on vectors and tensors has been shortened by moving the section on separation of variables to Chapter 8, where they are presented as a method of reducing partial to ordinary differential equations, and by deleting the sections on dyadics and elasticity, topics that are rarely taught in physics today.

Chapter 3 is now closely linked to Chapter 1 by using the geometrical aspects of scalar and vector products systematically to motivate the concept and definition of determinants and to solve linear equations. The concept of matrices and their multiplication is based on rotations as special linear transformations. The product theorem linking matrices and determinants is included. Dirac gamma matrices are now based on the metric and conventions of Bjorken and Drell, *Relativistic Quantum Mechanics,* which are becoming standard in the literature. A brief new subsection on commonly used functions of matrices is included.

The group theoretical sections have been collected and expanded into a separate Chapter 4. Generators are treated in more detail. Ladder operators have been adapted from Chapter 12 on spherical harmonics. Angular momentum coupling and Clebsch–Gordan coefficients are developed along with spherical tensor operators. Sections on the Lorentz group and Maxwell's equations are adapted to the Bjorken–Drell metric, in accord with Jackson's corresponding revision of *Classical Electrodynamics.*

Chapter 5 contains minor additions to Bernoulli numbers. The integral convergence test is extended to alternating and other series. An example from Fourier series in Chapter 14 is included.

Multivalent functions and their branch cuts are given more emphasis in Chapter 6, and this is continued in Chapter 7. The Mittag–Leffler pole expansion of meromorphic functions and product expansions of entire functions are included, along with applications such as Rouché's theorem.

In Chapter 8 characteristics are briefly introduced. A soliton solution of a nonlinear partial differential equation is included. The separation of variables from Chapter 1 has been moved here. Green's functions have been moved here from Chapter 16 and have been moved into Chapter 9 as well. The connections of Chapter 9 with the linear algebra of Chapters 1–4 are emphasized early on.

The product expansion of the Gamma function in Chapter 10 is now tied to product expansions of analytic functions discussed in Chapter 7.

In Chapter 12 vector spherical harmonics are adapted to the notation in some angular momentum texts.

As an application of Fourier series, Chapter 14 now contains the functional equation of the Riemann zeta function. The discussion expands the connections of this topic with Chapter 5 on Bernoulli numbers, and ties it in with Chapter 6 as an example of analytic continuation. The connection to analytic number theory is mentioned in more detail.

A new Chapter 18 on nonlinear methods takes into account a few of the major aspects of this vast and rapidly expanding field.

The problem sets have been examined closely. Some problems have been deleted and a number of new problems have been added.

The 4th edition is based on the advice and help of many people. Some of the additions and many revisions are in response to the comments of reviewers. We are grateful to them and to Senior Editor Robert Kaplan who organized the early stages of the revision. Dr. Michael Bozoian, Dr. Nelson Max, and Professor Philip A. Macklin have been most helpful with numerous suggestions and corrections in the text and problem sets. The final form of the 4th edition owes much to the expertise of Senior Editor Peter Renz and Production Editor Jacqueline Garrett.

INTRODUCTION

Many of the physical examples used to illustrate the applications of mathematics are taken from electromagnetism and quantum mechanics. For convenience the main equations are listed below and the symbols identified. References are also given.

ELECTROMAGNETIC THEORY

Maxwell's Equations (MKS Units—Vacuum)

$$\nabla \cdot \mathbf{D} = \rho \qquad \nabla \times \mathbf{E} = -\frac{\partial \mathbf{B}}{\partial t}$$

$$\nabla \cdot \mathbf{B} = 0 \qquad \nabla \times \mathbf{H} = \frac{\partial \mathbf{D}}{\partial t} + \mathbf{J}$$

Here $\mathbf{E}$ is the electric field defined in terms of force on a static charge and $\mathbf{B}$ the magnetic induction defined in terms of force on a moving charge. The related fields $\mathbf{D}$ and $\mathbf{H}$ are given (in vacuum) by

$$\mathbf{D} = \varepsilon_0 \mathbf{E} \qquad \text{and} \qquad \mathbf{B} = \mu_0 \mathbf{H}.$$

The quantity ρ represents free charge density while $\mathbf{J}$ is the corresponding current. The electric field $\mathbf{E}$ and the magnetic induction $\mathbf{B}$ are often expressed in terms of the scalar potential φ and the magnetic vector potential $\mathbf{A}$:

$$\mathbf{E} = \frac{\partial \mathbf{A}}{\partial t} - \nabla \varphi \qquad \mathbf{B} = \nabla \times \mathbf{A}.$$

For additional details see J. M. Marion, *Classical Electromagnetic Radiation.* New York: Academic Press (1965); J. D. Jackson, *Classical Electrodynamics,* 2nd ed. New York: Wiley (1975).

Note that Marion and Jackson prefer Gaussian units. A glance at the last two texts and the great demands they make upon the student's mathematical competence should provide considerable motivation for the study of this book.

QUANTUM MECHANICS

Schrödinger Wave Equation (Time Independent)

$$-\frac{\hbar^2}{2m}\nabla^2\psi + V\psi = E\psi$$

ψ is the (unknown) wave function. The potential energy, often a function of position, is denoted by V while E is the total energy of the system. The mass of the particle being described by ψ is m. $\hbar$ is Planck's constant h divided by 2π. Among the extremely large number of beginning or intermediate texts we might note: A. Messiah, *Quantum Mechanics*, 2 vols. New York; Wiley (1961); E. Merzbacher, *Quantum Mechanics*, 2nd ed. New York: Wiley (1970); G. Baym, *Lectures on Quantum Mechanics*, 2nd printing. Reading, MA: Benjamin (1973); J. J. Sakurai, *Modern Quantum Mechanics*, rev. ed. Reading, MA: Addison–Wesley (1994).

1 VECTOR ANALYSIS

1.1 DEFINITIONS, ELEMENTARY APPROACH

In science and engineering we frequently encounter quantities that have magnitude and magnitude only: mass, time, and temperature. These we label scalar quantities. In contrast, many interesting physical quantities have magnitude and, in addition, an associated direction. This second group includes displacement, velocity, acceleration, force, momentum, and angular momentum. Quantities with magnitude and direction are labeled vector quantities. Usually, in elementary treatments, a vector is defined as a quantity having magnitude and direction. To distinguish vectors from scalars, we identify vector quantities with boldface type, that is, $\mathbf{V}$.

Our vector may be conveniently represented by an arrow with length proportional to the magnitude. The direction of the arrow gives the direction of the vector, the positive sense of direction being indicated by the point. In this representation vector addition

$$\mathbf{C} = \mathbf{A} + \mathbf{B} \tag{1.1}$$

consists in placing the rear end of vector $\mathbf{B}$ at the point of vector $\mathbf{A}$. Vector $\mathbf{C}$ is then represented by an arrow drawn from the rear of $\mathbf{A}$ to the point of $\mathbf{B}$. This procedure, the triangle law of addition, assigns meaning to Eq. (1.1) and is illustrated in Fig. 1.1. By completing the parallelogram, we see that

$$\mathbf{C} = \mathbf{A} + \mathbf{B} = \mathbf{B} + \mathbf{A}, \tag{1.2}$$

as shown in Fig. 1.2. In words, vector addition in *commutative*.

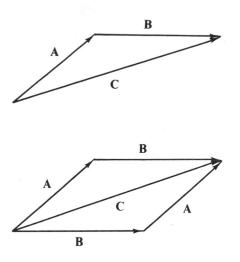

Figure 1.1 Triangle law of vector addition.

Figure 1.2 Parallelogram law of vector addition.

For the sum of three vectors

$$D = A + B + C,$$

Fig. 1.3, we may first add **A** and **B**

$$A + B = E.$$

Then this sum is added to **C**

$$D = E + C.$$

Similarly, we may first add **B** and **C**

$$B + C = F.$$

Then

$$D = A + F.$$

In terms of the original expression,

$$(A + B) + C = A + (B + C).$$

Vector addition is *associative*.

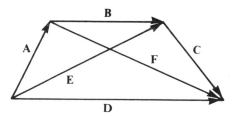

Figure 1.3 Vector addition is associative.

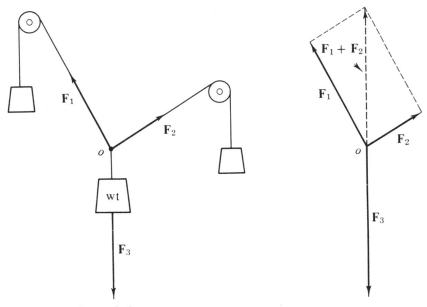

Figure 1.4 Equilibrium of forces. $\mathbf{F}_1 + \mathbf{F}_2 = -\mathbf{F}_3$.

A direct physical example of the parallelogram addition law is provided by a weight suspended by two cords. If the junction point (O in Fig. 1.4) is in equilibrium, the vector sum of the two forces $\mathbf{F}_1$ and $\mathbf{F}_2$ must just cancel the downward force of gravity, $\mathbf{F}_3$. Here the parallelogram addition law is subject to immediate experimental verification.[1]

Subtraction may be handled by defining the negative of a vector as a vector of the same magnitude but with reversed direction. Then

$$\mathbf{A} - \mathbf{B} = \mathbf{A} + (-\mathbf{B}).$$

In Fig. 1.3

$$\mathbf{A} = \mathbf{E} - \mathbf{B}.$$

Note that the vectors are treated as geometrical objects that are independent of any coordinate system. Indeed, we have not yet introduced a coordinate system. This concept of independence of a preferred coordinate system is developed in considerable detail in the next section.

The representation of vector $\mathbf{A}$ by an arrow suggests a second possibility. Arrow $\mathbf{A}$ (Fig. 1.5), starting from the origin,[2] terminates at the point (A_x, A_y, A_z). Thus, if we agree that the vector is to start at the origin, the

---

[1] Strictly speaking the parallologram addition was introduced as a definition. Experiments show that if we assume that the forces are vector quantities and we combine them by parallelogram addition the equilibrium condition of zero resultant force is satisfied.

[2] The reader will see that we could start from any point in our cartesian reference frame; we choose the origin for simplicity. This freedom of shifting the origin of the coordinate system without affecting the geometry is called **translation invariance**.

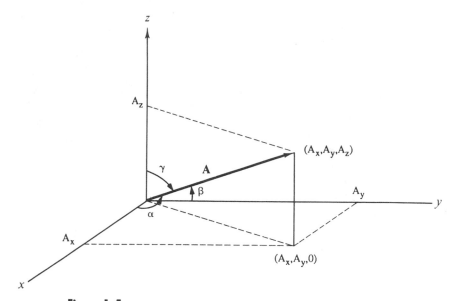

Figure 1.5   Cartesian components and direction cosines of **A**.

positive end may be specified by giving the cartesian coordinates (A_x, A_y, A_z) of the arrow head.

Although **A** could have represented any vector quantity (momentum, electric field, etc.), one particularly important vector quantity, the displacement from the origin to the point (x, y, z), is denoted by the special symbol **r**. We then have a choice of referring to the displacement as either the vector **r** or the collection (x, y, z), the coordinates of its end point:

$$\mathbf{r} \leftrightarrow (x, y, z). \tag{1.3}$$

Using r for the magnitude of vector **r**, we find that Fig. 1.5 shows that the end point coordinates and the magnitude are related by

$$x = r \cos \alpha, \qquad y = r \cos \beta, \qquad z = r \cos \gamma. \tag{1.4}$$

Cos α, cos β, and cos γ are called the *direction cosines*, α being the angle between the given vector and the positive x-axis, and so on. One further bit of vocabulary: The quantities A_x, A_y, and A_z are known as the (cartesian) *components* of **A** or the *projections* of **A**.

Thus, any vector **A** may be resolved into its components (or projected onto the coordinate axes) to yield $A_x = A \cos \alpha$, etc., as in Eq. (1.4). We may choose to refer to the vector as a single quantity **A** or to its components (A_x, A_y, A_z). Note that the subscript x in A_x denotes the x component and not a dependence on the variable x. A_x may be a function of x, y, and z as $A_x(x, y, z)$. The choice between using **A** or its components (A_x, A_y, A_z) is essentially a choice between a geometric or an algebraic representation. In the language of group theory (Chapter 4), the two representations are *isomorphic*.

Use either representation at your convenience. The geometric "arrow in space" may aid in visualization. The algebraic set of components is usually much more suitable for precise numerical or algebraic calculations.

Vectors enter physics in two distinct forms. (1) Vector **A** may represent a single force acting at a single point. The force of gravity acting at the center of gravity illustrates this form. (2) Vector **A** may be defined over some extended region; that is, **A** and its components may be functions of position: $A_x = A_x(x, y, z)$, and so on. Examples of this sort include the velocity of a fluid varying from point to point over a given volume and electric and magnetic fields. Some writers distinguish these two cases by referring to the vector defined over a region as a vector field. The concept of the vector defined over a region and being a function of position will be extremely important in Section 1.2 and in later sections where we differentiate and integrate vectors.

At this stage it is convenient to introduce unit vectors along each of the coordinate axes. Let $\hat{\mathbf{x}}$ be a vector of unit magnitude pointing in the positive x-direction, $\hat{\mathbf{y}}$, a vector of unit magnitude in the positive y-direction, and $\hat{\mathbf{z}}$, a vector of unit magnitude in the positive z-direction. Then $\hat{\mathbf{x}}A_x$ is a vector with magnitude equal to A_x and in the positive x-direction. By vector addition

$$\mathbf{A} = \hat{\mathbf{x}}A_x + \hat{\mathbf{y}}A_y + \hat{\mathbf{z}}A_z, \tag{1.5}$$

which states that a vector equals the vector sum of its components. Note that if **A** vanishes, all of its components must vanish individually; that is, if

$$\mathbf{A} = 0, \quad \text{then } A_x = A_y = A_z = 0.$$

Finally, by the Pythagorean theorem, the magnitude of vector **A** is

$$A = (A_x^2 + A_y^2 + A_z^2)^{1/2}. \tag{1.6}$$

This resolution of a vector into its components can be carried out in a variety of coordinate systems, as shown in Chapter 2. Here we restrict ourselves to cartesian coordinates.

Equation (1.5) is actually an assertion that the three unit vectors $\hat{\mathbf{x}}$, $\hat{\mathbf{y}}$, and $\hat{\mathbf{z}}$ *span* our real three-dimensional space: Any constant vector may be written as a linear combination of $\hat{\mathbf{x}}$, $\hat{\mathbf{y}}$, and $\hat{\mathbf{z}}$. Since $\hat{\mathbf{x}}$, $\hat{\mathbf{y}}$, and $\hat{\mathbf{z}}$ are linearly independent (no one is a linear combination of the other two), they form a *basis* for the real three-dimensional space.

As a replacement of the graphical technique, addition and subtraction of vectors may now be carried out in terms of their components. For $\mathbf{A} = \hat{\mathbf{x}}A_x + \hat{\mathbf{y}}A_y + \hat{\mathbf{z}}A_z$ and $\mathbf{B} = \hat{\mathbf{x}}B_x + \hat{\mathbf{y}}B_y + \hat{\mathbf{z}}B_z$,

$$\mathbf{A} \pm \mathbf{B} = \hat{\mathbf{x}}(A_x \pm B_x) + \hat{\mathbf{y}}(A_y \pm B_y) + \hat{\mathbf{z}}(A_z \pm B_z). \tag{1.7}$$

Example 1.1.1

Let

$$\mathbf{A} = 6\hat{\mathbf{x}} + 4\hat{\mathbf{y}} + 3\hat{\mathbf{z}}$$

$$\mathbf{B} = 2\hat{\mathbf{x}} - 3\hat{\mathbf{y}} - 3\hat{\mathbf{z}}.$$

Then by Eq. (1.7)

$$\mathbf{A} + \mathbf{B} = 8\hat{\mathbf{x}} + \hat{\mathbf{y}}$$

and

$$\mathbf{A} - \mathbf{B} = 4\hat{\mathbf{x}} + 7\hat{\mathbf{y}} + 6\hat{\mathbf{z}}.$$

It should be emphasized here that the unit vectors $\hat{\mathbf{x}}$, $\hat{\mathbf{y}}$, and $\hat{\mathbf{z}}$ are used for convenience. They are not essential; we can describe vectors and use them entirely in terms of their components: $\mathbf{A} \leftrightarrow (A_x, A_y, A_z)$. This is the approach of the two more powerful, more sophisticated definitions of vector discussed in the next section. However, $\hat{\mathbf{x}}$, $\hat{\mathbf{y}}$, and $\hat{\mathbf{z}}$ emphasize the *direction*, which will be useful in Chapter 2.

So far we have defined the operations of addition and subtraction of vectors. Three varieties of multiplication are defined on the basis of their applicability: a scalar or inner product in Section 1.3, a vector product peculiar to three-dimensional space in Section 1.4, and a direct or outer product yielding a second-rank tensor in Section 2.7. Division by a vector is not defined. See Exercises 3.2.21 and 3.2.22.

EXERCISES

1.1.1   Show how to find $\mathbf{A}$ and $\mathbf{B}$, given $\mathbf{A} + \mathbf{B}$ and $\mathbf{A} - \mathbf{B}$.

1.1.2   The vector $\mathbf{A}$ whose magnitude is 1.732 units makes equal angles with the coordinate axes. Find A_x, A_y, and A_z.

1.1.3   Calculate the components of a unit vector that lies in the xy-plane and makes equal angles with the positive directions of the x- and y-axes.

1.1.4   The velocity of sailboat A relative to sailboat B, $\mathbf{v}_{\text{rel}}$, is defined by the equation $\mathbf{v}_{\text{rel}} = \mathbf{v}_A - \mathbf{v}_B$, where $\mathbf{v}_A$ is the velocity of A and $\mathbf{v}_B$ is the velocity of B. Determine the velocity of A relative to B if

$$\mathbf{v}_A = 30 \text{ km/hr east}$$
$$\mathbf{v}_B = 40 \text{ km/hr north}.$$

ANS. $\mathbf{v}_{\text{rel}} = 50 \text{ km/hr}$, 53.1° south of east.

1.1.5   A sailboat sails for 1 hr at 4 km/hr (relative to the water) on a steady compass heading of 40° east of north. The sailboat is simultaneously carried along by a current. At the end of the hour the boat is 6.12 km from its starting point. The line from its starting point to its location lies 60° east of north. Find the x (easterly) and y (northerly) components of the water's velocity.

ANS. $v_{\text{east}} = 2.73 \text{ km/hr}$, $v_{\text{north}} \approx 0 \text{ km/hr}$.

1.1.6   A vector equation can be reduced to the form $\mathbf{A} = \mathbf{B}$. From this show that the one vector equation is equivalent to *three* scalar equations.
Assuming the validity of Newton's second law $\mathbf{F} = m\mathbf{a}$ as a *vector* equation, this means that a_x depends only on F_x and is independent of F_y and F_z.

1.1.7   The vertices of a triangle A, B, and C are given by the points $(-1, 0, 2)$, $(0, 1, 0)$, and $(1, -1, 0)$, respectively. Find point D so that the figure $ABCD$ forms a plane parallelogram.

ANS. $(2, 0, -2)$.

1.1.8   A triangle is defined by the vertices of three vectors, **A**, **B** and **C** that extend from the origin. In terms of **A**, **B**, and **C** show that the *vector* sum of the successive sides of the triangle $(AB + BC + CA)$ is zero.

1.1.9   A sphere of radius a is centered at a point $\mathbf{r}_1$.

(a) Write out the algebraic equation for the sphere.
(b) Write out a *vector* equation for the sphere.

ANS.   (a) $(x - x_1)^2 + (y - y_1)^2 + (z - z_1)^2 = a^2$.
(b) $\mathbf{r} = \mathbf{r}_1 + \mathbf{a}$.
($\mathbf{a}$ takes on all directions but has a fixed magnitude, a.)

1.1.10   A corner reflector is formed by three mutually perpendicular reflecting surfaces. Show that a ray of light incident upon the corner reflector (striking all three surfaces) is reflected back along a line parallel to the line of incidence.
Hint. Consider the effect of a reflection on the components of a vector describing the direction of the light ray.

1.1.11   Hubble's law. Hubble found that distant galaxies are receding with a velocity proportional to their distance from where we are on Earth. For the ith galaxy

$$\mathbf{v}_i = H_0 \mathbf{r}_i$$

with us at the origin. Show that this recession of the galaxies from us does *not* imply that we are at the center of the universe. Specifically, take the galaxy at $\mathbf{r}_1$ as a new origin and show that Hubble's law is still obeyed.

1.2 ROTATION OF THE COORDINATE AXES*

In the preceding section vectors were defined or represented in two equivalent ways: (1) geometrically by specifying magnitude and direction, as with an arrow, and (2) algebraically by specifying the components relative to cartesian coordinate axes. The second definition is adequate for the vector analysis of this chapter. In this section two more refined, sophisticated, and powerful definitions are presented. First, the vector field is defined in terms of the behavior of its components under rotation of the coordinate axes. This transformation theory approach leads into the tensor analysis of Chapter 2 and groups of transformations in Chapter 4. Second, the component definition of Section 1.1 is refined and generalized according to the mathematician's concepts of vector and vector space. This approach leads to function spaces including the Hilbert space—Section 9.4.

---

*This section is optional. It is not essential for the remaining sections of this chapter.

The definition of vector as a quantity with magnitude and direction breaks down in advanced work. On the one hand, we encounter quantities, such as elastic constants and index of refraction in anisotropic crystals, that have magnitude and direction *but* which are not vectors. On the other hand, our naïve approach is awkward to generalize, to extend to more complex quantities. We seek a new definition of vector field, using our displacement vector **r** as a prototype.

There is an important physical basis for our development of a new definition. We describe our physical world by mathematics, but it and any physical predictions we may make must be *independent* of our mathematical analysis. Some writers compare the physical system to a building and the mathematical analysis to the scaffolding used to construct the building. In the end the scaffolding is stripped off and the building stands.

In our specific case we assume that space is isotropic; that is, there is no preferred direction or all directions are equivalent. Then the physical system being analyzed or the physical law being enunciated cannot and must not depend on our choice or *orientation* of the coordinate axes.

Now we return to the concept of vector **r** as a geometric object independent of the coordinate system. Let us look at **r** in two different systems, one rotated in relation to the other.

For simplicity we consider first the two-dimensional case. If the x-, y-coordinates are rotated counterclockwise through an angle φ, *keeping* **r** *fixed* (Fig. 1.6), we get the following relations between the components resolved in the original system (unprimed) and those resolved in the new rotated system (primed):

$$x' = x \cos \varphi + y \sin \varphi,$$
$$y' = -x \sin \varphi + y \cos \varphi. \tag{1.8}$$

We saw in Section 1.1 that a vector could be represented by the coordinates of a point; that is, the coordinates were proportional to the vector components. Hence the components of a vector must transform under rotation as coordinates of a point (such as **r**). Therefore whenever any pair of quantities $A_x(x, y)$ and $A_y(x, y)$ in the xy-coordinate system is transformed in (A'_x, A'_y) by this rotation of the coordinate system with

$$A'_x = A_x \cos \varphi + A_y \sin \varphi$$
$$A'_y = -A_x \sin \varphi + A_y \cos \varphi, \tag{1.9}$$

we *define*[1] A_x and A_y as the components of a vector **A**. Our vector now is defined in terms of the transformation of its components under rotation of the coordinate system. If A_x and A_y transform in the same way as x and y, the components of the two-dimensional displacement vector, they are the

---

[1] The corresponding definition of a scalar quantity is $S' = S$, that is, invariant under rotation of the coordinates.

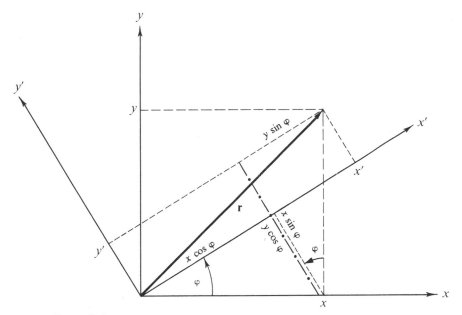

Figure 1.6   Rotation of cartesian coordinate axes about the z-axis.

components of a vector **A**. If A_x and A_y do not show this form invariance when the coordinates are rotated, they do not form a vector.

The vector field components A_x and A_y satisfying the defining equations, Eq. (1.9), associate a magnitude A and a direction with each point in space. The magnitude is a scalar quantity, invariant to the rotation of the coordinate system. The direction (relative to the unprimed system) is likewise invariant to the rotation of the coordinate system (see Exercise 1.2.1). The result of all this is that the components of a vector may vary according to the rotation of the primed coordinate system. This is what Eq. (1.9) says. But the variation with the angle is just such that the components in the rotated coordinate system A_x' and A_y' define a vector with the same magnitude and the same direction as the vector defined by the components A_x and A_y relative to the x-, y-coordinate axes. (Compare Exercise 1.2.1.) The components of **A** in a particular coordinate system constitute the *representation* of **A** in that coordinate system. Equation (1.9), the transformation relation, is a guarantee that the entity **A** is independent of the rotation of the coordinate system.

To go on to three and, later, four dimensions, we find it convenient to use a more compact notation. Let

$$
\begin{aligned}
x &\to x_1 \\
y &\to x_2
\end{aligned}
\tag{1.10}
$$

$$
\begin{aligned}
a_{11} &= \cos \varphi, & a_{12} &= \sin \varphi, \\
a_{21} &= -\sin \varphi, & a_{22} &= \cos \varphi.
\end{aligned}
\tag{1.11}
$$

Then Eq. (1.8) becomes

$$x_1' = a_{11}x_1 + a_{12}x_2,$$
$$x_2' = a_{21}x_1 + a_{22}x_2. \tag{1.12}$$

The coefficient a_{ij} may be interpreted as a direction cosine, the cosine of the angle between x_i' and x_j; that is,

$$a_{12} = \cos(x_1', x_2) = \sin \varphi,$$
$$a_{21} = \cos(x_2', x_1) = \cos\left(\varphi + \frac{\pi}{2}\right) = -\sin \varphi. \tag{1.13}$$

The advantage of the new notation[2] is that it permits us to use the summation symbol Σ and to rewrite Eqs. (1.12) as

$$x_i' = \sum_{j=1}^{2} a_{ij}x_j, \quad i = 1, 2. \tag{1.14}$$

Note that i remains as a parameter that gives rise to one equation when it is set equal to 1 and to a second equation when it is set equal to 2. The index j, of course, is a summation index, a dummy index, and as with a variable of integration, j may be replaced by any other convenient symbol.

The generalization to three, four, or N dimensions is now very simple. The set of N quantities, V_j, is said to be the components of an N-dimensional vector, **V**, if and only if their values relative to the rotated coordinate axes are given by

$$V_i' = \sum_{j=1}^{N} a_{ij}V_j, \quad i = 1, 2, \dots, N. \tag{1.15}$$

As before, a_{ij} is the cosine of the angle between x_i' and x_j. Often the upper limit N and the corresponding range of i will not be indicated. It is taken for granted that the reader knows how many dimensions his or her space has.

From the definition of a_{ij} as the cosine of the angle between the positive x_i' direction and the positive x_j direction we may write (cartesian coordinates)[3]

$$a_{ij} = \frac{\partial x_i'}{\partial x_j}. \tag{1.16a}$$

---

[2] The reader may wonder at the replacement of one parameter φ by four parameters a_{ij}. Clearly, the a_{ij} do not constitute a minimum set of parameters. For two dimensions the four a_{ij} are subject to the three constraints given in Eq. (1.18). The justification for the redundant set of direction cosines is the convenience it provides. Hopefully, this convenience will become more apparent in Chapters 2 and 3. For three-dimensional rotations (9 a_{ij} but only three independent) alternate descriptions are provided by: (1) the Euler angles discussed in Section 3.3, (2) quaternions, and (3) the Cayley–Klein parameters. These alternatives have their respective advantages and disadvantages.

[3] Differentiate $x_i' = \sum a_{ik}x_k$ with respect to x_j. See discussion following Eq. (1.21). Section 3.3 provides an alternate approach.

Using the inverse rotation ($\varphi \to -\varphi$) yields

$$x_j = \sum_{i=1}^{2} a_{ij} x_i' \qquad \text{or} \qquad \frac{\partial x_j}{\partial x_i'} = a_{ij}. \qquad (1.16b)$$

Note carefully that these are *partial derivatives*. By use of Eq. (1.16), Eq. (1.15) becomes

$$V_i' = \sum_{j=1}^{N} \frac{\partial x_i'}{\partial x_j} V_j = \sum_{j=1}^{N} \frac{\partial x_j}{\partial x_i'} V_j. \qquad (1.17)$$

The direction cosines a_{ij} satisfy an *orthogonality condition*

$$\sum_i a_{ij} a_{ik} = \delta_{jk} \qquad (1.18)$$

or, equivalently,

$$\sum_i a_{ji} a_{ki} = \delta_{jk}. \qquad (1.19)$$

The symbol δ_{jk} is the Kronecker delta defined by

$$\begin{aligned} \delta_{jk} &= 1 \qquad \text{for} \qquad j = k, \\ \delta_{jk} &= 0 \qquad \text{for} \qquad j \neq k. \end{aligned} \qquad (1.20)$$

The reader may easily verify that Eqs. (1.18) and (1.19) hold in the two-dimensional case by substituting in the specific a_{ij} from Eq. (1.11). The result is the well-known identity $\sin^2\varphi + \cos^2\varphi = 1$ for the nonvanishing case. To verify Eq. (1.18) in general form, we may use the partial derivative forms of Eqs. (1.16) to obtain

$$\sum_i \frac{\partial x_j}{\partial x_i'} \frac{\partial x_k}{\partial x_i'} = \sum_i \frac{\partial x_j}{\partial x_i'} \frac{\partial x_i'}{\partial x_k} = \frac{\partial x_j}{\partial x_k}. \qquad (1.21)$$

The last step follows by the standard rules for partial differentiation, assuming that x_j is a function of x_1', x_2', x_3', and so on. The final result, $\partial x_j / \partial x_k$, is equal to δ_{jk}, since x_j and x_k as coordinate lines ($j \neq k$) are assumed to be perpendicular (two or three dimensions) or orthogonal (for any number of dimensions). Equivalently, we may assume that x_j and x_k ($j \neq k$) are totally independent variables. If $j = k$, the partial derivative is clearly equal to 1.

In redefining a vector in terms of how its components transform under a rotation of the coordinate system, we should emphasize two points:

1. This definition is developed because it is useful and appropriate in describing our physical world. Our vector equations will be independent of any particular coordinate system. (The coordinate system need not even be cartesian.) The vector equation can always be expressed in some particular coordinate system and, to obtain numerical results, we must ultimately express the equation in some specific coordinate system.

2. This definition is subject to a generalization that will open up the branch of mathematics known as tensor analysis (Chapter 2).

A qualification is also in order. The behavior of the vector components under rotation of the coordinates is used in Section 1.3 to prove that a scalar product is a scalar, in Section 1.4 to prove that a vector product is a vector, and in Section 1.6 to show that the gradient of a scalar, $\nabla \psi$, is a vector. The remainder of this chapter proceeds on the basis of the less restrictive definitions of the vector given in Section 1.1.

VECTORS AND VECTOR SPACE

It is customary in mathematics to label an ordered triple of real numbers (x_1, x_2, x_3) a *vector* $\mathbf{x}$. The number x_n is called the nth component of vector $\mathbf{x}$. The collection of all such vectors (obeying the properties that follow) form a three-dimensional real *vector space*. We ascribe five properties to our vectors: If $\mathbf{x} = (x_1, x_2, x_3)$ and $\mathbf{y} = (y_1, y_2, y_3)$,

1. Vector equality: $\mathbf{x} = \mathbf{y}$ means $x_i = y_i$, $i = 1, 2, 3$.
2. Vector addition: $\mathbf{x} + \mathbf{y} = \mathbf{z}$ means $x_i + y_i = z_i$, $i = 1, 2, 3$.
3. Scalar multiplication: $a\mathbf{x} \leftrightarrow (ax_1, ax_2, ax_3)$ (with a real).
4. Negative of a vector: $-\mathbf{x} = (-1)\mathbf{x} \leftrightarrow (-x_1, -x_2, -x_3)$.
5. Null vector: There exists a null vector $\mathbf{0} \leftrightarrow (0, 0, 0)$.

Since our vector components are real numbers, the following properties also hold:

1. Addition of vectors is commutative: $\mathbf{x} + \mathbf{y} = \mathbf{y} + \mathbf{x}$.
2. Addition of vectors is associative: $(\mathbf{x} + \mathbf{y}) + \mathbf{z} = \mathbf{x} + (\mathbf{y} + \mathbf{z})$.
3. Scalar multiplication is distributive:

$$a(\mathbf{x} + \mathbf{y}) = a\mathbf{x} + a\mathbf{y}, \quad \text{also} \quad (a + b)\mathbf{x} = a\mathbf{x} + b\mathbf{x}.$$

4. Scalar multiplication is associative: $(ab)\mathbf{x} = a(b\mathbf{x})$.

Further, the null vector $\mathbf{0}$ is unique as is the negative of a given vector $\mathbf{x}$.

So far as the vectors themselves are concerned this approach merely formalizes the component discussion of Section 1.1. The importance lies in the extensions which will be considered in later chapters. In Chapter 4, we show that vectors form both an Abelian group under addition and a linear space with the transformations in the linear space described by matrices. Finally, and perhaps most important, for advanced physics the concept of vectors presented here may be generalized to (1) complex quantities,[4] (2) functions, and (3) an infinite number of components. This leads to infinite dimensional function spaces, the Hilbert spaces, which are important in modern quantum theory. A brief introduction to function expansions and Hilbert space appears in Section 9.4.

---

[4] The n-dimensional vector space of real n-tuples is often labeled $\mathbf{R}^n$ and the n-dimensional vector space of complex n-tuples is labeled $\mathbf{C}^n$.

EXERCISES

1.2.1   (a) Show that the magnitude of a vector **A**, $A = (A_x^2 + A_y^2)^{1/2}$ is independent of the orientation of the rotated coordinate system,
$$(A_x^2 + A_y^2)^{1/2} = (A_x'^2 + A_y'^2)^{1/2}$$
independent of the rotation angle φ.

This independence of angle is expressed by saying that A is *invariant* under rotations.

(b) At a given point (x, y), **A** defines an angle α relative to the positive x-axis and α' relative to the positive x'-axis. The angle from x to x' is φ. Show that **A** = **A**′ defines the *same* direction in space when expressed in terms of its primed components, as in terms of its unprimed components; that is,
$$\alpha' = \alpha - \varphi.$$

1.2.2   Prove the orthogonality condition $\sum_i a_{ji}a_{ki} = \delta_{jk}$. As a special case of this the direction cosines of Section 1.1 satisfy the relation
$$\cos^2\alpha + \cos^2\beta + \cos^2\gamma = 1,$$
a result that also follows from Eq. (1.6).

1.3 SCALAR OR DOT PRODUCT

Having defined vectors, we must now proceed to combine them. The laws for combining vectors must be mathematically consistent. From the possibilities that are consistent we select two that are both mathematically and physically interesting. A third possibility is introduced in Chapter 2, in which we form tensors.

The projection of a vector **A** onto a coordinate axis, which defines its cartesian components in Eq. (1.4), is a special case of the scalar product of **A** and the coordinate unit vectors,

$$A_x = A \cos \alpha \equiv \mathbf{A} \cdot \hat{\mathbf{x}}, \qquad A_y = A \cos \beta \equiv \mathbf{A} \cdot \hat{\mathbf{y}}, \qquad A_z = A \cos \gamma \equiv \mathbf{A} \cdot \hat{\mathbf{z}}.$$
$$(1.22)$$

Just as the projection is linear in **A**, we want the scalar product of two vectors to be linear in **A** and **B**, i.e., obey the distributive and associative laws

$$\mathbf{A} \cdot (\mathbf{B} + \mathbf{C}) = \mathbf{A} \cdot \mathbf{B} + \mathbf{A} \cdot \mathbf{C} \tag{1.23a}$$

$$\mathbf{A} \cdot (y\mathbf{B}) = (y\mathbf{A}) \cdot \mathbf{B} = y\mathbf{A} \cdot \mathbf{B}, \tag{1.23b}$$

where y is a number. Now we can use the decomposition of **B** into its cartesian components according to Eq. (1.5), $\mathbf{B} = B_x\hat{\mathbf{x}} + B_y\hat{\mathbf{y}} + B_z\hat{\mathbf{z}}$, to construct the

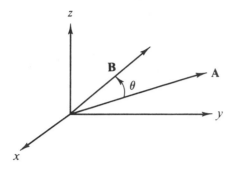

Figure 1.7 Scalar product $\mathbf{A} \cdot \mathbf{B} = AB \cos \theta$.

general scalar or dot product of the vectors **A** and **B** as

$$\mathbf{A} \cdot \mathbf{B} = \mathbf{A} \cdot (B_x \hat{\mathbf{x}} + B_y \hat{\mathbf{y}} + B_z \hat{\mathbf{z}})$$

$$= B_x \mathbf{A} \cdot \hat{\mathbf{x}} + B_y \mathbf{A} \cdot \hat{\mathbf{y}} + B_z \mathbf{A} \cdot \hat{\mathbf{z}}, \quad \text{upon applying Eq. (1.23)},$$

$$= B_x A_x + B_y A_y + B_z A_z, \quad \text{upon substituting Eq. (1.22)}.$$

Hence

$$\mathbf{A} \cdot \mathbf{B} \equiv \sum_i B_i A_i = \sum_i A_i B_i = \mathbf{B} \cdot \mathbf{A}. \tag{1.24}$$

If $\mathbf{A} = \mathbf{B}$ in Eq. (1.24), we recover the magnitude $A = (\sum A_i^2)^{1/2}$ of **A** in Eq. (1.6) from Eq. (1.24).

It is obvious from Eq. (1.24) that the scalar product treats **A** and **B** alike, or is symmetric in **A** and **B**, and is commutative. Thus, alternatively and equivalently, we can first generalize Eq. (1.22) to the projection A_B of **A** onto the direction of a vector $\mathbf{B} \neq 0$ as $A_B = A \cos \theta \equiv \mathbf{A} \cdot \hat{\mathbf{B}}$, where $\hat{\mathbf{B}} = \mathbf{B}/B$ is the unit vector in the direction of **B** and θ is the angle between **A** and **B** as shown in Fig. 1.7. Similarly, we project **B** onto **A** as $B_A = B \cos \theta \equiv \mathbf{B} \cdot \hat{\mathbf{A}}$. Second, we make these projections symmetric in **A** and **B**, which leads to the definition

$$\mathbf{A} \cdot \mathbf{B} \equiv A_B B = AB_A = AB \cos \theta. \tag{1.25}$$

The distributive law in Eq. (1.23a) is illustrated in Fig. 1.8, which shows that the sum of the projections of **B** and **C** onto **A**, $B_A + C_A$, is equal to the projection of $\mathbf{B} + \mathbf{C}$ onto **A**, $(\mathbf{B} + \mathbf{C})_A$.

It follows from Eqs. (1.22), (1.24), (1.25) that the coordinate unit vectors satisfy the relations

$$\hat{\mathbf{x}} \cdot \hat{\mathbf{x}} = \hat{\mathbf{y}} \cdot \hat{\mathbf{y}} = \hat{\mathbf{z}} \cdot \hat{\mathbf{z}} = 1, \tag{1.26a}$$

whereas

$$\hat{\mathbf{x}} \cdot \hat{\mathbf{y}} = \hat{\mathbf{x}} \cdot \hat{\mathbf{z}} = \hat{\mathbf{y}} \cdot \hat{\mathbf{z}} = 0. \tag{1.26b}$$

If the component definition, Eq. (1.24), is labeled an algebraic definition, then Eq. (1.25) is a geometric definition. One of the most common applications of

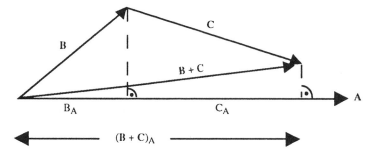

Figure 1.8 The distributive law $\mathbf{A} \cdot (\mathbf{B} + \mathbf{C}) = AB_A + AC_A = A(\mathbf{B} + \mathbf{C})_A$, Eq. (1.23a).

the scalar product in physics is in the calculation of work = force × displacement × cos θ, which is interpreted as displacement times the projection of the force along the displacement direction, i.e., the scalar product of force and displacement, $W = \mathbf{F} \cdot \mathbf{S}$.

Example 1.3.1

For the two vectors $\mathbf{A}$ and $\mathbf{B}$ of Example 1.1.1, $\mathbf{A} = 6\hat{\mathbf{x}} + 4\hat{\mathbf{y}} + 3\hat{\mathbf{z}}$, $\mathbf{B} = 2\hat{\mathbf{x}} - 3\hat{\mathbf{y}} - 3\hat{\mathbf{z}}$,

$$\mathbf{A} \cdot \mathbf{B} = (12 - 12 - 9) = -9$$

by Eq. (1.24). In this case the projection of $\mathbf{A}$ on $\mathbf{B}$ (or $\mathbf{B}$ on $\mathbf{A}$) is negative. Actually,

$$|\mathbf{A}| = (36 + 16 + 9)^{1/2} = (61)^{1/2} = 7.81,$$

$$|\mathbf{B}| = (4 + 9 + 9)^{1/2} = (22)^{1/2} = 4.69,$$

and cos $\theta = -0.246$, $\theta = 104.2°$.

If $\mathbf{A} \cdot \mathbf{B} = 0$ and we know that $\mathbf{A} \neq 0$ and $\mathbf{B} \neq 0$, then from Eq. (1.25), cos $\theta = 0$ or $\theta = 90°$, 270°, and so on. The vectors $\mathbf{A}$ and $\mathbf{B}$ must be perpendicular. Alternately, we may say $\mathbf{A}$ and $\mathbf{B}$ are orthogonal. The unit vectors $\hat{\mathbf{x}}$, $\hat{\mathbf{y}}$, and $\hat{\mathbf{z}}$ are mutually orthogonal. To develop this notion of orthogonality one more step, suppose that $\mathbf{n}$ is a unit vector and $\mathbf{r}$ is a nonzero vector in the xy-plane; that is, $\mathbf{r} = \hat{\mathbf{x}}x + \hat{\mathbf{y}}y$ (Fig. 1.9). If

$$\mathbf{n} \cdot \mathbf{r} = 0$$

for *all* choices of $\mathbf{r}$, then $\mathbf{n}$ must be perpendicular (orthogonal) to the xy-plane.

Often it is convenient to replace $\hat{\mathbf{x}}$, $\hat{\mathbf{y}}$, and $\hat{\mathbf{z}}$ by subscripted unit vectors $\mathbf{e}_m$, $m = 1, 2, 3$, with $\hat{\mathbf{x}} = \mathbf{e}_1$, and so on. Then Eqs. (1.26a) and (1.26b) become

$$\mathbf{e}_m \cdot \mathbf{e}_n = \delta_{mn}. \tag{1.26c}$$

For $m \neq n$ the unit vectors $\mathbf{e}_m$ and $\mathbf{e}_n$ are orthogonal. For $m = n$ each vector is normalized to unity, that is, has unit magnitude. The set $\mathbf{e}_m$ is said to be *orthonormal*. A major advantage of Eq. (1.26c) over Eqs. (1.26a) and (1.26b)

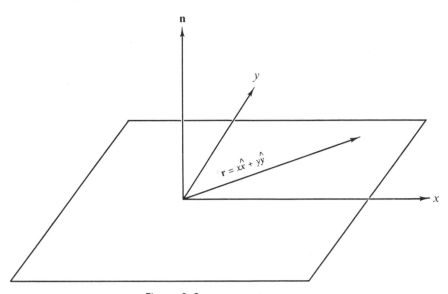

Figure 1.9   A normal vector.

is that Eq. (1.26c) may readily be generalized to N-dimensional space: $m, n = 1, 2, \ldots, N$. Finally, we are picking sets of unit vectors $\mathbf{e}_m$ that are orthonormal for convenience—a very great convenience.

Invariance of the Scalar Product under Rotations

We have not yet shown that the word scalar is justified or that the scalar product is indeed a scalar quantity. To do this, we investigate the behavior of $\mathbf{A} \cdot \mathbf{B}$ under a rotation of the coordinate system. By use of Eq. (1.15)

$$A'_x B'_x + A'_y B'_y + A'_z B'_z = \sum_i a_{xi} A_i \sum_j a_{xj} B_j + \sum_i a_{yi} A_i \sum_j a_{yj} B_j$$

$$+ \sum_i a_{zi} A_i \sum_j a_{zj} B_j. \tag{1.27}$$

Using the indices k and l to sum over x, y, and z, we obtain

$$\sum_k A'_k B'_k = \sum_l \sum_i \sum_j a_{li} A_i a_{lj} B_j, \tag{1.28}$$

and, by rearranging the terms on the right-hand side, we have

$$\sum_k A'_k B'_k = \sum_i \sum_j \sum_l (a_{li} a_{lj}) A_i B_j = \sum_i \sum_j \delta_{ij} A_i B_j = \sum_i A_i B_i. \tag{1.29}$$

The last two steps follow by using Eq. (1.18), the orthogonality condition of the direction cosines, and Eq. (1.20), which defines the Kronecker delta. The effect of the Kronecker delta is to cancel all terms in a summation over either index except the term for which the indices are equal. In Eq. (1.29) its effect

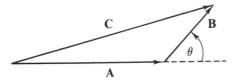

Figure 1.10 The law of cosines.

is to set $j = i$ and to eliminate the summation over j. Of course, we could equally well set $i = j$ and eliminate the summation over i. Equation (1.29) gives us

$$\sum_k A'_k B'_k = \sum_i A_i B_i, \tag{1.30}$$

which is just our definition of a scalar quantity, one that remains *invariant* under the rotation of the coordinate system.

In a similar approach which exploits this concept of invariance, we take $\mathbf{C} = \mathbf{A} + \mathbf{B}$ and dot it into itself.

$$\mathbf{C} \cdot \mathbf{C} = (\mathbf{A} + \mathbf{B}) \cdot (\mathbf{A} + \mathbf{B})$$
$$= \mathbf{A} \cdot \mathbf{A} + \mathbf{B} \cdot \mathbf{B} + 2\mathbf{A} \cdot \mathbf{B}. \tag{1.31}$$

Since

$$\mathbf{C} \cdot \mathbf{C} = C^2, \tag{1.32}$$

the square of the magnitude of vector $\mathbf{C}$ and thus an invariant quantity, we see that

$$\mathbf{A} \cdot \mathbf{B} = \tfrac{1}{2}(C^2 - A^2 - B^2), \qquad \text{invariant.} \tag{1.33}$$

Since the right-hand side of Eq. (1.33) is invariant—that is, a scalar quantity—the left-hand side, $\mathbf{A} \cdot \mathbf{B}$, must also be invariant under rotation of the coordinate system. Hence $\mathbf{A} \cdot \mathbf{B}$ is a scalar.

Equation (1.31) is really another form of the law of cosines which is

$$C^2 = A^2 + B^2 + 2AB \cos \theta. \tag{1.34}$$

Comparing Eqs. (1.31) and (1.34), we have another verification of Eq. (1.25), or, if preferred, a vector derivation of the law of cosines (Fig. 1.10).

The dot product, given by Eq. (1.24), may be generalized in two ways. The space need not be restricted to three dimensions. In n-dimensional space, Eq. (1.24) applies with the sum running from 1 to n. n may be infinity, with the sum then a convergent infinite series (Section 5.2). The other generalization extends the concept of vector to embrace functions. The function analog of a dot or inner product appears in Section 9.4.

EXERCISES

1.3.1   Two unit magnitude vectors $\mathbf{e}_i$ and $\mathbf{e}_j$ are required to be either parallel or perpendicular to each other. Show that $\mathbf{e}_i \cdot \mathbf{e}_j$ provides an interpretation of Eq. (1.18), the direction cosine orthogonality relation.

1.3.2   Given that (1) the dot product of a unit vector with itself is unity and (2) this relation is valid in all (rotated) coordinate systems, show that $\hat{\mathbf{x}}' \cdot \hat{\mathbf{x}}' = 1$ (with the primed system rotated 45° about the z-axis relative to the unprimed) implies that $\hat{\mathbf{x}} \cdot \hat{\mathbf{y}} = 0$.

1.3.3   The vector $\mathbf{r}$, starting at the origin, terminates at and specifies the point in space (x, y, z). Find the surface swept out by the tip of $\mathbf{r}$ if

(a) $(\mathbf{r} - \mathbf{a}) \cdot \mathbf{a} = 0$,

(b) $(\mathbf{r} - \mathbf{a}) \cdot \mathbf{r} = 0$.

The vector $\mathbf{a}$ is a constant (constant in magnitude and direction).

1.3.4

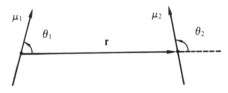

The interaction energy between two dipoles of moments $\boldsymbol{\mu}_1$ and $\boldsymbol{\mu}_2$ may be written in the vector form

$$V = -\frac{\boldsymbol{\mu}_1 \cdot \boldsymbol{\mu}_2}{r^3} + \frac{3(\boldsymbol{\mu}_1 \cdot \mathbf{r})(\boldsymbol{\mu}_2 \cdot \mathbf{r})}{r^5}$$

and in the scalar form

$$V = \frac{\mu_1 \mu_2}{r^3}(2 \cos \theta_1 \cos \theta_2 - \sin \theta_1 \sin \theta_2 \cos \varphi).$$

Here θ_1 and θ_2 are the angles of $\boldsymbol{\mu}_1$ and $\boldsymbol{\mu}_2$ relative to $\mathbf{r}$, while φ is the azimuth of $\boldsymbol{\mu}_2$ relative to the $\boldsymbol{\mu}_1 - \mathbf{r}$ plane. Show that these two forms are equivalent. *Hint.* Equation (12.198) will be helpful.

1.3.5   A pipe comes diagonally down the south wall of a building, making an angle of 45° with the horizontal. Coming into a corner, the pipe turns and continues diagonally down a west-facing wall, still making an angle of 45° with the horizontal. What is the angle between the south-wall and west-wall sections of the pipe?

ANS.   120°.

1.4 VECTOR OR CROSS PRODUCT

A second form of vector multiplication employs the sine of the included angle instead of the cosine. For instance, the angular momentum (Fig. 1.11) of a body is defined as

angular momentum = radius arm × linear momentum

= distance × linear momentum × sin θ.

For convenience in treating problems relating to quantities such as angular momentum, torque, and angular velocity, we define the vector or cross

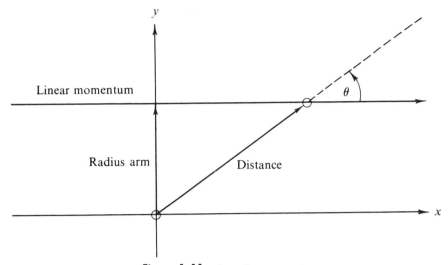

Figure 1.11   Angular momentum.

product as

$$C = A \times B,$$

with

$$C = AB \sin \theta. \tag{1.35}$$

Unlike the preceding case of the scalar product, **C** is now a vector, and we assign it a direction perpendicular to the plane of **A** and **B** such that **A**, **B**, and **C** form a right-handed system. With this choice of direction we have

$$A \times B = -B \times A, \qquad \text{anticommutation.} \tag{1.36a}$$

From this definition of cross product we have

$$\hat{x} \times \hat{x} = \hat{y} \times \hat{y} = \hat{z} \times \hat{z} = 0, \tag{1.36b}$$

whereas

$$\hat{x} \times \hat{y} = \hat{z}, \qquad \hat{y} \times \hat{z} = \hat{x}, \qquad \hat{z} \times \hat{x} = \hat{y}$$

and
$$\tag{1.36c}$$

$$\hat{y} \times \hat{x} = -\hat{z}, \qquad \hat{z} \times \hat{y} = -\hat{x}, \qquad \hat{x} \times \hat{z} = -\hat{y}.$$

Among the examples of the cross product in mathematical physics are the relation between linear momentum **p** and angular momentum **L** (defining angular momentum),

$$L = r \times p$$

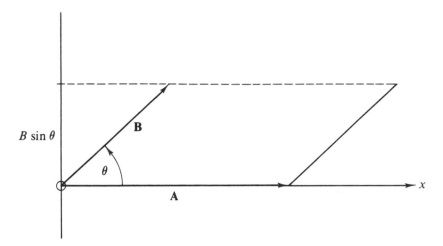

Figure 1.12   Parallelogram representation of the vector product.

and the relation between linear velocity $\mathbf{v}$ and angular velocity $\boldsymbol{\omega}$,

$$\mathbf{v} = \boldsymbol{\omega} \times \mathbf{r}.$$

Vectors $\mathbf{v}$ and $\mathbf{p}$ describe properties of the particle or physical system. However, the position vector $\mathbf{r}$ is determined by the choice of the origin of the coordinates. This means that $\boldsymbol{\omega}$ and $\mathbf{L}$ depend on the choice of the origin.

The familiar magnetic induction $\mathbf{B}$ is usually defined by the vector product force equation[1]

$$\mathbf{F}_M = q\mathbf{v} \times \mathbf{B}.$$

Here $\mathbf{v}$ is the velocity of the electric charge q and $\mathbf{F}_M$ is the resulting force on the moving charge.

The cross product has an important geometrical interpretation which we shall use in subsequent sections. In the parallelogram defined by $\mathbf{A}$ and $\mathbf{B}$ (Fig. 1.12), $B \sin \theta$ is the height if A is taken as the length of the base. Then $|\mathbf{A} \times \mathbf{B}| = AB \sin \theta$ is the *area* of the parallelogram. As a vector, $\mathbf{A} \times \mathbf{B}$ is the area of the parallelogram defined by $\mathbf{A}$ and $\mathbf{B}$, with the area vector normal to the plane of the parallelogram. This suggests that area may be treated as a vector quantity.

Parenthetically, it might be noted that Eq. (1.36c) and a modified Eq. (1.36b) form the starting point for the development of *quaternions*. Equation (1.36b) is replaced by $\hat{\mathbf{x}} \times \hat{\mathbf{x}} = \hat{\mathbf{y}} \times \hat{\mathbf{y}} = \hat{\mathbf{z}} \times \hat{\mathbf{z}} = -1$.

An alternate definition of the vector product can be derived from the special case of the coordinate unit vectors in Eq. (1.36) in conjunction with the linearity of the cross product in both vector arguments, in analogy with

[1] The electric field $\mathbf{E}$ is assumed here to be zero.

Eq. (1.23) for the dot product,

$$\mathbf{A} \times (\mathbf{B} + \mathbf{C}) = \mathbf{A} \times \mathbf{B} + \mathbf{A} \times \mathbf{C},$$

$$(\mathbf{A} + \mathbf{B}) \times \mathbf{C} = \mathbf{A} \times \mathbf{C} + \mathbf{B} \times \mathbf{C} \tag{1.37a}$$

$$\mathbf{A} \times (y\mathbf{B}) = y\mathbf{A} \times \mathbf{B} = (y\mathbf{A}) \times \mathbf{B}, \tag{1.37b}$$

where y is a number again. Using the decomposition of $\mathbf{A}$ and $\mathbf{B}$ into their cartesian components according to Eq. (1.5), we find

$$\mathbf{A} \times \mathbf{B} \equiv \mathbf{C} = (C_x, C_y, C_z) = (A_x\hat{\mathbf{x}} + A_y\hat{\mathbf{y}} + A_z\hat{\mathbf{z}}) \times (B_x\hat{\mathbf{x}} + B_y\hat{\mathbf{y}} + B_z\hat{\mathbf{z}})$$

$$= (A_xB_y - A_yB_x)\hat{\mathbf{x}} \times \hat{\mathbf{y}} + (A_xB_z - A_zB_x)\hat{\mathbf{x}} \times \hat{\mathbf{z}} + (A_yB_z - A_zB_y)\hat{\mathbf{y}} \times \hat{\mathbf{z}}$$

upon applying Eq. (1.37) and substituting Eq. (1.36), so that the cartesian components of $\mathbf{A} \times \mathbf{B}$ become

$$C_x = A_yB_z - A_zB_y, \quad C_y = A_zB_x - A_xB_z, \quad C_z = A_xB_y - A_yB_x, \tag{1.38}$$

or

$$C_i = A_jB_k - A_kB_j, \quad i, j, k \text{ all different}, \tag{1.39}$$

and with cyclic permutation of the indices i, j, and k. The vector product $\mathbf{C}$ may be conveniently represented by a determinant[2]

$$\mathbf{C} = \begin{vmatrix} \hat{\mathbf{x}} & \hat{\mathbf{y}} & \hat{\mathbf{z}} \\ A_x & A_y & A_z \\ B_x & B_y & B_z \end{vmatrix}. \tag{1.40}$$

Expansion of the determinant across the top row reproduces the three components of $\mathbf{C}$ listed in Eq. (1.38).

Equation (1.35) might be called a geometric definition of the vector product. Then Eq. (1.38) would be an algebraic definition.

To show the equivalence of Eq. (1.35) and the component definition, Eq. (1.38), let us form $\mathbf{A} \cdot \mathbf{C}$ and $\mathbf{B} \cdot \mathbf{C}$, using Eq. (1.38). We have

$$\mathbf{A} \cdot \mathbf{C} = \mathbf{A} \cdot (\mathbf{A} \times \mathbf{B})$$

$$= A_x(A_yB_z - A_zB_y) + A_y(A_zB_x - A_xB_z) + A_z(A_xB_y - A_yB_x)$$

$$= 0. \tag{1.41}$$

Similarly,

$$\mathbf{B} \cdot \mathbf{C} = \mathbf{B} \cdot (\mathbf{A} \times \mathbf{B}) = 0. \tag{1.42}$$

Equations (1.41) and (1.42) show that $\mathbf{C}$ is perpendicular to both $\mathbf{A}$ and $\mathbf{B}$ ($\cos \theta = 0$, $\theta = \pm90°$) and therefore perpendicular to the plane they determine. The positive direction is determined by considering special cases such as the unit vectors $\hat{\mathbf{x}} \times \hat{\mathbf{y}} = \hat{\mathbf{z}}$ ($C_z = +A_xB_y$).

[2] See Section 3.1 for a summary of determinants.

The magnitude is obtained from

$$(\mathbf{A} \times \mathbf{B}) \cdot (\mathbf{A} \times \mathbf{B}) = A^2 B^2 - (\mathbf{A} \cdot \mathbf{B})^2$$

$$= A^2 B^2 - A^2 B^2 \cos^2 \theta$$

$$= A^2 B^2 \sin^2 \theta. \tag{1.43}$$

Hence

$$C = AB \sin \theta. \tag{1.44}$$

The big first step in Eq. (1.43) may be verified by expanding out in component form, using Eq. (1.38) for $\mathbf{A} \times \mathbf{B}$ and Eq. (1.24) for the dot product. From Eqs. (1.41), (1.42), and (1.44) we see the equivalence of Eqs. (1.35) and (1.38), the two definitions of vector product.

There still remains the problem of verifying that $\mathbf{C} = \mathbf{A} \times \mathbf{B}$ is indeed a vector; that is, it obeys Eq. (1.15), the vector transformation law. Starting in a rotated (primed system)

$$C_i' = A_j' B_k' - A_k' B_j', \qquad i, j, \text{ and } k \text{ in cyclic order,}$$

$$= \sum_l a_{jl} A_l \sum_m a_{km} B_m - \sum_l a_{kl} A_l \sum_m a_{jm} B_m$$

$$= \sum_{l,m} (a_{jl} a_{km} - a_{kl} a_{jm}) A_l B_m. \tag{1.45}$$

The combination of direction cosines in parentheses vanishes for $m = l$. We therefore have j and k taking on fixed values, dependent on the choice of i, and six combinations of l and m. If $i = 3$, then $j = 1$, $k = 2$ (cyclic order), and we have the following direction cosine combinations[3]

$$a_{11} a_{22} - a_{21} a_{12} = a_{33},$$

$$a_{13} a_{21} - a_{23} a_{11} = a_{32}, \tag{1.46}$$

$$a_{12} a_{23} - a_{22} a_{13} = a_{31}$$

and their negatives. Equations (1.46) are identities satisfied by the direction cosines. They may be verified with the use of determinants and matrices (see Exercise 3.3.3). Substituting back into Eq. (1.45),

$$C_3' = a_{33} A_1 B_2 + a_{32} A_3 B_1 + a_{31} A_2 B_3 - a_{33} A_2 B_1 - a_{32} A_1 B_3 - a_{31} A_3 B_2$$

$$= a_{31} C_1 + a_{32} C_2 + a_{33} C_3$$

$$= \sum_n a_{3n} C_n. \tag{1.47}$$

By permuting indices to pick up C_1' and C_2', we see that Eq. (1.15) is satisfied and $\mathbf{C}$ is indeed a vector. It should be mentioned here that this vector nature of the cross product is an accident associated with the three-dimensional

[3] Equation (1.46) holds for rotations because they preserve volumes. For a more general orthogonal transformation the r.h.s. of Eq. (1.46) is multiplied by the determinant of the transformation matrix (see Chapter 3 for matrices and determinants).

nature of ordinary space.[4] It will be seen in Chapter 2 that the cross product may also be treated as a second-rank antisymmetric tensor!

If we define a vector as an ordered triple of numbers (or functions) as in the latter part of Section 1.2, then there is no problem identifying the cross product as a vector. The cross-product operation maps the two triples **A** and **B** into a third triple **C** which by definition is a vector.

We now have two ways of multiplying vectors; a third form appears in Chapter 2. But what about division by a vector? It turns out that the ratio **B/A** is not uniquely specified (Exercise 3.2.21) unless **A** and **B** are also required to be parallel. Hence division of one vector by another is not defined.

EXERCISES

1.4.1   Two vectors **A** and **B** are given by
$$\mathbf{A} = 2\hat{\mathbf{x}} + 4\hat{\mathbf{y}} + 6\hat{\mathbf{z}},$$
$$\mathbf{B} = 3\hat{\mathbf{x}} - 3\hat{\mathbf{y}} - 5\hat{\mathbf{z}}.$$
Compute the scalar and vector products $\mathbf{A} \cdot \mathbf{B}$ and $\mathbf{A} \times \mathbf{B}$.

1.4.2   Prove the law of cosines starting from $\mathbf{A}^2 = (\mathbf{B} - \mathbf{C})^2$.

1.4.3   Starting with $\mathbf{C} = \mathbf{A} + \mathbf{B}$, show that $\mathbf{C} \times \mathbf{C}$ leads to
$$\mathbf{A} \times \mathbf{B} = -\mathbf{B} \times \mathbf{A}.$$

1.4.4   Show that
(a) $(\mathbf{A} - \mathbf{B}) \cdot (\mathbf{A} + \mathbf{B}) = A^2 - B^2$,
(b) $(\mathbf{A} - \mathbf{B}) \times (\mathbf{A} + \mathbf{B}) = 2\mathbf{A} \times \mathbf{B}$.
The distributive laws needed here,
$$\mathbf{A} \cdot (\mathbf{B} + \mathbf{C}) = \mathbf{A} \cdot \mathbf{B} + \mathbf{A} \cdot \mathbf{C}$$
and
$$\mathbf{A} \times (\mathbf{B} + \mathbf{C}) = \mathbf{A} \times \mathbf{B} + \mathbf{A} \times \mathbf{C},$$
may easily be verified (if desired) by expansion in cartesian components.

1.4.5   Given the three vectors,
$$\mathbf{P} = 3\hat{\mathbf{x}} + 2\hat{\mathbf{y}} - \hat{\mathbf{z}},$$
$$\mathbf{Q} = -6\hat{\mathbf{x}} - 4\hat{\mathbf{y}} + 2\hat{\mathbf{z}},$$
$$\mathbf{R} = \hat{\mathbf{x}} - 2\hat{\mathbf{y}} - \hat{\mathbf{z}},$$
find two that are perpendicular and two that are parallel or antiparallel.

1.4.6   If $\mathbf{P} = \hat{\mathbf{x}}P_x + \hat{\mathbf{y}}P_y$ and $\mathbf{Q} = \hat{\mathbf{x}}Q_x + \hat{\mathbf{y}}Q_y$ are any two nonparallel (also non-antiparallel) vectors in the xy-plane, show that $\mathbf{P} \times \mathbf{Q}$ is in the z-direction.

1.4.7   Prove that $(\mathbf{A} \times \mathbf{B}) \cdot (\mathbf{A} \times \mathbf{B}) = (AB)^2 - (\mathbf{A} \cdot \mathbf{B})^2$.

---

[4] Specifically Eq. (1.46) holds only for three-dimensional space. Technically, it is also possible to define a cross product in R^7, seven-dimensional space, but the cross product turns out to have unacceptable (pathological) properties.

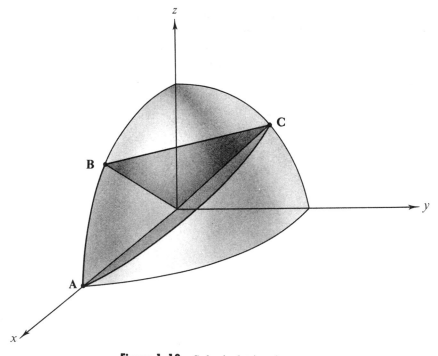

Figure 1.13 Spherical triangle.

1.4.8   Using the vectors
$$\mathbf{P} = \hat{\mathbf{x}} \cos \theta + \hat{\mathbf{y}} \sin \theta,$$
$$\mathbf{Q} = \hat{\mathbf{x}} \cos \varphi - \hat{\mathbf{y}} \sin \varphi,$$
$$\mathbf{R} = \hat{\mathbf{x}} \cos \varphi + \hat{\mathbf{y}} \sin \varphi,$$
prove the familiar trigonometric identities
$$\sin(\theta + \varphi) = \sin \theta \cos \varphi + \cos \theta \sin \varphi,$$
$$\cos(\theta + \varphi) = \cos \theta \cos \varphi - \sin \theta \sin \varphi.$$

1.4.9   (a) Find a vector $\mathbf{A}$ that is perpendicular to
$$\mathbf{U} = 2\hat{\mathbf{x}} + \hat{\mathbf{y}} - \hat{\mathbf{z}},$$
$$\mathbf{V} = \hat{\mathbf{x}} - \hat{\mathbf{y}} + \hat{\mathbf{z}}.$$

(b) What is $\mathbf{A}$ if, in addition to this requirement, we also demand that it have unit magnitude?

1.4.10   If four vectors $\mathbf{a}$, $\mathbf{b}$, $\mathbf{c}$, and $\mathbf{d}$ all lie in the same plane, show that
$$(\mathbf{a} \times \mathbf{b}) \times (\mathbf{c} \times \mathbf{d}) = 0.$$
Hint. Consider the directions of the cross-product vectors.

1.4.11   The coordinates of the three vertices of a triangle are $(2, 1, 5)$, $(5, 2, 8)$, and $(4, 8, 2)$. Compute its area by vector methods.

1.4.12 The vertices of parallelogram $ABCD$ are $(1, 0, 0)$, $(2, -1, 0)$, $(0, -1, 1)$, and $(-1, 0, 1)$ in order. Calculate the vector areas of triangle ABD and of triangle BCD. Are the two vector areas equal?

$$ANS. \quad \text{Area}_{ABD} = -\tfrac{1}{2}(\hat{\mathbf{x}} + \hat{\mathbf{y}} + 2\hat{\mathbf{z}}).$$

1.4.13 The origin and the three vectors $\mathbf{A}$, $\mathbf{B}$, and $\mathbf{C}$ (all of which start at the origin) define a tetrahedron. Taking the *outward* direction as positive, calculate the total vector area of the four tetrahedral surfaces.

Note. In Section 1.11 this result is generalized to any closed surface.

1.4.14 Find the sides and angles of the spherical triangle ABC defined by the three vectors

$$\mathbf{A} = (1, 0, 0),$$

$$\mathbf{B} = \left(\frac{1}{\sqrt{2}}, 0, \frac{1}{\sqrt{2}}\right),$$

and

$$\mathbf{C} = \left(0, \frac{1}{\sqrt{2}}, \frac{1}{\sqrt{2}}\right).$$

Each vector starts from the origin (Fig. 1.13).

1.4.15 Derive the law of sines:

$$\frac{\sin \alpha}{|\mathbf{A}|} = \frac{\sin \beta}{|\mathbf{B}|} = \frac{\sin \gamma}{|\mathbf{C}|}.$$

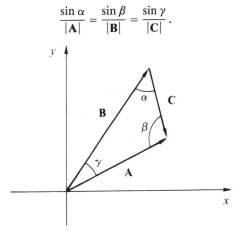

1.4.16 The magnetic induction $\mathbf{B}$ is *defined* by the Lorentz force equation

$$\mathbf{F} = q(\mathbf{v} \times \mathbf{B}).$$

Carrying out three experiments, we find that if

$$\mathbf{v} = \hat{\mathbf{x}}, \qquad \frac{\mathbf{F}}{q} = 2\hat{\mathbf{z}} - 4\hat{\mathbf{y}},$$

$$\mathbf{v} = \hat{\mathbf{y}}, \qquad \frac{\mathbf{F}}{q} = 4\hat{\mathbf{x}} - \hat{\mathbf{z}},$$

and

$$\mathbf{v} = \hat{\mathbf{z}}, \qquad \frac{\mathbf{F}}{q} = \hat{\mathbf{y}} - 2\hat{\mathbf{x}}.$$

From the results of these three separate experiments calculate the magnetic induction $\mathbf{B}$.

1.5 TRIPLE SCALAR PRODUCT, TRIPLE VECTOR PRODUCT

Triple Scalar Product

Sections 1.3 and 1.4 cover the two types of multiplication of interest here. However, there are combinations of three vectors, $\mathbf{A} \cdot (\mathbf{B} \times \mathbf{C})$ and $\mathbf{A} \times (\mathbf{B} \times \mathbf{C})$, which occur with sufficient frequency to deserve further attention. The combination

$$\mathbf{A} \cdot (\mathbf{B} \times \mathbf{C})$$

is known as the triple scalar product. $\mathbf{B} \times \mathbf{C}$ yields a vector which, dotted into $\mathbf{A}$, gives a scalar. We note that $(\mathbf{A} \cdot \mathbf{B}) \times \mathbf{C}$ represents a scalar crossed into a vector, an operation that is not defined. Hence, if we agree to exclude this undefined interpretation, the parentheses may be omitted and the triple scalar product written $\mathbf{A} \cdot \mathbf{B} \times \mathbf{C}$.

Using Eq. (1.38) for the cross product and Eq. (1.24) for the dot product, we obtain

$$\mathbf{A} \cdot \mathbf{B} \times \mathbf{C} = A_x(B_y C_z - B_z C_y) + A_y(B_z C_x - B_x C_z) + A_z(B_x C_y - B_y C_x)$$

$$= \mathbf{B} \cdot \mathbf{C} \times \mathbf{A} = \mathbf{C} \cdot \mathbf{A} \times \mathbf{B}$$

$$= -\mathbf{A} \cdot \mathbf{C} \times \mathbf{B} = -\mathbf{C} \cdot \mathbf{B} \times \mathbf{A} = -\mathbf{B} \cdot \mathbf{A} \times \mathbf{C}, \qquad \text{and so on.}$$

$$(1.48)$$

The high degree of symmetry present in the component expansion should be noted. Every term contains the factors A_i, B_j, and C_k. If i, j, and k are in cyclic order (x, y, z), the sign is positive. If the order is anticyclic, the sign is negative. Further, the dot and the cross may be interchanged,

$$\mathbf{A} \cdot \mathbf{B} \times \mathbf{C} = \mathbf{A} \times \mathbf{B} \cdot \mathbf{C}. \qquad (1.49)$$

A convenient representation of the component expansion of Eq. (1.48) is provided by the determinant

$$\mathbf{A} \cdot \mathbf{B} \times \mathbf{C} = \begin{vmatrix} A_x & A_y & A_z \\ B_x & B_y & B_z \\ C_x & C_y & C_z \end{vmatrix}. \qquad (1.50)$$

The rules for interchanging rows and columns of a determinant[1] provide an immediate verification of the permutations listed in Eq. (1.48), whereas the symmetry of $\mathbf{A}$, $\mathbf{B}$, and $\mathbf{C}$ in the determinant form suggests the relation given in Eq. (1.49).

The triple products encountered in Section 1.4, which showed that $\mathbf{A} \times \mathbf{B}$ was perpendicular to both $\mathbf{A}$ and $\mathbf{B}$, were special cases of the general result (Eq. (1.48)).

---

[1] See Section 3.1 for a summary of the properties of determinants.

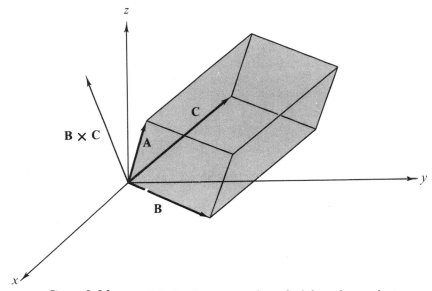

Figure 1.14   Parallelepiped representation of triple scalar product.

The triple scalar product has a direct geometrical interpretation. The three vectors **A, B,** and **C** may be interpreted as defining a parallelepiped (Fig. 1.14).

$$|\mathbf{B} \times \mathbf{C}| = BC \sin \theta$$

$$= \text{area of parallelogram base.} \qquad (1.51)$$

The direction, of course, is normal to the base. Dotting **A** into this means multiplying the base area by the projection of A onto the normal, or base times height. Therefore

$$\mathbf{A} \cdot \mathbf{B} \times \mathbf{C} = \text{volume of parallelepiped defined by } \mathbf{A}, \mathbf{B}, \text{ and } \mathbf{C}.$$

Example 1.5.1: A Parallelepiped

For

$$\mathbf{A} = \hat{\mathbf{x}} + 2\hat{\mathbf{y}} - \hat{\mathbf{z}},$$

$$\mathbf{B} = \hat{\mathbf{y}} + \hat{\mathbf{z}},$$

$$\mathbf{C} = \hat{\mathbf{x}} - \hat{\mathbf{y}},$$

$$\mathbf{A} \cdot \mathbf{B} \times \mathbf{C} = \begin{vmatrix} 1 & 2 & -1 \\ 0 & 1 & 1 \\ 1 & -1 & 0 \end{vmatrix}. \qquad (1.52)$$

By expansion by minors across the top row the determinant equals

$$1(0 + 1) - 2(0 - 1) - 1(0 - 1) = 4.$$

This is the volume of the parallelepiped defined by **A**, **B**, and **C**. The reader should note that $\mathbf{A} \cdot \mathbf{B} \times \mathbf{C}$ may sometimes turn out to be negative! This problem and its interpretation are considered in Chapter 2.

The triple scalar product finds an interesting and important application in the construction of a reciprocal crystal lattice. Let **a**, **b**, and **c** (not necessarily mutually perpendicular) represent the vectors that define a crystal lattice. The distance from one lattice point to another may then be written

$$\mathbf{r} = n_a \mathbf{a} + n_b \mathbf{b} + n_c \mathbf{c}, \tag{1.53}$$

with n_a, n_b, and n_c taking on integral values. With these vectors we may form

$$\mathbf{a}' = \frac{\mathbf{b} \times \mathbf{c}}{\mathbf{a} \cdot \mathbf{b} \times \mathbf{c}}, \qquad \mathbf{b}' = \frac{\mathbf{c} \times \mathbf{a}}{\mathbf{a} \cdot \mathbf{b} \times \mathbf{c}}, \qquad \mathbf{c}' = \frac{\mathbf{a} \times \mathbf{b}}{\mathbf{a} \cdot \mathbf{b} \times \mathbf{c}}. \tag{1.53a}$$

We see that $\mathbf{a}'$ is perpendicular to the plane containing **b** and **c** and has a magnitude proportional to a^{-1}. In fact, we can readily show that

$$\mathbf{a}' \cdot \mathbf{a} = \mathbf{b}' \cdot \mathbf{b} = \mathbf{c}' \cdot \mathbf{c} = 1, \tag{1.53b}$$

whereas

$$\mathbf{a}' \cdot \mathbf{b} = \mathbf{a}' \cdot \mathbf{c} = \mathbf{b}' \cdot \mathbf{a} = \mathbf{b}' \cdot \mathbf{c} = \mathbf{c}' \cdot \mathbf{a} = \mathbf{c}' \cdot \mathbf{b} = 0. \tag{1.53c}$$

It is from Eqs. (1.53b) and (1.53c) that the name reciprocal lattice is derived. The mathematical space in which this reciprocal lattice exists is sometimes called a Fourier space, on the basis of relations to the Fourier analysis of Chapters 14 and 15. This reciprocal lattice is useful in problems involving the scattering of waves from the various planes in a crystal. Further details may be found in R. B. Leighton's *Principles of Modern Physics*, pp. 440–448 [New York: McGraw–Hill (1959)].

Triple Vector Product

The second triple product of interest is $\mathbf{A} \times (\mathbf{B} \times \mathbf{C})$, which is a vector. Here the parentheses must be retained, as may be seen from a special case $(\hat{\mathbf{x}} \times \hat{\mathbf{x}}) \times \hat{\mathbf{y}} = 0$, while $\hat{\mathbf{x}} \times (\hat{\mathbf{x}} \times \hat{\mathbf{y}}) = \hat{\mathbf{x}} \times \hat{\mathbf{z}} = -\hat{\mathbf{y}}$. The triple product vector is perpendicular to **A** and to $\mathbf{B} \times \mathbf{C}$. The plane defined by **B** and **C** is perpendicular to $\mathbf{B} \times \mathbf{C}$ and so the triple product lies in this plane (see Fig. 1.15)

$$\mathbf{A} \times (\mathbf{B} \times \mathbf{C}) = x\mathbf{B} + y\mathbf{C}. \tag{1.54}$$

Multiplying Eq. (1.54) by **A** gives zero for the left-hand side, so that $x\mathbf{A} \cdot \mathbf{B} + y\mathbf{A} \cdot \mathbf{C} = 0$. Hence $x = z\mathbf{A} \cdot \mathbf{C}$ and $y = -z\mathbf{A} \cdot \mathbf{B}$ for a suitable z. Substituting these values into Eq. (1.54) gives

$$\mathbf{A} \times (\mathbf{B} \times \mathbf{C}) = z(\mathbf{B}\, \mathbf{A} \cdot \mathbf{C} - \mathbf{C}\, \mathbf{A} \cdot \mathbf{B}); \tag{1.55}$$

we want to show that $z = 1$ in Eq. (1.55), an important relation sometimes known as the *BAC–CAB* rule. Since Eq. (1.55) is linear in A, B, and C, z is

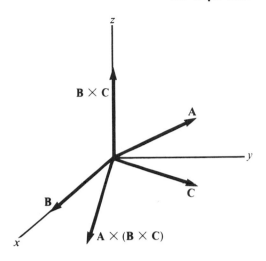

Figure 1.15 **B** and **C** are in the xy-plane. **B** × **C** is perpendicular to the xy-plane and is shown here along the z-axis. Then **A** × (**B** × **C**) is perpendicular to the z-axis and therefore is back in the xy-plane.

independent of these magnitudes. That is, we only need to show that $z = 1$ for unit vectors $\hat{\mathbf{A}}$, $\hat{\mathbf{B}}$, $\hat{\mathbf{C}}$. Let us denote $\hat{\mathbf{B}} \cdot \hat{\mathbf{C}} = \cos \alpha$, $\hat{\mathbf{C}} \cdot \hat{\mathbf{A}} = \cos \beta$, $\hat{\mathbf{A}} \cdot \hat{\mathbf{B}} = \cos \gamma$, and square Eq. (1.55) to obtain

$$[\hat{\mathbf{A}} \times (\hat{\mathbf{B}} \times \hat{\mathbf{C}})]^2 = \hat{\mathbf{A}}^2(\hat{\mathbf{B}} \times \hat{\mathbf{C}})^2 - [\hat{\mathbf{A}} \cdot (\hat{\mathbf{B}} \times \hat{\mathbf{C}})]^2 = 1 - \cos^2\alpha - [\hat{\mathbf{A}} \cdot (\hat{\mathbf{B}} \times \hat{\mathbf{C}})]^2$$

$$= z^2[(\hat{\mathbf{A}} \cdot \hat{\mathbf{C}})^2 + (\hat{\mathbf{A}} \cdot \hat{\mathbf{B}})^2 - 2\hat{\mathbf{A}} \cdot \hat{\mathbf{B}}\,\hat{\mathbf{A}} \cdot \hat{\mathbf{C}}\,\hat{\mathbf{B}} \cdot \hat{\mathbf{C}}]$$

$$= z^2(\cos^2\beta + \cos^2\gamma - 2\cos\alpha\cos\beta\cos\gamma), \qquad (1.56)$$

using $(\hat{\mathbf{A}} \times \hat{\mathbf{B}})^2 = \hat{\mathbf{A}}^2\hat{\mathbf{B}}^2 - (\hat{\mathbf{A}} \cdot \hat{\mathbf{B}})^2$ repeatedly. Consequently, the (squared) volume spanned by $\hat{\mathbf{A}}$, $\hat{\mathbf{B}}$, $\hat{\mathbf{C}}$ that occurs in Eq. (1.56) can be written as

$$[\hat{\mathbf{A}} \cdot \hat{\mathbf{B}} \times \hat{\mathbf{C}}]^2 = 1 - \cos^2\alpha - z^2\,(\cos^2\beta + \cos^2\gamma - 2\cos\alpha\cos\beta\cos\gamma).$$

Here $z^2 = 1$, as this volume is symmetric in α, β, γ. That is, $z = \pm 1$ and independent of $\hat{\mathbf{A}}, \hat{\mathbf{B}}, \hat{\mathbf{C}}$. Using again the special case $\hat{\mathbf{x}} \times (\hat{\mathbf{x}} \times \hat{\mathbf{y}}) = -\hat{\mathbf{y}}$ in Eq. (1.55) finally gives $z = 1$.

An alternate derivation using the Levi–Civita ε_{ijk} of Section 2 is the topic of Exercise 2.9.8.

It might be noted here that as vectors are independent of the coordinates so a vector equation is independent of the particular coordinate system. The coordinate system only determines the components. If the vector equation can be established in cartesian coordinates, it is established and valid in any of the coordinate systems to be introduced in Chapter 2. Thus, Eq. (1.55) may be verified by a direct though not very elegant method of expanding into cartesian components (see Exercise 1.5.2).

Example 1.5.2: A Triple Vector Product

By using the three vectors given in Example 1.5.1, we obtain

$$\mathbf{A} \times (\mathbf{B} \times \mathbf{C}) = (\hat{\mathbf{y}} + \hat{\mathbf{z}})(1 - 2) - (\hat{\mathbf{x}} - \hat{\mathbf{y}})(2 - 1)$$

$$= -\hat{\mathbf{x}} - \hat{\mathbf{z}}$$

by Eq. (1.55). In detail,

$$\mathbf{B} \times \mathbf{C} = \begin{vmatrix} \hat{\mathbf{x}} & \hat{\mathbf{y}} & \hat{\mathbf{z}} \\ 0 & 1 & 1 \\ 1 & -1 & 0 \end{vmatrix} = \hat{\mathbf{x}} + \hat{\mathbf{y}} - \hat{\mathbf{z}}$$

and

$$\mathbf{A} \times (\mathbf{B} \times \mathbf{C}) = \begin{vmatrix} \hat{\mathbf{x}} & \hat{\mathbf{y}} & \hat{\mathbf{z}} \\ 1 & 2 & -1 \\ 1 & 1 & -1 \end{vmatrix} = -\hat{\mathbf{x}} - \hat{\mathbf{z}}.$$

Other, more complicated, products may be simplified by using these forms of the triple scalar and triple vector products.

EXERCISES

1.5.1

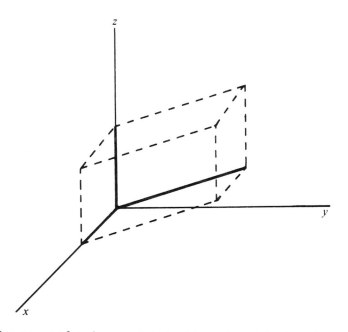

One vertex of a glass parallelepiped is at the origin. The three adjacent vertices are at (3, 0, 0), (0, 0, 2), and (0, 3, 1). All lengths are in centimeters. Calculate the number of cubic centimeters of glass in the parallelepiped by using the triple scalar product.

1.5.2   Verify the expansion of the triple vector product

$$\mathbf{A} \times (\mathbf{B} \times \mathbf{C}) = \mathbf{B}(\mathbf{A} \cdot \mathbf{C}) - \mathbf{C}(\mathbf{A} \cdot \mathbf{B})$$

by direct expansion in cartesian coordinates.

1.5.3 Show that the first step in Eq. (1.43), which is
$$(\mathbf{A} \times \mathbf{B}) \cdot (\mathbf{A} \times \mathbf{B}) = A^2 B^2 - (\mathbf{A} \cdot \mathbf{B})^2,$$
is consistent with the *BAC-CAB* rule for a triple vector product.

1.5.4 Given the three vectors **A**, **B**, and **C**,
$$\mathbf{A} = \hat{\mathbf{x}} + \hat{\mathbf{y}},$$
$$\mathbf{B} = \hat{\mathbf{y}} + \hat{\mathbf{z}},$$
$$\mathbf{C} = \hat{\mathbf{x}} - \hat{\mathbf{z}}.$$
(a) Compute the triple scalar product, $\mathbf{A} \cdot \mathbf{B} \times \mathbf{C}$. Noting that $\mathbf{A} = \mathbf{B} + \mathbf{C}$, give a geometric interpretation of your result for the triple scalar product.
(b) Compute $\mathbf{A} \times (\mathbf{B} \times \mathbf{C})$.

1.5.5 The angular momentum **L** of a particle is given by $\mathbf{L} = \mathbf{r} \times \mathbf{p} = m\mathbf{r} \times \mathbf{v}$, where **p** is the linear momentum. With linear and angular velocity related by $\mathbf{v} = \boldsymbol{\omega} \times \mathbf{r}$, show that
$$\mathbf{L} = mr^2[\boldsymbol{\omega} - \hat{\mathbf{r}}(\hat{\mathbf{r}} \cdot \boldsymbol{\omega})].$$
Here $\hat{\mathbf{r}}$ is a unit vector in the **r** direction. For $\mathbf{r} \cdot \boldsymbol{\omega} = 0$ this reduces to $\mathbf{L} = I\boldsymbol{\omega}$, with the moment of inertia I given by mr^2. In Section 3.5 this result is generalized to form an inertia tensor.

1.5.6 The kinetic energy of a single particle is given by $T = \frac{1}{2}mv^2$. For rotational motion this becomes $\frac{1}{2}m(\boldsymbol{\omega} \times \mathbf{r})^2$. Show that
$$T = \frac{1}{2}m[r^2\omega^2 - (\mathbf{r} \cdot \boldsymbol{\omega})^2].$$
For $\mathbf{r} \cdot \boldsymbol{\omega} = 0$ this reduces to $T = \frac{1}{2}I\omega^2$ with the moment of inertia I given by mr^2.

1.5.7 Show that[1]
$$\mathbf{a} \times (\mathbf{b} \times \mathbf{c}) + \mathbf{b} \times (\mathbf{c} \times \mathbf{a}) + \mathbf{c} \times (\mathbf{a} \times \mathbf{b}) = 0.$$

1.5.8 A vector **A** is decomposed into a radial vector $\mathbf{A}_r$ and a tangential vector $\mathbf{A}_t$. If $\hat{\mathbf{r}}$ is a unit vector in the radial direction, show that
(a) $\mathbf{A}_r = \hat{\mathbf{r}}(\mathbf{A} \cdot \hat{\mathbf{r}})$ and
(b) $\mathbf{A}_t = -\hat{\mathbf{r}} \times (\hat{\mathbf{r}} \times \mathbf{A})$.

1.5.9 Prove that a necessary and sufficient condition for the three (nonvanishing) vectors **A**, **B**, and **C** to be coplanar is the vanishing of the triple scalar product
$$\mathbf{A} \cdot \mathbf{B} \times \mathbf{C} = 0.$$

1.5.10 Three vectors **A**, **B**, and **C** are given by
$$\mathbf{A} = 3\hat{\mathbf{x}} - 2\hat{\mathbf{y}} + 2\hat{\mathbf{z}},$$
$$\mathbf{B} = 6\hat{\mathbf{x}} + 4\hat{\mathbf{y}} - 2\hat{\mathbf{z}},$$
$$\mathbf{C} = -3\hat{\mathbf{x}} - 2\hat{\mathbf{y}} - 4\hat{\mathbf{z}}.$$
Compute the values of $\mathbf{A} \cdot \mathbf{B} \times \mathbf{C}$ and $\mathbf{A} \times (\mathbf{B} \times \mathbf{C})$, $\mathbf{C} \times (\mathbf{A} \times \mathbf{B})$ and $\mathbf{B} \times (\mathbf{C} \times \mathbf{A})$.

---

[1] This is Jacobi's identity for vector products; for commutators it is important in the context of Lie algebras (see Eq. (4.16) in Section 4.2).

1.5.11   Vector D is a linear combination of three noncoplanar (and nonorthogonal) vectors:

$$D = aA + bB + cC.$$

Show that the coefficients are given by a ratio of triple scalar products,

$$a = \frac{D \cdot B \times C}{A \cdot B \times C}, \quad \text{and so on.}$$

1.5.12   Show that

$$(A \times B) \cdot (C \times D) = (A \cdot C)(B \cdot D) - (A \cdot D)(B \cdot C).$$

1.5.13   Show that

$$(A \times B) \times (C \times D) = (A \cdot B \times D)C - (A \cdot B \times C)D.$$

1.5.14   For a *spherical* triangle such as pictured in Fig. 1.13 show that

$$\frac{\sin A}{\sin \overline{BC}} = \frac{\sin B}{\sin \overline{CA}} = \frac{\sin C}{\sin \overline{AB}}.$$

Here $\sin A$ is the sine of the included angle at A while $\overline{BC}$ is the side opposite (in radians).
Hint. Exercise 1.5.13 will be useful.

1.5.15   Given

$$a' = \frac{b \times c}{a \cdot b \times c}, \quad b' = \frac{c \times a}{a \cdot b \times c}, \quad c' = \frac{a \times b}{a \cdot b \times c} \text{ and } a \cdot b \times c \neq 0,$$

show that

(a) $x' \cdot y = \delta_{xy}$, $(x, y = a, b, c)$,
(b) $a' \cdot b' \times c' = (a \cdot b \times c)^{-1}$,
(c) $a = \dfrac{b' \times c'}{a' \cdot b' \times c'}$.

1.5.16   If $x' \cdot y = \delta_{xy}$, $(x, y = a, b, c)$, prove that

$$a' = \frac{b \times c}{a \cdot b \times c}.$$

(This is the converse of Problem 1.5.15.)

1.5.17   Show that any vector V may be expressed in terms of the reciprocal vectors a', b', c' by

$$V = (V \cdot a)a' + (V \cdot b)b' + (V \cdot c)c'.$$

1.5.18   An electric charge q_1 moving with velocity v_1 produces a magnetic induction B given by

$$B = \frac{\mu_0}{4\pi} q_1 \frac{v_1 \times \hat{r}}{r^2} \quad \text{(mks units),}$$

where $\hat{r}$ points from q_1 to the point at which B is measured (Biot and Savart law).

(a) Show that the magnetic force on a second charge q_2, velocity $\mathbf{v}_2$, is given by the triple vector product

$$\mathbf{F}_2 = \frac{\mu_0}{4\pi} \frac{q_1 q_2}{r^2} \mathbf{v}_2 \times (\mathbf{v}_1 \times \hat{\mathbf{r}}).$$

(b) Write out the corresponding magnetic force $\mathbf{F}_1$ that q_2 exerts on q_1. Define your unit radial vector. How do $\mathbf{F}_1$ and $\mathbf{F}_2$ compare?

(c) Calculate $\mathbf{F}_1$ and $\mathbf{F}_2$ for the case of q_1 and q_2 moving along parallel trajectories side by side.

> *ANS.* (b) $\mathbf{F}_1 = -\dfrac{\mu_0}{4\pi} \dfrac{q_1 q_2}{r^2} \mathbf{v}_1 \times (\mathbf{v}_2 \times \hat{\mathbf{r}}).$
>
> In general, there is no simple relation between $\mathbf{F}_1$ and $\mathbf{F}_2$. Specifically, Newton's third law, $\mathbf{F}_1 = -\mathbf{F}_2$, does not hold.
>
> (c) $\mathbf{F}_1 = \dfrac{\mu_0}{4\pi} \dfrac{q_1 q_2}{r^2} v^2 \hat{\mathbf{r}} = -\mathbf{F}_2.$
>
> Mutual attraction.

1.6 GRADIENT, ∇

Suppose that $\varphi(x, y, z)$ is a scalar point function, that is, a function whose value depends on the values of the coordinates (x, y, z). As a scalar, it must have the same value at a given fixed point in space, independent of the rotation of our coordinate system, or

$$\varphi'(x_1', x_2', x_3') = \varphi(x_1, x_2, x_3). \tag{1.57}$$

By differentiating with respect to x_i' we obtain

$$\frac{\partial \varphi'(x_1', x_2', x_3')}{\partial x_i'} = \frac{\partial \varphi(x_1, x_2, x_3)}{\partial x_i'} = \sum_j \frac{\partial \varphi}{\partial x_j} \frac{\partial x_j}{\partial x_i'} = \sum_j a_{ij} \frac{\partial \varphi}{\partial x_j} \tag{1.58}$$

by the rules of partial differentiation and Eq. (1.16). But comparison with Eq. (1.17), the vector transformation law, now shows that we have *constructed* a vector with components $\partial \varphi / \partial x_j$. This vector we label the gradient of φ.

A convenient symbolism is

$$\nabla \varphi = \hat{\mathbf{x}} \frac{\partial \varphi}{\partial x} + \hat{\mathbf{y}} \frac{\partial \varphi}{\partial y} + \hat{\mathbf{z}} \frac{\partial \varphi}{\partial z} \tag{1.59}$$

or

$$\nabla = \hat{\mathbf{x}} \frac{\partial}{\partial x} + \hat{\mathbf{y}} \frac{\partial}{\partial y} + \hat{\mathbf{z}} \frac{\partial}{\partial z}. \tag{1.60}$$

$\nabla \varphi$ (or del φ) is our gradient of the scalar φ, whereas ∇ (del) itself is a vector differential operator (available to operate on or to differentiate a scalar φ). All the relationships for ∇ (del) can be derived from the hybrid nature of del in terms of both the partial derivatives and its vector nature.

Example 1.6.1 The Gradient of a Function of r

Let us calculate the gradient of $f(r) = f(\sqrt{x^2 + y^2 + z^2})$.

$$\nabla f(r) = \hat{x}\frac{\partial f(r)}{\partial x} + \hat{y}\frac{\partial f(r)}{\partial y} + \hat{z}\frac{\partial f(r)}{\partial z}.$$

Now $f(r)$ depends on x through the dependence of r on x. Therefore[1]

$$\frac{\partial f(r)}{\partial x} = \frac{df(r)}{dr} \cdot \frac{\partial r}{\partial x}.$$

From r as a function of x, y, z

$$\frac{\partial r}{\partial x} = \frac{\partial(x^2 + y^2 + z^2)^{1/2}}{\partial x} = \frac{x}{(x^2 + y^2 + z^2)^{1/2}} = \frac{x}{r}.$$

Therefore

$$\frac{\partial f(r)}{\partial x} = \frac{df(r)}{dr} \cdot \frac{x}{r}.$$

Permuting coordinates $(x \to y, y \to z, z \to x)$ to obtain the y and z derivatives, we get

$$\nabla f(r) = (\hat{x}x + \hat{y}y + \hat{z}z)\frac{1}{r}\frac{df}{dr}$$

$$= \frac{\mathbf{r}}{r}\frac{df}{dr}$$

$$= \hat{\mathbf{r}}\frac{df}{dr}.$$

Here $\hat{\mathbf{r}}$ is a unit vector $(\mathbf{r}/r)$ is the *positive* radial direction. The gradient of a function of r is a vector in the (positive or negative) radial direction. In Section 2.5, $\hat{\mathbf{r}}$ is seen as one of the three orthonormal unit vectors of spherical polar coordinates and $\hat{\mathbf{r}}\partial/\partial r$ as the radial component of ∇ (compare with Eq. (2.44)).

A Geometrical Interpretation

One immediate application of $\nabla\varphi$ is to dot it into an increment of length

$$d\mathbf{r} = \hat{x}\,dx + \hat{y}\,dy + \hat{z}\,dz. \qquad (1.61)$$

---

[1] This is a special case of the chain rule of partial differentiation:

$$\frac{\partial f(r, \theta, \varphi)}{\partial x} = \frac{\partial f}{\partial r}\frac{\partial r}{\partial x} + \frac{\partial f}{\partial \theta}\frac{\partial \theta}{\partial x} + \frac{\partial f}{\partial \varphi}\frac{\partial \varphi}{\partial x}.$$

Here $\partial f/\partial\theta = \partial f/\partial\varphi = 0$, $\partial f/\partial r \to df/dr$.

Thus we obtain

$$(\nabla\varphi) \cdot d\mathbf{r} = \frac{\partial\varphi}{\partial x}\,dx + \frac{\partial\varphi}{\partial y}\,dy + \frac{\partial\varphi}{\partial z}\,dz$$

$$= d\varphi, \tag{1.62}$$

the change in the scalar function φ corresponding to a change in position $d\mathbf{r}$. Now consider P and Q to be two points on a surface $\varphi(x, y, z) = C$, a constant. These points are chosen so that Q is a distance $d\mathbf{r}$ from P. Then moving from P to Q, the change in $\varphi(x, y, z) = C$ is given by

$$d\varphi = (\nabla\varphi) \cdot d\mathbf{r} = 0, \tag{1.63}$$

since we stay on the surface $\varphi(x, y, z) = C$. This shows that $\nabla\varphi$ is perpendicular to $d\mathbf{r}$. Since $d\mathbf{r}$ may have any direction from P *as long as it stays in the surface* φ, point Q being restricted to the surface, but having arbitrary direction, $\nabla\varphi$ is seen as normal to the surface $\varphi =$ constant (Fig. 1.16).

If we now permit $d\mathbf{r}$ to take us from one surface $\varphi = C_1$ to an adjacent surface $\varphi = C_2$ (Fig. 1.17a),

$$d\varphi = C_1 - C_1 = \Delta C = (\nabla\varphi) \cdot d\mathbf{r}. \tag{1.64}$$

For a given $d\varphi$, $|d\mathbf{r}|$ is a minimum when it is chosen parallel to $\nabla\varphi$ ($\cos\theta = 1$); or, for a given $|d\mathbf{r}|$, the change in the scalar function φ is maximized by choosing $d\mathbf{r}$ parallel to $\nabla\varphi$. *This identifies* $\nabla\varphi$ *as a vector having the direction*

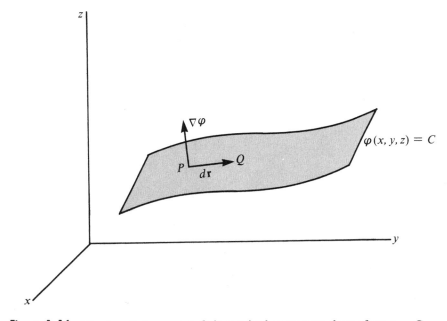

Figure 1.16   The length increment $d\mathbf{r}$ is required to stay on the surface $\varphi = C$.

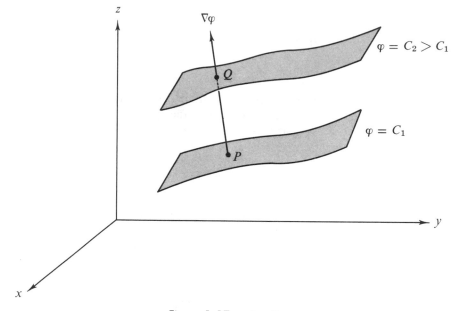

Figure 1.17a Gradient.

of the maximum space rate of change of φ, an identification that will be useful in Chapter 2 when we consider noncartesian coordinate systems.

This identification of $\nabla\varphi$ may also be developed by using the calculus of variations subject to a constraint, Exercise 17.6.9.

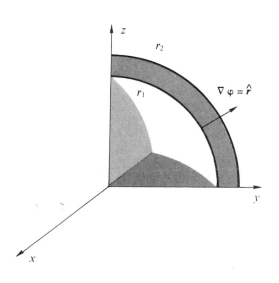

Figure 1.17b Gradient for $\varphi(x, y, z) = (x^2 + y^2 + z^2)^{1/2}$, spherical shells: $(x^2 + y^2 + z^2)^{1/2} = r_2 = C_2$, $(x^2 + y^2 + z^2)^{1/2} = r_1 = C_1$.

Example 1.6.2

As a specific example of the foregoing, and as an extension of Example 1.6.1, we consider the surfaces consisting of concentric spherical shells, Fig. 1.17b.

We have

$$\varphi(x, y, z) = (x^2 + y^2 + z^2)^{1/2} = r_i = C_i,$$

where r_i is the radius equal to C_i, our constant. $\Delta C = \Delta\varphi = \Delta r_i$, the distance between two shells. From Example 1.6.1

$$\nabla\varphi(r) = \hat{\mathbf{r}}\frac{d\varphi(r)}{dr} = \hat{\mathbf{r}}.$$

The gradient is in the radial direction and is normal to the spherical surface $\varphi = C$.

The gradient of a scalar is of extreme importance in physics in expressing the relation between a force field and a potential field.

$$\text{force} = -\nabla \text{ (potential)}. \tag{1.65}$$

This is illustrated by both gravitational and electrostatic fields, among others. Readers should note that the minus sign in Eq. (1.65) results in water flowing downhill rather than uphill! We reconsider Eq. (1.65) in a broader context in Section 1.13.

EXERCISES

1.6.1   If $S(x, y, z) = (x^2 + y^2 + z^2)^{-3/2}$, find

   (a) ∇S at the point $(1, 2, 3)$;
   (b) the magnitude of the gradient of S, $|\nabla S|$ at $(1, 2, 3)$; and
   (c) the direction cosines of ∇S at $(1, 2, 3)$.

1.6.2   (a) Find a unit vector perpendicular to the surface

$$x^2 + y^2 + z^2 = 3$$

   at the point $(1, 1, 1)$.
   (b) Derive the equation of the plane tangent to the surface at $(1, 1, 1)$.

ANS.   (a) $(\hat{\mathbf{x}} + \hat{\mathbf{y}} + \hat{\mathbf{z}})/\sqrt{3}$.
(b) $x + y + z = 3$.

1.6.3   Given a vector $\mathbf{r}_{12} = \hat{\mathbf{x}}(x_1 - x_2) + \hat{\mathbf{y}}(y_1 - y_2) + \hat{\mathbf{z}}(z_1 - z_2)$, show that $\nabla_1 r_{12}$ (gradient with respect to x_1, y_1, and z_1 of the magnitude r_{12}) is a unit vector in the direction of $\mathbf{r}_{12}$.

1.6.4   If a vector function $\mathbf{F}$ depends on both space coordinates (x, y, z) and time t, show that

$$d\mathbf{F} = (d\mathbf{r} \cdot \nabla)\mathbf{F} + \frac{\partial\mathbf{F}}{\partial t}\,dt.$$

1.6.5   Show that $\nabla(uv) = v\nabla u + u\nabla v$, where u and v are differentiable scalar functions of x, y, and z.

1.6.6   (a) Show that a necessary and sufficient condition that $u(x, y, z)$ and $v(x, y, z)$ are related by some function $f(u, v) = 0$ is that $(\nabla u) \times (\nabla v) = 0$.
        (b) If $u = u(x, y)$ and $v = v(x, y)$, show that the condition $(\nabla u) \times (\nabla v) = 0$ leads to the two-dimensional Jacobian

$$J\left(\frac{u, v}{x, y}\right) = \begin{vmatrix} \dfrac{\partial u}{\partial x} & \dfrac{\partial u}{\partial y} \\[2mm] \dfrac{\partial v}{\partial x} & \dfrac{\partial v}{\partial y} \end{vmatrix} = 0.$$

The functions u and v are assumed differentiable.

1.7 DIVERGENCE, $\nabla\cdot$

Differentiating a vector function is a simple extension of differentiating scalar quantities. Suppose $\mathbf{r}(t)$ describes the position of a satellite at some time t. Then, for differentiation with respect to time,

$$\frac{d\mathbf{r}(t)}{dt} = \lim_{\Delta t \to 0} \frac{\mathbf{r}(t + \Delta t) - \mathbf{r}(t)}{\Delta t}$$

$$= \mathbf{v}, \qquad \text{linear velocity.}$$

Graphically, we again have the slope of a curve, orbit, or trajectory, as shown in Fig. 1.18.

If we resolve $\mathbf{r}(t)$ into its cartesian components, $d\mathbf{r}/dt$ always reduces directly to a vector sum of not more than three (for three-dimensional space) scalar derivatives. In other coordinate systems (Chapter 2) the situation is a little more complicated, for the unit vectors are no longer constant in direction. Differentiation with respect to the space coordinates is handled in the same way as differentiation with respect to time, as seen in the following paragraphs.

In Section 1.6, ∇ was defined as a vector operator. Now, paying careful attention to both its vector and its differential properties, we let it operate on a vector. First, as a vector we dot it into a second vector to obtain

$$\nabla \cdot \mathbf{V} = \frac{\partial V_x}{\partial x} + \frac{\partial V_y}{\partial y} + \frac{\partial V_z}{\partial z}, \tag{1.66}$$

known as the divergence of $\mathbf{V}$. This is a scalar, as discussed in Section 1.3.

Example 1.7.1

Calculate $\nabla \cdot \mathbf{r}$

$$\nabla \cdot \mathbf{r} = \left(\hat{\mathbf{x}}\frac{\partial}{\partial x} + \hat{\mathbf{y}}\frac{\partial}{\partial y} + \hat{\mathbf{z}}\frac{\partial}{\partial z}\right) \cdot (\hat{\mathbf{x}}x + \hat{\mathbf{y}}y + \hat{\mathbf{z}}z) = \frac{\partial x}{\partial x} + \frac{\partial y}{\partial y} + \frac{\partial z}{\partial z},$$

or

$$\nabla \cdot \mathbf{r} = 3.$$

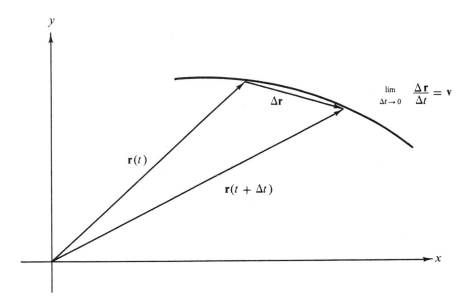

Figure 1.18   Differentiation of a vector.

Example 1.7.2
Generalizing Example 1.7.1,

$$\nabla \cdot \mathbf{r} f(r) = \frac{\partial}{\partial x}[xf(r)] + \frac{\partial}{\partial y}[yf(r)] + \frac{\partial}{\partial z}[zf(r)]$$

$$= 3f(r) + \frac{x^2}{r}\frac{df}{dr} + \frac{y^2}{r}\frac{df}{dr} + \frac{z^2}{r}\frac{df}{dr}$$

$$= 3f(r) + r\frac{df}{dr}.$$

The manipulation of the partial derivatives leading to the second equation in Example 1.7.2 is discussed in Example 1.6.1.

In particular, if $f(r) = r^{n-1}$,

$$\nabla \cdot \mathbf{r} r^{n-1} = \nabla \cdot \hat{\mathbf{r}} r^n$$

$$= 3r^{n-1} + (n-1)r^{n-1}$$

$$= (n+2)r^{n-1}. \tag{1.66a}$$

This divergence vanishes for $n = -2$, except at $r = 0$, an important fact in Section 1.14.

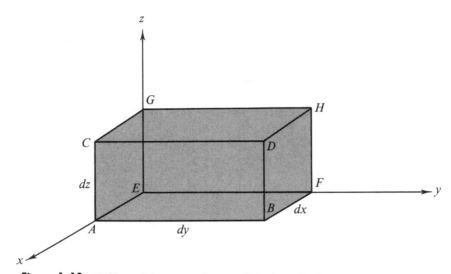

Figure 1.19 Differential rectangular parallelepiped (in first or positive octant).

A Physical Interpretation

To develop a feeling for the physical significance of the divergence, consider $\nabla \cdot (\rho \mathbf{v})$ with $\mathbf{v}(x, y, z)$, the velocity of a compressible fluid and $\rho(x, y, z)$, its density at point (x, y, z). If we consider a small volume $dx\,dy\,dz$ (Fig. 1.19), the fluid flowing into this volume per unit time (positive x-direction) through the face $EFGH$ is (rate of flow in)$_{EFGH} = \rho v_x|_{x=0}\,dy\,dz$. The components of the flow ρv_y and ρv_z tangential to this face contribute nothing to the flow through this face. The rate of flow out (still positive x-direction) through face $ABCD$ is $\rho v_x|_{x=dx}\,dy\,dz$. To compare these flows and to find the net flow out, we expand this last result in a Maclaurin series,[1] Section 5.6. This yields

$$\text{(rate of flow out)}_{ABCD} = \rho v_x|_{x=dx}\,dy\,dz$$

$$= \left[\rho v_x + \frac{\partial}{\partial x}(\rho v_x)\,dx\right]_{x=0} dy\,dz.$$

Here the derivative term is a first correction term allowing for the possibility of nonuniform density or velocity or both.[2] The zero-order term $\rho v_x|_{x=0}$ (corresponding to uniform flow) cancels out.

$$\text{Net rate of flow out}\,|_x = \frac{\partial}{\partial x}(\rho v_x)\,dx\,dy\,dz.$$

[1] A Maclaurin expansion for a single variable is given by Eq. (5.88), Section 5.6. Here we have the increment x of Eq. (5.88) replaced by dx. We show a partial derivative with respect to x since ρv_x may also depend on y and z.
[2] Strictly speaking, ρv_x is averaged over face $EFGH$ and the expression $\rho v_x + (\partial/\partial x)(\rho v_x)\,dx$ is similarly averaged over face $ABCD$. Using an arbitrarily small differential volume, we find that the averages reduce to the values employed here.

Equivalently, we can arrive at this result by

$$\lim_{\Delta x \to 0} \frac{\rho v_x(\Delta x, 0, 0) - \rho v_x(0, 0, 0)}{\Delta x} \equiv \left.\frac{\partial[\rho v_x(x, y, z)]}{\partial x}\right|_{0,0,0}.$$

Now the x-axis is not entitled to any preferred treatment. The preceding result for the two faces perpendicular to the x-axis must hold for the two faces perpendicular to the y-axis, with x replaced by y and the corresponding changes for y and z: $y \to z, z \to x$. This is a cyclic permutation of the coordinates. A further cyclic permutation yields the result for the remaining two faces of our parallelepiped. Adding the net rate of flow out for all three pairs of surfaces of our volume element, we have

$$\begin{array}{l} \text{net flow out} \\ \text{(per unit time)} \end{array} = \left[\frac{\partial}{\partial x}(\rho v_x) + \frac{\partial}{\partial y}(\rho v_y) + \frac{\partial}{\partial z}(\rho v_z)\right] dx\, dy\, dz$$

$$= \mathbf{\nabla} \cdot (\rho \mathbf{v})\, dx\, dy\, dz. \tag{1.67}$$

Therefore the net flow of our compressible fluid out of the volume element $dx\, dy\, dz$ per unit volume per unit time is $\mathbf{\nabla} \cdot (\rho \mathbf{v})$. Hence the name *divergence*. A direct application is in the continuity equation

$$\frac{\partial \rho}{\partial t} + \mathbf{\nabla} \cdot (\rho \mathbf{v}) = 0, \tag{1.68}$$

which simply states that a net flow out of the volume results in a decreased density inside the volume. Note that in Eq. (1.68), ρ is considered to be a possible function of time as well as of space: $\rho(x, y, z, t)$. The divergence appears in a wide variety of physical problems, ranging from a probability current density in quantum mechanics to neutron leakage in a nuclear reactor.

The combination $\mathbf{\nabla} \cdot (f\mathbf{V})$, in which f is a scalar function and $\mathbf{V}$ a vector function, may be written

$$\mathbf{\nabla} \cdot (f\mathbf{V}) = \frac{\partial}{\partial x}(fV_x) + \frac{\partial}{\partial y}(fV_y) + \frac{\partial}{\partial z}(fV_z)$$

$$= \frac{\partial f}{\partial x}V_x + f\frac{\partial V_x}{\partial x} + \frac{\partial f}{\partial y}V_y + f\frac{\partial V_y}{\partial y} + \frac{\partial f}{\partial z}V_z + f\frac{\partial V_z}{\partial z}$$

$$= (\mathbf{\nabla} f) \cdot \mathbf{V} + f\mathbf{\nabla} \cdot \mathbf{V}, \tag{1.68a}$$

which is just what we would expect for the derivative of a product. Notice that $\mathbf{\nabla}$ as a differential operator differentiates both f and $\mathbf{V}$; as a vector it is dotted into $\mathbf{V}$ (in each term).

If we have the special case of the divergence of a vector vanishing,

$$\mathbf{\nabla} \cdot \mathbf{B} = 0, \tag{1.69}$$

the vector $\mathbf{B}$ is said to be *solenoidal*, the term coming from the example in which $\mathbf{B}$ is the magnetic induction and Eq. (1.69) appears as one of Maxwell's equations. When a vector is solenoidal it may be written as the curl of another vector known as the vector potential. In Section 1.13 we shall calculate such a vector potential.

EXERCISES

1.7.1 For a particle moving in a circular orbit $\mathbf{r} = \hat{\mathbf{x}} r \cos \omega t + \hat{\mathbf{y}} r \sin \omega t$,

(a) evaluate $\mathbf{r} \times \dot{\mathbf{r}}$.

(b) Show that $\ddot{\mathbf{r}} + \omega^2 \mathbf{r} = 0$.

The radius r and the angular velocity ω are constant.

$\qquad\qquad\qquad\qquad\qquad\qquad$ *ANS.* (a) $\hat{\mathbf{z}} \omega r^2$.

Note. $\dot{\mathbf{r}} = d\mathbf{r}/dt$, $\ddot{\mathbf{r}} = d^2\mathbf{r}/dt^2$.

1.7.2 Vector $\mathbf{A}$ satisfies the vector transformation law, Eq. (1.15). Show directly that its time derivative $d\mathbf{A}/dt$ also satisfies Eq. (1.15) and is therefore a vector.

1.7.3 Show, by differentiating components, that

(a) $\dfrac{d}{dt}(\mathbf{A} \cdot \mathbf{B}) = \dfrac{d\mathbf{A}}{dt} \cdot \mathbf{B} + \mathbf{A} \cdot \dfrac{d\mathbf{B}}{dt}$,

(b) $\dfrac{d}{dt}(\mathbf{A} \times \mathbf{B}) = \dfrac{d\mathbf{A}}{dt} \times \mathbf{B} + \mathbf{A} \times \dfrac{d\mathbf{B}}{dt}$,

just like the derivative of the product of two algebraic functions.

1.7.4 In Chapter 2 it will be seen that the *unit* vectors in noncartesian coordinate systems are usually functions of the coordinate variables, $\mathbf{e}_i = \mathbf{e}_i(q_1, q_2, q_3)$ but $|\mathbf{e}_i| = 1$. Show that either $\partial \mathbf{e}_i/\partial q_j = 0$ or $\partial \mathbf{e}_i/\partial q_j$ is orthogonal to $\mathbf{e}_i$.
Hint. $\partial \mathbf{e}_i^2/\partial q_j = 0$.

1.7.5 Prove $\nabla \cdot (\mathbf{a} \times \mathbf{b}) = \mathbf{b} \cdot \nabla \times \mathbf{a} - \mathbf{a} \cdot \nabla \times \mathbf{b}$.
Hint. Treat as a triple scalar product.

1.7.6 The electrostatic field of a point charge q is

$$\mathbf{E} = \frac{q}{4\pi\varepsilon_0} \cdot \frac{\hat{\mathbf{r}}}{r^2}.$$

Calculate the divergence of $\mathbf{E}$. What happens at the origin?

1.8 CURL, $\nabla \times$

Another possible operation with the vector operator ∇ is to cross it into a vector. We obtain

$$\mathbf{\nabla} \times \mathbf{V} = \hat{\mathbf{x}}\left(\frac{\partial}{\partial y}V_z - \frac{\partial}{\partial z}V_y\right) + \hat{\mathbf{y}}\left(\frac{\partial}{\partial z}V_x - \frac{\partial}{\partial x}V_z\right) + \hat{\mathbf{z}}\left(\frac{\partial}{\partial x}V_y - \frac{\partial}{\partial y}V_x\right)$$

$$= \begin{vmatrix} \hat{\mathbf{x}} & \hat{\mathbf{y}} & \hat{\mathbf{z}} \\ \dfrac{\partial}{\partial x} & \dfrac{\partial}{\partial y} & \dfrac{\partial}{\partial z} \\ V_x & V_y & V_z \end{vmatrix}, \tag{1.70}$$

which is called the curl of **V**. In expanding this determinant we must consider the derivative nature of **∇**. Specifically, **V** × **V** is defined only as an operator, another vector differential operator. It is certainly not equal, in general, to −**V** × **V**.[1] In the case of Eq. (1.70) the determinant must be expanded *from the top down* so that we get the derivatives as shown in the middle portion of Eq. (1.70). If **∇** is crossed into the product of a scalar and a vector, we can show

$$\mathbf{\nabla} \times (f\mathbf{V})|_x = \left[\frac{\partial}{\partial y}(fV_z) - \frac{\partial}{\partial z}(fV_y)\right]$$

$$= \left(f\frac{\partial V_z}{\partial y} + \frac{\partial f}{\partial y}V_z - f\frac{\partial V_y}{\partial z} - \frac{\partial f}{\partial z}V_y\right)$$

$$= f\mathbf{\nabla} \times \mathbf{V}|_x + (\mathbf{\nabla}f) \times \mathbf{V}|_x. \tag{1.71}$$

If we permute the coordinates $x \to y$, $y \to z$, $z \to x$ to pick up the y-component and then permute them a second time to pick up the z-component,

$$\mathbf{\nabla} \times (f\mathbf{V}) = f\mathbf{\nabla} \times \mathbf{V} + (\mathbf{\nabla}f) \times \mathbf{V}, \tag{1.72}$$

which is the vector product analog of Eq. (1.68a). Again, as a differential operator **∇** differentiates both f and **V**. As a vector it is crossed into **V** (in each term).

Example 1.8.1

Calculate **∇** × **r**$f(r)$.
By Eq. (1.72),

$$\mathbf{\nabla} \times \mathbf{r}f(r) = f(r)\mathbf{\nabla} \times \mathbf{r} + [\mathbf{\nabla}f(r)] \times \mathbf{r}. \tag{1.73}$$

First,

$$\mathbf{\nabla} \times \mathbf{r} = \begin{vmatrix} \hat{\mathbf{x}} & \hat{\mathbf{y}} & \hat{\mathbf{z}} \\ \dfrac{\partial}{\partial x} & \dfrac{\partial}{\partial y} & \dfrac{\partial}{\partial z} \\ x & y & z \end{vmatrix} = 0. \tag{1.74}$$

---

[1] In this same spirit, if **A** is a differential operator, it is not necessarily true that **A** × **A** = 0. Specifically, for the quantum mechanical angular momentum *operator*, **L** = −i(**r** × **∇**), we find that **L** × **L** = i**L**.

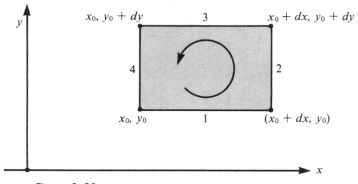

Figure 1.20   Circulation around a differential loop.

Second, using $\nabla f(r) = \hat{\mathbf{r}}(df/dr)$ (Example 1.6.1), we obtain

$$\nabla \times \mathbf{r}f(r) = \frac{\partial f}{\partial r}\hat{\mathbf{r}} \times \mathbf{r} = 0. \tag{1.75}$$

The vector product vanishes, since $\mathbf{r} = \hat{\mathbf{r}}r$ and $\hat{\mathbf{r}} \times \hat{\mathbf{r}} = 0$.

To develop a better feeling for the physical significance of the curl, we consider the circulation of fluid around a differential loop in the xy-plane, Fig. 1.20.

Although the circulation is technically given by a vector line integral $\int \mathbf{V} \cdot d\lambda$ (Section 1.10), we can set up the equivalent scalar integrals here. Let us take the circulation to be

$$\text{circulation}_{1234} = \int_1 V_x(x, y)\, d\lambda_x + \int_2 V_y(x, y)\, d\lambda_y$$

$$+ \int_3 V_x(x, y)\, d\lambda_x + \int_4 V_y(x, y)\, d\lambda_y. \tag{1.76}$$

The numbers 1, 2, 3, and 4 refer to the numbered line segments in Fig. 1.20. In the first integral $d\lambda_x = +dx$ but in the third integral $d\lambda_x = -dx$ because the third line segment is traversed in the negative x-direction. Similarly, $d\lambda_y = +dy$ for the second integral, $-dy$ for the fourth. Next, the integrands are referred to the point (x_0, y_0) with a Taylor expansion[2] taking into account the displacement of line segment 3 from 1 and 2 from 4. For our differential line

[2] $V_y(x_0 + dx, y_0) = V_y(x_0, y_0) + \left(\dfrac{\partial V_y}{\partial x}\right)_{x_0 y_0} dx + \cdots.$

The higher-order terms will drop out in the limit as $dx \to 0$. A correction term for the variation of V_y with y is canceled by the corresponding term in the fourth integral (see Section 5.6).

segments this leads to

$$\text{circulation}_{1234} = V_x(x_0, y_0)\, dx + \left[V_y(x_0, y_0) + \frac{\partial V_y}{\partial x}\, dx \right] dy$$

$$+ \left[V_x(x_0, y_0) + \frac{\partial V_x}{\partial y}\, dy \right](-dx) + V_y(x_0, y_0)(-dy)$$

$$= \left(\frac{\partial V_y}{\partial x} - \frac{\partial V_x}{\partial y} \right) dx\, dy. \tag{1.77}$$

Dividing by $dx\, dy$, we have

$$\text{circulation per unit area} = \nabla \times \mathbf{V}|_z. \tag{1.78}$$

The circulation[3] about our differential area in the xy-plane is given by the z-component of $\nabla \times \mathbf{V}$. In principle, the curl, $\nabla \times \mathbf{V}$ at (x_0, y_0), could be determined by inserting a (differential) paddle wheel into the moving fluid at point (x_0, y_0). The rotation of the little paddle wheel would be a measure of the curl, and its axis along the direction of $\nabla \times \mathbf{V}$ which is perpendicular to the plane of circulation.

We shall use the result, Eq. (1.77), in Section 1.13 to derive Stokes's theorem. Whenever the curl of a vector $\mathbf{V}$ vanishes,

$$\nabla \times \mathbf{V} = 0. \tag{1.79}$$

$\mathbf{V}$ is labeled irrotational. The most important physical examples of irrotational vectors are the gravitational and electrostatic forces. In each case

$$\mathbf{V} = C\frac{\hat{\mathbf{r}}}{r^2} = C\frac{\mathbf{r}}{r^3}, \tag{1.80}$$

where C is a constant and $\hat{\mathbf{r}}$ is the unit vector in the outward radial direction. For the gravitational case we have $C = -Gm_1 m_2$, given by Newton's law of universal gravitation. If $C = q_1 q_2 / 4\pi\varepsilon_0$, we have Coulomb's law of electrostatics (mks units). The force $\mathbf{V}$ given in Eq. (1.80) may be shown to be irrotational by direct expansion into cartesian components as we did in Example 1.8.1. Another approach is developed in Chapter 2, in which we express $\nabla \times$, the curl, in terms of spherical polar coordinates. In Section 1.13 we shall see that whenever a vector is irrotational, the vector may be written as the (negative) gradient of a scalar potential. In Section 1.15 we shall prove that a vector field may be resolved into an irrotational part and a solenoidal part (subject to conditions at inifinity). In terms of the electromagnetic field this corresponds to the resolution into an irrotational electric field and a solenoidal magnetic field.

For waves in an elastic medium, if the displacement $\mathbf{u}$ is irrotational, $\nabla \times \mathbf{u} = 0$, plane waves (or spherical waves at large distances) become

[3] In fluid dynamics $\nabla \times \mathbf{V}$ is called the "vorticity."

longitudinal. If **u** is solenoidal, $\nabla \cdot \mathbf{u} = 0$, then the waves become transverse. A seismic disturbance will produce a displacement that may be resolved into a solenoidal part and an irrotational part (compare Section 1.15). The irrotational part yields the longitudinal P (primary) earthquake waves. The solenoidal part gives rise to the slower transverse S (secondary) waves.

Using the gradient, divergence, and curl, and of course the *BAC-CAB* rule, we may construct or verify a large number of useful vector identities. For verification, complete expansion into cartesian components is always a possibility. Sometimes if we use insight instead of routine shuffling of cartesian components, the verification process can be shortened drastically.

Remember that ∇ is a vector operator, a hybrid creature satisfying two sets of rules:

1. vector rules, and
2. partial differentiation rules—including differentiation of a product.

Example 1.8.2 Gradient of a Dot Product
Verify that

$$\nabla(\mathbf{A} \cdot \mathbf{B}) = (\mathbf{B} \cdot \nabla)\mathbf{A} + (\mathbf{A} \cdot \nabla)\mathbf{B} + \mathbf{B} \times (\nabla \times \mathbf{A}) + \mathbf{A} \times (\nabla \times \mathbf{B}). \qquad (1.81)$$

This particular example hinges on the recognition that $\nabla(\mathbf{A} \cdot \mathbf{B})$ is the type of term that appears in the *BAC-CAB* expansion of a triple vector product, Eq. (1.55). For instance,

$$\mathbf{A} \times (\nabla \times \mathbf{B}) = \nabla(\mathbf{A} \cdot \mathbf{B}) - (\mathbf{A} \cdot \nabla)\mathbf{B},$$

with the ∇ differentiating only **B**, not **A**. From the commutativity of factors in a scalar product we may interchange **A** and **B** and write

$$\mathbf{B} \times (\nabla \times \mathbf{A}) = \nabla(\mathbf{A} \cdot \mathbf{B}) - (\mathbf{B} \cdot \nabla)\mathbf{A},$$

now with ∇ differentiating only **A**, not **B**. Adding these two equations, we obtain ∇ differentiating the product $\mathbf{A} \cdot \mathbf{B}$ and the identity, Eq. (1.81).

This identity is used frequently in advanced electromagnetic theory. Exercise 1.8.15 is one simple illustration.

EXERCISES

1.8.1   Show, by rotating the coordinates, that the components of the curl of a vector transform as a vector.
Hint. The direction cosine identities of Eq. (1.46) are available as needed.

1.8.2   Show that $\mathbf{u} \times \mathbf{v}$ is solenoidal if **u** and **v** are each irrotational.

1.8.3   If **A** is irrotational, show that $\mathbf{A} \times \mathbf{r}$ is solenoidal.

1.8.4   A rigid body is rotating with constant angular velocity $\boldsymbol{\omega}$. Show that the linear velocity **v** is solenoidal.

1.8.5 A vector function $\mathbf{f}(x, y, z)$ is not irrotational but the product of $\mathbf{f}$ and a scalar function $g(x, y, z)$ is irrotational. Show that

$$\mathbf{f} \cdot \nabla \times \mathbf{f} = 0.$$

1.8.6 If (a) $\mathbf{V} = \hat{\mathbf{x}} V_x(x, y) + \hat{\mathbf{y}} V_y(x, y)$ and (b) $\nabla \times \mathbf{V} \neq 0$, prove that $\nabla \times \mathbf{V}$ is perpendicular to $\mathbf{V}$.

1.8.7 Classically, angular momentum is given by $\mathbf{L} = \mathbf{r} \times \mathbf{p}$, where $\mathbf{p}$ is the linear momentum. To go from classical mechanics to quantum mechanics, replace $\mathbf{p}$ by the operator $-i\nabla$ (Section 15.6). Show that the quantum mechanical angular momentum operator has cartesian components

$$L_x = -i\left(y \frac{\partial}{\partial z} - z \frac{\partial}{\partial y} \right)$$

$$L_y = -i\left(z \frac{\partial}{\partial x} - x \frac{\partial}{\partial z} \right)$$

$$L_z = -i\left(x \frac{\partial}{\partial y} - y \frac{\partial}{\partial x} \right)$$

(in units of $\hbar$).

1.8.8 Using the angular momentum operators previously given, show that they satisfy commutation relations of the form

$$[L_x, L_y] \equiv L_x L_y - L_y L_x = iL_z$$

and hence

$$\mathbf{L} \times \mathbf{L} = i\mathbf{L}.$$

These commutation relations will be taken later as the defining relations as an angular momentum operator—Exercise 3.2.15 and the following one and Chapter 4.

1.8.9 With the commutator bracket notation $[L_x, L_y] = L_x L_y - L_y L_x$, the angular momentum vector $\mathbf{L}$ satisfies $[L_x, L_y] = iL_z$, etc., or $\mathbf{L} \times \mathbf{L} = i\mathbf{L}$.

Two other vectors $\mathbf{a}$ and $\mathbf{b}$ commute with each other and with $\mathbf{L}$, that is, $[\mathbf{a}, \mathbf{b}] = [\mathbf{a}, \mathbf{L}] = [\mathbf{b}, \mathbf{L}] = 0$. Show that

$$[\mathbf{a} \cdot \mathbf{L}, \mathbf{b} \cdot \mathbf{L}] = i(\mathbf{a} \times \mathbf{b}) \cdot \mathbf{L}.$$

1.8.10 For $\mathbf{A} = \hat{\mathbf{x}} A_x(x, y, z)$ and $\mathbf{B} = \hat{\mathbf{x}} B_x(x, y, z)$ evaluate each term in the vector identity

$$\nabla(\mathbf{A} \cdot \mathbf{B}) = (\mathbf{B} \cdot \nabla)\mathbf{A} + (\mathbf{A} \cdot \nabla)\mathbf{B} + \mathbf{B} \times (\nabla \times \mathbf{A}) + \mathbf{A} \times (\nabla \times \mathbf{B})$$

and verify that the identity is satisfied.

1.8.11 Verify the vector identity

$$\nabla \times (\mathbf{A} \times \mathbf{B}) = (\mathbf{B} \cdot \nabla)\mathbf{A} - (\mathbf{A} \cdot \nabla)\mathbf{B} - \mathbf{B}(\nabla \cdot \mathbf{A}) + \mathbf{A}(\nabla \cdot \mathbf{B}).$$

1.8.12 As an alternative to the vector identity of Example 1.8.2 show that

$$\nabla(\mathbf{A} \cdot \mathbf{B}) = (\mathbf{A} \times \nabla) \times \mathbf{B} + (\mathbf{B} \times \nabla) \times \mathbf{A} + \mathbf{A}(\nabla \cdot \mathbf{B}) + \mathbf{B}(\nabla \cdot \mathbf{A}).$$

1.8.13 Verify the identity

$$\mathbf{A} \times (\nabla \times \mathbf{A}) = \tfrac{1}{2}\nabla(A^2) - (\mathbf{A} \cdot \nabla)\mathbf{A}.$$

1.8.14   If **A** and **B** are constant vectors, show that

$$\nabla(\mathbf{A} \cdot \mathbf{B} \times \mathbf{r}) = \mathbf{A} \times \mathbf{B}.$$

1.8.15   A distribution of electric currents creates a constant magnetic moment **m**. The force on **m** in an external magnetic induction **B** is given by

$$\mathbf{F} = \nabla \times (\mathbf{B} \times \mathbf{m}).$$

Show that

$$\mathbf{F} = \nabla(\mathbf{m} \cdot \mathbf{B}).$$

Note. Assuming no time dependence of the fields, Maxwell's equations yield $\nabla \times \mathbf{B} = 0$. Also $\nabla \cdot \mathbf{B} = 0$.

1.8.16   An electric dipole of moment **p** is located at the origin. The dipole creates an electric potential at **r** given by

$$\psi(\mathbf{r}) = \frac{\mathbf{p} \cdot \mathbf{r}}{4\pi\varepsilon_0 r^3}.$$

Find the electric field, $\mathbf{E} = -\nabla\psi$ at **r**.

1.8.17   The vector potential **A** of a magnetic dipole, dipole moment **m**, is given by $\mathbf{A}(\mathbf{r}) = (\mu_0/4\pi)(\mathbf{m} \times \mathbf{r}/r^3)$. Show that the magnetic induction $\mathbf{B} = \nabla \times \mathbf{A}$ is given by

$$\mathbf{B} = \frac{\mu_0}{4\pi} \frac{3\hat{\mathbf{r}}(\hat{\mathbf{r}} \cdot \mathbf{m}) - \mathbf{m}}{r^3}.$$

Note. The limiting process leading to point dipoles is discussed in Section 12.1 for electric dipoles, Section 12.5 for magnetic dipoles.

1.8.18   The velocity of a two-dimensional flow of liquid is given by

$$\mathbf{V} = \hat{\mathbf{x}}u(x, y) - \hat{\mathbf{y}}v(x, y).$$

If the liquid is incompressible and the flow is irrotational show that

$$\frac{\partial u}{\partial x} = \frac{\partial v}{\partial y} \quad \text{and} \quad \frac{\partial u}{\partial y} = -\frac{\partial v}{\partial x}.$$

These are the Cauchy–Riemann conditions of Section 6.2.

1.8.19   The evaluation in this section of the four integrals for the circulation omitted Taylor series terms such as $\partial V_x/\partial x$, $\partial V_y/\partial y$ and all second derivatives. Show that $\partial V_x/\partial x$, $\partial V_y/\partial y$ cancel out when the four integrals are added and that the second derivative terms drop out in the limit as $dx \to 0$, $dy \to 0$. *Hint.* Calculate the circulation per unit area and then take the limit $dx \to 0$, $dy \to 0$.

1.9 SUCCESSIVE APPLICATIONS OF ∇

We have now defined gradient, divergence, and curl to obtain vector, scalar, and vector quantities, respectively. Letting ∇ operate on each of these quantities, we obtain

(a) $\nabla \cdot \nabla\varphi$ (b) $\nabla \times \nabla\varphi$ (c) $\nabla\nabla \cdot \mathbf{V}$

(d) $\nabla \cdot \nabla \times \mathbf{V}$ (e) $\nabla \times (\nabla \times \mathbf{V})$, all five expressions involving second derivatives and all five appearing in the second-order differential equations of mathematical physics, particularly in electromagnetic theory.

The first expression, $\mathbf{\nabla} \cdot \mathbf{\nabla}\varphi$, the divergence of the gradient, is named the Laplacian of φ. We have

$$\mathbf{\nabla} \cdot \mathbf{\nabla}\varphi = \left(\hat{\mathbf{x}}\frac{\partial}{\partial x} + \hat{\mathbf{y}}\frac{\partial}{\partial y} + \hat{\mathbf{z}}\frac{\partial}{\partial z}\right) \cdot \left(\hat{\mathbf{x}}\frac{\partial\varphi}{\partial x} + \hat{\mathbf{y}}\frac{\partial\varphi}{\partial y} + \hat{\mathbf{z}}\frac{\partial\varphi}{\partial z}\right)$$

$$= \frac{\partial^2\varphi}{\partial x^2} + \frac{\partial^2\varphi}{\partial y^2} + \frac{\partial^2\varphi}{\partial z^2}. \tag{1.82a}$$

When φ is the electrostatic potential, we have

$$\mathbf{\nabla} \cdot \mathbf{\nabla}\varphi = 0. \tag{1.82b}$$

which is Laplace's equation of electrostatics. Often the combination $\mathbf{\nabla} \cdot \mathbf{\nabla}$ is written ∇^2.

Example 1.9.1

Calculate $\mathbf{\nabla} \cdot \mathbf{\nabla}g(r)$.
Referring to Examples 1.6.1 and 1.7.2,

$$\mathbf{\nabla} \cdot \mathbf{\nabla}g(r) = \mathbf{\nabla} \cdot \hat{\mathbf{r}}\frac{dg}{dr} = \frac{2}{r}\frac{dg}{dr} + \frac{d^2g}{dr^2},$$

replacing $f(r)$ in Example 1.7.2 by $1/r \cdot dg/dr$. If $g(r) = r^n$, this reduces to

$$\mathbf{\nabla} \cdot \mathbf{\nabla}r^n = n(n + 1)r^{n-2}.$$

This vanishes for $n = 0$ [$g(r) = $ constant] and for $n = -1$; that is, $g(r) = 1/r$ is a solution of Laplace's equation, $\nabla^2 g(r) = 0$. This is for $r \neq 0$. At $r = 0$, a Dirac delta function is involved (see Eq. (1.168) and Section 8.7).

Expression (b) may be written

$$\mathbf{\nabla} \times \mathbf{\nabla}\varphi = \begin{vmatrix} \hat{\mathbf{x}} & \hat{\mathbf{y}} & \hat{\mathbf{z}} \\ \dfrac{\partial}{\partial x} & \dfrac{\partial}{\partial y} & \dfrac{\partial}{\partial z} \\ \dfrac{\partial\varphi}{\partial x} & \dfrac{\partial\varphi}{\partial y} & \dfrac{\partial\varphi}{\partial z} \end{vmatrix}.$$

By expanding the determinant, we obtain

$$\mathbf{\nabla} \times \mathbf{\nabla}\varphi = \hat{\mathbf{x}}\left(\frac{\partial^2\varphi}{\partial y\,\partial z} - \frac{\partial^2\varphi}{\partial z\,\partial y}\right) + \hat{\mathbf{y}}\left(\frac{\partial^2\varphi}{\partial z\,\partial x} - \frac{\partial^2\varphi}{\partial x\,\partial z}\right) + \hat{\mathbf{z}}\left(\frac{\partial^2\varphi}{\partial x\,\partial y} - \frac{\partial^2\varphi}{\partial y\,\partial x}\right)$$

$$= 0, \tag{1.83}$$

assuming that the order of partial differentiation may be interchanged. This is true as long as these second partial derivatives of φ are continuous functions.

Then, from Eq. (1.83), the curl of a gradient is identically zero. All gradients, therefore, are irrotational. Note carefully that the zero in Eq. (1.83) comes as a mathematical identity, independent of any physics. The zero in Eq. (1.82b) is a consequence of physics.

Expression (d) is a triple scalar product which may be written

$$\mathbf{\nabla} \cdot \mathbf{\nabla} \times \mathbf{V} = \begin{vmatrix} \dfrac{\partial}{\partial x} & \dfrac{\partial}{\partial y} & \dfrac{\partial}{\partial z} \\[2mm] \dfrac{\partial}{\partial x} & \dfrac{\partial}{\partial y} & \dfrac{\partial}{\partial z} \\[2mm] V_x & V_y & V_z \end{vmatrix}. \qquad (1.84)$$

Again, assuming continuity so that the order of differentiation is immaterial, we obtain

$$\mathbf{\nabla} \cdot \mathbf{\nabla} \times \mathbf{V} = 0. \qquad (1.85)$$

The divergence of a curl vanishes or all curls are solenoidal. In Section 1.15 we shall see that vectors may be resolved into solenoidal and irrotational parts by Helmholtz's theorem.

The two remaining expressions satisfy a relation

$$\mathbf{\nabla} \times (\mathbf{\nabla} \times \mathbf{V}) = \mathbf{\nabla}\mathbf{\nabla} \cdot \mathbf{V} - \mathbf{\nabla} \cdot \mathbf{\nabla}\mathbf{V}. \qquad (1.86)$$

This follows immediately from Eq. (1.55), the *BAC-CAB* rule, which we rewrite so that **C** appears at the extreme right of each term. The term $\mathbf{\nabla} \cdot \mathbf{\nabla}\mathbf{V}$ was not included in our list, but it may be *defined* by Eq. (1.86).

Example 1.9.2 Electromagnetic Wave Equation

One important application of this vector relation (Eq. (1.86)) is in the derivation of the electromagnetic wave equation. In vacuum Maxwell's equations become

$$\mathbf{\nabla} \cdot \mathbf{B} = 0, \qquad (1.87a)$$

$$\mathbf{\nabla} \cdot \mathbf{E} = 0, \qquad (1.87b)$$

$$\mathbf{\nabla} \times \mathbf{B} = \varepsilon_0 \mu_0 \frac{\partial \mathbf{E}}{\partial t}, \qquad (1.87c)$$

$$\mathbf{\nabla} \times \mathbf{E} = -\frac{\partial \mathbf{B}}{\partial t}. \qquad (1.87d)$$

Here **E** is the electric field, **B** the magnetic induction, ε_0 the electric permittivity, and μ_0 the magnetic permeability (mks or SI units). Suppose we eliminate **B** from Eqs. (1.87c) and (1.87d). We may do this by taking the curl of both sides of Eq. (1.87d) and the time derivative of both sides of Eq. (1.87c). Since the space and time derivatives commute,

$$\frac{\partial}{\partial t} \mathbf{V} \times \mathbf{B} = \mathbf{V} \times \frac{\partial \mathbf{B}}{\partial t}, \tag{1.88}$$

and we obtain

$$\mathbf{V} \times (\mathbf{V} \times \mathbf{E}) = -\varepsilon_0 \mu_0 \frac{\partial^2 \mathbf{E}}{\partial t^2}. \tag{1.89}$$

Application of Eqs. (1.86) and of (1.87b) yields

$$\mathbf{V} \cdot \mathbf{V} \mathbf{E} = \varepsilon_0 \mu_0 \frac{\partial^2 \mathbf{E}}{\partial t^2}, \tag{1.90}$$

the electromagnetic vector wave equation. Again, if **E** is expressed in cartesian coordinates, Eq. (1.90) separates into three scalar wave equations, each involving a scalar Laplacian.

EXERCISES

1.9.1   Verify Eq. (1.86)

$$\mathbf{V} \times (\mathbf{V} \times \mathbf{V}) = \mathbf{V} \mathbf{V} \cdot \mathbf{V} - \mathbf{V} \cdot \mathbf{V} \mathbf{V}$$

by direct expansion in cartesian coordinates.

1.9.2   Show that the identity

$$\mathbf{V} \times (\mathbf{V} \times \mathbf{V}) = \mathbf{V} \mathbf{V} \cdot \mathbf{V} - \mathbf{V} \cdot \mathbf{V} \mathbf{V}$$

follows from the *BAC-CAB* rule for a triple vector product. Justify any alteration of the order of factors in the *BAC* and *CAB* terms.

1.9.3   Prove that $\mathbf{V} \times (\varphi \mathbf{V} \varphi) = 0$.

1.9.4   You are given that the curl of **F** equals the curl of **G**. Show that **F** and **G** may differ by (a) a constant and (b) a gradient of a scalar function.

1.9.5   The Navier–Stokes equation of hydrodynamics contains a nonlinear term $(\mathbf{v} \cdot \mathbf{V})\mathbf{v}$. Show that the curl of this term may be written $-\mathbf{V} \times [\mathbf{v} \times (\mathbf{V} \times \mathbf{v})]$.

1.9.6   From the Navier–Stokes equation for the steady flow of an incompressible viscous fluid we have the term

$$\mathbf{V} \times [\mathbf{v} \times (\mathbf{V} \times \mathbf{v})],$$

where **v** is the fluid velocity. Show that this term vanishes for the special case

$$\mathbf{v} = \hat{\mathbf{x}} v(y, z).$$

1.9.7   Prove that $(\mathbf{V}u) \times (\mathbf{V}v)$ is solenoidal where u and v are differentiable scalar functions.

1.9.8 φ is a scalar satisfying Laplace's equation, $\mathbf{V}^2 \varphi = 0$. Show that $\mathbf{V}\varphi$ is *both* solenoidal and irrotational.

1.9.9   With ψ a scalar function, show that

$$(\mathbf{r} \times \boldsymbol{\nabla}) \cdot (\mathbf{r} \times \boldsymbol{\nabla})\psi = r^2 \nabla^2 \psi - r^2 \frac{\partial^2 \psi}{\partial r^2} - 2r \frac{\partial \psi}{\partial r}.$$

(This can actually be shown more easily in spherical polar coordinates, Section 2.5.)

1.9.10   In a (nonrotating) isolated mass such as a star, the condition for equilibrium is

$$\boldsymbol{\nabla} P + \rho \boldsymbol{\nabla} \varphi = 0.$$

Here P is the total pressure, ρ the density, and φ the gravitational potential. Show that at any given point the normals to the surfaces of constant pressure and constant gravitational potential are parallel.

1.9.11   In the Pauli theory of the electron one encounters the expression

$$(\mathbf{p} - e\mathbf{A}) \times (\mathbf{p} - e\mathbf{A})\psi,$$

where ψ is a scalar function. $\mathbf{A}$ is the magnetic vector potential related to the magnetic induction $\mathbf{B}$ by $\mathbf{B} = \boldsymbol{\nabla} \times \mathbf{A}$. Given that $\mathbf{p} = -i\boldsymbol{\nabla}$, show that this expression reduces to $ie\mathbf{B}\psi$.

1.9.12   Show that any solution of the equation

$$\boldsymbol{\nabla} \times \boldsymbol{\nabla} \times \mathbf{A} - k^2\mathbf{A} = 0$$

automatically satisfies the vector Helmholtz equation

$$\nabla^2\mathbf{A} + k^2\mathbf{A} = 0$$

and the solenoidal condition

$$\boldsymbol{\nabla} \cdot \mathbf{A} = 0.$$

Hint. Let $\boldsymbol{\nabla} \cdot$ operate on the first equation.

1.9.13   The theory of heat conduction leads to an equation

$$\nabla^2\Psi = k|\boldsymbol{\nabla}\Phi|^2,$$

where Φ is a potential satisfying Laplace's equation: $\nabla^2\Phi = 0$. Show that a solution of this equation is

$$\Psi = \tfrac{1}{2}k\Phi^2.$$

1.10 VECTOR INTEGRATION

The next step after differentiating vectors is to integrate them. Let us start with line integrals and then proceed to surface and volume integrals. In each case the method of attack will be to reduce the vector integral to scalar integrals with which the reader is assumed familiar.

LINE INTEGRALS

Using an increment of length $d\mathbf{r} = \hat{\mathbf{x}}\, dx + \hat{\mathbf{y}}\, dy + \hat{\mathbf{z}}\, dz$, we may encounter the line integrals

$$\int_c \varphi \, d\mathbf{r}, \tag{1.91a}$$

$$\int_c \mathbf{V} \cdot d\mathbf{r}, \tag{1.91b}$$

$$\int_c \mathbf{V} \times d\mathbf{r}, \tag{1.91c}$$

in each of which the integral is over some contour C that may be open (with starting point and ending point separated) or closed (forming a loop). Because of its physical interpretation that follows, the second form, Eq. (1.91b) is by far the most important of the three.

With φ, a scalar, the first integral reduces immediately to

$$\int_c \varphi \, d\mathbf{r} = \hat{\mathbf{x}} \int_c \varphi(x, y, z) \, dx + \hat{\mathbf{y}} \int_c \varphi(x, y, z) \, dy + \hat{\mathbf{z}} \int_c \varphi(x, y, z) \, dz. \tag{1.92}$$

This separation has employed the relation

$$\int \hat{\mathbf{x}} \varphi \, dx = \hat{\mathbf{x}} \int \varphi \, dx, \tag{1.93}$$

which is permissible because the cartesian unit vectors $\hat{\mathbf{x}}$, $\hat{\mathbf{y}}$, and $\hat{\mathbf{z}}$ are constant in both magnitude and direction. Perhaps this relation is obvious here, but it will not be true in the noncartesian systems encountered in Chapter 2.

The three integrals on the right side of Eq. (1.92) are ordinary scalar integrals and, to avoid complications, we assume that they are Riemann integrals. Note, however, that the integral with respect to x cannot be evaluated unless y and z are known in terms of x and similarly for the integrals with respect to y and z. This simply means that the path of integration C must be specified. Unless the integrand has special properties that lead the integral to depend only on the value of the end points, the value will depend on the particular choice of contour C. For instance, if we choose the very special case $\varphi = 1$, Eq. (1.91a) is just the vector distance from the start of contour C to the end point, in this case independent of the choice of path connecting fixed end points. With $d\mathbf{r} = \hat{\mathbf{x}} \, dx + \hat{\mathbf{y}} \, dy + \hat{\mathbf{z}} \, dz$, the second and third forms also reduce to scalar integrals and, like Eq. (1.91a), are dependent, in general, on the choice of path. The form (Eq. (1.91b)) is exactly the same as that encountered when we calculate the work done by a force that varies along the path,

$$W = \int \mathbf{F} \cdot d\mathbf{r}$$

$$= \int F_x(x, y, z) \, dx + \int F_y(x, y, z) \, dy + \int F_z(x, y, z) \, dz. \tag{1.94a}$$

In this expression $\mathbf{F}$ is the force exerted on a particle.

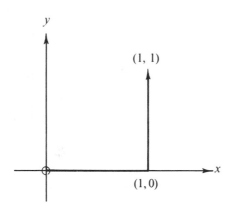

Figure 1.21   A path of integration.

Example 1.10.1

The force exerted on a body is $\mathbf{F} = -\hat{x}y + \hat{y}x$. The problem is to calculate the work done going from the origin to the point (1, 1).

$$W = \int_{0,0}^{1,1} \mathbf{F} \cdot d\mathbf{r} = \int_{0,0}^{1,1} (-y\,dx + x\,dy). \tag{1.94b}$$

Separating the two integrals, we obtain

$$W = -\int_{0}^{1} y\,dx + \int_{0}^{1} x\,dy. \tag{1.94c}$$

The first integral cannot be evaluated until we specify the values of y as x ranges from 0 to 1. Likewise, the second integral requires x as a function of y. Consider first the path shown in Fig. 1.21. Then

$$W = -\int_{0}^{1} 0\,dx + \int_{0}^{1} 1\,dy = 1, \tag{1.94d}$$

since $y = 0$ along the first segment of the path and $x = 1$ along the second.

If we select the path $[x = 0, 0 \le y \le 1]$ and $[0 \le x \le 1, y = 1]$, then Eq. (1.94c) gives $W = -1$. For this force the work done depends on the choice of path.

SURFACE INTEGRALS

Surface integrals appear in the same forms as line integrals, the element of area also being a vector, $d\boldsymbol{\sigma}$.[1] Often this area element is written $\mathbf{n}\,dA$ in which $\mathbf{n}$ is a unit (normal) vector to indicate the positive direction.[2] There are two conventions for choosing the positive direction. First, if the surface is a closed surface, we agree to take the outward normal as positive. Second, if the surface is an open surface, the positive normal depends on the direction in

---

[1] Recall that in Section 1.4 the area (of a parallelogram) is represented by a cross-product *vector*.
[2] Although $\mathbf{n}$ always has unit length, its direction may well be a function of position.

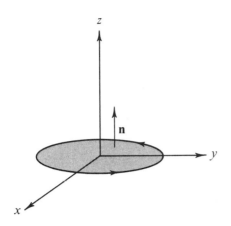

Figure 1.22   Right-hand rule for the positive normal.

which the perimeter of the open surface is traversed. If the right-hand fingers are placed in the direction of travel around the perimeter, the positive normal is indicated by the thumb of the right hand. As an illustration, a circle in the xy-plane (Fig. 1.22) mapped out from x to y to $-x$ to $-y$ and back to x will have its positive normal parallel to the positive z-axis (for the right-handed coordinate system). If readers should ever encounter one-sided surfaces, such as Moebius strips, it is suggested that they either cut the strips and form reasonable, well-behaved surfaces or label them pathological and send them to the nearest mathematics department.

Analogous to the line integrals, Eqs. (1.91a)–(1.91c) surface integrals may appear in the forms

$$\int \varphi \, d\boldsymbol{\sigma}$$

$$\int \mathbf{V} \cdot d\boldsymbol{\sigma}$$

$$\int \mathbf{V} \times d\boldsymbol{\sigma}.$$

Again, the dot product is by far the most commonly encountered form.

The surface integral $\int \mathbf{V} \cdot d\boldsymbol{\sigma}$ may be interpreted as a flow or flux through the given surface. This is really what we did in Section 1.7 to obtain the significance of the term divergence. This identification reappears in Section 1.11 as Gauss's theorem. Note that both physically and from the dot product the tangential components of the velocity contribute nothing to the flow through the surface.

VOLUME INTEGRALS

Volume integrals are somewhat simpler, for the volume element $d\tau$ is a scalar quantity.[3] We have

---

[3] Frequently the symbols d^3r and d^3x are used to denote a volume element in coordinate (xyz or $x_1x_2x_3$) space.

$$\int_V \mathbf{V}\, d\tau = \hat{\mathbf{x}} \int_V V_x\, d\tau + \hat{\mathbf{y}} \int_V V_y\, d\tau + \hat{\mathbf{z}} \int_V V_z\, d\tau, \qquad (1.95)$$

again reducing the vector integral to a vector sum of scalar integrals.

INTEGRAL DEFINITIONS OF GRADIENT, DIVERGENCE, AND CURL

One interesting and significant application of our surface and volume integrals is their use in developing alternate definitions of our differential relations. We find

$$\boldsymbol{\nabla}\varphi = \lim_{\int d\tau \to 0} \frac{\int \varphi\, d\boldsymbol{\sigma}}{\int d\tau}, \qquad (1.96)$$

$$\boldsymbol{\nabla}\cdot\mathbf{V} = \lim_{\int d\tau \to 0} \frac{\int \mathbf{V}\cdot d\boldsymbol{\sigma}}{\int d\tau}, \qquad (1.97)$$

$$\boldsymbol{\nabla}\times\mathbf{V} = \lim_{\int d\tau \to 0} \frac{\int d\boldsymbol{\sigma}\times\mathbf{V}}{\int d\tau}. \qquad (1.98)$$

In these three equations $\int d\tau$ is the volume of a small region of space and $d\boldsymbol{\sigma}$ is the vector area element of this volume. The identification of Eq. (1.97) as the divergence of $\mathbf{V}$ was carried out in Section 1.7. Here we show that Eq. (1.96) is consistent with our earlier definition of $\boldsymbol{\nabla}\varphi$ (Eq. (1.59)). For simplicity we choose $d\tau$ to be the differential volume $dx\, dy\, dz$ (Fig. 1.23). This time we place the origin at the geometric center of our volume element. The area integral leads to six integrals, one for each of the six faces. Remembering that $d\boldsymbol{\sigma}$ is outward, $d\boldsymbol{\sigma}\cdot\hat{\mathbf{x}} = -|d\boldsymbol{\sigma}|$ for surface $EFHG$, and $+|d\boldsymbol{\sigma}|$ for surface

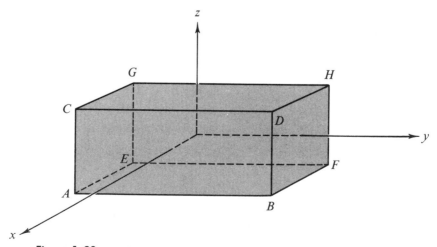

Figure 1.23 Differential rectangular parallelepiped (origin at center).

ABDC, we have

$$\int \varphi \, d\boldsymbol{\sigma} = -\hat{\mathbf{x}} \int_{EFHG} \left(\varphi - \frac{\partial \varphi}{\partial x} \frac{dx}{2} \right) dy \, dz + \hat{\mathbf{x}} \int_{ABDC} \left(\varphi + \frac{\partial \varphi}{\partial x} \frac{dx}{2} \right) dy \, dz$$

$$- \hat{\mathbf{y}} \int_{AEGC} \left(\varphi - \frac{\partial \varphi}{\partial y} \frac{dy}{2} \right) dx \, dz + \hat{\mathbf{y}} \int_{BFHD} \left(\varphi + \frac{\partial \varphi}{\partial y} \frac{dy}{2} \right) dx \, dz$$

$$- \hat{\mathbf{z}} \int_{ABFE} \left(\varphi - \frac{\partial \varphi}{\partial z} \frac{dz}{2} \right) dx \, dy + \hat{\mathbf{z}} \int_{CDHG} \left(\varphi + \frac{\partial \varphi}{\partial z} \frac{dz}{2} \right) dx \, dy.$$

Using the first two terms of a Maclaurin expansion, we evaluate each integrand at the origin with a correction included to correct for the displacement ($\pm dx/2$, etc.) of the center of the face from the origin.[4] Having chosen the total volume to be of differential size ($\int d\tau = dx \, dy \, dz$), we drop the integral signs on the right and obtain

$$\int \varphi \, d\boldsymbol{\sigma} = \left(\hat{\mathbf{x}} \frac{\partial \varphi}{\partial x} + \hat{\mathbf{y}} \frac{\partial \varphi}{\partial y} + \hat{\mathbf{z}} \frac{\partial \varphi}{\partial z} \right) dx \, dy \, dz. \tag{1.99}$$

Dividing by

$$\int d\tau = dx \, dy \, dz,$$

we verify Eq. (1.96).

This verification has been oversimplified in ignoring other correction terms beyond the first derivatives. These additional terms, which are introduced in Section 5.6 when the Taylor expansion is developed, vanish in the limit

$$\int d\tau \to 0 \, (dx \to 0, \, dy \to 0, \, dz \to 0).$$

This, of course, is the reason for specifying in Eqs. (1.96), (1.97), and (1.98) that this limit be taken.

Verification of Eq. (1.98) follows these same lines exactly, using a differential volume $dx \, dy \, dz$.

EXERCISES

1.10.1   The force field acting on a two-dimensional linear oscillator may be described by

$$\mathbf{F} = -\hat{\mathbf{x}} \, kx - \hat{\mathbf{y}} \, ky.$$

Compare the work done moving against this force field when going from (1, 1) to (4, 4) by the following straight-line paths:

[4] The origin has been placed at the geometric center.

(a) $(1, 1) \rightarrow (4, 1) \rightarrow (4, 4)$
(b) $(1, 1) \rightarrow (1, 4) \rightarrow (4, 4)$
(c) $(1, 1) \rightarrow (4, 4)$ along $x = y$.

This means evaluating

$$-\int_{(1,1)}^{(4,4)} \mathbf{F} \cdot d\mathbf{r}$$

along each path.

1.10.2   Find the work done going around a unit circle in the xy-plane:

(a) counterclockwise from 0 to π,
(b) clockwise from 0 to $-\pi$,

doing work *against* a force field given by

$$\mathbf{F} = \frac{-\hat{\mathbf{x}}y}{x^2 + y^2} + \frac{\hat{\mathbf{y}}x}{x^2 + y^2}.$$

Note that the work done depends on the path.

1.10.3   Calculate the work you do in going from point $(1, 1)$ to point $(3, 3)$. The force *you exert* is given by

$$\mathbf{F} = \hat{\mathbf{x}}(x - y) + \hat{\mathbf{y}}(x + y).$$

Specify clearly the path you choose. Note that this force field is non-conservative.

1.10.4   Evaluate $\oint \mathbf{r} \cdot d\mathbf{r}$.
Note. The symbol $\oint$ means that the path of integration is a closed loop.

1.10.5   Evaluate

$$\tfrac{1}{3} \int_s \mathbf{r} \cdot d\boldsymbol{\sigma}$$

over the unit cube defined by the point $(0, 0, 0)$ and the unit intercepts on the positive x-, y-, and z-axes. Note that (a) $\mathbf{r} \cdot d\boldsymbol{\sigma}$ is zero for three of the surfaces and (b) each of the three remaining surfaces contributes the same amount to the integral.

1.10.6   Show, by expansion of the surface integral, that

$$\lim_{\int d\tau \rightarrow 0} \frac{\int_s d\boldsymbol{\sigma} \times \mathbf{V}}{\int d\tau} = \nabla \times \mathbf{V}.$$

Hint. Choose the volume to be a differential volume, $dx \, dy \, dz$.

1.11 GAUSS'S THEOREM

Here we derive a useful relation between a surface integral of a vector and the volume integral of the divergence of that vector. Let us assume that the vector $\mathbf{V}$ and its first derivatives are continuous over the region of interest. Then Gauss's theorem states that

$$\int_S \mathbf{V} \cdot d\boldsymbol{\sigma} = \int_V \nabla \cdot \mathbf{V} \, d\tau. \qquad (1.100a)$$

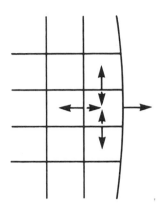

Figure 1.24 Exact cancellation of $d\sigma$'s on *interior* surfaces. No cancellation on the *exterior* surface.

In words, the surface integral of a vector over a closed surface equals the volume integral of the divergence of that vector integrated over the volume enclosed by the surface.

Imagine that volume V is subdivided into an arbitrarily large number of tiny (differential) parallelepipeds. For each parallelepiped

$$\sum_{\text{six surfaces}} \mathbf{V} \cdot d\boldsymbol{\sigma} = \boldsymbol{\nabla} \cdot \mathbf{V} \, d\tau \qquad (1.100b)$$

from the analysis of Section 1.7, Eq. (1.67), with $\rho\mathbf{v}$ replaced by $\mathbf{V}$. The summation is over the six faces of the parallelepiped. Summing over all parallelepipeds, we find that the $\mathbf{V} \cdot d\boldsymbol{\sigma}$ terms cancel (pairwise) for all *interior* faces; only the contributions of the *exterior* surfaces survive (Fig. 1.24). Analogous to the definition of a Riemann integral as the limit of a sum, we take the limit as the number of parallelepipeds approaches infinity ($\rightarrow\infty$) and the dimensions of each approach zero ($\rightarrow 0$).

$$\sum_{\text{exterior surfaces}} \mathbf{V} \cdot d\boldsymbol{\sigma} = \sum_{\text{volumes}} \boldsymbol{\nabla} \cdot \mathbf{V} \, d\tau$$

$$\downarrow \qquad\qquad \downarrow$$

$$\int_S \mathbf{V} \cdot d\boldsymbol{\sigma} = \int_V \boldsymbol{\nabla} \cdot \mathbf{V} \, d\tau.$$

The result is Eq. (1.100a), Gauss's theorem.

From a physical point of view Eq. (1.67) has established $\boldsymbol{\nabla} \cdot \mathbf{V}$ as the net outflow of fluid per unit volume. The volume integral then gives the total net outflow. But the surface integral $\int \mathbf{V} \cdot d\boldsymbol{\sigma}$ is just another way of expressing this same quantity, which is the equality, Gauss's theorem.

Green's Theorem

A frequently useful corollary of Gauss's theorem is a relation known as Green's theorem. If u and v are two scalar functions, we have the identities

$$\boldsymbol{\nabla} \cdot (u \, \boldsymbol{\nabla} v) = u \, \boldsymbol{\nabla} \cdot \boldsymbol{\nabla} v + (\boldsymbol{\nabla} u) \cdot (\boldsymbol{\nabla} v), \qquad (1.101)$$

$$\boldsymbol{\nabla} \cdot (v \, \boldsymbol{\nabla} u) = v \, \boldsymbol{\nabla} \cdot \boldsymbol{\nabla} u + (\boldsymbol{\nabla} v) \cdot (\boldsymbol{\nabla} u). \qquad (1.102)$$

Subtracting Eq. (1.102) from Eq. (1.101), integrating over a volume (u, v, and their derivatives, assumed continuous), and applying Eq. (1.100) (Gauss's theorem), we obtain

$$\int_V (u \, \nabla \cdot \nabla v - v \, \nabla \cdot \nabla u) \, d\tau = \int_S (u \, \nabla v - v \, \nabla u) \cdot d\boldsymbol{\sigma}. \qquad (1.103)$$

This is Green's theorem. We use it for developing Green's functions in Chapter 8. An alternate form of Green's theorem derived from Eq. (1.101) alone is

$$\int_S u \, \nabla v \cdot d\boldsymbol{\sigma} = \int_V u \, \nabla \cdot \nabla v \, d\tau + \int_V \nabla u \cdot \nabla v \, d\tau. \qquad (1.104)$$

This is the form of Green's theorem used in Section 1.16.

Alternate Forms of Gauss's Theorem

Although Eq. (1.100) involving the divergence is by far the most important form of Gauss's theorem, volume integrals involving the gradient and the curl may also appear. Suppose

$$\mathbf{V}(x, y, z) = V(x, y, z)\mathbf{a}, \qquad (1.105)$$

in which $\mathbf{a}$ is a vector with constant magnitude and constant but arbitrary direction. (You pick the direction, but once you have chosen it, hold it fixed.) Equation (1.100) becomes

$$\mathbf{a} \cdot \int_S V \, d\boldsymbol{\sigma} = \int_V \nabla \cdot \mathbf{a} V \, d\tau$$

$$= \mathbf{a} \cdot \int_V \nabla V \, d\tau \qquad (1.106)$$

by Eq. (1.68a). This may be rewritten

$$\mathbf{a} \cdot \left[\int_S V \, d\boldsymbol{\sigma} - \int_V \nabla V \, d\tau \right] = 0. \qquad (1.107)$$

Since $|\mathbf{a}| \neq 0$ and its direction is arbitrary, meaning that the cosine of the included angle cannot *always* vanish, the terms in brackets must be zero.[1] The result is

$$\int_S V \, d\boldsymbol{\sigma} = \int_V \nabla V \, d\tau. \qquad (1.108)$$

---

[1] This exploitation of the *arbitrary* nature of a part of a problem is a valuable and widely used technique. The arbitrary vector is used again in Sections 1.12 and 1.13. Other examples appear in Section 1.14 (integrands equated) and in Section 2.8, quotient rule.

In a similar manner, using $\mathbf{V} = \mathbf{a} \times \mathbf{P}$ in which $\mathbf{a}$ is a constant vector, we may show

$$\int_S d\boldsymbol{\sigma} \times \mathbf{P} = \int_V \boldsymbol{\nabla} \times \mathbf{P} \, d\tau. \tag{1.109}$$

These last two forms of Gauss's theorem are used in the vector form of Kirchoff diffraction theory. They may also be used to verify Eqs. (1.96) and (1.98).

Gauss's theorem may also be extended to tensors (see Section 2.10).

EXERCISES

1.11.1   Using Gauss's theorem prove that

$$\int_S d\boldsymbol{\sigma} = 0,$$

if S is a closed surface.

1.11.2   Show that

$$\tfrac{1}{3} \int_S \mathbf{r} \cdot d\boldsymbol{\sigma} = V,$$

where V is the volume enclosed by the closed surface S.
Note. This is a generalization of Exercise 1.10.5.

1.11.3   If $\mathbf{B} = \boldsymbol{\nabla} \times \mathbf{A}$, show that

$$\int_S \mathbf{B} \cdot d\boldsymbol{\sigma} = 0$$

for any closed surface S.

1.11.4   Over some volume V let ψ be a solution of Laplace's equation (with the derivatives appearing there continuous). Prove that the integral over any closed surface in V of the normal derivative of ψ, $(\partial \psi / \partial n$, or $\boldsymbol{\nabla} \psi \cdot \mathbf{n})$ will be zero.

1.11.5   In analogy to the integral definition of gradient, divergence, and curl of Section 1.10, show that

$$\nabla^2 \varphi = \lim_{\int d\tau \to 0} \frac{\int \boldsymbol{\nabla} \varphi \cdot d\boldsymbol{\sigma}}{\int d\tau}.$$

1.11.6   The electric displacement vector $\mathbf{D}$ satisfies the Maxwell equation $\boldsymbol{\nabla} \cdot \mathbf{D} = \rho$ where ρ is the charge density (per unit volume). At the boundary between two media there is a surface charge density σ (per unit area). Show that a boundary condition for $\mathbf{D}$ is

$$(\mathbf{D}_2 - \mathbf{D}_1) \cdot \mathbf{n} = \sigma.$$

$\mathbf{n}$ is a unit vector normal to the surface and out of medium 1.
Hint. Consider a *thin* pillbox as shown in the figure.

1.11.7   From Eq. (1.68a) with $\mathbf{V}$ the electric field $\mathbf{E}$, and f the electrostatic potential φ, show that

$$\int \rho\varphi \, d\tau = \varepsilon_0 \int E^2 \, d\tau.$$

This corresponds to a three-dimensional integration by parts.
Hint. $\mathbf{E} = -\nabla\varphi$, $\nabla \cdot \mathbf{E} = \rho/\varepsilon_0$. You may assume that φ vanishes at large r at least as fast as r^{-1}.

1.11.8   A particular steady-state electric current distribution is localized in space. Choosing a bounding surface far enough out so that the current density $\mathbf{J}$ is zero everywhere on the surface, show that

$$\int \mathbf{J} \, d\tau = 0.$$

Hint. Take one component of $\mathbf{J}$ at a time. With $\nabla \cdot \mathbf{J} = 0$, show that $J_i = \nabla \cdot x_i \mathbf{J}$ and apply Gauss's theorem.

1.11.9   The creation of a *localized* system of steady electric currents (current density $\mathbf{J}$) and magnetic fields may be shown to require an amount of work

$$W = \tfrac{1}{2} \int \mathbf{H} \cdot \mathbf{B} \, d\tau.$$

Transform this into

$$W = \tfrac{1}{2} \int \mathbf{J} \cdot \mathbf{A} \, d\tau.$$

Here $\mathbf{A}$ is the magnetic vector potential: $\nabla \times \mathbf{A} = \mathbf{B}$.
Hint. In Maxwell's equations take the displacement current term $\partial\mathbf{D}/\partial t = 0$. If the fields and currents are localized, a bounding surface may be taken far enough out so that the integrals of the fields and currents over the surface yield zero.

1.11.10   Prove the generalization of Green's theorem:

$$\int (v\mathcal{L}u - u\mathcal{L}v) \, d\tau = \int p(v\nabla u - u\nabla v) \cdot d\boldsymbol{\sigma}.$$

Here $\mathcal{L}$ is the self-adjoint operator (Section 9.1),

$$\mathcal{L} = \nabla \cdot [p(\mathbf{r})\nabla] + q(\mathbf{r})$$

and p, q, u, and v are functions of position, p and q having continuous first derivatives and u and v having continuous second derivatives.
Note. This generalized Green's theorem appears in Section 8.7.

1.12 STOKES'S THEOREM

Gauss's theorem relates the volume integral of a derivative of a function to an integral of the function over the closed surface bounding the volume. Here we consider an analogous relation between the surface integral of a derivative of a function and the line integral of the function, the path of integration being the perimeter bounding the surface.

Let us take the surface and subdivide it into a network of arbitrarily small rectangles. In Section 1.8 we showed that the circulation about such a differential rectangle (in the xy-plane) is $\nabla \times V|_2 \, dx \, dy$. From Eq. (1.77) applied to *one* differential rectangle

$$\sum_{\text{four sides}} \mathbf{V} \cdot d\lambda = \nabla \times \mathbf{V} \cdot d\sigma. \tag{1.110}$$

We sum over all the little rectangles as in the definition of a Riemann integral. The surface contributions (right-hand side of Eq. (1.110)) are added together. The line integrals (left-hand side of Eq. (1.110)) of all *interior* line segments cancel identically. Only the line integral around the perimeter survives (Fig. 1.25). Taking the usual limit as the number of rectangles approaches infinity while $dx \to 0$, $dy \to 0$, we have

$$\sum_{\substack{\text{exterior} \\ \text{line segments}}} \mathbf{V} \cdot d\lambda = \sum_{\text{rectangles}} \nabla \times \mathbf{V} \cdot d\sigma$$

$$\oint \mathbf{V} \cdot d\lambda \quad = \quad \int_s \nabla \times \mathbf{V} \cdot d\sigma. \tag{1.111}$$

This is Stokes's theorem. The surface integral on the right is over the surface bounded by the perimeter or contour for the line integral on the left. The direction of the vector representing the area is out of the paper plane toward the reader if the direction of traversal around the contour for the line integral is in the positive mathematical sense as shown in Fig. 1.25.

This demonstration of Stokes's theorem is limited by the fact that we used a Maclaurin expansion of $\mathbf{V}(x, y, z)$ in establishing Eq. (1.77) in Section 1.8. Actually we need only demand that the curl of $\mathbf{V}(x, y, z)$ exists and that it be integrable over the surface. A proof of the Cauchy integral theorem analogous to the development of Stokes's theorem here but using these less restrictive conditions appears in Section 6.3.

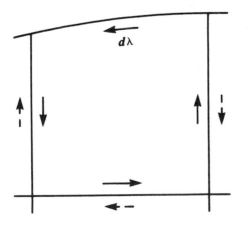

Figure 1.25 Exact cancellation on *interior* paths. No cancellation on the *exterior* path.

Stokes's theorem obviously applies to an open surface. It is possible to consider a closed surface as a limiting case of an open surface with the opening (and therefore the perimeter) shrinking to zero. This is the point of Exercise 1.12.7.

Alternate Forms of Stokes's Theorem

As with Gauss's theorem, other relations between surface and line integrals are possible. We find

$$\int_S d\boldsymbol{\sigma} \times \boldsymbol{\nabla}\varphi = \oint \varphi \, d\boldsymbol{\lambda} \tag{1.112}$$

and

$$\int_S (d\boldsymbol{\sigma} \times \boldsymbol{\nabla}) \times \mathbf{P} = \oint d\boldsymbol{\lambda} \times \mathbf{P}. \tag{1.113}$$

Equation (1.112) may readily be verified by the substitution $\mathbf{V} = \mathbf{a}\varphi$ in which $\mathbf{a}$ is a vector of constant magnitude and of constant direction, as in Section 1.11. Substituting into Stokes's theorem, Eq. (1.111)

$$\int_S (\boldsymbol{\nabla} \times \mathbf{a}\varphi) \cdot d\boldsymbol{\sigma} = -\int_S \mathbf{a} \times \boldsymbol{\nabla}\varphi \cdot d\boldsymbol{\sigma}$$

$$= -\mathbf{a} \cdot \int_S \boldsymbol{\nabla}\varphi \times d\boldsymbol{\sigma}. \tag{1.114}$$

For the line integral

$$\oint \mathbf{a}\varphi \cdot d\boldsymbol{\lambda} = \mathbf{a} \cdot \oint \varphi \, d\boldsymbol{\lambda}, \tag{1.115}$$

and we obtain

$$\mathbf{a} \cdot \left(\oint \varphi \, d\boldsymbol{\lambda} + \int_S \boldsymbol{\nabla}\varphi \times d\boldsymbol{\sigma} \right) = 0. \tag{1.116}$$

Since the choice of direction of $\mathbf{a}$ is arbitrary, the expression in parentheses must vanish, thus verifying Eq. (1.112). Equation (1.113) may be derived similarly by using $\mathbf{V} = \mathbf{a} \times \mathbf{P}$, in which $\mathbf{a}$ is again a constant vector.

Both Stokes's and Gauss's theorems are of tremendous importance in a wide variety of problems involving vector calculus. Some idea of their power and versatility may be obtained from the exercises of Sections 1.11 and 1.12 and the development of potential theory in Sections 1.13 and 1.14.

EXERCISES

1.12.1   Given a vector $\mathbf{t} = -\hat{x}y + \hat{y}x$, with the help of Stokes's theorem, show that the integral around a continuous closed curve in the xy-plane

$$\tfrac{1}{2} \oint \mathbf{t} \cdot d\lambda = \tfrac{1}{2} \oint (x\,dy - y\,dx) = A,$$

the area enclosed by the curve.

1.12.2   The calculation of the magnetic moment of a current loop leads to the line integral

$$\oint \mathbf{r} \times d\mathbf{r}.$$

(a) Integrate around the perimeter of a current loop (in the xy-plane) and show that the scalar magnitude of this line integral is twice the area of the enclosed surface.

(b) The perimeter of an ellipse is described by $\mathbf{r} = \hat{x}a \cos \theta + \hat{y}b \sin \theta$. From part (a) show that the area of the ellipse is πab.

1.12.3   Evaluate $\oint \mathbf{r} \times d\mathbf{r}$ by using the alternate form of Stokes's theorem given by Eq. (1.113):

$$\int_S (d\boldsymbol{\sigma} \times \boldsymbol{\nabla}) \times \mathbf{P} = \oint d\lambda \times \mathbf{P}.$$

Take the loop to be entirely in the xy-plane.

1.12.4   In steady state the magnetic field $\mathbf{H}$ satisfies the Maxwell equation $\boldsymbol{\nabla} \times \mathbf{H} = \mathbf{J}$, where $\mathbf{J}$ is the current density (per square meter). At the boundary between two media there is a surface current density $\mathbf{K}$ (perimeter). Show that a boundary condition on $\mathbf{H}$ is

$$\mathbf{n} \times (\mathbf{H}_2 - \mathbf{H}_1) = \mathbf{K}.$$

$\mathbf{n}$ is a unit vector normal to the surface and out of medium 1.
Hint. Consider a narrow loop perpendicular to the interface as shown in the figure.

1.12.5   From Maxwell's equations, $\boldsymbol{\nabla} \times \mathbf{H} = \mathbf{J}$ with $\mathbf{J}$ here the current density and $\mathbf{E} = 0$. Show from this that

$$\oint \mathbf{H} \cdot d\mathbf{r} = I,$$

where I is the net electric current enclosed by the loop integral. These are the differential and integral forms of Ampere's law of magnetism.

1.12.6   A magnetic induction $\mathbf{B}$ is generated by electric current in a ring of radius R. Show that the *magnitude* of the vector potential $\mathbf{A}$ ($\mathbf{B} = \boldsymbol{\nabla} \times \mathbf{A}$) at the ring is

$$|\mathbf{A}| = \frac{\varphi}{2\pi R},$$

where φ is the total magnetic flux passing through the ring.
Note. $\mathbf{A}$ is tangential to the ring.

1.12.7   Prove that

$$\int_S \mathbf{\nabla} \times \mathbf{V} \cdot d\mathbf{\sigma} = 0,$$

if S is a closed surface.

1.12.8   Evaluate $\oint \mathbf{r} \cdot d\mathbf{r}$ (Exercise 1.10.4) by Stokes's theorem.

1.12.9   Prove that

$$\oint u \, \mathbf{\nabla} v \cdot d\mathbf{\lambda} = -\oint v \, \mathbf{\nabla} u \cdot d\mathbf{\lambda}.$$

1.12.10   Prove that

$$\oint u \, \mathbf{\nabla} v \cdot d\mathbf{\lambda} = \int_S (\mathbf{\nabla} u) \times (\mathbf{\nabla} v) \cdot d\mathbf{\sigma}.$$

1.13 POTENTIAL THEORY

SCALAR POTENTIAL

If a force over a given region of space S can be expressed as the negative gradient of a scalar function φ,

$$\mathbf{F} = -\mathbf{\nabla}\varphi, \tag{1.117}$$

we call φ a scalar potential which describes the force by one function instead of three. A scalar potential is only determined up to an additive constant which can be used to adjust its origin. The force $\mathbf{F}$ appearing as the negative gradient of a single-valued scalar potential is labeled a *conservative* force. We want to know when a scalar potential function exists. To answer this question we establish two other relations as equivalent to Eq. (1.117). These are

$$\mathbf{\nabla} \times \mathbf{F} = 0 \tag{1.118}$$

and

$$\oint \mathbf{F} \cdot d\mathbf{r} = 0, \tag{1.119}$$

for every closed path in our region S. We proceed to show that each of these three equations implies the other two.

Let us start with

$$\mathbf{F} = -\mathbf{\nabla}\varphi. \tag{1.120}$$

Then

$$\mathbf{\nabla} \times \mathbf{F} = -\mathbf{\nabla} \times \mathbf{\nabla}\varphi = 0 \tag{1.121}$$

by Eq. (1.83) or Eq. (1.17) implies Eq. (1.118). Turning to the line integral, we have

$$\oint \mathbf{F} \cdot d\mathbf{r} = -\oint \mathbf{\nabla}\varphi \cdot d\mathbf{r} = -\oint d\varphi, \tag{1.122}$$

using Eq. (1.62). Now $d\varphi$ integrates to give φ. Since we have specified a closed loop, the end points coincide and we get zero for every closed path in our region S for which Eq. (1.117) holds. It is important to note the restriction here that the potential be single-valued and that Eq. (1.117) hold for *all* points in S. This problem may arise in using a scalar magnetic potential, a perfectly valid procedure as long as no net current is encircled. As soon as we choose a path in space that encircles a net current, the scalar magnetic potential ceases to be single-valued and our analysis no longer applies.

Continuing this demonstration of equivalence, let us assume that Eq. (1.119) holds. If $\oint \mathbf{F} \cdot d\mathbf{r} = 0$ for all paths in S, we see that the value of the integral joining two distinct points A and B is independent of the path (Fig. 1.26). Our premise is that

$$\oint_{ACBDA} \mathbf{F} \cdot d\mathbf{r} = 0. \tag{1.123}$$

Therefore

$$\int_{ACB} \mathbf{F} \cdot d\mathbf{r} = -\int_{BDA} \mathbf{F} \cdot d\mathbf{r} = \int_{ADB} \mathbf{F} \cdot d\mathbf{r}, \tag{1.124}$$

reversing the sign by reversing the direction of integration. Physically, this means that the work done in going from A to B is independent of the path and that the work done in going around a closed path is zero. This is the reason for labeling such a force conservative: Energy is conserved.

With the result shown in Eq. (1.124), we have the work done dependent only on the end points, A and B. That is,

$$\text{work done by force} = \int_A^B \mathbf{F} \cdot d\mathbf{r} = \varphi(A) - \varphi(B). \tag{1.25}$$

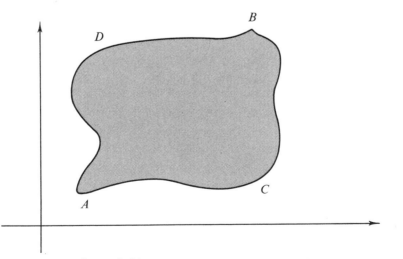

Figure 1.26   Possible paths for doing work.

Equation (1.125) defines a scalar potential (strictly speaking, the difference in potential between points A and B) and provides a means of calculating the potential. If point B is taken as a variable, say, (x, y, z), then differentiation with respect to x, y, and z will recover Eq. (1.117).

The choice of sign on the right-hand side is arbitrary. The choice here is made to achieve agreement with Eq. (1.117) and to ensure that water will run downhill rather than uphill. For points A and B separated by a length $d\mathbf{r}$, Eq. (1.125) becomes

$$\mathbf{F} \cdot d\mathbf{r} = -d\varphi$$

$$= -\nabla\varphi \cdot d\mathbf{r}. \tag{1.126}$$

This may be rewritten

$$(\mathbf{F} + \nabla\varphi) \cdot d\mathbf{r} = 0, \tag{1.127}$$

and since $d\mathbf{r}$ is arbitrary, Eq. (1.117) must follow.

If

$$\oint \mathbf{F} \cdot d\mathbf{r} = 0, \tag{1.128}$$

we may obtain Eq. (1.118) by using Stokes's theorem (Eq. (1.115)).

$$\oint \mathbf{F} \cdot d\mathbf{r} = \int \nabla \times \mathbf{F} \cdot d\boldsymbol{\sigma}. \tag{1.129}$$

If we take the path of integration to be the perimeter of an arbitrary differential area $d\boldsymbol{\sigma}$, the integrand in the surface integral must vanish. Hence Eq. (1.119) implies Eq. (1.118).

Finally, if $\nabla \times \mathbf{F} = 0$, we need only reverse our statement of Stokes's theorem (Eq. (1.129)) to derive Eq. (1.119). Then, by Eqs. (1.125)–(1.127) the initial statement $\mathbf{F} = -\nabla\varphi$ is derived. The triple equivalence is demonstrated (Fig. 1.27).

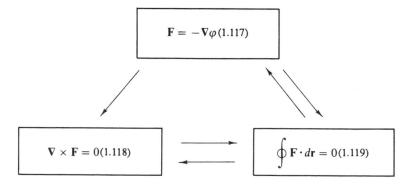

Figure 1.27   Equivalent formulations of a conservative force.

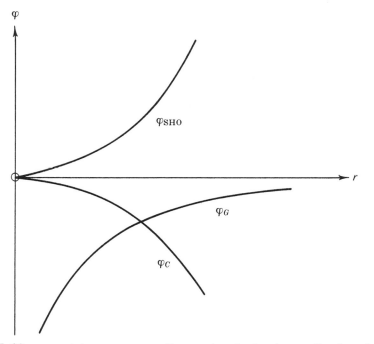

Figure 1.28 Potential energy versus distance (gravitational, centrifugal, and simple harmonic oscillator).

To summarize, a single-valued scalar potential function φ exists if and only if **F** is irrotational or the work done around every closed loop is zero. The gravitational and electrostatic force fields given by Eq. (1.81) are irrotational and therefore are conservative. Gravitational and electrostatic scalar potentials exist. Now, by calculating the work done (Eq. (1.125)), we proceed to determine three potentials, Fig. 1.28.

Example 1.13.1: Gravitational Potential
Find the scalar potential for the gravitational force on a unit mass m_1,

$$\mathbf{F}_G = -\frac{Gm_1 m_2 \hat{\mathbf{r}}}{r^2} = -\frac{k\hat{\mathbf{r}}}{r^2}, \qquad \text{radially } inward. \tag{1.130}$$

By integrating Eq. (1.117) from infinity into position **r**, we obtain

$$\varphi_G(r) - \varphi_G(\infty) = -\int_\infty^r \mathbf{F}_G \cdot d\mathbf{r} = +\int_r^\infty \mathbf{F}_G \cdot d\mathbf{r}. \tag{1.131}$$

By use of $\mathbf{F}_G = -\mathbf{F}_{\text{applied}}$, a comparison with Eq. (1.94a) shows that the potential is the work done in bringing the unit mass in from infinity. (We can define only potential difference. Here we arbitrarily assign infinity to be a zero of potential.) The integral on the right-hand side of Eq. (1.131) is negative, meaning that $\varphi_G(r)$ is negative. Since $\mathbf{F}_G$ is radial, we obtain a contribution to φ only when $d\mathbf{r}$ is radial or

$$\varphi_G(r) = -\int_r^\infty \frac{k\, dr}{r^2} = -\frac{k}{r} = -\frac{Gm_1 m_2}{r}.$$

The final negative sign is a consequence of the attractive force of gravity.

Example 1.13.2 Centrifugal Potential

Calculate the scalar potential for the *centrifugal* force per unit mass, $\mathbf{F}_C = \omega^2 r \hat{\mathbf{r}}$, radially *outward*. Physically, this might be you on a large horizontal spinning disk at an amusement park. Proceeding as in Example 1.13.1, but integrating from the origin outward and taking $\varphi_C(0) = 0$, we have

$$\varphi_C(r) = -\int_0^r \mathbf{F}_C \cdot d\mathbf{r} = -\frac{\omega^2 r^2}{2}.$$

If we reverse signs, taking $\mathbf{F}_{SHO} = -k\mathbf{r}$, we obtain $\varphi_{SHO} = \frac{1}{2}kr^2$, the simple harmonic oscillator potential.

The gravitational, centrifugal, and simple harmonic oscillator potentials are shown in Fig. 1.28. Clearly, the simple harmonic oscillator yields stability and describes a restoring force. The centrifugal potential describes an unstable situation.

Thermodynamics—Exact Differentials

In thermodynamics, which is sometimes called a search for exact differentials, we encounter equations of the form

$$df = P(x, y)\, dx + Q(x, y)\, dy. \tag{1.132}$$

The usual problem is to determine whether $\int (P(x, y)\, dx + Q(x, y)\, dy)$ depends only on the end points, that is, whether df is indeed an exact differential. The necessary and sufficient condition is that

$$df = \frac{\partial f}{\partial x}\, dx + \frac{\partial f}{\partial y}\, dy \tag{1.132a}$$

or that

$$P(x, y) = \frac{\partial f}{\partial x},$$

$$Q(x, y) = \frac{\partial f}{\partial y}. \tag{1.132b}$$

Equations (1.132b) depend on the relation

$$\frac{\partial P(x, y)}{\partial y} = \frac{\partial Q(x, y)}{\partial x} \tag{1.132c}$$

being satisfied. This, however, is exactly analogous to Eq. (1.118), the requirement that **F** be irrotational. Indeed, the z-component of Eq. (1.118) yields

$$\frac{\partial F_x}{\partial y} = \frac{\partial F_y}{\partial x},\qquad (1.132\text{d})$$

with

$$F_x = \frac{\partial f}{\partial x},\qquad F_y = \frac{\partial f}{\partial y}.$$

Vector Potential

In some branches of physics, especially electromagnetic theory, it is convenient to introduce a vector potential **A**, such that a (force) field **B** is given by

$$\mathbf{B} = \mathbf{\nabla} \times \mathbf{A}.\qquad (1.133)$$

Clearly, if Eq. (1.133) holds, $\mathbf{\nabla} \cdot \mathbf{B} = 0$ by Eq. (1.85) and **B** is solenoidal. Here we want to develop a converse, to show that when **B** is solenoidal a vector potential **A** exists. We demonstrate the existence of **A** by actually calculating it. Suppose $\mathbf{B} = \hat{\mathbf{x}}b_1 + \hat{\mathbf{y}}b_2 + \hat{\mathbf{z}}b_3$ and our unknown $\mathbf{A} = \hat{\mathbf{x}}a_1 + \hat{\mathbf{y}}a_2 + \hat{\mathbf{z}}a_3$. By Eq. (1.133)

$$\frac{\partial a_3}{\partial y} - \frac{\partial a_2}{\partial z} = b_1,\qquad (1.134\text{a})$$

$$\frac{\partial a_1}{\partial z} - \frac{\partial a_3}{\partial x} = b_2,\qquad (1.134\text{b})$$

$$\frac{\partial a_2}{\partial x} - \frac{\partial a_1}{\partial y} = b_3.\qquad (1.134\text{c})$$

Let us assume that the coordinates have been chosen so that **A** is parallel to the yz-plane; that is, $a_1 = 0$.[1] Then

$$b_2 = -\frac{\partial a_3}{\partial x}$$

$$\qquad\qquad (1.135)$$

$$b_3 = \frac{\partial a_2}{\partial x}.$$

Integrating, we obtain

$$a_2 = \int_{x_0}^{x} b_3 \, dx + f_2(y, z),$$

$$\qquad\qquad (1.136)$$

$$a_3 = -\int_{x_0}^{x} b_2 \, dx + f_3(y, z),$$

where f_2 and f_3 are arbitrary functions of y and z but are *not* functions of x. These two equations can be checked by differentiating and recovering

---

[1] Clearly, this can be done at any one point. It is not at all obvious that this assumption will hold at all points; that is, **A** will be two dimensional. The justification for the assumption is that it works; Eq. (1.140) satisfies Eq. (1.133).

Eq. (1.135). Equation (1.134a) becomes[2]

$$\frac{\partial a_3}{\partial y} - \frac{\partial a_2}{\partial z} = -\int_{x_0}^{x} \left(\frac{\partial b_2}{\partial y} + \frac{\partial b_3}{\partial z}\right) dx + \frac{\partial f_3}{\partial y} - \frac{\partial f_2}{\partial z}$$

$$= \int_{x_0}^{x} \frac{\partial b_1}{\partial x} dx + \frac{\partial f_3}{\partial y} - \frac{\partial f_2}{\partial z}, \tag{1.137}$$

using $\nabla \cdot \mathbf{B} = 0$. Integrating with respect to x, we obtain

$$\frac{\partial a_3}{\partial y} - \frac{\partial a_2}{\partial z} = b_1(x, y, z) - b_1(x_0, y, z) + \frac{\partial f_3}{\partial y} - \frac{\partial f_2}{\partial z}. \tag{1.138}$$

Remembering that f_3 and f_2 are arbitrary functions of y and z, we choose

$$f_2 = 0,$$

$$f_3 = \int_{y_0}^{y} b_1(x_0, y, z)\, dy, \tag{1.139}$$

so that the right-hand side of Eq. (1.138) reduces to $b_1(x, y, z)$ in agreement with Eq. (1.134a). With f_2 and f_3 given by Eq. (1.139), we can construct $\mathbf{A}$.

$$\mathbf{A} = \hat{\mathbf{y}} \int_{x_0}^{x} b_3(x, y, z)\, dx + \hat{\mathbf{z}} \left[\int_{y_0}^{y} b_1(x_0, y, z)\, dy - \int_{x_0}^{x} b_2(x, y, z)\, dx \right]. \tag{1.140}$$

This is not quite complete. We may add any constant since $\mathbf{B}$ is a derivative of $\mathbf{A}$. What is much more important, we may add any gradient of a scalar function $\nabla \varphi$ without affecting $\mathbf{B}$ at all. Finally, the functions f_2 and f_3 are not unique. Other choices could have been made. Instead of setting $a_1 = 0$ to get Eq. (1.134) any cyclic permutation of 1, 2, 3, x, y, z, x_0, y_0, z_0 would also work.

Example 1.13.3 A Magnetic Vector Potential for a Constant Magnetic Field

To illustrate the construction of a magnetic vector potential, we take the special but still important case of a constant magnetic induction

$$\mathbf{B} = \hat{\mathbf{z}} B_z, \tag{1.141}$$

in which B_z is a constant. Equation (1.134) becomes

$$\frac{\partial a_3}{\partial y} - \frac{\partial a_2}{\partial z} = 0,$$

$$\frac{\partial a_1}{\partial z} - \frac{\partial a_3}{\partial x} = 0, \tag{1.142}$$

$$\frac{\partial a_2}{\partial x} - \frac{\partial a_1}{\partial y} = B_z.$$

---

[2] Leibniz' formula in Exercise 8.6.13 is useful here.

If we assume that $a_1 = 0$, as before, then by Eq. (1.140)

$$\mathbf{A} = \hat{\mathbf{y}} \int^x B_z \, dx$$

$$= \hat{\mathbf{y}} x B_z,$$
(1.143)

setting a constant of integration equal to zero. It can readily be seen that this **A** satisfies Eq. (1.133).

To show that the choice $a_1 = 0$ was not sacred or at least not required, let us try setting $a_3 = 0$. From Eq. (1.142)

$$\frac{\partial a_2}{\partial z} = 0,$$
(1.144a)

$$\frac{\partial a_1}{\partial z} = 0,$$
(1.144b)

$$\frac{\partial a_2}{\partial x} - \frac{\partial a_1}{\partial y} = B_z.$$
(1.144c)

We see a_1 and a_2 are independent of z or

$$a_1 = a_1(x, y), \qquad a_2 = a_2(x, y).$$
(1.145)

Equation (1.144c) is satisfied if we take

$$a_2 = p \int^x B_z \, dx = p x B_z$$
(1.146)

and

$$a_1 = (p - 1) \int^y B_z \, dy = (p - 1) y B_z,$$
(1.147)

with p any constant. Then

$$\mathbf{A} = \hat{\mathbf{x}}(p - 1) y B_z + \hat{\mathbf{y}} p x B_z.$$
(1.148)

Again, Eqs. (1.133), (1.141), and (1.148) are seen to be consistent. Comparison of Eqs. (1.143) and (1.148) shows immediately that **A** is not unique. The difference between Eqs. (1.143) and (1.148) and the appearance of the parameter p in Eq. (1.148) may be accounted for by rewriting Eq. (1.148) as

$$\mathbf{A} = -\tfrac{1}{2}(\hat{\mathbf{x}} y - \hat{\mathbf{y}} x) B_z + (p - \tfrac{1}{2})(\hat{\mathbf{x}} y + \hat{\mathbf{y}} x) B_z$$

$$= -\tfrac{1}{2}(\hat{\mathbf{x}} y - \hat{\mathbf{y}} x) B_z + (p - \tfrac{1}{2}) B_z \nabla \varphi$$
(1.149)

with

$$\varphi = xy.$$
(1.150)

The first term in **A** corresponds to the usual form

$$\mathbf{A} = \tfrac{1}{2}(\mathbf{B} \times \mathbf{r})$$
(1.151)

for **B**, a constant.

Adding a gradient of a scalar function, Λ say, to the vector potential $\mathbf{A}$ does not affect $\mathbf{B}$, by Eq. (1.83), and is known as a gauge transformation (see Exercises 1.13.9 and 4.7.4)

$$\mathbf{A} \rightarrow \mathbf{A}' = \mathbf{A} + \nabla\Lambda. \tag{1.152}$$

Suppose now that the wave function ψ_0 solves the Schrödinger equation of quantum mechanics without magnetic induction field $\mathbf{B}$,

$$\left\{ \frac{1}{2m} (-i\hbar\nabla)^2 + V - E \right\} \psi_0 = 0, \tag{1.153}$$

describing a particle with mass m and charge e. When $\mathbf{B}$ is switched on the wave equation becomes

$$\left\{ \frac{1}{2m} \left(-i\hbar\nabla - \frac{e}{c}\mathbf{A} \right)^2 + V - E \right\} \psi = 0, \tag{1.154}$$

where c is the velocity of light. Its solution ψ picks up a phase factor that depends on the coordinates in general,

$$\psi(\mathbf{r}) = \exp\left[\frac{ie}{\hbar c} \int^{\mathbf{r}} \mathbf{A}(\mathbf{r}') \cdot d\mathbf{r}' \right] \psi_0(\mathbf{r}). \tag{1.155}$$

From the relation

$$\left(-i\hbar\nabla - \frac{e}{c}\mathbf{A} \right)\psi = \exp\left[\frac{ie}{\hbar c} \int \mathbf{A} \cdot d\mathbf{r}' \right] \left\{ \left(-i\hbar\nabla - \frac{e}{c}\mathbf{A} \right)\psi_0 - i\hbar\psi_0 \frac{ie}{\hbar c}\mathbf{A} \right\}$$

$$= \exp\left[\frac{ie}{\hbar c} \int \mathbf{A} \cdot d\mathbf{r}' \right] (-i\hbar\nabla\psi_0), \tag{1.156}$$

it is obvious that ψ solves Eq. (1.154), if ψ_0 solves Eq. (1.153). The *gauge covariant derivative* $\nabla - i(e/\hbar c)\mathbf{A}$ describes the coupling of a charged particle with the magnetic field. It is often called minimal substitution and plays a central role in quantum electromagnetism, the first and simplest gauge theory in physics.

To summarize this discussion of the vector potential, when a vector $\mathbf{B}$ is solenoidal, a vector potential $\mathbf{A}$ exists such that $\mathbf{B} = \nabla \times \mathbf{A}$. $\mathbf{A}$ is undetermined to within an additive gradient. This corresponds to the arbitrary zero of a potential, a constant of integration for the scalar potential.

In many problems the magnetic vector potential $\mathbf{A}$ will be obtained from the current distribution that produces the magnetic induction $\mathbf{B}$. This means solving Poisson's (vector) equation (see Exercise 1.14.4).

EXERCISES

1.13.1   If a force $\mathbf{F}$ is given by

$$\mathbf{F} = (x^2 + y^2 + z^2)^n (\hat{\mathbf{x}}x + \hat{\mathbf{y}}y + \hat{\mathbf{z}}z),$$

find

(a) $\mathbf{V} \cdot \mathbf{F}$.
(b) $\mathbf{V} \times \mathbf{F}$.
(c) A scalar potential $\varphi(x, y, z)$ so that $\mathbf{F} = -\mathbf{V}\varphi$.
(d) For what value of the exponent n does the scalar potential diverge at both the origin and infinity?

$$ANS. \quad (a) \; (2n + 3)r^{2n} \qquad (c) \; -\frac{1}{2n + 2}r^{2n+2}, \; n \neq -1$$

$$(b) \; 0 \qquad\qquad (d) \; n = -1, \; \varphi = -\ln r.$$

1.13.2   A sphere of radius a is uniformly charged (throughout its volume). Construct the electrostatic potential $\varphi(r)$ for $0 \le r < \infty$.
Hint. In Section 1.14 it is shown that the Coulomb force on a test charge at $r = r_0$ depends only on the charge at distances less than r_0 and is independent of the charge at distances greater than r_0. Note that this applies to a *spherically symmetric* charge distribution.

1.13.3   The usual problem in classical mechanics is to calculate the motion of a particle given the potential. For a uniform density (ρ_0), nonrotating massive sphere, Gauss's law of Section 1.14 leads to a gravitational force on a unit mass m_0 at a point r_0 produced by the attraction of the mass at $r \le r_0$. The mass at $r > r_0$ contributes nothing to the force.

(a) Show that $\mathbf{F}/m_0 = -(4\pi G\rho_0/3)\mathbf{r}$, $0 \le r \le a$, where a is the radius of the sphere.
(b) Find the corresponding gravitational potential, $0 \le r \le a$.
(c) Imagine a vertical hole running completely through the center of the earth and out to the far side. Neglecting the rotation of the earth and assuming a uniform density $\rho_0 = 5.5 \, \text{gm/cm}^3$, calculate the nature of the motion of a particle dropped into the hole. What is its period?

Note. $\mathbf{F} \propto \mathbf{r}$ is actually a very poor approximation. Because of varying density, the approximation $\mathbf{F} = \text{constant}$, along the outer half of a radial line and $\mathbf{F} \propto \mathbf{r}$ along the inner half is a much closer approximation.

1.13.4   The origin of the cartesian coordinates is at the Earth's center. The moon is on the z-axis, a fixed distance R away (center-to-center distance). The tidal force exerted by the moon on a particle at the earth's surface (point x, y, z) is given by

$$F_x = -GMm\,\frac{x}{R^3}, \qquad F_y = -GMm\,\frac{y}{R^3}, \qquad F_z = +GMm\,\frac{z}{R^3}.$$

Find the potential that yields this tidal force.

$$ANS. \quad \frac{-GMm}{R^3}(z^2 - \tfrac{1}{2}x^2 - \tfrac{1}{2}y^2)$$

In terms of the Legendre polynomials of Chapter 12 this becomes

$$\frac{-GMm}{R^3}r^2 P_2(\cos \theta).$$

1.13.5   A long straight wire carrying a current I produces a magnetic induction $\mathbf{B}$
with components

$$\mathbf{B} = \frac{\mu_0 I}{2\pi}\left(\frac{-y}{x^2+y^2}, \frac{x}{x^2+y^2}, 0\right).$$

Find a magnetic vector potential, $\mathbf{A}$

$$ANS. \quad \mathbf{A} = -\hat{\mathbf{z}}(\mu_0 I/4\pi)\ln(x^2+y^2).$$
(This solution is not unique.)

1.13.6   If

$$\mathbf{B} = \frac{\hat{\mathbf{r}}}{r^2} = \left(\frac{x}{r^3}, \frac{y}{r^3}, \frac{z}{r^3}\right).$$

find a vector $\mathbf{A}$ such that $\nabla \times \mathbf{A} = \mathbf{B}$. One possible solution is

$$\mathbf{A} = \frac{\hat{\mathbf{x}}yz}{r(x^2+y^2)} - \frac{\hat{\mathbf{y}}xz}{r(x^2+y^2)}.$$

1.13.7   Show that the pair of equations

$$\mathbf{A} = \tfrac{1}{2}(\mathbf{B} \times \mathbf{r}),$$
$$\mathbf{B} = \nabla \times \mathbf{A},$$

is satisfied by any constant vector $\mathbf{B}$ (any orientation).

1.13.8   Vector $\mathbf{B}$ is formed by the product of two gradients

$$\mathbf{B} = (\nabla u) \times (\nabla v),$$

where u and v are scalar functions.

(a) Show that $\mathbf{B}$ is solenoidal.
(b) Show that

$$\mathbf{A} = \tfrac{1}{2}(u\,\nabla v - v\,\nabla u)$$

is a vector potential for $\mathbf{B}$ in that

$$\mathbf{B} = \nabla \times \mathbf{A}.$$

1.13.9   The magnetic induction $\mathbf{B}$ is related to the magnetic vector potential $\mathbf{A}$ by
$\mathbf{B} = \nabla \times \mathbf{A}$. By Stokes's theorem

$$\int \mathbf{B} \cdot d\boldsymbol{\sigma} = \oint \mathbf{A} \cdot d\mathbf{r}.$$

Show that each side of this equation is invariant under the *gauge transformation*, $\mathbf{A} \to \mathbf{A} + \nabla\psi$.
Note. Take the function ψ to be single-valued. The complete gauge transformation is considered in Exercise 4.7.4.

1.13.10   With $\mathbf{E}$ the electric field and $\mathbf{A}$ the magnetic vector potential, show that
$[\mathbf{E} + \partial\mathbf{A}/\partial t]$ is irrotational and that therefore we may write

$$\mathbf{E} = -\nabla\varphi - \frac{\partial\mathbf{A}}{\partial t}.$$

1.13.11   The total force on a charge q moving with velocity $\mathbf{v}$ is

$$\mathbf{F} = q(\mathbf{E} + \mathbf{v} \times \mathbf{B}).$$

Using the scalar and vector potentials, show that

$$\mathbf{F} = q\left[-\nabla\varphi - \frac{d\mathbf{A}}{dt} + \nabla(\mathbf{A}\cdot\mathbf{v})\right].$$

Note that we now have a total time derivative of $\mathbf{A}$ in place of the partial derivative of Exercise 1.13.10.

1.14 GAUSS'S LAW, POISSON'S EQUATION

GAUSS'S LAW

Consider a point electric charge q at the origin of our coordinate system. This produces an electric field $\mathbf{E}$ given by[1]

$$\mathbf{E} = \frac{q\hat{\mathbf{r}}}{4\pi\varepsilon_0 r^2}. \tag{1.157}$$

We now derive Gauss's law which states that the surface integral in Fig. 1.29 is q/ε_0 if the closed surface S includes the origin (where q is located) and zero if the surface does not include the origin. The surface S is any closed surface; it need not be spherical.

Using Gauss's theorem, Eq. (1.100) (and neglecting the $q/4\pi\varepsilon_0$), we obtain

$$\int_S \frac{\hat{\mathbf{r}}\cdot d\boldsymbol{\sigma}}{r^2} = \int_V \nabla\cdot\left(\frac{\hat{\mathbf{r}}}{r^2}\right)d\tau = 0 \tag{1.158}$$

by Example 1.7.2, provided the surface S does not include the origin, where the integrands are not defined. This proves the second part of Gauss's law.

The first part, in which the surface S must include the origin, may be handled by surrounding the origin with a small sphere S' of radius δ (Fig. 1.30). So that there will be no question what is inside and what is outside, imagine the volume outside the outer surface S and the volume inside surface $S'(r < \delta)$ connected by a small hole. This joins surfaces S and S', combining them into one single simply connected closed surface. Because the radius of the imaginary hole may be made vanishingly small, there is no additional contribution to the surface integral. The inner surface is deliberately chosen to be spherical so that we will be able to integrate over it. Gauss's theorem now applies to the volume between S and S' without any difficulty. We have

$$\int_S \frac{\hat{\mathbf{r}}\cdot d\boldsymbol{\sigma}}{r^2} + \int_{S'} \frac{\hat{\mathbf{r}}\cdot d\boldsymbol{\sigma}'}{\delta^2} = 0. \tag{1.159}$$

We may evaluate the second integral, for $d\boldsymbol{\sigma}' = -\hat{\mathbf{r}}\delta^2\,d\Omega$, in which $d\Omega$ is an element of solid angle. The minus sign appears because we agreed in Section

---

[1] The electric field $\mathbf{E}$ is defined as the force per unit charge on a small stationary test charge q_t: $\mathbf{E} = \mathbf{F}/q_t$. From Coulomb's law the force on q_t due to q is $\mathbf{F} = (qq_t/4\pi\varepsilon_0)(\hat{\mathbf{r}}/r^2)$. When we divide by q_t, Eq. (1.157) follows.

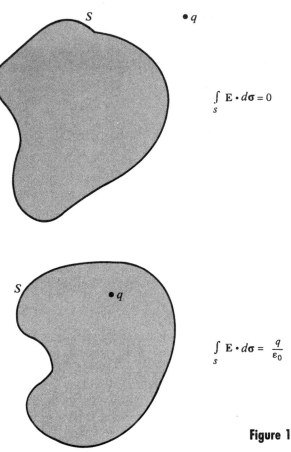

$$\int_s \mathbf{E} \cdot d\boldsymbol{\sigma} = 0$$

$$\int_s \mathbf{E} \cdot d\boldsymbol{\sigma} = \frac{q}{\varepsilon_0}$$

Figure 1.29 Gauss's law.

1.10 to have the positive normal $\hat{\mathbf{r}}'$ *outward* from the volume. In this case the outward $\hat{\mathbf{r}}'$ is in the negative radial direction, $\hat{\mathbf{r}}' = -\hat{\mathbf{r}}$. By integrating over all angles, we have

$$\int_{S'} \frac{\hat{\mathbf{r}} \cdot d\boldsymbol{\sigma}'}{\delta^2} = -\int_{S'} \frac{\hat{\mathbf{r}} \cdot \hat{\mathbf{r}} \delta^2 \, d\Omega}{\delta^2} = -4\pi, \qquad (1.160)$$

independent of the radius δ. With the constants from Eq. (1.157), this results in

$$\int_S \mathbf{E} \cdot d\boldsymbol{\sigma} = \frac{q}{4\pi\varepsilon_0} 4\pi = \frac{q}{\varepsilon_0}, \qquad (1.161)$$

completing the proof of Gauss's law. Notice carefully that although the surface S may be spherical, it *need not* be spherical.

Going just a bit further, we consider a distributed charge so that

$$q = \int_V \rho \, d\tau. \qquad (1.162)$$

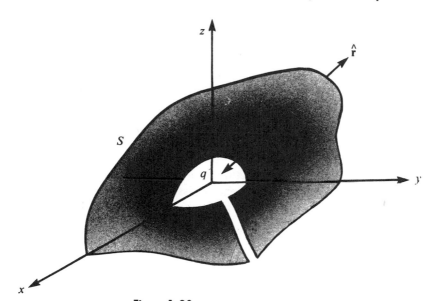

Figure 1.30   Exclusion of the origin.

Equation (1.161) still applies, with q now interpreted as the total distributed charge enclosed by surface S.

$$\int_S \mathbf{E} \cdot d\boldsymbol{\sigma} = \int_V \frac{\rho}{\varepsilon_0} \, d\tau. \tag{1.163}$$

Using Gauss's theorem, we have

$$\int_V \boldsymbol{\nabla} \cdot \mathbf{E} \, d\tau = \int_V \frac{\rho}{\varepsilon_0} \, d\tau. \tag{1.164}$$

Since our volume is completely arbitrary, the integrands must be equal or

$$\boldsymbol{\nabla} \cdot \mathbf{E} = \frac{\rho}{\varepsilon_0}, \tag{1.165}$$

one of Maxwell's equations. If we reverse the argument, Gauss's law follows immediately from Maxwell's equation.

POISSON'S EQUATION

Replacing $\mathbf{E}$ by $-\boldsymbol{\nabla}\varphi$, Eq. (1.165) becomes

$$\boldsymbol{\nabla} \cdot \boldsymbol{\nabla}\varphi = -\frac{\rho}{\varepsilon_0}, \tag{1.166}$$

which is Poisson's equation. For the condition $\rho = 0$ this reduces to an even more famous equation,

$$\boldsymbol{\nabla} \cdot \boldsymbol{\nabla}\varphi = 0, \tag{1.167}$$

Laplace's equation. We encounter Laplace's equation frequently in discussing various coordinate systems (Chapter 2) and the special functions of mathematical physics which appear as its solutions. Poisson's equation will be invaluable in developing the theory of Green's functions (Section 8.7).

From direct comparison of the Coulomb electrostatic force law and Newton's law of universal gravitation

$$\mathbf{F}_E = \frac{1}{4\pi\varepsilon_0}\frac{q_1 q_2}{r^2}\hat{\mathbf{r}}, \qquad \mathbf{F}_G = -G\frac{m_1 m_2}{r^2}\hat{\mathbf{r}}.$$

All of the potential theory of this section applies equally well to gravitational potentials. For example, the gravitational Poisson equation is

$$\mathbf{\nabla}\cdot\mathbf{\nabla}\varphi = +4\pi G\rho \tag{1.166a}$$

with ρ now a mass density.

EXERCISES

1.14.1   Develop Gauss's law for the two-dimensional case in which

$$\varphi = -q\,\frac{\ln\rho}{2\pi\varepsilon_0}, \qquad \mathbf{E} = -\mathbf{\nabla}\varphi = q\,\frac{\hat{\mathbf{\rho}}}{2\pi\varepsilon_0\rho}.$$

Here q is the charge at the origin or the line charge per unit length if the two-dimensional system is a unit thickness slice of a three-dimensional (circular cylindrical) system. The variable ρ is measured radially outward from the line charge. $\hat{\mathbf{\rho}}$ is the corresponding unit vector (see Section 2.4).

1.14.2   (a) Show that Gauss's law follows from Maxwell's equation

$$\mathbf{\nabla}\cdot\mathbf{E} = \frac{\rho}{\varepsilon_0}.$$

Here ρ is the usual charge density.

(b) Assuming that the electric field of a point charge q is spherically symmetric, show that Gauss's law implies the Coulomb inverse square expression

$$\mathbf{E} = \frac{q\hat{\mathbf{r}}}{4\pi\varepsilon_0 r^2}.$$

1.14.3   Show that the value of the electrostatic potential φ at any point P is equal to the *average* of the potential over any spherical surface centered on P. There are no electric charges on or within the sphere.

Hint. Use Green's theorem, Eq. (1.103), with $u^{-1} = r$, the distance from P, and $v = \varphi$. Also note Eq. (1.169) in Section 1.15.

1.14.4   Using Maxwell's equations, show that for a system (steady current) the magnetic vector potential $\mathbf{A}$ satisfies a vector Poisson equation,

$$\nabla^2\mathbf{A} = -\mu\mathbf{J},$$

provided we require $\mathbf{\nabla}\cdot\mathbf{A} = 0$.

1.15 DIRAC DELTA FUNCTION

From Example 1.6.1 and the development of Gauss's law in Section 1.14

$$\int \mathbf{\nabla} \cdot \mathbf{\nabla}\left(\frac{1}{r}\right) d\tau = -\int \mathbf{\nabla} \cdot \left(\frac{\hat{\mathbf{r}}}{r^2}\right) d\tau = \begin{cases} -4\pi \\ 0, \end{cases} \qquad (1.168)$$

depending on whether the integration includes the origin $\mathbf{r} = 0$ or not. This result may be conveniently expressed by introducing the Dirac delta function,

$$\mathbf{\nabla}^2\left(\frac{1}{r}\right) = -4\pi\delta(\mathbf{r}) = -4\pi\delta(x)\delta(y)\delta(z). \qquad (1.169)$$

This Dirac delta function is *defined* by its assigned properties

$$\delta(x) = 0, \qquad x \neq 0 \qquad (1.170a)$$

$$\int f(x)\delta(x)\,dx = f(0), \qquad (1.170b)$$

where $f(x)$ is any well-behaved function and the integration includes the origin. As a special case of Eq. (1.170b),

$$\int \delta(x)\,dx = 1. \qquad (1.170c)$$

From Eq. (1.170), $\delta(x)$ must be an infinitely high, infinitely thin spike at $x = 0$, as in the description of an impulsive force (Section 15.9) or the charge density for a point charge.[1] The problem is that *no such function exists* in the usual sense of function. However, the crucial property in Eq. (1.170b) can be developed rigorously as the limit of a *sequence* of functions, a distribution. For example, the delta function may be approximated by the sequences of functions, Eqs. (1.171) to (1.174) and Figs. 1.31 to 1.34:

$$\delta_n(x) = \begin{cases} 0, & x < -\dfrac{1}{2n} \\[2ex] n, & -\dfrac{1}{2n} < x < \dfrac{1}{2n} \\[2ex] 0, & x > \dfrac{1}{2n} \end{cases} \qquad (1.171)$$

$$\delta_n(x) = \frac{n}{\sqrt{\pi}}\exp(-n^2 x^2) \qquad (1.172)$$

---

[1] The delta function is frequently invoked to describe very short range forces such as nuclear forces. It also appears in the normalization of continuum wave functions of quantum mechanics. Compare Eq. (1.192c) for plane wave eigenfunctions.

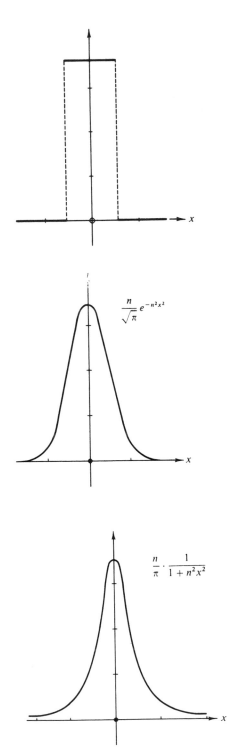

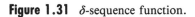

Figure 1.31 δ-sequence function.

$$\frac{n}{\sqrt{\pi}} e^{-n^2 x^2}$$

Figure 1.32 δ-sequence function.

$$\frac{n}{\pi} \cdot \frac{1}{1 + n^2 x^2}$$

Figure 1.33 δ-sequence function.

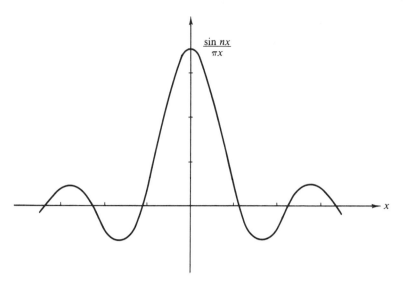

Figure 1.34 δ-sequence function.

$$\delta_n(x) = \frac{n}{\pi} \cdot \frac{1}{1 + n^2 x^2}$$ (1.173)

$$\delta_n(x) = \frac{\sin nx}{\pi x} = \frac{1}{2\pi} \int_{-n}^{n} e^{ixt}\, dt.$$ (1.174)

These approximations have varying degrees of usefulness. Equation (1.171) is useful in providing a simple derivation of the integral property, Eq. (1.170). Equation (1.172) is convenient to differentiate. Its derivatives lead to the Hermite polynomials. Equation (1.174) is particularly useful in Fourier analysis and in its applications to quantum mechanics. In the theory of Fourier series, Eq. (1.174) often appears (modified) as the Dirichlet kernel:

$$\delta_n(x) = \frac{1}{2\pi} \frac{\sin[(n + \frac{1}{2})x]}{\sin(\frac{1}{2}x)}.$$ (1.175)

In using these approximations in Eq. (1.170) and later, we assume that $f(x)$ is well behaved—it offers no problems at large x.

For most physical purposes such approximations are quite adequate. From a mathematical point of view the situation is still unsatisfactory: The limits

$$\lim_{n \to \infty} \delta_n(x)$$

do not exist.

A way out of this difficulty is provided by the theory of distributions. Recognizing that Eq. (1.170) is the fundamental property, we focus our attention on it rather than on $\delta(x)$ itself. Equations (1.171)–(1.174) with

$n = 1, 2, 3, \ldots$ may be interpreted as *sequences* of normalized functions:

$$\int_{-\infty}^{\infty} \delta_n(x)\, dx = 1. \tag{1.176}$$

The sequence of integrals has the limit

$$\lim_{n \to \infty} \int_{-\infty}^{\infty} \delta_n(x) f(x)\, dx = f(0). \tag{1.177}$$

Note that Eq. (1.177) is the limit of a sequence of integrals. Again, the limit of $\delta_n(x)$, $n \to \infty$, does not exist. (The limits for all four forms of $\delta_n(x)$ diverge at $x = 0$.)

We may treat $\delta(x)$ consistently in the form

$$\int_{-\infty}^{\infty} \delta(x) f(x)\, dx = \lim_{n \to \infty} \int_{-\infty}^{\infty} \delta_n(x) f(x)\, dx. \tag{1.178}$$

$\delta(x)$ is labeled a distribution (not a function) defined by the sequences $\delta_n(x)$ as indicated in Eq. (1.178). We might emphasize that the integral on the left-hand side of Eq. (1.178) is not a Riemann integral.[2] It is a limit.

This distribution $\delta(x)$ is only one of an infinity of possible distributions, but it is the one we are interested in because of Eq. (1.170).

From these sequences of functions we see that Dirac's delta function must be even in x, $\delta(-x) = \delta(x)$.

The integral property, Eq. (1.170b), is useful in cases, where the argument of the delta function is a function $g(x)$ with simple zeros on the real axis, which leads to the rules

$$\delta(ax) = \frac{1}{a}\,\delta(x), \qquad a > 0, \tag{1.179}$$

$$\delta(g(x)) = \sum_{\substack{a, \\ g(a)=0, \\ g'(a)\neq 0.}} \frac{\delta(x-a)}{|g'(a)|}. \tag{1.180}$$

To obtain Eq. (1.179) we change the integration variable in

$$\int_{-\infty}^{\infty} f(x)\delta(ax)\, dx = \frac{1}{a}\int_{-\infty}^{\infty} f\!\left(\frac{y}{a}\right)\delta(y)\, dy = \frac{1}{a} f(0),$$

and apply Eq. (1.170b). To prove Eq. (1.180) we decompose the integral

$$\int_{-\infty}^{\infty} f(x)\delta(g(x))\, dx = \sum_{a} \int_{a-\varepsilon}^{a+\varepsilon} f(x)\delta((x-a)g'(a))\, dx \tag{1.180a}$$

into a sum of integrals over small intervals containing the zeros of $g(x)$. In

[2] It can be treated as a Stieltjes integral if desired. $\delta(x)\,dx$ is replaced by $du(x)$, where $u(x)$ is the Heaviside step function (compare Exercise 1.15.13).

these intervals, $g(x) \approx g(a) + (x - a)g'(a) = (x - a)g'(a)$. Using Eq. (1.179) on the right-hand side of Eq. (1.180a) we obtain the integral of Eq. (1.180).

Using integration by parts we can also *define the derivative $\delta'(x)$ of the Dirac delta function* by the relation

$$\int_{-\infty}^{\infty} f(x)\delta'(x - x')\,dx = -\int_{-\infty}^{\infty} f'(x)\delta(x - x')\,dx = -f'(x'). \quad (1.181)$$

We use $\delta(x)$ frequently and call it the Dirac delta function[3]—for historical reasons. Remember that it is not really a function. It is essentially a shorthand notation, defined implicitly as the limit of integrals in a sequence, $\delta_n(x)$, according to Eq. (1.178). It should be understood that our Dirac delta function has significance only as part of an integrand. In this spirit the Dirac delta function is often regarded as an operator, a linear operator: $\delta(x - x_0)$ operates on $f(x)$ and yields $f(x_0)$.

$$\mathcal{L}(x_0)f(x) \equiv \int_{-\infty}^{\infty} \delta(x - x_0)f(x)\,dx = f(x_0). \quad (1.182)$$

It may also be classified as a linear mapping or simply as a generalized function. Shifting our singularity to the point $x = x'$, we write the Dirac delta function as $\delta(x - x')$. Equation (1.170) becomes

$$\int_{-\infty}^{\infty} f(x)\delta(x - x')\,dx = f(x'). \quad (1.183)$$

As a description of a singularity at $x = x'$, the Dirac delta function may be written as $\delta(x - x')$ or as $\delta(x' - x)$. Going to three dimensions and using spherical polar coordinates, we obtain

$$\int_0^{2\pi} \int_0^{\pi} \int_0^{\infty} \delta(\mathbf{r})r^2\,dr \sin\theta\,d\theta\,d\varphi = \iiint_{-\infty}^{\infty} \delta(x)\delta(y)\delta(z)\,dx\,dy\,dz = 1. \quad (1.184)$$

This corresponds to a singularity (or source) at the origin. Again, if our source is at $\mathbf{r} = \mathbf{r}_1$, Eq. (1.184) becomes

$$\iiint \delta(\mathbf{r}_2 - \mathbf{r}_1)r_2^2\,dr_2 \sin\theta_2\,d\theta_2\,d\varphi_2 = 1. \quad (1.185)$$

DELTA FUNCTION REPRESENTATION BY ORTHOGONAL FUNCTIONS*

Dirac's delta function can be expanded in terms of any basis of real orthogonal functions $\{\varphi_n(x),\ n = 0, 1, 2, \ldots\}$. Such functions will occur in Chapter 9 as solutions of ordinary differential equations of the Sturm–Liouville form.

---

[3] Dirac introduced the delta function to quantum mechanics. Actually the delta function can be traced back to Kirchhoff, 1882. For further details see M. Jammer, *The Conceptual Development of Quantum Mechanics*. New York: McGraw-Hill (1966) p. 301.

* This section is optional here. It is not needed until Chapter 9.

They satisfy the orthogonality relations

$$\int_a^b \varphi_m(x)\varphi_n(x)\, dx = \delta_{mn}, \tag{1.186}$$

where the interval (a, b) may be infinite at either end or both. [For convenience we assume that φ_n has been defined to include $(w(x))^{1/2}$ if the orthogonality relations contain an additional positive weight function $w(x)$.] We use the φ_n to expand the delta function as

$$\delta(x - t) = \sum_{n=0}^{\infty} a_n(t)\varphi_n(x), \tag{1.187}$$

where the coefficients a_n are functions of the variable t. Multiplying by $\varphi_m(x)$ and integrating over the orthogonality interval (Eq. (1.186)), we have

$$a_m(t) = \int_a^b \delta(x - t)\varphi_m(x)\, dx = \varphi_m(t) \tag{1.188}$$

or

$$\delta(x - t) = \sum_{n=0}^{\infty} \varphi_n(t)\varphi_n(x) = \delta(t - x). \tag{1.189}$$

This series is assuredly not uniformly convergent (see Chapter 5), but it may be used as part of an integrand in which the ensuing integration will make it convergent (compare Section 5.5).

Suppose we form the integral $\int F(t)\delta(t - x)\, dx$, where it is assumed that $F(t)$ can be expanded in a series of orthogonal functions $\varphi_p(t)$, a property called completeness. We then obtain

$$\int F(t)\delta(t - x)\, dt = \int \sum_{p=0}^{\infty} a_p\varphi_p(t) \sum_{n=0}^{\infty} \varphi_n(x)\varphi_n(t)\, dt$$

$$= \sum_{p=0}^{\infty} a_p\varphi_p(x) = F(x), \tag{1.190}$$

the cross products $\int \varphi_p\varphi_n\, dt$ $(n \neq p)$ vanishing by orthogonality (Eq. (1.186)). Referring back to the definition of the Dirac delta function, Eq. (1.170b), we see that our series representation, Eq. (1.189), satisfies the defining property of the Dirac delta function and therefore is a representation of it. This representation of the Dirac delta function is called *closure*. The assumption of completeness of a set of functions for expansion of $\delta(x - t)$ yields the closure relation. The converse, that closure implies completeness, is the topic of Exercise 1.15.16.

INTEGRAL REPRESENTATIONS FOR THE DELTA FUNCTION

Integral transforms, such as the Fourier integral

$$F(\omega) = \int_{-\infty}^{\infty} f(t)\exp(i\omega t)\, dt$$

of Chapter 15, lead to the corresponding integral representations of Dirac's delta function. For example, take

$$\delta_n(t - x) = \frac{\sin n(t - x)}{\pi(t - x)} = \frac{1}{2\pi} \int_{-n}^{n} \exp(i\omega(t - x)) \, d\omega, \qquad (1.191)$$

using Eq. (1.174). We have

$$f(x) = \lim_{n \to \infty} \int_{-\infty}^{\infty} f(t)\delta_n(t - x) \, dt, \qquad (1.192a)$$

where $\delta_n(t - x)$ is the sequence in Eq. (1.191) defining the distribution $\delta(t - x)$. Note that Eq. (1.192a) assumes that $f(t)$ is continuous at $t = x$. If we substitute Eq. (1.191) into Eq. (1.192a) we obtain

$$f(x) = \lim_{n \to \infty} \frac{1}{2\pi} \int_{-\infty}^{\infty} f(t) \int_{-n}^{n} \exp(i\omega(t - x)) \, d\omega \, dt. \qquad (1.192b)$$

Interchanging the order of integration and then taking the limit as $n \to \infty$, we have the Fourier integral theorem of Section 15.

With the understanding that it belongs under an integral sign as in Eq. (1.192a), the identification

$$\delta(t - x) = \frac{1}{2\pi} \int_{-\infty}^{\infty} \exp(i\omega(t - x)) \, d\omega \qquad (1.192c)$$

provides a very useful integral representation of the delta function. It is used to great advantage in Sections 15.5 and 15.6.

When the Laplace transform (see Sections 15.5 and 15.9)

$$L_\delta(s) = \int_0^{\infty} \exp(-st)\delta(t - t_0) = \exp(-st_0), \qquad t_0 > 0 \qquad (1.193)$$

is inverted, we obtain the complex representation

$$\delta(t - t_0) = \frac{1}{2\pi i} \int_{\gamma - i\infty}^{\gamma + i\infty} \exp(s(t - t_0)) \, ds, \qquad (1.194)$$

which is essentially equivalent to the previous Fourier representation of Dirac's delta function.

EXERCISES

1.15.1   Let

$$\delta_n(x) = \begin{cases} 0, & x < -\dfrac{1}{2n} \\[2mm] n, & -\dfrac{1}{2n} < x < \dfrac{1}{2n}, \\[2mm] 0, & \dfrac{1}{2n} < x. \end{cases}$$

Show that

$$\lim_{n \to \infty} \int_{-\infty}^{\infty} f(x)\delta_n(x) \, dx = f(0),$$

assuming that $f(x)$ is continuous at $x = 0$.

1.15.2   Verify that the sequence $\delta_n(x)$, based on the function

$$\delta_n = \begin{cases} 0, & x < 0 \\ ne^{-nx}, & x > 0, \end{cases}$$

is a delta sequence (satisfying Eq. (1.177)). Note that the singularity is at $+0$, the positive side of the origin.
Hint. Replace the upper limit (∞) by c/n, where c is large but finite and use the mean value theorem of integral calculus.

1.15.3   For

$$\delta_n(x) = \frac{n}{\pi} \cdot \frac{1}{1 + n^2 x^2},$$

(Eq. (1.173)), show that

$$\int_{-\infty}^{\infty} \delta_n(x) \, dx = 1.$$

1.15.4   Demonstrate that $\delta_n = \sin nx/\pi x$ is a delta distribution by showing that

$$\lim_{n \to \infty} \int_{-\infty}^{\infty} f(x) \frac{\sin nx}{\pi x} \, dx = f(0).$$

Assume that $f(x)$ is continuous at $x = 0$ and vanishes as $x \to \pm\infty$.
Hint. Replace x by y/n and take lim $n \to \infty$ *before* integrating. The needed integral is evaluated in Sections 7.2 and 15.7.

1.15.5   Fejér's method of summing series is associated with the function

$$\delta_n(t) = \frac{1}{2\pi n} \left[\frac{\sin(nt/2)}{\sin(t/2)} \right]^2.$$

Show that $\delta_n(t)$ is a delta distribution in the sense that

$$\lim_{n \to \infty} \frac{1}{2\pi n} \int_{-\infty}^{\infty} f(t) \left[\frac{\sin(nt/2)}{\sin(t/2)} \right]^2 dt = f(0).$$

1.15.6   Prove that

$$\delta[a(x - x_1)] = \frac{1}{a} \delta(x - x_1).$$

Note. If $\delta[a(x - x_1)]$ is considered even relative to x_1, the relation holds for negative a and $1/a$ may be replaced by $1/|a|$.

1.15.7   Show that

$$\delta[(x - x_1)(x - x_2)] = [\delta(x - x_1) + \delta(x - x_2)]/|x_1 - x_2|.$$

Hint. Try using Exercise 1.15.6.

1.15.8   Using the Gauss error curve delta sequence (δ_n), show that

$$x \frac{d}{dx} \delta(x) = -\delta(x),$$

treating $\delta(x)$ and its derivative as in Eq. (1.178).

1.15.9 Show that

$$\int_{-\infty}^{\infty} \delta'(x) f(x)\, dx = -f'(0).$$

Here we assume that $f'(x)$ is continuous at $x = 0$.

1.15.10 Prove that

$$\delta(f(x)) = \left|\frac{df(x)}{dx}\right|^{-1} \delta(x - x_0),$$

where x_0 is chosen so that $f(x_0) = 0$.
Hint. Note that $\delta(f)\, df = \delta(x)\, dx$.

1.15.11 Show that in spherical polar coordinates $(r, \cos\theta, \varphi)$ the delta function $\delta(\mathbf{r}_1 - \mathbf{r}_2)$ becomes

$$\frac{1}{r_1^2}\delta(r_1 - r_2)\delta(\cos\theta_1 - \cos\theta_2)\delta(\varphi_1 - \varphi_2).$$

Generalize this to the curvilinear coordinates (q_1, q_2, q_3) of Section 2.1 with scale factors h_1, h_2, and h_3.

1.15.12 A rigorous development of Fourier transforms (Sneddon, *Fourier Transforms*)[4] includes as a theorem the relations

$$\lim_{a\to\infty}\frac{2}{\pi}\int_{x_1}^{x_2} f(u + x)\frac{\sin ax}{x}\, dx$$

$$= \begin{cases} f(u+0) + f(u-0), & x_1 < 0 < x_2 \\ f(u+0), & x_1 = 0 < x_2 \\ f(u-0), & x_1 < 0 = x_2 \\ 0 & x_1 < x_2 < 0 \quad\text{or}\quad 0 < x_1 < x_2. \end{cases}$$

Verify these results using the Dirac delta function.

1.15.13 (a) If we define a sequence $\delta_n(x) = n/(2\cosh^2 nx)$, show that

$$\int_{-\infty}^{\infty}\delta_n(x)\, dx = 1, \qquad \text{independent of } n.$$

(b) Continuing this analysis, show that[*]

$$\int_{-\infty}^{x}\delta_n(x)\, dx = \tfrac{1}{2}[1 + \tanh nx] \equiv u_n(x)$$

and

$$\lim_{n\to\infty} u_n(x) = \begin{cases} 0, & x < 0, \\ 1, & x > 0. \end{cases}$$

This is the Heaviside unit step function (Fig. 1.35).

1.15.14 Show that the unit step function $u(x)$ may be represented by

$$u(x) = \frac{1}{2} + \frac{1}{2\pi i} P \int_{-\infty}^{\infty} e^{ixt}\frac{dt}{t},$$

where P means Cauchy principal value (Section 7.2).

[4] I. N. Sneddon, *Fourier Transforms*. New York: McGraw–Hill (1951).
[*] Many other symbols are used for this function. This is the AMS-55 notation: u for unit.

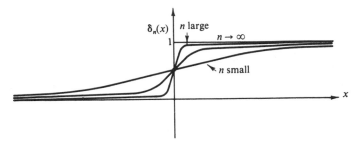

Figure 1.35 $\frac{1}{2}[1 + \tanh nx]$ and the Heaviside unit step function.

1.15.15 As a variation of Eq. (1.174), take

$$\delta_n(x) = \frac{1}{2\pi} \int_{-\infty}^{\infty} e^{ixt-|t|/n}\, dt.$$

Show that this reduces to $(n/\pi) \cdot 1/(1 + n^2x^2)$, Eq. (1.173), and that

$$\int_{-\infty}^{\infty} \delta_n(x)\, dx = 1.$$

Note. In terms of integral transforms, the initial equation here may be interpreted as either a Fourier exponential transform of $e^{-|t|/n}$ or a Laplace transform of e^{ixt}.

1.15.16 (a) The Dirac delta function representation given by Eq. (1.189)

$$\delta(x - t) = \sum_{n=0}^{n} \varphi_n(x)\varphi_n(t)$$

is often called the *closure relation*. For an orthonormal set of functions, φ_n, show that closure implies completeness, that is, Eq. (1.190) follows from Eq. (1.189).
Hint. One can take

$$F(x) = \int F(t)\delta(x - t)\, dt.$$

(b) Following the hint of part (a) you encounter the integral $\int F(t)\varphi_n(t)\, dt$. How do you know that this integral is finite?

1.15.17 For the finite interval $(-\pi, \pi)$ expand the Dirac delta function $\delta(x - t)$ in a series of sines and cosines: $\sin nx$, $\cos nx$, $n = 0, 1, 2, \ldots$. Note that although these functions are orthogonal, they are not normalized to unity.

1.15.18 In the interval $(-\pi, \pi)$, $\delta_n(x) = \dfrac{n}{\sqrt{\pi}} \exp(-n^2x^2)$.

(a) Expand $\delta_n(x)$ as a Fourier cosine series.
(b) Show that your Fourier series agrees with a Fourier expansion of $\delta(x)$ in the limit as $n \to \infty$.
(c) Confirm the delta function nature of your Fourier series by showing that for any $f(x)$ that is finite in the interval $[-\pi, \pi]$ and continuous at $x = 0$,

$$\int_{-\pi}^{\pi} f(x)[\text{Fourier expansion of } \delta_\infty(x)]\, dx = f(0).$$

1.15.19 In the interval $(-\infty, \infty)$,

(i) $\delta_n(x) = \dfrac{n}{\sqrt{\pi}} \exp(-n^2 x^2)$, and

(ii) $\delta_n(x) = 0$, for $x < 0$,
$n \exp(-nx)$, for $x > 0$.

(a) Expand $\delta_n(x)$ as a Fourier integral and compare the limit $n \to \infty$ with Eq. (1.192c).

(b) Expand $\delta_n(x)$ as a Laplace transform and compare the limit $n \to \infty$ with Eq. (1.194).

Hint. See Eqs. (15.22), (15.23) for (a), and Eq. (15.212) for (b).

1.15.20 (a) Show that the Dirac delta function $\delta(x - a)$, expanded in a Fourier sine series in the half interval $(0, L)$, $(0 < a < L)$, is given by

$$\delta(x - a) = \frac{2}{L} \sum_{n=1}^{\infty} \sin\left(\frac{n\pi a}{L}\right) \sin\left(\frac{n\pi x}{L}\right).$$

Note that this series actually describes
$$-\delta(x + a) + \delta(x - a) \text{ in the interval } (-L, L).$$

(b) By integrating both sides of the preceding equation from 0 to x, show that the cosine expansion of the square wave

$$f(x) = \begin{cases} 0, & 0 \le x < a \\ 1, & a < x < L, \end{cases}$$

is

$$f(x) = \frac{2}{\pi} \sum_{n=1}^{\infty} \frac{1}{n} \sin\left(\frac{n\pi a}{L}\right) - \frac{2}{\pi} \sum_{n=1}^{\infty} \frac{1}{n} \sin\left(\frac{n\pi a}{L}\right) \cos\left(\frac{n\pi x}{L}\right), \quad 0 \le x < L.$$

(c) Verify that the term

$$\frac{2}{\pi} \sum_{n=1}^{\infty} \frac{1}{n} \sin\left(\frac{n\pi a}{L}\right) \text{ is } \langle f(x) \rangle \equiv \frac{1}{L} \int_0^L f(x)\, dx.$$

1.15.21 Verify the Fourier cosine expansion of the square wave, Exercise 1.15.20(b), by direct calculation of the Fourier coefficients.

1.15.22 We may define a sequence

$$\delta_n(x) = \begin{cases} n, & |x| < 1/2n, \\ 0, & |x| > 1/2n. \end{cases}$$

(This is Eq. (1.171).) Express $\delta_n(x)$ as a Fourier integral (via the Fourier integral theorem, inverse transform, etc.). Finally, show that we may write

$$\delta(x) = \lim_{n \to \infty} \delta_n(x) = \frac{1}{2\pi} \int_{-\infty}^{\infty} e^{-ikx}\, dk.$$

1.15.23   Using the sequence

$$\delta_n(x) = \frac{n}{\sqrt{\pi}} \exp(-n^2 x^2),$$

show that

$$\delta(x) = \frac{1}{2\pi} \int_{-\infty}^{\infty} e^{-ikx}\, dk.$$

Note. Remember that $\delta(x)$ is defined in terms of its behavior as part of an integrand—especially Eqs. (1.177) and (1.188).

1.15.24   Derive sin and cosine representations of $\delta(t - x)$ that are comparable to the exponential representation, Eq. (1.192c).

$$ANS. \quad \frac{2}{\pi} \int_0^{\infty} \sin \omega t \sin \omega x\, d\omega$$

$$\frac{2}{\pi} \int_0^{\infty} \cos \omega t \cos \omega x\, d\omega.$$

1.16 HELMHOLTZ'S THEOREM

In Section 1.13 it was emphasized that the choice of a magnetic vector potential **A** was not unique. The divergence of **A** was still undetermined. In this section two theorems about the divergence and curl of a vector are developed. The first theorem is as follows.

A vector is uniquely specified by giving its divergence and its curl within a region and its normal component over the boundary.

Let us take

$$\mathbf{\nabla} \cdot \mathbf{V}_1 = s,$$
$$\mathbf{\nabla} \times \mathbf{V}_1 = \mathbf{c}, \tag{1.195}$$

where s may be interpreted as a source (charge) density and $\mathbf{c}$ as a circulation (current) density. Assuming also that the normal component V_{1n} on the boundary is given, we want to show that $\mathbf{V}_1$ is unique. We do this by assuming the existence of a second vector $\mathbf{V}_2$, which satisfies Eq. (1.195) and has the same normal component over the boundary, and then showing that $\mathbf{V}_1 - \mathbf{V}_2 = 0$. Let

$$\mathbf{W} = \mathbf{V}_1 - \mathbf{V}_2.$$

Then

$$\mathbf{\nabla} \cdot \mathbf{W} = 0 \tag{1.196}$$

and

$$\mathbf{\nabla} \times \mathbf{W} = 0. \tag{1.197}$$

Since **W** is irrotational we may write (by Section (1.13)

$$\mathbf{W} = -\nabla\varphi. \tag{1.198}$$

Substituting this into Eq. (1.196), we obtain

$$\nabla \cdot \nabla\varphi = 0, \tag{1.199}$$

Laplace's equation.

Now we draw upon Green's theorem in the form given in Eq. (1.104), letting u and v each equal φ. Since

$$W_n = V_{1n} - V_{2n} = 0 \tag{1.200}$$

on the boundary, Green's theorem reduces to

$$\int_V (\nabla\varphi) \cdot (\nabla\varphi)\, d\tau = \int_V \mathbf{W} \cdot \mathbf{W}\, d\tau = 0. \tag{1.201}$$

The quantity $\mathbf{W} \cdot \mathbf{W} = W^2$ is nonnegative and so we must have

$$\mathbf{W} = \mathbf{V}_1 - \mathbf{V}_2 = 0 \tag{1.202}$$

everywhere. Thus $\mathbf{V}_1$ is unique, proving the theorem.

For our magnetic vector potential **A** the relation $\mathbf{B} = \nabla \times \mathbf{A}$ specifies the curl of **A**. Often for convenience we set $\nabla \cdot \mathbf{A} = 0$ (compare Exercise 1.14.4). Then (with boundary conditions) **A** is fixed.

This theorem may be written as a uniqueness theorem for solutions of Laplace's equation, Exercise 1.16.1. In this form, this uniqueness theorem is of great importance in solving electrostatic and other Laplace equation boundary value problems. If we can find a solution of Laplace's equation that satisfies the necessary boundary conditions, then our solution is the complete solution. Such boundary value problems are taken up in Sections 12.3 and 12.5.

HELMHOLTZ'S THEOREM

The second theorem we shall prove is Helmholtz's theorem.

*A vector **V** satisfying Eq. (1.195)* with both source and circulation densities vanishing at infinity may be written as the sum of two parts, one of which is irrotational, the other solenoidal.

Helmholtz's theorem will clearly be satisfied if we may write **V** as

$$\mathbf{V} = -\nabla\varphi + \nabla \times \mathbf{A}, \tag{1.203}$$

$-\nabla\varphi$ being irrotational and $\nabla \times \mathbf{A}$ being solenoidal. We proceed to justify Eq. (1.203).

V is a known vector. Taking the divergence and curl

$$\nabla \cdot \mathbf{V} = s(\mathbf{r}) \tag{1.203a}$$

$$\nabla \times \mathbf{V} = \mathbf{c}(\mathbf{r}). \tag{1.203b}$$

with $s(\mathbf{r})$ and $\mathbf{c}(\mathbf{r})$ now known functions of position. From these two functions we construct a scalar potential $\varphi(\mathbf{r}_1)$,

$$\varphi(\mathbf{r}_1) = \frac{1}{4\pi} \int \frac{s(\mathbf{r}_2)}{r_{12}} d\tau_2, \tag{1.204a}$$

and a vector potential $\mathbf{A}(\mathbf{r}_1)$,

$$\mathbf{A}(\mathbf{r}_1) = \frac{1}{4\pi} \int \frac{\mathbf{c}(\mathbf{r}_2)}{r_{12}} d\tau_2. \tag{1.204b}$$

If $s = 0$ then $\mathbf{V}$ is solenoidal and Eq. (1.204a) implies $\varphi = 0$. From Eq. (1.203), $\mathbf{V} = \nabla \times \mathbf{A}$ with $\mathbf{A}$ as given in Eq. (1.140) is consistent with Section 1.13. Further, if $\mathbf{c} = 0$ then $\mathbf{V}$ is irrotational and Eq. (1.204b) implies $\mathbf{A} = 0$, and Eq. (1.203) implies $\mathbf{V} = -\nabla\varphi$, consistent with scalar potential theory of Section 1.13.

Here the argument $\mathbf{r}_1$ indicates (x_1, y_1, z_1), the field point; $\mathbf{r}_2$, the coordinates of the source point (x_2, y_2, z_2), whereas

$$r_{12} = [(x_1 - x_2)^2 + (y_1 - y_2)^2 + (z_1 - z_2)^2]^{1/2}. \tag{1.205}$$

When a direction is associated with r_{12}, the positive direction is taken to be away from the source toward the field point. Vectorially, $\mathbf{r}_{12} = \mathbf{r}_1 - \mathbf{r}_2$, as shown in Fig. 1.36. Of course, s and $\mathbf{c}$ must vanish sufficiently rapidly at large distance so that the integrals exist. The actual expansion and evaluation of integrals such as Eqs. (1.204a) and (1.204b) is treated in Section 12.1.

From the uniqueness theorem at the beginning of this section, $\mathbf{V}$ is uniquely specified by its divergence, s, and curl, $\mathbf{c}$ (and boundary conditions). Returning

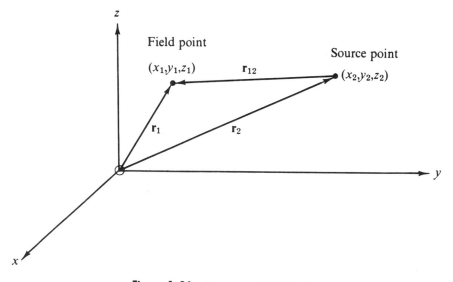

Figure 1.36   Source and field points.

to Eq. (1.203), we have

$$\mathbf{\nabla} \cdot \mathbf{V} = -\mathbf{\nabla} \cdot \mathbf{\nabla}\varphi, \tag{1.206a}$$

the divergence of the curl vanishing and

$$\mathbf{\nabla} \times \mathbf{V} = \mathbf{\nabla} \times \mathbf{\nabla} \times \mathbf{A}, \tag{1.206b}$$

the curl of the gradient vanishing. If we can show that

$$-\mathbf{\nabla} \cdot \mathbf{\nabla}\varphi(\mathbf{r}_1) = s(\mathbf{r}_1) \tag{1.206c}$$

and

$$\mathbf{\nabla} \times \mathbf{\nabla} \times \mathbf{A}(\mathbf{r}_1) = \mathbf{c}(\mathbf{r}_1), \tag{1.206d}$$

then $\mathbf{V}$ as given in Eq. (1.203) will have the proper divergence and curl. Our description will be internally consistent and Eq. (1.203) justified.[1]

First, we consider the divergence of $\mathbf{V}$:

$$\mathbf{\nabla} \cdot \mathbf{V} = -\mathbf{\nabla} \cdot \mathbf{\nabla}\varphi = -\frac{1}{4\pi} \mathbf{\nabla} \cdot \mathbf{\nabla} \int \frac{s(\mathbf{r}_2)}{r_{12}} \, d\tau_2. \tag{1.207}$$

The Laplacian operator, $\mathbf{\nabla} \cdot \mathbf{\nabla}$ or $\mathbf{\nabla}^2$, operates on the field coordinates (x_1, y_1, z_1) and so commutes with the integration with respect to (x_2, y_2, z_2). We have

$$\mathbf{\nabla} \cdot \mathbf{V} = -\frac{1}{4\pi} \int s(\mathbf{r}_2)\mathbf{\nabla}_1^2\left(\frac{1}{r_{12}}\right) d\tau_2. \tag{1.208}$$

We must make two minor modifications in Eq. (1.168) before applying it. First, our source is at r_2, not at the origin. This means that the 4π in Gauss's law appears if and only if the surface includes the point $\mathbf{r} = \mathbf{r}_2$. To show this, we rewrite Eq. (1.168):

$$\mathbf{\nabla}^2\left(\frac{1}{r_{12}}\right) = -4\pi \, \delta(\mathbf{r}_1 - \mathbf{r}_2). \tag{1.209}$$

This shift of the source to $\mathbf{r}_2$ may be incorporated in the defining equations (1.170) as

$$\delta(\mathbf{r}_1 - \mathbf{r}_2) = 0, \qquad \mathbf{r}_1 \neq \mathbf{r}_2, \tag{1.210a}$$

$$\int f(\mathbf{r}_1) \, \delta(\mathbf{r}_1 - \mathbf{r}_2) \, d\tau_1 = f(\mathbf{r}_2). \tag{1.210b}$$

Second, noting that differentiating r_{12}^{-1} twice with respect to x_2, y_2, z_2 is the same as differentiating *twice* with respect to x_1, y_1, z_1, we have

$$\mathbf{\nabla}_1^2\left(\frac{1}{r_{12}}\right) = \mathbf{\nabla}_2^2\left(\frac{1}{r_{12}}\right) = -4\pi \, \delta(\mathbf{r}_1 - \mathbf{r}_2)$$

$$= -4\pi \, \delta(\mathbf{r}_2 - \mathbf{r}_1). \tag{1.211}$$

[1] Alternatively, we could solve Eq. (1.206c), Poisson's equation, and compare the solution with the constructed potential, Eq. (1.204a). The solution of Poisson's equation is developed in Section 8.7.

Rewriting Eq. (1.208) and using the Dirac delta function, Eq. (1.211), we may integrate to obtain

$$\mathbf{\nabla} \cdot \mathbf{V} = -\frac{1}{4\pi} \int s(\mathbf{r}_2) \nabla_2^2 \left(\frac{1}{r_{12}}\right) d\tau_2$$

$$= -\frac{1}{4\pi} \int s(\mathbf{r}_2)(-4\pi)\delta(\mathbf{r}_2 - \mathbf{r}_1) \, d\tau_2$$

$$= s(\mathbf{r}_1). \tag{1.212}$$

The final step follows from Eq. (1.210b) with the subscripts 1 and 2 exchanged. Our result, Eq. (1.212), shows that the assumed form of $\mathbf{V}$ and of the scalar potential φ are in agreement with the given divergence (Eq. (1.203a)).

To complete the proof of Helmholtz's theorem, we need to show that our assumptions are consistent with Eq. (1.203b), that is, the curl of $\mathbf{V}$ is equal to $\mathbf{c}(\mathbf{r}_1)$. From Eq. (1.203),

$$\mathbf{\nabla} \times \mathbf{V} = \mathbf{\nabla} \times \mathbf{\nabla} \times \mathbf{A}$$

$$= \mathbf{\nabla}\mathbf{\nabla} \cdot \mathbf{A} - \nabla^2\mathbf{A}. \tag{1.213}$$

The first term, $\mathbf{\nabla}\mathbf{\nabla} \cdot \mathbf{A}$ leads to

$$4\pi\mathbf{\nabla}\mathbf{\nabla} \cdot \mathbf{A} = \int \mathbf{c}(\mathbf{r}_2) \cdot \mathbf{\nabla}_1\mathbf{\nabla}_1 \left(\frac{1}{r_{12}}\right) d\tau_2 \tag{1.214}$$

by Eq. (1.204b). Again replacing the second derivatives with respect to x_1, y_1, z_1 by second derivatives with respect to x_2, y_2, z_2, we integrate each component[2] of Eq. (1.214) by parts:

$$4\pi\mathbf{\nabla}\mathbf{\nabla} \cdot \mathbf{A}|_x = \int \mathbf{c}(\mathbf{r}_2) \cdot \mathbf{\nabla}_2 \frac{\partial}{\partial x_2}\left(\frac{1}{r_{12}}\right) d\tau_2$$

$$= \int \mathbf{\nabla}_2 \cdot \left[\mathbf{c}(\mathbf{r}_2)\frac{\partial}{\partial x_2}\left(\frac{1}{r_{12}}\right)\right] d\tau_2 - \int [\mathbf{\nabla}_2 \cdot \mathbf{c}(\mathbf{r}_2)]\frac{\partial}{\partial x_2}\left(\frac{1}{r_{12}}\right) d\tau_2.$$

$$\tag{1.215}$$

The second integral vanishes because the circulation density $\mathbf{c}$ is solenoidal.[3] The first integral may be transformed to a surface integral by Gauss's theorem. If $\mathbf{c}$ is bounded in space or vanishes faster that $1/r$ for large r, so that the integral in Eq. (1.204b) exists, then by choosing a sufficiently large surface the first integral on the right-hand side of Eq. (1.215) also vanishes.

With $\mathbf{\nabla}\mathbf{\nabla} \cdot \mathbf{A} = 0$, Eq. (1.213) now reduces to

$$\mathbf{\nabla} \times \mathbf{V} = -\nabla^2\mathbf{A} = -\frac{1}{4\pi} \int \mathbf{c}(\mathbf{r}_2)\nabla_1^2\left(\frac{1}{r_{12}}\right) d\tau_2. \tag{1.216}$$

[2] This avoids creating the *tensor* $\mathbf{c}(\mathbf{r}_2)\mathbf{\nabla}_2$.
[3] Remember $\mathbf{c} = \mathbf{\nabla} \times \mathbf{V}$ is known.

This is exactly like Eq. (1.208) except that the scalar $s(\mathbf{r}_2)$ is replaced by the vector circulation density $\mathbf{c}(\mathbf{r}_2)$. Introducing the Dirac delta function, as before, as a convenient way of carrying out the integration, we find that Eq. (1.216) reduces to Eq. (1.195). We see that our assumed form of $\mathbf{V}$, given by Eq. (1.203), and of the vector potential $\mathbf{A}$, given by Eq. (1.204b), are in agreement with Eq. (1.195) specifying the curl of $\mathbf{V}$.

This completes the proof of Helmholtz's theorem, showing that a vector may be resolved into irrotational and solenoidal parts. Applied to the electromagnetic field, we have resolved our field vector $\mathbf{V}$ into an irrotational electric field $\mathbf{E}$, derived from a scalar potential φ, and a solenoidal magnetic induction field $\mathbf{B}$, derived from a vector potential $\mathbf{A}$. The source density $s(\mathbf{r})$ may be interpreted as an electric charge density (divided by electric permittivity ε), whereas the circulation density $\mathbf{c}(\mathbf{r})$ becomes electric current density (times magnetic permeability μ).

EXERCISES

1.16.1 Implicit in this section is a proof that a function $\psi(\mathbf{r})$ is uniquely specified by requiring it to (1) satisfy Laplace's equation and (2) satisfy a complete set of boundary conditions. Develop this proof explicitly.

1.16.2 (a) Assuming that $\mathbf{P}$ is a solution of the vector Poisson equation, $\nabla_1^2 \mathbf{P}(\mathbf{r}_1) = -\mathbf{V}(\mathbf{r}_1)$, develop an alternate proof of Helmholtz's theorem, showing that $\mathbf{V}$ may be written as

$$\mathbf{V} = -\nabla\varphi + \nabla \times \mathbf{A},$$

where

$$\mathbf{A} = \nabla \times \mathbf{P},$$

and

$$\varphi = \nabla \cdot \mathbf{P}.$$

(b) Solving the vector Poisson equation, we find

$$\mathbf{P}(\mathbf{r}_1) = \frac{1}{4\pi} \int_V \frac{\mathbf{V}(\mathbf{r}_2)}{r_{12}} d\tau_2.$$

Show that this solution substituted into φ and $\mathbf{A}$ of part (a) leads to the expressions given for φ and $\mathbf{A}$ in Section 1.16.

ADDITIONAL READINGS

BORISENKO, A. I., and I. E. TARAPOV, *Vector and Tensor Analysis with Applications.* Englewood Cliffs, NJ: Prentice-Hall (1968).

DAVIS, H. F., and A. D. SNIDER, *Introduction to Vector Analysis*, 4th ed. Boston: Allyn & Bacon (1979).

KELLOGG, O. D., *Foundations of Potential Theory.* New York: Dover (1953). Originally published (1929). The classic text on potential theory.

LEWIS, P. E., *Vector Analysis for Engineers and Scientists*. Reading, MA: Addison-Wesley (1989).

MARION, J. B., *Principles of Vector Analysis*. New York: Academic Press (1965). A moderately advanced presentation of vector analysis oriented toward tensor analysis. Rotations and other transformations are described with the appropriate matrices.

WREDE, R. C., *Introduction to Vector and Tensor Analysis*. New York: Wiley (1963). Reprinted, New York: Dover (1972). Fine historical introduction. Excellent discussion of differentiation of vectors and applications to mechanics.

2 VECTOR ANALYSIS IN CURVED COORDINATES AND TENSORS

In Chapter 1 we restricted ourselves almost completely to cartesian coordinate systems. A cartesian coordinate system offers the unique advantage that all three unit vectors, $\hat{\mathbf{x}}$, $\hat{\mathbf{y}}$, and $\hat{\mathbf{z}}$, are constant in direction as well as in magnitude. We did introduce the radial distance $\mathbf{r}$ but even this was treated as a function of x, y, and z. Unfortunately, not all physical problems are well adapted to solution in cartesian coordinates. For instance, if we have a central force problem, $\mathbf{F} = \hat{\mathbf{r}}F(r)$, such as gravitational or electrostatic force, cartesian coordinates may be unusually inappropriate. Such a problem literally screams for the use of a coordinate system in which the radial distance is taken to be one of the coordinates, that is, spherical polar coordinates.

The point is that the coordinate system should be chosen to fit the problem, to exploit any constraint or symmetry present in it. Then, hopefully, it will be more readily soluble than if we had forced it into a cartesian framework.

Naturally, there is a price that must be paid for the use of a noncartesian coordinate system. We have not yet written expressions for gradient, divergence, or curl in any of the noncartesian coordinate systems. Such expressions are developed in very general form in Section 2.2. First, we must develop a system of curvilinear coordinates, a general system that may be specialized to any of the particular systems of interest. We shall specialize to circular cylindrical coordinates in Section 2.4 and to spherical polar coordinates in Section 2.5.

2.1 ORTHOGONAL COORDINATES

In cartesian coordinates we deal with three mutually perpendicular families of planes: $x =$ constant, $y =$ constant, and $z =$ constant. Imagine that we superimpose on this system three other families of surfaces. The surfaces of any one family need not be parallel to each other and they need not be planes. If this is difficult to visualize, the figure of a specific coordinate system such as Fig. 2.3 may be helpful. The three new families of surfaces need not be mutually perpendicular, but for simplicity we quickly impose this condition (Eq. (2.7)). We may describe any point (x, y, z) as the intersection of three planes in cartesian coordinates or as the intersection of the three surfaces that form our new, curvilinear coordinates. Describing the curvilinear coordinate surfaces by $q_1 =$ constant, $q_2 =$ constant, $q_3 =$ constant, we may identify our point by (q_1, q_2, q_3) as well as by (x, y, z). This means that in principle we may write

<table>
<tr><td>General curvilinear coordinates</td><td>Circular cylindrical coordinates</td><td></td></tr>
<tr><td>q_1, q_2, q_3</td><td>ρ, φ, z</td><td></td></tr>
<tr><td>$x = x(q_1, q_2, q_3)$</td><td>$-\infty < x = \rho \cos \varphi < \infty$</td><td></td></tr>
<tr><td>$y = y(q_1, q_2, q_3)$</td><td>$-\infty < y = \rho \sin \varphi < \infty$</td><td>(2.1)</td></tr>
<tr><td>$z = z(q_1, q_2, q_3)$</td><td>$-\infty < z = z < \infty$</td><td></td></tr>
</table>

specifying x, y, z in terms of the q's and the inverse relations,

<table>
<tr><td>$q_1 = q_1(x, y, z)$</td><td>$0 \le \rho = (x^2 + y^2)^{1/2} < \infty$</td><td></td></tr>
<tr><td>$q_2 = q_2(x, y, z)$</td><td>$0 \le \varphi = \arctan(y/x) < 2\pi$</td><td>(2.2)</td></tr>
<tr><td>$q_3 = q_3(x, y, z)$</td><td>$-\infty < z = z < \infty$</td><td></td></tr>
</table>

As a specific illustration of the general, abstract q_1, q_2, q_3 the transformation equations for circular cylindrical coordinates (Section 2.4) are included in Eqs. (2.1) and (2.2). With each family of surfaces $q_i =$ constant, we can associate a unit vector $\mathbf{e}_i$ normal to the surface $q_i =$ constant and in the direction of increasing q_i. Then a vector $\mathbf{V}$ may be written

$$\mathbf{V} = \mathbf{e}_1 V_1 + \mathbf{e}_2 V_2 + \mathbf{e}_3 V_3. \qquad (2.3)$$

The $\mathbf{e}_i$ are normalized to $\mathbf{e}_i^2 = 1$ and form a right-handed coordinate system with volume $\mathbf{e}_1 \cdot (\mathbf{e}_2 \times \mathbf{e}_3) > 0$.

Differentiation of x in Eq. (2.1) leads to

$$dx = \frac{\partial x}{\partial q_1} dq_1 + \frac{\partial x}{\partial q_2} dq_2 + \frac{\partial x}{\partial q_3} dq_3, \qquad (2.4)$$

and similarly for differentiation of y and z. From the Pythagorean theorem in cartesian coordinates the square of the distance between two neighboring points is

$$ds^2 = dx^2 + dy^2 + dz^2. \qquad (2.4a)$$

We assume that in our curvilinear coordinate space the square of the distance element can be written as a general quadratic form:

$$
\begin{aligned}
ds^2 &= g_{11}\, dq_1^2 + g_{12}\, dq_1\, dq_2 + g_{13}\, dq_1\, dq_3 \\
&\quad + g_{21}\, dq_2\, dq_1 + g_{22}\, dq_2^2 + g_{23}\, dq_2\, dq_3 \\
&\quad + g_{31}\, dq_3\, dq_1 + g_{32}\, dq_3\, dq_2 + g_{33}\, dq_3^2 \\
&= \sum_{ij} g_{ij}\, dq_i\, dq_j. \qquad (2.5)
\end{aligned}
$$

Spaces for which Eq. (2.5) is a legitimate expression are called metric or Riemannian. Substituting Eq. (2.4) (squared) and the corresponding results for dy^2 and dz^2 into Eq. (2.4a) and equating coefficients of $dq_i\, dq_j$,[1] we find

$$g_{ij} = \frac{\partial x}{\partial q_i}\frac{\partial x}{\partial q_j} + \frac{\partial y}{\partial q_i}\frac{\partial y}{\partial q_j} + \frac{\partial z}{\partial q_i}\frac{\partial z}{\partial q_j} = \sum_l \frac{\partial x_l}{\partial q_i}\frac{\partial x_l}{\partial q_j}. \qquad (2.6)$$

These coefficients g_{ij}, which we now proceed to investigate, may be viewed as specifying the nature of the coordinate system (q_1, q_2, q_3). Collectively these coefficients are referred to as the *metric* and in Section 2.8 will be shown to be a second-rank symmetric tensor.[2] In general relativity the metric components are determined by the properties of matter. Geometry is merged with physics.

At this point we limit ourselves to orthogonal (mutually perpendicular surfaces) coordinate systems, which means (see Exercise 2.1.1)[3]

$$g_{ij} = 0, \qquad i \neq j, \qquad (2.7)$$

and $e_i \cdot e_j = \delta_{ij}$.

(Nonorthogonal coordinate systems are considered in some detail in Sections 2.10 and 2.11 in the framework of tensor analysis.) Now, to simplify the notation, we write $g_{ii} = h_i^2$ so that

$$ds^2 = (h_1\, dq_1)^2 + (h_2\, dq_2)^2 + (h_3\, dq_3)^2 = \sum_i (h_i\, dq_i)^2. \qquad (2.8)$$

The specific orthogonal coordinate systems are described in subsequent sections by specifying these scale factors h_1, h_2, and h_3. Conversely, the scale factors may be conveniently identified by the relation

$$ds_i = h_i\, dq_i \qquad (2.9)$$

[1] The dq's are arbitrary. For instance, setting $dq_2 = dq_3 = 0$ isolates g_{11}. It might be noted that Eq. (2.6) can be derived from Eq. (2.4) more elegantly with the matrix notation of Chapter 3. Further, the matrix notation leads directly to the Jacobian determinant, Exercise 2.1.5.

[2] The tensor nature of the set of g_{ij}'s follows from the quotient rule (Section 2.8). Then the tensor transformation law yields Eq. (2.6).

[3] In relativistic cosmology the nondiagonal elements of the metric g_{ij} are usually set equal to zero as a consequence of physical assumptions such as no rotation.

for any given dq_i, holding the other q's constant. Note that the three curvilinear coordinates q_1, q_2, q_3 need not be lengths. The scale factors h_i may depend on the q's and they may have dimensions. The *product* $h_i\,dq_i$ must have dimensions of length and be positive. The differential distance vector $d\mathbf{r}$ may be written

$$d\mathbf{r} = h_1\,dq_1\,\mathbf{e}_1 + h_2\,dq_2\,\mathbf{e}_2 + h_3\,dq_3\,\mathbf{e}_3$$

$$= \sum_i h_i\,dq_i\,\mathbf{e}_i.$$

Using this curvilinear component form, we find that a line integral becomes

$$\int \mathbf{V} \cdot d\mathbf{r} = \sum_i \int V_i h_i\,dq_i.$$

From Eq. (2.9) we may immediately develop the area and volume elements

$$d\sigma_{ij} = ds_i\,ds_j = h_i h_j\,dq_i\,dq_j \tag{2.10}$$

and

$$d\tau = ds_1\,ds_2\,ds_3 = h_1 h_2 h_3\,dq_1\,dq_2\,dq_3. \tag{2.11}$$

The expressions in Eqs. (2.10) and (2.11) agree, of course, with the results of using the transformation equations, Eq. (2.1), and Jacobians.

From Eq. (2.10) an area element may be expanded:

$$d\boldsymbol{\sigma} = ds_2\,ds_3\,\mathbf{e}_1 + ds_3\,ds_1\,\mathbf{e}_2 + ds_1\,ds_2\,\mathbf{e}_3$$

$$= h_2 h_3\,dq_2\,dq_3\,\mathbf{e}_1 + h_3 h_1\,dq_3\,dq_1\,\mathbf{e}_2$$

$$+ h_1 h_2\,dq_1\,dq_2\,\mathbf{e}_3.$$

A surface integral becomes

$$\int \mathbf{V} \cdot d\boldsymbol{\sigma} = \int V_1 h_2 h_3\,dq_2\,dq_3 + \int V_2 h_3 h_1\,dq_3\,dq_1$$

$$+ \int V_3 h_1 h_2\,dq_1\,dq_2.$$

Examples of such line and surface integrals appear in Sections 2.4 and 2.5.

In anticipation of the new forms of equations for vector *calculus* that appear in the next section, the student should clearly understand that vector *algebra* is the same in orthogonal curvilinear coordinates as in cartesian coordinates. Specifically, for the dot product

$$\mathbf{A} \cdot \mathbf{B} = A_1 B_1 + A_2 B_2 + A_3 B_3, \tag{2.11a}$$

where the subscripts indicate curvilinear components. For the cross product

$$\mathbf{A} \times \mathbf{B} = \begin{vmatrix} \mathbf{e}_1 & \mathbf{e}_2 & \mathbf{e}_3 \\ A_1 & A_2 & A_3 \\ B_1 & B_2 & B_3 \end{vmatrix}, \tag{2.11b}$$

just like Eq. (1.40).

EXERCISES

2.1.1   Show that limiting our attention to orthogonal coordinate systems implies that $g_{ij} = 0$ for $i \neq j$ (Eq. (2.7)).

Hint. Construct a triangle with sides ds_1, ds_2, and ds. Equation (2.9) must hold regardless of whether $g_{ij} = 0$. Then compare ds^2 from Eq. (2.5) with a calculation using the law of cosines. Show that $\cos \theta_{12} = g_{12}/\sqrt{g_{11}g_{22}}$.

2.1.2   In the spherical polar coordinate system $q_1 = r$, $q_2 = \theta$, $q_3 = \varphi$. The transformation equations corresponding to Eq. (2.1) are

$$x = r \sin \theta \cos \varphi$$
$$y = r \sin \theta \sin \varphi$$
$$z = r \cos \theta.$$

(a) Calculate the spherical polar coordinate scale factors: h_r, h_θ, and h_φ.

(b) Check your calculated scale factors by the relation $ds_i = h_i dq_i$.

2.1.3   The u-, v-, z-coordinate system frequently used in electrostatics and in hydrodynamics is defined by

$$xy = u,$$
$$x^2 - y^2 = v,$$
$$z = z.$$

This u-, v-, z-system is orthogonal.

(a) In words, describe briefly the nature of each of the three families of coordinate surfaces.

(b) Sketch the system in the xy-plane showing the intersections of surfaces of constant u and surfaces of constant v with the xy-plane.

(c) Indicate the directions of the unit vector $\hat{u}$ and $\hat{v}$ in all four quadrants.

(d) Finally, is this u-, v-, z-system right-handed ($\hat{u} \times \hat{v} = +\hat{z}$) or left-handed ($\hat{u} \times \hat{v} = -\hat{z}$)?

2.1.4   The elliptic cylindrical coordinate system consists of three families of surfaces:

$$1. \quad \frac{x^2}{a^2 \cosh^2 u} + \frac{y^2}{a^2 \sinh^2 u} = 1$$

$$2. \quad \frac{x^2}{a^2 \cos^2 v} - \frac{y^2}{a^2 \sin^2 v} = 1$$

3. $z = z$.

Sketch the coordinate surfaces $u = \text{constant}$ and $v = \text{constant}$ as they intersect the first quadrant of the xy-plane. Show the unit vectors $\hat{u}$ and $\hat{v}$. The range of u is $0 \leq u < \infty$. The range of v is $0 \leq v \leq 2\pi$.

2.1.5   A *two*-dimensional orthogonal system is described by the coordinates q_1 and q_2. Show that the Jacobian

$$J\left(\frac{x, y}{q_1, q_2}\right) = h_1 h_2$$

is in agreement with Eq. (2.10).

Hint. It's easier to work with the square of each side of this equation.

2.1.6   In Minkowski space we define $x_1 = x$, $x_2 = y$, $x_3 = z$, and $x_0 = ct$. This is done so that the space–time interval $ds^2 = dx_0^2 - dx_1^2 - dx_2^2 - dx_3^2$ ($c =$ velocity of light). Show that the metric in Minkowski space is

$$(g_{ij}) = \begin{pmatrix} 1 & 0 & 0 & 0 \\ 0 & -1 & 0 & 0 \\ 0 & 0 & -1 & 0 \\ 0 & 0 & 0 & -1 \end{pmatrix}.$$

This indicates the advantage of using Minkowski space in a special relativity theory: It is a four-dimensional cartesian system. We use Minkowski space in Sections 4.6 and 4.2 for describing Lorentz transformations.

2.2 DIFFERENTIAL VECTOR OPERATORS

GRADIENT

The starting point for developing the gradient, divergence, and curl operators in curvilinear coordinates is our interpretation of the gradient as the vector having the magnitude and direction of the maximum space rate of change (compare Section 1.6). From this interpretation the component of $\nabla \psi(q_1, q_2, q_3)$ in the direction normal to the family of surfaces $q_1 =$ constant is given by[1]

$$\mathbf{e}_1 \cdot \nabla \psi = \nabla \psi|_1 = \frac{\partial \psi}{\partial s_1} = \frac{\partial \psi}{h_1 \, \partial q_1}, \tag{2.12}$$

since this is the rate of change of ψ for varying q_1, holding q_2 and q_3 fixed. The quantity ds_1 is a differential length in the direction of increasing q_1 (compare Eq. (2.9)). In Section 2.1 we introduced a unit vector $\mathbf{e}_1$ to indicate this direction. By repeating Eq. (2.12) for q_2 and again for q_3 and adding vectorially, we see that the gradient becomes

$$\nabla \psi(q_1, q_2, q_3) = \mathbf{e}_1 \frac{\partial \psi}{\partial s_1} + \mathbf{e}_2 \frac{\partial \psi}{\partial s_2} + \mathbf{e}_3 \frac{\partial \psi}{\partial s_3}$$

$$= \mathbf{e}_1 \frac{\partial \psi}{h_1 \, \partial q_1} + \mathbf{e}_2 \frac{\partial \psi}{h_2 \, \partial q_2} + \mathbf{e}_3 \frac{\partial \psi}{h_3 \, \partial q_3}$$

$$= \sum_i \mathbf{e}_i \frac{1}{h_i} \frac{\partial \psi}{\partial q_i}. \tag{2.13}$$

Exercise 2.2.4 offers a mathematical alternative independent of this physical interpretation of the gradient.

---

[1] Here the use of φ to label a function is avoided because it is conventional to use this symbol to denote an azimuthal coordinate.

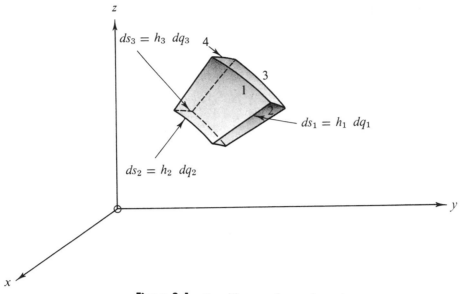

Figure 2.1 Curvilinear volume element.

DIVERGENCE

The divergence operator may be obtained from the second definition (Eq. (1.97)) of Chapter 1 or equivalently from Gauss's theorem, Section 1.11. Let us use Eq. (1.97),

$$\nabla \cdot \mathbf{V}(q_1, q_2, q_3) = \lim_{\int d\tau \to 0} \frac{\int \mathbf{V} \cdot d\boldsymbol{\sigma}}{\int d\tau}, \qquad (2.14)$$

with a differential volume $h_1 h_2 h_3 \, dq_1 \, dq_2 \, dq_3$ (Fig. 2.1). Note that the positive directions have been chosen so that (q_1, q_2, q_3) or $(\mathbf{e}_1, \mathbf{e}_2, \mathbf{e}_3)$ form a right-handed set, $\mathbf{e}_1 \times \mathbf{e}_2 = \mathbf{e}_3$.

The area integral for the two faces $q_1 = $ constant is given by

$$\left[V_1 h_2 h_3 + \frac{\partial}{\partial q_1} (V_1 h_2 h_3) \, dq_1 \right] dq_2 \, dq_3 - V_1 h_2 h_3 \, dq_2 \, dq_3$$

$$= \frac{\partial}{\partial q_1} (V_1 h_2 h_3) \, dq_1 \, dq_2 \, dq_3, \qquad (2.15)$$

exactly as in Sections 1.7 and 1.10.[2] Adding in the similar results for the other

[2] Since we take the limit $dq_1, dq_2, dq_3 \to 0$, the second- and higher-order derivatives will drop out.

two pairs of surfaces, we obtain

$$\int \mathbf{V}(q_1, q_2, q_3) \cdot d\boldsymbol{\sigma}$$
$$= \left[\frac{\partial}{\partial q_1} (V_1 h_2 h_3) + \frac{\partial}{\partial q_2} (V_2 h_3 h_1) + \frac{\partial}{\partial q_3} (V_3 h_1 h_2) \right] dq_1 \, dq_2 \, dq_3. \qquad (2.16)$$

Division by our differential volume (Eq. (2.14)) yields

$$\mathbf{V} \cdot \mathbf{V}(q_1, q_2, q_3) = \frac{1}{h_1 h_2 h_3} \left[\frac{\partial}{\partial q_1} (V_1 h_2 h_3) + \frac{\partial}{\partial q_2} (V_2 h_3 h_1) + \frac{\partial}{\partial q_3} (V_3 h_1 h_2) \right].$$
$$(2.17)$$

In Eq. (2.17), V_i is the component of $\mathbf{V}$ in the $\mathbf{e}_i$-direction, increasing q_i; that is, $V_i = \mathbf{e}_i \cdot \mathbf{V}$ is the projection of $\mathbf{V}$ onto the $\mathbf{e}_i$-direction.

We may obtain the Laplacian by combining Eqs. (2.13) and (2.17), using $\mathbf{V} = \boldsymbol{\nabla}\psi(q_1, q_2, q_3)$. This leads to

$$\boldsymbol{\nabla} \cdot \boldsymbol{\nabla}\psi(q_1, q_2, q_3)$$
$$= \frac{1}{h_1 h_2 h_3} \left[\frac{\partial}{\partial q_1} \left(\frac{h_2 h_3}{h_1} \frac{\partial \psi}{\partial q_1} \right) + \frac{\partial}{\partial q_2} \left(\frac{h_3 h_1}{h_2} \frac{\partial \psi}{\partial q_2} \right) + \frac{\partial}{\partial q_3} \left(\frac{h_1 h_2}{h_3} \frac{\partial \psi}{\partial q_3} \right) \right].$$
$$(2.18a)$$

CURL

Finally, to develop $\boldsymbol{\nabla} \times \mathbf{V}$, let us apply Stokes's theorem (Section 1.12) and, as with the divergence, take the limit as the surface area becomes vanishingly small. Working on one component at a time, we consider a differential surface element in the curvilinear surface $q_1 = $ constant. From

$$\int_s \boldsymbol{\nabla} \times \mathbf{V} \cdot d\boldsymbol{\sigma} = \mathbf{e}_1 \cdot (\boldsymbol{\nabla} \times \mathbf{V}) h_2 h_3 \, dq_2 \, dq_3 \qquad (2.18b)$$

(mean value theorem of integral calculus) Stokes's theorem yields

$$\mathbf{e}_1 \cdot (\boldsymbol{\nabla} \times \mathbf{V}) h_2 h_3 \, dq_2 \, dq_3 = \oint \mathbf{V} \cdot d\mathbf{r}, \qquad (2.19)$$

with the line integral lying in the surface $q_1 = $ constant. Following the loop $(1, 2, 3, 4)$ of Fig. 2.2,

$$\oint \mathbf{V}(q_1, q_2, q_3) \cdot d\mathbf{r} = V_2 h_2 \, dq_2 + \left[V_3 h_3 + \frac{\partial}{\partial q_2} (V_3 h_3) \, dq_2 \right] dq_3$$
$$- \left[V_2 h_2 + \frac{\partial}{\partial q_3} (V_2 h_2) \, dq_3 \right] dq_2 - V_3 h_3 \, dq_3$$
$$= \left[\frac{\partial}{\partial q_2} (h_3 V_3) - \frac{\partial}{\partial q_3} (h_2 V_2) \right] dq_2 \, dq_3. \qquad (2.20)$$

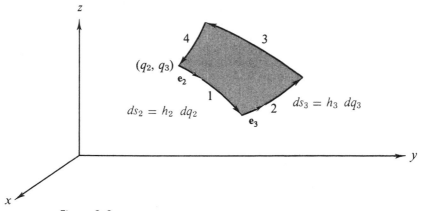

Figure 2.2 Curvilinear surface element with q_1 = constant.

We pick up a positive sign when going in the positive direction on parts 1 and 2 and a negative sign on parts 3 and 4 because here we are going in the negative direction. Higher-order terms in Maclaurin or Taylor expansion have been omitted. They will vanish in the limit as the surface becomes vanishingly small $(dq_2 \to 0, dq_3 \to 0)$.

From Eq. (2.19)

$$\mathbf{\nabla} \times \mathbf{V}|_1 = \frac{1}{h_2 h_3}\left[\frac{\partial}{\partial q_2}(h_3 V_3) - \frac{\partial}{\partial q_3}(h_2 V_2)\right]. \tag{2.21}$$

The remaining two components of $\mathbf{\nabla} \times \mathbf{V}$ may be picked up by cyclic permutation of the indices. As in Chapter 1, it is often convenient to write the curl in determinant form:

$$\mathbf{\nabla} \times \mathbf{V} = \frac{1}{h_1 h_2 h_3}\begin{vmatrix} \mathbf{e}_1 h_1 & \mathbf{e}_2 h_2 & \mathbf{e}_3 h_3 \\ \dfrac{\partial}{\partial q_1} & \dfrac{\partial}{\partial q_2} & \dfrac{\partial}{\partial q_3} \\ h_1 V_1 & h_2 V_2 & h_3 V_3 \end{vmatrix}. \tag{2.22}$$

Remember that because of the presence of the differential operators, this determinant must be expanded from the top down. Note that this equation is *not* identical with the form for the cross product of two vectors, Eq. (2.11b). $\mathbf{\nabla}$ is not an ordinary vector; it is a vector *operator*.

Our geometric interpretation of the gradient and the use of Gauss's and Stokes's theorems (or integral definitions of divergence and curl) have enabled us to obtain these quantities *without having to differentiate the unit vectors* $\mathbf{e}_i$. There exist alternate ways to determine grad, div, and curl based on direct differentiation of the $\mathbf{e}_i$. One approach resolves the $\mathbf{e}_i$ of a specific coordinate system into its cartesian components (Exercises 2.4.1 and 2.5.1) and differentiates this cartesian form (Exercises 2.4.3 and 2.5.2). The point here is that the derivatives of the cartesian $\hat{\mathbf{x}}$, $\hat{\mathbf{y}}$, and $\hat{\mathbf{z}}$ vanish since $\hat{\mathbf{x}}$, $\hat{\mathbf{y}}$, and $\hat{\mathbf{z}}$ are constant in direction as well as in magnitude. A second approach [L. J. Kijewski, *Am. J. Phys.* **33**, 816 (1965)] assumes the equality of $\partial^2 \mathbf{r}/\partial q_i\, \partial q_j$ and $\partial^2 \mathbf{r}/\partial q_j \partial q_i$ and develops the derivatives of $\mathbf{e}_i$ in a general curvilinear form. Exercises 2.2.3 and 2.2.4 are based on this method.

EXERCISES

2.2.1   Develop arguments to show that ordinary dot and cross products (not involving ∇) in orthogonal curvilinear coordinates proceed as in cartesian coordinates *with no involvement of scale factors.*

2.2.2   With $\mathbf{e}_1$ a unit vector in the direction of increasing q_1, show that

(a) $\nabla \cdot \mathbf{e}_1 = \dfrac{1}{h_1 h_2 h_3} \dfrac{\partial (h_2 h_3)}{\partial q_1}$

(b) $\nabla \times \mathbf{e}_1 = \dfrac{1}{h_1} \left[\mathbf{e}_2 \dfrac{\partial h_1}{h_3\, \partial q_3} - \mathbf{e}_3 \dfrac{\partial h_1}{h_2\, \partial q_2} \right].$

Note that even though $\mathbf{e}_1$ is a unit vector, its divergence and curl *do not necessarily vanish.*

2.2.3   Show that the orthogonal unit vectors $\mathbf{e}_i$ may be defined by

$$\mathbf{e}_i = \frac{1}{h_i} \frac{\partial \mathbf{r}}{\partial q_i}. \qquad (a)$$

In particular, show that $\mathbf{e}_i \cdot \mathbf{e}_i = 1$ leads to an expression for h_i in agreement with Eq. (2.6).

Equation (a) may be taken as a starting point for deriving

$$\frac{\partial \mathbf{e}_i}{\partial q_j} = \mathbf{e}_j \frac{\partial h_j}{h_i\, \partial q_i}, \qquad i \neq j$$

and

$$\frac{\partial \mathbf{e}_i}{\partial q_i} = - \sum_{j \neq i} \mathbf{e}_j \frac{\partial h_i}{h_j\, \partial q_j}.$$

2.2.4   Derive

$$\nabla \psi = \mathbf{e}_1 \frac{\partial \psi}{h_1\, \partial q_1} + \mathbf{e}_2 \frac{\partial \psi}{h_2\, \partial q_2} + \mathbf{e}_3 \frac{\partial \psi}{h_3\, \partial q_3}$$

by direct application of Eq. (1.96),

$$\nabla \psi = \lim_{\int d\tau \to 0} \frac{\int \psi\, d\boldsymbol{\sigma}}{\int d\tau}.$$

Hint. Evaluation of the surface integral will lead to terms like $(h_1 h_2 h_3)^{-1}$ $(\partial/\partial q_1)(\mathbf{e}_1 h_2 h_3)$. The results listed in Exercise 2.2.3 will be helpful. Cancellation of unwanted terms occurs when the contributions of all three pairs of surfaces are added together.

2.3 SPECIAL COORDINATE SYSTEMS: INTRODUCTION

As mentioned in Section 2.1, there are 11 coordinate systems in which the three-dimensional Helmholtz equation can be separated into three ordinary differential equations. Some of these coordinate systems have achieved prominence in the historical development of quantum mechanics. Other systems such as bipolar coordinates, satisfy special needs. Partly because the needs are rather infrequent, but mostly because the development of high-speed computing machines and efficient programming techniques reduces the need for these coordinate systems, the discussion in this chapter is limited to (1) cartesian coordinates, (2) spherical polar coordinates, and (3) circular cylindrical coordinates. Specifications and details of the other coordinate systems will be found in the first two editions of this work and in the references (Morse and Feshbach, Margenau and Murphy).

RECTANGULAR CARTESIAN COORDINATES

These are the cartesian coordinates on which Chapter 1 is based. In this simplest of all systems

$$h_1 = h_x = 1,$$

$$h_2 = h_y = 1, \qquad (2.23)$$

$$h_3 = h_z = 1.$$

The families of coordinate surfaces are three sets of parallel planes: $x = $ constant, $y = $ constant, and $z = $ constant. The cartesian coordinate system is unique in that all its h_i's are constant. This will be a significant advantage in treating tensors in Section 2.6. Note also that the unit vectors, $\mathbf{e}_1, \mathbf{e}_2, \mathbf{e}_3$ or $\hat{\mathbf{x}}, \hat{\mathbf{y}}, \hat{\mathbf{z}}$, have *fixed* directions.

From Eqs. (2.13), (2.17), (2.18), and (2.22) we reproduce the results of Chapter 1,

$$\nabla \psi = \hat{\mathbf{x}} \frac{\partial \psi}{\partial x} + \hat{\mathbf{y}} \frac{\partial \psi}{\partial y} + \hat{\mathbf{z}} \frac{\partial \psi}{\partial z}, \qquad (2.24)$$

$$\nabla \cdot \mathbf{V} = \frac{\partial V_x}{\partial x} + \frac{\partial V_y}{\partial y} + \frac{\partial V_z}{\partial z}, \qquad (2.25)$$

$$\nabla \cdot \nabla \psi = \frac{\partial^2 \psi}{\partial x^2} + \frac{\partial^2 \psi}{\partial y^2} + \frac{\partial^2 \psi}{\partial z^2}, \qquad (2.26)$$

$$\nabla \times \mathbf{V} = \begin{vmatrix} \hat{\mathbf{x}} & \hat{\mathbf{y}} & \hat{\mathbf{z}} \\ \dfrac{\partial}{\partial x} & \dfrac{\partial}{\partial y} & \dfrac{\partial}{\partial z} \\ V_x & V_y & V_z \end{vmatrix} \qquad (2.27)$$

2.4 CIRCULAR CYLINDRICAL COORDINATES (ρ, φ, z)

In the circular cylindrical coordinate system the three curvilinear coordinates (q_1, q_2, q_3) are relabeled (ρ, φ, z). The coordinate surfaces, shown in Fig. 2.3, are

1. Right circular cylinders having the z-axis as a common axis,

$$\rho = (x^2 + y^2)^{1/2} = \text{constant}.$$

2. Half planes through the z-axis,

$$\varphi = \tan^{-1}\left(\frac{y}{x}\right) = \text{constant}.$$

3. Planes parallel to the xy-plane, as in the cartesian system,

$$z = \text{constant}.$$

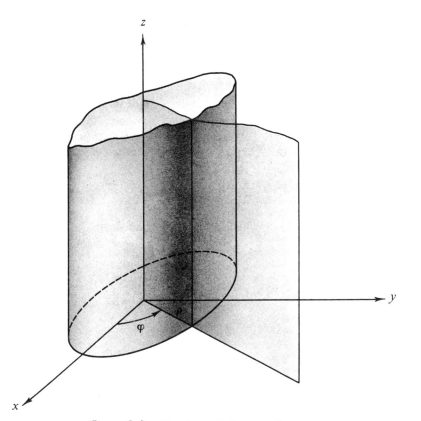

Figure 2.3 Circular cylinder coordinates.

The limits on ρ, φ, and z are

$$0 \leq \rho < \infty, \qquad 0 \leq \varphi \leq 2\pi, \qquad \text{and} \qquad -\infty < z < \infty.$$

Note that we are using ρ for the perpendicular distance from the z-axis and saving r for the distance from the origin.

Inverting the preceding equations for ρ and φ (or going directly to Fig. 2.3), we obtain the transformation relations

$$x = \rho \cos \varphi,$$

$$y = \rho \sin \varphi, \qquad (2.28)$$

$$z = z.$$

The z-axis remains unchanged. This is essentially a two-dimensional curvilinear system with a cartesian z-axis added on to form a three-dimensional system.

According to Eq. (2.28) or from the length elements ds_i, the scale factors are

$$h_1 = h_\rho = 1,$$

$$h_2 = h_\varphi = \rho, \qquad (2.29)$$

$$h_3 = h_z = 1.$$

The unit vectors $\mathbf{e}_1, \mathbf{e}_2, \mathbf{e}_3$ are relabeled $(\hat{\boldsymbol{\rho}}, \hat{\boldsymbol{\varphi}}, \hat{\mathbf{z}})$, Fig. 2.4. The unit vector $\hat{\boldsymbol{\rho}}$ is normal to the cylindrical surface pointing in the direction of increasing radius ρ. The unit vector $\hat{\boldsymbol{\varphi}}$ is tangential to the cylindrical surface, perpendicular to the half plane $\varphi =$ constant and pointing in the direction of increasing azimuth angle φ. The third unit vector, $\hat{\mathbf{z}}$, is the usual cartesian unit vector.

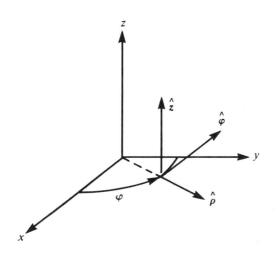

Figure 2.4 Circular cylindrical coordinate unit vectors.

A differential displacement $d\mathbf{r}$ may be written

$$d\mathbf{r} = \hat{\mathbf{\rho}}\, ds_\rho + \hat{\mathbf{\phi}}\, ds_\varphi + \hat{\mathbf{z}}\, dz$$

$$= \hat{\mathbf{\rho}}\, d\rho + \hat{\mathbf{\phi}}\rho\, d\varphi + \hat{\mathbf{z}}\, dz. \tag{2.30}$$

The differential operations involving $\mathbf{\nabla}$ follow from Eqs. (2.13), (2.17), (2.18), and (2.22),

$$\mathbf{\nabla}\psi(\rho, \varphi, z) = \hat{\mathbf{\rho}}\frac{\partial\psi}{\partial\rho} + \hat{\mathbf{\phi}}\frac{1}{\rho}\frac{\partial\psi}{\partial\varphi} + \hat{\mathbf{z}}\frac{\partial\psi}{\partial z}, \tag{2.31}$$

$$\mathbf{\nabla}\cdot\mathbf{V} = \frac{1}{\rho}\frac{\partial}{\partial\rho}(\rho V_\rho) + \frac{1}{\rho}\frac{\partial V_\varphi}{\partial\varphi} + \frac{\partial V_z}{\partial z}, \tag{2.32}$$

$$\mathbf{\nabla}^2\psi = \frac{1}{\rho}\frac{\partial}{\partial\rho}\left(\rho\frac{\partial\psi}{\partial\rho}\right) + \frac{1}{\rho^2}\frac{\partial^2\psi}{\partial\varphi^2} + \frac{\partial^2\psi}{\partial z^2}, \tag{2.33}$$

$$\mathbf{\nabla}\times\mathbf{V} = \frac{1}{\rho}\begin{vmatrix} \hat{\mathbf{\rho}} & \rho\hat{\mathbf{\phi}} & \hat{\mathbf{z}} \\ \dfrac{\partial}{\partial\rho} & \dfrac{\partial}{\partial\varphi} & \dfrac{\partial}{\partial z} \\ V_\rho & \rho V_\varphi & V_z \end{vmatrix}. \tag{2.34}$$

Finally, for problems such as circular wave guides or cylindrical cavity resonators the vector Laplacian $\mathbf{\nabla}^2\mathbf{V}$ resolved in circular cylindrical coordinates is

$$\mathbf{\nabla}^2\mathbf{V}\big|_\rho = \mathbf{\nabla}^2 V_\rho - \frac{1}{\rho^2}V_\rho - \frac{2}{\rho^2}\frac{\partial V_\varphi}{\partial\varphi},$$

$$\mathbf{\nabla}^2\mathbf{V}\big|_\varphi = \mathbf{\nabla}^2 V_\varphi - \frac{1}{\rho^2}V_\varphi + \frac{2}{\rho^2}\frac{\partial V_\rho}{\partial\varphi}, \tag{2.35}$$

$$\mathbf{\nabla}^2\mathbf{V}\big|_z = \mathbf{\nabla}^2 V_z.$$

The basic reason for the form of the z-component is that the z-axis is a cartesian axis; that is,

$$\mathbf{\nabla}^2(\hat{\mathbf{\rho}} V_\rho + \hat{\mathbf{\phi}} V_\varphi + \hat{\mathbf{z}} V_z) = \mathbf{\nabla}^2(\hat{\mathbf{\rho}} V_\rho + \hat{\mathbf{\phi}} V_\varphi) + \hat{\mathbf{z}}\mathbf{\nabla}^2 V_z$$

$$= \hat{\mathbf{\rho}} f(V_\rho, V_\varphi) + \hat{\mathbf{\phi}} g(V_\rho, V_\varphi) + \hat{\mathbf{z}}\mathbf{\nabla}^2 V_z.$$

The operator $\mathbf{\nabla}^2$ operating on the $\hat{\mathbf{\rho}}$, $\hat{\mathbf{\phi}}$ unit vectors stays in the $\hat{\mathbf{\rho}}\hat{\mathbf{\phi}}$-plane. This behavior holds in all such cylindrical systems.

Example 2.4.1 A Navier–Stokes Term

The Navier–Stokes equations of hydrodynamics contain a nonlinear term

$$\mathbf{\nabla}\times[\mathbf{v}\times(\mathbf{\nabla}\times\mathbf{v})],$$

where **v** is the fluid velocity. For fluid flowing through a cylindrical pipe in the z-direction

$$\mathbf{v} = \hat{\mathbf{z}} v(\rho).$$

From Eq. (2.34)

$$\nabla \times \mathbf{v} = \frac{1}{\rho} \begin{vmatrix} \hat{\boldsymbol{\rho}} & \rho\hat{\boldsymbol{\varphi}} & \hat{\mathbf{z}} \\ \dfrac{\partial}{\partial \rho} & \dfrac{\partial}{\partial \varphi} & \dfrac{\partial}{\partial z} \\ 0 & 0 & v(\rho) \end{vmatrix}$$

$$= -\hat{\boldsymbol{\varphi}} \frac{\partial v}{\partial \rho}$$

$$\mathbf{v} \times (\nabla \times \mathbf{v}) = \begin{vmatrix} \hat{\boldsymbol{\rho}} & \hat{\boldsymbol{\varphi}} & \hat{\mathbf{z}} \\ 0 & 0 & v \\ 0 & -\dfrac{\partial v}{\partial \rho} & 0 \end{vmatrix}$$

$$= \hat{\boldsymbol{\rho}} v(\rho) \frac{\partial v}{\partial \rho}.$$

Finally,

$$\nabla \times (\mathbf{v} \times (\nabla \times \mathbf{v})) = \frac{1}{\rho} \begin{vmatrix} \hat{\boldsymbol{\rho}} & \rho\hat{\boldsymbol{\varphi}} & \hat{\mathbf{z}} \\ \dfrac{\partial}{\partial \rho} & \dfrac{\partial}{\partial \varphi} & \dfrac{\partial}{\partial z} \\ v\dfrac{\partial v}{\partial \rho} & 0 & 0 \end{vmatrix}$$

$$= 0.$$

For this particular case the nonlinear term vanishes.

EXERCISES

2.4.1   Resolve the circular cylindrical unit vectors into their cartesian components (Fig. 2.5).

$ANS. \quad \hat{\boldsymbol{\rho}} = \hat{\mathbf{x}} \cos \varphi + \hat{\mathbf{y}} \sin \varphi,$
$\hat{\boldsymbol{\varphi}} = -\hat{\mathbf{x}} \sin \varphi + \hat{\mathbf{y}} \cos \varphi,$
$\hat{\mathbf{z}} = \hat{\mathbf{z}}.$

2.4.2   Resolve the cartesian unit vectors into their circular cylindrical components.

$ANS. \quad \hat{\mathbf{x}} = \hat{\boldsymbol{\rho}} \cos \varphi - \hat{\boldsymbol{\varphi}} \sin \varphi,$
$\hat{\mathbf{y}} = \hat{\boldsymbol{\rho}} \sin \varphi + \hat{\boldsymbol{\varphi}} \cos \varphi,$
$\hat{\mathbf{z}} = \hat{\mathbf{z}}.$

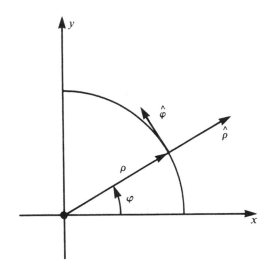

Figure 2.5

2.4.3   From the results of Exercise 2.4.1 show that

$$\frac{\partial \hat{\boldsymbol{\rho}}}{\partial \varphi} = \hat{\boldsymbol{\varphi}}, \qquad \frac{\partial \hat{\boldsymbol{\varphi}}}{\partial \varphi} = -\hat{\boldsymbol{\rho}}$$

and that all other first derivatives of the circular cylindrical unit vectors with respect to the circular cylindrical coordinates vanish.

2.4.4   Compare $\nabla \cdot \mathbf{V}$ (Eq. (2.32)) with the gradient operator

$$\nabla = \hat{\boldsymbol{\rho}}\frac{\partial}{\partial \rho} + \hat{\boldsymbol{\varphi}}\frac{1}{\rho}\frac{\partial}{\partial \varphi} + \hat{\mathbf{z}}\frac{\partial}{\partial z}$$

(Eq. (2.31)) dotted into $\mathbf{V}$. Note that the differential operators of ∇ differentiate *both* the unit vectors and the components of $\mathbf{V}$.

Hint. $\hat{\boldsymbol{\varphi}}(1/\rho)(\partial/\partial\varphi) \cdot \hat{\boldsymbol{\rho}}V_\rho$ becomes $\hat{\boldsymbol{\varphi}} \cdot \dfrac{1}{\rho}\dfrac{\partial}{\partial\varphi}(\hat{\boldsymbol{\rho}}V_\rho)$ and does *not* vanish.

2.4.5   (a) Show that $\mathbf{r} = \hat{\boldsymbol{\rho}}\rho + \hat{\mathbf{z}}z$.
(b) Working entirely in circular cylindrical coordinates, show that

$$\nabla \cdot \mathbf{r} = 3 \qquad \text{and} \qquad \nabla \times \mathbf{r} = 0.$$

2.4.6   (a) Show that the parity operation (reflection through the origin) on a point (ρ, φ, z) relative to *fixed* x-, y-, z-axes consists of the transformation

$$\rho \to \rho$$
$$\varphi \to \varphi \pm \pi$$
$$z \to -z.$$

(b) Show that $\hat{\boldsymbol{\rho}}$ and $\hat{\boldsymbol{\varphi}}$ have odd parity (reversal of direction) and that $\hat{\mathbf{z}}$ has even parity.
Note. The cartesian unit vectors $\hat{\mathbf{x}}$, $\hat{\mathbf{y}}$, and $\hat{\mathbf{z}}$ remain constant.

2.4.7 A rigid body is rotating about a fixed axis with a constant angular velocity ω. Take ω to lie along the z-axis. Express **r** in circular cylindrical coordinates and using circular cylindrical coordinates,

(a) calculate $\mathbf{v} = \boldsymbol{\omega} \times \mathbf{r}$.
(b) calculate $\boldsymbol{\nabla} \times \mathbf{v}$.

ANS. (a) $\mathbf{v} = \hat{\boldsymbol{\varphi}}\omega\rho$
(b) $\boldsymbol{\nabla} \times \mathbf{v} = 2\boldsymbol{\omega}$.

2.4.8 A particle is moving through space. Find the circular cylindrical components of its velocity and acceleration.

$$v_\rho = \dot{\rho}, \qquad a_\rho = \ddot{\rho} - \rho\dot{\varphi}^2,$$
$$v_\varphi = \rho\dot{\varphi}, \qquad a_\varphi = \rho\ddot{\varphi} + 2\dot{\rho}\dot{\varphi},$$
$$v_z = \dot{z}, \qquad a_z = \ddot{z}.$$

Hint.

$$\mathbf{r}(t) = \hat{\boldsymbol{\rho}}(t)\rho(t) + \hat{\mathbf{z}}z(t)$$
$$= [\hat{\mathbf{x}}\cos\varphi(t) + \hat{\mathbf{y}}\sin\varphi(t)]\rho(t) + \hat{\mathbf{z}}z(t).$$

Note. $\dot{\rho} = d\rho/dt$, $\ddot{\rho} = d^2\rho/dt^2$, and so on.

2.4.9 Solve Laplace's equation $\nabla^2\psi = 0$, in cylindrical coordinates for $\psi = \psi(\rho)$.

ANS. $\psi = k\ln\dfrac{\rho}{\rho_0}$.

2.4.10 In right circular cylindrical coordinates a particular vector function is given by

$$\mathbf{V}(\rho, \varphi) = \hat{\boldsymbol{\rho}}V_\rho(\rho, \varphi) + \hat{\boldsymbol{\varphi}}V_\varphi(\rho, \varphi).$$

Show that $\boldsymbol{\nabla} \times \mathbf{V}$ has only a z-component. Note that this result will hold for any vector confined to a surface $q_3 =$ constant as long as the products h_1V_1 and h_2V_2 are each independent of q_3.

2.4.11 For the flow of an incompressible viscous fluid the Navier–Stokes equations lead to

$$-\boldsymbol{\nabla} \times (\mathbf{v} \times (\boldsymbol{\nabla} \times \mathbf{v})) = \frac{\eta}{\rho_0}\nabla^2(\boldsymbol{\nabla} \times \mathbf{v}).$$

Here η is the viscosity and ρ_0 the density of the fluid. For axial flow in a cylindrical pipe we take the velocity **v** to be

$$\mathbf{v} = \hat{\mathbf{z}}v(\rho).$$

From Example 2.4.1

$$\boldsymbol{\nabla} \times (\mathbf{v} \times (\boldsymbol{\nabla} \times \mathbf{v})) = 0$$

for this choice of **v**.

Show that

$$\nabla^2(\boldsymbol{\nabla} \times \mathbf{v}) = 0$$

leads to the differential equation

$$\frac{1}{\rho}\frac{d}{d\rho}\left(\rho\frac{d^2v}{d\rho^2}\right) - \frac{1}{\rho^2}\frac{dv}{d\rho} = 0$$

and that this is satisfied by

$$v = v_0 + a_2\rho^2.$$

2.4.12   A conducting wire along the z-axis carries a current I. The resulting magnetic vector potential is given by

$$\mathbf{A} = \hat{\mathbf{z}} \frac{\mu I}{2\pi} \ln\left(\frac{1}{\rho}\right).$$

Show that the magnetic induction $\mathbf{B}$ is given by

$$\mathbf{B} = \hat{\boldsymbol{\varphi}} \frac{\mu I}{2\pi\rho}.$$

2.4.13   A force is described by

$$\mathbf{F} = -\hat{\mathbf{x}} \frac{y}{x^2 + y^2} + \hat{\mathbf{y}} \frac{x}{x^2 + y^2}.$$

(a) Express $\mathbf{F}$ in circular cylindrical coordinates.
   Operating entirely in circular cylindrical coordinates for (b) and (c),
(b) calculate the curl of $\mathbf{F}$ and
(c) calculate the work done by $\mathbf{F}$ in encircling the unit circle once counter-clockwise.
(d) How do you reconcile the results of (b) and (c)?

2.4.14   A transverse electromagnetic wave (TEM) in a coaxial wave guide has an electric field $\mathbf{E} = \mathbf{E}(\rho, \varphi)e^{i(kz-\omega t)}$ and a magnetic induction field of $\mathbf{B} = \mathbf{B}(\rho, \varphi)e^{i(kz-\omega t)}$. Since the wave is transverse neither $\mathbf{E}$ nor $\mathbf{B}$ has a z component. The two fields satisfy the *vector* Laplacian equation

$$\nabla^2 \mathbf{E}(\rho, \varphi) = 0$$
$$\nabla^2 \mathbf{B}(\rho, \varphi) = 0.$$

(a) Show that $\mathbf{E} = \hat{\boldsymbol{\rho}}E_0(a/\rho)e^{i(kz-\omega t)}$ and $\mathbf{B} = \hat{\boldsymbol{\varphi}}B_0(a/\rho)e^{i(kz-\omega t)}$ are solutions. Here a is the radius of the inner conductor and E_0 and B_0 are amplitudes.
(b) Assuming a vacuum inside the wave guide, verify that Maxwell's equations are satisfied with

$$B_0/E_0 = k/\omega = \mu_0\varepsilon_0(\omega/k) = 1/c.$$

2.4.15   A calculation of the magnetohydrodynamic pinch effect involves the evaluation of $(\mathbf{B} \cdot \nabla)\mathbf{B}$. If the magnetic induction $\mathbf{B}$ is taken to be $\mathbf{B} = \hat{\boldsymbol{\varphi}}B_\varphi(\rho)$, show that

$$(\mathbf{B} \cdot \nabla)\mathbf{B} = -\hat{\boldsymbol{\rho}}B_\varphi^2/\rho.$$

2.4.16   The linear velocity of particles in a rigid body rotating with angular velocity ω is given by

$$\mathbf{v} = \hat{\boldsymbol{\varphi}}\rho\omega.$$

Integrate $\oint \mathbf{v} \cdot d\boldsymbol{\lambda}$ around a circle in the xy-plane and verify that

$$\frac{\oint \mathbf{v} \cdot d\boldsymbol{\lambda}}{\text{area}} = \nabla \times \mathbf{v}|_z.$$

2.5 SPHERICAL POLAR COORDINATES (r, θ, φ)

Relabeling (q_1, q_2, q_3) as (r, θ, φ), we see that the spherical polar coordinate system consists of the following:

1. Concentric spheres centered at the origin,

$$r = (x^2 + y^2 + z^2)^{1/2} = \text{constant}.$$

2. Right circular cones centered on the z-(polar) axis, vertices at the origin,

$$\theta = \arccos \frac{z}{(x^2 + y^2 + z^2)^{1/2}} = \text{constant}.$$

3. Half planes through the z-(polar) axis,

$$\varphi = \arctan \frac{y}{x} = \text{constant}.$$

By our arbitrary choice of definitions of θ, the polar angle, and φ, the azimuth angle, the z-axis is singled out for special treatment. The transformation equations corresponding to Eq. (2.1) are

$$x = r \sin \theta \cos \varphi,$$

$$y = r \sin \theta \sin \varphi, \tag{2.36}$$

$$z = r \cos \theta,$$

measuring θ from the positive z-axis and φ in the xy-plane from the positive x-axis. The ranges of values are $0 \le r < \infty, 0 \le \theta \le \pi$, and $0 \le \varphi \le 2\pi$. From Eq. (2.6)

$$h_1 = h_r = 1,$$

$$h_2 = h_\theta = r, \tag{2.37}$$

$$h_3 = h_\varphi = r \sin \theta.$$

This gives a line element

$$d\mathbf{r} = \hat{\mathbf{r}} \, dr + \hat{\boldsymbol{\theta}} r \, d\theta + \hat{\boldsymbol{\varphi}} r \sin \theta \, d\varphi.$$

In this spherical coordinate system the area element (for $r = $ constant) is

$$dA = d\sigma_{\theta\varphi} = r^2 \sin \theta \, d\theta \, d\varphi, \tag{2.38}$$

the dark, shaded area in Fig. 2.6. Integrating over the aximuth φ, we find that the area element becomes a ring of width $d\theta$,

$$dA = 2\pi r^2 \sin \theta \, d\theta. \tag{2.39}$$

This form will appear repeatedly in problems in spherical polar coordinates with azimuthal symmetry—such as the scattering of an unpolarized beam of

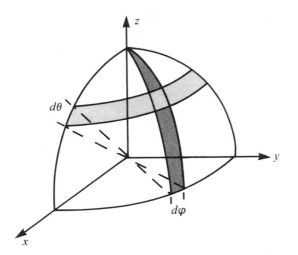

Figure 2.6 Spherical polar co-ordinate area elements.

nuclear particles. By definition of solid radians or steradians, an element of solid angle $d\Omega$ is given by

$$d\Omega = \frac{dA}{r^2} = \sin\theta\, d\theta\, d\varphi. \tag{2.40}$$

Integrating over the entire spherical surface, we obtain

$$\int d\Omega = 4\pi.$$

From Eq. (2.11) the volume element is

$$d\tau = r^2\, dr \sin\theta\, d\theta\, d\varphi$$
$$= r^2\, dr\, d\Omega. \tag{2.41}$$

The spherical polar coordinate unit vectors are shown in Fig. 2.7.

It must be emphasized that the unit vectors $\hat{\mathbf{r}}$, $\hat{\boldsymbol{\theta}}$, and $\hat{\boldsymbol{\varphi}}$ vary in direction as the angles θ and φ vary. Specifically, the θ and φ derivatives of these spherical polar coordinate unit vectors do not vanish (Exercise 2.5.2). When differentiating vectors in spherical polar (or in any noncartesian system) this variation of the unit vectors with position must not be neglected. In terms of the fixed direction cartesian unit vectors $\hat{\mathbf{x}}$, $\hat{\mathbf{y}}$, and $\hat{\mathbf{z}}$,

$$\hat{\mathbf{r}} = \hat{\mathbf{x}} \sin\theta \cos\varphi + \hat{\mathbf{y}} \sin\theta \sin\varphi + \hat{\mathbf{z}} \cos\theta,$$

$$\hat{\boldsymbol{\theta}} = \hat{\mathbf{x}} \cos\theta \cos\varphi + \hat{\mathbf{y}} \cos\theta \sin\varphi - \hat{\mathbf{z}} \sin\theta, \tag{2.42}$$

$$\hat{\boldsymbol{\varphi}} = -\hat{\mathbf{x}} \sin\varphi + \hat{\mathbf{y}} \cos\varphi.$$

Note that Exercise 2.5.5 gives the inverse transformation.

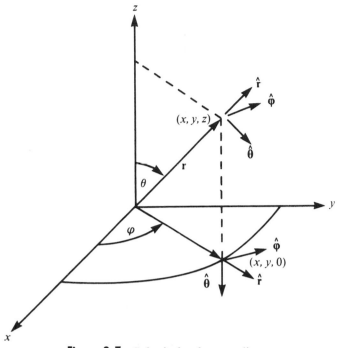

Figure 2.7   Spherical polar coordinates.

Note that a given vector can now be expressed in a number of different (but equivalent) ways. For instance, the position vector $\mathbf{r}$ may be written

$$\mathbf{r} = \hat{\mathbf{r}}r$$

$$= \hat{\mathbf{r}}(x^2 + y^2 + z^2)^{1/2}$$

$$= \hat{\mathbf{x}}x + \hat{\mathbf{y}}y + \hat{\mathbf{z}}z$$

$$= \hat{\mathbf{x}}r \sin \theta \cos \varphi + \hat{\mathbf{y}}r \sin \theta \sin \varphi + \hat{\mathbf{z}}r \cos \theta. \tag{2.43}$$

Select the form that is most useful for your particular problem.

From Section 2.2, relabeling the curvilinear coordinate unit vectors $\mathbf{e}_1$, $\mathbf{e}_2$, and $\mathbf{e}_3$ as $\hat{\mathbf{r}}$, $\hat{\boldsymbol{\theta}}$, and $\hat{\boldsymbol{\varphi}}$ gives

$$\nabla \psi = \hat{\mathbf{r}} \frac{\partial \psi}{\partial r} + \hat{\boldsymbol{\theta}} \frac{1}{r} \frac{\partial \psi}{\partial \theta} + \hat{\boldsymbol{\varphi}} \frac{1}{r \sin \theta} \frac{\partial \psi}{\partial \varphi}, \tag{2.44}$$

$$\nabla \cdot \mathbf{V} = \frac{1}{r^2 \sin \theta} \left[\sin \theta \frac{\partial}{\partial r} (r^2 V_r) + r \frac{\partial}{\partial \theta} (\sin \theta V_\theta) + r \frac{\partial V_\varphi}{\partial \varphi} \right], \tag{2.45}$$

$$\nabla \cdot \nabla \psi = \frac{1}{r^2 \sin \theta} \left[\sin \theta \frac{\partial}{\partial r} \left(r^2 \frac{\partial \psi}{\partial r} \right) + \frac{\partial}{\partial \theta} \left(\sin \theta \frac{\partial \psi}{\partial \theta} \right) + \frac{1}{\sin \theta} \frac{\partial^2 \psi}{\partial \varphi^2} \right], \tag{2.46}$$

$$\nabla \times \mathbf{V} = \frac{1}{r^2 \sin \theta} \begin{vmatrix} \hat{\mathbf{r}} & r\hat{\boldsymbol{\theta}} & r \sin \theta \hat{\boldsymbol{\varphi}} \\ \dfrac{\partial}{\partial r} & \dfrac{\partial}{\partial \theta} & \dfrac{\partial}{\partial \varphi} \\ V_r & rV_\theta & r \sin \theta V_\varphi \end{vmatrix}. \tag{2.47}$$

Occasionally, the vector Laplacian $\nabla^2 \mathbf{V}$ is needed in spherical polar coordinates. It is best obtained by using the vector identity (Eq. (1.86)) of Chapter 1. For future reference

$$\nabla^2 \mathbf{V}|_r = \left(-\frac{2}{r^2} + \frac{2}{r} \frac{\partial}{\partial r} + \frac{\partial^2}{\partial r^2} + \frac{\cos \theta}{r^2 \sin \theta} \frac{\partial}{\partial \theta} + \frac{1}{r^2} \frac{\partial^2}{\partial \theta^2} + \frac{1}{r^2 \sin^2 \theta} \frac{\partial^2}{\partial \varphi^2} \right) V_r$$

$$+ \left(-\frac{2}{r^2} \frac{\partial}{\partial \theta} - \frac{2 \cos \theta}{r^2 \sin \theta} \right) V_\theta + \left(-\frac{2}{r^2 \sin \theta} \frac{\partial}{\partial \varphi} \right) V_\varphi$$

$$= \nabla^2 V_r - \frac{2}{r^2} V_r - \frac{2}{r^2} \frac{\partial V_\theta}{\partial \theta} - \frac{2 \cos \theta}{r^2 \sin \theta} V_\theta - \frac{2}{r^2 \sin \theta} \frac{\partial V_\varphi}{\partial \varphi}, \tag{2.48}$$

$$\nabla^2 \mathbf{V}|_\theta = \nabla^2 V_\theta - \frac{1}{r^2 \sin^2 \theta} V_\theta + \frac{2}{r^2} \frac{\partial V_r}{\partial \theta} - \frac{2 \cos \theta}{r^2 \sin^2 \theta} \frac{\partial V_\varphi}{\partial \varphi}, \tag{2.49}$$

$$\nabla^2 \mathbf{V}|_\varphi = \nabla^2 V_\varphi - \frac{1}{r^2 \sin^2 \theta} V_\varphi + \frac{2}{r^2 \sin \theta} \frac{\partial V_r}{\partial \varphi} + \frac{2 \cos \theta}{r^2 \sin^2 \theta} \frac{\partial V_\theta}{\partial \varphi}. \tag{2.50}$$

These expressions for the components of $\nabla^2 \mathbf{V}$ are undeniably messy, but sometimes they are needed.

Example 2.5.1

Using Eqs. (2.44) to (2.47), we can reproduce by inspection some of the results derived in Chapter 1 by laborious application of cartesian coordinates.
From Eq. (2.44)

$$\nabla f(r) = \hat{\mathbf{r}} \frac{df}{dr}, \tag{2.51}$$

$$\nabla r^n = \hat{\mathbf{r}} n r^{n-1}.$$

From Eq. (2.45)

$$\nabla \cdot \hat{\mathbf{r}} f(r) = \frac{2}{r} f(r) + \frac{df}{dr}, \tag{2.52}$$

$$\nabla \cdot \hat{\mathbf{r}} r^n = (n + 2) r^{n-1}.$$

From Eq. (2.46)

$$\nabla^2 f(r) = \frac{2}{r} \frac{df}{dr} + \frac{d^2 f}{dr^2}, \tag{2.53}$$

$$\nabla^2 r^n = n(n + 1) r^{n-2}. \tag{2.54}$$

Finally, from Eq. (2.47)

$$\nabla \times \hat{\mathbf{r}}f(r) = 0. \tag{2.55}$$

Example 2.5.2: Magnetic Vector Potential

The computation of the magnetic vector potential of a single current loop in the xy-plane involves the evaluation of

$$\mathbf{V} = \nabla \times [\nabla \times \hat{\varphi}A_\varphi(r, \theta)]. \tag{2.56}$$

In spherical polar coordinates this reduces as

$$\mathbf{V} = \nabla \times \frac{1}{r^2 \sin\theta} \begin{vmatrix} \hat{\mathbf{r}} & r\hat{\theta} & r\sin\theta\hat{\varphi} \\ \dfrac{\partial}{\partial r} & \dfrac{\partial}{\partial\theta} & \dfrac{\partial}{\partial\varphi} \\ 0 & 0 & r\sin\theta A_\varphi(r,\theta) \end{vmatrix}$$

$$= \nabla \times \frac{1}{r^2 \sin\theta}\left[\hat{\mathbf{r}}\frac{\partial}{\partial\theta}(r\sin\theta A_\varphi) - r\hat{\theta}\frac{\partial}{\partial r}(r\sin\theta A_\varphi)\right]. \tag{2.57}$$

Taking the curl a second time, we obtain

$$\mathbf{V} = \frac{1}{r^2 \sin\theta}\begin{vmatrix} \hat{\mathbf{r}} & r\hat{\theta} & r\sin\theta\hat{\varphi} \\ \dfrac{\partial}{\partial r} & \dfrac{\partial}{\partial\theta} & \dfrac{\partial}{\partial\varphi} \\ \dfrac{1}{r^2\sin\theta}\dfrac{\partial}{\partial\theta}(r\sin\theta A_\varphi) & -\dfrac{1}{r\sin\theta}\dfrac{\partial}{\partial r}(r\sin\theta A_\varphi) & 0 \end{vmatrix}.$$

$$\tag{2.58}$$

By expanding the determinant, we have

$$\mathbf{V} = -\hat{\varphi}\left\{\frac{1}{r}\frac{\partial^2}{\partial r^2}(rA_\varphi) + \frac{1}{r^2}\frac{\partial}{\partial\theta}\left[\frac{1}{\sin\theta}\frac{\partial}{\partial\theta}(\sin\theta A_\varphi)\right]\right\}$$

$$= -\hat{\varphi}\left[\nabla^2 A_\varphi(r,\theta) - \frac{1}{r^2\sin^2\theta}A_\varphi(r,\theta)\right].$$

In Chapter 12 we shall see that $\mathbf{V}$ leads to the associated Legendre equation and that A_φ may be given by a series of associated Legendre polynomials.

EXERCISES

2.5.1   Resolve the spherical polar unit vectors into their cartesian components.

ANS. $\hat{\mathbf{r}} = \hat{\mathbf{x}}\sin\theta\cos\varphi + \hat{\mathbf{y}}\sin\theta\sin\varphi + \hat{\mathbf{z}}\cos\theta,$
$\hat{\theta} = \hat{\mathbf{x}}\cos\theta\cos\varphi + \hat{\mathbf{y}}\cos\theta\sin\varphi - \hat{\mathbf{z}}\sin\theta,$
$\hat{\varphi} = -\hat{\mathbf{x}}\sin\varphi + \hat{\mathbf{y}}\cos\varphi.$

2.5.2   (a) From the results of Exercise 2.5.1 calculate the partial derivatives of $\hat{\mathbf{r}}$, $\hat{\boldsymbol{\theta}}$, and $\hat{\boldsymbol{\varphi}}$ with respect to r, θ, and φ.

(b) With $\boldsymbol{\nabla}$ given by

$$\hat{\mathbf{r}}\frac{\partial}{\partial r} + \hat{\boldsymbol{\theta}}\frac{1}{r}\frac{\partial}{\partial \theta} + \hat{\boldsymbol{\varphi}}\frac{1}{r\sin\theta}\frac{\partial}{\partial \varphi}$$

(greatest space rate of change), use the results of part (a) to calculate $\boldsymbol{\nabla}\cdot\boldsymbol{\nabla}\psi$. This is an alternate derivation of the Laplacian.

Note. The derivatives of the left-hand $\boldsymbol{\nabla}$ operate on the unit vectors of the right-hand $\boldsymbol{\nabla}$ *before* the unit vectors are dotted together.

2.5.3   A rigid body is rotating about a fixed axis with a constant angular velocity $\boldsymbol{\omega}$. Take $\boldsymbol{\omega}$ to be along the z-axis. Using spherical polar coordinates,

(a) Calculate

$$\mathbf{v} = \boldsymbol{\omega} \times \mathbf{r}.$$

(b) Calculate

$$\boldsymbol{\nabla} \times \mathbf{v}.$$

ANS.   (a) $\mathbf{v} = \hat{\boldsymbol{\varphi}}\omega r \sin\theta$
(b) $\boldsymbol{\nabla} \times \mathbf{v} = 2\boldsymbol{\omega}$.

2.5.4   The coordinate system (x, y, z) is rotated through an angle $\boldsymbol{\Phi}$ counterclockwise about an axis defined by the unit vector $\mathbf{n}$ into system (x', y', z'). In terms of the new coordinates the radius vector becomes

$$\mathbf{r}' = \mathbf{r}\cos\boldsymbol{\Phi} + \mathbf{r}\times\mathbf{n}\sin\boldsymbol{\Phi} + \mathbf{n}(\mathbf{n}\cdot\mathbf{r})(1 - \cos\boldsymbol{\Phi}).$$

(a) Derive this expression from geometric considerations.
(b) Show that it reduces as expected for $\mathbf{n} = \hat{\mathbf{z}}$. The answer, in matrix form, appears in Section 3.3.
(c) Verify that $r'^2 = r^2$.

2.5.5   Resolve the cartesian unit vectors into their spherical polar components.

$$\hat{\mathbf{x}} = \hat{\mathbf{r}}\sin\theta\cos\varphi + \hat{\boldsymbol{\theta}}\cos\theta\cos\varphi - \hat{\boldsymbol{\varphi}}\sin\varphi,$$
$$\hat{\mathbf{y}} = \hat{\mathbf{r}}\sin\theta\sin\varphi + \hat{\boldsymbol{\theta}}\cos\theta\sin\varphi + \hat{\boldsymbol{\varphi}}\cos\varphi,$$
$$\hat{\mathbf{z}} = \hat{\mathbf{r}}\cos\theta - \hat{\boldsymbol{\theta}}\sin\theta.$$

2.5.6   The direction of one vector is given by the angles θ_1 and φ_1. For a second vector the corresponding angles are θ_2 and φ_2. Show that the cosine of the included angle γ is given by

$$\cos\gamma = \cos\theta_1\cos\theta_2 + \sin\theta_1\sin\theta_2\cos(\varphi_1 - \varphi_2).$$

See Fig. 12.15.

2.5.7   A certain vector $\mathbf{V}$ has no radial component. Its curl has no tangential components. What does this imply about the radial dependence of the tangential components of $\mathbf{V}$?

2.5.8   Modern physics lays great stress on the property of parity—whether a quantity remains invariant or changes sign under an inversion of the coordinate system. In cartesian coordinates this means $x \to -x$, $y \to -y$, and $z \to -z$.

(a) Show that the inversion (reflection through the origin) of a point (r, θ, φ) relative to *fixed* x-, y-, z-axes consists of the transformation

$$r \rightarrow r,$$
$$\theta \rightarrow \pi - \theta,$$
$$\varphi \rightarrow \varphi \pm \pi.$$

(b) Show that $\hat{\mathbf{r}}$ and $\hat{\boldsymbol{\varphi}}$ have odd parity (reversal of direction) and that $\hat{\boldsymbol{\theta}}$ has even parity.

2.5.9   With $\mathbf{A}$ any vector

$$\mathbf{A} \cdot \nabla \mathbf{r} - \mathbf{A}.$$

(a) Verify this result in cartesian coordinates.
(b) Verify this result using spherical polar coordinates. (Equation (2.44) provides ∇.)

2.5.10   A particle is moving through space. Find the spherical coordinate components of its velocity and acceleration:

$$v_r = \dot{r},$$
$$v_\theta = r\dot{\theta},$$
$$v_\varphi = r \sin \theta \dot{\varphi},$$
$$a_r = \ddot{r} - r\dot{\theta}^2 - r \sin^2\theta\dot{\varphi}^2,$$
$$a_\theta = r\ddot{\theta} + 2\dot{r}\dot{\theta} - r \sin \theta \cos \theta \dot{\varphi}^2,$$
$$a_\varphi = r \sin \theta \ddot{\varphi} + 2\dot{r} \sin \theta \dot{\varphi} + 2r \cos \theta \dot{\theta}\dot{\varphi}.$$

Hint.
$$\mathbf{r}(t) = \hat{\mathbf{r}}(t)r(t)$$
$$= [\hat{\mathbf{x}} \sin \theta(t) \cos \varphi(t) + \hat{\mathbf{y}} \sin \theta(t) \sin \varphi(t) + \hat{\mathbf{z}} \cos \theta(t)]r(t).$$

Note. Using the Lagrangian techniques of Section 17.3, we may obtain these results somewhat more elegantly. The dot in $\dot{r}$ means time derivative, $\dot{r} = dr/dt$. The notation was originated by Newton.

2.5.11   A particle m moves in response to a central force according to Newton's second law

$$m\ddot{\mathbf{r}} = \hat{\mathbf{r}}f(\mathbf{r}).$$

Show that $\mathbf{r} \times \dot{\mathbf{r}} = \mathbf{c}$, a constant and that the geometric interpretation of this leads to Kepler's second law.

2.5.12   Express $\partial/\partial x$, $\partial/\partial y$, $\partial/\partial z$ in spherical polar coordinates.

$$\text{ANS.} \quad \frac{\partial}{\partial x} = \sin \theta \cos \varphi \frac{\partial}{\partial r} + \cos \theta \cos \varphi \frac{1}{r} \frac{\partial}{\partial \theta} - \frac{\sin \varphi}{r \sin \theta} \frac{\partial}{\partial \varphi},$$
$$\frac{\partial}{\partial y} = \sin \theta \sin \varphi \frac{\partial}{\partial r} + \cos \theta \sin \varphi \frac{1}{r} \frac{\partial}{\partial \theta} + \frac{\cos \varphi}{r \sin \theta} \frac{\partial}{\partial \varphi},$$
$$\frac{\partial}{\partial z} = \cos \theta \frac{\partial}{\partial r} - \sin \theta \frac{1}{r} \frac{\partial}{\partial \theta}.$$

Hint. Equate ∇_{xyz} and $\nabla_{r\theta\varphi}$.

2.5.13   From Exercise 2.5.12 show that

$$-i\left(x\frac{\partial}{\partial y} - y\frac{\partial}{\partial x}\right) = -i\frac{\partial}{\partial\varphi}.$$

This is the quantum mechanical operator corresponding to the z-component of angular momentum.

2.5.14   With the quantum mechanical angular momentum operator defined as $\mathbf{L} = -i(\mathbf{r} \times \boldsymbol{\nabla})$, show that

(a) $L_x + iL_y = e^{i\varphi}\left(\dfrac{\partial}{\partial\theta} + i\cot\theta\dfrac{\partial}{\partial\varphi}\right)$,

(b) $L_x - iL_y = -e^{-i\varphi}\left(\dfrac{\partial}{\partial\theta} - i\cot\theta\dfrac{\partial}{\partial\varphi}\right)$.

These are the raising and lowering operators of Section 4.3.

2.5.15   Verify that $\mathbf{L} \times \mathbf{L} = i\mathbf{L}$ in spherical polar coordinates. $\mathbf{L} = -i(\mathbf{r} \times \boldsymbol{\nabla})$, the quantum mechanical angular momentum operator.
Hint. Use spherical polar coordinates for $\mathbf{L}$ but cartesian components for the cross product.

2.5.16   (a) From Eq. (2.44) show that

$$\mathbf{L} = -i(\mathbf{r} \times \boldsymbol{\nabla}) = i\left(\hat{\boldsymbol{\theta}}\frac{1}{\sin\theta}\frac{\partial}{\partial\varphi} - \hat{\boldsymbol{\varphi}}\frac{\partial}{\partial\theta}\right).$$

(b) Resolving $\hat{\boldsymbol{\theta}}$ and $\hat{\boldsymbol{\varphi}}$ into cartesian components, determine L_x, L_y, and L_z in terms of θ, φ, and their derivatives.

(c) From $L^2 = L_x^2 + L_y^2 + L_z^2$ show that

$$\mathbf{L}^2 = -\frac{1}{\sin\theta}\frac{\partial}{\partial\theta}\left(\sin\theta\frac{\partial}{\partial\theta}\right) - \frac{1}{\sin^2\theta}\frac{\partial^2}{\partial\varphi^2}$$

$$= -r^2\nabla^2 + \frac{\partial}{\partial r}\left(r^2\frac{\partial}{\partial r}\right).$$

This latter identity is useful in relating angular momentum and Legendre's differential equation, Exercise 8.3.1.

2.5.17   With $\mathbf{L} = -i\mathbf{r} \times \boldsymbol{\nabla}$ verify the operator identities

(a) $\boldsymbol{\nabla} = \hat{\mathbf{r}}\dfrac{\partial}{\partial r} - i\dfrac{\mathbf{r} \times \mathbf{L}}{r^2}$,

(b) $\mathbf{r}\nabla^2 - \boldsymbol{\nabla}\left(1 + r\dfrac{\partial}{\partial r}\right) = i\boldsymbol{\nabla} \times \mathbf{L}$.

2.5.18   Show that the following three forms (spherical coordinates) of $\nabla^2\psi(r)$ are equivalent.

(a) $\dfrac{1}{r^2}\dfrac{d}{dr}\left[r^2\dfrac{d\psi(r)}{dr}\right]$,

(υ) $\dfrac{1}{r}\dfrac{d^2}{dr^2}[r\psi(r)]$,

(c) $\dfrac{d^2\psi(r)}{dr^2} + \dfrac{2}{r}\dfrac{d\psi(r)}{dr}$.

The second form is particularly convenient in establishing a correspondence between spherical polar and cartesian descriptions of a problem. A generalization of this appears in Exercise 8.6.11.

2.5.19 One model of the solar corona assumes that the steady-state equation of heat flow

$$\nabla \cdot (k \nabla T) = 0$$

is satisfied. Here, k, the thermal conductivity, is proportional to $T^{5/2}$. Assuming that the temperature T is proportional to r^n, show that the heat flow equation is satisfied by $T = T_0 (r_0/r)^{2/7}$.

2.5.20 A certain force field is given by

$$\mathbf{F} = \hat{\mathbf{r}} \frac{2P \cos \theta}{r^3} + \hat{\boldsymbol{\theta}} \frac{P}{r^3} \sin \theta, \qquad r \geq P/2$$

(in spherical polar coordinates).

(a) Examine $\nabla \times \mathbf{F}$ to see if a potential exists.

(b) Calculate $\oint \mathbf{F} \cdot d\boldsymbol{\lambda}$ for a unit circle in the plane $\theta = \pi/2$.
What does this indicate about the force being conservative or nonconservative?

(c) If you believe that $\mathbf{F}$ may be described by $\mathbf{F} = -\nabla \psi$, find ψ. Otherwise simply state that no acceptable potential exists.

2.5.21 (a) Show that $\mathbf{A} = -\hat{\boldsymbol{\varphi}} \cot \theta / r$ is a solution of $\nabla \times \mathbf{A} = \hat{\mathbf{r}}/r^2$.

(b) Show that this spherical polar coordinate solution agrees with the solution given for Exercise 1.13.5:

$$\mathbf{A} = \hat{\mathbf{x}} \frac{yz}{r(x^2 + y^2)} - \hat{\mathbf{y}} \frac{xz}{r(x^2 + y^2)} .$$

Note that the solution diverges for $\theta = 0, \pi$ corresponding to $x, y = 0$.

(c) Finally, show that $\mathbf{A} = -\hat{\boldsymbol{\theta}} \varphi \sin \theta / r$ is a solution. Note that although this solution does not diverge $(r \neq 0)$ it is no longer single-valued for all possible azimuth angles.

2.5.22 A magnetic vector potential is given by

$$\mathbf{A} = \frac{\mu_0}{4\pi} \frac{\mathbf{m} \times \mathbf{r}}{r^3} .$$

Show that this leads to the magnetic induction $\mathbf{B}$ of a point magnetic dipole, dipole moment $\mathbf{m}$.

ANS. for $\mathbf{m} = \hat{\mathbf{z}}m$,

$$\nabla \times \mathbf{A} = \hat{\mathbf{r}} \frac{\mu_0}{4\pi} \frac{2m \cos \theta}{r^3} + \hat{\boldsymbol{\theta}} \frac{\mu_0}{4\pi} \frac{m \sin \theta}{r^3} .$$

Compare Eqs. (12.136) and (12.137).

2.5.23 At large distances from its source, electric dipole radiation has fields

$$\mathbf{E} = a_E \sin \theta \frac{e^{i(kr - \omega t)}}{r} \hat{\boldsymbol{\theta}}, \qquad \mathbf{B} = a_B \sin \theta \frac{e^{i(kr - \omega t)}}{r} \hat{\boldsymbol{\varphi}}.$$

Show that Maxwell's equations

$$\nabla \times \mathbf{E} = -\frac{\partial \mathbf{B}}{\partial t} \qquad \text{and} \qquad \nabla \times \mathbf{B} = \varepsilon_0 \mu_0 \frac{\partial \mathbf{E}}{\partial t}$$

are satisfied, if we take

$$\frac{a_E}{a_B} = \frac{\omega}{k} = c = (\varepsilon_0 \mu_0)^{-1/2}.$$

Hint. Since r is large, terms of order r^{-2} may be dropped.

2.5.24   The magnetic vector potential for a uniformly charged rotating spherical shell is

$$\mathbf{A} = \begin{cases} \hat{\varphi} \dfrac{\mu_0 a^4 \sigma \omega}{3} \cdot \dfrac{\sin\theta}{r^2}, & r > a \\[2ex] \hat{\varphi} \dfrac{\mu_0 a \sigma \omega}{3} \cdot r\cos\theta, & r < a. \end{cases}$$

(a = radius of spherical shell, σ surface charge density, and ω angular velocity.) Find the magnetic induction $\mathbf{B} = \nabla \times \mathbf{A}$.

$$ANS. \quad B_r(r,\theta) = \frac{2\mu_0 a^4 \sigma \omega}{3} \cdot \frac{\cos\theta}{r^3}, \qquad r > a$$

$$B_\theta(r,\theta) = \frac{\mu_0 a^4 \sigma \omega}{3} \cdot \frac{\sin\theta}{r^3}, \qquad r > a$$

$$\mathbf{B} = \hat{z}\frac{2\mu_0 a \sigma \omega}{3}, \qquad r < a.$$

2.5.25   (a) Explain why ∇^2 in plane polar coordinates follows from ∇^2 in circular cylindrical coordinates with z = constant.
   (b) Explain why taking ∇^2 in spherical polar coordinates and restricting θ to $\pi/2$ does *not* lead to the plane polar form of ∇^2.
   Note.

$$\nabla^2(\rho, \varphi) = \frac{\partial^2}{\partial\rho^2} + \frac{1}{\rho}\frac{\partial}{\partial\rho} + \frac{1}{\rho^2}\frac{\partial^2}{\partial\varphi^2}.$$

2.6 TENSOR ANALYSIS

INTRODUCTION, DEFINITIONS

Tensors are important in many areas of physics, including general relativity and electrodynamics. Scalars and vectors are special cases of tensors. In Chapter 1, a quantity that did not change under rotations of the coordinate system in three-dimensional space, an invariant, was labeled a scalar. A scalar is specified by one real number and is a *tensor of rank zero*. A quantity whose components transformed under rotations like those of the distance of a point from a chosen origin (Eq. (1.9), Section 1.2) was called a vector. The transformation of the components of the vector under a rotation of the coordinates

preserves the vector as a geometric entity (such as an arrow in space), independent of the orientation of the reference frame. In three-dimensional space, a
vector is specified by $3 = 3^1$ real numbers, for example, its cartesian components, and is a *tensor of rank one*. A *tensor of rank n* has 3^n components
which transform in a definite way.[1]

There is a possible ambiguity in the transformation law of a vector

$$A'_i = \sum_j a_{ij} A_j. \tag{2.59}$$

in which a_{ij} is the cosine of the angle between the x'_i-axis and the x_j-axis.

If we start with a differential distance vector $d\mathbf{r}$, then, taking dx'_i to be a
function of the unprimed variables,

$$dx'_i = \sum_j \frac{\partial x'_i}{\partial x_j} dx_j \tag{2.60}$$

by partial differentiation. If we set

$$a_{ij} = \frac{\partial x'_i}{\partial x_j}, \tag{2.61}$$

Eqs. (2.59) and (2.60) are consistent. Any set of quantities A_j transforming
according to

$$A'_i = \sum_j \frac{\partial x'_i}{\partial x_j} A_j \tag{2.62}$$

is defined as a contravariant vector.

However, we have already encountered a slightly different type of vector
transformation. The gradient of a scalar, $\nabla\varphi$, defined by

$$\nabla\varphi = \hat{\mathbf{x}} \frac{\partial \varphi}{\partial x_1} + \hat{\mathbf{y}} \frac{\partial \varphi}{\partial x_2} + \hat{\mathbf{z}} \frac{\partial \varphi}{\partial x_3} \tag{2.63}$$

(using x_1, x_2, x_3 for x, y, z), transforms as

$$\frac{\partial \varphi'}{\partial x'_i} = \sum_j \frac{\partial \varphi}{\partial x_j} \frac{\partial x_j}{\partial x'_i}, \tag{2.64}$$

using $\varphi = \varphi(x, y, z) = \varphi(x', y', z') = \varphi'$, φ defined as a scalar quantity. Notice
that this differs from Eq. (2.62) in that we have $\partial x_j / \partial x'_i$ instead of $\partial x'_i / \partial x_j$.
Equation (2.64) is taken as the definition of a covariant vector with the
gradient as the prototype.

In cartesian coordinates

$$\frac{\partial x_j}{\partial x'_i} = \frac{\partial x'_i}{\partial x_j} = a_{ij}, \tag{2.65}$$

and there is no difference between contravariant and covariant transformations. In other systems Eq. (2.65) in general does not apply, and the distinction

---

[1] In N-dimensional space a tensor of rank n has N^n components.

between contravariant and covariant is real and must be observed. This is of prime importance in the curved Riemannian space of general relativity.

In the remainder of this section the components of a contravariant vector are denoted by a superscript, A^i, whereas a subscript is used for the components of a covariant vector A_i.[2]

DEFINITION OF TENSORS OF RANK TWO

Now we proceed to define *contravariant, mixed, and covariant tensors of rank 2* by the following equations for their components under coordinate transformations:

$$A'^{ij} = \sum_{kl} \frac{\partial x_i'}{\partial x_k} \frac{\partial x_j'}{\partial x_l} A^{kl},$$

$$B'^{i}_{\ j} = \sum_{kl} \frac{\partial x_i'}{\partial x_k} \frac{\partial x_l}{\partial x_j'} B^k_{\ l}, \tag{2.66}$$

$$C'_{ij} = \sum_{kl} \frac{\partial x_k}{\partial x_i'} \frac{\partial x_l}{\partial x_j'} C_{kl}.$$

Clearly, the rank goes as the number of partial derivatives (or direction cosines) in the definition: zero for a scalar, one for a vector, two for a second-rank tensor, and so on. Each index (subscript or superscript) ranges over the number of dimensions of the space. The number of indices (rank of tensor) is independent of the dimensions of the space. We see that A^{kl} is contravariant with respect to both indices, C_{kl} is covariant with respect to both indices, and $B^k_{\ l}$ transforms contravariantly with respect to the first index k but covariantly with respect to the second index l. Once again, if we are using cartesian coordinates, all three forms of the tensors of second rank, contravariant, mixed, and covariant are the same.

As with the components of a vector, the transformation laws for the components of a tensor, Eq. (2.66), yield entities (and properties) that are independent of the choice of reference frame. This is what makes tensor analysis important in physics. The independence of reference frame (invariance) is ideal for expressing and investigating universal physical laws.

The second-rank tensor **A** (components A^{kl}) may be conveniently represented by writing out its components in a square array (3×3 if we are in three-dimensional space),

$$\mathbf{A} = \begin{pmatrix} A^{11} & A^{12} & A^{13} \\ A^{21} & A^{22} & A^{23} \\ A^{31} & A^{32} & A^{33} \end{pmatrix}. \tag{2.67}$$

[2] This means that the coordinates (x, y, z) should be written (x^1, x^2, x^3) since **r** transforms as a contravariant vector. Because we shall shortly restrict our attention to cartesian tensors (where the distinction between contravariance and covariance disappears) we continue to use subscripts on the coordinates. This avoids the ambiguity of x^2 representing both x squared and y.

This does not mean that any square array of numbers or functions forms a tensor. The essential condition is that the components transform according to Eq. (2.66).

In the context of matrix analysis the preceding transformation equations become (for cartesian coordinates) an orthogonal similarity transformation, Section 3.3. A geometrical interpretation of a second-rank tensor (the inertia tensor) is developed in Section 3.5.

ADDITION AND SUBTRACTION OF TENSORS

The addition and subtraction of tensors is defined in terms of the individual elements just as for vectors. To add or subtract two tensors, we add or subtract the corresponding elements. If

$$\mathbf{A} + \mathbf{B} = \mathbf{C}, \tag{2.68}$$

then

$$A^{ij} + B^{ij} = C^{ij}.$$

Of course, **A** and **B** must be tensors of the same rank and both expressed in a space of the same number of dimensions.

SUMMATION CONVENTION

In tensor analysis it is customary to adopt a summation convention to put Eq. (2.66) and subsequent tensor equations in a more compact form. As long as we are distinguishing between contravariance and covariance, let us agree that when an index appears on one side of an equation, once as a superscript and once as a subscript (except for the coordinates where both are subscripts), we automatically sum over that index. Then we may write the second expression in Eq. (2.66) as

$$B'^{i}_{\ j} = \frac{\partial x'_i}{\partial x_k} \frac{\partial x_l}{\partial x'_j} B^k_{\ l}; \tag{2.69}$$

with the summation of the right-hand side over k and i implied. This is the summation convention.[3]

To illustrate the use of the summation convention and some of the techniques of tensor analysis, let us show that the now familiar Kronecker delta, δ_{kl}, is really a mixed tensor of rank two, δ^k_l.[4] The question is: Does δ^k_l transform according to Eq. (2.66)? This is our criterion for calling it a tensor. We have, using the summation convention,

$$\delta^k_l \frac{\partial x'_i}{\partial x_k} \frac{\partial x_l}{\partial x'_j} = \frac{\partial x'_i}{\partial x_k} \frac{\partial x_k}{\partial x'_j} \tag{2.70}$$

---

[3] In this context $\partial x'_i / \partial x_k$ might better be written as $a^i_{\ k}$ and $\partial x_l / \partial x'_j$ as $b^l_{\ j}$.

[4] It is common practice to refer to a tensor **A** by specifying a typical component, A_{ij}. As long as the reader refrains from writing nonsense such as $\mathbf{A} = A_{ij}$, no harm is done.

by definition of the Kronecker delta. Now

$$\frac{\partial x_i'}{\partial x_k} \frac{\partial x_k}{\partial x_j'} = \frac{\partial x_i'}{\partial x_j'} \tag{2.71}$$

by direct partial differentiation of the right-hand side (chain rule). However, x_i' and x_j' are independent coordinates, and therefore the variation of one with respect to the other must be zero if they are different, unity if they coincide; that is,

$$\frac{\partial x_i'}{\partial x_j'} = \delta_j'^i. \tag{2.72}$$

Hence

$$\delta_j'^i = \frac{\partial x_i'}{\partial x_k} \frac{\partial x_l}{\partial x_j'} \delta_l^k,$$

showing that the δ_l^k are indeed the components of a mixed second-rank tensor. Notice that this result is independent of the number of dimensions of our space.

The Kronecker delta has one further interesting property. It has the same components in all of our rotated coordinate systems and is therefore called *isotropic*. In Section 2.9 we shall meet a third-rank isotropic tensor and three fourth-rank isotropic tensors. No isotropic first-rank tensor (vector) exists.

SYMMETRY—ANTISYMMETRY

The order in which the indices appear in our description of a tensor is important. In general, A^{mn} is independent of A^{nm}, but there are some cases of special interest. If, for all m and n,

$$A^{mn} = A^{nm}, \tag{2.73}$$

we call the tensor symmetric. If, on the other hand,

$$A^{mn} = -A^{nm}, \tag{2.74}$$

the tensor is antisymmetric. Clearly, every (second-rank) tensor can be resolved into symmetric and antisymmetric parts by the identity

$$A^{mn} = \tfrac{1}{2}(A^{mn} + A^{nm}) + \tfrac{1}{2}(A^{mn} - A^{nm}), \tag{2.75}$$

the first term on the right being a symmetric tensor, the second, an anti-symmetric tensor. A similar resolution of functions into symmetric and antisymmetric parts is of extreme importance to quantum mechanics.

SPINORS

It was once thought that the system of scalars, vectors, tensors (second-rank), and so on formed a complete mathematical system, one that is adequate for describing a physics independent of the choice of reference frame. But the universe and mathematical physics are not this simple. In the realm of elementary particles, for example, spin zero particles[5] (π mesons, α particles) may be described with scalars, spin 1 particles (deuterons) by vectors, and spin 2 particles (gravitons) by tensors. This listing omits the most common particles: electrons, protons, and neutrons, all with spin $\frac{1}{2}$. These particles are properly described by *spinors*. A spinor is not a scalar, vector, or tensor. A brief introduction to spinors in the context of group theory appears in Section 4.3.

EXERCISES

2.6.1   Show that if the components of any tensor of any rank vanish in one particular coordinate system they vanish in all coordinate systems.
Note. This point takes on especial importance in the four-dimensional curved space of general relativity. If a quantity, expressed as a tensor, exists in one coordinate system, it exists in all coordinate systems and is not just a consequence of a *choice* of a coordinate system (as are centrifugal and Coriolis forces in Newtonian mechanics).

2.6.2   The components of tensor **A** are equal to the corresponding components of tensor **B** in one particular coordinate system; that is,
$$A_{ij}^0 = B_{ij}^0.$$
Show that tensor **A** is equal to tensor **B**, $A_{ij} = B_{ij}$, in *all* coordinate systems.

2.6.3   The first three components of a four-dimensional vector vanish in each of two reference frames. If the second reference frame is not merely a rotation of the first about the x_4 axis, that is, if at least one of the coefficients a_{i4} $(i = 1, 2, 3) \neq 0$, show that the fourth component vanishes in all reference frames. Translated into relativistic mechanics this means that if momentum is conserved in two Lorentz frames, then energy is conserved in all Lorentz frames.

2.6.4   From an analysis of the behavior of a general second-rank tensor under 90° and 180° rotations about the coordinate axes, show that an *isotropic* second-rank tensor in three-dimensional space must be a multiple of δ_{ij}.

2.6.5   The four-dimensional fourth-rank Riemann–Christoffel curvature tensor of general relativity, R_{iklm}, satisfies the symmetry relations
$$R_{iklm} = -R_{ikml} = -R_{kilm}.$$
With the indices running from 1 to 4, show that the number of independent components is reduced from 256 to 36 and that the condition
$$R_{iklm} = R_{lmik}$$
further reduces the number of independent components to 21. Finally, if the components satisfy an identity $R_{iklm} + R_{ilmk} + R_{imkl} = 0$, show that the number of independent components is reduced to 20.

----

[5] The particle spin is intrinsic angular momentum (in units of $\hbar$). It is distinct from classical, orbital angular momentum due to motion.

Note. The final three-term identity furnishes new information only if all four indices are different. Then it reduces the number of independent components by one third.

2.6.6 T_{iklm} is antisymmetric with respect to all pairs of indices. How many independent components has it (three-dimensional space)?

2.7 CONTRACTION, DIRECT PRODUCT

CONTRACTION

When dealing with vectors, we formed a scalar product (Section 1.3) by summing products of corresponding components:

$$\mathbf{A} \cdot \mathbf{B} = A_i B_i \qquad \textbf{(summation convention)}. \tag{2.76}$$

The generalization of this expression in tensor analysis is a process known as contraction. Two indices, one covariant and the other contravariant, are set equal to each other and then (as implied by the summation convention) we sum over this repeated index. For example, let us contract the second-rank mixed tensor B'^i_j.

$$B'^i_j \rightarrow B'^i_i = \frac{\partial x'_i}{\partial x_k} \frac{\partial x_l}{\partial x'_i} B^k_l = \frac{\partial x_l}{\partial x_k} B^k_l \tag{2.77}$$

by Eq. (2.71) and then by Eq. (2.72)

$$B'^i_i = \delta^l_k B^k_l = B^k_k. \tag{2.78}$$

Our contracted second-rank mixed tensor is invariant and therefore a scalar.[1] This is exactly what we obtained in Section 1.3 for the dot product of two vectors and Section 1.7 for the divergence of a vector. In general, the operation of contraction reduces the rank of a tensor by 2. An example of the use of contraction appears in Chapter 4.

DIRECT PRODUCT

The components of a covariant vector (first-rank tensor) a_i and those of a contravariant vector (first-rank tensor) b^j may be multiplied component by component to give the general term $a_i b^j$. This, by Eq. (2.66) is actually a second-rank tensor, for

$$a'_i b'^j = \frac{\partial x_k}{\partial x'_i} a_k \frac{\partial x'_j}{\partial x_l} b^l = \frac{\partial x_k}{\partial x'_i} \frac{\partial x'_j}{\partial x_l} (a_k b^l). \tag{2.79}$$

Contracting, we obtain

$$a'_i b'^i = a_k b^k, \tag{2.80}$$

as in Eqs. (2.77) and (2.78) to give the regular scalar product.

---

[1] In matrix analysis this scalar is the *trace* of the matrix, Section 3.2.

The operation of adjoining two vectors a_i and b^j as in the last paragraph is known as forming the **direct product**. For the case of two vectors, the direct product is a tensor of second rank. In this sense we may attach meaning to **VE**, which was not defined within the framework of vector analysis. In general, the direct product of two tensors is a tensor of rank equal to the sum of the two initial ranks; that is,

$$A_j^i B^{kl} = C_j^{ikl}, \tag{2.81a}$$

where C_j^{ikl} is a tensor of fourth rank. From Eqs. (2.66)

$$C_j^{'ikl} = \frac{\partial x_i'}{\partial x_m} \frac{\partial x_n}{\partial x_j'} \frac{\partial x_k'}{\partial x_p} \frac{\partial x_l'}{\partial x_q} C_n^{mpq}. \tag{2.81b}$$

The direct product appears in mathematical physics as a technique for creating new higher-rank tensors. Exercise 2.7.1 is a form of the direct product in which the first factor is **V**. Applications appear in Section 4.6.

When **T** is an nth rank cartesian tensor, $(\partial/\partial x_i)T_{jkl\ldots}$, an element of **VT**, is a *cartesian* tensor of rank $n + 1$ (Exercise 2.7.1). However, $(\partial/\partial x_i)T_{jkl\ldots}$ is not a tensor under more general transformations. In noncartesian systems $\partial/\partial x_i'$ will act on the partial derivatives $\partial x_p/\partial x_q'$ and destroy the simple tensor transformation relation.

So far the distinction between a covariant transformation and a contravariant transformation has been maintained because it does exist in noncartesian space and because it is of great importance in general relativity. In Sections 2.10 and 2.11 we shall develop differential relations for noncartesian tensors. Now, however, because of the simplification achieved, we restrict ourselves to cartesian tensors. As noted in Section 2.6, the distinction between contravariance and covariance disappears and all indices are from now on shown as subscripts. We restate the summation convention and the operation of contraction.

SUMMATION CONVENTION
When a subscript (letter, not number) appears twice on one side of an equation, summation with respect to that subscript is implied.

CONTRACTION
Contraction consists of setting two unlike indices (subscripts) equal to each other and then summing as implied by the summation convention.

EXERCISES

2.7.1   If $T_{\ldots i}$ is a tensor of rank n, show that $\partial T_{\ldots i}/\partial x_j$ is a tensor of rank $n + 1$ (cartesian coordinates).

Note. In noncartesian coordinate systems the coefficients a_{ij} are, in general, functions of the coordinates, and the simple derivative of a tensor of rank

n is not a tensor except in the special case of $n = 0$. In this case the derivative does yield a covariant vector (tensor of rank 1) by Eq. (2.64).

2.7.2   If $T_{ijk...}$ is a tensor of rank n, show that $\sum \partial T_{ijk...}/\partial x_j$ is a tensor of rank $n - 1$ (cartesian coordinates).

2.7.3   The operator

$$\nabla^2 - \frac{1}{c^2}\frac{\partial^2}{\partial t^2}$$

may be written as

$$\sum_{i=1}^{4}\frac{\partial^2}{\partial x_i^2},$$

using $x_4 = ict$. This is the four-dimensional Laplacian, usually called the d'Alembertian and denoted by $\Box^2$. Show that it is a *scalar* operator.

2.8 QUOTIENT RULE

If A_i and B_j are vectors, as seen in Section 2.7, we can easily show that $A_i B_j$ is a second-rank tensor. Here we are concerned with a variety of inverse relations. Consider such equations as

$$K_i A_i = B \tag{2.82a}$$

$$K_{ij} A_j = B_i \tag{2.82b}$$

$$K_{ij} A_{jk} = B_{ik} \tag{2.82c}$$

$$K_{ijkl} A_{ij} = B_{kl} \tag{2.82d}$$

$$K_{ij} A_k = B_{ijk}. \tag{2.82e}$$

In each of these expressions **A** and **B** are known tensors of rank indicated by the number of indices and **A** is arbitrary. In each case K is an unknown quantity. We wish to establish the transformation properties of K. The quotient rule asserts that if the equation of interest holds in all (rotated) cartesian coordinate systems, K is a tensor of the indicated rank. The importance in physical theory is that the quotient rule can establish the tensor nature of quantities. Exercise 2.8.1 is a simple illustration of this. The quotient rule (Eq. (2.82b)) shows that the inertia matrix appearing in the angular momentum equation $\mathbf{L} = I\boldsymbol{\omega}$, Section 3.5, is a tensor.

In proving the quotient rule, we consider Eq. (2.82b) as a typical case. In our primed coordinate system

$$K'_{ij} A'_j = B'_i = a_{ik} B_k, \tag{2.83}$$

using the vector transformation properties of **B**. Since the equation holds in all rotated cartesian coordinate systems,

$$a_{ik} B_k = a_{ik}(K_{kl} A_l). \tag{2.84}$$

Now, transforming **A** back into the primed coordinate system[1] (compare Eq. (2.62)), we have

$$K'_{ij}A'_j = a_{ik}K_{kl}a_{jl}A'_j. \tag{2.85}$$

Rearranging, we obtain

$$(K'_{ij} - a_{ik}a_{jl}K_{kl})A'_j = 0. \tag{2.86}$$

This must hold for each value of the index i and for every primed coordinate system. Since the A'_j is arbitrary,[2] we conclude

$$K'_{ij} = a_{ik}a_{jl}K_{kl}, \tag{2.87}$$

which is our definition of second-rank tensor.

The other equations may be treated similarly, giving rise to other forms of the quotient rule. One minor pitfall should be noted—the quotient rule does not necessarily apply if B is zero. The transformation properties of zero are indeterminate.

EXERCISES

2.8.1   The double summation $K_{ij}A_iB_j$ is invariant for any two vectors A_i and B_j. Prove that K_{ij} is a second-rank tensor.
Note. In the form ds^2 (invariant) $= g_{ij} dx_i dx_j$, this result shows that g_{ij}, the "metric" is a tensor.

2.8.2   The equation $K_{ij}A_{jk} = B_{ik}$ holds for all orientations of the coordinate system. If **A** and **B** are second-rank tensors, show that **K** is a second-rank tensor also.

2.8.3   The exponential in a plane wave is $\exp[i(\mathbf{k} \cdot \mathbf{r} - \omega t)]$. We recognize $x^\mu = (ct; x_1, x_2, x_3)$ as a prototype vector in Minkowski space. If $\mathbf{k} \cdot \mathbf{r} - \omega t$ is a scalar under Lorentz transformations (Section 4.7), show that $k^\mu = (\omega/c; k_1, k_2, k_3)$ is a vector in Minkowski space.
Note. Multiplication by $\hbar$ yields $(E/c, \mathbf{p})$ as a vector in Minkowski space.

2.9 PSEUDOTENSORS, DUAL TENSORS

So far our coordinate transformations have been restricted to pure rotations. We now consider the effect of reflections or inversions. If we have

---

[1] Note carefully the order of the indices of the direction cosine a_{jl} in this *inverse* transformation. We have

$$A_l = \sum_j \frac{\partial x_l}{\partial x'_j} A'_j = \sum_j a_{jl}A'_j.$$

[2] We might, for instance, take $A'_1 = 1$ and $A'_m = 0$ for $m \neq 1$. Then the equation $K'_{i1} = a_{ik}a_{1l}K_{kl}$ follows immediately. The rest of Eq. (2.87) comes from other special choices of the arbitrary A'_j.

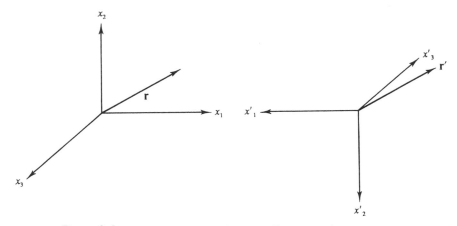

Figure 2.8 Inversion of cartesian coordinates—polar vector.

transformation coefficients $a_{ij} = -\delta_{ij}$, then by Eq. (2.60)

$$x_i = -x_i', \tag{2.88}$$

which is an inversion or parity transformation. Note carefully that this transformation changes our initial right-handed coordinate system into a left-handed coordinate system.[1] Our prototype vector $\mathbf{r}$ with components (x_1, x_2, x_3) transforms to $\mathbf{r}' = (x_1', x_2', x_3') = (-x_1, -x_2, -x_3)$. This new vector $\mathbf{r}'$ has negative components, relative to the new transformed set of axes. As shown in Fig. 2.8, reversing the directions of the coordinate axes and changing the signs of the components gives $\mathbf{r}' = \mathbf{r}$. The vector (an arrow in space) stays exactly as it was before the transformation was carried out. The position vector $\mathbf{r}$ and all other vectors whose components behave this way (reversing sign with a reversal of the coordinate axes) are called polar vectors and have odd parity.

A fundamental difference appears when we encounter a vector defined as the cross product of two polar vectors. Let $\mathbf{C} = \mathbf{A} \times \mathbf{B}$, where both $\mathbf{A}$ and $\mathbf{B}$ are polar vectors. From Eq. 1.33 of Section 1.4 the components of $\mathbf{C}$ are given by

$$C_1 = A_2 B_3 - A_3 B_2. \tag{2.89}$$

and so on. Now when the coordinate axes are inverted, $A_i \rightarrow -A_i'$, $B_j \rightarrow -B_j'$ but from its definition $C_k \rightarrow +C_k'$; that is, our cross-product vector, vector $\mathbf{C}$, does *not* behave like a polar vector under inversion. To distinguish, we label it a pseudovector or axial vector (see Fig. 2.9) that has even parity. The term axial vector is frequently used because these cross products often arise from a description of rotation.

---

[1] This is an inversion of the coordinate system or coordinate axes, objects in the physical world remaining fixed.

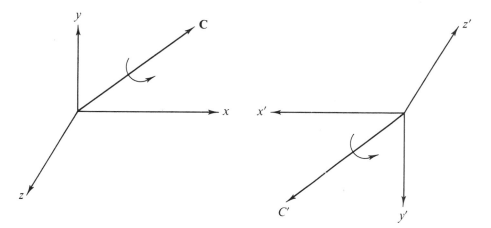

Figure 2.9   Inversion of cartesian coordinates—axial vector.

Examples are

$$\text{angular velocity,} \qquad \mathbf{v} = \boldsymbol{\omega} \times \mathbf{r}$$

$$\text{angular momentum,} \qquad \mathbf{L} = \mathbf{r} \times \mathbf{p},$$

$$\text{torque,} \qquad \mathbf{N} = \mathbf{r} \times \mathbf{f},$$

$$\text{magnetic induction field } \mathbf{B}, \qquad \frac{\partial \mathbf{B}}{\partial t} = -\boldsymbol{\nabla} \times \mathbf{E}.$$

In $\mathbf{v} = \boldsymbol{\omega} \times \mathbf{r}$, the axial vector is the angular velocity $\boldsymbol{\omega}$, and $\mathbf{r}$ and $\mathbf{v} = d\mathbf{r}/dt$ are polar vectors. Clearly, axial vectors occur frequently in elementary physics, although this fact is usually not pointed out. In a right-handed coordinate system an axial vector $\mathbf{C}$ has a sense of rotation associated with it given by a right-hand rule (compare Section 1.4). In the inverted left-handed system the sense of rotation is a left-handed rotation. This is indicated by the curved arrows in Fig. 2.9.

The distinction between polar and axial vectors may also be illustrated by a reflection. A polar vector reflects in a mirror like a real physical arrow, Fig. 2.10a. In Figs. 2.8 and 2.9 the coordinates are inverted; the physical world remains fixed. Here the coordinate axes remain fixed; the world is reflected— as in a mirror in the xz-plane. Specifically, in this representation we keep the axes fixed and associate a change of sign with the component of the vector. For a mirror in the xz-plane, $P_y \rightarrow -P_y$. We have

$$\mathbf{P} = (P_x, P_y, P_z)$$

$$\mathbf{P}' = (P_x, -P_y, P_z) \qquad \text{polar vector.}$$

An axial vector such as a magnetic field $\mathbf{H}$ or a magnetic moment $\boldsymbol{\mu}$ (= current × area of current loop) behaves quite differently under reflection.

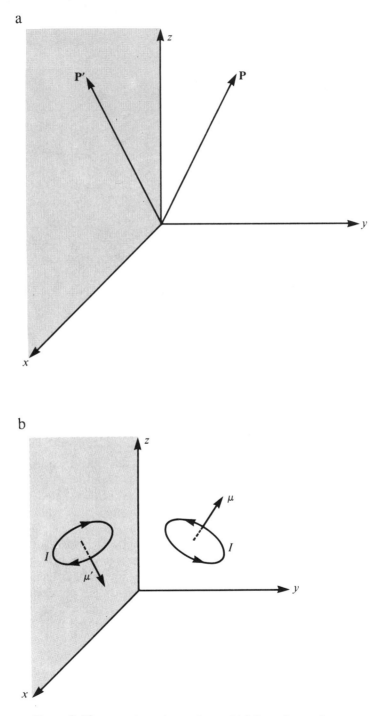

Figure 2.10 (a) Mirror in xz-plane; (b) Mirror in xz-plane.

Consider the magnetic field $\mathbf{H}$ and magnetic moment $\boldsymbol{\mu}$ to be produced by an electric charge moving in a circular path (Exercise 5.8.4 and Example 12.5.1). Reflection reverses the sense of rotation of the charge. The two current loops and the resulting magnetic moments are shown in Fig. 2.10b. We have

$$\boldsymbol{\mu} = (\mu_x, \mu_y, \mu_z)$$

$$\boldsymbol{\mu}' = (-\mu_x, \mu_y, -\mu_z) \quad \text{axial vector.}$$

If we agree that the universe does not care whether we use a right- or left-handed coordinate system, then it does not make sense to add an axial vector to a polar vector. In the vector equation $\mathbf{A} = \mathbf{B}$, both $\mathbf{A}$ and $\mathbf{B}$ are either polar vectors or axial vectors.[2] Similar restrictions apply to scalars and pseudoscalars and, in general, to the tensors and pseudotensors considered subsequently.

Usually, pseudoscalars, pseudovectors, and pseudotensors will transform as

$$S' = |a|S$$

$$C_i' = |a|a_{ij}C_j, \tag{2.90}$$

$$A_{ij}' = |a|a_{ik}a_{jl}A_{kl},$$

where $|a|$ is the determinant[3] of the array of coefficients a_{mn}. In our inversion the determinant is

$$|a| = \begin{vmatrix} -1 & 0 & 0 \\ 0 & -1 & 0 \\ 0 & 0 & -1 \end{vmatrix} = -1. \tag{2.91}$$

For a reflection of one axis, the x-axis,

$$|a| = \begin{vmatrix} -1 & 0 & 0 \\ 0 & 1 & 0 \\ 0 & 0 & 1 \end{vmatrix} = -1, \tag{2.92}$$

and again the determinant $|a| = -1$. On the other hand, for all pure rotations the determinant $|a|$ is always $+1$. This is discussed further in Section 3.3. Often quantities that transform according to Eq. (2.90) are known as tensor densities. They are regular tensors as far as rotations are concerned, differing from tensors only in reflections or inversions of the coordinates, and then the only difference is the appearance of an additional minus sign from the determinant $|a|$.

In Chapter 1 the triple scalar product $S = \mathbf{A} \times \mathbf{B} \cdot \mathbf{C}$ was shown to be a scalar (under rotations). Now by considering the parity transformation given

---

[2] The big exception to this is in beta decay, weak interactions. Here the universe distinguishes between right- and left-handed systems, and we add polar and axial vector interactions.

[3] Determinants are described in Section 3.1.

by Eq. (2.88), we see that $S \to -S$, proving that the triple scalar product is actually a pseudoscalar: This behavior was foreshadowed by the geometrical analogy of a volume. If all three parameters of the volume, length, depth, and height, change from positive distances to negative distances, the product of the three will be negative.

LEVI–CIVITA SYMBOL

For future use it is convenient to introduce the three-dimensional Levi–Civita symbol ε_{ijk} defined by

$$\varepsilon_{123} = \varepsilon_{231} = \varepsilon_{312} = 1,$$

$$\varepsilon_{132} = \varepsilon_{213} = \varepsilon_{321} = -1, \tag{2.93}$$

$$\text{all other } \varepsilon_{ijk} = 0.$$

Note that ε_{ijk} is totally antisymmetric with respect to all pairs of indices. Suppose now that we have a third-rank pseudotensor δ_{ijk}, which in one particular coordinate system is equal to ε_{ijk}. Then

$$\delta'_{ijk} = |a| a_{ip} a_{jq} a_{kr} \varepsilon_{pqr} \tag{2.94}$$

by definition of pseudotensor. Now

$$a_{1p} a_{2q} a_{3r} \varepsilon_{pqr} = |a| \tag{2.95}$$

by direct expansion of the determinant, showing that $\delta'_{123} = |a|^2 = 1 = \varepsilon_{123}$. Considering the other possibilities one by one, we find

$$\delta'_{ijk} = \varepsilon_{ijk} \tag{2.96}$$

for rotations and reflections. Hence ε_{ijk} is a pseudotensor.[4,5] Furthermore, it is seen to be an isotropic pseudotensor with the same components in all rotated cartesian coordinate systems.

DUAL TENSORS

With any *antisymmetric* second-rank tensor C_{ij} (in three-dimensional space) we may associate a dual pseudovector C_i defined by

$$C_i = \tfrac{1}{2} \varepsilon_{ijk} C_{jk}. \tag{2.97}$$

[4] The usefulness of ε_{ijk} extends far beyond this section. For instance, the matrices M_k of Exercise 3.2.16 are derived from $(M_k)_{ij} = -i\varepsilon_{ijk}$. Much of elementary vector analysis can be written in a very compact form by using ε_{ijk} and the identity of Exercise 2.9.4. See A. A. Evett, Permutation symbol approach to elementary vector analysis. *Am. J. Phys.* **34**, 503 (1966).

[5] The numerical value of ε_{ijk} is given by the triple scalar product of coordinate unit vectors:

$$\mathbf{e}_i \cdot \mathbf{e}_j \times \mathbf{e}_k.$$

From this point of view each element of ε_{ijk} is a pseudoscalar, but the ε_{ijk} collectively form a third-rank pseudotensor.

Here the antisymmetric C_{jk} may be written

$$C_{jk} = \begin{pmatrix} 0 & C_{12} & -C_{31} \\ -C_{12} & 0 & C_{23} \\ C_{31} & -C_{23} & 0 \end{pmatrix}. \tag{2.98}$$

We know that C_i must transform as a vector under rotations from the double contraction of the fifth-rank (pseudo) tensor $\varepsilon_{ijk} C_{mn}$ but that it is really a pseudovector from the pseudo nature of ε_{ijk}. Specifically, the components of **C** are given by

$$(C_1, C_2, C_3) = (C_{23}, C_{31}, C_{12}). \tag{2.99}$$

Notice the cyclic order of the indices that comes from the cyclic order of the components of ε_{ijk}. This duality, given by Eq. (2.99), means that our three-dimensional vector product may literally be taken to be either a pseudovector or an antisymmetric second-rank tensor, depending on how we chose to write it out.

If we take three (polar) vectors **A**, **B**, and **C**, we may define

$$V_{ijk} = \begin{vmatrix} A_i & B_i & C_i \\ A_j & B_j & C_j \\ A_k & B_k & C_k \end{vmatrix}$$

$$= A_i B_j C_k - A_i B_k C_j + \cdots. \tag{2.100}$$

By an extension of the analysis of Section 2.6 each term $A_p B_q C_r$ is seen to be a third-rank tensor, making V_{ijk} a tensor of third rank. From its definition as a determinant V_{ijk} is totally antisymmetric, reversing sign under the interchange of any two indices, that is, the interchange of any two rows of the determinant. The dual quantity is

$$V = \frac{1}{3!} \varepsilon_{ijk} V_{ijk}, \tag{2.101}$$

clearly a pseudoscalar. By expansion it is seen that

$$V = \begin{vmatrix} A_1 & B_1 & C_1 \\ A_2 & B_2 & C_2 \\ A_3 & B_3 & C_3 \end{vmatrix}, \tag{2.102}$$

our familiar triple scalar product.

For use in writing of Maxwell's equations in covariant form, Section 4.7, we want to extend this dual vector analysis to four-dimensional space and, in particular, to indicate that the four-dimensional volume element, $dx_1\, dx_2\, dx_3\, dx_4$, is a pseudoscalar.

We introduce the Levi–Civita symbol ε_{ijkl}, the four-dimensional analog of ε_{ijk}. This quantity ε_{ijkl} is defined as totally antisymmetric in all four indices. If $(ijkl)$ is an even permutation[6] of $(1, 2, 3, 4)$, then ε_{ijkl} is defined as $+1$; if it is an odd permutation, then ε_{ijkl} is -1. The Levi–Civita ε_{ijkl} may be proved a pseudotensor of rank 4 by analysis similar to that used for establishing the nature of ε_{ijk}. Introducing a fourth-rank tensor,

$$H_{ijkl} = \begin{vmatrix} A_i & B_i & C_i & D_i \\ A_j & B_j & C_j & D_j \\ A_k & B_k & C_k & D_k \\ A_l & B_l & C_l & D_l \end{vmatrix}, \tag{2.103}$$

built from the polar vectors **A**, **B**, **C**, and **D**, we may define the dual quantity

$$H = \frac{1}{4!} \varepsilon_{ijkl} H_{ijkl}. \tag{2.104}$$

We actually have a quadruple contraction which reduces the rank to zero. From the pseudo nature of ε_{ijkl}, H is a pseudoscalar. Now we let **A**, **B**, **C**, and **D** be infinitesimal displacements along the four coordinate axes (Minkowski space),

$$\mathbf{A} = (dx_1, 0, 0, 0)$$
$$\mathbf{B} = (0, dx_2, 0, 0), \quad \text{and so on,} \tag{2.105}$$

and

$$H = dx_1\, dx_2\, dx_3\, dx_4. \tag{2.106}$$

The four-dimensional volume element is now identified as a pseudoscalar. We use this result in Section 4.7. This result could have been expected from the results of the special theory of relativity. The Lorentz–Fitzgerald contraction of $dx_1\, dx_2\, dx_3$ just balances the time dilation of dx_4.

We slipped into this four-dimensional space as a simple mathematical extension of the three-dimensional space and, indeed, we could just as easily have discussed 5-, 6-, or N-dimensional space. This is typical of the power of the component analysis. Physically, this four-dimensional space may be taken as Minkowski space,

$$(x_1, x_2, x_3, x_4) = (x, y, z, ict), \tag{2.107}$$

where t is time. This is the merger of space and time achieved in special relativity. The transformations that describe the rotations in four-dimensional space are the Lorentz transformations of special relativity. We encounter these Lorentz transformations in Sections 4.7 and 4.13.

---

[6] A permutation is odd if it involves an odd number of interchanges of adjacent indices such as $(1\,2\,3\,4) \rightarrow (1\,3\,2\,4)$. Even permutations arise from an even number of transpositions of adjacent indices. (Actually the word "adjacent" is not necessary.)

IRREDUCIBLE TENSORS

For some applications, particularly in the quantum theory of angular momentum, our cartesian tensors are not particularly convenient. In mathematical language our general second-rank tensor A_{ij} is reducible, which means that it can be decomposed into parts of lower tensor rank. In fact, we have already done this. From Eq. (2.78)

$$A = A_{ii} \tag{2.108}$$

is a scalar quantity, the trace of A_{ij}.[7]

The antisymmetric portion

$$B_{ij} = \tfrac{1}{2}(A_{ij} - A_{ji}) \tag{2.109}$$

has just been shown to be equivalent to a (pseudo) vector, or

$$B_{ij} = C_k \qquad \text{cyclic permutation of } i, j, k. \tag{2.110}$$

By subtracting the scalar A and the vector C_k from our original tensor, we have an irreducible, symmetric, zero-trace second-rank tensor, S_{ij}, in which

$$S_{ij} = \tfrac{1}{2}(A_{ij} + A_{ji}) - \tfrac{1}{3}A\delta_{ij}, \tag{2.111}$$

with five independent components. Then, finally, our original cartesian tensor may be written

$$A_{ij} = \tfrac{1}{3}A\delta_{ij} + C_k + S_{ij}. \tag{2.112}$$

The three quantities A, C_k, and S_{ij} form spherical tensors of rank 0, 1, and 2, respectively, transforming like the spherical harmonics Y_L^M (Chapter 12) for $L = 0, 1,$ and 2. Further details of such spherical tensors and their uses will be found in the books by Rose and Edmonds cited in Chapter 4.

A specific example of the preceding reduction is furnished by the symmetric electric quadrupole tensor

$$Q_{ij} = \int (3x_i x_j - r^2\delta_{ij})\rho(x_1, x_2, x_3)\, d^3x.$$

The $-r^2\delta_{ij}$ term represents a subtraction of the scalar trace (the three $i = j$ terms). The resulting Q_{ij} has zero trace.

EXERCISES

2.9.1   An antisymmetric square array is given by

$$\begin{pmatrix} 0 & C_3 & -C_2 \\ -C_3 & 0 & C_1 \\ C_2 & -C_1 & 0 \end{pmatrix} = \begin{pmatrix} 0 & C_{12} & C_{13} \\ -C_{12} & 0 & C_{23} \\ -C_{13} & -C_{23} & 0 \end{pmatrix},$$

[7] An alternate approach, using matrices, is given in Section 3.3 (see Exercise 3.3.9).

where (C_1, C_2, C_3) form a pseudovector. Assuming that the relation

$$C_i = \frac{1}{2!}\varepsilon_{ijk}C_{jk}$$

holds in all coordinate systems, prove that C_{jk} is a tensor. (This is another form of the quotient theorem.)

2.9.2   Show that the vector product is unique to three-dimensional space, that is, only in three dimensions can we establish a one-to-one correspondence between the components of an antisymmetric tensor (second-rank) and the components of a vector.

2.9.3   Show that

(a) $\delta_{ii} = 3$,

(b) $\delta_{ij}\varepsilon_{ijk} = 0$,

(c) $\varepsilon_{ipq}\varepsilon_{jpq} = 2\delta_{ij}$,

(d) $\varepsilon_{ijk}\varepsilon_{ijk} = 6$.

2.9.4   Show that

$$\varepsilon_{ijk}\varepsilon_{pqk} = \delta_{ip}\delta_{jq} - \delta_{iq}\delta_{jp}.$$

2.9.5   (a) Express the components of a cross-product vector $\mathbf{C}$, $\mathbf{C} = \mathbf{A} \times \mathbf{B}$, in terms of ε_{ijk} and the components of $\mathbf{A}$ and $\mathbf{B}$.

(b) Use the antisymmetry of ε_{ijk} to show that $\mathbf{A} \cdot \mathbf{A} \times \mathbf{B} = 0$.

ANS.   (a) $C_i = \varepsilon_{ijk}A_jB_k$.

2.9.6   (a) Show that the inertia tensor (matrix) of Section 3.5 may be written

$$I_{ij} = m(x_nx_n\delta_{ij} - x_ix_j)$$

for a particle of mass m at (x_1, x_2, x_3).

(b) Show that

$$I_{ij} = -M_{il}M_{lj} = -m\varepsilon_{ilk}x_k\varepsilon_{ljm}x_m,$$

where $M_{il} = m^{1/2}\varepsilon_{ilk}x_k$. This is the contraction of two second-rank tensors and is identical with the matrix product of Section 3.2.

2.9.7   Write $\nabla \cdot \nabla \times \mathbf{A}$ and $\nabla \times \nabla\varphi$ in ε_{ijk} notation, so that it becomes obvious that each expression vanishes.

$$ANS.\quad \nabla \cdot \nabla \times \mathbf{A} = \varepsilon_{ijk}\frac{\partial}{\partial x_i}\frac{\partial}{\partial x_j}A_k$$

$$(\nabla \times \nabla\varphi)_i = \varepsilon_{ijk}\frac{\partial}{\partial x_j}\frac{\partial}{\partial x_k}\varphi.$$

2.9.8   Expressing cross products in terms of Levi–Civita symbols (ε_{ijk}), derive the *BAC-CAB* rule, Eq. (1.55).

Hint. The relation of Exercise 2.9.4 is helpful.

2.9.9   Verify that each of the following fourth-rank tensors is isotropic, that is, it has the same form independent of any rotation of the coordinate systems.

(a) $A_{ijkl} = \delta_{ij}\delta_{kl}$,

(b) $B_{ijkl} = \delta_{ik}\delta_{jl} + \delta_{il}\delta_{jk}$,

(c) $C_{ijkl} = \delta_{ik}\delta_{jl} - \delta_{il}\delta_{jk}$.

2.9.10   Show that the two index Levi–Civita symbol ε_{ij} is a second-rank pseudo-tensor (in two-dimensional space). Does this contradict the uniqueness of δ_{ij} (Exercise 2.6.4)?

2.9.11   (a) Represent ε_{ij} by a 2×2 matrix, and using the 2×2 rotation matrix of Section 3.3, show that ε_{ij} is invariant under orthogonal similarity transformations.

   (b) Demonstrate the pseudo nature of ε_{ij} by using $\begin{pmatrix} 1 & 0 \\ 0 & -1 \end{pmatrix}$ as the transforming matrix.

2.9.12   Given $A_k = \frac{1}{2}\varepsilon_{ijk}B_{ij}$ with $B_{ij} = -B_{ji}$, antisymmetric, show that

$$B_{mn} = \varepsilon_{mnk}A_k.$$

2.9.13   Show that the vector identity

$$(\mathbf{A} \times \mathbf{B}) \cdot (\mathbf{C} \times \mathbf{D}) = (\mathbf{A} \cdot \mathbf{C})(\mathbf{B} \cdot \mathbf{D}) - (\mathbf{A} \cdot \mathbf{D})(\mathbf{B} \cdot \mathbf{C})$$

(Exercise 1.5.12) follows directly from the description of a cross product with ε_{ijk} and the identity of Exercise 2.9.4.

2.10 NONCARTESIAN TENSORS, COVARIANT DIFFERENTIATION

The distinction between contravariant transformations and covariant transformations was established in Section 2.6. Then, for convenience, we restricted our attention to cartesian coordinates (in which the distinction disappears). Now in these two concluding sections we return to noncartesian coordinates and resurrect the contravariant and covariant dependence. As in Section 2.6, a superscript will be used for an index denoting contravariant and a subscript for an index denoting covariant dependence. The metric tensor of Section 2.1 will be used to relate contravariant and covariant indices.

The emphasis in this section is on differentiation, culminating in the construction of the covariant derivative. We saw in Section 2.7 that the derivative of a vector yields a second-rank tensor—in cartesian coordinates. The covariant derivative of a vector yields a second-rank tensor in noncartesian coordinate systems.

METRIC TENSOR, RAISING AND LOWERING INDICES

Let us start with a set of basis vectors ε_i such that an infinitesimal displacement $d\mathbf{r}$ would be given by

$$d\mathbf{r} = \varepsilon_1 \, dq^1 + \varepsilon_2 \, dq^2 + \varepsilon_3 \, dq^3. \tag{2.113}$$

For convenience we take ε_1, ε_2, and ε_3 to form a right-handed set. These vectors are not necessarily orthogonal. Also, a limitation to three-dimensional space will be required only for the discussions of cross products and curls. Otherwise these ε_i may be in N-dimensional space, including the four-dimensional space–time of special and general relativity. The basis vectors $\boldsymbol{\varepsilon}_i$ may be expressed by

$$\boldsymbol{\varepsilon}_i = \frac{\partial \mathbf{r}}{\partial q^i}, \tag{2.114}$$

as in Exercise 2.2.3. Note, however, that the $\boldsymbol{\varepsilon}_i$ here do *not* necessarily have unit magnitude. From Exercise 2.2.3, the unit vectors, are

$$\mathbf{e}_i = \frac{1}{h_i} \frac{\partial \mathbf{r}}{\partial q_i} \quad \text{(no summation)}$$

and therefore

$$\boldsymbol{\varepsilon}_i = h_i \mathbf{e}_i \quad \text{(no summation)}. \tag{2.115}$$

The $\boldsymbol{\varepsilon}_i$ are related to the unit vectors $\mathbf{e}_i$ by the scale factors h_i of Section 2.2. The $\mathbf{e}_i$ have no dimensions; the $\boldsymbol{\varepsilon}_i$ have the dimensions of h_i. In spherical polar coordinates, as a specific example,

$$\boldsymbol{\varepsilon}_r = \mathbf{e}_r$$

$$\boldsymbol{\varepsilon}_\theta = r\mathbf{e}_\theta \tag{2.116}$$

$$\boldsymbol{\varepsilon}_\varphi = r\sin\theta\,\mathbf{e}_\varphi.$$

As in Section 2.1, we construct the square of a differential displacement

$$(ds)^2 = d\mathbf{r} \cdot d\mathbf{r} = (\boldsymbol{\varepsilon}_i\,dq^i)^2 = \boldsymbol{\varepsilon}_i \cdot \boldsymbol{\varepsilon}_j\,dq^i\,dq^j. \tag{2.117}$$

Comparing this with $(ds)^2$ of Section 2.1, Eq. (2.5), we identify $\boldsymbol{\varepsilon}_i \cdot \boldsymbol{\varepsilon}_j$ as the covariant metric tensor

$$\boldsymbol{\varepsilon}_i \cdot \boldsymbol{\varepsilon}_j = g_{ij}. \tag{2.118}$$

Clearly, g_{ij} is symmetric. The tensor nature of g_{ij} follows from the quotient rule, Exercise 2.8.1. We take the relation

$$g^{ik}g_{kj} = \delta^i_j \tag{2.119}$$

to define the corresponding contravariant tensor g^{ik}. Contravariant g^{ik} enters as the inverse[1] of covariant g_{kj}. We use this contravariant g^{ik} to raise indices, converting a covariant index into a contravariant index, as shown subsequently. Likewise the covariant g_{kj} will be used to lower indices. The choice of g^{ik} and g_{kj} for this raising–lowering operation is arbitrary. Any second-rank tensor (and its inverse) would do. Specifically, we have

$$g^{ij}\boldsymbol{\varepsilon}_j = \boldsymbol{\varepsilon}^i \qquad \text{relating covariant and contravariant basis vectors,}$$

$$g^{ij}F_j = F^i \qquad \text{relating covariant and contravariant vector components.} \tag{2.120}$$

---

[1] If the tensor g_{kj} is written as a matrix, the tensor g^{ik} is given by the inverse matrix.

Then

$$g_{ij}\varepsilon^j = \varepsilon_i \qquad \text{as the corresponding index}$$
$$g_{ij}F^j = F_i \qquad \text{lowering relations.} \tag{2.121}$$

As an example of these transformations we start with the contravariant form of vector

$$\mathbf{F} = F^i\varepsilon_i. \tag{2.122}$$

From Eqs. (2.120) and (2.121)

$$\mathbf{F} = F_j g^{ji} g_{ik}\varepsilon^k$$

$$= F_j \varepsilon^j, \tag{2.123}$$

the final equality coming from Eq. (2.119). Equation (2.122) gives the contravariant representation of **F**. Equation (2.123) gives the corresponding covariant representation of the same **F**.

It should be emphasized again that the ε_i and ε^j do *not* have unit magnitude. This may be seen in Eqs. (2.116) and in the metric tensor g_{ij} for spherical polar coordinates and its inverse g^{ij}:

$$(g_{ij}) = \begin{pmatrix} 1 & 0 & 0 \\ 0 & r^2 & 0 \\ 0 & 0 & r^2\sin^2\theta \end{pmatrix} \qquad (g^{ij}) = \begin{pmatrix} 1 & 0 & 0 \\ 0 & \dfrac{1}{r^2} & 0 \\ 0 & 0 & \dfrac{1}{r^2\sin^2\theta} \end{pmatrix}.$$

DERIVATIVES, CHRISTOFFEL SYMBOLS
Let us form the differential of a scalar

$$d\psi = \frac{\partial \psi}{\partial q^i}\,dq^i. \tag{2.124}$$

Since the dq^i are the components of a contravariant vector, the partial derivatives $\partial\psi/\partial q^i$ must form a covariant vector—by the quotient rule. The gradient of a scalar becomes

$$\nabla\psi = \frac{\partial \psi}{\partial q^i}\,\varepsilon^i. \tag{2.125}$$

The reader should note that $\partial\psi/\partial q^i$ are not the gradient components of Section 2.2—because $\varepsilon^i \neq \mathbf{e}_i$ of Section 2.2.

Moving on to the derivatives of a vector, we find that the situation is much more complicated because the basis vectors ε_i are in general not constant. Remember we are no longer restricting ourselves to cartesian coordinates and the nice, convenient $\hat{\mathbf{x}}$, $\hat{\mathbf{y}}$, $\hat{\mathbf{z}}$! Direct differentiation yields

$$\frac{\partial \mathbf{V}}{\partial q^j} = \frac{\partial V^i}{\partial q^j} \boldsymbol{\varepsilon}_i + V^i \frac{\partial \boldsymbol{\varepsilon}_i}{\partial q^j}.$$ (2.126)

Now $\partial \boldsymbol{\varepsilon}_i / \partial q^j$ will be some linear combination of the $\boldsymbol{\varepsilon}_k$ with the coefficient depending on the indices i and j from the partial derivative and index k from the base vector. We write

$$\frac{\partial \boldsymbol{\varepsilon}_i}{\partial q^j} = \Gamma^k{}_{ij} \boldsymbol{\varepsilon}_k.$$ (2.127a)

Multiplying by $\boldsymbol{\varepsilon}^m$ and using $\boldsymbol{\varepsilon}^m \cdot \boldsymbol{\varepsilon}_k = \delta^m_k$ from Exercise 2.10.2, we have

$$\Gamma^m{}_{ij} = \boldsymbol{\varepsilon}^m \cdot \frac{\partial \boldsymbol{\varepsilon}_i}{\partial q^j}.$$ (2.127b)

The $\Gamma^k{}_{ij}$ is a Christoffel symbol (of the second kind). It is also called a "coefficient of connection." These $\Gamma^k{}_{ij}$ are *not* third-rank tensors and the $\partial V^i / \partial q^j$ of Eq. (2.126) are not second-rank tensors. Equation (2.127) should be compared with the results quoted in Exercise 2.2.3 (remembering that in general $\boldsymbol{\varepsilon}_i \neq \mathbf{e}_i$). In cartesian coordinates, $\Gamma^k{}_{ij} = 0$ for all values of the indices i, j, and k. These Christoffel three index symbols may be computed by the techniques of Chapter 2. This is the topic of Exercise 2.10.8. Equation (2.140) at the end of this section offers an easier method.

Using Eq. (2.114), we obtain

$$\frac{\partial \boldsymbol{\varepsilon}_i}{\partial q^j} = \frac{\partial^2 \mathbf{r}}{\partial q^j \, \partial q^i} = \frac{\partial \boldsymbol{\varepsilon}_j}{\partial q^i}$$

$$= \Gamma^k{}_{ji} \boldsymbol{\varepsilon}_k.$$ (2.128)

Hence these Christoffel symbols are symmetric in the two lower indices:

$$\Gamma^k{}_{ij} = \Gamma^k{}_{ji}.$$ (2.129)

COVARIANT DERIVATIVE

With the Christoffel symbols, Eq. (2.126) may be rewritten

$$\frac{\partial \mathbf{V}}{\partial q^j} = \frac{\partial V^i}{\partial q^j} \boldsymbol{\varepsilon}_i + V^i \Gamma^k{}_{ij} \boldsymbol{\varepsilon}_k.$$ (2.130)

Now i and k in the last term are dummy indices. Interchanging i and k (in this one term), we have

$$\frac{\partial \mathbf{V}}{\partial q^j} = \left(\frac{\partial V^i}{\partial q^j} + V^k \Gamma^i{}_{kj} \right) \boldsymbol{\varepsilon}_i.$$ (2.131)

The quantity in parenthesis is labeled a **covariant derivative**, $V^i{}_{;j}$. We have

$$V^i{}_{;j} \equiv \frac{\partial V^i}{\partial q^j} + V^k \Gamma^i{}_{kj}.$$ (2.132)

The $_{;j}$ subscript indicates differentiation with respect to q^j. The differential $d\mathbf{V}$ becomes

$$d\mathbf{V} = \frac{\partial \mathbf{V}}{\partial q^j} dq^j = [V^i_{;j} dq^j]\varepsilon_i. \qquad (2.133)$$

A comparison with Eq. (2.113) or (2.122) shows that the quantity in square brackets is the ith contravariant component of a vector. Since dq^j is the jth contravariant component of a vector (again, Eq. (2.113)), $V^i_{;j}$ must be the ijth component of a (mixed) second-rank tensor (quotient rule). The covariant derivatives of the contravariant components of a vector form a mixed second-rank tensor, $V^i_{;j}$.

Since the Christoffel symbols vanish in cartesian coordinates, the covariant derivative and the ordinary partial derivative coincide

$$\frac{\partial V^i}{\partial q^j} = V^i_{;j} \qquad \text{(cartesian coordinates)}. \qquad (2.134)$$

The covariant derivative of a covariant vector V_i is given by (Exercise 3.8.8)

$$V_{i;j} = \frac{\partial V_i}{\partial q^j} - V_k \Gamma^k{}_{ij}. \qquad (2.135)$$

Like $V^i_{;j}$, $V_{i;j}$ is a second-rank tensor.

The physical importance of the covariant derivative is that

> A consistent replacement of regular partial derivatives by covariant derivatives carries the laws of physics (in component form) from flat space time into the curved (Riemannian) space time of general relativity. Indeed, this substitution may be taken as a mathematical statement of Einstein's principle of equivalence.[2]

THE CHRISTOFFEL SYMBOLS AS DERIVATIVES OF THE METRIC TENSOR

It is often convenient to have an explicit expression for the Christoffel symbols in terms of derivatives of the metric tensor. As an initial step, we define the Christoffel symbol of the first kind $[ij, k]$ by

$$[ij, k] \equiv g_{mk} \Gamma^m{}_{ij}. \qquad (2.136)$$

This $[ij, k]$ is not a third-rank tensor. From Eq. (2.127b)

$$[ij, k] = g_{mk} \varepsilon^m \cdot \frac{\partial \varepsilon_i}{\partial q^j}$$

$$= \varepsilon_k \cdot \frac{\partial \varepsilon_i}{\partial q^j}. \qquad (2.137)$$

[2] C. W. Misner, K. S. Thorne, and J. A. Wheeler, *Gravitation*. San Francisco: W. H. Freeman (1973), p. 387.

Now we differentiate $g_{ij} = \varepsilon_i \cdot \varepsilon_j$, Eq. (2.118):

$$\frac{\partial g_{ij}}{\partial q^k} = \frac{\partial \varepsilon_i}{\partial q^k} \cdot \varepsilon_j + \varepsilon_i \cdot \frac{\partial \varepsilon_j}{\partial q^k}$$

$$= [ik, j] + [jk, i] \qquad (2.138)$$

by Eq. (2.137).
  Then

$$[ij, k] = \frac{1}{2} \left\{ \frac{\partial g_{ik}}{\partial q^j} + \frac{\partial g_{jk}}{\partial q^i} - \frac{\partial g_{ij}}{\partial q^k} \right\}, \qquad (2.139)$$

and

$$\Gamma^s_{ij} = g^{ks}[ij, k]$$

$$= \frac{1}{2} g^{ks} \left\{ \frac{\partial g_{ik}}{\partial q^j} + \frac{\partial g_{jk}}{\partial q^i} - \frac{\partial g_{ij}}{\partial q^k} \right\}. \qquad (2.140)$$

These Christoffel symbols and the covariant derivatives are applied in the next section.

EXERCISES

2.10.1 Equations (2.115) and (2.116) use the scale factor h_i, citing Exercise 2.2.3. In Section 2.2 we had restricted ourselves to orthogonal coordinate systems, yet Eq. (2.115) holds for nonorthogonal systems. Justify the use of Eq. (2.115) for nonorthogonal systems.

2.10.2 (a) Show that $\varepsilon^i \cdot \varepsilon_j = \delta^i_j$.
  (b) From the result of part (a) show that
$$F^i = \mathbf{F} \cdot \varepsilon^i \qquad \text{and} \qquad F_i = \mathbf{F} \cdot \varepsilon_i.$$

2.10.3 For the special case of three-dimensional space (ε_1, ε_2, ε_3 defining a right-handed coordinate system, not necessarily orthogonal) show that
$$\varepsilon^i = \frac{\varepsilon_j \times \varepsilon_k}{\varepsilon_j \times \varepsilon_k \cdot \varepsilon_i}, \qquad i, j, k = 1, 2, 3 \text{ and cyclic permutations.}$$

Note. These contravariant basis vectors, ε^i, define the reciprocal lattice space of Section 1.5.

2.10.4 Prove that the contravariant metric tensor is given by
$$g^{ij} = \varepsilon^i \cdot \varepsilon^j.$$

2.10.5 If the covariant vectors ε_i are orthogonal, show that
  (a) g_{ij} is diagonal,
  (b) $g^{ii} = 1/g_{ii}$ (no summation),
  (c) $|\varepsilon^i| = 1/|\varepsilon_i|$.

2.10.6 Derive the covariant and contravariant metric tensors for circular cylindrical coordinates.

2.10.7   Transform the right-hand side of Eq. (2.125)

$$\nabla\psi = \frac{\partial\psi}{\partial q^i}\, \varepsilon^i$$

into the e_i basis and verify that this expression agrees with the gradient developed in Section 2.2 (for orthogonal coordinates).

2.10.8   Evaluate $\partial\varepsilon_i/\partial q^j$ for the spherical polar coordinates, and from these results calculate the spherical polar coordinate $\Gamma^k{}_{ij}$.

Note. Exercise 2.5.1 offers a way of calculating the needed partial derivatives. Remember

$$\varepsilon_1 = \hat{\mathbf{r}} \qquad \text{but } \varepsilon_2 = r\hat{\theta} \qquad \text{and} \qquad \varepsilon_3 = r\sin\theta\hat{\phi}.$$

2.10.9   Show that the covariant derivative of a covariant vector is given by

$$V_{i;j} \equiv \frac{\partial V_i}{\partial q^j} - V_k\Gamma^k{}_{ij}.$$

Hint. Differentiate

$$\varepsilon^i \cdot \varepsilon_j = \delta^i_j.$$

2.10.10   Verify that $V_{i;j} = g_{ik}V^k{}_{;j}$ by showing that

$$\frac{\partial V_i}{\partial q^j} - V_s\Gamma^s{}_{ij} = g_{ik}\left\{\frac{\partial V^k}{\partial q^j} + V^m\Gamma^k{}_{mj}\right\}.$$

2.10.11   From the circular cylindrical metric tensor, g_{ij}, calculate the $\Gamma^k{}_{ij}$ for circular cylindrical coordinates.

Note. There are only three nonvanishing Γ's.

2.10.12   Using the $\Gamma^k{}_{ij}$ from Exercise 2.10.11, write out the covariant derivatives $V^i{}_{;j}$ of a vector $\mathbf{V}$ in circular cylindrical coordinates.

2.10.13   A triclinic crystal is described, using an oblique coordinate system. The three covariant base vectors are

$$\varepsilon_1 = 1.5\hat{\mathbf{x}},$$
$$\varepsilon_2 = 0.4\hat{\mathbf{x}} + 1.6\hat{\mathbf{y}},$$

and

$$\varepsilon_3 = 0.2\hat{\mathbf{x}} + 0.3\hat{\mathbf{y}} + 1.0\hat{\mathbf{z}}.$$

(a) Calculate the elements of the covariant metric tensor g_{ij}.

(b) Calculate the Christoffel three index symbols, $\Gamma^k{}_{ij}$. (This is a "by inspection" calculation.)

(c) From the cross-product form of Exercise 2.10.3 calculate the contravariant base vector ε^3.

(d) Using the explicit forms of ε^3 and ε_i, verify that $\varepsilon^3 \cdot \varepsilon_i = \delta^3_i$.

Note. If it were needed, the contravariant metric tensor could be determined by finding the inverse of g_{ij} or by finding the ε^i and using $g^{ij} = \varepsilon^i \cdot \varepsilon^j$.

2.10.14   Verify that

$$[ij, k] = \frac{1}{2}\left\{\frac{\partial g_{ik}}{\partial q^j} + \frac{\partial g_{jk}}{\partial q^i} - \frac{\partial g_{ij}}{\partial q^k}\right\}.$$

Hint. Substitute Eq. (2.138) into the right-hand side and show that an identity results.

2.11 TENSOR DIFFERENTIAL OPERATORS

In this section the covariant derivative of Section 2.10 is applied to rederive the vector differential operations of Section 2.2 in general tensor form.

DIVERGENCE

Replacing the partial derivative by the covariant derivative, we take the divergence to be

$$\nabla \cdot \mathbf{V} = V^i{}_{;i} = \frac{\partial V^i}{\partial q^i} + V^k \Gamma^i{}_{ik}. \tag{2.141}$$

Expressing $\Gamma^i{}_{ik}$ by Eq. (2.140), we have

$$\Gamma^i{}_{ik} = \frac{1}{2} g^{im} \left\{ \frac{\partial g_{im}}{\partial q^k} + \frac{\partial g_{km}}{\partial q^i} - \frac{\partial g_{ik}}{\partial q^m} \right\}. \tag{2.142}$$

When contracted with g^{im} the last two terms in the curly bracket cancel, since

$$g^{im} \frac{\partial g_{km}}{\partial q^i} = g^{mi} \frac{\partial g_{ki}}{\partial q^m} = g^{im} \frac{\partial g_{ik}}{\partial q^m}. \tag{2.143}$$

Then

$$\Gamma^i{}_{ik} = \frac{1}{2} g^{im} \frac{\partial g_{im}}{\partial q^k}. \tag{2.144}$$

From the theory of determinants Section 3.1,

$$\frac{\partial g}{\partial q^k} = g g^{im} \frac{\partial g_{im}}{\partial q^k}, \tag{2.145}$$

where g is the determinant of the metric, $g = \det(g_{ij})$. Substituting this result into Eq. (2.144), we obtain

$$\Gamma^i{}_{ik} = \frac{1}{2g} \frac{\partial g}{\partial q^k} = \frac{1}{g^{1/2}} \frac{\partial g^{1/2}}{\partial q^k}. \tag{2.146}$$

This yields

$$\nabla \cdot \mathbf{V} = V^i{}_{;i} = \frac{1}{g^{1/2}} \frac{\partial}{\partial q^k} (g^{1/2} V^k). \tag{2.147}$$

To compare this result with Eq. (2.17), note that $h_1 h_2 h_3 = g^{1/2}$ and V^i (contravariant coefficient of ε_i) $= V_i/h_i$ (no summation), where V_i is the Section 2.2 coefficient of $\mathbf{e}_i$.

LAPLACIAN

In Section 2.2 replacement of the vector $\mathbf{V}$ in $\nabla \cdot \mathbf{V}$ by $\nabla \psi$ led to the Laplacian $\nabla \cdot \nabla \psi$. Here we have a contravariant V^i. Using the metric tensor to create a contravariant $\nabla\psi$, we make the substitution

$$V^i \rightarrow g^{ik} \frac{\partial\psi}{\partial q^k}.$$

Then the Laplacian $\nabla \cdot \nabla\psi$ becomes

$$\nabla \cdot \nabla\psi = \frac{1}{g^{1/2}} \frac{\partial}{\partial q^i} \left(g^{1/2} g^{ik} \frac{\partial\psi}{\partial q^k} \right). \tag{2.148}$$

For the *orthogonal* systems of Section 2.2 the metric tensor is diagonal and the contravariant g^{ii} (no summation) becomes

$$g^{ii} = (h_i)^{-2}.$$

Equation (2.148) reduces to

$$\nabla \cdot \nabla\psi = \frac{1}{h_1 h_2 h_3} \frac{\partial}{\partial q^i} \left(\frac{h_1 h_2 h_3}{h_i^2} \frac{\partial\psi}{\partial q^i} \right)$$

in agreement with Eq. (2.18a).

CURL

The difference of derivatives that appears in the curl (Eq. (2.21)) will be written

$$\frac{\partial V_i}{\partial q^j} - \frac{\partial V_j}{\partial q^i}.$$

Again, remember that the components V_i here are coefficients of the contravariant (nonunit) base vectors ε^i. The V_i of Section 2.2 are coefficients of unit vectors $\mathbf{e}_i$. Adding and subtracting, we obtain

$$\Gamma^k_{ij} = \Gamma^k_{ji},$$

$$\frac{\partial V_i}{\partial q^j} - \frac{\partial V_j}{\partial q^i} = \frac{\partial V_i}{\partial q^j} - V_k \Gamma^k_{ij} - \frac{\partial V_j}{\partial q^i} + V_k \Gamma^k_{ji} \tag{2.149}$$

$$= V_{i;j} - V_{j;i}.$$

The characteristic difference of derivatives of the curl becomes a difference of covariant derivatives and therefore is a second-rank tensor (covariant in both indices). As emphasized in Section 2.9, the special vector form of the curl exists only in three-dimensional space.

From Eq. (2.140) it is clear that all the Christoffel three index symbols vanish in Minkowski space and in the real space–time of special relativity with

$$g_{\lambda\mu} = \begin{pmatrix} 1 & 0 & 0 & 0 \\ 0 & -1 & 0 & 0 \\ 0 & 0 & -1 & 0 \\ 0 & 0 & 0 & -1 \end{pmatrix}.$$

Here

$$x_0 = ct, \quad x_1 = x, \quad x_2 = y, \quad \text{and} \quad x_3 = z.$$

This completes the development of the differential operators in general tensor form. (The gradient was given in Section 2.10.) In addition to the fields of elasticity and electromagnetism, these differential forms find application in mechanics (Lagrangian mechanics, Hamiltonian mechanics, and the Euler equations for rotation of rigid body); fluid mechanics; and perhaps most important of all, in the curved space–time of modern theories of gravity.

EXERCISES

2.11.1   Verify Eq. (2.145)

$$\frac{\partial g}{\partial q^k} = g g^{im} \frac{\partial g_{im}}{\partial q^k}$$

for the specific case of spherical polar coordinates.

2.11.2   Starting with the divergence in tensor notation, Eq. (2.147), develop the divergence of a vector in spherical polar coordinates, Eq. (2.45).

2.11.3   The covariant vector A_i is the gradient of a scalar. Show that the difference of covariant derivatives $A_{i;j} - A_{j;i}$ vanishes.

ADDITIONAL READINGS

JEFFREYS, H., *Cartesian Tensors*. Cambridge: Cambridge University Press (1952). This is an excellent discussion of cartesian tensors and their application to a wide variety of fields of classical physics.

LAWDEN, D. F., *An Introduction to Tensor Calculus, Relativity and Cosmology*, 3rd ed. New York: Wiley (1982).

MARGENAU, H., and G. M. MURPHY, *The Mathematics of Physics and Chemistry*, 2nd ed. Princeton, NJ: Van Nostrand (1956). Chapter 5 covers curvilinear coordinates and 13 specific coordinate systems.

MISNER, C. W., K. S. THORNE, and J. A. WHEELER, *Gravitation*. San Francisco: W. H. Freeman (1973), p. 387.

MOLLER, C., *The Theory of Relativity*. Oxford: Oxford University Press (1955). Reprinted (1972). Most texts on general relativity include a discussion of tensor analysis. Chapter 4 develops tensor calculus, including the topic of dual tensors. The extension to non-cartesian systems, as required by general relativity, is presented in Chapter 9.

MORSE, P. M., and H. FESHBACH, *Methods of Theoretical Physics*. New York: McGraw-Hill (1953). Chapter 5 includes a description of several different coordinate systems. Note carefully that Morse and Feshbach are not above using left-handed coordinate systems even for cartesian coordinates. Elsewhere in this excellent (and difficult) book there are many examples of the use of the various coordinate systems in solving physical problems.

Eleven additional fascinating but seldom encountered orthogonal coordinate systems are discussed in the second (1970) edition of *Mathematical Methods for Physicists*.

OHANIAN, H. C., and R. RUFFINI, *Gravitation and Spacetime*, 2nd ed. New York: Norton & Co. (1994). A well-written introduction to Riemannian geometry.

SOKOLNIKOFF, I. S., *Tensor Analysis—Theory and Applications*, 2nd ed. New York: Wiley (1964). Particularly useful for its extension of tensor analysis to non-Euclidean geometries.

WEINBERG, S., *Gravitation and Cosmology; Principles and Applications of the General Theory of Relativity*. New York: Wiley (1972). This book and the one by Misner, Thorne, and Wheeler are the two leading texts on general relativity and cosmology (with tensors in noncartesian space).

YOUNG, E. C., *Vector and Tensor Analysis*, 2nd ed. New York: Dekker (1993).

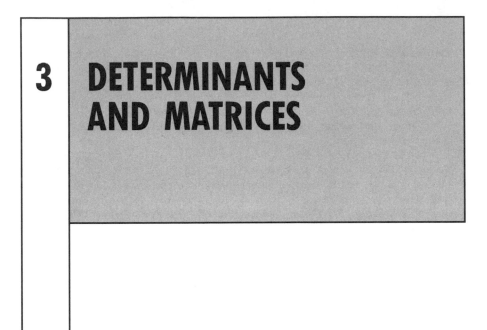

3 DETERMINANTS AND MATRICES

3.1 DETERMINANTS

We begin the study of matrices by solving linear equations which will lead us to determinants and matrices. The concept of "determinant" and the notation were introduced by Leibniz.

HOMOGENEOUS LINEAR EQUATIONS

One of the major applications of determinants is in the establishment of a condition for the existence of a nontrivial solution for a set of linear homogeneous algebraic equations. Suppose we have three unknowns x_1, x_2, x_3 (or n equations with n unknowns)

$$a_1 x_1 + a_2 x_2 + a_3 x_3 = 0,$$

$$b_1 x_1 + b_2 x_2 + b_3 x_3 = 0, \tag{3.1}$$

$$c_1 x_1 + c_2 x_2 + c_3 x_3 = 0.$$

The problem is to determine under what conditions there is any solution, apart from the trivial one $x_1 = 0$, $x_2 = 0$, $x_3 = 0$. If we use vector notation $\mathbf{x} = (x_1, x_2, x_3)$ for the solution and three rows $\mathbf{a} = (a_1, a_2, a_3)$, $\mathbf{b} = (b_1, b_2, b_3)$, $\mathbf{c} = (c_1, c_2, c_3)$ of coefficients, then the three equations, Eq. (3.1), become

$$\mathbf{a} \cdot \mathbf{x} = 0, \qquad \mathbf{b} \cdot \mathbf{x} = 0, \qquad \mathbf{c} \cdot \mathbf{x} = 0. \tag{3.2}$$

These three vector equations have the obvious geometrical interpretation that $\mathbf{x}$ is orthogonal to $\mathbf{a}$, $\mathbf{b}$ and $\mathbf{c}$. If the volume spanned by $\mathbf{a}, \mathbf{b}, \mathbf{c}$ given by the

156

determinant (or triple scalar product, see Eq. (1.50) of Section 1.5)

$$D_3 = (\mathbf{a} \times \mathbf{b}) \cdot \mathbf{c} = \det(\mathbf{a}, \mathbf{b}, \mathbf{c}) = \begin{vmatrix} a_1 & a_2 & a_3 \\ b_1 & b_2 & b_3 \\ c_1 & c_2 & c_3 \end{vmatrix} \tag{3.3}$$

is not zero, then clearly there is only the trivial solution $\mathbf{x} = 0$.

Conversely, if the above determinant of coefficients vanishes, then one of the row vectors is a linear combination of the other two. Let us assume that $\mathbf{c}$ lies in the plane spanned by $\mathbf{a}$ and $\mathbf{b}$, i.e., the third equation is a linear combination of the first two and not independent. Then $\mathbf{x}$ is orthogonal to that plane so that $\mathbf{x} \sim \mathbf{a} \times \mathbf{b}$. Since homogeneous equations can be multiplied by arbitrary numbers, only ratios of the x_i are relevant, for which we then obtain ratios of 2×2 determinants

$$x_1/x_3 = (a_2 b_3 - a_3 b_2)/(a_1 b_2 - a_2 b_1)$$
$$x_2/x_3 = -(a_1 b_3 - a_3 b_1)/(a_1 b_2 - a_2 b_1) \tag{3.4}$$

from the components of the cross product $\mathbf{a} \times \mathbf{b}$.

INHOMOGENEOUS LINEAR EQUATIONS

The simplest case of two equations with two unknowns

$$a_1 x_1 + a_2 x_2 = a_3,$$
$$b_1 x_1 + b_2 x_2 = b_3, \tag{3.5}$$

can be reduced to the previous case by imbedding it in three-dimensional space with a solution vector $\mathbf{x} = (x_1, x_2, -1)$ and row vectors $\mathbf{a} = (a_1, a_2, a_3)$, $\mathbf{b} = (b_1, b_2, b_3)$. As before, Eq. (3.5) in vector notation, $\mathbf{a} \cdot \mathbf{x} = 0$ and $\mathbf{b} \cdot \mathbf{x} = 0$, imply that $\mathbf{x} \sim \mathbf{a} \times \mathbf{b}$ so that the analog of Eq. (3.4) holds. For this to apply, though, the third component of $\mathbf{a} \times \mathbf{b}$ must not be zero, i.e., $a_1 b_2 - a_2 b_1 \neq 0$, because the third component of $\mathbf{x}$ is $-1 \neq 0$. This yields the x_i as

$$x_1 = (a_3 b_2 - b_3 a_2)/(a_1 b_2 - a_2 b_1) = \begin{vmatrix} a_3 & a_2 \\ b_3 & b_2 \end{vmatrix} \bigg/ \begin{vmatrix} a_1 & a_2 \\ b_1 & b_2 \end{vmatrix} \tag{3.6a}$$

$$x_2 = (a_1 b_3 - a_3 b_1)/(a_1 b_2 - a_2 b_1) = \begin{vmatrix} a_1 & a_3 \\ b_1 & b_3 \end{vmatrix} \bigg/ \begin{vmatrix} a_1 & a_2 \\ b_1 & b_2 \end{vmatrix} \tag{3.6b}$$

The determinant in the numerator of x_1 (x_2) is obtained from the determinant of the coefficients $\begin{vmatrix} a_1 & a_2 \\ b_2 & b_2 \end{vmatrix}$ by replacing the first (second) column vector by the vector $\begin{pmatrix} a_3 \\ b_3 \end{pmatrix}$ of the inhomogeneous side of Eq. (3.5).

These solutions of linear equations in terms of determinants can be generalized to n dimensions. The determinant is a square array

$$D_n = \begin{vmatrix} a_1 & a_2 & \cdots & a_n \\ b_1 & b_2 & \cdots & b_n \\ c_1 & c_2 & \cdots & c_n \\ \cdot & \cdot & \cdots & \cdot \end{vmatrix} \tag{3.7}$$

of numbers (or functions), the coefficients of n linear equations in our case here. The number n of columns (and of rows) in the array is sometimes called the *order* of the determinant. The generalization of the expansion in Eq. (1.48) of the triple scalar product (of row vectors of three linear equations) leads to the following value of the determinant D_n in n dimensions,

$$D_n = \sum_{i,j,k\ldots} \varepsilon_{ijk\ldots} a_i b_j c_k \ldots, \tag{3.8}$$

where $\varepsilon_{ijk\ldots}$, analogous to the Levi–Civita symbol of Section 2.9, is $+1$ for even permutations[1] $(ijk\ldots)$ of $(123\ldots n)$, -1 for odd permutations, and zero if any index is repeated.

Specifically, for the third-order determinant D_3 of Eq. (3.3), Eq. (3.8) leads to

$$D_3 = +a_1 b_2 c_3 - a_1 b_3 c_2 - a_2 b_1 c_3 + a_2 b_3 c_1 + a_3 b_1 c_2 - a_3 b_2 c_1. \tag{3.9}$$

The third-order determinant, then, is this particular linear combination of products. Each product contains one and only one element from each row and from each column. Each product is added if the columns (indices) represent an even permutation of (123) and subtracted if we have an odd permutation. Equation (3.3) may be considered short-hand notation for Eq. (3.9). The number of terms in the sum (Eq. (3.8)) is 24 for a fourth-order determinant, $n!$ for an nth-order determinant. Because of the appearance of the negative signs in Eq. (3.9) (and possibly in the individual elements as well), there may be considerable cancellation. It is quite possible that a determinant of large elements will have a very small value.

Several useful properties of the nth-order determinants follow from Eq. (3.8). Again, to be specific, Eq. (3.9) for third-order determinants is used to illustrate these properties.

LAPLACIAN DEVELOPMENT BY MINORS

Equation (3.9) may be written

$$D_3 = a_1(b_2 c_3 - b_3 c_2) - a_2(b_1 c_3 - b_3 c_1) + a_3(b_1 c_2 - b_2 c_1)$$

$$= a_1 \begin{vmatrix} b_2 & b_3 \\ c_2 & c_3 \end{vmatrix} - a_2 \begin{vmatrix} b_1 & b_3 \\ c_1 & c_3 \end{vmatrix} + a_3 \begin{vmatrix} b_1 & b_2 \\ c_1 & c_2 \end{vmatrix}. \tag{3.10}$$

---

[1] In a linear sequence *abcd*..., any single, simple transposition of adjacent elements yields an *odd* permutation of the original sequence: *abcd* → *bacd*. Two such transpositions yield an even permutation. In general, an odd number of such interchanges of adjacent elements results in an odd permutation; an even number of such transpositions yields an even permutation.

In general, the nth-order determinant may be expanded as a linear combination of the products of the elements of any row (or any column) and the $(n - 1)$-order determinants formed by striking out the row and column of the original determinant in which the element appears. This reduced array (2×2 in this specific example) is called a "minor." If the element is in the ith row and the jth column, the sign associated with the product is $(-1)^{i+j}$. The minor with this sign is called the "cofactor." If M_{ij} is used to designate the minor formed by omitting the ith row and the jth column and c_{ij} is the corresponding cofactor, Eq. (3.10) becomes

$$D_3 = \sum_{j=1}^{3} (-1)^{j+1} a_j M_{1j} = \sum_{j=1}^{3} a_j C_{1j}. \tag{3.11}$$

In this case, expanding along the first row, we have $i = 1$ and the summation over j, the columns.

This Laplace expansion may be used to advantage in the evaluation of high-order determinants in which a lot of the elements are zero. For example, to find the value of the determinant

$$D = \begin{vmatrix} 0 & 1 & 0 & 0 \\ -1 & 0 & 0 & 0 \\ 0 & 0 & 0 & 1 \\ 0 & 0 & -1 & 0 \end{vmatrix}, \tag{3.12}$$

we expand across the top row to obtain

$$D = (-1)^{1+2} \cdot (1) \begin{vmatrix} -1 & 0 & 0 \\ 0 & 0 & 1 \\ 0 & -1 & 0 \end{vmatrix}. \tag{3.13}$$

Again, expanding across the top row, we get

$$D = (-1) \cdot (-1)^{1+1} \cdot (-1) \begin{vmatrix} 0 & 1 \\ -1 & 0 \end{vmatrix}$$

$$= \begin{vmatrix} 0 & 1 \\ -1 & 0 \end{vmatrix} = 1. \tag{3.14}$$

This determinant D (Eq. (3.12)) is formed from one of the Dirac matrices appearing in Dirac's relativistic electron theory.

ANTISYMMETRY

The determinant changes sign if any two rows are interchanged or if any two columns are interchanged. This follows from the even–odd character of the Levi–Civita ε in Eq. (3.8) or explicitly from the form of Eqs. (3.9) and (3.10).[2]

---

[2] The sign reversal is reasonably obvious for the interchange of two adjacent rows (or columns), this clearly being an odd permutation. The reader may wish to show that the interchange of *any* two rows is still an odd permutation.

This property was used in Section 2.9 to develop a totally antisymmetric linear combination. It is also frequently used in quantum mechanics in the construction of a many particle wave function that, in accordance with the Pauli exclusion principle, will be antisymmetric under the interchange of any two identical spin $\frac{1}{2}$ particles (electrons, protons, neutrons, etc.).

As a special case of antisymmetry, any determinant with two rows equal or two columns equal equals zero.

If each element in a row or each element in a column is zero, the determinant is equal to zero.

If each element in a row or each element in a column is multiplied by a constant, the determinant is multiplied by that constant.

The value of a determinant is unchanged if a multiple of one row is added (column by column) to another row or if a multiple of one column is added (row by row) to another column.

We have

$$
\begin{vmatrix}
a_1 & a_2 & a_3 \\
b_1 & b_2 & b_3 \\
c_1 & c_2 & c_3
\end{vmatrix}
=
\begin{vmatrix}
a_1 + ka_2 & a_2 & a_3 \\
b_1 + kb_2 & b_2 & b_3 \\
c_1 + kc_2 & c_2 & c_3
\end{vmatrix}.
\tag{3.15}
$$

Using the Laplace development on the right-hand side, we obtain

$$
\begin{vmatrix}
a_1 + ka_2 & a_2 & a_3 \\
b_1 + kb_2 & b_2 & b_3 \\
c_1 + kc_2 & c_2 & c_3
\end{vmatrix}
=
\begin{vmatrix}
a_1 & a_2 & a_3 \\
b_1 & b_2 & b_3 \\
c_1 & c_2 & c_3
\end{vmatrix}
+ k
\begin{vmatrix}
a_2 & a_2 & a_3 \\
b_2 & b_2 & b_3 \\
c_2 & c_2 & c_3
\end{vmatrix},
\tag{3.16}
$$

then by the property of antisymmetry the second determinant on the right-hand side of Eq. (3.16) vanishes, verifying Eq. (3.15).

As a special case, a determinant is equal to zero if any two rows are proportional or any two columns are proportional.

Some useful relations involving determinants or matrices appear in Exercises 3.2 and 3.5.

Returning to the homogeneous Eqs. (3.1) and multiplying the determinant of the coefficients by x_1, then adding x_2 times the second column and x_3 times the third column, we can directly establish the condition for the presence of a nontrivial solution for Eq. (3.1):

$$
x_1
\begin{vmatrix}
a_1 & a_2 & a_3 \\
b_1 & b_2 & b_3 \\
c_1 & c_2 & c_3
\end{vmatrix}
=
\begin{vmatrix}
a_1 x_1 & a_2 & a_3 \\
b_1 x_1 & b_2 & b_3 \\
c_1 x_1 & c_2 & c_3
\end{vmatrix}
=
\begin{vmatrix}
a_1 x_1 + a_2 x_2 + a_3 x_3 & a_2 & a_3 \\
b_1 x_1 + b_2 x_2 + b_3 x_3 & b_2 & b_3 \\
c_1 x_1 + c_2 x_2 + c_3 x_3 & c_2 & c_3
\end{vmatrix}
$$

$$
=
\begin{vmatrix}
0 & a_2 & a_3 \\
0 & b_2 & b_3 \\
0 & c_2 & c_3
\end{vmatrix}
= 0.
\tag{3.17}
$$

Therefore x_1 (and x_2 and x_3) must be zero *unless the determinant of the coefficients vanishes*. Conversely (see text below Eq. (3.3)), we can show that if the determinant of the coefficients vanishes, a nontrivial solution does indeed exist. This is used in Section 8.6 to establish the linear dependence or independence of a set of functions.

If our linear equations are inhomogeneous, that is, as in Eq. (3.5) if the zeros on the right-hand side of Eq. (3.1) are replaced by a_4, b_4, and c_4, respectively, then from Eq. (3.17) we obtain,[3] instead,

$$x_1 = \begin{vmatrix} a_4 & a_2 & a_3 \\ b_4 & b_2 & b_3 \\ c_4 & c_2 & c_3 \end{vmatrix} \Bigg/ \begin{vmatrix} a_1 & a_2 & a_3 \\ b_1 & b_2 & b_3 \\ c_1 & c_2 & c_3 \end{vmatrix}, \tag{3.18}$$

which generalizes Eq. (3.6a) to $n = 3$ dimensions, etc. If the determinant of the coefficients vanishes, the inhomogeneous set of equations has no solution—unless the numerators also vanish. In this case solutions may exist but they are not unique (see Exercise 3.1.3 for a specific example).

For numerical work, this determinant solution, Eq. (3.18), is exceedingly unwieldy. The determinant may involve large numbers with alternate signs, and in the subtraction of two large numbers the relative error may soar to a point that makes the result worthless. Also, although the determinant method is illustrated here with 3 equations and 3 unknowns, we might easily have 200 equations with 200 unknowns which, involving up to 200! terms in each determinant, pose a challenge even to high speed electronic computers. There must be a better way.

In fact, there are better ways. One of the best is a straightforward process often called *Gauss elimination*. To illustrate this technique, consider the following set of equations.

Example 3.1.1: Gauss Elimination

Solve

$$3x + 2y + z = 11$$

$$2x + 3y + z = 13 \tag{3.19}$$

$$x + y + 4z = 12.$$

The determinant of the inhomogeneous linear equations (3.19) is 18, so that a solution exists.

For convenience and for the optimum numerical accuracy, the equations are rearranged so that the largest coefficients run along the main diagonal (upper left to lower right). This has already been done in the preceding set.

The Gauss technique is to use the first equation to eliminate the first unknown, x, from the remaining equations. Then the (new) second equation is

---

[3] Exercise 1.5.13 gives the vector analog of Eq. (3.18).

used to eliminate y from the last equation. In general, we work down through the set of equations, and then, with one unknown determined, we work back up to solve for each of the other unknowns in succession.

Dividing each row by its initial coefficient, we see that Eqs. (3.19) become

$$x + \tfrac{2}{3}y + \tfrac{1}{3}z = \tfrac{11}{3}$$
$$x + \tfrac{3}{2}y + \tfrac{1}{2}z = \tfrac{13}{2} \tag{3.20}$$
$$x + y + 4z = 12.$$

Now, using the first equation, we eliminate x from the second and third:

$$x + \tfrac{2}{3}y + \tfrac{1}{3}z = \tfrac{11}{3}$$
$$\tfrac{5}{6}y + \tfrac{1}{6}z = \tfrac{17}{6} \tag{3.21}$$
$$\tfrac{1}{3}y + \tfrac{11}{3}z = \tfrac{25}{3}$$

and

$$x + \tfrac{2}{3}y + \tfrac{1}{3}z = \tfrac{11}{3}$$
$$y + \tfrac{1}{5}z = \tfrac{17}{5} \tag{3.22}$$
$$y + 11z = 25.$$

Repeating the technique, we use the new second equation to eliminate y from the third equation:

$$x + \tfrac{2}{3}y + \tfrac{1}{3}z = \tfrac{11}{3}$$
$$y + \tfrac{1}{5}z = \tfrac{17}{5} \tag{3.23}$$
$$54z = 108,$$

or

$$z = 2.$$

Finally, working back up, we get

$$y + \tfrac{1}{5} \times 2 = \tfrac{17}{5},$$

or

$$y = 3.$$

Then with z and y determined,

$$x + \tfrac{2}{3} \times 3 + \tfrac{1}{3} \times 2 = \tfrac{11}{3},$$

and

$$x = 1.$$

The technique may not seem so elegant as Eq. (3.17), but it is well adapted to modern computing machines and is far faster than the time spent with determinants.

This Gauss technique may be used to convert a determinant into triangular form:

$$D = \begin{vmatrix} a_1 & b_1 & c_1 \\ 0 & b_2 & c_2 \\ 0 & 0 & c_3 \end{vmatrix}$$

for a third-order determinant whose elements are not to be confused with those in Eq. (3.3). In this form $D = a_1 b_2 c_3$. For an nth-order determinant the evaluation of the triangular form requires only $n - 1$ multiplications compared with the $n!$ required for the general case.

A variation of this progressive elimination is known as Gauss–Jordan elimination. We start as with the preceding Gauss elimination, but each new equation considered is used to eliminate a variable from *all* the other equations, not just those below it. If we had used this Gauss–Jordan elimination, Eq. (3.23) would become

$$x + \tfrac{1}{5}z = \tfrac{7}{5}$$
$$y + \tfrac{1}{5}z = \tfrac{17}{5} \tag{3.24}$$
$$z = 2,$$

using the second equation of Eq. (3.22) to eliminate y from both the first and third equations. Then the third equation of Eq. (3.24) is used to eliminate z from the first and second, giving

$$x = 1$$
$$y = 3 \tag{3.25}$$
$$z = 2.$$

We return to this Gauss–Jordan technique in Section 3.2 for inverting matrices.

Another technique suitable for computer use is the Gauss–Seidel iteration technique. Each technique has its advantages and disadvantages. The Gauss and Gauss–Jordan methods may have accuracy problems for large determinants. This is also a problem for matrix inversion (Section 3.2). The Gauss–Seidel method, as an iterative method, may have convergence problems. The IBM Scientific Subroutine Package (SSP) uses Gauss and Gauss–Jordan techniques. The Gauss–Seidel iterative method and the Gauss and Gauss–Jordan elimination methods are discussed in considerable detail by Ralston and Wilf and also by Pennington.[4] Computer codes in FORTRAN and other programming languages and extensive literature for the Gauss–Jordan elimination and others are also given by Press *et al.*[5]

[4] A. Ralston and H. Wilf, eds., *Mathematical Methods for Digital Computers*. New York: Wiley (1960); R. H. Pennington, *Introductory Computer Methods and Numerical Analysis*. New York: Macmillan (1970).

[5] W. H. Press, B. P. Flannery, S. A. Teukolsky, and W. T. Vetterling, *Numerical Recipes*, 2nd ed. Cambridge: Cambridge University Press (1992), Chapter 2.

EXERCISES

3.1.1   Evaluate the following determinants

(a) $\begin{vmatrix} 1 & 0 & 1 \\ 0 & 1 & 0 \\ 1 & 0 & 0 \end{vmatrix}$,   (b) $\begin{vmatrix} 1 & 2 & 0 \\ 3 & 1 & 2 \\ 0 & 3 & 1 \end{vmatrix}$,   (c) $\dfrac{1}{\sqrt{2}} \begin{vmatrix} 0 & \sqrt{3} & 0 & 0 \\ \sqrt{3} & 0 & 2 & 0 \\ 0 & 2 & 0 & \sqrt{3} \\ 0 & 0 & \sqrt{3} & 0 \end{vmatrix}$.

3.1.2   Test the set of linear homogeneous equations

$$x + 3y + 3z = 0,$$
$$x - y + z = 0,$$
$$2x + y + 3z = 0,$$

to see if it possesses a nontrivial solution.

3.1.3   Given the pair of equations

$$x + 2y = 3,$$
$$2x + 4y = 6,$$

(a) Show that the determinant of the coefficients vanishes.
(b) Show that the numerator determinants (Eq. (3.18)) also vanish.
(c) Find at least two solutions.

3.1.4   Express the *components* of $\mathbf{A} \times \mathbf{B}$ as 2×2 determinants. Show then that the dot product $\mathbf{A} \cdot (\mathbf{A} \times \mathbf{B})$ yields a Laplacian expansion of a 3×3 determinant. Finally, note that two rows of the 3×3 determinant are identical and hence $(\mathbf{A} \cdot (\mathbf{A} \times \mathbf{B})) = 0$.

3.1.5   If C_{ij} is the cofactor of element a_{ij} (formed by striking out the ith row and jth column and including a sign $(-1)^{i+j}$), show that

(a) $\sum_i a_{ij} C_{ij} = \sum_i a_{ji} C_{ji} = |A|$, where $|A|$ is the determinant with the elements a_{ij},

(b) $\sum_i a_{ij} C_{ik} = \sum_i a_{ji} C_{ki} = 0, j \neq k$.

3.1.6   A determinant with all elements of order unity may be surprisingly small. The Hilbert determinant $H_{ij} = (i + j - 1)^{-1}$, $i, j = 1, 2, \ldots, n$ is notorious for its small values.

(a) Calculate the value of the Hilbert determinants of order n for $n = 1, 2,$ and 3.
(b) If an appropriate subroutine is available, find the Hilbert determinants of order n for $n = 4, 5,$ and 6.

ANS.	n	Det(H_n)
	1	1.
	2	8.33333×10^{-2}
	3	4.62963×10^{-4}
	4	1.65344×10^{-7}
	5	3.74930×10^{-12}
	6	5.36730×10^{-18}

3.1.7 Solve the following set of linear simultaneous equations. Give the results to five decimal places.

$$1.0x_1 + 0.9x_2 + 0.8x_3 + 0.4x_4 + 0.1x_5 \qquad = 1.0$$
$$0.9x_1 + 1.0x_2 + 0.8x_3 + 0.5x_4 + 0.2x_5 + 0.1x_6 = 0.9$$
$$0.8x_1 + 0.8x_2 + 1.0x_3 + 0.7x_4 + 0.4x_5 + 0.2x_6 = 0.8$$
$$0.4x_1 + 0.5x_2 + 0.7x_3 + 1.0x_4 + 0.6x_5 + 0.3x_6 = 0.7$$
$$0.1x_1 + 0.2x_2 + 0.4x_3 + 0.6x_4 + 1.0x_5 + 0.5x_6 = 0.6$$
$$0.1x_2 + 0.2x_3 + 0.3x_4 + 0.5x_5 + 1.0x_6 = 0.5.$$

Note. These equations may also be solved by matrix inversion, Section 3.2.

3.2 MATRICES

Matrix analysis belongs to linear algebra because matrices are linear operators or maps such as rotations. Suppose, for instance, we rotate the cartesian coordinates of a two-dimensional space as in Section 1.2 so that, in vector notation,

$$\begin{pmatrix} x_1' \\ x_2' \end{pmatrix} = \begin{pmatrix} x_1 \cos \varphi + x_2 \sin \varphi \\ -x_2 \sin \varphi + x_2 \cos \varphi \end{pmatrix} = \left(\sum_j a_{ij} x_j \right). \tag{3.26}$$

We label the array of elements a_{ij} a 2×2 matrix A consisting of two rows and two columns and consider the vectors x, x' as 2×1 matrices. *We take the summation of products in Eq. (3.26) as a definition of matrix multiplication involving the scalar product of each row vector of A with the column vector x.* Thus, in matrix notation Eq. (3.26) becomes

$$x' = A x. \tag{3.27}$$

To extend this definition of multiplication of a matrix times a column vector to the product of two 2×2 matrices, let the coordinate rotation be followed by a second rotation given by matrix B such that

$$x'' = B x'. \tag{3.28}$$

In component form

$$x_i'' = \sum_j b_{ij} x_j' = \sum_j b_{ij} \sum_k a_{jk} x_k = \sum_k \left(\sum_j b_{ij} a_{jk} \right) x_k. \tag{3.29}$$

The summation over j is matrix multiplication defining a matrix $C = BA$ such that

$$x_i'' = \sum_k c_{ik} x_k, \tag{3.30}$$

or $x'' = Cx$ in matrix notation. Again, this definition involves the scalar products of row vectors of B with column vectors of A. This definition of matrix multiplication generalizes to $m \times n$ matrices and is found useful and indeed *this usefulness is the justification for its existence*. The physical interpretation is that the matrix product of the two matrices, BA, is the rotation that carries the unprimed system directly into the double-primed coordinate system. Before passing to formal definitions, the reader should note that operator A is described by its effect on the coordinates or basis vectors. The matrix elements a_{ij} constitute a *representation* of the operator, a representation that depends on the choice of a basis.

The special case where a matrix has one column and n rows is called a column vector, $|x\rangle$, with components x_i, $i = 1, 2, ..., n$. If A is an $n \times n$ matrix, $|x\rangle$ an n-component column vector, $A|x\rangle$ is defined as in Eqs. (3.27) and (3.26). Similarly, if a matrix has one row and n columns, it is called a row vector, $\langle x|$ with components x_i, $i = 1, 2, ..., n$. Clearly, $\langle x|$ results from $|x\rangle$ by interchanging rows and columns, a matrix operation called *transposition*, and for any matrix A, Ã is called[1] "A transpose" with matrix elements $(\tilde{A})_{ik} = A_{ki}$. Transposing a product of matrices AB reverses the order and gives B̃Ã; similarly $A|x\rangle$ transpose is $\langle x|\tilde{A}$. The scalar product takes the form $\langle x|y\rangle$.

BASIC DEFINITIONS

A matrix may be defined as a square or rectangular array of numbers or functions that obeys certain laws. This is a perfectly logical extension of familiar mathematical concepts. In arithmetic we deal with single numbers. In the theory of complex variables (Chapter 6) we deal with ordered pairs of numbers, $(1, 2) = 1 + 2i$, in which the ordering is important. We now consider numbers (or functions) ordered in a square or rectangular array. For convenience in later work the numbers are distinguished by two subscripts, the first indicating the row (horizontal) and the second indicating the column (vertical) in which the number appears. For instance, a_{13} is the matrix element in the first row, third column. Hence, if A is a matrix with m rows and n columns,

$$A = \begin{pmatrix} a_{11} & a_{12} & \cdots & a_{1n} \\ a_{21} & a_{33} & \cdots & a_{2n} \\ & \cdots & \cdots & \\ a_{m1} & a_{m2} & \cdots & a_{mn} \end{pmatrix}. \tag{3.31}$$

Perhaps the most important fact to note is that the elements a_{ij} are not combined with one another. A matrix is not a determinant. It is an ordered array of numbers, not a single number.

The matrix A, so far just an array of numbers, has the properties we assign to it. Literally, this means constructing a new form of mathematics. We postulate that matrices A, B, and C, with elements a_{ij}, b_{ij}, and c_{ij}, respectively, combine according to the following rules.

---

[1] Some texts denote A transpose by A^T.

EQUALITY

Matrix A = Matrix B if and only if $a_{ij} = b_{ij}$ for all values of i and j. This, of course, requires that A and B each be m by n arrays (m rows, n columns).

ADDITION

A + B = C if and only if $a_{ij} + b_{ij} = c_{ij}$ for all values of i and j, the elements combining according to the laws of ordinary algebra (or arithmetic if they are simple numbers). This means that A + B = B + A, commutation. Also, an associative law is satisfied (A + B) + C = A + (B + C). If all elements are zero, the matrix is called the null matrix and is denoted by O. For all A,

$$A + O = O + A = A,$$

with

$$O = \begin{pmatrix} 0 & 0 & 0 & \cdot & \cdot & \cdot \\ 0 & 0 & 0 & \cdot & \cdot & \cdot \\ 0 & 0 & 0 & \cdot & \cdot & \cdot \\ \cdot & \cdot & \cdot & \cdot & \cdot & \cdot \end{pmatrix}. \tag{3.32}$$

Such $m \times n$ matrices form a linear space with respect to addition and subtraction.

MULTIPLICATION (BY A SCALAR)

The multiplication of matrix A by the scalar quantity α is defined as

$$\alpha A = (\alpha A), \tag{3.33}$$

in which the elements of αA are αa_{ij}; that is, each element of matrix A is multiplied by the scalar factor. This is in striking contrast to the behavior of determinants in which the factor α multiplies only one column or one row and not every element of the entire determinant. A consequence of this scalar multiplication is that

$$\alpha A = A\alpha, \qquad \text{commutation.}$$

MULTIPLICATION (MATRIX MULTIPLICATION), INNER PRODUCT

$$AB = C \qquad \text{if and only if}^2 \; c_{ij} = \sum_k a_{ik} b_{kj}. \tag{3.34}$$

The ij element of C is formed as a scalar product of the ith row of A with the jth column of B (which demands that A have the same number of columns (n) as B has rows). The dummy index k takes on all values $1, 2, \ldots, n$ in succession, that is,

$$c_{ij} = a_{i1} b_{1j} + a_{i2} b_{2j} + a_{i3} b_{3j} \tag{3.35}$$

---

[2] Some authors follow the summation convention here (compare Section 2.6).

for $n = 3$. Obviously, the dummy index k may be replaced by any other symbol that is not already in use without altering Eq. (3.34). Perhaps the situation may be clarified by stating that Eq. (3.34) defines the method of combining certain matrices. This method of combination, to give it a label, is called matrix multiplication. To illustrate, consider two (so-called Pauli) matrices

$$\sigma_1 = \begin{pmatrix} 0 & 1 \\ 1 & 0 \end{pmatrix} \quad \text{and} \quad \sigma_3 = \begin{pmatrix} 1 & 0 \\ 0 & -1 \end{pmatrix}. \tag{3.36}$$

The $_{11}$ element of the product, $(\sigma_1\sigma_3)_{11}$ is given by the sum of the products of elements of the first *row* of σ_1 with the corresponding elements of the first *column* of σ_3:

$$\begin{pmatrix} 0 & 1 \\ 1 & 0 \end{pmatrix}\begin{pmatrix} 1 & 0 \\ 0 & -1 \end{pmatrix} \rightarrow 0 \cdot 1 + 1 \cdot 0 = 0.$$

Continuing, we have

$$\sigma_1\sigma_3 = \begin{pmatrix} 0 \cdot 1 + 1 \cdot 0 & 0 \cdot 0 + 1 \cdot (-1) \\ 1 \cdot 1 + 0 \cdot 0 & 1 \cdot 0 + 0 \cdot (-1) \end{pmatrix} = \begin{pmatrix} 0 & -1 \\ 1 & 0 \end{pmatrix}. \tag{3.37}$$

Here

$$(\sigma_1\sigma_3)_{ij} = \sigma_{1_{i1}}\sigma_{3_{1j}} + \sigma_{1_{i2}}\sigma_{3_{2j}}.$$

Direct application of the definition of matrix multiplication shows that

$$\sigma_3\sigma_1 = \begin{pmatrix} 0 & 1 \\ -1 & 0 \end{pmatrix} \tag{3.38}$$

and by Eq. (3.34)

$$\sigma_3\sigma_1 = -\sigma_1\sigma_3. \tag{3.39}$$

Except in special cases, matrix multiplication is not commutative.[3]

$$AB \neq BA. \tag{3.40}$$

However, from the definition of matrix multiplication we can show[4] that an associative law holds, $(AB)C = A(BC)$. There is also a distributive law, $A(B + C) = AB + AC$.

---

[3] Commutation or the lack of it is conveniently described by the commutator bracket symbol, $[A, B] = AB - BA$. Equation (3.40) becomes $[A, B] \neq 0$.

[4] The reader should note that the basic definitions of equality, addition, and multiplication are given in terms of the matrix elements, the a_{ij}'s, and so on. All our matrix operations can be carried out in terms of the matrix elements. However, we can also treat a matrix as a single algebraic operator, as in Eq. (3.40). Matrix elements and single operators each have their advantages as will be seen in the following section. We shall use both approaches.

The unit matrix 1 has elements δ_{ij}, Kronecker delta, and the property that $1A = A1 = A$ for all A,

$$1 = \begin{pmatrix} 1 & 0 & 0 & 0 & \cdot & \cdot & \cdot \\ 0 & 1 & 0 & 0 & \cdot & \cdot & \cdot \\ 0 & 0 & 1 & 0 & \cdot & \cdot & \cdot \\ 0 & 0 & 0 & 1 & \cdot & \cdot & \cdot \\ \cdot & \cdot & \cdot & \cdot & \cdot & \cdot & \cdot \end{pmatrix}. \tag{3.41}$$

It should be noted that it is possible for the product of two matrices to be the null matrix without either one being the null matrix. For example, if

$$A = \begin{pmatrix} 1 & 1 \\ 0 & 0 \end{pmatrix} \quad \text{and} \quad B = \begin{pmatrix} 1 & 0 \\ -1 & 0 \end{pmatrix},$$

AB = O. This differs from the multiplication of real or complex numbers which form a *field*, whereas the additive and multiplicative structure of matrices is called a *ring* by mathematicians. See also Exercise 3.2.6(a) from which it is evident that, if AB = 0, at least one of the matrices must have a zero determinant (i.e., be singular as defined after Eq. (3.50) in Section 3.2.

If A is an $n \times n$ matrix with determinant $|A| \neq 0$, then it has a unique inverse A^{-1} so that $AA^{-1} = A^{-1}A = 1$. If B is also an $n \times n$ matrix with inverse B^{-1}, then the product AB has the inverse

$$(AB)^{-1} = B^{-1}A^{-1} \tag{3.42}$$

because $ABB^{-1}A^{-1} = 1 = B^{-1}A^{-1}AB$ (see also Exercises 3.2.31 and 3.2.32).

The *product theorem* which says that the determinant of the product, $|AB|$, of two $n \times n$ matrices A and B is equal to the product of the determinants, $|A|\,|B|$, links matrices with determinants. To prove this consider the n column vectors $c_k = (\sum_j a_{ij} b_{jk},\ i = 1, 2, \ldots, n)$ of the product matrix $C = AB$ for $k = 1, 2, \ldots, n$. Each $c_k = \sum_{j_k} b_{j_k k} a_{j_k}$ is a sum of n column vectors $a_{j_k} = (a_{ij_k},\ i = 1, 2, \ldots, n)$. Note that we are now using a different product summation index j_k for each column c_k. Since any determinant $D(b_1 a_1 + b_2 a_2) = b_1 D(a_1) + b_2 D(a_2)$ is linear in its column vectors, we can pull out the summation sign in front of the determinant from each column vector in C together with the common column factor $b_{j_k k}$ so that

$$|C| = \sum_{j_k\text{'s}} b_{j_1 1} b_{j_2 2} \cdots b_{j_n n} \det(a_{j_1}, a_{j_2}, \ldots, a_{j_n}). \tag{3.43}$$

If we rearrange the column vectors a_{j_k} of the determinant factor in Eq. (3.43) in the proper order, then we can pull the common factor $\det(a_1, a_2, \ldots, a_n) = |A|$ in front of the n summation signs in Eq. (3.43). These column permutations generate just the right sign $\varepsilon_{j_1 j_2 \ldots j_n}$ to produce the expression in Eq. (3.8) for $|B|$ so that

$$|C| = |A| \sum_{j_k\text{'s}} \varepsilon_{j_1 j_2 \ldots j_n} b_{j_1 1} b_{j_2 2} \cdots b_{j_n n} = |A|\,|B|, \tag{3.44}$$

which proves the product theorem.

DIRECT PRODUCT

A second procedure for multiplying matrices, known as the *direct* tensor or Kronecker *product*, follows. If A is an $m \times m$ matrix and B an $n \times n$ matrix, then the direct product is

$$A \otimes B = C. \tag{3.45}$$

C is an $mn \times mn$ matrix with elements

$$C_{\alpha\beta} = A_{ij}B_{kl}, \tag{3.46}$$

with

$$\alpha = n(i-1) + k, \qquad \beta = n(j-1) + l.$$

For instance, if A and B are both 2×2 matrices,

$$A \otimes B = \begin{pmatrix} a_{11}B & a_{12}B \\ a_{21}B & a_{22}B \end{pmatrix}$$

$$= \begin{pmatrix} a_{11}b_{11} & a_{11}b_{12} & a_{12}b_{11} & a_{12}b_{12} \\ a_{11}b_{21} & a_{11}b_{22} & a_{12}b_{21} & a_{12}b_{22} \\ a_{21}b_{11} & a_{21}b_{12} & a_{22}b_{11} & a_{22}b_{12} \\ a_{21}b_{21} & a_{21}b_{22} & a_{22}b_{21} & a_{22}b_{22} \end{pmatrix}. \tag{3.47}$$

The direct product is associative but not commutative. As as example of the direct product, the Dirac matrices of Section 3.4 may be developed as direct products of the Pauli matrices and the unit matrix. Other examples appear in the construction of groups in group theory and in vector or Hilbert space in quantum theory.

The direct product defined here is sometimes called the *standard* form and denoted by $\otimes$. Three other types of direct products of matrices exist as mathematical possibilities or curiosities but have little or no application in mathematical physics.

DIAGONAL MATRICES

An important special type of matrix is the square matrix in which all the nondiagonal elements are zero. Specifically, if a 3×3 matrix A is diagonal,

$$A = \begin{pmatrix} a_{11} & 0 & 0 \\ 0 & a_{22} & 0 \\ 0 & 0 & a_{33} \end{pmatrix}.$$

A physical interpretation of such diagonal matrices and the method of reducing matrices to this diagonal form are considered in Section 3.5. Here we simply note a significant property of diagonal matrices—multiplication of diagonal matrices is commutative,

$$AB = BA, \qquad \text{if A and B are each diagonal.}$$

TRACE

In any square matrix the sum of the diagonal elements is called the *trace*. Clearly the trace is a linear operation:

$$\text{trace}(A - B) = \text{trace}(A) - \text{trace}(B).$$

One of its interesting and useful properties is that the trace of a product of two matrices A and B is independent of the order of multiplication:

$$\text{trace}(AB) = \sum_i (AB)_{ii} = \sum_i \sum_j a_{ij} b_{ji}$$

$$= \sum_j \sum_i b_{ji} a_{ij} = \sum_j (BA)_{jj}$$

$$= \text{trace}(BA). \tag{3.48}$$

This holds even though $AB \neq BA$. Equation (3.48) means that the trace of any commutator, $[A, B] = AB - BA$, is zero. From Eq. (3.48) we obtain

$$\text{trace}(ABC) = \text{trace}(BCA) = \text{trace}(CAB),$$

which shows that the trace is invariant under cyclic permutation of the matrices in a product.

For a real symmetric or a complex Hermitian matrix (see Section 3.5) the trace is the sum, and the determinant the product, of its eigenvalues, and both are coefficients of the characteristic polynomial. In Exercise 3.4.23 the operation of taking the trace selects one term out of a sum of 16 terms. The trace will serve a similar function relative to matrices as orthogonality serves for vectors and functions.

In terms of tensors (Section 2.7) the trace is a contraction and like the contracted second-rank tensor is a scalar (invariant).

Matrices are used extensively to represent the elements of groups (compare Exercise 3.2.7 and Chapter 4). The trace of the matrix representing the group element is known in group theory as the *character*. The reason for the special name and special attention is that while the matrices may vary the trace or character remains invariant (compare Exercise 3.3.9).

MATRIX INVERSION

At the beginning of this section matrix A is introduced as the representation of an operator that (linearly) transforms the coordinate axes. A rotation would be one example of such a linear transformation. Now we look for the inverse transformation A^{-1} that will restore the original coordinate axes. This means, as either a matrix or an operator equation,[5]

$$AA^{-1} = A^{-1}A = 1. \tag{3.49}$$

---

[5] Here and throughout this chapter our matrices have finite rank. If A is an infinite rank matrix ($n \times n$ with $n \to \infty$), then life is more difficult. For A^{-1} to be the inverse we must demand that *both*

$$AA^{-1} = 1 \quad and \quad A^{-1}A = 1.$$

One relation no longer implies the other.

From Exercise 3.2.32

$$a_{ij}^{-1} = \frac{C_{ji}}{|A|},\qquad(3.50)$$

with the assumption that the determinant of A ($|A|$) $\neq$ 0. If it is zero, we label A singular. No inverse exists. As explained at the end of Section 3.1, this determinant form is *totally unsuited for numerical work* with large matrices.

There is a wide variety of alternative techniques. One of the best and most commonly used is the Gauss–Jordan matrix inversion technique. The theory is based on the results of Exercises 3.2.34 and 3.2.35, which show that there exist matrices M_L such that the product $M_L A$ will be A but with

    a. one row multiplied by a constant, or
    b. one row replaced by the original row minus a multiple of another row, or
    c. rows interchanged.

Other matrices M_R operating on the right ($A M_R$) can carry out the same operations on the *columns* of A.

This means that the matrix rows and columns may be altered (by matrix multiplication) as though we were dealing with determinants, so we can apply the Gauss–Jordan elimination techniques of Section 3.1 to the matrix elements. Hence there exists a matrix M_L (or M_R) such that[6]

$$M_L A = 1.\qquad(3.51)$$

The $M_L = A^{-1}$. We determine M_L by carrying out the identical elimination operations on the unit matrix. Then

$$M_L 1 = M_L.\qquad(3.52)$$

To clarify this, we consider a specific example.

Example 3.2.1: Gauss–Jordan Matrix Inversion

We want to invert the matrix

$$A = \begin{pmatrix} 3 & 2 & 1 \\ 2 & 3 & 1 \\ 1 & 1 & 4 \end{pmatrix}.\qquad(3.53)$$

For convenience we write A and 1 side by side and carry out the identical operations on each:

$$\begin{pmatrix} 3 & 2 & 1 \\ 2 & 3 & 1 \\ 1 & 1 & 4 \end{pmatrix} \quad\text{and}\quad \begin{pmatrix} 1 & 0 & 0 \\ 0 & 1 & 0 \\ 0 & 0 & 1 \end{pmatrix}.\qquad(3.54)$$

---

[6] Remember that det(A) $\neq$ 0.

To be systematic, we multiply each row to get $a_{k1} = 1$,

$$\begin{pmatrix} 1 & \frac{2}{3} & \frac{1}{3} \\ 1 & \frac{3}{2} & \frac{1}{2} \\ 1 & 1 & 4 \end{pmatrix} \quad \text{and} \quad \begin{pmatrix} \frac{1}{3} & 0 & 0 \\ 0 & \frac{1}{2} & 0 \\ 0 & 0 & 1 \end{pmatrix}. \tag{3.55}$$

Subtracting the first row from the second and third, we obtain

$$\begin{pmatrix} 1 & \frac{2}{3} & \frac{1}{3} \\ 0 & \frac{5}{6} & \frac{1}{6} \\ 0 & \frac{1}{3} & \frac{11}{3} \end{pmatrix} \quad \text{and} \quad \begin{pmatrix} \frac{1}{3} & 0 & 0 \\ -\frac{1}{3} & \frac{1}{2} & 0 \\ -\frac{1}{3} & 0 & 1 \end{pmatrix}. \tag{3.56}$$

Then we divide the second row (of *both* matrices) by $\frac{5}{6}$ and subtract $\frac{2}{3}$ times it from the first row, and $\frac{1}{3}$ times it from the third row. The results for both matrices are

$$\begin{pmatrix} 1 & 0 & \frac{1}{5} \\ 0 & 1 & \frac{1}{5} \\ 0 & 0 & \frac{18}{5} \end{pmatrix} \quad \text{and} \quad \begin{pmatrix} \frac{3}{5} & -\frac{2}{5} & 0 \\ -\frac{2}{5} & \frac{3}{5} & 0 \\ -\frac{1}{5} & -\frac{1}{5} & 1 \end{pmatrix}. \tag{3.57}$$

We divide the third row (of *both* matrices) by $\frac{18}{5}$. Then as the last step $\frac{1}{5}$ times the third row is subtacted from each of the first two rows (of both matrices). Our final pair is

$$\begin{pmatrix} 1 & 0 & 0 \\ 0 & 1 & 0 \\ 0 & 0 & 1 \end{pmatrix} \quad \text{and} \quad \begin{pmatrix} \frac{11}{8} & -\frac{7}{18} & -\frac{1}{18} \\ -\frac{7}{18} & \frac{11}{18} & -\frac{1}{18} \\ -\frac{1}{18} & -\frac{1}{18} & \frac{5}{18} \end{pmatrix}. \tag{3.58}$$

The check is to multiply the original A by the calculated A^{-1} to see if we really do get the unit matrix 1.

As with the Gauss–Jordan solution of simultaneous linear algebraic equations, this technique is well adapted to large computing machines. Indeed, this Gauss–Jordan matrix inversion technique will probably be available in the program library as a subroutine (see Sections 2.3 and 2.4 of Press *et al.*, loc. cit.).

EXERCISES

3.2.1   Show that matrix multiplication is associative, $(AB)C = A(BC)$.

3.2.2   Show that

$$(A + B)(A - B) = A^2 - B^2$$

if and only if A and B commute,

$$[A, B] = 0.$$

3.2.3   Show that matrix A is a *linear operator* by showing that
$$A(c_1 \mathbf{r}_1 + c_2 \mathbf{r}_2) = c_1 A\mathbf{r}_1 + c_2 A\mathbf{r}_2.$$
It can be shown that an $n \times n$ matrix is the *most general* linear operator in an n-dimensional vector space. This means that every linear operator in this n-dimensional vector space is equivalent to a matrix.

3.2.4   (a) Complex numbers, $a + ib$, with a and b real, may be represented by (or, are isomorphic with) 2×2 matrices:
$$a + ib \leftrightarrow \begin{pmatrix} a & b \\ -b & a \end{pmatrix}.$$
Show that this matrix representation is valid for (i) addition and (ii) multiplication.
(b) Find the matrix corresponding to $(a + ib)^{-1}$.

3.2.5   If A is an $n \times n$ matrix, show that
$$\det(-A) = (-1)^n \det A.$$

3.2.6   (a) The matrix equation $A^2 = 0$ does not imply $A = 0$. Show that the most general 2×2 matrix whose square is zero may be written as
$$\begin{pmatrix} ab & b^2 \\ -a^2 & -ab \end{pmatrix},$$
where a and b are real or complex numbers.
(b) If $C = A + B$, in general
$$\det C \neq \det A + \det B.$$
Construct a specific numerical example to illustrate this inequality.

3.2.7   Given the three matrices
$$A = \begin{pmatrix} -1 & 0 \\ 0 & -1 \end{pmatrix}, \quad B = \begin{pmatrix} 0 & 1 \\ 1 & 0 \end{pmatrix}, \quad \text{and} \quad C = \begin{pmatrix} 0 & -1 \\ -1 & 0 \end{pmatrix},$$
find all possible products of A, B, and C, two at a time, including squares. Express your answers in terms of A, B, and C, and 1, the unit matrix. These three matrices together with the unit matrix form a representation of a mathematical group, the *vierergruppe* (see Chapter 4).

3.2.8   Given
$$K = \begin{pmatrix} 0 & 0 & i \\ -i & 0 & 0 \\ 0 & -1 & 0 \end{pmatrix},$$
show that
$$K^n = KKK \cdots (n \text{ factors}) = 1$$
(with the proper choice of n, $n \neq 0$).

3.2.9   Verify the Jacobi identity
$$[A, [B, C]] = [B, [A, C]] - [C, [A, B]].$$
This is useful in matrix descriptions of elementary particles (see Eq. (4.16)). As a mnemonic aid, the reader might note that the Jacobi identity has the same form as the *BAC-CAB* rule of Section 1.5.

3.2.10 Show that the matrices

$$A = \begin{pmatrix} 0 & 1 & 0 \\ 0 & 0 & 0 \\ 0 & 0 & 0 \end{pmatrix}, \quad B = \begin{pmatrix} 0 & 0 & 0 \\ 0 & 0 & 1 \\ 0 & 0 & 0 \end{pmatrix}, \quad C = \begin{pmatrix} 0 & 0 & 1 \\ 0 & 0 & 0 \\ 0 & 0 & 0 \end{pmatrix}$$

satisfy the communtation relations

$$[A, B] = C, \quad [A, C] = 0, \quad \text{and} \quad [B, C] = 0.$$

3.2.11 Let

$$i = \begin{pmatrix} 0 & 1 & 0 & 0 \\ -1 & 0 & 0 & 0 \\ 0 & 0 & 0 & 1 \\ 0 & 0 & -1 & 0 \end{pmatrix}, \quad j = \begin{pmatrix} 0 & 0 & 0 & -1 \\ 0 & 0 & -1 & 0 \\ 0 & 1 & 0 & 0 \\ 1 & 0 & 0 & 0 \end{pmatrix},$$

and

$$k = \begin{pmatrix} 0 & 0 & -1 & 0 \\ 0 & 0 & 0 & 1 \\ 1 & 0 & 0 & 0 \\ 0 & -1 & 0 & 0 \end{pmatrix}.$$

Show that

(a) $i^2 = j^2 = k^2 = -1$, where 1 is the unit matrix.
(b) $ij = -ji = k$,
    $jk = -kj = i$,
    $ki = -ik = j$.

These three matrices (i, j, and k) plus the unit matrix 1 form a basis for *quaternions*. An alternate basis is provided by the four 2×2 matrices, $i\sigma_1$, $i\sigma_2$, $-i\sigma_3$, and 1, where the σ's are the Pauli spin matrices of Exercise 3.2.13.

3.2.12 A matrix with elements $a_{ij} = 0$ for $j < i$ may be called upper right triangular. The elements in the lower left (below and to the left of the main diagonal) vanish. Examples are the matrices in Chapters 12 and 13 relating power series and eigenfunction expansions.
Show that the product of two upper right triangular matrices is an upper right triangular matrix.

3.2.13 The three Pauli spin matrices are

$$\sigma_1 = \begin{pmatrix} 0 & 1 \\ 1 & 0 \end{pmatrix}, \quad \sigma_2 = \begin{pmatrix} 0 & -i \\ i & 0 \end{pmatrix}, \quad \text{and} \quad \sigma_3 = \begin{pmatrix} 1 & 0 \\ 0 & -1 \end{pmatrix}.$$

Show that

(a) $\sigma_i^2 = 1$,
(b) $\sigma_i\sigma_j = i\sigma_k$, $(i, j, k) = (1, 2, 3), (2, 3, 1), (3, 1, 2)$ (cyclic permutation),
(c) $\sigma_i\sigma_j + \sigma_j\sigma_i = 2\delta_{ij}1$.

These matrices were used by Pauli in the nonrelativistic theory of electron spin.

3.2.14 Using the Pauli σ's of Exercise 3.2.13, show that

$$(\sigma \cdot a)(\sigma \cdot b) = a \cdot b1 + i\sigma \cdot (a \times b).$$

Here

$$\sigma \equiv \hat{x}\sigma_x + \hat{y}\sigma_y + \hat{z}\sigma_z$$

and **a** and **b** are ordinary vectors.

3.2.15   One description of spin 1 particles uses the matrices

$$M_x = \frac{1}{\sqrt{2}}\begin{pmatrix} 0 & 1 & 0 \\ 1 & 0 & 1 \\ 0 & 1 & 0 \end{pmatrix}, \qquad M_y = \frac{1}{\sqrt{2}}\begin{pmatrix} 0 & -i & 0 \\ i & 0 & -i \\ 0 & i & 0 \end{pmatrix},$$

and

$$M_z = \begin{pmatrix} 1 & 0 & 0 \\ 0 & 0 & 0 \\ 0 & 0 & -1 \end{pmatrix}.$$

Show that

(a) $[M_x, M_y] = iM_z$, and so on[7] (cyclic permutation of indices).
   Using the Levi–Civita symbol of Section 3.4, we may write

$$[M_i, M_j] = i\varepsilon_{ijk}M_k.$$

(b) $M^2 \equiv M^2_x + M^2_y + M^2_z = 2 \cdot 1$,
   where 1 is the unit matrix.

(c) $[M^2, M_i] = 0$,
   $[M_z, L^+] = L^+$,
   $[L^+, L^-] = 2M_z$,
   where
   $L^+ \equiv M_x + iM_y$,
   $L^- \equiv M_x - iM_y$.

3.2.16   Repeat Exercise 3.2.15 using an alternate representation,

$$M_x = \begin{pmatrix} 0 & 0 & 0 \\ 0 & 0 & -i \\ 0 & i & 0 \end{pmatrix}, \qquad M_y = \begin{pmatrix} 0 & 0 & i \\ 0 & 0 & 0 \\ -i & 0 & 0 \end{pmatrix},$$

and

$$M_z = \begin{pmatrix} 0 & -i & 0 \\ i & 0 & 0 \\ 0 & 0 & 0 \end{pmatrix}.$$

In Chapter 4 these matrices appear as the *generators* of the rotation matrices.

3.2.17   Show that the matrix-vector equation

$$\left(M \cdot \nabla + 1\frac{1}{c}\frac{\partial}{\partial t}\right)\psi = 0$$

reproduces Maxwell's equations in vacuum. Here ψ is a column vector with components $\psi_j = B_j - iE_j/c, j = x, y, z$. M is a vector whose elements are the angular momentum *matrices* of Exercise 3.2.16. Note that $\varepsilon_0\mu_0 = 1/c^2$. From Exercise 3.2.15(b)

$$M^2\psi = 2\psi.$$

[7] $[A, B] = AB - BA$.

A comparison with the Dirac relativistic electron equation suggests that the "particle" of electromagnetic radiation, the photon, has zero rest mass and a spin of 1 (in units of $\hbar$).

3.2.18 Repeat Exercise 3.2.15, using the matrices for a spin of $\frac{3}{2}$,

$$M_x = \frac{1}{2}\begin{pmatrix} 0 & \sqrt{3} & 0 & 0 \\ \sqrt{3} & 0 & 2 & 0 \\ 0 & 2 & 0 & \sqrt{3} \\ 0 & 0 & \sqrt{3} & 0 \end{pmatrix}, \quad M_y = \frac{i}{2}\begin{pmatrix} 0 & -\sqrt{3} & 0 & 0 \\ \sqrt{3} & 0 & -2 & 0 \\ 0 & 2 & 0 & -\sqrt{3} \\ 0 & 0 & \sqrt{3} & 0 \end{pmatrix},$$

and

$$M_z = \frac{1}{2}\begin{pmatrix} 3 & 0 & 0 & 0 \\ 0 & 1 & 0 & 0 \\ 0 & 0 & -1 & 0 \\ 0 & 0 & 0 & -3 \end{pmatrix}.$$

3.2.19 An operator P commutes with J_x and J_y, the x and y components of an angular momentum operator. Show that P commutes with the third component of angular momentum; that is,

$$[P, J_z] = 0.$$

Hint. The angular momentum components must satisfy the commutation relation of Exercise 3.2.15(a).

3.2.20 The L^+ and L^- matrices of Exercise 3.2.15 are "ladder operators" (see Chapter 4): L^+ operating on a system of spin projection m will raise the spin projection to $m + 1$ if m is below its maximum. L^+ operating on m_{max} yields zero. L^- reduces the spin projection in unit steps in a similar fashion. Dividing by $\sqrt{2}$, we have

$$L^+ = \begin{pmatrix} 0 & 1 & 0 \\ 0 & 0 & 1 \\ 0 & 0 & 0 \end{pmatrix}, \quad L^- = \begin{pmatrix} 0 & 0 & 0 \\ 1 & 0 & 0 \\ 0 & 1 & 0 \end{pmatrix}.$$

Show that

$$L^+|-1\rangle = |0\rangle, \quad L^-|-1\rangle = \text{null column vector,}$$
$$L^+|0\rangle = |1\rangle, \quad L^-|0\rangle = |-1\rangle,$$
$$L^+|1\rangle = \text{null column vector,} \quad L^-|1\rangle = |0\rangle,$$

where

$$|-1\rangle = \begin{pmatrix} 0 \\ 0 \\ 1 \end{pmatrix}, \quad |0\rangle = \begin{pmatrix} 0 \\ 1 \\ 0 \end{pmatrix}, \quad \text{and} \quad |1\rangle = \begin{pmatrix} 1 \\ 0 \\ 0 \end{pmatrix}$$

representing states of spin projection -1, 0, and 1, respectively.

Note. Differential operator analogs of these ladder operators appear in Exercise 12.6.7.

3.2.21 Vectors **A** and **B** are related by the tensor **T**

$$\mathbf{B} = \mathbf{TA}.$$

Given **A** and **B** show that there is *no unique solution* for the components of **T**. This is why vector division **B**/**A** is undefined (apart from the special case of **A** and **B** parallel and **T** then a scalar).

3.2.22   We might ask for a vector $\mathbf{A}^{-1}$, an inverse of a given vector $\mathbf{A}$ in the sense that

$$\mathbf{A} \cdot \mathbf{A}^{-1} = \mathbf{A}^{-1} \cdot \mathbf{A} = 1.$$

Show that this relation does not suffice to define $\mathbf{A}^{-1}$ uniquely. $\mathbf{A}$ has literally an infinite number of inverses.

3.2.23   If A is a diagonal, with all diagonal elements different, and A and B commute, show that B is diagonal.

3.2.24   If A and B are diagonal, show that A and B commute.

3.2.25   Show that trace(ABC) = trace(CBA) if any two of the three matrices commute.

3.2.26   Angular momentum matrices satisfy a commutation relation

$$[\mathsf{M}_i, \mathsf{M}_j] = i\mathsf{M}_k, \qquad i, j, k \text{ cyclic.}$$

Show that the trace of each angular momentum matrix vanishes.

3.2.27   (a) The operator Tr replaces a matrix A by its trace; that is,

$$\mathrm{Tr}(A) = \mathrm{trace}(A) = \sum_i a_{ii}.$$

Show that Tr is a *linear* operator.
(b) The operator det replaces a matrix A by its determinant; that is,

$$\det(A) = \text{determinant of A.}$$

Show that det is *not* a linear operator.

3.2.28   A and B anticommute. Also, $A^2 = 1$, $B^2 = 1$. Show that trace(A) = trace(B) = 0.
Note. The Pauli and Dirac (Section 3.4) matrices are specific examples.

3.2.29   With $|x\rangle$ an N-dimensional column vector and $\langle y|$ an N-dimensional row vector, show that

$$\mathrm{trace}(|x\rangle\langle y|) = \langle y|x\rangle.$$

Note. $|x\rangle\langle y|$ means column vector $|x\rangle$ multiplying row vector $\langle y|$. The result is a square matrix $N \times N$.

3.2.30   (a) If two nonsingular matrices anticommute, show that the trace of each one is zero. (Nonsingular means that the determinant of the matrix elements $\neq 0$.)
(b) For the conditions of part (a) to hold A and B must be $n \times n$ matrices with n even. Show that if n is *odd* a contradiction results.

3.2.31   If a matrix has an inverse, show that the inverse is unique.

3.2.32   If A^{-1} has elements

$$a_{ij}^{-1} = \frac{C_{ji}}{|A|},$$

where C_{ji} is the jith cofactor of $|A|$, show that

$$A^{-1}A = 1.$$

Hence A^{-1} is the inverse of A (if $|A| \neq 0$).
Note. In numerical work it sometimes happens that $|A|$ is almost equal to zero. Then there is trouble.

3.2.33   Show that det $A^{-1} = (\det A)^{-1}$.
Hint. Apply Exercise 3.2.6.
Note. If det A is zero, then A has no inverse. A is singular.

3.2.34   Find the matrices M_L such that the product $M_L A$ will be A but with:
(a) the ith row multiplied by a constant k $(a_{ij} \rightarrow k a_{ij}, j = 1, 2, 3, \ldots)$;
(b) the ith row replaced by the original ith row minus a multiple of the mth row $(a_{ij} \rightarrow a_{ij} - K a_{mj}, j = 1, 2, 3, \ldots)$;
(c) the ith and mth rows interchanged $(a_{ij} \rightarrow a_{mj}, a_{mj} \rightarrow a_{ij}, j = 1, 2, 3, \ldots)$.

3.2.35   Find the matrices M_R such that the product $A M_R$ will be A but with:
(a) the ith column multiplied by a constant k $(a_{ji} \rightarrow k a_{ji}, j = 1, 2, 3, \ldots)$;
(b) the ith column replaced by the original ith column minus a multiple of the mth column $(a_{ji} \rightarrow a_{ji} - k a_{jm}, j = 1, 2, 3, \ldots)$;
(c) the ith and mth columns interchanged $(a_{ji} \rightarrow a_{jm}, a_{jm} \rightarrow a_{ji}, j = 1, 2, 3, \ldots)$.

3.2.36   Find the inverse of
$$A = \begin{pmatrix} 3 & 2 & 1 \\ 2 & 2 & 1 \\ 1 & 1 & 4 \end{pmatrix}.$$

3.2.37   (a) Rewrite Eq. (2.4) of Chapter 2 (and the corresponding equations for dy and dz) as a single matrix equation
$$|dx_k\rangle = J|dq_j\rangle.$$
J is a matrix of derivatives, the Jacobian matrix. Show that
$$\langle dx_k | dx_k \rangle = \langle dq_i | G | dq_j \rangle$$
with the metric (matrix) G having elements g_{ij} given by Eq. (2.6).
(b) Show that
$$\det(J)\, dq_1\, dq_2\, dq_3 = dx\, dy\, dz.$$
Det(J) is the usual Jacobian.

3.2.38   Matrices are far too useful to remain the exclusive property of physicists. They may appear wherever there are linear relations. For instance, in a study of population movement the initial fraction of a fixed population in each of n areas (or industries or religions, etc.) is represented by an n-component column vector **P**. The movement of people from one area to another in a given time is described by an $n \times n$ (stochastic) matrix T. Here T_{ij} is the fraction of the popuation in the jth area that moves to the ith area. (Those not moving are covered by $i = j$.) With **P** describing the initial population distribution, the final population distribution is given by the matrix equation **TP** = **Q**.

From its definition $\sum_{i=1}^{n} P_i = 1$.

(a) Show that conservation of people requires that
$$\sum_{i=1}^{n} T_{ij} = 1, \quad j = 1, 2, \ldots, n.$$

(b) Prove that

$$\sum_{i=1}^{n} Q_i = 1$$

continues the conservation of people.

3.2.39 Given a 6×6 matrix A with elements $a_{ij} = 0.5^{|i-j|}$, $i = 0, 1, 2, \ldots, 5$; $j = 0, 1, 2, \ldots, 5$. Find A^{-1}. List the matrix elements a_{pq}^{-1} to five decimal places.

$$ANS. \quad A^{-1} = \frac{1}{3} \begin{pmatrix} 4 & -2 & 0 & 0 & 0 & 0 \\ -2 & 5 & -2 & 0 & 0 & 0 \\ 0 & -2 & 5 & -2 & 0 & 0 \\ 0 & 0 & -2 & 5 & -2 & 0 \\ 0 & 0 & 0 & -2 & 5 & -2 \\ 0 & 0 & 0 & 0 & -2 & 4 \end{pmatrix}.$$

3.2.40 Exercise 3.1.7 may be written in matrix form

$$AX = C.$$

Fine A^{-1} and calculate X as $A^{-1}C$.

3.2.41 (a) Write a *subroutine* that will multiply *complex* matrices. Assume that the complex matrices are in a general rectangular form.
(b) Test your subroutine by multiplying pairs of the Dirac 4×4 matrices, Section 3.4.

3.2.42 (a) Write a subroutine that will call the complex matrix multiplication subroutine of Exercise 3.2.41 and will calculate the commutator bracket of two complex matrices.
(b) Test your complex commutator bracket subroutine with the matrices of Exercise 3.2.16.

3.2.43 Interpolating polynomial is the name given to the $(n - 1)$-degree polynomial determined by (and passing through) n points, (x_i, y_i) with all the x_i's distinct. This interpolating polynomial forms the basis for the numerical quadrature developed in Appendix 2.

(a) Show that the requirement that an $(n - 1)$-degree polynomial in x passes through each of the n points (x_i, y_i) with all x_i distinct leads to n simultaneous equations of the form

$$\sum_{j=0}^{n-1} a_j x_i^j = y_i, \qquad i = 1, 2, \ldots, n.$$

(b) Write a computer program that will read in n data points and return the n coefficients a_j. Use a subroutine to solve the simultaneous equations if such a subroutine is available.
(c) Rewrite the set of simultaneous equations as a matrix equation

$$XA = Y.$$

(d) Repeat the computer calculation of part (b) but this time solve for vector A by inverting matrix X (again, using a subroutine).

3.2.44   A calculation of the values of electrostatic potential inside a cylinder leads to

$$V(0.0) = 52.640 \qquad V(0.6) = 25.844$$
$$V(0.2) = 48.292 \qquad V(0.8) = 12.648$$
$$V(0.4) = 38.270 \qquad V(1.0) = 0.0.$$

The problem is to determine the values of the argument for which $V = 10$, 20, 30, 40, and 50. Express $V(x)$ as a series $\sum_{n=0}^{5} a_{2n} x^{2n}$. (Symmetry requirements in the original problem require that $V(x)$ be an even function of x.)
Determine the coefficients a_{2n}. With $V(x)$ now a known function of x, find
the root of $V(x) - 10 = 0$, $0 \le x \le 1$. Repeat for $V(x) - 20$, and so on.

$$\begin{aligned} ANS. \quad a_0 &= 52.640 \\ a_2 &= -117.676 \\ V(0.6851) &= 20. \end{aligned}$$

3.3 ORTHOGONAL MATRICES

Ordinary three-dimensional space may be described with the cartesian coordinates (x_1, x_2, x_3). We consider a second set of cartesian coordinates (x'_1, x'_2, x'_3) whose origin and handedness coincides with that of the first set but whose orientation is different (Fig. 3.1). We can say that the primed coordinate axes have been *rotated* relative to the initial, unprimed coordinate axes. Since

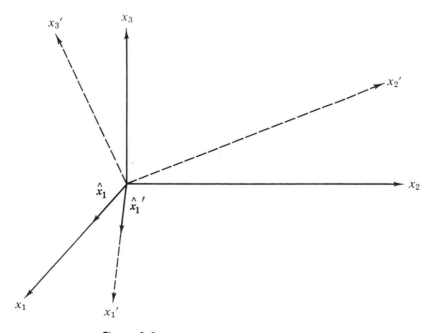

Figure 3.1   Cartesian coordinate systems.

this rotation is a *linear* operation, we expect a matrix equation relating the primed basis to the unprimed basis.

This section repeats portions of Chapters 1 and 2 in a slightly different context and with a different emphasis. Previously, attention was focused on the vector or tensor. In the case of the tensor, transformation properties were strongly stressed and were critical. Here emphasis is placed on the description of the coordinate rotation itself—the matrix. Transformation properties, the behavior of the matrix when the basis is changed, appear at the end of this section. Sections 3.4 and 3.5 continue with transformation properties in complex vector spaces.

DIRECTION COSINES

A unit vector along the x_1'-axis ($\hat{\mathbf{x}}_1'$) may be resolved into components along the x_1-, x_2-, and x_3-axes by the usual projection technique.

$$\hat{\mathbf{x}}_1' = \hat{\mathbf{x}}_1 \cos(x_1', x_1) + \hat{\mathbf{x}}_2 \cos(x_2', x_2) + \hat{\mathbf{x}}_3 \cos(x_3', x_3). \tag{3.59}$$

Equation (3.59) is a specific example of the linear relations discussed at the beginning of Section 3.2.

For convenience these cosines, which are the direction cosines, are labeled

$$\cos(x_1', x_1) = \hat{\mathbf{x}}_1' \cdot \hat{\mathbf{x}}_1 = a_{11},$$

$$\cos(x_1', x_2) = \hat{\mathbf{x}}_1' \cdot \hat{\mathbf{x}}_2 = a_{12}, \tag{3.60}$$

$$\cos(x_1', x_3) = \hat{\mathbf{x}}_1' \cdot \hat{\mathbf{x}}_3 = a_{13}.$$

Continuing, we have

$$\cos(x_2', x_1) = \hat{\mathbf{x}}_2' \cdot \hat{\mathbf{x}}_1 = a_{21}, \qquad (a_{21} \neq a_{12}),$$

$$\cos(x_2', x_2) = \hat{\mathbf{x}}_2' \cdot \hat{\mathbf{x}}_2 = a_{22}, \qquad \text{and so on.} \tag{3.61}$$

Now Eq. (3.59) may be rewritten

$$\hat{\mathbf{x}}_1' = \hat{\mathbf{x}}_1 a_{11} + \hat{\mathbf{x}}_2 a_{12} + \hat{\mathbf{x}}_3 a_{13}$$

and also

$$\hat{\mathbf{x}}_2' = \hat{\mathbf{x}}_1 a_{21} + \hat{\mathbf{x}}_2 a_{22} + \hat{\mathbf{x}}_3 a_{23},$$

$$\hat{\mathbf{x}}_3' = \hat{\mathbf{x}}_1 a_{31} + \hat{\mathbf{x}}_2 a_{32} + \hat{\mathbf{x}}_3 a_{33}. \tag{3.62}$$

We may also go the other way by resolving $\hat{\mathbf{x}}_1$, $\hat{\mathbf{x}}_2$, and $\hat{\mathbf{x}}_3$ into components in the primed system. Then

$$\hat{\mathbf{x}}_1 = \hat{\mathbf{x}}_1' a_{11} + \hat{\mathbf{x}}_2' a_{21} + \hat{\mathbf{x}}_3' a_{31},$$

$$\hat{\mathbf{x}}_2 = \hat{\mathbf{x}}_1' a_{12} + \hat{\mathbf{x}}_2' a_{22} + \hat{\mathbf{x}}_3' a_{32}, \tag{3.63}$$

$$\hat{\mathbf{x}}_3 = \hat{\mathbf{x}}_1' a_{13} + \hat{\mathbf{x}}_2' a_{23} + \hat{\mathbf{x}}_3' a_{33}.$$

Associating $\hat{\mathbf{x}}_1$ and $\hat{\mathbf{x}}_1'$ with the subscript 1, $\hat{\mathbf{x}}_2$ and $\hat{\mathbf{x}}_2'$ with the subscript 2, $\hat{\mathbf{x}}_3$ and $\hat{\mathbf{x}}_3'$ with the subscript 3, we see that in each case the first subscript of a_{ij}

refers to the primed unit vector $(\hat{\mathbf{x}}_1', \hat{\mathbf{x}}_2', \hat{\mathbf{x}}_3')$, whereas the second subscript refers to the unprimed unit vector $(\hat{\mathbf{x}}_1, \hat{\mathbf{x}}_2, \hat{\mathbf{x}}_3)$.

APPLICATIONS TO VECTORS

If we consider a vector whose components are functions of the position in space, then

$$
\begin{aligned}
\mathbf{V}(x_1, x_2, x_3) &= \hat{\mathbf{x}}_1 V_1 + \hat{\mathbf{x}}_2 V_2 + \hat{\mathbf{x}}_3 V_3 \\
&= \mathbf{V}'(x_1', x_2', x_3') = \hat{\mathbf{x}}_1' V_1' + \hat{\mathbf{x}}_2' V_2' + \hat{\mathbf{x}}_3' V_3',
\end{aligned}
\tag{3.64}
$$

since the point may be given both by the coordinates (x_1, x_2, x_3) and the coordinates (x_1', x_2', x_3'). Note that $\mathbf{V}$ and $\mathbf{V}'$ are geometrically the same vector (but with different components). The coordinate axes are being rotated; the vector stays fixed. Using Eq. (3.62) to eliminate $\hat{\mathbf{x}}_1$, $\hat{\mathbf{x}}_2$, and $\hat{\mathbf{x}}_3$, we may separate Eq. (3.64) into three scalar equations

$$
\begin{aligned}
V_1' &= a_{11} V_1 + a_{12} V_2 + a_{13} V_3 \\
V_2' &= a_{21} V_1 + a_{22} V_2 + a_{23} V_3
\end{aligned}
\tag{3.65}
$$

and

$$
V_3' = a_{31} V_1 + a_{32} V_2 + a_{33} V_3.
$$

In particular, these relations will hold for the coordinates of a point (x_1, x_2, x_3) and (x_1', x_2', x_3'), giving

$$
\begin{aligned}
x_1' &= a_{11} x_1 + a_{12} x_2 + a_{13} x_3, \\
x_2' &= a_{21} x_1 + a_{22} x_2 + a_{23} x_3,
\end{aligned}
\tag{3.66}
$$

and

$$
x_3' = a_{31} x_1 + a_{32} x_2 + a_{33} x_3,
$$

and similarly for the primed coordinates. In this notation the set of three equations (3.66) may be written as

$$
x_i' = \sum_{j=1}^{3} a_{ij} x_j,
\tag{3.67}
$$

where i takes on the values 1, 2, and 3 and the result is three *separate* equations.

Now let us set aside these results and try a different approach to the same problem. We consider two coordinate systems (x_1, x_2, x_3) and (x_1', x_2', x_3') with a common origin and one point (x_1, x_2, x_3) in the unprimed system, (x_1', x_2', x_3') in the primed system. Note the usual ambiguity. The same symbol x denotes both the coordinate axis and a particular distance along that axis. Since our system is linear, x_i' must be a linear combination of the x_i's. Let

$$
x_i' = \sum_{j=1}^{3} a_{ij} x_j.
\tag{3.68}
$$

The a_{ij} may be identified as our old friends, the direction cosines. This identification is carried out for the two-dimensional case later.

If we have two sets of quantities (V_1, V_2, V_3) in the unprimed system and (V_1', V_2', V_3') in the primed system, related in the same way as the coordinates of a point in the two different systems (Eq. (3.68)),

$$V_i' = \sum_{j=1}^{3} a_{ij} V_j, \qquad (3.69)$$

then, as in Section 1.2, the quantities (V_1, V_2, V_3) are defined as the components of a vector that stays fixed while the coordinates rotate; that is, a vector is defined in terms of transformation properties of its components under a rotation of the coordinate axes. In a sense the coordinates of a point have been taken as a prototype vector. The power and usefulness of this definition becomes apparent in Chapter 2, in which it is extended to define pseudovectors and tensors.

From Eq. (3.67) we can derive interesting information about the a_{ij}'s which describe the orientation of coordinate system (x_1', x_2', x_3') relative to the system (x_1, x_2, x_3). The length from the origin to the point is the same in both systems. Squaring, for convenience,[1]

$$\sum_i x_i^2 = \sum_i x_i'^2$$

$$= \sum_i \left(\sum_j a_{ij} x_j \right) \left(\sum_k a_{ik} x_k \right)$$

$$= \sum_{j,k} x_j x_k \sum_i a_{ij} a_{ik}. \qquad (3.70)$$

This can be true for all points if and only if

$$\sum_i a_{ij} a_{ik} = \delta_{jk}, \qquad j, k = 1, 2, 3. \qquad (3.71)$$

Note that Eq. (3.71) is equivalent to the matrix equation (3.83); see also Eqs. (3.87a)–(3.87d).

Verification of Eq. (3.71), if needed, may be obtained by returning to Eq. (3.70) and setting $\mathbf{r} = (x_1, x_2, x_3) = (1, 0, 0), (0, 1, 0), (0, 0, 1), (1, 1, 0)$, and so on to evaluate the nine relations given by Eq. (3.71). This process is valid, since Eq. (3.70) must hold for all $\mathbf{r}$ for a given set of a_{ij}. Equation (3.71), a consequence of requiring that the length remain constant (invariant) under rotation of the coordinate system, is called the *orthogonality condition*. The a_{ij}'s, written as a matrix A, form an orthogonal matrix. Note carefully that Eq. (3.71) is *not* matrix multiplication. Rather, it is interpreted later as a scalar product of two columns of A.

In matrix notation Eq. (3.67) becomes

$$|x'\rangle = A|x\rangle. \qquad (3.72)$$

---

[1] Note that *two* independent indices j and k are used.

ORTHOGONALITY CONDITIONS—TWO-DIMENSIONAL CASE

A better understanding of the a_{ij}'s and the orthogonality condition may be gained by considering rotation in two dimensions in detail. (This can be thought of as a three-dimensional system with the x_1-, x_2-axes rotated about x_3.) From Fig. 3.2,

$$x_1' = x_1 \cos \varphi + x_2 \sin \varphi,$$
$$x_2' = -x_1 \sin \varphi + x_2 \cos \varphi. \tag{3.73}$$

Therefore by Eq. (3.72)

$$A = \begin{pmatrix} \cos \varphi & \sin \varphi \\ -\sin \varphi & \cos \varphi \end{pmatrix}. \tag{3.74}$$

Notice that A reduces to the unit matrix for $\varphi = 0$. Zero rotation means nothing has changed. It is clear from Fig. 3.2 that

$$a_{11} = \cos \varphi = \cos(x_1', x_1),$$
$$a_{12} = \sin \varphi = \cos\left(\frac{\pi}{2} - \varphi\right) = \cos(x_1', x_2), \qquad \text{and so on,} \tag{3.75}$$

thus identifying the matrix elements a_{ij} as the direction cosines. Equation (3.71), the orthogonality condition, becomes

$$\sin^2 \varphi + \cos^2 \varphi = 1,$$
$$\sin \varphi \cos \varphi - \sin \varphi \cos \varphi = 0. \tag{3.76}$$

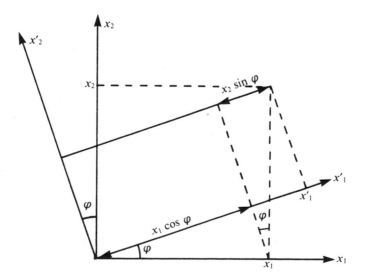

Figure 3.2

The extension to three dimensions (rotation of the coordinates through an angle φ counterclockwise about x_3) is simply

$$A = \begin{pmatrix} \cos\varphi & \sin\varphi & 0 \\ -\sin\varphi & \cos\varphi & 0 \\ 0 & 0 & 1 \end{pmatrix}. \tag{3.77}$$

The $a_{33} = 1$ expresses the fact that $x_3' = x_3$, since the rotation has been about the x_3-axis. The zeros guarantee that x_1' and x_2' do not depend on x_3 and that x_3' does not depend on x_1 and x_2. In more sophisticated language, x_1 and x_2 span an invariant *subspace*, whereas x_3 forms an invariant *subspace* alone. The general form of A is reducible. Equation (3.77) gives one possible decomposition.

INVERSE MATRIX, A^{-1}

Returning to the general transformation matrix A, the inverse matrix A^{-1} is defined such that

$$|x\rangle = A^{-1}|x'\rangle. \tag{3.78}$$

That is, A^{-1} describes the reverse of the rotation given by A and returns the coordinate system to its original position. Symbolically, Eqs. (3.72) and (3.78) combine to give

$$|x\rangle = A^{-1}A|x\rangle, \tag{3.79}$$

and since $|x\rangle$ is arbitrary,

$$A^{-1}A = 1, \tag{3.80}$$

the unit matrix. Similarly,

$$AA^{-1} = 1. \tag{3.81}$$

using Eqs. (3.72) and (3.78) and elimating $|x\rangle$ instead of $|x'\rangle$.

TRANSPOSE MATRIX, $\tilde{A}$

We can determine the elements of our postulated inverse matrix A^{-1} by employing the orthogonality condition. Equation (3.71), the orthogonality condition, does *not* conform to our definition of matrix multiplication, but it can be put in the required form by *defining* a new matrix $\tilde{A}$ such that

$$\tilde{a}_{ji} = a_{ij}. \tag{3.82}$$

Equation (3.71) becomes

$$\tilde{A}A = 1. \tag{3.83}$$

This is a restatement of the orthogonality condition and may be taken as a definition of orthogonality. Multiplying Eq. (3.83) by A^{-1} from the right and using Eq. (3.81), we have

$$\tilde{A} = A^{-1}. \tag{3.84}$$

This important result that the inverse equals the transpose holds only for orthogonal matrices and indeed may be taken as a further restatement of the orthogonality condition.

Multiplying Eq. (3.84) by A from the left, we obtain

$$A\tilde{A} = 1 \tag{3.85}$$

or

$$\sum_i a_{ji}a_{ki} = \delta_{jk}, \tag{3.86}$$

which is still another form of the orthogonality condition.

Summarizing, the orthogonality condition may be stated in several equivalent ways:

$$\sum_i a_{ij}a_{ik} = \delta_{jk} \tag{3.87a}$$

$$\sum_i a_{ji}a_{ki} = \delta_{jk} \tag{3.87b}$$

$$\tilde{A}A = A\tilde{A} = 1 \tag{3.87c}$$

$$\tilde{A} = A^{-1}. \tag{3.87d}$$

Any one of these relations is a necessary and a sufficient condition for A to be orthogonal.

It is now possible to see and understand why the term *orthogonal* is appropriate for these matrices. We have the general form

$$A = \begin{pmatrix} a_{11} & a_{12} & a_{13} \\ a_{21} & a_{22} & a_{23} \\ a_{31} & a_{32} & a_{33} \end{pmatrix}.$$

a matrix of direction cosines in which a_{ij} is the cosine of the angle between x_i' and x_j. Therefore a_{11}, a_{12}, a_{13} are the direction cosines of x_1' relative to x_1, x_2, x_3. These three elements of A *define* a unit length along x_1', that is, a unit vector $\hat{\mathbf{x}}_1'$,

$$\hat{\mathbf{x}}_1' = \hat{\mathbf{x}}_1 a_{11} + \hat{\mathbf{x}}_2 a_{12} + \hat{\mathbf{x}}_3 a_{13}.$$

The orthogonality relation (Eq. (3.86)) is simply a statement that the unit vectors $\hat{\mathbf{x}}_1'$, $\hat{\mathbf{x}}_2'$, and $\hat{\mathbf{x}}_3'$ are mutually perpendicular or orthogonal. Our orthogonal transformation matrix A transforms one orthogonal coordinate system into a second orthogonal coordinate system by rotation and/or reflection.

As an example of the use of matrices, the unit vectors in spherical polar coordinates may be written as

$$\begin{pmatrix} \hat{\mathbf{r}} \\ \hat{\boldsymbol{\theta}} \\ \hat{\boldsymbol{\phi}} \end{pmatrix} = C \begin{pmatrix} \hat{\mathbf{x}} \\ \hat{\mathbf{y}} \\ \hat{\mathbf{z}} \end{pmatrix}, \tag{3.88}$$

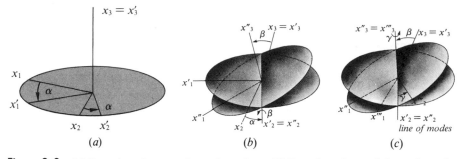

Figure 3.3   (a) Rotation about x_3 through angle α; (b) Rotation about x_2' through angle β; (c) Rotation about x_3'' through angle γ.

where C is given in Exercise 2.5.1. This is equivalent to Eq. (3.62) with $\mathbf{x}_1'$, $\mathbf{x}_2'$, and $\mathbf{x}_3'$ replaced by $\hat{\mathbf{r}}$, $\hat{\boldsymbol{\theta}}$, and $\hat{\boldsymbol{\phi}}$. From the preceding analysis C is orthogonal. Therefore the inverse relation becomes

$$
\begin{pmatrix} \hat{\mathbf{x}} \\ \hat{\mathbf{y}} \\ \hat{\mathbf{z}} \end{pmatrix} = \mathbf{C}^{-1} \begin{pmatrix} \hat{\mathbf{r}} \\ \hat{\boldsymbol{\theta}} \\ \hat{\boldsymbol{\phi}} \end{pmatrix} = \tilde{\mathbf{C}} \begin{pmatrix} \hat{\mathbf{r}} \\ \hat{\boldsymbol{\theta}} \\ \hat{\boldsymbol{\phi}} \end{pmatrix},
\tag{3.89}
$$

and Exercise 2.5.5 is solved by inspection. Similar applications of matrix inverses appear in connection with the transformation of a power series into a series of orthogonal functions (Gram–Schmidt orthogonalization) and the numerical solution of integral equations.

EULER ANGLES

Our transformation matrix A contains nine direction cosines. Clearly, only three of these are independent, Eq. (3.71) providing six constraints. Equivalently, we may say that two parameters (θ and φ in spherical polar coordinates) are required to fix the axis of rotation. Then one additional parameter describes the amount of rotation about the specified axis. In the Lagrangian formulation of mechanics (Section 17.3) it is necessary to describe A by using some set of three independent parameters rather than the redundant direction cosines. The usual choice of parameters is the Euler angles.[2]

The goal is to describe the orientation of a final rotated system (x_1''', x_2''', x_3''') relative to some initial coordinate system (x_1, x_2, x_3). The final system is developed in three steps—each step involving one rotation described by one Euler angle (Fig. 3.3):

---

[2] There are almost as many definitions of the Euler angles as there are authors. Here we follow the choice generally made by workers in the area of group theory and the quantum theory of angular momentum (compare Section 3.9).

1. The x_1'-, x_2'-, x_3'-axes are rotated about the x_3-axis through an angle α counterclockwise relative to x_1, x_2, x_3. (The x_3- and x_3'-axes coincide.)
2. The x_1''-, x_2''-, x_3''-axes are rotated about the x_2'-axis[3] through an angle β counterclockwise relative to x_1', x_2', x_3'. (The x_2'- and the x_2''-axes coincide.)
3. The third and final rotation is through an angle γ counterclockwise about the x_3''-axis, yielding the x_1''', x_2''', x_3''' system. (The x_3''- and x_3'''-axes coincide.)

The three matrices describing these rotations are

$$R_z(\alpha) = \begin{pmatrix} \cos\alpha & \sin\alpha & 0 \\ -\sin\alpha & \cos\alpha & 0 \\ 0 & 0 & 1 \end{pmatrix} \qquad (3.90)$$

exactly like Eq. (3.77),

$$R_y(\beta) = \begin{pmatrix} \cos\beta & 0 & -\sin\beta \\ 0 & 1 & 0 \\ \sin\beta & 0 & \cos\beta \end{pmatrix} \qquad (3.91)$$

and

$$R_z(\gamma) \begin{pmatrix} \cos\gamma & \sin\gamma & 0 \\ -\sin\gamma & \cos\gamma & 0 \\ 0 & 0 & 1 \end{pmatrix}. \qquad (3.92)$$

The total rotation is described by the triple matrix product.

$$A(\alpha, \beta, \gamma) = R_z(\gamma)R_y(\beta)R_z(\alpha). \qquad (3.93)$$

Note the order: $R_z(\alpha)$ operates first, then $R_y(\beta)$, and finally $R_z(\gamma)$. Direct multiplication gives

$$A(\alpha, \beta, \gamma) = \begin{pmatrix} \cos\gamma\cos\beta\cos\alpha - \sin\gamma\sin\alpha & \cos\gamma\cos\beta\sin\alpha + \sin\gamma\cos\alpha & -\cos\gamma\sin\beta \\ -\sin\gamma\cos\beta\cos\alpha - \cos\gamma\sin\alpha & -\sin\gamma\cos\beta\sin\alpha + \cos\gamma\cos\alpha & \sin\gamma\sin\beta \\ \sin\beta\cos\alpha & \sin\beta\sin\alpha & \cos\beta \end{pmatrix}. \qquad (3.94)$$

Equating $A(a_{ij})$ with $A(\alpha, \beta, \gamma)$, element by element, yields the direction cosines in terms of the three Euler angles. We could use this Euler angle

---

[3] Many authors choose this second rotation to be about the x_1'-axis.

identification to verify the direction cosine identities, Eq. (1.46) of Section 1.4, but the approach of Exercise 3.3.3 is much more elegant.

SYMMETRY PROPERTIES

Our matrix description leads to the rotation group $SO(3)$ in three-dimensional space $\mathbf{R}^3$, and the Euler angle description of rotations forms a basis for developing the rotation group in Chapter 4. Rotations may also be described by the unitary group $SU(2)$ in two-dimensional space $\mathbf{C}^2$ over the complex numbers. The concept of groups such as $SU(2)$ and its generalizations and group theoretical techniques are often encountered in modern particle physics where symmetry properties play an important role. The $SU(2)$ group is also considered in Chapter 4. The power and flexibility of matrices pushed quaternions into obscurity early in this century.[4]

It will be noted that matrices have been handled in two ways in the foregoing discussion: by their components and as single entities. Each technique has its own advantages and both are useful.

The transpose matrix is useful in a discussion of symmetry properties. If

$$A = \tilde{A}, \qquad a_{ij} = a_{ji}, \tag{3.95}$$

the matrix is called *symmetric*, whereas if

$$A = -\tilde{A}, \qquad a_{ij} = -a_{ji}, \tag{3.96}$$

it is called antisymmetric or skewsymmetric. The diagonal elements vanish. It is easy to show that any (square) matrix may be written as the sum of a symmetric matrix and an antisymmetric matrix. Consider the identity

$$A = \tfrac{1}{2}[A + \tilde{A}] + \tfrac{1}{2}[A - \tilde{A}]. \tag{3.97}$$

$[A + \tilde{A}]$ is clearly symmetric, whereas $[A - \tilde{A}]$ is clearly antisymmetric. This is the matrix analog of Eq. (2.75), Chapter 2, for tensors. Similarly, a function may be broken up into its even and odd function.

So far we have interpreted the orthogonal matrix as rotating the coordinate system. This changes the components of a fixed vector (not rotating with the coordinates) (Fig. 1.7, Chapter 1). However, an orthogonal matrix A may be interpreted equally well as a rotation of the *vector* in the *opposite* direction (Fig. 3.4).

These two possibilities, (1) rotating the vector keeping the basis fixed and (2) rotating the basis (in the opposite sense) keeping the vector fixed, have a direct analogy in quantum theory. Rotation (a time transformation) of the state vector gives the Schrödinger picture. Rotation of the basis keeping the state vector fixed yields the Heisenberg picture.

Suppose we interpret matrix A as rotating a *vector* **r** into the position shown by $\mathbf{r}_1$, i.e., in a particular coordinate system we have a relation

$$\mathbf{r}_1 = A\mathbf{r}. \tag{3.98}$$

[4] R. J. Stephenson, Development of vector analysis from quaternions. *Am. J. Phys.* **34**, 194 (1966).

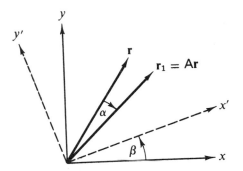

Figure 3.4 Fixed coordinates–rotated vector.

Now let us rotate the *coordinates* by applying matrix B, which rotates (x, y, z) into (x', y', z'),

$$\mathbf{r}_1' = \mathbf{Br}_1 = \mathbf{BAr} = (\mathbf{Ar})'$$

$$= \mathbf{BA}(\mathbf{B}^{-1}\mathbf{B})\mathbf{r}$$

$$= (\mathbf{BAB}^{-1})\mathbf{Br} = (\mathbf{BAB}^{-1})\mathbf{r}'. \tag{3.99}$$

$\mathbf{Br}_1$ is just $\mathbf{r}_1$ in the new coordinate system with a similar interpretation holding for $\mathbf{Br}$. Hence *in this new system* $(\mathbf{Br})$ is rotated into position $(\mathbf{Br}_1)$ by the matrix $\mathbf{BAB}^{-1}$.

$$\begin{array}{ccc} \mathbf{Br}_1 = & (\mathbf{BAB}^{-1}) & \mathbf{Br} \\ \downarrow & \downarrow & \downarrow \\ \mathbf{r}_1' = & \mathbf{A}' & \mathbf{r}' \end{array}$$

In the new system the coordinates have been rotated by matrix B, A has the form A′, in which

$$\mathbf{A}' = \mathbf{BAB}^{-1}. \tag{3.100}$$

A′ operates in the x', y', z' space as A operates in the x, y, z space.

The transformation defined by Eq. (3.100) with B any matrix, not necessarily orthogonal, is known as a similarity transformation. In component form Eq. (3.100) becomes

$$a_{ij}' = \sum_{k,l} b_{ik} a_{kl} (\mathbf{B}^{-1})_{lj}. \tag{3.101}$$

Now if B is orthogonal,

$$(\mathbf{B}^{-1})_{lj} = (\tilde{\mathbf{B}})_{lj} = b_{jl}, \tag{3.102}$$

and we have

$$a_{ij}' = \sum_{k,l} b_{ik} b_{jl} a_{kl}. \tag{3.103}$$

It may be helpful to think of A again as an operator, possibly as rotating coordinate axes, relating angular momentum and angular velocity of a rotating solid (Section 3.6). Matrix A is the representation in a given coordinate system—or basis. But there are directions associated with A—crystal axes, symmetry axes in the rotating solid, and so on—so that the representation A depends on the basis. The similarity transformation shows just how the representation changes with a change of basis.

RELATION TO TENSORS

Comparing Eq. (3.101) with the equations of Section 2.6, we see that it is the definition of a tensor of second rank. Hence a matrix that transforms by an *orthogonal* similarity transformation is, by definition, a tensor. Clearly, then, any *orthogonal* matrix A, interpreted as rotating a vector (Eq. (3.98)), may be called a tensor. If, however, we consider the orthogonal matrix as a collection of fixed direction cosines, giving the new orientation of a coordinate system, there is no tensor transformation involved.

The symmetry and antisymmetry properties defined earlier are preserved under *orthogonal* similarity transformations. Let A be a symmetric matrix, $A = \tilde{A}$, and

$$A' = BAB^{-1}. \tag{3.104}$$

Now

$$\tilde{A}' = B\tilde{A}B^{-1} = \tilde{B}^{-1}\tilde{A}\tilde{B} = B\tilde{A}B^{-1}, \tag{3.105}$$

since B is orthogonal. But $A = \tilde{A}$. Therefore

$$\tilde{A}' = BAB^{-1} = A', \tag{3.106}$$

showing that the property of symmetry is invariant under an orthogonal similarity transformation. In general, symmetry is *not* preserved under a nonorthogonal similarity transformation.

EXERCISES

Note. Assume all matrix elements are real.

3.3.1   Show that the product of two orthogonal matrices is orthogonal.
Note. This is a key step in showing that all $n \times n$ orthogonal matrices form a group (Section 4.1).

3.3.2   If A is orthogonal, show that its determinant $= \pm 1$.

3.3.3   If A is orthogonal and det A $= +1$, show that (det A)$a_{ij} = C_{ij}$, where C_{ij} is the *cofactor* of a_{ij}. This yields the identities of Eq. (1.46) used in Section 1.4 to show that a cross product of vectors (in three-space) is itself a vector. *Hint.* Note Exercise 3.2.32.

3.3.4   Another set of Euler rotations in common use is

1. a rotation about the x_3-axis through an angle φ, counterclockwise,
2. a rotation about the x_1'-axis through an angle θ, counterclockwise, and
3. a rotation about the x_3''-axis through an angle ψ, counterclockwise.

If
$$\alpha = \varphi - \pi/2 \qquad \varphi = \alpha + \pi/2$$
$$\beta = \theta \qquad \theta = \beta$$
$$\gamma = \psi + \pi/2 \qquad \psi = \gamma - \pi/2,$$

show that the final systems are identical.

3.3.5   Suppose the Earth is moved (rotated) so that the north pole goes to 30° north, 20° west (original latitude and longitude system) and the 10° west meridian points due south.

(a) What are the Euler angles describing this rotation?
(b) Find the corresponding direction cosines.

$$\text{ANS. (b) A} = \begin{pmatrix} 0.9551 & -0.2552 & -0.1504 \\ 0.0052 & 0.5221 & -0.8529 \\ 0.2962 & 0.8138 & 0.5000 \end{pmatrix}.$$

3.3.6   Verify that the Euler angle rotation matrix, Eq. (3.94), is invariant under the transformation

$$\alpha \to \alpha + \pi, \qquad \beta \to -\beta, \qquad \gamma \to \gamma - \pi.$$

3.3.7   Show that the Euler angle rotation matrix $A(\alpha, \beta, \gamma)$ satisfies the following relations:

(a) $A^{-1}(\alpha, \beta, \gamma) = \tilde{A}(\alpha, \beta, \gamma)$
(b) $A^{-1}(\alpha, \beta, \gamma) = A(-\gamma, -\beta, -\alpha)$.

3.3.8   Show that the trace of the product of a symmetric and an antisymmetric matrix is zero.

3.3.9   Show that the trace of a matrix remains invariant under similarity transformations.

3.3.10   Show that the determinant of a matrix remains invariant under similarity transformations.
Note. These two exercises (3.3.9) and 3.3.10) show that the trace and the determinant are independent of the basis. They are characteristics of the matrix (operator) itself.

3.3.11   Show that the property of antisymmetry is invariant under orthogonal similarity transformations.

3.3.12   A is 2 × 2 and orthogonal. Find the most general form of

$$A = \begin{pmatrix} a & b \\ c & d \end{pmatrix}.$$

Compare with two-dimensional rotation.

3.3.13 $|x\rangle$ and $|y\rangle$ are column vectors. Under an orthogonal transformation S, $|x'\rangle = S|x\rangle$, $|y'\rangle = S|y\rangle$. Show that the scalar product $\langle x | y \rangle$ is invariant under this orthogonal transformation.
Note. This is equivalent to the invariance of the dot product of two vectors, Section 1.3.

3.3.14   Show that the sum of the squares of the elements of a matrix remains invariant under orthogonal similarity transformations.

3.3.15   As a generalization of Exercise 3.3.14, show that

$$\sum_{jk} S_{jk} T_{jk} = \sum_{l,m} S'_{lm} T'_{lm},$$

where the primed and unprimed elements are related by an orthogonal similarity transformation. This result is useful in deriving invariants in electromagnetic theory (compare Section 4.6).

Note. This product $M_{jk} = \sum S_{jk} T_{jk}$ is sometimes called a Hadamard product. In the framework of tensor analysis, Chapter 2, this exercise becomes a double contraction of two second-rank tensors and therefore is clearly a scalar (invariant)!

3.3.16   A rotation $\varphi_1 + \varphi_2$ about the z-axis is carried out as two successive rotations φ_1 and φ_2, each about the z-axis. Use the matrix representation of the rotations to derive the trigonometric identities:

$$\cos(\varphi_1 + \varphi_2) = \cos \varphi_1 \cos \varphi_2 - \sin \varphi_1 \sin \varphi_2$$
$$\sin(\varphi_1 + \varphi_2) = \sin \varphi_1 \cos \varphi_2 + \cos \varphi_1 \sin \varphi_2.$$

3.3.17   A column vector **V** has components V_1 and V_2 in an initial (unprimed) system. Calculate V'_1 and V'_2 for a

(a) rotation of the coordinates through an angle of θ *counterclockwise*,
(b) rotation of the vector through an angle of θ *clockwise*.

The results for parts (a) and (b) should be identical.

3.3.18   Write a subroutine that will test whether a real $N \times N$ matrix is symmetric. Symmetry may be defined as

$$0 \le |a_{ij} - a_{ji}| \le \varepsilon,$$

where ε is some small tolerance (which allows for truncation error, and so on in the machine).

3.4 HERMITIAN MATRICES, UNITARY MATRICES

DEFINITIONS

Thus far it has generally been assumed that our linear vector space is a real space and that the matrix elements (the representations of the linear operators) are real. For many calculations in classical physics real matrix elements will suffice. However, in quantum mechanics complex variables are unavoidable because of the form of the basic commutation relations (or the form of the time-dependent Schrödinger equation). With this in mind, we generalize to the case of complex matrix elements. To handle these elements, let us define, or label, some new properties.

1. Complex conjugate, A*, formed by taking the complex conjugate $(i \rightarrow -i)$ of each element, where $i = \sqrt{-1}$.

2. Adjoint, $A^\dagger$, formed by transposing A^*,

$$A^\dagger = \widetilde{A^*} = \tilde{A}^*. \qquad (3.107)$$

3. Hermitian matrix: The matrix A is labeled Hermitian (or self-adjoint) if

$$A = A^\dagger. \qquad (3.108)$$

If A is real, then $A^\dagger = \tilde{A}$, and real Hermitian matrices are real symmetric matrices. In quantum mechanics (or matrix mechanics) matrices are usually constructed to be Hermitian, or unitary.

4. Unitary matrix: Matrix U is labeled unitary if

$$U^\dagger = U^{-1}. \qquad (3.109)$$

If U is real, then $U^{-1} = \tilde{U}$, so that real unitary matrices are orthogonal matrices. This represents a generalization of the concept of orthogonal matrix (compare Eq. (3.84)).

5. $(AB)^* = A^*B^*$, $(AB)^\dagger = B^\dagger A^\dagger$.

If the matrix elements are complex, the physicist is almost always concerned with adjoint matrices, Hermitian matrices, and unitary matrices. Unitary matrices are especially important in quantum mechanics because they leave the length of a (complex) vector unchanged—analogous to the operation of an orthogonal matrix on a real vector. It is for this reason that the S matrix of scattering theory is a unitary matrix. One important exception to this interest in unitary matrices is the group of Lorentz matrices, Chapter 4. Using Minkowski space, we see that these matrices are not *unitary*.

In a complex n-dimensional linear space the square of the length of a point $\tilde{x} = (x_1, x_2, \ldots, x_n)$, or the square of its distance from the origin 0, is defined as $x^\dagger x = \sum x_i^* x_i = \sum |x_i|^2$. If a coordinate transformation $y = Ux$ leaves the distance unchanged, then $x^\dagger x = y^\dagger y = (Ux)^\dagger Ux = x^\dagger U^\dagger Ux$. Since x is arbitrary it follows that $U^\dagger U = 1_n$, i.e., U is a unitary $n \times n$ matrix. If $x' = Ax$ is a linear map, then its matrix in the new coordinates becomes the unitary (analog of a similarity) transformation

$$A' = UAU^\dagger, \qquad (3.110)$$

because $Ux' = y' = UAx = UAU^{-1}y = UAU^\dagger y$.

PAULI AND DIRAC MATRICES

The set of three 2×2 Pauli matrices σ,

$$\sigma_1 = \begin{pmatrix} 0 & 1 \\ 1 & 0 \end{pmatrix}, \quad \sigma_2 = \begin{pmatrix} 0 & -i \\ i & 0 \end{pmatrix}, \quad \sigma_3 = \begin{pmatrix} 1 & 0 \\ 0 & -1 \end{pmatrix} \qquad (3.111)$$

were introduced by W. Pauli to describe a particle of spin $\frac{1}{2}$ in nonrelativistic quantum mechanics. It can readily be shown that (compare Exercises 3.2.13

and 3.2.14) the Pauli σ's satisfy

$$\sigma_i\sigma_j + \sigma_j\sigma_i = 2\delta_{ij}1_2, \qquad \text{anticommutation} \qquad (3.112)$$

$$\sigma_i\sigma_j = i\sigma_k, \qquad \text{cyclic permutation of indices} \qquad (3.113)$$

$$(\sigma_i)^2 = 1_2, \qquad (3.114)$$

where 1_2 is the 2×2 unit matrix. Thus, the vector $\sigma/2$ satisfies the same commutation relations

$$[\sigma_i,\sigma_j] \equiv \sigma_i\sigma_j - \sigma_j\sigma_i = 2i\varepsilon_{ijk}\sigma_k \qquad (3.115)$$

as the orbital angular momentum $\mathbf{L}$ ($\mathbf{L} \times \mathbf{L} = i\mathbf{L}$, see Exercise 2.5.15 and the $SO(3)$ and $SU(2)$ groups in Chapter 4).

The three Pauli matrices σ and the unit matrix form a complete set so that any 2×2 matrix M may be expanded as

$$\mathbf{M} = m_0 1 + m_1\sigma_1 + m_2\sigma_2 + m_3\sigma_3 = m_0 + \mathbf{m} \cdot \sigma, \qquad (3.116)$$

where the m_i are constants. Using $\sigma_i^2 = 1$ and $\mathrm{tr}(\sigma_i) = 0$ we obtain from Eq. (3.116) the expansion coefficients m_i by forming traces,

$$2m_0 = \mathrm{tr}(\mathbf{M}), \qquad 2m_i = \mathrm{tr}(\mathbf{M}\sigma_i), \qquad i = 1, 2, 3. \qquad (3.117)$$

Adding and multiplying such 2×2 matrices we generate the Pauli algebra.[1] Note that $\mathrm{tr}(\sigma_i) = 0$ for $i = 1, 2, 3$.

In 1927 P. M. Dirac extended this formalism to fast moving particles of spin $\frac{1}{2}$ such as electrons (and neutrinos). To include special relativity he started from Einstein's energy $E^2 = \mathbf{p}^2c^2 + m^2c^4$ instead of the nonrelativistic kinetic and potential energy $E = \mathbf{p}^2/2m + V$. The key to the Dirac equation is to factorize

$$E^2 - \mathbf{p}^2c^2 = E^2 - (c\sigma \cdot \mathbf{p})^2 = (E - c\sigma \cdot \mathbf{p})(E + c\sigma \cdot \mathbf{p}) = m^2c^4 \qquad (3.118)$$

using the 2×2 matrix identity

$$(\sigma \cdot \mathbf{p})^2 = \mathbf{p}^2 1_2. \qquad (3.119)$$

The 2×2 unit matrix 1_2 is not written explicitly in Eq. (3.118), and Eq. (3.119) follows from Exercise 3.2.14 for $\mathbf{a} = \mathbf{b} = \mathbf{p}$. Equivalently, we can introduce matrices γ_0 and γ to factorize $E^2 - \mathbf{p}^2c^2$ directly,

$$(\gamma_0 E - \gamma c\sigma \cdot \mathbf{p})^2 = \gamma_0^2 E^2 + \gamma^2 c^2(\sigma \cdot \mathbf{p})^2 - Ec\sigma \cdot \mathbf{p}(\gamma_0\gamma + \gamma\gamma_0)$$

$$= E^2 - \mathbf{p}^2c^2 = m^2c^4. \qquad (3.119')$$

If we recognize

$$\gamma_0 E - \gamma c\sigma \cdot \mathbf{p} = \gamma \cdot p = (\gamma_0, \gamma\sigma) \cdot (E, c\mathbf{p}) \qquad (3.120)$$

as a scalar product of two four-vectors γ^μ and p^μ (see Lorentz group in

[1] For its geometrical significance, see W. E. Baylis, J. Huschilt, and Jiansu Wei, *Am. J. Phys.* **60**, 788 (1992).

Chapter 4), then Eq. (3.119') with $p^2 = p \cdot p = E^2 - \mathbf{p}^2 c^2$ may be regarded as a four-vector generalization of Eq. (3.119). Clearly, for Eq. (3.119') to hold the conditions

$$\gamma_0^2 = 1 = -\gamma^2, \qquad \gamma_0\gamma + \gamma\gamma_0 = 0, \tag{3.121}$$

must be satisfied so that the four matrices γ^μ anticommute, just like the three Pauli matrices. Since the latter are a complete set of anticommuting 2×2 matrices, the conditions (3.121) cannot be met by 2×2 matrices, but they can be satisfied by 4×4 matrices

$$\gamma_0 = \gamma^0 = \begin{pmatrix} 1 & 0 & 0 & 0 \\ 0 & 1 & 0 & 0 \\ 0 & 0 & -1 & 0 \\ 0 & 0 & 0 & -1 \end{pmatrix} = \begin{pmatrix} 1_2 & 0 \\ 0 & -1_2 \end{pmatrix},$$

$$\gamma = \begin{pmatrix} 0 & 0 & 0 & 1 \\ 0 & 0 & 1 & 0 \\ 0 & -1 & 0 & 0 \\ -1 & 0 & 0 & 0 \end{pmatrix} = \begin{pmatrix} 0 & 1_2 \\ -1_2 & 0 \end{pmatrix}. \tag{3.122}$$

(The subscripts 2 and 4 will be omitted whenever it is clear what symbol is meant.) Alternatively, the vector of 4×4 matrices

$$\gamma = \begin{pmatrix} 0 & \sigma \\ -\sigma & 0 \end{pmatrix} = \gamma\sigma = i\sigma_2 \times \sigma \tag{3.123}$$

can be obtained as a direct product in the sense of Section 3.2 of Pauli 2×2 matrices. Similarly, $\gamma^0 = \sigma_3 \times 1_2$ and $1_4 = 1_2 \times 1_2$, etc.

Summarizing the relativistic treatment of a spin $\frac{1}{2}$ particle, it leads to 4×4 matrices, while the spin $\frac{1}{2}$ of a nonrelativistic particle is described by the 2×2 Pauli matrices σ.

In analogy with the Pauli algebra, we can form products of the basic γ^μ and linear combinations of them and the unit matrix $1 = 1_4$, thereby generating a 16 dimensional (so-called *Clifford*) algebra. A basis (with convenient Lorentz transformation properties, see Chapter 4) is given (in 2×2 matrix notation of Eq. (3.122)) by

$$1_4, \gamma_5 = i\gamma^0\gamma^1\gamma^2\gamma^3 = \begin{pmatrix} 0 & 1 \\ 1 & 0 \end{pmatrix}, \gamma^\mu, \gamma_5\gamma^\mu, \sigma^{\mu\nu} = i(\gamma^\mu\gamma^\nu - \gamma^\nu\gamma^\mu)/2. \tag{3.124}$$

Since the γ-matrices anticommute, their symmetric combinations

$$\gamma^\mu\gamma^\nu + \gamma^\nu\gamma^\mu = 2g^{\mu\nu}1_4, \tag{3.125}$$

where $g^{00} = 1 = -g^{11} = -g^{22} = -g^{33}$, and $g^{\mu\nu} = 0$ for $\mu \neq \nu$, are zero or proportional to the 4×4 unit matrix 1_4, while the six antisymmetric combinations in Eq. (3.124) give new basis elements which transform like a tensor under Lorentz transformations (see Chapter 4). Any 4×4 matrix can be expanded in terms of these 16 elements, and the expansion coefficients are given by forming traces similar to the 2×2 case in Eq. (3.117) using $\text{tr}(1_4) = 4$, $\text{tr}(\gamma_5) = 0$, $\text{tr}(\gamma^\mu) = 0 = \text{tr}(\gamma_5 \gamma^\mu)$, $\text{tr}(\sigma^{\mu\nu}) = 0$ for $\mu, \nu = 0, 1, 2, 3$ (see Exercise 3.4.23). In Chapter 4 we show that γ_5 is odd under parity so that $\gamma_5 \gamma^\mu$ transform like an axial vector that has even parity.

The discussion of orthogonal matrices in Section 3.3 and unitary matrices in this section is only a beginning. Further extensions are of vital concern in modern "elementary" particle physics. With the Pauli and Dirac matrices, we can develop *spinor* wave functions for electrons, protons, and other (relativistic) spin $\frac{1}{2}$ particles. The coordinate system rotations lead to $\mathbf{D}^j(\alpha, \beta, \gamma)$, the rotation group usually represented by matrices in which the elements are functions of the Euler angles describing the rotation. The special unitary group $SU(3)$ (composed of 3×3 unitary matrices with determinant $+1$) has been used with considerable success to describe mesons and baryons. These extensions are considered further in Chapter 4.

EXERCISES

3.4.1   Show that
$$\det(A^*) = (\det A)^* = \det(A^\dagger).$$

3.4.2   Three angular momentum matrices satisfy the basic commutation relation
$$[J_x, J_y] = iJ_z$$
(and cyclic permutation of indices). If two of the matrices have real elements, show that the elements of the third must be pure imaginary.

3.4.3   Show that $(AB)^\dagger = B^\dagger A^\dagger$.

3.4.4   Matrix $C = S^\dagger S$. Show that the trace is positive definite unless S is the null matrix in which case trace $(C) = 0$.

3.4.5   If A and B are Hermitian matrices, show that $(AB + BA)$ and $i(AB - BA)$ are also Hermitian.

3.4.6   Matrix C is *not* Hermitian. Show that $C + C^\dagger$ and $i(C - C^\dagger)$ are Hermitian. This means that a non-Hermitian matrix may be resolved into two Hermitian parts
$$C = \frac{1}{2}(C + C^\dagger) + \frac{1}{2i} i(C - C^\dagger).$$

This decomposition of a matrix into two Hermitian matrix parts parallels the decomposition of a complex number z into $x + iy$, where $x = (z + z^*)/2$ and $y = (z - z^*)/2i$.

3.4.7 A and B are two noncommuting Hermitian matrices:
$$AB - BA = iC.$$
Prove that C is Hermitian.

3.4.8 Show that a Hermitian matrix remains Hermitian under unitary similarity transformations.

3.4.9 Two matrices A and B are each Hermitian. Find a necessary and sufficient condition for their product AB to be Hermitian.

$ANS.$ [A, B] = 0.

3.4.10 Show that the reciprocal of a unitary matrix is unitary.

3.4.11 A particular similarity transformation yields

$$A' = UAU^{-1}$$
$$A^{\dagger\prime} = UA^{\dagger}U^{-1}.$$

If the adjoint relationship is preserved ($A^{\dagger\prime} = A'^{\dagger}$) and det U = 1, show that U must be unitary.

3.4.12 Two matrices U and H are related by

$$U = e^{iaH}$$

with a real. (The exponential function is defined by a Maclaurin expansion. This will be done in Section 5.11.)

(a) If H is Hermitian, show that U is unitary.
(b) If U is unitary, show that H is Hermitian. (H is independent of a.)

Note. With H the Hamiltonian,

$$\psi(x, t) = U(x, t)\psi(x, 0) = \exp(-itH/\hbar)\psi(x, 0)$$

is a solution of the time-dependent Schrödinger equation. $U(x, t) = \exp(-itH/\hbar)$ is the "evolution operator."

3.4.13 An operator $T(t + \varepsilon, t)$ describes the change in the wave function from t to $t + \varepsilon$. For ε real and small enough so that ε^2 may be neglected

$$T(t + \varepsilon, t) = 1 - \frac{i}{\hbar}\varepsilon H(t).$$

(a) If T is unitary, show that H is Hermitian.
(b) If H is Hermitian, show that T is unitary.

Note. When H(t) is independent of time, this relation may be put in exponential form—Exercise 3.4.12.

3.4.14 Show that an alternate form

$$T(t + \varepsilon, t) = \frac{1 - i\varepsilon H(t)/2\hbar}{1 + i\varepsilon H(t)/2\hbar}$$

agrees with the T of part (a) of Exercise 3.5.13 neglecting ε^2 and is exactly unitary (for H Hermitian).

3.4.15 Prove that the direct product of two unitary matrices is unitary.

3.4.16 Show that γ_5 anticommutes with all four γ^μ.

3.4.17 Use the four-dimensional Levi–Civita symbol $\varepsilon_{\lambda\mu\nu\rho}$ with $\varepsilon_{0123} = -1$ (generalizing Eq. (2.93) in Section 2.9 to four dimensions) and show that
(i) $2\gamma_5\sigma_{\mu\nu} = -i\varepsilon_{\mu\nu\alpha\beta}\sigma^{\alpha\beta}$ using the summation convention of Section 2.6 and
(ii) $\gamma_\lambda\gamma_\mu\gamma_\nu = g_{\lambda\mu}\gamma_\nu - g_{\lambda\nu}\gamma_\mu + g_{\mu\nu}\gamma_\lambda + i\varepsilon_{\lambda\mu\nu\rho}\gamma^\rho\gamma_5$. Define $\gamma_\mu = g_{\mu\nu}\gamma^\nu$ using $g^{\mu\nu} = g_{\mu\nu}$ to raise and lower indices.

3.4.18 Evaluate the following traces

$$\text{(i) } \text{tr}(\gamma \cdot a\gamma \cdot b) = 4a \cdot b, \qquad \text{(ii) } \text{tr}(\gamma \cdot a\gamma \cdot b\gamma \cdot c) = 0,$$

$$\text{(iii) } \text{tr}(\gamma \cdot a\gamma \cdot b\gamma \cdot c\gamma \cdot d) = 4(a \cdot bc \cdot d - a \cdot cb \cdot d + a \cdot db \cdot c),$$

$$\text{(iv) } \text{tr}(\gamma_5\gamma \cdot a\gamma \cdot b\gamma \cdot c\gamma \cdot d) = 4i\varepsilon_{\alpha\beta\mu\nu}a^\alpha b^\beta c^\mu d^\nu.$$

3.4.19 Show that (i) $\gamma_\mu\gamma^\alpha\gamma^\mu = -2\gamma^\alpha$, (ii) $\gamma_\mu\gamma^\alpha\gamma^\beta\gamma^\mu = 4g^{\alpha\beta}$, and (iii) $\gamma_\mu\gamma^\alpha\gamma^\beta\gamma^\nu\gamma^\mu = -2\gamma^\nu\gamma^\beta\gamma^\alpha$.

3.4.20 If $M = \frac{1}{2}(1 + \gamma_5)$, show that

$$M^2 = M.$$

Note that γ_5 may be replaced by any other Dirac matrix (any Γ_i of Eq. (3.124)). If M is Hermitian, then this result, $M^2 = M$, is the defining equation for a quantum mechanical projection operator.

3.4.21 Show that

$$\mathbf{\alpha} \times \mathbf{\alpha} = 2i\mathbf{\sigma},$$

where $\mathbf{\alpha} = \gamma_0\gamma$ is a vector whose components are the α matrices,

$$\mathbf{\alpha} = (\alpha_1, \alpha_2, \alpha_3).$$

Note that if $\mathbf{\alpha}$ is a polar vector (Section 2.4), then $\mathbf{\sigma}$ is an axial vector.

3.4.22 Prove that the 16 Dirac matrices form a linearly independent set.

3.4.23 If we assume that a given 4×4 matrix, A (with constant elements), can be written as a linear combination of the 16 Dirac matrices

$$A = \sum_{i=1}^{16} c_i\Gamma_i$$

show that

$$c_i \sim \text{trace}(A\Gamma_i).$$

3.4.24 If $C = i\gamma^2\gamma^0$ is the charge conjugation matrix, show that $C\gamma^\mu C^{-1} = -\tilde{\gamma}^\mu$, where $\tilde{\ }$ indicates transposition.

3.4.25 Let $x'_\mu = \Lambda_\mu{}^\nu x_\nu$ be a rotation by an angle θ about the 3-axis,

$$x'_0 = x_0, \qquad x'_1 = x_1 \cos \theta + x_2 \sin \theta,$$

$$x'_2 = -x_1 \sin \theta + x_2 \cos \theta, \qquad x'_3 = x_3.$$

Use $R = \exp(i\theta\sigma^{12}/2) = \cos \theta/2 + i\sigma^{12} \sin \theta/2$ (see Eq. (3.170)) and show that the γ's transform just like the coordinates x^μ, i.e., $\Lambda_\mu{}^\nu\gamma_\nu = R^{-1}\gamma_\mu R$. (Note that $\gamma_\mu = g_{\mu\nu}\gamma^\nu$, and that the γ^μ are well defined only up to a similarity transformation.) Similarly, if $x' = \Lambda x$ is a boost (pure Lorentz transformation) along the 1-axis, i.e.,

$$x'_0 = x_0 \cosh \zeta - x_1 \sinh \zeta, \qquad x'_1 = -x_0\sinh \zeta + x_1 \cosh \zeta,$$

$$x'_2 = x_2, \qquad x'_3 = x_3,$$

with $\cosh \zeta = 1/\sqrt{(1 - v^2/c^2)}$, $\sinh \zeta = v/c/\sqrt{(1 - v^2/c^2)}$, and $B = \exp(-i\zeta\sigma^{01}/2) = \cosh \zeta/2 - i\sigma^{01} \sinh \zeta/2$ (see Eq. (3.170)), show that $\Lambda_\mu{}^\nu\gamma_\nu = B\gamma_\mu B^{-1}$.

3.4.26 (a) Given $\mathbf{r}' = U\mathbf{r}$, with U a unitary matrix and $\mathbf{r}$ a (column) vector with complex elements, show that the norm (magnitude) of $\mathbf{r}$ is invariant under this operation.

(b) The matrix U transforms any column vector **r** with complex elements into **r′** leaving the magnitude invariant: $\mathbf{r}^\dagger\mathbf{r} = \mathbf{r}'^\dagger\mathbf{r}'$. Show that U is unitary.

3.4.27 Write a subroutine that will test whether a complex $N \times N$ matrix is self-adjoint. In demanding equality of matrix elements $a_{ij} = a_{ij}^\dagger$, allow some small tolerance ε to compensate for truncation error, and so on in the machine.

3.4.28 Write a subroutine that will form the adjoint of a complex $M \times N$ matrix.

3.4.29 (a) Write a subroutine that will take a complex $M \times N$ matrix A and will yield the product $\mathsf{A}^\dagger\mathsf{A}$.
Hint. This subroutine can call the subroutines of Exercises 3.2.41 and 3.4.28.
(b) Test your subroutine by taking A to be one or more of the Dirac matrices, Eq. (3.124).

3.5 DIAGONALIZATION OF MATRICES

MOMENT OF INERTIA MATRIX

In many physical problems involving real symmetric or complex Hermitian matrices it is desirable to carry out a (real) orthogonal similarity transformation or a unitary transformation (corresponding to a rotation of the coordinate system) to reduce the matrix to a diagonal form, nondiagonal elements all equal to zero. One particularly direct example of this is the moment of inertia matrix I of a rigid body. From the definition of angular momentum **L** we have

$$\mathbf{L} = \mathsf{I}\boldsymbol{\omega} \tag{3.126}$$

$\boldsymbol{\omega}$ being the angular velocity.[1] The inertia matrix I is found to have diagonal components

$$I_{xx} = \sum_i m_i(r_i^2 - x_i^2), \quad \text{and so on,} \tag{3.127}$$

the subscript i referring to mass m_i located at $\mathbf{r}_i = (x_i, y_i, z_i)$. For the non-diagonal components we have

$$I_{xy} = -\sum_i m_i x_i y_i = I_{yx}. \tag{3.128}$$

By inspection, matrix I is symmetric. Also, since I appears in a physical equation of the form (3.126), which holds for all orientations of the coordinate system, it may be considered to be a tensor (quotient rule, Section 2.3).

The key now is to orient the coordinate axes (along a body-fixed frame) so that the I_{xy} and the other nondiagonal elements will vanish. As a consequence of this orientation and an indication of it, if the angular velocity is along one

---

[1] The moment of inertia matrix may also be developed from the kinetic energy of a rotating body, $T = \frac{1}{2}\langle\boldsymbol{\omega}|\mathsf{I}|\boldsymbol{\omega}\rangle$.

such realigned axis, the angular velocity and the angular momentum will be parallel.

EIGENVECTORS, EIGENVALUES

It is perhaps instructive to consider a geometrical picture of this problem. If the inertia matrix I is multiplied from each side by a unit vector of variable direction, $\hat{\mathbf{n}} = (\alpha, \beta, \gamma)$, then in the Dirac bra-ket notation of Section 3.2,

$$\langle \hat{n} | \mathbf{I} | \hat{n} \rangle = I, \tag{3.129}$$

where I is the moment of inertia about the direction $\hat{n}$ and a positive number (scalar). Carrying out the multiplication, we obtain

$$I = I_{xx}\alpha^2 + I_{yy}\beta^2 + I_{zz}\gamma^2 + 2I_{xy}\alpha\beta + 2I_{xz}\alpha\gamma + 2I_{yz}\beta\gamma. \tag{3.130}$$

If we introduce

$$\mathbf{n} = \frac{\hat{\mathbf{n}}}{\sqrt{I}} = (n_1, n_2, n_3), \tag{3.131}$$

which is variable in direction *and* magnitude then Eq. (3.130) becomes

$$1 = I_{xx}n_1^2 + I_{yy}n_2^2 + I_{zz}n_3^2 + 2I_{xy}n_1 n_2 + 2I_{xz}n_1 n_3 + 2I_{yz}n_2 n_3, \tag{3.132}$$

a positive definite quadratic form which must be an ellipsoid (see Fig. 3.5).

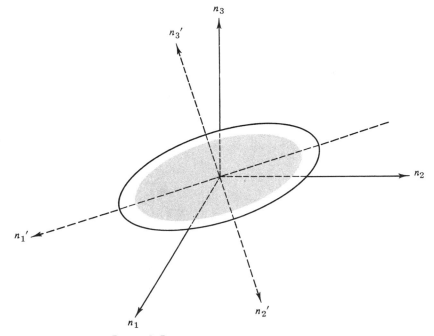

Figure 3.5 Moment of inertia ellipsoid.

From analytic geometry it is known that the coordinate axes can always be rotated to coincide with the axes of our ellipsoid. In many elementary cases, especially when symmetry is present, these new axes, called the *principal axes*, can be found by inspection. We now proceed to develop a general method of finding the diagonal elements and the principal axes.

If $R^{-1} = \tilde{R}$ is the corresponding real orthogonal matrix so that $\mathbf{n}' = R\mathbf{n}$, or $|\mathbf{n}'\rangle = R|\mathbf{n}\rangle$ in Dirac notation, are the new coordinates, then we obtain using $\langle\mathbf{n}'|R = \langle\mathbf{n}|$ in Eq. (3.132),

$$\langle\mathbf{n}|I|\mathbf{n}\rangle = \langle\mathbf{n}'|R I \tilde{R}|\mathbf{n}'\rangle = I_1' n_1'^2 + I_2' n_2'^2 + I_3' n_3'^2 = 1, \qquad (3.133)$$

where the $I_i' > 0$ are the principal moments of inertia. The inertia matrix I' in Eq. (3.133) is diagonal in the new coordinates,

$$I' = R I \tilde{R} = \begin{pmatrix} I_1' & 0 & 0 \\ 0 & I_2' & 0 \\ 0 & 0 & I_3' \end{pmatrix}. \qquad (3.134)$$

If we rewrite Eq. (3.134) using $R^{-1} = \tilde{R}$

$$\tilde{R} I' = I\tilde{R}, \qquad (3.135)$$

and take $\tilde{R} = (\mathbf{v}_1, \mathbf{v}_2, \mathbf{v}_3)$ to consist of three column vectors, then Eq. (3.135) splits up into three *eigenvalue* equations

$$I\mathbf{v}_i = I_i'\mathbf{v}_i, \qquad i = 1, 2, 3 \qquad (3.136)$$

with *eigenvalues* I_i' and *eigenvectors* $\mathbf{v}_i$. The terms were introduced from the early German literature on quantum mechanics. As these equations are linear and homogeneous (for fixed i), by Section 3.1 their determinants have to vanish:

$$\begin{vmatrix} I_{11} - I_i' & I_{12} & I_{13} \\ I_{12} & I_{22} - I_i' & I_{23} \\ I_{13} & I_{23} & I_{33} - I_i' \end{vmatrix} = 0. \qquad (3.137)$$

Replacing the eigenvalue I_i' by a variable λ times the unit matrix 1, we may rewrite Eq. (3.136) as

$$(I - \lambda 1)|\mathbf{v}\rangle = 0, \qquad (3.136')$$

whose determinant

$$|I - \lambda 1| = 0, \qquad (3.137')$$

is a cubic polynomial in λ; its three roots, of course, are the I_i'. Substituting one root at a time back into Eq. (3.136) (or (3.136')), we can find the corresponding eigenvectors. Because of its applications in astronomical theories Eq. (3.137) (or (3.137')) is known as the *secular equation*.[2] The same treatment

---

[2] Equation (3.126) will take on this form when ω is along one of the principal axes. Then $L = \lambda\omega$ and $I\omega = \lambda\omega$. In the mathematics literature λ is usually called a *characteristic value*, ω a *characteristic vector*.

applies to any real symmetric matrix I, except that its eigenvalues need not all be positive. Also, the orthogonality condition in Eq. (3.87) for R say that, in geometric terms, the eigenvectors v_i are mutually orthogonal unit vectors. Indeed they form the new coordinate system. The fact that any two eigenvectors v_i, v_j are orthogonal if $I_i' \neq I_j'$ follows from Eq. (3.136) in conjunction with the symmetry of I by multiplying with v_i and v_j, respectively,

$$\langle v_j | I | v_i \rangle = I_i' v_j \cdot v_i = \langle v_i | I | v_j \rangle = I_j' v_i \cdot v_j. \tag{3.138}$$

Since $I_i' \neq I_j'$ and Eq. (3.138) implies that $(I_j' - I_i')v_i \cdot v_j = 0$, so $v_i \cdot v_j = 0$.

HERMITIAN MATRICES

For complex vector spaces Hermitian and unitary matrices play the same role as symmetric and orthogonal matrices over real vector spaces, respectively. First, let us generalize the important theorem about the diagonal elements and the principal axes for the eigenvalue equation

$$A|r\rangle = \lambda|r\rangle. \tag{3.139}$$

We now show that if A is a Hermitian matrix,[3] its eigenvalues are real and its eigenvectors orthogonal.

Let λ_i and λ_j be two eigenvalues and $|r_i\rangle$ and $|r_j\rangle$, the corresponding eigenvectors of A, a Hermitian matrix. Then

$$A|r_i\rangle = \lambda_i|r_i\rangle \tag{3.140}$$

$$A|r_j\rangle = \lambda_j|r_j\rangle. \tag{3.141}$$

Equation (4.140) is multiplied by $\langle r_j|$

$$\langle r_j|A|r_i\rangle = \lambda_i\langle r_j|r_i\rangle. \tag{3.142}$$

Equation (4.141) is multiplied by $\langle r_i|$ to give

$$\langle r_i|A|r_j\rangle = \lambda_j\langle r_i|r_j\rangle. \tag{3.143}$$

Taking the adjoint* of this equation, we have

$$\langle r_j|A^\dagger|r_i\rangle = \lambda_j^*\langle r_j|r_i\rangle \tag{3.144}$$

or

$$\langle r_j|A|r_i\rangle = \lambda_j^*\langle r_j|r_i\rangle \tag{3.145}$$

since A is Hermitian. Subtracting Eq. (3.145) from Eq. (3.142), we obtain

$$(\lambda_i - \lambda_j^*)\langle r_j|r_i\rangle = 0. \tag{3.146}$$

This is a general result for all possible combinations of i and j. First, let $j = i$. Then Eq. (3.146) becomes

$$(\lambda_i - \lambda_i^*)\langle r_i|r_i\rangle = 0. \tag{3.147}$$

---

[3] If A is real, the Hermitian requirement is replaced by a requirement of symmetry.
* Note $\langle r_j| = |r_j\rangle^\dagger$ for complex vectors.

Since $\langle r_i | r_i \rangle = 0$ would be a trivial solution of Eq. (3.147), we conclude that

$$\lambda_i = \lambda_i^*, \tag{3.148}$$

or λ_i is real, for all i.

Second, for $i \neq j$, and $\lambda_i \neq \lambda_j$,

$$(\lambda_i - \lambda_j)\langle r_j | r_i \rangle = 0 \tag{3.149}$$

or

$$\langle r_j | r_i \rangle = 0, \tag{3.150}$$

which means that the eigenvectors of *distinct* eigenvalues are orthogonal, Eq. (3.150) being our generalization of orthogonality in this complex space.[4]

If $\lambda_i = \lambda_j$ (degenerate case), $|r_i\rangle$ is not automatically orthogonal to $|r_j\rangle$, but it may be *made* orthogonal.[5] Consider the physical problem of the moment of inertia matrix again. If x_1 is an axis of rotational symmetry, then we will find that $\lambda_2 = \lambda_3$. Eigenvectors $|r_2\rangle$ and $|r_3\rangle$ are each perpendicular to the symmetry axis, $|r_1\rangle$, but they lie anywhere in the plane perpendicular to $|r_1\rangle$; that is, any linear combination of $|r_2\rangle$ and $|r_3\rangle$ is also an eigenvector. Consider $(a_2|r_2\rangle + a_3|r_3\rangle)$ with a_2 and a_3 constants. Then

$$A(a_2|r_2\rangle + a_3|r_3\rangle) = a_2\lambda_2|r_2\rangle + a_3\lambda_3|r_3\rangle$$

$$= \lambda_2(a_2|r_2\rangle + a_3|r_3\rangle), \tag{3.151}$$

as is to be expected, for x_1 is an axis of rotational symmetry. Therefore, if $|r_1\rangle$ and $|r_2\rangle$ are fixed, $|r_3\rangle$ may simply be chosen to lie in the plane perpendicular to $|r_1\rangle$ and also perpendicular to $|r_2\rangle$. A general method of orthogonalizing solutions, the Gram–Schmidt process, is applied to functions in Section 9.3.

The set of n orthogonal eigenvectors of our $n \times n$ Hermitian matrix forms a *complete* set, spanning the n-dimensional (complex) space. This fact is useful in a variational calculation of the eigenvalues, Section 17.8 (Exercise 3.7.19). Eigenvalues and eigenvectors are not limited to Hermitian matrices. All matrices have eigenvalues and eigenvectors. For instance, the stochastic population matrix T satisfies an eigenvalue equation

$$Tp_{\text{equilibrium}} = \lambda p_{\text{equilibrium}},$$

with $\lambda = 1$. However, only Hermitian matrices have all eigenvectors orthogonal *and* all eigenvalues real.

---

[4] The corresponding theory for differential operators (Sturm–Liouville theory) appears in Section 9.2. The integral equation analog (Hilbert–Schmidt theory) is given in Section 16.4.

[5] We are assuming here that the eigenvectors of the n-fold degenerate λ_i span the corresponding n-dimensional space. This may be shown by including a parameter ε in the original matrix to remove the degeneracy and then letting ε approach zero (compare Exercise 3.6.30). This is analogous to breaking a degeneracy in atomic spectroscopy by applying an external magnetic field (Zeeman effect).

ANTIHERMITIAN MATRICES

Occasionally, in quantum theory we encounter antihermitian matrices:

$$A^\dagger = -A.$$

Following the analysis of the first portion of this section, we can show that

a. The eigenvalues are pure imaginary (or zero).
b. The eigenvectors corresponding to distinct eigenvalues are orthogonal.

The matrix R formed from the normalized eigenvectors is unitary. This antihermitian property is preserved under *unitary* transformations.

Example 3.5.1 Eigenvalues and Eigenvectors of a Real Symmetric Matrix

Let

$$A = \begin{pmatrix} 0 & 1 & 0 \\ 1 & 0 & 0 \\ 0 & 0 & 0 \end{pmatrix}. \tag{3.152}$$

The secular equation is

$$\begin{vmatrix} -\lambda & 1 & 0 \\ 1 & -\lambda & 0 \\ 0 & 0 & -\lambda \end{vmatrix} = 0, \tag{3.153}$$

or

$$-\lambda(\lambda^2 - 1) = 0, \tag{3.154}$$

expanding by minors. The roots are $\lambda = -1, 0, 1$. To find the eigenvector corresponding to $\lambda = -1$, we substitute this value back into the eigenvalue equation, Eq. (3.139),

$$\begin{pmatrix} -\lambda & 1 & 0 \\ 1 & -\lambda & 0 \\ 0 & 0 & -\lambda \end{pmatrix} \begin{pmatrix} x \\ y \\ z \end{pmatrix} = \begin{pmatrix} 0 \\ 0 \\ 0 \end{pmatrix}. \tag{3.155}$$

With $\lambda = -1$, this yields

$$x + y = 0, \qquad z = 0. \tag{3.156}$$

Within an arbitrary scale factor, and an arbitrary sign (or phase factor), $\langle r_1| = (1, -1, 0)$. Note carefully that (for real $|r\rangle$ in ordinary space) the eigenvector singles out a line in space. The positive or negative sense is not determined. This indeterminancy could be expected if we noted that Eq. (3.139) is homogeneous in $|r\rangle$. For convenience we will require that the eigenvectors be normalized to unity, $\langle r_1|r_1\rangle = 1$. With this choice of sign,

$$\langle r_1| \text{ or } r_1 = \left(\frac{1}{\sqrt{2}}, \frac{-1}{\sqrt{2}}, 0 \right) \tag{3.157}$$

is fixed. For $\lambda = 0$, Eq. (3.139) yields

$$y = 0, \qquad x = 0, \tag{3.158}$$

$\langle \mathbf{r}_2|$ or $\mathbf{r}_2 = (0, 0, 1)$ is a suitable eigenvector. Finally, for $\lambda = 1$, we get

$$-x + y = 0, \qquad z = 0, \tag{3.159}$$

or

$$\langle \mathbf{r}_3| \text{ or } \mathbf{r}_3 = \left(\frac{1}{\sqrt{2}}, \frac{1}{\sqrt{2}}, 0\right). \tag{3.160}$$

The orthogonality of $\mathbf{r}_1$, $\mathbf{r}_2$, and $\mathbf{r}_3$, corresponding to three distinct eigenvalues, may be easily verified.

Example 3.5.2 Degenerate Eigenvalues
Consider

$$A = \begin{pmatrix} 1 & 0 & 0 \\ 0 & 0 & 1 \\ 0 & 1 & 0 \end{pmatrix}. \tag{3.161}$$

The secular equation is

$$\begin{vmatrix} 1 - \lambda & 0 & 0 \\ 0 & -\lambda & 1 \\ 0 & 1 & -\lambda \end{vmatrix} = 0 \tag{3.162}$$

or

$$(1 - \lambda)(\lambda^2 - 1) = 0, \qquad \lambda = -1, 1, 1, \tag{3.163}$$

a degenerate case. If $\lambda = -1$, the eigenvalue equation (3.139) yields

$$2x = 0, \qquad y + z = 0. \tag{3.164}$$

A suitable normalized eigenvector is

$$\langle \mathbf{r}_1| \text{ or } \mathbf{r}_1 = \left(0, \frac{1}{\sqrt{2}}, \frac{-1}{\sqrt{2}}\right). \tag{3.165}$$

For $\lambda = 1$, we get

$$-y + z = 0. \tag{3.166}$$

Any eigenvector satisfying Eq. (3.166) is perpendicular to $\mathbf{r}_1$. We have an infinite number of choices. Suppose, as one possible choice, $\mathbf{r}_2$ is taken as

$$\langle \mathbf{r}_2| \text{ or } \mathbf{r}_2 = \left(0, \frac{1}{\sqrt{2}}, \frac{1}{\sqrt{2}}\right), \tag{3.167}$$

which clearly satisfies Eq. (3.166). Then $\mathbf{r}_3$ must be perpendicular to $\mathbf{r}_1$ and may be made perpendicular to $\mathbf{r}_2$ by[6]

$$\mathbf{r}_3 = \mathbf{r}_1 \times \mathbf{r}_2 = (1, 0, 0). \tag{3.168}$$

---

[6] The use of the cross product is limited to three-dimensional space (see Section 1.4).

FUNCTIONS OF MATRICES

Polynomials with one or more matrix arguments are well defined and occur often. Power series of a matrix may also be defined provided the series converge (see Chapter 5) for each matrix element. For example, if A is any $n \times n$ matrix then the power series

$$\exp(A) = \sum_{i=0}^{\infty} A^i/i!, \tag{3.169a}$$

$$\sin(A) = \sum_{i=0}^{\infty} (-1)^i A^{2i+1}/(2i+1)!, \tag{3.169b}$$

$$\cos(A) = \sum_{i=0}^{\infty} (-1)^i A^{2i}/(2i)! \tag{3.169c}$$

are well defined $n \times n$ matrices. For the Pauli matrices σ_k the Euler identity for real θ and $k = 1, 2,$ or 3

$$\exp(i\sigma_k \theta) = 1_2 \cos \theta + i\sigma_k \sin \theta, \tag{3.170a}$$

follows from collecting all even and odd powers of θ in separate series using $\sigma_k^2 = 1$. For the 4×4 Dirac matrices σ^{jk} with $(\sigma^{jk})^2 = 1$ if $j \neq k = 1, 2,$ or 3 we obtain similarly (without writing the obvious unit matrix 1_4 anymore)

$$\exp(i\sigma^{jk}\theta) = \cos \theta + i\sigma^{jk} \sin \theta, \tag{3.170b}$$

while

$$\exp(i\sigma^{0k}\zeta) = \cosh \zeta + i\sigma^{0k} \sinh \zeta \tag{3.170c}$$

holds for real ζ because $(i\sigma^{0k})^2 = 1$ for $k = 1, 2,$ or 3.

For a Hermitian matrix A there is a unitary matrix U that diagonalizes it, i.e., $UAU^\dagger = [a_1, a_2, \ldots, a_n]$. Then the *trace formula*

$$\det(\exp(A)) = \exp(\operatorname{tr}(A)) \tag{3.171}$$

is obtained (see Exercises 3.5.2 and 3.5.9) from

$$\det(\exp(A)) = \det(U \exp(A)U^\dagger) = \det(\exp(UAU^\dagger))$$

$$= \det(\exp[a_1, a_2, \ldots, a_n] = \det[e^{a_1}, e^{a_2}, \ldots, e^{a_n}]$$

$$= \prod e^{a_i} = \exp(\sum a_i) = \exp(\operatorname{tr}(A)),$$

using $UA^iU^\dagger = (UAU^\dagger)^i$ in the power series Eq. (3.169a) for $\exp(UAU^\dagger)$.

Another important relation is the *Baker–Hausdorff formula*

$$\exp(iG) H \exp(-iG) = H + [iG, H] + [iG, [iG, H]]/2 + \cdots \tag{3.172}$$

which follows from multiplying the power series for $\exp(iG)$ and collecting the terms with the same powers of iG. Here we define

$$[G, H] = GH - HG$$

as the *commutator* of G and H.

The preceding analysis has the advantage of exhibiting and clarifying conceptual relationships in the diagonalization of matrices. However, for matrices larger that 3×3, or perhaps 4×4, the process rapidly becomes so cumbersome that we turn gratefully to high-speed computers and iterative techniques.[7] One such technique is the Jacobi method for determining eigenvalues and eigenvectors of real symmetric matrices. This Jacobi technique for determining eigenvalues and eigenvectors and the Gauss–Seidel method of solving systems of simultaneous linear equations are examples of relaxation methods. They are iterative techniques in which hopefully the errors will decrease or relax as the iterations continue. Relaxation methods are used extensively for the solution of partial differential equations.

EXERCISES

3.5.1   (a) Starting with the angular momentum of the ith element of mass,

$$\mathbf{L}_i = \mathbf{r}_i \times \mathbf{p}_i = m_i \mathbf{r}_i \times (\boldsymbol{\omega} \times \mathbf{r}_i),$$

derive the inertia matrix such that $\mathbf{L} = \mathbf{I}\boldsymbol{\omega}$, $|\mathbf{L}\rangle = \mathbf{I}|\boldsymbol{\omega}\rangle$.

(b) Repeat the derivation starting with kinetic energy

$$T_i = \tfrac{1}{2}m_i(\boldsymbol{\omega} \times \mathbf{r}_i)^2 \qquad (T = \tfrac{1}{2}\langle\boldsymbol{\omega}|\mathbf{I}|\boldsymbol{\omega}\rangle).$$

3.5.2   Show that the eigenvalues of a matrix are unaltered if the matrix is transformed by a similarity transformation. This property is not limited to symmetric or Hermitian matrices. It holds for any matrix satisfying the eigenvalue equation, Eq. (3.139). If our matrix can be brought into diagonal form by a similarity transformation, then two immediate consequences are

1. The trace (sum of eigenvalues) is invariant under a similarity transformation.
2. The determinant (product of eigenvalues) is invariant under a similarity transformation.

Note. Prove this *separately* (for matrices that cannot be diagonalized). The invariance of the trace and determinant are often demonstrated by using the Cayley–Hamilton theorem: A matrix satisfies its own characteristic (secular) equation.

3.5.3   As a converse of the theorem that Hermitian matrices have real eigenvalues and that eigenvectors corresponding to distinct eigenvalues are orthogonal, show that if

(a) the eigenvalues of a matrix are real and
(b) the eigenvectors satisfy $\mathbf{r}_i^\dagger \mathbf{r}_j = \delta_{ij}$ or $\langle \mathbf{r}_i | \mathbf{r}_j \rangle = \delta_{ij}$, then the matrix is Hermitian.

---

[7] In higher-dimensional systems the secular equation may be strongly ill-conditioned with respect to the determination of its roots (the eigenvalues). Direct solution by machine may be very inaccurate. Iterative techniques for diagonalizing the original matrix are usually preferred. See Sections 2.7 and 2.9 of Press *et al.*, loc. cit.

3.5.4   Show that a real matrix that is not symmetric cannot be diagonalized by an orthogonal similarity transformation.
Hint. Assume that the nonsymmetric real matrix can be diagonalized and develop a contradiction.

3.5.5   The matrices representing the angular momentum components J_x, J_y, and J_z are all Hermitian. Show that the eigenvalues of J^2 where $J^2 = J_x^2 + J_y^2 + J_z^2$ are real and nonnegative.

3.5.6   A has eigenvalues λ_i and corresponding eigenvectors $|x_i\rangle$. Show that A^{-1} has the *same* eigenvectors but with eigenvalues λ_i^{-1}.

3.5.7   A square matrix with zero determinant is labeled *singular*.
(a) If A is singular, show that there is at least one nonzero column vector **v** such that
$$A|v\rangle = 0.$$
(b) If there is a nonzero vector $|v\rangle$ such that
$$A|v\rangle = 0,$$
show that A is a singular matrix. This means that if a matrix (or operator) has zero as an eigenvalue, the matrix (or operator) has no inverse.

3.5.8   The same similarity transformation diagonalizes each of two matrices. Show that the original matrices must commute. (This is particularly important in the matrix (Heisenberg) formulation of quantum mechanics.)

3.5.9   Two Hermitian matrices A and B have the same eigenvalues. Show that A and B are related by a unitary similarity transformation.

3.5.10   Find the eigenvalues and an orthonormal (orthogonal and normalized) set of eigenvectors for the matrices of Exercise 3.2.15.

3.5.11   Show that the inertia matrix for a single particle of mass m at (x, y, z) has a zero determinant. Explain this result in terms of the invariance of the determinant of a matrix under similarity transformations (Exercise 3.3.10) and a possible rotation of the coordinate system.

3.5.12   A certain rigid body may be represented by three point masses:
$$m_1 = 1 \quad \text{at } (1, 1, -2)$$
$$m_2 = 2 \quad \text{at } (-1, -1, 0)$$
$$m_3 = 1 \quad \text{at } (1, 1, 2).$$
(a) Find the inertia matrix.
(b) Diagonalize the inertia matrix obtaining the eigenvalues and the principal axes (as orthonormal eigenvectors).

3.5.13

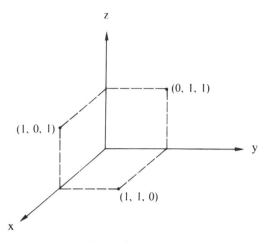

Unit masses are placed as shown in the figure.

(a) Find the moment of inertia matrix.

(b) Find the eigenvalues and a set of orthonormal eigenvectors.

(c) Explain the degeneracy in terms of the symmetry of the system.

$$ANS. \quad | = \begin{pmatrix} 4 & -1 & -1 \\ -1 & 4 & -1 \\ -1 & -1 & 4 \end{pmatrix} \quad \begin{aligned} \lambda_1 &= 2 \\ r_1 &= (1/\sqrt{3}, 1/\sqrt{3}, 1/\sqrt{3}) \\ \lambda_2 &= \lambda_3 = 5. \end{aligned}$$

3.5.14   A mass $m_1 = \frac{1}{2}$ kg is located at $(1, 1, 1)$ (meters), a mass $m_2 = \frac{1}{2}$ kg is at $(-1, -1, -1)$. The two masses are held together by an ideal (weightless, rigid) rod.

(a) Find the moment of inertia tensor of this pair of masses.

(b) Find the eigenvalues and eigenvectors of this inertia matrix.

(c) Explain the meaning, the physical significance of the $\lambda = 0$ eigenvalue. What is the significance of the corresponding eigenvector?

(d) Now that you have solved this problem by rather sophisticated matrix tensor techniques, explain how you could obtain

(1) $\lambda = 0$ and $\lambda = $?—by inspection.

(2) $r_{\lambda = 0} = $?—by inspection.

(By inspection means using freshman physics.)

3.5.15   Unit masses are at the eight corners of a cube $(\pm 1, \pm 1, \pm 1)$. Find the moment of inertia matrix and show that there is a triple degeneracy. This means that so far as moments of inertia are concerned, the cubic structure exhibits spherical symmetry.

3.5.16   Find the eigenvalues and corresponding orthonormal eigenvectors of the following matrices (as a numerical check, note that the sum of the eigenvalues equals the sum of the diagonal elements of the original matrix— Exercise 3.3.9). Note also the correspondence between det A = 0 and the existence of $\lambda = 0$—as required by Exercises 3.5.2 and 3.5.7.

$$A = \begin{pmatrix} 1 & 0 & 1 \\ 0 & 1 & 0 \\ 1 & 0 & 1 \end{pmatrix}. \qquad ANS. \quad \lambda = 0, 1, 2.$$

3.5.17
$$A = \begin{pmatrix} 1 & \sqrt{2} & 0 \\ \sqrt{2} & 0 & 0 \\ 0 & 0 & 0 \end{pmatrix}. \qquad ANS. \quad \lambda = -1, 0, 2.$$

3.5.18
$$A = \begin{pmatrix} 1 & 1 & 0 \\ 1 & 0 & 1 \\ 0 & 1 & 1 \end{pmatrix}. \qquad ANS. \quad \lambda = -1, 1, 2.$$

3.5.19
$$A = \begin{pmatrix} 1 & \sqrt{8} & 0 \\ \sqrt{8} & 1 & \sqrt{8} \\ 0 & \sqrt{8} & 1 \end{pmatrix}. \qquad ANS. \quad \lambda = -3, 1, 5.$$

3.5.20
$$A = \begin{pmatrix} 1 & 0 & 0 \\ 0 & 1 & 1 \\ 0 & 1 & 1 \end{pmatrix}. \qquad ANS. \quad \lambda = 0, 1, 2.$$

3.5.21
$$A = \begin{pmatrix} 1 & 0 & 0 \\ 0 & 1 & \sqrt{2} \\ 0 & \sqrt{2} & 0 \end{pmatrix}. \qquad ANS. \quad \lambda = -1, 1, 2.$$

3.5.22
$$A = \begin{pmatrix} 0 & 1 & 0 \\ 1 & 0 & 1 \\ 0 & 1 & 0 \end{pmatrix}. \qquad ANS. \quad \lambda = -\sqrt{2}, 0, \sqrt{2}.$$

3.5.23
$$A = \begin{pmatrix} 2 & 0 & 0 \\ 0 & 1 & 1 \\ 0 & 1 & 1 \end{pmatrix}. \qquad ANS. \quad \lambda = 0, 2, 2.$$

3.5.24
$$A = \begin{pmatrix} 0 & 1 & 1 \\ 1 & 0 & 1 \\ 1 & 1 & 0 \end{pmatrix}. \qquad ANS. \quad \lambda = -1, -1, 2.$$

3.5.25
$$A = \begin{pmatrix} 1 & -1 & -1 \\ -1 & 1 & -1 \\ -1 & -1 & 1 \end{pmatrix}. \qquad ANS. \quad \lambda = -1, 2, 2.$$

3.5.26
$$A = \begin{pmatrix} 1 & 1 & 1 \\ 1 & 1 & 1 \\ 1 & 1 & 1 \end{pmatrix}. \qquad ANS. \quad \lambda = 0, 0, 3.$$

3.5.27
$$A = \begin{pmatrix} 5 & 0 & 2 \\ 0 & 1 & 0 \\ 2 & 0 & 2 \end{pmatrix}. \qquad ANS. \quad \lambda = 1, 1, 6.$$

3.5.28
$$A = \begin{pmatrix} 1 & 1 & 0 \\ 1 & 1 & 0 \\ 0 & 0 & 0 \end{pmatrix}.$$ *ANS.* $\lambda = 0, 0, 2.$

3.5.29
$$A = \begin{pmatrix} 5 & 0 & \sqrt{3} \\ 0 & 3 & 0 \\ \sqrt{3} & 0 & 3 \end{pmatrix}.$$ *ANS.* $\lambda = 2, 3, 6.$

3.5.30   (a) Determine the eigenvalues and eigenvectors of
$$\begin{pmatrix} 1 & \varepsilon \\ \varepsilon & 1 \end{pmatrix}.$$

Note that the eigenvalues are degenerate for $\varepsilon = 0$ but the eigenvectors are orthogonal for all $\varepsilon \neq 0$ and $\varepsilon \to 0$.

(b) Determine the eigenvalues and eigenvectors of
$$\begin{pmatrix} 1 & 1 \\ \varepsilon^2 & 1 \end{pmatrix}.$$

Note that the eigenvalues are degenerate for $\varepsilon = 0$ and for this (non-symmetric) matrix the eigenvectors ($\varepsilon = 0$) do *not* span the space.

(c) Find the cosine of the angle between the two eigenvectors as a function of ε for $0 \leq \varepsilon \leq 1$.

3.5.31   (a) Take the coefficients of the simultaneous linear equations of Exercise 3.1.7 to be the matrix elements a_{ij} of matrix A (symmetric). Calculate the eigenvalues and eigenvectors.

(b) Form a matrix R whose columns are the eigenvectors of A and calculate the triple matrix product $\tilde{R}AR$.

ANS. $\lambda = 3.33163$

3.5.32   Repeat Exercise 3.5.31 by using the matrix of Exercise 3.2.39.

3.6 NORMAL MATRICES[1]

In Section 3.5 we concentrate primarily on Hermitian or real symmetric matrices and on the actual process of finding the eigenvalues and eigenvectors. In this section we generalize to normal matrices with Hermitian and unitary matrices as special cases. The physically important problem of normal modes of vibration and the numerically important problem of ill-conditioned matrices are also considered.

A normal matrix is a matrix that commutes with its adjoint,

$$[A, A^{\dagger}] = 0.$$

[1] Normal matrices are the largest class of matrices that can be diagonalized by unitary transformations. For an extensive discussion of normal matrices, see P. A. Macklin, Normal matrices for physicists. *Am. J. Phys.* **52**, 513 (1984).

Table 3.1

Matrix	Eigenvalues	Eigenvectors (for different eigenvalues)
Hermitian	Real	Orthogonal
Antihermitian	Pure imaginary (or zero)	Orthogonal
Unitary	Unit magnitude	Orthogonal
Normal	If A has eigenvalue λ $A^{\dagger}$ has eigenvalue $\lambda*$	Orthogonal A and $A^{\dagger}$ have the same eigenvectors

Obvious and important examples are Hermitian and unitary matrices. We will show that normal matrices have orthogonal eigenvectors (see Table 3.1). We proceed in two steps.

I. Let A have an eigenvector $|x\rangle$ and corresponding eigenvalue λ. Then

$$A|x\rangle = \lambda|x\rangle \tag{3.173}$$

or

$$(A - \lambda 1)|x\rangle = 0. \tag{3.174}$$

For convenience the combination $A - \lambda 1$ will be labeled B. Taking the adjoint of Eq. (3.174), we obtain

$$\langle x|(A - \lambda 1)^{\dagger} = 0 = \langle x|B^{\dagger}. \tag{3.175}$$

Because

$$[(A - \lambda 1)^{\dagger}, (A - \lambda 1)] = [A, A^{\dagger}] = 0,$$

we have

$$[B, B^{\dagger}] = 0. \tag{3.176}$$

The matrix B is also normal.

From Eqs. (3.174) and (3.175) we form

$$\langle x|B^{\dagger}B|x\rangle = 0. \tag{3.177}$$

This equals

$$\langle x|BB^{\dagger}|x\rangle = 0 \tag{3.178}$$

by Eq. (3.176). Now Eq. (3.178) may be rewritten as

$$(B^{\dagger}|x\rangle)^{\dagger}(B^{\dagger}|x\rangle) = 0. \tag{3.179}$$

Thus

$$B^{\dagger}|x\rangle = (A^{\dagger} - \lambda*1)|x\rangle = 0. \tag{3.180}$$

We see that for normal matrices, $A^{\dagger}$ has the same eigenvectors as A but the complex conjugate eigenvalues.

II. Now, considering more than one eigenvector–eigenvalue, we have

$$A|\mathbf{x}_i\rangle = \lambda_i|\mathbf{x}_i\rangle \tag{3.181}$$

$$A|\mathbf{x}_j\rangle = \lambda_j|\mathbf{x}_j\rangle. \tag{3.182}$$

Multiplying Eq. (3.182) from the left by $\langle\mathbf{x}_i|$ yields

$$\langle\mathbf{x}_i|A|\mathbf{x}_j\rangle = \lambda_j\langle\mathbf{x}_i|\mathbf{x}_j\rangle. \tag{3.183}$$

Operating on the left side of Eq. (3.183), we obtain

$$\langle\mathbf{x}_i|A = (A^\dagger|\mathbf{x}_i\rangle)^\dagger. \tag{3.184}$$

From Eq. (3.180) with $A^\dagger$ having the same eigenvectors as A but the complex conjugate eigenvalues

$$(A^\dagger|\mathbf{x}_i\rangle)^\dagger = (\lambda_i^*|\mathbf{x}_i\rangle)^\dagger = \lambda_i\langle\mathbf{x}_i|. \tag{3.185}$$

Substituting into Eq. (3.183) we have

$$\lambda_i\langle\mathbf{x}_i|\mathbf{x}_j\rangle = \lambda_j\langle\mathbf{x}_i|\mathbf{x}_j\rangle$$

or

$$(\lambda_i - \lambda_j)\langle\mathbf{x}_i|\mathbf{x}_j\rangle = 0. \tag{3.186}$$

This is the same as Eq. (3.149).

For $\lambda_i \neq \lambda_j$

$$\langle\mathbf{x}_j|\mathbf{x}_i\rangle = 0.$$

The eigenvectors corresponding to different eigenvalues of a normal matrix are *orthogonal*. This means that a normal matrix may be diagonalized by a unitary transformation. The required unitary matrix may be constructed from the orthonormal eigenvectors as shown earlier in Section 3.5.

The converse of this result is also valid. If A can be diagonalized by a unitary transformation, then A is normal.

NORMAL MODES OF VIBRATION

We consider the vibrations of a classical model of the CO_2 molecule. It is an illustration of the application of matrix techniques to a problem that does not start as a matrix problem. It also provides an example of the eigenvalues and eigenvectors of an asymmetric real matrix.

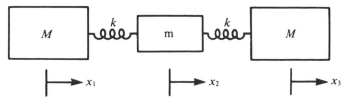

Figure 3.6

Example 3.6.1 Normal Modes

Consider three masses on the x-axis joined by springs as shown in Fig. 3.6. The spring forces are assumed to be linear (small displacements, Hooke's law) and the mass is constrained to stay on the x-axis.

Using a different coordinate for each mass Newton's second law yields the set of equations

$$\ddot{x}_1 = -\frac{k}{M}(x_1 - x_2)$$

$$\ddot{x}_2 = -\frac{k}{m}(x_2 - x_1) - \frac{k}{m}(x_2 - x_3) \qquad (3.187)$$

$$\ddot{x}_3 = -\frac{k}{M}(x_3 - x_2).$$

The system of masses is vibrating. We seek the common frequencies, ω, such that all masses vibrate at this same frequency. These are the normal modes. Let

$$x_i = x_{i0}e^{i\omega t}, \qquad i = 1, 2, 3.$$

Substituting into Eq. (3.187), we may rewrite this set as

$$\begin{pmatrix} \frac{k}{M} & -\frac{k}{M} & 0 \\ -\frac{k}{m} & \frac{2k}{m} & -\frac{k}{m} \\ 0 & -\frac{k}{M} & \frac{k}{M} \end{pmatrix} \begin{pmatrix} x_1 \\ x_2 \\ x_3 \end{pmatrix} = +\omega^2 \begin{pmatrix} x_1 \\ x_2 \\ x_3 \end{pmatrix}, \qquad (3.188)$$

with the common factor $e^{i\omega t}$ divided out. We have a matrix–eigenvalue equation with the matrix asymmetric. The secular equation is

$$\begin{vmatrix} \frac{k}{M} - \omega^2 & -\frac{k}{M} & 0 \\ -\frac{k}{m} & \frac{2k}{m} - \omega^2 & -\frac{k}{m} \\ 0 & -\frac{k}{M} & \frac{k}{M} - \omega^2 \end{vmatrix} = 0. \qquad (3.189)$$

This leads to

$$\omega^2 \left(\frac{k}{M} - \omega^2 \right)\left(\omega^2 - \frac{2k}{m} - \frac{k}{M} \right) = 0.$$

The eigenvalues are

$$\omega^2 = 0, \qquad \frac{k}{M}, \qquad \text{and} \qquad \frac{k}{M} + \frac{2k}{m},$$

all real.

The corresponding eigenvectors are determined by substituting the eigenvalues back into Eq. (3.188) one eigenvalue at a time. For $\omega^2 = 0$, Eq. (3.188) yields

$$x_1 - x_2 = 0$$

$$-x_1 + 2x_2 - x_3 = 0$$

$$-x_2 + x_3 = 0.$$

Then, we get

$$x_1 = x_2 = x_3.$$

This describes pure translation with no relative motion of the masses and no vibration.

For $\omega^2 = k/M$, Eq. (3.188) yields

$$x_1 = -x_3, \qquad x_2 = 0.$$

The two outer masses are moving in opposite direction. The center mass is stationary.

For $\omega^2 = k/M + 2k/m$ the eigenvector components are

$$x_1 = x_3, \qquad x_2 = -\frac{2M}{m}x_1.$$

The two outer masses are moving together. The center mass is moving opposite to the two outer ones. The net momentum is zero.

Any displacement of the three masses along the x-axis can be described as a linear combination of these three types of motion: translation plus two forms of vibration.

ILL-CONDITIONED SYSTEMS

A system of simultaneous linear equations may be written as

$$A|x\rangle = |y\rangle \qquad \text{or} \qquad A^{-1}|y\rangle = |x\rangle, \tag{3.190}$$

with A and $|y\rangle$ known and $|x\rangle$ unknown. The reader may encounter examples in which a small error in $|y\rangle$ results in a larger error in $|x\rangle$. In this case the matrix A is called "ill-conditioned." With $|\delta x\rangle$ and error in $|x\rangle$ and $|\delta y\rangle$ an error in $|y\rangle$, the *relative* errors may be written as

$$\left[\frac{\langle \delta x|\delta x\rangle}{\langle x|x\rangle}\right]^{1/2} \leq K(A)\left[\frac{\langle \delta y|\delta y\rangle}{\langle y|y\rangle}\right]^{1/2}. \tag{3.191}$$

Here $K(A)$, a property of matrix A, is labeled the *condition number*. For A Hermitian one form of the condition number is given by[1]

$$K(A) = \frac{|\lambda|_{\max}}{|\lambda|_{\min}}. \tag{3.192}$$

An approximate form due to Turing[2] is

$$K(A) = n[A_{ij}]_{\max}[A_{ij}^{-1}]_{\max}, \tag{3.193}$$

in which n is the order of the matrix and $[A_{ij}]_{\max}$ is the maximum element in A.

Example 3.6.1 An Ill-Conditioned Matrix

A common example of an ill-conditioned matrix is the Hilbert matrix, $H_{ij} = (i + j - 1)^{-1}$. The Hilbert matrix of order 4, H_4, is encountered in a least-squares fit of data to a third-degree polynomial. We have

$$H_4 = \begin{pmatrix} 1 & \frac{1}{2} & \frac{1}{3} & \frac{1}{4} \\ \frac{1}{2} & \frac{1}{3} & \frac{1}{4} & \frac{1}{5} \\ \frac{1}{3} & \frac{1}{4} & \frac{1}{5} & \frac{1}{6} \\ \frac{1}{4} & \frac{1}{5} & \frac{1}{6} & \frac{1}{7} \end{pmatrix}. \tag{3.194}$$

The elements of the inverse matrix (order n) are given by

$$(H_n^{-1})_{ij} = \frac{(-1)^{i+j}}{i + j - 1} \cdot \frac{(n + i - 1)!(n + j - 1)!}{[(i - 1)!(j - 1)!]^2(n - i)!(n - j)!}. \tag{3.195}$$

For $n = 4$

$$H_4^{-1} = \begin{pmatrix} 16 & -120 & 240 & -140 \\ -120 & 1200 & -2700 & 1680 \\ 240 & -2700 & 6480 & -4200 \\ -140 & 1680 & -4200 & 2800 \end{pmatrix}. \tag{3.196}$$

From Eq. (3.193) the Turing estimate of the condition number for H_4 becomes

$$K_{\text{Turing}} = 4 \times 1 \times 6480$$

$$= 2.59 \times 10^4.$$

This is a warning that an input error may be multiplied by 26,000 in the calculation of the output result. It is a statement that H_4 is ill-conditioned. If you encounter a highly ill-conditioned system you have two alternatives (besides abandoning the problem).

---

[1] G. E. Forsythe, and C. B. Moler, *Computer Solution of Linear Algebraic Systems.* Englewood Cliffs, NJ, Prentice Hall (1967).

[2] Compare J. Todd, *The Condition of the Finite Segments of the Hilbert Matrix,* Applied Mathematics Series No. 313. Washington, DC: National Bureau of Standards.

a. Try a different mathematical attack.

b. Arrange to carry more significant figures and push through by brute force.

As previously seen, matrix eigenvector–eigenvalue techniques are not limited to the solution of strictly matrix problems. A further example of the transfer of techniques from one area to another is seen in the application of matrix techniques to the solution of Fredholm eigenvalue integral equations, Section 16.3. In turn, these matrix techniques are strengthened by a variational calculation of Section 17.8.

EXERCISES

3.6.1   Show that every 2×2 matrix has two eigenvectors and corresponding eigenvalues. The eigenvectors are not necessarily orthogonal or different. The eigenvalues are not necessarily real.

3.6.2   As an illustration of Exercise 3.6.1, find the eigenvalues and corresponding eigenvectors for

$$\begin{pmatrix} 2 & 4 \\ 1 & 2 \end{pmatrix}.$$

Note that the eigenvectors are *not* orthogonal.

$$ANS. \quad \lambda_1 = 0, \mathbf{r}_1 = (2, -1);$$
$$\lambda_2 = 4, \mathbf{r}_2 = (2, 1).$$

3.6.3   If A is a 2×2 matrix show that its eigenvalues λ satisfy the equation

$$\lambda^2 - \lambda \, \text{trace}(A) + \det A = 0.$$

3.6.4   Assuming a unitary matric U to satisfy an eigenvalue equation $U\mathbf{r} = \lambda \mathbf{r}$, show that the eigenvalues of the unitary matrix have unit magnitude. This same result holds for real orthogonal matrices.

3.6.5   Since an orthogonal matrix describing a rotation in real three-dimensional space is a special case of a unitary matrix, such an orthogonal matrix can be diagonalized by a unitary transformation.

(a) Show that the sum of the three eigenvalues is $1 + 2 \cos \varphi$; where φ is the net angle of rotation about a single fixed axis.

(b) Given that one eigenvalue is 1, show that the other two eigenvalues must be $e^{i\varphi}$ and $e^{-i\varphi}$.

Our orthogonal rotation matrix (real elements) has complex eigenvalues.

3.6.6   A is an nth order Hermitian matrix with orthonormal eigenvectors $|x_i\rangle$ and real eigenvalues $\lambda_1 \leq \lambda_2 \leq \lambda_3 \leq \cdots \leq \lambda_n$. Show that for a unit magnitude vector $|\mathbf{y}\rangle$,

$$\lambda_1 \leq \langle \mathbf{y}|A|\mathbf{y}\rangle \leq \lambda_n.$$

3.6.7   A particular matrix is both Hermitian and unitary. Show that its eigenvalues are all ± 1.

Note. The Pauli and Dirac matrices are specific examples.

3.6.8   For his relativistic electron theory Dirac required a set of *four* anticommuting matrices. Assume that these matrices are to be Hermitian and unitary. If these are $n \times n$ matrices, show that n must be *even*. With 2×2 matrices inadequate (why?), this demonstrates that the smallest possible matrices forming a set of four anticommuting, Hermitian, unitary matrices are 4×4.

3.6.9   A is a normal matrix with eigenvalues λ_n and orthonormal eigenvectors $|x_n\rangle$. Show that A may be written as

$$A = \sum_n \lambda_n |x_n\rangle\langle x_n|.$$

Hint. Show that both this eigenvector form of A and the original A give the same result acting on an arbitrary vector $|y\rangle$.

3.6.10   A has eigenvalues 1 and -1 and corresponding eigenfunctions $\binom{1}{0}$ and $\binom{0}{1}$. Construct A.

$$ANS. \quad A = \begin{pmatrix} 1 & 0 \\ 0 & -1 \end{pmatrix}.$$

3.6.11   A non-Hermitian matrix A has eigenvalues λ_i and corresponding eigenvectors $|u_i\rangle$. The adjoint matrix $A^\dagger$ has the same set of eigenvalues but *different* corresponding eigenvectors, $|v_i\rangle$. Show that the eigenvectors form a *biorthogonal* set in the sense that

$$\langle v_i | u_j \rangle = 0 \quad \text{for } \lambda_i^* \neq \lambda_j.$$

3.6.12   You are given a pair of equations:

$$A|f_n\rangle = \lambda_n |g_n\rangle$$
$$\tilde{A}|g_n\rangle = \lambda_n |f_n\rangle \quad \text{with A real.}$$

(a) Prove that $|f_n\rangle$ is an eigenvector of $(\tilde{A}A)$ with eigenvalue λ_n^2.
(b) Prove that $|g_n\rangle$ is an eigenvector of $(A\tilde{A})$ with eigenvalue λ_n^2.
(c) State how you know that
   1. The $|f_n\rangle$ form an orthogonal set.
   2. The $|g_n\rangle$ form an orthogonal set.
   3. λ_n^2 is real.

3.6.13   Prove that A of the preceding problem may be written as

$$A = A_1 = \sum_n \lambda_n |g_n\rangle\langle f_n|,$$

with the $|g_n\rangle$ and $\langle f_n|$ normalized to unity.
Hint.   (a) Show that A_1 operating on an arbitrary vector yields the same result as A operating on that vector.
   (b) Expand your arbitrary vector as a linear combination of $|f_n\rangle$.

3.6.14   Given

$$A = \frac{1}{\sqrt{5}} \begin{pmatrix} 2 & 2 \\ 1 & -4 \end{pmatrix}$$

(a) Construct the transpose $\tilde{A}$ and the symmetric forms $\tilde{A}A$ and $A\tilde{A}$.
(b) From $A\tilde{A}|g_n\rangle = \lambda_n^2|g_n\rangle$ find λ_n and $|g_n\rangle$. Normalize the $|g_n\rangle$'s.
(c) From $\tilde{A}A|f_n\rangle = \lambda_n^2|f_n\rangle$ find λ_n [same as (b)] and $|f_n\rangle$. Normalize the $|f_n\rangle$'s.
(d) Verify that $A|f_n\rangle = \lambda_n|g_n\rangle$ and $\tilde{A}|g_n\rangle = \lambda_n|f_n\rangle$.
(e) Verify that $A = \sum_n \lambda_n|g_n\rangle\langle f_n|$.

3.6.15   Given the eigenvalues $\lambda_1 = 1$, $\lambda_2 = -1$, and the corresponding eigenvectors

$$|f_1\rangle = \begin{pmatrix} 1 \\ 0 \end{pmatrix}, \qquad |g_1\rangle = \frac{1}{\sqrt{2}}\begin{pmatrix} 1 \\ 1 \end{pmatrix}, \qquad |f_2\rangle = \begin{pmatrix} 0 \\ 1 \end{pmatrix}, \qquad \text{and}$$

$$|g_2\rangle = \frac{1}{\sqrt{2}}\begin{pmatrix} 1 \\ -1 \end{pmatrix}.$$

(a) Construct A.
(b) Verify that $A|f_n\rangle = \lambda_n|g_n\rangle$.
(c) Verify that $\tilde{A}|g_n\rangle = \lambda_n|f_n\rangle$.

$$ANS. \quad A = \frac{1}{\sqrt{2}}\begin{pmatrix} 1 & -1 \\ 1 & 1 \end{pmatrix}.$$

3.6.16   This is a continuation of Exercise 3.4.12, where the unitary matrix U and the Hermitian matrix H are related by

$$U = e^{iaH}.$$

(a) If trace $H = 0$, show that det $U = +1$.
(b) If det $U = +1$, show that trace $H = 0$.

Hint. H may be diagonalized by a similarity transformation. Then, interpreting the exponential by a Maclaurin expansion, U is also diagonal. The corresponding eigenvalues are given by $u_j = \exp(iah_j)$.

Note. These properties, and those of Exercise 3.4.12, are vital in the development of the concept of *generators* in group theory—Section 4.2.

3.6.17   An $n \times n$ matrix A has n eigenvalues A_i. If $B = e^A$ show that B has the same eigenvectors as A with the corresponding eigenvalues B_i given by $B_i = \exp(A_i)$.

Note. e^A is defined by the Maclaurin expansion of the exponential:

$$e^A = 1 + A + \frac{A^2}{2!} + \frac{A^3}{3!} + \cdots.$$

3.6.18   A matrix P is a projection operator satisfying the condition

$$P^2 = P.$$

Show that the corresponding eigenvalues $(p^2)_\lambda$ and p_λ satisfy the relation

$$(p^2)_\lambda = (p_\lambda)^2 = p_\lambda.$$

This means that the eigenvalues of P are 0 and 1.

3.6.19   In the *matrix* eigenvector, eigenvalue equation

$$A|r_i\rangle = \lambda_i|r_i\rangle,$$

A is an $n \times n$ Hermitian matrix. For simplicity assume that its n real eigenvalues are distinct, λ_1 being the largest. If $|r\rangle$ is an approximation to $|r_1\rangle$,

$$|r\rangle = |r_1\rangle + \sum_{i=2}^{n} \delta_i|r_i\rangle,$$

show that

$$\frac{\langle r|A|r\rangle}{\langle r|r\rangle} \leq \lambda_1$$

and that the error in λ_1 is of the order $|\delta_i|^2$. Take $|\delta_i| \ll 1$.

Hint. The $n|r_i\rangle$ for a *complete* orthogonal set spanning the n-dimensional (complex) space.

3.6.20   Two equal masses are connected to each other and to walls by springs as shown in the figure. The masses are constrained to stay on a horizontal line.

(a) Set up the Newtonian acceleration equation for each mass.
(b) Solve the secular equation for the eigenvectors.
(c) Determine the eigenvectors and thus the normal modes of motion.

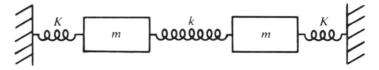

ADDITIONAL READINGS

AITKEN, A. C., *Determinants and Matrices*. New York: Interscience (1956). Reprinted, Greenwood (1983). A readable introduction to determinants and matrices.

BARNETT, S., *Matrices: Methods and Applications*. Oxford: Clarendon Press (1990).

BICKLEY, W. G., and R. S. H. G. THOMPSON, *Matrices—Their Meaning and Manipulation*. Princeton, NJ: Van Nostrand (1964). A comprehensive account of the occurrence of matrices in physical problems, their analytic properties, and numerical techniques.

BROWN, W. C., *Matrices and Vector Spaces*. New York: Dekker (1991).

HEADING, J. *Matrix Theory for Physicists*. London: Longmans, Green and Co. (1958). A readable introduction to determinants and matrices with applications to mechanics, electromagnetism, special relativity, and quantum mechanics.

4 GROUP THEORY

Disciplined judgment about what is neat and symmetrical and elegant has time and time again proved an excellent guide to how nature works.

MURRAY GELL-MANN

4.1 INTRODUCTION TO GROUP THEORY

In classical mechanics the *symmetry* of a physical system leads to *conservation laws*. Conservation of angular momentum is a direct consequence of rotational symmetry, which means *invariance* under spatial rotations. In the first third of this century, Wigner and others realized that invariance was a key concept in understanding the new phenomena and in developing appropriate theories. Thus, in quantum mechanics the concept of angular momentum and spin has become even more central. Its generalizations, *isospin* in nuclear physics and the *flavor* symmetry in particle physics, are indispensable tools in building and solving theories. Generalizations of the concept of *gauge* invariance of classical electrodynamics to the isospin symmetry lead to the electroweak gauge theory.

In each case the set of these symmetry operations forms a *group*. Group theory is the mathematical tool to treat invariants and symmetries. It brings unification and formalization of principles such as spatial reflections, or parity, angular momentum, and geometry that are widely used by physicists.

In geometry the fundamental role of group theory was recognized more than a century ago by mathematicians (e.g., Felix Klein's Erlanger Programm). In euclidean geometry the distance between two points, the scalar product of two vectors or metric, does not change under rotations or translations. These symmetries are characteristic of this geometry. In special relativity the metric, or scalar product of four vectors, differs from that of euclidean geometry in that it is no longer positive definite and is invariant under Lorentz transformations.

For a crystal the symmetry group contains only a finite number of rotations at discrete values of angles or reflections. The theory of such *discrete* or *finite* groups, developed originally as a branch of pure mathematics, now is a useful tool for the development of crystallography and condensed matter physics. A brief introduction to this area appears in Section 4.7. When the rotations depend on continuously varying angles (the Euler angles of Section 3.3) the rotation groups have an infinite number of elements. Such *continuous* (or *Lie*[1]) *groups* are the topic of Sections 4.2–4.6.

DEFINITION OF GROUP

A group G may be defined as a set of objects or operations, called the elements of G, that may be combined or "multiplied" to form a well-defined product in G which satisfy the following four conditions.

1. If a and b are any two elements of G, then the product ab is also an element of G; or $(a, b) \rightarrow ab$ maps $G \times G$ onto G.
2. This multiplication is associative, $(ab)c = a(bc)$.
3. There is a unit element I in G such that $Ia = aI = a$ for every element a in G.[2]
4. There must be an inverse or reciprocal of each element a of G, labeled a^{-1}, such that $aa^{-1} = a^{-1}a = I$.

An example for a group is the set of counterclockwise coordinate rotations

$$R(\varphi) = \begin{pmatrix} \cos \varphi & \sin \varphi \\ -\sin \varphi & \cos \varphi \end{pmatrix} \tag{4.1}$$

through an angle φ of the xy-coordinate system to a new orientation in Eq. (1.8), Fig. 1.6, and Section 3.3. The product of two rotations $R(\varphi_1)R(\varphi_2)$ is defined by rotating first by the angle φ_2, then by φ_1. According to Eq. (3.29), this corresponds to the product of the orthogonal 2×2 matrices

$$\begin{pmatrix} \cos \varphi_1 & \sin \varphi_1 \\ -\sin \varphi_1 & \cos \varphi_1 \end{pmatrix}\begin{pmatrix} \cos \varphi_2 & \sin \varphi_2 \\ -\sin \varphi_2 & \cos \varphi_2 \end{pmatrix}$$
$$= \begin{pmatrix} \cos(\varphi_1 + \varphi_2) & \sin(\varphi_1 + \varphi_2) \\ -\sin(\varphi_1 + \varphi_2) & \cos(\varphi_1 + \varphi_2) \end{pmatrix}, \tag{4.2}$$

using the addition formulas for the trigonometric functions. The product is clearly a rotation represented by the orthogonal matrix with angle $\varphi_1 + \varphi_2$. The product is the associative matrix multiplication. It is *commutative* or *abelian* because the order in which these rotations are performed does not matter. The inverse of the rotation with angle φ is that with angle $-\varphi$. The unit corresponds to the angle $\varphi = 0$. The group's name is SO(2), if the angle φ varies continuously from 0 to 2π. Clearly, SO(2) has infinitely many elements. The

---

[1] After the Norwegian mathematician Sophus Lie.
[2] Following E. Wigner, the unit element of a group is often labeled E, from the German *Einheit*, i.e., unit, or just 1, or I for identity.

unity with angle $\varphi = 0$ and the rotation with $\varphi = \pi$ form a finite *subgroup*. A subgroup G' of a group G consists of elements of G so that the product of any of its elements is again in the subgroup G', i.e., G' is closed under the multiplication of G. If $gg'g^{-1}$ is an element of G' for any g of G and g' of G', then G' is called an *invariant subgroup* of G.

Orthogonal $n \times n$ matrices form the group $O(n)$, and also $SO(n)$ if their determinants are $+1$ (S stands for "special"). If $\tilde{O}_i = O_i^{-1}$ for $i = 1$ and 2 (see Chapter 3.3 for orthogonal matrices), then the product

$$\overline{O_1 O_2} = \tilde{O}_2 \tilde{O}_1 = O_2^{-1} O_1^{-1} = (O_1 O_2)^{-1}$$

is also an orthogonal matrix in $SO(n)$. The inverse is the transpose (orthogonal) matrix. The unit of the group is 1_n. A real orthogonal $n \times n$ matrix has $n(n-1)/2$ *independent* parameters. For $n = 2$, there is only one parameter: one angle in Eq. (4.1). For $n = 3$, there are three independent parameters: the three Euler angles of Section 3.3.

Likewise, unitary $n \times n$ matrices form the group $U(n)$, and also $SU(n)$ if their determinants are $+1$. If $U_i^\dagger = U_i^{-1}$ (see Chapter 3.4 for unitary matrices), then

$$(U_1 U_2)^\dagger = U_2^\dagger U_1^\dagger = U_2^{-1} U_1^{-1} = (U_1 U_2)^{-1},$$

so that the product is unitary and an element of $SU(n)$. Each unitary matrix obviously has an inverse which again is unitary.

In the following we shall discuss only the rotation groups $SO(n)$ and unitary groups $SU(n)$ among the classical Lie groups. More examples for finite groups will be given in Section 4.7.

HOMOMORPHISM, ISOMORPHISM

There may be a correspondence between the elements of two groups (or between two representations), one-to-one, two-to-one, or many-to-one. If this correspondence preserves the group multiplication, we say that the two groups are *homomorphic*. A most important homomorphic correspondence between the rotation group $SO(3)$ and the unitary group $SU(2)$ is developed in Section 4.2. If the correspondence is one-to-one, still preserving the group multiplication,[3] then the groups are *isomorphic*. An example is the *rotations of the coordinates* through a finite angle φ counterclockwise about the z-axis in three-dimensional space described by

$$R_z(\varphi) = \begin{pmatrix} \cos\varphi & \sin\varphi & 0 \\ -\sin\varphi & \cos\varphi & 0 \\ 0 & 0 & 1 \end{pmatrix}. \tag{4.3}$$

[3] Suppose the elements of one group are labeled g_i, the elements of a second group h_i. Then $g_i \leftrightarrow h_i$ is a one-to-one correspondence for all values of i. If $g_i g_j = g_k$ and $h_i h_j = h_k$, then g_k and h_k must be the corresponding group elements.

The group of rotations R_z is obviously isomorphic to the group of rotations in Eq. (4.1). (See Exercise 4.1.4.)

MATRIX REPRESENTATIONS—REDUCIBLE AND IRREDUCIBLE

The representation of group elements by matrices is a very powerful technique and has been almost universally adopted by physicists. The use of matrices imposes no significant restriction. It can be shown that the elements of any finite group and of the continuous groups of Sections 4.2–4.4 may be represented by matrices. Examples are the rotations described in Eqs. (4.1) and (4.3).

To illustrate how matrix representations arise from a symmetry, consider the stationary Schrödinger equation (or some other eigenvalue equation such as $Iv_i = I_i v_i$ for the principal moments of inertia of a rigid body in classical mechanics, say)

$$H\psi = E\psi. \tag{4.4}$$

Let us assume that Eq. (4.4) stays invariant under a group G of transformations R in G (coordinate rotations, for example, for a central potential $V(r)$ in the Hamiltonian H), i.e.,

$$H_R = \mathsf{R}H\mathsf{R}^{-1} = H. \tag{4.5}$$

Now take a solution ψ of Eq. (4.4) and "rotate" it: $\psi \rightarrow \mathsf{R}\psi$. Then $\mathsf{R}\psi$ has the *same energy* E because multiplying Eq. (4.4) by R and using Eq. (4.5) yields

$$\mathsf{R}H\psi = E(\mathsf{R}\psi) = (\mathsf{R}H\mathsf{R}^{-1})\mathsf{R}\psi = H(\mathsf{R}\psi). \tag{4.6}$$

In other words, all rotated solutions $\mathsf{R}\psi$ are *degenerate* in energy or form what physicists call a *multiplet*. Let us assume that this vector space V_ψ of transformed solutions has a finite dimension n. Let $\psi_1, \psi_2, \ldots, \psi_n$ be a basis. Since $\mathsf{R}\psi_j$ is a member of the multiplet, we can expand it in terms of its basis,

$$\mathsf{R}\psi_j = \sum_k r_{jk}\psi_k. \tag{4.7}$$

Thus, with each R in G we can associate a matrix (r_{jk}), and this map $\mathsf{R} \rightarrow (r_{jk})$ is called a representation of G. If we can take any element of V_ψ and by rotating with *all* elements R of G transform it into all other elements of V_ψ then the representation is *irreducible*. If all elements of V_ψ are not reached, then V_ψ splits into a direct sum of two or more vector subspaces, $V_\psi = V_1 + V_2 + \ldots$, which are mapped into themselves by rotating their elements. In this case the representation is called *reducible*. Then we can find a basis in V_ψ (i.e., there is a unitary matrix U) so that

$$\mathsf{U}(r_{jk})\mathsf{U}^\dagger = \begin{pmatrix} \mathbf{r}_1 & \mathbf{0} & \cdots \\ \mathbf{0} & \mathbf{r}_2 & \cdots \\ \vdots & \vdots & \end{pmatrix} \tag{4.8}$$

for *all* R of *G* and *all* matrices (r_{jk}). Here $\mathbf{r}_1$, $\mathbf{r}_2$, ..., are matrices of lower dimension than (r_{jk}) that are lined up along the diagonal and the **0**'s are matrices made up of zeros. We may say that *R* has been decomposed into $\mathbf{r}_1 + \mathbf{r}_2 + \ldots$ along with $V_\psi = V_1 \oplus V_2 \oplus \ldots$.

The irreducible representations play a role in group theory that is roughly analogous to the unit vectors of vector analysis. They are the simplest representations—all others can be built from them. (See Section 4.4 on Clebsch–Gordan coefficients and Young tableaux.)

EXERCISES

4.1.1   Show that an $n \times n$ orthogonal matrix has $n(n - 1)/2$ *independent* parameters.

Hint. The orthogonality condition, Eq. (3.71), provides constraints.

4.1.2   Show that an $n \times n$ unitary matrix has $n^2 - 1$ *independent* parameters.

Hint. Each element may be complex, doubling the number of possible parameters. Some of the constraint equations are likewise complex—and count as two constraints.

4.1.3   The special linear group SL(2) consists of all 2×2 matrices (with complex elements) having a determinant of $+1$. Show that such matrices form a group.

Note. The SL(2) group can be related to the full Lorentz group in Section 4.4 much as the SU(2) group is related to SO(3).

4.1.4   Show that the rotations about the *z*-axis form a subgroup of SO(3). Is it an invariant subgroup?

4.1.5   Show that if *R*, *S*, *T* are elements of a group *G* so that $RS = T$, and $R \rightarrow (r_{ik})$, $S \rightarrow (s_{ik})$ is a representation according to Eq. (4.7), then

$$(r_{ik})(s_{ik}) = \left(t_{ik} = \sum_n r_{in} s_{nk} \right),$$

i.e., group multiplication translates into matrix multiplication for any group representation.

4.2 GENERATORS OF CONTINUOUS GROUPS

A characteristic of continuous groups known as Lie groups is that the parameters of a product element are analytic functions[1] of the parameters of the factors. The analytic nature of the functions (differentiability) allows us to develop the concept of *generator* and reduce the study of the whole group to a study of the group elements in the neighborhood of the identity element.

---

[1] Analytic (see Section 6.2) means having derivatives of all orders.

Lie's essential idea was to study elements R in a group G that are infinitesimally close to the unity of G. Let us consider the SO(2) group as a simple example. The 2×2 rotation matrices in Eq. (4.1) can be written in *exponential* form using the Euler identity Eq. (3.170) as

$$R(\varphi) = \begin{pmatrix} \cos\varphi & \sin\varphi \\ -\sin\varphi & \cos\varphi \end{pmatrix} = 1_2 \cos\varphi + i\sigma_2 \sin\varphi = \exp(i\sigma_2\varphi). \quad (4.9)$$

From the exponential form it is obvious that multiplication of these matrices is equivalent to addition of the arguments

$$R(\varphi_2)R(\varphi_1) = \exp(i\sigma_2\varphi_2)\exp(i\sigma_2\varphi_1) = \exp(i\sigma_2(\varphi_1 + \varphi_2)) = R(\varphi_1 + \varphi_2).$$

Of course, the rotations close to 1 have small angle $\varphi \cong 0$.

This suggests that we look for an *exponential* representation

$$R = \exp(i\varepsilon S), \qquad \varepsilon \to 0, \qquad (4.10)$$

for group elements $R\varepsilon G$ close to the unity 1. The infinitesimal transformations S are called *generators* of G. They form a linear space whose dimension is the *order* of G because multiplication of the elements R of the group translates into addition of generators S.

If R does not change the volume element, i.e., $\det(R) = 1$, we use Eq. (3.171) to see that

$$\det(R) = \exp(\text{tr}(\ln R)) = \exp(i\varepsilon\text{tr}(S)) = 1$$

implies that *generators are traceless*,

$$\text{tr}(S) = 0. \qquad (4.11)$$

This is the case for the rotation groups SO(n) and unitary groups SU(n), as we shall see below.

If R of G in Eq. (4.10) is unitary, then $S^\dagger = S$ is hermitian, which is also the case for SO(n) and SU(n). Hence the extra i in Eq. (4.10).

Next we go around the unity in four steps, similar to parallel transport in differential geometry. We expand the group elements

$$R_i = \exp(i\varepsilon_i S_i) = 1 + i\varepsilon_i S_i - 0.5\varepsilon_i^2 S_i^2 + \dots,$$
$$R_i^{-1} = \exp(-i\varepsilon_i S_i) = 1 - i\varepsilon_i S_i - 0.5\varepsilon_i^2 S_i^2 + \dots, \qquad (4.12)$$

to second order in the small group parameter ε_i because the linear terms and several quadratic terms all cancel in the product (Fig. 4.1)

$$R_i^{-1}R_j^{-1}R_iR_j = 1 + \varepsilon_i\varepsilon_j[S_j, S_i] + \dots,$$
$$= 1 + \varepsilon_i\varepsilon_j \sum_k c_{ji}^k S_k + \dots, \qquad (4.13)$$

when Eqs. (4.12) are substituted into Eq. (4.13). The last line holds because the product in Eq. (4.13) is again an element, R_{ij}, close to the unity in the

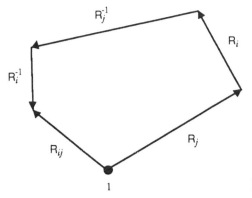

Figure 4.1 Illustration of Eq. (4.13).

group G. Hence its exponent must be a linear combination of the generators S_k and its infinitesimal group parameter has to be proportional to the product $\varepsilon_i \varepsilon_j$. Comparing both lines in Eq. (4.13) we find the *closure* relation of the generators of the Lie group G,

$$[S_i, S_j] = \sum_k c_{ij}^k S_k. \tag{4.14}$$

The coefficients c_{ij}^k are the *structure constants* of the group G. Since the commutator in Eq. (4.14) is antisymmetric in i and j, so are the structure constants in the lower indices,

$$c_{ij}^k = -c_{ji}^k. \tag{4.15}$$

If the *commutator* in Eq. (4.14) is taken as a *multiplication law of generators*, we see that the vector space of generators becomes an algebra, the *Lie algebra G* of the group G. For SU($l + 1$) the Lie algebra is called A_l, for SO($2l + 1$) it is B_l, and for SO($2l$) it is D_l, where $l = 1, 2, \ldots$ is a positive integer, later called the *rank* of the Lie group G or of its algebra G.

Finally, the *Jacobi identity* holds for all double commutators

$$[[S_i, S_j], S_k] + [[S_j, S_k], S_i] + [[S_k, S_i], S_j] = 0, \tag{4.16}$$

which is easily verified using the definition of any commutator $[A, B] = AB - BA$. When Eq. (4.14) is substituted into Eq. (4.16) we find another constraint on structure constants,

$$\sum_m \{c_{ij}^m [S_m, S_k] + c_{jk}^m [S_m, S_i] + c_{ki}^m [S_m, S_j]\} = 0. \tag{4.17}$$

Upon inserting Eq. (4.14) again, Eq. (4.17) implies that

$$\sum_{mn} \{c_{ij}^m c_{mk}^n S_n + c_{jk}^m c_{mi}^n S_n + c_{ki}^m c_{mj}^n S_n\} = 0, \tag{4.18}$$

where the common factor S_n (and the sum over n) may be dropped because the generators are linearly independent. Hence

$$\sum_m \{c_{ij}^m c_{mk}^n + c_{jk}^m c_{mi}^n + c_{ki}^m c_{mj}^n\} = 0. \qquad (4.19)$$

The relations (4.14), (4.15), and (4.19) form the basis of Lie algebras from which finite elements of the Lie group near its unity can be reconstructed.

Returning to Eq. (4.5), the inverse of R is just $R^{-1} = \exp(-i\varepsilon S)$. We expand H_R according to the Baker–Hausdorff formula Eq. (3.172),

$$H = H_R = \exp(i\varepsilon S)H \exp(-i\varepsilon S) = H + i\varepsilon[S, H] - \varepsilon^2[S[S, H]]/2 + \dots. \qquad (4.20)$$

We drop H from Eq. (4.20), divide by ε, and let $\varepsilon \to 0$. Then Eq. (4.20) implies that for any rotation close to 1 in G the commutator

$$[S, H] = 0. \qquad (4.21)$$

If S and H are Hermitian matrices, Eq. (4.21) says that S and H can be simultaneously diagonalized. If S and H are differential operators like the Hamiltonian and orbital angular momentum in quantum mechanics, then Eq. (4.21) says that S and H have common eigenfunctions, and that the degenerate eigenvalues of H can be distinguished by the eigenvalues of the generators S. This is by far the most important application of group theory in quantum mechanics.

In the next subsections we shall study orthogonal and unitary groups as examples to understand better the general concepts of this section.

ROTATION GROUPS SO(2) AND SO(3)

For SO(2) as defined by Eq. (4.1) there is only one linearly independent generator, σ_2, and the order of SO(2) is 1. We get σ_2 from Eq. (4.9) by differentiation at the unity of SO(2), i.e., $\varphi = 0$,

$$-idR(\varphi)/d\varphi|_{\varphi=0} = -i \begin{pmatrix} -\sin\varphi & \cos\varphi \\ -\cos\varphi & -\sin\varphi \end{pmatrix}\bigg|_{\varphi=0} = -i \begin{pmatrix} 0 & 1 \\ -1 & 0 \end{pmatrix} = \sigma_2. \qquad (4.22)$$

For the rotations $R_z(\varphi)$ about the z-axis described by Eq. (4.3), the generator is given by

$$-idR_z(\varphi)/d\varphi|_{\varphi=0} = S_z = \begin{pmatrix} 0 & -i & 0 \\ i & 0 & 0 \\ 0 & 0 & 0 \end{pmatrix}, \qquad (4.23)$$

where the extra factor i is inserted to make S_z Hermitian. The rotation $R_z(\delta\varphi)$ through an infinitesimal angle $\delta\varphi$ may then be written as

$$R_z(\delta\varphi) = 1_3 + i\delta\varphi S_z, \qquad (4.24)$$

a Maclaurin–Taylor expansion (see Chapter 5) of R_z near the unity ($\varphi = 0$) with terms of order $(\delta\varphi)^2$ and higher omitted. A finite rotation $R(\varphi)$ may be compounded of successive infinitesimal rotations

$$R_z(\delta\varphi_1 + \delta\varphi_2) = (1 + i\delta\varphi_1 S_z)(1 + i\delta\varphi_2 S_z). \tag{4.25}$$

Let $\delta\varphi = \varphi/N$ for N rotations, with $N \rightarrow \infty$. Then,

$$R_z(\varphi) = \lim_{N\to\infty} [1 + (i\varphi/N)S_z]^N = \exp(iS_z). \tag{4.26}$$

This form identifies S_z as the generator of the group R_z, an abelian subgroup of SO(3), the group of rotations in three dimensions with determinant $+1$. Each 3×3 matrix $R_z(\varphi)$ is orthogonal, hence unitary, and tr $(S_z) = 0$ in accord with Eq. (4.11).

By differentiation of the coordinate rotations

$$R_x(\psi) = \begin{pmatrix} 1 & 0 & 0 \\ 0 & \cos\psi & \sin\psi \\ 0 & -\sin\psi & \cos\psi \end{pmatrix}, \quad R_y(\theta) = \begin{pmatrix} \cos\theta & 0 & -\sin\theta \\ 0 & 1 & 0 \\ \sin\theta & 0 & \cos\theta \end{pmatrix}, \tag{4.27}$$

we get the generators

$$S_x = \begin{pmatrix} 0 & 0 & 0 \\ 0 & 0 & -i \\ 0 & i & 0 \end{pmatrix}, \quad S_y = \begin{pmatrix} 0 & 0 & -i \\ 0 & 0 & 0 \\ i & 0 & 0 \end{pmatrix}, \tag{4.28}$$

of $R_x(R_y)$, the subgroup of rotations about the x- (y-) axis.

ROTATION OF FUNCTIONS AND ORBITAL ANGULAR MOMENTUM

In the foregoing discussion the group elements are matrices that rotate the coordinates. Any physical system being described is held fixed. Now let us hold the coordinates fixed and rotate a function $\psi(x, y, z)$ relative to our fixed coordinates. With R to rotate the coordinates,

$$\mathbf{x}' = R\mathbf{x}, \tag{4.29}$$

we define R by

$$R\psi(x, y, z) = \psi'(x, y, z) \rightarrow \psi(\mathbf{x}'). \tag{4.30}$$

In words, R operates on the function ψ, creating a *new function* ψ' that is numerically equal to $\psi(\mathbf{x}')$, where $\mathbf{x}'$ are the coordinates rotated by R. If R rotates the coordinates counterclockwise, the effect of R is to rotate the pattern of the function ψ counterclockwise.

Returning to Eqs. (4.3) and (4.28), consider an infinitesimal rotation again, $\varphi \rightarrow \delta\varphi$. Then, using R_z, Eq. (4.3), we obtain

$$R_z(\delta\varphi)\psi(x, y, z) = \psi(x + y\delta\varphi, y - x\delta\varphi, z). \tag{4.31}$$

The right side may be expanded as a Taylor series (Section 5.6) to first order in $\delta\varphi$ to give

$$R_z(\delta\varphi)\psi(x, y, z) = \psi(x, y, z) - \delta\varphi\{x\partial\psi/\partial y - y\partial\psi/\partial x\} + O(\delta\varphi)^2$$

$$= (1 - i\delta\varphi L_z)\psi(x, y, z), \tag{4.32}$$

the differential expression in curly brackets being iL_z (Exercise 1.8.7). Since a rotation of first φ and then $\delta\varphi$ about the z-axis is given by

$$R_z(\varphi + \delta\varphi)\psi = R_z(\delta\varphi)R_z(\varphi)\psi = (1 - i\delta\varphi L_z)R_z(\varphi)\psi, \tag{4.33}$$

we have (as an operator equation)

$$(R_z(\varphi + \delta\varphi) - R_z(\varphi))/\delta\varphi = -iL_z R_z(\varphi). \tag{4.34}$$

The left side is just $dR_z(\varphi)/d\varphi$ (for $\delta\varphi \to 0$). In this form Eq. (4.34) integrates immediately to

$$R_z(\varphi) = \exp(-i\varphi L_z). \tag{4.35}$$

Note carefully that $R_z(\varphi)$ rotates functions (counterclockwise) relative to fixed coordinates and that L_z is the z-component of the orbital angular momentum **L**. The constant of integration is fixed by the boundary condition $R_z(0) = 1$.

If we recognize that the matrix element

$$L_z = (x, y, z)S_z\begin{pmatrix} \partial/\partial x \\ \partial/\partial y \\ \partial/\partial z \end{pmatrix}, \tag{4.36}$$

it becomes clear why L_x, L_y, and L_z satisfy the same commutation relation

$$[L_i, L_j] = i\varepsilon_{ijk}L_k \tag{4.37}$$

as S_x, S_y, and S_z and yield the same structure constants $i\varepsilon_{ijk}$ of SO(3).

SU(2)–SO(3) HOMOMORPHISM

The *special unitary* group SU(2) of unitary 2×2 matrices with determinant $+1$ has all three Pauli matrices σ_i as generators (while the rotations of Eq. (4.3) form a one-dimensional abelian subgroup). So SU(2) is of order 3 and depends on three real continuous parameters ξ, η, ζ which are often called the *Caley–Klein* parameters. Its general element has the form

$$U_2(\xi, \eta, \zeta) = \begin{pmatrix} e^{i\xi}\cos\eta & e^{i\zeta}\sin\eta \\ -e^{-i\zeta}\sin\eta & e^{-i\xi}\cos\eta \end{pmatrix} = \begin{pmatrix} a & b \\ -b^* & a^* \end{pmatrix}. \tag{4.38}$$

It is easy to check that the determinant $\text{Det}(U_2) = 1$ and that $U_2^\dagger U_2 = 1 = U_2 U_2^\dagger$ hold.

To get the generators we differentiate (and drop irrelevant overall factors)

$$-i\partial U_2/\partial\xi|_{\xi=0,\eta=0} = \begin{pmatrix} 1 & 0 \\ 0 & -1 \end{pmatrix} = \sigma_3,$$

$$[-i\partial U_2/\partial\zeta|_{\zeta=0}]/\sin\eta = \begin{pmatrix} 0 & 1 \\ 1 & 0 \end{pmatrix} = \sigma_1,$$ (4.39)

$$-i\partial U_2/\partial\eta|_{\eta=0,\zeta=0} = \begin{pmatrix} 0 & -i \\ i & 0 \end{pmatrix} = \sigma_2.$$

Of course, the Pauli matrices are all traceless and Hermitian.

With the Pauli matrices as generators the elements (U_1, U_2, U_3) of SU(2) may be generated by

$$U_1 = \exp(ia_1\sigma_1/2), \qquad U_2 = \exp(ia_2\sigma_2/2), \qquad U_3 = \exp(ia_3\sigma_3/2).$$ (4.40)

The three parameters a_i are real. The extra factor $1/2$ is present in the exponents because $s_i = \sigma_i/2$ satisfy the same commutation relations[2]

$$[s_i, s_j] = i\varepsilon_{ijk}s_k$$ (4.41)

as the angular momentum in Eq. (4.37).

Equation (4.3) gives the rotation operator for rotating the cartesian coordinates in the three-space. Using the angular momentum matrix s_3, we have as the corresponding rotation operator in two-dimensional (complex) space $R_z(\varphi) = \exp(i\varphi\sigma_3/2)$.

For rotating the two-component vector wave function (spinor) or a spin $1/2$ particle relative to fixed coordinates, the rotation operator is $R_z(\varphi) = \exp(-i\varphi\sigma_3/2)$ according to Eq. (4.35).

Using in Eq. (4.40) the Euler identity, Eq. (3.170), we obtain

$$U_j = \cos(a_j/2) + i\sigma_j\sin(a_j/2).$$

Here the parameter a_j appears as an angle, the coefficient of an angular momentum matrix—like φ in Eq. (4.26). With this identification of the exponentials, the general form of the SU(2) matrix (for rotating functions rather than coordinates) may be written as

$$U(\alpha, \beta, \gamma) = \exp(-i\gamma\sigma_3/2)\exp(-i\beta\sigma_2/2)\exp(-i\alpha\sigma_3/2).$$

This reproduces Eq. (4.57) below. With the rotation matrix convention

---

[2] The structure constants $(i\varepsilon_{ijk})$ lead to the SU(2) representations of dimension $2J + 1$ for generators of dimension $2J + 1$, $J = 0, 1/2, 1, \ldots$. The *integral J* cases also lead to the representations of SO(3).

$$D(\alpha, \beta, \gamma) = U^{\dagger}(\alpha, \beta, \gamma),$$

$$D^{1/2}(\alpha, \beta, \gamma) = \exp(i\alpha\sigma_3/2)\exp(i\beta\sigma_2/2)\exp(i\gamma\sigma_3/2)$$

$$D^{L}(\alpha, \beta, \gamma) = \exp(i\alpha L_z)\exp(i\beta L_y)\exp(i\gamma L_z) \tag{4.42}$$

$$D^{J}_{m'm}(\alpha, \beta, \gamma) = \langle Jm|\exp(i\alpha J_z)\exp(i\beta J_y)\exp(i\gamma J_z)|Jm'\rangle$$

leads to Eq. (4.58) below. The selection of Pauli matrices corresponds to the Euler angle rotations described in Chapter 3.3.

As just seen, the elements of SU(2) describe rotations in a two-dimensional complex space that leave $|z_1|^2 + |z_2|^2$ invariant. The determinant is $+1$. There are three independent parameters. Our real orthogonal group SO(3) determinant $+1$, clearly describes rotations in ordinary three-dimensional space with the important characteristic of leaving $x^2 + y^2 + z^2$ *invariant*. Also, there are three independent parameters. The rotation interpretations and the equality of numbers of parameters suggest the existence of some sort of correspondence between the groups SU(2) and SO(3). Here we develop this correspondence.

The operation of SU(2) on a matrix is given by a unitary transformation, Eq. (4.5), with $R = U$ and Fig. 4.2

$$M' = UMU^{\dagger}. \tag{4.43}$$

Taking M to be a 2×2 matrix, we note that any 2×2 matrix may be written as a linear combination of the unit matrix and the three Pauli matrices of Section 3.2. Let M be the zero-trace matrix,

$$M = x\sigma_1 + y\sigma_2 + z\sigma_3 = \begin{pmatrix} z & x - iy \\ x + iy & -z \end{pmatrix}, \tag{4.44}$$

the unit matrix not entering. Since the trace is invariant under a unitary transformation (Exercise 3.3.9), M' must have the same form,

$$M' = x'\sigma_1 + y'\sigma_2 + z'\sigma_3 = \begin{pmatrix} z' & x' - iy' \\ x' + iy' & -z' \end{pmatrix}. \tag{4.45}$$

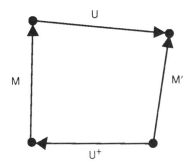

Figure 4.2 Illustration of $M' = UMU^{\dagger}$ in Eq. (4.43).

The determinant is also invariant under a unitary transformation (Exercise 3.3.10). Therefore

$$-(x^2 + y^2 + z^2) = -(x'^2 + y'^2 + z'^2),$$ (4.46)

or $x^2 + y^2 + z^2$ is invariant under this operation of SU(2), just as with SO(3). SU(2) must, therefore, describe a rotation. This suggests that SU(2) and SO(3) may be isomorphic or homomorphic.

We approach the problem of what rotation SU(2) describes by considering special cases. Returning to Eq. (4.38) let $a = e^{i\xi}$ and $b = 0$, or

$$U_z = \begin{pmatrix} e^{i\xi} & 0 \\ 0 & e^{-i\xi} \end{pmatrix}.$$ (4.47)

In anticipation of Eq. (4.51), this U is given a subscript z.

Carrying out a unitary transformation on each of the three Pauli σ's, we have

$$U_z \sigma_1 U_z^\dagger = \begin{pmatrix} e^{i\xi} & 0 \\ 0 & e^{-i\xi} \end{pmatrix} \begin{pmatrix} 0 & 1 \\ 1 & 0 \end{pmatrix} \begin{pmatrix} e^{-i\xi} & 0 \\ 0 & e^{i\xi} \end{pmatrix}$$

$$= \begin{pmatrix} 0 & e^{2i\xi} \\ e^{-2i\xi} & 0 \end{pmatrix}.$$ (4.48)

We reexpress this result in terms of the Pauli σ's to obtain

$$U_z x \sigma_1 U_z^\dagger = x \cos 2\xi \sigma_1 - x \sin 2\xi \sigma_2.$$ (4.49)

Similarly,

$$U_z y \sigma_2 U_z^\dagger = y \sin 2\xi \sigma_1 + y \cos 2\xi \sigma_2$$
$$U_z z \sigma_3 U_z^\dagger = z \sigma_3.$$ (4.50)

From these double angle expressions we see that we should start with a *half-angle*: $\xi = \alpha/2$. Then, from Eqs. (4.43)–(4.45), (4.49), and (4.50),

$$x' = x \cos \alpha + y \sin \alpha$$
$$y' = -x \sin \alpha + y \cos \alpha$$ (4.51)
$$z' = z.$$

The 2×2 unitary transformation using $U_z(\alpha/2)$ is equivalent to the rotation operator $R(\alpha)$ of Eq. (4.3).

The establishment of the correspondence of

$$U_y(\beta/2) = \begin{pmatrix} \cos \beta/2 & \sin \beta/2 \\ -\sin \beta/2 & \cos \beta/2 \end{pmatrix}$$ (4.52)

and $R_y(\beta)$ and of

$$U_x(\varphi/2) = \begin{pmatrix} \cos \varphi/2 & i \sin \varphi/2 \\ i \sin \varphi/2 & \cos \varphi/2 \end{pmatrix}$$ (4.53)

and $R_x(\varphi)$ are left as Exercise 4.2.7. The reader might note that $U_k(\psi/2)$ has the general form

$$U_k(\psi/2) = 1 \cos \psi/2 + i\sigma_k \sin \psi/2, \qquad (4.54)$$

where $k = x, y, z$. We made this point in Section 4.2.

The correspondence

$$U_z(\alpha/2) = \begin{pmatrix} e^{i\alpha/2} & 0 \\ 0 & e^{-i\alpha/2} \end{pmatrix} \leftrightarrow \begin{pmatrix} \cos \alpha & \sin \alpha & 0 \\ -\sin \alpha & \cos \alpha & 0 \\ 0 & 0 & 1 \end{pmatrix} = R_z(\alpha) \qquad (4.55)$$

is not a simple one-to-one correspondence. Specifically, as α in R_z ranges from 0 to 2π, the parameter in U_z, $\alpha/2$, goes from 0 to π. We find

$$R_z(\alpha + 2\pi) = R_z(\alpha)$$

$$U_z(\alpha/2 + \pi) = \begin{pmatrix} -e^{i\alpha/2} & 0 \\ 0 & -e^{-i\alpha/2} \end{pmatrix} = -U_z(\alpha/2). \qquad (4.56)$$

Therefore *both* $U_z(\alpha/2)$ and $U_z(\alpha/2 + \pi) = -U_z(\alpha/2)$ correspond to $R_z(\alpha)$. The correspondence is 2 to 1, or SU(2) and SO(3) are *homomorphic*. This establishment of the correspondence between the representations of SU(2) and those of SO(3) means that the known representations of SU(2) automatically provide us with the representations[3] of SO(3).

Combining the various rotations, we find that a unitary transformation using

$$U(\alpha, \beta, \gamma) = U_z(\gamma/2)U_y(\beta/2)U_z(\alpha/2) \qquad (4.57)$$

corresponds to the general Euler rotation $R_z(\gamma)R_y(\beta)R_z(\alpha)$. By direct multiplication,

$$U(\alpha, \beta, \gamma) = \begin{pmatrix} e^{i\gamma/2} & 0 \\ 0 & e^{-i\gamma/2} \end{pmatrix} \begin{pmatrix} \cos \beta/2 & \sin \beta/2 \\ -\sin \beta/2 & \cos \beta/2 \end{pmatrix} \begin{pmatrix} e^{i\alpha/2} & 0 \\ 0 & e^{-i\alpha/2} \end{pmatrix}$$

$$= \begin{pmatrix} e^{i(\gamma+\alpha)/2} \cos \beta/2 & e^{i(\gamma-\alpha)/2} \sin \beta/2 \\ -e^{-i(\gamma-\alpha)/2} \sin \beta/2 & e^{-i(\gamma+\alpha)/2} \cos \beta/2 \end{pmatrix}. \qquad (4.58)$$

This is our alternate general form, Eq. (4.38), with

$$\varepsilon = (\gamma + \alpha)/2, \qquad \eta = \beta/2, \qquad \zeta = (\gamma - \alpha)/2. \qquad (4.59)$$

From Eq. (4.58) we may identify the parameters of Eq. (4.38) as

$$a = e^{i(\gamma+\alpha)/2} \cos \beta/2$$

$$b = e^{i(\gamma-\alpha)/2} \sin \beta/2. \qquad (4.60)$$

[3] Whereas SU(2) has representations for integral and half odd integral values of j ($j = 0, \frac{1}{2}, 1, \frac{3}{2}, \ldots$), SO(3) is limited to integral values of j ($j = 0, 1, 2, \ldots$). Further discussion of this point—the relation between SO(3) and orbital angular momentum—appears in Section 12.7.

SU(2)-ISOSPIN AND SU(3)-EIGHTFOLD WAY

The application of group theory to "elementary" particles has been labeled by Wigner the third stage of group theory and physics. The first stage was the search for the 32 point groups and the 230 space groups giving crystal symmetries—Section 4.7. The second stage was a search for representations such as of SO(3) and SU(2)—Section 4.2. Now in this stage, physicists are back to a search for groups.

In discussing the strongly interacting particles of nuclear and high energy physics that lead to the SU(2) isospin group and the SU(3) flavor symmetry, we should look to angular momentum and the rotation group SO(3) for an analogy. Suppose we have an electron in the spherically symmetric attractive potential of some atomic nucleus. The electron's Schrödinger wave function may be characterized by three quantum numbers n, l, and m, that are related to the eigenvalues of conserved operators H, $\mathbf{L}^2$, and L_z. The energy,[4] however, is $2l + 1$-fold degenerate, depending only on n and l. The reason for this degeneracy may be stated in two equivalent ways:

1. The potential is spherically symmetric, independent of θ, φ.
2. The Schrödinger Hamiltonian $-(\hbar^2/2m_e)\nabla^2 + V(r)$ is *invariant* under ordinary spatial rotations [SO(3)].

As a consequence of the spherical symmetry of the potential $V(r)$, the orbital angular momentum $\mathbf{L}$ is conserved. In Section 4.2 the cartesian components of $\mathbf{L}$ are identified as the generators of the rotation group SO(3). Instead of representing L_x, L_y, and L_z by operators, let us use matrices. The exercises at the end of Section 3.2 provide examples for $l = \frac{1}{2}$, 1, and $\frac{3}{2}$. The L_i matrices are $(2l + 1) \times (2l + 1)$ matrices with the dimension the same as the number of the degenerate states.[5] These L_i matrices generate the $(2l + 1) \times (2l + 1)$ irreducible representations of SO(3). The dimension $2l + 1$ is identified with the $2l + 1$ degenerate states.

This degeneracy is lifted by a constant magnetic induction field $\mathbf{B}$ leading to the Zeeman effect. This magnetic interaction adds a term to the Schrödinger Hamiltonian that is *not* invariant under SO(3). It is a symmetry-breaking term. So much for the analogy.

In the case of strongly interacting particles (protons, neutrons, etc.) we cannot follow the analogy directly, because we do not yet fully understand the nuclear interaction. The strong force is described by a Yang–Mills gauge theory based on the SU(3) color symmetry called quantum chromodynamics or QCD for short. However, QCD is a nonlinear theory and so complicated at long distances and low energy that it remains unsolved there. Therefore, we do not know the Hamiltonian. So instead, let us run the analogy backward.

---

[4] For a pure Coulomb potential the energy depends only on n (Section 13.2).

[5] With L_i a matrix, the Schrödinger wave function $\psi(r, \theta, \varphi)$ is replaced by a state vector with $2l + 1$ components. Angular momentum and the $(2l + 1)$-fold degeneracy are discussed at some length in Section 4.3.

Table 4.1 Baryons with Spin $\frac{1}{2}$ Even Parity

		Mass (MeV)	Y	I	I_3
Ξ	Ξ^-	1321.32			$-\frac{1}{2}$
			-1	$\frac{1}{2}$	
	Ξ^0	1314.9			$+\frac{1}{2}$
Σ	Σ^-	1197.43			-1
	Σ^0	1192.55	0	1	0
	Σ^+	1189.37			$+1$
Λ	Λ	1115.63	0	0	0
N	n	939.566			$-\frac{1}{2}$
			1	$\frac{1}{2}$	
	p	938.272			$+\frac{1}{2}$

In the 1930s, after the neutron was discovered, Heisenberg proposed that the nuclear forces were charge independent. The neutron mass differs from that of the proton by only 1.6%. If this tiny mass difference is ignored, the neutron and proton may be considered as two charge (or isospin) states of a doublet, called the nucleon. The isospin $\mathbf{I}$ has z-projection $I_3 = \frac{1}{2}$ for the proton and $I_3 = -\frac{1}{2}$ for the neutron. Isospin has nothing to do with spin (the particle's intrinsic angular momentum) but the two-component isospin state obeys the same mathematical relations as the spin $\frac{1}{2}$ state. For the nucleon, $\mathbf{I} = \tau/2$ are the usual Pauli matrices and the $\pm\frac{1}{2}$ isospin states are eigenvectors of the Pauli matrix $\tau_3 = \begin{pmatrix} 1 & 0 \\ 0 & -1 \end{pmatrix}$. Similarly, the three charge states of the pion—π^+, π^0, π^-—form a triplet. The pion is the lightest of all strongly interacting particles and is the carrier of the nuclear force at long distances, much like the photon is that of the electromagnetic force. The strong interaction treats alike members of these particle families, or multiplets, and conserves isospin. The symmetry is the SU(2) isospin group.

By the 1960s particles produced as resonances by accelerators had proliferated. The eight shown in Table 4.1 attracted particular attention.[6] The relevant conserved quantum numbers that are analogs and generalizations of L_z and $\mathbf{L}^2$ from SO(3) are I_3 and I^2 for isospin, and Y for *hypercharge*. Particles may be grouped into charge or isospin multiplets. Then the hypercharge may be taken as twice the average charge of the multiplet. For the nucleon, i.e., the neutron–proton doublet, $Y = 2 \cdot \frac{1}{2}(0 + 1) = 1$. The hypercharge and isospin values are listed in Table 4.1 for baryons like the nucleon and its (approximately degenerate) partners. They form an octet as shown in Fig. 4.3 after which the corresponding symmetry is called the *eightfold way*. In 1961 Gell–Mann, and independently Ne'eman, suggested that the strong interaction should be (approximately) invariant under a three-dimensional special unitary group, SU(3), that is, have SU(3) *flavor symmetry*.

[6] All masses are given in energy units, 1 MeV = 10^6 e-Volts.

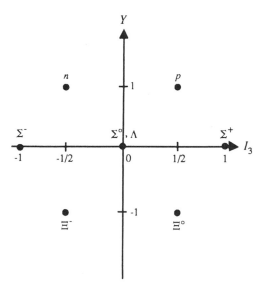

Figure 4.3 Baryon octet weight diagram for SU(3).

The choice of SU(3) was based first on the two conserved and independent quantum numbers $H_1 = I_3$ and $H_2 = Y$ (i.e., generators with $[I_3, Y] = 0$, see the summary in Section 4.3, not Casimir invariants) that call for a group of rank 2. Second, the group had to have an eight-dimensional representation to account for the nearly degenerate baryons and four similar octets for the mesons. In a sense SU(3) is the simplest generalization of SU(2) isospin. Three of its generators are zero-trace Hermitian 3×3 matrices that contain the 2×2 isospin Pauli matrices τ_i in the upper left corner,

$$\lambda_i = \begin{pmatrix} \tau_i & & 0 \\ & & 0 \\ 0 & 0 & 0 \end{pmatrix}, \qquad i = 1, 2, 3. \tag{4.61a}$$

Thus, the SU(2)–isospin group is a subgroup of SU(3)–flavor with $I_3 = \lambda_3/2$. Four other generators have the off-diagonal 1's of τ_1 and $-i$, i of τ_2 in all other possible locations to form zero-trace Hermitian 3×3 matrices,

$$\lambda_4 = \begin{pmatrix} 0 & 0 & 1 \\ 0 & 0 & 0 \\ 1 & 0 & 0 \end{pmatrix}, \qquad \lambda_5 = \begin{pmatrix} 0 & 0 & -i \\ 0 & 0 & 0 \\ i & 0 & 0 \end{pmatrix},$$

$$\lambda_6 = \begin{pmatrix} 0 & 0 & 0 \\ 0 & 0 & 1 \\ 0 & 1 & 0 \end{pmatrix}, \qquad \lambda_7 = \begin{pmatrix} 0 & 0 & 0 \\ 0 & 0 & -i \\ 0 & i & 0 \end{pmatrix}. \tag{4.61b}$$

The second diagonal generator has the two-dimensional unit matrix 1_2 in the upper left corner, which makes it clearly independent of the SU(2)–isospin subgroup because of its nonzero trace in that subspace, and -2 in the third diagonal place to make it traceless,

$$\lambda_8 = \frac{1}{\sqrt{3}} \begin{pmatrix} 1 & 0 & 0 \\ 0 & 1 & 0 \\ 0 & 0 & -2 \end{pmatrix}. \tag{4.61c}$$

Altogether there are $3^2 - 1 = 8$ generators for SU(3) which has order 8. From the commutators of these generators the structure constants of SU(3) can be easily obtained.

Returning to the SU(3) flavor symmetry we imagine the Hamiltonian for our eight baryons to be composed of three parts

$$H = H_{strong} + H_{medium} + H_{electromagnetic}. \tag{4.62}$$

The first part, H_{strong}, has the SU(3) symmetry and leads to the eightfold degeneracy. Introduction of the symmetry breaking term, H_{medium}, removes part of the degeneracy giving the four isospin multiplets (Ξ^-, Ξ^0), $(\Sigma^-, \Sigma^0, \Sigma^+)$, Λ, and $N = (p, n)$ different masses. These are still multiplets because H_{medium} has SU(2)–isospin symmetry. Finally, the presence of charge dependent forces splits the isospin multiplets and removes the last degeneracy. This imagined sequence is shown in Fig. 4.4.

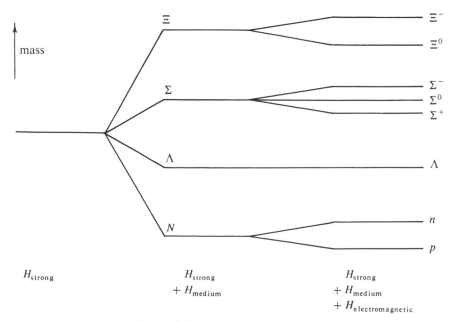

Figure 4.4 Baryon mass splitting.

Applying first-order perturbation theory of quantum mechanics, simple relations about baryon masses may be calculated. Perhaps the most spectacular success of this SU(3) model has been its prediction of new particles. In 1961 four K and three π mesons (all pseudoscalar; spin 0, odd parity) suggested another octet, similar to the baryon octet. SU(3) predicted an eighth meson η, mass 563 MeV. The η meson with experimentally determined mass 548 MeV was found soon after. Groupings of nine of the heavier baryons (all with spin $\frac{3}{2}$, even parity) suggested a 10-member multiplet or decuplet of SU(3). The missing tenth baryon was predicted to have a mass of about 1680 MeV and a negative charge. In 1964 the negatively charged Ω^- with mass (1675 ± 12) MeV was discovered. Since the completion of this $\frac{3}{2}^+$ decuplet, octets for scalar, vector, and axial vector mesons have been established, among others.

The octet representation is not the simplest SU(3) representation. The simplest representations are the triangular ones shown in Fig. 4.5 from which all others can be generated by generalized angular momentum coupling (see Section 4.4 on Young Tableaux). The *fundamental* representation in Fig. 4.5a contains the u (up), d (down), and s (strange) quarks, and Fig. 4.5b the corresponding antiquarks. Since the meson octets can be obtained from the quark representations as $q\bar{q}$, $3^2 = 8 + 1$, this suggests that mesons contain quarks (and antiquarks) as their constituents. The resulting quark model gives a successful description of hadronic spectroscopy. The resolution of its problem with the Pauli exclusion principle eventually led to the SU(3)–color gauge theory of the strong interaction called quantum chromodynamics or QCD for short.

To keep group theory and its very real accomplishment in proper perspective, we should emphasize that group theory identifies and formalizes symmetries. It classifies (and sometimes predicts) particles. But aside from saying that one

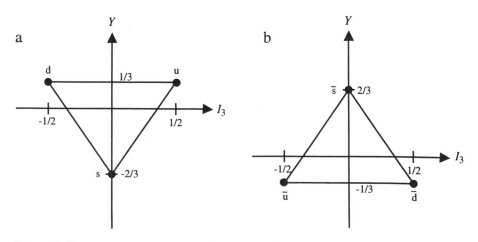

Figure 4.5 (a) Fundamental representation of SU(3), the weight diagram for the u, d, s quarks; (b) weight diagram for the antiquarks ū, d̄, s̄.

part of the Hamiltonian has SU(2) symmetry and another part has SU(3) symmetry, group theory says *nothing* about the particle interaction. Remember that the statement that the atomic potential is spherically symmetric tells us nothing about the radial dependence of the potential or of the wave function. In contrast, in a gauge theory the interaction is mediated by vector bosons (like the photon in quantum electrodynamics) and uniquely determined by the gauge covariant derivative (see Chapter 1.13).

EXERCISES

4.2.1   (i) Show that the Pauli matrices are the generators of SU(2) without using the parameterization of the general unitary 2×2 matrix in Eq. (4.38). (ii) Derive the eight independent generators λ_i of SU(3) similarly. Normalize them so that $\mathrm{tr}(\lambda_i \lambda_j) = 2\delta_{ij}$. Then determine the structure constants of SU(3).

Hint. The λ_i are traceless and Hermitian 3×3 matrices.

(iii) Construct the quadratic Casimir invariant of SU(3).

Hint. Work by analogy with $\sigma_1^2 + \sigma_2^2 + \sigma_3^2$ of SU(2) or $\mathbf{L}^2$ of SO(3).

4.2.2   Prove that the general form of a 2×2 unitary, unimodular matrix is

$$U = \begin{pmatrix} a & b \\ -b^* & a^* \end{pmatrix}$$

with $a^*a + b^*b = 1$.

4.2.3   Determine three SU(2) subgroups of SU(3).

4.2.4   A *translation* operator $T(a)$ converts $\psi(x)$ to $\psi(x + a)$,

$$T(a)\psi(x) = \psi(x + a).$$

In terms of the (quantum mechanical) linear momentum operator $p_x = -id/dx$, show that $T(a) = \exp(iap_x)$.

Hint. Expand $\psi(x + a)$ as a Taylor series.

4.2.5   Consider the general SU(2) element Eq. (4.38) to be built up of three Euler rotations: (i) a rotation of $a/2$ about the z-axis, (ii) a rotation of $b/2$ about the new x-axis, and (iii) a rotation of $c/2$ about the new z-axis. (All rotations are counterclockwise.) Using the Pauli σ generators, show that these rotation angles are determined by

$$a = \xi - \zeta + \pi/2 = \alpha + \pi/2$$
$$b = 2\eta \qquad\qquad = \beta$$
$$c = \xi + \zeta - \pi/2 = \gamma - \pi/2.$$

Note. The angles a and b here are not the a and b of Eq. (4.38).

4.3 ORBITAL ANGULAR MOMENTUM

The classical concept of angular momentum $\mathbf{L}_{\text{class}} = \mathbf{r} \times \mathbf{p}$ is presented in Section 1.4 to introduce the cross product. Following the usual Schrödinger representation of quantum mechanics, the classical linear momentum $\mathbf{p}$ is replaced by the operator $-i\nabla$. The quantum mechanical angular momentum *operator* becomes[1]

$$\mathbf{L}_{QM} = -i\mathbf{r} \times \nabla. \tag{4.63}$$

This is used repeatedly in Sections 1.8, 1.9, and 2.4 to illustrate vector differential operators. From Exercise 1.8.6 the angular momentum components satisfy the commutation relations

$$[L_i, L_j] = i\varepsilon_{ijk}L_k. \tag{4.64}$$

The ε_{ijk} is the Levi–Civita symbol of Section 2.9. A summation over the index k is understood.

The differential operator corresponding to the square of the angular momentum

$$\mathbf{L}^2 = \mathbf{L} \cdot \mathbf{L} = L_x^2 + L_y^2 + L_z^2 \tag{4.65}$$

may be determined from

$$\mathbf{L} \cdot \mathbf{L} = (\mathbf{r} \times \mathbf{p}) \cdot (\mathbf{r} \times \mathbf{p}), \tag{4.66}$$

which is the subject of Exercises 1.9.9 and 2.5.17(b). Since $\mathbf{L}^2$ is a rotational scalar, $[\mathbf{L}^2, L_i] = 0$, which can also be verified directly.

Equation (4.64) presents the basic commutation relations of the components of the quantum mechanical angular momentum. Indeed, within the framework of quantum mechanics and group theory, these commutation relations *define* an angular momentum operator.

LADDER OPERATOR APPROACH

Let us start with a more general approach, where the angular momentum $\mathbf{J}$ we consider may represent an orbital angular momentum $\mathbf{L}$, a spin $\sigma/2$, or a total angular momentum $\mathbf{L} + \sigma/2$, etc. We assume that

1. $\mathbf{J}$ is an Hermitian operator whose components satisfy the commutation relations

$$[J_i, J_j] = i\varepsilon_{ijk}J_k, \qquad [\mathbf{J}^2, J_i] = 0. \tag{4.67}$$

   Otherwise $\mathbf{J}$ is arbitrary. (See Exercise 4.3.1.)
2. $|\lambda M\rangle$ is simultaneously a normalized eigenfunction (or eigenvector) of J_z with eigenvalue M and an eigenfunction[2] of $\mathbf{J}^2$,

$$J_z|\lambda M\rangle = M|\lambda M\rangle, \qquad \mathbf{J}^2|\lambda M\rangle = \lambda|\lambda M\rangle. \tag{4.68}$$

---

[1] For simplicity, the $\hbar$ is set equal to 1. This means that the angular momentum is measured in units of $\hbar$.

[2] That $|\lambda M\rangle$ is an eigenfunction of *both* J_z and $\mathbf{J}^2$ follows from $[J_z, \mathbf{J}^2] = 0$ in Eq. (4.67).

We shall show that $\lambda = J(J + 1)$. Otherwise $|\lambda M\rangle$ is *unknown*. This treatment will illustrate the generality and power of operator techniques—particularly the use of *ladder operators*.[3]

The *ladder operators* are defined as

$$J_+ = J_x + iJ_y, \qquad J_- = J_x - iJ_y. \tag{4.69}$$

In terms of these operators $\mathbf{J}^2$ may be rewritten as

$$\mathbf{J}^2 = \tfrac{1}{2}(J_+J_- + J_-J_+) + J_z^2. \tag{4.70}$$

From the commutation relations, Eq. (4.67), we find

$$[J_z, J_+] = +J_+, \qquad [J_z, J_-] = -J_-, \qquad [J_+, J_-] = 2J_z. \tag{4.71}$$

Since J_+ commutes with $\mathbf{J}^2$ (Exercise 4.3.1),

$$\mathbf{J}^2(J_+|\lambda M\rangle) = J_+(\mathbf{J}^2|\lambda M\rangle) = \lambda(J_+|\lambda M\rangle). \tag{4.72}$$

Therefore, $J_+|\lambda M\rangle$ is still an eigenfunction of $\mathbf{J}^2$ with eigenvalue λ, and similarly for $J_-|\lambda M\rangle$. But from Eq. (4.71)

$$J_zJ_+ = J_+(J_z + 1), \tag{4.73}$$

or

$$J_z(J_+|\lambda M\rangle) = J_+(J_z + 1)|\lambda M\rangle = (M + 1)J_+|\lambda M\rangle. \tag{4.74}$$

Therefore, $J_+|\lambda M\rangle$ is still an eigenfunction of J_z but with eigenvalue $M + 1$. J_+ has raised the eigenvalue by 1 and so is often called a *raising operator*. Similarly, J_- lowers the eigenvalue by 1 and is often called a *lowering operator*.

Taking expectation values and using $J_x^\dagger = J_x$, $J_y^\dagger = J_y$,

$$\langle\lambda M|\mathbf{J}^2 - J_z^2|\lambda M\rangle = \langle\lambda M|J_x^2 + J_y^2|\lambda M\rangle = |J_x|\lambda M\rangle|^2 + |J_y|\lambda M\rangle|^2$$

we see that $\lambda - M^2 \geq 0$, so M is bounded. Let J be the *largest* M. Then $J_+|\lambda M\rangle = 0$, which implies $J_-J_+|\lambda M\rangle = 0$. Hence, combining Eqs. (4.70) and (4.71) to get

$$\mathbf{J}^2 = J_-J_+ + J_z(J_z + 1), \tag{4.75}$$

we find from Eq. (4.75) that

$$0 = J_-J_+|\lambda M\rangle = (\mathbf{J}^2 - J_z^2 - J_z)|\lambda M\rangle = (\lambda - J^2 - J)|\lambda M\rangle.$$

Therefore

$$\lambda = J(J + 1) \geq 0, \tag{4.76}$$

with nonnegative J. We now relabel the states $|\lambda M\rangle = |JM\rangle$. Similarly, let J' be the *smallest* M. Then $J_-|JJ'\rangle = 0$. From

$$\mathbf{J}^2 = J_+J_- - J_z(J_z + 1), \tag{4.77}$$

---

[3] Ladder operators can be developed for other mathematical functions. Compare the next subsection on other Lie groups and Section 13.1 for Hermite polynomials.

we see that

$$0 = J_+ J_- |JJ'\rangle = (\mathbf{J}^2 + J_z - J_z^2)|JJ'\rangle = (\lambda + J' - J'^2)|JJ'\rangle. \quad (4.78)$$

Hence

$$\lambda = J(J + 1) = J'(J' - 1) = (-J)(-J - 1).$$

So $J' = -J$, and M runs in integer steps from $-J$ to $+J$,

$$-J \le M \le J. \quad (4.79)$$

Starting from $|JJ\rangle$ and applying J_- repeatedly, we reach all other states $|JM\rangle$. Hence the $|JM\rangle$ form an irreducible representation; M varies and J is fixed.
Then using Eqs. (4.67), (4.75), and (4.77) we obtain

$$J_- J_+ |JM\rangle = [J(J + 1) - M(M + 1)]|JM\rangle = (J - M)(J + M + 1)|JM\rangle,$$

$$J_+ J_- |JM\rangle = [J(J + 1) - M(M - 1)]|JM\rangle = (J + M)(J - M + 1)|JM\rangle. \quad (4.80)$$

As J_+ and J_- are Hermitian conjugates,[4]

$$J_+^\dagger = J_-, \qquad J_-^\dagger = J_+, \quad (4.81)$$

the eigenvalues or expectation values in Eq. (4.80) must be positive or zero.[5] Examples of Eq. (4.81) are provided by the matrices of Exercise 3.2.13 (spin $\frac{1}{2}$), 3.2.15 (spin 1), and 3.2.18 (spin $\frac{3}{2}$). For the orbital angular momentum ladder operators, L_+ and L_-, explicit forms are given in Exercises 2.5.14 and 12.6.7. The reader can now show (see also Exercise 12.7.2) that

$$\langle JM|J_-(J_+|JM\rangle) = (J_+|JM\rangle)^\dagger J_+|JM\rangle. \quad (4.82)$$

Since J_+ raises the eigenvalue M to $M + 1$, we relabel the resultant eigenfunction $|JM + 1\rangle$. The normalization is given by Eq. (4.80) as

$$J_+|JM\rangle = \sqrt{(J - M)(J + M + 1)}|JM + 1\rangle, \quad (4.83)$$

taking the positive square root and *not* introducing any phase factor. By the same arguments

$$J_-|JM\rangle = \sqrt{(J + M)(J - M + 1)}|JM - 1\rangle. \quad (4.84)$$

Finally, since M ranges from $-J$ to $+J$ in unit steps, $2J$ *must be an integer. J is either an integer or half of an odd integer. As seen later, orbital angular momentum is described with integral J. From the spins of some of the fundamental particles and of some nuclei, we get $J = \frac{1}{2}, \frac{3}{2}, \frac{5}{2}, \dots$. Our angular momentum is quantized—essentially as a result of the commutation relations.
  In spherical polar coordinates θ, φ the functions $\langle\theta, \varphi|lm\rangle = Y_l^m(\theta, \varphi)$ are the spherical harmonics of Section 12.6.

---

[4] The Hermitian conjugation or *adjoint* operation is defined for matrices in Section 3.5, and for operators in general in Section 9.1.
  [5] For an excellent discussion of adjoint operators and Hilbert space see A. Messiah, *Quantum Mechanics*. New York: Wiley (1961), Chapter 7.

SUMMARY OF LIE GROUPS AND LIE ALGEBRAS

The general commutation relations, Eq. (4.14) in Section 4.2, for a classical Lie group [SO(n) and SU(n) in particular] can be simplified to look more like Eq. (4.71) for SO(3) and SU(2) in Section 4.3. Here we merely review and, as a rule, do not provide proofs for various theorems that are quoted.

First we choose linearly independent and mutually commuting generators H_i which are generalizations of J_z for SO(3) and SU(2). Let l be the maximum number of such H_i's with

$$[H_i, H_k] = 0. \tag{4.85}$$

Then l is called the *rank* of the Lie group G or its Lie algebra G. The rank and dimension or order of some Lie groups are given in Table 4.2. All other generators E_α can be shown to be raising and lowering operators with respect to all the H_i, so that

$$[H_i, E_\alpha] = \alpha_i E_\alpha. \tag{4.86}$$

The set of so-called *root vectors* $(\alpha_1, \alpha_2, \ldots, \alpha_l)$ form the *root diagram of G*.

Since the H_i commute, they can be simultaneously diagonalized. They provide us with a set of eigenvalues $m_1, m_2, \ldots, m_l$ [projection or additive quantum numbers generalizing M of J_z in SO(3) and SU(2)]. The set of so-called *weight vectors* $(m_1, m_2, \ldots, m_l)$ for an irreducible representation form a *weight diagram*.

There are l invariant operators C_i, called *Casimir* operators, which commute with all generators and are generalizations of $\mathbf{J}^2$,

$$[C_i, H_j] = 0, \qquad [C_i, E_\alpha] = 0. \tag{4.87}$$

The first one, C_1, is a quadratic function of the generators, the others are more complicated. Since the C_i commute with all H_j, they can be simultaneously diagonalized with the H_j's. Their eigenvalues $c_1, c_2, \ldots, c_l$ characterize irreducible representations and stay constant while the weight vector varies over any particular irreducible representation. Thus the general eigenfunction may be written as

$$|(c_1, c_2, \ldots, c_l)\, m_1, m_2, \ldots, m_l\rangle, \tag{4.88}$$

generalizing $|JM\rangle$ of SO(3) and SU(2). Their eigenvalue equations are

$$H_i|(c_1, c_2, \ldots, c_l)m_1, m_2, \ldots, m_l\rangle = m_i|(c_1, c_2, \ldots, c_l)m_1, m_2, \ldots, m_l\rangle \tag{4.89a}$$

$$C_i|(c_1, c_2, \ldots, c_l)m_1, m_2, \ldots, m_l\rangle = c_i|(c_1, c_2, \ldots, c_l)m_1, m_2, \ldots, m_l\rangle. \tag{4.89b}$$

Table 4.2 Rank and Order of Unitary and Rotational Groups

Lie algebra	A_l	B_l	D_l
Lie group	SU($l + 1$)	SO($2l + 1$)	SO($2l$)
Rank	l	l	l
Order	$l(l + 2)$	$l(2l + 1)$	$l(2l - 1)$

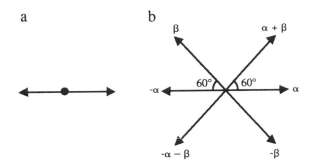

Figure 4.6   Root diagram for (a) SU(2) and (b) SU(3).

We can now show that $E_\alpha|(c_1, c_2, \ldots, c_l)m_1, m_2, \ldots, m_l\rangle$ has the weight vector $(m_1 + \alpha_1, m_2 + \alpha_2, \ldots, m_l + \alpha_l)$ using the commutation relations, Eq. (4.86), in conjunction with Eqs. (4.89a), (4.89b),

$$H_i E_\alpha|(c_1, c_2, \ldots, c_l)m_1, m_2, \ldots, m_l\rangle$$
$$= (E_\alpha H_i + [H_i, E_\alpha])|(c_1, c_2, \ldots, c_l)m_1, m_2, \ldots, m_l\rangle$$
$$= (m_i + \alpha_i)E_\alpha|(c_1, c_2, \ldots, c_l)m_1, m_2, \ldots, m_l\rangle. \qquad (4.90)$$

Therefore

$$E_\alpha|(c_1, c_2, \ldots, c_l)m_1, m_2, \ldots, m_l\rangle \sim |(c_1, \ldots, c_l)m_1 + \alpha_1, \ldots, m_l + \alpha_l\rangle,$$

the generalization of Eqs. (4.83) and (4.84) from SO(3). These changes of eigenvalues by the operator E_α are called its *selection rules* in quantum mechanics. They are displayed in the root diagram of a Lie algebra.

Examples of root diagrams are given in Fig. 4.6 for SU(2) and SU(3). If we attach the roots denoted by arrows in Fig. 4.6b to a weight in Figs. 4.3 or 4.5a, b we can reach any other state (represented by a dot in the weight diagram).

EXERCISES

4.3.1   Show that (a) $[J_+, \mathbf{J}^2] = 0$, (b) $[J_-, \mathbf{J}^2] = 0$.

4.3.2   Derive the root diagram of SU(3) in Fig. 4.6b from the generators λ_i in Eq. (4.61).
Hint. Work out first the SU(2) case in Fig. 4.6a from the Pauli matrices.

4.4 ANGULAR MOMENTUM COUPLING

In many-body systems of classical mechanics the total angular momentum is the sum, $\mathbf{L} = \Sigma \mathbf{L}_i$, of the individual orbital angular momenta. In quantum mechanics conserved angular momentum arises when particles move in a central potential, such as the Coulomb potential in atomic physics, a shell model potential in nuclear physics, or a confinement potential of a quark model in particle physics. In the relativistic Dirac equation, orbital angular momentum is no longer conserved, but $\mathbf{J} = \mathbf{L} + \mathbf{s}$ is conserved, the total angular momentum of a particle consisting of its orbital and intrinsic angular momentum, called spin $\mathbf{s} = \boldsymbol{\sigma}/2$ in units of $\hbar$.

It is readily shown that the sum of angular momentum operators obeys the same commutation relations in Eq. (4.37) or (4.41) as the individual angular momentum operators provided those from different particles commute.

CLEBSCH–GORDAN COEFFICIENTS FOR SU(2)–SO(3)

Clearly, combining two commuting angular momenta $\mathbf{J}_i$ to form their sum

$$\mathbf{J} = \mathbf{J}_1 + \mathbf{J}_2, \qquad [J_{1i}, J_{2k}] = 0, \qquad (4.91)$$

occurs often in applications. For a single particle with spin $\frac{1}{2}$, e.g. an electron or a quark, the total angular momentum is a sum of orbital angular momentum and spin. For two spinless particles their total orbital angular momentum $\mathbf{L} = \mathbf{L}_1 + \mathbf{L}_2$. For $\mathbf{J}^2$ and J_z of Eq. (4.91) to be both diagonal, $[\mathbf{J}^2, J_z] = 0$ has to hold. To show this we use the obvious commutation relations $[J_{iz}, \mathbf{J}_j^2] = 0$, etc., and

$$\mathbf{J}^2 = \mathbf{J}_1^2 + \mathbf{J}_2^2 + 2\mathbf{J}_1 \cdot \mathbf{J}_2 = \mathbf{J}_1^2 + \mathbf{J}_2^2 + J_{1+}J_{2-} + J_{1-}J_{2+} + 2J_{1z}J_{2z} \qquad (4.91')$$

in conjunction with Eq. (4.71) for both $\mathbf{J}_i$'s, to obtain

$$[\mathbf{J}^2, J_z] = [J_{1-}J_{2+} + J_{1+}J_{2-}, J_{1z} + J_{2z}]$$

$$= [J_{1-}, J_{1z}]J_{2+} + J_{1-}[J_{2+}, J_{2z}] + [J_{1+}, J_{1z}]J_{2-} + J_{1+}[J_{2-}, J_{2z}]$$

$$= J_{1-}J_{2+} - J_{1-}J_{2+} - J_{1+}J_{2-} + J_{1+}J_{2-} = 0.$$

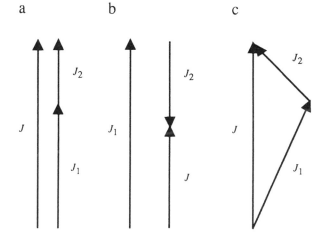

Figure 4.7 Coupling of two angular momenta; (a) parallel stretched, (b) antiparallel, (c) general case.

Similarly $[\mathbf{J}^2, \mathbf{J}_i^2] = 0$ are proved. Hence the eigenvalues of $\mathbf{J}_i^2, \mathbf{J}^2, J_z$ can be used to label the total angular momentum states $|J_1 J_2 JM\rangle$.

The product states $|J_1 m_1\rangle|J_2 m_2\rangle$ obviously satisfy the eigenvalue equations

$$J_z|J_1 m_1\rangle|J_2 m_2\rangle = (J_{1z} + J_{2z})|J_1 m_1\rangle|J_2 m_2\rangle = (m_1 + m_2)|J_1 m_1\rangle|J_2 m_2\rangle$$

$$= M|J_1 m_1\rangle|J_2 m_2\rangle, \tag{4.92}$$

$$\mathbf{J}_i^2|J_1 m_1\rangle|J_2 m_2\rangle = J_i(J_i + 1)|J_1 m_1\rangle|J_2 m_2\rangle,$$

but will not have good $\mathbf{J}^2$ except for the maximally stretched states with $M = \pm(J_1 + J_2)$ and $J = J_1 + J_2$ (see Fig. 4.7a). To see this we use Eq. (4.91') again in conjunction with Eqs. (4.83) and (4.84) in

$$\mathbf{J}^2|J_1 m_1\rangle|J_2 m_2\rangle = \{J_1(J_1 + 1) + J_2(J_2 + 1) + 2m_1 m_2\}|J_1 m_1\rangle|J_2 m_2\rangle +$$

$$\{J_1(J_1 + 1) - m_1(m_1 + 1)\}^{1/2}\{J_2(J_2 + 1) - m_2(m_2 - 1)\}^{1/2}|J_1 m_1 + 1\rangle|J_2 m_2 - 1\rangle +$$

$$\{J_1(J_1 + 1) - m_1(m_1 - 1)\}^{1/2}\{J_2(J_2 + 1) - m_2(m_2 + 1)\}^{1/2}|J_1 m_1 - 1\rangle|J_2 m_2 + 1\rangle. \tag{4.93}$$

The last two terms in Eq. (4.93) vanish only when $m_1 = J_1$ and $m_2 = J_2$, or $m_1 = -J_1$ and $m_2 = -J_2$. In both cases $J = J_1 + J_2$ follows from the first line of Eq. (4.93). In general, therefore, we have to form appropriate linear combinations of product states

$$|J_1 J_2 JM\rangle = \sum_{m_1,m_2} C(J_1 J_2 J|m_1 m_2 M)|J_1 m_1\rangle|J_2 m_2\rangle, \tag{4.94}$$

so that $\mathbf{J}^2$ has eigenvalue $J(J + 1)$. The quantities $C(J_1 J_2 J|m_1 m_2 M)$ in Eq. (4.94) are called *Clebsch-Gordan coefficients*. From Eq. (4.92) we see that they vanish unless $M = m_1 + m_2$, reducing the double sum to a single sum.

Clearly, the maximal $J_{max} = J_1 + J_2$ (see Fig. 4.7a). In this case Eq. (4.93) reduces to a pure product state

$$|J_1 J_2 J = J_1 + J_2\, M = J_1 + J_2\rangle = |J_1 J_1\rangle|J_2 J_2\rangle, \tag{4.95a}$$

so that the Clebsch-Gordan coefficient

$$C(J_1 J_2 J = J_1 + J_2|J_1 J_2 J_1 + J_2) = 1. \tag{4.95b}$$

The minimal $J = J_1 - J_2$ (if $J_1 > J_2$, see Fig. 4.7b) and $J = J_2 - J_1$ for $J_2 > J_1$ follow if we keep in mind that there are just as many product states as $|JM\rangle$ states,

$$\sum_{J=J_{min}}^{J_{max}} (2J + 1) = (J_{max} - J_{min} + 1)(J_{max} + J_{min} + 1)$$

$$= (2J_1 + 1)(2J_2 + 1). \tag{4.96}$$

This condition holds because the $|J_1 J_2 JM\rangle$ states merely rearrange all product states into irreducible representations of total angular momentum. It is equivalent to the *triangle rule*

$$\Delta(J_1 J_2 J) = 1, \qquad \text{if } |J_1 - J_2| \le J \le J_1 + J_2;$$
$$\Delta(J_1 J_2 J) = 0, \qquad \text{else.} \tag{4.97}$$

In other words, Eq. (4.94) defines a unitary transformation from the orthogonal basis set of products of single-particle states $|J_1 M_1; J_2 m_2\rangle = |J_1 m_1\rangle |J_2 m_2\rangle$ to the two-particle states $|J_1 J_2 JM\rangle$. The Clebsch–Gordan coefficients are just the overlap matrix elements

$$C(J_1 J_2 J | m_1 m_2 M) = \langle J_1 J_2 JM | J_1 m_1; J_2 m_2 \rangle. \tag{4.98}$$

The explicit construction below shows that they are all real. From the orthonormality of both bases we have the immediate constraints

$$\sum_{\substack{m_1, m_2 \\ m_1 + m_2 = M}} C(J_1 J_2 J | m_1 m_2 M) C(J_1 J_2 J' | m_1 m_2 M')$$

$$= \langle J_1 J_2 JM | J_1 J_2 J'M' \rangle = \delta_{JJ'MM'} \tag{4.99a}$$

$$\sum_{J,M} C(J_1 J_2 J | m_1 m_2 M) C(J_1 J_2 J | m_1' m_2' M)$$

$$= \langle J_1 m_1 | J_1 m_1' \rangle \langle J_2 m_2 | J_2 m_2' \rangle = \delta_{m_1 m_1'} \delta_{m_2 m_2'}. \tag{4.99b}$$

Now we are ready to construct more directly the total angular momentum states starting from $|J_{\max} = J_1 + J_2 M = J_1 + J_2\rangle$ in Eq. (4.95a) and using the lowering operator $J_- = J_{1-} + J_{2-}$ repeatedly. In the first step we use Eq. (4.84) for

$$J_{i-}|J_i J_i\rangle = \{J_i(J_i + 1) - J_i(J_i - 1)\}^{1/2}|J_i J_i - 1\rangle = (2J_i)^{1/2}|J_i J_i - 1\rangle,$$

which we substitute into $(J_{1-} + J_{2-})|J_1 J_1\rangle |J_2 J_2\rangle$. Normalizing the resulting state with $M = J_1 + J_2 - 1$ properly to 1 we obtain

$$|J_1 J_2 J_1 + J_2 J_1 + J_2 - 1\rangle = \{J_1/(J_1 + J_2)]^{1/2}|J_1 J_1 - 1\rangle |J_2 J_2\rangle$$

$$+ \{J_2/(J_1 + J_2)\}^{1/2}|J_1 J_1\rangle |J_2 J_2 - 1\rangle. \tag{4.100}$$

Equation (4.100) yields the Clebsch–Gordan coefficients

$$C(J_1 J_2 J_1 + J_2 | J_1 - 1 J_2 J_1 + J_2 - 1) = \{J_1/(J_1 + J_2)\}^{1/2},$$
$$C(J_1 J_2 J_1 + J_2 | J_1 J_2 - 1 J_1 + J_2 - 1) = \{J_2/(J_1 + J_2)\}^{1/2}. \tag{4.101}$$

Then we apply J_- again and normalize the states obtained until we reach $|J_1 J_2 J_1 + J_2 M\rangle$ with $M = -(J_1 + J_2)$. The Clebsch–Gordan coefficients $C(J_1 J_2 J_1 + J_2 | m_1 m_2 M)$ may thus be calculated step by step, and they are all real.

The next step is to realize that the only other state with $M = J_1 + J_2 - 1$ is the top of the next lower tower of $|J_1 + J_2 - 1M\rangle$ states. Since $|J_1 + J_2 - 1 J_1 + J_2 - 1\rangle$ is orthogonal to $|J_1 + J_2 J_1 + J_2 - 1\rangle$ in Eq. (4.100), it must be the other linear combination with a relative minus sign,

$$|J_1 + J_2 - 1 J_1 + J_2 - 1\rangle = -\{J_2/(J_1 + J_2)\}^{1/2}|J_1 J_1 - 1\rangle|J_2 J_2\rangle$$
$$+ \{J_1/(J_1 + J_2)\}^{1/2}|J_1 J_1\rangle|J_2 J_2 - 1\rangle. \qquad (4.102)$$

Hence we have determined the Clebsch–Gordan coefficients (for $J_2 \geq J_1$)

$$C(J_1 J_2 J_1 + J_2 - 1|J_1 - 1 J_2 J_1 + J_2 - 1) = -\{J_2/(J_1 + J_2)\}^{1/2},$$
$$\qquad (4.103)$$
$$C(J_1 J_2 J_1 + J_2 - 1|J_1 J_2 - 1 J_1 + J_2 - 1) = \{J_1/(J_1 + J_2)\}^{1/2}.$$

Again we continue using J_- until we reach $M = -(J_1 + J_2 - 1)$, and we keep normalizing the resulting states $|J_1 + J_2 - 1\,M\rangle$ of the $J_1 + J_2 - 1$ tower.

In order to get to the top of the next tower $|J_1 + J_2 - 2\,M\rangle$ with $M = J_1 + J_2 - 2$, we remember that we have already constructed two states with that M. Both $|J_1 + J_2 J_1 + J_2 - 2\rangle$ and $|J_1 + J_2 - 1 J_1 + J_2 - 2\rangle$ are known linear combinations of the three product states $|J_1 J_1\rangle|J_2 J_2 - 2\rangle$, $|J_1 J_1 - 1\rangle|J_2 J_2 - 1\rangle$, and $|J_1 J_1 - 2\rangle|J_2 J_2\rangle$. The third linear combination is easy to find from orthogonality to these two states, up to an overall phase, which is chosen by the *Condon–Shortley phase conventions*[1] so that the coefficient $C(J_1 J_2 J_1 + J_2 - 2|J_1 J_2 - 2 J_1 + J_2 - 2)$ of the last product state is positive for $|J_1 J_2 J_1 + J_2 - 2 J_1 + J_2 - 2\rangle$. It is straightforward, though a bit tedious, to determine the rest of the Clebsch–Gordan coefficients.

Numerous recursion relations can be derived from matrix elements of various angular momentum operators, for which we refer to the literature.[2]

The symmetry properties of Clebsch–Gordan coefficients are best displayed in Wigner's $3j$-symbols which are tabulated,[2]

$$\begin{pmatrix} J_1 J_2 J_3 \\ m_1 m_2 m_3 \end{pmatrix} = (-1)^{J_1 - J_2 - m_3} C(J_1 J_2 J_3 | m_1 m_2, -m_3)/(2J_3 + 1)^{1/2}. \qquad (4.104)$$

One of the most important places where Clebsch–Gordan coefficients occur is in matrix elements of tensor operators, which are governed by the Wigner–Eckart theorem discussed in the next section on spherical tensors. Another is coupling of operators or state vectors to total angular momentum, such as spin-orbit coupling. Recoupling of operators and states in matrix elements leads to $6j$- and $9j$-symbols.[2] Clebsch–Gordan coefficients can and have been calculated for other Lie groups, such as SU(3).

[1] E. U. Condon and G. H. Shortley, *Theory of Atomic Spectra*. Cambridge: Cambridge University Press (1935).

[2] There is a rich literature on this subject, e.g., A. R. Edmonds, *Angular Momentum in Quantum Mechanics*. Princeton, NJ: Princeton University Press (1957); M. E. Rose, *Elementary Theory of Angular Momentum*. New York: Wiley (1957); A. de-Shalit and I. Talmi, *Nuclear Shell Model*. New York: Academic Press (1963). Clebsch–Gordan coefficients are tabulated in M. Rotenberg, R. Bivins, N. Metropolis, and J. K. Wooten, Jr., *The 3j- and 6j-Symbols*. Cambridge, MA: Massachusetts Institute of Technology (1959).

SPHERICAL TENSORS

The properties of cartesian tensors in Chapter 2 are defined using the group of nonsingular (linear infinitesimal) transformations, which contains the three-dimensional rotations as a subgroup. A tensor of a given rank that is irreducible with respect to the full group may well become reducible for the rotation group SO(3). To explain this point, consider the second rank tensor with components $T_{jk} = x_j y_k$ for $j, k = 1, 2, 3$. It contains the symmetric tensor $S_{jk} = (x_j y_k + x_k y_j)/2$ and the antisymmetric tensor $A_{jk} = (x_j y_k - x_k y_j)/2$ so that $T_{jk} = S_{jk} + A_{jk}$. Under rotations the scalar product $\mathbf{x} \cdot \mathbf{y}$ is invariant and may therefore be subtracted from S_{jk} leading to the SO(3)-irreducible tensor

$$S'_{jk} = (x_j y_k + x_k y_j)/2 - \mathbf{x} \cdot \mathbf{y} \delta_{yk}/3.$$

Tensors of higher rank may be treated similarly. When we form tensors from products of the components of the coordinate vector $\mathbf{r}$ then, in polar coordinates that are tailored to SO(3) symmetry, we end up with the spherical harmonics of Chapter 12.

The form of the ladder operators for SO(3) in Section 4.3 leads us to introduce the *spherical components* (note the different normalization and signs, though, prescribed by the Y_{lm}) of a vector $\mathbf{A}$,

$$A_{+1} = -(A_x + iA_y)/\sqrt{2}, \qquad A_{-1} = (A_x - iA_y)/\sqrt{2}, \qquad A_0 = A_z. \tag{4.105}$$

Then we have for the coordinate vector $\mathbf{r}$ in polar coordinates

$$r_{+1} = -\frac{r}{\sqrt{2}} \sin \theta e^{i\varphi} = r\sqrt{\frac{4\pi}{3}} Y_{11}, \qquad r_{-1} = \frac{r}{\sqrt{2}} \sin \theta e^{-i\varphi} = r\sqrt{\frac{4\pi}{3}} Y_{1,-1},$$

$$r_0 = r\sqrt{\frac{4\pi}{3}} Y_{10}, \tag{4.106}$$

where $Y_{lm}(\theta, \varphi)$ are the spherical harmonics of Section 9. Again, the spherical jm components of tensors T_{jm} of higher rank j may be introduced similarly.

An irreducible *spherical tensor operator* T_{jm} of rank j has $2j + 1$ components, just as for spherical harmonics Y_{lm}, m runs from $-l$ to $+l$. Under a rotation $R(\alpha)$, where α stands for the Euler angles, the Y_{lm} transforms like

$$Y_{lm}(\hat{\mathbf{r}}') = \sum_{m'} Y_{lm'}(\hat{\mathbf{r}}) D^l_{m'm}(R), \tag{4.107a}$$

where $\hat{\mathbf{r}}' = (\theta', \varphi')$ are obtained from $\hat{\mathbf{r}} = (\theta, \varphi)$ by the rotation R and are the angles of the same point in the rotated frame. So, for the operator T_{jm} we define

$$RT_{jm}R^{-1} = \sum_{m'} T_{jm'} D^j_{m'm}(\alpha), \tag{4.107b}$$

with the rotation matrices of Eq. (4.42). For an infinitesimal rotation (see Eq. (4.20) in Section 4.2 on generators) the left side of Eq. (4.107b) simplifies to a commutator and the right side to the matrix elements of $\mathbf{J}$, the infinitesimal generator of the rotation R,

$$[J_n, T_{jm}] = \sum_{m'} T_{jm'} \langle jm' | J_n | jm \rangle. \tag{4.108}$$

If we substitute Eqs. (4.83) and (4.84) for the matrix elements of J_m we obtain the alternative transformation laws of a tensor operator,

$$[J_0, T_{jm}] = m T_{jm}, \qquad [J_\pm, T_{jm}] = T_{jm\pm1}\{(j - m)(j \pm m + 1)\}^{1/2}. \qquad (4.109)$$

We can use the Clebsch–Gordan coefficients of the previous subsection to couple two tensors of given rank to another rank. An example is the cross or vector product of two vectors $\mathbf{a}$ and $\mathbf{b}$ from Chapter 1. Let us write both vectors in spherical components, a_m and b_m. Then it is easy to verify that the tensor C_m of rank 1 defined as

$$C_m \equiv \sum_{m_1 m_2} C(111|m_1 m_2 m) a_{m_1} b_{m_2} = \frac{i}{\sqrt{2}} (\mathbf{a} \times \mathbf{b})_m. \qquad (4.110)$$

Even simpler is the usual scalar product of two vectors in Chapter 1, in which $\mathbf{a}$ and $\mathbf{b}$ are coupled to zero angular momentum:

$$\mathbf{a} \cdot \mathbf{b} = -(\mathbf{ab})_0\sqrt{3} \equiv -\sqrt{3} \sum_m C(110|m, -m0) a_m b_{-m}. \qquad (4.111)$$

Another often used application of tensors is the *recoupling* that involves *6j-symbols* for three operators and *9j*'s for four operators.[2] An example is the following scalar product, for which it can be shown[2] that

$$\boldsymbol{\sigma}_1 \cdot \mathbf{r} \boldsymbol{\sigma}_2 \cdot \mathbf{r} = \tfrac{1}{3} r^2 \boldsymbol{\sigma}_1 \cdot \boldsymbol{\sigma}_2 + (\boldsymbol{\sigma}_1 \boldsymbol{\sigma}_2)_2 \cdot (\mathbf{rr})_2, \qquad (4.112)$$

but which can also be rearranged by elementary means. Here the tensor operators are defined as

$$(\boldsymbol{\sigma}_1 \boldsymbol{\sigma}_2)_{2m} = \sum_{m_1 m_2} C(112|m_1 m_2 m) \sigma_{1m_1} \sigma_{2m_2} \qquad (4.113)$$

$$(\mathbf{rr})_{2m} = \sum_{m_1 m_2} C(112|m_1 m_2 m) r_{m_1} r_{m_2} = \sqrt{\frac{8\pi}{15}} r^2 Y_{2m}(\hat{\mathbf{r}}), \qquad (4.114)$$

and the scalar product of tensors of rank 2 as

$$(\boldsymbol{\sigma}_1 \boldsymbol{\sigma}_2)_2 \cdot (\mathbf{rr})_2 = \sum_m (-1)^m (\boldsymbol{\sigma}_1 \boldsymbol{\sigma}_2)_{2m} (\mathbf{rr})_{2,-m} = \sqrt{5}((\boldsymbol{\sigma}_1 \boldsymbol{\sigma}_2)_2 (\mathbf{rr})_2)_0. \qquad (4.115)$$

One of the most important applications of spherical tensor operators is the *Wigner–Eckart theorem*. It says that a matrix element of a spherical tensor operator T_{km} of rank k between states of angular momentum j and j' factorizes into a Clebsch–Gordan coefficient and a so-called reduced matrix element denoted by double bars that no longer has any dependence on the projection quantum numbers m, m', n:

$$\langle j'm'|T_{kn}|jm\rangle = C(kjj'|nmm')(-1)^{k-j+j'}\langle j'||T_k||j\rangle/\sqrt{(2j' + 1)}. \qquad (4.116)$$

In other words, such a matrix element splits into a dynamic part, the reduced matrix element, and a geometric part, the Clebsch–Gordan coefficient that contains the rotational properties (expressed by the projection quantum numbers) from the SO(3) invariance. To see this we couple T_{kn} with the initial state to total angular momentum j'

$$|j'm'\rangle_0 \equiv \sum_{nm} C(kjj'|nmm') T_{kn}|jm\rangle. \qquad (4.117)$$

Under rotations the state $|j'm'\rangle_0$ transforms just like $|j'm'\rangle$. Thus, the overlap matrix element $\langle j'm'|j'm'\rangle_0$ is a rotational scalar that has no m' dependence so that we can average over the projections,

$$\langle JM|j'm'\rangle_0 = \delta_{Jj'}\delta_{Mm'} \sum_\mu \langle j'\mu|j'\mu\rangle_0/(2j'+1). \tag{4.118}$$

Next we substitute our definition, Eq. (4.116), into Eq. (4.118) and invert the relation using orthogonality, Eq. (4.99b), to find that

$$\langle JM|T_{kn}|jm\rangle = \sum_{j'm'} C(kjj'|nmm')\delta_{Jj'}\delta_{Mm'} \sum_\mu \langle J\mu|J\mu\rangle_0/(2J+1), \tag{4.119}$$

which proves the Wigner–Eckart theorem, Eq. (4.116).[3]

As an application, we can write the Pauli matrix elements in terms of Clebsch–Gordan coefficients. We apply the Wigner–Eckart theorem to

$$\langle \tfrac{1}{2}\gamma|\sigma_\alpha|\tfrac{1}{2}\beta\rangle = (\sigma_\alpha)_{\gamma\beta} = -C(1\tfrac{1}{2}\tfrac{1}{2}|\alpha\beta\gamma)\langle \tfrac{1}{2}||\sigma||\tfrac{1}{2}\rangle/\sqrt{2}. \tag{4.120}$$

Since $\langle \tfrac{1}{2}\tfrac{1}{2}|\sigma_0|\tfrac{1}{2}\tfrac{1}{2}\rangle = 1$ and $C(1\tfrac{1}{2}\tfrac{1}{2}|0\tfrac{1}{2}\tfrac{1}{2}) = -1/\sqrt{3}$, we find

$$\langle \tfrac{1}{2}||\sigma||\tfrac{1}{2}\rangle = \sqrt{6}, \tag{4.121}$$

which substituted into Eq. (4.120) yields

$$(\sigma_\alpha)_{\gamma\beta} = -\sqrt{3}C(1\tfrac{1}{2}\tfrac{1}{2}|\alpha\beta\gamma). \tag{4.122}$$

Note carefully that the $\alpha = \pm 1, 0$ denote the spherical components of the Pauli matrices.

YOUNG TABLEAUX FOR SU(n)

Young tableaux (YT) provide a powerful and elegant method for decomposing products of SU(n) group representations into sums of irreducible representations. YTs provide the dimensions and symmetry types of the irreducible representations in this so-called *Clebsch–Gordan series*, though not the Clebsch–Gordan coefficients by which the product states are coupled to the quantum numbers of each irreducible representation of the series (see Eq. (4.94)).

Products of representations correspond to multiparticle states. In this context permutations of particles are important when we deal with several identical particles. Permutations of n identical objects form the *symmetric group* S_n. A close connection between irreducible representations of S_n which are the YTs and those of SU(n) is provided by the *theorem*: Every N-particle state of S_n that is made up of single-particle states of the fundamental n-dimensional SU(n) multiplet belongs to an irreducible SU(n) representation. A proof is in Chapter 22.[4]

---

[3] The extra factor $(-1)^{k-j+j'}/\sqrt{(2j'+1)}$ in Eq. (4.116) is just a convention that varies in the literature.

[4] B. G. Wyborne, *Classical Groups for Physicists*. New York: Wiley (1974).

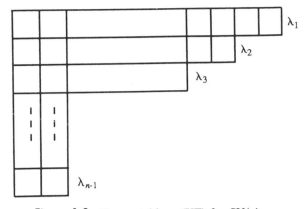

Figure 4.8 Young tableau (YT) for SU(n).

For SU(2) the fundamental representation is a box that stands for the spin $+\frac{1}{2}$ (up) and $-\frac{1}{2}$ (down) states and has dimension two. For SU(3) the box comprises the three quark states in the triangle of Fig. 4.5a; it has dimension 3.

An array of boxes shown in Fig. 4.8 with λ_1 boxes in the first row, λ_2 boxes in the second row, ..., and λ_{n-1} boxes in the last row is called a Young tableau (YT) and represents an irreducible representation of SU(n) if and only if

$$\lambda_1 \geq \lambda_2 \geq \cdots \geq \lambda_{n-1}. \tag{4.123}$$

Boxes in the same row are symmetric representations; those in the same column are antisymmetric. A YT consisting of one row is totally symmetric. A YT consisting of a single column is totally antisymmetric.

There are at most $n - 1$ rows for SU(n) YTs because a column of n boxes is the totally antisymmetric (Slater determinant of single-particle states) singlet representation that may be struck from the YT.

An array of N boxes is an N-particle state whose boxes may be labeled by positive integers so that the (particle labels or) numbers in one row of the YT do not decrease from left to right and those in any one column increase from top to bottom. In contrast to the possible repetitions of row numbers, the numbers in any column must be different because of the antisymmetry of these states.

The product of a YT with a single box is the sum of YTs formed when the box is put at the end of each row of the YT, provided the resulting YT is legitimate, i.e., obeys Eq. (4.123). For SU(2) the product of two boxes, spin $\frac{1}{2}$ representations, generates

$$\square \times \square = \square\square \oplus \begin{array}{c}\square\\\square\end{array} \tag{4.124}$$

the symmetric spin 1 representation of dimension 3 and the antisymmetric singlet of dimension 1.

a

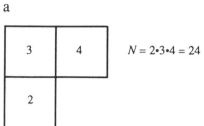

$N = 2 \cdot 3 \cdot 4 = 24$

b

Figure 4.9 Illustration of (a) N and (b) D in Eq. (4.125) for the octet Young tableaux of SU(3).

The column of $n - 1$ boxes is the conjugate representation of the fundamental representation; its product with a single box contains the column of n boxes which is the singlet. For SU(3) the conjugate representation is the inverted triangle in Fig. 4.5b that represents the three antiquarks $\bar{u}, \bar{d}, \bar{s}$.

The dimension of a YT is given by the ratio

$$\dim \text{YT} = N/D. \tag{4.125}$$

The numerator N is obtained by writing an n in all boxes of the YT along the diagonal, $(n + 1)$ in all boxes immediately above the diagonal, $(n - 1)$ immediately below the diagonal, etc. N is the product of all the numbers in the YT. An example is shown in Fig. 4.9a for the octet representation of SU(3), where $N = 2 \cdot 3 \cdot 4 = 24$. There is a closed formula that is equivalent to Eq. (4.125).[5] The denominator D is the product of all *hooks*.[6] A hook is drawn through each box of the YT by starting a horizontal line from the right to the box in question, then continuing it vertically out of the YT. The number of boxes encountered by the hook-line is the hook-number of the box. D is the product

[5] See, e.g., M. Hamermesh, *Group Theory and Its Application to Physical Problems*. Reading, MA: Addison–Wesley (1962).

[6] F. Close, *Introduction to Quarks and Partons*. New York: Academic Press (1979).

of all hook-numbers of the YT. An example is shown in Fig. 4.9b for the octet of SU(3) whose hook-number is $D = 1 \cdot 3 \cdot 1 = 3$. Hence the dimension of the SU(3) octet is $24/3 = 8$, whence its name.

Now we can calculate the dimensions of the YTs in Eq. (4.124). For SU(2) they are $2 \times 2 = 3 + 1 = 4$. For SU(3) they are $3 \cdot 3 = 3 \cdot 4/(1 \cdot 2) + 3 \cdot 2/(2 \cdot 1)$ $= 6 + 3 = 9$. For the product of the quark times antiquark YTs of SU(3) we get

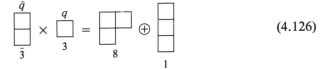

$$ (4.126) $$

which are precisely the meson multiplets considered in the subsection on the eightfold way, the SU(3) flavor symmetry, that suggest mesons are bound states of a quark and antiquark, $q\bar{q}$ configurations. For the product of three quarks we get

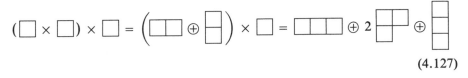

$$ (4.127) $$

which are the observed multiplets for the baryons that suggest they are bound states of three quarks, q^3 configurations.

As we have seen, YTs describe the decomposition of a product of SU(n) irreducible representations into irreducible representations of SU(n), which is called the Clebsch–Gordan series, while the Clebsch–Gordan coefficients considered above allow constructing the individual states in this series.

EXERCISES

4.4.1 Derive recursion relations for Clebsch–Gordan coefficients. Use them to calculate $C(11J|m_1 m_2 M)$ for $J = 0, 1, 2$.
Hint. Use the known matrix elements of $J_+ = J_{1+} + J_{2+}$, J_{i+}, and $\mathbf{J}^2 = (\mathbf{J}_1 + \mathbf{J}_2)^2$, etc.

4.4.2 Show that $(Y_l \chi)^J_M = \sum C(l\frac{1}{2}J|m_l m_s M) Y_{lm_l} \chi_{m_s}$, where $\chi_{\pm 1/2}$ are the spin up and down eigenfunctions of $\sigma_3 = \sigma_z$, transforms like a spherical tensor of rank J.

4.4.3 When the spin of quarks is taken into account, the SU(3) flavor symmetry is replaced by the SU(6) symmetry. Why? Obtain the Young tableau for the antiquark configuration $\bar{q}$. Then decompose the product $q\bar{q}$. Which SU(3) representations are contained in the nontrivial SU(6) representation for mesons?
Hint. Determine the dimensions of all YTs.

4.4.4   For $l = 1$, Eq. (4.107a) becomes

$$Y_1^m(\theta', \varphi') = \sum_{m' = -1}^{1} D_{m'm}^1(\alpha, \beta, \gamma) Y_1^{m'}(\theta, \varphi).$$

Rewrite these spherical harmonics in *cartesian* form. Show that the resulting cartesian coordinate equations are equivalent to the Euler rotation matrix $A(\alpha, \beta, \gamma)$, Eq. (3.94) rotating the coordinates.

4.4.5   Assuming that $D^j(\alpha, \beta, \gamma)$ is unitary, show that

$$\sum_{m = -l}^{l} Y_l^{m*}(\theta_1, \varphi_1) Y_l^m(\theta_2, \varphi_2)$$

is a scalar quantity (invariant under rotations). This is a function analog of a scalar product of vectors.

4.4.6   (a) Show that the α and γ dependence of $D^j(\alpha, \beta, \gamma)$ may be factored out such that

$$D^j(\alpha, \beta, \gamma) = A^j(\alpha) d^j(\beta) C^j(\gamma).$$

(b) Show that $A^j(\alpha)$ and $C^j(\gamma)$ are diagonal. Find the explicit forms.
(c) Show that $d^j(\beta) = D^j(0, \beta, 0)$.

Hint. Exercises 3.2.28 and 3.2.29 may be helpful.

4.4.7   The angular momentum-exponential form of the Euler angle rotation operators is

$$\mathcal{R} = \mathcal{R}_{z''}(\gamma) \mathcal{R}_{y'}(\beta) \mathcal{R}_z(\alpha)$$
$$= \exp(-i\gamma J_{z''}) \exp(-i\beta J_{y'}) \exp(-i\alpha J_z).$$

Show that in terms of the *original axes*

$$\mathcal{R} = \exp(-i\alpha J_z) \exp(-i\beta J_y) \exp(-i\gamma J_z).$$

Hint. The $\mathcal{R}$ operators transform as matrices. The rotation about the y'-axis (second Euler rotation) may be referred to the original y-axis by

$$\exp(-i\beta J_{y'}) = \exp(-i\alpha J_z) \exp(-i\beta J_y) \exp(i\alpha J_z).$$

4.5 HOMOGENEOUS LORENTZ GROUP

Generalizing the approach to vectors of Section 1.2, in special relativity we demand that our physical laws be covariant[1] under

a. space and time translations,
b. rotations in real, three-dimensional space, and
c. Lorentz transformations.

The demand for covariance under translations is based on the homogeneity of space and time. Covariance under rotations is an assertion of the isotropy of space. The requirement of Lorentz covariance follows from special relativity. All three of these transformations together form the inhomogeneous Lorentz

[1] To be covariant means to have the same form in different coordinate systems so that there is *no preferred reference* system (compare Sections 1.2 and 2.6).

group or the Poincaré group. Here we exclude translations. The space rotations and the Lorentz transformations together form a group—the homogeneous Lorentz group.

We first generate a subgroup, the Lorentz transformations in which the relative velocity $\mathbf{v}$ is along the $x = x_1$ axis. The generator may be determined by considering space–time reference frames moving with a relative velocity δv, an infinitesimal.[2] The relations are similar to those for rotations in real space, Sections 1.2, 2.6, and 3.3, except that here the angle of rotation is pure imaginary (compare Section 4.6).

Lorentz transformations are linear not only in the space coordinates x_i but in the time t as well. They originate from Maxwell's equations of electrodynamics which are invariant under Lorentz transformations, as we shall see later. Lorentz transformations leave the quadratic form $c^2 t^2 - x_1^2 - x_2^2 - x_3^2 = x_0^2 - x_1^2 - x_2^2 - x_3^2$ invariant, where $x_0 = ct$. We see this if we switch on a light source at the origin of the coordinate system. At time t light has traveled the distance $ct = \sqrt{\Sigma x_i^2}$, so that $c^2 t^2 - x_1^2 - x_2^2 - x_3^2 = 0$. Special relativity requires that in all (inertial) frames that move with velocity $v \le c$ in any direction relative to the x_i-system and have the same origin at time $t = 0$, $c^2 t'^2 - x_1'^2 - x_2'^2 - x_3'^2 = 0$ holds also. Four-dimensional space–time with the metric $x_0^2 - x_1^2 - x_2^2 - x_3^2$ is called Minkowski space with the scalar product of two four-vectors defined as $a \cdot b = a_0 b_0 - \mathbf{a} \cdot \mathbf{b}$. Using the metric tensor

$$(g_{\mu\nu}) = (g^{\mu\nu}) = \begin{pmatrix} 1 & 0 & 0 & 0 \\ 0 & -1 & 0 & 0 \\ 0 & 0 & -1 & 0 \\ 0 & 0 & 0 & -1 \end{pmatrix} \tag{4.128}$$

we can raise and lower the indices of a four-vector such as the coordinates $x^\mu = (x_0, \mathbf{x})$ so that $x_\mu = g_{\mu\nu} x^\nu = (x_0, -\mathbf{x})$ and $x^\mu g_{\mu\nu} x^\nu = x_0^2 - \mathbf{x}^2$, Einstein's summation convention being understood. For the gradient $\partial^\mu = (\partial/\partial x_0, -\boldsymbol{\nabla}) = \partial/\partial x_\mu$ and $\partial_\mu = (\partial/\partial x_0, \boldsymbol{\nabla})$ so that $\partial^2 = \partial^2/\partial x_0^2 - \boldsymbol{\nabla}^2$ is a Lorentz scalar just like the metric $x_0^2 - \mathbf{x}^2$.

For $v \ll c$, in the nonrelativistic limit, a Lorentz transformation must be Galilean. Hence, to derive the form of a Lorentz transformation along the x_1-axis, we start with a Galilean transformation for infinitesimal relative velocity δv:

$$x_1' = x_1 - \delta v t = x_1 - x_0 \delta\beta. \tag{4.129}$$

Here, as usual, $\beta = v/c$. By symmetry we also write

$$x_0' = x_0 + a \delta\beta x_1 \tag{4.129'}$$

with a a parameter that is fixed by the requirement that $x_0^2 - x_1^2$ be invariant,

$$x_0'^2 - x_1'^2 = x_0^2 - x_1^2. \tag{4.130}$$

[2] This derivation, with a slightly different metric, appears in an article by J. L. Strecker, *Am. J. Phys.* **35**, 12 (1967).

Remember $x = (x_0; x_1, x_2, x_3)$ is the prototype four-dimensional vector in Minkowski space. Thus Eq. (4.130) is simply a statement of the invariance of the square of the magnitude of the "distance" vector under rotation in Minkowski space. Here is where the special relativity is brought into our transformation. Squaring and subtracting Eqs. (4.129) and (4.129′) and discarding terms of order $(\delta\beta)^2$, we find $a = -1$. Equations (4.129) and (4.129′) may be combined as a matrix equation

$$\begin{pmatrix} x_0' \\ x_1' \end{pmatrix} = (1 - \delta\beta\sigma_1)\begin{pmatrix} x_0 \\ x_1 \end{pmatrix}, \tag{4.131}$$

σ_1 happens to be the Pauli matrix, σ_1, and the parameter $\delta\beta$ represents an infinitesimal change. Using the same techniques as in Section 4.2, we repeat the transformation N times to develop a *finite* transformation with the velocity parameter $\rho = N\delta\beta$. Then

$$\begin{pmatrix} x_0' \\ x_1' \end{pmatrix} = \left(1 - \frac{\rho\sigma_1}{N}\right)^N\begin{pmatrix} x_0 \\ x_1 \end{pmatrix}. \tag{4.132}$$

In the limit as $N \to \infty$,

$$\lim_{N \to \infty}\left(1 - \frac{\rho\sigma_1}{N}\right)^N = \exp(-\rho\sigma_1). \tag{4.133}$$

As in Section 4.2, the exponential is interpreted by a Maclaurin expansion

$$\exp(-\rho\sigma_1) = 1 - \rho\sigma_1 + (\rho\sigma_1)^2/2! - (\rho\sigma_1)^3/3! + \cdots. \tag{4.134}$$

Noting that $\sigma_1^2 = 1$,

$$\exp(-\rho\sigma_1) = 1 \cosh\rho - \sigma_1 \sinh\rho. \tag{4.135}$$

Hence our finite Lorentz transformation is

$$\begin{pmatrix} x_0' \\ x_1' \end{pmatrix} = \begin{pmatrix} \cosh\rho & -\sinh\rho \\ -\sinh\rho & \cosh\rho \end{pmatrix}\begin{pmatrix} x_0 \\ x_1 \end{pmatrix}. \tag{4.136}$$

σ_1 has *generated* the representations of this special Lorentz transformation.

Cosh ρ and sinh ρ may be identified by considering the origin of the primed coordinate system, $x_1' = 0$, or $x_1 = vt$. Substituting into Eq. (4.136), we have

$$0 = x_1 \cosh\rho - x_0 \sinh\rho. \tag{4.137}$$

With $x_1 = vt$ and $x_0 = ct$.

$$\tanh\rho = \beta = v/c.$$

Note that the rapidity $\rho \neq v/c$ except in the limit as $v \to 0$.
Using $1 - \tanh^2\rho = (\cosh^2\rho)^{-1}$,

$$\cosh\rho = (1 - \beta^2)^{-1/2} \equiv \gamma, \qquad \sinh\rho = \beta\gamma. \tag{4.138}$$

The preceding special case of the velocity parallel to one space axis is easy, but it illustrates the infinitesimal velocity—exponentiation—generator technique. Now this exact technique may be applied to derive the Lorentz transformation for the relative velocity **v** *not* parallel to any space axis. The matrices given by Eq. (4.136) for the case of $\mathbf{v} = \hat{\mathbf{x}}v_x$ form a subgroup. The matrices in the general case do *not*. The product of two Lorentz transformation matrices, $L(\mathbf{v}_1)$ and $L(\mathbf{v}_2)$, yields a third Lorentz matrix $L(\mathbf{v}_3)$—if the two velocities $\mathbf{v}_1$ and $\mathbf{v}_2$ are parallel. The resultant velocity $\mathbf{v}_3$ is related to $\mathbf{v}_1$ and $\mathbf{v}_2$ by the Einstein velocity addition law, Exercise 4.5.4. If $\mathbf{v}_1$ and $\mathbf{v}_2$ are not parallel, no such simple relation exists. Specifically, consider three reference frames S, S', and S'', with S and S' related by $L(\mathbf{v}_1)$, and S' and S'' related by $L(\mathbf{v}_2)$. If the velocity of S'' relative to the original system S is $\mathbf{v}_3$, S'' is *not* obtained from S by $L(\mathbf{v}_3) = L(\mathbf{v}_2)L(\mathbf{v}_1)$. Rather, we find that

$$L(\mathbf{v}_3) = RL(\mathbf{v}_2)L(\mathbf{v}_1), \tag{4.139}$$

where R is a 3×3 space rotation matrix embedded in our four-dimensional space–time. With $\mathbf{v}_1$ and $\mathbf{v}_2$ not parallel, the final system S'' is *rotated* relative to S. This rotation is the origin of the Thomas precession involved in spin-orbit coupling terms in atomic and nuclear physics. Because of its presence, the $L(\mathbf{v})$ by themselves do not form a group.

EXERCISES

4.5.1   Obtain $\sigma(\lambda, \mu, \nu)$ in Exercise 4.5.3 by differentiating the final matrix.

4.5.2   Two Lorentz transformations are carried out in succession: v_1 along the x-axis, then v_2 along the y-axis. Show that the resultant transformation (given by the product of these two successive transformations) *cannot* be put in the form of a single Lorentz transformation.
Note. The discrepancy corresponds to a rotation.

4.5.3   Rederive the Lorentz transformation working entirely in the real space (x_0, x_1, x_2, x_3) with $x_0 = ct$. Show that the Lorentz transformation may be written $L(\mathbf{v}) = \exp(\rho\sigma)$, with

$$\sigma = \begin{pmatrix} 0 & -\lambda & -\mu & -\nu \\ -\lambda & 0 & 0 & 0 \\ -\mu & 0 & 0 & 0 \\ -\nu & 0 & 0 & 0 \end{pmatrix}.$$

4.5.4   Using the matrix relation, Eq. (4.136), let the velocity parameter ρ_1 relate the Lorentz reference frames (x_0', x_1') and (x_0, x_1). Let ρ_2 relate (x_0'', x_1'') and (x_0', x_1'). Finally, let ρ relate (x_0'', x_1'') and (x_0, x_1). From $\rho = \rho_1 + \rho_2$ derive the Einstein velocity addition law

$$v = \frac{v_1 + v_2}{1 + v_1 v_2/c^2}.$$

4.6 LORENTZ COVARIANCE OF MAXWELL'S EQUATIONS

If a physical law is to hold for all orientations of our (real) special coordinates (i.e., to be invariant under rotations), the terms of the equation must be covariant under rotations (Sections 1.2 and 2.6). This means that we write the physical laws in the mathematical form scalar = scalar, vector = vector, second-rank tensor = second-rank tensor, and so on. Similarly, if a physical law is to hold for all inertial systems, the terms of the equation must be covariant under Lorentz transformations.

Using Minkowski space ($x = x_1$, $y = x_2$, $z = x_3$, $ct = x_0$), we have a four-dimensional cartesian space with the metric $g_{\mu\nu}$ (Eq. (4.128), Section 4.5). The Lorentz transformations are linear in space and time in this four-dimensional real space.[1]

Here we consider Maxwell's equations

$$\nabla \times \mathbf{E} = -\frac{\partial \mathbf{B}}{\partial t}, \tag{4.140a}$$

$$\nabla \times \mathbf{H} = \frac{\partial \mathbf{D}}{\partial t} + \rho \mathbf{v}, \tag{4.140b}$$

$$\nabla \cdot \mathbf{D} = \rho, \tag{4.140c}$$

$$\nabla \cdot \mathbf{B} = 0, \tag{4.140d}$$

and the relations

$$\mathbf{D} = \varepsilon_0 \mathbf{E}, \qquad \mathbf{B} = \mu_0 \mathbf{H}. \tag{4.141}$$

The symbols have their usual meanings as given in the introduction. For simplicity we assume vacuum ($\varepsilon = \varepsilon_0$, $\mu = \mu_0$).

We assume that Maxwell's equations hold in all inertial systems; that is, Maxwell's equations are consistent with special relativity. (The covariance of Maxwell's equations under Lorentz transformations was actually shown by Lorentz and Poincaré before Einstein proposed his theory of special relativity.) Our immediate goal is to rewrite Maxwell's equations as tensor equations in Minkowski space. This will make the Lorentz covariance explicit.

In terms of scalar and magnetic vector potentials, we may write[2]

$$\mathbf{B} = \nabla \times \mathbf{A}$$
$$\mathbf{E} = -\frac{\partial \mathbf{A}}{\partial t} - \nabla \varphi. \tag{4.142}$$

---

[1] A group theoretic derivation of the Lorentz transformation in Minkowski space appears in Section 4.5. See also H. Goldstein, *Classical Mechanics*. Cambridge, MA: Addison–Wesley (1951), Chapter 6. The metric equation $x_0^2 - x^2 = 0$, independent of reference frame, leads to the Lorentz transformations.

[2] Compare Section 1.13, especially Exercise 1.13.10.

Equation (4.142) specifies the curl of **A**; the divergence of **A** is still undefined (compare Sections 1.13 and 1.16). We may, and for future convenience we do, impose the further restriction on the vector potential **A**,

$$\nabla \cdot \mathbf{A} + \varepsilon_0 \mu_0 \frac{\partial \varphi}{\partial t} = 0. \tag{4.143}$$

This is the Lorentz gauge relation. It will serve the purpose of uncoupling the differential equations for **A** and φ that follow. The potentials **A** and φ are not yet completely fixed. The freedom remaining is the topic of Exercise 4.6.4.

Now we rewrite the Maxwell equations in terms of the potentials **A** and φ. From Eqs. (4.140c) for $\nabla \cdot \mathbf{D}$ and (4.142)

$$\nabla^2 \varphi + \nabla \cdot \frac{\partial \mathbf{A}}{\partial t} = -\frac{\rho}{\varepsilon_0}, \tag{4.144}$$

whereas Eqs. (4.140b) for $\nabla \times \mathbf{H}$ and (4.142) and Eq. (1.86) of Chapter 1 yield

$$\frac{\partial^2 \mathbf{A}}{\partial t^2} + \nabla \frac{\partial \varphi}{\partial t} + \frac{1}{\varepsilon_0 \mu_0} \{\nabla\nabla \cdot \mathbf{A} - \nabla^2 \mathbf{A}\} = \frac{\rho \mathbf{v}}{\varepsilon_0}. \tag{4.145}$$

Using the Lorentz relation, Eq. (4.143), and the relation $\varepsilon_0 \mu_0 = 1/c^2$, we obtain

$$\left[\nabla^2 - \frac{1}{c^2} \frac{\partial^2}{\partial t^2} \right] \mathbf{A} = -\mu_0 \rho \mathbf{v},$$

$$\left[\nabla^2 - \frac{1}{c^2} \frac{\partial^2}{\partial t^2} \right] \varphi = -\frac{\rho}{\varepsilon_0}. \tag{4.146}$$

Now the differential operator (see also Exercise 2.7.3)

$$\nabla^2 - \frac{1}{c^2} \frac{\partial^2}{\partial t^2} = -\partial^2 = -\partial^\mu \partial_\mu$$

is a four-dimensional Laplacian, usually called the d'Alembertian and denoted also by $\Box^2$. It may readily be proved a scalar (see Exercise 2.7.3).

For convenience we define

$$A^1 \equiv \frac{A_x}{\mu_0 c} = c\varepsilon_0 A_x, \qquad A^3 \equiv \frac{A_z}{\mu_0 c} = c\varepsilon_0 A_z,$$

$$A^2 \equiv \frac{A_y}{\mu_0 c} = c\varepsilon_0 A_y, \qquad A_0 \equiv \varepsilon_0 \varphi = A^0. \tag{4.147}$$

If we further put

$$\frac{\rho v_x}{c} \equiv i^1, \qquad \frac{\rho v_y}{c} \equiv i^2, \qquad \frac{\rho v_z}{c} \equiv i^3, \qquad \rho \equiv i_0 = i^0, \tag{4.148}$$

then Eq. (4.146) may be written in the form

$$\partial^2 A^\mu = i^\mu. \tag{4.149}$$

Equation (4.149) looks like a tensor equation, but looks do not constitute proof. To prove that it is a tensor equation, we start by investigating the transformation properties of the generalized current i^{μ}.

Since an electric charge element de is an invariant quantity, we have

$$de = \rho \, dx_1 \, dx_2 \, dx_3, \qquad \text{invariant.} \qquad (4.150)$$

We saw in Section 2.9 that the four-dimensional volume element, $dx_1 \, dx_2 \, dx_3 \, dx_0$, was also invariant. Comparing this result, Eq. (2.106) with Eq. (4.150), we see that the charge density ρ must transform the same way as dx_0, the zeroth component of a four-dimensional vector dx_λ. We put $\rho = i^0$ with i^0 now established as the zeroth component of a four-dimensional vector. The other parts of Eq. (4.148) may be expanded as

$$i^1 = \frac{\rho v_x}{c} = \frac{\rho \, dx_1}{c \, dt}$$

$$= i^0 \frac{dx_1}{dx_0}. \qquad (4.151)$$

Since we have just shown that i_0 transforms as dx_0, this means that i_1 transforms as dx_1. With similar results for i_2 and i_3, We have i^λ transforming as dx^λ, proving that i^λ is a vector, a four-dimensional vector in Minkowski space.

Equation (4.149), which follows directly from Maxwell's equations, Eq. (4.140), is assumed to hold in all cartesian systems (all Lorentz frames). Then, by the quotient rule, Section 2.8, A_μ is also a vector and Eq. (4.149) is a legitimate tensor equation.

Now, working backward, Eq. (4.142) may be written

$$\varepsilon_0 E_j = -\frac{\partial A^j}{\partial x_0} + \frac{\partial A^0}{\partial x_j}, \qquad j = 1, 2, 3,$$

$$\frac{1}{\mu_0 c} B_i = \frac{\partial A^k}{\partial x_j} - \frac{\partial A^j}{\partial x_k}, \qquad (i, j, k) = (1, 2, 3) \qquad (4.152)$$

and *cyclic* permutations.

We define a new tensor

$$^\mu A^\lambda - \partial^\lambda A^\mu = \frac{\partial A^\lambda}{\partial x_\mu} - \frac{\partial A^\mu}{\partial x_\lambda} \equiv f^{\mu\lambda} = -f^{\lambda\mu} \qquad (\mu, \lambda = 0, 1, 2, 3)$$

symmetric second-rank tensor, since A^λ is a vector. Written out ly,

$$= \varepsilon_0 \begin{pmatrix} 0 & E_x & E_y & E_z \\ -E_x & 0 & -cB_z & cB_y \\ -E_y & cB_z & 0 & -cB_x \\ -E_z & -cB_y & cB_x & 0 \end{pmatrix}, f^{\mu\lambda} = \varepsilon_0 \begin{pmatrix} 0 & -E_x & -E_y & -E_z \\ E_x & 0 & -cB_z & cB_y \\ E_y & cB_z & 0 & -cB_x \\ E_z & -cB_y & cB_x & 0 \end{pmatrix}.$$

$$(4.153)$$

Notice that in our four-dimensional Minkowski space $\mathbf{E}$ and $\mathbf{B}$ are no longer vectors but together form a second-rank tensor. With this tensor we may write the two nonhomogeneous Maxwell equations ((4.140b) and (4.140c)) combined as a tensor equation

$$\frac{\partial f_{\lambda\mu}}{\partial x_\mu} = i_\lambda. \tag{4.154}$$

The left-hand side of Eq. (4.154) is a four-dimensional divergence of a tensor and therefore a vector. This, of course, is equivalent to contracting a third-rank tensor $\partial f^{\lambda\mu}/\partial x_\nu$ (compare Exercises 2.7.1 and 2.7.2). The two homogeneous Maxwell equations—(4.140a) for $\nabla \times \mathbf{E}$ and (4.140d) for $\nabla \cdot \mathbf{B}$—may be expressed in the tensor form

$$\frac{\partial f_{23}}{\partial x_1} + \frac{\partial f_{31}}{\partial x_2} + \frac{\partial f_{12}}{\partial x_3} = 0 \tag{4.155}$$

for Eq. (4.140d) and three equations of the form

$$-\frac{\partial f_{30}}{\partial x_2} - \frac{\partial f_{02}}{\partial x_3} + \frac{\partial f_{23}}{\partial x_0} = 0 \tag{4.156}$$

for Eq. (4.140a). (A second equation permutes 120, a third permutes 130.) Since

$$\partial^\lambda f^{\mu\nu} = \frac{\partial f^{\mu\nu}}{\partial x_\lambda} \equiv t^{\lambda\mu\nu}$$

is a tensor (of third rank), Eqs. (4.140a) and (4.140d) are given by the tensor equation

$$t^{\lambda\mu\nu} + t^{\nu\lambda\mu} + t^{\mu\nu\lambda} = 0. \tag{4.157}$$

From Eqs. (4.155) and (4.156) the reader will understand that the indices λ, μ, and ν are supposed to be different. Actually Eq. (4.157) automatically reduces to $0 = 0$ if any two indices coincide. An alternate form of Eq. (4.157) appears in Exercise 4.6.14.

LORENTZ TRANSFORMATION OF E AND B

The construction of the tensor equations ((4.154) and (4.157)) completes our initial goal of rewriting Maxwell's equations in tensor form.[3] Now we exploit the tensor properties of our four vectors and the tensor $f_{\mu\nu}$.

For the Lorentz transformation corresponding to motion along the $z(x_3)$-axis with velocity v, the "direction cosines" are given by[4]

$$x_0' = \gamma(x_0 - \beta x_3)$$
$$x_3' = \gamma(x_3 - \beta x_0), \tag{4.158}$$

---

[3] Modern theories of quantum electrodynamics and elementary particles are often written in this "manifestly covariant" form to guarantee consistency with special relativity. Conversely, the insistence on such tensor form has been a useful guide in the construction of these theories.

[4] A group theoretic derivation of the Lorentz transformation appears in Section 4.5. See also Goldstein, Chapter 6.

where

$$\beta = \frac{v}{c}$$

and

$$\gamma = (1 - \beta^2)^{-1/2}. \tag{4.159}$$

Using the tensor transformation properties, we may calculate the electric and magnetic fields in the moving system in terms of the values in the original reference frame. From Eqs. (2.66), (4.153), and (4.158) we obtain

$$E_x' = \frac{1}{\sqrt{1 - \beta^2}} \left(E_x - \frac{v}{c^2} B_y \right),$$

$$E_y' = \frac{1}{\sqrt{1 - \beta^2}} \left(E_y + \frac{v}{c^2} B_x \right), \tag{4.160}$$

$$E_z' = E_z,$$

and

$$B_x' = \frac{1}{\sqrt{1 - \beta^2}} \left(B_x + \frac{v}{c^2} E_y \right),$$

$$B_y' = \frac{1}{\sqrt{1 - \beta^2}} \left(B_y - \frac{v}{c^2} E_x \right), \tag{4.161}$$

$$B_z' = B_z.$$

This coupling of **E** and **B** is to be expected. Consider, for instance, the case of zero electric field in the unprimed system

$$E_x = E_y = E_z = 0.$$

Clearly, there will be no force on a stationary charged particle. When the particle is in motion with a *small* velocity **v** along the z-axis,[5] an observer on the particle sees fields (exerting a force on his charged particle) given by

$$E_x' = -vB_y,$$

$$E_y' = vB_x,$$

where **B** is a magnetic induction field in the unprimed system. These equations may be put in vector form

$$\mathbf{E}' = \mathbf{v} \times \mathbf{B}$$

or   (4.162)

$$\mathbf{F} = q\mathbf{v} \times \mathbf{B},$$

which is usually taken as the operational definition of the magnetic induction **B**.

---

[5] If the velocity is not small (so that v^2/c^2 is negligible), a relativistic transformation of force is needed.

ELECTROMAGNETIC INVARIANTS

Finally, the tensor (or vector) properties allow us to construct a multitude of invariant quantities. A more important one is the scalar product of the two four-dimensional vectors or four vectors A_λ and i_λ. We have

$$A^\lambda i_\lambda = -c\varepsilon_0 A_x \frac{\rho v_x}{c} - c\varepsilon_0 A_y \frac{\rho v_y}{c} - c\varepsilon_0 A_z \frac{\rho v_z}{c} + \varepsilon_0 \varphi \rho$$

$$= \varepsilon_0(\rho\varphi - \mathbf{A} \cdot \mathbf{J}), \quad \text{invariant,} \quad (4.163)$$

with $\mathbf{A}$ the usual magnetic vector potential and $\mathbf{J}$ the ordinary current density. The first term $\rho\varphi$ is the ordinary static electric coupling with dimensions of energy per unit volume. Hence our newly constructed scalar invariant is an energy density. The dynamic interaction of field and current is given by the product $\mathbf{A} \cdot \mathbf{J}$. This invariant $A^\lambda i_\lambda$ appears in the electromagnetic Lagrangians of Exercise 17.3.6 and 17.5.1.

Other possible electromagnetic invariants appear in Exercises 4.6.9 and 4.6.11.

EXERCISES

4.6.1 (a) Show that every four-vector in Minkowski space may be decomposed into an ordinary three-space vector and a three-space scalar. Examples: $(ct, \mathbf{r})$, $(\rho, \rho\mathbf{v}/c)$, $(\varepsilon_0\varphi, c\varepsilon_0\mathbf{A})$, $(E/c, \mathbf{p})$, $(\omega/c, \mathbf{k})$.
Hint. Consider a rotation of the three-space coordinates with time fixed.
(b) Show that the converse of (a) is *not* true—every three-vector plus scalar does *not* form a Minkowski four-vector.

4.6.2 (a) Show that

$$\partial^\mu i_\mu = \partial \cdot i = \frac{\partial i_\mu}{\partial x_\mu} = 0.$$

(b) Show how the previous tensor equation may be interpreted as a statement of continuity of charge and current in ordinary three-dimensional space and time.
(c) If this equation is known to hold in all Lorentz reference frames, why can we not conclude that i_μ is a vector?

4.6.3 Write the Lorentz condition (Eq. (4.143)), as a tensor equation in Minkowski space.

4.6.4 A gauge transformation consists of varying the scalar potential φ_1 and the vector potential $\mathbf{A}_1$ according to the relation

$$\varphi_2 = \varphi_1 + \frac{\partial\chi}{\partial t},$$

$$\mathbf{A}_2 = \mathbf{A}_1 - \nabla\chi.$$

The new function χ is required to satisfy the homogeneous wave equation

$$\nabla^2\chi - \varepsilon_0\mu_0 \frac{\partial^2\chi}{\partial t^2} = 0.$$

Show the following:

(a) The Lorentz relation is unchanged.

(b) The new potentials satisfy the same inhomogeneous wave equations as did the original potentials.

(c) The fields $\mathbf{E}$ and $\mathbf{B}$ are unaltered.

The invariance of our electromagnetic theory under this transformation is called *gauge invariance*.

4.6.5   A charged particle, charge q, mass m, obeys the Lorentz covariant equation

$$dp^{\mu}/d\tau = (q/\varepsilon_0 m_0 c)f^{\mu\nu}p_{\nu}.$$

p_{ν} is the four-dimensional momentum vector $(E/c; p_1, p_2, p_3)$. τ is the proper time; $d\tau = dt\sqrt{1 - v^2/c^2}$, a Lorentz scalar. Show that the explicit space–time forms are

$$d\mathbf{p}/dt = q(\mathbf{E} + \mathbf{v} \times \mathbf{B})$$

$$dE/dt = q\mathbf{v} \cdot \mathbf{E}.$$

4.6.6   From the Lorentz transformation matrix elements (Eq. (4.158)) derive the Einstein velocity addition law

$$u' = \frac{u - v}{1 - (uv/c^2)} \quad \text{or} \quad u = \frac{u' + v}{1 + (u'v/c^2)},$$

where $u = c \, dx_3/dx_0$ and $u' = c \, dx_3'/dx_0'$.

Hint. If $L_{12}(v)$ is the matrix transforming system 1 into system 2, $L_{23}(u')$ the matrix transforming system 2 into system 3, $L_{13}(u)$ the matrix transforming system 1 directly into system 3, then $L_{13}(u) = L_{23}(u')L_{12}(v)$. From this matrix relation extract the Einstein velocity addition law.

4.6.7   The dual of a four-dimensional second-rank tensor $\mathbf{B}$ may be defined by $\tilde{\mathbf{B}}$, where the elements of the dual tensor are given by

$$\tilde{B}^{ij} = \frac{1}{2!} \varepsilon^{ijkl} B_{kl}.$$

Show that $\tilde{\mathbf{B}}$ transforms as

(a) a second-rank tensor under rotations,

(b) a pseudotensor under inversions.

Note. The tilde here does *not* mean transpose.

4.6.8   Construct $\tilde{\mathbf{f}}$, the dual of $\mathbf{f}$, where $\mathbf{f}$ is the electromagnetic tensor given by Eq. (4.153).

$$ANS. \quad \tilde{f}^{\mu\nu} = \varepsilon_0 \begin{pmatrix} 0 & -cB_x & -cB_y & -cB_z \\ cB_x & 0 & E_z & -E_y \\ cB_y & -E_z & 0 & E_x \\ cB_z & E_y & -E_x & 0 \end{pmatrix}.$$

This corresponds to

$$c\mathbf{B} \rightarrow -\mathbf{E},$$

$$\mathbf{E} \rightarrow c\mathbf{B}.$$

This transformation, sometimes called a "dual transformation," leaves Maxwell's equations in vacuum ($\rho = 0$) invariant.

4.6.9   As the quadruple contraction of a fourth-rank pseudotensor and two second-rank tensors $\varepsilon_{\mu\lambda\nu\sigma} f^{\mu\lambda} f^{\nu\sigma}$ is clearly a pseudoscalar, evaluate it.

$$ANS. \quad -8\varepsilon_0^2 c\mathbf{B} \cdot \mathbf{E}.$$

4.6.10   (a) If an electromagnetic field is purely electric (or purely magnetic) in one particular Lorentz frame, show that $\mathbf{E}$ and $\mathbf{B}$ will be orthogonal in other Lorentz reference systems.

(b) Conversely, if $\mathbf{E}$ and $\mathbf{B}$ are orthogonal in one particular Lorentz frame, there exists a Lorentz reference system in which $\mathbf{E}$ (or $\mathbf{B}$) vanishes. Find that reference system.

4.6.11   Show that $c^2 B^2 - E^2$ is a scalar invariant.

4.6.12   Since $(dx_0; dx_1, dx_2, dx_3)$ is a vector, $dx_\mu\, dx^\mu$ is a scalar. Evaluate this scalar for a moving particle in two different coordinate systems: (a) a coordinate system fixed relative to you (lab system), and (b) a coordinate system moving with a moving particle (velocity v relative to you). With the time increment labeled $d\tau$ in the particle system and dt in the lab system, show that

$$d\tau = dt\sqrt{1 - v^2/c^2}.$$

$d\tau$ or τ is the proper time of the particle, a Lorentz invariant quantity.

4.6.13   Expand the scalar expression

$$-\frac{1}{4\varepsilon_0} f_{\mu\nu} f^{\mu\nu} + \frac{1}{\varepsilon_0} i_\mu A^\mu$$

in terms of the fields and potentials. The resulting expression is the Lagrangian density used in Exercise 17.5.1.

4.6.14   Show that Eq. (4.157) may be written

$$\varepsilon_{\alpha\beta\gamma\delta} \frac{\partial f^{\alpha\beta}}{\partial x_\gamma} = 0.$$

4.7 DISCRETE GROUPS

Here we return to groups with a finite number of elements. In physics, groups usually appear as a set of operations that leave a system unchanged, invariant. This is an expression of symmetry. Indeed, a symmetry may be defined as the invariance of the Hamiltonian of a system under a group of transformations. Symmetry in this sense is important in classical mechanics, but it becomes even more important and more profound in quantum mechanics. In this section we investigate the symmetry properties of sets of objects (atoms in a molecule or crystal). This provides additional illustrations of the group concepts of Section 4.1 and leads directly to dihedral groups. The dihedral groups in turn open up the study of the 32 point groups and 230 space groups that are of such importance in crystallography and solid state physics. It might be noted that it was through the study of crystal symmetries that the concepts of symmetry and group theory entered physics. In physics, the abstract group conditions often take on direct physical meaning in terms of transformations of vectors, spinors, and tensors.

As a very simple, but not trivial, example of a finite group, consider the set $1, a, b, c$ that combine according to the group multiplication table[1]

	1	a	b	c
1	1	a	b	c
a	a	b	c	1
b	b	c	1	a
c	c	1	a	b

Clearly, the four conditions of the definition of "group" are satisfied. The elements a, b, c, and 1 are abstract mathematical entities, completely unrestricted except for the preceding multiplication table.

Now, for a specific *representation* of these group elements, let

$$1 \to 1, \qquad a \to i, \qquad b \to -1, \qquad c \to -i, \qquad (4.164)$$

combining by ordinary multiplication. Again, the four group conditions are satisfied, and these four elements form a group. We label this group C_4. Since the multiplication of the group elements is commutative, the group is labeled *commutative* or *abelian*. Our group is also a *cyclic group* in that the elements may be written as successive powers of one element, in this case i^n, $n = 0, 1, 2, 3$. Note that in writing out Eq. (4.164) we have selected a specific *representation* for this group of four objects, C_4.

We recognize that the group elements $1, i, -1, -i$ may be interpreted as successive 90° rotations in the complex plane. Then, from Eq. (3.74) we create the set of four 2×2 matrices (replacing φ by $-\varphi$ in Eq. (3.74) to rotate a vector rather than rotate the coordinates.)

$$R(\varphi) = \begin{pmatrix} \cos\varphi & -\sin\varphi \\ \sin\varphi & \cos\varphi \end{pmatrix},$$

and for $\varphi = 0, \pi/2, \pi$, and $3\pi/2$ we have

$$1 = \begin{pmatrix} 1 & 0 \\ 0 & 1 \end{pmatrix} \qquad A = \begin{pmatrix} 0 & -1 \\ 1 & 0 \end{pmatrix}$$

$$B = \begin{pmatrix} -1 & 0 \\ 0 & -1 \end{pmatrix} \qquad C = \begin{pmatrix} 0 & 1 \\ -1 & 0 \end{pmatrix}. \qquad (4.165)$$

This set of four matrices forms a group with the law of combination being matrix multiplication. Here is a second representation, now in terms of matrices. A little matrix multiplication verifies that this representation is also abelian and cyclic. Clearly, there is a one-to-one correspondence of the two

[1] The order of the factors is row–column: $ab = c$ in the indicated previous example.

Table 4.3

	I	V_1	V_2	V_3
I	I	V_1	V_2	V_3
V_1	V_1	I	V_3	V_2
V_2	V_2	V_3	I	V_1
V_3	V_3	V_2	V_1	I

representations

$$1 \leftrightarrow 1 \leftrightarrow 1 \qquad a \leftrightarrow i \leftrightarrow A \qquad b \leftrightarrow -1 \leftrightarrow B \qquad c \leftrightarrow -i \leftrightarrow C. \qquad (4.166)$$

In the group C_4 the two representations $(1, i, -1, -i)$ and $(1, A, B, C)$ are isomorphic.

In contrast to this, there is no such correspondence between either of these representations of group C_4 and another group of four objects, the *vierergruppe* (Exercise 3.2.7). The *vierergruppe* has the multiplication table shown in Table 4.3. Confirming the lack of correspondence between the group represented by $(1, i, -1, -i)$ or the matrices $(1, A, B, C)$ of Eq. (4.165), note that although the *vierergruppe* is abelian, it is not cyclic. The cyclic group C_4 and the *vierergruppe* are not isomorphic.

CLASSES AND CHARACTER

Consider a group element x transformed into a group element y by a similarity transform with respect to g_i, an element of the group

$$g_i x g_i^{-1} = y. \qquad (4.167)$$

The group element y is conjugate to x. A *class* is a set of mutually conjugate group elements. In general, this set of elements forming a class does not satisfy the group postulates and is not a group. Indeed, the unit element 1 which is always in a class by itself is the only class that is also a subgroup. All members of a given class are equivalent in the sense that any one element is a similarity transform of any other element. Clearly, if a group is abelian, every element is a class by itself. We find that

1. Every element of the original group belongs to one and only one class.
2. The number of elements in a class is a factor of the order of the group.

We get a possible physical interpretation of the concept of class by noting that y is a similarity transform of x. If g_i represents a rotation of the coordinate system, then y is the same operation as x but relative to the new, related coordinates.

In Section 3.3 we see that a real matrix transforms under rotation of the coordinates by an orthogonal similarity transformation. Depending on the choice of reference frame, essentially the same matrix may take on an infinity of different forms. Likewise, our group representations may be put in an infinity of different forms by using unitary transformations. But each such transformed representation is isomorphic with the original. From Exercise 3.3.9 the trace of each element (each matrix of our representation) is invariant under unitary transformations. Just because it is invariant, the trace (relabeled the *character*) assumes a role of some importance in group theory, particularly in applications to solid state physics. Clearly, all members of a given class (in a given representation) have the same character. Elements of different classes may have the same character but elements with different characters cannot be in the same class.

The concept of class is important (1) because of the trace or character and (2) because *the number of nonequivalent irreducible representations of a group is equal to the number of classes*.

SUBGROUPS AND COSETS

Frequently a subset of the group elements (including the unit element I) will by itself satisfy the four group requirements and therefore is a group. Such a subset is called a *subgroup*. Every group has two trivial subgroups: the unit element alone and the group itself. The elements 1 and b of the four element group C_4 discussed earlier form a nontrivial subgroup. In Section 4.1 we consider SO(3), the (continuous) group of all rotations in ordinary space. The rotations about any single axis form a subgroup of SO(3). Numerous other examples of subgroups appear in the following sections.

Consider a subgroup H with elements h_i and a group element x not in H. Then xh_i and $h_i x$ are not in subgroup H. The sets generated by

$$xh_i, i = 1, 2, \ldots \quad \text{and} \quad h_i x, i = 1, 2, \ldots$$

are called *cosets*, respectively, the left and right cosets of subgroup H with respect to x. It can be shown (assume the contrary and prove a contradiction) that the coset of a subgroup has the same number of distinct elements as the subgroup. Extending this result we may express the original group G as the sum of H and cosets:

$$G = H + x_1 H + x_2 H + \cdots.$$

Then the order of any subgroup is a divisor of the order of the group. It is this result that makes the concept of coset significant. In the next section the six-element group D_3 (order 6) has subgroups of order 1, 2, and 3. D_3 cannot (and does not) have subgroups of order 4 or 5.

The similarity transform of a subgroup H by a fixed group element x *not* in H, xHx^{-1} yields a subgroup—Exercise 4.7.8. If this new subgroup is identical with H for all x,

$$xHx^{-1} = H,$$

then H is called an *invariant, normal,* or *self-conjugate subgroup*. Such subgroups are involved in the analysis of multiplets of atomic and nuclear spectra and the particles discussed in Section 4.2. All subgroups of a commutative (abelian) group are automatically invariant.

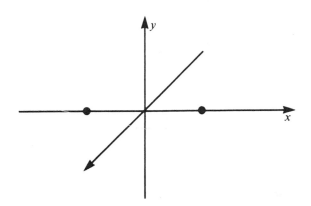

Figure 4.10 Diatomic molecule H_2, N_2, O_2, Cl_2, and so on.

TWO OBJECTS—TWOFOLD SYMMETRY AXIS

Consider first the two-dimensional system of two identical atoms in the xy-plane at $(1, 0)$ and $(-1, 0)$, Fig. 4.10. What rotations[2] can be carried out (keeping both atoms in the xy-plane) that will leave this system invariant? The first candidate is, of course, the unit operator 1. A rotation of π radians about the z-axis completes the list. So we have a rather uninteresting group of two members $(1, -1)$. The z-axis is labeled a twofold symmetry axis—corresponding to the two rotation angles 0 and π that leave the system invariant.

Our system becomes more interesting in three dimensions. Now imagine a molecule (or part of a crystal) with atoms of element X at $\pm a$ on the x-axis, atoms of element Y at $\pm b$ on the y-axis, and atoms of element Z at $\pm c$ on the z-axis as show in Fig. 4.11. Clearly, each axis is now a twofold symmetry axis. Using $R_x(\pi)$ to designate a rotation of π radians about the x-axis, we may

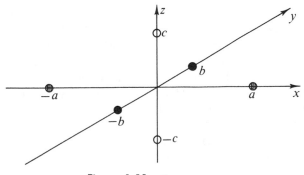

Figure 4.11 D_2 symmetry.

---

[2] Here we deliberately exclude reflections and inversions. They must be brought in to develop the full set of 32 point groups.

Table 4.4

	1	$R_x(\pi)$	$R_y(\pi)$	$R_z(\pi)$
1	1	R_x	R_y	R_x
$R_x(\pi)$	R_x	1	R_z	R_y
$R_y(\pi)$	R_y	R_z	1	R_x
$R_z(\pi)$	R_z	R_y	R_x	1

set up a matrix representation of the rotations as in Section 3.3:

$$R_x(\pi) = \begin{pmatrix} 1 & 0 & 0 \\ 0 & -1 & 0 \\ 0 & 0 & -1 \end{pmatrix} \qquad R_y(\pi) = \begin{pmatrix} -1 & 0 & 0 \\ 0 & 1 & 0 \\ 0 & 0 & -1 \end{pmatrix}$$

$$R_z(\pi) = \begin{pmatrix} -1 & 0 & 0 \\ 0 & -1 & 0 \\ 0 & 0 & 1 \end{pmatrix} \qquad 1 = \begin{pmatrix} 1 & 0 & 0 \\ 0 & 1 & 0 \\ 0 & 0 & 1 \end{pmatrix}.$$

$$(4.168)$$

These four elements $[1, R_x(\pi), R_y(\pi), R_z(\pi)]$ form an abelian group with the group multiplication table shown in Table 4.4.

The products shown in Table 4.4 can be obtained in either of two distinct ways: (1) We may analyze the operations themselves—a rotation of π about the x-axis followed by a rotation of π about the y-axis is equivalent to a rotation of π about the z-axis: $R_y(\pi)R_x(\pi) = R_z(\pi)$. (2) Alternatively, once the matrix representation is established, we can obtain the products by matrix multiplication. This is where the power of mathematics is shown—when the system is too complex for a direct physical interpretation.

Comparison with Exercises 3.2.7, 4.7.2, or 4.7.3 shows immediately that this group is the *vierergruppe*. The matrices of Eq. (4.168) are isomorphic with those of Exercise 3.2.7. Also, they are obviously reducible—being diagonal. The subgroups are $(1, R_x)$, $(1, R_y)$, and $(1, R_z)$. They are invariant. It should be noted that a rotation of π about the y-axis and a rotation of π about the z-axis is equivalent to a rotation of π about the x-axis: $R_z(\pi)R_y(\pi) = R_x(\pi)$. In symmetry terms, if y and z are twofold symmetry axes, x is automatically a twofold symmetry axis.

This symmetry group,[3] the *vierergruppe*, is often labeled D_2, the D signifying a dihedral group and the subscript 2 signifying a twofold symmetry axis (and no higher symmetry axis).

---

[3] A symmetry group is a group of symmetry-preserving operations, that is, rotations, reflections, and inversions. A symmetric group is the group of permutations of n distinct objects—of order n!

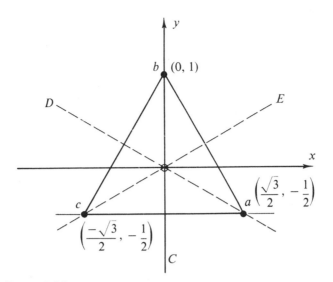

Figure 4.12   Symmetry operations on an equilateral triangle.

THREE OBJECTS—THREEFOLD SYMMETRY AXIS

Consider now three identical atoms at the vertices of an equilateral triangle, Fig. 4.12. Rotations *of the triangle* of 0, $2\pi/3$, and $4\pi/3$ leave the triangle invariant. In matrix form, we have[4]

$$1 = R_z(0) = \begin{pmatrix} 1 & 0 \\ 0 & 1 \end{pmatrix}$$

$$A = R_z(2\pi/3) = \begin{pmatrix} \cos 2\pi/3 & -\sin 2\pi/3 \\ \sin 2\pi/3 & \cos 2\pi/3 \end{pmatrix} = \begin{pmatrix} -1/2 & -\sqrt{3}/2 \\ \sqrt{3}/2 & -1/2 \end{pmatrix} \qquad (4.169)$$

$$B = R_z(4\pi/3) = \begin{pmatrix} -1/2 & \sqrt{3}/2 \\ -\sqrt{3}/2 & -1/2 \end{pmatrix}.$$

The z-axis is a threefold symmetry axis. (1, A, B) form a cyclic group, a subgroup of the complete six-element group that follows.

In the xy-plane there are three additional axes of symmetry—each atom (vertex) and the geometric center defining an axis. Each of these is a twofold symmetry axis. These rotations may most easily be described within our two-dimensional framework by introducing reflections. The rotation of π about the C or y-axis, which means the interchanging of atoms a and c, is just a reflection of the x-axis:

$$C = R_C(\pi) = \begin{pmatrix} -1 & 0 \\ 0 & 1 \end{pmatrix}. \qquad (4.170)$$

---

[4] Note that here we are rotating the *triangle* counterclockwise relative to fixed coordinates.

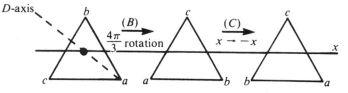

Figure 4.13 The triangle on the right is the triangle on the left rotated 180° about the D-axis. $D = CB$.

We may replace the rotation about the D-axis by a rotation of $4\pi/3$ (about our z-axis) followed by a reflection of the x-axis ($x \to -x$) (Fig. 4.13):

$$D = R_D(\pi) = CB$$

$$= \begin{pmatrix} -1 & 0 \\ 0 & 1 \end{pmatrix} \begin{pmatrix} -1/2 & \sqrt{3}/2 \\ -\sqrt{3}/2 & -1/2 \end{pmatrix}$$

$$= \begin{pmatrix} 1/2 & -\sqrt{3}/2 \\ -\sqrt{3}/2 & -1/2 \end{pmatrix}. \tag{4.171}$$

In a similar manner; the rotation of π about the E-axis interchanging a and b is replaced by a rotation of $2\pi/3$ (A) and then a reflection[5] of the x-axis ($x \to -x$):

$$E = R_E(\pi) = CA$$

$$= \begin{pmatrix} -1 & 0 \\ 0 & 1 \end{pmatrix} \begin{pmatrix} -1/2 & -\sqrt{3}/2 \\ \sqrt{3}/2 & -1/2 \end{pmatrix}$$

$$= \begin{pmatrix} 1/2 & \sqrt{3}/2 \\ \sqrt{3}/2 & -1/2 \end{pmatrix}. \tag{4.172}$$

The complete group multiplication table is

	1	A	B	C	D	E
1	1	A	B	C	D	E
A	A	B	1	D	E	C
B	B	1	A	E	C	D
C	C	E	D	1	B	A
D	D	C	E	A	1	B
E	E	D	C	B	A	1

Notice that each element of the group appears only once in each row and in each column—as required by the rearrangement theorem, Exercise 4.7.4.

[5] Note that, as a consequence of these reflections, $\det(C) = \det(D) = \det(E) = -1$. The rotations A and B, of course, have a determinant of $+1$.

Also, from the multiplication table the group is not abelian. We have constructed a six-element group and a 2×2 irreducible matrix representation of it. The only other distinct six-element group is the cyclic group $[1, R, R^2, R^3, R^4, R^5]$ with

$$R = \begin{pmatrix} \cos \pi/3 & -\sin \pi/3 \\ \sin \pi/3 & \cos \pi/3 \end{pmatrix} = \begin{pmatrix} 1/2 & -\sqrt{3}/2 \\ \sqrt{3}/2 & 1/2 \end{pmatrix}. \tag{4.173}$$

Our group [1, A, B, C, D, E] is labeled D_3 in crystallography, the dihedral group with a threefold axis of symmetry. The three axes (C, D, and E) in the xy-plane automatically become twofold symmetry axes. As a consequence, (1, C), (1, D), and (1, E) all form two-element subgroups. None of these two-element subgroups of D_3 is invariant.

There are two other irreducible representation of the symmetry group of the equilateral triangle: (1) the trivial (1, 1, 1, 1, 1, 1), and (2) the almost as trivial (1, 1, 1, -1, -1, -1), the positive signs corresponding to proper rotations and the negative signs to improper rotations (involving a reflection). Both of these representations are homomorphic with D_3.

A general and most important result for *finite* groups of h elements is that

$$\sum_i n_i^2 = h, \tag{4.174}$$

where n_i is the dimension of the matrices of the ith irreducible representation. This equality, sometimes called the dimensionality theorem, is very useful in establishing the irreducible representation of a group. Here for D_3 we have $1^2 + 1^2 + 2^2 = 6$ for our three representations. No other irreducible representations of the symmetry group of three objects exist.

DIHEDRAL GROUPS, D_n

A dihedral group D_n with an n-fold symmetry axis implies n axes with angular separation of $2\pi/n$ radians. n is a positive integer, but otherwise unrestricted. If we apply the symmetry arguments to *crystal lattices*, then n is limited to 1, 2, 3, 4, and 6. The requirement of invariance of the crystal lattice under translations in the plane perpendicular to the n-fold axis excludes $n = 5, 7$, and higher values. Try to cover a plane completely with identical regular pentagons and with no overlapping.[6] For individual molecules, this constraint does not exist, although the examples with $n > 6$ are rare. $n = 5$ is a real possibility. As an example, the symmetry group for ruthenocene, $(C_5H_5)_2Ru$, illustrated in Fig. 4.14, is D_5.[7]

CRYSTALLOGRAPHIC POINT AND SPACE GROUPS

The dihedral groups just considered are examples of the crystallographic point groups. A point group is composed of combinations of rotations and

---

[6] For D_6 imagine a plane covered with regular hexagons and the axis of rotation through the geometric center of one of them.

[7] Actually the full technical label is D_{5h}, the h, indicating invariance under a *reflection* of the fivefold axis.

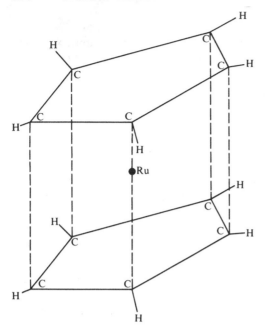

Figure 4.14 Ruthenocene.

reflections (including inversions) that will leave some crystal lattice unchanged. Limiting the operations to rotations and reflections (including inversions) means that one point—the origin—remains *fixed*, hence the term *point group*. Including the cyclic groups, two cubic groups (tetrahedon and octahedron symmetries), and the improper forms (involving reflections), we come to a total of 32-point groups.

If, to the rotation and reflection operations that produced the point groups, we add the possibility of translations and still demand that some crystal lattice remain invariant, we come to the space groups. There are 230 distinct space groups, a number that is appalling except, possibly, to specialists in the field. For details (which can cover hundreds of pages) see the references.

EXERCISES

4.7.1   Show that the matrices 1, A, B, and C of Eq. (4.165) are reducible. Reduce them.

Note. This means transforming A and C to diagonal form (by the *same* unitary transformation).

Hint. A and C are anti-Hermitian. Their eigenvectors will be orthogonal.

4.7.2   Possible operations on a crystal lattice include A_π (rotation by π), m (reflection), and i (inversion). These three operations combine as

$$A_\pi^2 = m^2 = i^2 = 1,$$

$$A_\pi \cdot m = i, \qquad m \cdot i = A_\pi, \qquad \text{and} \qquad i \cdot A_\pi = m.$$

Show that the group $(1, A_\pi, m, i)$ is isomorphic with the *vierergruppe*.

4.7.3   Four possible operations in the xy-plane are:

1. no change $\begin{cases} x \to x \\ y \to y \end{cases}$

2. inversion $\begin{cases} x \to -x \\ y \to -y \end{cases}$

3. reflection $\begin{cases} x \to -x \\ y \to y \end{cases}$

4. reflection $\begin{cases} x \to x \\ y \to -y. \end{cases}$

(a) Show that these four operations form a group.
(b) Show that this group is isomorphic with the *vierergruppe*.
(c) Set up a 2×2 matrix representation.

4.7.4   Rearrangement theorem: Given a group of n distinct elements $(I, a, b, c, \ldots, n)$, show that the set of products $(aI, a^2, ab, ac \ldots, an)$ reproduces the n distinct elements in a new order.

4.7.5   Using the 2×2 matrix representation of Exercise 3.2.7 for the *vierergruppe*,

(a) Show that there are four classes, each with one element.
(b) Calculate the character (trace) of each class. Note that two different classes may have the same character.
(c) Show that there are three two-element subgroups. (The unit element by itself always forms a subgroup.)
(d) For any one of the two-element subgroups show that the subgroup and a single coset reproduce the original *vierergruppe*.

Note that subgroups, classes, and cosets are entirely different.

4.7.6   Using the 2×2 matrix representation, Eq. (4.165), of C_4,

(a) Show that there are four classes, each with one element.
(b) Calculate the character (trace) of each class.
(c) Show that there is one two-element subgroup.
(d) Show that the subgroup and a single coset reproduce the original group.

4.7.7   Prove that the number of distinct elements in a coset of a subgroup is the same as the number of elements in the subgroup.

4.7.8   A subgroup H has elements h_i. x is a fixed element of the original group G and is *not* a member of H. The transform

$$xh_ix^{-1}, \qquad i = 1, 2, \ldots$$

generates a *conjugate subgroup* xHx^{-1}. Show that this conjugate subgroup satisfies each of the four group postulates and therefore is a group.

4.7.9   (a) A particular group is abelian. A second group is created by replacing g_i by g_i^{-1} for each element in the original group. Show that the two groups are isomorphic.
Note. This means showing that if $a_ib_i = c_i$, then, $a_i^{-1}b_i^{-1} = c_i^{-1}$.
(b) Continuing part (a), if the two groups are isomorphic, show that each must be abelian.

4.7.10   (a) Once you have a matrix representation of any group, a one-dimensional representation can be obtained by taking the determinants of the matrices. Show that the multiplicative relations are preserved in this determinant representation.

(b) Use determinants to obtain a one-dimensional representative of D_3.

4.7.11   Explain how the relation

$$\sum_i n_i^2 = h$$

applies to the *vierergruppe* ($h = 4$) and to the dihedral group, D_3 ($h = 6$).

4.7.12   Show that the subgroup (1, A, B) of D_3 is an *invariant* subgroup.

4.7.13   The group D_3 may be discussed as a *permutation* group of three objects. Matrix B, for instance, rotates vertex a (originally in location 1) to the position formerly occupied by c (location 3). Vertex b moves from location 2 to location 1, and so on. As a permutation $(a\ b\ c) \rightarrow (b\ c\ a)$. In three dimensions

$$\begin{pmatrix} 0 & 1 & 0 \\ 0 & 0 & 1 \\ 1 & 0 & 0 \end{pmatrix} \begin{pmatrix} a \\ b \\ c \end{pmatrix} = \begin{pmatrix} b \\ c \\ a \end{pmatrix}.$$

(a) Develop analogous 3×3 representations for the other elements of D_3.

(b) Reduce your 3×3 representation to the 2×2 representation of this section.

(This 3×3 representation *must* be reducible or Eq. (4.174) would be violated.)

Note. The actual reduction of a reducible representation may be awkward. It is often easier to develop directly a new representation of the required dimension.

4.7.14   (a) The permutation group of four objects, P_4, has $4! = 24$ elements. Treating the four elements of the cyclic group, C_4, as permutations, set up a 4×4 matrix representation of C_4. C_4 becomes a subgroup of P_4.

(b) How do you know that this 4×4 matrix representation of C_4 *must* be reducible?

Note. C_4 is abelian and every abelian group of h objects has only h one-dimensional irreducible representations.

4.7.15   (a) The objects $(a\ b\ c\ d)$ are permuted to $(d\ a\ c\ b)$. Write out a 4×4 matrix representation of this one permutation.

(b) Is permutation, $(a\ b\ d\ c) \rightarrow (d\ a\ c\ b)$, odd or even?

(c) Is this permutation a possible member of the D_4 group? Why or why not?

4.7.16   The elements of the dihedral group D_n may be written in the form

$$S^\lambda R_z^\mu(2\pi/n), \qquad \lambda = 0, 1$$

$$\mu = 0, 1, \ldots, n - 1,$$

where $R_z(2\pi/n)$ represents a rotation of $2\pi/n$ about the n-fold symmetry axis, whereas S represents a rotation of π about an axis through the center of the regular polygon and one of its vertices.

For S = E show that this form may describe the matrices A, B, C, and D of D_3.
Note. The elements R_z and S are called the *generators* of this finite group.
Similarly, i is the generator of the group given by Eq. (4.164).

4.7.17   Show that the cyclic group of n objects, C_n, may be represented by r^m,
$m = 0, 1, 2, ..., n - 1$. Here r is a generator given by

$$r = \exp(2\pi is/n).$$

The parameter s takes on the values $s = 1, 2, 3, ..., n$, each value of s
yielding a different one-dimensional (irreducible) representation of C_n.

4.7.18   Develop the irreducible 2×2 matrix representation of the group of
operations (rotations and reflections) that transform a square into itself.
Give the group multiplication table.
Note. This is the symmetry group of a square and also the dihedral group, D_4.

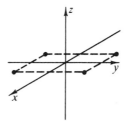

4.7.19   The permutation group of four objects contains $4! = 24$ elements. From
Exercise 4.7.18, D_4, the symmetry group for a square, has far less than 24
elements. Explain the relation between D_4 and the permutation group of
four objects.

4.7.20   A plane is covered with *regular* hexagons, as shown.
(a) Determine the dihedral symmetry of an axis perpendicular to the plane
through the common vertex of three hexagons (A). That is, if the axis
has n-fold symmetry, show (with careful explanation) what n is. Write
out the 2×2 matrix describing the minimum (nonzero) positive rotation
of the array of hexagons that is a member of your D_n group.
(b) Repeat part (a) for an axis perpendicular to the plane through the
geometric center of one hexagon (B).

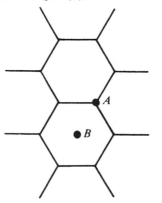

4.7.21   In a simple cubic crystal, we might have identical atoms at $r = (la, ma, na)$, l, m, and n taking on all integral values.

(a) Show that each cartesian axis is a fourfold symmetry axis.

(b) The cubic group will consist of all operations (rotations, reflections, inversion) that leave the simple cubic crystal invariant. From a consideration of the permutation of the positive and negative coordinate axes, predict how many elements this cubic group will contain.

4.7.22   (a) From the D_3 multiplication table construct a similarity transform table showing xyx^{-1}, where x and y each range over all six elements of D_3:

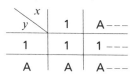

(b) Divide the elements of D_3 into classes. Using the 2×2 matrix representation of Eqs. (4.169)–(4.172) note the trace (character) of each class.

ADDITIONAL READINGS

BUERGER, M. J., *Elementary Crystallography*. New York: Wiley (1956). A comprehensive discussion of crystal symmetries. Buerger develops all 32 point groups and all 230 space groups. Related books by this author include *Contemporary Crystallography*. New York: McGraw-Hill (1970), *Crystal Structure Analysis*. New York: Krieger (1979) (reprint, 1960), and *Introduction to Crystal Geometry*. New York: Krieger (1977) (reprint, 1971).

BURNS G., and A. M. GLAZER, *Space Groups for Solid State Scientists*. New York: Academic Press (1978). A well-organized, readable treatment of groups and their application to the solid state.

DE-SHALIT, A., and I. TALMI, *Nuclear Shell Model*. New York: Academic Press (1963). We adopt the Condon-Shortley phase conventions of this text.

EDMONDS, A. R., *Angular Momentum in Quantum Mechanics*. Princeton, NJ: Princeton University Press (1957).

FALICOV, L. M., *Group Theory and Its Physical Applications*. Notes compiled by A. Luehrmann. Chicago: University of Chicago Press (1966). Group theory with an emphasis on applications to crystal symmetries and solid state physics.

GELL-MANN, M., and Y. NE'EMAN, *The Eightfold Way*. New York: Benjamin (1965). A collection of reprints of significant papers on SU(3) and the particles of high-energy physics. The several introductory sections by Gell-Mann and Ne'eman are especially helpful.

GREINER, W., and B. MÜLLER, *Quantum Mechanics Symmetries*. Berlin: Springer (1989). We refer to this textbook for more details and numerous exercises that are worked out in detail.

HAMERMESH, M., *Group Theory and Its Application to Physical Problems*. Reading, MA: Addison-Wesley (1962). A detailed, rigorous account of both finite and continuous groups. The 32 point groups are developed. The continuous groups are treated with Lie algebra included. A wealth of applications to atomic and nuclear physics.

HEITLER, W., *The Quantum Theory of Radiation*, 2nd ed. Oxford: Oxford University Press (1947). Reprinted, New York: Dover (1983).

HIGMAN, B., *Applied Group-Theoretic and Matrix Methods*. Oxford: Clarendon Press (1955). A rather complete and unusually intelligible development of matrix analysis and group theory.

JACKSON, J. D., *Classical Electrodynamics*, 2nd ed. New York: Wiley (1975).

MESSIAH, A., *Quantum Mechanics*, Vol. II. Amsterdam: North-Holland (1961).

PANOFSKY, W. K. H., and M. PHILLIPS, *Classical Electricity and Magnetism*, 2nd ed. Reading, MA: Addison-Wesley (1962). The Lorentz covariance of Maxwell's equations is developed for both vacuum and material media. Panofsky and Phillips use contravariant and covariant tensors rather than Minkowski space. Discussions using Minkowski space are given by Heitler and Stratton.

PARK, D., Resource letter SP-1 on symmetry in physics. *Am. J. Phys.* **36**, 577–584 (1968). Includes a large selection of basic references on group theory and its applications to physics: atoms, molecules, nuclei, solids, and elementary particles.

RAM, B., Physics of the SU(3) symmetry model. *Am. J. Phys.* **35**, 16 (1967). An excellent discussion of the applications of SU(3) to the strongly interacting particles (baryons). For a sequel to this see R. D. Young, Physics of the Quark model. *Am. J. Phys.* **41**, 472 (1973).

ROSE, M. E., *Elementary Theory of Angular Momentum*. New York: Wiley (1957). As part of the development of the quantum theory of angular momentum, Rose includes a detailed and readable account of the rotation group.

WIGNER, E. P., *Group Theory and Its Application to the Quantum Mechanics of Atomic Spectra* (translated by J. J. Griffin). New York and London: Academic Press (1959). This is the classic reference on group theory for the physicist. The rotation group is treated in considerable detail. There are a wealth of applications to atomic physics.

5 INFINITE SERIES

5.1 FUNDAMENTAL CONCEPTS

Infinite series, literally summations of an infinite number of terms, occur frequently in both pure and applied mathematics. They may be used by the pure mathematician to define functions as a fundamental approach to the theory of functions, as well as for calculating accurate values of transcendental constants and transcendental functions. In the mathematics of science and engineering infinite series are ubiquitous, for they appear in the evaluation of integrals (Sections 5.6 and 5.7), in the solution of differential equations (Sections 8.5 and 8.6), and as Fourier series (Chapter 14) and compete with integral representations for the description of a host of special functions (Chapters 11, 12, and 13). In Section 16.3 the Neumann series solution for integral equations provides one more example of the occurrence and use of infinite series.

Right at the start we face the problem of attaching meaning to the sum of an infinite number of terms. The usual approach is by partial sums. If we have a sequence of infinite terms u_1, u_2, u_3, u_4, u_5, ..., we define the ith partial sum as

$$s_i = \sum_{n=1}^{i} u_n. \tag{5.1}$$

This is a finite summation and offers no difficulties. If the partial sums s_i converge to a (finite) limit as $i \to \infty$,

$$\lim_{i \to \infty} s_i = S, \tag{5.2}$$

the infinite series $\sum_{n=1}^{\infty} u_n$ is said to be *convergent* and to have the value S. Note carefully that we reasonably, plausibly, but still arbitrarily *define* the infinite series as equal to S. The reader should also note that a necessary condition for this convergence to a limit is that $\lim_{n \to \infty} u_n = 0$. This condition, however, is *not* sufficient to guarantee convergence. Equation (5.2) is usually written in formal mathematical notation:

The condition for the existence of a limit S is that for each $\varepsilon > 0$, there is a fixed N such that

$$|S - s_i| < \varepsilon, \qquad \text{for } i > N.$$

This condition is often derived from the *Cauchy criterion* applied to the partial sums s_i. The Cauchy criterion is:

A necessary and sufficient condition that a sequence (s_i) converge is that for each $\varepsilon > 0$ there is a fixed number N such that

$$|s_j - s_i| < \varepsilon \qquad \text{for all } i, j > N.$$

This means that the individual partial sums must cluster together as we move far out in the sequence.

The Cauchy criterion may easily be extended to sequences of functions. We see it in this form in Section 5.5 in the definition of uniform convergence and in Section 9.4 in the development of Hilbert space.

Our partial sums s_i may not converge to a single limit but may oscillate, as in the case

$$\sum_{n=1}^{\infty} u_n = 1 - 1 + 1 - 1 + 1 + \cdots - (-1)^n + \cdots.$$

Clearly, $s_i = 1$ for i odd but 0 for i even. There is no convergence to a limit, and series such as this one are labeled oscillatory.

For the series

$$1 + 2 + 3 + \cdots + n + \cdots$$

we have

$$s_n = \frac{n(n + 1)}{2}.$$

As $n \to \infty$,

$$\lim_{n \to \infty} s_n = \infty.$$

Whenever the sequence of partial sums diverges (approaches $\pm \infty$), the infinite series is said to *diverge*. Often the term divergent is extended to include oscillatory series as well.

Because we evaluate the partial sums by ordinary arithmetic, the convergent series, defined in terms of a limit of the partial sums, assumes a position of supreme importance. Two examples may clarify the nature of convergence or divergence of a series and will also serve as a basis for a further detailed investigation in the next section.

Example 5.1.1 The Geometric Series

The geometrical sequence, starting with a and with a ratio r $(r \geq 0)$, is given by

$$a + ar + ar^2 + ar^3 + \cdots + ar^{n-1} + \cdots .$$

The nth partial sum is given by[1]

$$S_n = a \frac{1 - r^n}{1 - r}. \tag{5.3}$$

Taking the limit as $n \to \infty$,

$$\lim_{n \to \infty} s_n = \frac{a}{1 - r}, \qquad \text{for } r < 1. \tag{5.4}$$

Hence, by definition, the infinite geometric series converges for $r < 1$ and is given by

$$\sum_{n=1}^{\infty} ar^{n-1} = \frac{a}{1 - r}. \tag{5.5}$$

On the other hand, if $r \geq 1$, the necessary condition $u_n \to 0$ is not satisfied and the infinite series diverges.

Example 5.1.2 The Harmonic Series

As a second and more involved example, we consider the harmonic series

$$\sum_{n=1}^{\infty} n^{-1} = 1 + \frac{1}{2} + \frac{1}{3} + \frac{1}{4} + \cdots + \frac{1}{n} + \cdots . \tag{5.6}$$

We have the $\lim_{n \to \infty} u_n = \lim_{n \to \infty} 1/n = 0$, but this is not sufficient to guarantee convergence. If we group the terms (no change in order) as

$$1 + \tfrac{1}{2} + (\tfrac{1}{3} + \tfrac{1}{4}) + (\tfrac{1}{5} + \tfrac{1}{6} + \tfrac{1}{7} + \tfrac{1}{8}) + (\tfrac{1}{9} + \cdots + \tfrac{1}{16}) + \cdots , \tag{5.7}$$

it will be seen that each pair of parentheses encloses p terms of the form

$$\frac{1}{p + 1} + \frac{1}{p + 2} + \cdots + \frac{1}{p + p} > \frac{p}{2p} = \frac{1}{2}. \tag{5.8}$$

Forming partial sums by adding the parenthetical groups one by one, we

---

[1] Multiply and divide $s_n = \sum_{m=0}^{n-1} ar^m$ by $1 - r$.

obtain

$$s_1 = 1, \quad s_4 > \frac{5}{2},$$

$$s_2 = \frac{3}{2}, \quad s_5 > \frac{6}{2}, \tag{5.9}$$

$$s_3 > \frac{4}{2}, \quad s_n > \frac{n+1}{2}.$$

The harmonic series considered in this way is certainly divergent.[2] An alternate and independent demonstration of its divergence appears in Section 5.2.

Using the binomial theorem[3] (Section 5.6), we may expand the function $(1 + x)^{-1}$:

$$\frac{1}{1+x} = 1 - x + x^2 - x^3 + \cdots + (-x)^{n-1} + \cdots. \tag{5.10}$$

If we let $x \to 1$, this series becomes

$$1 - 1 + 1 - 1 + 1 - 1 + \cdots, \tag{5.11}$$

a series that we labeled oscillatory earlier in this section. Although it does not converge in the usual sense, meaning can be attached to this series. Euler, for example, assigned a value of $\frac{1}{2}$ to this oscillatory sequence on the basis of the correspondence between this series and the well-defined function $(1 + x)^{-1}$. Unfortunately, such correspondence between series and function is not unique and this approach must be refined. Other methods of assigning a meaning to a divergent or oscillatory series, methods of defining a sum, have been developed. In general, however, this aspect of infinite series is of relatively little interest to the scientist or the engineer. An exception to this statement, the very important asymptotic or semiconvergent series, is considered in Section 5.10.

EXERCISES

5.1.1   Show that

$$\sum_{n=1}^{\infty} \frac{1}{(2n-1)(2n+1)} = \frac{1}{2}.$$

Hint. Show (by mathematical induction) that $s_m = m/(2m + 1)$.

5.1.2   Show that

$$\sum_{n=1}^{\infty} \frac{1}{n(n+1)} = 1.$$

---

[2] The (finite) harmonic series appears in an interesting note on the maximum stable displacement of a stack of coins. P. R. Johnson, The Leaning Tower of Lire. *Am. J. Phys.* **23**, 240 (1955).

[3] Actually Eq. (5.10) may be taken as an identity and verified by multiplying both sides by $1 + x$.

Find the partial sum s_m and verify its correctness by mathematical induction. *Note.* The method of expansion in partial fractions, Section 15.8, offers an alternative way of solving Exercises 5.1.1 and 5.1.2.

5.2 CONVERGENCE TESTS

Although nonconvergent series may be useful in certain special cases, (compare Section 5.10), we usually insist, as a matter of convenience if not necessity, that our series be convergent. It therefore becomes a matter of extreme importance to be able to tell whether a given series is convergent. We shall develop a number of possible tests, starting with the simple and relatively insensitive tests and working up to the more complicated but quite sensitive tests.

For the present let us consider a series of positive terms, $a_n > 0$, postponing negative terms until the next section

COMPARISON TEST

If term by term a series of terms $u_n \le a_n$, in which the a_n form a convergent series, the series $\sum_n u_n$ is also convergent. Symbolically, we have

$$\sum_n a_n = a_1 + a_2 + a_3 + \cdots, \qquad \text{convergent,}$$

$$\sum_n u_n = u_1 + u_2 + u_3 + \cdots.$$

If $u_n \le a_n$ for all n, then $\sum_n u_n \le \sum_n a_n$ and $\sum_n u_n$ therefore is convergent.

If term by term a series of terms $v_n \ge b_n$, in which the b_n form a divergent series, the series $\sum_n v_n$ is also divergent. Note that comparisons of u_n with b_n or v_n with a_n yield no information. Here we have

$$\sum_n b_n = b_1 + b_2 + b_3 + \cdots, \qquad \text{divergent,}$$

$$\sum_n v_n = v_1 + v_2 + v_3 + \cdots.$$

If $v_n \ge b_n$ for all n, then $\sum_n v_n \ge \sum_n b_n$ and $\sum_n v_n$ therefore is divergent.

For the convergent series a_n we already have the geometric series, whereas the harmonic series will serve as the divergent series b_n. As other series are identified as either convergent or divergent, they may be used for the known series in this comparison test.

All tests developed in this section are essentially comparison tests. Figure 5.1 exhibits these tests and the interrelationships.

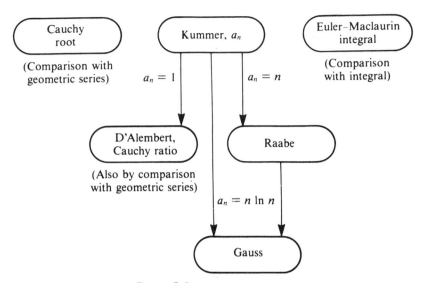

Figure 5.1 Comparison tests.

Example 5.2.1 The p Series

Test $\sum_n n^{-p}$, $p = 0.999$, for convergence. Since $n^{-0.999} > n^{-1}$, and $b_n = n^{-1}$ forms the divergent harmonic series, the comparison test shows that $\sum_n n^{-0.999}$ is divergent. Generalizing, $\sum_n n^{-p}$ is seen to be divergent for all $p \leq 1$.

CAUCHY ROOT TEST

If $(a_n)^{1/n} \leq r < 1$ for all sufficiently large n, with r independent of n, then $\sum_n a_n$ is convergent. If $(a_n)^{1/n} \geq 1$ for all sufficiently large n, then $\sum_n a_n$ is divergent.

The first part of this test is verified easily by raising $(a_n)^{1/n} \leq r$ to the nth power. We get

$$a_n \leq r^n < 1.$$

Since r^n is just the nth term in a convergent geometric series, $\sum_n a_n$ is convergent by the comparison test. Conversely, if $(a_n)^{1/n} \geq 1$, then $a_n \geq 1$ and the series must diverge. This root test is particularly useful in establishing the properties of power series (Section 5.7).

D'ALEMBERT OR CAUCHY RATIO TEST

If $a_{n+1}/a_n \leq r < 1$ for all sufficiently large n, and r is independent of n, then $\sum_n a_n$ is convergent. If $a_{n+1}/a_n \geq 1$ for all sufficiently large n, then $\sum_n a_n$ is divergent.

Convergence is proved by direct comparison with the geometric series $(1 + r + r^2 + \cdots)$. In the second part $a_{n+1} \geq a_n$ and divergence should be reasonably obvious. Although not quite so sensitive as the Cauchy root test, this D'Alembert ratio test is one of the easiest to apply and is widely used. An alternate statement of the ratio test is in the form of a limit: If

$$\lim_{n \to \infty} \frac{a_{n+1}}{a_n} < 1, \qquad \text{convergence,}$$

$$> 1, \qquad \text{divergence,} \qquad (5.12)$$

$$= 1, \qquad \text{indeterminant.}$$

Because of this final indeterminant possibility, the ratio test is likely to fail at crucial points, and more delicate, sensitive tests are necessary.

The alert reader may wonder how this indeterminacy arose. Actually it was concealed in the first statement $a_{n+1}/a_n \leq r < 1$. We might encounter $a_{n+1}/a_n < 1$ for all *finite* n but be unable to choose an $r < 1$ *and independent of* n such that $a_{n+1}/a_n \leq r$ for all sufficiently large n. An example is provided by the harmonic series

$$\frac{a_{n+1}}{a_n} = \frac{n}{n+1} < 1. \qquad (5.13)$$

Since

$$\lim_{n \to \infty} \frac{a_{n+1}}{a_n} = 1, \qquad (5.14)$$

no fixed ratio $r < 1$ exists and the ratio test fails.

Example 5.2.2 D'Alembert Ratio Test

Test $\sum_n n/2^n$ for convergence.

$$\frac{a_{n+1}}{a_n} = \frac{(n+1)/2^{n+1}}{n/2^n} = \frac{1}{2} \cdot \frac{n+1}{n}. \qquad (5.15)$$

Since

$$\frac{a_{n+1}}{a_n} \leq \frac{3}{4} \qquad \text{for } n \geq 2, \qquad (5.16)$$

we have convergence. Alternatively,

$$\lim_{n \to \infty} \frac{a_{n+1}}{a_n} = \frac{1}{2} \qquad (5.17)$$

and again—convergence.

CAUCHY OR MACLAURIN INTEGRAL TEST

This is another sort of comparison test in which we compare a series with an integral. Geometrically, we compare the area of a series of unit-width rectangles with the area under a curve.

Let $f(x)$ be a continuous, monotonic decreasing function in which $f(n) = a_n$.

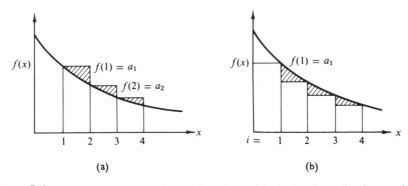

Figure 5.2 (a) Comparison of integral and sum-blocks leading. (b) Comparison of integral and sum-blocks lagging.

Then $\sum_n a_n$ converges if $\int_1^\infty f(x)\,dx$ is finite and diverges if the integral is infinite. For the ith partial sum

$$S_i = \sum_{n=1}^{i} a_n = \sum_{n=1}^{i} f(n). \tag{5.18}$$

But

$$S_i > \int_1^{i+1} f(x)\,dx \tag{5.19}$$

by Fig. 5.2a, $f(x)$ being monotonic decreasing. On the other hand, from Fig. 5.2b,

$$S_i - a_1 < \int_1^{i} f(x)\,dx, \tag{5.20}$$

in which the series is represented by the inscribed rectangles. Taking the limit as $i \to \infty$, we have

$$\int_1^\infty f(x)\,dx < \sum_{n=1}^{\infty} a_n < \int_1^\infty f(x)\,dx + a_1. \tag{5.21}$$

Hence the infinite series converges or diverges as the corresponding integral converges or diverges.

This integral test is particularly useful in setting upper and lower bounds on the remainder of a series after some number of initial terms have been summed. That is,

$$\sum_{n=1}^{\infty} a_n = \sum_{n=1}^{N} a_n + \sum_{n=N+1}^{\infty} a_n,$$

where

$$\int_{N+1}^{\infty} f(x)\,dx < \sum_{n=N+1}^{\infty} a_n < \int_{N+1}^{\infty} f(x)\,dx + a_{N+1}.$$

To free the integral test from the quite restrictive requirement that the interpolating function $f(x)$ be positive and monotonic, we show for any function $f(x)$ with a continuous derivative that

$$\sum_{n=N_i+1}^{N_f} f(n) = \int_{N_i}^{N_f} f(x)\, dx + \int_{N_i}^{N_f} (x - [x]) f'(x)\, dx \qquad (5.22)$$

holds. Here $[x]$ denotes the largest integer below x, so that $x - [x]$ varies saw-toothlike between 0 and 1. To get Eq. (5.22) we observe that

$$\int_{N_i}^{N_f} x f'(x)\, dx = N_f f(N_f) - N_i f(N_i) - \int_{N_i}^{N_f} f(x)\, dx, \qquad (5.23)$$

using integration by parts. Next we evaluate the integral

$$\int_{N_i}^{N_f} [x] f'(x)\, dx = \sum_{n=N_i}^{N_f-1} n \int_{n}^{n+1} f'(x)\, dx = \sum_{n=N_i}^{N_f-1} n\{f(n+1) - f(n)\}$$

$$= -\sum_{n=N_i+1}^{N_f} f(n) - N_i f(N_i) + N_f f(N_f). \qquad (5.24)$$

Subtracting Eq. (5.24) from (5.23) we arrive at Eq. (5.22). Note carefully that $f(x)$ may go up or down and even change sign, so that Eq. (5.22) applies to alternating series (see Section 5.3) as well. Usually $f'(x)$ falls faster than $f(x)$ for $x \to \infty$ so that the remainder term in Eq. (5.22) converges better. It is easy to improve Eq. (5.22) replacing $x - [x]$ by $x - [x] - \frac{1}{2}$ which varies between $-\frac{1}{2}$ and $\frac{1}{2}$:

$$\sum_{N_i < n \le N_f} f(n) = \int_{N_i}^{N_f} f(x)\, dx + \int_{N_i}^{N_f} (x - [x] - \tfrac{1}{2}) f'(x)\, dx + \tfrac{1}{2}\{f(N_f) - f(N_i)\}$$

$$(5.25)$$

Then the $f'(x)$-integral becomes even smaller, if $f'(x)$ does not change sign too often.

Example 5.2.3 Riemann Zeta Function

The Riemann zeta function is defined by

$$\zeta(p) = \sum_{n=1}^{\infty} n^{-p}. \qquad (5.26)$$

We may take $f(x) = x^{-p}$ and then

$$\int_{1}^{\infty} x^{-p}\, dx = \left. \frac{x^{-p+1}}{-p+1} \right|_{1}^{\infty}, \qquad p \ne 1$$

$$= \ln x \big|_{1}^{\infty}, \qquad p = 1. \qquad (5.27)$$

The integral and therefore the series are divergent for $p \le 1$, convergent for $p > 1$. Hence Eq. (5.26) should carry the condition $p > 1$. This, incidentally, is an independent proof that the harmonic series ($p = 1$) diverges and diverges logarithmically. The sum of the first million terms $\sum^{1,000,000} n^{-1}$, is only 14.392 726 ....

This integral comparison may also be used to set an upper limit to the Euler–Mascheroni constant[1] defined by

$$\gamma = \lim_{n \to \infty} \left(\sum_{m=1}^{n} m^{-1} - \ln n \right). \tag{5.28}$$

Returning to partial sums,

$$s_n = \sum_{m=1}^{n} m^{-1} - \ln n < \int_1^n \frac{dx}{x} - \ln n + 1. \tag{5.29}$$

Evaluating the integral on the right, $s_n < 1$ for all n and therefore $\gamma < 1$. Exercise 5.2.12 leads to more restrictive bounds. Actually the Euler–Mascheroni constant is 0.577 215 66 ....

KUMMER'S TEST

This is the first of three tests that are somewhat more difficult to apply than the preceding tests. Their importance lies in their power and sensitivity. Frequently, at least one of the three will work when the simpler easier tests are indecisive. It must be remembered, however, that these tests, like those previously discussed, are ultimately based on comparisons. It can be shown that there is no most slowly converging convergent series and no most slowly diverging divergent series. This means that all convergence tests given here, including Kummer's, may fail sometime.

We consider a series of positive terms u_i and a sequence of finite positive constants a_i. If

$$a_n \frac{u_n}{u_{n+1}} - a_{n+1} \geq C > 0 \tag{5.30}$$

for all $n \geq N$, some fixed number,[2] then $\sum_{i=1}^{\infty} u_i$ converges. If

$$a_n \frac{u_n}{u_{n+1}} - a_{n+1} \leq 0 \tag{5.31}$$

and $\sum_{i=1}^{\infty} a_i^{-1}$ diverges, then $\sum_{i=1}^{\infty} u_i$ diverges.

The proof of this powerful test is remarkably simple. From Eq. (5.30), with C some positive constant,

$$\begin{aligned}
Cu_{N+1} &\leq a_N u_N & - a_{N+1} u_{N+1} \\
Cu_{N+2} &\leq a_{N+1} u_{N+1} - a_{N+2} u_{N+2} \\
&\cdots\cdots\cdots\cdots\cdots\cdots\cdots \\
Cu_n &\leq a_{n-1} u_{n-1} - a_n u_n.
\end{aligned} \tag{5.32}$$

---

[1] This is the notation of National Bureau of Standards, *Handbook of Mathematical Functions*, Applied Mathematics Series-55 (AMS-55). New York: Dover (1972).

[2] With u_m finite, the partial sum s_N will always be finite for N finite. The convergence or divergence of a series depends on the behavior of the last infinity of terms, *not* on the first N terms.

Adding and dividing by C $(C \neq 0)$, we obtain

$$\sum_{i=N+1}^{n} u_i \leq \frac{a_N u_N}{C} - \frac{a_n u_n}{C}. \tag{5.33}$$

Hence for the partial sum, s_n,

$$s_n \leq \sum_{i=1}^{N} u_i + \frac{a_N u_N}{C} - \frac{a_n u_n}{C}$$

$$< \sum_{i=1}^{N} u_i + \frac{a_N u_N}{C}, \qquad \text{a constant, independent of } n, \tag{5.34}$$

The partial sums therefore have an upper bound. With zero as an obvious lower bound, the series $\sum u_i$ must converge.

Divergence is shown as follows. From Eq. (5.31)

$$a_n u_n \geq a_{n-1} u_{n-1} \geq \cdots \geq a_N u_N, \qquad n > N. \tag{5.35}$$

Thus

$$u_n \geq \frac{a_N u_N}{a_n} \tag{5.36}$$

and

$$\sum_{i=N+1}^{\infty} u_i \geq a_N u_N \sum_{i=N+1}^{\infty} a_i^{-1}. \tag{5.37}$$

If $\sum_{i=1}^{\infty} a_i^{-1}$ diverges, then by the comparison test $\sum_i u_i$ diverges.

Equations (5.30) and (5.31) are often given in a limit form:

$$\lim_{n \to \infty} \left(a_n \frac{u_n}{u_{n+1}} - a_{n+1} \right) = C. \tag{5.38}$$

Thus for $C > 0$ we have convergence, whereas for $C < 0$ (and $\sum_i a_i^{-1}$ divergent) we have divergence. It is perhaps useful to show the equivalence of Eq. (5.38) and Eqs. (5.30) and (5.31) and to show why indeterminacy creeps in when the limit $C = 0$. From the definition of limit

$$\left| a_n \frac{u_n}{u_{n+1}} - a_{n+1} - C \right| < \varepsilon \tag{5.39}$$

for all $n \geq N$ and all $\varepsilon > 0$, no matter how small ε may be. When the absolute value signs are removed,

$$C - \varepsilon < a_n \frac{u_n}{u_{n+1}} - a_{n+1} < C + \varepsilon. \tag{5.40}$$

Now if $C > 0$, Eq. (5.30) follows from ε sufficiently small. On the other hand, if $C < 0$, Eq. (5.31) follows. However, if $C = 0$, the center term $a_n(u_n/u_{n+1}) - a_{n+1}$ may be either positive or negative and the proof fails. The primary use of Kummer's test is to prove other tests such as Raabe's (compare also Exercise 5.2.3).

If the positive constants a_n of Kummer's test are chosen $a_n = n$, we have Raabe's test.

RAABE'S TEST

If $u_n > 0$ and if

$$n\left(\frac{u_n}{u_{n+1}} - 1\right) \geq P > 1 \tag{5.41}$$

for all $n \geq N$, where N is a positive integer independent of n, then $\sum_i u_i$ converges. If

$$n\left(\frac{u_n}{u_{n+1}} - 1\right) \leq 1, \tag{5.42}$$

then $\sum_i u_i$ diverges ($\sum n^{-1}$ diverges).

The limit form of Raabe's test is

$$\lim_{n \to \infty} n\left(\frac{u_n}{u_{n+1}} - 1\right) = P. \tag{5.43}$$

We have convergence for $P > 1$, divergence for $P < 1$, and no test for $P = 1$ exactly as with the Kummer test. This indeterminacy is pointed up by Exercise 5.2.4, which presents a convergent series and a divergent series with both series yielding $P = 1$ in Eq. (5.43).

Raabe's test is more sensitive than the d'Alembert ratio test because $\sum_{n=1}^{\infty} n^{-1}$ diverges more slowly than $\sum_{n=1}^{\infty} 1$. We obtain a still more sensitive test (and one that is relatively easy to apply) by choosing $a_n = n \ln n$. This is Gauss's test.

GAUSS'S TEST

If $u_n > 0$ for all finite n and

$$\frac{u_n}{u_{n+1}} = 1 + \frac{h}{n} + \frac{B(n)}{n^2}, \tag{5.44a}$$

in which $B(n)$ is a bounded function of n for $n \to \infty$, then $\sum_i u_i$ converges for $h > 1$ and diverges for $h \leq 1$.

The ratio u_n/u_{n+1} of Eq. (5.44a) often comes as the ratio of two quadratic forms:

$$\frac{u_n}{u_{n+1}} = \frac{n^2 + a_1 n + a_0}{n^2 + b_1 n + b_0}. \tag{5.44b}$$

It may be shown (Exercise 5.2.5) that we have convergence for $a_1 > b_1 + 1$ and divergence for $a_1 \leq b_1 + 1$.

The Gauss test is an extremely sensitive test of series convergence. It will work for all series the physicist is likely to encounter. For $h > 1$ or $h < 1$ the proof follows directly from Raabe's test

$$\lim_{n \to \infty} n \left[1 + \frac{h}{n} + \frac{B(n)}{n^2} - 1 \right] = \lim_{n \to \infty} \left[h + \frac{B(b)}{n} \right]$$

$$= h. \tag{5.45}$$

If $h = 1$, Raabe's test fails. However, if we return to Kummer's test and use $a_n = n \ln n$, Eq. (5.38) leads to

$$\lim_{n \to \infty} \left\{ n \ln n \left[1 + \frac{1}{n} + \frac{B(n)}{n^2} \right] - (n + 1) \ln(n + 1) \right\}$$

$$= \lim_{n \to \infty} \left[n \ln n \cdot \frac{(n + 1)}{n} - (n + 1) \ln(n + 1) \right]$$

$$= \lim_{n \to \infty} (n + 1) \left[\ln n - \ln n - \ln\left(1 + \frac{1}{n}\right) \right]. \tag{5.46}$$

Borrowing a result from Section 5.6 (which is not dependent on Gauss's test), we have

$$\lim_{n \to \infty} - (n + 1) \ln\left(1 + \frac{1}{n}\right) = \lim_{n \to \infty} - (n + 1)\left(\frac{1}{n} - \frac{1}{2n^2} + \frac{1}{3n^3} \cdots\right)$$

$$= -1 < 0. \tag{5.47}$$

Hence we have divergence for $h = 1$. This is an example of a successful application of Kummer's test in which Raabe's test had failed.

Example 5.2.4 Legendre Series

The recurrence relation for the series solution of Legendre's equation (Section 8.5) may be put in the form

$$\frac{a_{2j+2}}{a_{2j}} = \frac{2j(2j + 1) - l(l + 1)}{(2j + 1)(2j + 2)}. \tag{5.48}$$

This is equivalent to u_{2j+2}/u_{2j} for $x = +1$. For $j \gg l$,[3]

$$\frac{u_{2j}}{u_{2j+2}} \to \frac{(2j + 1)(2j + 2)}{2j(2j + 1)} = \frac{2j + 2}{2j}$$

$$= 1 + \frac{1}{j}. \tag{5.49}$$

By Eq. (5.44b) the series is divergent. Later we shall demand that the Legendre series be finite at $x = 1$. We shall eliminate the divergence by setting the parameter $n = 2j_0$, an even integer. This will truncate the series, converting the infinite series into a polynomial.

---

[3] The l dependence enters $B(2j)$ but does not affect h in Eq. (5.44a).

IMPROVEMENT OF CONVERGENCE

This section so far has been concerned with establishing convergence as an abstract mathematical property. In practice, the *rate* of convergence may be of considerable importance. Here we present one method of improving the rate of convergence of a convergent series. Other techniques are given in Sections 5.4 and 5.9.

The basic principle of this method, due to Kummer, is to form a linear combination of our slowly converging series and one or more series whose sum is known. For the known series the collection

$$\alpha_1 = \sum_{n=1}^{\infty} \frac{1}{n(n+1)} = 1$$

$$\alpha_2 = \sum_{n=1}^{\infty} \frac{1}{n(n+1)(n+2)} = \frac{1}{4}$$

$$\alpha_3 = \sum_{n=1}^{\infty} \frac{1}{n(n+1)(n+2)(n+3)} = \frac{1}{18}$$

$$\vdots \qquad \vdots \qquad \qquad \vdots$$

$$\alpha_p = \sum_{n=1}^{\infty} \frac{1}{n(n+1)\cdots(n+p)} = \frac{1}{p \cdot p!}$$

is particularly useful.[4] The series are combined term by term and the coefficients in the linear combination chosen to cancel the most slowly converging terms.

Example 5.2.5 Riemann Zeta Function, $\zeta(3)$

Let the series to be summed be $\sum_{n=1}^{\infty} n^{-3}$. In Section 5.9 this is identified as a Riemann zeta function, $\zeta(3)$. We form a linear combination

$$\sum_{n=1}^{\infty} n^{-3} + a_2 \alpha_2 = \sum_{n=1}^{\infty} n^{-3} + \frac{a_2}{4}.$$

α_1 is not included since it converges more slowly than $\zeta(3)$. Combining terms, we obtain on the left-hand side

$$\sum_{n=1}^{\infty} \left\{ \frac{1}{n^3} + \frac{a_2}{n(n+1)(n+2)} \right\} = \sum_{n=1}^{\infty} \frac{n^2(1+a_2) + 3n + 2}{n^3(n+1)(n+2)}.$$

If we choose $a_2 = -1$, the preceding equations yield

$$\zeta(3) = \sum_{n=1}^{\infty} n^{-3} = \frac{1}{4} + \sum_{n=1}^{\infty} \frac{3n+2}{n^3(n+1)(n+2)}. \qquad (5.50)$$

The resulting series may not be beautiful but it does converge as n^{-4}, appreciably faster than n^{-3}. A more convenient form comes from Exercise 5.2.21. There, the symmetry leads to convergence as n^{-5}.

---

[4] These series sums may be verified by expanding the forms by partial fractions, writing out the initial terms and inspecting the pattern of cancellation of positive and negative terms.

The method can be extended including $a_3 \alpha_3$ to get convergence as n^{-5}, $a_4 \alpha_4$ to get convergence as n^{-6}, and so on. Eventually, you have to reach a compromise between how much algebra you do and how much arithmetic the computing machine does. As computing machines get faster, the balance is steadily shifting to less algebra for you and more arithmetic for the machine.

EXERCISES

5.2.1 (a) Prove that if
$$\lim_{n \to \infty} n^p u_n \to A < \infty, \qquad p > 1,$$
the series $\sum_{n=1}^{\infty} u_n$ converges.

(b) Prove that if
$$\lim_{n \to \infty} n u_n = A > 0,$$
the series diverges. (The test fails for $A = 0$.)

These two tests, known as limit tests, are often convenient for establishing the convergence of a series. They may be treated as comparison tests, comparing with
$$\sum n^{-q}, \qquad 1 \le q < p.$$

5.2.2 If
$$\lim_{n \to \infty} \frac{b_n}{a_n} = K,$$
a constant with $0 < K < \infty$, show that $\sum b_n$ converges or diverges with $\sum a_n$. *Hint.* If $\sum a_n$ converges, use
$$b'_n = \frac{1}{2K} b_n.$$

If $\sum a_n$ diverges, use
$$b''_n = \frac{2}{K} b_n.$$

5.2.3 Show that the complete d'Alembert ratio test follows directly from Kummer's test with $a_i = 1$.

5.2.4 Show that Raabe's test is indecisive for $P = 1$ by establishing that $P = 1$ for the series

(a) $u_n = \dfrac{1}{n \ln n}$ and that this series diverges.

(b) $u_n = \dfrac{1}{n(\ln n)^2}$ and that this series converges.

Note. By direct addition $\sum_{2}^{100,000} [n(\ln n)^2]^{-1} = 2.02288$. The remainder of the series $n > 10^5$ yields 0.08686 by the integral comparison test. The total, then, 2 to ∞, is 2.1097.

5.2.5 Gauss's test is often given in the form of a test of the ratio

$$\frac{u_n}{u_{n+1}} = \frac{n^2 + a_1 n + a_0}{n^2 + b_1 n + b_0}.$$

For what values of the parameters a_1 and b_1 is there convergence? Divergence?

> ANS. Convergent for $a_1 - b_1 > 1$, divergent for $a_1 - b_1 \le 1$.

5.2.6 Test for convergence

(a) $\sum\limits_{n=2}^{\infty} (\ln n)^{-1}$

(d) $\sum\limits_{n=1}^{\infty} [n(n+1)]^{-1/2}$

(b) $\sum\limits_{n=1}^{\infty} \dfrac{n!}{10^n}$

(e) $\sum\limits_{n=0}^{\infty} \dfrac{1}{2n+1}$.

(c) $\sum\limits_{n=1}^{\infty} \dfrac{1}{2n(2n-1)}$

5.2.7 Test for convergence

(a) $\sum\limits_{n=1}^{\infty} \dfrac{1}{n(n+1)}$

(d) $\sum\limits_{n=1}^{\infty} \ln\left(1 + \dfrac{1}{n}\right)$

(b) $\sum\limits_{n=2}^{\infty} \dfrac{1}{n \ln n}$

(e) $\sum\limits_{n=1}^{\infty} \dfrac{1}{n \cdot n^{1/n}}$.

(c) $\sum\limits_{n=1}^{\infty} \dfrac{1}{n2^n}$

5.2.8 For what values of p and q will the following series converge?

$$\sum_{n=2}^{\infty} \frac{1}{n^p (\ln n)^q}.$$

> ANS. Convergent for $\begin{cases} p > 1, & \text{all } q, \\ p = 1, & q > 1, \end{cases}$
>
> divergent for $\begin{cases} p < 1, & \text{all } q, \\ p = 1, & q \le 1. \end{cases}$

5.2.9 Determine the range of convergence for Gauss's hypergeometric series

$$F(\alpha, \beta, \gamma; x) = 1 + \frac{\alpha\beta}{1!\gamma} x + \frac{\alpha(\alpha+1)\beta(\beta+1)}{2!\gamma(\gamma+1)} x^2 + \cdots.$$

Hint. Gauss developed Gauss's test for the specific purpose of establishing the convergence of this series.

> ANS. Convergent for $-1 < x < 1$ and $x = \pm 1$ if $\gamma > \alpha + \beta$.

5.2.10 A simple machine calculation yields

$$\sum_{n=1}^{100} n^{-3} = 1.202\,007.$$

Show that

$$1.202\,056 \le \sum_{n=1}^{\infty} n^{-3} \le 1.202\,057.$$

Hint. Use integrals to set upper and lower bounds on $\sum_{n=101}^{\infty} n^{-3}$.
Comment. A more exact value for summation $\sum_{1}^{\infty} n^{-3}$ is $1.202\,056\,903\,\ldots$.

5.2.11   Set upper and lower bounds on $\sum_{n=1}^{1,000,000} n^{-1}$, assuming that

(a) the Euler–Mascheroni constant is known.
$$ANS. \quad 14.392\,726 < \sum_{n=1}^{1,000,000} n^{-1} < 14.392\,727.$$
(b) The Euler–Mascheroni constant is unknown.

5.2.12   Given $\sum_{n=1}^{1,000} n^{-1} = 7.485\,470\,\ldots$, set upper and lower bounds on the Euler–Mascheroni constant.
$$ANS. \quad 0.5767 < \gamma < 0.5778.$$

5.2.13   (From Olbers's paradox.) Assume a static universe in which the stars are uniformly distributed. Divide all space into shells of constant thickness; the stars in any one shell by themselves subtend a solid angle of ω_0. *Allowing for the blocking out of distant stars by nearer stars*, show that the total net solid angle subtended by all stars, shells extending to infinity, is *exactly* 4π. (Therefore the night sky should be ablaze with light.)

5.2.14   Test for convergence
$$\sum_{n=1}^{\infty} \left[\frac{1 \cdot 3 \cdot 5 \cdots (2n-1)}{2 \cdot 4 \cdot 6 \cdots (2n)} \right]^2 = \frac{1}{4} + \frac{9}{64} + \frac{25}{256} + \cdots.$$

5.2.15   The Legendre series, $\sum_{j\,\text{even}} u_j(x)$, satisfies the recurrence relations
$$u_{j+2}(x) = \frac{(j+1)(j+2) - l(l+1)}{(j+2)(j+3)} x^2 u_j(x),$$
in which the index j is even and l is some constant (but, in this problem, *not* a nonnegative odd integer). Find the range of values of x for which this Legendre series is convergent. Test the end points carefully.
$$ANS. \quad -1 < x < 1.$$

5.2.16   A series solution (Section 8.5) of the Chebyshev equation leads to successive terms having the ratio
$$\frac{u_{j+2}(x)}{u_j(x)} = \frac{(k+j)^2 - n^2}{(k+j+1)(k+j+2)} x^2,$$
with $k = 0$ and $k = 1$. Test for convergence at $x = \pm 1$.
$$ANS. \quad \text{Convergent.}$$

5.2.17   A series solution for the ultraspherical (Gegenbauer) function $C_n^{\alpha}(x)$ leads to the recurrence
$$a_{j+2} = a_j \frac{(k+j)(k+j+2\alpha) - n(n+2\alpha)}{(k+j+1)(k+j+2)}.$$
Investigate the convergence of each of these series at $x = \pm 1$ as a function of the parameter α.
$$ANS. \quad \text{Convergent for } \alpha < 1,$$
$$\text{divergent for } \alpha \geq 1.$$

5.2.18 A series expansion of the incomplete beta function (Section 10.4) yields

$$B_x(p, q) = x^p \left\{ \frac{1}{p} + \frac{1 - q}{p + 1} x + \frac{(1 - q)(2 - q)}{2!(p + 2)} x^2 + \cdots \right.$$

$$\left. + \frac{(1 - q)(2 - q) \cdots (n - q)}{n!(p + n)} x^n + \cdots \right\}.$$

Given that $0 \le x \le 1$, $p > 0$, and $q > 0$, test this series for convergence. What happens at $x = 1$?

5.2.19 Show that the following series is convergent.

$$\sum_{s=0}^{\infty} \frac{(2s - 1)!!}{(2s)!!(2s + 1)}.$$

Note. $(2s - 1)!! = (2s - 1)(2s - 3) \cdots 3 \cdot 1$ with $(-1)!! = 1$. $(2s)!! = (2s)(2s - 2) \cdots 4 \cdot 2$ with $0!! = 1$. The series appears as a series expansion of $\sin^{-1}(1)$ and equals $\pi/2$.

5.2.20 Show how to combine $\zeta(2) = \sum_{n=1}^{\infty} n^{-2}$ with α_1 and α_2 to obtain a series converging as n^{-4}.
Note. $\zeta(2)$ is actually available in closed form: $\zeta(2) = \pi^2/6$ (see Section 5.9).

5.2.21 The convergence improvement of Example 5.2.5 may be carried out more expediently (in this special case) by putting α_2 into a more symmetric form: Replacing n by $n - 1$, we have

$$\alpha_2' = \sum_{n=2}^{\infty} \frac{1}{(n - 1)n(n + 1)} = \frac{1}{4}.$$

(a) Combine $\zeta(3)$ and α_2' to obtain convergence as n^{-5}.
(b) Let α_4' be α_4 with $n \to n - 2$. Combine $\zeta(3)$, α_2', and α_4' to obtain convergence as n^{-7}.
(c) If $\zeta(3)$ is to be calculated to 6 decimal accuracy (error 5×10^{-7}), how many terms are required for $\zeta(3)$ alone? Combined as in part (a)? Combined as in part (b)?

Note. The error may be estimated using the corresponding integral.

$$ANS. \quad (a) \ \zeta(3) = \frac{5}{4} - \sum_{n=2}^{\infty} \frac{1}{n^3(n^2 - 1)}.$$

5.2.22 Catalan's constant ($\beta(2)$ of AMS-55, Chapter 23) is defined by

$$\beta(2) = \sum_{k=0}^{\infty} (-1)^k (2k + 1)^{-2} = \frac{1}{1^2} - \frac{1}{3^2} + \frac{1}{5^2} \cdots.$$

Calculate $\beta(2)$ to six-digit accuracy.
Hint. The rate of convergence is enhanced by pairing the terms:

$$\frac{1}{(4k - 1)^2} - \frac{1}{(4k + 1)^2} = \frac{16k}{(16k^2 - 1)^2}.$$

If you have carried enough digits in your series summation, $\sum_{k=1}^{N} 16k/(16k^2 - 1)^2$, additional significant figures may be obtained by setting upper and lower bounds on the tail of the series, $\sum_{k=N+1}^{\infty}$. These bounds may be set by comparison with integrals as in the Maclaurin integral test.

$$ANS. \quad \beta(2) = 0.9159 \ 6559 \ 4177 \ldots.$$

5.3 ALTERNATING SERIES

In Section 5.2 we limited ourselves to series of positive terms. Now, in contrast, we consider infinite series in which the signs alternate. The partial cancellation due to alternating signs makes convergence more rapid and much easier to identify. We shall prove the Leibniz criterion, a general condition for the convergence of an alternating series.

LEIBNIZ CRITERION

Consider the series $\sum_{n=1}^{\infty}(-1)^{n+1}a_n$ with $a_n > 0$. If a_n is monotonic decreasing (for sufficiently large n) and $\lim_{n\to\infty} a_n = 0$, then the series converges.

To prove this, we examine the even partial sums

$$s_{2n} = a_1 - a_2 + a_3 - \cdots - a_{2n},$$

$$s_{2n+2} = s_{2n} + (a_{2n+1} - a_{2n+2}).$$

(5.51)

Since $a_{2n+1} > a_{2n+2}$, we have

$$s_{2n+2} > s_{2n}.$$ (5.52)

On the other hand,

$$s_{2n+2} = a_1 - (a_2 - a_3) - (a_4 - a_5) - \cdots - a_{2n+2}.$$ (5.53)

Hence, with each pair of terms $a_{2p} - a_{2p+1} > 0$,

$$s_{2n+2} < a_1.$$ (5.54)

With the even partial sums bounded $s_{2n} < s_{2n+2} < a_1$ and the terms a_n decreasing monotonically and approaching zero, this alternating series converges.

One further important result can be extracted from the partial sums. From the difference between the series limit S and the partial sum s_n

$$S - s_n = a_{n+1} - a_{n+2} + a_{n+3} - a_{n+4} + \cdots$$

$$= a_{n+1} - (a_{n+2} - a_{n+3}) - (a_{n+4} - a_{n+5}) - \cdots$$ (5.55)

or

$$S - s_n < a_{n+1}.$$ (5.56)

Equation (5.56) says that the error in cutting off an alternating series after n terms is less than a_{n+1}, the first term dropped. A knowledge of the error obtained this way may be of great practical importance.

ABSOLUTE CONVERGENCE

Given a series of terms u_n in which u_n may vary in sign, if $\sum|u_n|$ converges, then $\sum u_n$ is said to be absolutely convergent. If $\sum u_n$ converges but $\sum|u_n|$ diverges, the convergence is called conditional.

The alternating harmonic series is a simple example of this conditional convergence. We have

$$\sum_{n=1}^{\infty} (-1)^{n-1} n^{-1} = 1 - \frac{1}{2} + \frac{1}{3} - \frac{1}{4} + \cdots + \frac{1}{n} - \cdots, \tag{5.57}$$

convergent by the Leibniz criterion; but

$$\sum_{n=1}^{\infty} n^{-1} = 1 + \frac{1}{2} + \frac{1}{3} + \frac{1}{4} + \cdots + \frac{1}{n} + \cdots$$

has been shown to be divergent in Sections 5.1 and 5.2.

The reader will note that all the tests developed in Section 5.2 assume a series of positive terms. Therefore all the tests in that section guarantee absolute convergence.

Example 5.3.1

For $0 < x < 2\pi$ the Fourier series (see Chapter 14.1)

$$\sum_{n=1}^{\infty} \frac{\cos(nx)}{n} = -\ln\left(2 \sin \frac{x}{2}\right) \tag{5.58}$$

converges having coefficients that change sign often, but not so that Leibniz's convergence criterion applies easily. Let us apply the integral test of Eq. (5.22). Using integration by parts we see immediately that

$$\int_{1}^{\infty} \frac{\cos(nx)}{n} \, dn = \left[\frac{\sin(nx)}{nx} \right]_{1}^{\infty} - \frac{1}{x} \int_{1}^{\infty} \frac{\sin(nx)}{n^2} \, dn$$

converges for $n \to \infty$, and the integral on the right-hand side even converges absolutely. The derivative term in Eq. (5.22) has the form

$$\int_{1}^{\infty} (n - [n]) \left\{ -\frac{x}{n} \sin(nx) - \frac{\cos(nx)}{n^2} \right\} dn,$$

where the second term converges absolutely and need not be considered further. Next we observe that $g(N) = \int_{1}^{N} (n - [n]) \sin(nx) \, dn$ is bounded for $N \to \infty$, just as $\int^N \sin(nx) \, dn$ is bounded because of the periodic nature of $\sin(nx)$ and its regular sign changes. Using integration by parts again

$$\int_{1}^{\infty} \frac{g'(n)}{n} \, dn = \left[\frac{g(n)}{n} \right]_{1}^{\infty} + \int_{1}^{\infty} \frac{g(n)}{n^2} \, dn$$

we see that the second term is absolutely convergent, and the first goes to zero at the upper limit. Hence the series in Eq. (5.58) converges, which is hard to see from other convergence tests.

Alternatively, we may apply the $q = 1$ case of the Euler–Maclaurin integration formula in Eq. (5.168b),

$$\sum_{v=1}^{n} f(v) = \int_{1}^{n} f(x)\,dx + \{f(n) + f(1)\}/2 + \{f'(n) - f'(1)\}/12$$

$$- \frac{1}{2}\int_{0}^{1}\left(x^2 - x + \frac{1}{6}\right)\sum_{v=1}^{n-1} f''(x + v)\,dx,$$

which is straightforward but more tedious because of the second derivative.

EXERCISES

5.3.1   (a) From the electrostatic two hemisphere problem (Exercise 12.3.20) we obtain the series

$$\sum_{s=0}^{\infty} (-1)^s (4s + 3) \frac{(2s - 1)!!}{(2s + 2)!!}.$$

Test for convergence.

(b) The corresponding series for the surface charge density is

$$\sum_{s=0}^{\infty} (-1)^s (4s + 3) \frac{(2s - 1)!!}{(2s)!!}.$$

Test for convergence. The !! notation is explained in Section 10.1 and Exercise 5.2.19.

5.3.2   Show by direct numerical computation that the sum of the first 10 terms of

$$\lim_{x \to 1} \ln(1 + x) = \ln 2 = \sum_{n=1}^{\infty} (-1)^{n-1} n^{-1}$$

differs from ln 2 by less than the eleventh term: $\ln 2 = 0.69314\,71806\ldots$.

5.3.3   In Exercise 5.2.9 the hypergeometric series is shown convergent for $x = \pm 1$, if $\gamma > \alpha + \beta$. Show that there is conditional convergence for $x = -1$ for γ down to $\gamma > \alpha + \beta - 1$.
Hint. The asymptotic behavior of the factorial function is given by Stirling's series, Section 10.3.

5.4 ALGEBRA OF SERIES

The establishment of absolute convergence is important because it can be proved that absolutely convergent series may be handled according to the ordinary familiar rules of algebra or arithmetic.

1. If an infinite series is absolutely convergent, the series sum is independent of the order in which the terms are added.
2. The series may be multiplied with another absolutely convergent series. The limit of the product will be the product of the individual series limits. The product series, a double series, will also converge absolutely.

No such guarantees can be given for conditionally convergent series. Again consider the alternating harmonic series. If we write

$$1 - \tfrac{1}{2} + \tfrac{1}{3} - \tfrac{1}{4} + \cdots = 1 - (\tfrac{1}{2} - \tfrac{1}{3}) - (\tfrac{1}{4} - \tfrac{1}{5}) - \cdots, \qquad (5.59)$$

it is clear that the sum

$$\sum_{n=1}^{\infty} (-1)^{n-1} n^{-1} < 1. \qquad (5.60)$$

However, if we rearrange the terms slightly, we may make the alternating harmonic series converge to $\tfrac{3}{2}$. We regroup the terms of Eq. (5.59), taking

$$(1 + \tfrac{1}{3} + \tfrac{1}{5}) - (\tfrac{1}{2}) + (\tfrac{1}{7} + \tfrac{1}{9} + \tfrac{1}{11} + \tfrac{1}{13} + \tfrac{1}{15}) - (\tfrac{1}{4})$$

$$+ (\tfrac{1}{17} + \cdots + \tfrac{1}{25}) - (\tfrac{1}{6}) + (\tfrac{1}{27} + \cdots + \tfrac{1}{35}) - (\tfrac{1}{8}) + \cdots. \qquad (5.61)$$

Treating the terms grouped in parentheses as single terms for convenience, we obtain the partial sums

$$s_1 = 1.5333 \qquad s_2 = 1.0333$$
$$s_3 = 1.5218 \qquad s_4 = 1.2718$$
$$s_5 = 1.5143 \qquad s_6 = 1.3476$$
$$s_7 = 1.5103 \qquad s_8 = 1.3853$$
$$s_9 = 1.5078 \qquad s_{10} = 1.4078.$$

From this tabulation of s_n and the plot of s_n versus n in Fig. 5.3 the convergence to $\tfrac{3}{2}$ is fairly clear. We have rearranged the terms, taking positive terms until the partial sum was equal to or greater than $\tfrac{3}{2}$, then adding in

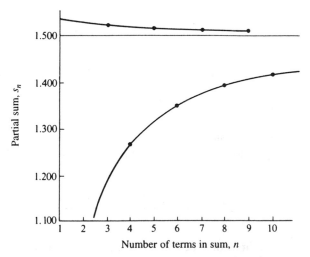

Figure 5.3   Alternating harmonic series—terms rearranged to give convergence to 1.5.

negative terms until the partial sum just fell below $\frac{3}{2}$, and so on. As the series extends to infinity, all original terms will eventually appear, but the partial sums of this rearranged alternating harmonic series converge to $\frac{3}{2}$.

By a suitable rearrangement of terms a conditionally convergent series may be made to converge to any desired value or even to diverge. This statement is sometimes given as Riemann's theorem. Obviously, conditionally convergent series must be treated with caution.

IMPROVEMENT OF CONVERGENCE, RATIONAL APPROXIMATIONS

The series

$$\ln(1 + x) = \sum_{n=1}^{\infty} (-1)^{n-1} x^n/n, \qquad -1 < x \le 1, \qquad (5.61a)$$

converges very slowly as x approaches $+1$. The *rate* of convergence may be improved substantially by multiplying both sides of Eq. (5.61a) by a polynomial and adjusting the polynomial coefficients to cancel the more slowly converging portions of the series. Consider the simplest possibility: Multiply $\ln(1 + x)$ by $1 + a_1 x$.

$$(1 + a_1 x)\ln(1 + x) = \sum_{n=1}^{\infty} (-1)^{n-1} x^n/n + a_1 \sum_{n=1}^{\infty} (-1)^{n-1} x^{n+1}/n.$$

Combining the two series on the right term by term, we obtain

$$(1 + a_1 x)\ln(1 + x) = x + \sum_{n=2}^{\infty} (-1)^{n-1}\left(\frac{1}{n} - \frac{a_1}{n-1}\right) x^n$$

$$= x + \sum_{n=2}^{\infty} (-1)^{n-1} \frac{n(1 - a_1) - 1}{n(n-1)} x^n.$$

Clearly, if we take $a_1 = 1$, the n in the numerator disappears and our combined series converges as n^{-2}.

Continuing this process, we find that $(1 + 2x + x^2)\ln(1 + x)$ vanishes as n^{-3}, $(1 + 3x + 3x^2 + x^3)\ln(1 + x)$ vanishes as n^{-4}. In effect we are shifting from a simple series expansion of Eq. (5.61a) to a rational fraction representation in which the function $\ln(1 + x)$ is represented by the ratio of a series and a polynomial:

$$\ln(1 + x) = \frac{x + \sum_{n=2}^{\infty}(-1)^n x^n/[n(n-1)]}{1 + x}.$$

Such rational approximations may be both compact and accurate. Computer subroutines make extensive use of such approximations.

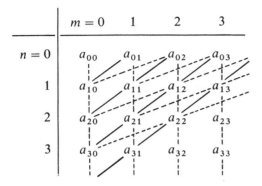

	$m = 0$	1	2	3
$n = 0$	a_{00}	a_{01}	a_{02}	a_{03}
1	a_{10}	a_{11}	a_{12}	a_{13}
2	a_{20}	a_{21}	a_{22}	a_{23}
3	a_{30}	a_{31}	a_{32}	a_{33}

Figure 5.4 Double series—summation over n indicated by vertical dashed lines.

REARRANGEMENT OF DOUBLE SERIES

Another aspect of the rearrangement of series appears in the treatment of double series (Fig. 5.4):

$$\sum_{m=0}^{\infty} \sum_{n=0}^{\infty} a_{n,m}.$$

Let us substitute

$$n = q \geq 0,$$

$$m = p - q \geq 0,$$

$$(q \leq p).$$

This results in the identity

$$\sum_{m=0}^{\infty} \sum_{n=0}^{\infty} a_{n,m} = \sum_{p=0}^{\infty} \sum_{q=0}^{p} a_{q,p-q}. \tag{5.62}$$

The summation over p and q of Eq. (5.62) is illustrated in Fig. 5.5. The substitution

$$n = s \geq 0, \qquad m = r - 2s \geq 0, \qquad \left(s \leq \frac{r}{2}\right)$$

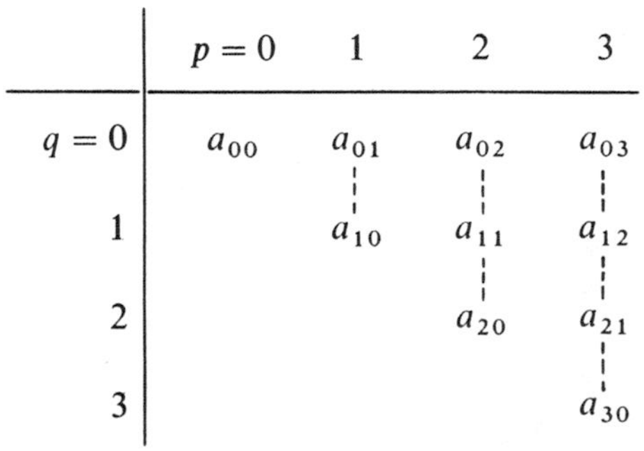

	$p = 0$	1	2	3
$q = 0$	a_{00}	a_{01}	a_{02}	a_{03}
1		a_{10}	a_{11}	a_{12}
2			a_{20}	a_{21}
3				a_{30}

Figure 5.5 Double series—again, the first summation is represented by vertical dashed lines but these vertical lines correspond to diagonals in Fig. 5.4.

	$r = 0$	1	2	3	4
$s = 0$	a_{00}	a_{01}	a_{02}	a_{03}	a_{04}
			$\vdots$	$\vdots$	$\vdots$
1			a_{10}	a_{11}	a_{12}
					$\vdots$
2				a_{20}	

Figure 5.6 Double series. The summation over s corresponds to a summation along the almost horizontal slanted lines in Fig. 5.4.

leads to

$$\sum_{m=0}^{\infty} \sum_{n=0}^{\infty} a_{n,m} = \sum_{r=0}^{\infty} \sum_{s=0}^{[r/2]} a_{s,r-2s} \tag{5.63}$$

with $[r/2] = r/2$ for r even, $(r-1)/2$ for r odd. The summation over r and s of Eq. (5.63) is shown in Fig. 5.6. Equations (5.62) and (5.63) are clearly rearrangements of the array of coefficients a_{nm}, rearrangements that are valid as long as we have absolute convergence.

The combination of Eqs. (5.62) and (5.63),

$$\sum_{p=0}^{\infty} \sum_{q=0}^{p} a_{q,p-q} = \sum_{r=0}^{\infty} \sum_{s=0}^{[r/2]} a_{s,r-2s} \tag{5.64}$$

is used in Section 12.1 in the determination of the series form of the Legendre polynomials.

EXERCISES

5.4.1 Given the series (derived in Section 5.6)

$$\ln(1 + x) = x - \frac{x^2}{2} + \frac{x^3}{3} - \frac{x^4}{4} \cdots, \qquad -1 < x \le 1,$$

show that

(a) $\ln(1 - x) = -x - \dfrac{x^2}{2} - \dfrac{x^3}{3} - \dfrac{x^4}{4} - \cdots, \qquad -1 \le x < 1.$

(b) $\ln\left(\dfrac{1 + x}{1 - x}\right) = 2\left(x + \dfrac{x^3}{3} + \dfrac{x^5}{5} + \cdots\right), \qquad -1 < x < 1.$

The original series, $\ln(1 + x)$, appears in an analysis of binding energy in crystals. It is $\frac{1}{2}$ the Madelung constant (2 ln 2) for a chain of atoms. The second series (b) is useful in normalizing the Legendre polynomials (Section 12.3) and in developing a second solution for Legendre's differential equation (Section 12.10).

5.4.2 Determine the values of the coefficients a_1, a_2, and a_3 that will make $(1 + a_1x + a_2x^2 + a_3x^3) \ln(1 + x)$ converge as n^{-4}. Find the resulting series.

5.4.3   Show that

(a) $\displaystyle\sum_{n=2}^{\infty} [\zeta(n) - 1] = 1$.

(b) $\displaystyle\sum_{n=2}^{\infty} (-1)^n[\zeta(n) - 1] = \frac{1}{2}$.

where $\zeta(n)$ is the Riemann zeta function.

5.4.4   Write a program that will rearrange the terms of the alternating harmonic series to make the series converge to 1.5. Group your terms as indicated in Eq. (5.61). List the first 100 successive partial sums that just climb above 1.5 or just drop below 1.5, and list the new terms included in each such partial sum.

ANS.

n	1	2	3	4	5
s_n	1.5333	1.0333	1.5218	1.2718	1.5143

5.5 SERIES OF FUNCTIONS

We extend our concept of infinite series to include the possibility that each term u_n may be a function of some variable, $u_n = u_n(x)$. Numerous illustrations of such series of functions appear in Chapters 11–14. The partial sums become functions of the variable x

$$s_n(x) = u_1(x) + u_2(x) + \cdots + u_n(x), \tag{5.65}$$

as does the series sum, defined as the limit of the partial sums

$$\sum_{n=1}^{\infty} u_n(x) = S(x) = \lim_{n \to \infty} s_n(x). \tag{5.66}$$

So far we have concerned ourselves with the behavior of the partial sums as a function of n. Now we consider how the foregoing quantities depend on x. The key concept here is that of uniform convergence.

UNIFORM CONVERGENCE

If for any small $\varepsilon > 0$ there exists a number N, *independent of x* in the interval $[a, b]$ ($a \le x \le b$) such that

$$|S(x) - s_n(x)| < \varepsilon, \qquad \text{for all } n \ge N, \tag{5.67}$$

the series is said to be uniformly convergent in the interval $[a, b]$. This says that for our series to be uniformly convergent, it must be possible to find a finite N so that the tail of the infinite series, $|\sum_{i=N+1}^{\infty} u_i(x)|$, will be less than an arbitrarily small ε for all x in the given interval.

This condition, Eq. (5.67), which defines uniform convergence, is illustrated in Fig. 5.7. The point is that no matter how small ε is taken to be we can always choose n large enough so that the absolute magnitude of the difference

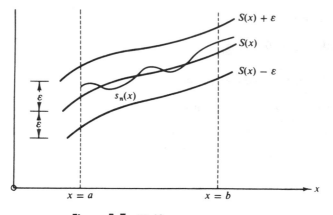

Figure 5.7 Uniform convergence.

between $S(x)$ and $s_n(x)$ is less than ε for all x, $a \le x \le b$. If this cannot be done, then $\sum u_n(x)$ is *not* uniformly convergent in $[a, b]$.

Example 5.5.1

$$\sum_{n=1}^{\infty} u_n(x) = \sum_{n=1}^{\infty} \frac{x}{[(n-1)x + 1][nx + 1]}. \tag{5.68}$$

The partial sum $s_n(x) = nx(nx + 1)^{-1}$ as may be verified by mathematical induction. By inspection this expression for $s_n(x)$ holds for $n = 1, 2$. We assume it holds for n terms and then prove it holds for $n + 1$ terms.

$$s_{n+1}(x) = s_n(x) + \frac{x}{[nx + 1][(n + 1)x + 1]}$$

$$= \frac{nx}{[nx + 1]} + \frac{x}{[nx + 1][(n + 1)x + 1]}$$

$$= \frac{(n + 1)x}{(n + 1)x + 1},$$

completing the proof.

Letting n approach infinity, we obtain

$$S(0) = \lim_{n \to \infty} s_n(0) = 0,$$

$$S(x \ne 0) = \lim_{n \to \infty} s_n(x \ne 0) = 1.$$

We have a discontinuity in our series limit at $x = 0$. However, $s_n(x)$ is a continuous function of x, $0 \le x \le 1$, for all finite n. Equation (5.67) with ε sufficiently small, will be violated for all *finite* n. Our series does not converge uniformly.

WEIERSTRASS M (MAJORANT) TEST

The most commonly encountered test for uniform convergence is the Weierstrass M test. If we can construct a series of numbers $\sum_1^\infty M_i$, in which $M_i \geq |u_i(x)|$ for all x in the interval $[a, b]$ and $\sum_1^\infty M_i$ is convergent, our series $\sum_1^\infty u_i(x)$ will be *uniformly* convergent in $[a, b]$.

The proof of this Weierstrass M test is direct and simple. Since $\sum_i M_i$ converges, some number N exists such that for $n + 1 \geq N$,

$$\sum_{i = n+1}^\infty M_i < \varepsilon. \tag{5.69}$$

This follows from our definition of convergence. Then, with $|u_i(x)| \leq M_i$ for all x in the interval $a \leq x \leq b$,

$$\sum_{i = n+1}^\infty |u_i(x)| < \varepsilon. \tag{5.70}$$

Hence

$$|S(x) - s_n(x)| = \left| \sum_{i = n+1}^\infty u_i(x) \right| < \varepsilon, \tag{5.71}$$

and by definition $\sum_{i=1}^\infty u_i(x)$ is uniformly convergent in $[a, b]$. Since we have specified absolute values in the statement of the Weierstrass M test, the series $\sum_{i=1}^\infty u_i(x)$ is also seen to be *absolutely* convergent.

The reader should note carefully that uniform convergence and absolute convergence are independent properties. Neither implies the other. For specific examples,

$$\sum_{n = 1}^\infty \frac{(-1)^n}{n + x^2}, \qquad -\infty < x < \infty \tag{5.72}$$

and

$$\sum_{n = 1}^\infty (-1)^{n-1} \frac{x^n}{n} = \ln(1 + x), \qquad 0 \leq x \leq 1 \tag{5.73}$$

converge uniformly in the indicated intervals but do not converge absolutely. On the other hand,

$$\sum_{n = 0}^\infty (1 - x)x^n = 1, \qquad 0 \leq x < 1$$
$$= 0, \qquad x = 1 \tag{5.74}$$

converges absolutely but does not converge uniformly in $[0, 1]$.

From the definition of uniform convergence we may show that any series

$$f(x) = \sum_{n = 1}^\infty u_n(x) \tag{5.75}$$

cannot converge uniformly in any interval that includes a discontinuity of $f(x)$.

Since the Weierstrass M test establishes both uniform and absolute convergence, it will necessarily fail for series that are uniformly but conditionally convergent.

ABEL'S TEST

A somewhat more delicate test for uniform convergence has been given by Abel. If

$$u_n(x) = a_n f_n(x),$$

$$\sum a_n = A, \qquad \text{convergent,}$$

and the functions $f_n(x)$ are monotonic $[f_{n+1}(x) \le f_n(x)]$ and bounded, $0 \le f_n(x) \le M$, for all x in $[a, b]$, then $\sum u_n(x)$ converges uniformly in $[a, b]$.

This test is especially useful in analyzing power series (compare Section 5.7). Details of the proof of Abel's test and other tests for uniform convergence are given in the references listed at the end of this chapter.

Uniformly convergent series have three particularly useful properties.

1. If the individual terms $u_n(x)$ are continuous, the series sum

$$f(x) = \sum_{n=1}^{\infty} u_n(x) \tag{5.76}$$

is also continuous.

2. If the individual terms $u_n(x)$ are continuous, the series may be integrated term by term. The sum of the integrals is equal to the integral of the sum.

$$\int_a^b f(x)\, dx = \sum_{n=1}^{\infty} \int_a^b u_n(x)\, dx. \tag{5.77}$$

3. The derivative of the series sum $f(x)$ equals the sum of the individual term derivatives,

$$\frac{d}{dx} f(x) = \sum_{n=1}^{\infty} \frac{d}{dx} u_n(x), \tag{5.78}$$

provided the following conditions are satisfied:

$$u_n(x) \text{ and } \frac{du_n(x)}{dx} \text{ are continuous in } [a, b].$$

$$\sum_{n=1}^{\infty} \frac{du_n(x)}{dx} \text{ is uniformly convergent in } [a, b].$$

Term-by-term integration of a uniformly convergent series[1] requires only continuity of the individual terms. This condition is almost always satisfied in physical applications. Term-by-term differentiation of a series is often not valid because more restrictive conditions must be satisfied. Indeed, we shall encounter cases in Chapter 14, Fourier Series, in which term-by term differentiation of a uniformly convergent series leads to a divergent series.

---

[1] Term-by-term integration *may* also be valid in the absence of uniform convergence.

EXERCISES

5.5.1   Find the range of *uniform* convergence of

(a) $\displaystyle\sum_{n=1}^{\infty} \frac{(-1)^{n-1}}{n^x}$

(b) $\displaystyle\sum_{n=1}^{\infty} \frac{1}{n^x}$.

> ANS.   (a) $1 \le x < \infty$.
> (b) $1 < s \le x < \infty$.

5.5.2   For what range of x is the geometric series $\sum_{n=0}^{\infty} x^n$ uniformly convergent?
> ANS. $-1 < -s \le x \le s < 1$.

5.5.3   For what range of positive values of x is $\sum_{n=0}^{\infty} 1/(1 + x^n)$

(a) convergent?

(b) uniformly convergent?

5.5.4   If the series of the coefficients $\sum a_n$ and $\sum b_n$ are absolutely convergent, show that the Fourier series

$$\sum (a_n \cos nx + b_n \sin nx)$$

is *uniformly* convergent for $-\infty < x < \infty$.

5.6 TAYLOR'S EXPANSION

This is an expansion of a function into an infinite series or into a finite series plus a remainder term. The coefficients of the successive terms of the series involve the successive derivatives of the function. We have already used Taylor's expansion in the establishment of a physical interpretation of divergence (Section 1.7) and in other sections of Chapters 1 and 2. Now we derive the Taylor expansion.

We assume that our function $f(x)$ has a continuous nth derivative[1] in the interval $a \le x \le b$. Then, integrating this nth derivative n times,

$$\int_a^x f^{(n)}(x)\, dx = f^{(n-1)}(x)\Big|_a^x = f^{(n-1)}(x) - f^{(n-1)}(a)$$

$$\int_a^x \left(\int_a^x f^{(n)}(x)\, dx \right) dx = \int_a^x [f^{(n-1)}(x) - f^{(n-1)}(a)]\, dx \qquad (5.79)$$

$$= f^{(n-2)}(x) - f^{(n-2)}(a) - (x - a)f^{(n-1)}(a).$$

---

[1] Taylor's expansion may be derived under slightly less restrictive conditions, compare H. Jeffreys and B. S. Jeffreys, *Methods of Mathematical Physics*, 3rd ed. Cambridge: Cambridge University Press (1956), Section 1.133.

Continuing, we obtain

$$\int\int\int_a^x f^{(n)}(x)(dx)^3 = f^{(n-3)}(x) - f^{(n-3)}(a) - (x-a)f^{(n-2)}(a)$$

$$- \frac{(x-a)^2}{2} f^{(n-1)}(a). \tag{5.80}$$

Finally, on integrating for the nth time,

$$\int_a^x \cdots \int f^{(n)}(x)(dx)^n = f(x) - f(a) - (x-a)f'(a)$$

$$- \frac{(x-a)^2}{2!} f''(a) - \cdots - \frac{(x-a)^{n-1}}{(n-1)!} f^{(n-1)}(a). \tag{5.81}$$

Note that this expression is exact. No terms have been dropped, no approximations made. Now, solving for $f(x)$, we have

$$f(x) = f(a) + (x-a)f'(a)$$

$$+ \frac{(x-a)^2}{2!} f''(a) + \cdots + \frac{(x-a)^{n-1}}{(n-1)!} f^{(n-1)}(a) + R_n. \tag{5.82}$$

The remainder, R_n, is given by the n-fold integral

$$R_n = \int_a^x \cdots \int f^{(n)}(x)(dx)^n. \tag{5.83}$$

This remainder, Eq. (5.83), may be put into perhaps more intelligible form by using the mean value theorem of integral calculus

$$\int_a^x g(x)\, dx = (x-a)g(\xi), \tag{5.84}$$

with $a \le \xi \le x$. By integrating n times we get the Lagrangian form[2] of the remainder:

$$R_n = \frac{(x-a)^n}{n!} f^{(n)}(\xi). \tag{5.85}$$

With Taylor's expansion in this form we are not concerned with any questions of infinite series convergence. This series is finite, and the only questions concern the magnitude of the remainder.

When the function $f(x)$ is such that

$$\lim_{n \to \infty} R_n = 0, \tag{5.86}$$

---

[2] An alternate form derived by Cauchy is

$$R_n = \frac{(x-\zeta)^{n-1}(x-a)}{(n-1)!} f^{(n)}(\zeta).$$

with $a \le \zeta \le x$.

Eq. (5.82) becomes Taylor's series

$$f(x) = f(a) + (x - a)f'(a) + \frac{(x - a)^2}{2!}f''(a) + \cdots$$

$$= \sum_{n=0}^{\infty} \frac{(x - a)^n}{n!}f^{(n)}(a).^3 \tag{5.87}$$

Our Taylor series specifies the value of a function at one point, x, in terms of the value of the function and its derivatives at a reference point, a. It is an expansion in powers of the *change* in the variable, $\Delta x = x - a$ in this case. The notation may be varied at the user's convenience. With the substitution $x \rightarrow x + h$ and $a \rightarrow x$ we have an alternate form

$$f(x + h) = \sum_{n=0}^{\infty} \frac{h^n}{n!}f^{(n)}(x).$$

When we use the operator $D = d/dx$ the Taylor expansion becomes

$$f(x + h) = \sum_{n=0}^{\infty} \frac{h^n D^n}{n!}f(x) = e^{hD}f(x).$$

(The transition to the exponential form anticipates Eq. (5.90) that follows.) An equivalent operator form of this Taylor expansion appears in Exercise 4.2.4. A derivation of the Taylor expansion in the context of complex variable theory appears in Section 6.5.

MACLAURIN THEOREM

If we expand about the origin ($a = 0$), Eq. (5.87) is known as Maclaurin's series

$$f(x) = f(0) + xf'(0) + \frac{x^2}{2!}f''(0) + \cdots$$

$$= \sum_{n=0}^{\infty} \frac{x^n}{n!}f^{(n)}(0). \tag{5.88}$$

An immediate application of the Maclaurin series (or the Taylor series) is in the expansion of various transcendental functions into infinite series.

Example 5.6.1

Let $f(x) = e^x$. Differentiating, we have

$$f^{(n)}(0) = 1 \tag{5.89}$$

for all n, $n = 1, 2, 3, \ldots$. Then, Eq. (5.88), we have

$$e^x = 1 + x + \frac{x^2}{2!} + \frac{x^3}{3!} + \cdots = \sum_{n=0}^{\infty} \frac{x^n}{n!}. \tag{5.90}$$

---

^{3} Note that $0! = 1$ (compare Section 10.1).

This is the series expansion of the exponential function. Some authors use this series to define the exponential function.

Although this series is clearly convergent for all x, we should check the remainder term, R_n. By Eq. (5.85) we have

$$R_n = \frac{x^n}{n!} f^{(n)}(\xi)$$

$$= \frac{x^n}{n!} e^\xi, \qquad 0 \le |\xi| \le x. \tag{5.91}$$

Therefore

$$|R_n| \le \frac{x^n e^x}{n!} \tag{5.92}$$

and

$$\lim_{n \to \infty} R_n = 0 \tag{5.93}$$

for all *finite* values of x, which indicates that this Maclaurin expansion of e^x is valid over the range $-\infty < x < \infty$.

Example 5.6.2

Let $f(x) = \ln(1 + x)$. By differentiating, we obtain

$$f'(x) = (1 + x)^{-1},$$
$$f^{(n)}(x) = (-1)^{n-1}(n - 1)!(1 + x)^{-n}. \tag{5.94}$$

The Maclaurin expansion (Eq. (5.88)) yields

$$\ln(1 + x) = x - \frac{x^2}{2} + \frac{x^3}{3} - \frac{x^4}{4} + \cdots + R_n$$

$$= \sum_{p=1}^{n} (-1)^{p-1} \frac{(x)^p}{p} + R_n. \tag{5.95}$$

In this case our remainder is given by

$$R_n = \frac{x^n}{n!} f^{(n)}(\xi), \qquad 0 \le \xi \le x$$

$$\le \frac{x^n}{n}, \qquad 0 \le \xi \le x \le 1. \tag{5.96}$$

Now the remainder approaches zero as n is increased indefinitely, provided $0 \le x \le 1$.[4] As an infinite series

$$\ln(1 + x) = \sum_{n=1}^{\infty} (-1)^{n-1} \frac{x^n}{n}, \tag{5.97}$$

[4] This range can easily be extended to $-1 < x \le 1$ but not to $x = -1$.

which converges for $-1 < x \leq 1$. The range $-1 < x < 1$ is easily established by the d'Alembert ratio test (Section 5.2). Convergence at $x = 1$ follows by the Leibniz criterion (Section 5.3). In particular, at $x = 1$, we have

$$\ln 2 = 1 - \tfrac{1}{2} + \tfrac{1}{3} - \tfrac{1}{4} + \tfrac{1}{5} - \cdots$$

$$= \sum_{n=1}^{\infty} (-1)^{n-1} n^{-1}, \tag{5.98}$$

the conditionally convergent alternating harmonic series.

BINOMIAL THEOREM

A second, extremely important application of the Taylor and Maclaurin expansions is the derivation of the binomial theorem for negative and/or nonintegral powers.

Let $f(x) = (1 + x)^m$, in which m may be negative and is not limited to integral values. Direct application of Eq. (5.88) gives

$$(1 + x)^m = 1 + mx + \frac{m(m-1)}{2!} x^2 + \cdots + R_n. \tag{5.99}$$

For this function the remainder is

$$R_n = \frac{x^n}{n!} (1 + \xi)^{m-n} \times m(m-1) \cdots (m-n+1) \tag{5.100}$$

and ξ lies between 0 and x, $0 \leq \xi \leq x$. Now, for $n > m$, $(1 + \xi)^{m-n}$ is a maximum for $\xi = 0$. Therefore

$$R_n \leq \frac{x^n}{n!} \times m(m-1) \cdots (m-n+1). \tag{5.101}$$

Note that the m dependent factors do not yield a zero unless m is a nonnegative integer; R_n tends to zero as $n \to \infty$ if x is restricted to the range $0 \leq x < 1$.
The binomial expansion therefore is shown to be

$$(1 + x)^m = 1 + mx + \frac{m(m-1)}{2!} x^2 + \frac{m(m-1)(m-2)}{3!} x^3 + \cdots. \tag{5.102}$$

In other, equivalent notation

$$(1 + x)^m = \sum_{n=0}^{\infty} \frac{m!}{n!(m-n)!} x^n$$

$$= \sum_{n=0}^{\infty} \binom{m}{n} x^n. \tag{5.103}$$

The quantity $\binom{m}{n}$, which equals $m!/n!(m-n)!$ is called a *binomial coefficient*. Although we have only shown that the remainder vanishes,

$$\lim_{n \to \infty} R_n = 0,$$

for $0 \le x < 1$, the series in Eq. (5.102) actually may be shown to be convergent for the extended range $-1 < x < 1$. For m an integer, $(m - n)! = \pm\infty$ if $n > m$ (Section 10.1) and the series automatically terminates at $n = m$.

Example 5.6.3 Relativistic Energy

The total relativistic energy of a particle is

$$E = mc^2\left(1 - \frac{v^2}{c^2}\right)^{-1/2}. \tag{5.104}$$

Compare this equation with the classical kinetic energy, $\frac{1}{2}mv^2$.

By Eq. (5.102) with $x = -v^2/c^2$ and $m = -\frac{1}{2}$ we have

$$E = mc^2\left[1 - \frac{1}{2}\left(-\frac{v^2}{c^2}\right) + \frac{(-1/2)(-3/2)}{2!}\left(-\frac{v^2}{c^2}\right)^2 \right.$$
$$\left. + \frac{(-1/2)(-3/2)(-5/2)}{3!}\left(-\frac{v^2}{c^2}\right)^3 + \cdots\right]$$

or

$$E = mc^2 + \frac{1}{2}mv^2 + \frac{3}{8}mv^2 \cdot \frac{v^2}{c^2} + \frac{5}{16}mv^2 \cdot \left(\frac{v^2}{c^2}\right)^2 + \cdots. \tag{5.105}$$

The first term, mc^2, is identified as the rest mass energy. Then

$$E_{\text{kinetic}} = \frac{1}{2}mv^2\left[1 + \frac{3}{4}\frac{v^2}{c^2} + \frac{5}{8}\left(\frac{v^2}{c^2}\right)^2 + \cdots\right]. \tag{5.106}$$

For particle velocity $v \ll c$, the velocity of light, the expression in the brackets reduces to unity and we see that the kinetic portion of the total relativistic energy agrees with the classical result.

For polynomials we can generalize the binomial expansion to

$$(a_1 + a_2 + \cdots + a_m)^n = \sum \frac{n!}{n_1!n_2!\cdots n_m!}a_1^{n_1}a_2^{n_2}\cdots a_m^{n_m},$$

where the summation includes all different combinations of $n_1, n_2, \ldots, n_m$ with $\sum_{i=1}^m n_i = n$. Here n_i and n are all integral. This generalization finds considerable use in statistical mechanics.

Maclaurin series may sometimes appear indirectly rather than by direct use of Eq. (5.88). For instance, the most convenient way to obtain the series expansion

$$\sin^{-1}x = \sum_{n=0}^{\infty} \frac{(2n-1)!!}{(2n)!!} \cdot \frac{x^{2n+1}}{(2n+1)} = x + \frac{x^3}{6} + \frac{3x^5}{40} + \cdots, \tag{5.106a}$$

is to make use of the relation

$$\sin^{-1}x = \int_0^x \frac{dt}{(1-t^2)^{1/2}}.$$

We expand $(1 - t^2)^{-1/2}$ (binomial theorem) and then integrate term by term. This term-by-term integration is discussed in Section 5.7. The result is Eq. (5.106a). Finally, we may take the limit as $x \to 1$. The series converges by Gauss's test, Exercise 5.2.5.

TAYLOR EXPANSION—MORE THAN ONE VARIABLE

If the function f has more than one independent variable, say, $f = f(x, y)$, the Taylor expansion becomes

$$f(x, y) = f(a, b) + (x - a)\frac{\partial f}{\partial x} + (y - b)\frac{\partial f}{\partial y}$$

$$+ \frac{1}{2!}\left[(x - a)^2 \frac{\partial^2 f}{\partial x^2} + 2(x - a)(y - b)\frac{\partial^2 f}{\partial x \, \partial y} + (y - b)^2 \frac{\partial^2 f}{\partial y^2} \right]$$

$$+ \frac{1}{3!}\left[(x - a)^3 \frac{\partial^3 f}{\partial x^3} + 3(x - a)^2 (y - b)\frac{\partial^3 f}{\partial x^2 \, \partial y} \right.$$

$$\left. + 3(x - a)(y - b)^2 \frac{\partial^3 f}{\partial x \, \partial y^2} + (y - b)^3 \frac{\partial^3 f}{\partial y^3} \right] + \cdots, \tag{5.107}$$

with all derivatives evaluated at the point (a, b). Using $\alpha_j t = x_j - x_{j0}$, we may write the Taylor expansion for m independent variables in the symbolic form

$$f(x_j) = \sum_{n=0}^{\infty} \frac{t^n}{n!}\left(\sum_{i=1}^{m} \alpha_i \frac{\partial}{\partial x_i} \right)^n f(x_k)\bigg|_{x_k = x_{k0}}. \tag{5.108}$$

A convenient vector form is

$$\psi(\mathbf{r} + \mathbf{a}) = \sum_{n=0}^{\infty} \frac{1}{n!}(\mathbf{a} \cdot \nabla)^n \psi(\mathbf{r}). \tag{5.109}$$

EXERCISES

5.6.1   Show that

(a) $\sin x = \sum_{n=0}^{\infty} (-1)^n \frac{x^{2n+1}}{(2n + 1)!}$,

(b) $\cos x = \sum_{n=0}^{\infty} (-1)^n \frac{x^{2n}}{(2n)!}$.

In Section 6.1, e^{ix} is *defined* by a series expansion such that

$$e^{ix} = \cos x + i \sin x.$$

This is the basis for the polar representation of complex quantities. As a special case we find, with $x = \pi$,

$$e^{i\pi} = -1.$$

5.6.2   Derive a series expansion of cot x in increasing powers of x by dividing cos x by sin x.

Note. The resultant series that starts with $1/x$ is actually a Laurent series (Section 6.5). Although the two series for sin x and cos x were valid for all x, the convergence of the series for cot x is limited by the zeros of the denominator, sin x.

5.6.3   (a) Expand $(1 + x)\ln(1 + x)$ in a Maclaurin series. Find the limits on x for convergence.

(b) From the results for part (a) show that

$$\ln 2 = \frac{1}{2} + \frac{1}{2}\sum_{n=1}^{\infty}\frac{(-1)^{n+1}}{n(n + 1)}.$$

ANS.   (a) $(1 + x)\ln(1 + x) = x + \sum_{n=2}^{\infty}(-1)^n\frac{x^n}{n(n - 1)}$, $-1 \le x \le 1$.

5.6.4   The Raabe test for $\sum(n \ln n)^{-1}$ leads to

$$\lim_{n\to\infty} n\left[\frac{(n + 1)\ln(n + 1)}{n \ln n} - 1\right].$$

Show that this limit is unity (which means that the Raabe test here is indeterminate).

5.6.5   Show by series expansion that

$$\frac{1}{2}\ln\frac{\eta_0 + 1}{\eta_0 - 1} = \coth^{-1}\eta_0,\qquad |\eta_0| > 1.$$

This identity may be used to obtain a second solution for Legendre's equation.

5.6.6   Show that $f(x) = x^{1/2}$ (a) has no Maclaurin expansion but (b) has a Taylor expansion about any point $x_0 \ne 0$. Find the range of convergence of the Taylor expansion about $x = x_0$.

5.6.7   Let x be an approximation for a zero of $f(x)$ and Δx, the correction. Show that by neglecting terms of order $(\Delta x)^2$

$$\Delta x = -\frac{f(x)}{f'(x)}.$$

This is Newton's formula for finding a root. Newton's method has the virtues of illustrating series expansions and elementary calculus but is very treacherous. See Appendix A1 for details and an alternative.

5.6.8   Expand a function $\Phi(x, y, z)$ by Taylor's expansion. Evaluate $\bar{\Phi}$, the average value of Φ, averaged over a small cube of side a centered on the origin and show that the Laplacian of Φ is a measure of deviation of Φ from $\Phi(0, 0, 0)$.

5.6.9   The ratio of two differentiable functions $f(x)$ and $g(x)$ takes on the indeterminate form $0/0$ at $x = x_0$. Using Taylor expansions prove L'Hospital's rule

$$\lim_{x\to x_0}\frac{f(x)}{g(x)} = \lim_{x\to x_0}\frac{f'(x)}{g'(x)}.$$

5.6.10   With $n > 1$, show that

(a) $\dfrac{1}{n} - \ln\left(\dfrac{n}{n-1}\right) < 0,$ 　　　　　　　(b) $\dfrac{1}{n} - \ln\left(\dfrac{n+1}{n}\right) > 0.$

Use these inequalities to show that the limit defining the Euler–Mascheroni constant is finite.

5.6.11   Expand $(1 - 2tz + t^2)^{-1/2}$ in powers of t. Assume that t is small. Collect the coefficients of t^0, t^1, and t^2.

$$\begin{aligned} ANS. \quad a_0 &= P_0(z) = 1, \\ a_1 &= P_1(z) = z, \\ a_2 &= P_2(z) = \tfrac{1}{2}(3z^2 - 1), \end{aligned}$$

where $a_n = P_n(z)$, the nth Legendre polynomial.

5.6.12   Using the double factorial notation of Section 10.1, show that

$$(1 + x)^{-m/2} = \sum_{n=0}^{\infty} (-1)^n \frac{(m + 2n - 2)!!}{2^n n!(m - 2)!!} x^n,$$

for $m = 1, 2, 3, \ldots$.

5.6.13   Using binomial expansions, compare the three Doppler shift formulas:

(a) $v' = v\left(1 \mp \dfrac{v}{c}\right)^{-1}$ 　　　moving source;

(b) $v' = v\left(1 \pm \dfrac{v}{c}\right),$ 　　　moving observer;

(c) $v' = v\left(1 \pm \dfrac{v}{c}\right)\left(1 - \dfrac{v^2}{c^2}\right)^{-1/2},$ 　　　relativistic.

Note. The relativistic formula agrees with the classical formulas if terms of order v^2/c^2 can be neglected.

5.6.14   In the theory of general relativity there are various ways of relating (defining) a velocity of recession of a galaxy to its red shift, δ. Milne's model (kinematic relativity) gives

(a) $v_1 = c\delta(1 + \tfrac{1}{2}\delta),$

(b) $v_2 = c\delta(1 + \tfrac{1}{2}\delta)(1 + \delta)^{-2},$

(c) $1 + \delta = \left[\dfrac{1 + v_3/c}{1 - v_3/c}\right]^{1/2}.$

1. Show that for $\delta \ll 1$ (and $v_3/c \ll 1$) all three formulas reduce to $v = c\delta$.
2. Compare the three velocities through terms of order δ^2.
Note. In special relativity (with δ replaced by z), the ratio of observed wavelength λ to emitted wavelength λ_0 is given by

$$\frac{\lambda}{\lambda_0} = 1 + z = \left(\frac{c + v}{c - v}\right)^{1/2}.$$

5.6.15   The relativistic sum w of two velocities u and v is given by

$$\frac{w}{c} = \frac{u/c + v/c}{1 + uv/c^2}.$$

If

$$\frac{v}{c} = \frac{u}{c} = 1 - \alpha,$$

where $0 \le \alpha \le 1$, find w/c in powers of α through terms in α^3.

5.6.16   The displacement x of a particle of rest mass m_0, resulting from a constant force $m_0 g$ along the x-axis, is

$$x = \frac{c^2}{g}\left\{\left[1 + \left(g\frac{t}{c}\right)^2\right]^{1/2} - 1\right\},$$

including relativistic effects. Find the displacement x as a power series in time t. Compare with the classical result

$$x = \tfrac{1}{2}gt^2.$$

5.6.17   By use of Dirac's relativistic theory the fine structure formula of atomic spectroscopy is given by

$$E = mc^2\left[1 + \frac{\gamma^2}{(s + n - |k|)^2}\right]^{-1/2},$$

where

$$s = (|k|^2 - \gamma^2)^{1/2}, \qquad k = \pm 1, \pm 2, \pm 3, \dots .$$

Expand in powers of γ^2 through order γ^4 ($\gamma^2 = Ze^2/\hbar c$, with Z the atomic number). This expansion is useful in comparing the predictions of the Dirac electron theory with those of a relativistic Schrödinger electron theory. Experimental results support the Dirac theory.

5.6.18   In a head-on proton–proton collision, the ratio of the kinetic energy in the center of mass system to the incident kinetic energy is

$$R = \frac{\sqrt{2mc^2(E_k + 2mc^2)} - 2mc^2}{E_k}.$$

Find the value of this ratio of kinetic energies for

(a) $E_k \ll mc^2$ (nonrelativistic)
(b) $E_k \gg mc^2$ (extreme-relativistic).

> ANS. (a) $\tfrac{1}{2}$, (b) 0. The latter answer is a sort of law of diminishing returns for high energy particle accelerators (with stationary targets).

5.6.19   With binomial expansions

$$\frac{x}{1 - x} = \sum_{n=1}^{\infty} x^n, \qquad \frac{x}{x - 1} = \frac{1}{1 - x^{-1}} = \sum_{n=0}^{\infty} x^{-n}.$$

Adding these two series yields $\sum_{n=-\infty}^{\infty} x^n = 0$.
Hopefully, we can agree that this is nonsense but what has gone wrong?

5.6.20   (a) Planck's theory of quantized oscillators led to an average energy

$$\langle \varepsilon \rangle = \frac{\displaystyle\sum_{n=1}^{\infty} n\varepsilon_0 \exp(-n\varepsilon_0/kT)}{\displaystyle\sum_{n=0}^{\infty} \exp(-n\varepsilon_0/kT)},$$

where ε_0 was a fixed energy. Identify the numerator and denominator as binomial expansions and show that the ratio is

$$\langle \varepsilon \rangle = \frac{\varepsilon_0}{\exp(\varepsilon_0/kT) - 1}.$$

(b) Show that the $\langle \varepsilon \rangle$ of part (a) reduces to kT, the classical result, for $kT \gg \varepsilon_0$.

5.6.21   (a) Expand by the binomial theorem and integrate term by term to obtain the Gregory series for $\tan^{-1} x$:

$$\tan^{-1} x = \int_0^x \frac{dt}{1 + t^2} = \int_0^x \{1 - t^2 + t^4 - t^6 + \cdots\} \, dt$$

$$= \sum_{n=0}^{\infty} (-1)^n \frac{x^{2n+1}}{2n + 1}, \qquad -1 \le x \le 1.$$

(b) By comparing series expansions, show that

$$\tan^{-1} x = \frac{i}{2} \ln\left(\frac{1 - ix}{1 + ix}\right).$$

Hint. Compare Exercise 5.4.1.

5.6.22   In numerical analysis it is often convenient to approximate $d^2\psi(x)/dx^2$ by

$$\frac{d^2}{dx^2} \psi(x) \approx \frac{1}{h^2} [\psi(x + h) - 2\psi(x) + \psi(x - h)].$$

Find the error in this approximation.

$$ANS. \quad \text{Error} = \frac{h^2}{12} \psi^{(4)}(x).$$

5.6.23   You have a function $y(x)$ tabulated at equally spaced values of the argument

$$\begin{cases} y_n = y(x_n) \\ x_n = x + nh. \end{cases}$$

Show that the linear combination

$$\frac{1}{12h}\{-y_2 + 8y_1 - 8y_{-1} + y_{-2}\}$$

yields

$$y_0' - \frac{h^4}{30} y_0^{(5)} + \cdots.$$

Hence this linear combination yields y_0' if $(h^4/30)y_0^{(5)}$ and higher powers of h and higher derivatives of $y(x)$ are negligible.

5.6.24   In a numerical integration of a partial differential equation the three-dimensional Laplacian is replaced by

$$\nabla^2\psi(x, y, z) \to h^{-2}[\psi(x + h, y, z) + \psi(x - h, y, z)$$
$$+ \psi(x, y + h, z) + \psi(x, y - h, z) + \psi(x, y, z + h)$$
$$+ \psi(x, y, z - h) - 6\psi(x, y, z)].$$

Determine the error in this approximation. Here h is the step size, the distance between adjacent points in the x-, y-, or z-direction.

5.6.25   Using double precision, calculate e from its Maclaurin series.
Note. This simple, direct approach is the best way of calculating e to high accuracy. Sixteen terms give e to 16 significant figures. The reciprocal factorials give very rapid convergence.

5.7 POWER SERIES

The power series is a special and extremely useful type of infinite series of the form

$$f(x) = a_0 + a_1 x + a_2 x^2 + a_3 x^3 + \cdots$$

$$= \sum_{n=0}^{\infty} a_n x^n, \qquad (5.110)$$

where the coefficients a_i are constants, independent of x.[1]

CONVERGENCE

Equation (5.110) may readily be tested for convergence by either the Cauchy root test or the d'Alembert ratio test (Section 5.2). If

$$\lim_{n \to \infty} \frac{a_{n+1}}{a_n} = R^{-1}, \qquad (5.111)$$

the series converges for $-R < x < R$. This is the interval or radius of convergence. Since the root and ratio tests fail when the limit is unity, the end points of the interval require special attention.

For instance, if $a_n = n^{-1}$, then $R = 1$ and, from Sections 5.1, 5.2, and 5.3, the series converges for $x = -1$ but diverges for $x = +1$. If $a_n = n!$, then $R = 0$ and the series diverges for all $x \neq 0$.

UNIFORM AND ABSOLUTE CONVERGENCE

Suppose our power series (Eq. (5.110)) has been found convergent for $-R < x < R$; then it will be uniformly and absolutely convergent in any *interior* interval, $-S \leq x \leq S$, where $0 < S < R$.

This may be proved directly by the Weierstrass M test (Section 5.5) by using $M_i = |a_i| S^i$.

CONTINUITY

Since each of the terms $u_n(x) = a_n x^n$ is a continuous function of x and $f(x) = \sum a_n x^n$ converges uniformly for $-S \leq x \leq S$, $f(x)$ must be a continuous function in the interval of uniform convergence.

---

[1] Equation (5.110) may be rewritten with $z = x + iy$, replacing x. The following sections will then yield uniform convergence, integrability, and differentiability in a region of a complex plane in place of an interval on the x-axis.

This behavior is to be contrasted with the strikingly different behavior of the Fourier series (Chapter 14), in which the Fourier series is used frequently to represent discontinuous functions such as sawtooth and square waves.

DIFFERENTIATION AND INTEGRATION

With $u_n(x)$ continuous and $\sum a_n x^n$ uniformly convergent, we find that the differentiated series is a power series with continuous functions and the same radius of convergence as the original series. The new factors introduced by differentiation (or integration) do not affect either the root or the ratio test. Therefore our power series may be differentiated or integrated as often as desired within the interval of uniform convergence (Exercise 5.7.13).

In view of the rather severe restrictions placed on differentiation (Section 5.5), this is a remarkable and valuable result.

UNIQUENESS THEOREM

In the preceding section, using the Maclaurin series, we expanded e^x and $\ln(1 + x)$ into infinite series. In the succeeding chapters functions are frequently represented or perhaps defined by infinite series. We now establish that the power-series representation is unique.

If

$$f(x) = \sum_{n=0}^{\infty} a_n x^n, \qquad -R_a < x < R_a$$

$$= \sum_{n=0}^{\infty} b_n x^n, \qquad -R_b < x < R_b, \tag{5.112}$$

with overlapping intervals of convergence, including the origin, then

$$a_n = b_n \tag{5.113}$$

for all n; that is, we assume two (different) power-series representations and then proceed to show that the two are actually identical.

From Eq. (5.112)

$$\sum_{n=0}^{\infty} a_n x^n = \sum_{n=0}^{\infty} b_n x^n, \qquad -R < x < R, \tag{5.114}$$

where R is the smaller of R_a, R_b. By setting $x = 0$ to eliminate all but the constant terms, we obtain

$$a_0 = b_0. \tag{5.115}$$

Now, exploiting the differentiability of our power series, we differentiate Eq. (5.114), getting

$$\sum_{n=1}^{\infty} n a_n x^{n-1} = \sum_{n=1}^{\infty} n b_n x^{n-1}. \tag{5.116}$$

We again set $x = 0$ to isolate the new constant terms and find

$$a_1 = b_1. \tag{5.117}$$

By repeating this process n times, we get

$$a_n = b_n, \tag{5.118}$$

which shows that the two series coincide. Therefore our power-series representation is unique.

This will be a crucial point in Section 8.5, in which we use a power series to develop solutions of differential equations. This uniqueness of power series appears frequently in theoretical physics. The establishment of perturbation theory in quantum mechanics is one example. The power-series representation of functions is often useful in evaluating indeterminate forms, particularly when l'Hospital's rule may be awkward to apply (Exercise 5.7.9).

Example 5.7.1
Evaluate

$$\lim_{x \to 0} \frac{1 - \cos x}{x^2}. \tag{5.119}$$

Replacing $\cos x$ by its Maclaurin series expansion, we obtain

$$\frac{1 - \cos x}{x^2} = \frac{1 - (1 - x^2/2! + x^4/4! - \cdots)}{x^2}$$

$$= \frac{x^2/2! - x^4/4! + \cdots}{x^2}$$

$$= \frac{1}{2!} - \frac{x^2}{4!} + \cdots.$$

Letting $x \to 0$, we have

$$\lim_{x \to 0} \frac{1 - \cos x}{x^2} = \frac{1}{2}. \tag{5.120}$$

The uniqueness of power series means that the coefficients a_n may be identified with the derivatives in a Maclaurin series. From

$$f(x) = \sum_{n=0}^{\infty} a_n x^n = \sum_{n=0}^{\infty} \frac{1}{n!} f^{(n)}(0) x^n$$

we have

$$a_n = \frac{1}{n!} f^{(n)}(0).$$

REVERSION (INVERSION) OF POWER SERIES

Suppose we are given a series

$$y - y_0 = a_1(x - x_0) + a_2(x - x_0)^2 + \cdots$$

$$= \sum_{n=1}^{\infty} a_n(x - x_0)^n. \tag{5.121}$$

This gives $(y - y_0)$ in terms of $(x - x_0)$. However, it may be desirable to have an explicit expression for $(x - x_0)$ in terms of $(y - y_0)$. We may solve Eq. (5.121) for $x - x_0$ by reversion (or inversion) of our series. Assume that

$$x - x_0 = \sum_{n=1}^{\infty} b_n(y - y_0)^n, \tag{5.122}$$

with the b_n to be determined in terms of the assumed known a_n. A brute-force approach, which is perfectly adequate for the first few coefficients, is simply to substitute Eq. (5.121) into Eq. (5.122). By equating coefficients of $(x - x_0)^n$ on both sides of Eq. (5.122), since the power series is unique, we obtain

$$b_1 = \frac{1}{a_1},$$

$$b_2 = -\frac{a_2}{a_1^3},$$

$$b_3 = \frac{1}{a_1^5}(2a_2^2 - a_1 a_3), \tag{5.123}$$

$$b_4 = \frac{1}{a_1^7}(5a_1 a_2 a_3 - a_1^2 a_4 - 5a_2^3), \quad \text{and so on.}$$

Some of the higher coefficients are listed by Dwight.[2] A more general and much more elegant approach is developed by the use of complex variables in the first and second editions of *Mathematical Methods for Physicists*.

EXERCISES

5.7.1   The classical Langevin theory of paramagnetism leads to an expression for the magnetic polarization

$$P(x) = c\left(\frac{\cosh x}{\sinh x} - \frac{1}{x}\right).$$

Expand $P(x)$ as a power series for small x (low fields, high temperature).

5.7.2   The depolarizing factor L for an oblate ellipsoid in a uniform electric field parallel to the axis of rotation is

$$L = \frac{1}{\varepsilon_0}(1 + \zeta_0^2)(1 - \zeta_0 \cot^{-1}\zeta_0),$$

where ζ_0 defines an oblate ellipsoid in oblate spheroidal coordinates (ξ, ζ, φ). Show that

$$\lim_{\zeta_0 \to \infty} L = \frac{1}{3\varepsilon_0} \quad \text{(sphere)},$$

$$\lim_{\zeta_0 \to 0} L = \frac{1}{\varepsilon_0} \quad \text{(thin sheet)}.$$

[2] H. B. Dwight, *Tables of Integrals and Other Mathematical Data*, 4th ed. New York: Macmillan (1961). (Compare Formula No. 50.)

5.7.3   The corresponding depolarizing factor (Exercise 5.7.2) for a prolate ellipsoid is

$$L = \frac{1}{\varepsilon_0}(\eta_0^2 - 1)\left(\frac{1}{2}\eta_0 \ln\frac{\eta_0 + 1}{\eta_0 - 1} - 1\right).$$

Show that

$$\lim_{\eta_0 \to \infty} L = \frac{1}{3\varepsilon_0} \qquad \text{(sphere)},$$

$$\lim_{\eta_0 \to 0} L = 0 \qquad \text{(long needle)}.$$

5.7.4   The analysis of the diffraction pattern of a circular opening involves

$$\int_0^{2\pi} \cos(c \cos \varphi)\, d\varphi.$$

Expand the integrand in a series and integrate by using

$$\int_0^{2\pi} \cos^{2n}\varphi\, d\varphi = \frac{(2n)!}{2^{2n}(n!)^2} \cdot 2\pi,$$

$$\int_0^{2\pi} \cos^{2n+1}\varphi\, d\varphi = 0.$$

The result is 2π times the Bessel function $J_0(c)$.

5.7.5   Neutrons are created (by a nuclear reaction) inside a hollow sphere of radius R. The newly created neutrons are uniformly distributed over the spherical volume. Assuming that all directions are equally probable (isotropy), what is the average distance a neutron will travel before striking the surface of the sphere? Assume straight line motion, no collisions.

(a) Show that

$$\bar{r} = \frac{3}{2}R \int_0^1 \int_0^{\pi} \sqrt{1 - k^2 \sin^2\theta}\, k^2\, dk\, \sin\theta\, d\theta.$$

(b) Expand the integrand as a series and integrate to obtain

$$\bar{r} = R\left[1 - 3 \sum_{n=1}^{\infty} \frac{1}{(2n - 1)(2n + 1)(2n + 3)}\right].$$

(c) Show that the sum of this infinite series is $\frac{1}{12}$, giving $\bar{r} = \frac{3}{4}R$.
Hint. Show that $s_n = \frac{1}{12} - [4(2n + 1)(2n + 3)]^{-1}$ by mathematical induction. Then let $n \to \infty$.

5.7.6   Given that

$$\int_0^1 \frac{dx}{1 + x^2} = \tan^{-1}x \Big|_0^1 = \frac{\pi}{4},$$

expand the integrand into a series and integrate term by term obtaining[3]

$$\frac{\pi}{4} = 1 - \frac{1}{3} + \frac{1}{5} - \frac{1}{7} + \frac{1}{9} - \cdots + (-1)^n\frac{1}{2n + 1} + \cdots,$$

which is Leibniz's formula for π. Compare the convergence (or lack of it) of the integrand series and the integrated series at $x = 1$.

---

[3] The series expansion of $\tan^{-1}x$ (upper limit 1 replaced by x) was discovered by James Gregory in 1671, 3 years before Leibniz. See Peter Beckmann's entertaining and informative book, *A History of Pi*, 2nd ed. Boulder, CO: Golem Press (1971).

Leibniz's formula converges so slowly that it is quite useless for numerical work; π has been computed to 100,000 decimals[4] by using expressions such as

$$\pi = 24 \tan^{-1}\tfrac{1}{8} + 8 \tan^{-1}\tfrac{1}{57} + 4 \tan^{-1}\tfrac{1}{239},$$

$$\pi = 48 \tan^{-1}\tfrac{1}{18} + 32 \tan^{-1}\tfrac{1}{57} - 20 \tan^{-1}\tfrac{1}{239}.$$

These expressions may be verified by the use of Exercise 5.6.2.

5.7.7 Expand the incomplete factorial function

$$\int_0^x e^{-t} t^n \, dt$$

in a series of powers of x for small values of x. What is the range of convergence of the resulting series? Why was x specified to be small?

$$\text{ANS.} \quad \int_0^x e^{-t} t^n \, dt = x^{n+1} \left[\frac{1}{(n+1)} - \frac{x}{(n+2)} + \frac{x^2}{2!(n+3)} \right.$$

$$\left. - \cdots \frac{(-1)^p x^p}{p!(n+p+1)} + \cdots \right].$$

5.7.8 Derive the series expansion of the incomplete beta function

$$B_x(p, q) = \int_0^x t^{p-1}(1 - t)^{q-1} \, dt$$

$$= x^p \left\{ \frac{1}{p} + \frac{1-q}{p+1} x + \cdots + \frac{(1-q)\cdots(n-q)}{n!(p+n)} x^n + \cdots \right\}$$

for $0 \le x \le 1$, $p > 0$ and $q > 0$ (if $x = 1$).

5.7.9 Evaluate

(a) $\displaystyle\lim_{x \to 0} \frac{\sin(\tan x) - \tan(\sin x)}{x^7}$,

(b) $\displaystyle\lim_{x \to 0} x^{-n} j_n(x)$ for $n = 3$,

where $j_n(x)$ is a spherical Bessel function (Section 11.7) defined by

$$j_n(x) = (-1)^n x^n \left(\frac{d}{x\,dx} \right)^n \left(\frac{\sin x}{x} \right).$$

$$\text{ANS.} \quad \text{(a)} \; -\frac{1}{30}, \quad \text{(b)} \; \frac{1}{1 \cdot 3 \cdot 5 \cdots (2n+1)} \to \frac{1}{105} \; \text{for } n = 3.$$

5.7.10 Neutron transport theory gives the following expression for the inverse neutron diffusion length of k:

$$\frac{a - b}{k} \tanh^{-1}\left(\frac{k}{a} \right) = 1.$$

By series inversion or otherwise, determine k^2 as a series of powers of b/a. Give the first two terms of the series.

$$\text{ANS.} \quad k^2 = 3ab\left(1 - \frac{4}{5}\frac{b}{a} \right).$$

[4] D. Shanks and J. W. Wrench, Jr., Computation of π to 100,000 decimals. *Math. Comput.* **16**, 76 (1962).

5.7.11   Develop a series expansion of $\sinh^{-1}x$ in powers of x by

(a) reversion of the series for $\sinh y$,

(b) a direct Maclaurin expansion.

5.7.12   A function $f(z)$ is represented by a *descending* power series

$$f(z) = \sum_{n=0}^{\infty} a_n z^{-n}, \qquad R \le z < \infty.$$

Show that this series expansion is unique; that is, if $f(z) = \sum_{n=0}^{\infty} b_n z^{-n}$, $R \le z < \infty$, then $a_n = b_n$ for all n.

5.7.13   A power series given by

$$f(x) = \sum_{n=0}^{\infty} a_n x^n$$

converges for $-R < x < R$. Show that the differentiated series and the integrated series have the same interval of convergence. (Do not bother about the end points $x = \pm R$.)

5.7.14   Assuming that $f(x)$ may be expanded in a power series about the origin, $f(x) = \sum_{n=0}^{\infty} a_n x^n$, with some nonzero range of convergence. Use the techniques employed in proving uniqueness of series to show that your assumed series is a Maclaurin series:

$$a_n = \frac{1}{n!} f^{(n)}(0).$$

5.7.15   The Klein–Nishina formula for the scattering of photons by electrons contains a term of the form

$$f(\varepsilon) = \frac{(1 + \varepsilon)}{\varepsilon^2} \left[\frac{2 + 2\varepsilon}{1 + 2\varepsilon} - \frac{\ln(1 + 2\varepsilon)}{\varepsilon} \right].$$

Here $\varepsilon = h\nu/mc^2$, the ratio of the photon energy to the electron rest mass energy. Find

$$\lim_{\varepsilon \to 0} f(\varepsilon).$$

ANS. $\frac{4}{3}$.

5.7.16   The behavior of a neutron losing energy by colliding elastically with nuclei of mass A is described by a parameter ξ_1,

$$\xi_1 = 1 + \frac{(A - 1)^2}{2A} \ln \frac{A - 1}{A + 1}.$$

An approximation, good for large A, is

$$\xi_2 = \frac{2}{A + \frac{2}{3}}.$$

Expand ξ_1 and ξ_2 in powers of A^{-1}. Show that ξ_2 agrees with ξ_1 through $(A^{-1})^2$. Find the difference in the coefficients of the $(A^{-1})^3$ term.

5.7.17   Show that each of these two integrals equals Catalan's constant

(a) $\int_0^1 \arctan t \, \dfrac{dt}{t}$,

(b) $-\int_0^1 \ln x \, \dfrac{dx}{1 + x^2}$.

5.7.18   Calculate π (double precision) by each of the following arc tangent expressions:

$$\pi = 16\tan^{-1}(1/5) - 4\tan^{-1}(1/239)$$
$$\pi = 24\tan^{-1}(1/8) + 8\tan^{-1}(1/57) + 4\tan^{-1}(1/239)$$
$$\pi = 48\tan^{-1}(1/18) + 32\tan^{-1}(1/57) - 20\tan^{-1}(1/239).$$

You should obtain 16 significant figures.
Note. These formulas have been used in some of the more accurate calculations of π.[5]

5.7.19   An analysis of the Gibbs phenomenon of Section 14.5 leads to the expression

$$\frac{2}{\pi}\int_0^\pi \frac{\sin\xi}{\xi}\,d\xi.$$

(a) Expand the integrand in a series and integrate term by term. Find the numerical value of this expression to four significant figures.
(b) Evaluate this expression by the Gaussian quadrature (Appendix A2).
ANS.   1.178980.

5.8 ELLIPTIC INTEGRALS

Elliptic integrals are included here partly as an illustration of the use of power series and partly for their own intrinsic interest. This interest includes the occurrence of elliptic integrals in physical problems (Example 5.81 and Exercise 5.8.4) and applications in mathematical problems.

Example 5.8.1 Period of a Simple Pendulum

For small amplitude oscillations our pendulum (Fig. 5.8) has simple harmonic motion with a period $T = 2\pi(l/g)^{1/2}$. For a maximum amplitude θ_M large enough so that $\sin\theta_M \neq \theta_M$, Newton's second law of motion and Lagrange's equation (Section 17.7) lead to a nonlinear differential equation ($\sin\theta$ is a nonlinear function of θ), so we turn to a different approach.

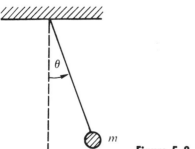

Figure 5.8 Simple pendulum.

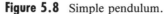

[5] D. Shanks and J. W. Wrench, Computation of π to 100,000 decimals. *Math. Comput.* **16**, 76 (1962).

The swinging mass m has a kinetic energy of $\frac{1}{2}ml^2(d\theta/dt)^2$ and a potential energy of $-mgl\cos\theta$ ($\theta = \frac{1}{2}\pi$ taken for the arbitrary zero of potential energy). Since $d\theta/dt = 0$ at $\theta = \theta_M$, the conservation of energy principle gives

$$\frac{1}{2}ml^2\left(\frac{d\theta}{dt}\right)^2 - mgl\cos\theta = -mgl\cos\theta_M. \tag{5.124}$$

Solving for $d\theta/dt$ we obtain

$$\frac{d\theta}{dt} = \pm\left(\frac{2g}{l}\right)^{1/2}(\cos\theta - \cos\theta_M)^{1/2} \tag{5.125}$$

with the mass m canceling out. We take t to be zero when $\theta = 0$ and $d\theta/dt > 0$. An integration from $\theta = 0$ to $\theta = \theta_M$ yields

$$\int_0^{\theta_M}(\cos\theta - \cos\theta_M)^{-1/2}\,d\theta = \left(\frac{2g}{l}\right)^{1/2}\int_0^t dt = \left(\frac{2g}{l}\right)^{1/2}t. \tag{5.126}$$

This is $\frac{1}{4}$ of a cycle, and therefore the time t is $\frac{1}{4}$ of the period, T. We note that $\theta \le \theta_M$, and with a bit of clairvoyance we try the half-angle substitution

$$\sin\left(\frac{\theta}{2}\right) = \sin\left(\frac{\theta_M}{2}\right)\sin\varphi. \tag{5.127}$$

With this, Eq. (5.126) becomes

$$T = 4\left(\frac{l}{g}\right)^{1/2}\int_0^{\pi/2}\left(1 - \sin^2\left(\frac{\theta_M}{2}\right)\sin^2\varphi\right)^{-1/2}d\varphi. \tag{5.128}$$

Although not an obvious improvement over Eq. (5.126), the integral now defines the complete elliptic integral of the first kind, $K(\sin\theta_M/2)$. From the series expansion, the period of our pendulum may be developed as a power series—powers of $\sin\theta_M/2$:

$$T = 2\pi\left(\frac{l}{g}\right)^{1/2}\left\{1 + \frac{1}{4}\sin^2\frac{\theta_M}{2} + \frac{9}{64}\sin^4\frac{\theta_M}{2} + \cdots\right\}. \tag{5.129}$$

DEFINITIONS

Generalizing Example 5.8.1 to include the upper limit as a variable, the *elliptic integral of the first kind* is defined as

$$F(\varphi\backslash\alpha) = \int_0^\varphi(1 - \sin^2\alpha\sin^2\theta)^{-1/2}\,d\theta \tag{5.130a}$$

or

$$F(x|m) = \int_0^x[(1 - t^2)(1 - mt^2)]^{-1/2}\,dt, \qquad 0 \le m < 1. \tag{5.130b}$$

(This is the notation of AMS-55.) For $\varphi = \pi/2$, $x = 1$, we have the *complete* elliptic integral of the first kind,

$$K(m) = \int_0^{\pi/2} (1 - m \sin^2\theta)^{-1/2} \, d\theta$$

$$= \int_0^1 [(1 - t^2)(1 - mt^2)]^{-1/2} \, dt, \tag{5.131}$$

with $m = \sin^2\alpha$, $0 \le m < 1$.

The *elliptic integral of the second kind* is defined by

$$E(\varphi\backslash\alpha) = \int_0^\varphi (1 - \sin^2\alpha \sin^2\theta)^{1/2} \, d\theta \tag{5.132a}$$

or

$$E(x|m) = \int_0^x \left(\frac{1 - mt^2}{1 - t^2}\right)^{1/2} dt, \qquad 0 \le m \le 1. \tag{5.132b}$$

Again, for the case $\varphi = \pi/2$, $x = 1$, we have the *complete elliptic integral of the second kind*:

$$E(m) = \int_0^{\pi/2} (1 - m \sin^2\theta)^{1/2} \, d\theta$$

$$= \int_0^1 \left(\frac{1 - mt^2}{1 - t^2}\right)^{1/2} dt, \qquad 0 \le m \le 1. \tag{5.133}$$

Exercise 5.8.1 is an example of its occurrence. Figure 5.9 shows the behavior of $K(m)$ and $E(m)$. Extensive tables are available in AMS-55.

SERIES EXPANSION

For our range $0 \le m < 1$, the denominator of $K(m)$ may be expanded by the binomial series

$$(1 - m \sin^2\theta)^{-1/2} = 1 + \frac{1}{2} m \sin^2\theta + \frac{3}{8} m^2 \sin^4\theta + \cdots$$

$$= \sum_{n=0}^\infty \frac{(2n - 1)!!}{(2n)!!} m^n \sin^{2n}\theta. \tag{5.134}$$

For any closed interval $[0, m_{max}]$, $m_{max} < 1$ this series is uniformly convergent and may be integrated term by term. From Exercise 10.4.9

$$\int_0^{\pi/2} \sin^{2n}\theta \, d\theta = \frac{(2n - 1)!!}{(2n)!!} \cdot \frac{\pi}{2}. \tag{5.135}$$

Hence

$$K(m) = \frac{\pi}{2}\left\{1 + \left(\frac{1}{2}\right)^2 m + \left(\frac{1 \cdot 3}{2 \cdot 4}\right)^2 m^2 + \left(\frac{1 \cdot 3 \cdot 5}{2 \cdot 4 \cdot 6}\right)^2 m^3 + \cdots\right\}. \tag{5.136}$$

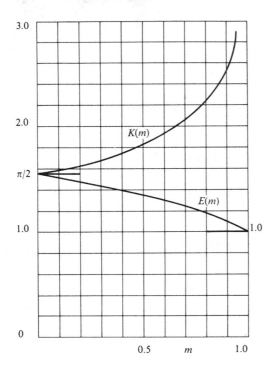

Figure 5.9 Complete elliptic integrals, $K(m)$ and $E(m)$.

Similarly,

$$E(m) = \frac{\pi}{2}\left\{1 - \left(\frac{1}{2}\right)^2\frac{m}{1} - \left(\frac{1 \cdot 3}{2 \cdot 4}\right)^2\frac{m^2}{3} - \left(\frac{1 \cdot 3 \cdot 5}{2 \cdot 4 \cdot 6}\right)^2\frac{m^3}{5} - \cdots\right\}. \qquad (5.137)$$

(Exercise 5.8.2). In Section 13.5 these series are identified as hypergeometric functions, and we have

$$K(m) = \frac{\pi}{2}\,_2F_1(\tfrac{1}{2}, \tfrac{1}{2}, 1; m) \qquad (5.138)$$

$$E(m) = \frac{\pi}{2}\,_2F_1(-\tfrac{1}{2}, \tfrac{1}{2}, 1; m). \qquad (5.139)$$

LIMITING VALUES

From the series Eqs. (5.136) and (5.137), or from the defining integrals,

$$\lim_{m \to 0} K(m) = \frac{\pi}{2}, \qquad (5.140)$$

$$\lim_{m \to 0} E(m) = \frac{\pi}{2}. \qquad (5.141)$$

For $m \to 1$ the series expansions are of little use. However, the integrals yield

$$\lim_{m \to 1} K(m) = \infty, \qquad (5.142)$$

the integral diverging logarithmically, and

$$\lim_{m \to 1} E(m) = 1. \qquad (5.143)$$

The elliptic integrals have been used extensively in the past for evaluating integrals. For instance, integrals of the form

$$I = \int_0^x R(t, \sqrt{a_4 t^4 + a_3 t^3 + a_2 t^2 + a_1 t + a_0}) \, dt,$$

where R is a rational function of t and of the radical, may be expressed in terms of elliptic integrals. Jahnke and Emde, Chapter 5, give pages of such transformations. With high-speed computers available for direct numerical evaluation, interest in these elliptic integral techniques has declined. However, elliptic integrals still remain of interest because of their appearance in physical problems—Exercises 5.8.4 and 5.8.5.

EXERCISES

5.8.1 The ellipse $x^2/a^2 + y^2/b^2 = 1$ may be represented parametrically by $x = a \sin \theta$, $y = b \cos \theta$. Show that the length of arc within the first quadrant is

$$a \int_0^{\pi/2} (1 - m \sin^2\theta)^{1/2} \, d\theta = aE(m).$$

Here $0 \le m = (a^2 - b^2)/a^2 \le 1$.

5.8.2 Derive the series expansion

$$E(m) = \frac{\pi}{2} \left\{ 1 - \left(\frac{1}{1}\right)^2 \frac{m}{1} - \left(\frac{1 \cdot 3}{2 \cdot 4}\right)^2 \frac{m^2}{3} - \cdots \right\}$$

$$= \frac{\pi}{2} \left\{ 1 - \sum_{n=1}^{\infty} \left[\frac{(2n - 1)!!}{(2n)!!} \right]^2 \frac{m^n}{(2n - 1)} \right\}.$$

5.8.3 Show that

$$\lim_{m \to 0} \frac{(K - E)}{m} = \frac{\pi}{4}.$$

5.8.4 A circular loop of wire in the xy-plane, as shown, carries a current I. Given that the vector potential is

$$A_\varphi(\rho, \varphi, z) = \frac{a\mu_0 I}{2\pi} \int_0^{\pi} \frac{\cos \alpha \, d\alpha}{(a^2 + \rho^2 + z^2 - 2a\rho \cos \alpha)^{1/2}},$$

show that

$$A_\varphi(\rho, \varphi, z) = \frac{\mu_0 I}{\pi k} \left(\frac{a}{\rho}\right)^{1/2} \left[\left(1 - \frac{k^2}{2}\right) K - E \right],$$

where

$$k^2 = \frac{4a\rho}{(a + \rho)^2 + z^2}.$$

Note. For extension of Exercise 5.8.4 to **B** see Smythe, p. 270.[1]

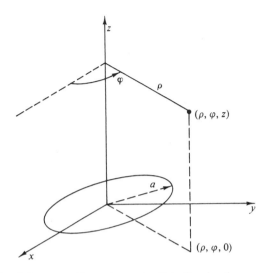

5.8.5.   An analysis of the magnetic vector potential of a circular current loop leads to the expression

$$f(k^2) = k^{-2}[(2 - k^2)K(k^2) - 2E(k^2)],$$

where $K(k^2)$ and $E(k^2)$ are the complete elliptic integrals of the first and second kinds. Show that for $k^2 \ll 1$ ($r \gg$ radius of loop)

$$f(k^2) \approx \frac{\pi k^2}{16}.$$

5.8.6   Show that

(a) $\dfrac{dE(k^2)}{dk} = \dfrac{1}{k}(E - K),$

(b) $\dfrac{dK(k^2)}{dk} = \dfrac{E}{k(1 - k^2)} - \dfrac{K}{k}.$

Hint. For part (b) show that

$$E(k^2) = (1 - k^2) \int_0^{\pi/2} (1 - k \sin^2 \theta)^{-3/2} \, d\theta$$

by comparing series expansions.

5.8.7   (a) Write a function subroutine that will compute $E(m)$ from the series expansion, Eq. (5.137).

(b) Test your function subroutine by using it to calculate $E(m)$ over the range $m = 0.0(0.1)0.9$ and comparing the result with the values given by AMS-55.

[1] W. R. Smythe, *Static and Dynamic Electricity*, 3rd ed. New York: McGraw-Hill (1969).

5.8.8   Repeat Exercise 5.8.7 for $K(m)$, written out as in Exercise 5.8.7.
Note. These series for $E(m)$, Eq. (5.137), and $K(m)$, Eq. (5.136), converge
only very slowly for m near 1. More rapidly converging series for $E(m)$ and
$K(m)$ exist. See Dwight's Tables of Integrals:[2] No. 773.2 and 774.2. Your
computer subroutine for computing E and K probably uses polynomial
approximations: AMS-55, Chapter 17.

5.8.9   A simple pendulum is swinging with a maximum amplitude of θ_M. In the
limit as $\theta_M \to 0$, the period is 1 sec. Using the elliptic integral, $K(k^2)$,
$k = \sin(\theta_M/2)$ calculate the period T for $\theta_M = 0$ (10°) 90°.
Caution. Some elliptic integral subroutines require $k = m^{1/2}$ as an input
parameter, not m itself. *Check values.*

θ_M	10°	50°	90°
T (sec)	1.00193	1.05033	1.18258

5.8.10   Calculate the magnetic vector potential $\mathbf{A}(\rho, \varphi, z) = \hat{\varphi} A_\varphi(\rho, \varphi, z)$ of a
circular current loop (Exercise 5.8.4) for the ranges $\rho/a = 2, 3, 4$, and
$z/a = 0, 1, 2, 3, 4$.
Note. This elliptic integral calculation of the magnetic vector potential may
be checked by an associated Legendre function calculation, Example 12.5.1.
       Check value.   For $\rho/a = 3$ and $z/a = 0$; $A_\varphi = 0.029023 \mu_0 I$.

5.9 BERNOULLI NUMBERS, EULER–MACLAURIN FORMULA

The Bernoulli numbers were introduced by Jacques (James, Jacob) Bernoulli.
There are several equivalent definitions, but extreme care must be taken, for
some authors introduce variations in numbering or in algebraic signs. One
relatively simple approach is to define the Bernoulli numbers by the series[1]

$$\frac{x}{e^x - 1} = \sum_{n=0}^{\infty} \frac{B_n x^n}{n!}, \tag{5.144}$$

which converges for $|x| < 2\pi$ by the ratio test using Eq. (5.153) (see also
Example 7.2.6). By differentiating this power series repeatedly and then setting
$x = 0$, we obtain

$$B_n = \left[\frac{d^n}{dx^n} \left(\frac{x}{e^x - 1} \right) \right]_{x=0}. \tag{5.145}$$

Specifically,

$$B_1 = \frac{d}{dx} \left(\frac{x}{e^x - 1} \right) \bigg|_{x=0} = \frac{1}{e^x - 1} - \frac{xe^x}{(e^x - 1)^2} \bigg|_{x=0}$$

$$= -\frac{1}{2}, \tag{5.146}$$

---

[2] H. B. Dwight, *Tables of Integrals and Other Mathematical Data.* New York: Macmillan
(1947).
[1] The function $x/(e^x - 1)$ may be considered a *generating function* since it generates the
Bernoulli numbers. Generating functions that generate the special functions of mathematical
physics appear in Chapters 11, 12, and 13.

as may be seen by series expansion of the denominators. Using $B_0 = 1$ and $B_1 = -\frac{1}{2}$, it is easy to verify that the function

$$\frac{x}{e^x - 1} - 1 + x/2 = \sum_{n=2}^{\infty} B_n x^n/n! = -x/(e^{-x} - 1) - 1 - x/2 \quad (5.147)$$

is even in x, so that all $B_{2n+1} = 0$.

To derive a recursion relation for the Bernoulli numbers, we multiply

$$\frac{e^x - 1}{x} \cdot \frac{x}{e^x - 1} = 1 = \left\{ \sum_{m=0}^{\infty} x^m/(m+1)! \right\} \left\{ 1 - x/2 + \sum_{n=1}^{\infty} B_{2n} x^{2n}/(2n)! \right\}$$

$$= 1 + \sum_{m=1}^{\infty} x^m \{1/(m+1)! - 1/m!/2\}$$

$$+ \sum_{N=2}^{\infty} x^N \sum_{1 \le n \le N/2} B_{2n}/[(2n)!(N - 2n + 1)!]. \quad (5.148)$$

Equation (5.148) yields

$$\tfrac{1}{2}(N+1) - 1 = \sum_{1 \le n \le N/2} B_{2n} \binom{N+1}{2n} = \tfrac{1}{2}(N-1), \quad (5.149)$$

which is equivalent to

$$N - 1/2 = \sum_{n=1}^{N} B_{2n} \binom{2N+1}{2n},$$

$$\quad (5.150)$$

$$N - 1 = \sum_{n=1}^{N-1} B_{2n} \binom{2N}{2n}.$$

From Eq. (5.150) the Bernoulli numbers in Table 5.1 are readily obtained. If the variable x in Eq. (5.144) is replaced by $2ix$ (and B_1 set equal to $-\frac{1}{2}$), we obtain an alternate (and equivalent) definition of B_{2n} by the expression

$$x \cot x = \sum_{n=0}^{\infty} (-1)^n B_{2n} \frac{(2x)^{2n}}{(2n)!}, \quad -\pi < x < \pi. \quad (5.151)$$

Table 5.1 Bernoulli Numbers

n	B_n	B_n
0	1	1.0000 00000
1	$-\frac{1}{2}$	$-0.5000\,00000$
2	$\frac{1}{6}$	0.1666 66667
4	$-\frac{1}{30}$	$-0.0333\,33333$
6	$\frac{1}{42}$	0.0238 09524
8	$-\frac{1}{30}$	$-0.0333\,33333$
10	$\frac{5}{66}$	0.0757 57576

Note. Further values are given in National Bureau of Standards, *Handbook of Mathematical Functions* (AMS-55).

Using the method of residues (Section 7.2) or working from the infinite product representation of $\sin x$ (Section 5.11), we find that

$$B_{2n} = \frac{(-1)^{n-1}2(2n)!}{(2\pi)^{2n}} \sum_{p=1}^{\infty} p^{-2n}, \qquad n = 1, 2, 3, \ldots. \qquad (5.152)$$

This representation of the Bernoulli numbers was discovered by Euler. It is readily seen from Eq. (5.152) that $|B_{2n}|$ increases without limit as $n \to \infty$. Numerical values have been calculated by Glaisher.[2] Illustrating the divergent behavior of the Bernoulli numbers, we have

$$B_{20} = -5.291 \times 10^2$$

$$B_{200} = -3.647 \times 10^{215}.$$

Some authors prefer to define the Bernoulli numbers with a modified version of Eq. (5.152) by using

$$B_n = \frac{2(2n)!}{(2\pi)^{2n}} \sum_{p=1}^{\infty} p^{-2n}, \qquad (5.153)$$

the subscript being just half of our subscript and all signs are positive. Again, when using other texts or references the reader must check carefully to see exactly how the Bernoulli numbers are defined.

The Bernoulli numbers occur frequently in number theory. The von Standt–Clausen theorem states that

$$B_{2n} = A_n - \frac{1}{p_1} - \frac{1}{p_2} - \frac{1}{p_3} - \cdots - \frac{1}{p_k}, \qquad (5.154)$$

in which A_n is an integer and $p_1, p_2, \ldots, p_k$ are prime numbers so that $p_i - 1$ is a divisor of $2n$. It may readily be verified that this holds for

$$B_6(A_3 = 1, \quad p = 2, 3, 7),$$

$$B_8(A_4 = 1, \quad p = 2, 3, 5), \qquad (5.155)$$

$$B_{10}(A_5 = 1, \quad p = 2, 3, 11),$$

and other special cases.

The Bernoulli numbers appear in the summation of integral powers of the integers,

$$\sum_{j=1}^{N} j^p, \qquad p \text{ integral,}$$

and in numerous series expansions of the transcendental functions, including $\tan x$, $\cot x$, $(\sin x)^{-1}$, $\ln|\sin x|$, $\ln|\cos x|$, $\ln|\tan x|$, $\tanh x$, $\coth x$, and $(\cosh x)^{-1}$.

---

[2] J. W. L. Glaisher, Table of the first 250 Bernoulli's numbers (to nine figures) and their logarithms (to ten figures). *Trans. Cambridge Philos. Soc.* **12**, 390 (1871–1879).

For example,

$$\tan x = x + \frac{x^3}{3} + \frac{2}{15}x^5 + \cdots + \frac{(-1)^{n-1}2^{2n}(2^{2n} - 1)B_{2n}}{(2n)!}x^{2n-1} + \cdots .$$

(5.156)

The Bernoulli numbers are likely to come in such series expansions because of the defining equations (5.144) and (5.150) and because of their relation to the Riemann zeta function

$$\zeta(2n) = \sum_{p=1}^{\infty} p^{-2n}.$$

(5.157)

BERNOULLI FUNCTIONS

If Eq. (5.144) is generalized slightly, we have

$$\frac{xe^{xs}}{e^x - 1} = \sum_{n=0}^{\infty} B_n(s)\frac{x^n}{n!}$$

(5.158)

defining the *Bernoulli functions*, $B_n(s)$. The first seven Bernoulli functions are given in Table 5.2.

From the generating function, Eq. (5.158),

$$B_n(0) = B_n, \qquad n = 0, 1, 2, \ldots,$$

(5.159)

the Bernoulli function evaluated at zero equals the corresponding Bernoulli number. Two particularly important properties of the Bernoulli functions follow from the defining relation: a differentiation relation

$$B_n'(s) = nB_{n-1}(s), \qquad n = 1, 2, 3, \ldots,$$

(5.160)

and a symmetry relation

$$B_n(1) = (-1)^n B_n(0), \qquad n = 0, 1, 2, \ldots .$$

(5.161)

These relations are used in the development of the Euler–Maclaurin integration formula.

Table 5.2   Bernoulli Functions

$B_0 = 1$
$B_1 = x - \frac{1}{2}$
$B_2 = x^2 - x + \frac{1}{6}$
$B_3 = x^3 - \frac{3}{2}x^2 + \frac{1}{2}x$
$B_4 = x^4 - 2x^3 + x^2 - \frac{1}{30}$
$B_5 = x^5 - \frac{5}{2}x^4 + \frac{5}{3}x^3 - \frac{1}{6}x$
$B_6 = x^6 - 3x^5 + \frac{5}{2}x^4 - \frac{1}{2}x^2 + \frac{1}{42}$
$B_n(0) = B_n, \qquad$ Bernoulli number

EULER–MACLAURIN INTEGRATION FORMULA

One use of the Bernoulli functions is in the derivation of the Euler–Maclaurin integration formula. This formula is used in Section 10.3 for the development of an asymptotic expression for the factorial function—Stirling's series.

The technique is repeated integration by parts using Eq. (5.160) to create new derivatives. We start with

$$\int_0^1 f(x)\,dx = \int_0^1 f(x)B_0(x)\,dx. \tag{5.162}$$

From Eq. (5.160) and Exercise 5.9.2

$$B_1'(x) = B_0(x) = 1. \tag{5.163}$$

Substituting $B_1'(x)$ into eq. (5.162) and integrating by parts, we obtain

$$\int_0^1 f(x)\,dx = f(1)B_1(1) - f(0)B_1(0) - \int_0^1 f'(x)B_1(x)\,dx$$

$$= \frac{1}{2}[f(1) + f(0)] - \int_0^1 f'(x)B_1(x)\,dx. \tag{5.164}$$

Again, using Eq. (5.160), we have

$$B_1(x) = \tfrac{1}{2}B_2'(x), \tag{5.165}$$

and integrating by parts

$$\int_0^1 f(x)\,dx = \frac{1}{2}[f(1) + f(0)] - \frac{1}{2!}[f'(1)B_2(1) - f'(0)B_2(0)]$$

$$+ \frac{1}{2!}\int_0^1 f^{(2)}(x)B_2(x)\,dx. \tag{5.166}$$

Using the relations,

$$B_{2n}(1) = B_{2n}(0) = B_{2n}, \qquad n = 0, 1, 2, \ldots$$

$$B_{2n+1}(1) = B_{2n+1}(0) = 0, \qquad n = 1, 2, 3, \ldots \tag{5.167}$$

and continuing this process, we have

$$\int_0^1 f(x)\,dx = \frac{1}{2}[f(1) + f(0)] - \sum_{p=1}^{q} \frac{1}{(2p)!} B_{2p}[f^{(2p-1)}(1) - f^{(2p-1)}(0)]$$

$$+ \frac{1}{(2q)!}\int_0^1 f^{(2q)}(x)B_{2q}(x)\,dx. \tag{5.168a}$$

This is the Euler–Maclaurin integration formula. It assumes that the function $f(x)$ has the required derivatives.

Table 5.3 Riemann Zeta Function

s	$\zeta(s)$
2	1.64493 40668
3	1.20205 69032
4	1.08232 32337
5	1.03692 77551
6	1.01734 30620
7	1.00834 92774
8	1.00407 73562
9	1.00200 83928
10	1.00099 45751

The range of integration in Eq. (5.168a) may be shifted from $[0, 1]$ to $[1, 2]$ by replacing $f(x)$ by $f(x + 1)$. Adding such results up to $[n - 1, n]$,

$$\int_0^n f(x)\,dx = \tfrac{1}{2}f(0) + f(1) + f(2) + \cdots + f(n - 1) + \tfrac{1}{2}f(n)$$

$$- \sum_{p=1}^{q} \frac{1}{(2p)!} B_{2p}[f^{(2p-1)}(n) - f^{(2p-1)}(0)]$$

$$+ \frac{1}{(2q)!} \int_0^1 B_{2q}(x) \sum_{v=0}^{n-1} f^{(2q)}(x + v)\,dx. \qquad (5.168b)$$

The terms $\tfrac{1}{2}f(0) + f(1) + \cdots + \tfrac{1}{2}f(n)$ appear exactly as in trapezoidal integration or quadrature. The summation over p may be interpreted as a correction to the trapezoidal approximation. Equation (5.168b) is the form used in Exercise 5.9.5 for summing positive powers of integers and in Section 10.3 for the derivation of Stirling's formula.

The Euler–Maclaurin formula is often useful in summing series by converting them to integrals.[3]

RIEMANN ZETA FUNCTION

This series $\sum_{p=1}^{\infty} p^{-2n}$ was used as a comparison series for testing convergence (Section 5.2) and in Eq. (5.151) as one definition of the Bernoulli numbers, B_{2n}. It also serves to define the Riemann zeta function by

$$\zeta(s) \equiv \sum_{n=1}^{\infty} n^{-s}, \qquad s > 1. \qquad (5.169)$$

Table 5.3 lists the values of $\zeta(s)$ for integral s, $s = 2, 3, \ldots, 10$. Closed forms for even s appear in Exercise 5.9.6. Figure 5.10 is a plot of $\zeta(s) - 1$. An

---

[3]Compare R. P. Boas and C. Stutz, Estimating sums with integrals. *Am. J. Phys.* **39**, 745 (1971), for a number of examples.

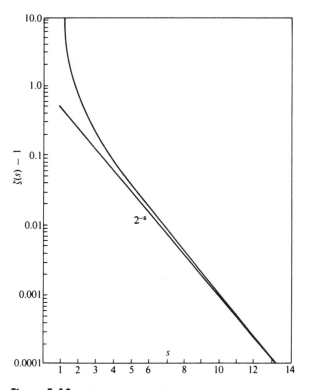

Figure 5.10   Riemann zeta function, $\zeta(s) - 1$ versus s.

integral expression for this Riemann zeta function appears in Section 10.2 as part of the development of the gamma function.

Another interesting expression for the Riemann zeta function may be derived as

$$\zeta(s)(1 - 2^{-s}) = 1 + \frac{1}{2^s} + \frac{1}{3^s} + \cdots - \left(\frac{1}{2^s} + \frac{1}{4^s} + \frac{1}{6^s} + \cdots\right), \qquad (5.170)$$

eliminating all the n^{-s}, where n is a multiple of 2. Then

$$\zeta(s)(1 - 2^{-s})(1 - 3^{-s}) = 1 + \frac{1}{3^s} + \frac{1}{5^s} + \frac{1}{7^s} + \frac{1}{9^s} + \cdots$$

$$- \left(\frac{1}{3^s} + \frac{1}{9^s} + \frac{1}{15^s} + \cdots\right), \qquad (5.171)$$

eliminating all the remaining terms in which n is a multiple of 3. Continuing, we have $\zeta(s)(1 - 2^{-s})(1 - 3^{-s})(1 - 5^{-s}) \cdots (1 - P^{-s})$, where P is a prime number, and all terms n^{-s}, in which n is a multiple of any integer up through P, are canceled out. As $P \to \infty$,

$$\zeta(s)(1 - 2^{-s})(1 - 3^{-s}) \cdots (1 - P^{-s}) = \zeta(s) \prod_{P(\text{prime}) = 2}^{\infty} (1 - P^{-s}) = 1. \quad (5.172)$$

Therefore

$$\zeta(s) = \left[\prod_{P(\text{prime}) = 2}^{\infty} (1 - P^{-s}) \right]^{-1}, \quad (5.173)$$

giving $\zeta(s)$ as an infinite product.[4]

This cancellation procedure has a clear application in numerical computation. Equation (5.170) will give $\zeta(s)(1 - 2^{-s})$ to the same accuracy as
Eq. (5.169) gives $\zeta(s)$, but with only half as many terms. (In either case, a
correction would be made for the neglected tail of the series by the Maclaurin
integral test technique—replacing the series by an integral, Section 5.2.)

Along with the Riemann zeta function, AMS-55 (Chapter 23) defines three
other functions of sums of reciprocal powers:

$$\eta(s) = \sum_{n = 1}^{\infty} (-1)^{n-1} n^{-s} = (1 - 2^{1-s})\zeta(s),$$

$$\lambda(s) = \sum_{n = 0}^{\infty} (2n + 1)^{-s} = (1 - 2^{-s})\zeta(s),$$

and

$$\beta(s) = \sum_{n = 0}^{\infty} (-1)^n (2n + 1)^{-s}.$$

From the Bernoulli numbers (Exercise 5.9.6) or Fourier series (Example 14.3.3
and Exercise 14.3.13) special values are

$$\zeta(2) = 1 + \frac{1}{2^2} + \frac{1}{3^2} + \cdots = \frac{\pi^2}{6}$$

$$\zeta(4) = 1 + \frac{1}{2^4} + \frac{1}{3^4} + \cdots = \frac{\pi^4}{90}$$

$$\eta(2) = 1 - \frac{1}{2^2} + \frac{1}{3^2} - \cdots = \frac{\pi^2}{12}$$

$$\eta(4) = 1 - \frac{1}{2^4} + \frac{1}{3^4} - \cdots = \frac{7\pi^4}{720}$$

[4]This is the starting point for the extensive applications of the Riemann zeta function to analytic
number theory. See H. M. Edwards, *Riemann's Zeta Function*. New York: Academic Press (1974);
A. Ivić, *The Riemann Zeta Function*. New York: Wiley (1985); S. J. Patterson, *Introduction to the
Theory of the Riemann Zeta Function*. Cambridge: Cambridge University Press (1988).

$$\lambda(2) = 1 + \frac{1}{3^2} + \frac{1}{5^2} + \cdots = \frac{\pi^2}{8}$$

$$\lambda(4) = 1 + \frac{1}{3^4} + \frac{1}{5^4} + \cdots = \frac{\pi^4}{96}$$

$$\beta(1) = 1 - \frac{1}{3} + \frac{1}{5} - \cdots = \frac{\pi}{4}$$

$$\beta(3) = 1 - \frac{1}{3^3} + \frac{1}{5^3} - \cdots = \frac{\pi^3}{32}$$

Catalan's constant,

$$\beta(2) = 1 - \frac{1}{3^2} + \frac{1}{5^2} - \cdots = 0.9159\,6559\ldots,$$

is the topic of Exercise 5.2.22.

IMPROVEMENT OF CONVERGENCE

If we are required to sum a convergent series $\sum_{n=1}^{\infty} a_n$ whose terms are rational functions of n, the convergence may be improved dramatically by introducing the Riemann zeta function.

Example 5.9.1   Improvement of Convergence

The problem is to evaluate the series $\sum_{n=1}^{\infty} 1/(1+n^2)$. Expanding $(1+n^2)^{-1} = n^{-2}(1+n^{-2})^{-1}$ by direct division, we have

$$(1 + n^2)^{-1} = n^{-2}\left(1 - n^{-2} + n^{-4} - \frac{n^{-6}}{1 + n^{-2}}\right)$$

$$= \frac{1}{n^2} - \frac{1}{n^4} + \frac{1}{n^6} - \frac{1}{n^8 + n^6}.$$

Therefore

$$\sum_{n=1}^{\infty} \frac{1}{1 + n^2} = \zeta(2) - \zeta(4) + \zeta(6) - \sum_{n=1}^{\infty} \frac{1}{n^8 + n^6}.$$

The ζ functions are tabulated and the remainder series converges as n^{-8}. Clearly, the process can be continued as desired. You make a choice between how much algebra you will do and how much arithmetic the computing machine will do.

Other methods for improving computational effectiveness are given at the end of Sections 5.2 and 5.4.

EXERCISES

5.9.1   Show that

$$\tan x = \sum_{n=1}^{\infty} \frac{(-1)^{n-1} 2^{2n}(2^{2n} - 1)B_{2n}}{(2n)!} x^{2n-1}, \qquad -\frac{\pi}{2} < x < \frac{\pi}{2}.$$

Hint. $\tan x = \cot x - 2 \cot 2x$.

5.9.2   The Bernoulli numbers generated in Eq. (5.144) may be generalized to Bernoulli polynomials,

$$\frac{xe^{xs}}{e^x - 1} = \sum_{n=0}^{\infty} B_n(s) \frac{x^n}{n!}.$$

Show that

$$B_0(s) = 1$$
$$B_1(s) = s - \tfrac{1}{2}$$
$$B_2(s) = s^2 - s + \tfrac{1}{6}.$$

Note that $B_n(0) = B_n$, the Bernoulli number.

5.9.3   Show that $B_n'(s) = nB_{n-1}(s), \quad n = 1, 2, 3, \dots$.
Hint. Differentiate the equation in Exercise 5.9.2.

5.9.4   Show that

$$B_n(1) = (-1)^n B_n(0).$$

Hint. Go back to the generating function, Eq. (5.158) or Exercise 5.9.2.

5.9.5   The Euler–Maclaurin integration formula may be used for the evaluation of finite series:

$$\sum_{m=1}^{n} f(m) = \int_0^n f(x)\, dx + \frac{1}{2} f(1) + \frac{1}{2} f(n) + \frac{B_2}{2!} [f'(n) - f'(1)] + \cdots.$$

Show that

(a) $\displaystyle\sum_{m=1}^{n} m = \tfrac{1}{2} n(n + 1)$.

(b) $\displaystyle\sum_{m=1}^{n} m^2 = \tfrac{1}{6} n(n + 1)(2n + 1)$.

(c) $\displaystyle\sum_{m=1}^{n} m^3 = \tfrac{1}{4} n^2(n + 1)^2$.

(d) $\displaystyle\sum_{m=1}^{n} m^4 = \tfrac{1}{30} n(n + 1)(2n + 1)(3n^2 + 3n - 1)$.

5.9.6   From

$$B_{2n} = (-1)^{n-1} \frac{2(2n)!}{(2\pi)^{2n}} \zeta(2n),$$

show that

(a) $\zeta(2) = \dfrac{\pi^2}{6}$

(d) $\zeta(8) = \dfrac{\pi^8}{9450}$

(b) $\zeta(4) = \dfrac{\pi^4}{90}$

(e) $\zeta(10) = \dfrac{\pi^{10}}{93,555}$.

(c) $\zeta(6) = \dfrac{\pi^6}{945}$

5.9.7   Planck's black-body radiation law involves the integral

$$\int_0^\infty \frac{x^3\,dx}{e^x - 1}.$$

Show that this equals $6\,\zeta(4)$. From Exercise 5.9.6

$$\zeta(4) = \frac{\pi^4}{90}.$$

Hint. Make use of the gamma function, Chapter 10.

5.9.8   Prove that

$$\int_0^\infty \frac{x^n e^x\,dx}{(e^x - 1)^2} = n!\,\zeta(n).$$

Assuming n to be real, show that each side of the equation diverges if $n = 1$. Hence the preceding equation carries the condition $n > 1$. Integrals such as this appear in the quantum theory of transport effects—thermal and electrical conductivity.

5.9.9   The Bloch–Gruneissen approximation for the resistance in a monovalent metal is

$$\rho = C\frac{T^5}{\Theta^6}\int_0^{\Theta/T} \frac{x^5\,dx}{(e^x - 1)(1 - e^{-x})},$$

where Θ is the Debye temperature characteristic of the metal.

(a) For $T \to \infty$ show that

$$\rho \approx \frac{C}{4}\cdot\frac{T}{\Theta^2}.$$

(b) For $T \to 0$, show that

$$\rho \approx 5!\,\zeta(5)C\,\frac{T^5}{\Theta^6}.$$

5.9.10   Show that

(a) $\displaystyle\int_0^1 \frac{\ln(1 + x)}{x}\,dx = \frac{1}{2}\,\zeta(2),$

(b) $\displaystyle\lim_{a \to 1}\int_0^a \frac{\ln(1 - x)}{x}\,dx = \zeta(2).$

From Exercise 5.9.6, $\zeta(2) = \pi^2/6$. Note that the integrand in part (b) diverges for $a = 1$ but that the *integrated* series is convergent.

5.9.11   The integral

$$\int_0^1 [\ln(1-x)]^2 \frac{dx}{x}$$

appears in the fourth-order correction to the magnetic moment of the electron. Show that it equals 2 $\zeta(3)$.

Hint. Let $1 - x = e^{-t}$.

5.9.12   Show that

$$\int_0^\infty \frac{(\ln z)^2}{1+z^2}\, dz = 4\left(1 - \frac{1}{3^3} + \frac{1}{5^3} - \frac{1}{7^3} + \cdots\right).$$

By contour integration (Exercises 7.2.17), this may be shown equal to $\pi^3/8$.

5.9.13   For "small" values of x

$$\ln(x!) = -\gamma x + \sum_{n=2}^\infty (-1)^n \frac{\zeta(n)}{n} x^n,$$

where γ is the Euler–Mascheroni constant and $\zeta(n)$ the Riemann zeta function. For what values of x does this series converge?

                                                            ANS. $-1 < x \le 1$.

Note that if $x = 1$, we obtain

$$\gamma = \sum_{n=2}^\infty (-1)^n \frac{\zeta(n)}{n},$$

a series for the Euler–Mascheroni constant,. The convergence of this series is exceedingly slow. For actual computation of γ, other, indirect approaches are far superior (see Exercises 5.9.17, 5.10,11, and 10.5.16).

5.9.14   Show that the series expansion of $\ln(x!)$ (Exercise 5.9.13) may be written as

(a) $\ln(x!) = \dfrac{1}{2}\ln\left(\dfrac{\pi x}{\sin \pi x}\right) - \gamma x - \displaystyle\sum_{n=1}^\infty \dfrac{\zeta(2n+1)}{2n+1} x^{2n+1}$,

(b) $\ln(x!) = \dfrac{1}{2}\ln\left(\dfrac{\pi x}{\sin \pi x}\right) - \dfrac{1}{2}\ln\left(\dfrac{1+x}{1-x}\right) + (1-\gamma)x$

    $- \displaystyle\sum_{n=1}^\infty [\zeta(2n+1) - 1]\dfrac{x^{2n+1}}{2n+1}$.

Determine the range of convergence of each of these expressions.

5.9.15   Show that Catalan's constant, $\beta(2)$, may be written as

$$\beta(2) = 2 \sum_{k=1}^\infty (4k-3)^{-2} - \frac{\pi^2}{8}.$$

Hint. $\pi^2 = 6\zeta(2)$.

5.9.16   Derive the following expansions of the Debye functions

(a) $\displaystyle\int_0^x \frac{t^n\, dt}{e^t - 1} = x^n\left[\frac{1}{n} - \frac{x}{2(n+1)} + \sum_{k=1}^\infty \frac{B_{2k}x^{2k}}{(2k+n)(2k)!}\right],$ $|x| < 2\pi, n \ge 1,$

(b) $\displaystyle\int_x^\infty \frac{t^n\, dt}{e^t - 1} = \sum_{k=1}^\infty e^{-kx}\left[\frac{x^n}{k} + \frac{nx^{n-1}}{k^2} + \frac{n(n-1)x^{n-2}}{k^3} + \cdots + \frac{n!}{k^{n+1}}\right],$

                                                            $x > 0, n \ge 1.$

The complete integral $(0, \infty)$ equals $n!\, \zeta(n+1)$, Exercise 10.2.15.

5.9.17   Derive the following Bernoulli number series for the Euler–Mascheroni constant.

$$\gamma = \sum_{s=1}^{n} s^{-1} - \ln n - \frac{1}{2n} + \sum_{k=1}^{\infty} \frac{B_{2k}}{(2k)n^{2k}}.$$

Hint. Apply the Euler–Maclaurin integration formula to $f(x) = x^{-1}$ over the range $[n, N]$.

5.9.18   (a) Show that the equation $\ln 2 = \sum_{s=1}^{\infty} (-1)^{s+1} s^{-1}$ (Exercise 5.4.1) may be rewritten as

$$\ln 2 = \sum_{s=2}^{\infty} 2^{-s}\zeta(s) + \sum_{p=1}^{\infty} (2p)^{-n-1}\left[1 - \frac{1}{2p}\right]^{-1}.$$

Hint. Take the terms in pairs.
(b) Calculate $\ln 2$ to six significant figures.

5.9.19   (a) Show that the equation $\pi/4 = \sum_{n=1}^{\infty} (-1)^{n+1}(2n-1)^{-1}$ (Exercise 5.7.6) may be rewritten as

$$\frac{\pi}{4} = 1 - 2\sum_{s=1}^{n} 4^{-2s}\zeta(2s)$$

$$- 2\sum_{p=1}^{\infty} (4p)^{-2n-2}\left[1 - \frac{1}{(4p)^2}\right]^{-1}.$$

(b) Calculate $\pi/4$ to six significant figures.

5.9.20   Write a function subprogram ZETA(N) that will calculate the Riemann zeta function for integer argument. Tabulate $\zeta(s)$ for $s = 2, 3, 4, \ldots, 20$. Check your values against Table 5.3 and AMS-55, Chapter 23.
Hint. If you simply supply the function subprogram with the values of $\zeta(2)$, $\zeta(3)$, and $\zeta(4)$, you avoid the more slowly converging series. Calculation time may be further shortened by using Eq. (5.170).

5.9.21   Calculate the logarithm (base 10) of $|B_{2n}|$, $n = 10, 20, \ldots, 100$.
Hint. Program the zeta function as a function subprogram, Exercise 5.9.20.
    Check values. $\log|B_{100}| = 78.45$
                       $\log|B_{200}| = 215.56.$

5.10 ASYMPTOTIC OR SEMICONVERGENT SERIES

   Asymptotic series frequently occur in physics. In numerical computations they are employed for the accurate computation of a variety of functions. We consider here two types of integrals that lead to asymptotic series: first, an integral of the form

$$I_1(x) = \int_x^{\infty} e^{-u} f(u)\, du,$$

where the variable x appears as the lower limit of an integral. Second, we consider the form

$$I_2(x) = \int_0^\infty e^{-u} f\left(\frac{u}{x}\right) du,$$

with the function f to be expanded as a Taylor series (binomial series). Asymptotic series often occur as solutions of differential equations. An example of this appears in Section 11.6 as a solution of Bessel's equation.

INCOMPLETE GAMMA FUNCTION

The nature of an asymptotic series is perhaps best illustrated by a specific example. Suppose that we have the exponential integral function[1]

$$Ei(x) = \int_{-\infty}^x \frac{e^u}{u} du, \tag{5.174}$$

or

$$-Ei(-x) = \int_x^\infty \frac{e^{-u}}{u} du = E_1(x), \tag{5.175}$$

to be evaluated for large values of x. Better still, let us take a generalization of the incomplete factorial function (incomplete gamma function),[2]

$$I(x, p) = \int_x^\infty e^{-u} u^{-p} du = \Gamma(1 - p, x), \tag{5.176}$$

in which x and p are positive. Again, we seek to evaluate it for large values of x.
Integrating by parts, we obtain

$$I(x, p) = \frac{e^{-x}}{x^p} - p \int_x^\infty e^{-u} u^{-p-1} du$$

$$= \frac{e^{-x}}{x^p} - \frac{pe^{-x}}{x^{p+1}} + p(p + 1) \int_x^\infty e^{-u} u^{-p-2} du. \tag{5.177}$$

Continuing to integrate by parts, we develop the series

$$I(x, p) = e^{-x}\left(\frac{1}{x^p} - \frac{p}{x^{p+1}} + \frac{p(p + 1)}{x^{p+2}} - \cdots + (-1)^{n-1}\frac{(p + n - 2)!}{(p - 1)!x^{p+n-1}}\right)$$

$$+ (-1)^n \frac{(p + n - 1)!}{(p - 1)!} \int_x^\infty e^{-u} u^{-p-n} du. \tag{5.178}$$

[1] This function occurs frequently in astrophysical problems involving gas with a Maxwell–Boltzmann energy distribution.
[2] See also Section 10.5.

This is a remarkable series. Checking the convergence by the d'Alembert ratio test, we find

$$\lim_{n \to \infty} \frac{|u_{n+1}|}{|u_n|} = \lim_{n \to \infty} \frac{(p + n)!}{(p + n - 1)!} \cdot \frac{1}{x}$$

$$= \lim_{n \to \infty} \frac{p + n}{x}$$

$$= \infty \qquad (5.179)$$

for all finite values of x. Therefore our series as an infinite series diverges everywhere! Before discarding Eq. (5.178) as worthless, let us see how well a given partial sum approximates the incomplete factorial function, $I(x, p)$.

$$I(x, p) - s_n(x, p) = (-1)^{n+1} \frac{(p + n)!}{(p - 1)!} \int_x^{\infty} e^{-u} u^{-p-n-1} \, du = R_n(x, p). \qquad (5.180)$$

In absolute value

$$|I(x, p) - s_n(x, p)| \le \frac{(p + n)!}{(p - 1)!} \int_x^{\infty} e^{-u} u^{-p-n-1} \, du.$$

When we substitute $u = v + x$ the integral becomes

$$\int_x^{\infty} e^{-u} u^{-p-n-1} \, du = e^{-x} \int_0^{\infty} e^{-v} (v + x)^{-p-n-1} \, dv$$

$$= \frac{e^{-x}}{x^{p+n+1}} \int_0^{\infty} e^{-v} \left(1 + \frac{v}{x}\right)^{-p-n-1} \, dv.$$

For large x the final integral approaches 1 and

$$|I(x, p) - s_n(x, p)| \approx \frac{(p + n)!}{(p - 1)!} \cdot \frac{e^{-x}}{x^{p+n+1}}. \qquad (5.181)$$

This means that if we take x large enough, our partial sum s_n is an arbitrarily good approximation to the desired function $I(x, p)$. Our divergent series (Eq. (5.178)) therefore is perfectly good for computations of partial sums. For this reason it is sometimes called a semiconvergent series. Note that the power of x in the denominator of the remainder $(p + n + 1)$ is higher than the power of x in the last term included in $s_n(x, p)$, $(p + n)$.

Since the remainder $R_n(x, p)$ alternates in sign, the successive partial sums give alternatively upper and lower bounds for $I(x, p)$. The behavior of the series (with $p = 1$) as a function of the number of terms included is shown in Fig. 5.11. We have

$$e^x E_1(x) = e^x \int_x^{\infty} \frac{e^{-u}}{u} \, du$$

$$\approx s_n(x) = \frac{1}{x} - \frac{1!}{x^2} + \frac{2!}{x^3} - \frac{3!}{x^4} + \cdots + (-1)^n \frac{n!}{x^{n+1}}, \qquad (5.182)$$

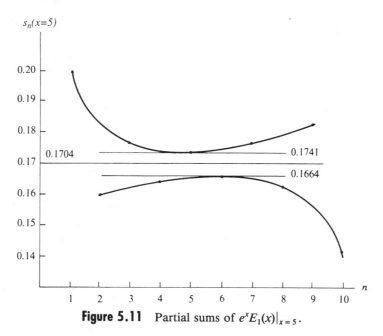

Figure 5.11   Partial sums of $e^x E_1(x)|_{x=5}$.

which is evaluated at $x = 5$. For a given value of x the successive upper and lower bounds given by the partial sums first converge and then diverge. The optimum determination of $e^x E_1(x)$ is then given by the closest approach of the upper and lower bounds, that is, between $s_4 = s_6 = 0.1664$ and $s_5 = 0.1741$ for $x = 5$. Therefore

$$0.1664 \leq e^x E_1(x)|_{x=5} \leq 0.1741. \tag{5.183}$$

Actually, from tables,

$$e^x E_1(x)|_{x=5} = 0.1704, \tag{5.184}$$

within the limits established by our asymptotic expansion. Note carefully that inclusion of additional terms in the series expansion beyond the optimum point literally reduces the accuracy of the representation.

As x is increased, the spread between the lowest upper bound and the highest lower bound will diminish. By taking x large enough, one may compute $e^x E_1(x)$ to any desired degree of accuracy. Other properties of $E_1(x)$ are derived and discussed in Section 10.5.

COSINE AND SINE INTEGRALS

Asymptotic series may also be developed from definite integrals— if the integrand has the required behavior. As an example, the cosine and sine integrals (Section 10.5) are defined by

$$Ci(x) = -\int_x^\infty \frac{\cos t}{t}\, dt, \tag{5.185}$$

$$si(x) = -\int_x^\infty \frac{\sin t}{t} \, dt. \tag{5.186}$$

Combining these with regular trigonometric functions, we may define

$$f(x) = Ci(x) \sin x - si(x) \cos x = \int_0^\infty \frac{\sin y}{y + x} \, dy,$$

$$g(x) = -Ci(x) \cos x - si(x) \sin x = \int_0^\infty \frac{\cos y}{y + x} \, dy, \tag{5.187}$$

with the new variable $y = t - x$. Going to complex variables, Section 6.1, we have

$$g(x) + if(x) = \int_0^\infty \frac{e^{iy}}{y + x} \, dy$$

$$= \int_0^\infty \frac{ie^{-xu}}{1 + iu} \, du, \tag{5.188}$$

in which $u = -iy/x$. The limits of integration, 0 to ∞, rather than 0 to $-i\infty$, may be justified by Cauchy's theorem, Section 6.3. Rationalizing the denominator and equating real part to real part and imaginary part to imaginary part, we obtain

$$g(x) = \int_0^\infty \frac{ue^{-xu}}{1 + u^2} \, du,$$

$$f(x) = \int_0^\infty \frac{e^{-xu}}{1 + u^2} \, du. \tag{5.189}$$

For convergence of the intergrals we must require that $\Re(x) > 0$.[3]

Now, to develop the asymptotic expansions, let $v = xu$ and expand the factor $[1 + (v/x)^2]^{-1}$ by the binomial theorem.[4] We have

$$f(x) \approx \frac{1}{x} \int_0^\infty e^{-v} \sum_{0 \le n \le N} (-1)^n \frac{v^{2n}}{x^{2n}} \, dv = \frac{1}{x} \sum_{0 \le n \le N} (-1)^n \frac{(2n)!}{x^{2n}}$$

$$\tag{5.190}$$

$$g(x) \approx \frac{1}{x^2} \int_0^\infty e^{-v} \sum_{0 \le n \le N} (-1)^n \frac{v^{2n+1}}{x^{2n}} \, dv = \frac{1}{x^2} \sum_{0 \le n \le N} (-1)^n \frac{(2n + 1)!}{x^{2n}}.$$

[3] $\Re(x)$ = real part of (complex) x (compare Section 6.1).
[4] This step is valid for $v \le x$. The contributions from $v \ge x$ will be negligible (for large x) because of the negative exponential. It is because the binomial expansion does not converge for $v \ge x$ that our final series is asymptotic rather than convergent.

From Eqs. (5.187) and (5.190)

$$Ci(x) \approx \frac{\sin x}{x} \sum_{0 \le n \le N} (-1)^n \frac{(2n)!}{x^{2n}} - \frac{\cos x}{x^2} \sum_{0 \le n \le N} (-1)^n \frac{(2n+1)!}{x^{2n}}$$

$$si(x) \approx -\frac{\cos x}{x} \sum_{0 \le n \le N} (-1)^n \frac{(2n)!}{x^{2n}} - \frac{\sin x}{x^2} \sum_{0 \le n \le N} (-1)^n \frac{(2n+1)!}{x^{2n}},$$

(5.191)

the desired asymptotic expansions.

This technique of expanding the integrand of a definite integral and integrating term by term is applied in Section 11.6 to develop an asymptotic expansion of the modified Bessel function K_ν and in Section 13.6 for expansions of the two confluent hypergeometric functions $M(a, c; x)$ and $U(a, c; x)$.

DEFINITION OF ASYMPTOTIC SERIES

The behavior of these series (Eqs. (5.178) and (5.191) in consistent with the defining properties of an asymptotic series.[5] Following Poincaré, we take[6]

$$x^n R_n(x) = x^n [f(x) - s_n(x)],$$

(5.192)

where

$$s_n(x) = a_0 + \frac{a_1}{x} + \frac{a_2}{x^2} + \cdots + \frac{a_n}{x^n}.$$

(5.193)

The asymptotic expansion of $f(x)$ has the properties that

$$\lim_{x \to \infty} x^n R_n(x) = 0, \qquad \text{for fixed } n,$$

(5.194)

and

$$\lim_{n \to \infty} x^n R_n(x) = \infty, \qquad \text{for fixed } x.[7]$$

(5.195)

See Eqs. (5.178), (5.179) as an example of these properties. For power series, as assumed in the form of $s_n(x)$, $R_n(x) \sim x^{-n-1}$. With conditions (5.194) and (5.195) satisfied, we write

$$f(x) \approx \sum_{n=0}^{\infty} a_n x^{-n}.$$

(5.196)

Note the use of $\approx$ in place of $=$. The function $f(x)$ is equal to the series only in the limit as $x \to \infty$.

---

[5] It is not necessary that the asymptotic series be a power series. The required property is that the remainder $R_n(x)$ be of higher order than the last term kept—as in Eq. (5.194).

[6] Poincaré's definition allows (or neglects) exponentially decreasing functions. The refinement of Poincaré's definition is of considerable importance for the advanced theory of asymptotic expansions, particularly for extensions into the complex plane. However, for purposes of an introductory treatment and especially for numerical computation with x real and positive, Poincaré's approach is perfectly satisfactory.

[7] This excludes convergent series of inverse powers of x. Some writers feel that this distribution, this exclusion, is artificial and unnecessary.

Asymptotic expansions of two functions may be multiplied together and the result will be an asymptotic expansion of the product of the two functions.

The asymptotic expansion of a given function $f(t)$ may be integrated term by term (just as in a uniformly convergent series of continuous functions) from $x \leq t < \infty$ and the result will be an asymptotic expansion of $\int_x^\infty f(t)\, dt$. Term-by-term differentiation, however, is valid only under very special conditions.

Some functions do not possess an asymptotic expansion; e^x is an example of such a function. However, if a function has an asymptotic expansion, it has only one. The correspondence is not one to one; many functions may have the same asymptotic expansion.

One of the most useful and powerful methods of generating asymptotic expansions, the method of steepest descents, will be developed in Section 7.4. Applications include the derivation of Stirling's formula for the (complete) factorial function (Section 10.3) and the asymptotic forms of the various Bessel functions (Section 11.6). Asymptotic series occur fairly often in mathematical physics. One of the earliest and still important approximations of quantum mechanics, the *WKB* expansion, is an asymptotic series.

APPLICATIONS TO COMPUTING

Asymptotic series are frequently used in the computations of functions by modern high-speed electronic computers. This is the case for the Neumann functions $N_0(x)$ and $N_1(x)$, and the modified Bessel functions $I_n(x)$ and $K_n(x)$. The relevant asymptotic series are given as Eqs. (11.127), (11.134), and (11.136). A further discussion of these functions is included in Section 11.6. The asymptotic series for the exponential integral, Eq. (5.182), for the Fresnel integrals, Exercise 5.10.2, and for the Gauss error function, Exercise 5.10.4, are used for the evaluation of these integrals for large values of the argument. How large the argument should be depends on what accuracy is required. In actual practice, a finite portion of the asymptotic series is telescoped by using Chebyshev techniques to optimize the accuracy as discussed in Section 13.4.

EXERCISES

5.10.1 Stirling's formula for the logarithm of the factorial function is

$$\ln(x!) = \frac{1}{2}\ln 2\pi + \left(x + \frac{1}{2}\right)\ln x - x - \sum_{n=1}^\infty \frac{B_{2n}}{(2n)(2n-1)}x^{1-2n}.$$

The B_{2n} are the Bernoulli numbers (Section 5.9). Show that Stirling's formula is an *asymptotic* expansion.

5.10.2 Integrating by parts, develop asymptotic expansions of the Fresnel integrals.

(a) $C(x) = \displaystyle\int_0^x \cos\frac{\pi u^2}{2}\, du$ (b) $S(x) = \displaystyle\int_0^x \sin\frac{\pi u^2}{2}\, du.$

These integrals appear in the analysis of a knife-edge diffraction pattern.

5.10.3   Rederive the asymptotic expansions of $Ci(x)$ and $si(x)$ by repeated integration by parts.

Hint. $Ci(x) + i\, si(x) = -\displaystyle\int_x^\infty \frac{e^{it}}{t}\, dt.$

5.10.4   Derive the asymptotic expansion of the Gauss error function

$$\text{erf}(x) = \frac{2}{\sqrt{\pi}} \int_0^x e^{-t^2}\, dt$$

$$\approx 1 - \frac{e^{-x^2}}{\sqrt{\pi}\, x} \left(1 - \frac{1}{2x^2} + \frac{1\cdot 3}{2^2 x^4} - \frac{1\cdot 3\cdot 5}{2^3 x^6} + \cdots + (-1)^n \frac{(2n-1)!!}{2^n x^{2n}} \right).$$

Hint. $\text{erf}(x) = 1 - \text{erfc}(x) = 1 - \dfrac{2}{\sqrt{\pi}} \displaystyle\int_x^\infty e^{-t^2}\, dt.$

Normalized so that $\text{erf}(\infty) = 1$, this function plays an important role in probability theory. It may be expressed in terms of the Fresnel integrals (Exercise 5.10.2), the incomplete gamma functions (Section 10.5), and the confluent hypergeometric functions (Section 13.6).

5.10.5   The asymptotic expressions for the various Bessel functions, Section 11.6, contain the series

$$P_\nu(z) \sim 1 + \sum_{n=1}^{\infty} (-1)^n \frac{\prod_{s=1}^{2n} [4\nu^2 - (2s-1)^2]}{(2n)!\,(8z)^{2n}},$$

$$Q_\nu(z) \sim \sum_{n=1}^{\infty} (-1)^{n+1} \frac{\prod_{s=1}^{2n-1} [4\nu^2 - (2s-1)^2]}{(2n-1)!\,(8z)^{2n-1}}.$$

Show that these two series are indeed asymptotic series.

5.10.6   For $x > 1$

$$\frac{1}{1+x} = \sum_{n=0}^{\infty} (-1)^n \frac{1}{x^{n+1}}.$$

Test this series to see if it is an asymptotic series.

5.10.7   In Exercise 5.9.17 the Euler–Mascheroni constant γ is expressed with a Bernoulli number series:

$$\sum_{k=1}^{N} \frac{B_{2k}}{(2k)n^{2k}}.$$

Show that this is an *asymptotic* series.

5.10.8   Develop an asymptotic series for

$$\int_0^\infty e^{-xv}(1+v^2)^{-2}\, dv.$$

Take x to be real and positive.

$$ANS. \quad \frac{1}{x} - \frac{2!}{x^3} + \frac{4!}{x^5} + \cdots + \frac{(-1)^n (2n)!}{x^{2n+1}}.$$

5.10.9 Calculate partial sums of $e^x E_1(x)$ for $x = 5, 10$, and 15 to exhibit the behavior shown in Fig. 5.11. Determine the width of the throat for $x = 10$ and 15 analogous to Eq. (5.183).

ANS. Throat width: $n = 10$, 0.000051
$n = 15$, 0.0000002.

5.10.10 The knife-edge diffraction pattern is described by

$$I = 0.5I_0\{[C(u_0) + 0.5]^2 + [S(u_0) + 0.5]^2\},$$

where $C(u_0)$ and $S(u_0)$ are the Fresnel integrals of Exercise 5.10.2. Here I_0 is the incident intensity and I the diffracted intensity. u_0 is proportional to the distance away from the knife edge (measured at right angles to the incident beam). Calculate I/I_0 for u_0 varying from -1.0 to $+4.0$ in steps of 0.1. Tabulate your results and, if a plotting routine is available, plot them.

Check value. $u_0 = 1.0$, $I/I_0 = 1.259226$.

5.10.11 The Euler–Maclaurin integration formula of Section 5.9 provides a way of calculating the Euler–Mascheroni constant γ to high accuracy. Using $f(x) = 1/x$ in Eq. (5.168b) (with interval $[1, n]$) and the definition of γ (Eq. 5.28), we obtain

$$\gamma = \sum_{s=1}^{n} s^{-1} - \ln n - \frac{1}{2n} + \sum_{k=1}^{\infty} \frac{B_{2k}}{(2k)n^{2k}}.$$

Using double precision arithmetic, calculate γ.

Note. D. E. Knuth, Euler's constant to 1271 places. *Math. Comput.* **16**, 275 (1962). An even more precise calculation appears in Exercise. 10.5.16.

ANS. For $n = 1000$,
$\gamma = 0.5772\,1566\,4901$.

5.11 INFINITE PRODUCTS

Consider a succession of positive factors $f_1 \cdot f_2 \cdot f_3 \cdot f_4 \cdots f_n \, (f_i > 0)$. Using capital pi to indicate product, as capital sigma indicates a sum, we have

$$f_1 \cdot f_2 \cdot f_3 \cdots f_n = \prod_{i=1}^{n} f_i. \tag{5.197}$$

We define p_n, a partial product, in analogy with s_n the partial sum,

$$p_n = \prod_{i=1}^{n} f_i \tag{5.198}$$

and then investigate the limit

$$\lim_{n \to \infty} p_n = P. \tag{5.199}$$

If P is finite (but not zero), we say the infinite product is convergent. If P is infinite *or zero*, the infinite product is labeled divergent.

Since the product will diverge to infinity if

$$\lim_{n=\infty} f_n > 1 \tag{5.200}$$

or to zero for

$$\lim_{n=\infty} f_n < 1 \quad \text{(and} >0), \tag{5.201}$$

it is convenient to write our infinite products as

$$\prod_{n=1}^{\infty} (1 + a_n).$$

The condition $a_n \to 0$ is then a necessary (but not sufficient) condition for convergence.

The infinite product may be related to an infinite series by the obvious method of taking the logarithm

$$\ln \prod_{n=1}^{\infty} (1 + a_n) = \sum_{n=1}^{\infty} \ln(1 + a_n). \tag{5.202}$$

A more useful relationship is stated by the following theorem.

CONVERGENCE OF INFINITE PRODUCT

If $0 \le a_n < 1$, the infinite products $\prod_{n=1}^{\infty} (1 + a_n)$ and $\prod_{n=1}^{\infty} (1 - a_n)$ converge if $\sum_{n=1}^{\infty} a_n$ converges and diverge if $\sum_{n=1}^{\infty} a_n$ diverges.

Considering the term $1 + a_n$, we see from Eq. (5.90)

$$1 + a_n \le e^{a_n}. \tag{5.203}$$

Therefore for the partial product p_n

$$p_n \le e^{s_n}, \tag{5.204}$$

and letting $n \to \infty$,

$$\prod_{n=1}^{\infty} (1 + a_n) \le \exp \sum_{n=1}^{\infty} a_n, \tag{5.205}$$

thus establishing an upper bound for the infinite product.

To develop a lower bound, we note that

$$p_n = 1 + \sum_{i=1}^{n} a_i + \sum_{i=1}^{n} \sum_{j=1}^{n} a_i a_j + \cdots > s_n, \tag{5.206}$$

since $a_i \ge 0$. Hence

$$\prod_{n=1}^{\infty} (1 + a_n) \ge \sum_{n=1}^{\infty} a_n. \tag{5.207}$$

If the infinite sum remains finite, the infinite product will also. If the infinite sum diverges, so will the infinite product.

The case of $\prod (1 - a_n)$ is complicated by the negative signs, but a proof that depends on the foregoing proof may be developed by noting that for $a_n < \frac{1}{2}$ (remember $a_n \to 0$ for convergence)

$$(1 - a_n) \le (1 + a_n)^{-1}$$

and

$$(1 - a_n) \ge (1 + 2a_n)^{-1}. \tag{5.208}$$

SINE, COSINE, AND GAMMA FUNCTIONS

The reader will recognize that an nth-order polynomial $P_n(x)$ with n real roots may be written as a product of n factors:

$$P_n(x) = (x - x_1)(x - x_2) \cdots (x - x_n) = \prod_{i=1}^{n} (x - x_i). \tag{5.209}$$

In much the same way we may expect that a function with an infinite number of roots may be written as an infinite product, one factor for each root. This is indeed the case for the trigonometric functions. We have two very useful infinite product representations,

$$\sin x = x \prod_{n=1}^{\infty} \left(1 - \frac{x^2}{n^2 \pi^2}\right) \tag{5.210}$$

$$\cos x = \prod_{n=1}^{\infty} \left[1 - \frac{4x^2}{(2n - 1)^2 \pi^2}\right]. \tag{5.211}$$

The most convenient and perhaps most elegant derivation of these two expressions is by the use of complex variables.[1] By our theorem of convergence, Eqs. (5.210) and (5.211) are convergent for all finite values of x. Specifically, for the infinite product for $\sin x$, $a_n = x^2/n^2\pi^2$,

$$\sum_{n=1}^{\infty} a_n = \frac{x^2}{\pi^2} \sum_{n=1}^{\infty} n^{-2} = \frac{x^2}{\pi^2} \zeta(2)$$

$$= \frac{x^2}{6} \tag{5.212}$$

by Exercise 5.9.6. The series corresponding to Eq. (5.211) behaves in a similar manner.

Equation (5.210) leads to two interesting results. First, if we set $x = \pi/2$, we obtain

$$1 = \frac{\pi}{2} \prod_{n=1}^{\infty} \left[1 - \frac{1}{(2n)^2}\right] = \frac{\pi}{2} \prod_{n=1}^{\infty} \left[\frac{(2n)^2 - 1}{(2n)^2}\right]. \tag{5.213}$$

[1]The derivation appears in Chapter 7.

Solving for $\pi/2$, we have

$$\frac{\pi}{2} = \prod_{n=1}^{\infty} \left[\frac{(2n)^2}{(2n-1)(2n+1)} \right] = \frac{2 \cdot 2}{1 \cdot 3} \cdot \frac{4 \cdot 4}{3 \cdot 5} \cdot \frac{6 \cdot 6}{5 \cdot 7} \cdots, \tag{5.214}$$

which is Wallis's famous formula for $\pi/2$.

The second result involves the gamma or factorial function (Section 10.1). One definition of the gamma function is

$$\Gamma(x) = \left[xe^{\gamma x} \prod_{r=1}^{\infty} \left(1 + \frac{x}{r} \right) e^{-x/r} \right]^{-1}, \tag{5.215}$$

where γ is the usual Euler–Mascheroni constant (compare Section 5.2). If we take the product of $\Gamma(x)$ and $\Gamma(-x)$, Eq. (5.215) leads to

$$\Gamma(x)\Gamma(-x) = -\left[xe^{\gamma x} \prod_{r=1}^{\infty} \left(1 + \frac{x}{r} \right) e^{-x/r} xe^{-\gamma x} \prod \left(1 - \frac{x}{r} \right) e^{x/r} \right]^{-1}$$

$$= -\left[x^2 \prod_{r=1}^{\infty} \left(1 - \frac{x^2}{r^2} \right) \right]^{-1}. \tag{5.216}$$

Using Eq. (5.210) with x replaced by πx, we obtain

$$\Gamma(x)\Gamma(-x) = -\frac{\pi}{x \sin \pi x}. \tag{5.217}$$

Anticipating a recurrence relation developed in Section 10.1, we have $-x\Gamma(-x) = \Gamma(1-x)$. Equation (5.217) may be written as

$$\Gamma(x)\Gamma(1-x) = \frac{\pi}{\sin \pi x}. \tag{5.218}$$

This will be useful in treating the gamma function (Chapter 10).

Strictly speaking, we should check the range of x for which Eq. (5.215) is convergent. Clearly, individual factors will vanish for $x = 0, -1, -2, \ldots$. The proof that the infinite product converges for all other (finite) values of x is left as Exercise 5.11.9.

These infinite products have a variety of uses in analytical mathematics. However, because of rather slow convergence, they are not suitable for precise numerical work.

EXERCISES

5.11.1   Using

$$\ln \prod_{n=1}^{\infty} (1 \pm a_n) = \sum_{n=1}^{\infty} \ln(1 \pm a_n)$$

and the Maclaurin expansion of $\ln(1 \pm a_n)$, show that the infinite product $\prod_{n=1}^{\infty} (1 \pm a_n)$ converges or diverges with the infinite series $\sum_{n=1}^{\infty} a_n$.

5.11.2   An infinite product appears in the form

$$\prod_{n=1}^{\infty} \left(\frac{1 + a/n}{1 + b/n}\right),$$

where a and b are constants. Show that this infinite product converges only if $a = b$.

5.11.3   Show that the infinite product representations of $\sin x$ and $\cos x$ are consistent with the identity $2 \sin x \cos x = \sin 2x$.

5.11.4   Determine the limit to which

$$\prod_{n=2}^{\infty} \left(1 + \frac{(-1)^n}{n}\right)$$

converges.

5.11.5   Show that

$$\prod_{n=2}^{\infty} \left[1 - \frac{2}{n(n+1)}\right] = \frac{1}{3}.$$

5.11.6   Prove that

$$\prod_{n=2}^{\infty} \left(1 - \frac{1}{n^2}\right) = \frac{1}{2}.$$

5.11.7   Using the infinite product representations of $\sin x$, show that

$$x \cot x = 1 - 2 \sum_{m,n=1}^{\infty} \left(\frac{x}{n\pi}\right)^{2m},$$

hence that the Bernoulli number

$$B_{2n} = (-1)^{n-1} \frac{2(2n)!}{(2\pi)^{2n}} \zeta(2n).$$

5.11.8   Verify the Euler identity

$$\prod_{p=1}^{\infty} (1 + z^p) = \prod_{q=1}^{\infty} (1 - z^{2q-1})^{-1}, \qquad |z| < 1.$$

5.11.9   Show that $\prod_{r=1}^{\infty} (1 + x/r)e^{-x/r}$ converges for all finite x (except for the zeros of $1 + x/r$).
Hint. Write the nth factor as $1 + a_n$.

5.11.10   Calculate $\cos x$ from its infinite product representation, Eq. (5.2.11), using (a) 10, (b) 100, and (c) 1000 factors in the product. Calculate the absolute error. Note how slowly the partial products converge—making the infinite product quite unsuitable for precise numerical work.
ANS.   For 1000 factors $\cos \pi = -1.00051$.

ADDITIONAL READINGS

BENDER, C. M., and S. ORSZAG, *Advanced Mathematical Methods for Scientists and Engineers*. New York: McGraw-Hill (1978). Particularly recommended for methods of accelerating convergence.

DAVIS, H. T., *Tables of Higher Mathematical Functions*. Bloomington, IN: Principia Press (1935). Volume II contains extensive information on Bernoulli numbers and polynomials.

DINGLE, R. B., *Asymptotic Expansions: Their Derivation and Interpretation*. London and New York: Academic Press (1973).

GALAMBOS, J., *Representations of Real Numbers by Infinite Series*. Berlin: Springer (1976).

GRADSHTEYN, I. S., and I. N. RYZHIK, *Table of Integrals, Series and Products*. Corrected and enlarged edition prepared by Alan Jeffrey. New York: Academic Press (1980).

HANSEN, E., *A Table of Series and Products*. Englewood Cliffs, NJ: Prentice-Hall (1975). A tremendous compilation of series and products.

HARDY, G. H., *Divergent Series*. Oxford: Clarendon Press (1956). A standard, comprehensive work on methods of treating divergent series. Hardy includes an instructive account of the gradual development of the concepts of convergence and divergence.

KNOPP, K., *Theory and Application of Infinite Series*. London: Blackie and Son (2nd ed.); New York: Hafner (1971). This is a thorough, comprehensive, and authoritative work, which covers infinite series and products. Proofs of almost all of the statements not proved in Chapter 5 will be found in this book.

MANGULIS, V., *Handbook of Series for Scientists and Engineers*. New York and London: Academic Press (1965). A most convenient and useful collection of series. Includes algebraic functions, Fourier series, and series of the special functions: Bessel, Legendre, and so on.

OLVER, F. W. J., *Asymptotics and Special Functions*. New York: Academic Press (1974). A detailed, readable development of asymptotic theory. Considerable attention is paid to error bounds for use in computation.

RAINVILLE, E. D., *Infinite Series*. New York: Macmillan (1967). A readable and useful account of series—constants and functions.

SOKOLNIKOFF, I. S., and R. M. REDHEFFER, *Mathematics of Physics and Modern Engineering*, 2nd ed. New York: McGraw-Hill (1966). A long Chapter 2 (101 pages) presents infinite series in a thorough but very readable form. Extensions to the solutions of differential equations, to complex series, and to Fourier series are included.

The topic of infinite series is treated in many texts on advanced calculus.

6 FUNCTIONS OF A COMPLEX VARIABLE I

Analytic Properties
Mapping

The imaginary numbers are a wonderful flight of God's spirit; they are almost an amphibian between being and not being.

GOTTFRIED WILHELM VON LEIBNIZ, 1702

We turn now to a study of functions of a complex variable. In this area we develop some of the most powerful and widely useful tools in all of mathematical analysis. To indicate, at least partly, why complex variables are important, we mention briefly several areas of application.

1. For many pairs of functions u and v, both u and v satisfy Laplace's equation

$$\nabla^2 \psi = \frac{\partial^2 \psi(x, y)}{\partial x^2} + \frac{\partial^2 \psi(x, y)}{\partial y^2} = 0.$$

Hence either u or v may be used to describe a two-dimensional electrostatic potential. The other function that gives a family of curves orthogonal to those of the first function, may then be used to describe the electric field **E**. A similar situation holds for the hydrodynamics of an ideal fluid in irrotational motion. The function u might describe the velocity potential, whereas the function v would then be the stream function.

In many cases in which the functions u and v are unknown, mapping or transforming in the complex plane permits us to create a coordinate system tailored to the particular problem.

2. In Chapter 8 we shall see that the second-order differential equations of interest in physics may be solved by power series. The same power series may be used in the complex plane to replace x by the complex variable z. The dependence of the solution $f(z)$ at a given z_0 on the behavior of $f(z)$ elsewhere gives us greater insight into the behavior of our solution and a powerful tool (analytic continuation) for extending the region in which the solution is valid.

3. The change of a parameter k from real to imaginary, $k \to ik$, transforms the Helmholtz equation into the diffusion equation. The same change transforms the Helmholtz equation solutions (Bessel and spherical Bessel functions) into the diffusion equation solutions (modified Bessel and modified spherical Bessel functions).

4. Integrals in the complex plane have a wide variety of useful applications.

    a. Evaluating definite integrals.
    b. Inverting power series.
    c. Forming infinite products.
    d. Obtaining solutions of differential equations for large values of the variable (asymptotic solutions).
    e. Investigating the stability of potentially oscillatory systems.
    f. Inverting integral transforms.

5. Many physical quantities that were originally real become complex as a simple physical theory is made more general. The real index of refraction of light becomes a complex quantity when absorption is included. The real energy associated with an energy level becomes complex when the finite lifetime of the level is considered.

6.1 COMPLEX ALGEBRA

A complex number is nothing more than an ordered pair of two ordinary numbers, (a, b) or $a + ib$, in which i is $(-1)^{1/2}$. Similarly, a complex variable is an ordered pair of two real variables,

$$z = (x, y) = x + iy. \tag{6.1}$$

The reader will see that the ordering is significant, that in general $a + ib$ is not equal to $b + ia$ and $x + iy$ is not equal to $y + ix$.[1]

It is frequently convenient to employ a graphical representation of the complex variable. By plotting x—the real part of z—as the abscissa and y— the imaginary part of z—as the ordinate, we have the complex plane or Argand plane shown in Fig. 6.1. If we assign specific values to x and y, then z corresponds to a point (x, y) in the plane. In terms of the ordering mentioned before, it is obvious that the point (x, y) does not coincide with the point (y, x) except for the special case of $x = y$.

---

[1] The algebra of complex numbers, $a + ib$, is isomorphic with that of matrices of the form

$$\begin{pmatrix} a & b \\ -b & a \end{pmatrix}.$$

(compare Exercise 3.2.4).

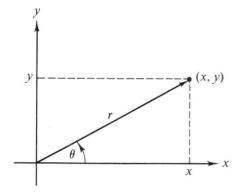

Figure 6.1 Complex plane—Argand diagram.

All our complex variable analyses can be developed in terms of ordered pairs[2] of numbers (a, b), variables (x, y), and functions $(u(x, y), v(x, y))$. The i is not necessary but it is convenient. It serves to keep pairs in order—somewhat like the unit vectors of Chapter 1.

Hence addition and multiplication of complex numbers may be defined in terms of their Cartesian components as

$$z_1 + z_2 = (x_1, y_1) + (x_2, y_2) = (x_1 + x_2, y_1 + y_2)$$

$$= x_1 + x_2 + i(y_1 + y_2), \qquad (6.2a)$$

$$z_1 z_2 = (x_1, y_1) \cdot (x_2, y_2) = (x_1 x_2 - y_1 y_2, x_1 y_2 + x_2 y_1). \qquad (6.2b)$$

In Chapter 1 the points in the xy-plane are identified with the two-dimensional displacement vector $\mathbf{r} = \hat{\mathbf{x}}x + \hat{\mathbf{y}}y$. As a result, two-dimensional vector analogs can be developed for much of our complex analysis. Exercise 6.1.2 is one simple example; Cauchy's theorem, Section 6.3, is another.

Further, from Fig. 6.1 we may write

$$x = r \cos \theta$$
$$\qquad (6.3)$$
$$y = r \sin \theta$$

and

$$z = r(\cos \theta + i \sin \theta). \qquad (6.4a)$$

Using a result that is suggested (but not rigorously proved)[3] by Section 5.6, we have the very useful polar representation

$$z = r e^{i\theta}. \qquad (6.4b)$$

---

[2] This is how a computer would do complex arithmetic.

[3] Strictly speaking, Chapter 5 was limited to real variables. However, we can define e^z as $\sum_{n=0}^{\infty} z^n/n!$ for complex z. The development of power-series expansions for complex functions is taken up in Section 6.5 (Laurent expansion).

In this representation r is called the modulus or magnitude of z ($r = |z| = (x^2 + y^2)^{1/2}$) and the angle θ ($= \tan^{-1}(y/x)$) is labeled the argument or phase of z.

The choice of polar representation, Eq. (6.4), or Cartesian representation, Eq. (6.1), is a matter of convenience. Addition and subtraction of complex variables are easier in the Cartesian representation, Eq. (6.2a). Multiplication, division, powers, and roots are easier to handle in polar form, Eq. (6.4b).

Analytically or graphically, using the vector analogy, we may show that the modulus of the sum of two complex numbers is no greater than the sum of the moduli and no less than the difference, Exercise 6.1.3,

$$|z_1| - |z_2| \le |z_1 + z_2| \le |z_1| + |z_2|. \tag{6.5}$$

Because of the vector analogy, these are called the triangle inequalities.

Using the polar form, Eq. (6.4b), we find that the magnitude of a product is the product of the magnitudes,

$$|z_1 \cdot z_2| = |z_1| \cdot |z_2|. \tag{6.6}$$

Also,

$$\arg(z_1 \cdot z_2) = \arg z_1 + \arg z_2. \tag{6.7}$$

From our complex variable z complex functions $f(z)$ or $w(z)$ may be constructed. These complex functions may then be resolved into real and imaginary parts

$$w(z) = u(x, y) + iv(x, y), \tag{6.8}$$

in which the separate functions $u(x, y)$ and $v(x, y)$ are pure real. For example, if $f(z) = z^2$, we have

$$f(z) = (x + iy)^2$$

$$= (x^2 - y^2) + i2xy.$$

The real part of a function $f(z)$ will be labeled $\mathcal{R}f(z)$, whereas the imaginary part will be labeled $\mathcal{I}f(z)$. In Eq. (6.8)

$$\mathcal{R}w(z) = u(x, y),$$

$$\mathcal{I}w(z) = v(x, y).$$

The relationship between the independent variable z and the dependent variable w is perhaps best pictured as a mapping operation. A given $z = x + iy$ means a given point in the z-plane. The complex value of $w(z)$ is then a point in the w-plane. Points in the z-plane map into points in the w-plane and curves in the z-plane map into curves in the w-plane as indicated in Fig. 6.2.

COMPLEX CONJUGATION

In all these steps, complex number, variable, and function, the operation of replacing i by $-i$ is called "taking the complex conjugate." The complex

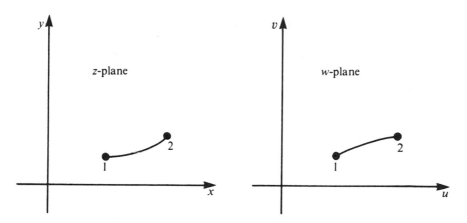

Figure 6.2 The function $w(z) = u(x, y) + iv(x, y)$ maps points in the xy-plane into points in the uv plane.

conjugate of z is denoted by z^*, where[4]

$$z^* = x - iy. \tag{6.9}$$

The complex variable z and its complex conjugate z^* are mirror images of each other reflected in the x-axis, that is, inversion of the y-axis (compare Fig. 6.3). The product zz^* leads to

$$zz^* = (x + iy)(x - iy) = x^2 + y^2$$

$$= r^2. \tag{6.10}$$

Hence

$$(zz^*)^{1/2} = |z|,$$

the magnitude of z.

FUNCTIONS OF A COMPLEX VARIABLE

All the elementary functions of real variables may be extended into the complex plane—replacing the real variable x by the complex variable z.

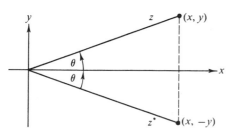

Figure 6.3 Complex conjugate points.

---

[4] The complex conjugate is often denoted by $\bar{z}$ in the mathematical literature.

This is an example of the analytic continuation mentioned in Section 6.5. The extremely important relation, Eq. (6.4), is an illustration of this. Moving into the complex plane opens up new opportunities for analysis.

Example 6.1.1 De Moivre's Formula

If Eq. (6.4) is raised to the nth power, we have

$$e^{in\theta} = (\cos \theta + i \sin \theta)^n. \tag{6.11}$$

Expanding the exponential now with argument $n\theta$, we obtain

$$\cos n\theta + i \sin n\theta = (\cos \theta + i \sin \theta)^n. \tag{6.12}$$

This is De Moivre's formula.

Now if the right-hand side of Eq. (6.12) is expanded by the binomial theorem, we obtain $\cos n\theta$ as a series of powers of $\cos \theta$ and $\sin \theta$, Exercise 6.16.

Numerous other examples of relations among the exponential, hyperbolic, and trigonometric functions in the complex plane appear in the exercises.

Occasionally there are complications. The logarithm of a complex variable may be expanded using the polar representation

$$\ln z = \ln re^{i\theta}$$

$$= \ln r + i\theta. \tag{6.13a}$$

This is not complete. To the phase angle, θ, we may add any integral multiple of 2π without changing z. Hence Eq. (6.13a) should read

$$\ln z = \ln re^{i(\theta + 2n\pi)}$$

$$= \ln r + i(\theta + 2n\pi). \tag{6.13b}$$

The parameter n may be any integer. This means that $\ln z$ is a multivalued function having an infinite number of values for a single pair of real values r and θ. To avoid ambiguity, we usually agree to set $n = 0$ and limit the phase to an interval of length 2π such as $(-\pi, \pi)$.[5] The line in the z-plane that is not crossed, the negative real axis in this case, is labeled a *cut line*. The value of $\ln z$ with $n = 0$ is called the *principal value* of $\ln z$. Further discussion of these functions, including the logarithm, appears in Section 6.6.

EXERCISES

6.1.1   (a) Find the reciprocal of $x + iy$, working entirely in the Cartesian representation.
         (b) Repeat part (a), working in polar form but expressing the final result in Cartesian form.

[5] There is no standard choice of phase: The appropriate phase depends on each problem.

6.1.2   The complex quantities $a = u + iv$ and $b = x + iy$ may also be represented as two-dimensional vectors, $\mathbf{a} = \hat{x}u + \hat{y}v$, $\mathbf{b} = \hat{x}x + \hat{y}y$. Show that

$$a^*b = \mathbf{a} \cdot \mathbf{b} + i\hat{z} \cdot \mathbf{a} \times \mathbf{b}.$$

6.1.3   Prove algebraically that

$$|z_1| - |z_2| \le |z_1 + z_2| \le |z_1| + |z_2|.$$

Interpret this result in terms of vectors. Prove that

$$|z - 1| < |\sqrt{z^2 - 1}| < |z + 1|, \qquad \text{for } \mathcal{R}(z) > 0.$$

6.1.4   We may define a complex conjugation operator K such that $Kz = z^*$. Show that K is *not* a linear operator.

6.1.5   Show that complex numbers have square roots and that the square roots are contained in the complex plane. What are the square roots of i?

6.1.6   Show that

(a) $\cos n\theta = \cos^n \theta - \binom{n}{2} \cos^{n-2} \theta \sin^2 \theta + \binom{n}{4} \cos^{n-4} \theta \sin^4 \theta - \cdots$.

(b) $\sin n\theta = \binom{n}{1} \cos^{n-1} \theta \sin \theta - \binom{n}{3} \cos^{n-3} \theta \sin^3 \theta + \binom{n}{5} \cos^{n-5} \theta \sin^5 \theta - \cdots$.

Note. The quantities $\binom{n}{m}$ are binomial coefficients: $\binom{n}{m} = n!/[(n-m)!m!]$.

6.1.7   Prove that

(a) $\displaystyle\sum_{n=0}^{N-1} \cos nx = \frac{\sin N(x/2)}{\sin x/2} \cos(N-1)\frac{x}{2}$,

(b) $\displaystyle\sum_{n=0}^{N-1} \sin nx = \frac{\sin N(x/2)}{\sin x/2} \sin(N-1)\frac{x}{2}$.

These series occur in the analysis of the multiple-slit diffraction pattern. Another application is the analysis of the Gibbs phenomenon, Section 14.5. *Hint.* Parts (a) and (b) may be combined to form a *geometric* series (compare Section 5.1).

6.1.8   For $-1 < p < 1$ prove that

(a) $\displaystyle\sum_{n=0}^{\infty} p^n \cos nx = \frac{1 - p \cos x}{1 - 2p \cos x + p^2}$,

(b) $\displaystyle\sum_{n=0}^{\infty} p^n \sin nx = \frac{p \sin x}{1 - 2p \cos x + p^2}$.

These series occur in the theory of the Fabry–Perot interferometer.

6.1.9   Assume that the trigonometric functions and the hyperbolic functions are defined for complex argument by the appropriate power series

$$\sin z = \sum_{n=1,\text{odd}}^{\infty} (-1)^{(n-1)/2} \frac{z^n}{n!} = \sum_{s=0}^{\infty} (-1)^s \frac{z^{2s+1}}{(2s+1)!},$$

$$\cos z = \sum_{n=0,\text{even}}^{\infty} (-1)^{n/2} \frac{z^n}{n!} = \sum_{s=0}^{\infty} (-1)^s \frac{z^{2s}}{(2s)!},$$

$$\sinh z = \sum_{n=1,\text{odd}}^{\infty} \frac{z^n}{n!} = \sum_{s=0}^{\infty} \frac{z^{2s+1}}{(2s+1)!},$$

$$\cosh z = \sum_{n=0,\text{even}}^{\infty} \frac{z^n}{n!} = \sum_{s=0}^{\infty} \frac{z^{2s}}{(2s)!}.$$

(a) Show that

$$i \sin z = \sinh iz, \qquad \sin iz = i \sinh z,$$
$$\cos z = \cosh iz, \qquad \cos iz = \cosh z.$$

(b) Verify that familiar functional relations such as

$$\cosh z = \frac{e^z + e^{-z}}{2},$$

$$\sin(z_1 + z_2) = \sin z_1 \cos z_2 + \sin z_2 \cos z_1,$$

still hold in the complex plane.

6.1.10   Using the identities

$$\cos z = \frac{e^{iz} + e^{-iz}}{2},$$

$$\sin z = \frac{e^{iz} - e^{-iz}}{2i},$$

established from comparison of power series, show that

(a) $\sin(x + iy) = \sin x \cosh y + i \cos x \sinh y$,
    $\cos(x + iy) = \cos x \cosh y - i \sin x \sinh y$,
(b) $|\sin z|^2 = \sin^2 x + \sinh^2 y$,
    $|\cos z|^2 = \cos^2 x + \sinh^2 y$.

This demonstrates that we may have $|\sin z|, |\cos z| > 1$ in the complex plane.

6.1.11   From the identities in Exercises 6.1.9 and 6.1.10 show that

(a) $\sinh(x + iy) = \sinh x \cos y + i \cosh x \sin y$.
    $\cosh(x + iy) = \cosh x \cos y + i \sinh x \sin y$,
(b) $|\sinh z|^2 = \sinh^2 x + \sin^2 y$.
    $|\cosh z|^2 = \sinh^2 x + \cos^2 y$.

6.1.12   Prove that

(a) $|\sin z| \geq |\sin x|$
(b) $|\cos z| \geq |\cos x|$.

6.1.13   Show that the exponential function e^z is periodic with a pure imaginary period of $2\pi i$.

6.1.14   Show that

(a) $\tanh(z/2) = \dfrac{\sinh x + i \sin y}{\cosh x + \cos y}$,

(b) $\coth(z/2) = \dfrac{\sinh x - i \sin y}{\cosh x - \cos y}$.

6.1.15   Find all the zeros of

(a) $\sin z$,
(b) $\cos z$,
(c) $\sinh z$,
(d) $\cosh z$.

6.1.16   Show that

(a) $\sin^{-1} z = -i \ln(iz \pm \sqrt{1 - z^2})$,   (d) $\sinh^{-1} z = \ln(z + \sqrt{z^2 + 1})$,

(b) $\cos^{-1} z = -i \ln(z \pm \sqrt{z^2 - 1})$,   (e) $\cosh^{-1} z = \ln(z + \sqrt{z^2 - 1})$,

(c) $\tan^{-1} z = \dfrac{i}{2} \ln\left(\dfrac{i + z}{i - z}\right)$,   (f) $\tanh^{-1} z = \dfrac{1}{2} \ln\left(\dfrac{1 + z}{1 - z}\right)$.

Hint. 1. Express the trigonometric and hyperbolic functions in terms of exponentials. 2. Solve for the exponential and then for the exponent.

6.1.17   In the quantum theory of the photoionization we encounter the identity

$$\left(\frac{ia - 1}{ia + 1}\right)^{ib} = \exp(-2b \cot^{-1} a),$$

in which a and b are real. Verify this identity.

6.1.18   A plane wave of light of angular frequency ω is represented by

$$e^{i\omega(t - nx/c)}$$

In a certain substance the simple real index of refraction n is replaced by the complex quantity $n - ik$. What is the effect of k on the wave? What does k correspond to physically? The generalization of a quantity from real to complex form occurs frequently in physics. Examples range from the complex Young's modulus of viscoelastic materials to the complex (optical) potential of the "cloudy crystal ball" model of the atomic nucleus.

6.1.19   We see that for the angular momentum components defined in Exercise 2.5.14

$$(L_x - iL_y) \neq (L_x + iL_y)^*.$$

Explain why this occurs.

6.1.20   Show that the *phase* of $f(z) = u + iv$ is equal to the imaginary part of the logarithm of $f(z)$. Exercise 10.2.13 depends on this result.

6.1.21   (a) Show that $e^{\ln z}$ always equals z.
(b) Show that $\ln e^z$ does *not* always equal z.

6.1.22   The infinite product representations of Section 5.11 hold when the real variable x is replaced by the complex variable z. From this, develop infinite product representations for

(a) $\sinh z$
(b) $\cosh z$.

6.1.23   The equation of motion of a mass m *relative to a rotating coordinate system* is

$$m \frac{d^2 \mathbf{r}}{dt^2} = \mathbf{F} - m\boldsymbol{\omega} \times (\boldsymbol{\omega} \times \mathbf{r}) - 2m\left(\boldsymbol{\omega} \times \frac{d\mathbf{r}}{dt}\right) - m\left(\frac{d\boldsymbol{\omega}}{dy} \times \mathbf{r}\right).$$

Consider the case $\mathbf{F} = 0$, $\mathbf{r} = \hat{\mathbf{x}}x + \hat{\mathbf{y}}y$, and $\boldsymbol{\omega} = \omega\hat{\mathbf{z}}$, with ω constant. Show that the replacement of $\mathbf{r} = \hat{\mathbf{x}}x + \hat{\mathbf{y}}y$ by $z = x + iy$ leads to

$$\frac{d^2 z}{dt^2} + i2\omega \frac{dz}{dt} - \omega^2 z = 0.$$

Note. This differential equation may be solved by the substitution $z = fe^{-i\omega t}$.

6.1.24   Using the complex arithmetic available in FORTRAN IV, write a program that will calculate the complex exponential e^z from its series expansion (definition). Calculate e^z for $z = e^{in\pi/6}$, $n = 0, 1, 2, ..., 12$. Tabulate the phase angle ($\theta = n\pi/6$), $\mathcal{R}(z)$, $\mathcal{I}(z)$, $\mathcal{R}(e^z)$, $\mathcal{I}(e^z)$, $|e^z|$, and the phase of e^z.

Check value. $n = 5$, $\theta = 2.61799$, $\mathcal{R}(z) = -0.86602$,
$\mathcal{I}(z) = 0.50000$, $\mathcal{R}(e^z) = 0.36913$, $\mathcal{I}(e^z) = 0.20166$,
$|e^z| = 0.42062$, phase$(e^z) = 0.50000$.

6.1.25   Using the complex arithmetic available in FORTRAN IV, calculate and tabulate $\mathcal{R}(\sinh z)$, $\mathcal{I}(\sinh z)$, $|\sinh z|$, and phase$(\sinh z)$ for $x = 0.0 \ (0.1) \ 1.0$ and $y = 0.0 \ (0.1) \ 1.0$.

Hint. Beware of dividing by zero if calculating an angle as an arc tangent.

Check value. $z = 0.2 + 0.1i$, $\mathcal{R}(\sinh z) = 0.20033$,
$\mathcal{I}(\sinh z) = 0.10184$, $|\sinh z| = 0.22473$,
phase$(\sinh z) = 0.47030$.

6.1.26   Repeat Exercise 6.1.25 for $\cosh z$.

6.2 CAUCHY–RIEMANN CONDITIONS

Having established complex functions of a complex variable, we now proceed to differentiate them. The derivative of $f(z)$, like that of a real function, is defined by

$$\lim_{\delta z \to 0} \frac{f(z + \delta z) - f(z)}{z + \delta z - z} = \lim_{\delta z \to 0} \frac{\delta f(z)}{\delta z}$$

$$= \frac{df}{dz} \quad \text{or} \quad f'(z), \tag{6.14}$$

provided that the limit is *independent* of the particular approach to the point z. For real variables we require that the right-hand limit ($x \to x_0$ from above) and the left-hand limit ($x \to x_0$ from below) be equal for the derivative $df(x)/dx$ to exist at $x = x_0$. Now, with z (or z_0) some point in a *plane*, our requirement that the limit be independent of the direction of approach is very restrictive.

Consider increments δx and δy of the variables x and y, respectively. Then

$$\delta z = \delta x + i \, \delta y. \tag{6.15}$$

Also,

$$\delta f = \delta u + i \, \delta v, \tag{6.16}$$

so that

$$\frac{\delta f}{\delta z} = \frac{\delta u + i \, \delta v}{\delta x + i \, \delta y}. \tag{6.17}$$

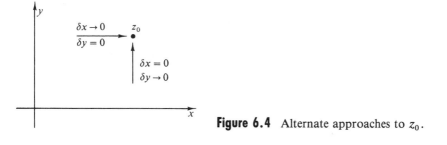

Figure 6.4 Alternate approaches to z_0.

Let us take the limit indicated by Eq. (6.14) by two different approaches as shown in Fig. 6.4. First, with $\delta y = 0$, we let $\delta x \to 0$. Equation (6.14) yields

$$\lim_{\delta z \to 0} \frac{\delta f}{\delta z} = \lim_{\delta x \to 0} \left(\frac{\delta u}{\delta x} + i \frac{\delta v}{\delta x} \right)$$

$$= \frac{\partial u}{\partial x} + i \frac{\partial v}{\partial x}, \tag{6.18}$$

assuming the partial derivatives exist. For a second approach, we set $\delta x = 0$ and then let $\delta y \to 0$. This leads to

$$\lim_{\delta z \to 0} \frac{\delta f}{\delta z} = \lim_{\delta y \to 0} \left(-i \frac{\delta u}{\delta y} + \frac{\delta v}{\delta y} \right)$$

$$= -i \frac{\partial u}{\partial y} + \frac{\partial v}{\partial y}. \tag{6.19}$$

If we are to have a derivative df/dz, Eqs. (6.18) and (6.19) must be identical. Equating real parts to real parts and imaginary parts to imaginary parts (like components of vectors), we obtain

$$\frac{\partial u}{\partial x} = \frac{\partial v}{\partial y}, \qquad \frac{\partial u}{\partial y} = -\frac{\partial v}{\partial x}. \tag{6.20}$$

These are the famous Cauchy–Riemann conditions. They were discovered by Cauchy and used extensively by Riemann in his theory of analytic functions. These Cauchy–Riemann conditions are necessary for the existence of a derivative of $f(z)$, that is, if df/dz exists, the Cauchy–Riemann conditions must hold.

Conversely, if the Cauchy–Riemann conditions are satisfied and the partial derivatives of $u(x, y)$ and $v(x, y)$ are continuous, the derivative df/dz exists. This may be shown by writing

$$\delta f = \left(\frac{\partial u}{\partial x} + i \frac{\partial v}{\partial x} \right) \delta x + \left(\frac{\partial u}{\partial y} + i \frac{\partial v}{\partial y} \right) \delta y. \tag{6.21}$$

The justification for this expression depends on the continuity of the partial derivatives of u and v. Dividing by δz, we have

$$\frac{\delta f}{\delta z} = \frac{(\partial u/\partial x + i(\partial v/\partial x))\, \delta x + (\partial u/\partial y + i(\partial u/\partial y))\, \delta y}{\delta x + i\, \delta y}$$

$$= \frac{(\partial u/\partial x + i(\partial v/\partial x)) + (\partial u/\partial y + i(\partial v/\partial y))\, \delta y/\delta x}{1 + i(\delta y/\delta x)}. \tag{6.22}$$

If $\delta f/\delta z$ is to have a unique value, the dependence on $\delta y/\delta x$ must be eliminated. Applying the Cauchy–Riemann conditions to the y derivatives, we obtain

$$\frac{\partial u}{\partial y} + i\frac{\partial v}{\partial y} = -\frac{\partial v}{\partial x} + i\frac{\partial u}{\partial x}. \tag{6.23}$$

Substituting Eq. (6.23) into Eq. (6.22), we may cancel out the $\delta y/\delta x$ dependence and

$$\frac{\delta f}{\delta z} = \frac{\partial u}{\partial x} + i\frac{\partial v}{\partial x}, \tag{6.24}$$

which shows that $\lim \delta f/\delta z$ is independent of the direction of approach in the complex plane as long as the partial derivatives are continuous.

It is worthwhile noting that the Cauchy–Riemann conditions guarantee that the curves $u = c_1$ will be orthogonal to the curves $v = c_2$ (compare Section 2.1). This is fundamental in application to potential problems in a variety of areas of physics. If $u = c_1$ is a line of electric force, then $v = c_2$ is an equipotential line (surface), and vice versa. A further implication for potential theory is developed in Exercise 6.2.1.

ANALYTIC FUNCTIONS

Finally, if $f(z)$ is differentiable at $z = z_0$ and in some small region around z_0, we say that $f(z)$ is *analytic*[1] at $z = z_0$. If $f(z)$ is analytic everywhere in the (finite) complex plane, we call it an *entire* function. Our theory of complex variables here is essentially one of analytic functions of complex variables, which points up the crucial importance of the Cauchy–Riemann conditions. The concept of analyticity carried on in advanced theories of modern physics plays a crucial role in dispersion theory (of elementary particles). If $f'(z)$ does not exist at $z = z_0$, then z_0 is labeled a singular point and consideration of it is postponed until Section 7.1.

To illustrate the Cauchy–Riemann conditions, consider two very simple examples.

----

[1] Some writers use the term holomorphic or regular.

Example 6.2.1

Let $f(z) = z^2$. Then the real part $u(x, y) = x^2 - y^2$ and the imaginary part $v(x, y) = 2xy$. Following Eq. (6.20),

$$\frac{\partial u}{\partial x} = 2x = \frac{\partial v}{\partial y}, \qquad \frac{\partial u}{\partial y} = -2y = -\frac{\partial v}{\partial x}.$$

We see that $f(z) = z^2$ satisfies the Cauchy–Riemann conditions throughout the complex plane. Since the partial derivatives are clearly continuous, we conclude that $f(z) = z^2$ is analytic.

Example 6.2.2

Let $f(z) = z^*$. Now $u = x$ and $v = -y$. Applying the Cauchy–Riemann conditions, we obtain

$$\frac{\partial u}{\partial x} = 1 \neq \frac{\partial v}{\partial y}.$$

The Cauchy–Riemann conditions are *not* satisfied and $f(z) = z^*$ is not an analytic function of z. It is interesting to note that $f(z) = z^*$ is continuous, thus providing an example of a function that is everywhere continuous but nowhere differentiable.

The derivative of a real function of a real variable is essentially a local characteristic, in that it provides information about the function only in a local neighborhood—for instance, as a truncated Taylor expansion. The existence of a derivative of a function of a complex variable has much more far-reaching implications. The real and imaginary parts of our analytic function must separately satisfy Laplace's equation. This is Exercise 6.2.1. Further, our analytic function is guaranteed derivatives of all orders, Section 6.4. In this sense the derivative not only governs the local behavior of the complex function, but controls the distant behavior as well.

EXERCISES

6.2.1   The functions $u(x, y)$ and $v(x, y)$ are the real and imaginary parts, respectively, of an analytic function $w(z)$.

(a) Assuming that the required derivatives exist, show that

$$\nabla^2 u = \nabla^2 v = 0.$$

Solutions of Laplace's equation such as $u(x, y)$ and $v(x, y)$ are called *harmonic functions*.

(b) Show that

$$\frac{\partial u}{\partial x}\frac{\partial u}{\partial y} + \frac{\partial v}{\partial x}\frac{\partial v}{\partial y} = 0.$$

and give a geometric interpretation.

Hint. The technique of Section 1.6 allows you to construct vectors *normal* to the curve $u(x, y) = c_i$ and $v(x, y) = c_j$.

6.2.2   Show whether or not the function $f(z) = \Re(z) = x$ is analytic.

6.2.3   Having shown that the real part $u(x, y)$ and the imaginary part $v(x, y)$ of an analytic function $w(z)$ each satisfy Laplace's equation, show that $u(x, y)$ and $v(x, y)$ cannot have either a maximum or a minimum in the interior of any region in which $w(z)$ is analytic. (They can have saddle points.)

6.2.4   Let $A = \partial^2 w/\partial x^2$, $B = \partial^2 w/\partial x\,\partial y$, $C = \partial^2 w/\partial y^2$. From the calculus of functions of *two* variables, $w(x, y)$, we have a saddle point if
$$B^2 - AC > 0.$$
With $f(z) = u(x, y) + iv(x, y)$, apply the Cauchy–Riemann conditions and show that neither $u(x, y)$ nor $v(x, y)$ has a maximum or a minimum in a finite region of the complex plane.

6.2.5   Find the analytic functions
$$w(z) = u(x, y) + iv(x, y)$$
if
(a) $u(x, y) = x^3 - 3xy^2$,
(b) $v(x, y) = e^{-y} \sin x$.

6.2.6   If there is some common region in which $w_1 = u(x, y) + iv(x, y)$ and $w_2 = w_1^* = u(x, y) - iv(x, y)$ are both analytic, prove that $u(x, y)$ and $v(x, y)$ are constants.

6.2.7   The function $f(z) = u(x, y) + iv(x, y)$ is analytic. Show that $f^*(z^*)$ is also analytic.

6.2.8   Using $f(re^{i\theta}) = R(r, \theta)e^{i\Theta(r, \theta)}$, in which $R(r, \theta)$ and $\Theta(r, \theta)$ are differentiable real functions of r and θ, show that the Cauchy–Riemann conditions in polar coordinates become

(a) $\dfrac{\partial R}{\partial r} = \dfrac{R}{r}\dfrac{\partial \Theta}{\partial \theta}$,

(b) $\dfrac{\partial R}{r\,\partial \theta} = -R\dfrac{\partial \Theta}{\partial r}$.

Hint. Set up the derivative first with δz radial and then with δz tangential.

6.2.9   As an extension of Exercise 6.2.8 show that $\Theta(r, \theta)$ satisfies Laplace's equation in polar coordinates, Eq. (2.33) (without the final term).

6.2.10   Two-dimensional irrotational fluid flow is conveniently described by a complex potential $f(z) = u(x, v) + iv(x, y)$. We label the real part $u(x, y)$, the velocity potential and the imaginary part $v(x, y)$, the stream function. The fluid velocity $\mathbf{V}$ is given by $\mathbf{V} = \nabla u$. If $f(z)$ is analytic,
(a) Show that $df/dz = V_x - iV_y$;
(b) Show that $\nabla \cdot \mathbf{V} = 0$ (no sources or sinks);
(c) Show that $\nabla \times \mathbf{V} = 0$ (irrotational, nonturbulent flow).

6.2.11   A proof of the Schwarz inequality (Section 9.4) involves minimizing an expression
$$f = \psi_{aa} + \lambda\psi_{ab} + \lambda^*\psi_{ab}^* + \lambda\lambda^*\psi_{bb} \geq 0.$$
The ψ's are integrals of products of functions; ψ_{aa} and ψ_{bb} are real, ψ_{ab} is complex. λ is a parameter, possibly complex.

(a) Differentiate the preceding expression with respect to λ^*, treating λ as an independent parameter, independent of λ^*. Show that setting the derivative $\partial f/\partial \lambda^*$ equal to zero yields

$$\lambda = -\psi_{ab}^*/\psi_{bb}.$$

(b) Show that $\partial f/\partial \lambda = 0$ leads to the same result.

(c) Let $\lambda = x + iy$, $\lambda^* = x - iy$. Set the x and y derivatives equal to zero and show that again

$$\lambda = -\psi_{ab}^*/\psi_{bb}.$$

This independence of λ and λ^* appears again in Section 17.7.

6.2.12   The function $f(z)$ is analytic. Show that the derivative of $f(z)$ with respect to z^* does not exist unless $f(z)$ is a constant.

Hint. Use the chain rule and take $x = (z + z^*)/2$, $y = (z - z^*)/2i$.

Note. This result emphasizes that our analytic function $f(z)$ is not just a complex function of two real variables x and y. It is a function of the complex variable $x + iy$.

6.3 CAUCHY'S INTEGRAL THEOREM

CONTOUR INTEGRALS

With differentiation under control, we turn to integration. The integral of a complex variable over a contour in the complex plane may be defined in close analogy to the (Riemann) integral of a real function integrated along the real x-axis.

We divide the contour $z_0 z_0'$ into n intervals by picking $n - 1$ intermediate points $z_1, z_2, \ldots$, on the contour (Fig. 6.5). Consider the sum

$$S_n = \sum_{j=1}^{n} f(\zeta_j)(z_j - z_{j-1}), \tag{6.25}$$

where ζ_j is a point on the curve between z_j and z_{j-1}. Now let $n \to \infty$ with

$$|z_j - z_{j-1}| \to 0$$

for all j. If the $\lim_{n \to \infty} S_n$ exists and is independent of the details of choosing the points z_j and ζ_j, then

$$\lim_{n \to \infty} \sum_{j=1}^{n} f(\zeta_j)(z_j - z_{j-1}) = \int_{z_0}^{z_0'} f(z)\, dz. \tag{6.26}$$

The right-hand side of Eq. (6.26) is called the contour integral of $f(z)$ (along the specified contour C from $z = z_0$ to $z = z_0'$).

The preceding development of the contour integral is closely analogous to the Riemann integral of a real function of a real variable. As an alternative,

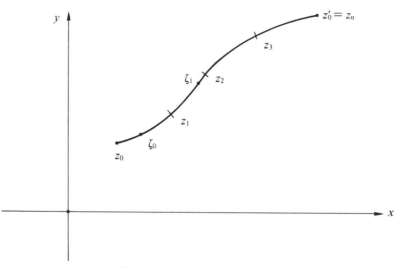

Figure 6.5 Integration path.

the contour integral may be *defined* by

$$\int_{z_1}^{z_2} f(z)\, dz = \int_{x_1,y_1}^{x_2,y_2} [u(x, y) + iv(x, y)][dx + i\, dy]$$

$$= \int_{x_1,y_1}^{x_2,y_2} [u(x, y)\, dx - v(x, y)\, dy] + i \int_{x_1,y_1}^{x_2,y_2} [v(x, y)\, dx + u(x, y)\, dy],$$

with the path joining (x_1, y_1) and (x_2, y_2) *specified*. This reduces the complex integral to the complex sum of real integrals. It's somewhat analogous to the replacement of a vector integral by the vector sum of scalar integrals, Section 1.10.

An important example is the contour integral $\int_C z^n\, dz$, where C is a circle of radius $r > 0$ around the origin $z = 0$ in the positive mathematical sense (counterclockwise). In polar coordinates of Eq. (6.4b) we parameterize $z = re^{i\theta}$ and $dz = ire^{i\theta}\, d\theta$. For $n \neq -1$ we obtain

$$(2\pi i)^{-1} \int_C z^n\, dz = (2\pi)^{-1} r^{n+1} \int_0^{2\pi} \exp[i(n + 1)\theta]\, d\theta$$

$$= [2\pi i(n + 1)]^{-1} r^{n+1} [e^{i(n+1)\theta}]_0^{2\pi} = 0 \qquad (6.27a)$$

because 2π is a period of $e^{i(n+1)\theta}$, while for $n = -1$

$$(2\pi i)^{-1} \int_C dz/z = (2\pi)^{-1} \int_0^{2\pi} d\theta = 1, \qquad (6.27b)$$

again independent of r. These integrals are examples of Cauchy's integral theorem which we consider in the next section.

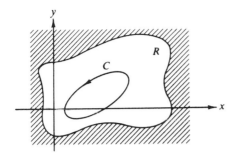

Figure 6.6 A closed contour C within a simply connected region R.

STOKES'S THEOREM PROOF

Cauchy's integral theorem is the first of two basic theorems in the theory of the behavior of functions of a complex variable. First, a proof under relatively restrictive conditions—conditions that are intolerable to the mathematician developing a beautiful abstract theory but that are usually satisfied in physical problems.

If a function $f(z)$ is analytic (therefore single-valued) and its partial derivatives are continuous throughout some simply connected region R,[1] for every closed path C (Fig. 6.6) in R the line integral of $f(z)$ around C is zero or

$$\int_C f(z)\,dz = \oint_C f(z)\,dz = 0. \tag{6.27c}$$

The symbol $\oint$ is used to emphasize that the path is closed. The reader will recall that in Section 1.13 such a function $f(z)$, identified as a force, was labeled conservative.

In this form the Cauchy integral theorem may be proved by direct application of Stokes's theorem (Section 1.12). With $f(z) = u(x, y) + iv(x, y)$ and $dz = dx + i\,dy$,

$$\oint_C f(z)\,dz = \oint_C (u + iv)(dx + i\,dy)$$

$$= \oint_C (u\,dx - v\,dy) + i \oint (v\,dx + u\,dy). \tag{6.28}$$

These two line integrals may be converted to surface integrals by Stokes's theorem, a procedure that is justified if the partial derivatives are continuous within C. In applying Stokes's theorem, the reader might note that the final two integrals of Eq. (6.28) are completely real. Using

$$\mathbf{V} = \hat{\mathbf{x}}V_x + \hat{\mathbf{y}}V_y,$$

---

[1] A simply connected region or domain is one in which every closed contour in that region encloses only the points contained in it. If a region is not simply connected, it is called multiply connected. As an example of a multiply connected region, consider the z-plane with the interior of the unit circle *excluded*.

we have

$$\oint_C (V_x\, dx + V_y\, dy) = \int \left(\frac{\partial V_y}{\partial x} - \frac{\partial V_x}{\partial y}\right) dx\, dy. \tag{6.29}$$

For the first integral in the last part of Eq. (6.28) let $u = V_x$ and $v = -V_y$.[2]
Then

$$\oint_C (u\, dx - v\, dy) = \oint_C (V_x\, dx + V_y\, dy)$$

$$= \int \left(\frac{\partial V_y}{\partial x} - \frac{\partial V_x}{\partial y}\right) dx\, dy$$

$$= -\int \left(\frac{\partial v}{\partial x} + \frac{\partial u}{\partial y}\right) dx\, dy. \tag{6.30}$$

For the second integral on the right side of Eq. (6.28) we let $u = V_y$ and $v = V_x$.
Using Stokes's theorem again, we obtain

$$\oint (v\, dx + u\, dy) = \int \left(\frac{\partial u}{\partial x} - \frac{\partial v}{\partial y}\right) dx\, dy. \tag{6.31}$$

On application of the Cauchy–Riemann conditions that must hold, since $f(z)$
is assumed analytic, each integrand vanishes and

$$\oint f(z)\, dz = -\int \left(\frac{\partial v}{\partial x} + \frac{\partial u}{\partial y}\right) dx\, dy + i \int \left(\frac{\partial u}{\partial x} - \frac{\partial v}{\partial y}\right) dx\, dy$$

$$= 0. \tag{6.32}$$

CAUCHY–GOURSAT PROOF

This completes the proof of Cauchy's integral theorem. However, the proof
is marred from a theoretical point of view by the need for continuity of the
first partial derivatives. Actually, as shown by Goursat, this condition is not
essential. An outline of the Goursat proof is as follows. We subdivide the
region inside the contour C into a network of small squares as indicated in
Fig. 6.7. Then

$$\oint_C f(z)\, dz = \sum_j \oint_{C_j} f(z)\, dz, \tag{6.33}$$

all integrals along interior lines canceling out. To attack the $\oint_{C_j} f(z)\, dz$, we
construct the function

$$\delta_j(z, z_j) = \frac{f(z) - f(z_j)}{z - z_j} - \frac{df(z)}{dz}\bigg|_{z = z_j}, \tag{6.34}$$

---

[2] In the proof of Stokes's theorem, Section 1.12, V_x and V_y are any two functions (with
continuous partial derivatives).

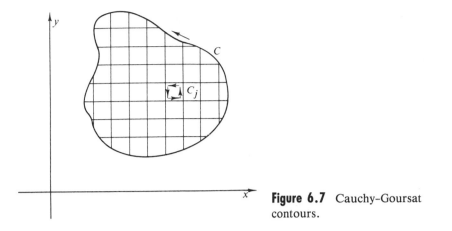

Figure 6.7 Cauchy-Goursat contours.

with z_j an interior point of the jth subregion. Note that $[f(z) - f(z_j)]/(z - z_j)$ is an approximation to the derivative at $z = z_j$. Equivalently, we may note that if $f(z)$ had a Taylor expansion (which we have not yet proved), then $\delta_j(z, z_j)$ would be of order $z - z_j$, approaching zero as the network was made finer. We may make

$$|\delta_j(z, z_j)| < \varepsilon, \tag{6.35}$$

where ε is an arbitrarily chosen small positive quantity.

Solving Eq. (6.34) for $f(z)$ and integrating around C_j, we obtain

$$\oint_{C_j} f(z)\, dz = \oint_{C_j} (z - z_j)\delta_j(z, z_j)\, dz, \tag{6.36}$$

the integrals of the other terms vanishing.[3] When Eqs. (6.35) and (6.36) are combined, one may show that

$$\left| \sum_j \oint_{C_j} f(z)\, dz \right| < A\varepsilon, \tag{6.37}$$

where A is a term of the order of the area of the enclosed region. Since ε is arbitrary, we let $\varepsilon \to 0$ and conclude that if a function $f(z)$ is analytic on and within a closed path C,

$$\oint_C f(z)\, dz = 0. \tag{6.38}$$

Details of the proof of this significantly more general and more powerful form can be found in Churchill and in the other references cited. Actually we can still prove the theorem for $f(z)$ analytic within the interior of C and only continuous on C.

[3] $\oint dz$ and $\oint z\, dz = 0$ by Eq. (6.27a).

The consequence of the Cauchy integral theorem is that for analytic functions the line integral is a function only of its end points, independent of the path of integration,

$$\int_{z_1}^{z_2} f(z)\, dz = F(z_2) - F(z_1) = -\int_{z_2}^{z_1} f(z)\, dz, \qquad (6.39)$$

again exactly like the case of a conservative force, Section 1.13.

MULTIPLY CONNECTED REGIONS

The original statement of our theorem demanded a simply connected region. This restriction may easily be relaxed by the creation of a barrier, a contour line. Consider the multiply connected region of Fig. 6.8, in which $f(z)$ is not defined for the interior R'. Cauchy's integral theorem is not valid for the contour C, as shown, but we can construct a contour C' for which the theorem holds. We draw a line from the interior forbidden region R' to the forbidden region exterior to R and then run a new contour C', as shown in Fig. 6.9.

The new contour C' through $ABDEFGA$ never crosses the contour line that literally converts R into a simply connected region. The three-dimensional analog of this technique was used in Section 1.14 to prove Gauss's law. By Eq. (6.39)

$$\int_{G}^{A} f(z)\, dz = -\int_{E}^{D} f(z)\, dz, \qquad (6.40)$$

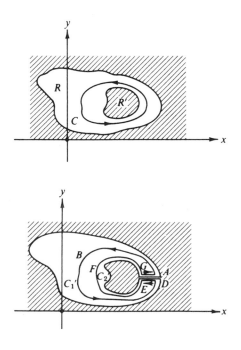

Figure 6.8 A closed contour C in a multiply connected region.

Figure 6.9 Conversion of a multiply connected region into a simply connected region.

$f(z)$ having been continuous across the contour line and line segments DE and GA arbitrarily close together. Then

$$\oint_{C'} f(z)\,dz = \int_{ABD} f(z)\,dz + \int_{EFG} f(z)\,dz$$

$$= 0 \tag{6.41}$$

by Cauchy's integral theorem, with region R now simply connected. Applying Eq. (6.39) once again with $ABD \to C_1'$ and $EFG \to -C_2'$, we obtain

$$\oint_{C_1'} f(z)\,dz = \oint_{C_2'} f(z)\,dz, \tag{6.42}$$

in which C_1' and C_2' are both traversed in the same (counterclockwise) direction.

It should be emphasized that the contour line here is a matter of mathematical convenience, to permit the application of Cauchy's integral theorem. Since $f(z)$ is analytic in the annular region, it is necessarily single-valued and continuous across any such contour line. When we consider branch points (Sections 6.6 and 7.1) our functions will not be single-valued and a cut line will be required to make them single-valued.

EXERCISES

6.3.1   Show that $\int_{z_1}^{z_2} f(z)\,dz = -\int_{z_2}^{z_1} f(z)\,dz$.

6.3.2   Prove that

$$\left| \int_C f(z)\,dz \right| \le |f|_{\max} \cdot L,$$

where $|f|_{\max}$ is the maximum value of $|f(z)|$ along the contour C and L is the length of the contour.

6.3.3   Verify that

$$\int_{0,0}^{1,1} z^* \, dz$$

depends on the path by evaluating the integral for the two paths shown in Fig. 6.10. Recall that $f(z) = z^*$ is not an analytic function of z and that Cauchy's integral theorem therefore does not apply.

6.3.4   Show that

$$\oint \frac{dz}{z^2 + z} = 0,$$

in which the contour C is a circle defined by $|z| = R > 1$.
Hint. Direct use of the Cauchy integral theorem is illegal. Why? The integral may be evaluated by transforming to polar coordinates and using tables. The preferred technique would be the calculus of residues, Section 7.2. This yields 0 for $R > 1$ and $2\pi i$ for $R < 1$.

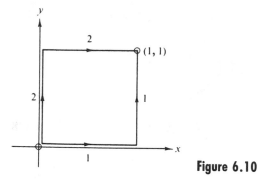

Figure 6.10

6.4 CAUCHY'S INTEGRAL FORMULA

As in the preceding section, we consider a function $f(z)$ that is analytic on a closed contour C and within the interior region bounded by C. We seek to prove that

$$\oint_C \frac{f(z)}{z - z_0}\, dz = 2\pi i f(z_0), \tag{6.43}$$

in which z_0 is some point in the interior region bounded by C. This is the second of the two basic theorems mentioned in Section 6.2. Note carefully that since z is on the contour C while z_0 is in the interior, $z - z_0 \neq 0$ and the integral Eq. (6.4.3) is well defined.

Although $f(z)$ is assumed analytic, the integrand is $f(z)/(z - z_0)$ and is *not* analytic at $z = z_0$ unless $f(z_0) = 0$. If the contour is deformed as shown in Fig. 6.11 (or Fig. 6.9, Section 6.3), Cauchy's integral theorem applies. By Eq. (6.42)

$$\oint_C \frac{f(z)}{z - z_0}\, dz - \oint_{C_2} \frac{f(z)}{z - z_0}\, dz = 0, \tag{6.44}$$

where C is the original outer contour and C_2 is the circle surrounding the point z_0 traversed in a *counterclockwise* direction. Let $z = z_0 + re^{i\theta}$, using the polar representation because of the circular shape of the path around z_0. Here r is

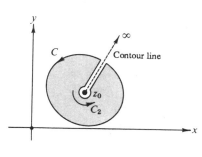

Figure 6.11   Exclusion of a singular point.

small and will eventually be made to approach zero. We have

$$\oint_{C_2} \frac{f(z)}{z - z_0}\, dz = \oint_{C_2} \frac{f(z_0 + re^{i\theta})}{re^{i\theta}} rie^{i\theta} d\theta.$$

Taking the limit as $r \to 0$, we obtain

$$\oint_{C_2} \frac{f(z)}{z - z_0}\, dz = if(z_0) \int_{C_2} d\theta$$

$$= 2\pi i f(z_0), \tag{6.45}$$

since $f(z)$ is analytic and therefore continuous at $z = z_0$. This proves the Cauchy integral formula.

Here is a remarkable result. The value of an analytic function $f(z)$ is given at an interior point $z = z_0$ once the values on the boundary C are specified. This is closely analogous to a two-dimensional form of Gauss's law (Section 1.14) in which the magnitude of an interior line charge would be given in terms of the cylindrical surface integral of the electric field $\mathbf{E}$.

A further analogy is the determination of a function in real space by an integral of the function and the corresponding Green's function (and their derivatives) over the bounding surface. Kirchhoff diffraction theory is an example of this.

It has been emphasized that z_0 is an interior point. What happens if z_0 is exterior to C? In this case the entire integrand is analytic on and within C. Cauchy's integral theorem, Section 6.3, applies and the integral vanishes. We have

$$\frac{1}{2\pi i} \oint_C \frac{f(z)\, dz}{z - z_0} = \begin{cases} f(z_0), & z_0 \text{ interior} \\ 0, & z_0 \text{ exterior.} \end{cases}$$

DERIVATIVES

Cauchy's integral formula may be used to obtain an expression for the derivative of $f(z)$. From Eq. (6.43), with $f(z)$ analytic,

$$\frac{f(z_0 + \delta z_0) - f(z_0)}{\delta z_0} = \frac{1}{2\pi i \delta z_0} \left(\oint \frac{f(z)}{z - z_0 - \delta z_0}\, dz - \oint \frac{f(z)}{z - z_0}\, dz \right).$$

Then, by definition of derivative (Eq. (6.14)),

$$f'(z_0) = \lim_{\delta z_0 \to 0} \frac{1}{2\pi i \delta z_0} \oint \frac{\delta z_0 f(z)}{(z - z_0 - \delta z_0)(z - z_0)}\, dz$$

$$= \frac{1}{2\pi i} \oint \frac{f(z)}{(z - z_0)^2}\, dz. \tag{6.46}$$

The alert reader will see that this result could have been obtained by differentiating Eq. (6.43) under the integral sign with respect to z_0. This formal or turning-the-crank approach is valid, but the justification for it is contained in the preceding analysis.

This technique for constructing derivatives may be repeated. We write $f'(z_0 + \delta z_0)$ and $f'(z_0)$, using Eq. (6.46). Subtracting, dividing by δz_0, and finally taking the limit as $\delta z_0 \to 0$, we have

$$f^{(2)}(z_0) = \frac{2}{2\pi i} \oint \frac{f(z)\, dz}{(z - z_0)^3}.$$

Note that $f^{(2)}(z_0)$ is independent of the direction of δz_0 as it must be. Continuing, we get

$$f^{(n)}(z_0) = \frac{n!}{2\pi i} \oint \frac{f(z)\, dz}{(z - z_0)^{n+1}};^1 \tag{6.47}$$

that is, the requirement that $f(z)$ be analytic not only guarantees a first derivative but derivatives of *all* orders as well! The derivatives of $f(z)$ are automatically analytic. The reader should notice that this statement assumes the Goursat version of the Cauchy integral theorem. This is why Goursat's contribution is so significant in the development of the theory of complex variables.

MORERA'S THEOREM

A further application of Cauchy's integral formula is in the proof of Morera's theorem, which is the converse of Cauchy's integral theorem. The theorem states the following:

If a function $f(z)$ is continuous in a simply connected region R and $\oint_C f(z)\, dz = 0$ for every closed contour C within R, then $f(z)$ is analytic throughout R.

Let us integrate $f(z)$ from z_1 to z_2. Since every closed path integral of $f(z)$ vanishes, the integral is independent of path and depends only on its end points. We label the result of the integration $F(z)$, with

$$F(z_2) - F(z_1) = \int_{z_1}^{z_2} f(z)\, dz. \tag{6.48}$$

As an identity,

$$\frac{F(z_2) - F(z_1)}{z_2 - z_1} - f(z_1) = \frac{\int_{z_1}^{z_2} [f(t) - f(z_1)]\, dt}{z_2 - z_1}, \tag{6.49}$$

using t as another complex variable. Now we take the limit as $z_2 \to z_1$.

$$\lim_{z_2 \to z_1} \frac{\int_{z_1}^{z_2} [f(t) - f(z_1)]\, dt}{z_2 - z_1} = 0, \tag{6.50}$$

---

[1] This expression is the starting point for defining derivatives of *fractional order*. See A. Erdelyi (Ed.), *Tables of Integral Transforms*, Vol. 2. New York: McGraw-Hill (1954). For recent applications to mathematical analysis see T. J. Osler, An integral analogue of Taylor's series and its use in computing Fourier transforms. *Math. Comput.* **26**, 449 (1972), and his references.

since $f(t)$ is continuous.[2] Therefore

$$\lim_{z_2 \to z_1} \frac{F(z_2) - F(z_1)}{z_2 - z_1} = F'(z) \bigg|_{z = z_1} = f(z_1) \tag{6.51}$$

by definition of derivative (Eq. (6.14)). We have proved that $F'(z)$ at $z = z_1$ exists and equals $f(z_1)$. Since z_1 is any point in R, we see that $F(z)$ is analytic. Then by Cauchy's integral formula (compare Eq. (6.47)) $F'(z) = f(z)$ is also analytic, proving Morcra's theorem.

Drawing once more on our electrostatic analog, we might use $f(z)$ to represent the electrostatic field **E**. If the net charge within every closed region in R is zero (Gauss's law), the charge density is everywhere zero in R. Alternatively, in terms of the analysis of Section 1.13, $f(z)$ represents a conservative force (by definition of conservative), and then we find that it is always possible to express it as the derivative of a potential function $F(z)$.

If $f(z) = \sum a_n z^n$ is analytic and bounded, $|f(z)| \le M$ on a circle of radius r about the origin, then

$$|a_n| r^n \le M \qquad \text{(Cauchy's inequality)} \tag{6.52}$$

gives upper bounds for the coefficients of its Taylor expansion. To prove Eq. (6.52) let us define $M(r) = \max_{|z| = r} |f(z)|$ and use the Cauchy integral for a_n

$$|a_n| = (2\pi)^{-1} \left| \int_{|z| = r} \frac{f(z)}{z^{n+1}} \, dz \right| \le M(r) \frac{2\pi r}{2\pi r^{n+1}}.$$

An immediate consequence of the inequality (6.52) is *Liouville's theorem*: If $f(z)$ is analytic and bounded in the complex plane it is a constant. In fact, if $|f(z)| \le M$ for any z, then Cauchy's inequality (6.52) gives $|a_n| \le M r^{-n} \to 0$ as $r \to \infty$ for $n > 0$. Hence $f(z) = a_0$.

Conversely, the slightest deviation of an analytic function from a constant value implies that there must be at least one singularity somewhere in the infinite complex plane. Apart from the trivial constant functions, then, singularities are a fact of life, and we must learn to live with them. But we shall do more than that. We shall next expand a function in a Laurent series at a singularity, and we shall use singularities to develop the powerful and useful calculus of residues in Chapter 7.

The *fundamental theorem of algebra* (due to C. F. Gauss), which says that any polynomial $P(z) = \sum_{\nu=0}^{n} a_\nu z^\nu$ with $n > 0$ and $a_n \ne 0$ has n roots, also follows from Liouville's theorem. Suppose $P(z)$ has no zero. Then $1/P(z)$ is analytic and bounded as $|z| \to \infty$. Hence $f(z)$ is a constant by Liouville's theorem, q.e.a. Thus $P(z)$ has at least one root which we can divide out. Then we repeat the process for the resulting polynomial of degree $n - 1$. This leads to the conclusion that $P(z)$ has exactly n roots.

---

[2] We can quote the mean value theorem of calculus here.

EXERCISES

6.4.1   Show that

$$\oint_C (z - z_0)^n \, dz = \begin{cases} 2\pi i, & n = -1, \\ 0, & n \neq -1, \end{cases}$$

where the contour C encircles the point $z = z_0$ in a positive (counterclockwise) sense. The exponent n is an integer. See also Eq. (6.27).
The calculus of residues, Chapter 7, is based on this result.

6.4.2   Show that

$$\frac{1}{2\pi i} \oint z^{m-n-1} \, dz, \qquad m \text{ and } n \text{ integers}$$

(with the contour encircling the origin once counterclockwise), is a representation of the Kronecker delta $\delta_{m,n}$.

6.4.3   Solve Exercise 6.3.4 by separating the integrand into partial fractions and then applying Cauchy's integral theorem for multiply connected regions.
Note. Partial fractions are explained in Section 15.7 in connection with Laplace transforms.

6.4.4   Evaluate

$$\oint_C \frac{dz}{z^2 - 1},$$

where C is the circle $|z| = 2$.

6.4.5   Assuming that $f(z)$ is analytic on and within a closed contour C and that the point z_0 is within C, show that

$$\oint_C \frac{f'(z)}{(z - z_0)} \, dz = \oint_C \frac{f(z)}{(z - z_0)^2} \, dz.$$

6.4.6   You know that $f(z)$ is analytic on and within a closed contour C. You suspect that the nth derivative $f^{(n)}(z_0)$ is given by

$$f^{(n)}(z_0) = \frac{n!}{2\pi i} \oint_C \frac{f(z)}{(z - z_0)^{n+1}} \, dz.$$

Using mathematical induction, prove that this expression is correct.

6.4.7   (a) A function $f(z)$ is analytic within a closed contour C (and continuous on C). If $f(z) \neq 0$ within C and $|f(z)| \geq M$ on C, show that

$$|f(z)| \geq M$$

for all points within C.
Hint. Consider $w(z) = 1/f(z)$.

(b) If $f(z) = 0$ within the contour C, show that the foregoing result does not hold, that it is possible to have $|f(z)| = 0$ at one or more points in the interior with $|f(z)| > 0$ over the entire bounding contour. Cite a specific example of an analytic function that behaves this way.

6.4.8   Using the Cauchy integral formula for the nth derivative, convert the following Rodrigues formulas into the corresponding Schlaefli integrals.

(a) Legendre

$$P_n(x) = \frac{1}{2^n n!} \frac{d^n}{dx^n} (x^2 - 1)^n.$$

$$ANS. \quad \frac{(-1)^n}{2^n} \cdot \frac{1}{2\pi i} \oint \frac{(1 - z^2)^n}{(z - x)^{n+1}} dz.$$

(b) Hermite

$$H_n(x) = (-1)^n e^{x^2} \frac{d^n}{dx^n} e^{-x^2}.$$

(c) Laguerre

$$L_n(x) = \frac{e^x}{n!} \frac{d^n}{dx^n} (x^n e^{-x}).$$

Note. From the Schlaefli integral representations one can develop generating functions for these special functions. Compare Sections 12.4, 13.1,
and 13.2.

6.5 LAURENT EXPANSION

TAYLOR EXPANSION

The Cauchy integral formula of the preceding section opens up the way for
another derivation of Taylor's series (Section 5.6), but this time for functions
of a complex variable. Suppose we are trying to expand $f(z)$ about $z = z_0$
and we have $z = z_1$ as the nearest point on the Argand diagram for which
$f(z)$ is not analytic. We construct a circle C centered at $z = z_0$ with radius
$|z' - z_0| < |z_1 - z_0|$ (Fig. 6.12). Since z_1 was assumed to be the nearest point
at which $f(z)$ was not analytic, $f(z)$ is necessarily analytic on and within C.

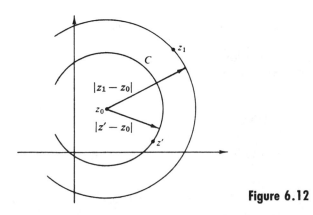

Figure 6.12

From Eq. (6.43), the Cauchy integral formula,

$$f(z) = \frac{1}{2\pi i} \oint_C \frac{f(z')\,dz'}{z' - z}$$

$$= \frac{1}{2\pi i} \oint_C \frac{f(z')\,dz'}{(z' - z_0) - (z - z_0)}$$

$$= \frac{1}{2\pi i} \oint_C \frac{f(z')\,dz'}{(z' - z_0)[1 - (z - z_0)/(z' - z_0)]} . \tag{6.53}$$

Here z' is a point on the contour C and z is any point interior to C. It is not quite rigorously legal to expand the denominator of the integrand in Eq. (6.53) by the binomial theorem, for we have not yet proved the binomial theorem for complex variables. Instead, we note the identity

$$\frac{1}{1 - t} = 1 + t + t^2 + t^3 + \cdots = \sum_{n=0}^{\infty} t^n, \tag{6.54}$$

which may easily be verified by multiplying both sides by $1 - t$. The infinite series, following the methods of Section 5.2, is convergent for $|t| < 1$.

Now for point z interior to C, $|z - z_0| < |z' - z_0|$, and, using Eq. (6.5.4), Eq. (6.53) becomes

$$f(z) = \frac{1}{2\pi i} \oint_C \sum_{n=0}^{\infty} \frac{(z - z_0)^n f(z')\,dz'}{(z' - z_0)^{n+1}} . \tag{6.55}$$

Interchanging the order of integration and summation (valid since Eq. (6.54) is uniformly convergent for $|t| < 1$), we obtain

$$f(z) = \frac{1}{2\pi i} \sum_{n=0}^{\infty} (z - z_0)^n \oint_C \frac{f(z')\,dz'}{(z' - z_0)^{n+1}} . \tag{6.56}$$

Referring to Eq. (6.47), we get

$$f(z) = \sum_{n=0}^{\infty} (z - z_0)^n \frac{f^{(n)}(z_0)}{n!} , \tag{6.57}$$

which is our desired Taylor expansion. Note that it is based only on the assumption that $f(z)$ is analytic for $|z - z_0| < |z_1 - z_0|$. Just as for real variable power series (Section 5.7), this expansion is unique for a given z_0.

From the Taylor expansion for $f(z)$ a binomial theorem may be derived—Exercise 6.5.2.

SCHWARZ REFLECTION PRINCIPLE

From the binomial expansion of $g(z) = (z - x_0)^n$ for integral n it is easy to see that the complex conjugate of the function is the function of the complex conjugate, for real x_0

$$g^*(z) = (z - x_0)^{n*} = (z^* - x_0)^n = g(z^*). \tag{6.58}$$

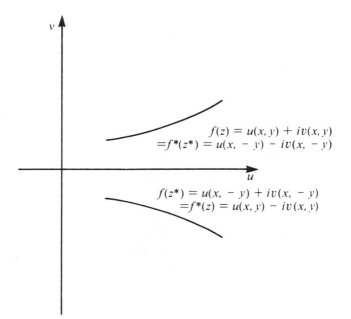

Figure 6.13 Schwarz reflection.

This leads us to the Schwarz reflection principle:

If a function $f(z)$ is (1) analytic over some region including the real axis and (2) real when z is real, then

$$f^*(z) = f(z^*). \qquad (6.59)$$

(See Fig. 6.13).

Expanding $f(z)$ about some (nonsingular) point x_0 on the real axis,

$$f(z) = \sum_{n=0}^{\infty} (z - x_0)^n \frac{f^{(n)}(x_0)}{n!} \qquad (6.60)$$

by Eq. (6.56). Since $f(z)$ is analytic at $z = x_0$, this Taylor expansion exists. Since $f(z)$ is real when z is real, $f^{(n)}(x_0)$ must be real for all n. Then when we use Eq. (6.58), Eq. (6.59), the Schwarz reflection principle, follows immediately. Exercise 6.5.6 is another form of this principle.

ANALYTIC CONTINUATION

It is natural to think of the values $f(z)$ of an analytic function f as a single entity that is usually defined in some restricted region S_1 of the complex plane, for example by a Taylor series (see Fig. 6.14). Then f is analytic inside the circle of convergence C_1, whose radius is given by the distance r_1 from the center of C_1 to the nearest singularity of f at z_1 (in Fig. 6.14). If we choose a point inside C_1 that is farther than r_1 from the singularity z_1 and make a Taylor

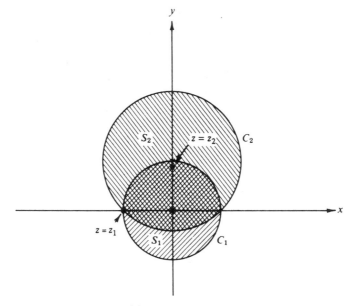

Figure 6.14 Analytic continuation.

expansion of f about it (z_2 in Fig. 6.14), then the circle of convergence C_2 will usually extend beyond the first circle C_1. In the overlap region of both circles C_1, C_2 the function f is uniquely defined. In the region of the circle C_2 that extends beyond C_1, $f(z)$ is uniquely defined by the Taylor series about the center of C_2 and analytic there, although the Taylor series about the center of C_1 is no longer convergent there. After Weierstrass this process is called *analytic continuation*. It defines the analytic function f in terms of its original definition (in C_1, say) and all its continuations.

A specific example is the *meromorphic* function

$$f(z) = \frac{1}{1 + z},\qquad(6.61)$$

which has a simple pole at $z = -1$ and is analytic elsewhere. The geometric series expansion

$$\frac{1}{1 + z} = 1 - z + z^2 + \cdots = \sum_{n=0}^{\infty} (-z)^n \qquad(6.62)$$

converges for $|z| < 1$, i.e., inside the circle C_1 in Fig. 6.14.

Suppose we expand $f(z)$ about $z = i$, so that

$$f(z) = \frac{1}{1 + z} = \frac{1}{1 + i + (z - i)} = \frac{1}{(1 + i)(1 + (z - i)/(1 + i))}$$

$$= \left[1 - \frac{z - i}{1 + i} + \frac{(z - i)^2}{(1 + i)^2} - \cdots \right] \frac{1}{1 + i} \qquad(6.63)$$

converges for $|z - i| < |1 + i| = \sqrt{2}$. Our circle of convergence is C_2 in Fig. 6.14. Now $f(z)$ is defined by the expansion (6.63) in S_2 which overlaps S_1 and extends further out in the complex plane.[1] This extension is an analytic continuation, and when we have only isolated singular points to contend with, the function can be extended indefinitely. Equations (6.61), (6.62), and (6.63) are three different representations of the *same* function. Each representation has its own domain of convergence. Equation (6.62) is a Maclaurin series. Equation (6.63) is a Taylor expansion about $z = i$ and from the following paragraphs Eq. (6.61) is seen to be a one-term Laurent series.

Analytic continuation may take many forms and the series expansion just considered is not necessarily the most convenient technique. As an alternate technique we shall use a recurrence relation in Section 10.1 to extend the factorial function around the isolated singular points, $z = -n$, $n = 1, 2, 3, \ldots$. As another example, the hypergeometric equation is satisfied by the hypergeometric function defined by the series, Eq. (13.114), for $|z| < 1$. The integral representation given in Exercise 13.5.8 permits a continuation over the entire complex plane.

PERMANENCE OF ALGEBRAIC FORM

All our elementary functions, e^z, $\sin z$, and so on can be extended into the complex plane (compare Exercise 6.19). For instance, they can be *defined* by power-series expansions such as

$$e^z = 1 + \frac{z}{1!} + \frac{z^2}{2!} + \cdots = \sum_{n=0}^{\infty} \frac{z^n}{n!} \qquad (6.64)$$

for the exponential. Such definitions agree with the real variable definitions along the real x-axis and literally constitute an analytic continuation of the corresponding real functions into the complex plane. This result is often called *permanence of the algebraic form*.

LAURENT SERIES

We frequently encounter functions that are analytic in an annular region, say, of inner radius r and outer radius R, as shown in Fig. 6.15. Drawing an imaginary contour line to convert our region into a simply connected region,

---

[1] One of the most powerful and beautiful results of the more abstract theory of functions of a complex variable is that if two analytic functions coincide in any region, such as the overlap of S_1 and S_2, or coincide on any line segment, they are the same function in the sense that they will coincide everywhere as long as they are both well defined. In this case the agreement of the expansions (Eqs. (6.62) and (6.63)) over the region common to S_1 and S_2 would establish the identity of the functions these expansions represent. Then Eq. (6.63) would represent an analytic continuation or extension of $f(z)$ into regions not covered by Eq. (6.62). We could equally well say that $f(z) = 1/(1 + z)$ is itself an analytic continuation of either of the series given by Eqs. (6.62) and (6.63).

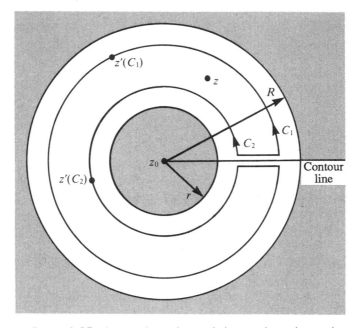

Figure 6.15 $|z' - z_0|_{C_1} > |z - z_0|$; $|z' - z_0|_{C_2} < |z - z_0|$.

we apply Cauchy's integral formula, and for two circles, C_2 and C_1, centered at $z = z_0$ and with radii r_2 and r_1, respectively, where $r < r_2 < r_1 < R$, we have[2]

$$f(z) = \frac{1}{2\pi i} \oint_{C_1} \frac{f(z')\,dz'}{z' - z} - \frac{1}{2\pi i} \oint_{C_2} \frac{f(z')\,dz'}{z' - z}. \tag{6.65}$$

Note carefully that in Eq. (6.65) an explicit minus sign has been introduced so that contour C_2 (like C_1) is to be traversed in the *positive* (counterclockwise) sense. The treatment of Eq. (6.65) now proceeds exactly like that of Eq. (6.53) in the development of the Taylor series. Each denominator is written as $(z' - z_0) - (z - z_0)$ and expanded by the binomial theorem which now follows from the Taylor series (Eq. (6.57)).

Noting that for C_1, $|z' - z_0| > |z - z_0|$ while for C_2, $|z' - z_0| < |z - z_0|$, we find

$$f(z) = \frac{1}{2\pi i} \sum_{n=0}^{\infty} (z - z_0)^n \oint_{C_1} \frac{f(z')\,dz'}{(z' - z_0)^{n+1}}$$

$$+ \frac{1}{2\pi i} \sum_{n=1}^{\infty} (z - z_0)^{-n} \oint_{C_2} (z' - z_0)^{n-1} f(z')\,dz'. \tag{6.66}$$

[2] We may take r_2 arbitrarily close to r and r_1 arbitrarily close to R, maximizing the area enclosed between C_1 and C_2.

The minus sign of Eq. (6.65) has been absorbed by the binomial expansion. Labeling the first series S_1 and the second S_2,

$$S_1 = \frac{1}{2\pi i} \sum_{n=0}^{\infty} (z - z_0)^n \oint_{C_1} \frac{f(z')\, dz'}{(z' - z_0)^{n+1}}, \qquad (6.67)$$

which is the regular Taylor expansion, convergent for $|z - z_0| < |z' - z_0| = r_1$, that is, for all z *interior* to the larger circle, C_1. For the second series in Eq. (6.66) we have

$$S_2 = \frac{1}{2\pi i} \sum_{n=1}^{\infty} (z - z_0)^{-n} \oint_{C_2} (z' - z_0)^{n-1} f(z')\, dz' \qquad (6.68)$$

convergent for $|z - z_0| > |z' - z_0| = r_2$, that is, for all z *exterior* to the smaller circle C_2. Remember, C_2 now goes counterclockwise.

These two series may be combined into one series[3] (a Laurent series) by

$$f(z) = \sum_{n=-\infty}^{\infty} a_n (z - z_0)^n, \qquad (6.69)$$

where

$$a_n = \frac{1}{2\pi i} \oint_C \frac{f(z')\, dz'}{(z' - z_0)^{n+1}}. \qquad (6.70)$$

Since, in Eq. (6.70), convergence of a binomial expansion is no longer a problem, C may be any contour within the annular region $r < |z - z_0| < R$ encircling z_0 once in a counterclockwise sense. If we assume that such an annular region of convergence does exist, Eq. (6.69) is the Laurent series or Laurent expansion of $f(z)$.

The use of the contour line (Fig. 6.15) is convenient in converting the annular region into a simply connected region. Since our function is analytic in this annular region (and therefore single-valued), the contour line is not essential and, indeed, does not appear in the final result, Eq. (6.70). In contrast to this, functions with branch points must have cut lines—Section 7.1.

Laurent series coefficients need not come from evaluation of contour integrals (which may be very intractable). Other techniques such as ordinary series expansions may provide the coefficients.

Numerous examples of Laurent series appear in Chapter 7. We limit ourselves here to one simple example to illustrate the application of Eq. (6.69).

Example 6.5.1

Let $f(z) = [z(z - 1)]^{-1}$. If we choose $z_0 = 0$, then $r = 0$ and $R = 1$, $f(z)$ diverging at $z = 1$. From Eqs. (6.70) and (6.69)

$$a_n = \frac{1}{2\pi i} \oint \frac{dz'}{(z')^{n+2}(z' - 1)}$$

$$= \frac{-1}{2\pi i} \oint \sum_{m=0}^{\infty} (z')^m \frac{dz'}{(z')^{n+2}}. \qquad (6.71)$$

---

[3] Replace n by $-n$ in S_2 and add.

Again, interchanging the order of summation and integration (uniformly convergent series), we have

$$a_n = -\frac{1}{2\pi i} \sum_{m=0}^{\infty} \oint \frac{dz'}{(z')^{n+2-m}}. \tag{6.72}$$

If we employ the polar form, as in Eq. (6.47) (or compare Exercise 6.4.1),

$$a_n = -\frac{1}{2\pi i} \sum_{m=0}^{\infty} \oint \frac{rie^{i\theta}\, d\theta}{r^{n+2-m}e^{i(n+2-m)\theta}}$$

$$= -\frac{1}{2\pi i} \cdot 2\pi i \sum_{m=0}^{\infty} \delta_{n+2-m,1}. \tag{6.73}$$

In other words,

$$a_n = \begin{cases} -1 & \text{for } n \geq -1, \\ 0 & \text{for } n < -1. \end{cases} \tag{6.74}$$

The Laurent expansion (Eq. (6.69)) becomes

$$\frac{1}{z(z-1)} = -\frac{1}{z} - 1 - z - z^2 - z^3 - \cdots$$

$$= -\sum_{n=-1}^{\infty} z^n. \tag{6.75}$$

For this simple function the Laurent series can, of course, be obtained by a direct binomial expansion.

The Laurent series differs from the Taylor series by the obvious feature of negative powers of $(z - z_0)$. For this reason the Laurent series will always diverge at least at $z = z_0$ and perhaps as far out as some distance r (Fig. 6.15).

EXERCISES

6.5.1   Develop the Taylor expansion of $\ln(1 + z)$.   *ANS.* $\sum_{n=1}^{\infty} (-1)^{n-1} \dfrac{z^n}{n}$.

6.5.2   Derive the binomial expansion

$$(1 + z)^m = 1 + mz + \frac{m(m-1)}{1 \cdot 2} z^2 + \cdots$$

$$= \sum_{n=0}^{\infty} \binom{m}{n} z^n$$

for m any real number. The expansion is convergent for $|z| < 1$.

6.5.3   A function $f(z)$ is analytic on and within the unit circle. Also, $|f(z)| \leq 1$ for $|z| \leq 1$ and $f(0) = 0$. Show that $|f(z)| \leq |z|$ for $|z| \leq 1$.
Hint. One approach is to show that $f(z)/z$ is analytic and then express $[f(z_0)/z_0]^n$ by the Cauchy integral formula. Finally, consider absolute magnitudes and take the nth root. This exercise is sometimes called Schwarz's theorem.

6.5.4   If $f(z)$ is a real function of the complex variable $z = x + iy$, i.e., $f(x) = f^*(x)$, and the Laurent expansion about the origin, $f(z) = \sum a_n z^n$, has $a_n = 0$ for $n < -N$, show that all of the coefficients, a_n, are real.
Hint. Show that $z^N f(z)$ is analytic (via Morera's theorem, Section 6.4).

6.5.5   A function $f(z) = u(x, y) + iv(x, y)$ satisfies the conditions for the Schwarz reflection principle. Show that

(a) u is an even function of y.
(b) v is an odd function of y.

6.5.6   A function $f(z)$ can be expanded in a Laurent series about the origin with the coefficients a_n real. Show that the complex conjugate of this function of z is the same function of the complex conjugate of z; that is,

$$f^*(z) = f(z^*).$$

Verify this explicitly for

(a) $f(z) = z^n$, n an integer,
(b) $f(z) = \sin z$.

If $f(z) = iz$, $(a_1 = i)$, show that the foregoing statement does *not* hold.

6.5.7   The function $f(z)$ is analytic in a domain that includes the real axis. When z is real $(z = x)$, $f(x)$ is pure imaginary.
(a) Show that

$$f(z^*) = -[f(z)]^*.$$

(b) For the specific case $f(z) = iz$, develop the Cartesian forms of $f(z)$, $f(z^*)$, and $f^*(z)$. Do *not* quote the general result of part (a).

6.5.8   Develop the first three nonzero terms of the Laurent expansion of

$$f(z) = (e^z - 1)^{-1}$$

about the origin. Notice the resemblance to the Bernoulli number generating function, Eq. (5.144) of Section 5.9.

6.5.9   Prove that the Laurent expansion of a given function about a given point is unique; that is, if

$$f(z) = \sum_{n=-N}^{\infty} a_n(z - z_0)^n = \sum_{n=-N}^{\infty} b_n(z - z_0)^n,$$

show that $a_n = b_n$ for all n.
Hint. Use the Cauchy integral formula.

6.5.10   (a) Develop a Laurent expansion of $f(z) = [z(z - 1)]^{-1}$ about the point $z = 1$ valid for small values of $|z - 1|$. Specify the exact range over which your expansion holds. This is an analytic continuation of Eq. (6.75).
(b) Determine the Laurent expansion of $f(z)$ about $z = 1$ but for $|z - 1|$ large.
Hint. Partial fraction this function and use the geometric series.

6.5.11   (a) Given $f_1(z) = \int_0^{\infty} e^{-zt}\, dt$ (with t real), show that the domain in which $f_1(z)$ exists (and is analytic) is $\Re(z) > 0$.
(b) Show that $f_2(z) = 1/z$ equals $f_1(z)$ over $\Re(z) > 0$ and is therefore an analytic continuation of $f_1(z)$ over the entire z-plane *except for* $z = 0$.
(c) Expand $1/z$ about the point $z = i$. You will have $f_3(z) = \sum_{n=0}^{\infty} a_n(z - i)^n$. What is the domain of $f_3(z)$?

$$ANS. \quad \frac{1}{z} = -i \sum_{n=0}^{\infty} (i)^n(z - i)^n, \qquad |z - i| < 1$$

6.6 MAPPING

In the preceding sections we have defined analytic functions and developed some of their main features. From these developments the integral relations of Chapter 7 follow directly. Here we introduce some of the more geometric aspects of functions of complex variables, aspects that will be useful in visualizing the integral operations in Chapter 7 and that are valuable in their own right in solving Laplace's equation in two-dimensional systems.

In ordinary analytic geometry we may take $y = f(x)$ and then plot y versus x. Our problem here is more complicated, for z is a function of two variables x and y. We use the notation

$$w = f(z) = u(x, y) + iv(x, y). \tag{6.76}$$

Then for a point in the z-plane (specific values for x *and* y) there may correspond specific values for $u(x, y)$ and $v(x, y)$ which then yield a point in the w-plane. As points in the z-plane transform or are mapped into points in the w-plane, lines or areas in the z-plane will be mapped into lines or areas in the w-plane. Our immediate purpose is to see how lines and areas map from the z-plane to the w-plane for a number of simple functions.

TRANSLATION

$$w = z + z_0. \tag{6.77}$$

The function w is equal to the variable z plus a constant, $z_0 = x_0 + iy_0$. By Eqs. (6.1) and (6.76)

$$u = x + x_0,$$
$$v = y + y_0, \tag{6.78}$$

representing a pure translation of the coordinate axes as shown in Fig. 6.16.

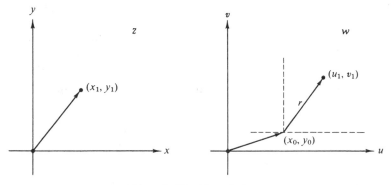

Figure 6.16   Translation.

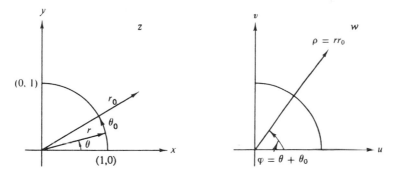

Figure 6.17 Rotation.

ROTATION

$$w = zz_0. \tag{6.79}$$

Here it is convenient to return to the polar representation, using

$$w = \rho e^{i\varphi}, \quad z = r e^{i\theta}, \quad \text{and} \quad z_0 = r_0 e^{i\theta_0}, \tag{6.80}$$

then

$$\rho e^{i\varphi} = r r_0 e^{i(\theta + \theta_0)} \tag{6.81}$$

or

$$\rho = r r_0, \\ \varphi = \theta + \theta_0. \tag{6.82}$$

Two things have occurred. First, the modulus r has been modified, either expanded or contracted, by the factor r_0. Second, the argument θ has been increased by the additive constant θ_0 (Fig. 6.17). This represents a *rotation* of the complex variable through an angle θ_0. For the special case of $z_0 = i$, we have a pure rotation through $\pi/2$ radians.

INVERSION

$$w = \frac{1}{z}. \tag{6.83}$$

Again, using the polar form, we have

$$\rho e^{i\varphi} = \frac{1}{r e^{i\theta}} = \frac{1}{r} e^{-i\theta}, \tag{6.84}$$

which shows that

$$\rho = \frac{1}{r}, \quad \varphi = -\theta. \tag{6.85}$$

The first part of Eq. (6.85) shows that inversion clearly. The interior of the unit circle is mapped onto the exterior and vice versa (Fig. 6.18). In addition,

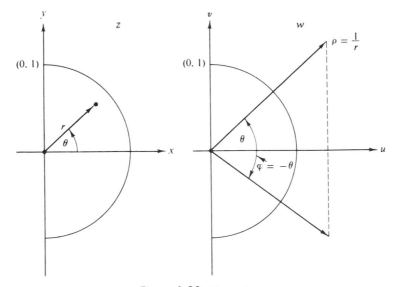

Figure 6.18   Inversion.

the second part of Eq. (6.85) shows that the polar angle is reversed in sign. Equation (6.83) therefore also involves a reflection of the y-axis exactly like the complex conjugate equation.

To see how lines in the z-plane transform into the w-plane, we simply return to the Cartesian form:

$$u + iv = \frac{1}{x + iy}.$$

(6.86)

Rationalizing the right-hand side by multiplying numerator and denominator by z^* and then equating the real parts and the imaginary parts, we have

$$u = \frac{x}{x^2 + y^2}, \qquad x = \frac{u}{u^2 + v^2},$$

$$v = -\frac{y}{x^2 + y^2}, \qquad y = -\frac{v}{u^2 + v^2}.$$

(6.87)

A circle centered at the origin in the z-plane has the form

$$x^2 + y^2 = r^2$$

(6.88)

and by Eq. (6.87) transforms into

$$\frac{u^2}{(u^2 + v^2)^2} + \frac{v^2}{(u^2 + v^2)^2} = r^2.$$

(6.89)

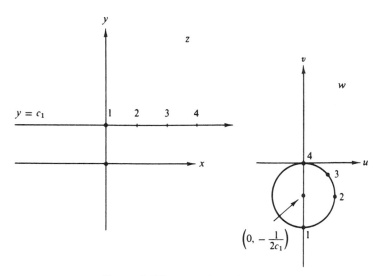

Figure 6.19 Inversion, line ↔ circle.

Simplifying Eq. (6.89), we obtain

$$u^2 + v^2 = \frac{1}{r^2} = \rho^2, \tag{6.90}$$

which describes a circle in the w-plane also centered at the origin.

The horizontal line $y = c_1$ transforms into

$$\frac{-v}{u^2 + v^2} = c_1 \tag{6.91}$$

or

$$u^2 + v^2 + \frac{v}{c_1} + \frac{1}{(2c_1)^2} = \frac{1}{(2c_1)^2}, \tag{6.92}$$

which describes a circle in the w-plane of radius $(1/2)c_1$ and centered at $u = 0$, $v = -\frac{1}{2}c_1$ (Fig. 6.19).

The reader may pick up the other three possibilities, $x = \pm c_1$, $y = -c_1$, by rotating the xy-axes. In general, any straight line or circle in the z-plane will transform into a straight line or a circle in the w-plane (compare Exercise 6.6.1).

BRANCH POINTS AND MULTIVALENT FUNCTIONS

The three transformations just discussed have all involved one-to-one correspondence of points in the z-plane to points in the w-plane. Now to illustrate the variety of transformations that are possible and the problems that can arise, we introduce first a two-to-one correspondence and then a many-to-one correspondence. Finally, we take up the inverses of these two transformations.

Consider first the transformation

$$w = z^2,$$ (6.93)

which leads to

$$\rho = r^2, \qquad \varphi = 2\theta.$$ (6.94)

Clearly, our transformation is nonlinear, for the modulus is squared, but the significant feature of Eq. (6.94) is that the phase angle or argument is doubled. This means that the

first quadrant of z, $0 \le \theta < \dfrac{\pi}{2} \to$ upper half-plane of w, $0 \le \varphi < \pi$,

upper half-plane of z, $0 \le \theta < \pi \to$ whole plane of w, $0 \le \varphi < 2\pi$.

The lower half-plane of z maps into the already covered entire plane of w, thus covering the w-plane a *second* time. This is our two-to-one correspondence, two distinct points in the z-plane, z_0 and $z_0 e^{i\pi} = -z_0$, corresponding to the single point $w = z_0^2$.

In Cartesian representation

$$u + iv = (x + iy)^2$$
$$= x^2 - y^2 + i2xy,$$ (6.95)

leading to

$$u = x^2 - y^2,$$
$$v = 2xy.$$ (6.96)

Hence the lines $u = c_1$, $v = c_2$ in the w-plane correspond to $x^2 - y^2 = c_1$, $2xy = c_2$, rectangular (and orthogonal) hyperbolas in the z-plane (Fig. 6.20). To every point on the hyperbola $x^2 - y^2 = c_1$ in the right half-plane, $x > 0$, one point on the line $u = c_1$ corresponds and vice versa. However, every point on the line $u = c_1$ also corresponds to a point on the hyperbola $x^2 - y^2 = c_1$ in the left half-plane, $x < 0$, as already explained.

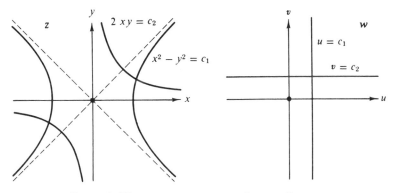

Figure 6.20 Mapping—hyperbolic coordinates.

It will be shown in Section 6.7 that if lines in the w-plane are orthogonal the corresponding lines in the z-plane are also orthogonal, as long as the transformation is analytic. Since $u = c_1$ and $v = c_2$ are constructed perpendicular to each other, the corresponding hyperbolas in the z-plane are orthogonal. We have literally constructed a new orthogonal system of hyperbolic lines (or surfaces if we add an axis perpendicular to x and y). Exercise 2.1.3 was an analysis of this system. It might be noted that if the hyperbolic lines are electric or magnetic lines of force, then we have a quadrupole lens useful in focusing beams of high energy particles.

The inverse of the fourth transformation (Eq. (6.93)) is

$$w = z^{1/2}. \tag{6.97}$$

From the relation

$$\rho e^{i\varphi} = r^{1/2} e^{i\theta/2}, \tag{6.98}$$

and

$$2\varphi = \theta, \tag{6.99}$$

we now have two points in the w-plane (arguments φ and $\varphi + \pi$) corresponding to one point in the z-plane (except for the point $z = 0$). Or, to put it another way, θ and $\theta + 2\pi$ correspond to φ and $\varphi + \pi$, two distinct points in the w-plane. This is the complex variable analog of the simple real variable equation $y^2 = x$, in which two values of y, plus and minus, correspond to each value of x.

The important point here is that we can make the function w of Eq. (6.97) a single-valued function instead of a double-valued function if we agree to restrict θ to a range such as $0 \le \theta < 2\pi$. This may be done by agreeing never to cross the line $\theta = 0$ in the z-plane (Fig. 6.21). Such a line of demarcation is called a cut line.

The *cut line* joins the two branch point singularities at 0 and ∞, where the function is clearly not analytic. Any line from $z = 0$ to infinity would serve equally well. The purpose of the cut line is to restrict the argument of z. The points z and $z \exp(2\pi i)$ coincide in the z-plane but yield different points w

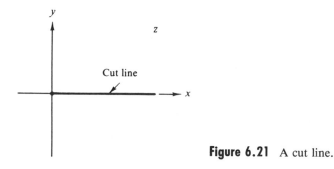

Figure 6.21 A cut line.

and $-w = w \exp(\pi i)$ in the w-plane. Hence in the absence of a cut line the function $w = z^{1/2}$ is ambiguous. Alternatively, since the function $w = z^{1/2}$ is double-valued, we can also glue two sheets of the complex z-plane together along the cut line so that $\arg(z)$ increases beyond 2π along the cut line and steps down from 4π on the second sheet to the start on the first sheet. This construction is called the *Riemann surface* of $w = z^{1/2}$. We shall encounter branch points and cut lines frequently in Chapter 7.

The transformation

$$w = e^z \tag{6.100}$$

leads to

$$\rho e^{i\varphi} = e^{x+iy} \tag{6.101}$$

or

$$\rho = e^x, \tag{6.102}$$
$$\varphi = y.$$

If y ranges from $0 \le y < 2\pi$ (or $-\pi \le y < \pi$), then φ covers the same range. But this is the whole w-plane. In other words, a horizontal strip in the z-plane of width 2π maps into the entire w-plane. Further, any point $x + i(y + 2n\pi)$, in which n is any integer, maps into the same point (by Eq. (6.102)), in the w-plane. We have a many-(infinitely many)-to-one correspondence.

Finally, as the inverse of the fifth transformation (Eq. (6.100)), we have

$$w = \ln z. \tag{6.103}$$

By expanding it, we obtain

$$u + iv = \ln re^{i\theta}$$
$$= \ln r + i\theta. \tag{6.104}$$

For a given point z_0 in the z-plane the argument θ is unspecified within an integral multiple of 2π. This means that

$$v = \theta + 2n\pi, \tag{6.105}$$

and as in the exponential transformation, we have an infinitely many-to-one correspondence.

Equation (6.103) has a nice physical representation. If we go around the unit circle in the z-plane, $r = 1$, and by Eq. (6.104), $u = \ln r = 0$; but $v = \theta$, and θ is steadily increasing and continues to increase as θ continues, past 2π.

The cut line joins the branch point at the origin with infinity. As θ increases past 2π we glue a new sheet of the complex z-plane along the cut-line, etc. Going around the unit circle in the z-plane is like the advance of a screw as it is rotated or the ascent of a person walking up a spiral staircase (Fig. 6.22), which is the Riemann surface of $w = \ln z$.

As in the preceding example, we can also make the correspondence unique (and Eq. (6.103) unambiguous) by restricting θ to a range such as $0 \le \theta < 2\pi$

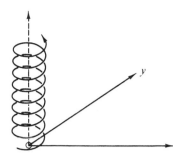

Figure 6.22   This is the Riemann surface for $\ln z$, a multivalued function.

by taking the line $\theta = 0$ (positive real axis) as a cut line. This is equivalent to taking one and only one complete turn of the spiral staircase.

It is because of the multivalued nature of $\ln z$ that the contour integral

$$\oint \frac{dz}{z} = 2\pi i \neq 0,$$

integrating about the origin. This property appears in Exercises 6.4.1 and 6.4.2 and is the basis for the entire calculus of residues (Chapter 7).

The concept of mapping is a very broad and useful one in mathematics. Our mapping from a complex z-plane to a complex w-plane is a simple generalization of one definition of function: a mapping of x (from one set) into y in a second set. A more sophisticated form of mapping appears in Section 1.15 where we use the Dirac delta function $\delta(x - a)$ to map a function $f(x)$ into its value at the point a. Then in Chapter 15 integral transforms are used to map one function $f(x)$ in x-space into a second (related) function $F(t)$ in t-space.

EXERCISES

6.6.1   How do circles centered on the origin in the z-plane transform for

(a) $w_1(z) = z + \dfrac{1}{z}$,   (b) $w_2(z) = z - \dfrac{1}{z}$,   for $z \neq 0$?

What happens when $|z| \to 1$?

6.6.2   What part of the z-plane corresponds to the interior of the unit circle in the w-plane if

(a) $w = \dfrac{z - 1}{z + 1}$,   (b) $w = \dfrac{z - i}{z + i}$?

6.6.3   Discuss the transformations

(a) $w(z) = \sin z$,   (c) $w(z) = \sinh z$,

(b) $w(z) = \cos z$,   (d) $w(z) = \cosh z$.

Show how the lines $x = c_1$, $y = c_2$ map into the w-plane. Note that the last three transformations can be obtained from the first one by appropriate translation and/or rotation.

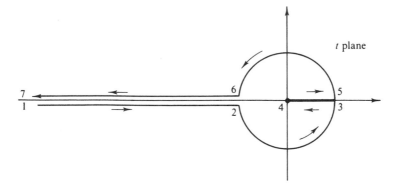

Figure 6.23 Bessel function integration contour.

6.6.4   Show that the function
$$w(z) = (z^2 - 1)^{1/2}$$
is single-valued if we take $-1 \le x \le 1$, $y = 0$ as a cut line.

6.6.5   Show that negative numbers have logarithms in the complex plane. In particular, find $\ln(-1)$.   *ANS.* $\ln(-1) = i\pi$.

6.6.6   An integral representation of the Bessel function follows the contour in the t-plane shown in Fig. 6.23. Map this contour into the θ-plane with $t = e^\theta$. Many additional examples of mapping are given in Chapters 11, 12, and 13.

6.6.7   For noninteger m, show that the binomial expansion of Exercise 6.5.2 holds only for a suitably defined branch of the function $(1 + z)^m$. Show how the z-plane is cut. Explain why $|z| < 1$ may be taken as the circle of convergence for the expansion of this branch, in light of the cut you have chosen.

6.6.8   The Taylor expansion of Exercises 6.5.2 and 6.6.7 is *not* suitable for branches other than the one suitably defined branch of the function $(1 + z)^m$ for noninteger m. [Note that other branches cannot have the same Taylor expansion since they must be distinguishable.] Using the same branch cut of the earlier exercises for all other branches, find the corresponding Taylor expansions detailing the phase assignments and Taylor coefficients.

6.7 CONFORMAL MAPPING

In Section 6.6 hyperbolas were mapped into straight lines and straight lines were mapped into circles. Yet in all these transformations one feature stayed constant. This constancy was a result of the fact that all the transformations of Section 6.6 were analytic.

As long as $w = f(z)$ is an analytic function, we have

$$\frac{df}{dz} = \frac{dw}{dz} = \lim_{\Delta z \to 0} \frac{\Delta w}{\Delta z}. \tag{6.106}$$

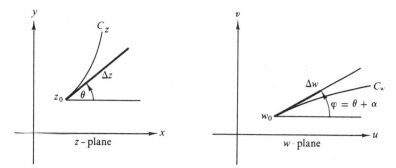

Figure 6.24   Conformal mapping—preservation of angles.

Assuming that this equation is in polar form, we may equate modulus to modulus and argument to argument. For the latter (assuming that $df/dz \neq 0$)

$$\arg \lim_{\Delta z \to 0} \frac{\Delta w}{\Delta z} = \lim_{\Delta z \to 0} \arg \frac{\Delta w}{\Delta z}$$

$$= \lim_{\Delta z \to 0} \arg \Delta w - \lim_{\Delta z \to 0} \arg \Delta z$$

$$= \arg \frac{df}{dz} = \alpha, \tag{6.107}$$

where α, the argument of the derivative, may depend on z but is a constant for a fixed z, independent of the direction of approach. To see the significance of this, consider two curves, C_z in the z-plane and the corresponding curve C_w in the w-plane (Fig. 6.24). The increment Δz is shown at an angle of θ relative to the real (x) axis whereas the corresponding increment Δw forms an angle of φ with the real (u) axis. From Eq. (6.107)

$$\varphi = \theta + \alpha, \tag{6.108}$$

or any line in the z-plane is rotated through an angle α in the w-plane as long as w is an analytic transformation and the derivative is not zero.[1]

Since this result holds for any line through z_0, it will hold for a pair of lines. Then for the angle between these two lines

$$\varphi_2 - \varphi_2 = (\theta_2 + \alpha) - (\theta_1 + \alpha) = \theta_2 - \theta_1, \tag{6.109}$$

which shows that the included angle is preserved under an analytic transformation. Such angle-preserving transformations are called *conformal*. The rotation angle α will, in general, depend on z. In addition $|f'(z)|$ will, usually, be a function of z.

---

[1] If $df/dz = 0$, its argument or phase is undefined and the (analytic) transformation will not necessarily preserve angles.

Historically, these conformal transformations have been of great importance to scientists and engineers in solving Laplace's equation for problems of electrostatics, hydrodynamics, heat flow, and so on. Unfortunately, the conformal transformation approach, however elegant, is limited to problems that can be reduced to two dimensions. The method is often beautiful if there is a high degree of symmetry present but often impossible if the symmetry is broken or absent. Because of these limitations and primarily because high-speed electronic computers offer a useful alternative (iterative solution of the partial differential equation), the details and applications of conformal mappings are omitted.

EXERCISES

6.7.1   Expand $w(x)$ in a Talyor series about the point $z = z_0$ where $f'(z_0) = 0$. (Angles are not preserved.) Show that if the first $n - 1$ derivatives vanish but $f^{(n)}(z_0) \neq 0$, then angles in the z-plane with vertices at $z = z_0$ appear in the w-plane multiplied by n.

6.7.2   Develop the transformations that create each of the four cylindrical coordinate systems:

(a) Circular cylindrical $x = \rho \cos \varphi,$
$y = \rho \sin \varphi.$

(b) Elliptic cylindrical $x = a \cosh u \cos v,$
$y = a \sinh u \sin v.$

(c) Parabolic cylindrical $x = \xi \eta,$
$y = \frac{1}{2}(\eta^2 - \xi^2).$

(d) Bipolar $x = \dfrac{a \sinh \eta}{\cosh \eta - \cos \xi},$

$y = \dfrac{a \sin \xi}{\cosh \eta - \cos \xi}.$

Note. These transformations are not necessarily analytic.

6.7.3   In the transformation

$$e^z = \frac{a - w}{a + w}$$

how do the coordinate lines in the z-plane transform? What coordinate system have you constructed?

ADDITIONAL READINGS

AHLFORS, L. V., *Complex Analysis*, 3rd ed. New York: McGraw-Hill (1979). This text is detailed, thorough, rigorous, and extensive.

CHURCHILL, R. V., J. W. BROWN, and R. F. VERKEY, *Complex Variables and Applications*, 3rd ed. New York: McGraw-Hill (1974). This is an excellent text for both the beginning and advanced student. It is readable and quite complete. A detailed proof of the Cauchy–Goursat theorem is given in Chapter 5.

GREENLEAF, F. P., *Introduction to Complex Variables*. Philadelphia: Saunders (1972). This very readable book has detailed, careful explanations.

KYRALA, A., *Applied Functions of a Complex Variable*. New York: Wiley (Interscience) (1972). An intermediate level text designed for scientists and engineers. Includes many physical applications.

LEVINSON, N., and R. M. REDHEFFER, *Complex Variables*. San Francisco: Holden-Day (1970). This text is written for scientists and engineers who are interested in applications.

MORSE, P. M., and H. FESHBACH, *Methods of Theoretical Physics*. New York: McGraw-Hill (1953). Chapter 4 is a presentation of portions of the theory of functions of a complex variable of interest to theoretical physicists.

REMMERT, R., *Theory of Complex Functions*. New York: Springer (1991).

SOKOLNIKOFF, I. S., and R. M. REDHEFFER, *Mathematics of Physics and Modern Engineering*, 2nd ed. New York: McGraw-Hill (1966). Chapter 7 covers complex variables.

SPIEGEL, M. R., *Theory and Problems of Complex Variables*. New York: Schaum (1964). An excellent summary of the theory of complex variables for scientists.

TITCHMARSH, E. C., *The Theory of Functions*, 2nd ed. New York: Oxford University Press (1958).

WATSON, G. N., *Complex Integration and Cauchy's Theorem*. New York: Hafner (orig. 1917, reprinted, 1960). A short work containing a rigorous development of the Cauchy integral theorem and integral formula. Applications to the calculus of residues are included. *Cambridge Tracts in Mathematics, and Mathematical Physics*, No. 15.

Other references are given at the end of Chapter 15.

7 FUNCTIONS OF A COMPLEX VARIABLE II

Calculus of Residues

7.1 SINGULARITIES

In this chapter we return to the line of analysis that started with the Cauchy–Riemann conditions in Chapter 6 and led on through the Laurent expansion (Section 6.5). The Laurent expansion represents a generalization of the Taylor series in the presence of singularities. We define the point z_0 as an isolated singular point of the function $f(z)$ if $f(z)$ is not analytic at $z = z_0$ but is analytic at neighboring points. A function that is analytic throughout the entire finite complex plane *except* for isolated poles is called *meromorphic*.

POLES

In the Laurent expansion of $f(z)$ about z_0

$$f(z) = \sum_{n=-\infty}^{\infty} a_n(z - z_0)^n. \tag{7.1}$$

If $a_n = 0$ for $n < -m < 0$ and $a_{-m} \neq 0$, we say that z_0 is a pole of order m. For instance, if $m = 1$; that is, if $a_{-1}/(z - z_0)$ is the first nonvanishing term in the Laurent series, we have a pole of order one, often called a simple pole.

If, on the other hand, the summation continues to $n = -\infty$, then z_0 is a pole of infinite order and is called an essential singularity. These essential singularities have many pathological features. For instance, we can show that in any small neighborhood of an essential singularity of $f(z)$ the function $f(z)$ comes arbitrarily close to any (and therefore every) preselected complex quantity w_0.[1] Literally, the entire w-plane is mapped into the neighborhood

---

[1] This theorem is due to Picard. A proof is given by E. C. Titchmarsh, *The Theory of Functions*, 2nd ed. New York: Oxford University Press (1939).

of the point z_0. One point of fundamental difference between a pole of finite order and an essential singularity is that a pole of order m can be removed by multiplying $f(z)$ by $(z - z_0)^m$. This obviously cannot be done for an essential singularity.

The behavior of $f(z)$ as $z \to \infty$ is defined in terms of the behavior of $f(1/t)$ as $t \to 0$. Consider the function

$$\sin z = \sum_{n=0}^{\infty} \frac{(-1)^n z^{2n+1}}{(2n + 1)!}. \tag{7.2}$$

As $z \to \infty$, we replace the z by $1/t$ to obtain

$$\sin\left(\frac{1}{t}\right) = \sum_{n=0}^{\infty} \frac{(-1)^n}{(2n + 1)! t^{2n+1}}. \tag{7.3}$$

Clearly, from the definition, $\sin z$ has an essential singularity at infinity. This result could be anticipated from Exercise 6.1.9 since

$$\sin z = \sin iy, \qquad \text{when } x = 0,$$

$$= i \sinh y,$$

which approaches infinity exponentially as $y \to \infty$. Thus, although the absolute value of $\sin x$ for real x is equal to or less than one, the absolute value of $\sin z$ is not bounded.

BRANCH POINTS

There is another sort of singularity that will be important in the later sections of this chapter. Consider

$$f(z) = z^a,$$

in which a is not an interger.[2] As z moves around the unit circle from e^0 to $e^{2\pi i}$,

$$f(z) \to e^{2\pi a i} \neq e^{0i},$$

for nonintegral a. As in Section 6.6, we have a branch point at the origin and another at infinity. The points e^{0i} and $e^{2\pi i}$ in the z-plane coincide but these coincident points lead to *different* values of $f(z)$; that is, $f(z)$ is a multivalued function. The problem is resolved by constructing a cut line joining both branch points so that $f(z)$ will be uniquely specified for a given point in the z-plane.

Note carefully that a function with a branch point and a required cut line will not be continuous across the cut line. In general, there will be a phase

---

[2] $z = 0$ is technically a singular point, for z^a has only a finite number of derivatives, whereas an analytic function is guaranteed an infinite number of derivatives (Section 6.4). The problem is that $f(z)$ is *not* single-valued as we encircle the origin. The Cauchy integral formula may not be applied.

difference on opposite sides of this cut line. Exercises 7.2.18, 7.2.19, and 7.2.23 are examples for this situation. Hence line integrals on opposite sides of this branch point cut line will not generally cancel each other. Numerous examples of this appear in the exercises.

The contour line used to convert a multiply connected region into a simply connected region (Section 6.3) is completely different. Our function is continuous across this contour line, and no phase difference exists.

Example 7.11

Consider the function

$$f(z) = (z^2 - 1)^{1/2} = (z + 1)^{1/2}(z - 1)^{1/2}. \qquad (7.4)$$

The first factor on the right-hand side, $(z + 1)^{1/2}$, has a branch point at $z = -1$. The second factor has a branch point at $z = +1$. At infinity $f(z)$ has a simple pole. The cut line has to connect both branch points. To check on the possibility of taking the line segment joining $z = +1$ and $z = -1$ as a cut line, let us follow the phases of these two factors as we move along the contour shown in Fig. 7.1.

For convenience in following the changes of phase let $z + 1 = re^{i\theta}$ and $z - 1 = \rho e^{i\varphi}$. Then the phase of $f(z)$ is $(\theta + \varphi)/2$. We start at point 1 where both $z + 1$ and $z - 1$ have a phase of zero. Moving from point 1 to point 2, φ, the phase of $z - 1 = \rho e^{i\varphi}$ increases by π. ($z - 1$ becomes negative.) φ then stays constant until the circle is completed, moving from 6 to 7. θ, the phase of $z + 1 = re^{i\theta}$ shows a similar behavior increasing by 2π as we move from 3 to 5. The phase of the function $f(z) = (z + 1)^{1/2}(z - 1)^{1/2} = r^{1/2}\rho^{1/2}e^{i(\theta+\varphi)/2}$ is $(\theta + \varphi)/2$. This is tabulated in the final column of Table 7.1.

Two features emerge;

1. The phase at points 5 and 6 is not the same as the phase at points 2 and 3. This behavior can be expected at a branch point cut line.

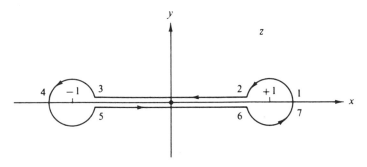

Figure 7.1

Table 7.1 Phase Angle

Point	θ	φ	$(\theta + \varphi)/2$
1	0	0	0
2	0	π	$\pi/2$
3	0	π	$\pi/2$
4	π	π	π
5	2π	π	$3\pi/2$
6	2π	π	$3\pi/2$
7	2π	2π	2π

2. The phase at point 7 exceeds that at point 1 by 2π and the function $f(z) = (z^2 - 1)^{1/2}$ is therefore *single-valued* for the contour shown, encircling *both* branch points.

If we take the x-axis $-1 \leq x \leq 1$ as a cut line, $f(z)$ is uniquely specified. Alternatively, the positive x-axis for $x > 1$ and the negative x-axis for $x < -1$ may be taken as cut lines. The branch points cannot be encircled and the function remains single-valued.

Generalizing from this example, we have that the phase of a function

$$f(z) = f_1(z) \cdot f_2(z) \cdot f_3(z) \cdots$$

is the algebraic sum of the phase of its individual factors:

$$\arg f(z) = \arg f_1(z) + \arg f_2(z) + \arg f_3(z) + \cdots.$$

The phase of an individual factor may be taken as the arctangent of the ratio of its imaginary part to its real part,

$$\arg f_i(z) = \tan^{-1}(v_i/u_i).$$

For the case of a factor of the form

$$f_i(z) = (z - z_0)$$

the phase corresponds to the phase angle of a two-dimensional vector from $+z_0$ to z, the phase increasing by 2π as the point $+z_0$ is encircled. Conversely, the traversal of any closed loop not encircling z_0 does *not* change the phase of $z - z_0$.

EXERCISES

7.1.1   The function $f(z)$ expanded in a Laurent series exhibits a pole of order m at $z = z_0$. Show that the coefficient of $(z - z_0)^{-1}$, a_{-1}, is given by

$$a_{-1} = \frac{1}{(m - 1)!} \frac{d^{m-1}}{dz^{m-1}} \left[(z - z_0)^m f(z) \right]_{z = z_0},$$

with

$$a_{-1} = [(z - z_0)f(z)]_{z=z_0},$$

when the pole is a simple pole ($m = 1$). These equations for a_{-1} are extremely useful in determining the residue to be used in the residue theorem of the next section.

Hint. The technique that was so successful in proving the uniqueness of power series, Section 5.7, will work here also.

7.1.2   A function $f(z)$ can be represented by

$$f(z) = \frac{f_1(z)}{f_2(z)}.$$

in which $f_1(z)$ and $f_2(z)$ are analytic. The denominator $f_2(z)$ vanishes at $z = z_0$ showing that $f(z)$ has a pole at $z = z_0$. However, $f_1(z_0) \neq 0$, $f_2'(z_0) \neq 0$. Show that a_{-1}, the coefficient of $(z - z_0)^{-1}$ in a Laurent expansion of $f(z)$ at $z = z_0$, is given by

$$a_{-1} = \frac{f_1(z_0)}{f_2'(z_0)}.$$

This result leads to the Heaviside expansion theorem, Section 15.12.

7.1.3   In analogy with Example 7.1.1 consider in detail the phase of each factor and the resultant overall phase of $f(z) = (z^2 + 1)^{1/2}$ following a contour similar to that of Fig. 7.1, but encircling the new branch points.

7.1.4   The Legendre function of the second kind, $Q_\nu(z)$, has branch points at $z = \pm 1$. The branch points are joined by a cut line along the real (x) axis.

(a) Show that $Q_0(z) = \frac{1}{2} \ln((z + 1)/(z - 1))$ is single-valued (with the real axis $-1 \leq x \leq 1$ taken as a cut line).

(b) For real argument x and $|x| < 1$ it is convenient to take

$$Q_0(x) = \frac{1}{2} \ln[(1 + x)/(1 - x)].$$

Show that

$$Q_0(x) = \frac{1}{2}[Q_0(x + i0) + Q_0(x - i0)].$$

Here $x + i0$ indicates z approaches the real axis from above, $x - i0$ indicates an approach from below.

7.1.5   As an example of an essential singularity consider $e^{1/z}$ as z approaches zero. For any complex number z_0, $z_0 \neq 0$, show that

$$e^{1/z} = z_0$$

has an infinite number of solutions.

7.2 CALCULUS OF RESIDUES

RESIDUE THEOREM

If the Laurent expansion of a function $f(z) = \sum_{n=-\infty}^{\infty} a_n(z - z_0)^n$ is integrated term by term by using a closed contour that encircles one isolated singular point z_0 once in a counterclockwise sense, we obtain (Exercise 6.4.1)

$$a_n \oint (z - z_0)^n \, dz = a_n \frac{(z - z_0)^{n+1}}{n + 1} \Big|_{z_1}^{z_1}$$

$$= 0 \quad \text{for all } n \neq -1. \tag{7.5}$$

However, if $n = -1$,

$$a_{-1} \oint (z - z_0)^{-1} \, dz = a_{-1} \oint \frac{ire^{i\theta} \, d\theta}{re^{i\theta}} = 2\pi i a_{-1}. \tag{7.6}$$

Summarizing Eqs. (7.5) and (7.6), we have

$$\frac{1}{2\pi i} \oint f(z) \, dz = a_{-1}. \tag{7.7}$$

The constant a_{-1}, the coefficient of $(z - z_0)^{-1}$ in the Laurent expansion, is called the *residue* of $f(z)$ at $z = z_0$.

A set of isolated singularities can be handled very nicely by deforming our contour as shown in Fig. 7.2. Cauchy's integral theorem (Section 6.3) leads to

$$\oint_C f(z) \, dz + \oint_{C_0} f(z) \, dz + \oint_{C_1} f(z) \, dz + \oint_{C_2} f(z) \, dz + \cdots = 0. \tag{7.8}$$

The circular integral around any given singular point is given by Eq. (7.7).

$$\oint_{C_i} f(z) \, dz = -2\pi i a_{-1z_i} \tag{7.9}$$

assuming a Laurent expansion about the singular point, $z = z_i$. The negative sign comes from the clockwise integration as shown in Fig. 7.2. Combining Eqs. (7.8) and (7.9), we have

$$\oint_C f(z) \, dz = 2\pi i (a_{-1z_0} + a_{-1z_1} + a_{-1z_2} + \cdots)$$

$$= 2\pi i (\text{sum of enclosed residues}). \tag{7.10}$$

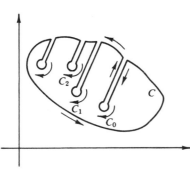

Figure 7.2 Excluding isolated singularities.

This is the *residue theorem*. The problem of evaluating one or more contour integrals is replaced by the algebraic problem of computing residues at the enclosed singular points.

We first use this residue theorem to develop the concept of the Cauchy principal value. Then in the remainder of this section we apply the residue theorem to a wide variety of definite integrals of mathematical and physical interest. In Section 7.3 the concept of Cauchy principal value is used to obtain the important dispersion relations. The residue theorem will also be needed in Chapter 16 for a variety of integral transforms, particularly the inverse Laplace transform.

Using the transformation $z = 1/w$ for $w \to 0$, we can find the nature of a singularity at $z \to \infty$ and the residue of a function $f(z)$ with just isolated singularities and no branch points. In such cases we know that

$$\sum \{\text{residues in the finite } z\text{-plane}\} + \{\text{residue at } z \to \infty\} = 0.$$

CAUCHY PRINCIPAL VALUE

Occasionally an isolated first-order pole will be directly on the contour of integration. In this case we may deform the contour to include or exclude the residue as desired by including a semicircular detour of *infinitesimal radius*. This is shown in Fig. 7.3. The integration over the semicircle then gives with $z - x_0 = \delta e^{i\varphi}$, $dz = i\delta e^{i\varphi}\, d\varphi$,

$$\int \frac{dz}{z - x_0} = i \int_{\pi}^{2\pi} d\varphi = i\pi, \text{ i.e., } \pi i a_{-1} \qquad \text{if counterclockwise,}$$

$$\int \frac{dz}{z - x_0} = i \int_{\pi}^{0} d\varphi = -i\pi, \text{ i.e., } -\pi i a_{-1} \qquad \text{if clockwise.}$$

This contribution, $+$ or $-$, appears on the left-hand side of Eq. (7.10). If our detour were clockwise, the residue would not be enclosed and there would be no corresponding term on the right-hand side of Eq. (7.10). However, if our detour were counterclockwise, this residue would be enclosed by the contour C and a term $2\pi i a_{-1}$ would appear on the right-hand side of Eq. (7.10). The net result for either clockwise or counterclockwise detour is that a simple pole on the contour is counted as one half what it would be if it were within the contour. This corresponds to taking the Cauchy principal value.

For instance, let us suppose that $f(z)$ with a simple pole at $z = x_0$ is integrated over the entire real axis. The contour is closed with an infinite

x_0

Figure 7.3 Bypassing singular points.

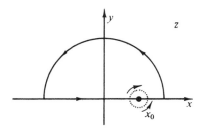

Figure 7.4 Closing the contour with an infinite radius semicircle.

semicircle in the upper half-plane (Fig. 7.4). Then

$$\oint f(z)\, dz = \int_{-\infty}^{x_0-\delta} f(x)\, dx + \int_{C_{x_0}} f(z)\, dz$$

$$+ \int_{x_0+\delta}^{\infty} f(x)\, dx + \int_{C \text{ infinite semicircle}}$$

$$= 2\pi i \sum enclosed \text{ residues.} \tag{7.11}$$

If the small semicircle C_{x_0} includes x_0 (by going below the x-axis, counter-clockwise), x_0 is enclosed, and its contribution appears *twice*—as $\pi i a_{-1}$ in $\int_{C_{x_0}}$ and as $2\pi i a_{-1}$ in the term $2\pi i \sum$ enclosed residues—for a net contribution of $\pi i a_{-1}$. If the upper small semicircle is elected, x_0 is excluded. The only contribution is from the *clockwise* integration over C_{x_0} which yields $-\pi i a_{-1}$. Moving this to the extreme right of Eq. (7.11), we have $+\pi i a_{-1}$, as before.

The integrals along the x-axis may be combined and the semicircle radius permitted to approach zero. We therefore define

$$\lim_{\delta \to 0} \left\{ \int_{-\infty}^{x_0-\delta} f(x)\, dx + \int_{x_0+\delta}^{\infty} f(x)\, dx \right\} = P \int_{-\infty}^{\infty} f(x)\, dx. \tag{7.12}$$

P indicates the Cauchy principal value and represents the preceding limiting process. Note carefully that the Cauchy principal value is a balancing or canceling process. In the vicinity of our singularity at $z = x_0$,

$$f(x) \approx \frac{a_{-1}}{x - x_0}. \tag{7.13}$$

This is odd, relative to x_0. The symmetric or even interval (relative to x_0) provides cancellation of the shaded areas, Fig. 7.5. The contribution of the singularity is in the integration about the semicircle.

Sometimes, this same limiting technique is applied to the integration limits $\pm\infty$. We may define

$$P \int_{-\infty}^{\infty} f(x)\, dx = \lim_{a \to \infty} \int_{-a}^{a} f(x)\, dx. \tag{7.14}$$

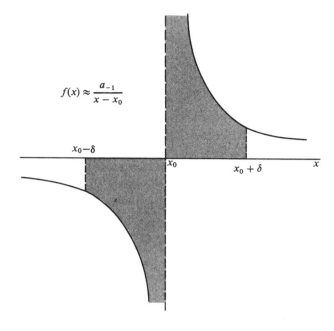

$$f(x) \approx \frac{a_{-1}}{x - x_0}$$

Figure 7.5

An alternate treatment moves the pole off the contour and then considers the limiting behavior as it is brought back. This technique is illustrated in Example 7.2.4, in which the singular points are moved off the contour in such a way that the solution is forced into the form desired to satisfy the boundary conditions of the physical problem.

POLE EXPANSION OF MEROMORPHIC FUNCTIONS

Analytic functions $f(z)$ that have only well separated poles as singularities are called *meromorphic*. For simplicity we assume that these poles at finite $z = a_n$ with $0 < |a_1| < |a_2| < \cdots$ are all simple with residues b_n. Then an expansion of $f(z)$ in terms of $b_n(z - a_n)^{-1}$ depends only on intrinsic properties of $f(z)$, in contrast to the Taylor expansion about an arbitrary analytic point z_0 of $f(z)$ or the Laurent expansion about some singular point of $f(z)$.

Let us consider a series of concentric circles C_n about the origin so that C_n includes $a_1, a_2, \ldots, a_n$ but no other poles, its radius $R_n \to \infty$ as $n \to \infty$. To guarantee convergence we assume that $|f(z)| < \varepsilon R_n$ for any small positive constant ε and all z on C_n. Then the series

$$f(z) = f(0) + \sum_{n=1}^{\infty} b_n \{(z - a_n)^{-1} + a_n^{-1}\} \qquad (7.15)$$

converges to $f(z)$. To prove this theorem (due to Mittag–Leffler) we use the residue theorem to evaluate the contour integral for z inside C_n:

$$I_n = (2\pi i)^{-1} \int_{C_n} \frac{f(w)}{w(w-z)}\, dw$$

$$= \sum_{m=1}^{n} \frac{b_m}{a_m(a_m - z)} + \frac{f(z) - f(0)}{z}. \tag{7.16}$$

On C_n we have for $n \to \infty$

$$|I_n| \le 2\pi R_n \max_{w \text{ on } C_n} \frac{|f(w)|}{2\pi R_n(R_n - |z|)} < \frac{\varepsilon R_n}{R_n - |z|} \le \varepsilon.$$

Using $I_n \to 0$ in Eq. (7.16) proves Eq. (7.15).

If $|f(z)| < \varepsilon R_n^{p+1}$, then we evaluate similarly the integral

$$I_n = (2\pi i)^{-1} \int \frac{f(w)}{w^{p+1}(w-z)}\, dw \to 0 \qquad \text{as } n \to \infty$$

and obtain the analogous pole expansion

$$f(z) = f(0) + zf'(0) + \cdots + \frac{z^p f^{(p)}(0)}{p!} + \sum_{n=1}^{\infty} \frac{b_n z^{p+1}/a_n^{p+1}}{z - a_n}. \tag{7.17}$$

PRODUCT EXPANSION OF ENTIRE FUNCTIONS

A function $f(z)$ that is analytic for all finite z is called an *entire* function. The logarithmic derivative f'/f is a meromorphic function with a pole expansion.

If $f(z)$ has a simple zero at $z = a_n$, then $f(z) = (z - a_n)g(z)$ with analytic $g(z)$ and $g(a_n) \ne 0$. Hence the logarithmic derivative

$$\frac{f'(z)}{f(z)} = (z - a_n)^{-1} + \frac{g'(z)}{g(z)} \tag{7.18}$$

has a simple pole at $z = a_n$ with residue 1, and g'/g is analytic there. If f'/f satisfies the conditions that lead to the pole expansion in Eq. (7.15), then

$$\frac{f'(z)}{f(z)} = \frac{f'(0)}{f(0)} + \sum_{n=1}^{\infty} \left[\frac{1}{a_n} + \frac{1}{z - a_n} \right] \tag{7.19}$$

holds. Integrating Eq. (7.19) yields

$$\int_0^z \frac{f'(z)}{f(z)}\, dz = \ln f(z) - \ln f(0)$$

$$= \frac{zf'(0)}{f(0)} + \sum_{n=1}^{\infty} \left\{ \ln(z - a_n) - \ln(-a_n) + \frac{z}{a_n} \right\},$$

and exponentiating we obtain the product expansion

$$f(z) = f(0) \exp \frac{zf'(0)}{f(0)} \prod_{1}^{\infty} \left(1 - \frac{z}{a_n}\right) \exp \frac{z}{a_n}. \qquad (7.20)$$

Examples are the product expansions (see Chapter 5) for

$$\sin z = z \prod_{\substack{n = -\infty \\ n \neq 0}}^{\infty} \left(1 - \frac{z}{n\pi}\right) e^{z/n\pi} = z \prod_{n=1}^{\infty} \left(1 - \frac{z^2}{n^2\pi^2}\right),$$

$$\cos z = \prod_{n=1}^{\infty} \left\{1 - \frac{z^2}{(n - 1/2)^2\pi^2}\right\}. \qquad (7.21)$$

Another example is the product expansion of the gamma function which will be discussed in Chapter 10.

As a consequence of Eq. (7.18) the contour integral of the logarithmic derivative may be used to count the number N_f of zeros (including their multiplicities) of the function $f(z)$ inside the contour C:

$$(2\pi i)^{-1} \int_C \frac{f'(z)}{f(z)} dz = N_f. \qquad (7.22)$$

Moreover, using

$$\int \frac{f'(z)}{f(z)} dz = \ln f(z) = \ln|f(z)| + i \arg f(z), \qquad (7.23)$$

we see that the real part in Eq. (7.23) does not change as z moves once around the contour, while the corresponding change in $\arg f$ must be

$$\Delta_C \arg(f) = 2\pi N_f. \qquad (7.24)$$

This leads to *Rouché's theorem*: If $f(z)$ and $g(z)$ are analytic inside and on a closed contour C, and $|g(z)| < |f(z)|$ on C, then $f(z)$ and $f(z) + g(z)$ have the same number of zeros inside C.

To show this we use

$$2\pi N_{f+g} = \Delta_C \arg(f + g) = \Delta_C \arg(f) + \Delta_C \arg\left(1 + \frac{g}{f}\right).$$

Since $|g| < |f|$ on C, the point $w = 1 + g(z)/f(z)$ is always an interior point of the circle in the w-plane with center at 1 and radius 1. Hence $\arg(1 + g/f)$ must return to its original value when z moves around C; it cannot decrease or increase by a multiple of 2π so that $\Delta_C \arg(1 + g/f) = 0$.

Rouché's theorem may be used for an alternative proof of the fundamental theorem of algebra: A polynomial $\sum_{m=0}^{n} a_m z^m$ with $a_n \neq 0$ has n zeros. We define $f(z) = a_n z^n$. Then f has an n-fold zero at the origin and no other zeros. Let $g(z) = \sum_{0}^{n-1} a_m z^m$. We apply Rouché's theorem to a circle C with center at the origin and radius $R > 1$. On C, $|f(z)| = |a_n| R^n$ and

$$|g(z)| \leq |a_0| + |a_1| R + \cdots + |a_{n-1}| R^{n-1} \leq \left(\sum_0^{n-1} |a_m| \right) R^{n-1}.$$

Hence $|g(z)| < |f(z)|$ for z on C provided $R > (\sum_0^{n-1} |a_m|)/|a_n|$. For all sufficiently large circles C, therefore, $f + g = \sum_{m=0}^{n} a_m z^m$ has n zeros inside C according to Rouché's theorem.

EVALUATION OF DEFINITE INTEGRALS

Definite integrals appear repeatedly in problems of mathematical physics as well as in pure mathematics. Three moderately general techniques are useful in evaluating definite integrals: (1) contour integration, (2) conversion to gamma or beta functions (Chapter 10), and (3) numerical quadrature (Appendix A2). Other approaches include series expansion with term-by-term integration and integral transforms. As will be seen subsequently, the method of contour integration is perhaps the most versatile of these methods, since it is applicable to a wide variety of integrals.

EVALUATION OF DEFINITE INTEGRALS: $\int_0^{2\pi} f(\sin \theta, \cos \theta) \, d\theta$

The calculus of residues is useful in evaluating a wide variety of definite integrals in both physical and purely mathematical problems. We consider, first, integrals of the form

$$I = \int_0^{2\pi} f(\sin \theta, \cos \theta) \, d\theta, \tag{7.25}$$

where f is finite for all values of θ. We also require f to be a rational function of $\sin \theta$ and $\cos \theta$ so that it will be single-valued. Let

$$z = e^{i\theta}, \qquad dz = ie^{i\theta} \, d\theta.$$

From this,

$$d\theta = -i \frac{dz}{z}, \qquad \sin \theta = \frac{z - z^{-1}}{2i}, \qquad \cos \theta = \frac{z + z^{-1}}{2}. \tag{7.26}$$

Our integral becomes

$$I = -i \oint f\left(\frac{z - z^{-1}}{2i}, \frac{z + z^{-1}}{2} \right) \frac{dz}{z}, \tag{7.27}$$

with the path of integration the unit circle. By the residue theorem Eq. (7.20),

$$I = (-i)2\pi i \sum \text{ residues within the unit circle.} \tag{7.28}$$

Note that we are after the residues of $f(z)/z$. Illustrations of integrals of this type are provided by Exercises 7.2.7–7.2.10.

Example 7.2.1

Our problem is to evaluate the definite integral

$$I = \int_0^{2\pi} \frac{d\theta}{1 + \varepsilon \cos \theta}, \qquad |\varepsilon| < 1.$$

By Eq. (7.27) this becomes

$$I = -i \oint_{\text{unit circle}} \frac{dz}{z[1 + (\varepsilon/2)(z + z^{-1})]}$$

$$= -i \frac{2}{\varepsilon} \oint \frac{dz}{z^2 + (2/\varepsilon)z + 1}.$$

The denominator has roots

$$z_- = -\frac{1}{\varepsilon} - \frac{1}{\varepsilon}\sqrt{1 - \varepsilon^2} \qquad \text{and} \qquad z_+ = -\frac{1}{\varepsilon} + \frac{1}{\varepsilon}\sqrt{1 - \varepsilon^2}.$$

z_+ is within the unit circle; z_- is outside. Then by Eq. (7.28) and Exercise 7.11

$$I = -i \frac{2}{\varepsilon} \cdot 2\pi i \frac{1}{z + 1/\varepsilon + (1/\varepsilon)\sqrt{1 - \varepsilon^2}} \bigg|_{z = -1/\varepsilon + (1/\varepsilon)\sqrt{1-\varepsilon^2}}.$$

We obtain

$$\int_0^{2\pi} \frac{d\theta}{1 + \varepsilon \cos \theta} = \frac{2\pi}{\sqrt{1 - \varepsilon^2}}, \qquad |\varepsilon| < 1.$$

EVALUATION OF DEFINITE INTEGRALS: $\int_{-\infty}^{\infty} f(x)\, dx$

Suppose that our definite integral has the form

$$I = \int_{-\infty}^{\infty} f(x)\, dx \tag{7.29}$$

and satisfies the two conditions:

a. $f(z)$ is analytic in the upper half-plane except for a finite number of poles. (It will be assumed that there are no poles on the real axis. If poles are present on the real axis, they may be included or excluded as discussed earlier in this section.)

b. $f(z)$ vanishes as strongly[1] as $1/z^2$ for $|z| \to \infty$, $0 \le \arg z \le \pi$.

With these conditions, we may take as a contour of integration the real axis and a semicircle in the upper half-plane as shown in Fig. 7.6. We let the

---

[1] We could use $f(z)$ vanishes faster than $1/z$, but we wish to have $f(z)$ single-valued.

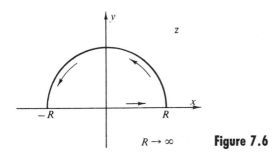

$R \to \infty$ Figure 7.6

radius R of the semicircle become infinitely large. Then

$$\oint f(z)\, dz = \lim_{R \to \infty} \int_{-R}^{R} f(x)\, dx + \lim_{R \to \infty} \int_{0}^{\pi} f(Re^{i\theta})iRe^{i\theta}\, d\theta$$

$$= 2\pi i \sum \text{residues (upper half-plane).} \qquad (7.30)$$

From the second condition the second integral (over the semicircle) vanishes and

$$\int_{-\infty}^{\infty} f(x)\, dx = 2\pi i \sum \text{residues (upper half-plane).} \qquad (7.31)$$

Example 7.2.2

Evaluate

$$I = \int_{-\infty}^{\infty} \frac{dx}{1 + x^2}. \qquad (7.32)$$

From Eq. (7.31)

$$\int_{-\infty}^{\infty} \frac{dx}{1 + x^2} = 2\pi i \sum \text{residues (upper half-plane).}$$

Here and in every other similar problem we have the question: Where are the poles? Rewriting the integrand as

$$\frac{1}{z^2 + 1} = \frac{1}{z + i} \cdot \frac{1}{z - i}, \qquad (7.33)$$

we see that there are simple poles (order 1) at $z = i$ and $z = -i$.

A simple pole at $z = z_0$ indicates (and is indicated by) a Laurent expansion of the form

$$f(z) = \frac{a_{-1}}{z - z_0} + a_0 + \sum_{n=1}^{\infty} a_n(z - z_0)^n. \qquad (7.34)$$

The residue a_{-1} is easily isolated as (Exercise 7.1.1)

$$a_{-1} = (z - z_0)f(z)|_{z = z_0}. \qquad (7.35)$$

Using Eq. (7.35), we find that the residue at $z = i$ is $1/2i$, whereas that at $z = -i$ is $-1/2i$.

Then

$$\int_{-\infty}^{\infty} \frac{dx}{1 + x^2} = 2\pi i \cdot \frac{1}{2i} = \pi. \tag{7.36}$$

Here we have used $a_{-1} = 1/2i$ for the residue of the one included pole at $z = i$. Readers should satisfy themselves that it is possible to use the lower semicircle and that this choice will lead to the same result, $I = \pi$. A somewhat more delicate problem is provided by the next example.

EVALUATION OF DEFINITE INTEGRALS: $\int_{-\infty}^{\infty} f(x)e^{iax} dx$

Consider the definite integral

$$I = \int_{-\infty}^{\infty} f(x)e^{iax} dx, \tag{7.37}$$

with a real and positive. This is a Fourier transform, Chapter 15. We assume the two conditions:

a. $f(z)$ is analytic in the upper half-plane except for a finite number of poles.
b.

$$\lim_{|z| \to \infty} f(z) = 0, \qquad 0 \le \arg z \le \pi. \tag{7.38}$$

Note that this is a less restrictive condition than the second condition imposed on $f(z)$ for integrating $\int_{-\infty}^{\infty} f(x) dx$ previously.

We employ the contour shown in Fig. 7.6. The application of the calculus of residues is the same as the one just considered, but here we have to work a little harder to show that the integral over the (infinite) semicircle goes to zero. This integral becomes

$$I_R = \int_0^{\pi} f(Re^{i\theta})e^{iaR \cos\theta - aR \sin\theta} iRe^{i\theta} d\theta. \tag{7.39}$$

Let R be so large that $|f(z)| = |f(Re^{i\theta})| < \varepsilon$. Then

$$|I_R| \le \varepsilon R \int_0^{\pi} e^{-aR \sin\theta} d\theta$$

$$= 2\varepsilon R \int_0^{\pi/2} e^{-aR \sin\theta} d\theta. \tag{7.40}$$

In the range $[0, \pi/2]$

$$\frac{2}{\pi}\theta \le \sin\theta$$

Therefore (Fig.7.7)

$$|I_R| \le 2\varepsilon R \int_0^{\pi/2} e^{-aR2\theta/\pi} d\theta. \tag{7.41}$$

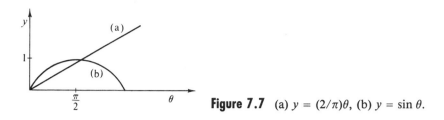

Figure 7.7 (a) $y = (2/\pi)\theta$, (b) $y = \sin \theta$.

Now, integrating by inspection, we obtain

$$|I_R| \le 2\varepsilon R \, \frac{1 - e^{-aR}}{aR2/\pi}.$$

Finally,

$$\lim_{R \to \infty} |I_R| \le \frac{\pi}{a} \varepsilon. \tag{7.42}$$

From Eq. (7.38), $\varepsilon \to 0$ as $R \to \infty$ and

$$\lim_{R \to \infty} |I_R| = 0. \tag{7.43}$$

This useful result is sometimes called *Jordan's lemma*. With it, we are prepared to tackle Fourier integrals of the form shown in Eq. (7.37).
   Using the contour shown in Fig. 7.6, we have

$$\int_{-\infty}^{\infty} f(x)e^{iax}\, dx + \lim_{R \to \infty} I_R = 2\pi i \sum \text{residues (upper half-plane)}.$$

Since the integral over the upper semicircle I_R vanishes as $R \to \infty$ (Jordan's lemma),

$$\int_{-\infty}^{\infty} f(x)e^{iax}\, dx = 2\pi i \sum \text{residues (upper half-plane)} \quad (a > 0). \tag{7.44}$$

Example 7.2.3 Singularity on Contour of Integration
The problem is to evaluate

$$I = \int_{0}^{\infty} \frac{\sin x}{x}\, dx. \tag{7.45}$$

This may be taken as the imaginary part[2] of

$$I_z = P \int_{-\infty}^{\infty} \frac{e^{iz}\, dz}{z}. \tag{7.46}$$

---

[2] One can use $\int [(e^{iz} - e^{-iz})/2iz]\, dz$, but then two different contours will be needed for the two exponentials (compare Example 7.2.4).

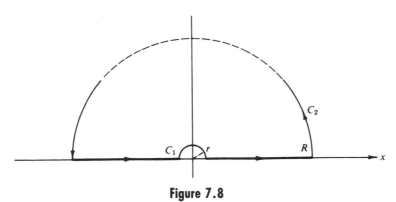

Figure 7.8

Now the only pole is a simple pole at $z = 0$ and the residue there by Eq. (7.35) is $a_{-1} = 1$. We choose the contour shown in Fig. 7.8 (1) to avoid the pole, (2) to include the real axis, and (3) to yield a vanishingly small integrand for $z = iy$, $y \to \infty$. Note that in this case a large (infinite) semicircle in the lower half-plane would be disastrous. We have

$$\oint \frac{e^{iz}\, dz}{z} = \int_{-R}^{-r} e^{ix}\frac{dx}{x} + \int_{C_1} \frac{e^{iz}\, dz}{z} + \int_{r}^{R} \frac{e^{ix}\, dx}{x} + \int_{C_2} \frac{e^{iz}\, dz}{z} = 0, \qquad (7.47)$$

the final zero coming from the residue theorem (Eq. (7.20)). By Jordan's lemma

$$\int_{C_2} \frac{e^{iz}\, dz}{z} = 0, \qquad (7.48)$$

and

$$\oint \frac{e^{iz}\, dz}{z} = \int_{C_1} \frac{e^{iz}\, dz}{z} + P\int_{-\infty}^{\infty} \frac{e^{ix}\, dx}{x} = 0. \qquad (7.49)$$

The integral over the small semicircle yields $(-)\,\pi i$ times the residue of 1, minus, as a result of going clockwise. Taking the imaginary part,[3] we have

$$\int_{-\infty}^{\infty} \frac{\sin x}{x}\, dx = \pi \qquad (7.50)$$

or

$$\int_{0}^{\infty} \frac{\sin x}{x}\, dx = \frac{\pi}{2}. \qquad (7.51)$$

The contour of Fig. 7.8, although convenient, is not at all unique. Another choice of contour for evaluating Eq. (7.45) is presented as Exercise 7.2.15.

[3] Alternatively, we may combine the integrals of Eq. (7.47) as

$$\int_{-R}^{-r} e^{ix}\frac{dx}{x} + \int_{r}^{R} e^{ix}\frac{dx}{x} = \int_{r}^{R} (e^{ix} - e^{-ix})\frac{dx}{x} = 2i\int_{r}^{R} \frac{\sin x}{x}\, dx.$$

Example 7.2.4   Quantum Mechanical Scattering

The quantum mechanical analysis of scattering leads to the function

$$I(\sigma) = \int_{-\infty}^{\infty} \frac{x \sin x \, dx}{x^2 - \sigma^2},\tag{7.52}$$

where σ is real and positive. From the physical conditions of the problem there is a further requirement: $I(\sigma)$ is to have the form $e^{i\sigma}$ so that it will represent an outgoing scattered wave.

Using

$$\sin z = \frac{1}{i}\sinh iz$$

$$= \frac{1}{2i}e^{iz} - \frac{1}{2i}e^{-iz},\tag{7.53}$$

we write Eq. (7.52) in the complex plane as

$$I(\sigma) = I_1 + I_2,\tag{7.54}$$

with

$$I_1 = \frac{1}{2i}\int_{-\infty}^{\infty} \frac{ze^{iz}}{z^2 - \sigma^2}dz,$$

$$I_2 = -\frac{1}{2i}\int_{-\infty}^{\infty} \frac{ze^{-iz}}{z^2 - \sigma^2}dz.\tag{7.55}$$

Integral I_1 is similar to Example 7.2.3 and, as in that case, we may complete the contour by an infinite semicircle in the upper half-plane as shown in Fig. 7.9a. For I_2 the exponential is negative and we complete the contour by an infinite semicircle in the lower half-plane, as shown in Fig. 7.9b. As in Example 7.2.3, neither semicircle contributes anything to the integral—Jordan's lemma.

There is still the problem of locating the poles and evaluating the residues. We find poles at $z = +\sigma$ and $z = -\sigma$ *on the contour of integration.* The residues are (Exercises 7.1.1, 7.2.1)

	$z = \sigma$	$z = -\sigma$
I_1	$\dfrac{e^{i\sigma}}{2}$	$\dfrac{e^{-i\sigma}}{2}$
I_2	$\dfrac{e^{-i\sigma}}{2}$	$\dfrac{e^{i\sigma}}{2}$

Detouring around the poles, as shown in Fig. 7.9 (it matters little whether we go above or below), we find that the residue theorem leads to

$$PI_1 - \pi i\left(\frac{1}{2i}\right)\frac{e^{-i\sigma}}{2} + \pi i\left(\frac{1}{2i}\right)\frac{e^{i\sigma}}{2} = 2\pi i\left(\frac{1}{2i}\right)\frac{e^{i\sigma}}{2},\tag{7.56}$$

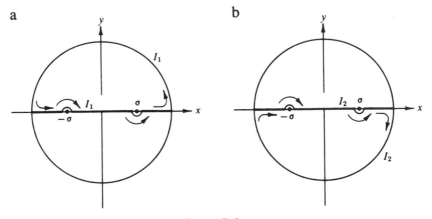

Figure 7.9

for we have enclosed the singularity at $z = \sigma$ but excluded the one at $z = -\sigma$. In similar fashion, but noting that the contour for I_2 is *clockwise*,

$$PI_2 - \pi i \left(\frac{-1}{2i}\right)\frac{e^{i\sigma}}{2} + \pi i \left(\frac{-1}{2i}\right)\frac{e^{-i\sigma}}{2} = -2\pi i \left(\frac{-1}{2i}\right)\frac{e^{i\sigma}}{2}. \qquad (7.57)$$

Adding Eqs. (7.56) and (7.57), we have

$$PI(\sigma) = PI_1 + PI_2 = \frac{\pi}{2}(e^{i\sigma} + e^{-i\sigma}) = \pi \cosh i\sigma$$

$$= \pi \cos \sigma. \qquad (7.58)$$

This is a perfectly good evaluation of Eq. (7.52), but unfortunately the cosine dependence is appropriate for a standing wave and not for the outgoing scattered wave as specified.

To obtain the desired form, we try a different technique. Instead of dodging around the singular points, let us move them off the real axis. Specifically, let $\sigma \to \sigma + i\gamma$, $-\sigma \to -\sigma - i\gamma$, where γ is positive but small and will eventually be made to approach zero, that is,

$$I_+(\sigma) = \lim_{\gamma \to 0} I(\sigma + i\gamma). \qquad (7.59)$$

With this simple substitution, the first integral I_1 becomes

$$I_1(\sigma + i\gamma) = 2\pi i \left(\frac{1}{2i}\right)\frac{e^{i(\sigma + i\gamma)}}{2} \qquad (7.60)$$

by direct application of the residue theorem. Also,

$$I_2(\sigma + i\gamma) = -2\pi i \left(\frac{-1}{2i}\right)\frac{e^{i(\sigma + i\gamma)}}{2}. \qquad (7.61)$$

Adding Eqs. (7.60) and (7.61) and then letting $\gamma \to 0$, we obtain

$$I_+(\sigma) = \lim_{\gamma \to 0} [I_1(\sigma + i\gamma) + I_2(\sigma + i\gamma)]$$

$$= \lim_{\gamma \to 0} \pi e^{i(\sigma + i\gamma)} = \pi e^{i\sigma}, \tag{7.62}$$

a result that does fit the boundary conditions of our scattering problem.

It is interesting to note that the substitution $\sigma \to \sigma - i\gamma$ would have led to

$$I_-(\sigma) = \pi e^{-i\sigma}, \tag{7.63}$$

which could represent an incoming wave. Our earlier result (Eq. (7.58)) is seen to be the arithmetic average of Eqs. (7.62) and (7.63). This average is the Cauchy principal value of the integral. Note that we have these possibilities (Eqs. (7.58), (7.62), and (7.63)) because our integral is not uniquely defined until we specify the particular limiting process (or average) to be used.

EVALUATION OF DEFINITE INTEGRALS: EXPONENTIAL FORMS

With exponential or hyperbolic functions present in the integrand, life gets somewhat more complicated than before. Instead of a general overall prescription, the contour must be chosen to fit the specific integral. These cases are also opportunities to illustrate the versatility and power of contour integration.

As an example, we consider an integral that will be quite useful in developing a relation between $z!$ and $(-z)!$. Notice how the periodicity along the imaginary axis is exploited.

Example 7.2.5   Factorial Function

We wish to evaluate

$$I = \int_{-\infty}^{\infty} \frac{e^{ax}}{1 + e^x} dx, \qquad 0 < a < 1. \tag{7.64}$$

The limits on a are necessary (and sufficient) to prevent the integral from diverging as $x \to \pm\infty$. This integral (Eq. (7.64) may be handled by replacing the real variable x by the complex variable z and integrating around the contour shown in Fig. 7.10. If we take the limit as $R \to \infty$, the real axis, of course, leads to the integral we want. The return path along $y = 2\pi$ is chosen to leave the denominator of the integral invariant, at the same time introducing a constant factor $e^{i2\pi a}$ in the numerator. We have, in the complex plane,

$$\oint \frac{e^{az}}{1 + e^z} dz = \lim_{R \to \infty} \left(\int_{-R}^{R} \frac{e^{ax}}{1 + e^x} dx - e^{i2\pi a} \int_{-R}^{R} \frac{e^{ax}}{1 + e^x} dx \right)$$

$$= (1 - e^{i2\pi a}) \int_{-\infty}^{\infty} \frac{e^{ax}}{1 + e^x} dx. \tag{7.65}$$

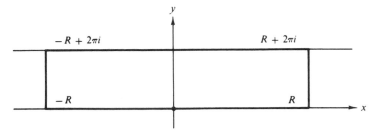

Figure 7.10

In addition there are two vertical sections $(0 \leq y \leq 2\pi)$, which vanish (exponentially) as $R \to \infty$.

Now where are the poles and what are the residues? We have a pole when

$$e^z = e^x e^{iy} = -1. \tag{7.66}$$

Equation (7.66) is satisfied at $z = 0 + i\pi$. By a Laurent expansion[4] in powers of $(z - i\pi)$ the pole is seen to be a simple pole with a residue of $-e^{i\pi a}$. Then, applying the residue theorem,

$$(1 - e^{i2\pi a}) \int_{-\infty}^{\infty} \frac{e^{ax}}{1 + e^x} dx = 2\pi i (-e^{i\pi a}). \tag{7.67}$$

This quickly reduces to

$$\int_{-\infty}^{\infty} \frac{e^{ax}}{1 + e^x} dx = \frac{\pi}{\sin a\pi}, \qquad 0 < a < 1. \tag{7.68}$$

Using the beta function (Section 10.4), we can show the integral to be equal to the product $(a - 1)! \, (-a)!$. This results in the interesting and useful factorial function relation

$$a!(-a)! = \frac{\pi a}{\sin \pi a}. \tag{7.69}$$

Although Eq. (7.68) holds for real a, $0 < a < 1$, Eq. (7.69) may be extended by analytic continuation to all values of a, real and complex, excluding only real integral values.

As a final example of contour integrals of exponential functions, we consider Bernoulli numbers again.

[4] $1 + e^z = 1 + e^{z - i\pi} e^{i\pi} = 1 - e^{z - i\pi} = -(z - i\pi)\left(1 + \frac{z - i\pi}{2!} + \frac{(z - i\pi)^2}{3!} + \cdots\right).$

Example 7.2.6   Bernoulli Numbers

In Section 5.9 the Bernoulli numbers were defined by the expansion

$$\frac{x}{e^x - 1} = \sum_{n=0}^{\infty} \frac{B_n}{n!} x^n. \tag{7.70}$$

Replacing x with z (analytic continuation), we have a Taylor series (compare Eq. (6.47) with

$$B_n = \frac{n!}{2\pi i} \oint_{C_0} \frac{z}{e^z - 1} \frac{dz}{z^{n+1}}, \tag{7.71}$$

where the contour C_0 is around the origin counterclockwise with $|z| < 2\pi$ to avoid the poles at $\pm 2\pi i$.

For $n = 0$ we have a simple pole at $z = 0$ with a residue of $+1$. Hence by Eq. (7.20)

$$B_0 = \frac{0!}{2\pi i} \cdot 2\pi i(1) = 1. \tag{7.72}$$

For $n = 1$ the singularity at $z = 0$ becomes a second-order pole. The residue may be shown to be $-\frac{1}{2}$ by series expansion of the exponential, followed by a binomial expansion. This results in

$$B_1 = \frac{1!}{2\pi i} \cdot 2\pi i\left(-\frac{1}{2}\right) = -\frac{1}{2}. \tag{7.73}$$

For $n \geq 2$ this procedure becomes rather tedious, and we resort to a different means of evaluating Eq. (7.71). The contour is deformed, as shown in Fig. 7.11.

The new contour C still encircles the origin, as required, but now it also encircles (in a negative direction) and infinite series of singular points along the imaginary axis at $z = \pm p2\pi i$, $p = 1, 2, 3, \ldots$. The integration back and forth along the x-axis cancels out, and for $R \to \infty$ the integration over the infinite

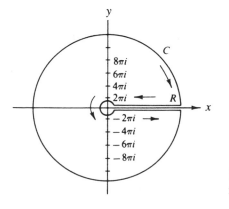

Figure 7.11   Contour of integration for Bernoulli numbers.

circle yields zero. Remember that $n \geq 2$. Therefore

$$\oint_{C_0} \frac{z}{e^z - 1} \frac{dz}{z^{n+1}} = -2\pi i \sum_{p=1}^{\infty} \text{residues} \qquad (z = \pm p2\pi i). \qquad (7.74)$$

At $z = p2\pi i$ we have a simple pole with a residue $(p2\pi i)^{-n}$. When n is odd, the residue from $z = p2\pi i$ exactly cancels that from $z = -p2\pi i$ and $B_n = 0$, $n = 3, 5, 7$, and so on. For n even the residues add, giving

$$B_n = \frac{n!}{2\pi i} (-2\pi i)2 \sum_{p=1}^{\infty} \frac{1}{p^n(2\pi i)^n}$$

$$= -\frac{(-1)^{n/2}2n!}{(2\pi)^n} \sum_{p=1}^{\infty} p^{-n}$$

$$= -\frac{(-1)^{n/2}2n!}{(2\pi)^n} \zeta(n) \qquad (n \text{ even}), \qquad (7.75)$$

where $\zeta(n)$ is the Riemann zeta function introduced in Section 5.9. Equation (7.75) corresponds to Eq. (5.152) of Section 5.9.

EXERCISES

7.2.1 Determine the nature of the singularities of each of the following functions and evaluate the residues $(a > 0)$.

(a) $\dfrac{1}{z^2 + a^2}$.

(b) $\dfrac{1}{(z^2 + a^2)^2}$.

(c) $\dfrac{z^2}{(z^2 + a^2)^2}$.

(d) $\dfrac{\sin 1/z}{z^2 + a^2}$.

(e) $\dfrac{ze^{+iz}}{z^2 + a^2}$.

(f) $\dfrac{ze^{+iz}}{z^2 - a^2}$.

(g) $\dfrac{e^{+iz}}{z^2 - a^2}$.

(h) $\dfrac{z^{-k}}{z + 1}$, $\quad 0 < k < 1$.

Hint. For the point at infinity, use the transformation $w = 1/z$ for $|z| \to 0$. For the residue transform $f(z)\,dz$ into $g(w)\,dw$ and look at the behavior of $g(w)$.

7.2.2 Locate the singularities and evaluate the residues of each of the following functions

(a) $z^{-n}(e^z - 1)^{-1}$, $\quad z \neq 0$,

(b) $\dfrac{z^2 e^z}{1 + e^{2z}}$.

(c) Find a closed form expression (i.e., not a sum) for the sum of the finite-plane singularities.

(d) Using the result in part (c), what is the residue at $|z| \to \infty$?

Hint. See Section 5.9 for expressions involving Bernoulli numbers. Note that Eq. (5.144) can *not* be used to investigate the singularity at $z \to \infty$, since this series is only valid for $|z| < 2\pi$.

7.2.3   The statement that the integral halfway around a singular point is equal to one half the integral all the way around was limited to simple poles. Show, by a specific example, that

$$\int_{\text{Semicircle}} f(z) \, dz = \frac{1}{2} \oint_{\text{Circle}} f(z) \, dz$$

does not necessarily hold if the integral encircles a pole of higher order. *Hint.* Try $f(z) = z^{-2}$.

7.2.4   A function $f(z)$ is analytic along the real axis except for a third-order pole at $z = x_0$. The Laurent expansion about $z = x_0$ has the form

$$f(z) = \frac{a_{-3}}{(z - x_0)^3} + \frac{a_{-1}}{(z - x_0)} + g(z),$$

with $g(z)$ analytic at $z = x_0$. Show that the Cauchy principal value technique is applicable in the sense that

(a) $\displaystyle \lim_{\delta \to 0} \left\{ \int_{-\infty}^{x_0 - \delta} f(x) \, dx + \int_{x_0 + \delta}^{\infty} f(x) \, dx \right\}$

is well behaved.

(b) $\displaystyle \int_{C_{x_0}} f(z) \, dz = \pm i\pi a_{-1},$

where C_{x_0} denotes a *small semicircle* about $z = x_0$.

7.2.5   The unit step function is defined as (compare Exercise 8.7.13)

$$u(s - a) = \begin{cases} 0, & s < a \\ 1, & s > a. \end{cases}$$

Show that $u(s)$ has the integral representations

(a) $\displaystyle u(s) = \lim_{\varepsilon \to 0^+} \frac{1}{2\pi i} \int_{-\infty}^{\infty} \frac{e^{ixs}}{x - i\varepsilon} \, dx,$

(b) $\displaystyle u(s) = \frac{1}{2} + \frac{1}{2\pi i} P \int_{-\infty}^{\infty} \frac{e^{ixs}}{x} \, dx.$

Note. The parameter s is real.

7.2.6   Most of the special functions of mathematical physics may be generated (defined) by a generating function of the form

$$g(t, x) = \sum_{n} f_n(x) t^n.$$

Given the following integral representations, derive the corresponding generating function:

(a) Bessel

$$J_n(x) = \frac{1}{2\pi i} \oint e^{(x/2)(t - 1/t)} t^{-n-1} \, dt.$$

(b) Modified Bessel

$$I_n(x) = \frac{1}{2\pi i} \oint e^{(x/2)(t+1/t)} t^{-n-1} \, dt.$$

(c) Legendre

$$P_n(x) = \frac{1}{2\pi i} \oint (1 - 2tx + t^2)^{-1/2} t^{-n-1} \, dt.$$

(d) Hermite

$$H_n(x) = \frac{n!}{2\pi i} \oint e^{-t^2 + 2tx} t^{-n-1} \, dt.$$

(e) Laguerre

$$L_n(x) = \frac{1}{2\pi i} \oint \frac{e^{-xt/(1-t)}}{(1 - t)t^{n+1}} \, dt.$$

(f) Chebyshev

$$T_n(x) = \frac{1}{4\pi i} \oint \frac{(1 - t^2)t^{-n-1}}{(1 - 2tx + t^2)} \, dt.$$

Each of the contours encircles the origin and no other singular points.

7.2.7   Generalizing Example 7.2.1, show that

$$\int_0^{2\pi} \frac{d\theta}{a \pm b \cos \theta} = \int_0^{2\pi} \frac{d\theta}{a \pm b \sin \theta} = \frac{2\pi}{(a^2 - b^2)^{1/2}}, \qquad \text{for } a > |b|.$$

What happens if $|b| > |a|$?

7.2.8   Show that

$$\int_0^{\pi} \frac{d\theta}{(a + \cos \theta)^2} = \frac{\pi a}{(a^2 - 1)^{3/2}}, \qquad a > 1.$$

7.2.9   Show that

$$\int_0^{2\pi} \frac{d\theta}{1 - 2t \cos \theta + t^2} = \frac{2\pi}{1 - t^2}, \qquad \text{for } |t| < 1.$$

What happens if $|t| > 1$?
What happens if $|t| = 1$?

7.2.10   With the calculus of residues show that

$$\int_0^{\pi} \cos^{2n} \theta \, d\theta = \pi \frac{(2n)!}{2^{2n}(n!)^2} = \pi \frac{(2n - 1)!!}{(2n)!!}, \qquad n = 0, 1, 2, \ldots.$$

(The double factorial notation is defined in Section 10.1.)
Hint. $\cos \theta = \frac{1}{2}(e^{i\theta} + e^{-i\theta}) = \frac{1}{2}(z + z^{-1})$, $\qquad |z| = 1$.

7.2.11   Evaluate

$$\int_{-\infty}^{\infty} \frac{\cos bx - \cos ax}{x^2} \, dx, \qquad a > b > 0.$$

ANS. $\pi(a - b)$.

7.2.12   Prove that

$$\int_0^\infty \frac{\sin^2 x}{x^2}\,dx = \frac{\pi}{2}.$$

Hint. $\sin^2 x = \frac{1}{2}(1 - \cos 2x).$

7.2.13   A quantum mechanical calculation of a transition probability leads to the function $f(t, \omega) = 2(1 - \cos \omega t)/\omega^2$. Show that

$$\int_{-\infty}^\infty f(t, \omega)\,d\omega = 2\pi t.$$

7.2.14   Show that $(a > 0)$

(a) $\displaystyle\int_{-\infty}^\infty \frac{\cos x}{x^2 + a^2}\,dx = \frac{\pi}{a}e^{-a}.$

How is the right side modified if $\cos x$ is replaced by $\cos kx$?

(b) $\displaystyle\int_{-\infty}^\infty \frac{x \sin x}{x^2 + a^2}\,dx = \pi e^{-a}.$

How is the right side modified if $\sin x$ is replaced by $\sin kx$?

These integrals may also be interpreted as Fourier cosine and sine transforms—Chapter 15.

7.2.15   Use the contour shown (Fig. 7.12) with $R \to \infty$ to prove that

$$\int_{-\infty}^\infty \frac{\sin x}{x}\,dx = \pi.$$

7.2.16   In the quantum theory of atomic collisions we encounter the integral

$$I = \int_{-\infty}^\infty \frac{\sin t}{t}e^{ipt}\,dt$$

in which p is real. Show that

$$I = 0, \qquad |p| > 1$$
$$I = \pi, \qquad |p| < 1.$$

What happens if $p = \pm 1$?

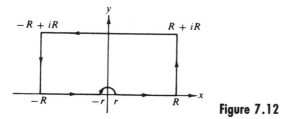

Figure 7.12

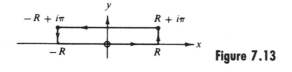

Figure 7.13

7.2.17   Evaluate

$$\int_0^\infty \frac{(\ln x)^2}{1 + x^2} \, dx.$$

(a) by appropriate series expansion of the integrand to obtain

$$4 \sum_{n=0}^\infty (-1)^n (2n + 1)^{-3},$$

(b) by contour integration to obtain

$$\frac{\pi^3}{8}.$$

Hint. $x \rightarrow z = e^t$. Try the contour shown in Fig. 7.13, letting $R \rightarrow \infty$.

7.2.18   Show that

$$\int_0^\infty \frac{x^a}{(x + 1)^2} \, dx = \frac{\pi a}{\sin \pi a},$$

where $-1 < a < 1$. Here is still another way of deriving Eq. (7.69).
Hint. Use the contour shown in Fig. 7.14, noting that $z = 0$ is a branch point and the positive x-axis is a cut line. Note also the comments on phases following Example 7.1.1.

7.2.19   Show that

$$\int_0^\infty \frac{x^{-a}}{x + 1} \, dx = \frac{\pi}{\sin a\pi},$$

where $0 < a < 1$. This opens up another way of deriving the factorial function relation given by Eq. (7.69).

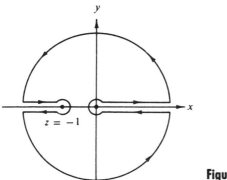

Figure 7.14

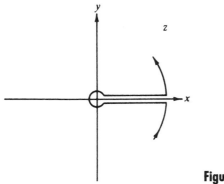

Figure 7.15

Hint. You have a branch point and you will need a cut line. Recall that $z^{-a} = w$ in polar form is

$$[re^{i(\theta + 2\pi n)}]^{-a} = \rho e^{i\varphi},$$

which leads to $-a\theta - 2an\pi = \varphi$.
You must restrict n to zero (or any other single integer) in order that φ may be uniquely specified. Try the contour shown in Fig. 7.15.

7.2.20   Show that

$$\int_0^\infty \frac{dx}{(x^2 + a^2)^2} = \frac{\pi}{4a^3}, \qquad a > 0.$$

7.2.21   Evaluate

$$\int_{-\infty}^\infty \frac{x^2}{1 + x^4}\, dx.$$

ANS. $\pi/\sqrt{2}$.

7.2.22   Show that

$$\int_0^\infty \cos(t^2)\, dt = \int_0^\infty \sin(t^2)\, dt = \frac{\sqrt{\pi}}{2\sqrt{2}}.$$

Hint. Try the contour shown in Fig. 7.16.
Note. These are the Fresnel integrals for the special case of infinity as the upper limit. For the general care of a varying upper limit, asymptotic expansions of the Fresnel integrals are the topic of Exercise 5.10.2. Spherical Bessel expansions are the subject of Exercise 11.7.13.

7.2.23   Several of the Bromwich integrals, Section 15.12, involve a portion that may be approximated by

$$I(y) = \int_{a-iy}^{a+iy} \frac{e^{zt}}{z^{1/2}}\, dz.$$

Here a and t are positive and finite. Show that

$$\lim_{y \to \infty} I(y) = 0.$$

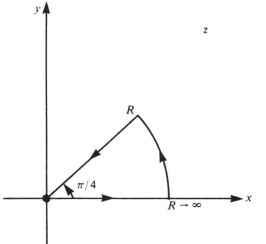

Figure 7.16

7.2.24   Show that

$$\int_0^\infty \frac{1}{1 + x^n}\, dx = \frac{\pi/n}{\sin(\pi/n)}.$$

Hint. Try the contour shown in Fig. 7.17.

7.2.25   (a) Show that

$$f(z) = z^4 - 2\cos 2\theta z^2 + 1$$

has zeros at $e^{i\theta}$, $e^{-i\theta}$, $-e^{i\theta}$, and $-e^{-i\theta}$.

(b) Show that

$$\int_{-\infty}^\infty \frac{dx}{x^4 - 2\cos 2\theta\, x^2 + 1} = \frac{\pi}{2\sin\theta} = \frac{\pi}{2^{1/2}(1 - \cos 2\theta)^{1/2}}.$$

Exercise 7.2.24 ($n = 4$) is a special case of this result.

7.2.26   Show that

$$\int_{-\infty}^\infty \frac{x^2\, dx}{x^4 - 2\cos 2\theta\, x^2 + 1} = \frac{\pi}{2\sin\theta} = \frac{\pi}{2^{1/2}(1 - \cos 2\theta)^{1/2}}.$$

Exercise 7.2.21 is a special case of this result.

7.2.27   Apply the techniques of Example 7.2.4 to the evaluation of the improper integral

$$I = \int_{-\infty}^\infty \frac{dx}{x^2 - \sigma^2}.$$

(a) Let $\sigma \to \sigma + i\gamma$.
(b) Let $\sigma \to \sigma - i\gamma$.
(c) Take the Cauchy principal value.

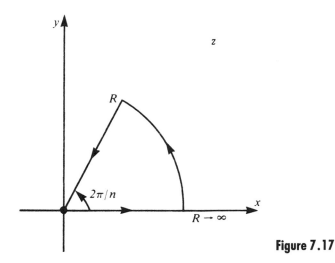

Figure 7.17

7.2.28   The integral in Exercise 7.2.17 may be transformed into

$$\int_0^\infty e^{-y}\frac{y^2}{1+e^{-2y}}\,dy = \frac{\pi^3}{16}.$$

Evaluate this integral by the Gauss-Laguerre quadrature, Appendix A2, and compare your result with $\pi^3/16$.

ANS.   Integral $= 1.93775$ (10 points).

7.3 DISPERSION RELATIONS

The concept of dispersion relations entered physics with the work of Kronig and Kramers in optics. The name dispersion comes from optical dispersion, a result of the dependence of the index of refraction on wavelength or angular frequency. The index of refraction n may have a real part determined by the phase velocity and a (negative) imaginary part determined by the absorption—see Eq. (7.89). Kronig and Kramers showed in 1926-1927 that the real part of $(n^2 - 1)$ could be expressed as an integral of the imaginary part. Generalizing this, we shall apply the label dispersion relations to any pair of equations giving the real part of a function as an integral of its imaginary part and the imaginary part as an integral of its real part—Eqs. (7.81a) and (7.81b) that follow. The existence of such integral relations might be suspected as an integral analog of the Cauchy-Riemann differential relations, Section 6.2.

The applications in modern physics are widespread. For instance, the real part of the function might describe the forward scattering of a gamma ray in a nuclear Coulomb field (a dispersive process). Then the imaginary part would describe the electron–positron pair production in that same Coulomb field

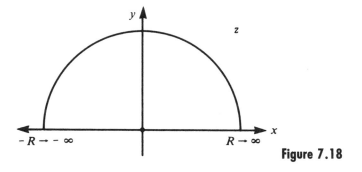

Figure 7.18

(the absorptive process). As will be seen later, the dispersion relations may be taken as a consequence of causality and therefore are independent of the details of the particular interaction.

We consider a complex function $f(z)$ that is analytic in the upper half-plane and on the real axis. We also require that

$$\lim_{|z| \to \infty} |f(z)| = 0, \qquad 0 \le \arg z \le \pi, \tag{7.76}$$

in order that the integral over an infinite semicircle will vanish. The point of these conditions is that we may express $f(z)$ by the Cauchy integral formula, Eq. 6.43,

$$f(z_0) = \frac{1}{2\pi i} \oint \frac{f(z)}{z - z_0}\, dz. \tag{7.77}$$

The integral over the upper semicircle[1] vanishes and we have

$$f(z_0) = \frac{1}{2\pi i} \int_{-\infty}^{\infty} \frac{f(x)}{x - z_0}\, dx. \tag{7.78}$$

The integral over the contour shown in Fig. 7.18 has become an integral along the x-axis.

Equation (7.78) assumes that z_0 is in the upper half-plane—interior to the closed contour. If z_0 were in the lower half-plane, the integral would yield zero by the Cauchy integral theorem, Section 6.3. Now, either letting z_0 approach the real axis from above ($z_0 \to x_0$), or placing it on the real axis and taking an average of Eq. (7.78) and zero, we find that Eq. (7.78) becomes

$$f(x_0) = \frac{1}{\pi i} P \int_{-\infty}^{\infty} \frac{f(x)}{x - x_0}\, dx, \tag{7.79}$$

where P indicates the Cauchy principal value.

---

[1] The use of a semicircle to close the path of integration is convenient, not mandatory. Other paths are possible.

Splitting Eq. (7.79) into real and imaginary parts[2] yields

$$f(x_0) = u(x_0) + iv(x_0)$$

$$= \frac{1}{\pi} P \int_{-\infty}^{\infty} \frac{v(x)}{x - x_0} dx - \frac{i}{\pi} P \int_{-\infty}^{\infty} \frac{u(x)}{x - x_0} dx. \tag{7.80}$$

Finally, equating real part to real part and imaginary part to imaginary part, we obtain

$$u(x_0) = \frac{1}{\pi} P \int_{-\infty}^{\infty} \frac{v(x)}{x - x_0} dx \tag{7.81a}$$

$$v(x_0) = -\frac{1}{\pi} P \int_{-\infty}^{\infty} \frac{u(x)}{x - x_0} dx. \tag{7.81b}$$

These are the dispersion relations. The real part of our complex function is expressed as an integral over the imaginary part. The imaginary part is expressed as an integral over the real part. The real and imaginary parts are *Hilbert transforms* of each other. Note that these relations are meaningful only when $f(x)$ is a *complex* function of the real variable x. Compare Exercise 7.3.1.

From a physical point of view $u(x)$ and/or $v(x)$ represent some physical measurements. Then $f(z) = u(z) + iv(z)$ is an analytic continuation over the upper half-plane, with the value on the real axis serving as a boundary condition.

SYMMETRY RELATIONS

On occasion $f(x)$ will satisfy a symmetry relation and the integral from $-\infty$ to $+\infty$ may be replaced by an integral over positive values only. This is of considerable physical importance because the variable x might represent a frequency and only zero and positive frequencies are available for physical measurements. Suppose[3]

$$f(-x) = f^*(x). \tag{7.82}$$

Then

$$u(-x) + iv(-x) = u(x) - iv(x). \tag{7.83}$$

The real part of $f(x)$ is even and the imaginary part is odd.[4] In quantum mechanical scattering problems these relations (Eq. (7.83)) are called *crossing conditions*. To exploit these crossing conditions, we rewrite Eq. (7.81a) as

$$u(x_0) = \frac{1}{\pi} P \int_{-\infty}^{0} \frac{v(x)}{x - x_0} dx + \frac{1}{\pi} P \int_{0}^{\infty} \frac{v(x)}{x - x_0} dx. \tag{7.84}$$

---

[2] The second argument, $y = 0$, is dropped. $u(x_0, 0) \to u(x_0)$.

[3] This is not just a happy coincidence. It ensures that the Fourier transform of $f(x)$ will be real. In turn, Eq. (7.82) is a consequence of obtaining $f(x)$ as the Fourier transform of a real function.

[4] $u(x, 0) = u(-x, 0)$, $v(x, 0) = -v(-x, 0)$. Compare these symmetry conditions with those that follow from the Schwarz reflection principle, Section 6.5.

Letting $x \rightarrow -x$ in the first integral on the right-hand side of Eq. (7.84) and substituting $v(-x) = -v(x)$ from Eq. (7.83), we obtain

$$u(x_0) = \frac{1}{\pi} P \int_0^\infty v(x) \left\{ \frac{1}{x + x_0} + \frac{1}{x - x_0} \right\} dx$$

$$= \frac{2}{\pi} P \int_0^\infty \frac{x v(x)}{x^2 - x_0^2} dx. \tag{7.85}$$

Similarly,

$$v(x_0) = -\frac{2}{\pi} P \int_0^\infty \frac{x_0 u(x)}{x^2 - x_0^2} dx. \tag{7.86}$$

The original Kronig–Kramers optical dispersion relations were in this form. The asymptotic behavior ($x_0 \rightarrow \infty$) of Eqs. (7.85) and (7.86) lead to quantum mechanical *sum rules*, Exercise 7.3.4.

OPTICAL DISPERSION

The function $\exp[i(kx - \omega t)]$ describes a wave moving along the x-axis in the positive direction with velocity $v = \omega/k$. ω is the angular frequency, k the wave number or propagation vector, and $n = ck/\omega$ the index of refraction. From Maxwell's equations, electric permittivity ε, and Ohm's law with conductivity σ the propagation vector k for a dielectric becomes[5]

$$k^2 = \varepsilon \frac{\omega^2}{c^2} \left(1 + i \frac{4\pi\sigma}{\omega\varepsilon} \right) \tag{7.87}$$

(with μ, the magnetic permeability taken to be unity). The presence of the conductivity (which means absorption) gives rise to an imaginary part. The propagation vector k (and therefore the index of refraction n) have become complex.

Conversely, the (positive) imaginary part implies absorption. For poor conductivity ($4\pi\sigma/\omega\varepsilon \ll 1$) a binomial expansion yields

$$k = \sqrt{\varepsilon} \frac{\omega}{c} + i \frac{2\pi\sigma}{c\sqrt{\varepsilon}}$$

and

$$e^{i(kx - \omega t)} = e^{i\omega(x\sqrt{\varepsilon}/c - t)} e^{-2\pi\sigma x/c\sqrt{\varepsilon}},$$

an attenuated wave.

Returning to the general expression for k^2, we find that Eq. (7.87) the index of refraction becomes

$$n^2 = \frac{c^2 k^2}{\omega^2} = \varepsilon + i \frac{4\pi\sigma}{\omega}. \tag{7.88}$$

[5] See J. D. Jackson, *Classical Electrodynamics*, 2nd ed. New York: Wiley (1975), Sections 7.7 and 7.10. Equation (7.87) follows Jackson in the use of Gaussian units.

We take n^2 to be a function of the *complex* variable ω (with ε and σ depending on ω). However, n^2 does not vanish as $\omega \to \infty$ but instead approaches unity. So to satisfy the condition, Eq. (7.76), one works with $f(\omega) = n^2(\omega) - 1$. The original Kronig–Kramers optical dispersion relations were in the form of

$$\Re[n^2(\omega_0) - 1] = \frac{2}{\pi} P \int_0^\infty \frac{\omega \Im[n^2(\omega) - 1]}{\omega^2 - \omega_0^2} \, d\omega,$$

$$\Im(n^2(\omega_0) - 1] = -\frac{2}{\pi} P \int_0^\infty \frac{\omega_0 \Re[n^2(\omega) - 1]}{\omega^2 - \omega_0^2} \, d\omega.$$

(7.89)

Knowledge of the absorption coefficient at all frequencies specifies the real part of the index of refraction and vice versa.

THE PARSEVAL RELATION

When the functions $u(x)$ and $v(x)$ are Hilbert transforms of each other and each is square integrable,[6] the two functions are related by

$$\int_{-\infty}^\infty |u(x)|^2 \, dx = \int_{-\infty}^\infty |v(x)|^2 \, dx.$$

(7.90)

This is the Parseval relation.

To derive Eq. (7.90), we start with

$$\int_{-\infty}^\infty |u(x)|^2 \, dx = \int_{-\infty}^\infty \frac{1}{\pi} \int_{-\infty}^\infty \frac{v(s) \, ds}{s - x} \frac{1}{\pi} \int_{-\infty}^\infty \frac{v(t) \, dt}{t - x} \, dx,$$

using Eq. (7.81a) twice.

Integrating first with respect to x, we have

$$\int_{-\infty}^\infty |u(x)|^2 \, dx = \int_{-\infty}^\infty \int_{-\infty}^\infty \frac{1}{\pi^2} \int_{-\infty}^\infty \frac{dx}{(s - x)(t - x)} v(s) \, ds \, v(t) \, dt. \quad (7.91)$$

From Exercise 7.3.8 the x integration yields a delta function:

$$\frac{1}{\pi^2} \int_{-\infty}^\infty \frac{dx}{(s - x)(t - x)} = \delta(s - t).$$

We have

$$\int_{-\infty}^\infty |u(x)|^2 \, dx = \int_{-\infty}^\infty \int_{-\infty}^\infty v(s)\delta(s - t) \, ds \, v(t) \, dt. \quad (7.92)$$

Then the s integration is carried out by inspection, using the defining property of the delta function.

$$\int_{-\infty}^\infty v(s)\delta(s - t) \, ds = v(t). \quad (7.93)$$

---

[6] This means that $\int_{-\infty}^\infty |u(x)|^2 \, dx$ and $\int_{-\infty}^\infty |v(x)|^2 \, dx$ are finite.

Substituting Eq. (7.93) into Eq. (7.92), we have Eq. (7.90), the Parseval relation. Again, in terms of optics, the presence of refraction over some frequency range ($n \neq 1$) implies the existence of absorption and vice versa.

CAUSALITY

The real significance of the dispersion relations in physics is that they are a direct consequence of assuming that the particular physical system obeys causality. Causality is awkward to define precisely but the general meaning is that the effect cannot precede the cause. A scattered wave cannot be emitted by the scattering center before the incident wave has arrived. For linear systems the most general relation between an input function G (the cause) and an output function H (the effect) may be written as

$$H(t) = \int_{-\infty}^{\infty} F(t - t')G(t')\,dt'. \tag{7.94}$$

Causality is imposed by requiring that

$$F(t - t') = 0 \qquad \text{for } t - t' < 0.$$

Equation (7.94) gives the time dependence. The frequency dependence is obtained by taking Fourier transforms. By the Fourier convolution theorem, Section 15.5,

$$h(\omega) = f(\omega)g(\omega),$$

where $f(\omega)$ is the Fourier transform of $F(t)$, and so on. Conversely, $F(t)$ is the Fourier transform of $f(\omega)$.

The connection with the dispersion relations is provided by the Titchmarsh theorem.[7] This states that if $f(\omega)$ is square integrable over the real ω-axis, then any one of the following three statements implies the other two.

1. The Fourier transform of $f(\omega)$ is zero for $t < 0$: Eq. (7.94).
2. Replacing ω by z, the function $f(z)$ is analytic in the complex z plane for $y > 0$ and approaches $f(x)$ almost everywhere as $y \to 0$. Further,

$$\int_{-\infty}^{\infty} |f(x + iy)|^2\,dx < K \qquad \text{for } y > 0,$$

   that is, the integral is bounded.
3. The real and imaginary parts of $f(z)$ are Hilbert transforms of each other: Eqs. (7.81a) and (7.81b).

[7] Refer to E. C. Titchmarsh, *Introduction to the Theory of Fourier Integrals*, 2nd ed. New York: Oxford University Press (1937). For a more informal discussion of the Titchmarsh theorem and further details on causality see J. Hilgevoord, *Dispersion Relations and Causal Description*. Amsterdam: North-Holland (1962).

The assumption that the relationship between the input and the output of our linear system is causal (Eq. (7.94)) means that the first statement is satisfied. If $f(\omega)$ is square integrable, then the Titchmarsh theorem has the third statement as a consequence and we have dispersion relations.

EXERCISES

7.3.1   The function $f(z)$ satisfies the conditions for the dispersion relations. In addition, $f(z) = f^*(z^*)$, the Schwarz reflection principle, Section 6.5. Show that $f(z)$ is identically zero.

7.3.2   For $f(z)$ such that we may replace the closed contour of the Cauchy integral formula by an integral over the real axis we have

$$f(x_0) = \frac{1}{2\pi i} \left\{ \int_{-\infty}^{x_0-\delta} \frac{f(x)}{x - x_0} \, dx + \int_{x_0+\delta}^{\infty} \frac{f(x)}{x - x_0} \, dx \right\} + \frac{1}{2\pi i} \int_{C_{x_0}} \frac{f(x)}{x - x_0} \, dx.$$

Here C_{x_0} designates a small semicircle about x_0 in the lower half-plane. Show that this reduces to

$$f(x_0) = \frac{1}{\pi i} P \int_{-\infty}^{\infty} \frac{f(x)}{x - x_0} \, dx,$$

which is Eq. (7.79).

7.3.3   (a) For $f(z) = e^{iz}$, Eq. (7.76) does not hold at the end points, arg $z = 0, \pi$. Show, with the help of Jordan's lemma, Section 7.2, that Eq. (7.77) still holds.

(b) For $f(z) = e^{iz}$ verify the dispersion relations, Eq. (7.81) or Eqs. (7.85) and (7.86), by direct integration.

7.3.4   With $f(x) = u(x) + iv(x)$ and $f(x) = f^*(-x)$, show that as $x_0 \to \infty$,

(a) $u(x_0) \sim -\dfrac{2}{\pi x_0^2} \displaystyle\int_0^{\infty} xv(x) \, dx$,

(b) $v(x_0) \sim \dfrac{2}{\pi x_0} \displaystyle\int_0^{\infty} u(x) \, dx$.

In quantum mechanics relations of this form are often called *sum rules*.

7.3.5   (a) Given the integral equation

$$\frac{1}{1 + x_0^2} = \frac{1}{\pi} P \int_{-\infty}^{\infty} \frac{u(x)}{x - x_0} \, dx,$$

use Hilbert transforms to determine $u(x_0)$.

(b) Verify that the integral equation of part (a) is satisfied.

(c) From $f(z)|_{y=0} = u(x) + iv(x)$, replace x by z and determine $f(z)$. Verify that the conditions for the Hilbert transforms are satisfied.

(d) Are the crossing conditions satisfied?

$$ANS. \quad (a) \; u(x_0) = \frac{x_0}{(1 + x_0^2)},$$

$$(c) \; f(z) = (z + i)^{-1}.$$

7.3.6   (a) If the real part of the complex index of refraction (squared) is constant (no optical dispersion), show that the imaginary part is zero (no absorption).

(b) Conversely, if there is absorption, show that there must be dispersion. In other words, if the imaginary part of $n^2 - 1$ is not zero, show that the real part of $n^2 - 1$ is not constant.

7.3.7   Given $u(x) = x/(x^2 + 1)$ and $v(x) = -1/(x^2 + 1)$, show by direct evaluation of each integral that

$$\int_{-\infty}^{\infty} |u(x)|^2 \, dx = \int_{-\infty}^{\infty} |v(x)|^2 \, dx.$$

$$ANS. \quad \int_{-\infty}^{\infty} |u(x)|^2 \, dx = \int_{-\infty}^{\infty} |v(x)|^2 \, dx = \frac{\pi}{2}.$$

7.3.8   Take $u(x) = \delta(x)$, a delta function, and *assume* that the Hilbert transform equations hold.

(a) Show that

$$\delta(w) = \frac{1}{\pi^2} \int_{-\infty}^{\infty} \frac{dy}{y(y - w)}.$$

(b) With changes of variables $w = s - t$ and $x = s - y$, transform the δ representation of part (a) into

$$\delta(s - t) = \frac{1}{\pi^2} \int_{-\infty}^{\infty} \frac{dx}{(x - s)(x - t)}.$$

Note. The δ function is discussed in Section 1.15.

7.3.9   Show that

$$\delta(x) = \frac{1}{\pi^2} \int_{-\infty}^{\infty} \frac{dt}{t(t - x)}$$

is a valid representation of the delta function in the sense that

$$\int_{-\infty}^{\infty} f(x)\delta(x) \, dx = f(0).$$

Assume that $f(x)$ satisfies the condition for the existence of a Hilbert transform.

Hint. Apply Eq. (7.79) twice.

7.4 THE METHOD OF STEEPEST DESCENTS

In analyzing problems in mathematical physics, one often finds it desirable to know the behavior of a function for large values of the variable, that is, the asymptotic behavior of the function. Specific examples are furnished by the gamma function (Chapter 10) and the various Bessel functions (Chapter 11). The method of steepest descents is a method of determining such asymptotic behavior when the function can be expressed as an integral of the general form

$$I(s) = \int_C g(z)e^{sf(z)} \, dz.$$
(7.95)

For the present, let us take s to be real. The contour of integration C is then chosen so that the real part of $f(z)$ approaches minus infinity at both limits and that the integrand will vanish at the limits, or is chosen as a closed contour. It is further assumed that the factor $g(z)$ in the integrand is dominated by the exponential in the region of interest.

If the parameter s is large and positive, the value of the integrand will become large when the real part of $f(z)$ is large and small when the real part of $f(z)$ is small or negative. In particular, as s is permitted to increase indefinitely (leading to the asymptotic dependence), the entire contribution of the integrand to the integral will come from the region in which the real part of $f(z)$ takes on a positive maximum value. Away from this positive maximum the integrand will become negligibly small in comparison. This is seen by expressing $f(z)$ as

$$f(z) = u(x, y) + iv(x, y).$$

Then the integral may be written as

$$I(s) = \int_C g(z)e^{su(x,y)}e^{isv(x,y)} \, dz.$$
(7.96)

If now, in addition, we impose the condition that the imaginary part of the exponent, $iv(x, y)$, be constant in the region in which the real part takes on its maximum value, that is, $v(x, y) = v(x_0, y_0) = v_0$, we may approximate the integral by

$$I(s) \approx e^{isv_0} \int_C g(z)e^{su(x,y)} \, dz.$$
(7.97)

Away from the maximum of the real part, the imaginary part may be permitted to oscillate as it wishes, for the integrand is negligibly small and the varying phase factor is therefore irrelevant.

The real part of $sf(z)$ is a maximum for a given s when the real part of $f(z)$, $u(x, y)$, is a maximum. This implies that

$$\frac{\partial u}{\partial x} = \frac{\partial u}{\partial y} = 0,$$

and therefore, by the use of the Cauchy–Riemann conditions of Section 6.2

$$\frac{df(z)}{dz} = 0.$$
(7.98)

We proceed to search for such zeros of the derivative.

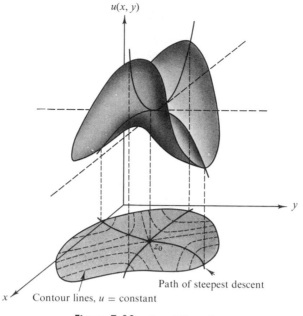

$u(x, y)$

y

Path of steepest descent

x Contour lines, u = constant

Figure 7.19 A saddle point.

It is essential to note that the maximum value of $u(x, y)$ is the maximum only along a given contour. In the finite plane neither the real nor the imaginary part of our analytic function possesses an absolute maximum. This may be seen by recalling that both u and v satisfy Laplace's equation

$$\frac{\partial^2 u}{\partial x^2} + \frac{\partial^2 u}{\partial y^2} = 0. \tag{7.99}$$

From this, if the second derivative with respect to x is positive, the second derivative with respect to y must be negative, and therefore neither u nor v can possess an absolute maximum or minimum. Since the function $f(z)$ was taken to be analytic, singular points are clearly excluded. The vanishing of the derivative (Eq. (7.98)) then implies that we have a saddle point, a stationary value, which may be a maximum of $u(x, y)$ for one contour and a minimum for another (Fig. 7.19).

Our problem, then, is to choose the contour of integration to satisfy two conditions. (1) The contour must be chosen so that $u(x, y)$ has a maximum at the saddle point. (2) The contour must pass through the saddle in such a way that the imaginary part, $v(x, y)$, is a constant. This second condition leads to the path of steepest descent and gives the method its name. From Section 6.2, especially Exercise 6.2.1, we know that the curves corresponding to u = constant and v = constant form an orthogonal system. This means that a curve $v = c_i$, constant, is everywhere tangential to the gradient of u, ∇u.

Hence the curve $v = $ constant is the curve that gives the line of steepest descent from the saddle point.[1]

At the saddle point the function $f(z)$ can be expanded in a Taylor series to give

$$f(z) = f(z_0) + \tfrac{1}{2}(z - z_0)^2 f''(z_0) + \cdots. \tag{7.100}$$

The first derivative is absent, since obviously Eq. (7.98) is satisfied. The first correction term, $\tfrac{1}{2}(z - z_0)^2 f''(z_0)$, is real and negative. It is real, for we have specified that the imaginary part shall be constant along our contour and negative because we are moving down from the saddle point or mountain pass. Then, assuming that $f''(z_0) \neq 0$, we have

$$f(z) - f(z_0) \approx \tfrac{1}{2}(z - z_0)^2 f''(z_0) = -\frac{1}{2s}t^2, \tag{7.101}$$

which serves to define a new variable t. If $(z - z_0)$ is written in polar form

$$(z - z_0) = \delta e^{i\alpha}, \tag{7.102}$$

(with the phase α held constant), we have

$$t^2 = -sf''(z_0)\delta^2 e^{2i\alpha}. \tag{7.103}$$

Since t is real,[2] it may be written as

$$t = \pm\delta|sf''(z_0)|^{1/2}. \tag{7.104}$$

Substituting Eq. (7.101) into Eq. (7.95), we obtain

$$I(s) \approx g(z_0)e^{sf(z_0)} \int_{-\infty}^{\infty} e^{-t^2/2}\frac{dz}{dt}\,dt. \tag{7.105}$$

We have

$$\frac{dz}{dt} = \left(\frac{dt}{dz}\right)^{-1} = \left(\frac{dt}{d\delta}\frac{d\delta}{dz}\right)^{-1} = |sf''(z_0)|^{-1/2}e^{i\alpha}, \tag{7.106}$$

from Eqs. (7.102) and (7.104). Equation (7.105) becomes

$$I(s) \approx \frac{g(z_0)e^{sf(z_0)}e^{i\alpha}}{|sf''(z_0)|^{1/2}} \int_{-\infty}^{\infty} e^{-t^2/2}\,dt. \tag{7.107}$$

It will be noted that the limits have been set as minus infinity to plus infinity. This is permissible, for the integrand is essentially zero when t departs appreciably from the origin. Noting that the remaining integral is just a Gauss

[1] The line of steepest ascent is also characterized by constant v. The saddle point must be inspected carefully to distinguish the line of steepest descent from the line of steepest ascent. This is discussed later in two examples.

[2] The phase of the contour (specified by α) at the saddle point is chosen so that $\Im[f(z) - f(z_0)] = 0$, that is, $\tfrac{1}{2}(z - z_0)^2 f''(z_0)$ must be real.

error integral equal to $\sqrt{2\pi}$, we finally obtain

$$I(s) \approx \frac{\sqrt{2\pi}\, g(z_0) e^{sf(z_0)} e^{i\alpha}}{|sf''(z_0)|^{1/2}}.\qquad(7.108)$$

The phase α was introduced in Eq. (7.102) as the phase of the contour as it passed through the saddle point. It is chosen so that the two conditions given $[\alpha = \text{constant}; \Re f(z) = \text{maximum}]$ are satisfied. It sometimes happens that the contour passes through two or more saddle points in succession. If this is the case, we need only add the contribution made by Eq. (7.108) from each of the saddle points in order to get an approximation for the total integral.

One note of warning: We assumed that the only significant contribution to the integral came from the immediate vicinity of the saddle point(s) $z = z_0$, that is,

$$\Re[f(z)] = u(x, y) \ll u(x_0, y_0)$$

over the entire contour away from $z_0 = x_0 + iy_0$. This condition must be checked for each new problem (Exercise 7.4.5).

Example 7.4.1   Asymptotic Form of the Hankel Function, $H_\nu^{(1)}(s)$

In Section 11.4 it is shown that the Hankel functions, which satisfy Bessel's equation, may be defined by

$$H_\nu^{(1)}(s) = \frac{1}{\pi i} \int_{0\,C_1}^{-\infty} e^{(s/2)(z-1/z)} \frac{dz}{z^{\nu+1}},\qquad(7.109)$$

$$H_\nu^{(2)}(s) = \frac{1}{\pi i} \int_{-\infty\,C_2}^{\infty} e^{(s/2)(z-1/z)} \frac{dz}{z^{\nu+1}}.\qquad(7.110)$$

The contour C_1 is the curve in the upper half-plane of Fig. 7.20. The contour C_2 is in the lower half-plane. We apply the method of steepest descents to the first Hankel function, $H_\nu^{(1)}(s)$, which is conveniently in the form specified by Eq. (7.95), with $f(z)$ given by

$$f(z) = \frac{1}{2}\left(z - \frac{1}{z}\right).\qquad(7.111)$$

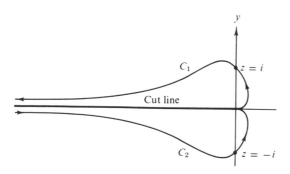

Figure 7.20   Hankel function contours.

By differentiating, we obtain

$$f'(z) = \frac{1}{2} + \frac{1}{2z^2}. \qquad (7.112)$$

Setting $f'(z) = 0$ in accordance with Eq. (7.88), we obtain

$$z = i, -i. \qquad (7.113)$$

Hence there are saddle points at $z = +i$ and $z = -i$. The integral for $H_\nu^{(1)}(s)$ is chosen so that it starts at the origin, moves out tangentially to the positive real axis, and then moves around through the saddle point at $z = +i$ and on out to minus infinity, asymptotic with the negative real axis. We must choose the contour through the point $z = +i$ in such a way that the real part of $(z - 1/z)$ will be a maximum and the phase will be constant in the vicinity of the saddle point. We have

$$\Re\left(z - \frac{1}{z}\right) = 0 \qquad \text{for } z = i.$$

We require $\Re(z - 1/z) < 0$ for the rest of C_1 ($z \neq i$).

In the vicinity of the saddle point at $z_0 = +i$ we have

$$z - i = \delta e^{i\alpha}, \qquad (7.114)$$

where δ is a small number. Then

$$2f(z) = z - \frac{1}{z} = \delta e^{i\alpha} + i - \frac{1}{\delta e^{i\alpha} + i}$$

$$= \delta \cos \alpha + i(\delta \sin \alpha + 1) - \frac{1}{\delta \cos \alpha + i(\delta \sin \alpha + 1)}$$

$$= \delta \cos \alpha + i(\delta \sin \alpha + 1) - \frac{\delta \cos \alpha - i(\delta \sin \alpha + 1)}{1 + 2\delta \sin \alpha + \delta^2}. \qquad (7.115)$$

Therefore our real part becomes

$$\Re\left(z - \frac{1}{z}\right) = \delta \cos \alpha - \delta \cos \alpha (1 + 2\delta \sin \alpha + \delta^2)^{-1}. \qquad (7.116)$$

Recalling that δ is small, we expanded by the binomial theorem and neglect terms of order δ^3 and higher.

$$\Re\left(z - \frac{1}{z}\right) = 2\delta^2 \cos \alpha \sin \alpha + O(\delta^3) \approx \delta^2 \sin 2\alpha. \qquad (7.117)$$

We see that the real part of $(z - 1/z)$ will take on an extreme value if $\sin 2\alpha$ is an extremum, that is, if 2α is $\pi/2$ or $3\pi/2$. Hence the phase of the contour α should be chosen to be $\pi/4$ or $3\pi/4$. One choice will represent the path of steepest descent that we want. The other choice will represent a path of steepest ascent that we must avoid. We distinguish the two possibilities by substituting in the specific values of α. For $\alpha = \pi/4$

$$\Re\left(z - \frac{1}{z}\right) = \delta^2. \tag{7.118}$$

For this choice $z = i$ is a minimum.
  For $\alpha = 3\pi/4$

$$\Re\left(z - \frac{1}{z}\right) = -\delta^2 \tag{7.119}$$

and $z = i$ is a maximum. This is the phase we want.
  Direct substitution into Eq. (7.108) with $\alpha = 3\pi/4$ now yields

$$H_\nu^{(1)}(s) = \frac{1}{\pi i} \frac{\sqrt{2\pi i}^{-\nu-1} e^{(s/2)(i-1/i)} e^{i(3\pi/4)}}{|(s/2)(-2/i^3)|^{1/2}}$$

$$= \sqrt{\frac{2}{\pi s}} e^{(i\pi/2)(-\nu-2)} e^{is} e^{i(3\pi/4)}. \tag{7.120}$$

By combining terms, we finally obtain

$$H_\nu^{(1)}(s) \approx \sqrt{\frac{2}{\pi s}} e^{i(s - \nu(\pi/2) - \pi/4)} \tag{7.121}$$

as the leading term of the asymptotic expansion of the Hankel function $H_\nu^{(1)}(s)$. Additional terms, if desired, may be picked up by assuming a series of descending powers and substituting back into Bessel's equation.

Example 7.4.2 *Asymptotic Form of the Factorial Function, s!*

In many physical problems, particularly in the field of statistical mechanics, it is desirable to have an accurate approximation of the gamma or factorial function of very large numbers. As developed in Section 10.1, the factorial function may be defined by the integral

$$s! = \int_0^\infty \rho^s e^{-\rho}\, d\rho = s^{s+1} \int_{0C}^\infty e^{s(\ln z - z)}\, dz. \tag{7.122}$$

Here we have made the substitution $\rho = zs$ in order to throw the integral into the form required by Eq. (7.95). As before, we assume that s is real and positive, from which it follows that the integrand vanishes at the limits 0 and ∞. By differentiating the z-dependence appearing in the exponent, we obtain

$$\frac{df(z)}{dz} = \frac{d}{dz}(\ln z - z) = \frac{1}{z} - 1, \tag{7.123}$$

which shows that the point $z = 1$ is a saddle point. We let

$$z - 1 = \delta e^{i\alpha}, \tag{7.124}$$

with δ small to describe the contour in the vicinity of the saddle point. Substituting into $f(z) = \ln z - z$, we develop a series expansion

$$f(z) = \ln(1 + \delta e^{i\alpha}) - (1 + \delta e^{i\alpha})$$
$$= \delta e^{i\alpha} - \tfrac{1}{2}\delta^2 e^{2i\alpha} + \cdots - 1 - \delta e^{i\alpha}$$
$$= -1 - \tfrac{1}{2}\delta^2 e^{2i\alpha} + \cdots. \tag{7.125}$$

From this we see that the integrand takes on a maximum value (e^{-s}) at the saddle point if we choose our contour C to follow the real axis, a conclusion that the reader may well have reached more or less intuitively.

Direct substitution into Eq. (7.108) with $\alpha = 0$ now gives

$$s! \approx \frac{\sqrt{2\pi}\, s^{s+1} e^{-s}}{|s(-1^{-2})|^{1/2}}. \tag{7.126}$$

Thus the first term in the asymptotic expansion of the factorial function is

$$s! \approx \sqrt{2\pi s}\, s^s e^{-s}. \tag{7.127}$$

This result is the first term in Stirling's expansion of the factorial function. The method of steepest descent is probably the easiest way of obtaining this first term. If more terms in the expansion are desired, then the method of Section 10.3 is preferable.

In the foregoing example the calculation was carried out by assuming s to be real. This assumption is not necessary. The student may show (Exercise 7.4.6) that Eq. (7.127) also holds when s is replaced by the complex variable w, provided only that the real part of w is required to be large and positive.

EXERCISES

7.4.1   Using the method of steepest descents, evaluate the second Hankel function given by

$$H_\nu^{(2)}(s) = \frac{1}{\pi i} \int_{-\infty C_2}^{\infty} e^{(s/2)(z-1/z)}\,\frac{dz}{z^{\nu+1}},$$

with contour C_2 as shown in Fig. 7.20.

$$\text{ANS.}\quad H_\nu^{(2)}(s) \approx \sqrt{\frac{2}{\pi s}}\, e^{-i(s-\pi/4-\nu\pi/2)}.$$

7.4.2   The negative square root in Eq. (7.104) does not appear in Eq. (7.107). What is the justification for dropping it? Illustrate your argument by detailed reference to $H_\nu^{(1)}(z)$, Example 7.4.1.

7.4.3   (a) In applying the method of steepest descent to the Hankel function $H_\nu^{(1)}(s)$, show that

$$\Re[f(z)] < \Re[f(z_0)] = 0$$

for z on the contour C_1 but away from the point $z = z_0 = i$.

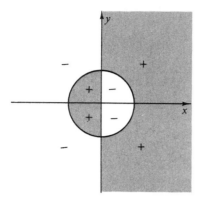

Figure 7.21

(b) Show that

$$\Re[f(z)] > 0 \quad \text{for} \quad 0 < r < 1, \quad \begin{cases} \dfrac{\pi}{2} < \theta \le \pi \\ \\ -\pi \le \theta < \dfrac{\pi}{2} \end{cases}$$

and

$$\text{for} \quad r > 1, \quad -\frac{\pi}{2} < \theta < \frac{\pi}{2}$$

(Fig. 7.21).
This is why C_1 may *not* be deformed to pass through the second saddle point $z = -i$.

7.4.4 Determine the asymptotic dependence of the modified Bessel functions $I_\nu(x)$, given

$$I_\nu(x) = \frac{1}{2\pi i} \int_C e^{(x/2)(t+1/t)} \frac{dt}{t^{\nu+1}}.$$

The contour starts and ends at $t = -\infty$, encircling the origin in a positive sense. There are two saddle points. Only the one at $z = +1$ contributes significantly to the asymptotic form.

7.4.5 Determine the asymptotic dependence of the modified Bessel function of the second kind, $K_\nu(x)$, by using

$$K_\nu(x) = \frac{1}{2} \int_0^\infty e^{(-x/2)(s+1/s)} \frac{ds}{s^{1-\nu}}.$$

7.4.6 Show that Stirling's formula

$$s! \approx \sqrt{2\pi s} \, s^s e^{-s}$$

holds for complex values of s (with $\Re(s)$ large and positive).
Hint. This involves assigning a phase to s and then demanding that $\Im[sf(z)] = \text{constant}$ in the vicinity of the saddle point.

7.4.7   Assume $H_\nu^{(1)}(s)$ to have a negative power-series expansion of the form

$$H_\nu^{(1)}(s) = \sqrt{\frac{2}{\pi s}}\, e^{i(s - \nu(\pi/2) - \pi/4)} \sum_{n=0}^{\infty} a_{-n} s^{-n},$$

with the coefficient of the summation obtained by the method of steepest descent. Substitute into Bessel's equation and show that you reproduce the asymptotic series for $H_\nu^{(1)}(s)$ given in Section 11.6.

ADDITIONAL READINGS

NUSSENZVEIG, H. M., *Causality and Dispersion Relations*, Mathematics in Science and Engineering Series, Vol. 95. New York: Academic Press (1972). This is an advanced text covering causality and dispersion relations in the first chapter and then moving on to develop the implications in a variety of areas of theoretical physics.

WYLD, H. W., *Mathematical Methods for Physics*. Reading, MA: Benjamin/Cummings (1976). This is a relatively advanced text that contains an extensive discussion of the dispersion relations.

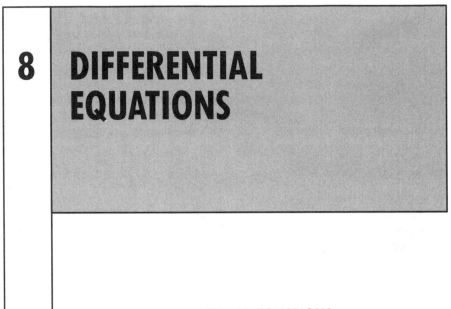

8 DIFFERENTIAL EQUATIONS

8.1 PARTIAL DIFFERENTIAL EQUATIONS, CHARACTERISTICS, AND BOUNDARY CONDITIONS

In physics the knowledge of the force in an equation of motion usually leads to a differential equation. Thus, almost all the elementary and numerous advanced parts of theoretical physics are formulated in terms of differential equations. Sometimes these are ordinary differential equations in one variable (abbreviated ODE). More often the equations are partial differential equations (PDE) in two or more variables.

Let us recall from calculus that the operation of taking an ordinary or partial derivative is a *linear operation* $(\mathcal{L})$[1]

$$\frac{d(a\varphi(x) + b\psi(x))}{dx} = a\frac{d\varphi}{dx} + b\frac{d\psi}{dx}$$

for ODEs involving derivatives in one variable x only and no quadratic, $(d\psi/dx)^2$, or higher powers. Similarly, for partial derivations,

$$\frac{\partial(a\varphi(x, y) + b\psi(x, y))}{\partial x} = a\frac{\partial\varphi(x, y)}{\partial x} + b\frac{\partial\psi(x, y)}{\partial x}.$$

In general

$$\mathcal{L}(a\varphi + b\psi) = a\mathcal{L}(\varphi) + b\mathcal{L}(\psi). \tag{8.1}$$

---

[1] We are especially interested in linear operators because in quantum mechanics physical quantities are represented by linear operators operating in a complex, infinite dimensional Hilbert space.

456

Thus, ODEs and PDEs appear as linear operator equations

$$\mathcal{L}\psi = F,$$

where F is a known function of one (for ODEs) or more variables (for PDEs), $\mathcal{L}$ is a linear combination of derivatives, and ψ is the unknown function or solution. Any linear combination of solutions is again a solution; this is the *superposition principle*.

Since the dynamics of many physical systems involve just two derivatives, e.g., acceleration in classical mechanics and the kinetic energy operator, $\sim\nabla^2$, in quantum mechanics, differential equations of second order occur most frequently in physics. [Maxwell's and Dirac's equations are first-order but involve two unknown functions. Eliminating one unknown yields a second-order differential equation for the other (compare Section 1.9).]

EXAMPLES OF PDEs

Among the most frequently encountered PDEs are the following:

1. Laplace's equation, $\nabla^2\psi = 0$.
   This very common and very important equation occurs in studies of

   a. electromagnetic phenomena including electrostatics, dielectrics, steady currents, and magnetostatics,
   b. hydrodynamics (irrotational flow of perfect fluid and surface waves),
   c. heat flow,
   d. gravitation.

2. Poisson's equation, $\nabla^2\psi = -\rho/\varepsilon_0$.
   In contrast to the homogeneous Laplace equation, Poisson's equation is nonhomogeneous with a source term $-\rho/\varepsilon_0$.

3. The wave (Helmholtz) and time-independent diffusion equations, $\nabla^2\psi \pm k^2\psi = 0$.
   These equations appear in such diverse phenomena as

   a. elastic waves in solids including vibrating strings, bars, membranes,
   b. sound or acoustics,
   c. electromagnetic waves,
   d. nuclear reactors.

4. The time-dependent diffusion equation

   $$\nabla^2\psi = \frac{1}{a^2}\frac{\partial\psi}{\partial t}$$

   and the corresponding four-dimensional forms involving the d'Alembertian, a four-dimensional analog of the Laplacian in Minkowski space,

   $$\partial^\mu\partial_\mu = \partial^2 = \frac{1}{c^2}\frac{\partial^2}{\partial t^2} - \nabla^2.$$

5. The time-dependent wave equation, $\partial^2\psi = 0$.

6. The scalar potential equation, $\partial^2 \psi = \rho/\varepsilon_0$.
   Like Poisson's equation this equation is nonhomogeneous with a source term ρ/ε_0.
7. The Klein–Gordon equation, $\partial^2 \psi = -\mu^2 \psi$, and the corresponding vector equations in which the scalar function ψ is replaced by a vector function. Other more complicated forms are common.
8. The Schrödinger wave equation,

$$-\frac{\hbar^2}{2m} \nabla^2 \psi + V\psi = i\hbar \frac{\partial \psi}{\partial t}$$

and

$$-\frac{\hbar^2}{2m} \nabla^2 \psi + V\psi = E\psi$$

for the time-independent case.
9. The equations for elastic waves and viscous fluids and the telegraphy equation.
10. Maxwell's coupled partial differential equations for electric and magnetic fields are those of Dirac for relativistic electron wave functions. For Maxwell's equations see the Introduction and also Section 1.9.

Some general techniques for solving second-order PDEs are discussed in this section:

1. Separation of variables, where the PDE is split into ODEs that are related by common constants which appear as eigenvalues of linear operators, $\mathcal{L}\psi = l\psi$, usually in one variable. The Helmholtz equation given as example 3 above has this form, where the eigenvalue k^2 may arise by separation of the time t from the spatial variables. Likewise in example 8, the energy E is the eigenvalue that arises in the separation of t from $\mathbf{r}$ in the Schrödinger equation. This is pursued in Chapter 9 in greater detail, while Section 8.2 serves as introduction. ODEs may be attacked by Frobenius's power series method in Section 8.3. It does not always work but is often the simplest method when it does.
2. Conversion of a PDE into an integral equation using *Green's functions* applies to *inhomogeneous PDEs* such as examples 2 and 6 given above. An introduction to the Green's function technique is given in Section 8.7.
3. Other analytical methods such as the use of integral transforms are developed and applied in Chapter 15.
4. Numerical calculations. The development of modern high-speed computing machines has opened up a wealth of possibilities based on the calculus of finite differences. Here we have the relaxation methods. In Section 8.8 two numerical methods, Runge–Kutta and a predictor–corrector, are applied to ODEs.[2]

---

[2] For further details of numerical computation the reader could start with R. W. Hamming's *Numerical Methods for Scientists and Engineers* [New York: McGraw-Hill (1973)] and proceed to specialized references.

Occasionally, we encounter equations of higher order. In both the theory of the slow motion of a viscous fluid and the theory of an elastic body we find the equation

$$(\nabla^2)^2 \psi = 0.$$

Fortunately, these higher-order differential equations are relatively rare and are not discussed in introductory texts such as this one.

Although not so frequently encountered and perhaps not so important as second-order differential equations, first-order differential equations do appear in theoretical physics and are sometimes intermediate steps for second-order differential equations. The solutions of some more important types of first-order ODEs are developed in Section 8.2. First-order PDEs can always be reduced to ODEs. This is a straightforward but lengthy process and involves a search for characteristics that are briefly introduced below; for more details we refer to the literature.

CLASSES OF PDEs AND CHARACTERISTICS *

Second order PDEs form three classes: (i) Elliptic PDEs involve ∇^2 or $c^{-2}\partial^2/\partial t^2 + \nabla^2$; (ii) parabolic PDEs, $a\partial/\partial t - \nabla^2$; (iii) hyperbolic PDEs, $c^{-2}\partial^2/\partial t^2 - \nabla^2$. These canonical operators come about by a change of variables $\xi = \xi(x, y)$, $\eta = \eta(x, y)$ in a linear operator (for two variables just for simplicity)

$$\mathcal{L} = a\frac{\partial^2}{\partial x^2} + 2b\frac{\partial^2}{\partial x \partial y} + c\frac{\partial^2}{\partial y^2} + d\frac{\partial}{\partial x} + e\frac{\partial}{\partial y} + f, \qquad (8.2)$$

which can be reduced to the canonical forms (i), (ii), (iii) according to whether the discriminant $D = ac - b^2 > 0$, $=0$, or <0. If $\xi(x, y)$ is determined from the first-order, but nonlinear, PDE

$$a\left(\frac{\partial \xi}{\partial x}\right)^2 + 2b\left(\frac{\partial \xi}{\partial x}\right)\left(\frac{\partial \xi}{\partial y}\right) + c\left(\frac{\partial \xi}{\partial y}\right)^2 = 0, \qquad (8.3)$$

where the lower-order terms in $\mathcal{L}$ are ignored, then the coefficient of $\partial^2/\partial \xi^2$ in $\mathcal{L}$ (i.e., Eq. (8.3)) is zero. If η is an independent solution of the same Eq. (8.3), then the coefficient of $\partial^2/\partial \eta^2$ is also zero. The remaining operator $\partial^2/\partial \xi \partial \eta$ in $\mathcal{L}$ is characteristic of the hyperbolic case (iii) with $D < 0$, where the quadratic form $a\lambda^2 + 2b\lambda + c$ factorizes and, therefore, Eq. (8.3) has two independent solutions $\xi(x, y)$, $\eta(x, y)$. In the elliptic case (i) with $D > 0$ the two solutions ξ, η are complex conjugate which, when substituted into Eq. (8.2), remove the mixed second-order derivative instead of the other second-order terms yielding the canonical form (i). In the parabolic case (ii) with $D = 0$, only $\partial^2/\partial \xi^2$ remains in $\mathcal{L}$, while the coefficients of the other two second-order derivatives vanish.

If the coefficients a, b, c in $\mathcal{L}$ are functions of the coordinates, then this classification is only local, i.e., its type may change as the coordinates vary.

---

* This section is optional. It is not essential for a first reading of this chapter.

Let us illustrate the physics underlying the hyperbolic case by looking at the wave equation (in $1 + 1$ dimensions for simplicity)

$$\left(\frac{1}{c^2}\frac{\partial^2}{\partial t^2} - \frac{\partial^2}{\partial x^2}\right)\psi = 0. \tag{8.4}$$

Since Eq. (8.3) becomes

$$\left(\frac{\partial\xi}{\partial t}\right)^2 - c^2\left(\frac{\partial\xi}{\partial x}\right)^2 = \left(\frac{\partial\xi}{\partial t} - c\frac{\partial\xi}{\partial x}\right)\left(\frac{\partial\xi}{\partial t} + c\frac{\partial\xi}{\partial x}\right) = 0 \tag{8.5}$$

and factorizes, we determine the solution of $\partial\xi/\partial t - c\partial\xi/\partial x = 0$. This is an arbitrary function $\xi = F(x + ct)$, and $\xi = G(x - ct)$ solves $\partial\xi/\partial t + c\partial\xi/\partial x = 0$, which is readily verified. By linear superposition a general solution of Eq. (8.4) is $\psi = F(x + ct) + G(x - ct)$. For periodic functions F, G we recognize the arguments $x + ct$ and $x - ct$ as the phases of plane waves or wave fronts, where the solutions of the wave equation (8.4) change abruptly (from zero to their actual values) and are not uniquely determined. Normal to the wave fronts are the rays of geometric optics. Thus, the solutions of Eq. (8.5), or (8.3) more generally, that are called *characteristics* or sometimes *bicharacteristics* (for second-order PDEs) in the mathematical literature correspond to the wave fronts of the geometric optics solution of the full wave equation.

For the elliptic case let us consider Laplace's equation

$$\frac{\partial^2\psi}{\partial x^2} + \frac{\partial^2\psi}{\partial y^2} = 0, \tag{8.6}$$

for a potential ψ of two variables. Here the characteristic equation

$$\left(\frac{\partial\xi}{\partial x}\right)^2 + \left(\frac{\partial\xi}{\partial y}\right)^2 = \left(\frac{\partial\xi}{\partial x} + i\frac{\partial\xi}{\partial y}\right)\left(\frac{\partial\xi}{\partial x} - i\frac{\partial\xi}{\partial y}\right) = 0 \tag{8.7}$$

has complex conjugate solutions: $\xi = F(x + iy)$ for $\partial\xi/\partial x + i\partial\xi/\partial y = 0$ and $\xi = G(x - iy)$ for $\partial\xi/\partial x - i\partial\xi/\partial y = 0$. A general solution of the potential equation (8.6) is therefore $\psi = F(x + iy) + G(x - iy)$, as well as the real and imaginary parts of ψ, which are called *harmonic* functions, while polynomial solutions are called *harmonic polynomials*.

In quantum mechanics the Wentzel–Kramers–Brillouin (WKB) form $\psi = \exp(-iS/\hbar)$ for the solution of the Schrödinger equation

$$\left(-\frac{\hbar^2}{2m}\nabla^2 + V\right)\psi = i\hbar\frac{\partial\psi}{\partial t} \tag{8.8}$$

leads to the Hamilton–Jacobi equation of classical mechanics,

$$\frac{1}{2m}(\nabla S)^2 + V = \frac{\partial S}{\partial t}, \tag{8.9}$$

in the limit $\hbar \to 0$. The classical action S then becomes the characteristic of the Schrödinger equation. Substituting $\nabla\psi = -i\psi\,\nabla S/\hbar$, $\partial\psi/\partial t = -i\psi\,\partial S/\partial t/\hbar$ into Eq. (8.8), dropping the overall nonvanishing factor ψ, and approximating $\nabla^2\psi = -i\psi\nabla^2 S/\hbar - \psi(\nabla S)^2/\hbar^2 \simeq -\psi(\nabla S)^2/\hbar^2$, i.e., neglecting $-i\nabla^2\psi/\hbar$, we indeed obtain Eq. (8.9).

Finding solutions of PDEs by solving for the characteristics is one of several general techniques. For more examples and a detailed treatment of characteristics, which we will not pursue further here, we refer to H. Bateman, *Partial Differential Equations of Mathematical Physics*. New York: Dover (1944); K. E. Gustafson, *Partial Differential Equations and Hilbert Space Methods*, 2nd ed. New York: Wiley (1987).

NONLINEAR PDEs

Nonlinear ODEs and PDEs are a rapidly growing and important field. We encountered above the simplest linear wave equation

$$\frac{\partial\psi}{\partial t} + c\frac{\partial\psi}{\partial x} = 0$$

as the first-order PDE from the characteristics of the wave equation. The simplest nonlinear wave equation

$$\frac{\partial\psi}{\partial t} + c(\psi)\frac{\partial\psi}{\partial x} = 0 \tag{8.10}$$

results if the local speed of propagation, c, is not constant but depends on the wave ψ. When a nonlinear equation has a solution of the form $\psi(x, t) = A\cos(kx - \omega t)$ where $\omega(k)$ varies with k so that $\omega''(k) \neq 0$, then it is called *dispersive*. Perhaps the best known nonlinear dispersive equation of second order is the *Korteweg–deVries* equation

$$\frac{\partial\psi}{\partial t} + \psi\frac{\partial\psi}{\partial x} + \frac{\partial^3\psi}{\partial x^3} = 0, \tag{8.11}$$

which models the lossless propagation of shallow water waves and other phenomena. It is widely known for its *soliton* solutions. A soliton is a traveling wave with the property of persisting through an interaction with another soliton: After they pass through each other, they emerge in the same shape and with the same velocity and acquire no more than a phase shift. Let $\psi(\xi = x - ct)$ be such a traveling wave. When substituted into Eq. (8.11) this yields the nonlinear ODE

$$(\psi - c)\frac{d\psi}{d\xi} + \frac{d^3\psi}{d\xi^3} = 0, \tag{8.12}$$

which can be integrated to yield

$$\frac{d^2\psi}{d\xi^2} = c\psi - \frac{\psi^2}{2}. \tag{8.13}$$

There is no additive integration constant in Eq. (8.13) to ensure that $d^2\psi/d\xi^2 \to 0$ with $\psi \to 0$ for large ξ, so that ψ is localized at the characteristic $\xi = 0$, or $x = ct$. Multiplying Eq. (8.13) by $d\psi/d\xi$ and integrating again yields

$$\left(\frac{d\psi}{d\xi}\right)^2 = c\psi^2 - \frac{\psi^3}{3}, \tag{8.14}$$

where $d\psi/d\xi \to 0$ for large ξ. Taking the root of Eq. (8.14) and integrating once more yields the soliton solution

$$\psi(x - ct) = \frac{3c}{\cosh^2\left(\sqrt{c}\,\dfrac{x - ct}{2}\right)}. \tag{8.15}$$

Some nonlinear topics, for example, the logistic equation and the onset of chaos, are reviewed in Chapter 18. For more details and literature, see J. Guckenheimer, *Nonlinear Oscillations, Dynamical Systems and Bifurcations of Vector Fields*. New York: Springer-Verlag (1983).

BOUNDARY CONDITIONS

Usually, when we know a physical system at some time and the law governing the physical process, then we are able to predict the subsequent development. Such initial values are the most common boundary conditions associated with ODEs and PDEs. Finding solutions that match given points, curves, or surfaces corresponds to boundary value problems. Eigenfunctions usually are required to satisfy certain imposed (e.g., asymptotic) boundary conditions. These boundary conditions may take three forms:

1. Cauchy boundary conditions. The value of a function and normal derivative specified on the boundary. In electrostatics this would mean φ, the potential, and E_n the normal components of the electric field.
2. Dirichlet boundary conditions. The value of a function specified on the boundary.
3. Neumann boundary conditions. The normal derivative (normal gradient) of a function specified on the boundary. In the electrostatic case this would be E_n and therefore σ, the surface charge density.

A summary of the relation of these three types of boundary conditions to the three types of two-dimensional partial differential equations is given in Table 8.1. For extended discussions of these partial differential equations the reader may consult Sommerfeld, Chapter 2, or Morse and Feshbach, Chapter 6 (see Additional Readings).

Parts of Table 8.1 are simply a matter of maintaining internal consistency, or common sense. For instance, for Poisson's equation with a closed surface, Dirichlet conditions lead to a unique, stable solution. Neumann conditions, independent of the Dirichlet conditions, likewise lead to a unique stable solution independent of the Dirichlet solution. Therefore Cauchy boundary conditions (meaning Dirichlet plus Neumann) could lead to an inconsistency.

Table 8.1

Boundary conditions	Type of partial differential equation		
	Elliptic	Hyperbolic	Parabolic
	Laplace, Poisson in (x, y)	Wave equation in (x, t)	Diffusion equation in (x, t)
Cauchy			
Open surface	Unphysical results (instability)	*Unique, stable solution*	Too restrictive
Closed surface	Too restrictive	Too restrictive	Too restrictive
Dirichlet			
Open surface	Insufficient	Insufficient	*Unique, stable solution in one direction*
Closed surface	*Unique, stable solution*	Solution not unique	Too restrictive
Neumann			
Open surface	Insufficient	Insufficient	*Unique, stable solution in one direction*
Closed surface	*Unique, stable solution*	Solution not unique	Too restrictive

The term boundary conditions includes as a special case the concept of initial conditions. For instance, specifying the initial position x_0 and the initial velocity v_0 in some dynamical problem would correspond to the Cauchy boundary conditions. The only difference in the present usage of boundary conditions in these one-dimensional problems is that we are going to apply the conditions on *both* ends of the allowed range of the variable.

8.2 FIRST-ORDER DIFFERENTIAL EQUATIONS

Physics involves some first-order differential equations. For completeness (and possible review) it seems desirable to touch on them briefly.

We consider here differential equations of the general form

$$\frac{dy}{dx} = f(x, y) = -\frac{P(x, y)}{Q(x, y)}. \tag{8.16}$$

Equation (8.16) is clearly a first-order, ordinary differential equation. It is *first-order* because it contains the first and no higher derivatives. *Ordinary* because the only derivative dy/dx is an ordinary or total derivative. Equation (8.16) may or may not be *linear*, although we shall treat the linear case explicitly later, Eq. (8.125).

SEPARABLE VARIABLES

Frequently Eq. (8.16) will have the special form

$$\frac{dy}{dx} = f(x, y) = -\frac{P(x)}{Q(y)}. \tag{8.17}$$

Then it may be rewritten as

$$P(x)\,dx + Q(y)\,dy = 0.$$

Integrating from (x_0, y_0) to (x, y) yields

$$\int_{x_0}^{x} P(x)\,dx + \int_{y_0}^{y} Q(y)\,dy = 0. \tag{8.18}$$

Since the lower limits x_0 and y_0 contribute constants, we may ignore the lower limits of integration and simply add a constant of integration. Note that this separation of variables technique does *not* require that the differential equation be linear.

Example 8.2.1 Boyle's Law

In differential form Boyle's gas law is

$$\frac{dV}{dP} = -\frac{V}{P}$$

for the volume V of a fixed quantity of gas at pressure P (and constant temperature). Separating variables, we have

$$\frac{dV}{V} = -\frac{dP}{P}$$

or

$$\ln V = -\ln P + C.$$

With two logarithms already, it is most convenient to rewrite the constant of integration C as $\ln k$. Then

$$\ln V + \ln P = \ln PV = \ln k$$

and

$$PV = k.$$

EXACT DIFFERENTIAL EQUATIONS

We rewrite Eq. (8.16) as

$$P(x, y)\,dx + Q(x, y)\,dy = 0. \tag{8.19}$$

This equation is said to be *exact* if we can match the left-hand side of it to a differential $d\varphi$,

$$d\varphi = \frac{\partial \varphi}{\partial x} dx + \frac{\partial \varphi}{\partial y} dy. \tag{8.20}$$

Since Eq. (8.19) has a zero on the right, we look for an unknown function $\varphi(x, y) =$ constant and $d\varphi = 0$.

We have (if such a function $\varphi(x, y)$ exists)

$$P(x, y) dx + Q(x, y) dy = \frac{\partial \varphi}{\partial x} dx + \frac{\partial \varphi}{\partial y} dy \tag{8.21}$$

and

$$\frac{\partial \varphi}{\partial x} = P(x, y), \qquad \frac{\partial \varphi}{\partial y} = Q(x, y). \tag{8.22}$$

The necessary and sufficient condition for our equation to be exact is that the second, mixed partial derivatives of $\varphi(x, y)$ (assumed continuous) are independent of the order of differentiation:

$$\frac{\partial^2 \varphi}{\partial y \, \partial x} = \frac{\partial P(x, y)}{\partial y} = \frac{\partial Q(x, y)}{\partial x} = \frac{\partial^2 \varphi}{\partial x \, \partial y}. \tag{8.23}$$

Note the resemblance to Eqs. (1.132) of Section 1.13, "Potential Theory." If Eq. (8.19) corresponds to a curl (equal to zero), then a potential, $\varphi(x, y)$, must exist.

If $\varphi(x, y)$ exists then from Eqs. (8.19) and (8.21) our solution is

$$\varphi(x, y) = C. \tag{8.24}$$

We may construct $\varphi(x, y)$ from its partial derivatives just as we constructed a magnetic vector potential in Section 1.13 from its curl. See Exercises 8.2.7 and 8.2.8.

It may well turn out that Eq. (8.19) is not exact, that Eq. (8.23) is not satisfied. However, there always exists at least one and perhaps an infinity of integrating factors, $\alpha(x, y)$, such that

$$\alpha(x, y)P(x, y) dx + \alpha(x, y)Q(x, y) dy = 0$$

is exact. Unfortunately, an integrating factor is not always obvious or easy to find. Unlike the case of the *linear* first-order differential equation to be considered next, there is no systematic way to develop an integrating factor for Eq. (8.19).

A differential equation in which the variables have been separated is automatically exact. An exact differential equation is *not* necessarily separable.

LINEAR FIRST-ORDER ORDINARY DIFFERENTIAL EQUATIONS

If $f(x, y)$ in Eq. (8.16) has the form $-p(x)y + q(x)$, then Eq. (8.16) becomes

$$\frac{dy}{dx} + p(x)y = q(x). \tag{8.25}$$

Equation (8.25) is the most general *linear* first-order ODE. If $q(x) = 0$, Eq. (8.25) is homogeneous (in y). A nonzero $q(x)$ may represent a source or a driving term. Equation (8.25) is *linear*; each term is linear in y or dy/dx. There are no higher powers; that is, y^2, and no products, $y(dy/dx)$. Note that the linearity refers to the y and dy/dx; $p(x)$ and $q(x)$ need not be linear in x. Equation (8.25), the most important of these first-order differential equations for physics, may be solved exactly.

Let us look for an *integrating factor* $\alpha(x)$ so that

$$\alpha(x)\frac{dy}{dx} + \alpha(x)p(x)y = \alpha(x)q(x) \tag{8.26}$$

may be rewritten as

$$\frac{d}{dx}[\alpha(x)y] = \alpha(x)q(x). \tag{8.27}$$

The purpose of this is to make the left-hand side of Eq. (8.25) a derivative so that it can be integrated—by inspection. It also, incidentally, makes Eq. (8.25) exact. Expanding Eq. (8.27), we obtain

$$\alpha(x)\frac{dy}{dx} + \frac{d\alpha}{dx}y = \alpha(x)q(x).$$

Comparison with Eq. (8.26) shows that we must require

$$\frac{d\alpha(x)}{dx} = \alpha(x)p(x). \tag{8.28}$$

Here is a differential equation for $\alpha(x)$, with the variables α and x *separable*. We separate variables, integrate, and obtain

$$\alpha(x) = \exp\left[\int^x p(x)\,dx\right] \tag{8.29}$$

as our integrating factor.

With $\alpha(x)$ known we proceed to integrate Eq. (8.27). This, of course, was the point of introducing α in the first place. We have

$$\int^x \frac{d}{dx}[\alpha(x)y(x)]\,dx = \int^x \alpha(x)q(x)\,dx.$$

Now integrating by inspection, we have

$$\alpha(x)y(x) = \int^x \alpha(x)q(x)\,dx + C.$$

The constants from a constant lower limit of integration are lumped into the constant C. Dividing by $\alpha(x)$, we obtain

$$y(x) = [\alpha(x)]^{-1}\left\{\int^x \alpha(x)q(x)\,dx + C\right\}.$$

Finally, substituting in Eq. (8.29) for α yields

$$y(x) = \exp\left[-\int^x p(t)\,dt\right]\left\{\int^x \exp\left[\int^s p(t)\,dt\right] q(s)\,ds + C\right\}. \quad (8.30)$$

Here the (dummy) variables of integration have been rewritten to make them unambiguous. Equation (8.30) is the complete general solution of the linear, first-order differential equation, Eq. (8.25). The portion

$$y_1(x) = C\exp\left[-\int^x p(t)\,dt\right] \quad (8.31)$$

corresponds to the case $q(x) = 0$ and is a general solution of the homogeneous differential equation. The other term in Eq. (8.30),

$$y_2(x) = \exp\left[-\int^x p(t)\,dt\right]\int^x \exp\left[\int^s p(t)\,dt\right] q(s)\,ds, \quad (8.32)$$

is a particular solution corresponding to the specific source term $q(x)$.

The reader might note that if our linear first-order differential equation is homogeneous ($q = 0$), then it is separable. Otherwise, apart from special cases such as $p = $ constant, $q = $ constant, or $q(x) = ap(x)$, Eq. (8.25) is not separable.

Example 8.2.2 RL Circuit

For a resistance–inductance circuit Kirchhoff's law leads to

$$L\frac{dI(t)}{dt} + RI(t) = V(t)$$

for the current $I(t)$, where L is the inductance and R the resistance, both constant. $V(t)$ is the time-dependent impressed voltage.

From Eq. (8.29) our integrating factor $\alpha(t)$ is

$$\alpha(t) = \exp\int^t \frac{R}{L}\,dt$$

$$= e^{Rt/L}.$$

Then by Eq. (8.30)

$$I(t) = e^{-Rt/L}\left[\int^t e^{Rt/L}\frac{V(t)}{L}\,dt + C\right],$$

with the constant C to be determined by an initial condition (a boundary condition).

For the special case $V(t) = V_0$, a constant,

$$I(t) = e^{-Rt/L}\left[\frac{V_0}{L}\cdot\frac{L}{R}e^{Rt/L} + C\right]$$

$$= \frac{V_0}{R} + Ce^{-Rt/L}.$$

If the initial condition is $I(0) = 0$, then $C = -V_0/R$ and

$$I(t) = \frac{V_0}{R}[1 - e^{-Rt/L}].$$

CONVERSION TO INTEGRAL EQUATION

Our first-order differential equation, Eq. (8.16), may be converted to an integral equation by direct integration:

$$y(x) - y(x_0) = \int_{x_0}^{x} f[x, y(x)]\, dx. \qquad (8.33)$$

As an integral equation there is a possibility of a Neumann series solution (Section 16.3) with the initial approximation $y(x) \approx y(x_0)$. In the differential equation literature this is called the "Picard method of successive approximations."

First-order differential equations will be encountered again in Chapter 15 in connection with Laplace transforms and in Chapter 17 from the Euler equation of the calculus of variations. Numerical techniques for solving first-order differential equations are examined in Section 8.8.

EXERCISES

8.2.1   From Kirchhoff's law the current I in an RC (resistance–capacitance) circuit (Fig. 8.1) obeys the equation

$$R\frac{dI}{dt} + \frac{1}{C}I = 0.$$

(a) Find $I(t)$.
(b) For a capacitance of 10,000 microfarads charged to 100 volts and discharging through a resistance of 1 megohm, find the current I for $t = 0$ and for $t = 100$ seconds.

Note. The initial voltage is $I_0 R$ or Q/C, where $Q = \int_0^\infty I(t)\, dt$.

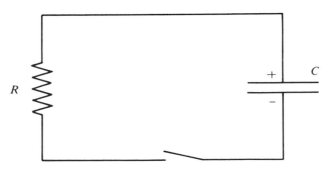

Figure 8.1   RC circuit.

8.2.2   The Laplace transform of Bessel's equation ($n = 0$) leads to
$$(s^2 + 1)f'(s) + sf(s) = 0.$$
Solve for $f(s)$.

8.2.3   The decay of a population by catastrophic two body collisions is described by
$$\frac{dN}{dt} = -kN^2.$$
This is a first-order, *nonlinear* differential equation. Derive the solution
$$N(t) = N_0\left(1 + \frac{t}{\tau_0}\right)^{-1},$$
where $\tau_0 = (kN_0)^{-1}$. This implies an infinite population at $t = -\tau_0$.

8.2.4   The rate of a particular chemical reaction $A + B \to C$ is proportional to the concentrations of the reactants A and B:
$$\frac{dC(t)}{dt} = \alpha[A(0) - C(t)][B(0) - C(t)].$$
(a) Find $C(t)$ for $A(0) \neq B(0)$.
(b) Find $C(t)$ for $A(0) = B(0)$.
The initial condition is that $C(0) = 0$.

8.2.5   A boat, coasting through the water, experiences a resisting force proportional to v^n, v being the boat's instantaneous velocity. Newton's second law leads to
$$m\frac{dv}{dt} = -kv^n.$$
With $v(t = 0) = v_0, x(t = 0) = 0$, integrate to find v as a function of time and v as a function of distance.

8.2.6   In the first-order differential equation $dy/dx = f(x, y)$ the function $f(x, y)$ is a function of the ratio y/x:
$$\frac{dy}{dx} = g(y/x).$$
Show that the substitution of $u = y/x$ leads to a *separable* equation in u and x.

8.2.7   The differential equation
$$P(x, y)\, dx + Q(x, y)\, dy = 0$$
is *exact*. Construct a solution
$$\varphi(x, y) = \int_{x_0}^{x} P(x, y)\, dx + \int_{y_0}^{y} Q(x_0, y)\, dy = \text{constant}.$$

8.2.8   The differential equation
$$P(x, y)\, dx + Q(x, y)\, dy = 0$$
is *exact*. If
$$\varphi(x, y) = \int_{x_0}^{x} P(x, y)\, dx + \int_{y_0}^{y} Q(x_0, y)\, dy,$$

show that

$$\frac{\partial \varphi}{\partial x} = P(x, y), \qquad \frac{\partial \varphi}{\partial y} = Q(x, y).$$

Hence $\varphi(x, y) = $ constant is a solution of the original differential equation.

8.2.9   Prove that Eq. (8.26) is exact in the sense of Eq. (8.23), provided that $\alpha(x)$ satisfies Eq. (8.28).

8.2.10   A certain differential equation has the form

$$f(x)\, dx + g(x)h(y)\, dy = 0,$$

with none of the functions $f(x), g(x), h(y)$ identically zero. Show that a necessary and sufficient condition for this equation to be exact is that $g(x) = $ constant.

8.2.11   Show that

$$y(x) = \exp\left[-\int^x p(t)\, dt \right] \left\{ \int^x \exp\left[\int^s p(t)\, dt \right] q(s)\, ds + C \right\}$$

is a solution of

$$\frac{dy}{dx} + p(x)y(x) = q(x)$$

by differentiating the expression for $y(x)$ and substituting into the differential equation.

8.2.12   The motion of a body falling in a resisting medium may be described by

$$m\frac{dv}{dt} = mg - bv$$

when the retarding force is proportional to the velocity, v. Find the velocity. Evaluate the constant of integration by demanding that $v(0) = 0$.

8.2.13   Radioactive nuclei decay according to the law

$$\frac{dN}{dt} = -\lambda N,$$

N being the concentration of a given nuclide and λ, the particular decay constant. In a radioactive series of n different nuclides, starting with N_1,

$$\frac{dN_1}{dt} = -\lambda_1 N_1,$$

$$\frac{dN_2}{dt} = \lambda_1 N_1 - \lambda_2 N_2, \qquad \text{and so on.}$$

Find $N_2(t)$ for the conditions $N_1(0) = N_0$ and $N_2(0) = 0$.

8.2.14   The rate of evaporation from a particular spherical drop of liquid (constant density) is proportional to its surface area. Assuming this to be the sole mechanism of mass loss, find the radius of the drop as a function of time.

8.2.15 In the linear homogeneous differential equation

$$\frac{dv}{dt} = -av$$

the variables are separable. When the variables are separated the equation is exact. Solve this differential equation subject to $v(0) = v_0$ by the following three methods:

(a) Separating variables and integrating.
(b) Treating the separated variable equation as exact.
(c) Using the result for a linear homogeneous differential equation.

$$ANS. \quad v(t) = v_0 e^{-at}.$$

8.2.16 Bernoulli's equation,

$$\frac{dy}{dx} + f(x)y = g(x)y^n$$

is nonlinear for $n \neq 0$ or 1. Show that the substitution $u = y^{1-n}$ reduces Bernoulli's equation to a linear equation.

(See Chapter 18.4.) $\qquad ANS. \quad \dfrac{du}{dx} + (1 - n)f(x)u = (1 - n)g(x).$

8.2.17 Solve the linear, first-order equation, Eq. (8.25), by assuming $y(x) = u(x)v(x)$, where $v(x)$ is a solution of the corresponding homogeneous equation $[q(x) = 0]$. This is the method of *variation of parameters* due to Lagrange. We apply it to second-order equations in Exercise 8.6.25.

8.3 SEPARATION OF VARIABLES

The equations of mathematical physics listed in Section 8.1 are all partial differential equations. Our first technique for their solution splits the partial differential equation of n variables into n ordinary differential equations. Each separation introduces an arbitrary constant of separation. If we have n variables, we have to introduce $n - 1$ constants, determined by the conditions imposed in the problem being solved.

Cartesian Coordinates

In Cartesian coordinates the Helmholtz equation becomes

$$\frac{\partial^2 \psi}{\partial x^2} + \frac{\partial^2 \psi}{\partial y^2} + \frac{\partial^2 \psi}{\partial z^2} + k^2 \psi = 0, \tag{8.34}$$

using Eq. (2.26) for the Laplacian. For the present let k^2 be a constant. Perhaps the simplest way of treating a partial differential equation such as (8.34) is to split it into a set of ordinary differential equations. This may be done as follows. Let

$$\psi(x, y, z) = X(x)Y(y)Z(z) \tag{8.35}$$

and substitute back into Eq. (8.34). How do we know Eq. (8.35) is valid? The answer is very simple. We do *not* know it is valid! Rather, we are proceeding in the spirit of let's try and see if it works. If our attempt succeeds, then Eq. (8.35) will be justified. If it does not succeed, we shall find out soon enough and then we shall try another attack such as Green's functions, integral transforms, or brute force numerical analysis. With ψ assumed given by Eq. (8.35), Eq. (8.34) becomes

$$YZ\frac{d^2X}{dx^2} + XY\frac{d^2Y}{dy^2} + XY\frac{d^2Z}{dz^2} + k^2XYZ = 0. \tag{8.36}$$

Dividing by $\psi = XYZ$ and rearranging terms, we obtain

$$\frac{1}{X}\frac{d^2X}{dx^2} = -k^2 - \frac{1}{Y}\frac{d^2Y}{dy^2} - \frac{1}{Z}\frac{d^2Z}{dz^2}. \tag{8.37}$$

Equation (8.37) exhibits one separation of variables. The left-hand side is a function of x alone, whereas the right-hand side depends only on y and z. So Eq. (8.37) is a sort of paradox. A function of x is equated to a function of y and z, but x, y, and z are all independent coordinates. This independence means that the behavior of x as an independent variable is not determined by y and z. The paradox is resolved by setting each side equal to a constant, a constant of separation. We choose[1]

$$\frac{1}{X}\frac{d^2X}{dx^2} = -l^2, \tag{8.38}$$

$$-k^2 - \frac{1}{Y}\frac{d^2Y}{dy^2} - \frac{1}{Z}\frac{d^2Z}{dz^2} = -l^2. \tag{8.39}$$

Now, turning our attention to Eq. (8.39), we obtain

$$\frac{1}{Y}\frac{d^2Y}{dy^2} = -k^2 + l^2 - \frac{1}{Z}\frac{d^2Z}{dz^2}, \tag{8.40}$$

and a second separation has been achieved. Here we have a function of y equated to a function of z and the same paradox appears. We resolve it as before by equating each side to another constant of separation, $-m^2$,

$$\frac{1}{Y}\frac{d^2Y}{dy^2} = -m^2, \tag{8.41}$$

$$\frac{1}{Z}\frac{d^2Z}{dz^2} = -k^2 + l^2 + m^2 = -n^2, \tag{8.42}$$

introducing a constant n^2 by $k^2 = l^2 + m^2 + n^2$ to produce a symmetric set of equations. Now we have three ordinary differential equations ((8.38), (8.41),

---

[1] The choice of sign, completely arbitrary here, will be fixed in specific problems by the need to satisfy specific boundary conditions.

and (8.42)) to replace Eq. (8.34). Our assumption (Eq. (8.35)) has succeeded and is thereby justified.

Our solution should be labeled according to the choice of our constants l, m, and n, that is,

$$\psi_{lmn}(x, y, z) = X_l(x)Y_m(y)Z_n(z). \tag{8.43}$$

Subject to the conditions of the problem being solved and to the condition $k^2 = l^2 + m^2 + n^2$, we may choose l, m, and n as we like, and Eq. (8.43) will still be a solution of Eq. (8.34), provided $X_l(x)$ is a solution of Eq. (8.38), and so on. We may develop the most general solution of Eq. (8.34) by taking a linear combination of solutions ψ_{lmn},

$$\Psi = \sum_{l, m, n} a_{lmn}\psi_{lmn}. \tag{8.44}$$

The constant coefficients a_{lmn} are finally chosen to permit Ψ to satisfy the boundary conditions of the problem.

Circular Cylindrical Coordinates

With our unknown function ψ dependent on ρ, φ, and z, the Helmholtz equation becomes

$$\nabla^2\psi(\rho, \varphi, z) + k^2\psi(\rho, \varphi, z) = 0, \tag{8.45}$$

or

$$\frac{1}{\rho}\frac{\partial}{\partial\rho}\left(\rho\frac{\partial\psi}{\partial\rho}\right) + \frac{1}{\rho^2}\frac{\partial^2\psi}{\partial\varphi^2} + \frac{\partial^2\psi}{\partial z^2} + k^2\psi = 0. \tag{8.46}$$

As before, we assume a factored form for ψ,

$$\psi(\rho, \varphi, z) = P(\rho)\Phi(\varphi)Z(z). \tag{8.47}$$

Substituting into Eq. (8.46), we have

$$\frac{\Phi Z}{\rho}\frac{d}{d\rho}\left(\rho\frac{dP}{d\rho}\right) + \frac{PZ}{\rho^2}\frac{d^2\Phi}{d\varphi^2} + P\Phi\frac{d^2Z}{dz^2} + k^2P\Phi Z = 0. \tag{8.48}$$

All the partial derivatives have become ordinary derivatives. Dividing by $P\Phi Z$ and moving the z derivative to the right-hand side yields

$$\frac{1}{\rho P}\frac{d}{d\rho}\left(\rho\frac{dP}{d\rho}\right) + \frac{1}{\rho^2\Phi}\frac{d^2\Phi}{d\varphi^2} + k^2 = -\frac{1}{Z}\frac{d^2Z}{dz^2}. \tag{8.49}$$

Again, we have the paradox. A function of z on the right appears to depend on a function of ρ and φ on the left. We resolve the paradox by setting each side of Eq. (8.49) equal to a constant, the same constant. Let us choose[2] $-l^2$.

---

[2] The choice of sign of the separation constant is arbitrary. However, a minus sign is chosen for the axial coordinate z in expectation of a possible exponential dependence on z (from Eq. (8.50)). A positive sign is chosen for the azimuthal coordinate φ in expectation of a periodic dependence on φ (from Eq. (8.53)).

Then

$$\frac{d^2Z}{dz^2} = l^2 Z, \tag{8.50}$$

and

$$\frac{1}{\rho P}\frac{d}{d\rho}\left(\rho\frac{dP}{d\rho}\right) + \frac{1}{\rho^2\Phi}\frac{d^2\Phi}{d\varphi^2} + k^2 = -l^2. \tag{8.51}$$

Setting $k^2 + l^2 = n^2$, multiplying by ρ^2, and rearranging terms, we obtain

$$\frac{\rho}{P}\frac{d}{d\rho}\left(\rho\frac{dP}{d\rho}\right) + n^2\rho^2 = -\frac{1}{\Phi}\frac{d^2\Phi}{d\varphi^2}. \tag{8.52}$$

We may set the right-hand side to m^2 and

$$\frac{d^2\Phi}{d\varphi^2} = -m^2\Phi. \tag{8.53}$$

Finally, for the ρ dependence we have

$$\rho\frac{d}{d\rho}\left(\rho\frac{dP}{d\rho}\right) + (n^2\rho^2 - m^2)P = 0. \tag{8.54}$$

This is Bessel's differential equation. The solutions and their properties are presented in Chapter 11. The separation of variables of Laplace's equation in parabolic coordinates also gives rise to Bessel's equation. It may be noted that the Bessel equation is notorious for the variety of disguises it may assume. For an extensive tabulation of possible forms the reader is referred to *Tables of Functions* by Jahnke and Emde.[3]

The original Helmholtz equation, a three-dimensional partial differential equation, has been replaced by three ordinary differential equations, Eqs. (8.50), (8.53), and (8.54). A solution of the Helmholtz equation is

$$\psi(\rho, \varphi, z) = P(\rho)\Phi(\varphi)Z(z). \tag{8.55}$$

Identifying the specific P, Φ, Z solutions by subscripts, we see that the most general solution of the Helmholtz equation is a linear combination of the product solutions:

$$\Psi(\rho, \varphi, z) = \sum_{m,n} a_{mn} P_{mn}(\rho)\Phi_m(\varphi)Z_n(z). \tag{8.56}$$

---

[3] E. Jahnke and F. Emde, *Tables of Functions*, 4th rev. ed. New York: Dover (1945), p. 146; also, E. Jahnke, F. Emde, and F. Lösch, *Tables of Higher Functions*, 6th ed. New York: McGraw–Hill (1960).

Spherical Polar Coordinates

Let us try to separate the Helmholtz equation, again with k^2 constant, in spherical polar coordinates. Using Eq. (2.46), we obtain

$$\frac{1}{r^2 \sin \theta} \left[\sin \theta \frac{\partial}{\partial r} \left(r^2 \frac{\partial \psi}{\partial r} \right) + \frac{\partial}{\partial \theta} \left(\sin \theta \frac{\partial \psi}{\partial \theta} \right) + \frac{1}{\sin \theta} \frac{\partial^2 \psi}{\partial \varphi^2} \right] = -k^2 \psi. \quad (8.57)$$

Now, in analogy with Eq. (8.35) we try

$$\psi(r, \theta, \varphi) = R(r)\Theta(\theta)\Phi(\varphi). \quad (8.58)$$

By substituting back into Eq. (8.57) and dividing by $R\Theta\Phi$, we have

$$\frac{1}{Rr^2} \frac{d}{dr} \left(r^2 \frac{dR}{dr} \right) + \frac{1}{\Theta r^2 \sin \theta} \frac{d}{d\theta} \left(\sin \theta \frac{d\Theta}{d\theta} \right) + \frac{1}{\Phi r^2 \sin^2 \theta} \frac{d^2 \Phi}{d\varphi^2} = -k^2. \quad (8.59)$$

Note that all derivatives are now ordinary derivatives rather than partials. By multiplying by $r^2 \sin^2 \theta$, we can isolate $(1/\Phi)(d^2\Phi/d\varphi^2)$ to obtain[4]

$$\frac{1}{\Phi} \frac{d^2 \Phi}{d\varphi^2} = r^2 \sin^2 \theta \left[-k^2 - \frac{1}{r^2 R} \frac{d}{dr} \left(r^2 \frac{dR}{dr} \right) - \frac{1}{r^2 \sin \theta \Theta} \frac{d}{d\theta} \left(\sin \theta \frac{d\Theta}{d\theta} \right) \right]. \quad (8.60)$$

Equation (8.60) relates a function of φ alone to a function of r and θ alone. Since r, θ, and φ are independent variables, we equate each side of Eq. (8.60) to a constant. Here a little consideration can simplify the later analysis. In almost all physical problems φ will appear as an azimuth angle. This suggests a periodic solution rather than an exponential. With this in mind, let us use $-m^2$ as the separation constant. Any constant will do, but this one will make life a little easier. Then

$$\frac{1}{\Phi} \frac{d^2 \Phi(\varphi)}{d\varphi^2} = -m^2 \quad (8.61)$$

and

$$\frac{1}{r^2 R} \frac{d}{dr} \left(r^2 \frac{dR}{dr} \right) + \frac{1}{r^2 \sin \theta \Theta} \frac{d}{d\theta} \left(\sin \theta \frac{d\Theta}{d\theta} \right) - \frac{m^2}{r^2 \sin^2 \theta} = -k^2. \quad (8.62)$$

Multiplying Eq. (8.62) by r^2 and rearranging terms, we obtain

$$\frac{1}{R} \frac{d}{dr} \left(r^2 \frac{dR}{dr} \right) + r^2 k^2 = - \frac{1}{\sin \theta \Theta} \frac{d}{d\theta} \left(\sin \theta \frac{d\Theta}{d\theta} \right) + \frac{m^2}{\sin^2 \theta}. \quad (8.63)$$

Again, the variables are separated. We equate each side to a constant Q and finally obtain

$$\frac{1}{\sin \theta} \frac{d}{d\theta} \left(\sin \theta \frac{d\Theta}{d\theta} \right) - \frac{m^2}{\sin^2 \theta} \Theta + Q\Theta = 0, \quad (8.64)$$

---

[4] The order in which the variables are separated here is not unique. Many quantum mechanics texts show the r dependence split off first.

$$\frac{1}{r^2}\frac{d}{dr}\left(r^2\frac{dR}{dr}\right) + k^2 R - \frac{QR}{r^2} = 0. \tag{8.65}$$

Once more we have replaced a partial differential equation of three variables by three ordinary differential equations. The solutions of these ordinary differential equations are discussed in Chapters 11 and 12. In Chapter 12, for example, Eq. (8.64) is identified as the associated Legendre equation in which the constant Q becomes $l(l + 1)$; l is an integer. If k^2 is a (positive) constant, Eq. (8.65) becomes the spherical Bessel equation of Section 11.7.

Again, our most general solution may be written

$$\psi_{Qm}(r, \theta, \varphi) = \sum_{Q, m} R_Q(r)\Theta_{Qm}(\theta)\Phi_m(\varphi). \tag{8.66}$$

The restriction that k^2 be a constant is unnecessarily severe. The separation process will still be possible for k^2 as general as

$$k^2 = f(r) + \frac{1}{r^2}g(\theta) + \frac{1}{r^2\sin^2\theta}h(\varphi) + k'^2. \tag{8.67}$$

In the hydrogen atom problem, one of the most important examples of the Schrödinger wave equation with a closed form solution is $k^2 = f(r)$. Equation (8.65) for the hydrogen atom becomes the associated Laguerre equation.

The great importance of this separation of variables in spherical polar coordinates stems from the fact that the case $k^2 = k^2(r)$ covers a tremendous amount of physics: a great deal of the theories of gravitation, electrostatics, atomic physics, and nuclear physics. And, with $k^2 = k^2(r)$, the angular dependence is isolated in Eqs. (8.61) and (8.64), *which can be solved exactly*.

Finally, as an illustration of how the constant m in Eq. (8.61) is restricted, we note that φ in cylindrical and spherical polar coordinates is an azimuth angle. If this is a classical problem, we shall certainly require that the azimuthal solution $\Phi(\varphi)$ be single-valued, that is,

$$\Phi(\varphi + 2\pi) = \Phi(\varphi). \tag{8.68}$$

This is equivalent to requiring the azimuthal solution to have a period of 2π or some integral multiple of it.[5] Therefore m must be an integer. Which integer it is depends on the details of the problem. This is discussed in Chapter 9. Whenever a coordinate corresponds to an axis of translation or to an azimuth angle the separated equation always has the form

$$\frac{d^2\Phi(\varphi)}{d\varphi^2} = -m^2\Phi(\varphi)$$

---

[5] This also applies in most quantum mechanical problems but the argument is much more involved. If m is not an integer, rotation group relations and ladder operator relations (Section 4.3) are disrupted. Compare E. Merzbacher, Single valuedness of wave functions. *Am. J. Phys.* **30**, 237 (1962).

for φ, the azimuth angle, and

$$\frac{d^2Z(z)}{dz^2} = \pm a^2 Z(z) \tag{8.69}$$

for z, an axis of translation in one of the cylindrical coordinate systems. The solutions, of course, are $\sin az$ and $\cos az$ for $-a^2$ and the corresponding hyperbolic function (or exponentials) $\sinh az$ and $\cosh az$ for $+a^2$.

Other occasionally encountered ordinary differential equations include the Laguerre and associated Laguerre equations from the supremely important hydrogen atom problem in quantum mechanics:

$$x\frac{d^2y}{dx^2} + (1 - x)\frac{dy}{dx} + \alpha y = 0, \tag{8.70}$$

$$x\frac{d^2y}{dx^2} + (1 + k - x)\frac{dy}{dx} + \alpha y = 0. \tag{8.71}$$

From the quantum mechanical theory of the linear oscillator we have Hermite's equation,

$$\frac{d^2y}{dx^2} - 2x\frac{dy}{dx} + 2\alpha y = 0. \tag{8.72}$$

Finally, from time to time we find the Chebyshev differential equation

$$(1 - x^2)\frac{d^2y}{dx^2} - x\frac{dy}{dx} + n^2 y = 0. \tag{8.73}$$

For convenient reference, the forms of the solutions of Laplace's equation, Helmholtz's equation, and the diffusion equation for spherical polar coordinates are collected in Table 8.2. The solutions of Laplace's equation in circular cylindrical coordinates are presented in Table 8.3.

Table 8.2 Solutions in Spherical Polar Coordinates*

$$\psi = \sum_{l,m} a_{lm} \psi_{lm}$$

1.	$\nabla^2 \psi = 0$	$\psi_{lm} = \begin{Bmatrix} r^l \\ r^{-l-1} \end{Bmatrix} \begin{Bmatrix} P_l^m(\cos\theta) \\ Q_l^m(\cos\theta) \end{Bmatrix} \begin{Bmatrix} \cos m\varphi \\ \sin m\varphi \end{Bmatrix}^\dagger$	
2.	$\nabla^2 \psi + k^2 \psi = 0$	$\psi_{lm} = \begin{Bmatrix} j_l(kr) \\ n_l(kr) \end{Bmatrix} \begin{Bmatrix} P_l^m(\cos\theta) \\ Q_l^m(\cos\theta) \end{Bmatrix} \begin{Bmatrix} \cos m\varphi \\ \sin m\varphi \end{Bmatrix}^\dagger$	
3.	$\nabla^2 \psi - k^2 \psi = 0$	$\psi_{lm} = \begin{Bmatrix} i_l(kr) \\ k_l(kr) \end{Bmatrix} \begin{Bmatrix} P_l^m(\cos\theta) \\ Q_l^m(\cos\theta) \end{Bmatrix} \begin{Bmatrix} \cos m\varphi \\ \sin m\varphi \end{Bmatrix}^\dagger$	

* References for some of the functions are $P_l^m(\cos\theta)$, $m = 0$, Section 12.1; $m \neq 0$, Section 12.5; $Q_l^m(\cos\theta)$, Section 12.10; $j_l(kr)$, $n_l(kr)$, $i_l(kr)$, and $k_l(kr)$, Section 11.7.

† $\cos m\varphi$ and $\sin m\varphi$ may be replaced by $e^{\pm im\varphi}$.

Table 8.3 Solutions in Circular Cylindrical Coordinates*

$$\psi = \sum_{m,\,\alpha} a_{m\alpha}\psi_{m\alpha}, \qquad \nabla^2\psi = 0$$

a. $\qquad \psi_{m\alpha} = \begin{Bmatrix} J_m(\alpha\rho) \\ N_m(\alpha\rho) \end{Bmatrix}\begin{Bmatrix} \cos m\varphi \\ \sin m\varphi \end{Bmatrix}\begin{Bmatrix} e^{-\alpha z} \\ e^{\alpha z} \end{Bmatrix}$

b. $\qquad \psi_{m\alpha} = \begin{Bmatrix} I_m(\alpha\rho) \\ K_m(\alpha\rho) \end{Bmatrix}\begin{Bmatrix} \cos m\varphi \\ \sin m\varphi \end{Bmatrix}\begin{Bmatrix} \cos \alpha z \\ \sin \alpha z \end{Bmatrix}$

c. If $\alpha = 0$ (no z-dependence) $\qquad \psi_m = \begin{Bmatrix} \rho^m \\ \rho^{-m} \end{Bmatrix}\begin{Bmatrix} \cos m\varphi \\ \sin m\varphi \end{Bmatrix}$

* References for the radial functions are $J_m(\alpha\rho)$, Section 11.1; $N_m(\alpha\rho)$, Section 11.3; $I_m(\alpha\rho)$ and $K_m(\alpha\rho)$, Section 11.5.

For the Helmholtz and the diffusion equation the constant $\pm k^2$ is added to the separation constant $\pm\alpha^2$ to define a new parameter γ^2 or $-\gamma^2$. For the choice $+\gamma^2$ (with $\gamma^2 > 0$) we get $J_m(\gamma\rho)$ and $N_m(\gamma\rho)$. For the choice $-\gamma^2$ (with $\gamma^2 > 0$) we get $I_m(\gamma\rho)$ and $K_m(\gamma\rho)$ as previously.

These ordinary differential equations and two generalizations of them will be examined and systematized in the next section. General properties following from the form of the differential equations are discussed in Chapter 9. The individual solutions are developed and applied in Chapters 10–13.

The practicing physicist may and probably will meet other second-order ordinary differential equations, some of which may possibly be transformed into the examples studied here. Some of these differential equations may be solved by the techniques of Sections 8.5 and 8.6. Others may require a calculating machine for a numerical solution.

We refer the reader to the second edition of this text for other important coordinate systems.

EXERCISES

8.3.1 By letting the operator $\nabla^2 + k^2$ act on the general form $a_1\psi_1(x,y,z) + a_2\psi_2(x,y,z)$, show that it is linear, that is, $(\nabla^2 + k^2)(a_1\psi_1 + a_2\psi_2) = a_1(\nabla^2 + k^2)\psi_1 + a_2(\nabla^2 + k^2)\psi_2$.

8.3.2 Show that the Helmholtz equation

$$\nabla^2\psi + k^2\psi = 0$$

is still separable in circular cylindrical coordinates if k^2 is generalized to $k^2 + f(\rho) + (1/\rho^2)g(\varphi) + h(z)$.

8.3.3 Separate variables in the Helmholtz equation in spherical polar coordinates splitting off the radial dependence *first*. Show that your separated equations have the same form as Eqs. (8.61), (8.64), and (8.65).

8.3.4   Verify that

$$\nabla^2\psi(r, \theta, \varphi) + \left[k^2 + f(r) + \frac{1}{r^2}g(\theta) + \frac{1}{r^2\sin^2\theta}h(\varphi)\right]\psi(r, \theta, \varphi) = 0$$

is separable (in spherical polar coordinates). The functions f, g, and h are functions only of the variables indicated; k^2 is a constant.

8.3.5   An atomic (quantum mechanical) particle is confined inside a rectangular box of sides a, b, and c. The particle is described by a wave function ψ which satisfies the Schrödinger wave equation

$$-\frac{\hbar^2}{2m}\nabla^2\psi = E\psi.$$

The wave function is required to vanish at each surface of the box (but not to be identically zero). This condition imposes constraints on the separation constants and therefore on the energy E. What is the smallest value of E for which such a solution can be obtained?

$$ANS. \quad E = \frac{\pi^2\hbar^2}{2m}\left(\frac{1}{a^2} + \frac{1}{b^2} + \frac{1}{c^2}\right).$$

8.3.6   For a homogeneous spherical solid with constant thermal diffusivity, K, and no heat sources the equation of heat conduction becomes

$$\frac{\partial T(r, t)}{\partial t} = K\nabla^2 T(r, t).$$

Assume a solution of the form

$$T = R(r)T(t)$$

and separate variables. Show that the radial equation may take on the standard form

$$r^2\frac{d^2R}{dr^2} + 2r\frac{dR}{dr} + [\alpha^2r^2 - n(n + 1)]R = 0; \quad n = \text{integer}.$$

The solutions of this equation are called spherical Bessel functions.

8.3.7   Separate variables in the thermal diffusion equation of Exercise 8.3.6 in circular cylindrical coordinates. Assume that you can neglect end effects and take $T = T(\rho, t)$.

8.3.8   The quantum mechanical angular momentum operator is given by $\mathbf{L} = -i(\mathbf{r} \times \nabla)$. Show that

$$\mathbf{L} \cdot \mathbf{L}\psi = l(l + 1)\psi$$

leads to the associated Legendre equation.
Hint. Exercises 1.9.9 and 2.5.16 may be helpful.

8.3.9   The one-dimensional Schrödinger wave equation for a particle in a potential field $V = \frac{1}{2}kx^2$ is

$$-\frac{\hbar^2}{2m}\frac{d^2\psi}{dx^2} + \frac{1}{2}kx^2\psi = E\psi(x).$$

(a) Using $\xi = ax$ and a constant λ, we have

$$a = \left(\frac{mk}{\hbar^2}\right)^{1/4}$$

$$\lambda = \frac{2E}{\hbar}\left(\frac{m}{k}\right)^{1/2};$$

show that

$$\frac{d^2\psi(\xi)}{d\xi^2} + (\lambda - \xi^2)\psi(\xi) = 0.$$

(b) Substituting

$$\psi(\xi) = y(\xi)e^{-\xi^2/2},$$

show that $y(\xi)$ satisfies the Hermite differential equation.

8.3.10 Verify that the following are solutions of Laplace's equation:

(a) $\psi_1 = 1/r$,

(b) $\psi_2 = \dfrac{1}{2r}\ln\dfrac{r+z}{r-z}$.

Note. The z derivatives of $1/r$ generate the Legendre polynomials, $P_n(\cos\theta)$, Exercise 12.1.7. The z derivatives of $(1/2r)\ln[(r+z)/(r-z)]$ generate the Legendre functions, $Q_n(\cos\theta)$.

8.3.11 If Ψ is a solution of Laplace's equation, $\nabla^2\Psi = 0$, show that $\partial\Psi/\partial z$ is also a solution.

8.4 SINGULAR POINTS

In this section the concept of a singular point or singularity (as applied to a differential equation) is introduced. The interest in this concept stems from its usefulness in (1) classifying differential equations and (2) investigating the feasibility of a series solution. This feasibility is the topic of Fuchs's theorem, Sections 8.5 and 8.6. First we give a definition.

All the ordinary differential equations listed in Section 8.3 may be solved for d^2y/dx^2. Using the notation $d^2y/dx^2 = y''$, we have[1]

$$y'' = f(x, y, y'). \tag{8.74}$$

Now, if in Eq. (8.74), y and y' can take on all finite values at $x = x_0$ and y'' remains finite, point $x = x_0$ is an ordinary point. On the other hand, if y'' becomes infinite for *any* finite choice of y and y', point $x = x_0$ is labeled a singular point.

Another way of presenting this definition of singular point is to write our second order homogeneous differential equation (in y) as

$$y'' + P(x)y' + Q(x)y = 0. \tag{8.75}$$

---

[1] This prime notation, $y' = dy/dx$, was introduced by Lagrange in the late eighteenth century as an abbreviation for Leibniz's more explicit but more cumbersome dy/dx.

Now, if the functions $P(x)$ and $Q(x)$ remain finite at $x = x_0$, point $x = x_0$ is an ordinary point. However, if either $P(x)$ or $Q(x)$ (or both) diverges as $x \to x_0$, point x_0 is a singular point.

Using Eq. (8.75), we may distinguish between two kinds of singular points.

1. If either $P(x)$ or $Q(x)$ diverges as $x \to x_0$ but $(x - x_0) P(x)$ and $(x - x_0)^2 Q(x)$ remain finite as $x \to x_0$, then $x = x_0$ is called a regular or non-essential singular point.

2. If $P(x)$ diverges faster than $1/(x - x_0)$, so that $(x - x_0) P(x)$ goes to infinity as $x \to x_0$, or $Q(x)$ diverges faster than $1/(x - x_0)^2$ so that $(x - x_0)^2 Q(x)$ goes to infinity as $x \to x_0$, then point $x = x_0$ is labeled an irregular or essential singularity.

These definitions hold for all finite values of x_0. The analysis of point $x \to \infty$ is similar to the treatment of functions of a complex variable (Section 6.6). We set $x = 1/z$, substitute into the differential equation, and then let $z \to 0$. By changing variables in the derivatives, we have

$$\frac{dy(x)}{dx} = \frac{dy(z^{-1})}{dz}\frac{dz}{dx} = -\frac{1}{x^2}\frac{dy(z^{-1})}{dz} = -z^2\frac{dy(z^{-1})}{dz} \qquad (8.76)$$

$$\frac{d^2y(x)}{dx^2} = \frac{d}{dz}\left[\frac{dy(x)}{dx}\right]\frac{dz}{dx} = (-z^2)\left[-2z\frac{dy(z^{-1})}{dz} - z^2\frac{d^2y(z^{-1})}{dz^2}\right]$$

$$= 2z^3\frac{dy(z^{-1})}{dz} + z^4\frac{d^2y(z^{-1})}{dz^2}. \qquad (8.77)$$

Using these results, we transform Eq. (8.75) into

$$z^4\frac{d^2y}{dz^2} + [2z^3 - z^2 P(z^{-1})]\frac{dy}{dz} + Q(z^{-1})y = 0. \qquad (8.78)$$

The behavior at $x = \infty$ ($z = 0$) then depends on the behavior of the new coefficients

$$\frac{2z - P(z^{-1})}{z^2} \qquad \text{and} \qquad \frac{Q(z^{-1})}{z^4}$$

as $z \to 0$. If these two expressions remain finite, point $x = \infty$ is an ordinary point. If they diverge no more rapidly than $1/z$ and $1/z^2$, respectively, point $x = \infty$ is a regular singular point, otherwise an irregular singular point (an essential singularity).

Example 8.4.1

Bessel's equation is

$$x^2 y'' + xy' + (x^2 - n^2)y = 0. \qquad (8.79)$$

Comparing it with Eq. (8.75) we have

$$P(x) = \frac{1}{x}, \qquad Q(x) = 1 - \frac{n^2}{x^2},$$

Table 8.4

Equation	Regular singularity $x =$	Irregular singularity $x =$
1. Hypergeometric $x(x-1)y'' + [(1+a+b)x - c]y' + aby = 0.$	0, 1, ∞	—
2. Legendre* $(1-x^2)y'' - 2xy' + l(l+1)y = 0.$	−1, 1, ∞	—
3. Chebyshev $(1-x^2)y'' - xy' + n^2y = 0.$	−1, 1, ∞	—
4. Confluent hypergeometric $xy'' + (c-x)y' - ay = 0.$	0	∞
5. Bessel $x^2y'' + xy' + (x^2 - n^2)y = 0.$	0	∞
6. Laguerre* $xy'' + (1-x)y' + ay = 0.$	0	∞
7. Simple harmonic oscillator $y'' + \omega^2 y = 0.$	—	∞
8. Hermite $y'' - 2xy' + 2\alpha y = 0.$	—	∞

* The associated equations have the same singular points.

which shows that point $x = 0$ is a regular singularity. By inspection we see that there are no other singular points in the finite range. As $x \to \infty$ ($z \to 0$), from Eq. (8.78) we have the coefficients

$$\frac{2z - z}{z^2} \quad \text{and} \quad \frac{1 - n^2 z^2}{z^4}.$$

Since the latter expression diverges as z^4, point $x = \infty$ is an irregular or essential singularity.

The ordinary differential equations of Section 8.3, plus two others, the hypergeometric and the confluent hypergeometric, have singular points, as shown in Table 8.4.

It will be seen that the first three equations in the preceding tabulation, hypergeometric, Legendre, and Chebyshev, all have three regular singular points. The hypergeometric equation with regular singularities at 0, 1, and ∞ is taken as the standard, the canonical form. The solutions of the other two may then be expressed in terms of its solutions, the hypergeometric functions. This is done in Chapter 13.

In a similar manner, the confluent hypergeometric equation is taken as the canonical form of a linear second-order differential equation with one regular and one irregular singular point.

EXERCISES

8.4.1   Show that Legendre's equation has regular singularities at $x = -1$, 1, and ∞.

8.4.2   Show that Laguerre's equation, like the Bessel equation, has a regular singularity at $x = 0$ and an irregular singularity at $x = \infty$.

8.4.3   Show that the substitution

$$x \rightarrow \frac{1 - x}{2}, \qquad a = -l, \qquad b = l + 1, \qquad c = 1$$

converts the hypergeometric equation into Legendre's equation.

8.5 SERIES SOLUTIONS—FROBENIUS' METHOD

In this section we develop a method of obtaining one solution of the linear, second-order, homogeneous differential equation. The method, a series expansion, will always work, provided the point of expansion is no worse than a regular singular point. In physics this very gentle condition is almost always satisfied.

A *linear, second-order, homogeneous* ODE may be put in the form

$$\frac{d^2y}{dx^2} + P(x)\frac{dy}{dx} + Q(x)y = 0. \tag{8.80}$$

The equation is *homogeneous* because each term contains $y(x)$ or a derivative; *linear* because each y, dy/dx, or d^2y/dx^2 appears as the first power—and no products. In this section we develop (at least) one solution of Eq. (8.80). In Section 8.6 we develop a second, independent solution and prove that no third, independent solution exists. Therefore the most general solution of Eq. (8.80) may be written as

$$y(x) = c_1 y_1(x) + c_2 y_2(x). \tag{8.81}$$

Our physical problem may lead to a *nonhomogeneous*, linear, second-order differential equation

$$\frac{d^2y}{dx^2} + P(x)\frac{dy}{dx} + Q(x)y = F(x). \tag{8.82}$$

The function on the right, $F(x)$, represents a source (such as electrostatic charge) or a driving force (as in a driven oscillator). Specific solutions of this nonhomogeneous equation are touched on in Exercise 8.6.25. They are explored in some detail, using Green's function techniques, in Section 8.7 and 9.5, and with a Laplace transform technique in Section 15.11. Calling this solution y_p, we may add to it any solution of the corresponding homogeneous equation (Eq. (8.80)). Hence the most general solution of Eq. (8.82) is

$$y(x) = c_1 y_1(x) + c_2 y_2(x) + y_p(x). \tag{8.83}$$

The constants c_1 and c_2 will eventually be fixed by boundary conditions.

For the present, we assume that $F(x) = 0$, that our differential equation is homogeneous. We shall attempt to develop a solution of our linear, second-order, homogeneous differential equation, Eq. (8.80), by substituting in a power series with undetermined coefficients. Also available as a parameter is the power of the lowest nonvanishing term of the series. To illustrate, we apply the method to two important differential equations. First, the linear oscillator equation

$$\frac{d^2 y}{dx^2} + \omega^2 y = 0, \tag{8.84}$$

with known solutions $y = \sin \omega x$, $\cos \omega x$.

We try

$$y(x) = x^k(a_0 + a_1 x + a_2 x^2 + a_3 x^3 + \cdots)$$

$$= \sum_{\lambda=0}^{\infty} a_\lambda x^{k+\lambda}, \qquad a_0 \neq 0, \tag{8.85}$$

with the exponent k and all the coefficients a_λ still undetermined. Note that k need *not* be an integer. By differentiating twice, we obtain

$$\frac{dy}{dx} = \sum_{\lambda=0}^{\infty} a_\lambda(k + \lambda)x^{k+\lambda-1},$$

$$\frac{d^2 y}{dx^2} = \sum_{\lambda=0}^{\infty} a_\lambda(k + \lambda)(k + \lambda - 1)x^{k+\lambda-2}.$$

By substituting into Eq. (8.84), we have

$$\sum_{\lambda=0}^{\infty} a_\lambda(k + \lambda)(k + \lambda - 1)x^{k+\lambda-2} + \omega^2 \sum_{\lambda=0}^{\infty} a_\lambda x^{k+\lambda} = 0. \tag{8.86}$$

From our analysis of the uniqueness of power series (Chapter 5) the coefficients of each power of x on the left-hand side of Eq. (8.86) must vanish individually.

The lowest power of x appearing in Eq. (8.86) is x^{k-2}, for $\lambda = 0$ in the first summation. The requirement that the coefficient vanish[1] yields

$$a_0 k(k - 1) = 0.$$

We had chosen a_0 as the coefficient of the lowest nonvanishing terms of the series (Eq. (8.85)), hence, by definition, $a_0 \neq 0$. Therefore we have

$$k(k - 1) = 0. \tag{8.87}$$

---

[1] See the uniqueness of power series, Section 5.7.

This equation, coming from the coefficient of the lowest power of x, we call the *indicial equation*. The indicial equation and its roots are of critical importance to our analysis. If $k = 1$, the coefficient $a_1(k + 1)k$ of x^{k-1} must vanish so that $a_1 = 0$. Clearly, in this example we must require either that $k = 0$ or $k = 1$.

Before considering these two possibilities for k, we return to Eq. (8.86) and demand that the remaining net coefficients, say, the coefficient of x^{k+j} ($j \geq 0$), vanish. We set $\lambda = j + 2$ in the first summation and $\lambda = j$ in the second. (They are independent summations and λ is a dummy index.) This results in

$$a_{j+2}(k + j + 2)(k + j + 1) + \omega^2 a_j = 0$$

or

$$a_{j+2} = -a_j \frac{\omega^2}{(k + j + 2)(k + j + 1)}. \tag{8.88}$$

This is a two-term *recurrence relation*.[2] Given a_j, we may compute a_{j+2} and then a_{j+4}, a_{j+6}, and so on up as far as desired. The reader will note that for this example, if we start with a_0, Eq. (8.88) leads to the even coefficients a_2, a_4, and so on, and ignores a_1, a_3, a_5, and so on. Since a_1 is arbitrary if $k = 0$ and necessarily zero if $k = 1$, let us set it equal to zero (compare Exercises 8.5.3 and 8.5.4) and then by Eq. (8.88)

$$a_3 = a_5 = a_7 = \cdots = 0,$$

and all the odd numbered coefficients vanish. Do not worry about the lost terms; the object here is to get a solution. The rejected powers of x will actually reappear when the *second* root of the indicial equation is used.

Returning to Eq. (8.87) our indicial equation, we first try the solution $k = 0$. The recurrence relation (Eq. (8.88)) becomes

$$a_{j+2} = -a_j \frac{\omega^2}{(j + 2)(j + 1)}, \tag{8.89}$$

which leads to

$$a_2 = -a_0 \frac{\omega^2}{1 \cdot 2} = -\frac{\omega^2}{2!} a_0,$$

$$a_4 = -a_2 \frac{\omega^2}{3 \cdot 4} = +\frac{\omega^4}{4!} a_0,$$

$$a_6 = -a_4 \frac{\omega^2}{5 \cdot 6} = -\frac{\omega^6}{6!} a_0, \qquad \text{and so on.}$$

---

[2] The recurrence relation may involve three terms; that is, a_{j+2}, depending on a_j and a_{j-2}. Equation (13.12) for the Hermite functions provides an example of this behavior.

By inspection (and mathematical induction)

$$a_{2n} = (-1)^n \frac{\omega^{2n}}{(2n)!} a_0, \qquad (8.90)$$

and our solution is

$$y(x)_{k=0} = a_0 \left[1 - \frac{(\omega x)^2}{2!} + \frac{(\omega x)^4}{4!} - \frac{(\omega x)^6}{6!} + \cdots \right]$$

$$= a_0 \cos \omega x. \qquad (8.91)$$

If we choose the indicial equation root $k = 1$ (Eq. (8.88)), the recurrence relation becomes

$$a_{j+2} = -a_j \frac{\omega^2}{(j + 3)(j + 2)}. \qquad (8.92)$$

Substituting in $j = 0, 2, 4$, successively, we obtain

$$a_2 = -a_0 \frac{\omega^2}{2 \cdot 3} = -\frac{\omega^2}{3!} a_0,$$

$$a_4 = -a_2 \frac{\omega^2}{4 \cdot 5} = +\frac{\omega^4}{5!} a_0,$$

$$a_6 = -a_4 \frac{\omega^2}{6 \cdot 7} = -\frac{\omega^6}{7!} a_0, \qquad \text{and so on.}$$

Again, by inspection and mathematical induction,

$$a_{2n} = (-1)^n \frac{\omega^{2n}}{(2n + 1)!} a_0. \qquad (8.93)$$

For this choice, $k = 1$, we obtain

$$y(x)_{k=1} = a_0 x \left[1 - \frac{(\omega x)^2}{3!} + \frac{(\omega x)^4}{5!} - \frac{(\omega x)^6}{7!} + \cdots \right]$$

$$= \frac{a_0}{\omega} \left[(\omega x) - \frac{(\omega x)^3}{3!} + \frac{(\omega x)^5}{5!} - \frac{(\omega x)^7}{7!} + \cdots \right]$$

$$= \frac{a_0}{\omega} \sin \omega x. \qquad (8.94)$$

To summarize this approach, we may write Eq. (8.86) schematically as shown in Fig. 8.2. From the uniqueness of power series (Section 5.7), the total coefficient of each power of x must vanish—all by itself. The requirement that the first coefficient (1) vanish leads to the indicial equation, Eq. (8.87). The second coefficient is handled by setting $a_1 = 0$. The vanishing of the coefficient of x^k (and higher powers, taken one at a time) leads to the recurrence relation Eq. (8.88).

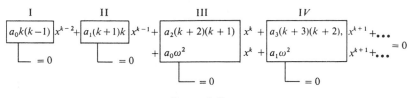

I   II   III   IV

Figure 8.2

This series substitution, known as Frobenius' method, has given us two series solutions of the linear oscillator equation. However, there are two points about such series solutions that must be strongly emphasized:

1. The series solution should always be substituted back into the differential equation, to see if it works, as a precaution against algebraic and logical errors. Conversely, if it works, it is a solution.
2. The acceptability of a series solution depends on its convergence (including asymptotic convergence). It is quite possible for Frobenius' method to give a series solution that satisfies the original differential equation when substituted in the equation but that does *not* converge over the region of interest. Legendre's differential equation illustrates this situation.

EXPANSION ABOUT x_0

Equation (8.85) is an expansion about the origin, $x_0 = 0$. It is perfectly possible to replace Eq. (8.85) with

$$y(x) = \sum_{\lambda = 0}^{\infty} a_\lambda (x - x_0)^{k+\lambda}, \qquad a_0 \neq 0. \tag{8.95}$$

Indeed, for the Legendre, Chebyshev, and hypergeometric equations the choice $x_0 = 1$ has some advantages. The point x_0 should not be chosen at an essential singularity—or our Frobenius method will probably fail. The resultant series (x_0 an ordinary point or regular singular point) will be valid where it converges. You can expect a divergence of some sort when $|x - x_0| = |z_s - x_0|$, where z_s is the closest singularity to x_0 in the complex plane.

SYMMETRY OF SOLUTIONS

The alert reader will note that we obtained one solution of even symmetry, $y_1(x) = y_1(-x)$, and one of odd symmetry, $y_2(x) = -y_2(-x)$. This is not just an accident but a direct consequence of the form of the differential equation. Writing a general differential equation as

$$\mathcal{L}(x)\,y(x) = 0, \tag{8.96}$$

in which $\mathcal{L}(x)$ is the differential operator, we see that for the linear oscillator equation (Eq. (8.84)), $\mathcal{L}(x)$ is even; that is,

$$\mathcal{L}(x) = \mathcal{L}(-x). \tag{8.97}$$

Often this is described as even parity.

Whenever the differential operator has a specific parity or symmetry, either even or odd, we may interchange $+x$ and $-x$, and Eq. (8.96) becomes

$$\pm \mathcal{L}(x)y(-x) = 0, \tag{8.98}$$

$+$ if $\mathcal{L}(x)$ is even, $-$ if $\mathcal{L}(x)$ is odd. Clearly, if $y(x)$ is a solution of the differential equation, $y(-x)$ is also a solution. Then any solution may be resolved into even and odd parts,

$$y(x) = \tfrac{1}{2}[y(x) + y(-x)] + \tfrac{1}{2}[y(x) - y(-x)], \tag{8.99}$$

the first bracket on the right giving an even solution, the second an odd solution.

If we refer back to Section 8.4, we can see that Legendre, Chebyshev, Bessel, simple harmonic oscillator, and Hermite equations (or differential operators) all exhibit this even parity, i.e., their $P(x)$ in Eq. (8.80) is odd and $Q(x)$ even. Solutions of all of them may be presented as series of even powers of x and separate series of odd powers of x. The Laguerre differential operator has neither even nor odd symmetry; hence its solutions cannot be expected to exhibit even or odd parity. Our emphasis on parity stems primarily from the importance of parity in quantum mechanics. We find that wave functions usually are either even or odd, meaning that they have a definite parity. Most interactions (beta decay is the big exception) are also even or odd and the result is that parity is conserved.

LIMITATIONS OF SERIES APPROACH—BESSEL'S EQUATION

This attack on the linear oscillator equation was perhaps a bit too easy. By substituting the power series (Eq. (8.85)) into the differential equation (Eq. (8.84)), we obtained two independent solutions with no trouble at all.

To get some idea of what can happen we try to solve Bessel's equation,

$$x^2 y'' + xy' + (x^2 - n^2)y = 0, \tag{8.100}$$

using y' for dy/dx and y'' for d^2y/dx^2. Again, assuming a solution of the form

$$y(x) = \sum_{\lambda=0}^{\infty} a_\lambda x^{k+\lambda},$$

we differentiate and substitute into Eq. (8.100). The result is

$$\sum_{\lambda=0}^{\infty} a_\lambda (k+\lambda)(k+\lambda-1)x^{k+\lambda} + \sum_{\lambda=0}^{\infty} a_\lambda (k+\lambda)x^{k+\lambda}$$

$$+ \sum_{\lambda=0}^{\infty} a_\lambda x^{k+\lambda+2} - \sum_{\lambda=0}^{\infty} a_\lambda n^2 x^{k+\lambda} = 0. \tag{8.101}$$

By setting $\lambda = 0$, we get the coefficient of x^k, the lowest power of x appearing on the left-hand side,

$$a_0[k(k - 1) + k - n^2] = 0, \qquad (8.102)$$

and again $a_0 \neq 0$ by definition. Equation (8.102) therefore yields the *indicial equation*

$$k^2 - n^2 = 0 \qquad (8.103)$$

with solutions $k = \pm n$.

It is of some interest to examine the coefficients of x^{k+1} also. Here we obtain

$$a_1[(k + 1)k + k + 1 - n^2] = 0$$

or

$$a_1(k + 1 - n)(k + 1 + n) = 0. \qquad (8.104)$$

For $k = \pm n$ neither $k + 1 - n$ nor $k + 1 + n$ vanishes and we *must* require $a_1 = 0.$[3]

Proceeding to the coefficient of x^{k+j} for $k = n$, we set $\lambda = j$ in the first, second, and fourth terms of Eq. (8.101) and $\lambda = j - 2$ in the third term. By requiring the resultant coefficient of x^{k+j} to vanish, we obtain

$$a_j[(n + j)(n + j - 1) + (n + j) - n^2] + a_{j-2} = 0.$$

When j is replaced by $j + 2$, this can be rewritten for $j \geq 0$ as

$$a_{j+2} = -a_j \frac{1}{(j + 2)(2n + j + 2)}, \qquad (8.105)$$

which is the desired recurrence relation. Repeated application of this recurrence relation leads to

$$a_2 = -a_0 \frac{1}{2(2n + 2)} = -\frac{a_0 n!}{2^2 1!(n + 1)!},$$

$$a_4 = -a_2 \frac{1}{4(2n + 4)} = +\frac{a_0 n!}{2^4 2!(n + 2)!},$$

$$a_6 = -a_4 \frac{1}{6(2n + 6)} = -\frac{a_0 n!}{2^6 3!(n + 3)!}, \qquad \text{and so on,}$$

and in general,

$$a_{2p} = (-1)^p \frac{a_0 n!}{2^{2p} p!(n + p)!}. \qquad (8.106)$$

Inserting these coefficients in our assumed series solution, we have

$$y(x) = a_0 x^n \left[1 - \frac{n! x^2}{2^2 1!(n + 1)!} + \frac{n! x^4}{2^4 2!(n + 2)!} - \cdots \right]. \qquad (8.107)$$

[3] $k = \pm n = -\frac{1}{2}$ are exceptions.

In summation form

$$y(x) = a_0 \sum_{j=0}^{\infty} (-1)^j \frac{n! \, x^{n+2j}}{2^{2j} j! (n+j)!}$$

$$= a_0 2^n n! \sum_{j=0}^{\infty} (-1)^j \frac{1}{j!(n+j)!} \left(\frac{x}{2}\right)^{n+2j}. \qquad (8.108)$$

In Chapter 11 the final summation is identified as the Bessel function $J_n(x)$. Notice that this solution $J_n(x)$ has either even or odd symmetry[4] as might be expected from the form of Bessel's equation.

When $k = -n$ and n is not an integer, we may generate a second distinct series to be labeled $J_{-n}(x)$. However, when $-n$ is a negative integer, trouble develops. The recurrence relation for the coefficients a_j is still given by Eq. (8.105), but with $2n$ replaced by $-2n$. Then, when $j + 2 = 2n$ or $j = 2(n-1)$, the coefficient a_{j+2} blows up and we have no series solution. This castastrophe can be remedied in Eq. (8.108), as it is done in Chapter 11, with the result that

$$J_{-n}(x) = (-1)^n J_n(x), \qquad n \text{ an integer.} \qquad (8.109)$$

The second solution simply reproduces the first. We have failed to construct a second independent solution for Bessel's equation by this series technique when n is an integer.

By substituting in an infinite series, we have obtained two solutions for the linear oscillator equation and one for Bessel's equation (two if n is not an integer). To the questions "Can we always do this? Will this method always work?" the answer is no, we cannot always do this. This method of series solution will not always work.

REGULAR AND IRREGULAR SINGULARITIES

The success of the series substitution method depends on the roots of the indicial equation and the degree of singularity of the coefficients in the differential equation. To understand better the effect of the equation coefficients on this naïve series substitution approach, consider four simple equations:

$$y'' - \frac{6}{x^2} y = 0, \qquad (8.110a)$$

$$y'' - \frac{6}{x^3} y = 0, \qquad (8.110b)$$

$$y'' + \frac{1}{x} y' - \frac{a^2}{x^2} y = 0, \qquad (8.110c)$$

$$y'' + \frac{1}{x^2} y' - \frac{a^2}{x^2} y = 0. \qquad (8.110d)$$

---

[4] $J_n(x)$ is an even function if n is an even integer, an odd function if n is an odd integer. For nonintegral n the x^n has no such simple symmetry.

The reader may show easily that for Eq. (8.110a) the indicial equation is

$$k^2 - k - 6 = 0,$$

giving $k = 3, -2$. Since the equation is homogeneous in x (counting d^2/dx^2 as x^{-2}), there is no recurrence relation. However, we are left with two perfectly good solutions, x^3 and x^{-2}.

Equation (8.110b) differs from Eq. (8.110a) by only one power of x, but this sends the indicial equation to

$$-6a_0 = 0,$$

with no solution at all, for we have agreed that $a_0 \neq 0$. Our series substitution worked for Eq. (8.110a), which had only a regular singularity, but broke down at Eq. (8.110b) which has an irregular singular point at the origin.

Continuing with Eq. (8.110c), we have added a term y'/x. The indicial equation is

$$k^2 - a^2 = 0,$$

but again, there is no recurrence relation. The solutions are $y = x^a, x^{-a}$, both perfectly acceptable one term series.

When we change the power of x in the coefficient of y' from -1 to -2, Eq. (8.110d), there is a drastic change in the solution. The indicial equation (with only the y' term contributing) becomes

$$k = 0.$$

There is a recurrence relation

$$a_{j+1} = +a_j \frac{a^2 - j(j-1)}{j+1}.$$

Unless the parameter a is selected to make the series terminate, we have

$$\lim_{j \to \infty} \left| \frac{a_{j+1}}{a_j} \right| = \lim_{j \to \infty} \frac{j(j+1)}{j+1}$$

$$= \lim_{j \to \infty} \frac{j^2}{j} = \infty.$$

Hence our series solution diverges for all $x \neq 0$. Again, our method worked for Eq. (8.110c) with a regular singularity but failed when we had the irregular singularity of (8.110d).

FUCHS'S THEOREM

The answer to the basic question when the method of series substitution can be expected to work is given by Fuchs's theorem, which asserts that we can always obtain at least one power-series solution, provided we are expanding about a point that is an ordinary point or at worst a regular singular point.

If we attempt an expansion about an irregular or essential singularity, our method may fail as it did for Eqs. (8.110b) and (8.110d). Fortunately, the more important equations of mathematical physics listed in Section 8.4 have no irregular singularities in the finite plane. Further discussion of Fuchs's theorem appears in Section 8.6.

From Table 8.3, Section 8.4, infinity is seen to be a singular point for all equations considered. As a further illustration of Fuchs's theorem, Legendre's equation (with infinity as a regular singularity) has a convergent series solution in negative powers of the argument (Section 12.10). In contrast, Bessel's equation (with an irregular singularity at infinity) yields asymptotic series (Sections 5.10 and 11.6). Although extremely useful, these asymptotic solutions are technically divergent.

SUMMARY

If we are expanding about an ordinary point or at worst about a regular singularity, the series substitution approach will yield at least one solution (Fuchs's theorem).

Whether we get one or two distinct solutions depends on the roots of the indicial equation.

1. If the two roots of the indicial equation are equal, we can obtain only one solution by this series substitution method.
2. If the two roots differ by a nonintegral number, two independent solutions may be obtained.
3. If the two roots differ by an integer, the larger of the two will yield a solution.

The smaller may or may not give a solution, depending on the behavior of the coefficients. In the linear oscillator equation we obtain two solutions; for Bessel's equation, only one solution.

The usefulness of the series solution in terms of what is the solution (i.e., numbers) depends on the rapidity of convergence of the series and the availability of the coefficients. Many, probably most, differential equations will not yield nice simple recurrence relations for the coefficients. In general, the available series will probably be useful for $|x|$ (or $|x - x_0|$) very small. Computers can be used to determine additional series coefficients using a symbolic language such as Mathematica,[5] Maple,[6] or Reduce.[7] Often, however, for numerical work a direct numerical integration will be preferred— Section 8.8.

[5] S. Wolfram, *Mathematica, A System for Doing Mathematics by Computer*. New York: Addison–Wesley (1991).
[6] A. Heck, *Introduction to Maple*. New York: Springer (1993).
[7] G. Rayna, *Reduce Software for Algebraic Computation*. New York: Springer (1987).

EXERCISES

8.5.1   Uniqueness theorem. The function $y(x)$ satisfies a second-order, linear, homogeneous differential equation. At $x = x_0$, $y(x) = y_0$, and $dy/dx = y_0'$. Show that $y(x)$ is unique in that no other solution of this differential equation passes through the points (x_0, y_0) with a slope of y_0'.
Hint. Assume a second solution satisfying these conditions and compare the Taylor series expansions.

8.5.2   A series solution of Eq. (8.80) is attempted, expanding about the point $x = x_0$. If x_0 is an ordinary point show that the indicial equation has roots $k = 0, 1$.

8.5.3   In the development of a series solution of the simple harmonic oscillator equation the second series coefficient a_1 was neglected except to set it equal to zero. From the coefficient of the next to the lowest power of x, x^{k-1}, develop a second indicial type equation.

(a) (SHO equation with $k = 0$). Show that a_1 may be assigned any finite value (including zero).

(b) (SHO equation with $k = 1$). Show that a_1 *must* be set equal to zero.

8.5.4   Analyze the series solutions of the following differential equations to see when a_1 *may* be set equal to zero without irrevocably losing anything and when a_1 *must* be set equal to zero.

(a) Legendre, (b) Chebyshev, (c) Bessel, (d) Hermite.
   ANS.   (a) Legendre, (b) Chebyshev, and (d) Hermite: For $k = 0$, a_1 *may* be set equal to zero; For $k = 1$, a_1 *must* be set equal to zero.
   (c) Bessel: a_1 *must* be set equal to zero (except for $k = \pm n = -\frac{1}{2}$).

8.5.5   Solve the Legendre equation

$$(1 - x^2)y'' - 2xy' + n(n + 1)y = 0$$

by direct series substitution.

(a) Verify that the indicial equation is

$$k(k - 1) = 0.$$

(b) Using $k = 0$, obtain a series of even powers of x, $(a_1 = 0)$.

$$y_{\text{even}} = a_0 \left[1 - \frac{n(n + 1)}{2!} x^2 + \frac{n(n - 2)(n + 1)(n + 3)}{4!} x^4 + \cdots \right],$$

where

$$a_{j+2} = \frac{j(j + 1) - n(n + 1)}{(j + 1)(j + 2)} a_j.$$

(c) Using $k = 1$, develop a series of odd powers of x $(a_1 = 0)$.

$$y_{\text{odd}} = a_0 \left[x - \frac{(n - 1)(n + 2)}{3!} x^3 + \frac{(n - 1)(n - 3)(n + 2)(n + 4)}{5!} x^5 + \cdots \right],$$

where

$$a_{j+2} = \frac{(j + 1)(j + 2) - n(n + 1)}{(j + 2)(j + 3)} a_j.$$

(d) Show that both solutions, y_{even} and y_{odd}, diverge for $x = \pm 1$ *if the series continue to infinity.*

(e) Finally, show that by an appropriate choice of n, one series at a time may be converted into a polynomial, thereby avoiding the divergence catastrophe. In quantum mechanics this restriction of n to integral values corresponds to *quantization* of angular momentum.

8.5.6   Develop series solutions for Hermite's differential equation

(a) $y'' - 2xy' + 2\alpha y = 0.$

ANS. $k(k - 1) = 0$, indicial equation.

For $k = 0$

$$a_{j+2} = 2a_j \frac{j - \alpha}{(j + 1)(j + 2)} \qquad (j \text{ even}),$$

$$y_{even} = a_0 \left[1 + \frac{2(-\alpha)x^2}{2!} + \frac{2^2(-\alpha)(2 - \alpha)x^4}{4!} + \cdots \right].$$

For $k = 1$

$$a_{j+2} = 2a_j \frac{j + 1 - \alpha}{(j + 2)(j + 3)} \qquad (j \text{ even}),$$

$$y_{odd} = a_0 \left[x + \frac{2(1 - \alpha)x^3}{3!} + \frac{2^2(1 - \alpha)(3 - \alpha)x^5}{5!} + \cdots \right].$$

(b) Show that both series solutions are convergent for all x, the ratio of successive coefficients behaving, for large index, like the corresponding ratio in the expansion of $\exp(2x^2)$.

(c) Show that by appropriate choice of α the series solutions may be cut off and converted to finite polynomials. (These polynomials, properly normalized, become the Hermite polynomials in Section 13.1.)

8.5.7   Laguerre's differential equation is

$$xL_n''(x) + (1 - x)L_n'(x) + nL_n(x) = 0.$$

Develop a series solution selecting the parameter n to make your series a polynomial.

8.5.8   Solve the Chebyshev equation

$$(1 - x^2)T_n'' - xT_n' + n^2 T_n = 0,$$

by series substitution. What restrictions are imposed on n if you demand that the series solution converge for $x = \pm 1$?

ANS.   The infinite series does converge for $x = \pm 1$. Therefore no restriction on n exists (compare Exercise 5.2.16).

8.5.9   Solve

$$(1 - x^2)U_n''(x) - 3xU_n'(x) + n(n + 2)U_n(x) = 0,$$

choosing the root of the indicial equation to obtain a series of *odd* powers of x. Since the series will diverge for $x = 1$, choose n to convert it into a polynomial.

$$k(k - 1) = 0.$$

For $k = 1$

$$a_{j+2} = \frac{(j + 1)(j + 3) - n(n + 2)}{(j + 2)(j + 3)} a_j.$$

8.5.10   Obtain a series solution of the hypergeometric equation

$$x(x - 1)y'' + [(1 + a + b)x - c]y' + aby = 0.$$

Test your solution for convergence.

8.5.11   Obtain two series solutions of the confluent hypergeometric equation

$$xy'' + (c - x)y' - ay = 0.$$

Test your solutions for convergence.

8.5.12   A quantum mechanical analysis of the Stark effect (parabolic coordinates) leads to the differential equation

$$\frac{d}{d\xi}\left(\xi\frac{du}{d\xi}\right) + \left(\frac{1}{2}E\xi + \alpha - \frac{m^2}{4\xi} - \frac{1}{4}F\xi^2\right)u = 0.$$

Here α is a separation constant, E is the total energy, and F is a constant, where Fz is the potential energy added to the system by the introduction of an electric field.

Using the larger root of the indicial equation, develop a power series solution about $\xi = 0$. Evaluate the first three coefficients in terms of a_0.

$$\text{Indicial equation} \qquad k^2 - \frac{m^2}{4} = 0,$$

$$u(\xi) = a_0\xi^{m/2}\left\{1 - \frac{\alpha}{m + 1}\xi + \left[\frac{\alpha^2}{2(m + 1)(m + 2)} - \frac{E}{4(m + 2)}\right]\xi^2 + \cdots\right\}.$$

Note that the perturbation E does not appear until a_3 is included.

8.5.13   For the special case of no azimuthal dependence, the quantum mechanical analysis of the hydrogen molecular ion leads to the equation

$$\frac{d}{d\eta}\left[(1 - \eta^2)\frac{du}{d\eta}\right] + \alpha u + \beta\eta^2 u = 0.$$

Develop a power-series solution for $u(\eta)$. Evaluate the first three non-vanishing coefficients in terms of a_0.

$$\text{Indicial equation} \qquad k(k - 1) = 0,$$

$$u_{k=1} = a_0\eta\left\{1 + \frac{2 - \alpha}{6}\eta^2 + \left[\frac{(2 - \alpha)(12 - \alpha)}{120} - \frac{\beta}{20}\right]\eta^4 + \cdots\right\}.$$

8.5.14   To a good approximation, the interaction of two nucleons may be described by a meson potential

$$V = \frac{Ae^{-ax}}{x},$$

attractive for A negative. Develop a series solution of the resultant Schrödinger wave equation

$$\frac{\hbar^2}{2m}\frac{d^2\psi}{dx^2} + (E - V)\psi = 0.$$

through the first three nonvanishing coefficients.

$$\psi_{k=1} = a_0\{x + \frac{1}{2}A'x^2 + \frac{1}{6}[\frac{1}{2}A'^2 - E' - aA']x^3 + \cdots\},$$

where the prime indicates multiplication by $2m/\hbar^2$.

8.5.15   Near the nucleus of a complex atom the potential energy of one electron is given by

$$V = \frac{Ze^2}{r}(1 + b_1 r + b_2 r^2),$$

where the coefficients b_1 and b_2 arise from screening effects. For the case of zero angular momentum show that the first three terms of the solution of the Schrödinger equation have the same form as those of Exercise 8.5.14. By appropriate translation of coefficients or parameters, write out the first three terms in a series expansion of the wave function.

8.5.16   If the parameter a^2 in Eq. (8.110d) is equal to 2, Eq. (8.110d) becomes

$$y'' + \frac{1}{x^2} y' - \frac{2}{x^2} y = 0.$$

From the indicial equation and the recurrence relation *derive* a solution $y = 1 + 2x + 2x^2$. Verify that this is indeed a solution by substituting back into the differential equation.

8.5.17   The modified Bessel function $I_0(x)$ satisfies the differential equation

$$x^2 \frac{d^2}{dx^2} I_0(x) + x \frac{d}{dx} I_0(x) - x^2 I_0(x) = 0.$$

From Exercise 7.4.4 the leading term in an asymptotic expansion is found to be

$$I_0(x) \sim \frac{e^x}{\sqrt{2\pi x}}.$$

Assume a series of the form

$$I_0(x) \sim \frac{e^x}{\sqrt{2\pi x}}\{1 + b_1 x^{-1} + b_2 x^{-2} + \cdots\}.$$

Determine the coefficients b_1 and b_2.

$$ANS. \quad b_1 = \tfrac{1}{8},$$
$$b_2 = \tfrac{9}{128}.$$

8.5.18   The even power-series solution of Legendre's equation is given by Exercise 8.5.5. Take $a_0 = 1$ and n *not* an even integer, say $n = 0.5$. Calculate the partial sums of the series through x^{200}, x^{400}, x^{600}, ..., x^{2000} for $x = 0.95(0.01)1.00$. Also, write out the individual term corresponding to each of these powers.
Note. This calculation does *not* constitute proof of convergence at $x = 0.99$ or divergence at $x = 1.00$, but perhaps you can see the difference in the behavior of the sequence of partial sums for these two values of x.

8.5.19   (a) The odd power-series solution of Hermite's equation is given by Exercise 8.5.6. Take $a_0 = 1$. Evaluate this series for $\alpha = 0$, $x = 1, 2, 3$. Cut off your calculation after the last term calculated has dropped below the maximum term by a factor of 10^6 or more. Set an upper bound to the error made in ignoring the remaining terms in the infinite series.
   (b) As a check on the calculation of part (a), show that the Hermite series $y_{odd}(\alpha = 0)$ corresponds to $\int_0^x \exp(x^2)\, dx$.
   (c) Calculate this integral for $x = 1, 2, 3$.

8.6 A SECOND SOLUTION

In Section 8.5 a solution of a second-order homogeneous differential equation was developed by substituting in a power series. By Fuchs's theorem this is possible, provided the power series is an expansion about an ordinary point or a nonessential singularity.[1] There is no guarantee that this approach will yield the two independent solutions we expect from a linear second-order differential equation. Indeed, the technique gave only one solution for Bessel's equation (n an integer). In this section we develop two methods of obtaining a second independent solution: an integral method and a power series containing a logarithmic term. First, however, we consider the question of independence of a set of functions.

LINEAR INDEPENDENCE OF SOLUTIONS

Given a set of functions, φ_λ, the criterion for linear *dependence* is the existence of a relation of the form

$$\sum_\lambda k_\lambda \varphi_\lambda = 0, \tag{8.111}$$

in which not all the coefficients k_λ are zero. On the other hand, if the only solution of Eq. (8.111) is $k_\lambda = 0$ for all λ, the set of functions φ_λ is said to be linearly *independent*.

It may be helpful to think of linear dependence of vectors. Consider $\mathbf{A}$, $\mathbf{B}$ and $\mathbf{C}$ in three-dimensional space with $\mathbf{A} \cdot \mathbf{B} \times \mathbf{C} \neq 0$. Then no nontrivial relation of the form

$$a\mathbf{A} + b\mathbf{B} + c\mathbf{C} = 0 \tag{8.112}$$

exists. $\mathbf{A}$, $\mathbf{B}$, and $\mathbf{C}$ are linearly independent. On the other hand, any fourth vector $\mathbf{D}$ may be expressed as a linear combination of $\mathbf{A}$, $\mathbf{B}$, and $\mathbf{C}$ (see Section 3.4). We can always write an equation of the form

$$\mathbf{D} - a\mathbf{A} - b\mathbf{B} - c\mathbf{C} = 0, \tag{8.113}$$

and the four vectors are *not* linearly independent. The three noncoplanar vectors $\mathbf{A}$, $\mathbf{B}$, and $\mathbf{C}$ span our real three-dimensional space.

If a set of vectors or functions are mutually orthogonal, then they are automatically linearly independent. Orthogonality implies linear independence. This can easily be demonstrated by taking inner products (scalar or dot product for vectors, orthogonality integral of Section 9.2 for functions).

Let us assume that the functions φ_λ are differentiable as needed. Then, differentiating Eq. (8.111) repeatedly, we generate a set of equations

$$\sum_\lambda k_\lambda \varphi_\lambda' = 0 \tag{8.114}$$

$$\sum_\lambda k_\lambda \varphi_\lambda'' = 0, \qquad \text{and so on.} \tag{8.115}$$

[1] This is why the classification of singularities in Section 8.4 is of vital importance.

This gives us a set of homogeneous linear equations in which k_λ are the unknown quantities. By Section 3.1 there is a solution $k_\lambda \neq 0$ only if the determinant of the coefficients of the k_λ's vanishes. This means

$$\begin{vmatrix} \varphi_1 & \varphi_2 & \cdots & \varphi_n \\ \varphi_1' & \varphi_2' & \cdots & \varphi_n' \\ \cdots & \cdots & \cdots & \cdots \\ \varphi_1^{(n-1)} & \varphi_2^{(n-1)} & \cdots & \varphi_n^{(n-1)} \end{vmatrix} = 0. \qquad (8.116)$$

This determinant is called the *Wronskian*.

1. If the Wronskian is not equal to zero, then Eq. (8.111) has no solution other than $k_\lambda = 0$. The set of functions φ_λ is therefore independent.
2. If the Wronskian vanishes at isolated values of the argument, this does not necessarily prove linear dependence (unless the set of functions has only two functions). However, if the Wronskian is zero over the entire range of the variable, the functions φ_λ are linearly dependent over this range[2] (compare Exercise 8.5.2 for the simple case of two functions).

Example 8.6.1 Linear Independence

The solutions of the linear oscillator equation (8.84) are $\varphi_1 = \sin \omega x$, $\varphi_2 = \cos \omega x$. The Wronskian becomes

$$\begin{vmatrix} \sin \omega x & \cos \omega x \\ \omega \cos \omega x & -\omega \sin \omega x \end{vmatrix} = -\omega \neq 0.$$

These two solutions, φ_1 and φ_2, are therefore linearly independent. For just two functions this means that one is not a multiple of the other, which is obviously true in this case.

You know that

$$\sin \omega x = \pm (1 - \cos^2 \omega x)^{1/2},$$

but this is *not* a *linear* relation, of the form of (8.111).

Examples 8.6.2 Linear Dependence

For an illustration of linear dependence, consider the solutions of the one-dimensional diffusion equation. We have $\varphi_1 = e^x$ and $\varphi_2 = e^{-x}$, and we add $\varphi_3 = \cosh x$, also a solution. The Wronskian is

$$\begin{vmatrix} e^x & e^{-x} & \cosh x \\ e^x & -e^{-x} & \sinh x \\ e^x & e^{-x} & \cosh x \end{vmatrix} = 0.$$

---

[2] Compare H. Lass, *Elements of Pure and Applied Mathematics*. New York: McGraw-Hill (1957), p. 187, for proof of this assertion. It is assumed that the functions have continuous derivatives and that at least one of the minors of the bottom row of Eq. (8.116) (Laplace expansion) does not vanish in $[a, b]$, the interval under consideration.

The determinant vanishes for all x because the first and third rows are identical. Hence e^x, e^{-x}, and $\cosh x$ are linearly dependent, and indeed, we have a relation of the form of Eq. (8.111):

$$e^x + e^{-x} - 2\cosh x = 0 \qquad \text{with } k_\lambda \neq 0.$$

A SECOND SOLUTION

Returning to our linear, second-order, homogeneous, differential equation of the general form

$$y'' + P(x)y' + Q(x)y = 0, \tag{8.117}$$

let y_1 and y_2 be two independent solutions. Then the Wronskian, by definition, is

$$W = y_1 y_2' - y_1' y_2. \tag{8.118}$$

By differentiating the Wronskian, we obtain

$$\begin{aligned} W' &= y_1' y_2' + y_1 y_2'' - y_1'' y_2 - y_1' y_2' \\ &= y_1[-P(x)y_2' - Q(x)y_2] - y_2[-P(x)y_1' - Q(x)y_1] \\ &= -P(x)(y_1 y_2' - y_1' y_2). \end{aligned} \tag{8.119}$$

The expression in parentheses is just W, the Wronskian, and we have

$$W' = -P(x)W. \tag{8.120}$$

In the special case that $P(x) = 0$, i.e.,

$$y'' + Q(x)y = 0, \tag{8.121}$$

the Wronskian

$$W = y_1 y_2' - y_1' y_2 = \text{constant}. \tag{8.122}$$

Since our original differential equation is homogeneous, we may multiply the solutions y_1 and y_2 by whatever constants we wish and arrange to have the Wronskian equal to unity (or -1). This case, $P(x) = 0$, appears more frequently than might be expected. The reader will recall that ∇^2 in Cartesian coordinates contains no first derivative. Similarly, the radial dependence of $\nabla^2(r\psi)$ in spherical polar coordinates lacks a first derivative. Finally, every linear second-order differential equation can be transformed into an equation of the form of Eq. (8.121) (compare Exercise 8.6.11).

For the general case, let us now assume that we have one solution of Eq. (8.117) by a series substitution (or by guessing). We now proceed to develop a second, independent solution for which $W \neq 0$. Rewriting Eq. (8.120) as

$$\frac{dW}{W} = -P\,dx_1,$$

we integrate, from $x_1 = a$ to $x_1 = x$ to obtain

$$\ln \frac{W(x)}{W(a)} = -\int_a^x P(x_1)\, dx_1,$$

or[3]

$$W(x) = W(a) \exp\left[-\int_a^x P(x_1)\, dx_1\right]. \tag{8.123}$$

But

$$W(x) = y_1 y_2' - y_1' y_2$$

$$= y_1^2 \frac{d}{dx}\left(\frac{y_2}{y_1}\right). \tag{8.124}$$

By combining Eqs. (8.123) and (8.124), we have

$$\frac{d}{dx}\left(\frac{y_2}{y_1}\right) = W(a)\frac{\exp[-\int_a^x P(x_1)\, dx_1]}{y_1^2}. \tag{8.125}$$

Finally, by integrating Eq. (8.125) from $x_2 = b$ to $x_2 = x$ we get

$$y_2(x) = y_1(x) W(a) \int_b^x \frac{\exp[-\int_a^{x_2} P(x_1)\, dx_1]}{[y_1(x_2)]^2}\, dx_2. \tag{8.126}$$

Here a and b are arbitrary constants and a term $y_1(x)y_2(b)/y_1(b)$ has been dropped, for it leads to nothing new. Since $W(a)$, the Wronskian evaluated at $x = a$, is a constant and our solutions for the homogeneous differential equation always contain an unknown normalizing factor, we set $W(a) = 1$ and write

$$y_2(x) = y_1(x) \int^x \frac{\exp[-\int^{x_2} P(x_1)\, dx_1]}{[y_1(x_2)]^2}\, dx_2. \tag{8.127}$$

Note that the lower limits $x_1 = a$ and $x_2 = b$ have been omitted. If they are retained, they simply make a contribution equal to a constant times the known first solution, $y_1(x)$, hence add nothing new.

If we have the important special case of $P(x) = 0$, Eq. (8.127) reduces to

$$y_2(x) = y_1(x) \int^x \frac{dx_2}{[y_1(x_2)]^2}. \tag{8.128}$$

This means that by using either Eq. (8.127) or (8.128) we can take one known solution and by integrating can generate a second independent solution of Eq. (8.117). This technique is used in Section 12.10 to generate a second solution of Legendre's differential equation.

---

[3] If $P(x_1)$ remains finite, $a \le x_1 \le x$, $W(x) \ne 0$ unless $W(a) = 0$. That is, the Wronskian *of our two solutions* is either identically zero or never zero.

Example 8.6.3 A Second Solution for the
Linear Oscillator Equation

From $d^2y/dx^2 + y = 0$ with $P(x) = 0$ let one solution be $y_1 = \sin x$. By applying Eq. (8.128), we obtain

$$y_2(x) = \sin x \int^x \frac{dx_2}{\sin^2 x_2}$$

$$= \sin x(-\cot x) = -\cos x,$$

which is clearly independent (not a linear multiple) of $\sin x$.

SERIES FORM OF THE SECOND SOLUTION

Further insight into the nature of the second solution of our differential equation may be obtained by the following sequence of operations:

1. Express $P(x)$ and $Q(x)$ in Eq. (8.117) as

$$P(x) = \sum_{i=-1}^{\infty} p_i x^i, \qquad Q(x) = \sum_{j=-2}^{\infty} q_j x^j. \tag{8.129}$$

The lower limits of the summations are selected to create the strongest possible *regular* singularity (at the origin). These conditions just satisfy Fuchs's theorem and thus help us gain a better understanding of Fuchs's theorem.

2. Develop the first few terms of a power-series solution, as in Section 8.5.
3. Using this solution as y_1, obtain a second series type solution, y_2, with Eq. (8.127), integrating term by term.

Proceeding with Step 1, we have

$$y'' + (p_{-1}x^{-1} + p_0 + p_1 x + \cdots)y' + (q_{-2}x^{-2} + q_{-1}x^{-1} + \cdots)y = 0, \tag{8.130}$$

in which point $x = 0$ is at worst a regular singular point. If $p_{-1} = q_{-1} = q_{-2} = 0$, it reduces to an ordinary point. Substituting

$$y = \sum_{\lambda=0}^{\infty} a_\lambda x^{k+\lambda}$$

(Step 2), we obtain

$$\sum_{\lambda=0}^{\infty} (k + \lambda)(k + \lambda - 1)a_\lambda x^{k+\lambda-2} + \sum_{i=-1}^{\infty} p_i x^i \sum_{\lambda=0}^{\infty} (k + \lambda)a_\lambda x^{k+\lambda-1}$$

$$+ \sum_{j=-2}^{\infty} q_j x^j \sum_{\lambda=0}^{\infty} a_\lambda x^{k+\lambda} = 0. \tag{8.131}$$

Assuming that $p_{-1} \neq 0$, $q_{-2} \neq 0$, our indicial equation is

$$k(k - 1) + p_{-1}k + q_{-2} = 0,$$

which sets the net coefficient of x^{k-2} equal to zero. This reduces to

$$k^2 + (p_{-1} - 1)k + q_{-2} = 0. \tag{8.132}$$

We denote the two roots of this indicial equation by $k = \alpha$ and $k = \alpha - n$, where n is zero or a positive integer. (If n is not an integer, we expect two independent series solutions by the methods of Section 8.5 and there is no problem.) Then

$$(k - \alpha)(k - \alpha + n) = 0, \tag{8.133}$$

or

$$k^2 + (n - 2\alpha)k + \alpha(a - n) = 0,$$

and equating coefficients of k in Eqs. (8.132) and (8.133), we have

$$p_{-1} - 1 = n - 2\alpha. \tag{8.134}$$

The known series solution corresponding to the larger root $k = \alpha$ may be written as

$$y_1 = x^\alpha \sum_{\lambda = 0}^{\infty} a_\lambda x^\lambda.$$

Substituting this series solution into Eq. (8.127) (Step 3), we are faced with

$$y_2(x) = y_1(x) \int^x \frac{\exp(-\int_a^{x_2} \sum_{i=-1}^\infty p_i x_1^i \, dx_1)}{x_2^{2\alpha}(\sum_{\lambda=0}^\infty a_\lambda x_2^\lambda)^2} \, dx_2, \tag{8.135}$$

where the solutions y_1 and y_2 have been normalized so that the Wronskian, $W(a) = 1$. Tackling the exponential factor first, we have

$$\int_a^{x_2} \sum_{i=-1}^\infty p_i x_1^i \, dx_1 = p_{-1} \ln x_2 + \sum_{k=0}^\infty \frac{p_k}{k+1} x_2^{k+1} + f(a). \tag{8.136}$$

Hence

$$\exp\left(-\int_a^{x_2} \sum_i p_i x_1^i \, dx_1\right)$$

$$= \exp[-f(a)] x_2^{-p_{-1}} \exp\left(-\sum_{k=0}^\infty \frac{p_k}{k+1} x_2^{k+1}\right)$$

$$= \exp[-f(a)] x_2^{-p_{-1}} \left[1 - \sum_{k=0}^\infty \frac{p^k}{k+1} x_2^{k+1} + \frac{1}{2!}\left(\sum_{k=0}^\infty \frac{p_k}{k+1} x_2^{k+1}\right)^2 + \cdots\right]. \tag{8.137}$$

This final series expansion of the exponential is certainly convergent if the original expansion of the coefficient $P(x)$ was convergent.

The denominator in Eq. (8.135) may be handled by writing

$$\left[x_2^{2\alpha} \left(\sum_{\lambda=0}^{\infty} a_\lambda x_2^\lambda \right)^2 \right]^{-1} = x_2^{-2\alpha} \left(\sum_{\lambda=0}^{\infty} a_\lambda x_2^\lambda \right)^{-2}$$

$$= x_2^{-2\alpha} \sum_{\lambda=0}^{\infty} b_\lambda x_2^\lambda. \tag{8.138}$$

Neglecting constant factors that will be picked up anyway by the requirement that $W(a) = 1$, we obtain

$$y_2(x) = y_1(x) \int^x x_2^{-p_{-1}-2\alpha} \left(\sum_{\lambda=0}^{\infty} c_\lambda x_2^\lambda \right) dx_2. \tag{8.139}$$

By Eq. (8.134),

$$x_2^{-p_{-1}-2\alpha} = x_2^{-n-1}, \tag{8.140}$$

and we have assumed here that n is an integer. Substituting this result into Eq. (8.139), we obtain

$$y_2(x) = y_1(x) \int^x (c_0 x_2^{-n-1} + c_1 x_2^{-n} + c_2 x_2^{n+1} + \cdots + c_n x_2^{-1} + \cdots) \, dx_2. \tag{8.141}$$

The integration indicated in Eq. (8.141) leads to a coefficient of $y_1(x)$ consisting of two parts:

1. A power series starting with x^{-n}.
2. A logarithm term from the integration of x^{-1} (when $\lambda = n$). This term always appears when n is an integer *unless* c_n fortuitously happens to vanish.[4]

Example 8.6.4 *A Second Solution of Bessel's Equation*
From Bessel's equation, Eq. (8.100) (divided by x^2 to agree with Eq. (8.117)), we have

$$P(x) = x^{-1} \qquad Q(x) = 1 \qquad \text{for the case } n = 0.$$

Hence $p_{-1} = 1$, $q_0 = 1$; all other p_i's and q_j's vanish. The Bessel indicial equation is

$$k^2 = 0$$

(Eq. (8.103) with $n = 0$). Hence we verify Eqs. (8.132)–(8.134) with n and $\alpha = 0$.

Our first solution is available from Eq. (8.108). Relabeling it to agree with Chapter 11 (and using $a_0 = 1$), we obtain[5]

$$y_1(x) = J_0(x) = 1 - \frac{x^2}{4} + \frac{x^4}{64} - O(x^6). \tag{8.142a}$$

---

[4] For parity considerations, $\ln x$ is taken to be $\ln|x|$, even.
[5] The capital O (order of) as written here means terms proportional to x^6 and possibly higher powers of x.

Now, substituting all this into Eq. (8.127), we have the specific case corresponding to Eq. (8.135):

$$y_2(x) = J_0(x) \int^x \frac{\exp[-\int^{x_2} x_1^{-1}\, dx_1]}{[1 - x_2^2/4 + x_2^4/64 - \cdots]^2}\, dx_2. \qquad (8.142b)$$

From the numerator of the integrand

$$\exp\left[-\int^{x_2} \frac{dx_1}{x_1} \right] = \exp[-\ln x_2] = \frac{1}{x_2}.$$

This corresponds to the x_2^{-p-1} in Eq. (8.137). From the denominator of the integrand, using a binomial expansion, we obtain

$$\left[1 - \frac{x_2^2}{4} + \frac{x_2^4}{64} \right]^{-2} = 1 + \frac{x_2^2}{2} + \frac{5x_2^4}{32} + \cdots.$$

Corresponding to Eq. (8.139), we have

$$y_2(x) = J_0(x) \int^x \frac{1}{x_2}\left[1 + \frac{x_2^2}{2} + \frac{5x_2^4}{32} + \cdots \right] dx_2$$

$$= J_0(x)\left\{ \ln x + \frac{x^2}{4} + \frac{5x^4}{128} + \cdots \right\}. \qquad (8.142c)$$

Let us check this result. From Eqs. (11.62) and (11.64), which give the standard form of the second solution,

$$N_0(x) = \frac{2}{\pi}[\ln x - \ln 2 + \gamma]J_0(x) + \frac{2}{\pi}\left\{ \frac{x^2}{4} - \frac{3x^4}{128} + \cdots \right\}. \qquad (8.142d)$$

Two points arise: (1) Since Bessel's equation is homogeneous, we may multiply $y_2(x)$ by any constant. To match $N_0(x)$, we multiply our $y_2(x)$ by $2/\pi$. (2) To our second solution $(2/\pi)y_2(x)$, we may add any constant multiple of the first solution. Again, to match $N_0(x)$ we add

$$\frac{2}{\pi}[-\ln 2 + \gamma]J_0(x),$$

where γ is the usual Euler–Mascheroni constant (Section 5.2).[6] Our new, modified second solution is

$$y_2(x) = \frac{2}{\pi}[\ln x - \ln 2 + \gamma]J_0(x) + \frac{2}{\pi}J_0(x)\left\{ \frac{x^2}{4} + \frac{5x^4}{128} + \cdots \right\}. \qquad (8.142e)$$

Now the comparison with $N_0(x)$ becomes a simple multiplication of $J_0(x)$ from Eq. (8.142a) and the curly bracket of Eq. (8.142c). The multiplication checks—through terms of order x^2 and x^4, which is all we carried. Our second solution

[6] The Neumann function N_0 is defined as it is in order to achieve convenient *asymptotic* properties, Section 11.6.

from Eqs. (8.127) and (8.135) agrees with the standard second solution, the Neumann function, $N_0(x)$.

From the preceding analysis, the second solution of Eq. (8.117), $y_2(x)$, may be written as

$$y_2(x) = y_1(x) \ln x + \sum_{j=-n}^{\infty} d_j x^{j+\alpha}, \qquad (8.142f)$$

the first solution times $\ln x$ and another power series, this one starting with $x^{\alpha-n}$, which means that we may look for a logarithmic term when the indicial equation of Section 8.5 gives only one series solution. With the form of the second solution specified by Eq. (8.142f), we can substitute Eq. (8.142f) into the original differential equation and determine the coefficients d_j exactly as in Section 8.5. It may be worth noting that no series expansion of $\ln x$ is needed. In the substitution $\ln x$ will drop out; its derivatives will survive.

The second solution will usually diverge at the origin because of the logarithmic factor and the negative powers of x in the series. For this reason $y_2(x)$ is often referred to as the irregular solution. The first series solution, $y_1(x)$, which usually converges at the origin, is called the regular solution. The question of behavior at the origin is discussed in more detail in Chapters 11 and 12 in which we take up Bessel functions, modified Bessel functions, and Legendre functions.

SUMMARY

The two sections (together with the exercises) provide a complete solution of our linear, homogeneous, second-order differential equation—assuming that the point of expansion is no worse than a regular singularity. At least one solution can always be obtained by series substitution (Section 8.5). A second, linearly independent solution can be constructed by the Wronskian double integral, Eq. (8.127). This is all there are: no third, linearly independent solution exists (compare Exercise 8.6.10).

The *nonhomogeneous*, linear, second-order differential equation will have an additional solution: the particular solution. This particular solution may be obtained by the method of variation of parameters, Exercise 8.6.25, or by techniques such as Green's functions, Section 8.7.

EXERCISES

8.6.1   You know that the three unit vectors $\hat{x}$, $\hat{y}$, and $\hat{z}$ are mutually perpendicular (orthogonal). Show that $\hat{x}$, $\hat{y}$, and $\hat{z}$ are linearly independent. Specifically, show that no relation of the form of Eq. (8.111) exists for $\hat{x}$, $\hat{y}$, and $\hat{z}$.

8.6.2   The criterion for the linear *independence* of three vectors $\mathbf{A}$, $\mathbf{B}$, and $\mathbf{C}$ is that the equation

$$a\mathbf{A} + b\mathbf{B} + c\mathbf{C} = 0$$

(analogous to Eq. (8.111)) has no solution other than the trivial $a = b = c = 0$. Using components $\mathbf{A} = (A_1, A_2, A_3)$, and so on, set up the determinant criterion for the existence or nonexistence of a nontrivial solution for the coefficients a, b, and c. Show that your criterion is equivalent to the scalar product $\mathbf{A} \cdot \mathbf{B} \times \mathbf{C}$.

8.6.3   Using the Wronski determinant, show that the set of functions

$$\left\{1, \frac{x^n}{n!} (n = 1, 2, ..., N)\right\}$$

is linearly independent.

8.6.4   If the Wronskian of two functions y_1 and y_2 is identically zero, show by direct integration that

$$y_1 = c y_2;$$

That is, y_1 and y_2 are dependent. Assume the functions have continuous derivatives and that at least one of the functions does not vanish in the interval under consideration.

8.6.5   The Wronskian of two functions is found to be zero at $x = x_0$. Show that this Wronskian vanishes for all x and that the functions are linearly dependent.

8.6.6   The three functions $\sin x$, e^x, and e^{-x} are linearly independent. No one function can be written as a linear combination of the other two. Show that the Wronskian of $\sin x$, e^x, and e^{-x} vanishes but only at isolated points.

ANS. $W = 4 \sin x$,
$W = 0$ for $x = \pm n\pi$, $n = 0, 1, 2, \ldots$.

8.6.7   Consider two functions $\varphi_1 = x$ and $\varphi_2 = |x| = x \operatorname{sgn} x$ (Fig. 8.3). The function $\operatorname{sgn} x$ is just the *sign* of x. Since $\varphi_1' = 1$ and $\varphi_2' = \operatorname{sgn} x$, $W(\varphi_1, \varphi_2) = 0$ for any interval including $[-1, +1]$. Does the vanishing of the Wronskian over $[-1, +1]$ prove that φ_1 and φ_2 are linearly dependent? Clearly, they are not. What is wrong?

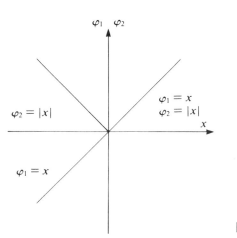

Figure 8.3 x and $|x|$.

8.6.8   Explain that *linear independence* does not mean the absence of any dependence. Illustrate your argument with $\cosh x$ and e^x.

8.6.9   Legendre's differential equation
$$(1 - x^2)y'' - 2xy' + n(n + 1)y = 0$$
has a regular solution $P_n(x)$ and an irregular solution $Q_n(x)$. Show that the Wronskian of P_n and Q_n is given by
$$P_n(x)Q_n'(x) - P_n'(x)Q_n(x) = \frac{A_n}{1 - x^2},$$
with A_n *independent* of x.

8.6.10   Show, by means of the Wronskian, that a linear, second-order, homogeneous, differential equation of the form
$$y''(x) + P(x)y'(x) + Q(x)y(x) = 0$$
cannot have three independent solutions. (Assume a third solution and show that the Wronskian vanishes for all x.)

8.6.11   Transform our linear, second-order, differential equation
$$y'' + P(x)y' + Q(x)y = 0$$
by the substitution
$$y = z \exp\left[-\frac{1}{2} \int^x P(t)\, dt \right]$$
and show that the resulting differential equation for z is
$$z'' + q(x)z = 0,$$
where
$$q(x) = Q(x) - \tfrac{1}{2}P'(x) - \tfrac{1}{4}P^2(x).$$
Note. This substitution can be *derived* by the technique of Exercise 8.6.24.

8.6.12   Use the result of Exercise 8.6.11 to show that the replacement of $\varphi(r)$ by $r\varphi(r)$ may be expected to eliminate the first derivative from the Laplacian in spherical polar coordinates. See also Exercise 2.5.18(b).

8.6.13   By direct differentiation and substitution show that
$$y_2(x) = y_1(x) \int^x \frac{\exp[-\int^s P(t)\, dt]}{[y_1(s)]^2}\, ds$$
satisfies
$$y_2''(x) + P(x)y_2'(x) + Q(x)y_2(x) = 0.$$
Note. The Leibniz formula for the derivative of an integral is
$$\frac{d}{d\alpha} \int_{g(\alpha)}^{h(\alpha)} f(x, \alpha)\, dx = \int_{g(\alpha)}^{h(\alpha)} \frac{\partial f(x, \alpha)}{\partial \alpha}\, dx$$
$$+ f[h(\alpha), \alpha]\frac{dh(\alpha)}{d\alpha} - f[g(\alpha), \alpha]\frac{dg(\alpha)}{d\alpha}.$$

8.6.14   In the equation
$$y_2(x) = y_1(x) \int^x \frac{\exp[-\int^s P(t)\, dt]}{[y_1(s)]^2}\, ds$$

$y_1(x)$ satisfies

$$y_1'' + P(x)y_1' + Q(x)y_1 = 0.$$

The function $y_2(x)$ is a linearly *independent* second solution of the same equation. Show that the inclusion of lower limits on the two integrals leads to nothing new; that is, it throws in only overall factors and/or a multiple of the known solution $y_1(x)$.

8.6.15   Given that one solution of

$$R'' + \frac{1}{r}R' - \frac{m^2}{r^2}R = 0$$

is $R = r^m$, show that Eq. (8.127) predicts a second solution, $R = r^{-m}$.

8.6.16   Using $y_1(x) = \sum_{n=0}^{\infty}(-1)^n x^{2n+1}/(2n+1)!$ as a solution of the linear oscillator equation, follow the analysis culminating in Eq. (8.142f) and show that $c_1 = 0$ so that the second solution does not, in this case, contain a logarithmic term.

8.6.17   Show that when n is *not* an integer the second solution of Bessel's equation, obtained from Eq. (8.127), does *not* contain a logarithmic term.

8.6.18   (a) One solution of Hermite's differential equation

$$y'' - 2xy' + 2\alpha y = 0$$

for $\alpha = 0$ is $y_1(x) = 1$. Find a second solution $y_2(x)$, using Eq. (8.127). Show that your second solution is equivalent to y_{odd} (Exercise 8.5.6).

(b) Find a second solution for $\alpha = 1$, where $y_1(x) = x$, using Eq. (8.127). Show that your second solution is equivalent to y_{even} (Exercise 8.5.6).

8.6.19   One solution of Laguerre's differential equation

$$xy'' + (1 - x)y' + ny = 0$$

for $n = 0$ is $y_1(x) = 1$. Using Eq. (8.127), develop a second, linearly independent solution. Exhibit the logarithmic term explicitly.

8.6.20   For Laguerre's equation with $n = 0$

$$y_2(x) = \int^x \frac{e^s}{s}\,ds.$$

(a) Write $y_2(x)$ as a logarithm plus a power series.

(b) Verify that the integral form of $y_2(x)$, previously given, is a solution of Laguerre's equation ($n = 0$) by direct differentiation of the integral and substitution into the differential equation.

(c) Verify that the series form of $y_2(x)$, part (a), is a solution by differentiating the series and substituting back into Laguerre's equation.

8.6.21   One solution of the Chebyshev equation

$$(1 - x^2)y'' - xy' + n^2 y = 0$$

for $n = 0$ *is* $y_1 = 1$.

(a) Using Eq. (8.127), develop a second, linearly independent solution.

(b) Find a second solution by direct integration of the Chebyshev equation.

Hint. Let $v = y'$ and integrate. Compare your result with the second solution given in Section 13.3.

$$\text{ANS.} \quad \text{(a)} \quad y_2 = \sin^{-1} x.$$
$$\text{(b) The second solution, } V_n(x), \text{ is not defined for } n = 0.$$

8.6.22 One solution of the Chebyshev equation

$$(1 - x^2)y'' - xy' + n^2 y = 0$$

for $n = 1$ is $y_1(x) = x$. Set up the Wronskian double integral solution and derive a second solution, $y_2(x)$.

$$\text{ANS.} \quad y_2 = -(1 - x^2)^{1/2}$$

8.6.23 The radial Schrödinger wave equation has the form

$$\left\{ -\frac{\hbar^2}{2m}\frac{d^2}{dr^2} + l(l+1)\frac{\hbar^2}{2mr^2} + V(r) \right\} y(r) = E y(r).$$

The potential energy $V(r)$ may be expanded about the origin as

$$V(r) = \frac{b_{-1}}{r} + b_0 + b_1 r + \cdots.$$

(a) Show that there is one (regular) solution starting with r^{l+1}.
(b) From Eq. (8.128) show that the irregular solution diverges at the origin as r^{-l}.

8.6.24 Show that if a second solution, y_2, is assumed to have the form $y_2(x) = y_1(x)f(x)$, substitution back into the original equation

$$y_2'' + P(x)y_2' + Q(x)y_2 = 0$$

leads to

$$f(x) = \int^x \frac{\exp[-\int^s P(t)\, dt]}{[y_1(s)]^2}\, ds$$

in agreement with Eq. (8.127).

8.6.25 If our linear, second-order differential equation is nonhomogeneous, that is, of the form of Eq. (8.82), the most general solution is

$$y(x) = y_1(x) + y_2(x) + y_p(x).$$

(y_1 and y_2 are solutions of the homogeneous equation.)
Show that

$$y_p(x) = y_2(x) \int^x \frac{y_1(s)F(s)\, ds}{W\{y_1(s), y_2(s)\}} - y_1(x) \int^x \frac{y_2(s)F(s)\, ds}{W\{y_1(s), y_2(s)\}},$$

with $W\{y_1(s), y_2(s)\}$ the Wronskian of $y_1(s)$ and $y_2(s)$.
Hint. As in Exercise 8.6.24, let $y_p(x) = y_1(x)v(x)$ and develop a first-order differential equation for $v'(x)$.

8.6.26 (a) Show that

$$y'' + \frac{1 - \alpha^2}{4x^2} y = 0$$

has two solutions:

$$y_1(x) = a_0 x^{(1+\alpha)/2}$$
$$y_2(x) = a_0 x^{(1-\alpha)/2}.$$

(b) For $\alpha = 0$ the two linearly independent solutions of part (a) reduce to $y_{10} = a_0 x^{1/2}$. Using Eq. (8.128) derive a second solution

$$y_{20}(x) = a_0 x^{1/2} \ln x.$$

Verify that y_{20} is indeed a solution.

(c) Show that the second solution from part (b) may be obtained as a limiting case from the two solutions of part (a):

$$y_{20}(x) = \lim_{\alpha \to 0} \left(\frac{y_1 - y_2}{\alpha} \right).$$

8.7 NONHOMOGENEOUS EQUATION—GREEN'S FUNCTION

The series substitution of Section 8.5 and the Wronskian double integral of Section 8.6 provide the most general solution of the *homogeneous*, linear, second-order differential equation. The specific solution, y_p, linearly dependent on the source term ($F(x)$ of Eq. (8.122)) may be cranked out by the variation of parameters method, Exercise 8.6.25. In this section we turn to a different method of solution—Green's functions.

For a brief introduction to Green's function method, as applied to the solution of a nonhomogeneous partial differential equation, it is helpful to use the electrostatic analog. In the presence of charges the electrostatic potential ψ satisfies Poisson's nonhomogeneous equation (compare Section 1.14)

$$\nabla^2 \psi = -\frac{\rho}{\varepsilon_0} \quad \text{(mks units)} \tag{8.143}$$

and Laplace's homogeneous equation,

$$\nabla^2 \psi = 0, \tag{8.144}$$

in the absence of electric charge ($\rho = 0$). If the charges are point charges q_i, we know that the solution is

$$\psi = \frac{1}{4\pi\varepsilon_0} \sum_i \frac{q_i}{r_i}, \tag{8.145}$$

a superposition of single-point charge solutions obtained from Coulomb's law for the force between two point charges q_1 and q_2,

$$\mathbf{F} = \frac{q_1 q_2 \hat{\mathbf{r}}}{4\pi\varepsilon_0 r^2}. \tag{8.146}$$

By replacement of the discrete point charges with a smeared out distributed charge, charge density ρ, Eq. (8.145) becomes

$$\psi(r = 0) = \frac{1}{4\pi\varepsilon_0} \int \frac{\rho(\mathbf{r})}{r} d\tau \tag{8.147}$$

or, for the potential at $\mathbf{r} = \mathbf{r}_1$ away from the origin and the charge at $\mathbf{r} = \mathbf{r}_2$,

$$\psi(\mathbf{r}_1) = \frac{1}{4\pi\varepsilon_0} \int \frac{\rho(\mathbf{r}_2)}{|\mathbf{r}_1 - \mathbf{r}_2|} \, d\tau_2. \tag{8.148}$$

We use ψ as the potential corresponding to the given distribution of charge and therefore satisfying Poisson's equation (8.143), whereas a function G, which we label Green's function, is required to satisfy Poisson's equation with a point source at the point defined by $\mathbf{r}_2$:

$$\nabla^2 G = -\delta(\mathbf{r}_1 - \mathbf{r}_2). \tag{8.149}$$

Physically, then, G is the potential at $\mathbf{r}_1$ corresponding to a unit source (ε_0) at $\mathbf{r}_2$. By Green's theorem (Section 1.11)

$$\int (\psi\nabla^2 G - G\nabla^2\psi) \, d\tau_2 = \int (\psi\,\nabla G - G\,\nabla\psi) \cdot d\boldsymbol{\sigma}. \tag{8.150}$$

Assuming that the integrand falls off faster than r^{-2}, we may simplify our problem by taking the volume so large that the surface integral vanishes, leaving

$$\int \psi\nabla^2 G \, d\tau_2 = \int G\nabla^2\psi \, d\tau_2 \tag{8.151}$$

or by substituting in Eqs. (8.143) and (8.149), we have

$$-\int \psi(\mathbf{r}_2)\delta(\mathbf{r}_1 - \mathbf{r}_2) \, d\tau_2 = -\int \frac{G(\mathbf{r}_1, \mathbf{r}_2)\rho(\mathbf{r}_2)}{\varepsilon_0} \, d\tau_2. \tag{8.152}$$

Integration by employing the defining property of the Dirac delta function (Eq. (1.170b)) produces

$$\psi(\mathbf{r}_1) = \frac{1}{\varepsilon_0} \int G(\mathbf{r}_1, \mathbf{r}_2)\rho(\mathbf{r}_2) \, d\tau_2. \tag{8.153}$$

Note that we have used Eq. (8.149) to eliminate $\nabla^2 G$ but that the function G itself is still unknown. In Section 1.14, Gauss's law, we found that

$$\int \nabla^2 \left(\frac{1}{r}\right) d\tau = \begin{cases} 0, \\ -4\pi, \end{cases} \tag{8.154}$$

0 if the volume did not include the origin and -4π if the origin were included. This result from Section 1.14 may be rewritten as in Eq. (1.169), or

$$\nabla^2 \left(\frac{1}{4\pi r}\right) = -\delta(\mathbf{r}), \quad \text{or} \quad \nabla^2 \left(\frac{1}{4\pi r_{12}}\right) = -\delta(\mathbf{r}_1 - \mathbf{r}_2), \tag{8.155}$$

corresponding to a shift of the electrostatic charge from the origin to the position $\mathbf{r} = \mathbf{r}_2$. Here $r_{12} = |\mathbf{r}_1 - \mathbf{r}_2|$, and the Dirac delta function $\delta(\mathbf{r}_1 - \mathbf{r}_2)$ vanishes unless $\mathbf{r}_1 = \mathbf{r}_2$. Therefore in a comparison of Eqs. (8.149) and (8.155) the function G (Green's function) is given by

$$G(\mathbf{r}_1, \mathbf{r}_2) = \frac{1}{4\pi|\mathbf{r}_1 - \mathbf{r}_2|}. \tag{8.156}$$

The solution of our differential equation (Poisson's equation) is

$$\psi(\mathbf{r}_1) = \frac{1}{4\pi\varepsilon_0} \int \frac{\rho(\mathbf{r}_2)}{|\mathbf{r}_1 - \mathbf{r}_2|} \, d\tau_2 \tag{8.157}$$

in complete agreement with Eq. (8.148). Actually $\psi(\mathbf{r}_1)$, Eq. (8.157), is the particular solution of Poisson's equation. We may add solutions of Laplace's equation (compare Eq. (8.83)). Such solutions could describe an external field.

These results will be generalized to the second-order linear but nonhomogeneous, differential equation

$$\mathcal{L}y(\mathbf{r}_1) = -f(\mathbf{r}_1), \tag{8.158}$$

where $\mathcal{L}$ is a linear differential operator. The Green's function is taken to be a solution of

$$\mathcal{L}G(\mathbf{r}_1, \mathbf{r}_2) = -\delta(\mathbf{r}_1 - \mathbf{r}_2) \tag{8.159}$$

(analogous to Eq. (8.149)). Then the particular solution $y(\mathbf{r}_1)$ becomes

$$y(\mathbf{r}_1) = \int G(\mathbf{r}_1, \mathbf{r}_2) f(\mathbf{r}_2) \, d\tau_2. \tag{8.160}$$

(There may also be an integral over a bounding surface depending on the conditions specified.)

In summary, Green's function, often written $G(\mathbf{r}_1, \mathbf{r}_2)$ as a reminder of the name, is a solution of Eq. (8.149). It enters in an integral solution of our differential equation, as in Eq. (8.148). For the simple, but important, electrostatic case we obtain Green's function $G(\mathbf{r}_1, \mathbf{r}_2)$ by Gauss's law, comparing Eqs. (8.149) and (8.155). Finally, from the final solution (Eq. (8.157)) it is possible to develop a physical interpretation of Green's function. It occurs as a weighting function or propagator function that enhances or reduces the effect of the charge element $\rho(\mathbf{r}_2) \, d\tau_2$ according to its distance from the field point $\mathbf{r}_1$. Green's function, $G(\mathbf{r}_1, \mathbf{r}_2)$, gives the effect of a unit point source at $\mathbf{r}_2$ in producing a potential at $\mathbf{r}_1$. This is how it was introduced in Eq. (8.149); this is how it appears in Eq. (8.157).

SYMMETRY OF GREEN'S FUNCTION

An important property of Green's function is the symmetry of its two variables, that is,

$$G(\mathbf{r}_1, \mathbf{r}_2) = G(\mathbf{r}_2, \mathbf{r}_1). \tag{8.161}$$

Although this is obvious in the electrostatic case just considered, it can be proved under much more general conditions. In place of Eq. (8.149), let us require that $G(\mathbf{r}, \mathbf{r}_1)$ satisfy[1]

$$\nabla \cdot [p(\mathbf{r}) \nabla G(\mathbf{r}, \mathbf{r}_1)] + \lambda q(\mathbf{r}) G(\mathbf{r}, \mathbf{r}_1) = -\delta(\mathbf{r} - \mathbf{r}_1), \qquad (8.162)$$

corresponding to a mathematical point source at $\mathbf{r} = \mathbf{r}_1$. Here the functions $p(\mathbf{r})$ and $q(\mathbf{r})$ are well behaved but otherwise arbitrary functions of $\mathbf{r}$. Green's function, $G(\mathbf{r}, \mathbf{r}_2)$, satisfies the same equation but the subscript 1 is replaced by subscript 2. Then $G(\mathbf{r}, \mathbf{r}_2)$ is a sort of potential at $\mathbf{r}$, created by a unit point source at $\mathbf{r}_2$. We multiply the equation for $G(\mathbf{r}, \mathbf{r}_1)$ by $G(\mathbf{r}, \mathbf{r}_2)$ and the equation for $G(\mathbf{r}, \mathbf{r}_2)$ by $G(\mathbf{r}, \mathbf{r}_1)$ and then subtract the two:

$$G(\mathbf{r}, \mathbf{r}_2)\nabla \cdot [p(\mathbf{r}) \nabla G(\mathbf{r}, \mathbf{r}_1)] - G(\mathbf{r}, \mathbf{r}_1)\nabla \cdot [p(\mathbf{r}) \nabla G(\mathbf{r}, \mathbf{r}_2)]$$

$$= -G(\mathbf{r}, \mathbf{r}_2)\delta(\mathbf{r} - \mathbf{r}_1) + G(\mathbf{r}, \mathbf{r}_1)\delta(\mathbf{r} - \mathbf{r}_2). \qquad (8.163)$$

The first term in Eq. (8.163),

$$G(\mathbf{r}, \mathbf{r}_2)\nabla \cdot [p(\mathbf{r}) \nabla G(\mathbf{r}, \mathbf{r}_1)]$$

may be replaced by

$$\nabla \cdot [G(\mathbf{r}, \mathbf{r}_2)p(\mathbf{r}) \nabla G(\mathbf{r}, \mathbf{r}_1)] - \nabla G(\mathbf{r}, \mathbf{r}_2) \cdot p(\mathbf{r}) \nabla G(\mathbf{r}, \mathbf{r}_1).$$

A similar transformation is carried out on the second term. Then integrating over whatever volume is involved and using Green's theorem, we obtain a surface integral:

$$\int_S [G(\mathbf{r}, \mathbf{r}_2)p(\mathbf{r}) \nabla G(\mathbf{r}, \mathbf{r}_1) - G(\mathbf{r}, \mathbf{r}_1)p(\mathbf{r}) \nabla G(\mathbf{r}, \mathbf{r}_2)] \cdot d\boldsymbol{\sigma}$$

$$= -G(\mathbf{r}_1, \mathbf{r}_2) + G(\mathbf{r}_2, \mathbf{r}_1). \qquad (8.164)$$

The terms on the right-hand side appear when we use the Dirac delta functions and carry out the volume integration. Under the requirement that Green's functions, $G(\mathbf{r}, \mathbf{r}_1)$ and $G(\mathbf{r}, \mathbf{r}_2)$, have the same values over the surface S and that their normal derivatives have the same values over the surfaces S, or that the Green's functions vanish (Dirichlet boundary conditions, Section 9.1)[2] over the surface S, the surface integral vanishes and

$$G(\mathbf{r}_1, \mathbf{r}_2) = G(\mathbf{r}_2, \mathbf{r}_1), \qquad (8.165)$$

which shows that Green's function is symmetric. If the eigenfunctions are complex, boundary conditions corresponding to Eqs. (9.19)–(9.20) are appropriate. Equation (8.165) becomes

$$G(\mathbf{r}_1, \mathbf{r}_2) = G^*(\mathbf{r}_2, \mathbf{r}_1). \qquad (8.166)$$

---

[1] Equation (8.162) is a three-dimensional version of the *self-adjoint* eigenvalue equation, Eq. (9.4).

[2] Any attempt to demand that the normal derivatives vanish at the surface (Neumann's conditions, Section 9.1) leads to trouble with Gauss's Law. It is like demanding that $\int \mathbf{E} \cdot d\boldsymbol{\sigma} = 0$ when you know perfectly well that there is some electric charge inside the surface.

Note that this symmetry property holds for Green's functions in every equation in the form of Eq. (8.162). In Chapter 9 we shall call equations in this form self-adjoint. The symmetry is the basis of various reciprocity theorems; the effect of a charge at r_2 on the potential at r_1 is the same as the effect of a charge at r_1 on the potential at r_2.

This use of Green's functions is a powerful technique for solving many of the more difficult problems of mathematical physics.

FORM OF GREEN'S FUNCTIONS

Let us assume that $\mathcal{L}$ is a self-adjoint differential operator of the general form[3]

$$\mathcal{L}_1 = \nabla_1 \cdot [p(\mathbf{r}_1)\nabla_1] + q(\mathbf{r}_1). \tag{8.167}$$

Here the subscript 1 on $\mathcal{L}$ emphasizes that $\mathcal{L}$ operates on $\mathbf{r}_1$. Then, as a simple generalization of Green's theorem, Eq. (1.103), we have

$$\int (v\mathcal{L}_2 u - u\mathcal{L}_2 v)\, d\tau_2 = \int p(v\nabla_2 u - u\nabla_2 v) \cdot d\sigma_2, \tag{8.168}$$

in which all quantities have $\mathbf{r}_2$ as their argument. (To verify Eq. (8.168), take the divergence of the integrand of the surface integral.) We let $u(\mathbf{r}_2) = y(\mathbf{r}_2)$ so that Eq. (8.158) applies and $v(\mathbf{r}_2) = G(\mathbf{r}_1, \mathbf{r}_2)$ so that Eq. (8.159) applies. (Remember $G(\mathbf{r}_1, \mathbf{r}_2) = G(\mathbf{r}_2, \mathbf{r}_1)$.) Substituting into Green's theorem

$$\int \{-G(\mathbf{r}_1, \mathbf{r}_2)f(\mathbf{r}_2) + y(\mathbf{r}_2)\delta(\mathbf{r}_1 - \mathbf{r}_2)\}\, d\tau_2$$

$$= \int p(\mathbf{r}_2)\{G(\mathbf{r}_1, \mathbf{r}_2)\nabla_2 y(\mathbf{r}_2) - y(\mathbf{r}_2)\nabla_2 G(\mathbf{r}_1, \mathbf{r}_2)\} \cdot d\sigma_2. \tag{8.169}$$

Integrating over the Dirac delta function

$$y(\mathbf{r}_1) = \int G(\mathbf{r}_1, \mathbf{r}_2)f(\mathbf{r}_2)\, d\tau_2$$

$$+ \int p(\mathbf{r}_2)\{G(\mathbf{r}_1, \mathbf{r}_2)\nabla_2 y(\mathbf{r}_2) - y(\mathbf{r}_2)\nabla_2 G(\mathbf{r}_1, \mathbf{r}_2)\} \cdot d\sigma_2, \tag{8.170}$$

our solution to Eq. (8.158) appears as a volume integral plus a surface integral. If y and G both satisfy Dirichlet boundary conditions, or if both satisfy Neumann boundary conditions, the surface integral vanishes and we regain Eq. (8.160). The volume integral is a weighted integral over the source term $f(\mathbf{r}_2)$ with our Green's function $G(\mathbf{r}_1, \mathbf{r}_2)$ as the weighting function.

For the special case of $p(\mathbf{r}_1) = 1$ and $q(\mathbf{r}_1) = 0$, $\mathcal{L}$ is ∇^2, the Laplacian. Let us integrate

$$\nabla_1^2 G(\mathbf{r}_1, \mathbf{r}_2) = -\delta(\mathbf{r}_1 - \mathbf{r}_2) \tag{8.171}$$

[3] $\mathcal{L}_1$ may be in 1, 2, or 3 dimensions (with appropriate interpretation of ∇_1).

over a small volume including the point source. Then

$$\int \nabla_1 \cdot \nabla_1 G(\mathbf{r}_1, \mathbf{r}_2) \, d\tau_1 = - \int \delta(\mathbf{r}_1 - \mathbf{r}_2) \, d\tau_1$$

$$= -1. \tag{8.172}$$

The volume integral on the left may be transformed by Gauss's theorem as in the development of Gauss's law—Section 1.14. We find that

$$\int \nabla_1 G(\mathbf{r}_1, \mathbf{r}_2) \cdot d\boldsymbol{\sigma}_1 = -1. \tag{8.173}$$

This shows, incidentally, that it may not be possible to impose a Neumann boundary condition, that the normal derivative of the Green's function, $\partial G/\partial n$, vanishes over the entire surface.

If we are in three-dimensional space, Eq. (8.173) is satisfied by taking

$$\frac{\partial}{\partial r_{12}} G(\mathbf{r}_1, \mathbf{r}_2) = -\frac{1}{4\pi} \cdot \frac{1}{|\mathbf{r}_1 - \mathbf{r}_2|^2}, \qquad r_{12} = |\mathbf{r}_1 - \mathbf{r}_2|. \tag{8.174}$$

The integration is over the surface of a sphere centered at $\mathbf{r}_2$. The integral of Eq. (8.174) is

$$G(\mathbf{r}_1, \mathbf{r}_2) = \frac{1}{4\pi} \cdot \frac{1}{|\mathbf{r}_1 - \mathbf{r}_2|}, \tag{8.175}$$

in agreement with Section 1.14.

If we are in two-dimensional space Eq. (8.173) is satisfied by taking

$$\frac{\partial}{\partial \rho_{12}} G(\boldsymbol{\rho}_1, \boldsymbol{\rho}_2) = -\frac{1}{2\pi} \cdot \frac{1}{|\boldsymbol{\rho}_1 - \boldsymbol{\rho}_2|}, \tag{8.176}$$

with $\mathbf{r}$ being replaced by $\boldsymbol{\rho}$, $\rho = (x^2 + y^2)^{1/2}$ and the integration being over the circumference of a circle centered on $\boldsymbol{\rho}_2$. Here $\rho_{12} = |\boldsymbol{\rho}_1 - \boldsymbol{\rho}_2|$. Integrating Eq. (8.176), we obtain

$$G(\boldsymbol{\rho}_1, \boldsymbol{\rho}_2) = -\frac{1}{2\pi} \ln |\boldsymbol{\rho}_1 - \boldsymbol{\rho}_2|. \tag{8.177}$$

To $G(\boldsymbol{\rho}_1, \boldsymbol{\rho}_2)$ (and to $G(\mathbf{r}_1, \mathbf{r}_2)$) we may add any multiple of the regular solution of the homogeneous equation as needed to satisfy boundary conditions.

The behavior of the Laplace operator Green's function in the vicinity of the source point $\mathbf{r}_1 = \mathbf{r}_2$ shown by Eqs. (8.175) and (8.177) facilitates the identification of the Green's functions for the other cases, such as the Helmholtz and modified Helmholtz equations.

1. For $\mathbf{r}_1 \neq \mathbf{r}_2$. $G(\mathbf{r}_1, \mathbf{r}_2)$ must satisfy the *homogeneous* differential equation

$$\mathcal{L}_1 G(\mathbf{r}_1, \mathbf{r}_2) = 0, \qquad \mathbf{r}_1 \neq \mathbf{r}_2. \tag{8.178}$$

Table 8.5 Green's Functions[a]

	Laplace ∇^2	Helmholtz $\nabla^2 + k^2$	Modified Helmholtz $\nabla^2 - k^2$
One-dimensional space	No solution for $(-\infty, \infty)$	$\dfrac{i}{2k}\exp(ik\lvert x_1 - x_2\rvert)$	$\dfrac{1}{2k}\exp(-k\lvert x_1 - x_2\rvert)$
Two-dimensional space	$-\dfrac{1}{2\pi}\ln\lvert\boldsymbol{\rho}_1 - \boldsymbol{\rho}_2\rvert$	$\dfrac{i}{4}H_0^{(1)}(k\lvert\boldsymbol{\rho}_1 - \boldsymbol{\rho}_2\rvert)$	$\dfrac{1}{2\pi}K_0(k\lvert\boldsymbol{\rho}_1 - \boldsymbol{\rho}_2\rvert)$
Three-dimensional space	$\dfrac{1}{4\pi}\cdot\dfrac{1}{\lvert\mathbf{r}_1 - \mathbf{r}_2\rvert}$	$\dfrac{\exp(ik\lvert\mathbf{r}_1 - \mathbf{r}_2\rvert)}{4\pi\lvert\mathbf{r}_1 - \mathbf{r}_2\rvert}$	$\dfrac{\exp(-k\lvert\mathbf{r}_1 - \mathbf{r}_2\rvert)}{4\pi\lvert\mathbf{r}_1 - \mathbf{r}_2\rvert}$

[a] These are the Green's functions satisfying the boundary condition $G(\mathbf{r}_1, \mathbf{r}_2) = 0$ as $r_1 \to \infty$ for the Laplace and modified Helmholtz operators. For the Helmholtz operator, $G(\mathbf{r}_1, \mathbf{r}_2)$ corresponds to an outgoing wave. $H_0^{(1)}$ is the Hankel function of Section 11.4. K_0 is the modified Bessel function of Section 11.5.

2. As $\mathbf{r}_1 \to \mathbf{r}_2$ (or $\boldsymbol{\rho}_1 \to \boldsymbol{\rho}_2$),

$$G(\boldsymbol{\rho}_1, \boldsymbol{\rho}_2) \approx -\frac{1}{2\pi}\ln\lvert\boldsymbol{\rho}_1 - \boldsymbol{\rho}_2\rvert, \qquad \text{two-dimensional space,} \qquad (8.179)$$

$$G(\mathbf{r}_1, \mathbf{r}_2) \approx \frac{1}{4\pi}\cdot\frac{1}{\lvert\mathbf{r}_1 - \mathbf{r}_2\rvert}, \qquad \text{three-dimensional space.} \qquad (8.180)$$

The term $\pm k^2$ in the operator does not affect the behavior of G near the singular point $\mathbf{r}_1 = \mathbf{r}_2$. For convenience, the Green's functions for the Laplace, Helmholtz, and modified Helmholtz operators are listed in Table 8.5.

SPHERICAL POLAR COORDINATE EXPANSION

As an alternate determination of the Green's function of the Laplace operator, let us assume a spherical harmonic expansion of the form

$$G(\mathbf{r}_1, \mathbf{r}_2) = \sum_{l=0}^{\infty}\sum_{m=-l}^{l} g_l(r_1, r_2) Y_l^m(\theta_1, \varphi_1) Y_l^{m*}(\theta_2, \varphi_2). \qquad (8.181)$$

We will determine $g_l(r_1, r_2)$. From Exercises 8.6.7 and 12.6.6,

$$\delta(\mathbf{r}_1 - \mathbf{r}_2) = \frac{1}{r_1^2}\delta(r_1 - r_2)\delta(\cos\theta_1 - \cos\theta_2)\delta(\varphi_1 - \varphi_2)$$

$$= \frac{1}{r_1^2}\delta(r_1 - r_2)\sum_{l=0}^{\infty}\sum_{m=-l}^{l} Y_l^m(\theta_1, \varphi_1) Y_l^{m*}(\theta_2, \varphi_2). \qquad (8.182)$$

Substituting Eqs. (8.181) and (8.182) into the Green's function differential equation, Eq. (8.171), and making use of the orthogonality of the spherical harmonics, we obtain a radial equation:

$$r_1 \frac{d^2}{dr_1^2}[r_1 g_1(r_1, r_2)] - l(l+1)g_l(r_1, r_2) = -\delta(r_1 - r_2). \quad (8.183)$$

This is now a one-dimensional problem. The solutions[4] of the corresponding homogeneous equation are r_1^l and r_1^{-l-1}. If we demand that g_l remain finite as $r_1 \to 0$ and vanish as $r_1 \to \infty$, the technique of Section 16.5 leads to

$$g_l(r_1, r_2) = \frac{1}{2l+1} \begin{cases} \dfrac{r_1^l}{r_2^{l+1}}, & r_1 < r_2, \\[2mm] \dfrac{r_2^l}{r_1^{l+1}}, & r_1 > r_2, \end{cases} \quad (8.184)$$

or

$$g_l(r_1, r_2) = \frac{1}{2l+1} \cdot \frac{r_<^l}{r_<^{l+1}}. \quad (8.185)$$

Hence our Green's function is

$$G(\mathbf{r}_1, \mathbf{r}_2) = \sum_{l=0}^{\infty} \sum_{m=-l}^{l} \frac{1}{2l+1} \frac{r_<^l}{r_>^{l+1}} Y_l^m(\theta_1, \varphi_1) Y_l^{m*}(\theta_2, \varphi_2). \quad (8.186)$$

Since we already have $G(\mathbf{r}_1, \mathbf{r}_2)$ in closed form, Eq. (8.175), we may write

$$\frac{1}{4\pi} \cdot \frac{1}{|\mathbf{r}_1 - \mathbf{r}_2|} = \sum_{l=0}^{\infty} \sum_{m=-l}^{l} \frac{1}{2l+1} \frac{r_<^l}{r_>^{l+1}} Y_l^m(\theta_1, \varphi_1) Y_l^{m*}(\theta_2, \varphi_2). \quad (8.187)$$

One immediate use for this spherical harmonic expansion of the Green's function is in the development of an electrostatic multipole expansion. The potential for an arbitrary charge distribution is

$$\psi(\mathbf{r}_1) = \frac{1}{4\pi\varepsilon_0} \int \frac{\rho(\mathbf{r}_2)}{|\mathbf{r}_1 - \mathbf{r}_2|} d\tau_2$$

(which is Eq. (8.148)). Substituting Eq. (8.187), we get

$$\psi(\mathbf{r}_1) = \frac{1}{\varepsilon_0} \sum_{l=0}^{\infty} \sum_{m=-l}^{l} \left\{ \frac{1}{2l+1} \frac{Y_l^m(\theta_1, \varphi_1)}{r_1^{l+1}} \right.$$

$$\left. \cdot \int \rho(\mathbf{r}_2) Y_l^{m*}(\theta_2, \varphi_2) r_2^l \, d\varphi_2 \sin\theta_2 \, d\theta_2 \, r_2^2 \, dr_2 \right\}, \quad \text{for } r_1 > r_2.$$

This is the multipole expansion. The relative importance of the various terms in the double sum depends on the form of the source $\rho(\mathbf{r}_2)$.

---

[4] Compare Table 8.2.

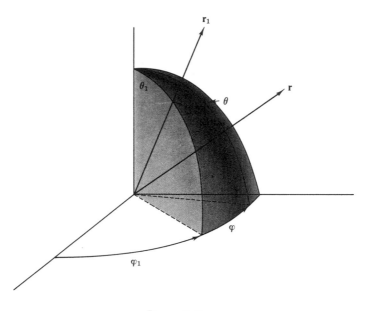

Figure 8.4

LEGENDRE POLYNOMIAL ADDITION THEOREM

From the generating expression for Legendre polynomials, Eq. (12.4a),

$$\frac{1}{4\pi} \cdot \frac{1}{|\mathbf{r}_1 - \mathbf{r}_2|} = \frac{1}{4\pi} \sum_{l=0}^{\infty} \frac{r_<^l}{r_>^{l+1}} P_l(\cos \gamma), \tag{8.188}$$

where γ is the angle included between vectors $\mathbf{r}_1$ and $\mathbf{r}_2$, Fig. 8.4. Equating Eqs. (8.187) and (8.188), we have the Legendre polynomial addition theorem

$$P_l(\cos \gamma) = \frac{4\pi}{2l + 1} \sum_{m=-l}^{l} Y_l^m(\theta_1, \varphi_1) Y_l^{m*}(\theta_2, \varphi_2). \tag{8.189}$$

Compare the simplicity (once Green's functions are understood) of this derivation with the relatively cumbersome derivation of Section 12.8.

CIRCULAR CYLINDRICAL COORDINATE EXPANSION

In analogy with the preceding spherical polar coordinate expansion, we write

$$\delta(\mathbf{r}_1 - \mathbf{r}_2) = \frac{1}{\rho_1} \delta(\rho_1 - \rho_2)\delta(\varphi_1 - \varphi_2)\delta(z_1 - z_2)$$

$$= \frac{1}{\rho_1} \delta(\rho_1 - \rho_2)\frac{1}{2\pi} \sum_{m=-\infty}^{\infty} e^{im(\varphi_1-\varphi_2)} \frac{1}{2\pi} \int_{-\infty}^{\infty} e^{ik(z_1-z_2)} \, dk, \tag{8.190}$$

using Exercise 12.6.5 and Eq. (1.192c). But why this choice? Why a summation for the φ-dependence and an integration for the z-dependence? The requirement that the azimuthal dependence be single-valued quantizes m, hence the summation. No such restriction applies to k.

To avoid problems later with negative values of k, we rewrite Eq. (8.190) as

$$\delta(\mathbf{r}_1 - \mathbf{r}_2) = \frac{1}{\rho_1}\delta(\rho_1 - \rho_2)\frac{1}{2\pi}\sum_{m=-\infty}^{\infty} e^{im(\varphi_1-\varphi_2)}\frac{1}{\pi}\int_0^\infty \cos k(z_1 - z_2)\,dk, \quad (8.191)$$

using the Cauchy principal value. We assume a similar expansion of the Green's function

$$G(\mathbf{r}_1, \mathbf{r}_2) = \frac{1}{2\pi^2}\sum_{m=-\infty}^{\infty} g_m(\rho_1,\rho_2)e^{im(\varphi_1-\varphi_2)}\int_0^\infty \cos k(z_1 - z_2)\,dk, \quad (8.192)$$

with the ρ-dependent coefficients $g_m(\rho_1,\rho_2)$ to be determined. Substituting into Eq. (8.171), now in circular cylindrical coordinates, we find that if $g_m(\rho_1,\rho_2)$ satisfies

$$\frac{d}{d\rho_1}\left[\rho_1\frac{dg_m}{d\rho_1}\right] - \left[k^2\rho_1 + \frac{m^2}{\rho_1}\right]g_m = -\delta(\rho_1 - \rho_2), \quad (8.193)$$

then Eq. (8.171) is satisfied.

The operator in Eq. (8.193) is identified as the modified Bessel operator (in self-adjoint form). Hence the solutions of the corresponding homogeneous equation are $u_1 = I_m(k\rho)$, $u_2 = K_m(k\rho)$. As in the spherical polar coordinate case, we demand that G be finite at $\rho_1 = 0$ and vanish as $\rho_1 \to \infty$. Then the technique yields

$$g_m(\rho_1,\rho_2) = -\frac{1}{A}I_m(k\rho_<)K_m(k\rho_>). \quad (8.194)$$

This corresponds to Eq. (8.155). The constant A comes from the Wronskian:

$$I_m(k\rho)K_m'(k\rho) - I_m'(k\rho)K_m(k\rho) = \frac{A}{p(k\rho)}. \quad (8.195)$$

From Exercise 11.5.10, $A = -1$ and

$$g_m(\rho_1,\rho_2) = I_m(k\rho_<)K_m(k\rho_>). \quad (8.196)$$

Therefore our circular cylindrical coordinate Green's function is

$$G(\mathbf{r}_1, \mathbf{r}_2) = \frac{1}{4\pi}\cdot\frac{1}{|\mathbf{r}_1 - \mathbf{r}_2|}$$

$$= \frac{1}{2\pi^2}\sum_{m=-\infty}^{\infty}\int_0^\infty I_m(k\rho_<)K_m(k\rho_>)e^{im(\varphi_1-\varphi_2)}\cos k(z_1 - z_2)\,dk.$$

$$(8.197)$$

Exercise 8.7.1 is a special case of this result.

Example 8.7.1 Quantum Mechanical Scattering—
Neumann Series Solution

The quantum theory of scattering provides a nice illustration of integral equation techniques and an application of a Green's function. Our physical picture of scattering is as follows. A beam of particles moves along the negative z-axis toward the origin. A small fraction of the particles is scattered by the potential $V(\mathbf{r})$ and goes off as an outgoing spherical wave. Our wave function $\psi(\mathbf{r})$ must satisfy the time-independent Schrödinger equation

$$-\frac{\hbar^2}{2m} \nabla^2 \psi(\mathbf{r}) + V(\mathbf{r})\psi(\mathbf{r}) = E\psi(\mathbf{r}) \qquad (8.198a)$$

or

$$\nabla^2 \psi(\mathbf{r}) + k^2 \psi(\mathbf{r}) = -\left[-\frac{2m}{\hbar^2} V(\mathbf{r})\psi(\mathbf{r}) \right] \qquad (8.198b)$$

with

$$k^2 = 2mE/\hbar^2.$$

From the physical picture just presented we look for a solution having an *asymptotic* form

$$\psi(\mathbf{r}) \sim e^{i\mathbf{k}_0 \cdot \mathbf{r}} + f_k(\theta, \varphi) \frac{e^{ikr}}{r}. \qquad (8.199)$$

Here $e^{i\mathbf{k}_0 \cdot \mathbf{r}}$ is the incident plane wave[5] with $\mathbf{k}_0$ the propagation vector carrying the subscript 0 to indicate that it is in the $\theta = 0$ (z-axis) direction. The magnitudes k_0 and k are equal (ignoring recoil) e^{ikr}/r is the outgoing spherical wave with an angular- (and energy) dependent amplitude factor $f_k(\theta, \varphi)$.[6] Vector $\mathbf{k}$ has the direction of the outgoing scattered wave. In quantum mechanics texts it is shown that the differential probability of scattering, $d\sigma/d\Omega$, the scattering cross section per unit solid angle, is given by $|f_k(\theta, \varphi)|^2$.

Identifying $[-(2m/\hbar^2)V(\mathbf{r})\psi(\mathbf{r})]$ with $f(\mathbf{r})$ of Eq. (8.158), we have

$$\psi(\mathbf{r}_1) = -\int \frac{2m}{\hbar^2} V(\mathbf{r}_2)\psi(\mathbf{r}_2)G(\mathbf{r}_1, \mathbf{r}_2)d^3r_2 \qquad (8.200)$$

by Eq. (8.170). This does not have the desired asymptotic form Eq. (8.199), but we may add to Eq. (8.200), $e^{i\mathbf{k}_0 \cdot \mathbf{r}_1}$, a solution of the homogeneous equation and put $\psi(r)$ into the desired form:

$$\psi(\mathbf{r}_1) = e^{i\mathbf{k}_0 \cdot \mathbf{r}_1} - \int \frac{2m}{\hbar^2} V(\mathbf{r}_2)\psi(\mathbf{r}_2)G(\mathbf{r}_1, \mathbf{r}_2)d^3r_2. \qquad (8.201)$$

---

[5] For simplicity we assume a continuous incident beam. In a more sophisticated and more realistic treatment Eq. (8.199) would be one component of a Fourier wave packet.

[6] If $V(\mathbf{r})$ represents a central force, f_k will be a function of θ only, independent of azimuth.

Our Green's function is the Green's function of the operator $\mathcal{L} = \nabla^2 + k^2$ (Eq. (8.198)) satisfying the boundary condition that it describe an outgoing wave. Then, from Table 8.5, $G(\mathbf{r}_1, \mathbf{r}_2) = \exp(ik|\mathbf{r}_1 - \mathbf{r}_2|)/(4\pi|\mathbf{r}_1 - \mathbf{r}_2|)$ and

$$\psi(\mathbf{r}_1) = e^{i\mathbf{k}_0 \cdot \mathbf{r}_1} - \int \frac{2m}{\hbar^2} V(\mathbf{r}_2)\psi(\mathbf{r}_2) \frac{e^{ik|\mathbf{r}_1-\mathbf{r}_2|}}{4\pi|\mathbf{r}_1 - \mathbf{r}_2|} d^3r_2. \tag{8.202}$$

This integral equation analog of the original Schrödinger wave equation is *exact*.

Employing the Neumann series technique of Section 16.3 (remember, the scattering probability is very small), we have

$$\psi_0(\mathbf{r}_1) = e^{i\mathbf{k}_0 \cdot \mathbf{r}_1}, \tag{8.203a}$$

which has the physical interpretation of no scattering.

Substituting $\psi_0(\mathbf{r}_2) = e^{i\mathbf{k}_0 \cdot \mathbf{r}_2}$ into the integral, we obtain the first correction term

$$\psi_1(\mathbf{r}_1) = e^{i\mathbf{k}_0 \cdot \mathbf{r}_1} - \int \frac{2m}{\hbar^2} V(\mathbf{r}_2) \frac{e^{ik|\mathbf{r}_1-\mathbf{r}_2|}}{4\pi|\mathbf{r}_1 - \mathbf{r}_2|} e^{i\mathbf{k}_0 \cdot \mathbf{r}_2} d^3r_2. \tag{8.203b}$$

This is the famous Born approximation. It is expected to be most accurate for weak potentials and high incident energy. If a more accurate approximation is desired the Neumann series may be continued.[7]

Example 8.7.2 Quantum Mechanical Scattering— Green's Function

Again, we consider the Schrödinger wave equation (Eq. (8.198b)) for the scattering problem. This time we use Fourier transform techniques and derive the desired form of the Green's function by contour integration. Substituting the desired asymptotic form of the solution (with k replaced by k_0)

$$\psi(\mathbf{r}) \sim e^{ik_0 z} + f_{k_0}(\theta, \varphi) \frac{e^{ik_0 r}}{r} = e^{ik_0 z} + \Phi(\mathbf{r}) \tag{8.204}$$

into the Schrödinger wave equation, Eq. (8.198b), yields

$$(\nabla^2 + k_0^2)\Phi(\mathbf{r}) = U(\mathbf{r})e^{ik_0 z} + U(\mathbf{r})\Phi(\mathbf{r}). \tag{8.205a}$$

Here

$$\frac{\hbar^2}{2m} U(\mathbf{r}) = V(\mathbf{r}),$$

the scattering (perturbing) potential. Since the probability of scattering is much less than one, the second term on the right-hand side of Eq. (8.205a) is expected to be negligible (relative to the first term on the right-hand side) and

---

[7] This assumes the Neumann series is convergent. In some physical situations it is not convergent and then other techniques are needed.

thus we drop it. Note that we are *approximating* our differential equation with

$$(\nabla^2 + k_0^2)\Phi(\mathbf{r}) = U(\mathbf{r})e^{ik_0z}. \tag{8.205b}$$

We now proceed to solve Eq. (8.205b), a nonhomogeneous differential equation. The differential operator ∇^2 generates a continuous set of eigenfunctions

$$\nabla^2\psi_\mathbf{k}(\mathbf{r}) = -k^2\psi_\mathbf{k}(\mathbf{r}), \tag{8.206}$$

where

$$\psi_\mathbf{k}(\mathbf{r}) = (2\pi)^{-3/2}e^{i\mathbf{k}\cdot\mathbf{r}}.$$

These plane wave eigenfunctions form a continuous but orthonormal set in the sense that

$$\int \psi_{\mathbf{k}_1}^*(\mathbf{r})\psi_{\mathbf{k}_2}(\mathbf{r})d^3r = \delta(\mathbf{k}_1 - \mathbf{k}_2)$$

(compare Eq. (15.21d).[8] We use these eigenfunctions to derive a Green's function.

We expand the unknown function $\Phi(\mathbf{r}_1)$ in these eigenfunctions,

$$\Phi(\mathbf{r}_1) = \int A_{\mathbf{k}_1}\psi_{\mathbf{k}_1}(\mathbf{r}_1)d^3k_1 \tag{8.207}$$

a Fourier integral with $A_{\mathbf{k}_1}$ the unknown coefficients. Substituting Eq. (8.207) into Eq. (8.205b) and using Eq. (8.206), we obtain

$$\int A_\mathbf{k}(k_0^2 - k^2)\psi_\mathbf{k}(\mathbf{r})d^3k = U(\mathbf{r})e^{ik_0z}. \tag{8.208}$$

Using the now familiar technique of multiplying by $\psi_{\mathbf{k}_2}^*(\mathbf{r})$ and integrating over the space coordinates, we have

$$\int A_{\mathbf{k}_1}(k_0^2 - k_1^2)d^3k_1 \int \psi_{\mathbf{k}_2}^*(\mathbf{r})\psi_{\mathbf{k}_1}(\mathbf{r})d^3r = A_{\mathbf{k}_2}(k_0^2 - k_2^2)$$

$$= \int \psi_{\mathbf{k}_2}^*(\mathbf{r})U(\mathbf{r})e^{ik_0z}d^3r. \tag{8.209}$$

Solving for $A_{\mathbf{k}_2}$ and substituting into Eq. (8.207) we have

$$\Phi(\mathbf{r}_2) = \int \left[(k_0^2 - k_2^2)^{-1} \int \psi_{\mathbf{k}_2}^*(\mathbf{r}_1)U(\mathbf{r}_1)e^{ik_0z_1}d^3r_1 \right]\psi_{\mathbf{k}_2}(\mathbf{r}_2)d^3k_2. \tag{8.210}$$

Hence

$$\Phi(\mathbf{r}_1) = \int \psi_{\mathbf{k}_1}(\mathbf{r}_1)(k_0^2 - k_1^2)^{-1}d^3k_1 \int \psi_{\mathbf{k}_1}^*(\mathbf{r}_2)U(\mathbf{r}_2)e^{ik_0z_2}d^3r_2, \tag{8.211}$$

---

[8] $d^3r = dx\,dy\,dz$, a (three-dimensional) volume element in **r**-space.

replacing $\mathbf{k}_2$ by $\mathbf{k}_1$ and $\mathbf{r}_1$ by $\mathbf{r}_2$ to agree with Eq. (8.207). Reversing the order of integration, we have

$$\Phi(\mathbf{r}_1) = -\int G_{k_0}(\mathbf{r}_1, \mathbf{r}_2) U(\mathbf{r}_2) e^{ik_0 z_2} d^3 r_2, \tag{8.212}$$

where $G_{k_0}(\mathbf{r}_1, \mathbf{r}_2)$, our Green's function, is given by

$$G_{k_0}(\mathbf{r}_1, \mathbf{r}_2) = \int \frac{\psi_{\mathbf{k}_1}^*(\mathbf{r}_2) \psi_{\mathbf{k}_1}(\mathbf{r}_1)}{k_1^2 - k_0^2} d^3 k_1 \tag{8.213}$$

analogous to Eq. (9.91) of Section 9.4 for *discrete* eigenfunctions. Equation (8.212) should be compared with the Green's function solution of Poisson's equation (8.157).

It is perhaps worth evaluating this integral to emphasize once more the vital role played by the boundary conditions. Using the eigenfunctions from Eq. (8.206) and

$$d^3 k = k^2 \, dk \sin \theta \, d\theta \, d\varphi,$$

we obtain

$$G_{k_0}(\mathbf{r}_1, \mathbf{r}_2) = \frac{1}{(2\pi)^3} \int_0^\infty \int_0^\pi \int_0^{2\pi} \frac{e^{ik\rho \cos \theta}}{k^2 - k_0^2} d\varphi \sin \theta \, d\theta \, k^2 \, dk. \tag{8.214}$$

Here $k\rho \cos \theta$ has replaced $\mathbf{k} \cdot (\mathbf{r}_1 - \mathbf{r}_2)$, with $\boldsymbol{\rho} = \mathbf{r}_1 - \mathbf{r}_2$ indicating the polar axis in k-space. Integrating over φ by inspection, we pick up a 2π. The θ-integration then leads to

$$G_{k_0}(\mathbf{r}_1, \mathbf{r}_2) = \frac{1}{4\pi^2 \rho i} \int_0^\infty \frac{(e^{ik\rho} - e^{-ik\rho})}{k^2 - k_0^2} k \, dk, \tag{8.215}$$

and since the integrand is an *even* function of k, we may set

$$G_{k_0}(\mathbf{r}_1, \mathbf{r}_2) = \frac{1}{8\pi^2 \rho i} \int_{-\infty}^\infty \frac{(e^{i\kappa} - e^{-i\kappa})}{\kappa^2 - \sigma^2} \kappa \, d\kappa. \tag{8.216}$$

The latter step is taken in anticipation of the evaluation of $G_k(\mathbf{r}_1, \mathbf{r}_2)$ as a contour integral. The symbols κ and σ $(\sigma > 0)$ represent $k\rho$ and $k_0\rho$, respectively.

If the integral in Eq. (8.216) is interpreted as a Riemann integral, the integral *does not exist*. This implies that $\mathcal{L}^{-1}$ does not exist, and in a literal sense it *does not*. $\mathcal{L} = \nabla^2 + k^2$ is singular since there exist nontrivial solutions ψ for which the homogeneous equation $\mathcal{L}\psi = 0$ (compare Exercise 3.5.6). We avoid this problem by introducing a parameter γ, defining a different operator $\mathcal{L}_\gamma^{-1}$ and taking the limit as $\gamma \to 0$.

Splitting the integral into two parts so each part may be written as a suitable contour integral gives us

$$G(\mathbf{r}_1, \mathbf{r}_2) = \frac{1}{8\pi^2 \rho i} \oint_{C_1} \frac{\kappa e^{i\kappa} \, d\kappa}{\kappa^2 - \sigma^2} + \frac{1}{8\pi^2 \rho i} \oint_{C_2} \frac{\kappa e^{-i\kappa} \, d\kappa}{\kappa^2 - \sigma^2}. \tag{8.217}$$

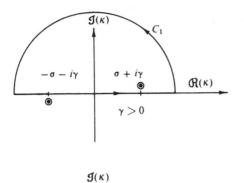

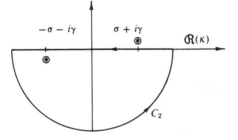

Figure 8.5 Possible Green's function contours of integration.

Contour C_1 is closed by a semicircle in the upper half-plane, C_2 by a semicircle in the lower half-plane. These integrals were evaluated in Chapter 7 by using appropriately chosen infinitesimal semicircles to go around the singular points $\kappa = \pm\sigma$. As an alternative procedure, let us first displace the singular points from the real axis by replacing σ by $\sigma + i\gamma$ and then, after evaluation, taking the limit as $\gamma \to 0$ (Fig. 8.5).

For γ *positive*, contour C_1 encloses the singular point $\kappa = \sigma + i\gamma$ and the first integral contributes

$$2\pi i \cdot \tfrac{1}{2} e^{i(\sigma + i\gamma)}.$$

From the second integral we also obtain

$$2\pi i \cdot \tfrac{1}{2} e^{i(\sigma + i\gamma)},$$

the enclosed singularity being $\kappa = -(\sigma + i\gamma)$. Returning to Eq. (8.217) and letting $\gamma \to 0$, we have

$$G(\mathbf{r}_1, \mathbf{r}_2) = \frac{1}{4\pi\rho} e^{i\sigma}$$

$$= \frac{e^{ik_0|\mathbf{r}_1 - \mathbf{r}_2|}}{4\pi|\mathbf{r}_1 - \mathbf{r}_2|}, \tag{8.218}$$

in full agreement with Exercise 8.7.16. This result depends on starting with γ positive. Had we chosen γ negative, our Green's function would have included

$e^{-i\sigma}$, which corresponds to an *incoming* wave. The choice of positive γ is dictated by the boundary conditions we wish to satisfy.

Equations (8.212) and (8.218) reproduce the scattered wave in Eq. (8.203b) and constitute an exact solution of the approximate Eq. (8.205b). Exercises 8.7.18 and 8.7.20 extend these results.

EXERCISES

8.7.1 Verify Eq. (8.168),

$$\int (v\mathcal{L}_2 u - u\mathcal{L}_2 v)\, d\tau_2 = \int p(v\nabla_2 u - u\nabla_2 v)\cdot d\boldsymbol{\sigma}_2.$$

8.7.2 Show that the terms $+k^2$ in the Helmholtz operator and $-k^2$ in the modified Helmholtz operator do not affect the behavior of $G(\mathbf{r}_1, \mathbf{r}_2)$ in the immediate vicinity of the singular point $\mathbf{r}_1 = \mathbf{r}_2$. Specifically, show that

$$\lim_{|\mathbf{r}_1 - \mathbf{r}_2| \to 0} \int k^2 G(\mathbf{r}_1, \mathbf{r}_2)\, d\tau_2 = 0.$$

8.7.3 Show that

$$\frac{\exp(ik|\mathbf{r}_1 - \mathbf{r}_2|)}{4\pi|\mathbf{r}_1 - \mathbf{r}_2|}$$

satisfies the two appropriate criteria and therefore is a Green's function for the Helmholtz equation.

8.7.4 (a) Find the Green's function for the three-dimensional Helmholtz equation, Exercise 8.7.3, when the wave is a standing wave.
(b) How is this Green's function related to the spherical Bessel functions?

8.7.5 The homogeneous Helmholtz equation

$$\nabla^2\varphi + \lambda^2\varphi = 0$$

has eigenvalues λ_i^2 and eigenfunctions φ_i. Show that the corresponding Green's function that satisfies

$$\nabla^2 G(\mathbf{r}_1, \mathbf{r}_2) + \lambda^2 G(\mathbf{r}_1, \mathbf{r}_2) = -\delta(\mathbf{r}_1 - \mathbf{r}_2)$$

may be written as

$$G(\mathbf{r}_1, \mathbf{r}_2) = \sum_{i=1}^{\infty} \frac{\varphi_i(\mathbf{r}_1)\varphi_i(\mathbf{r}_2)}{\lambda_i^2 - \lambda^2}.$$

An expansion of this form is called a bilinear expansion. If the Green's function is available in *closed form*, this provides a means of generating functions.

8.7.6 An electrostatic potential (mks units) is

$$\varphi(r) = \frac{Z}{4\pi\varepsilon_0}\cdot\frac{e^{-ar}}{r}.$$

Reconstruct the electrical charge distribution that will produce this potential.

Note that $\varphi(r)$ vanishes exponentially for large r, showing that the net charge is zero.

$$ANS. \quad \rho(r) = Z\delta(r) - \frac{Za^2}{4\pi}\frac{e^{-ar}}{r}.$$

8.7.7   Transform the differential equation

$$\frac{d^2y(r)}{dr^2} - k^2y(r) + V_0\frac{e^{-r}}{r}y(r) = 0,$$

and the boundary conditions $y(0) = y(\infty) = 0$ into a Fredholm integral equation of the form

$$y(r) = \lambda\int_0^\infty G(r, t)\frac{e^{-t}}{t}y(t)\, dt.$$

The quantities V_0 and k^2 are constants. The differential equation is derived from the Schrödinger wave equation with a meson potential.

$$\lambda = V_0$$

$$G(r, t) = \begin{cases} \dfrac{1}{k}e^{-kt}\sinh kr, & 0 \le r < t, \\[2mm] \dfrac{1}{k}e^{-kr}\sinh kt, & t < r < \infty. \end{cases}$$

8.7.8   A charged conducting ring of radius a (Example 12.3.3) may be described by

$$\rho(\mathbf{r}) = \frac{q}{2\pi a^2}\delta(r - a)\delta(\cos\theta).$$

Using the known Green's function for this system, find the electrostatic potential.
Hint. Exercise 12.6.3 will be helpful.

8.7.9   Changing a separation constant from k^2 to $-k^2$ and putting the discontinuity of the first derivative into the z-dependence, show that

$$\frac{1}{4\pi|\mathbf{r}_1 - \mathbf{r}_2|} = \frac{1}{4\pi}\sum_{m=-\infty}^{\infty}\int_0^\infty e^{im(\varphi_1-\varphi_2)}J_m(k\rho_1)J_m(k\rho_2)e^{-k|z_1-z_2|}\, dk.$$

Hint. The required $\delta(\rho_1 - \rho_2)$ may be obtained from Exercise 15.1.2.

8.7.10   Derive the expansion

$$\frac{\exp[ik|\mathbf{r}_1 - \mathbf{r}_2|]}{4\pi|\mathbf{r}_1 - \mathbf{r}_2|} = ik\sum_{l=0}^{\infty}\begin{cases} j_l(kr_1)h_l^{(1)}(kr_2) \\ j_l(kr_2)h_l^{(1)}(kr_1) \end{cases}.$$

$$\sum_{m=-l}^{l}Y_l^m(\theta_1\varphi_1)Y_l^{m*}(\theta_2, \varphi_2), \qquad \begin{matrix} r_1 < r_2 \\ r_1 > r_2. \end{matrix}$$

Hint. The left side is a known Green's function. Assume a spherical harmonic expansion and work on the remaining radial dependence. The spherical harmonic closure relation, Exercise 12.6.6, covers the angular dependence.

8.7.11 Show that the modified Helmholtz operator Green's function,

$$\frac{\exp{(-k|\mathbf{r}_1 - \mathbf{r}_2|)}}{4\pi|\mathbf{r}_1 - \mathbf{r}_2|}$$

has the spherical polar coordinate expansion

$$\frac{\exp(-k|\mathbf{r}_1 - \mathbf{r}_2|)}{4\pi|\mathbf{r}_1 - \mathbf{r}_2|} = k \sum_{l=0}^{\infty} i_l(kr_<)k_l(kr_>) \sum_{m=-l}^{l} Y_l^m(\theta_1, \varphi_1)Y_l^{m*}(\theta_2, \varphi_2).$$

Note. The modified spherical Bessel functions $i_l(kr)$ and $k_l(kr)$ are defined in Exercise 11.7.15.

8.7.12 From the spherical Green's function of Exercise 8.7.10, derive the plane wave expansion

$$e^{i\mathbf{k}\cdot\mathbf{r}} = \sum_{l=0}^{\infty} i^l(2l + 1)j_l(kr)P_l(\cos \gamma),$$

where γ is the angle included between $\mathbf{k}$ and $\mathbf{r}$. This is the Rayleigh equation of Exercise 12.4.7.

Hint. Take $r_2 \gg r_1$ so that

$$|\mathbf{r}_1 - \mathbf{r}_2| \to r_2 - \mathbf{r}_{20} \cdot \mathbf{r}_1 = r_2 - \frac{\mathbf{k} \cdot \mathbf{r}_1}{k}.$$

Let $r_2 \to \infty$ and cancel a factor of e^{ikr_2}/r_2.

8.7.13 From the results of Exercises 8.7.10 and 8.7.12, show that

$$e^{ix} = \sum_{l=0}^{\infty} i^l(2l + 1)j_l(x).$$

8.7.14 (a) From the circular cylindrical coordinate expansion of the Laplace Green's function (Eq. (8.197)), show that

$$\frac{1}{(\rho^2 + z^2)^{1/2}} = \frac{2}{\pi} \int_0^{\infty} K_0(k\rho) \cos kz \, dk.$$

This same result is obtained directly in Exercise 15.3.11.

(b) As a special case of part (a) show that

$$\int_0^{\infty} K_0(k) \, dk = \frac{\pi}{2}.$$

8.7.15 Noting that

$$\psi_k(\mathbf{r}) = \frac{1}{(2\pi)^{3/2}} e^{i\mathbf{k}\cdot\mathbf{r}}$$

is an eigenfunction of

$$(\nabla^2 + k^2)\psi_k(\mathbf{r}) = 0$$

(Eqs. (8.203) and (8.205)), show that the Green's function of $\mathcal{L} = \nabla^2$ may be expanded as

$$\frac{1}{4\pi|\mathbf{r}_1 - \mathbf{r}_2|} = \frac{1}{(2\pi)^3} \int e^{i\mathbf{k}\cdot(\mathbf{r}_1 - \mathbf{r}_2)} \frac{d^3k}{k^2}.$$

8.7.16   Using Fourier transforms, show that the Green's function satisfying the nonhomogeneous Helmholtz equation

$$(\nabla^2 + k_0^2)G(\mathbf{r}_1, \mathbf{r}_2) = -\delta(\mathbf{r}_1 - \mathbf{r}_2)$$

is

$$G(\mathbf{r}_1, \mathbf{r}_2) = \frac{1}{(2\pi)^3} \int \frac{e^{i\mathbf{k}\cdot(\mathbf{r}_1-\mathbf{r}_2)}}{k^2 - k_0^2} \, d^3k,$$

in agreement with Eq. (8.213).

8.7.17   The basic equation of the scalar Kirchhoff diffraction theory is

$$\psi(\mathbf{r}_1) = \frac{1}{4\pi} \int_{S_2} \left[\frac{e^{ikr}}{r} \nabla\psi(\mathbf{r}_2) - \psi(\mathbf{r}_2)\nabla\left(\frac{e^{ikr}}{r}\right) \right] \cdot d\boldsymbol{\sigma}_2,$$

where ψ satisfies the homogeneous Helmholtz equation and $r = |\mathbf{r}_1 - \mathbf{r}_2|$. Derive this equation. Assume that $\mathbf{r}_1$ is interior to the closed surface S_2.
Hint. Use Green's theorem.

8.7.18   The Born approximation for the scattered wave is given by Eq. (8.203b) (and Eq. (8.211)). From the asymptotic form, Eq. (8.199),

$$f_k(\theta, \varphi)\frac{e^{ikr}}{r} = -\frac{2m}{\hbar^2} \int V(\mathbf{r}_2) \frac{e^{ik|\mathbf{r}_1-\mathbf{r}_2|}}{4\pi|\mathbf{r} - \mathbf{r}_2|} e^{i\mathbf{k}_0\cdot\mathbf{r}_2} d^3r_2.$$

For the scattering potential $V(\mathbf{r}_2)$ independent of angles and for $r \gg r_2$, show that

$$f_k(\theta, \varphi) = -\frac{2m}{\hbar^2} \int_0^\infty r_2 V(r_2) \frac{\sin(|\mathbf{k}_0 - \mathbf{k}|r_2)}{|\mathbf{k}_0 - \mathbf{k}|} dr_2.$$

Here $\mathbf{k}_0$ is in the $\theta = 0$ (original z-axis) direction, whereas $\mathbf{k}$ is in the (θ, φ) direction. The magnitudes are equal: $|\mathbf{k}_0| = |\mathbf{k}|$. m is the reduced mass.
Hint. You have Exercise 8.7.12 to simplify the exponential and Exercise 15.3.20 to transform the three-dimensional Fourier exponential transform into a one-dimensional Fourier sine transform.

8.7.19   Calculate the scattering amplitude $f_k(\theta, \varphi)$ for a meson potential $V(r) = V_0(e^{-\alpha r}/\alpha r)$.
Hint. This particular potential permits the Born integral, Exercise 8.7.18 to be evaluated as a Laplace transform.

$$ANS. \quad f_k(\theta, \varphi) = -\frac{2mV_0}{\hbar^2\alpha}\frac{1}{\alpha^2 + (\mathbf{k}_0 - \mathbf{k})^2}.$$

8.7.20   The meson potential $V(r) = V_0(e^{-\alpha r}/\alpha r)$ may be used to describe the Coulomb scattering of two charges q_1 and q_2. We let $\alpha \to 0$ and $V_0 \to 0$ but take the ratio V_0/α to be $q_1 q_2/4\pi\varepsilon_0$. (For Gaussian units omit the $4\pi\varepsilon_0$.) Show that the differential scattering cross section $d\sigma/d\Omega = |f_k(\theta, \varphi)|^2$ is given by

$$\frac{d\sigma}{d\Omega} = \left(\frac{q_1 q_2}{4\pi\varepsilon_0}\right)^2 \frac{1}{16E^2 \sin^4(\theta/2)}, \qquad E = \frac{p^2}{2m} = \frac{\hbar^2 k^2}{2m}.$$

It happens (coincidentally) that this Born approximation is in exact agreement with both the exact quantum mechanical calculations and the classical Rutherford calculation.

8.8 NUMERICAL SOLUTIONS

The analytic solutions and approximate solutions to differential equations in this chapter and in succeeding chapters may suffice to solve the problem at hand—particularly if there is some symmetry present. The power-series solutions show how the solution behaves at small values of x. The asymptotic solutions (compare Sections 11.6 and 12.10) show how the solution behaves at large values of x. These limiting cases and also the possible resemblance of our differential equation to the standard forms with known solutions (Chapters 11 to 13) are invaluable in helping us gain an understanding of the general behavior of our solution.

However, the usual situation is that we have a *different* equation, perhaps a different potential in the Schrödinger wave equation, and we want a reasonably exact solution. So we turn to numerical techniques.

FIRST-ORDER DIFFERENTIAL EQUATIONS

The differential equation involves a continuity of points. The independent variable x is continuous. The (unknown) dependent variable $y(x)$ is assumed continuous. The concept of differentiation demands continuity. Our numerical processes replace these continua by discrete sets. We consider x at

$$x_0, \quad x_0 + h, \quad x_0 + 2h, \quad x_0 + 3h, \quad \text{and so on,}$$

where h is some small interval. The smaller h is, the better the approximation is—in principle. But if h is made too small, the demands on machine time will be excessive, and accuracy may actually decline because of accumulated round-off errors. We refer to the successive discrete values of x as x_n, x_{n+1}, and so on, and the corresponding values of $y(x)$ as $y(x_n) = y_n$. If x_0 and y_0 are given, the problem is to find y_1, then to find y_2, and so on.

TAYLOR SERIES SOLUTION

Consider the ordinary (possibly nonlinear) first-order differential equation

$$\frac{d}{dx} y(x) = f(x, y) \tag{8.219}$$

with the initial condition $y(x_0) = y_0$. In principle, a step-by-step solution of the first-order equation, Eq. (8.219), may be developed to any degree of accuracy by a Taylor expansion

$$y(x_0 + h) = y(x_0) + hy'(x_0) + \frac{h^2}{2!} y''(x_0) + \cdots + \frac{h^n}{n!} y^{(n)}(x_0) + \cdots, \tag{8.220}$$

(assuming the derivatives exist and the series is convergent). The initial value $y(x_0)$ is known and $y'(x_0)$ is given as $f(x_0, y_0)$. In principle, the higher derivatives may be obtained by differentiating $y'(x) = f(x, y)$. In practice, this differentiation may be tedious. Now, however, this differentiation can be done by computer, using symbolic languages such as Mathematica, Maple, or Reduce. For equations of the form encountered in this chapter a large computer has no trouble generating and evaluating ten or more derivatives.

The Taylor series solution is a form of analytic continuation, Section 6.5. If the right-hand side of Eq. (8.220) is truncated after two terms, we have

$$y_1 = y_0 + hy_0'$$

$$= y_0 + hf(x_0, y_0), \qquad (8.221)$$

neglecting the terms of order h^2. Equation (8.221) is often called the Euler solution. Clearly, it is subject to serious error with the neglect of terms of order h^2.

RUNGE–KUTTA METHOD

The Runge–Kutta method is a refinement of this, with an error of order h^5. The relevant formulas are

$$y_{n+1} = y_n + [k_0 + 2k_1 + 2k_2 + k_3]/6, \qquad (8.222)$$

where

$$k_0 = hf(x_n, y_n)$$

$$k_1 = hf(x_n + \tfrac{1}{2}h, y_n + \tfrac{1}{2}k_0),$$

$$k_2 = hf(x_n + \tfrac{1}{2}h, y_n + \tfrac{1}{2}k_1), \qquad (8.223)$$

$$k_3 = hf(x_n + h, y_n + k_2).$$

A derivation of these equations appears in Ralston and Wilf[1] (see Chapter 9 by M. J. Romanelli) and in Press et al.[2]

Equations (8.222) and (8.223) define what might be called the classic fourth-order Runge–Kutta method (accurate through terms of order h^4). This is the form followed in Sections 15.1 and 15.2 of Press et al. Many other Runge-Kutta methods exist. Lapidus and Seinfeld (see Additional Readings) analyze and compare other possibilities and recommend a fifth-order form due to Butcher as slightly superior to the classic method. But, for applications not demanding high precision and for not so smooth ODEs the fourth-order Runge-Kutta method *with adaptive step-size control* is and remains the method of choice for numerical solutions of ODEs.

The form of Eqs. (8.222) and (8.223) is assumed and the parameters adjusted to fit a Taylor expansion through h^4. From this Taylor expansion viewpoint the Runge–Kutta method is also an example of analytic continuation.

[1] A. Ralston and H. S. Wilf, Eds., *Mathematical Methods for Digital Computers*. New York: Wiley (1960).
[2] W. H. Press, B. P. Flannery, S. A. Teukolsky, and W. T. Vetterling, *Numerical Recipes*, 2nd ed. Cambridge: Cambridge University Press (1992).

For the special case in which dy/dx is a function of x alone [$f(x, y)$ in Eq. (8.219) $\rightarrow f(x)$], the last term in Eq. (8.222) reduces to a Simpson rule numerical integration from x_n to x_{n+1}.

The Runge–Kutta method is stable, meaning that small errors do not get amplified. It is self-starting, meaning that we just take the x_0 and y_0 and away we go. But it has disadvantages. Four separate calculations of $f(x, y)$ are required at each step. The errors, although of order h^5 per step, are not known. One checks the numerical solution by cutting h in half and repeating the calculation. If the second result agrees with the first, then h was small enough.

Finally, the Runge–Kutta method can be extended to a set of *coupled* first-order equations:

$$\frac{du}{dx} = f_1(x, u, v)$$

$$\frac{dv}{dx} = f_2(x, u, v), \qquad \text{and so on,}$$
(8.224)

with as many *dependent* variables as desired. Again, Eq. (8.224) may be nonlinear, an advantage of the numerical solution.

For high precision applications one can use either Richardson's extrapolation in conjuction with the Burlish–Stoer method,[3] or the predictor–corrector method to be described below. Richardson's extrapolation is based on approximating the numerical solution by a rational function which can then be evaluated in the limit of stepsize $h \rightarrow 0$. This often allows for a large actual stepsize in applications.

PREDICTOR–CORRECTOR METHODS

As an alternate attack on Eq. (8.219), we might estimate or *predict* a tentative value of y_{n+1} by

$$\bar{y}_{n+1} = y_{n-1} + 2hy'_n$$

$$= y_{n-1} + 2hf(x_n, y_n).$$
(8.225)

This is not quite the same as Eq. (8.221). Rather, it may be interpreted as

$$y'_n \approx \frac{\Delta y}{\Delta x} = \frac{y_{n+1} - y_{n-1}}{2h},$$
(8.226)

the derivative as a tangent being replaced by a chord. Next we *calculate*

$$y'_{n+1} = f(x_{n+1}, \bar{y}_{n+1}).$$
(8.227)

[3] See Section 15.4 of Press *et al.*, loc. cit., and J. Stoer and R. Bulirsch, *Introduction to Numerical Analysis*. New York: Springer-Verlag (1980), Chapter 7.

Then to *correct* for the crudeness of Eq. (8.225), we take

$$y_{n+1} = y_n + \frac{h}{2}(\bar{y}_{n+1} + y'_n). \tag{8.228}$$

Here the finite difference ratio $\Delta y/h$ is approximated by the average of the two derivatives. This technique—a prediction followed by a correction (and iteration until agreement is reached)—is the heart of the predictor–corrector method. It should be emphasized that the preceding set of equations is intended only to illustrate the predictor–corrector method. The accuracy of this set (to order h^3) is usually inadequate.

The iteration (substituting y_{n+1} from Eq. (8.228) back into Eq. (8.227) and recycling until y_{n+1} settles down to some limit) is time-consuming in a computing machine operation. Consequently, the iteration is usually replaced by an intermediate step (the *modifier*) between Eqs. (8.225) and (8.227).

This modified predictor–corrector method has the major advantage over the Runge–Kutta method of requiring only two computations of $f(x, y)$ per step, instead of four. Unfortunately, the method as originally developed was unstable—small errors (round-off and truncation) tended to propagate and become amplified.

This very serious problem of instability has been overcome in a version of the predictor–corrector method devised by Hamming. The formulas (which are moderately involved), a partial derivation, and detailed instructions for starting the solution are all given by Ralston (Chapter 8 of Ralston and Wilf). Hamming's method is accurate to order h^4. It is stable for all reasonable values of h and provides an estimate of the error. Unlike the Runge–Kutta method, it is *not* self-starting. For example, Eq. (8.225) requires both y_{n-1} and y_n. Starting values (y_0, y_1, y_2, y_3) for the Hamming predictor–corrector method may be computed by series solution (power series for small x, asymptotic series for large x) or by the Runge–Kutta method.

The Hamming predictor–corrector method may be extended to cover a set of coupled first-order differential equations, that is, Eq. (8.224).

SECOND-ORDER DIFFERENTIAL EQUATIONS

Any second-order differential equation

$$y''(x) + P(x)y'(x) + Q(x)y(x) = F(x), \tag{8.229}$$

may be split into two first-order differential equations by writing

$$y'(x) = z(x), \tag{8.230}$$

and then

$$z'(x) + P(x)z(x) + Q(x)y(x) = F(x). \tag{8.231}$$

These coupled first-order differential equations may be solved by either the Runge–Kutta or Hamming predictor–corrector techniques previously described.

As a final note, a thoughtless "turning the crank" application of these powerful numerical techniques is an invitation to disaster. The solution of a new and different differential equation will usually involve a mixture of analysis and numerical calculation. There is little point in trying to force a Runge–Kutta solution through a singular point where the solution is going to blow up. For a more extensive treatment of computational methods we refer the reader to Garcia in Additional Readings.

EXERCISES

8.8.1   The Runge–Kutta method, Eq. (8.222), is applied to a first-order differential equation $dy/dx = f(x)$. Note that this function $f(x)$ is independent of y. Show that in this special case the Runge–Kutta method reduces to Simpson's rule for numerical quadrature, Appendix A2.

8.8.2   (a) A body falling through a resisting medium is described by

$$\frac{dv}{dt} = g - av$$

(for a retarding force proportional to the velocity). Take the constants to be $g = 9.80 \, (\text{m/sec}^2)$ and $a = 0.2 \, (\text{sec}^{-1})$. The initial conditions are $t = 0$, $v = 0$. Integrate this equation out to $t = 20.0$ in steps of 0.1 sec. Tabulate the value of the velocity for each whole second, $v(1.0)$, $v(2.0)$, and so on. If a plotting routine is available, plot $v(t)$ versus t.

(b) Calculate the ratio of $v(20.0)$ to the terminal velocity $v(\infty)$.

Check value. $v(10) = 42.369 \, \text{m/sec}$.

ANS.   (b) 0.9817.

8.8.3   The differential equation for the population of a radioactive daughter element is

$$\frac{dN_2(t)}{dt} = \lambda_1 \exp(-\lambda_1 t) - \lambda_2 N_2,$$

$\lambda_1 \exp(-\lambda_1 t)$ being the rate of production resulting from the decay of the parent element, $\lambda_1 = 0.10 \, \text{sec}^{-1}$, $\lambda_2 = 0.08 \, \text{sec}^{-1}$. Integrate this differential equation from $t = 0$ out to $t = 40$ seconds for the initial condition $N_2(0) = 0$. Tabulate and plot $N_2(t)$ vs t.

8.8.4   The time-reversed asteroid depletion equation is

$$\frac{dN}{dt} = dN^2.$$

Solve this equation by using a Runge–Kutta or equivalent subroutine. The initial conditions are

$$t_0 = 0 \text{ (years)}$$
$$N_0 = 100 \text{ (asteroids)}$$
$$k = 0.25 \times 10^{-11} \text{ (years)}^{-1} \text{ (asteroid)}^{-1}.$$

Carry out your solution as far as you can. (There will be trouble as you approach $t = 5 \times 10^9$ years.) Tabulate $N(t)$ versus t, with $\Delta t = 5 \times 10^7$ years. *Note.* Exercise 8.2.3 (with k replaced by $-k$) gives the analytic solution.

8.8.5   Integrate Legendre's differential equation, Exercise 8.5.5, from $x = 0$ to $x = 1$ with the initial conditions $y(0) = 1$, $y'(0) = 0$ (even solution). Tabulate $y(x)$ and dy/dx at intervals of 0.05. Take $n = 2$.

8.8.6   The Lane–Emden equation of astrophysics is

$$\frac{d^2y}{dx^2} + \frac{2}{x}\frac{dy}{dx} + y^s = 0.$$

Take $y(0) = 1$, $y'(0) = 0$, and investigate the behavior of $y(x)$ for $s = 0, 1, 2, 3, 4, 5$, and 6. In particular, locate the first zero of $y(x)$.
Hint. From a power-series solution $y''(0) = -\frac{1}{3}$.
Note. For $s = 0$, $y(x)$ is a parabola, for $s = 1$, a spherical Bessel function, $j_0(x)$. As $s \to 5$, the first zero moves out to ∞, and for $s > 5$, $y(x)$ never crosses the positive x-axis.

$$\begin{aligned} ANS. \quad &\text{For } y(x_s) = 0, &x_0 &= 2.45(\sqrt{6}), \\ &x_1 = 3.14(\pi), &x_2 &= 4.35, \\ &x_3 = 6.90. \end{aligned}$$

8.8.7   As a check on Exercise 8.6.18(a), integrate Hermite's equation

$$\frac{d^2y}{dx^2} - 2x\frac{dy}{dx} = 0$$

from $x = 0$ out to $x = 3$. The initial conditions are $y(0) = 0$, $y'(0) = 1$. Tabulate $y(1)$, $y(2)$, and $y(3)$.

$$\begin{aligned} ANS \quad y(1) &= 1.463 \\ y(2) &= 16.45 \\ y(3) &= 1445. \end{aligned}$$

ADDITIONAL READINGS

BATEMAN, H. *Partial Differential Equations of Mathematical Physics.* New York: Dover (1944). 1st ed. (1932). A wealth of applications of various partial differential equations in classical physics. Excellent examples of the use of different coordinate systems—ellipsoidal, paraboloidal, toroidal coordinates, and so on.

COHEN, H., *Mathematics for Scientists and Engineers.* Englewood Cliffs, NJ: Prentice-Hall (1992).

COURANT, R., and D. HILBERT, *Methods of Mathematical Physics*, Vol. 1 (English edition). New York: Interscience (1953). This is one of the classic works of mathematical physics. Originally published in German in 1924, the revised English edition is an excellent reference for a rigorous treatment of Green's functions, and a wide variety of other topics on mathematical physics.

DAVIS, P. J., and P. RABINOWITZ, *Numerical Integration*. Waltham, MA: Blaisdell (1967). This book covers a great deal of material in a relatively easy-to-read form. Appendix 1 (*On the Practical Evaluation of Integrals* by M. Abramowitz) is excellent as an overall view.

GARCIA, A. L., *Numerical Methods for Physics*. Englewood Cliffs, NJ: Prentice-Hall (1994).

HAMMING, R. W., *Numerical Methods for Scientists and Engineers*, 2nd ed. New York: McGraw-Hill (1973). This well-written text discusses a wide variety of numerical methods from zeros of functions to the fast Fourier transform. All topics are selected and developed with a modern high-speed computer in mind.

HUBBARD, J., *Applied Differential Equations for Scientists and Engineers*. Boston: Computational Mechanics Publishers (1991).

INCE, E. L., *Ordinary Differential Equations*. New York: Dover (1926). The classic work in the theory of ordinary differential equations.

LAPIDUS, L., and J. H. SEINFELD, *Numerical Solutions of Ordinary Differential Equations*. New York: Academic Press (1971). A detailed and comprehensive discussion of numerical techniques with emphasis on the Runge–Kutta and predictor-corrector methods. Recent work on the improvement of characteristics such as stability is clearly presented.

MARGENAU, H., and G. M. MURPHY, *The Mathematics of Physics and Chemistry*, 2nd ed. Princeton, NJ: Van Nostrand (1956). Chapter 5 covers curvilinear coordinates and 13 specific coordinate systems.

MILLER, R. K., and A. N. MICHEL, *Ordinary Differential Equations*. New York: Academic Press (1982).

MORSE, P. M., and H. FESHBACH, *Methods of Theoretical Physics*. New York: McGraw-Hill (1953). Chapter 5 includes a description of several different coordinate systems. Note carefully that Morse and Feshbach are not above using left-handed coordinate systems even for Cartesian coordinates. Elsewhere in this excellent (and difficult) book there are many examples of the use of the various coordinate systems in solving physical problems. Eleven additional fascinating but seldom encountered orthogonal coordinate systems are discussed in the second edition (1970) of *Mathematical Methods for Physicists*. Chapter 7 is a particularly detailed, complete discussion of Green's functions from the point of view of mathematical physics. Note, however, that Morse and Feshbach frequently choose a source of $4\pi\delta(\mathbf{r} - \mathbf{r}')$ in place of our $\delta(\mathbf{r} - \mathbf{r}')$. Considerable attention is devoted to bounded regions.

MURPHY, G. M., *Ordinary Differential Equations and Their Solutions*. Princeton, NJ: Van Nostrand (1960). A thorough, relatively readable treatment of ordinary differential equations, both linear and nonlinear.

PRESS, W. H., B. P. FLANNERY, S. A. TEUCHOLSKY, and W. T. VETTERLING, *Numerical Recipes*, 2nd ed. Cambridge: Cambridge University Press (1992).

RALSTON, A., and H. WILF, eds., *Mathematical Methods for Digital Computers*. New York: Wiley (1960).

RITGER, P. D., and N. J. ROSE, *Differential Equations with Applications*. New York: McGraw-Hill (1968).

STAKGOLD, I., *Green's Functions and Boundary Value Problems*. New York: Wiley (1979).

STOER, J., and R. BURLIRSCH, *Introduction to Numerical Analysis*. New York: Springer-Verlag (1980).

STROUD, A. H., *Numerical Quadrature and Solution of Ordinary Differential Equations*, Applied Mathematics Series, Vol. 10. New York: Springer-Verlag (1974). A balanced, readable, and very helpful discussion of various methods of integrating differential equations. Stroud is familiar with recent work in this field and provides numerous current references.

9 STURM–LIOUVILLE THEORY—ORTHOGONAL FUNCTIONS

In the preceding chapter we developed two linearly independent solutions of the second-order linear homogeneous differential equation and proved that no third, linearly independent solution existed. In this chapter the emphasis shifts from solving the differential equation to developing and understanding general properties of the solutions. There is a close analogy between the concepts in this chapter and those of linear algebra in Chapter 3. Functions here play the role of vectors there, and linear operators that of matrices in Chapter 3. The diagonalization of a real symmetric matrix in Chapter 3 corresponds here to the solution of an ODE defined by an *adjoint* operator $\mathcal{L}$ in terms of its eigenfunctions, which are the ''continuous'' analog of the eigenvectors in Chapter 3.

In Section 9.1 the concepts of self-adjoint operator, eigenfunction, eigenvalue, and Hermitian operator are presented. The concept of adjoint operator, given first in terms of differential equations is then redefined in accordance with usage in quantum mechanics. The vital properties of reality of eigenvalues and orthogonality of eigenfunctions are derived in Section 9.2. In Section 9.3 we discuss the Gram–Schmidt procedure for systematically constructing sets of *orthogonal* functions. Finally, the general property of the completeness of a set of eigenfunctions is explored in Section 9.4, and Green's functions are continued in Section 9.5.

9.1 SELF-ADJOINT DIFFERENTIAL EQUATIONS

In Chapter 8 we studied, classified, and solved linear, second-order, differential equations corresponding to linear, second-order, differential operators of the general form

$$\mathcal{L}u(x) = p_0(x)\frac{d^2}{dx^2}u(x) + p_1(x)\frac{d}{dx}u(x) + p_2(x)u(x). \tag{9.1}$$

The *functions* $p_0(x)$, $p_1(x)$, and $p_2(x)$ are not to be confused with the constants p_i of Section 8.6. Reference to Eq. (8.117) shows that $P(x) = p_1(x)/p_0(x)$ and $Q(x) = p_2(x)/p_0(x)$.

These coefficients, $p_0(x)$, $p_1(x)$, and $p_2(x)$ are real functions of x and over the region of interest, $a \le x \le b$, the first $2 - i$ derivatives of $p_i(x)$ are continuous. Further, $p_0(x)$ does not vanish for $a < x < b$. Now, the zeros of $p_0(x)$ are singular points (Section 8.4), and the preceding statement simply means that we choose our interval $[a, b]$ so that there are no singular points in the interior of the interval. There may be and often are singular points on the boundaries.

For a linear operator $\mathcal{L}$ the analog of a quadratic form for a matrix in Chapter 3 is the integral

$$\langle u|\mathcal{L}|u\rangle \equiv \langle u|\mathcal{L}u\rangle \equiv \int_a^b u(x)\mathcal{L}u(x)dx$$

$$= \int_a^b u\{p_0 u'' + p_1 u' + p_2 u\}dx, \tag{9.2}$$

where the primes on the real function $u(x)$ denote derivatives as usual. If we shift the derivatives to the first factor u in Eq. (9.2) by integrating by parts once or twice, we are led to the equivalent expression

$$\langle u|\mathcal{L}|u\rangle = [u(x)(p_1 - p_0')u(x)]_{x=a}^b$$

$$+ \int_a^b \left\{\frac{d^2}{dx^2}[p_0 u] - \frac{d}{dx}[p_1 u] + p_2 u\right\}udx. \tag{9.3}$$

Comparing the integrands in Eqs. (9.2) and (9.3) yields

$$u(p_0'' - p_1')u + 2u(p_0' - p_1)u' = 0,$$

or

$$p_0'(x) = p_1(x) \tag{9.4}$$

and, as a bonus, the terms at the boundary $x = a$ and $x = b$ in Eq. (9.3) also vanish then.

Because of the analogy with the transposed matrix in Chapter 3, it is convenient to define the linear operator in Eq. (9.3)

$$\bar{\mathcal{L}}u = \frac{d^2}{dx^2}[p_0 u] - \frac{d}{dx}[p_1 u] + p_2 u$$

$$= p_0\frac{d^2 u}{dx^2} + (2p_0' - p_1)\frac{du}{dx} + (p_0'' - p_1' + p_2)u \tag{9.5}$$

as the *adjoint*[1] *operator* $\bar{\mathfrak{L}}$. The necessary and sufficient condition that $\bar{\mathfrak{L}} = \mathfrak{L}$, or $\langle u|\mathfrak{L}u\rangle = \langle\mathfrak{L}u|u\rangle$ upon rewriting Eq. (9.3), is that Eq. (9.4) be satisfied. When this condition is satisfied,

$$\bar{\mathfrak{L}}u = \mathfrak{L}u = \frac{d}{dx}\left[p(x)\frac{du(x)}{dx}\right] + q(x)u(x) \tag{9.6}$$

and the operator $\mathfrak{L}$ is said to be *self-adjoint*. Here, for the self-adjoint case, $p_0(x)$ is replaced by $p(x)$ and $p_2(x)$ by $q(x)$ to avoid unnecessary subscripts. The importance of the form of Eq. (9.6) is that we will be able to carry out two integrations by parts in Eq. (9.3) and Eq. (9.22) and following.[2]

In a survey of the differential equations introduced in Section 8.3, Legendre's equation and the linear oscillator equation are self-adjoint, but others, such as the Laguerre and Hermite equations, are not. However, the theory of linear, second-order, self-adjoint differential equations is perfectly general because we can *always* transform the non-self-adjoint operator into the required self-adjoint form. Consider Eq. (9.1) with $p_0' \neq p_1$. If we multiply $\mathfrak{L}$ by[3]

$$\frac{1}{p_0(x)}\exp\left[\int^x \frac{p_1(t)}{p_0(t)}\,dt\right],$$

we obtain

$$\frac{1}{p_0(x)}\exp\left[\int^x \frac{p_1(t)}{p_0(t)}\,dt\right]\mathfrak{L}u(x) = \frac{d}{dx}\left\{\exp\left[\int^x \frac{p_1(t)}{p_0(t)}\,dt\right]\frac{du(x)}{dx}\right\}$$

$$+ \frac{p_2(x)}{p_0(x)}\cdot\exp\left[\int^x \frac{p_1(t)}{p_0(t)}\,dt\right]u, \tag{9.7}$$

which is clearly self-adjoint (see Eq. (9.6)). Notice the $p_0(x)$ in the denominator. This is why we require $p_0(x) \neq 0$, $a < x < b$. In the following development we assume that $\mathfrak{L}$ has been put into self-adjoint form.

---

[1] The *adjoint* operator bears a somewhat forced relationship to the *adjoint* matrix. A better justification for the nomenclature is found in a comparison of the *self-adjoint* operator (plus appropriate boundary conditions) with the *self-adjoint* matrix. The significant properties are developed in Section 9.2. Because of these properties, we are interested in *self-adjoint* operators.

[2] The full importance of the self-adjoint form (plus boundary conditions) will become apparent in Section 9.2. In addition, self-adjoint forms will be required for developing Green's functions in Section 9.5.

[3] If we multiply $\mathfrak{L}$ by $f(x)/p_0(x)$ and then demand that

$$f'(x) = \frac{fp_1}{p_0},$$

so that the new operator will be self-adjoint, we obtain

$$f(x) = \exp\left[\int^x \frac{p_1(t)}{p_0(t)}\,dt\right].$$

EIGENFUNCTIONS, EIGENVALUES

Schrödinger's wave equation

$$H\psi(x) = E\psi(x)$$

is the major example of an eigenvalue equation in physics; here the differential operator $\mathcal{L}$ is defined by the Hamiltonian H and the eigenvalue becomes the total energy E of the system. The eigenfunction $\psi(x)$ is usually called a wave function. A variational derivation of this Schrödinger equation appears in Section 17.7. Based on spherical, cylindrical, or some other symmetry properties, a three- or four-dimensional PDE or eigenvalue equation such as the Schrödinger equation usually separates into eigenvalue equations in a single variable. Examples are Eqs. (8.41), (8.42), (8.50), and (8.53). However, sometimes an eigenvalue equation takes the more general self-adjoint form

$$\mathcal{L}u(x) + \lambda w(x)u(x) = 0, \tag{9.8}$$

where the constant λ is the eigenvalue and $w(x)$ is a known weight or density function; $w(x) > 0$ except possibly at isolated points at which $w(x) = 0$. For a given choice of the parameter λ, a function $u_\lambda(x)$, which satisfies Eq. (9.8) *and the imposed boundary conditions*, is called an *eigenfunction* corresponding to λ. The constant λ is then called an *eigenvalue*. There is no guarantee that an eigenfunction $u_\lambda(x)$ will exist for an arbitrary choice of the parameter λ. Indeed, the requirement that there be an eigenfunction often restricts the acceptable values of λ to a discrete set. Examples of this for the Legendre, Hermite, and Chebyshev equations appear in the excercises of Section 8.5. Here we have the mathematical approach to the process of quantization in quantum mechanics.

The extra weight function $w(x)$ appears sometimes as an asymptotic wave function ψ_∞ that is a common factor in all solutions of a PDE such as the Schrödinger equation, for example, when the potential $V(x) \to 0$ as $x \to \infty$ in $H = T + V$. We can find ψ_∞ when we set $V = 0$ in the Schrödinger equation. Another source for $w(x)$ may be a nonzero angular momentum barrier $l(l + 1)/x^2$ in a PDE or (8.65) that has a regular singularity and dominates at $x \to 0$. In such a case the indicial equation such as Eq. (8.87) or (8.103) shows that the wave function has x^l as an overall factor. Since the wave function enters twice in matrix elements and orthogonality relations, the weight functions in Table 9.1 come from these common factors in both radial wave functions. This is how the $\exp(-x)$ for Laguerre polynomials arises and $x^k \exp(-x)$ for associated Laguerre polynomials in Table 9.1.

Example 9.1.1 Legendre's Equation

Legendre's equation is given by

$$(1 - x^2)y'' - 2xy' + n(n + 1)y = 0. \tag{9.9}$$

Table 9.1

Equation	$p(x)$	$q(x)$	λ	$w(x)$
Legendre Shifted Legendre	$1 - x^2$	0	$l(l + 1)$	1
Shifted Legendre	$x(1 - x)$	0	$l(l + 1)$	1
Associated Legendre	$1 - x^2$	$-m^2/(1 - x^2)$	$l(l + 1)$	1
Chebyshev I	$(1 - x^2)^{1/2}$	0	n^2	$(1 - x^2)^{-1/2}$
Shifted Chebyshev I	$[x(1 - x)]^{1/2}$	0	n^2	$[x(1 - x)]^{-1/2}$
Chebyshev II	$(1 - x^2)^{3/2}$	0	$n(n + 2)$	$(1 - x^2)^{1/2}$
Ultraspherical (Gegenbauer)	$(1 - x^2)^{\alpha + 1/2}$	0	$n(n + 2\alpha)$	$(1 - x^2)^{\alpha - 1/2}$
Bessel*	x	$-\dfrac{n^2}{x}$	a^2	x
Laguerre	xe^{-x}	0	α	e^{-x}
Associated Laguerre	$x^{k+1}e^{-x}$	0	$\alpha - k$	$x^k e^{-x}$
Hermite	e^{-x^2}	0	2α	e^{-x^2}
Simple harmonic oscillator[†]	1	0	n^2	1

* Orthogonality of Bessel functions is rather special. Compare Section 11.2 for details. A second type of orthogonality is developed in Section 11.7.
[†] This will form the basis for Chapter 14, Fourier series.

From Eqs. (9.1) and (9.8)

$$p_0(x) = 1 - x^2 = p, \qquad w(x) = 1,$$

$$p_1(x) = -2x = p', \qquad \lambda = n(n + 1),$$

$$p_2(x) = 0 = q.$$

The reader will recall that our series solutions of Legendre's equation (Section 8.5)[4] diverged unless n was restricted to one of the integers. This represents a quantization of the eigenvalue λ.

When the equations of Chapter 8 are transformed into the self-adjoint form, we find the following values of the coefficients and parameters (Table 9.1).

The coefficient $p(x)$ is the coefficient of the second derivative of the eigenfunction and hopefully can be identified with no difficulty. The eigenvalue λ is the parameter that is available in a term of the form $\lambda w(x)y(x)$. Any x dependence apart from the eigenfunction becomes the weighting function $w(x)$. If there is another term containing the eigenfunction (not the derivatives), the coefficient of the eigenfunction in this additional term is identified as $q(x)$. If no such term is present, $q(x)$ is simply zero.

[4] Compare also Sections 5.2 and 12.10.

Example 9.1.2 Deuteron

Further insight into the concepts of eigenfunction and eigenvalue may be provided by an extremely simple model of the deuteron. The neutron–proton nuclear interaction is represented by a square well potential: $V = V_0 < 0$ for $0 \le r < a$, $V = 0$ for $r > a$. The Schrödinger wave equation is

$$-\frac{\hbar^2}{2M} \nabla^2 \psi + V\psi = E\psi. \tag{9.10}$$

With $\psi = \psi(r)$, we may write $u(r) = r\psi(r)$, and using Exercise 2.5.18, the wave equation becomes

$$\frac{d^2u}{dr^2} + k_1^2 u = 0, \tag{9.11}$$

with

$$k_1^2 = \frac{2M}{\hbar^2}(E - V_0) > 0 \tag{9.12}$$

for the interior range, $0 \le r < a$. Here M is the reduced mass of the neutron–proton system. For $a < r < \infty$, we have

$$\frac{d^2u}{dr^2} - k_2^2 u = 0, \tag{9.13}$$

with

$$k_2^2 = -\frac{2ME}{\hbar^2} > 0. \tag{9.14}$$

From the boundary condition that ψ remain finite, $u(0) = 0$ and

$$u_1(r) = \sin k_1 r, \qquad 0 \le r < a. \tag{9.15}$$

In the range outside the potential well, we have a linear combination of the two exponentials,

$$u_2(r) = A \exp k_2 r + B \exp(-k_2 r), \qquad a < r < \infty. \tag{9.16}$$

Continuity of particle density and current demand that $u_1(a) = u_2(a)$ and that $u_1'(a) = u_2'(a)$. These *joining conditions* give

$$\sin k_1 a = A \exp k_2 a + B \exp(-k_2 a),$$
$$k_1 \cos k_1 a = k_2 A \exp k_2 a - k_2 B \exp(-k_2 a). \tag{9.17}$$

The condition that we actually have one proton–neutron combination is that $\int \psi^* \psi \, d\tau = 1$. This constraint can be met if we impose a boundary condition that $\psi(r)$ remain finite as $r \to \infty$. And this, in turn, means that $A = 0$. Dividing the preceding pair of equations (to cancel B), we obtain

$$\tan k_1 a = -\frac{k_1}{k_2} = -\sqrt{\frac{E - V_0}{-E}}, \tag{9.18}$$

a transcendental equation for the energy E with only certain *discrete* solutions. If E is such that Eq. (9.18) can be satisfied, our solutions $u_1(r)$ and $u_2(r)$ can satisfy the boundary conditions. If Eq. (9.18) is not satisfied, *no acceptable solution exists*. The values of E for which Eq. (9.18) is satisfied are the eigenvalues; the corresponding functions u_1 and u_2 (or ψ) are the eigenfunctions. For the actual deuteron problem there is one (and only one) negative value of E satisfying Eq. (9.18), that is, the deuteron has one and only one bound state.

Now, what happens if E does *not* satisfy Eq. (9.18), if E is *not* an eigenvalue? In graphical form, imagine that E and therefore k_1 are varied slightly.

For $E = E_1 < E_0$, k_1 is reduced, and $\sin k_1 a$ has not turned down as much. The joining conditions, Eq. (9.17), require $A > 0$ and the wave function goes to $+\infty$, exponentially. For $E = E_2 > E_0$, k_1 is larger, $\sin k_1 a$ peaks sooner and is descending more rapidly at $r = a$. The joining conditions demand $A < 0$, and the wave function goes to $-\infty$, exponentially. Only for $E = E_0$, an eigenvalue, will the wave function have the required negative exponential asymptotic behavior (see Fig. 9.1).

BOUNDARY CONDITIONS

In the foregoing definition of eigenfunction, it was noted that the eigenfunction $u_\lambda(x)$ was required to satisfy certain imposed boundary conditions.

The term boundary conditions includes as a special case the concept of initial conditions. For instance, specifying the initial position x_0 and the initial

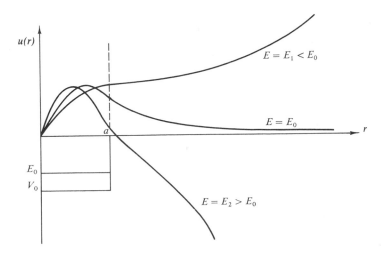

Figure 9.1 A deuteron eigenfunction.

velocity v_0 in some dynamical problem would correspond to the Cauchy boundary conditions. The only difference in the present usage of boundary conditions in these one-dimensional problems is that we are going to apply the conditions on *both* ends of the allowed range of the variable.

Usually the form of the differential equation or the boundary conditions on the solutions will guarantee that at the ends of our interval (that is, at the boundary) the following products will vanish:

$$p(x)v^*(x)\frac{du(x)}{dx}\bigg|_{x=a} = 0.$$

and   (9.19)

$$p(x)v^*(x)\frac{du(x)}{dx}\bigg|_{x=b} = 0.$$

Here $u(x)$ and $v(x)$ are solutions of the particular differential equation (Eq. (9.8)) being considered. We can, however, work with a somewhat less restrictive set of boundary conditions,

$$v^*pu'|_{x=a} = v^*pu'|_{x=b},$$ (9.20)

in which $u(x)$ and $v(x)$ are solutions of the differential equation corresponding to the same or to different eigenvalues. Equation (9.20) might well be satisfied if we were dealing with a periodic physical system such as a crystal lattice.

Equations (9.19) and (9.20) are written in terms of v^*, complex conjugate. When the solutions are real, $v = v^*$ and the asterisk may be ignored. However, in Fourier exponential expansions and in quantum mechanics the functions will be complex and the complex conjugate will be needed.

These properties (Eq. (9.19) or (9.20)) are so important for the concept of Hermitian operator (which follows) and the consequences (Section 9.2) that literally the interval (a, b) will be chosen to ensure that Eq. (9.19) or (9.20) is *satisfied*. If our solutions are polynomials, the coefficient $p(x)$ will determine the range of integration. Note that $p(x)$ also determines the singular points of the differential equation, Section 8.3. For nonpolynomial solutions, for example, $\sin nx$, $\cos nx$; ($p = 1$), the range of integration is determined by properties of the solutions—as in Example 9.1.3.

Example 9.1.3 Choice of Integration Interval, $[a, b]$

For $\mathcal{L} = d^2/dx^2$ a possible eigenvalue equation is

$$\frac{d^2}{dx^2}y(x) + n^2 y(x) = 0,$$ (9.21)

with eigenfunctions

$$u_n = \cos nx$$

$$v_m = \sin mx.$$

Equation (9.20) becomes

$$-n \sin mx \sin nx \, \big|_a^b = 0$$

or

$$m \cos mx \cos nx \, \big|_a^b = 0,$$

interchanging u_n and v_m. Since $\sin mx$ and $\cos nx$ are periodic with period 2π (for n and m integral), Eq. (9.20) is clearly satisfied if $a = x_0$ and $b = x_0 + 2\pi$.

The interval is chosen so that the boundary conditions (Eq. (9.20) etc.) are satisfied. For this case (Fourier series) the usual choices are $x_0 = 0$ leading to $(0, 2\pi)$ and $x_0 = -\pi$ leading to $(-\pi, \pi)$. Here and throughout the following several chapters *the integration interval is chosen so that the boundary conditions* (Eq. (9.20)) *will be satisfied.* The interval $[a, b]$ and the weighting factor $w(x)$ for the most commonly encountered second-order differential equations are listed in Table 9.2.

HERMITIAN OPERATORS

We now prove an important property of the combination self-adjoint, second-order differential operator (Eq. (9.8)), plus solutions $u(x)$ and $v(x)$ that satisfy boundary conditions given by Eq. (9.20).

By integrating v^* (complex conjugate) times the second-order self-adjoint differential operator $\mathcal{L}$ (operating on u) over the range $a \leq x \leq b$, we obtain

$$\int_a^b v^* \mathcal{L} u \, dx = \int_a^b v^* (pu')' \, dx + \int_a^b v^* qu \, dx \tag{9.22}$$

Table 9.2

Equation	a	b	$w(x)$
Legendre	-1	1	1
Shifted Legendre	0	1	1
Associated Legendre	-1	1	1
Chebyshev I	-1	1	$(1 - x^2)^{-1/2}$
Shifted Chebyshev I	0	1	$[x(1 - x)]^{-1/2}$
Chebyshev II	-1	1	$(1 - x^2)^{1/2}$
Laguerre	0	∞	e^{-x}
Associated Laguerre	0	∞	$x^k e^{-x}$
Hermite	$-\infty$	∞	e^{-x^2}
Simple harmonic oscillator	0	2π	1
	$-\pi$	π	1

Note.
1. The orthogonality interval $[a, b]$ is determined by the boundary conditions of Section 9.1.
2. The weighting function is established by putting the differential equation in self-adjoint form.

using Eq. (9.6). Integrating by parts, we have

$$\int_a^b v^*(pu')'\,dx = v^*pu'\Big|_a^b - \int_a^b v^{*'}pu'\,dx. \tag{9.23}$$

The integrated part vanishes on application of the boundary conditions (Eq. (9.20)). Integrating the remaining integral by parts a second time, we have

$$-\int_a^b v^{*'}pu'\,dx = -v^{*'}pu\Big|_a^b + \int_a^b u(pv^{*'})'\,dx. \tag{9.24}$$

Again, the integrated part vanishes in an application of Eq. (9.20). A combination of Eqs. (9.22)–(9.24) gives us

$$\int_a^b v^*\mathcal{L}u\,dx = \int_a^b u\mathcal{L}v^*\,dx. \tag{9.25}$$

This property, given by Eq. (9.25), is expressed by saying that the operator $\mathcal{L}$ is Hermitian with respect to the functions $u(x)$ and $v(x)$ which satisfy the boundary conditions specified by Eq. (9.20). Note carefully that this Hermitian property follows from self-adjointness plus boundary conditions.

HERMITIAN OPERATORS IN QUANTUM MECHANICS

The preceding development in this section has focused on the classical second-order differential operators of mathematical physics. Generalizing our Hermitian operator theory as required in quantum mechanics, we have an extension: The operators need be neither second-order differential operators nor real. $p_x = -i\hbar(\partial/\partial x)$ will be an Hermitian operator. We simply assume (as is customary in quantum mechanics) that the wave functions satisfy appropriate boundary conditions: vanishing sufficiently strongly at infinity or having periodic behavior (as in a crystal lattice, or unit intensity for waves). The operator $\mathcal{L}$ is called *Hermitian* if

$$\int \psi_1^*\mathcal{L}\psi_2\,d\tau = \int (\mathcal{L}\psi_1)^*\psi_2\,d\tau. \tag{9.26}$$

Apart from the simple extension to complex quantities, this definition is identical with Eq. (9.25).

The *adjoint* $A^\dagger$ of an operator A is defined by

$$\int \psi_1^*A^\dagger\psi_2\,d\tau \equiv \int (A\psi_1)^*\psi_2\,d\tau. \tag{9.27}$$

This generalizes our classical, second derivative operator-oriented definition, Eq. (9.5). Here the adjoint is defined in terms of the resultant integral, with the $A^\dagger$ as part of the integrand. Clearly, if $A = A^\dagger$ (self-adjoint) and satisfies the above mentioned boundary conditions, then A is Hermitian.

The *expectation value* of an operator $\mathcal{L}$ is defined as

$$\langle \mathcal{L} \rangle = \int \psi^* \mathcal{L} \psi \, d\tau. \tag{9.28a}$$

In the framework of quantum mechanics $\langle \mathcal{L} \rangle$ corresponds to the result of a measurement of the physical quantity represented by $\mathcal{L}$ when the physical system is in a state described by the wave function ψ. If we require $\mathcal{L}$ to be Hermitian, it is easy to show that $\langle \mathcal{L} \rangle$ is real (as would be expected from a measurement in a physical theory). Taking the complex conjugate of Eq. (9.28a), we obtain

$$\langle \mathcal{L} \rangle^* = \left[\int \psi^* \mathcal{L} \psi \, d\tau \right]^*$$

$$= \int \psi \mathcal{L}^* \psi^* \, d\tau.$$

Rearranging the factors in the integrand, we have

$$\langle \mathcal{L} \rangle^* = \int (\mathcal{L}\psi)^* \psi \, d\tau.$$

Then, applying our definition of Hermitian operator, Eq. (9.26), we get

$$\langle \mathcal{L} \rangle^* = \int \psi^* \mathcal{L} \psi \, d\tau = \langle \mathcal{L} \rangle, \tag{9.28b}$$

or $\langle \mathcal{L} \rangle$ is real. It is worth noting that ψ is not necessarily an eigenfunction of $\mathcal{L}$.

EXERCISES

9.1.1   Show that Laguerre's equation may be put into self-adjoint form by multiplying by e^{-x} and that $w(x) = e^{-x}$ is the weighting function.

9.1.2   Show that the Hermite equation may be put into self-adjoint form by multiplying by e^{-x^2} and that this gives $w(x) = e^{-x^2}$ as the appropriate density function.

9.1.3   Show that the Chebyshev equation (type I) may be put into self-adjoint form by multiplying by $(1 - x^2)^{-1/2}$ and that this gives $w(x) = (1 - x^2)^{-1/2}$ as the appropriate density function.

9.1.4   Show the following when the linear second-order differential equation is expressed in self-adjoint form:

(a) The Wronskian is equal to a constant divided by the initial coefficient p.

$$W[y_1, y_2] = \frac{C}{p(x)}.$$

(b) A second solution is given by

$$y_2(x) = Cy_1(x) \int^x \frac{dt}{p[y_1(t)]^2}.$$

9.1.5 $U_n(x)$, the Chebyshev polynomial (type II) satisfies the differential equation

$$(1 - x^2)U_n''(x) - 3xU_n'(x) + n(n + 2)U_n(x) = 0.$$

(a) Locate the singular points that appear in the *finite* plane and show whether they are regular or irregular.
(b) Put this equation in self-adjoint form.
(c) Identify the complete eigenvalue.
(d) Identify the weighting function.

9.1.6   For the very special case $\lambda = 0$ and $q(x) = 0$ the self-adjoint eigenvalue equation becomes

$$\frac{d}{dx}\left[p(x)\frac{du(x)}{dx} \right] = 0,$$

satisfied by

$$\frac{du}{dx} = \frac{1}{p(x)}.$$

Use this to obtain a "second" solution of the following:

(a) Legendre's equation,
(b) Laguerre's equation,
(e) Hermite's equation.

$$ANS \quad (a) \quad u_2(x) = \frac{1}{2}\ln\frac{1 + x}{1 - x},$$

$$(b) \quad u_2(x) - u_2(x_0) = \int_{x_0}^x e^t \frac{dt}{t},$$

$$(c) \quad u_2(x) = \int_0^x e^{t^2}\, dt.$$

These second solutions illustrate the divergent behavior usually found in a second solution.
Note. In all three cases $u_1(x) = 1$.

9.1.7   Given that $\mathcal{L}u = 0$ and $g\mathcal{L}u$ is self-adjoint, show that for the adjoint operator $\bar{\mathcal{L}}$, $\bar{\mathcal{L}}(gu) = 0$.

9.1.8   For a second-order differential operator $\mathcal{L}$ that is self-adjoint show that

$$\int_a^b [y_2\mathcal{L}y_1 - y_1\mathcal{L}y_2]\, dx = p(y_1'y_2 - y_1y_2')|_a^b.$$

9.1.9   Show that if a function ψ is required to satisfy Laplace's equation in a finite region of space and to satisfy Dirichlet boundary conditions over the entire closed bounding surface, then ψ is unique.
Hint. One of the forms of Green's theorem, Section 1.11, will be helpful.

9.1.10   Consider the solutions of the Legendre, Chebyshev, Hermite, and Laguerre equations to be polynomials. Show that the ranges of integration that guarantee that the Hermitian operator boundary conditions will be satisfied are

(a) Legendre $[-1, 1]$,
(b) Chebyshev $[-1, 1]$,
(c) Hermite $(-\infty, \infty)$,
(d) Laguerre $[0, \infty)$.

9.1.11 Within the framework of quantum mechanics (Eqs. (9.26) and following), show that the following are Hermitian operators:

(a) momentum $\mathbf{p} = -i\hbar \nabla \equiv -i\dfrac{h}{2\pi} \nabla$

(b) angular momentum $\mathbf{L} = -i\hbar \mathbf{r} \times \nabla \equiv -i\dfrac{h}{2\pi} \mathbf{r} \times \nabla$.

Hint. In Cartesian form $\mathbf{L}$ is a linear combination of noncommuting Hermitian operators.

9.1.12 (a) A is a non-Hermitian operator. In the sense of Eqs. (9.26) and (9.27), show that

$$A + A^\dagger \quad \text{and} \quad i(A - A^\dagger)$$

are Hermitian operators.

(b) Using the preceding result, show that every non-Hermitian operator may be written as a linear combination of two Hermitian operators.

9.1.13 U and V are two arbitrary operators, not necessarily Hermitian. In the sense of Eq. (9.27), show that

$$(UV)^\dagger = V^\dagger U^\dagger.$$

Note the resemblance to adjoint *matrices*.
Hint. Apply the definition of adjoint operator—Eq. (9.27).

9.1.14 Prove that the product of two Hermitian operators is Hermitian (Eq. (9.26)) if and only if the two operators commute.

9.1.15 A and B are noncommuting quantum mechanical operators:

$$AB - BA = iC.$$

Show that C is Hermitian. Assume that appropriate boundary conditions are satisfied.

9.1.16 The operator $\mathcal{L}$ is Hermitian. Show that $\langle \mathcal{L}^2 \rangle \geq 0$.

9.1.17 A quantum mechanical expectation value is defined by

$$\langle A \rangle = \int \psi^*(x) A \psi(x)\, dx,$$

where A is a linear operator. Show that demanding that $\langle A \rangle$ be real means that A must be Hermitian—with respect to $\psi(x)$.

9.1.18 From the definition of adjoint, Eq. (9.27), show that $A^{\dagger\dagger} = A$ in the sense that $\int \psi_1^* A^{\dagger\dagger} \psi_2\, d\tau = \int \psi_1^* A \psi_2\, d\tau$. The adjoint of the adjoint is the original operator.

Hint. The function ψ_1 and ψ_2 of Eq. (9.27) represent a class of functions. The subscripts 1 and 2 may be interchanged or replaced by other subscripts.

9.1.19 The Schrödinger wave equation for the deuteron (with a Woods–Saxon potential) is

$$-\frac{\hbar^2}{2M}\nabla^2\psi + \frac{V_0}{1 + \exp[(r - r_0)/a]}\psi = E\psi.$$

Here $E = -2.224$ MeV. a is a "thickness parameter," 0.4×10^{-13} cm. Expressing lengths in fermis (10^{-13} cm) and energies in million electron volts (MeV), we may rewrite the wave equation as

$$\frac{d^2}{dr^2}(r\psi) + \frac{1}{41.47}\left[E - \frac{V_0}{1 + \exp\left((r - r_0)/a\right)}\right](r\psi) = 0.$$

E is assumed known from experiment. The game is to find V_0 for a specified value of r_0 (say, $r_0 = 2.1$). If we let $y(r) = r\psi(r)$, then $y(0) = 0$ and we take $y'(0) = 1$. Find V_0 such that $y(20.0) = 0$. (This should be $y(\infty)$, but $r = 20$ is far enough beyond the range of nuclear forces to approximate infinity.)

ANS.   For $a = 0.4$ and $r_0 = 2.1$ fm, $V_0 = -34.159$ MeV.

9.1.20 Determine the nuclear potential well parameter V_0 of Exercise 9.1.19 as a function of r_0 for $r = 2.00(0.05)\ 2.25$ fermis.

Express your results as a power law

$$|V_0|r_0^{\nu} = k.$$

Determine the exponent ν and the constant k. This power law formulation is useful for accurate interpolation.

9.1.21 In Exercise 9.1.19 it was assumed that 20 fermis was a good approximation to infinity. Check on this by calculating V_0 for $r\psi(r) = 0$ at (a) $r = 15$, (b) $r = 20$, (c) $r = 25$, and (d) $r = 30$. Sketch your results. Take $r_0 = 2.10$ and $a = 0.4$ (fermis).

9.1.22 For a quantum particle moving in a potential well, $V(x) = \frac{1}{2}m\omega^2x^2$, the Schrödinger wave equation is

$$-\frac{\hbar^2}{2m}\frac{d^2\psi(x)}{dx^2} + \frac{1}{2}m\omega^2x^2\psi(x) = E\psi(x),$$

or

$$\frac{d^2\psi(z)}{dz^2} - z^2\psi(z) = -\frac{2E}{\hbar\omega}\psi(z),$$

where $z = (m\omega/\hbar)^{1/2}x$. Since this operator is even, we expect solutions of definite parity. For the initial conditions that follow integrate out from the origin and determine the minimum constant $2E/\hbar\omega$ that will lead to $\psi(\infty) = 0$ in each case. (You may take $z = 6$ as an approximation of infinity.)

(a) For an even eigenfunction,

$$\psi(0) = 1, \qquad \psi'(0) = 0.$$

(b) For an odd eigenfunction

$$\psi(0) = 0, \qquad \psi'(0) = 1.$$

Note. Analytical solutions appear in Section 13.1.

9.2 HERMITIAN OPERATORS

Hermitian or self-adjoint operators with appropriate boundary conditions have three properties that are of extreme importance in physics, both classical and quantum.

1. The eigenvalues of an Hermitian operator are real.
2. The eigenfunctions of an Hermitian operator are orthogonal.
3. The eigenfunctions of an Hermitian operator form a complete set.[1]

REAL EIGENVALUES

We proceed to prove the first two of these three properties. Let

$$\mathcal{L}u_i + \lambda_i w u_i = 0. \tag{9.29}$$

Assuming the existence of a second eigenvalue and eigenfunction

$$\mathcal{L}u_j + \lambda_j w u_j = 0. \tag{9.30}$$

Then, taking the complex conjugate, we obtain

$$\mathcal{L}u_j^* + \lambda_j^* w u_j^* = 0. \tag{9.31}$$

Here $\mathcal{L}$ is a real operator (p and q are real functions of x) and $w(x)$ is a real function. But we permit λ_k, the eigenvalues, and u_k, the eigenfunctions, to be complex. Multiplying Eq. (9.29) by u_j^* and Eq. (9.31) by u_i and then subtracting, we have

$$u_j^* \mathcal{L}u_i - u_i \mathcal{L}u_j^* = (\lambda_j^* - \lambda_i) w u_i u_j^*. \tag{9.32}$$

We integrate over the range $a \leq x \leq b$,

$$\int_a^b u_j^* \mathcal{L}u_i \, dx - \int_a^b u_i \mathcal{L}u_j^* \, dx = (\lambda_j^* - \lambda_i) \int_a^b u_i u_j^* w \, dx. \tag{9.33}$$

Since $\mathcal{L}$ is Hermitian, the left-hand side vanishes by Eq. (9.27) and

$$(\lambda_j^* - \lambda_i) \int_a^b u_i u_j^* w \, dx = 0. \tag{9.34}$$

If $i = j$, the integral cannot vanish [$w(x) > 0$, apart from isolated points], except in the trivial case $u_i = 0$. Hence the coefficient $(\lambda_i^* - \lambda_i)$ must be zero,

$$\lambda_i^* = \lambda_i, \tag{9.35}$$

which is a mathematical statement that the eigenvalue is real. Since λ_i can represent any one of the eigenvalues, this proves the first property. This is an

---

[1] This third property is not universal. It *does* hold for our linear, second-order differential operators in Sturm–Liouville (self-adjoint) form. Completeness is defined and discussed in Section 9.4. A proof that the eigenfunctions of our linear, second-order, self-adjoint, differential equations form a complete set may be developed from the calculus of variations of Section 17.8.

exact analog of the nature of the eigenvalues of real symmetric (and of Hermitian) matrices (compare Section 3.4).

This reality of the eigenvalues of Hermitian operators has a fundamental significance in quantum mechanics. In quantum mechanics the eigenvalues correspond to precisely measurable quantities, such as energy and angular momentum. With the theory formulated in terms of Hermitian operators, this proof of the reality of the eigenvalues guarantees that the theory will predict real numbers for these measurable physical quantities. In Section 17.8 it will be seen that the set of real eigenvalues has a lower bound.

ORTHOGONAL EIGENFUNCTIONS

If we now take $i \neq j$ and if $\lambda_i \neq \lambda_j$, the integral of the product of the two different eigenfunctions must vanish.

$$\int_a^b u_i u_j^* w \, dx = 0. \tag{9.36}$$

This condition, called orthogonality, is the continuum analog of the vanishing of a scalar product of two vectors.[2] We say that the eigenfunctions $u_i(x)$ and $u_j(x)$ are orthogonal with respect to the weighting function $w(x)$ over the interval $[a, b]$. Equation (9.36) constitutes a partial proof of the second property of our Hermitian operators. Again, the precise analogy with matrix analysis should be noted. Indeed, we can establish a one-to-one correspondence between this Sturm–Liouville theory of differential equations and the treatment of Hermitian matrices. Historically, this correspondence has been significant in establishing the mathematical equivalence of matrix mechanics developed by Heisenberg and wave mechanics developed by Schrödinger. Today, the two diverse approaches are merged into the theory of quantum mechanics and the mathematical formulation that is more convenient for a particular problem is used for that problem. Actually the mathematical alternatives do not end here. Integral equations, Chapter 16, form a third equivalent and sometimes more convenient or more powerful approach.

This proof of orthogonality is not quite complete. There is a loophole, because we may have $i \neq j$ but still have $\lambda_i = \lambda_j$. Such a case is labeled degenerate. Illustrations of degeneracy are given at the end of this section. If $\lambda_i = \lambda_j$, the integral in Eq. (9.34) need not vanish. This means that linearly independent eigenfunctions corresponding to the same eigenvalue are not automatically orthogonal and that some other method must be sought to

---

[2] From the definition of Riemann integral

$$\int_a^b f(x)g(x) \, dx = \lim_{N \to \infty} \left(\sum_{i=1}^N f(x_i)g(x_i) \, \Delta x \right),$$

where $x_0 = a$, $x_N = b$, and $x_i - x_{i-1} = \Delta x$. If we interpret $f(x_i)$ and $g(x_i)$ as the ith components of an N component vector, then this sum (and therefore this integral) corresponds directly to a scalar product of vectors, Eq. (1.24). The vanishing of the scalar product is the condition for *orthogonality* of the vectors—or functions.

obtain an orthogonal set. Although the eigenfunctions in this degenerate case may not be orthogonal, they can always be made orthogonal. One method is developed in the next section.

We shall see in succeeding chapters that it is just as desirable to have a given set of functions orthogonal as it is to have an orthogonal coordinate system. We can work with nonorthogonal functions, but they are likely to prove as messy as an oblique coordinate system.

Example 9.2.1 Fourier Series—Orthogonality

Continuing Example 9.1.3, the eigenvalue equation, Eq. (9.21),

$$\frac{d^2}{dx^2} y(x) + n^2 y(x) = 0,$$

perhaps describes a quantum mechanical particle in a box, perhaps a vibrating violin string with (degenerate) eigenfunctions—$\cos nx$, $\sin nx$.

With n real (here taken to be integral), the orthogonality integrals become

(a) $\displaystyle\int_{x_0}^{x_0+2\pi} \sin mx \sin nx \, dx = C_n \delta_{nm},$

(b) $\displaystyle\int_{x_0}^{x_0+2\pi} \cos mx \cos nx \, dx = D_n \delta_{nm},$

(c) $\displaystyle\int_{x_0}^{x_0+2\pi} \sin mx \cos nx \, dx = 0.$

For an interval of 2π the preceding analysis guarantees the Kronecker delta in (a) and (b) but not the zero in (c) because (c) involves degenerate eigenfunctions. However, inspection shows that (c) always vanishes for all integral m and n.

Our Sturm–Liouville theory says nothing about the values of C_n and D_n. Actual calculation yields

$$C_n = \begin{cases} \pi, & n \neq 0, \\ 0, & n = 0, \end{cases}$$

$$D_n = \begin{cases} \pi, & n \neq 0, \\ 2\pi, & n = 0. \end{cases}$$

These orthogonality integrals form the basis of the Fourier series developed in Chapter 14.

Example 9.2.2 Expansion in Orthogonal Eigenfunctions—Square Wave

The property of completeness means that certain classes of function (i.e., sectionally or piecewise continuous) may be represented by a series of orthogonal eigenfunctions to any desired degree of accuracy. Consider the square wave

$$f(x) = \begin{cases} \dfrac{h}{2}, & 0 < x < \pi, \\[2mm] -\dfrac{h}{2}, & -\pi < x < 0. \end{cases} \tag{9.37}$$

This function may be expanded in any of a variety of eigenfunctions— Legendre, Hermite, Chebyshev, and so on. The choice of eigenfunction is made on the basis of convenience. To illustrate the expansion technique, let us choose the eigenfunctions of Example 9.2.1, $\cos nx$ and $\sin nx$.

The eigenfunction series is conveniently (and conventionally) written as

$$f(x) = \frac{a_0}{2} + \sum_{n=1}^{\infty} (a_n \cos nx + b_n \sin nx).$$

From the orthogonality integrals of Example 9.2.1 the coefficients are given by

$$a_n = \frac{1}{\pi} \int_{-\pi}^{\pi} f(t) \cos nt \, dt,$$

$$b_n = \frac{1}{\pi} \int_{-\pi}^{\pi} f(t) \sin nt \, dt, \qquad n = 0, 1, 2, \ldots.$$

Direct substitution of $\pm h/2$ for $f(t)$ yields

$$a_n = 0,$$

which is expected here because of the antisymmetry, and

$$b_n = \frac{h}{n\pi} (1 - \cos n\pi) = \begin{cases} 0, & n \text{ even}, \\[2mm] \dfrac{2h}{n\pi}, & n \text{ odd}. \end{cases}$$

Hence the eigenfunction (Fourier) expansion of the square wave is

$$f(x) = \frac{2h}{\pi} \sum_{n=0}^{\infty} \frac{\sin(2n + 1)x}{(2n + 1)}. \tag{9.38}$$

Additional examples, using other eigenfunctions, appear in Chapters 11 and 12.

Degeneracy

The concept of degeneracy was introduced earlier. If N linearly independent eigenfunctions correspond to the same eigenvalue, the eigenvalue is said to be N-fold degenerate. A particularly simple illustration is provided by the eigenvalues and eigenfunctions of the linear oscillator equation, Example 9.2.1. For each value of the eigenvalue n, there are two possible solutions: $\sin nx$ and $\cos nx$ (and any linear combination). We may say the eigenfunctions are degenerate or the eigenvalue is degenerate.

A more involved example is furnished by the physical system of an electron in an atom (nonrelativistic treatment, spin neglected). From the Schrödinger equation, Eq. (13.53) for hydrogen, the total energy of the electron is our eigenvalue. We may label it E_{nLM} by using the quantum numbers n, L, and M as subscripts. For each distinct set of quantum numbers (n, L, M) there is a distinct, linearly independent eigenfunction $\psi_{nLM}(r, \theta, \varphi)$. For hydrogen, the energy E_{nLM} is independent of L and M. With $0 \le L \le n - 1$ and $-L \le M \le L$, the eigenvalue is n^2-fold degenerate (including the electron spin would raise this to $2n^2$). In atoms with more than one electron the electrostatic potential is no longer a simple r^{-1} potential. The energy depends on L as well as on n, although *not* on M. E_{nLM} is still $(2L + 1)$-fold degenerate. This degeneracy may be removed by applying an external magnetic field, giving rise to the Zeeman effect.

Often an underlying symmetry, such as rotational invariance, is causing the degeneracies. States belonging to the same energy eigenvalue then will form a multiplet or representation of the symmetry group. The powerful group-theoretical methods are treated in Chapter 4 in some detail.

EXERCISES

9.2.1 The functions $u_1(x)$ and $u_2(x)$ are eigenfunctions of the same Hermitian operator but for distinct eigenvalues λ_1 and λ_2. Prove that $u_1(x)$ and $u_2(x)$ are linearly independent.

9.2.2 (a) The vectors $\mathbf{e}_n$ are orthogonal to each other: $\mathbf{e}_n \cdot \mathbf{e}_m = 0$ for $n \ne m$. Show that they are linearly independent.

(b) The functions $\psi_n(x)$ are orthogonal to each other over the interval $[a, b]$ and with respect to the weighting function $w(x)$. Show that the $\psi_n(x)$ are linearly independent.

9.2.3 Given that

$$P_1(x) = x \quad \text{and} \quad Q_0(x) = \tfrac{1}{2} \ln\left(\frac{1 + x}{1 - x}\right)$$

are solutions of Legendre's differential equation corresponding to *different* eigenvalues:

(a) Evaluate their orthogonality integral

$$\int_{-1}^{1} \frac{x}{2} \ln\left(\frac{1 + x}{1 - x}\right) dx.$$

(b) Explain why these two functions are *not* orthogonal, why the proof of orthogonality does not apply.

9.2.4 $T_0(x) = 1$ and $V_1(x) = (1 - x^2)^{1/2}$ are solutions of the Chebyshev differential equation corresponding to *different* eigenvalues. Explain, in terms of the boundary conditions, why these two functions are *not* orthogonal.

9.2.5   (a) Show that the first derivatives of the Legendre polynomials satisfy a self-adjoint differential equation with eigenvalue $\lambda = n(n + 1) - 2$.

(b) Show that these Legendre polynomial derivatives satisfy an orthogonality relation

$$\int_{-1}^{1} P_m'(x)P_n'(x)(1 - x^2)\, dx = 0, \qquad m \neq n.$$

Note. In Section 12.5, $(1 - x^2)^{1/2}P_n'(x)$ will be labeled an associated Legendre polynomial, $P_n^1(x)$.

9.2.6   A set of functions $u_n(x)$ satisfies the Sturm–Liouville equation

$$\frac{d}{dx}\left[p(x)\frac{d}{dx}u_n(x) \right] + \lambda_n w(x)u_n(x) = 0.$$

The functions $u_m(x)$ and $u_n(x)$ satisfy boundary conditions that lead to orthogonality. The corresponding eigenvalues λ_m and λ_n are distinct. Prove that for appropriate boundary conditions $u_m'(x)$ and $u_n'(x)$ are orthogonal with $p(x)$ as a weighting function.

9.2.7   A linear operator A has n distinct eigenvalues and n corresponding eigenfunctions. $A\psi_i = \lambda_i \psi_i$. Show that the n eigenfunctions are linearly independent. A is not necessarily Hermitian.

Hint. Assume linear dependence, that $\psi_n = \sum_{i=1}^{n-1}a_i\psi_i$. Use this relation and the operator–eigenfunction equation first in one order, then in the reverse order. Show that a contradiction results.

9.2.8   (a) Show that the Liouville substitution

$$u(x) = v(\xi)[p(x)w(x)]^{-1/4}, \qquad \xi = \int_a^x [w(t)/p(t)]^{1/2}\, dt$$

transforms

$$\frac{d}{dx}\left[p(x)\frac{d}{dx}u \right] + [\lambda w(x) - q(x)]u(x) = 0$$

into

$$\frac{d^2v}{d\xi^2} + [\lambda - Q(\xi)]v(\xi) = 0,$$

where

$$Q(\xi) = \frac{q(x(\xi))}{w(x(\xi))} + [p(x(\xi))w(x(\xi))]^{-1/4}\frac{d^2}{d\xi^2}(pw)^{1/4}.$$

(b) If $v_1(\xi)$ and $v_2(\xi)$ are obtained from $u_1(x)$ and $u_2(x)$, respectively, by a Liouville substitution, show that $\int_a^b w(x)u_1 u_2\, dx$ is transformed into $\int_0^c v_1(\xi)v_2(\xi)\, d\xi$ with $c = \int_a^b [w/p]^{1/2}\, dx$.

9.2.9   The ultraspherical polynomials $C_n^{(\alpha)}(x)$ are solutions of the differential equation

$$\left\{ (1 - x^2)\frac{d^2}{dx^2} - (2\alpha + 1)x\frac{d}{dx} + n(n + 2\alpha) \right\}C_n^{(\alpha)}(x) = 0.$$

(a) Transform this differential equation into self-adjoint form.

(b) Show that the $C_n^{(\alpha)}(x)$ are orthogonal for different n. Specify the interval of integration and the weighting factor.

Note. Assume that your solutions are polynomials.

9.2.10   With $\mathcal{L}$ *not* self-adjoint,

$$\mathcal{L}u_i + \lambda_i w u_i = 0$$

and

$$\bar{\mathcal{L}}v_j + \lambda_j w v_j = 0.$$

(a) Show that

$$\int_a^b v_j \mathcal{L}u_i \, dx = \int_a^b u_i \bar{\mathcal{L}}v_j \, dx,$$

provided

$$u_i p_0 v_j' \bigg|_a^b = v_j p_0 u_i' \bigg|_a^b$$

and

$$u_i (p_1 - p_0') v_j \bigg|_a^b = 0.$$

(b) Show that the orthogonality integral for the eigenfunctions u_i and v_j becomes

$$\int_a^b u_i v_j w \, dx = 0 \qquad (\lambda_i \neq \lambda_j).$$

9.2.11   In Exercise 8.5.8 the series solution of the Chebyshev equation is found to be convergent for all n. Therefore n is *not* quantized by the argument used for Legendre's (Exercise 8.5.5). Calculate the sum of the $k = 0$ Chebyshev series for $n = v = 0.8$, 0.9, and 1.0 and for $x = 0.0(0.1)0.9$.
Note. The Chebyshev series recurrence relation is given in Exercise 5.2.16.

9.2.12   (a) Evaluate the $n = v = 0.9$, $k = 0$ Chebyshev series for $x = 0.98$, 0.99, and 1.00. The series converges very slowly at $x = 1.00$. You may wish to use double precision. Upper bounds to the error in your calculation can be set by comparison with the $v = 1.0$ case which corresponds to $(1 - x^2)^{1/2}$.
(b) These series solutions for $v = 0.9$ and for $v = 1.0$ are obviously *not* orthogonal despite the fact that they satisfy a self-adjoint eigenvalue equation with different eigenvalues. From the behavior of the solutions in the vicinity of $x = 1.00$ try to formulate a hypothesis as to why the proof of orthogonality does not apply.

9.2.13   The Fourier expansion of the (asymmetric) square wave is given by Eq. (9.38). With $h = 2$, evaluate this series for $x = 0(\pi/18)\pi/2$, using the first (a) 10 terms, (b) 100 terms of the series.
Note. For 10 terms and $x = \pi/18$ or $10°$ your Fourier representation has a sharp hump. This is the Gibbs phenomenon of Section 14.5. For 100 terms this hump has been shifted over to about $1°$.

9.2.14   The *symmetric* square wave

$$f(x) = \begin{cases} 1, & |x| < \dfrac{\pi}{2} \\ -1, & \dfrac{\pi}{2} < |x| < \pi \end{cases}$$

has a Fourier expansion

$$f(x) = \frac{4}{\pi} \sum_{n=0}^{\infty} (-1)^n \frac{\cos(2n+1)x}{2n+1}.$$

Evaluate this series for $x = 0(\pi/18)\pi/2$ using the first

(a) 10 terms,
(b) 100 terms of the series.

Note. As in Exercise 9.2.13, the Gibbs phenomenon appears at the discontinuity. This means that a Fourier series is not suitable for precise numerical work in the vicinity of a discontinuity.

9.3 GRAM–SCHMIDT ORTHOGONALIZATION

The Gram–Schmidt orthogonalization is a method that takes a non-orthogonal set of linearly independent functions[1] and literally constructs an orthogonal set over an arbitrary interval and with respect to an arbitrary weight or density factor. In the language of linear algebra the process is equivalent to a matrix transformation relating an orthogonal set of basis vectors (functions) to a nonorthogonal set. A specific example of this matrix transformation appears in Exercise 12.2.1. The functions involved may be real or complex. Here for convenience they are assumed to be real. The generalization to the complex case should offer little difficulty.

Before taking up orthogonalization, we should consider normalization of functions. So far no normalization has been specified. This means that

$$\int_a^b \varphi_i^2 w \, dx = N_i^2,$$

but no attention has been paid to the value of N_i. Since our basic equation (Eq. (9.8)), is linear and homogeneous, we may multiply our solution by any constant and it will still be a solution. We now demand that each solution $\varphi_i(x)$ be multiplied by N_i^{-1} so that the new (normalized) φ_i will satisfy

$$\int_a^b \varphi_i^2(x) w(x) \, dx = 1 \tag{9.39}$$

or

$$\int_a^b \varphi_i(x)\varphi_j(x) w(x) \, dx = \delta_{ij}. \tag{9.40}$$

[1] Such a set of functions might well arise from the solutions of a (partial) differential equation in which the eigenvalue was independent of one or more of the constants of separation. As an example, we have the hydrogen atom problem (Sections 9.2 and 13.2). The eigenvalue (energy) is independent of both the electron orbital angular momentum and its projection on the z-axis, m. The student should note, however, that the origin of the set of functions is irrelevant to the Gram–Schmidt orthogonalization procedure.

Equation (9.39) says that we have normalized to unity. Including the property of orthogonality, we have Eq. (9.40). Functions satisfying this equation are said to be orthonormal (orthogonal plus unit normalization). It should be emphasized that other normalizations (or a minus sign) are possible, and indeed, by historical convention, each of the special functions of mathematical physics treated in Chapters 12 and 13 will be normalized differently!

We consider three sets of functions: an original, linearly independent given set $u_n(x)$, $n = 0, 1, 2, \ldots$; an orthogonalized set $\psi_n(x)$ to be constructed; and a final set of functions $\varphi_n(x)$ which are the normalized ψ_n's. The original u_n's may be degenerate eigenfunctions, but this is not necessary. We shall have

$u_n(x)$	$\psi_n(x)$	$\varphi_n(x)$
linearly independent	*linearly independent*	*linearly independent*
nonorthogonal	*orthogonal*	*orthogonal*
unnormalized	unnormalized	*normalized*
		(orthonormal)

The Gram–Schmidt procedure is to take the nth ψ function (ψ_n) to be $u_n(x)$ plus an unknown linear combination of the previous φ's. The presence of the new $u_n(x)$ will guarantee linear independence. The requirement that $\psi_n(x)$ be orthogonal to each of the previous φ's yields just enough constraints to determine each of the unknown coefficients. Then the fully determined ψ_n will be normalized to unity, yielding $\varphi_n(x)$. Then the sequence of steps is repeated for $\psi_{n+1}(x)$.

Starting with $n = 0$, let

$$\psi_0(x) = u_0(x) \tag{9.41}$$

with no "previous" φ's to worry about. Normalizing,

$$\varphi_0(x) = \frac{\psi_0(x)}{[\int \psi_0^2 w \, dx]^{1/2}}. \tag{9.42}$$

For $n = 1$, let

$$\psi_1(x) = u_1(x) + a_{10}\varphi_0(x). \tag{9.43}$$

We demand that $\psi_1(x)$ be orthogonal to $\varphi_0(x)$. (At this stage the normalization of $\psi_1(x)$ is irrelevant.) This demand of orthogonality leads to

$$\int \psi_1 \varphi_0 w \, dx = \int u_1 \varphi_0 w \, dx + a_{10} \int \varphi_0^2 w \, dx.$$
$$= 0. \tag{9.44}$$

Since φ_0 is normalized to unity (Eq. (9.42)), we have

$$a_{10} = -\int u_1 \varphi_0 w \, dx, \tag{9.45}$$

fixing the value of a_{10}. Normalizing, we define

$$\varphi_1(x) = \frac{\psi_1(x)}{(\int \psi_1^2 w \, dx)^{1/2}}. \tag{9.46}$$

Generalizing, we have

$$\varphi_i(x) = \frac{\psi_i(x)}{(\int \psi_i^2(x) w(x) \, dx)^{1/2}}, \tag{9.47}$$

where

$$\psi_i(x) = u_i + a_{i0} \varphi_0 + a_{i1} \varphi_1 + \cdots + a_{ii-1} \varphi_{i-1}. \tag{9.48}$$

The coefficients a_{ij} are given by

$$a_{ij} = -\int u_i \varphi_j w \, dx. \tag{9.49}$$

Equation (9.49) is for unit normalization. If some other normalization is selected, then

$$\int_a^b [\varphi_j(x)]^2 w(x) \, dx = N_j^2.$$

Equation (9.47) is replaced by

$$\varphi_i(x) = N_i \frac{\psi_i(x)}{(\int \psi_i^2 w \, dx)^{1/2}} \tag{9.47a}$$

and a_{ij} becomes

$$a_{ij} = -\frac{\int u_i \varphi_j w \, dx}{N_j^2}. \tag{9.49a}$$

Equations (9.48) and (9.49) may be rewritten in terms of projection operators, P_j. If we consider the $\varphi_n(x)$ to form a linear vector space, then the integral in Eq. (9.49) may be interpreted as the projection of u_i into the φ_j "coordinate" or the jth component of u_i. With

$$P_j u_i(x) = \left\{ \int u_i(t) \varphi_j(t) w(t) \, dt \right\} \varphi_j(x),$$

Eq. (9.48) becomes

$$\psi_i(x) = \left\{ 1 - \sum_{j=1}^{i-1} P_j \right\} u_i(x). \tag{9.48a}$$

Subtracting off the jth components, $j = 1$ to $i - 1$ leaves $\psi_i(x)$ orthogonal to all the $\varphi_j(x)$.

It will be noticed that although this Gram–Schmidt procedure is one possible way of constructing an orthogonal or orthonormal set, the functions $\varphi_i(x)$ are not unique. There is an infinite number of possible orthonormal sets for a given interval and a given density function. As an illustration of the freedom involved, consider two (nonparallel) vectors $\mathbf{A}$ and $\mathbf{B}$ in the xy-plane. We may normalize $\mathbf{A}$ to unit magnitude and then form $\mathbf{B}' = a\mathbf{A} + \mathbf{B}$ so that $\mathbf{B}'$ is perpendicular to $\mathbf{A}$. By normalizing $\mathbf{B}'$ we have completed the Gram–Schmidt orthogonalization for two vectors. But any two perpendicular unit vectors such as $\hat{\mathbf{x}}$ and $\hat{\mathbf{y}}$ could have been chosen as our orthonormal set. Again, with an infinite number of possible rotations of $\hat{\mathbf{x}}$ and $\hat{\mathbf{y}}$ about the z-axis, we have an infinite number of possible orthonormal sets.

Example 9.3.1 Legendre Polynomials by Gram–Schmidt Orthogonalization

Let us form an orthonormal set from the set of functions $u_n(x) = x^n$, $n = 0, 1, 2, \ldots$. The interval is $-1 \le x \le 1$ and the density function is $w(x) = 1$.

In accordance with the Gram–Schmidt orthogonalization process described,

$$u_0 = 1 \quad \text{and} \quad \varphi_0 = \frac{1}{\sqrt{2}}. \tag{9.50}$$

Then

$$\psi_1(x) = x + a_{10}\frac{1}{\sqrt{2}} \tag{9.51}$$

and

$$a_{10} = -\int_{-1}^{1}\frac{x}{\sqrt{2}}\,dx = 0 \tag{9.52}$$

by symmetry. Normalizing ψ_1, we obtain

$$\varphi_1(x) = \sqrt{\frac{3}{2}}x. \tag{9.53}$$

Continuing the Gram–Schmidt process, we define

$$\psi_2(x) = x^2 + a_{20}\frac{1}{\sqrt{2}} + a_{21}\sqrt{\frac{3}{2}}x, \tag{9.54}$$

where

$$a_{20} = -\int_{-1}^{1}\frac{x^2}{\sqrt{2}}\,dx = -\frac{\sqrt{2}}{3}, \tag{9.55}$$

$$a_{21} = -\int_{-1}^{1}\sqrt{\frac{3}{2}}x^3\,dx = 0, \tag{9.56}$$

again by symmetry. Therefore

$$\psi_2(x) = x^2 - \frac{1}{3}, \tag{9.57}$$

and, on normalizing to unity, we have

$$\varphi_2(x) = \sqrt{\frac{5}{2}} \cdot \frac{1}{2}(3x^2 - 1). \tag{9.58}$$

The next function $\varphi_3(x)$ is

$$\varphi_3(x) = \sqrt{\frac{7}{2}} \cdot \frac{1}{2}(5x^3 - 3x). \tag{9.59}$$

Reference to Chapter 12 will show that

$$\varphi_n(x) = \sqrt{\frac{2n+1}{2}} P_n(x), \tag{9.60}$$

where $P_n(x)$ is the nth-order Legendre polynomial. Our Gram–Schmidt process provides a possible but very cumbersome method of generating the Legendre polynomials. It illustrates how a power series expansion in $u_n(x) = x^n$, which is not orthogonal, can be converted into an orthogonal series.

The equations for Gram–Schmidt orthogonalization tend to be ill-conditioned because of the subtractions. A technique for avoiding this difficulty using the polynomial recurrence relation is discussed by Hamming.[2]

In Example 9.3.1 we have specified an orthogonality interval $[-1, 1]$, a unit weighting function, and a set of functions, x^n, to be taken one at a time in increasing order. Given all these specifications the Gram–Schmidt procedure is unique (to within a normalization factor and an overall sign as discussed subsequently). Our resulting orthogonal set, the Legendre polynomials, P_0 up through P_n, form a complete set for the description of polynomials of order $\leq n$ over $[-1, 1]$. This concept of completeness is taken up in detail in Section 9.4. Expansions of functions in series of Legendre polynomials are found in Section 12.3.

ORTHOGONAL POLYNOMIALS

This particular example has been chosen strictly to illustrate the Gram–Schmidt procedure. Although it has the advantage of introducing the Legendre polynomials, the initial functions $u_n = x^n$ are not degenerate eigenfunctions and are not solutions of Legendre's equation. They are simply a set of functions that we have here rearranged to create an orthonormal set for the given interval and given weighting function. The fact that we obtained the Legendre polynomials is not quite black magic but a direct consequence of the choice of interval and weighting function. The use of $u_n(x) = x^n$ but with other choices of interval and weighting function leads to other sets of orthogonal polynomials as shown in Table 9.3. We consider these polynomials in detail in Chapters 12 and 13 as solutions of particular differential equations.

An examination of this orthogonalization process will reveal two arbitrary

---

[2] R. W. Hamming, *Numerical Methods for Scientists and Engineers,* 2nd ed. New York: McGraw-Hill (1973). See Section 27.2 and references given there.

Table 9.3 Orthogonal Polynomials Generated by Gram–Schmidt Orthogonalization of $u_n(x) = x^n$, $n = 0, 1, 2, \ldots$

Polynomials	Interval	Weighting function $w(x)$	Standard normalization
Legendre	$-1 \le x \le 1$	1	$\int_{-1}^{1} [P_n(x)]^2\, dx = \dfrac{2}{2n+1}$
Shifted Legendre	$0 \le x \le 1$	1	$\int_{0}^{1} [P_n^*(x)]^2\, dx = \dfrac{1}{2n+1}$
Chebyshev I	$-1 \le x \le 1$	$(1-x^2)^{-1/2}$	$\int_{-1}^{1} [T_n(x)]^2(1-x^2)^{-1/2}\, dx = \begin{cases} \pi/2, & n \ne 0 \\ \pi & n = 0 \end{cases}$
Shifted Chebyshev I	$0 \le x \le 1$	$[x(1-x)]^{-1/2}$	$\int_{0}^{1} [T_n^*(x)]^2[x(1-x)]^{-1/2}\, dx = \begin{cases} \pi/2, & n > 0 \\ \pi, & n = 0 \end{cases}$
Chebyshev II	$-1 \le x \le 1$	$(1-x^2)^{1/2}$	$\int_{-1}^{1} [U_n(x)]^2(1-x^2)^{1/2}\, dx = \dfrac{\pi}{2}$
Laguerre	$0 \le x < \infty$	e^{-x}	$\int_{0}^{\infty} [L_n(x)]^2\, e^{-x}\, dx = 1$
Associated Laguerre	$0 \le x < \infty$	$x^k e^{-x}$	$\int_{0}^{\infty} [L_n^k(x)]^2 x^k\, e^{-x}\, dx = \dfrac{(n+k)!}{n!}$
Hermite	$-\infty < x < \infty$	e^{-x^2}	$\int_{-\infty}^{\infty} [H_n(x)]^2\, e^{-x^2}\, dx = 2^n \pi^{1/2} n!$

features. First, as emphasized before, it is not necessary to normalize the functions to unity. In the example just given we could have required

$$\int_{-1}^{1} \varphi_n(x)\varphi_m(x)\, dx = \frac{2}{2n+1}\, \delta_{nm}, \tag{9.61}$$

and the resulting set would have been the actual Legendre polynomials. Second, the sign of φ_n is always indeterminate. In the example we chose the sign by requiring the coefficient of the highest power of x in the polynomial to be positive. For the Laguerre polynomials, on the other hand, we would require the coefficient of the highest power to be $(-1)^n/n!$

EXERCISES

9.3.1 Rework Example 9.3.1 by replacing $\varphi_n(x)$ by the conventional Legendre polynomial, $P_n(x)$.

$$\int_{-1}^{1} [P_n(x)]^2\, dx = \frac{2}{2n+1}.$$

Using Eqs. (9.38a), (9.47a), and (9.49a), construct P_0, $P_1(x)$, and $P_2(x)$.

ANS. $P_0 = 1$,
$P_1 = x$,
$P_2 = \frac{3}{2}x^2 - \frac{1}{2}$.

9.3.2   Following the Gram–Schmidt procedure, construct a set of polynomials $P_n^*(x)$ orthogonal (unit weighting factor) over the range [0, 1] from the set $\{1, x\}$. Normalize so that $P_n^*(1) = 1$.

$$ANS. \quad P_0^*(x) = 1,$$
$$P_1^*(x) = 2x - 1,$$
$$P_2^*(x) = 6x^2 - 6x + 1,$$
$$P_3^*(x) = 20x^3 - 30x^2 + 12x - 1.$$

These are the first four *shifted* Legendre polynomials.

Note. The "*" is the standard notation for "shifted": [0, 1] instead of $[-1, 1]$. It does *not* mean complex conjugate.

9.3.3   Apply the Gram–Schmidt procedure to form the first three Laguerre polynomials

$$u_n(x) = x^n, \quad n = 0, 1, 2, \ldots,$$
$$0 \le x < \infty,$$
$$w(x) = e^{-x}.$$

The conventional normalization is

$$\int_0^\infty L_m(x)L_n(x)e^{-x}\,dx = \delta_{mn}.$$

$$ANS. \quad L_0 = 1,$$
$$L_1 = 1 - x,$$
$$L_2 = \frac{(2 - 4x + x^2)}{2}.$$

9.3.4   You are given

(a) a set of functions $u_n(x) = x^n, n = 0, 1, 2, \ldots,$
(b) an interval $(0, \infty)$,
(c) a weighting function $w(x) = xe^{-x}$.

Use the Gram–Schmidt procedure to construct the first *three orthonormal* functions from the set $u_n(x)$ for this interval and this weighting function.

$$ANS. \quad \varphi_0(x) = 1,$$
$$\varphi_1(x) = (x - 2)/\sqrt{2},$$
$$\varphi_2(x) = (x^2 - 6x + 6)/2\sqrt{3}.$$

9.3.5   Using the Gram–Schmidt orthogonalization procedure, construct the lowest three Hermite polynomials:

$$u_n(x) = x^n, \quad n = 0, 1, 2, \ldots, \quad -\infty < x < \infty, \quad w(x) = e^{-x^2}.$$

For this set of polynomials the usual normalization is

$$\int_{-\infty}^\infty H_m(x)H_n(x)w(x)\,dx = \delta_{mn}2^m m!\pi^{1/2}.$$

$$ANS. \quad H_0 = 1,$$
$$H_1 = 2x,$$
$$H_2 = 4x^2 - 2.$$

9.3.6   Use the Gram–Schmidt orthogonalization scheme to construct the first three Chebyshev polynomials (type I):

$$u_n(x) = x^n, \quad n = 0, 1, 2, \ldots, \quad -1 \le x \le 1, \quad w(x) = (1 - x^2)^{-1/2}.$$

Take the normalization

$$\int_{-1}^{1} T_m(x)T_n(x)w(x)\, dx = \delta_{mn} \begin{cases} \pi, & m = n = 0 \\ \dfrac{\pi}{2}, & m = n \geq 1. \end{cases}$$

Hint. The needed integrals are given in Exercise 10.4.3.

$$\begin{aligned} ANS. \quad T_0 &= 1, \\ T_1 &= x, \\ T_2 &= 2x^2 - 1, \\ (T_3 &= 4x^3 - 3x). \end{aligned}$$

9.3.7 Use the Gram–Schmidt orthogonalization scheme to construct the first three Chebyshev polynomials (type II):

$$u_n(x) = x^n, \qquad n = 0, 1, 2, \ldots, \qquad -1 \leq x \leq 1, \qquad w(x) = (1 - x^2)^{+1/2}.$$

Take the normalization to be

$$\int_{-1}^{1} U_m(x)U_n(x)w(x)\, dx = \delta_{mn}\frac{\pi}{2}.$$

Hint.

$$\int_{-1}^{1} (1 - x^2)^{1/2}x^{2n}\, dx = \frac{\pi}{2} \times \frac{1 \cdot 3 \cdot 5 \cdots (2n-1)}{4 \cdot 6 \cdot 8 \cdots (2n+2)}, \qquad n = 1, 2, 3, \ldots$$

$$= \frac{\pi}{2}, \qquad n = 0.$$

$$\begin{aligned} ANS. \quad U_0 &= 1, \\ U_1 &= 2x, \\ U_2 &= 4x^2 - 1. \end{aligned}$$

9.3.8 As a modification of Exercise 9.3.5, apply the Gram–Schmidt orthogonalization procedure to the set $u_n(x) = x^n$, $n = 0, 1, 2, \ldots$, $0 \leq x < \infty$. Take $w(x)$ to be $\exp[-x^2]$. Find the first two nonvanishing polynomials. Normalize so that the coefficient of the highest power of x is unity. In Exercise 9.3.5 the interval $(-\infty, \infty)$ led to the Hermite polynomials. These are certainly not the Hermite polynomials.

$$\begin{aligned} ANS. \quad \varphi_0 &= 1, \\ \varphi_1 &= x - \pi^{-1/2}. \end{aligned}$$

9.3.9 Form an orthogonal set over the interval $0 \leq x < \infty$, using $u_n(x) = e^{-nx}$, $n = 1, 2, 3, \ldots$. Take the weighting factor, $w(x)$, to be unity. These functions are solutions of $u_n'' - n^2 u_n = 0$, which is clearly already in Sturm–Liouville (self-adjoint) form. Why doesn't the Sturm–Liouville theory guarantee the orthogonality of these functions?

9.4 COMPLETENESS OF EIGENFUNCTIONS

The third important property of an Hermitian operator is that its eigenfunctions form a complete set. This completeness means that any well-behaved (at least piecewise continuous) function $F(x)$ can be approximated by a series

$$F(x) = \sum_{n=0}^{\infty} a_n \varphi_n(x) \tag{9.62}$$

to any desired degree of accuracy.[1] More precisely, the set $\varphi_n(x)$ is called complete[2] if the limit of the mean square error vanishes;

$$\lim_{m \to \infty} \int_a^b \left[F(x) - \sum_{n=0}^{m} a_n \varphi_n(x) \right]^2 w(x)\, dx = 0. \tag{9.63}$$

Technicially, the integral here is a Lebesgue integral. We have not required that the error vanish identically in $[a, b]$ but only that the integral of the error squared go to zero.

This convergence in the mean, Eq. (9.63), should be compared with uniform convergence, Section 5.5, Eq. (5.67). Clearly, uniform convergence implies convergence in the mean but the converse does not hold; convergence in the mean is less restrictive. Specifically, Eq. (9.63) is not upset by piecewise continuous functions, a finite number of finite discontinuities. A relevant example is the Gibbs phenomenon of discontinuous Fourier series discussed in Section 14.5, which occurs for other eigenfunction series as well. Equation (9.63) is perfectly adequate for our purposes and is far more convenient than Eq. (5.67). Indeed, since we frequently use eigenfunctions to describe discontinuous functions, convergence in the mean is all we can expect.

In the language of linear algebra, we have a linear space, a function space. The linearly independent, orthonormal functions $\varphi_n(x)$, form the basis for this (infinite-dimensional) space. Equation (9.62) is a statement that the functions $\varphi_n(x)$ span this linear space. With an inner product defined by Eq. (9.65), our linear space is a Hilbert space.

The question of completeness of a set of functions is often determined by comparison with a Laurent series, Section 6.5. In Section 14.1 this is done for Fourier series, thus establishing the completeness of Fourier series. For all orthogonal polynomials mentioned in Section 9.3 it is possible to find a polynomial expansion of each power of z,

$$z^n = \sum_{i=0}^{n} a_i P_i(z), \tag{9.64}$$

where $P_i(z)$ is the ith polynomial. Exercises 12.4.6, 13.1.8, 13.2.5, and 13.3.22 are specific examples of Eq. (9.64). Using Eq. (9.64), we may reexpress the Laurent expansion of $f(z)$ in terms of the polynomials, showing that the polynomial expansion exists (and existing, it is unique, Exercise 9.4.1). The limitation of this Laurent series development is that it requires the function to be analytic. Equations (9.62) and (9.63) are more general. $F(x)$ may be only piecewise continuous. Numerous examples of the representation of such

---

[1] If we have a finite set, as with vectors, the summation is over the number of linearly independent members of the set.

[2] Many authors use the term closed here.

piecewise continuous functions appear in Chapter 14 (Fourier series). A proof that our Sturm–Liouville eigenfunctions form complete sets appears in Courant and Hilbert.[3]

In Eq. (9.6.2) the expansion coefficients a_m may be determined by

$$a_m = \int_a^b F(x)\varphi_m(x)w(x)\,dx. \tag{9.65}$$

This follows from multiplying Eq. (9.62) by $\varphi_m(x)w(x)$ and integrating. From the orthogonality of the eigenfunctions, $\varphi_n(x)$, only the mth term survives. Here we see the value of orthogonality. Equation (9.65) may be compared with the dot or inner product of vectors, Section 1.3, and a_m interpreted as the mth projection of the function $F(x)$. Often the coefficient a_m is called a generalized Fourier coefficient.

For a known function, $F(x)$, Eq. (9.65) gives a_m as a *definite* integral which can always be evaluated, by machine if not analytically.

For examples of particular eigenfunction expansions, see the following: Fourier series, Section 9.2 and Chapter 14; Bessel and Fourier–Bessel expansions, Section 11.2; Legendre series, Section 12.3; Laplace series, Section 12.6; Hermite series, Section 13.1; Laguerre series, Section 13.2; and Chebyshev series, Section 13.3.

It may also happen that the eigenfunction expansion, Eq. (9.62), is the expansion of an unknown $F(x)$ in a series of known eigenfunctions $\varphi_n(x)$ with unknown coefficients a_n. An example would be the quantum chemist's attempt to describe an (unknown) molecular wave function as a linear combination of known atomic wave functions. The unknown coefficients a_n would be determined by a variational technique—Rayleigh–Ritz, Section 17.8.

BESSEL'S INEQUALITY

If the set of functions $\varphi_n(x)$ does not form a complete set, possibly because we simply have not included the required infinite number of members of an infinite set, we are led to Bessel's inequality. First, consider the finite case. Let A be an n component vector,

$$\mathbf{A} = \mathbf{e}_1 a_1 + \mathbf{e}_2 a_2 + \cdots + \mathbf{e}_n a_n, \tag{9.66}$$

in which $\mathbf{e}_i$ is a unit vector and a_i is the corresponding component (projection) of $\mathbf{A}$, that is,

$$a_i = \mathbf{A} \cdot \mathbf{e}_i. \tag{9.67}$$

Then

$$\left(\mathbf{A} - \sum_i \mathbf{e}_i a_i\right)^2 \geq 0. \tag{9.68}$$

[3] R. Courant and D. Hilbert, *Methods of Mathematical Physics* (English translation), Vol. 1. New York: Interscience (1953), Chapter 6, Section 3.

If we sum over all n components, clearly, the summation equals $\mathbf{A}$ by Eq. (9.66) and the equality holds. If, however, the summation does not include all n components, the inequality results. By expanding Eq. (9.68) and remembering that the unit vectors satisfy an orthogonality relation,

$$\mathbf{e}_i \cdot \mathbf{e}_j = \delta_{ij}, \tag{9.69}$$

we have

$$A^2 \geq \sum_i a_i^2. \tag{9.70}$$

This is Bessel's inequality.

For functions we consider the integral

$$\int_a^b \left[f(x) - \sum_i a_i \varphi_i(x) \right]^2 w(x)\, dx \geq 0. \tag{9.71}$$

This is the continuum analog of Eq. (9.68), letting $n \to \infty$ and replacing the summation by an integration. Again, with the weighting factor $w(x) > 0$, the integrand is nonnegative. The integral vanishes by Eq. (9.62) if we have a complete set. Otherwise it is positive. Expanding the squared term, we obtain

$$\int_a^b [f(x)]^2 w(x)\, dx - 2 \sum_i a_i \int_a^b f(x)\varphi_i(x) w(x)\, dx + \sum_i a_i^2 \geq 0. \tag{9.72}$$

Applying Eq. (9.65), we have

$$\int_a^b [f(x)]^2 w(x)\, dx \geq \sum_i a_i^2. \tag{9.73}$$

Hence the sum of the squares of the expansion coefficients a_i is less than or equal to the weighted integral of $[f(x)]^2$, the equality holding if and only if the expansion is exact, that is if the set of functions $\varphi_n(x)$ is a complete set.

In later chapters when we consider eigenfunctions that form complete sets (such as Legendre polynomials), Eq. (9.73) with the equal sign holding will be called a Parseval relation.

Bessel's inequality has a variety of uses, including proof of convergence of the Fourier series.

SCHWARZ INEQUALITY

The frequently used Schwarz inequality is similar to the Bessel inequality. Consider the quadratic equation

$$\sum_{i=1}^n (a_i x + b_i)^2 = \sum_{i=1}^n a_i^2 (x + b_i/a_i)^2 = 0. \tag{9.74}$$

If $b_i/a_i = $ constant, c, then the solution is $x = -c$. If b_i/a_i is *not* a constant, all terms cannot vanish simultaneously for real x. So the solution must be complex. Expanding, we find that

$$x^2 \sum_i^n a_i^2 + 2x \sum_i^n a_i b_i + \sum_i^n b_i^2 = 0, \tag{9.75}$$

and since x is complex (or $= -b_i/a_i$), the quadratic formula[4] for x leads to

$$\left(\sum_{i=1}^n a_i b_i \right)^2 \leq \left(\sum_{i=1}^n a_i^2 \right) \left(\sum_{i=1}^n b_i^2 \right), \tag{9.76}$$

the equality holding when b_i/a_i equals a constant.

Once more, in terms of vectors, we have

$$(\mathbf{a} \cdot \mathbf{b})^2 = a^2 b^2 \cos^2 \theta \leq a^2 b^2, \tag{9.77}$$

where θ is the included angle.

The Schwarz inequality for functions has the form

$$\left| \int_a^b f^*(x)g(x)\, dx \right|^2 \leq \int_a^b f^*(x)f(x)\, dx \int_a^b g^*(x)g(x)\, dx, \tag{9.78}$$

the equality holding if and only if $g(x) = \alpha f(x)$, α being a constant. To prove this function form of the Schwarz inequality,[5] consider a complex function $\psi(x) = f(x) + \lambda g(x)$ with λ a complex constant. The functions $f(x)$ and $g(x)$ are any two functions (for which the integrals exist). Multiplying by the complex conjugate and integrating, we obtain

$$\int_a^b \psi^* \psi\, dx \equiv \int_a^b f^* f\, dx + \lambda \int_a^b f^* g\, dx + \lambda^* \int_a^b g^* f\, dx$$

$$+ \lambda \lambda^* \int_a^b g^* g\, dx \geq 0. \tag{9.79}$$

The ≥ 0 appears since $\psi^* \psi$ is nonegative, the equal ($=$) sign holding only if $\psi(x)$ is identically zero. Noting that λ and λ^* are linearly independent, we differentiate with respect to one of them and set the derivative equal to zero to minimize $\int_a^b \psi^* \psi\, dx$:

$$\frac{\partial}{\partial \lambda^*} \int_a^b \psi^* \psi\, dx = \int_a^b g^* f\, dx + \lambda \int_a^b g^* g\, dx = 0.$$

This yields

$$\lambda = - \frac{\int_a^b g^* f\, dx}{\int_a^b g^* g\, dx}. \tag{9.80a}$$

---

[4] With discriminant $b^2 - 4ac$ negative (or zero).

[5] An alternate derivation is provided by the inequality $\iint [f(x)g(y) - f(y)g(x)]^* [f(x)g(y) - f(y)g(x)]\, dx\, dy \geq 0$.

Taking the complex conjugate, we obtain

$$\lambda^* = -\frac{\int_a^b f^*g\, dx}{\int_a^b g^*g\, dx}. \tag{9.80b}$$

Substituting these values of λ and λ^* back into Eq. (9.79), we obtain Eq. (9.78), the Schwarz inequality.

In quantum mechanics $f(x)$ and $g(x)$ might each represent a state or configuration of a physical system. Then the Schwarz inequality guarantees that the inner product $\int_a^b f^*(x)g(x)\, dx$ exists. In some texts the Schwarz inequality is a key step in the derivation of the Heisenberg uncertainty principle.

The function notation of Eqs. (9.78) and (9.79) is relatively cumbersome. In advanced mathematical physics and especially in quantum mechanics it is common to use a different notation:

$$\langle f|g\rangle \equiv \int_a^b f^*(x)g(x)\, dx.$$

Using this new notation, we simply understand the range of integration, (a, b), and any weighting function. In this notation the Schwarz inequality becomes

$$|\langle f|g\rangle|^2 \le \langle f|f\rangle\langle g|g\rangle. \tag{9.78a}$$

If $g(x)$ is a normalized eigenfunction, $\varphi_i(x)$, Eq. (9.78) yields [here $w(x) = 1$]

$$a_i^* a_i \le \int_a^b f^*(x)f(x)\, dx, \tag{9.81}$$

a result that also follows from Eq. (9.73).

For useful representations of Dirac's delta function in terms of orthogonal sets of functions and the relation between closure and completeness we refer to the relevant subsection of Section 1.15 including Exercise 1.15.16.

SUMMARY—VECTOR SPACES—COMPLETENESS

Here we summarize some properties of vector space, first with the vectors taken to be the familiar real vectors of Chapter 1 and then with the vectors taken to be ordinary functions—polynomials. The concept of completeness is developed for finite vector spaces and carried over into infinite vector spaces.

1v.   We shall describe our vector space with a set of n linearly independent vectors $\mathbf{e}_i$, $i = 1, 2, \ldots, n$. If $n = 3$, $\mathbf{e}_1 = \hat{\mathbf{x}}$, $\mathbf{e}_2 = \hat{\mathbf{y}}$, and $\mathbf{e}_3 = \hat{\mathbf{z}}$. The n $\mathbf{e}_i$ *span* the linear vector space.

1f.   We shall describe our vector (function) space with a set of n linearly independent functions, $\varphi_i(x)$, $i = 0, 1, \ldots, n - 1$. The index i starts with 0 to agree with the labeling of the classical polynomials. Here $\varphi_i(x)$ is assumed to be a polynomial of degree i. The n $\varphi_i(x)$ *span* the linear vector (function) space.

2v.   The vectors in our vector space satisfy the following relations (Section 1.2; the vector components are numbers):

a. Vector addition is commutative $\mathbf{u} + \mathbf{v} = \mathbf{v} + \mathbf{u}$
b. Vector addition is associative $[\mathbf{u} + \mathbf{v}] + \mathbf{w} = \mathbf{u} + [\mathbf{v} + \mathbf{w}]$
c. There is a null vector $\mathbf{0} + \mathbf{v} = \mathbf{v}$
d. Multiplication by a scalar
    Distributive $a[\mathbf{u} + \mathbf{v}] = a\mathbf{u} + a\mathbf{v}$
    Distributive $(a + b)\mathbf{u} = a\mathbf{u} + b\mathbf{u}$
    Associative $a[b\mathbf{u}] = (ab)\mathbf{u}$
e. Multiplication
    By unit scalar $1\mathbf{u} = \mathbf{u}$
    By zero $0\mathbf{u} = 0$
f. Negative vector $(-1)\mathbf{u} = -\mathbf{u}.$

2f. The functions in our linear function space satisfy the properties listed for vectors (substitute "function" for "vector"):

$$f(x) + g(x) = g(x) + f(x)$$

$$[f(x) + g(x)] + h(x) = f(x) + [g(x) + h(x)]$$

$$0 + f(x) = f(x)$$

$$a[f(x) + g(x)] = af(x) + ag(x)$$

$$(a + b)f(x) = af(x) + bf(x)$$

$$a[bf(x)] = (ab)f(x)$$

$$1 \cdot f(x) = f(x)$$

$$0 \cdot f(x) = 0$$

$$(-1) \cdot f(x) = -f(x).$$

3v. In n-dimensional vector space an arbitrary vector $\mathbf{c}$ is described by its n components $(c_1, c_2, \ldots, c_n)$ or

$$\mathbf{c} = \sum_{i=1}^{n} c_i \mathbf{e}_i.$$

When (1) $n\mathbf{e}_i$ are linearly independent and (2) span the n-dimensional vector space, then the $\mathbf{e}_i$ form a *basis* and constitute a *complete* set.

3f. In n-dimensional function space a polynomial of degree $m \le n - 1$ is described by

$$f(x) = \sum_{i=0}^{n-1} c_i \varphi_i(x).$$

When (1) the $n\varphi_i(x)$ are linearly independent and (2) span the n-dimensional function space, then the $\varphi_i(x)$ form a *basis* and constitute a *complete* set (for describing polynomials of degree $m \le n - 1$).

4v. An inner product (scalar, dot product) is defined by

$$\mathbf{c} \cdot \mathbf{d} = \sum_{i=1}^{n} c_i d_i.$$

(If $\mathbf{c}$ and $\mathbf{d}$ have complex components, the inner product is defined as $\sum_{i=1}^{n} c_i^* d_i$.)

The inner product has the properties of

a. Distributive law of addition $\mathbf{c} \cdot (\mathbf{d} + \mathbf{e}) = \mathbf{c} \cdot \mathbf{d} + \mathbf{c} \cdot \mathbf{e}$
b. Scalar multiplication $\mathbf{c} \cdot a\mathbf{d} = a\mathbf{c} \cdot \mathbf{d}$
c. Complex conjugation $\mathbf{c} \cdot \mathbf{d} = (\mathbf{d} \cdot \mathbf{c})^*.$

4f. An inner product is defined by

$$\langle f | g \rangle = \int_a^b f^*(x) g(x) w(x)\, dx.$$

The choice of the weighting function $w(x)$ and the interval (a, b) follows from the differential equation satisfied by $\varphi_i(x)$ and the boundary conditions— Section 9.1. In matrix terminology, Section 3.2, $|g\rangle$ is a column vector and $\langle f|$ is a row vector, the adjoint of $|f\rangle$.

The inner product has the properties listed for vectors:

a. $\langle f | g + h \rangle = \langle f | g \rangle + \langle f | h \rangle$
b. $\langle f | ag \rangle = a\langle f | g \rangle$
c. $\langle f | g \rangle = \langle g | f \rangle^*.$

5v. Orthogonality:

$$\mathbf{e}_i \cdot \mathbf{e}_j = 0, \qquad i \neq j.$$

If the $n\mathbf{e}_i$ are not already orthogonal, the Gram–Schmidt process may be used to create an orthogonal set.

5f. Orthogonality:

$$\langle \varphi_i | \varphi_j \rangle = \int_a^b \varphi_i^*(x) \varphi_j(x) w(x)\, dx = 0, \qquad i \neq j.$$

If the $n\varphi_i(x)$ are not already orthogonal, the Gram–Schmidt process (Section 9.3) may be used to create an orthogonal set.

6v. Definition of norm:

$$|\mathbf{c}| = (\mathbf{c} \cdot \mathbf{c})^{1/2} = \left(\sum_{i=1}^{n} c_i^2 \right)^{1/2}.$$

The basis vectors $\mathbf{e}_i$ are taken to have unit norm (length) $\mathbf{e}_i \cdot \mathbf{e}_i = 1$. The components of $\mathbf{c}$ are given by

$$c_i = \mathbf{e}_i \cdot \mathbf{c}, \qquad i = 1, 2, \ldots, n.$$

6f. Definition of norm:

$$\|f\| = \langle f|f\rangle^{1/2} = \left[\int_a^b |f(x)|^2 w(x)\, dx\right]^{1/2}$$

$$= \left[\sum_{i=0}^{n-1} |c_i|^2\right]^{1/2},$$

Parseval's identity. $\|f\| > 0$ unless $f(x)$ is identically zero. The basis functions $\varphi_i(x)$ may be taken to have unit norm (unit normalization),

$$\|\varphi_i\| = 1.$$

The expansion coefficients of our polynomial $f(x)$ are given by

$$c_i = \langle \varphi_i|f\rangle, \qquad i = 0, 1, \ldots, n-1.$$

7v. Bessel's inequality:

$$\mathbf{c}\cdot\mathbf{c} \geq \sum_i c_i^2.$$

If the equal sign holds for all $\mathbf{c}$, it indicates that the $\mathbf{e}_i$ span the vector space; that is, they are complete.

7f. Bessel's inequality:

$$\langle f|f\rangle = \int_a^b |f(x)|^2 w(x)\, dx \geq \sum_i |c_i|^2.$$

If the equal sign holds for all allowable f's, it indicates that the $\varphi_i(x)$ span the function space, that is, they are complete.

8v. Schwarz inequality:

$$\mathbf{c}\cdot\mathbf{d} \leq |\mathbf{c}|\cdot|\mathbf{d}|.$$

The equal sign holds when $\mathbf{c}$ is a multiple of $\mathbf{d}$. If the angle included between $\mathbf{c}$ and $\mathbf{d}$ is θ, then $|\cos\theta| \leq 1$.

8f. Schwarz inequality:

$$|\langle f|g\rangle| \leq \langle f|f\rangle^{1/2}\langle g|g\rangle^{1/2} = \|f\|\cdot\|g\|.$$

The equals sign holds when $f(x)$ and $g(x)$ are linearly dependent, that is, when $f(x)$ is a multiple of $g(x)$.

Now, let $n \to \infty$, forming an infinite-dimensional linear vector space, l^2.

9v. In an infinite-dimensional space our vector $\mathbf{c}$ is

$$\mathbf{c} = \sum_{i=1}^{\infty} c_i \mathbf{e}_i.$$

We require that

$$\sum_{i=1}^{\infty} c_i^2 < \infty.$$

The components of $\mathbf{c}$ are given by

$$c_i = \mathbf{e}_i \cdot \mathbf{c}, \qquad i = 1, 2, \ldots, \infty,$$

exactly as in a finite-dimensional space.

Then let $n \to \infty$, forming an infinite-dimensional linear vector (function) space, L^2. Then L stands for Lebesgue, the superscript 2 for the 2 in $|f(x)|^2$. Our functions need no longer be polynomials but we do require that $f(x)$ be at least piecewise continuous (Dirichlet conditions for Fourier series) and that $\langle f | f \rangle = \int_a^b |f(x)|^2 w(x)\, dx$ exist. This latter condition is often stated as a requirement that $f(x)$ be *square integrable*.

9f.   Cauchy sequence: Let

$$f_n(x) = \sum_{i=0}^{n} c_i \varphi_i(x).$$

If

$$\| f(x) - f_n(x) \| \to 0 \qquad \text{as} \quad n \to \infty$$

or

$$\lim_{n \to \infty} \int \left| f(x) - \sum_{i=0}^{n} c_i \varphi_i(x) \right|^2 w(x)\, dx = 0,$$

then we have convergence in the mean. This is analogous to the partial sum—Cauchy sequence criterion for the convergence of an infinite series, Section 5.1.

If every Cauchy sequence of allowable vectors (square integrable, piecewise continuous functions) converges to a limit vector in our linear space, the space is said to be complete. Then

$$f(x) = \sum_{i=0}^{\infty} c_i \varphi_i(x) \qquad \text{(almost everywhere)}$$

in the sense of convergence in the mean. As noted before, this is a weaker requirement than pointwise convergence (fixed value of x) or uniform convergence.

EXPANSION (FOURIER) COEFFICIENTS

$$c_i = \langle \varphi_i | f \rangle, \qquad i = 0, 1, \ldots, \infty,$$

exactly as in a finite-dimensional space. Then

$$f(x) = \sum_i \langle \varphi_i | f \rangle \varphi_i(x).$$

A linear space (finite- or infinite-dimensional) that (1) has an inner product defined ($\langle f | g \rangle$) and (2) is complete is a *Hilbert space*.

Infinite-dimensional Hilbert space provides a natural mathematical framework for modern quantum mechanics. Away from quantum mechanics, Hilbert space retains its abstract mathematical power and beauty but the necessity for its use is reduced.

EXERCISES

9.4.1 A function $f(x)$ is expanded in a series of orthonormal eigenfunctions

$$f(x) = \sum_{n=0}^{\infty} a_n \varphi_n(x).$$

Show that the series expansion is unique for a given set of $\varphi_n(x)$. The functions $\varphi_n(x)$ are being taken here as the *basis* vectors in an infinite dimensional Hilbert space.

9.4.2 A function $f(x)$ is represented by a finite set of basis functions $\varphi_i(x)$,

$$f(x) = \sum_{i=1}^{N} c_i \varphi_i(x)$$

Show that the components c_i are unique, that no different set c_i' exists.
Note. Your basis functions are automatically linearly independent. They are not necessarily orthogonal.

9.4.3 A function $f(x)$ is approximated by a power series $\sum_{i=0}^{n-1} c_i x^i$ over the interval $[0, 1]$. Show that minimizing the mean square error leads to a set of linear equations

$$A\mathbf{c} = \mathbf{b},$$

where

$$A_{ij} = \int_0^1 x^{i+j} \, dx = \frac{1}{i+j+1}, \qquad i, j = 0, 1, 2, \ldots, n-1$$

and

$$b_i = \int_0^1 x^i f(x) \, dx, \qquad i = 0, 1, 2, \ldots, n-1.$$

Note. The A_{ij} are the elements of the Hilbert matrix of order n. The determinant of this Hilbert matrix is a rapidly decreasing function of n. For $n = 5$, $\det A = 3.7 \times 10^{-12}$ and the set of equations $A\mathbf{c} = \mathbf{b}$ is becoming ill-conditioned and unstable.

9.4.4 In place of the expansion of a function $F(x)$ given by

$$F(x) = \sum_{n=0}^{\infty} a_n \varphi_n(x),$$

with

$$a_n = \int_a^b F(x) \varphi_n(x) w(x) \, dx,$$

take the *finite* series approximation

$$F(x) \approx \sum_{n=0}^{m} c_n \varphi_n(x).$$

Show that the mean square error

$$\int_a^b \left[F(x) - \sum_{n=0}^{m} c_n \varphi_n(x) \right]^2 w(x) \, dx$$

is minimized by taking $c_n = a_n$.

Note. The values of the coefficients are independent of the number of terms in the finite series. This independence is a consequence of orthogonality and would not hold for a least-squares fit using powers of x.

9.4.5   From Example 9.2.2

$$f(x) = \begin{cases} h/2, & 0 < x < \pi \\ -h/2, & -\pi < x < 0 \end{cases} = \frac{2h}{\pi} \sum_{n=0}^{\infty} \frac{\sin(2n+1)x}{2n+1}.$$

(a) Show that

$$\int_{-\pi}^{\pi} [f(x)]^2 \, dx = \frac{\pi}{2} h^2$$

$$= \frac{4h^2}{\pi} \sum_{n=0}^{\infty} (2n+1)^{-2}.$$

For a finite upper limit this would be Bessel's inequality. For the upper limit, ∞, as shown, this is Parseval's identity.

(b) Verify that

$$\frac{\pi}{2} h^2 = \frac{4h^2}{\pi} \sum_{n=0}^{\infty} (2n+1)^{-2}$$

by evaluating the series.

Hint. The series can be expressed as a Riemann zeta function.

9.4.6   Differentiate Eq. (9.79),

$$\langle \psi | \psi \rangle = \langle f | f \rangle + \lambda \langle f | g \rangle + \lambda^* \langle g | f \rangle + \lambda \lambda^* \langle g | g \rangle$$

with respect to λ^* and show that you get the Schwarz inequality, Eq. (9.78).

9.4.7   Derive the Schwarz inequality from the identity

$$\left[\int_a^b f(x)g(x) \, dx \right]^2 = \int_a^b [f(x)]^2 \, dx \int_a^b [g(x)]^2 \, dx$$
$$- \frac{1}{2} \int_a^b \int_a^b [f(x)g(y) - f(y)g(x)]^2 dx \, dy.$$

9.4.8   If the functions $f(x)$ and $g(x)$ of the Schwarz inequality, Eq. (9.78), may be expanded in a series of eigenfunctions $\varphi_i(x)$, show that Eq. (9.78) reduces to Eq. (9.76) (with n possibly infinite).

Note the description of $f(x)$ as a vector in a function space in which $\varphi_i(x)$ corresponds to the unit vector $\mathbf{e}_i$.

9.4.9   The operator H is Hermitian and positive definite, that is,

$$\int_a^b f^* Hf \, dx > 0.$$

Prove the generalized Schwarz inequality:

$$\left| \int_a^b f^* Hg \, dx \right|^2 \leq \int_a^b f^* Hf \, dx \int_a^b g^* Hg \, dx.$$

9.4.10   A normalized wave function $\psi(x) = \sum_{n=0}^{\infty} a_n \varphi_n(x)$. The expansion coefficients a_n are known as probability amplitudes. We may define a density matrix ρ with elements $\rho_{ij} = a_i a_j^*$. Show that

$$(\rho^2)_{ij} = \rho_{ij}$$

or

$$\rho^2 = \rho.$$

This result, by definition, makes ρ a projection operator.
Hint:

$$\int \psi^* \psi \, dx = 1.$$

9.4.11   Show that
(a) the operator

$$|\varphi_i(x)\rangle\langle\varphi_i(t)|$$

operating on

$$f(t) = \sum_j c_j |\varphi_j(t)\rangle$$

yields

$$c_i |\varphi_i(x)\rangle.$$

(b) $\sum_i |\varphi_i(x)\rangle\langle\varphi_i(x)| = 1.$

This operator is a *projection operator* projecting $f(x)$ onto the ith coordinate, selectively picking out the ith component $c_i|\varphi_i(x)\rangle$ of $f(x)$. *Hint.* The operator operates via the defined inner product.

9.5 GREEN'S FUNCTION—EIGENFUNCTION EXPANSION

A series somewhat similar to that representing $\delta(x - t)$ results when we expand the Green's function in the eigenfunctions of the corresponding homogeneous equation. In the inhomogeneous Helmholtz equation we have

$$\nabla^2 \psi(\mathbf{r}) + k^2 \psi(\mathbf{r}) = -\rho(\mathbf{r}). \tag{9.82}$$

The homogeneous Helmholtz equation is satisfied by its eigenfunctions φ_n,

$$\nabla^2 \varphi_n(\mathbf{r}) + k_n^2 \varphi_n(\mathbf{r}) = 0. \tag{9.83}$$

As outlined in Section 8.7, the Green's function $G(\mathbf{r}_1, \mathbf{r}_2)$ satisfies the point source equation

$$\nabla^2 G(\mathbf{r}_1, \mathbf{r}_2) + k^2 G(\mathbf{r}_1 - \mathbf{r}_2) = -\delta(\mathbf{r}_1 - \mathbf{r}_2). \tag{9.84}$$

We expand the Green's function in a series of eigenfunctions of the homogeneous equation (9.83), that is,

$$G(\mathbf{r}_1, \mathbf{r}_2) = \sum_{n=0}^{\infty} a_n(\mathbf{r}_2)\varphi_n(\mathbf{r}_1), \tag{9.85}$$

and by substituting into Eq. (9.84) obtain

$$- \sum_{n=0}^{\infty} a_n(\mathbf{r}_2)k_n^2\varphi_n(\mathbf{r}_1) + k^2 \sum_{n=0}^{\infty} a_n(\mathbf{r}_2)\varphi_n(\mathbf{r}_1) = - \sum_{n=0}^{\infty} \varphi_n(\mathbf{r}_1)\varphi_n(\mathbf{r}_2). \qquad (9.86)$$

Here $\delta(\mathbf{r}_1 - \mathbf{r}_2)$ has been replaced by its eigenfunction expansion, Eq. (1.189). When we employ the orthogonality of $\varphi_n(\mathbf{r}_1)$ to isolate a_n and then substitute into Eq. (9.85), the Green's function becomes

$$G(\mathbf{r}_1, \mathbf{r}_2) = \sum_{n=0}^{\infty} \frac{\varphi_n(\mathbf{r}_1)\varphi_n(\mathbf{r}_2)}{k_n^2 - k^2}, \qquad (9.87)$$

a bilinear expansion, symmetric with respect to $\mathbf{r}_1$ and $\mathbf{r}_2$ as expected. Finally, $\psi(\mathbf{r}_1)$, the desired solution of the inhomogeneous equation, is given by

$$\psi(\mathbf{r}_1) = \int G(\mathbf{r}_1, \mathbf{r}_2)\rho(\mathbf{r}_2)\, d\tau_2. \qquad (9.88)$$

If we generalize our inhomogeneous differential equation to

$$\mathcal{L}\psi + \lambda\psi = -\rho, \qquad (9.89)$$

where $\mathcal{L}$ is an Hermitian operator, we find that

$$G(\mathbf{r}_1, \mathbf{r}_2) = \sum_{n=0}^{\infty} \frac{\varphi_n(\mathbf{r}_1)\varphi_n(\mathbf{r}_2)}{\lambda_n - \lambda}, \qquad (9.90)$$

where λ_n is the nth eigenvalue and φ_n, the corresponding orthonormal eigenfunction of the homogeneous differential equation

$$\mathcal{L}\psi + \lambda\psi = 0. \qquad (9.91)$$

The eigenfunction expansion of the Green's function in Eq. (9.90) makes the symmetry property $G(\mathbf{r}_1, \mathbf{r}_2) = G(\mathbf{r}_2, \mathbf{r}_1)$ explicit.

GREEN'S FUNCTIONS—ONE-DIMENSIONAL

The development of the Green's function for two-and three-dimensional systems was the topic above and of Section 8.7. Here, for simplicity, we restrict ourselves to one-dimensional cases and follow a somewhat different approach.

Defining Properties

In our one-dimensional analysis we consider first the nonhomogeneous Sturm–Liouville equation

$$\mathcal{L}y(x) + f(x) = 0, \qquad (9.92)$$

in which $\mathcal{L}$ is the *self-adjoint* differential operator

$$\mathcal{L} = \frac{d}{dx}\left(p(x)\frac{d}{dx}\right) + q(x). \qquad (9.93)$$

As in Section 9.1, $y(x)$ is required to satisfy certain boundary conditions at the end points a and b of our interval $[a, b]$. Indeed, the interval may well be chosen so that appropriate boundary conditions can be satisfied. We now proceed to define a rather strange and arbitrary function G over the interval $[a, b]$. At this stage the most that can be said in defense of G is that the defining properties are legitimate, or mathematically acceptable. Later, it is hoped, G may appear reasonable if not obvious.

1. The interval $a \leq x \leq b$ is divided by a parameter t. We label $G(x) = G_1(x)$ for $a \leq x < t$ and $G(x) = G_2(x)$ for $t < x \leq b$.
2. The functions $G_1(x)$ and $G_2(x)$ each satisfy the homogeneous[1] Sturm–Liouville equation; that is,

$$\mathcal{L}G_1(x) = 0, \qquad a \leq x < t,$$
$$\mathcal{L}G_2(x) = 0, \qquad t < x \leq b. \tag{9.94}$$

3. At $x = a$, $G_1(x)$ satisfies the boundary conditions we impose on $y(x)$. At $x = b$, $G_2(x)$ satisfies the boundary conditions imposed on $y(x)$ at this end point of the interval. For convenience in renormalizing the boundary conditions are taken to be homogeneous; that is, at $x = a$

$$y(a) = 0,$$

or

$$y'(a) = 0,$$

or

$$\alpha y(a) + \beta y'(a) = 0$$

and similarly for $x = b$.

4. We demand that $G(x)$ be *continuous*,[2]

$$\lim_{x \to t_-} G_1(x) = \lim_{x \to t_+} G_2(x). \tag{9.95}$$

5. We require that $G'(x)$ be *discontinuous*, specifically that[2]

$$\left. \frac{d}{dx} G_2(x) \right|_t - \left. \frac{d}{dx} G_1(x) \right|_t = -\frac{1}{p(t)}, \tag{9.96}$$

where $p(t)$ comes from the self-adjoint operator, Eq. (9.93). Note that with the first derivative discontinuous the second derivative does not exist.

These requirements, in effect, make G a function of two variables, $G(x, t)$. Also, we note that $G(x, t)$ depends on both the form of the differential operator $\mathcal{L}$ *and* the boundary conditions that $y(x)$ must satisfy.

---

[1] Homogeneous with respect to the unknown function. The function $f(x)$ in Eq. (9.92) is set equal to zero.
[2] Strictly speaking, this is the limit as $x \to t$.

Now, assuming that we can find a function $G(x, t)$ that has these properties, we label it a Green's function and proceed to show that a solution of Eq. (9.92) is

$$y(x) = \int_a^b G(x, t) f(t) \, dt. \tag{9.97}$$

To do this we first construct the Green's function, $G(x, t)$. Let $u(x)$ be a solution of the homogeneous Sturm–Liouville equation that satisfies the boundary conditions at $x = a$ and $v(x)$ is a solution that satisfies the boundary conditions at $x = b$. Then we may take[3]

$$G(x, t) = \begin{cases} c_1 u(x), & a \le x < t, \\ c_2 v(x), & t < x \le b. \end{cases} \tag{9.98}$$

Continuity at $x = t$ (Eq. (9.95)) requires

$$c_2 v(t) - c_1 u(t) = 0. \tag{9.99}$$

Finally, the discontinuity in the first derivative (Eq. (9.96)) becomes

$$c_2 v'(t) - c_1 u'(t) = -\frac{1}{p(t)}. \tag{9.100}$$

There will be a unique solution for our unknown coefficients c_1 and c_2 if the Wronskian determinant

$$\begin{vmatrix} u(t) & v(t) \\ u'(t) & v'(t) \end{vmatrix} = u(t)v'(t) - v(t)u'(t)$$

does not vanish. We have seen in Section 8.6 that the nonvanishing of this determinant is a necessary condition for linear independence. Let us consider $u(x)$ and $v(x)$ to be independent. The contrary, which occurs when $u(x)$ satisfies the boundary conditions at both end points, requires a generalized Green's function. Strictly speaking, no Green's function exists when $u(x)$ and $v(x)$ are linearly dependent. This is also true when $\lambda = 0$ is an eigenvalue of the homogenous equation. However, a "generalized Green's function" may be defined. This situation, which occurs with Legendre's equation, is discussed in Courant and Hilbert and other references. For independent $u(x)$ and $v(x)$ we have the Wronskian (again from Section 8.6 or Exercise 9.1.4)

$$u(t)v'(t) - v(t)u'(t) = \frac{A}{p(t)}, \tag{9.101}$$

in which A is a constant. Equation (9.101) is sometimes called Abel's formula.

---

[3] The "constants" c_1 and c_2 are independent of x, but they may (and do) depend on the other variable, t.

Numerous examples have appeared in connection with Bessel and Legendre functions. Now, from Eq. (9.100), we identify

$$c_1 = -\frac{v(t)}{A},$$
$$\tag{9.102}$$
$$c_2 = -\frac{u(t)}{A}.$$

Equation (9.99) is clearly satisfied. Substitution into Eq. (9.98) yields our Green's function

$$G(x, t) = \begin{cases} -\dfrac{1}{A} u(x)v(t), & a \leq x < t, \\[2mm] -\dfrac{1}{A} u(t)v(x), & t < x \leq b. \end{cases} \tag{9.103}$$

Note carefully that $G(x, t) = G(t, x)$. This is the symmetry property that was proved earlier in Section 8.7. Its physical interpretation is given by the reciprocity principle (via our propagator function)—a cause at t yields the same effect at x as a cause at x produces at t. In terms of our electrostatic analogy this is obvious, the propagator function depending only on the magnitude of the distance between the two points

$$|\mathbf{r}_1 - \mathbf{r}_2| = \mathbf{r}_2 - \mathbf{r}_1|.$$

GREEN'S FUNCTION INTEGRAL—DIFFERENTIAL EQUATION

We have constructed $G(x, t)$, but there still remains the task of showing that the integral (Eq. (9.97)) with our new Green's function is indeed a solution of the original differential equation (9.92). This we do by direct substitution. With $G(x, t)$ given by Eq. (9.103),[4] Eq. (9.97) becomes

$$y(x) = -\frac{1}{A} \int_a^x v(x)u(t)f(t)\, dt - \frac{1}{A} \int_x^b u(x)v(t)f(t)\, dt. \tag{9.104}$$

Differentiating, we obtain

$$y'(x) = -\frac{1}{A} \int_a^x v'(x)u(t)f(t)\, dt - \frac{1}{A} \int_x^b u'(x)v(t)f(t)\, dt, \tag{9.105}$$

the derivatives of the limits canceling. A second differentiation yields

$$y''(x) = -\frac{1}{A} \int_a^x v''(x)u(t)f(t)\, dt - \frac{1}{A} \int_x^b u''(x)v(t)f(t)\, dt$$
$$- \frac{1}{A} [u(x)v'(x) - v(x)u'(x)]f(x). \tag{9.106}$$

---

[4] In the first integral $a \leq t \leq x$. Hence $G(x, t) = G_2(x, t) = -(1/A)u(t)v(x)$. Similarly, the second integral requires $G = G_1$.

By Eqs. (9.100) and (9.102) this may be rewritten as

$$y''(x) = -\frac{v''(x)}{A} \int_a^x u(t)f(t)\,dt - \frac{u''(x)}{A} \int_x^b v(t)f(t)\,dt - \frac{f(x)}{p(x)}. \qquad (9.107)$$

Now, by substituting into Eq. (9.93), we have

$$\mathcal{L}y(x) = -\frac{[\mathcal{L}v(x)]}{A} \int_a^x u(t)f(t)\,dt - \frac{[\mathcal{L}u(x)]}{A} \int_x^b v(t)f(t)\,dt - f(x). \qquad (9.108)$$

Since $u(x)$ and $v(x)$ were chosen to satisfy the homogeneous Sturm–Liouville equation, the factors in brackets are zero and the integral terms vanish. Transposing $f(x)$, we see that Eq. (9.92) is satisfied.

We must also check that $y(x)$ satisfies the required boundary conditions. At point $x = a$

$$y(a) = -\frac{u(a)}{A} \int_a^b v(t)f(t)\,dt = cu(a), \qquad (9.109)$$

$$y'(a) = -\frac{u'(a)}{A} \int_a^b v(t)f(t)\,dt = cu'(a), \qquad (9.110)$$

since the definite integral is a constant. We chose $u(x)$ to satisfy

$$\alpha u(a) + \beta u'(a) = 0. \qquad (9.111)$$

Multiplying by the constant c, we verify that $y(x)$ also satisfies Eq. (9.111). This illustrates the utility of the *homogeneous* boundary conditions: The normalization does not matter. In quantum mechanical problems the boundary condition on the wave function is often expressed in terms of the ratio

$$\frac{\psi'(x)}{\psi(x)} = \frac{d}{dx}\ln\psi(x),$$

equivalent to Eq. (9.111). The advantage is that the wave function need not be normalized.

Summarizing, we have Eq. (9.97)

$$y(x) = \int_a^b G(x,t)f(t)\,dt,$$

which satisfies the differential equation (Eq. (9.92))

$$\mathcal{L}y(x) + f(x) = 0$$

and the boundary conditions, these boundary conditions having been built into the Green's function $G(x, t)$.

Basically, what we have done is to use the solutions of the homogeneous Sturm–Liouville equation to construct a solution of the nonhomogeneous equation. Again, Poisson's equation is an illustration. The solution (Eq. (8.148)) represents a weighted $[\rho(\mathbf{r}_2)]$ combination of solutions of the corresponding homogeneous Laplace's equation. (We did this same sort of thing early in Section 9.5.)

It should be noted that our $y(x)$, Eq. (9.97), is actually the *particular* solution of the differential equation, Eq. (9.92). Our boundary conditions exclude the addition of solutions of the homogeneous equation. In an actual physical problem we may well have both types of solutions. In electrostatics, for instance (compare Section 8.7), the Green's function solution of Poisson's equation gives the potential created by the given charge distribution. In addition, there may be external fields superimposed. These would be described by solutions of the homogeneous equation, Laplace.

EIGENFUNCTION, EIGENVALUE EQUATION

The preceding analysis placed no special restrictions on our $f(x)$. Let us now assume that $f(x) = \lambda \rho(x) y(x)$.[5] Then we have

$$y(x) = \lambda \int_a^b G(x, t)\rho(t)y(t)\, dt \tag{9.112}$$

as a solution of

$$\mathcal{L}y(x) + \lambda \rho(x)y(x) = 0 \tag{9.113}$$

and its boundary conditions. Equation (9.112) is a homogeneous Fredholm integral equation of the second kind and Eq. (9.113) is the Sturm–Liouville eigenvalue equation of Chapter 9 [with the weighting function $w(x)$ replaced by $\rho(x)$].

There is a change in the interpretation of our Green's function. It started as a propagator function, a weighting function giving the importance of the charge $\rho(\mathbf{r}_2)$ in producing the potential $\varphi(\mathbf{r}_1)$. The charge ρ was the nonhomogeneous term in the nonhomogeneous differential equation (9.92). Now the differential equation and the integral equation are both *homogeneous*. $G(x, t)$ has become a link relating the two equations, differential and integral.

To complete the discussion of this differential equation—integral equation equivalence—let us now show that Eq. (9.113) implies Eq. (9.112); that is, a solution of our differential equation (9.113) with its boundary conditions satisfies the integral equation (9.112). We multiply Eq. (9.113) by $G(x, t)$, the appropriate Green's function, and integrate from $x = a$ to $x = b$ to obtain

$$\int_a^b G(x, t)\mathcal{L}y(x)\, dx + \lambda \int_a^b G(x, t)\rho(x)y(x)\, dx = 0. \tag{9.114}$$

[5] The function $\rho(x)$ is a weighting function, not a charge density.

The first integral is split in two ($x < t$, $x > t$), according to the construction of our Green's function, giving

$$-\int_a^t G_1(x, t)\mathcal{L}y(x)\,dx - \int_t^b G_2(x, t)\mathcal{L}y(x)\,dx = \lambda \int_a^b G(x, t)p(x)y(x)\,dx.$$

(9.115)

Note that t is the upper limit for the G_1 integrals and the lower limit for the G_2 integrals. We are going to reduce the left-hand side of Eq. (9.115) to $y(t)$. Then, with $G(x, t) = G(t, x)$, we have Eq. (9.112) (with x and t interchanged).

Applying Green's theorem to the left-hand side or, equivalently, integrating by parts, we obtain

$$-\int_a^t G_1(x, t)\left[\frac{d}{dx}\left(p(x)\frac{d}{dx}y(x)\right) + q(x)y(x)\right]dx$$

$$= -\left.|G_1(x, t)p(x)y'(x)|_a^t + \int_a^t G_1'(x, t)p(x)y'(x)\,dx - \int_a^t G_1(x, t)q(x)y(x)\,dx,\right.$$

(9.116)

with an equivalent expression for the second integral. A second integration by parts yields

$$-\int_a^t G_1(x, t)\mathcal{L}y(x)\,dx = -\int_a^t y(x)\mathcal{L}G_1(x, t)\,dx$$

$$-|G_1(x, t)p(x)y'(x)|_a^t + |G_1'(x, t)p(x)y(x)|_a^t. \quad (9.117)$$

The integral on the right vanishes because $\mathcal{L}G_1 = 0$. By combining the integrated terms with those from integrating G_2, we have

$$-p(t)[G_1(t, t)y'(t) - G_1'(t, t)y(t) - G_2(t, t)y'(t) + G_2'(t, t)y(t)]$$

$$+ p(a)[G_1(a, t)y'(a) - G_1'(a, t)y(a)] - p(b)[G_2(b, t)y'(b) - G_2'(b, t)y(b)].$$

(9.118)

Each of the last two expressions vanishes, for $G(x, t)$ and $y(x)$ satisfy the same boundary conditions. The first expression, with the help of Eqs. (9.95) and (9.96), reduces to $y(t)$. Substituting into Eq. (9.115), we have Eq. (9.112), thus completing the demonstration of the equivalence of the integral equation and the differential equation plus boundary conditions.

Example 9.5.1 Linear Oscillator

As a simple example, consider the linear oscillator equation (for a vibrating string)

$$y''(x) + \lambda y(x) = 0.$$

(9.119)

We impose the conditions $y(0) = y(1) = 0$, which correspond to a string clamped at both ends. Now, to construct our Green's function, we need solutions of the homogeneous Sturm–Liouville equation, $\mathcal{L}y(x) = 0$, which is $y''(x) = 0$. To satisfy the boundary conditions, we must have one solution vanish at $x = 0$, the other at $x = 1$. Such solutions (unnormalized) are

$$u(x) = x,$$
$$v(x) = 1 - x.$$
(9.120)

We find that

$$uv' - vu' = -1 \qquad (9.121)$$

or, by Eq. (9.101) with $p(x) = 1$, $A = -1$. Our Green's function becomes

$$G(x, t) = \begin{cases} x(1 - t), & 0 \le x < t, \\ t(1 - x), & t < x \le 1. \end{cases} \qquad (9.122)$$

Hence by Eq. (9.112) our clamped vibrating string satisfies

$$y(x) = \lambda \int_0^1 G(x, t)y(t)\, dt. \qquad (9.123)$$

The reader may show that the known solutions of Eq. (9.119).

$$y = \sin n\pi x, \qquad \lambda = n^2\pi^2,$$

do indeed satisfy Eq. (9.123). Note that our eigenvalue λ is *not* the wavelength.

GREEN'S FUNCTION AND THE DIRAC DELTA FUNCTION

One more approach to the Green's function may shed additional light on our formulation and particularly on its relation to physical problems. Let us refer once more to Poisson's equation, this time for a point charge

$$\nabla^2\varphi(\mathbf{r}) = -\frac{\rho_{point}}{\varepsilon_0}. \qquad (9.124)$$

The Green's function solution of this equation was developed in Section 8.7. This time let us take a one-dimensional analog

$$\mathcal{L}y(x) + f(x)_{point} = 0. \qquad (9.125)$$

Here $f(x)_{point}$ refers to a unit point "charge" or a point force. We may represent it by a number of forms, but perhaps the most convenient is

$$f(x)_{point} = \begin{cases} \dfrac{1}{2\varepsilon}, & t - \varepsilon < x < t + \varepsilon, \\ 0, & \text{elsewhere,} \end{cases} \qquad (9.126)$$

which is essentially the same as Eq. (1.171). Then, integrating Eq. (9.125), we have

$$\int_{t-\varepsilon}^{t+\varepsilon} \mathscr{L}y(x)\, dx = -\int_{t-\varepsilon}^{t+\varepsilon} f(x)_{\text{point}}\, dx = -1 \qquad (9.127)$$

from the definition of $f(x)$. Let us examine $\mathscr{L}y(x)$ more closely. We have

$$\int_{t-\varepsilon}^{t+\varepsilon} \frac{d}{dx}[p(x)y'(x)]\, dx + \int_{t-\varepsilon}^{t+\varepsilon} q(x)y(x)\, dx$$

$$= |p(x)y'(x)|_{t-\varepsilon}^{t+\varepsilon} + \int_{t-\varepsilon}^{t+\varepsilon} q(x)y(x)\, dx = -1. \qquad (9.128)$$

In the limit $\varepsilon \to 0$ we may satisfy this relation by *permitting* $y'(x)$ to have a discontinuity of $-1/p(x)$ at $x = t$, $y(x)$ itself remaining continuous.[6] These, however, are just the properties used to define our Green's function, $G(x, t)$. In addition, we note that in the limit $\varepsilon \to 0$

$$f(x)_{\text{point}} = \delta(x - t), \qquad (9.129)$$

in which $\delta(x - t)$ is our Dirac delta function, defined in this manner in Section 1.15. Hence Eq. (9.125) has become

$$\mathscr{L}G(x, t) = -\delta(x - t). \qquad (9.130)$$

This is Eq. (8.159), which we exploit for the development of Green's functions in two and three dimensions—Section 8.7. It will be recalled that we used this relation in Section 8.7 to determine our Green's functions.

Equation (9.130) could have been expected since it is actually a consequence of our differential equation, Eq. (9.92), and Green's function integral solution, Eq. (9.97). If we let $\mathscr{L}_x$ (subscript to emphasize that it operates on the x-dependence) operate on both sides of Eq. (9.97), then

$$\mathscr{L}_x y(x) = \mathscr{L}_x \int_a^b G(x, t)f(t)\, dt.$$

By Eq. (9.92) the left-hand side is just $-f(x)$. On the right $\mathscr{L}_x$ is independent of the variable of integration t, so we may write

$$-f(x) = \int_a^b \{\mathscr{L}_x G(x, t)\}f(t)\, dt.$$

By definition of Dirac's delta function, Eqs. (1.170b) and (1.183), we have Eq. (9.130).

---

[6] The functions $p(x)$ and $q(x)$ appearing in the operator $\mathscr{L}$ are continuous functions. With $y(x)$ remaining continuous. $\int q(x)y(x)\, dx$ is certainly continuous. Hence this integral over an interval 2ε (Eq. (9.128)) vanishes as ε vanishes.

EXERCISES

9.5.1   Show that

$$G(x, t) = \begin{cases} x, & 0 \le x < t, \\ t, & t < x \le 1, \end{cases}$$

is the Green's function for the operator $\mathcal{L} = d^2/dx^2$ and the boundary conditions

$$y(0) = 0,$$
$$y'(1) = 0.$$

9.5.2   Find the Green's function for

(a) $\mathcal{L}y(x) = \dfrac{d^2y(x)}{dx^2} + y(x), \qquad \begin{cases} y(0) = 0, \\ y'(1) = 0, \end{cases}$

(b) $\mathcal{L}y(x) = \dfrac{d^2y(x)}{dx^2} - y(x), \qquad y(x)$ finite for $-\infty < x < \infty.$

9.5.3   Find the Green's function for the operators

(a) $\mathcal{L}y(x) = \dfrac{d}{dx}\left(x\dfrac{dy(x)}{dx} \right).$

$$ANS. \quad (a) \ \ G(x, t) = \begin{cases} -\ln t, & 0 \le x < t, \\ -\ln x, & t < x \le 1. \end{cases}$$

(b) $\mathcal{L}y(x) = \dfrac{d}{dx}\left(x\dfrac{dy(x)}{dx} \right) - \dfrac{n^2}{x}y(x),$

with $y(0)$ finite and $y(1) = 0.$

$$ANS. \quad (b) \ \ G(x, t) = \begin{cases} \dfrac{1}{2n}\left[\left(\dfrac{x}{t}\right)^n - (xt)^n \right], & 0 \le x < t, \\ \dfrac{1}{2n}\left[\left(\dfrac{t}{x}\right)^n - (xt)^n \right], & t < x \le 1. \end{cases}$$

The combination of operator and interval specified in Exercise 9.5.3(a) is pathological in that one of the end points of the interval (zero) is a singular point of the operator. As a consequence, the integrated part (the surface integral of Green's theorem) does not vanish. The next four exercises explore this situation.

9.5.4   (a) Show that the particular solution of

$$\frac{d}{dx}\left[x\frac{d}{dx}y(x) \right] = -1$$

is $y_P(x) = -x.$

(b) Show that

$$y_P(x) = -x \ne \int_0^1 G(x, t)(-1)\, dt,$$

where $G(x, t)$ is the Green's function of Exercise 9.5.3(a).

9.5.5   Show that Green's theorem, Eq. (1.103) in one dimension with a Sturm–Liouville type operator $(d/dt)p(t)(d/dt)$ replacing $\nabla \cdot \nabla$, may be rewritten as

$$\int_a^b \left[u(t) \frac{d}{dt}\left(p(t)\frac{dv(t)}{dt} \right) - v(t)\frac{d}{dt}\left(p(t)\frac{du(t)}{dt} \right) \right] dt$$

$$= \left| u(t)p(t)\frac{dv(t)}{dt} - v(t)p(t)\frac{du(t)}{dt} \right|_a^b.$$

9.5.6   Using the one-dimensional form of Green's theorem of Exercise 9.5.5, let

$$v(t) = y(t) \qquad \text{and} \qquad \frac{d}{dt}\left(p(t)\frac{dy(t)}{dt} \right) = -f(t)$$

$$u(t) = G(x, t) \qquad \text{and} \qquad \frac{d}{dt}\left(p(t)\frac{dG(x, t)}{dt} \right) = -\delta(x - t).$$

Show that Green's theorem yields

$$y(x) = \int_a^b G(x, t) f(t)\, dt$$

$$+ \left| G(x, t)p(t)\frac{dy(t)}{dt} - y(t)p(t)\frac{d}{dt} G(x, t) \right|_a^b.$$

9.5.7   For $p(t) = t$, $y(t) = -t$,

$$G(x, t) = \begin{cases} -\ln t & 0 \le x < t \\ -\ln x & t < x \le 1, \end{cases}$$

verify that the integrated part does not vanish.

9.5.8   Construct the Green's function for

$$x^2 \frac{d^2y}{dx^2} + x\frac{dy}{dx} + (k^2x^2 - 1)y = 0,$$

subject to the boundary conditions

$$y(0) = 0,$$
$$y(1) = 0.$$

9.5.9   Given that

$$\mathcal{L} = (1 - x^2)\frac{d^2}{dx^2} - 2x\frac{d}{dx}$$

and

$$G(\pm 1, t) \text{ remains finite,}$$

show that no Green's function can be constructed by the techniques of this section. ($u(x)$ and $v(x)$ are linearly dependent.)

9.5.10   Construct the infinite one-dimensional Green's function for the Helmholtz equation

$$(\nabla^2 + k^2)\psi(x) = g(x).$$

The boundary conditions are those for a wave advancing in the positive x direction—assuming a time dependence $e^{-i\omega t}$.

$$ANS. \quad G(x_1, x_2) = \frac{i}{2k} \exp(ik|x_1 - x_2|).$$

9.5.11 Construct the infinite one-dimensional Green's function for the modified Helmholtz equation

$$(\nabla^2 - k^2)\psi(x) = f(x).$$

The boundary conditions are that the Green's function must vanish for $x \to \infty$ and $x \to -\infty$.

$$ANS. \quad G(x_1, x_2) = \frac{1}{2k} \exp(-k|x_1 - x_2|).$$

9.5.12 From the eigenfunction expansion of the Green's function show that

(a) $\dfrac{2}{\pi^2} \displaystyle\sum_{n=1}^{\infty} \dfrac{\sin n\pi x \sin n\pi t}{n^2} = \begin{cases} x(1-t), & 0 \le x < t, \\ t(1-x), & t < x \le 1. \end{cases}$

(b) $\dfrac{2}{\pi^2} \displaystyle\sum_{n=0}^{\infty} \dfrac{\sin(n+\frac{1}{2})\pi x \sin(n+\frac{1}{2})\pi t}{(n+\frac{1}{2})^2} = \begin{cases} x, & 0 \le x < t, \\ t, & t < x \le 1. \end{cases}$

Note. In Section 9.4 the Green's function of $\mathcal{L} + \lambda$ is expanded in eigenfunctions. The λ there is an adjustable parameter, not an eigenvalue.

9.5.13 In the Fredholm equation,

$$f(x) = \lambda^2 \int_a^b G(x, t)\varphi(t)\, dt,$$

$G(x, t)$ is a Green's function given by

$$G(x, t) = \sum_{n=1}^{\infty} \frac{\varphi_n(x)\varphi_n(t)}{\lambda_n^2 - \lambda^2}.$$

Show that the solution is

$$\varphi(x) = \sum_{n=1}^{\infty} \frac{\lambda_n^2 - \lambda^2}{\lambda^2} \varphi_n(x) \int_a^b f(t)\varphi_n(t)\, dt.$$

9.5.14 Show that the Green's function integral transform operator

$$\int_a^b G(x, t)[\ ]\, dt$$

is equal to $-\mathcal{L}^{-1}$ in the sense that

(a) $\mathcal{L}_x \displaystyle\int_a^b G(x, t)y(t)\, dt = -y(x),$

(b) $\displaystyle\int_a^b G(x, t)\mathcal{L}_t y(t)\, dt = -y(x).$

Note. Take $\mathcal{L}y(x) + f(x) = 0$, Eq. (9.92).

9.5.15 Substitute Eq. (9.87), the eigenfunction expansion of Green's function, into Eq. (9.88) and then show that Eq. (9.88) is indeed a solution of the nonhomogenous Helmholtz equation (9.82).

9.5.16   (a) Starting with a one-dimensional nonhomogeneous differential equation, (Eq. (9.89)), assume that $\psi(x)$ and $\rho(x)$ may be represented by eigenfunction expansions. Without any use of the Dirac delta function or its representations, show that

$$\psi(x) = \sum_{n=0}^{\infty} \frac{\int_a^b \rho(t)\varphi_n(t)\,dt}{\lambda_n - \lambda}\varphi_n(x).$$

Note that (1) if $\rho = 0$, no solution exists unless $\lambda = \lambda_n$ and (2) if $\lambda = \lambda_n$, no solution exists unless ρ is orthogonal to φ_n. This same behavior will reappear with integral equations in Section 16.4.

(b) Interchanging summation and integration, show that you have constructed the Green's function corresponding to Eq. (9.90).

9.5.17   The eigenfunctions of the Schrödinger equation are often complex. In this case the orthogonality integral, Eq. (9.40), is replaced by

$$\int_a^b \varphi_i^*(x)\varphi_j(x)w(x)\,dx = \delta_{ij}.$$

Instead of Eq. (1.189), we have

$$\delta(\mathbf{r}_1 - \mathbf{r}_2) = \sum_{n=0}^{\infty} \varphi_n(\mathbf{r}_1)\varphi_n^*(\mathbf{r}_2).$$

Show that the Green's function, Eq. (9.87), becomes

$$G(\mathbf{r}_1, \mathbf{r}_2) = \sum_{n=0}^{\infty} \frac{\varphi_n(\mathbf{r}_1)\varphi_n^*(\mathbf{r}_2)}{k_n^2 - k^2}$$

$$= G^*(\mathbf{r}_2, \mathbf{r}_1).$$

ADDITIONAL READINGS

BYRON, F. W., JR., and R. W. FULLER, *Mathematics of Classical and Quantum Physics.* Reading, MA: Addison-Wesley (1969).

MILLER, K. S., *Linear Differential Equations in the Real Domain.* New York: Norton (1963).

TITCHMARSH, E. C., *Eigenfunction Expansions Associated with Second Order Differential Equations*, 2nd ed., Vol. I. London: Oxford University Press (1962), Vol. II (1958).

10 THE GAMMA FUNCTION (FACTORIAL FUNCTION)

The gamma function appears occasionally in physical problems such as the normalization of Coulomb wave functions and the computation of probabilities in statistical mechanics. In general, however, it has less direct physical application and interpretation than, say, the Legendre and Bessel functions of Chapters 11 and 12. Rather, its importance stems from its usefulness in developing other functions that have direct physical application. The gamma function, therefore, is included here. A discussion of the numerical evaluation of the gamma function appears in Section 10.3.

10.1 DEFINITIONS, SIMPLE PROPERTIES

At least three different, convenient definitions of the gamma function are in common use. Our first task is to state these definitions, to develop some simple, direct consequences, and to show the equivalence of the three forms.

INFINITE LIMIT (EULER)
The first definition, named after Euler, is

$$\Gamma(z) \equiv \lim_{n \to \infty} \frac{1 \cdot 2 \cdot 3 \cdots n}{z(z + 1)(z + 2) \cdots (z + n)} n^z, \qquad z \neq 0, -1, -2, -3, \ldots.$$

$$(10.1)$$

This definition of $\Gamma(z)$ is useful in developing the Weierstrass infinite-product form of $\Gamma(z)$ and Eq. (10.16) and in obtaining the derivative of $\ln \Gamma(z)$

(Section 10.2). Here and elsewhere in this chapter z may be either real or complex. Replacing z with $z + 1$, we have

$$\Gamma(z + 1) = \lim_{n \to \infty} \frac{1 \cdot 2 \cdot 3 \cdots n}{(z + 1)(z + 2)(z + 3) \cdots (z + n + 1)} n^{z+1}$$

$$= \lim_{n \to \infty} \frac{nz}{z + n + 1} \cdot \frac{1 \cdot 2 \cdot 3 \cdots n}{z(z + 1)(z + 2) \cdots (z + n)} n^z$$

$$= z\Gamma(z). \tag{10.2}$$

This is the basic functional relation for the gamma function. It should be noted that it is a *difference* equation. It has been shown that the gamma function is one of a general class of functions that do not satisfy any differential equation with rational coefficients. Specifically, the gamma function is one of the very few functions of mathematical physics that does not satisfy either the hypergeometric differential equation (Section 13.5) or the confluent hypergeometric equation (Section 13.6).

Also, from the definition

$$\Gamma(1) = \lim_{n \to \infty} \frac{1 \cdot 2 \cdot 3 \cdots n}{1 \cdot 2 \cdot 3 \cdots n(n + 1)} n$$

$$= 1. \tag{10.3}$$

Now, application of Eq. (10.2) gives

$$\Gamma(2) = 1,$$

$$\Gamma(3) = 2\Gamma(2) = 2, \tag{10.4}$$

$$\Gamma(n) = 1 \cdot 2 \cdot 3 \cdots (n - 1) = (n - 1)!.$$

DEFINITE INTEGRAL (EULER)

A second definition, also frequently called Euler's form, is

$$\Gamma(z) \equiv \int_0^\infty e^{-t} t^{z-1} \, dt, \qquad \Re(z) > 0. \tag{10.5}$$

The restriction on z is necessary to avoid divergence of the integral. When the gamma function does appear in physical problems, it is often in this form or some variation such as

$$\Gamma(z) = 2 \int_0^\infty e^{-t^2} t^{2z-1} \, dt, \qquad \Re(z) > 0, \tag{10.6}$$

$$\Gamma(z) = \int_0^1 \left[\ln\left(\frac{1}{t}\right) \right]^{z-1} dt, \qquad \Re(z) > 0. \tag{10.7}$$

When $z = \frac{1}{2}$, Eq. (10.6) is just the Gauss error function, and we have the interesting result

$$\Gamma(\tfrac{1}{2}) = \sqrt{\pi}.$$ (10.8)

Generalizations of Eq. (10.6), the Gaussian integrals, are considered in Exercise 10.1.11. This definite integral form of $\Gamma(z)$, Eq. (10.5), leads to the beta function, Section 10.4.

To show the equivalence of these two definitions, Eqs. (10.1) and (10.5), consider the function of two variables

$$F(z, n) = \int_0^n \left(1 - \frac{t}{n}\right)^n t^{z-1} \, dt, \qquad \mathfrak{R}(z) > 0,$$ (10.9)

with n a positive integer.[1] Since

$$\lim_{n \to \infty} \left(1 - \frac{t}{n}\right)^n \equiv e^{-t},$$ (10.10)

from the definition of the exponential

$$\lim_{n \to \infty} F(z, n) = F(z, \infty) = \int_0^\infty e^{-t} t^{z-1} \, dt$$

$$\equiv \Gamma(z)$$ (10.11)

by Eq. (10.5).

Returning to $F(z, n)$, we evaluate it in successive integrations by parts. For convenience let $u = t/n$. Then

$$F(z, n) = n^z \int_0^1 (1 - u)^n u^{z-1} \, du.$$ (10.12)

Integrating by parts, we obtain

$$\frac{F(z, n)}{n^z} = (1 - u)^n \frac{u^z}{z} \Big|_0^1 + \frac{n}{z} \int_0^1 (1 - u)^{n-1} u^z \, du.$$ (10.13)

Repeating this with the integrated part vanishing at both end points each time, we finally get

$$F(z, n) = n^z \frac{n(n - 1) \cdots 1}{z(z + 1) \cdots (z + n - 1)} \int_0^1 u^{z+n-1} \, du$$

$$= \frac{1 \cdot 2 \cdot 3 \cdots n}{z(z + 1)(z + 2) \cdots (z + n)} n^z.$$ (10.14)

This is identical with the expression on the right side of Eq. (10.1). Hence

$$\lim_{n \to \infty} F(z, n) = F(z, \infty) \equiv \Gamma(z),$$ (10.15)

by Eq. (10.1), completing the proof.

[1] The form of $F(z, n)$ is suggested by the beta function (compare Eq. (10.60)).

INFINITE PRODUCT (WEIERSTRASS)

The third definition (Weierstrass's form) is

$$\frac{1}{\Gamma(z)} \equiv z e^{\gamma z} \prod_{n=1}^{\infty} \left(1 + \frac{z}{n}\right) e^{-z/n}, \tag{10.16}$$

where γ is the usual Euler–Mascheroni constant,

$$\gamma = 0.577\,216\ldots. \tag{10.17}$$

This infinite-product form may be used to develop the reflection identity, Eq. (10.23), and applied in the exercises, such as Exercise 10.1.19. This form can be derived from the original definition (Eq. (10.1)) by rewriting it as

$$\Gamma(z) = \lim_{n \to \infty} \frac{1 \cdot 2 \cdot 3 \cdots n}{z(z+1) \cdots (z+n)} n^z$$

$$= \lim_{n \to \infty} \frac{1}{z} \prod_{m=1}^{n} \left(1 + \frac{z}{m}\right)^{-1} n^z. \tag{10.18}$$

Inverting and using

$$n^{-z} = e^{(-\ln n)z}, \tag{10.19}$$

we obtain

$$\frac{1}{\Gamma(z)} = z \lim_{n \to \infty} e^{(-\ln n)z} \prod_{m=1}^{n} \left(1 + \frac{z}{m}\right). \tag{10.20}$$

Multiplying and dividing by

$$\exp\left[\left(1 + \frac{1}{2} + \frac{1}{3} + \cdots + \frac{1}{n}\right)z\right] = \prod_{m=1}^{n} e^{z/m}, \tag{10.21}$$

we get

$$\frac{1}{\Gamma(z)} = z\left\{ \lim_{n \to \infty} \exp\left[\left(1 + \frac{1}{2} + \frac{1}{3} + \cdots + \frac{1}{n} - \ln n\right)z\right]\right\}$$

$$\times \left[\lim_{n \to \infty} \prod_{m=1}^{n} \left(1 + \frac{z}{m}\right) e^{-z/m}\right]. \tag{10.22}$$

As shown in Section 5.2, the infinite series in the exponent converges and defines γ, the Euler–Mascheroni constant. Hence Eq. (10.16) follows.

It was shown in Section 5.11 that the Weierstrass infinite-product definition of $\Gamma(z)$ led directly to an important identity,

$$\Gamma(z)\Gamma(1-z) = \frac{\pi}{\sin z\pi}. \tag{10.23}$$

This identity may also be derived by contour integration (Example 7.2.5 and Exercises 7.2.18 and 7.2.19) and the beta function, Section 10.4. Setting $z = \frac{1}{2}$

in Eq. (10.23), we obtain

$$\Gamma(\tfrac{1}{2}) = \sqrt{\pi} \tag{10.24}$$

(taking the positive square root) in agreement with Eq. (10.8).

The Weierstrass definition shows immediately that $\Gamma(z)$ has simple poles at $z = 0, -1, -2, -3, \ldots$, and that $[\Gamma(z)]^{-1}$ has no poles in the finite complex plane, which means that $\Gamma(z)$ has no zeros. This behavior may also be seen in Eq. (10.23), in which we note that $\pi/(\sin \pi z)$ is never equal to zero.

Actually the infinite-product definition of $\Gamma(z)$ may be derived from the Weierstrass factorization theorem with the specification that $[\Gamma(z)]^{-1}$ have simple zeros at $z = 0, -1, -2 -3, \ldots$. The Euler–Masheroni constant is fixed by requiring $\Gamma(1) = 1$. See also the products expansions of entire functions in Chapter 7.

In probability theory the gamma distribution (probability density) is given by

$$f(x) = \begin{cases} \dfrac{1}{\beta^{\alpha}\Gamma(\alpha)} x^{\alpha-1} e^{-x/\beta}, & x > 0 \\ 0, & x \leq 0. \end{cases} \tag{10.24a}$$

The constant $[\beta^{\alpha}\Gamma(\alpha)]^{-1}$ is chosen so that the total (integrated) probability will be unity. For $x \to E$, kinetic energy, $\alpha \to \tfrac{3}{2}$ and $\beta \to kT$, Eq. (10.24a) yields the classical Maxwell–Boltzmann statistics.

FACTORIAL NOTATION

So far this discussion has been presented in terms of the classical notation. As pointed out by Jeffreys and others, the -1 of the $z - 1$ exponent in our second definition (Eq. (10.5) is a continual nuisance. Accordingly, Eq. (10.5) is rewritten as

$$\int_0^\infty e^{-t} t^z \, dt \equiv z!, \qquad \mathfrak{R}(z) > -1, \tag{10.25}$$

to *define* a factorial function $z!$. Occasionally we may still encounter Gauss's notation, $\Pi(z)$, for the factorial function

$$\Pi(z) = z!. \tag{10.26}$$

The Γ notation is due to Legendre. The factorial function of Eq. (10.25) is, of course, related to the gamma function by

$$\Gamma(z) = (z - 1)!$$

or $\tag{10.27}$

$$\Gamma(z + 1) = z!.$$

If $z = n$, a positive integer (Eq. (10.4)) shows that

$$z! = n! = 1 \cdot 2 \cdot 3 \cdots n, \tag{10.28}$$

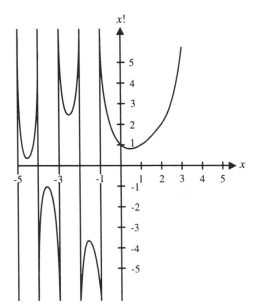

Figure 10.1 The factorial function—extension to negative arguments.

the familiar factorial. However, it should be noted carefully that since $z!$ is now defined by Eq. (10.25) (or equivalently by Eq. (10.27)) *the factorial function is no longer limited to positive integral values of the argument* (Fig. 10.1). The difference relation (Eq. (10.2)) becomes

$$(z - 1)! = \frac{z!}{z}.$$ (10.29)

This shows immediately that

$$0! = 1$$ (10.30)

and

$$n! = \pm\infty \qquad \text{for } n, \text{ a } \textit{negative} \text{ integer.}$$ (10.31)

In terms of the factorial Eq. (10.23) becomes

$$z!(-z)! = \frac{\pi z}{\sin \pi z}.$$ (10.32)

By restricting ourselves to the real values of the argument, we find that $x!$ defines the curve shown in Fig. 10.1. The minimum of the curve is

$$x! = (0.46163 \cdots)! = 0.88560 \cdots.$$ (10.33a)

DOUBLE FACTORIAL NOTATION

In many problems of mathematical physics, particularly in connection with Legendre polynomials (Chapter 12), we encounter products of the odd positive integers and products of the even positive integers. For convenience these are

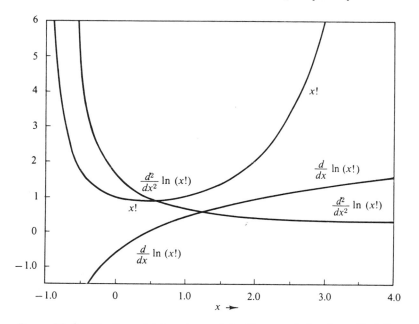

Figure 10.2   The factorial function and the first two derivatives of $\ln(x!)$.

given special labels as double factorials:

$$1 \cdot 3 \cdot 5 \cdots (2n + 1) = (2n + 1)!!$$

$$2 \cdot 4 \cdot 6 \cdots (2n) = (2n)!!.$$

(10.33b)

Clearly, these are related to the regular factorial functions by

$$(2n)!! = 2^n n! \quad \text{and} \quad (2n + 1)!! = \frac{(2n + 1)!}{2^n n!}.$$

(10.33c)

INTEGRAL REPRESENTATION

An integral representation that is useful in developing asymptotic series for the Bessel functions is

$$\int_C e^{-z} z^\nu \, dz = (e^{2\pi i \nu} - 1)\nu!,$$

(10.34)

where C is the contour shown in Fig. 10.3. This contour integral representation is particularly useful when ν is not an integer, $z = 0$ then being a *branch point*. Equation (10.34) may be readily verified for $\nu > -1$ by deforming the contour as shown in Fig. 10.4. The integral from ∞ into the origin yields $-(\nu!)$, placing the phase of z at 0. The integral out to ∞ (in the fourth quadrant) then yields $e^{2\pi i \nu}\nu!$, the phase of z having increased to 2π. Since the circle around the origin contributes nothing when $\nu > -1$, Eq. (10.34) follows.

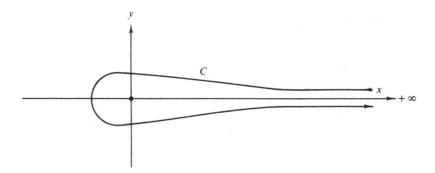

Figure 10.3 Factorial function contour.

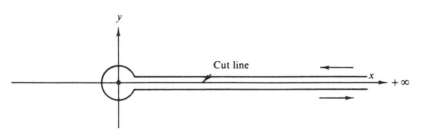

Figure 10.4 The contour of Fig. 10.3 deformed.

It is often convenient to throw this result into a more symmetrical form

$$\int_C e^{-z}(-z)^v \, dz = 2i \sin v\pi \, v!. \tag{10.35}$$

This corresponds to choosing the phase of z to have a range of $-\pi$ to $+\pi$ in Eq. (10.34).

This analysis establishes Eqs. (10.34) and (10.35) for $v > -1$. It is relatively simple to extend the range to include all nonintegral v. First, we note that the integral exists for $v < -1$ as long as we stay away from the origin. Second, integrating by parts we find that Eq. (10.35) yields the familiar difference relation (Eq. (10.29)). If we take the difference relation to define the factorial function of $v < -1$, then Eqs. (10.34) and (10.35) are verified for all v (except negative integers).

EXERCISES

10.1.1   Derive the recurrence relations

$$\Gamma(z + 1) = z\Gamma(z)$$

from the Euler integral form (Eq. (10.5)),

$$\Gamma(z) = \int_0^\infty e^{-t} t^{z-1} \, dt.$$

10.1.2   In a power-series solution for the Legendre functions of the second kind we encounter the expression

$$\frac{(n + 1)(n + 2)(n + 3) \cdots (n + 2s - 1)(n + 2s)}{2 \cdot 4 \cdot 6 \cdot 8 \cdots (2s - 2)(2s) \cdot (2n + 3)(2n + 5)(2n + 7) \cdots (2n + 2s + 1)},$$

in which s is a positive integer. Rewrite this expression in terms of factorials.

10.1.3   Show that

$$\frac{(s - n)!}{(2s - 2n)!} = \frac{(-1)^{n-s}(2n - 2s)!}{(n - s)!}.$$

Here s and n are integers with $s < n$. This result can be used to avoid negative factorials such as in the series representations of the spherical Neumann functions and the Legendre functions of the second kind.

10.1.4   Show that $\Gamma(z)$ may be written

$$\Gamma(z) = 2 \int_0^\infty e^{-t^2} t^{2z-1} \, dz, \qquad \mathcal{R}(z) > 0,$$

$$\Gamma(z) = \int_0^1 \left[\ln\left(\frac{1}{t}\right) \right]^{z-1} dt, \qquad \mathcal{R}(z) > 0.$$

10.1.5   In a Maxwellian distribution the fraction of particles between the speed v and $v + dv$ is

$$\frac{dN}{N} = 4\pi \left(\frac{m}{2\pi kT}\right)^{3/2} \exp(-mv^2/2kT) v^2 \, dv,$$

N being the total number of particles. The average or expectation value of v^n is defined as $\langle v^n \rangle = N^{-1} \int v^n \, dN$. Show that

$$\langle v^n \rangle = \left(\frac{2kT}{m}\right)^{n/2} \left(\frac{n + 1}{2}\right)! \bigg/ \frac{1}{2}!.$$

10.1.6   By transforming the integral into a gamma function, show that

$$-\int_0^1 x^k \ln x \, dx = \frac{1}{(k + 1)^2}, \qquad k > -1.$$

10.1.7   Show that

$$\int_0^\infty e^{-x^4} \, dx = (\tfrac{1}{4})!$$

10.1.8   Show that

$$\lim_{x \to 0} \frac{(ax - 1)!}{(x - 1)!} = \frac{1}{a}.$$

10.1.9   Locate the poles of $\Gamma(z)$. Show that they are simple poles and determine the residues.

10.1.10   Show that the equation $x! = k$, $k \neq 0$, has an infinite number of real roots.

10.1.11   Show that

(a) $\displaystyle \int_0^\infty x^{2s+1} \exp(-ax^2) \, dx = \frac{s!}{2a^{s+1}}.$

(b) $\displaystyle\int_0^\infty x^{2s} \exp(-ax^2)\,dx = \frac{(s - \frac{1}{2})!}{2a^{s+1/2}}$

$$= \frac{(2s - 1)!!}{2^{s+1}a^s}\sqrt{\frac{\pi}{a}}.$$

These Gaussian integrals are of major importance in statistical mechanics.

10.1.12   (a) Develop recurrence relations for $(2n)!!$ and for $(2n + 1)!!$.
  (b) Use these recurrence relations to calculate (or to define) $0!!$ and $(-1)!!$.

  ANS. $0!! = 1$
  $(-1)!! = 1$.

10.1.13   For s a nonnegative integer, show that

$$(-2s - 1)!! = \frac{(-1)^s}{(2s - 1)!!} = \frac{(-1)^s 2^s s!}{(2s)!}.$$

10.1.14   Express the coefficient of the nth term of the expansion of $(1 + x)^{1/2}$

  (a) in terms of factorials of integers,
  (b) in terms of the double factorial $(!!)$ functions.

$$ANS.\quad a_n = (-1)^{n+1}\frac{(2n - 3)!}{2^{2n-2}n!(n - 2)!} = (-1)^{n+1}\frac{(2n - 3)!!}{(2n)!!},$$

$$n = 2, 3, 4, \ldots.$$

10.1.15   Express the coefficient of the nth term of the expansion of $(1 + x)^{-1/2}$

  (a) in terms of the factorials of integers,
  (b) in terms of the double factorial $(!!)$ functions.

$$ANS.\quad a_n = (-1)^n\frac{(2n)!}{2^{2n}(n!)^2} = (-1)^n\frac{(2n - 1)!!}{(2n)!!},\quad n = 1, 2, 3, \ldots.$$

10.1.16   The Legendre polynomial may be written as

$$P_n(\cos\theta) = 2\frac{(2n - 1)!!}{(2n)!!}\Bigg\{\cos n\theta + \frac{1}{1}\cdot\frac{n}{2n - 1}\cos(n - 2)\theta$$

$$+ \frac{1\cdot 3}{1\cdot 2}\frac{n(n - 1)}{(2n - 1)(2n - 3)}\cos(n - 4)\theta$$

$$+ \frac{1\cdot 3\cdot 5}{1\cdot 2\cdot 3}\frac{n(n - 1)(n - 2)}{(2n - 1)(2n - 3)(2n - 5)}\cos(n - 6)\theta + \cdots\Bigg\}.$$

For n we let $n = 2s + 1$. Then

$$P_n(\cos\theta) = P_{2s+1}(\cos\theta) = \sum_{m=0}^{s} a_m \cos(2m + 1)\theta.$$

Find a_m in terms of factorials and double factorials.

10.1.17   (a) Show that

$$\Gamma(\tfrac{1}{2} - n)\Gamma(\tfrac{1}{2} + n) = (-1)^n\pi,$$

  where n is an integer.
  (b) Express $\Gamma(\tfrac{1}{2} + n)$ and $\Gamma(\tfrac{1}{2} - n)$ separately in terms of $\pi^{1/2}$ and a $!!$ function.

$$ANS.\quad \Gamma(\tfrac{1}{2} + n) = \frac{(2n - 1)!!}{2^n}\pi^{1/2}.$$

10.1.18 From one of the definitions of the factorial or gamma function, show that

$$|(ix)!|^2 = \frac{\pi x}{\sinh \pi x}.$$

10.1.19 Prove that

$$|\Gamma(\alpha + i\beta)| = \Gamma \frac{(1 + \alpha)}{\alpha} \prod_{n=0}^{\infty} \left[1 + \frac{\beta^2}{(\alpha + n)^2} \right]^{-1/2}.$$

This equation has been useful in calculations in the theory of beta decay.

10.1.20 Show that

$$|(n + ib)!| = \left(\frac{\pi b}{\sinh \pi b} \right)^{1/2} \prod_{s=1}^{n} (s^2 + b^2)^{1/2}$$

for n, a positive integer.

10.1.21 Show that

$$|x!| \geq |(x + iy)!|$$

for all x. The variables x and y are real.

10.1.22 Show that

$$|(-\tfrac{1}{2} + iy)!|^2 = \frac{\pi}{\cosh \pi y}.$$

10.1.23 The probability density associated with the normal distribution of statistics is given by

$$f(x) = \frac{1}{\sigma(2\pi)^{1/2}} \exp[-(x - \mu)^2/2\sigma^2]$$

with $(-\infty, \infty)$ for the range of x. Show that

(a) the mean value of x, $\langle x \rangle$ is equal to μ,
(b) the standard deviation $(\langle x^2 \rangle - \langle x \rangle^2)^{1/2}$ is given by σ.

10.1.24 From the gamma distribution

$$f(x) = \begin{cases} \dfrac{1}{\beta^\alpha \Gamma(\alpha)} x^{\alpha-1} e^{-x/\beta}, & x > 0 \\ 0, & x \leq 0, \end{cases}$$

show that

(a) $\langle x \rangle$ (mean) $= \alpha\beta$,
(b) σ^2 (variance) $\equiv \langle x^2 \rangle - \langle x \rangle^2 = \alpha\beta^2$.

10.1.25 The wave function of a particle scattered by a pure Coulomb potential is $\psi(r, \theta)$. At the origin the wave function becomes

$$\psi(0) = e^{-\pi\gamma/2} \Gamma(1 + i\gamma),$$

where $\gamma = Z_1 Z_2 e^2/\hbar v$. Show that

$$\psi^*(0)\psi(0) = \frac{2\pi\gamma}{e^{2\pi\gamma} - 1}.$$

10.1.26   Derive the contour integral representation of Eq. (10.34)

$$2i \sin v\pi \, v! = \int_C e^{-z}(-z)^v \, dz.$$

10.1.27   Write a function subprogram $FACT(N)$ (fixed point independent variable) that will calculate $N!$. Include provision for rejection and appropriate error message if N is negative.

Note. For small N direct multiplication is simplest. For large N, if large N are considered, Eq. (10.55), Stirling's series would be appropriate.

10.1.28   (a) Write a function subprogram to calculate the double factorial ratio $(2N - 1)!!/(2N)!!$. Include provision for $N = 0$ and for rejection and an error message if N is negative. Calculate and tabulate this ratio for $N = 1(1)100$.

(b) Check your function subprogram calculation of $199!!/200!!$ against the value obtained from Stirling's series (Section 10.3).

$$ANS. \quad \frac{199!!}{200!!} = 0.056348.$$

10.1.29   Using either the Fortran supplied GAMMA or a library supplied subroutine for $x!$ or $\Gamma(x)$, determine the value of x for which $\Gamma(x)$ is a minimum $(1 \le x \le 2)$ and this minimum value of $\Gamma(x)$. Notice that although the minimum value of $\Gamma(x)$ may be obtained to about six significant figures (single precision), the corresponding value of x is much less accurate. Why this relatively low accuracy?

10.1.30   The factorial function expressed in integral form can be evaluated by the Gauss–Laguerre quadrature. For a 10-point formula Appendix 2 guarantees the resultant $x!$ theoretically exact for x an integer, 0 up through 19. What happens if x is not an integer? Use the Gauss–Laguerre quadrature to evaluate $x!$, $x = 0.0(0.1)2.0$. Tabulate the absolute error as a function of x.

Check value. $\quad x!_{\text{exact}} - x!_{\text{quadrature}} = 0.00034 \qquad$ for $x = 1.3$.

10.2 DIGAMMA AND POLYGAMMA FUNCTIONS

DIGAMMA FUNCTIONS

As may be noted from the three definitions in Section 10.1, it is inconvenient to deal with the derivatives of the gamma or factorial function directly. Instead, it is customary to take the natural logarithm of the factorial function (Eq. (10.1)), convert the product to a sum, and then differentiate, that is,

$$z! = z\Gamma(z) = \lim_{n \to \infty} \frac{n!}{(z + 1)(z + 2) \cdots (z + n)} \, n^z \tag{10.36}$$

and

$$\ln(z!) = \lim_{n \to \infty} [\ln(n!) + z \ln n - \ln(z + 1)$$

$$- \ln(z + 2) - \cdots - \ln(z + n)], \tag{10.37}$$

in which the logarithm of the limit is equal to the limit of the logarithm. Differentiating with respect to z, we obtain

$$\frac{d}{dz}\ln(z!) \equiv F(z) = \lim_{n \to \infty} \left(\ln n - \frac{1}{z+1} - \frac{1}{z+2} - \cdots - \frac{1}{z+n} \right), \quad (10.38)$$

which defines $F(z)$, the digamma function. From the definition of the Euler–Mascheroni constant[1] Eq. (10.38) may be rewritten as

$$F(z) = -\gamma - \sum_{n=1}^{\infty} \left(\frac{1}{z+n} - \frac{1}{n} \right)$$

$$= -\gamma + \sum_{n=1}^{\infty} \frac{z}{n(n+z)}. \quad (10.39)$$

One application of Eq. (10.39) is in the derivation of the series form of the Neumann function (Section 11.3). Clearly,

$$F(0) = -\gamma = -0.577\,215\,664\,901 \cdots .[2] \quad (10.40)$$

Another, perhaps more useful, expression for $F(z)$ is derived in Section 10.3.

POLYGAMMA FUNCTION

The digamma function may be differentiated repeatedly, giving rise to the polygamma function:

$$F^{(m)}(z) \equiv \frac{d^{m+1}}{dz^{m+1}} \ln(z!)$$

$$= (-1)^{m+1} m! \sum_{n=1}^{\infty} \frac{1}{(z+n)^{m+1}}, \quad m = 1, 2, 3, \ldots . \quad (10.41)$$

A plot of $F(x)$ and $F'(x)$ is included in Fig. 10.2. Since the series in Eq. (10.41) defines the Riemann zeta function[3] (with $z = 0$),

$$\zeta(m) \equiv \sum_{n=1}^{\infty} \frac{1}{n^m}, \quad (10.42)$$

we have

$$F^{(m)}(0) = (-1)^{m+1} m! \zeta(m+1), \quad m = 1, 2, 3, \ldots . \quad (10.43)$$

The values of the polygamma functions of positive integral argument, $F^{(m)}(n)$, may be calculated by using Exercise 10.2.6.

---

[1] Compare Sections 5.2 and 5.6. We add and subtract $\sum_{s=1}^{n} s^{-1}$.

[2] γ has been computed to 1271 places by D. E. Knuth, *Math. Comput.* **16**, 275 (1962) and to 3566 decimal places by D. W. Sweeney, *ibid.* **17**, 170 (1963). It may be of interest that the fraction 228/395 gives γ accurate to six places.

[3] See Section 5.9. For $z \neq 0$ this series may be used to define a generalized zeta function.

In terms of the perhaps more common Γ notation,

$$\frac{d^{n+1}}{dz^{n+1}} \ln \Gamma(z) = \frac{d^n}{dz^n} \psi(z) = \psi^{(n)}(z). \tag{10.44a}$$

From Eq. (10.27)

$$\psi^{(n)}(z) = \mathbf{F}^{(n)}(z - 1). \tag{10.44b}$$

MACLAURIN EXPANSION, COMPUTATION

It is now possible to write a Maclaurin expansion for $\ln(z!)$.

$$\ln(z!) = \sum_{n=1}^{\infty} \frac{z^n}{n!} \mathbf{F}^{(n-1)}(0)$$

$$= -\gamma z + \sum_{n=2}^{\infty} (-1)^n \frac{z^n}{n} \zeta(n) \tag{10.44c}$$

convergent for $|z| < 1$; for $z = x$, the range is $-1 < x \le 1$. Alternate forms of this series appear in Exercise 5.9.14. Equation (10.44c) is a possible means of computing $z!$ for real or complex z, but Stirling's series (Section 10.3) is usually better, and in addition, an excellent table of values of the gamma function for complex arguments based on the use of Stirling's series and the recurrence relation (Eq. (10.29)) is now available.[4]

SERIES SUMMATION

The digamma and polygamma functions may also be used in summing series. If the general term of the series has the form of a rational fraction (with the highest power of the index in the numerator at least two less than the highest power of the index in the denominator), it may be transformed by the method of partial fractions (compare Section 15.8). The infinite series may then be expressed as a finite sum of digamma and polygamma functions. The usefulness of this method depends on the availability of tables of digamma and polygamma functions. Such tables and examples of series summation are given in AMS-55, Chapter 6.

Example 10.2.1 Catalan's Constant
Catalan's constant, Exercise 5.2.22, or $\beta(2)$ of Section 5.9 is given by

$$K = \beta(2) = \sum_{k=0}^{\infty} \frac{(-1)^k}{(2k + 1)^2}. \tag{10.44d}$$

Grouping the positive and negative terms separately and starting with unit index (to match the form of $\mathbf{F}^{(1)}$, Eq. (10.41)), we obtain

$$K = 1 + \sum_{n=1}^{\infty} \frac{1}{(4n + 1)^2} - \frac{1}{9} - \sum_{n=1}^{\infty} \frac{1}{(4n + 3)^2}.$$

[4] *Table of the Gamma Function for Complex Arguments*, Applied Mathematics Series No. 34. Washington, DC: National Bureau of Standards (1954).

Now, quoting Eqs. (10.41) and (10.44b), we get

$$K = \tfrac{8}{9} + \tfrac{1}{16}\mathbf{F}^{(1)}(\tfrac{1}{4}) - \tfrac{1}{16}\mathbf{F}^{(1)}(\tfrac{3}{4})$$

$$= \tfrac{8}{9} + \tfrac{1}{16}\psi^{(1)}(1 + \tfrac{1}{4}) - \tfrac{1}{16}\psi^{(1)}(1 + \tfrac{3}{4}). \qquad (10.44e)$$

Using the values of $\psi^{(1)}$ from Table 6.1 of AMS-55, we obtain

$$K = 0.9159\,6559\ldots.$$

Compare this calculation of Catalan's constant with the calculations of Chapter 5, either direct summation by machine or a modification using Riemann zeta functions and then a (shorter) machine computation.

EXERCISES

10.2.1   Verify that the following two forms of the digamma function,

$$\mathbf{F}(x) = \sum_{r=1}^{x} \frac{1}{r} - \gamma$$

and

$$\mathbf{F}(x) = \sum_{r=1}^{\infty} \frac{x}{r(r+x)} - \gamma,$$

are equal to each other (for x a positive integer).

10.2.2   Show that $\mathbf{F}(z)$ has the series expansion

$$\mathbf{F}(z) = -\gamma + \sum_{n=2}^{\infty} (-1)^n \zeta(n) z^{n-1}.$$

10.2.3   For a power series expansion of $\ln(z!)$, AMS-55 lists

$$\ln(z!) = -\ln(1 + z) + z(1 - \gamma) + \sum_{n=2}^{\infty} (-1)^n [\zeta(n) - 1] z^n / n.$$

(a) Show that this agrees with Eq. (10.44c) for $|z| < 1$.
(b) What is the range of convergence of this new expression?

10.2.4   Show that

$$\frac{1}{2}\ln\left(\frac{\pi z}{\sin \pi z}\right) = \sum_{n=1}^{\infty} \frac{\zeta(2n)}{2n} z^{2n}, \qquad |z| < 1.$$

Hint. Try Eq. (10.32).

10.2.5   Write out a Weierstrass infinite product definition of $\ln(z!)$. Without differentiating, show that this leads directly to the Maclaurin expansion of $\ln(z!)$, Eq. (10.44c).

10.2.6   Derive the difference relation for the polygamma function

$$\mathbf{F}^{(m)}(z + 1) = \mathbf{F}^{(m)}(z) + (-1)^m \frac{m!}{(z+1)^{m+1}}, \qquad m = 0, 1, 2, \ldots.$$

10.2.7   Show that if
$$\Gamma(x + iy) = u + iv$$
then
$$\Gamma(x - iy) = u - iv.$$
This is a special case of the Schwarz reflection principle, Section 6.5.

10.2.8   The Pochhammer symbol $(a)_n$ is defined as
$$(a)_n = a(a + 1) \cdots (a + n - 1)$$
$$(a)_0 = 1$$
(for integral n).

(a) Express $(a)_n$ in terms of factorials.
(b) Find $(d/da)(a)_n$ in terms of $(a)_n$ and digamma functions.

$$ANS. \quad \frac{d}{da}(a)_n = (a)_n[\mathbf{F}(a + n - 1) - \mathbf{F}(a - 1)].$$

(c) Show that
$$(a)_{n+k} = (a + n)_k \cdot (a)_n.$$

10.2.9   Verify the following special values of the ψ form of the di- and polygamma functions
$$\psi(1) = -\gamma$$
$$\psi^{(1)}(1) = \zeta(2)$$
$$\psi^{(2)}(1) = -2\zeta(3).$$

10.2.10   Derive the polygamma function recurrence relation
$$\psi^{(m)}(1 + z) = \psi^{(m)}(z) + (-1)^m m!/z^{m+1}, \quad m = 0, 1, 2, \ldots.$$

10.2.11   Verify

(a) $\displaystyle\int_0^\infty e^{-r} \ln r \, dr = -\gamma.$

(b) $\displaystyle\int_0^\infty re^{-r} \ln r \, dr = 1 - \gamma.$

(c) $\displaystyle\int_0^\infty r^n e^{-r} \ln r \, dr = (n - 1)! + n \int_0^\infty r^{n-1} e^{-r} \ln r \, dr, \quad n = 1, 2, 3, \ldots.$

Hint. These may be verified by integration by parts, three parts, or differentiating the integral form of $n!$ with respect to n.

10.2.12   Dirac relativistic wave functions for hydrogen involve factors such as $[2(1 - \alpha^2 Z^2)^{1/2}]!$ where α, the fine structure constant, is $\frac{1}{137}$ and Z is the atomic number. Expand $[2(1 - \alpha^2 Z^2)^{1/2}]!$ in a series of powers of $\alpha^2 Z^2$.

10.2.13   The quantum mechanical description of a particle in a coulomb field requires a knowledge of the phase of the complex factorial function. Determine the phase of $(1 + ib)!$ for small b.

10.2.14   The total energy radiated by a black body is given by

$$u = \frac{8\pi k^4 T^4}{c^3 h^3} \int_0^\infty \frac{x^3}{e^x - 1} \, dx.$$

Show that the integral in this expression is equal to 3! $\zeta(4)$.
[$\zeta(4) = \pi^4/90 = 1.0823\ldots$.] The final result is the Stefan–Boltzmann law.

10.2.15   As a generalization of the result in Exercise 10.2.14, show that

$$\int_0^\infty \frac{x^s \, dx}{e^x - 1} = s! \, \zeta(s + 1), \qquad \Re(s) > 0.$$

10.2.16   The neutrino energy density (Fermi distribution) in the early history of the universe is given by

$$\rho_\nu = \frac{4\pi}{h^3} \int_0^\infty \frac{x^3}{\exp(x/kT) + 1} \, dx.$$

Show that

$$\rho_\nu = \frac{7\pi^5}{30 h^3} (kT)^4.$$

10.2.17   Prove that

$$\int_0^\infty \frac{x^s \, dx}{e^x + 1} = s!(1 - 2^{-s})\zeta(s + 1), \qquad \Re(s) > 0.$$

Exercises 10.2.15 and 10.2.17 actually constitute Mellin integral transforms (compare Section 15.1).

10.2.18   Prove that

$$\psi^{(n)}(z) = (-1)^{n+1} \int_0^\infty \frac{t^n e^{-zt}}{1 - e^{-t}} \, dt, \qquad \Re(z) > 0.$$

10.2.19   Using di- and polygamma functions sum the series

(a) $\displaystyle\sum_{n=1}^\infty \frac{1}{n(n+1)}$

(b) $\displaystyle\sum_{n=2}^\infty \frac{1}{n^2 - 1}$.

Note. You can use Exercise 10.2.6 to calculate the needed digamma functions.

10.2.20   Show that

$$\sum_{n=1}^\infty \frac{1}{(n+a)(n+b)} = \frac{1}{(b-a)} \{F(b) - F(a)\}$$

$$= \frac{1}{(b-a)} \{\psi(1+b) - \psi(1+a)\}.$$

$a \neq b$, and neither a nor b is a negative integer. It is of some interest to compare this summation with the corresponding integral

$$\int_1^\infty \frac{dx}{(x+a)(x+b)} = \frac{1}{b-a} \{\ln(1+b) - \ln(1+a)\}.$$

The relation between $\psi(x)$ (or $F(x)$) and $\ln x$ is made explicit in Eq. (10.51) in the next section.

10.2.21   Verify the contour integral representation of $\zeta(s)$,

$$\zeta(s) = -\frac{(-s)!}{2\pi i} \int_C \frac{(-z)^{s-1}}{e^z - 1} dz.$$

The contour C is the same as that for Eq. (10.35). The points $z = \pm 2n\pi i$, $n = 1, 2, 3 \ldots$ are all excluded.

10.2.22   Show that $\zeta(s)$ is analytic in the entire finite complex plane except at $s = 1$ where it has a simple pole with a residue of $+1$.
Hint. The contour integral representation will be useful.

10.2.23   Using the complex variable capability of FORTRAN IV calculate $\mathfrak{R}(1 + ib)!$, $\mathfrak{I}(1 + ib)!$, $|(1 + ib)!|$ and phase $(1 + ib)!$ for $b = 0.0(0.1)1.0$. Plot the phase of $(1 + ib)!$ versus b.
Hint. Exercise 10.2.3 offers a convenient approach. You will need to calculate $\zeta(n)$.

10.3 STIRLING'S SERIES

For computation of $\ln(z!)$ for very large z (statistical mechanics) and for numerical computations at nonintegral values of z a series expansion of $\ln(z!)$ in negative powers of z is desirable. Perhaps the most elegant way of deriving such an expansion is by the method of steepest descents (Section 7.4). The following method, starting with a numerical integration formula, does not require knowledge of contour integration and is particularly direct.

DERIVATION FROM EULER–MACLAURIN INTEGRATION FORMULA

The Euler–Maclaurin formula for evaluating a definite integral[1] is

$$\int_0^n f(x) \, dx = \tfrac{1}{2}f(0) + f(1) + f(2) + \cdots + \tfrac{1}{2}f(n)$$

$$- b_2[f'(n) - f'(0)] - b_4[f'''(n) - f'''(0)] - \cdots, \quad (10.45)$$

in which the b_{2n} are related to the Bernoulli numbers B_{2n} (compare Section 5.9) by

$$(2n)! \, b_{2n} = B_{2n}, \tag{10.46}$$

$$B_0 = 1, \qquad B_6 = \tfrac{1}{42},$$
$$B_2 = \tfrac{1}{6}, \qquad B_8 = -\tfrac{1}{30}, \tag{10.47}$$
$$B_4 = -\tfrac{1}{30}, \qquad B_{10} = \tfrac{5}{66}, \text{ and so on.}$$

By applying Eq. (10.45) to the definite integral, we have

$$\int_0^\infty \frac{dx}{(z + x)^2} = \frac{1}{z} \tag{10.48}$$

---

[1] This is obtained by repeated integration by parts, Section 5.9.

(for z not on the negative real axis), we obtain

$$\frac{1}{z} = \frac{1}{2z^2} + \mathbf{F}^{(1)}(z) - \frac{2!b_2}{z^3} - \frac{4!b_4}{z^5} - \cdots . \tag{10.49}$$

This is the reason for using Eq. (10.48). The Euler–Maclaurin evaluation yields $\mathbf{F}^{(1)}(z)$, which is $d^2 \ln(z!)/dz^2$.

Using Eq. (10.46) and solving for $\mathbf{F}^{(1)}(z)$, we have

$$\mathbf{F}^{(1)}(z) = \frac{d}{dz} \mathbf{F}(z) = \frac{1}{z} - \frac{1}{2z^2} + \frac{B_2}{z^3} + \frac{B_4}{z^5} + \cdots$$

$$= \frac{1}{z} - \frac{1}{2z^2} + \sum_{n=1}^{\infty} \frac{B_{2n}}{z^{2n+1}} . \tag{10.50}$$

Since the Bernoulli numbers diverge strongly, this series does not converge! It is a semiconvergent or asymptotic series, useful for computation despite its divergence (compare Section 5.10).

Integrating once, we get the digamma function

$$\mathbf{F}(z) = C_1 + \ln z + \frac{1}{2z} - \frac{B_2}{2z^2} - \frac{B_4}{4z^4} - \cdots$$

$$= C_1 + \ln z + \frac{1}{2z} - \sum_{n=1}^{\infty} \frac{B_{2n}}{2nz^{2n}} . \tag{10.51}$$

Integrating Eq. (10.51) with respect to z from $z - 1$ to z and then letting z approach infinity, C_1, the constant of integration may be shown to vanish. This gives us a second expression for the digamma function, often more useful than Eq. (10.38).

STIRLING'S SERIES

The indefinite integral of the digamma function (Eq. (10.51)) is

$$\ln(z!) = C_2 + \left(z + \frac{1}{2} \right) \ln z - z + \frac{B_2}{2z} + \cdots + \frac{B_{2n}}{2n(2n - 1)z^{2n-1}} + \cdots , \tag{10.52}$$

in which C_2 is another constant of integration. To fix C_2 we find it convenient to use the doubling or Legendre duplication formula derived in Section 10.4,

$$z!(z - \tfrac{1}{2})! = 2^{-2z} \pi^{1/2} (2z)! . \tag{10.53}$$

This may be proved directly when z is a positive integer by writing $(2z)!$ as a product of even terms times a product of odd terms and extracting a factor of two from each term (Exercise 10.3.5). Substituting Eq. (10.52) into the logarithm of the doubling formula, we find that C_2 is

$$C_2 = \tfrac{1}{2} \ln 2\pi , \tag{10.54}$$

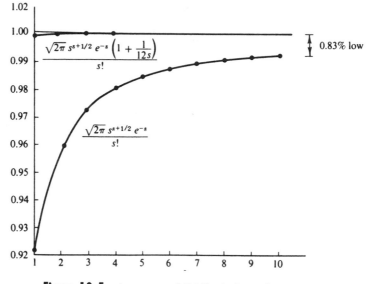

Figure 10.5   Accuracy of Stirling's formula.

giving

$$\ln(z!) = \frac{1}{2}\ln 2\pi + \left(z + \frac{1}{2}\right)\ln z - z + \frac{1}{12z} - \frac{1}{360z^3} + \frac{1}{1260z^5} - \cdots .$$

(10.55)

This is Stirling's series, an asymptotic expansion. The absolute value of the error is less than the absolute value of the first term neglected.

The constants of integration C_1 and C_2 may also be evaluated by comparison with the first term of the series expansion obtained by the method of "steepest descent." This is carried out in Section 7.4.

To help convey a feeling of the remarkable precision of Stirling's series for $s!$ the ratio of the first term of Stirling's approximation to $s!$ is plotted in Fig. 10.5. A tabulation gives the ratio of the first term in the expansion to $s!$ and the ratio of the first two terms in the expansion to $s!$ (Table 10.1). The derivation of these forms in Exercise 10.3.1.

NUMERICAL COMPUTATION

The possibility of using the Maclaurin expansion, Eq. (10.44c), for the numerical evaluation of the factorial function is mentioned in Section 10.2. However, for large x, Stirling's series, Eq. (10.55), gives much more rapid convergence. The *Table of the Gamma Function for Complex Arguments*, Applied Mathematics Series No. 34, National Bureau of Standards, is based on the use of Stirling's series for $z = x + iy$, $9 \le x \le 10$. Lower values of x

Table 10.1

s	$\dfrac{\sqrt{2\pi}\,s^{s+1/2}e^{-s}}{s!}$	$\dfrac{\sqrt{2\pi}\,s^{s+1/2}e^{-s}\left[1+\dfrac{1}{12s}\right]}{s!}$
1	0.92213	0.99898
2	0.95950	0.99949
3	0.97270	0.99972
4	0.97942	0.99983
5	0.98349	0.99988
6	0.98621	0.99992
7	0.98817	0.99994
8	0.98964	0.99995
9	0.99078	0.99996
10	0.99170	0.99998

are reached with the recurrence relation, Eq. (10.29). Now suppose the numerical value of $x!$ is needed for some particular value of x in a program in a large, high-speed digital computer. How shall we instruct the computer to compute $x!$? Stirling's series followed by the recurrence relation is a good possibility. An even better possibility is to fit $x!$, $0 \le x \le 1$, by a short power series (polynomial) and then calculate $x!$ directly from this empirical fit. Presumably, the computing machine has been told the values of the coefficients of the polynomial. Such polynomial fits have been made by Hastings[2] for various accuracy requirements. For example,

$$x! = 1 + \sum_{n=1}^{8} b_n x^n + \varepsilon(x), \qquad (10.56a)$$

with

$$b_1 = -0.57719\ 1652 \qquad b_5 = -0.75670\ 4078$$
$$b_2 = 0.98820\ 5891 \qquad b_6 = 0.48219\ 9394$$
$$\qquad\qquad\qquad\qquad\qquad\qquad\qquad\qquad (10.56b)$$
$$b_3 = -0.89705\ 6937 \qquad b_7 = -0.19352\ 7818$$
$$b_4 = 0.91820\ 6857 \qquad b_8 = 0.03586\ 8343$$

with the magnitude of the error $|\varepsilon(x)| < 3 \times 10^{-7}$, $0 \le x \le 1$.

This is *not* a least-squares fit. Hastings employed a Chebyshev polynomial technique similar to that described in Section 13.4 to minimize the *maximum* value of $|\varepsilon(x)|$.

[2] C. Hastings, Jr., *Approximations for Digitial Computers*. Princeton, NJ: Princeton University Press (1955).

EXERCISES

10.3.1   Rewrite Stirling's series to give $z!$ instead of $\ln(z!)$.

$$ANS. \quad z! = \sqrt{2\pi}\, z^{z+1/2} e^{-z} \left(1 + \frac{1}{12z} + \frac{1}{288z^2} - \frac{139}{51{,}840z^3} + \cdots \right).$$

10.3.2   Use Stirling's formula to estimate 52!, the number of possible rearrangements of cards in a standard deck of playing cards.

10.3.3   By integrating Eq. (10.51) from $z - 1$ to z and then letting $z \to \infty$, evaluate the constant C_1 in the asymptotic series for the digamma function $\mathbf{F}(z)$.

10.3.4   Show that the constant C_2 in Stirling's formula equals $\frac{1}{2} \ln 2\pi$ by using the logarithm of the doubling formula.

10.3.5   By direct expansion verify the doubling formula for $z = n + \frac{1}{2}$; n is an integer.

10.3.6   Without using Stirling's series show that

(a) $\ln(n!) < \displaystyle\int_1^{n+1} \ln x \, dx,$

(b) $\ln(n!) > \displaystyle\int_1^{n} \ln x \, dx; \qquad n$ is an integer ≥ 2.

Notice that the arithmetic mean of these two integrals gives a good approximation for Stirling's series.

10.3.7   Test for convergence

$$\sum_{p=0}^{\infty} \left[\frac{(p - \frac{1}{2})!}{p!} \right]^2 \times \frac{2p + 1}{2p + 2} = \pi \sum_{p=0}^{\infty} \frac{(2p - 1)!!(2p + 1)!!}{(2p)!!(2p + 2)!!}.$$

This series arises in an attempt to describe the magnetic field created by and enclosed by a current loop.

10.3.8   Show that

$$\lim_{x \to \infty} x^{b-a} \frac{(x + a)!}{(x + b)!} = 1.$$

10.3.9   Show that

$$\lim_{n \to \infty} \frac{(2n - 1)!!}{(2n)!!} n^{1/2} = \pi^{-1/2}.$$

10.3.10   Calculate the binomial coefficient $\binom{2n}{n}$ to six significant figures for $n = 10$, 20, and 30. Check your values by

(a) a Stirling series approximation through terms in n^{-1},
(b) a double precision calculation.

$$ANS. \quad \binom{20}{10} = 1.84756 \times 10^5$$
$$\binom{40}{20} = 1.37846 \times 10^{11}$$
$$\binom{60}{30} = 1.18264 \times 10^{17}.$$

10.3.11 Write a program (or subprogram) that will calculate $\log_{10}(x!)$ directly from Stirling's series. Assume that $x \geq 10$. (Smaller values could be calculated via the factorial recurrence relation.) Tabulate $\log_{10}(x!)$ versus x for $x = 10(10)300$. Check your results against AMS-55 or by direct multiplication (for $n = 10, 20$, and 30).

Check value. $\log_{10}(100!) = 157.97$.

10.3.12 Using the complex capability of FORTRAN IV, write a subroutine that will calculate $\ln(z!)$ for complex z based on Stirling's series. Include a test and an appropriate error message if z is too close to a negative real integer. Check your subroutine against alternate calculations for z real, z pure imaginary, and $z = 1 + ib$ (Exercise 10.2.23).

Check values. $|(i0.5)!| = 0.82618$
phase $(i0.5)! = -0.24406$.

10.4 THE BETA FUNCTION

Using the integral definition (Eq. (10.25)), we write the product of two factorials as the product of two integrals. To facilitate a change in variables, we take the integrals over a finite range.

$$m!\,n! = \lim_{a^2 \to \infty} \int_0^{a^2} e^{-u} u^m \, du \int_0^{a^2} e^{-v} v^n \, dv, \qquad \begin{matrix} \mathfrak{R}(m) > -1, \\ \mathfrak{R}(n) > -1. \end{matrix} \qquad (10.57a)$$

Replacing u with x^2 and v with y^2, we obtain

$$m!\,n! = \lim_{a = \infty} 4 \int_0^a e^{-x^2} x^{2m+1} \, dx \int_0^a e^{-y^2} y^{2n+1} \, dy. \qquad (10.57b)$$

Transforming to polar coordinates gives us

$$m!\,n! = \lim_{a \to \infty} 4 \int_0^a e^{-r^2} r^{2m+2n+3} \, dr \int_0^{\pi/2} \cos^{2m+1} \theta \sin^{2n+1} \theta \, d\theta$$

$$= (m + n + 1)!\,2 \int_0^{\pi/2} \cos^{2m+1} \theta \sin^{2n+1} \theta \, d\theta. \qquad (10.58)$$

Here the Cartesian area element $dx\,dy$ has been replaced by $r\,dr\,d\theta$ (Fig. 10.6). The last equality in Eq. (10.58) follows from Exercise 10.1.11.

The definite integral, together with the factor 2, has been named the beta function

$$B(m + 1, n + 1) \equiv 2 \int_0^{\pi/2} \cos^{2m+1} \theta \sin^{2n+1} \theta \, d\theta$$

$$= \frac{m!\,n!}{(m + n + 1)!} = B(n + 1, m + 1). \qquad (10.59a)$$

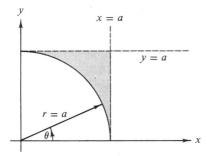

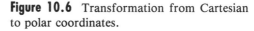

Figure 10.6 Transformation from Cartesian to polar coordinates.

Equivalently, in terms of the gamma function

$$B(p, q) = \frac{\Gamma(p)\Gamma(q)}{\Gamma(p + q)}. \tag{10.59b}$$

The only reason for choosing $m + 1$ and $n + 1$, rather than m and n, as the arguments of B is to be in agreement with the conventional, historical beta function.

In this manipulation the transformation from Cartesian to polar coordinates needs some justification. As seen in Fig. 10.6, the shaded area is being neglected. However, the maximum value of the integrand in this region is $e^{-a^2}a^{2m+2n+3}$ which vanishes so strongly as a approaches infinity that the integral over the neglected region vanishes.

DEFINITE INTEGRALS, ALTERNATE FORMS

The beta function is useful in the evaluation of a wide variety of definite integrals. The substitution $t = \cos^2 \theta$ converts Eq. (10.59a) to[1]

$$B(m + 1, n + 1) = \frac{m!n!}{(m + n + 1)!} = \int_0^1 t^m(1 - t)^n \, dt. \tag{10.60a}$$

Replacing t by x^2, we obtain

$$\frac{m!n!}{2(m + n + 1)!} = \int_0^1 x^{2m+1}(1 - x^2)^n \, dx. \tag{10.60b}$$

The substitution $t = u/(1 + u)$ in Eq. (10.60a) yields still another useful form,

$$\frac{m!n!}{(m + n + 1)!} = \int_0^\infty \frac{u^m}{(1 + u)^{m+n+2}} \, du. \tag{10.61}$$

The beta function as a definite integral is useful in establishing integral representations of the Bessel function (Exercise 11.1.18) and the hypergeometric function (Exercise 13.5.7).

---

[1] The Laplace transform convolution theorem provides an alternate derivation of Eq. (10.60a), compare Exercise 15.11.2.

VERIFICATION OF $\pi a/\sin \pi a$ RELATION

If we take $m = a$, $n = -a$, $-1 < a < 1$, then

$$\int_0^\infty \frac{u^a}{(1 + u)^2}\, du = a!(-a)!. \tag{10.62}$$

By contour integration this integral may be shown to be equal to $\pi a/\sin \pi a$ (Exercise 7.2.18), thus providing another method of obtaining Eq. (10.32).

DERIVATION OF LEGENDRE DUPLICATION FORMULA

The form of Eq. (10.59) suggests that the beta function may be useful in deriving the doubling formula used in the preceding section. From Eq. (10.60a) with $m = n = z$ and $\Re(z) > -1$,

$$\frac{z!\,z!}{(2z + 1)!} = \int_0^1 t^z(1 - t)^z\, dt. \tag{10.63}$$

By substituting $t = (1 + s)/2$, we have

$$\frac{z!\,z!}{(2z + 1)!} = 2^{-2z-1} \int_{-1}^1 (1 - s^2)^z\, ds$$

$$= 2^{-2z} \int_0^1 (1 - s^2)^z\, ds. \tag{10.64}$$

The last equality holds because the integrand is even. Evaluating this integral as a beta function (Eq. (10.60b)), we obtain

$$\frac{z!\,z!}{(2z + 1)!} = 2^{-2z-1} \frac{z!(-\tfrac{1}{2})!}{(z + \tfrac{1}{2})!}. \tag{10.65}$$

Rearranging terms and recalling that $(-\tfrac{1}{2})! = \pi^{1/2}$, we quickly reduce these equations to one form of the Legendre duplication formula,

$$z!(z + \tfrac{1}{2})! = 2^{-2z-1}\pi^{1/2}(2z + 1)!. \tag{10.66a}$$

Dividing by $(z + \tfrac{1}{2})$, we obtain an alternate form of the duplication formula.

$$z!(z - \tfrac{1}{2})! = 2^{-2z}\pi^{1/2}(2z)!. \tag{10.66b}$$

Although the integrals used in this derivation are defined only for $\Re(z) > -1$, the results (Eqs. (10.66a) and (10.66b)) hold for all z by analytic continuation.[2]

Using the double factorial notation (Section 10.1), we may rewrite Eq. (10.66a) (with $z = n$, an integer) as

$$(n + \tfrac{1}{2})! = \pi^{1/2}(2n + 1)!!/2^{n+1}. \tag{10.66c}$$

This is often convenient for eliminating factorials of fractions.

---

[2] If $2z$ is a negative integer, we get the valid but unilluminating result $\infty = \infty$.

INCOMPLETE BETA FUNCTION

Just as there is an incomplete gamma function (Section 10.5), there is also an incomplete beta function,

$$B_x(p, q) = \int_0^x t^{p-1}(1 - t)^{q-1}\, dt, \qquad 0 \le x \le 1, p > 0, q > 0 \,(\text{if } x = 1).$$

$$(10.67)$$

Clearly, $B_{x=1}(p, q)$ becomes the regular (complete) beta function, Eq. (10.60). A power-series expansion of $B_x(p, q)$ is the subject of Exercises 5.2.18 and 5.7.8. The relation to hypergeometric functions appears in Section 13.5.

The incomplete beta function makes an appearance in probability theory in calculating the probability of at most k successes in n independent trials.[3]

EXERCISES

10.4.1   Derive the doubling formula for the factorial function by integrating $(\sin 2\theta)^{2n+1} = (2 \sin \theta \cos \theta)^{2n+1}$ (and using the beta function).

10.4.2   Verify the following beta function identities:

(a) $B(a, b) = B(a + 1, b) + B(a, b + 1)$,

(b) $B(a, b) = \dfrac{a + b}{b} B(a, b + 1)$,

(c) $B(a, b) = \dfrac{b - 1}{a} B(a + 1, b - 1)$,

(d) $B(a, b)B(a + b, c) = B(b, c)B(a, b + c)$.

10.4.3   (a) Show that

$$\int_{-1}^1 (1 - x^2)^{1/2} x^{2n}\, dx = \begin{cases} \dfrac{\pi}{2} & n = 0 \\ \pi \dfrac{(2n - 1)!!}{(2n + 2)!!}, & n = 1, 2, 3, \ldots. \end{cases}$$

(b) Show that

$$\int_{-1}^1 (1 - x^2)^{-1/2} x^{2n}\, dx = \begin{cases} \pi & n = 0 \\ \pi \dfrac{(2n - 1)!!}{(2n)!!}, & n = 1, 2, 3, \ldots. \end{cases}$$

10.4.4   Show that

$$\int_{-1}^1 (1 - x^2)^n\, dx = \begin{cases} 2^{2n+1} \dfrac{n!\,n!}{(2n + 1)!}, & n > -1 \\ 2 \dfrac{(2n)!!}{(2n + 1)!!}, & n = 0, 1, 2, \ldots. \end{cases}$$

[3] W. Feller, *An Introduction to Probability Theory and Its Applications*, 3rd ed. New York: Wiley (1968), Section VI.10.

10.4.5 Evaluate $\int_{-1}^{1} (1 + x)^a(1 - x)^b \, dx$ in terms of the beta function.

ANS. $2^{a+b+1}B(a + 1, b + 1)$.

10.4.6 Show, by means of the beta function, that

$$\int_t^z \frac{dx}{(z - x)^{1-\alpha}(x - t)^\alpha} = \frac{\pi}{\sin \pi\alpha}, \qquad 0 < \alpha < 1.$$

This result is used in Section 16.2 to solve Abel's generalized integral equation.

10.4.7 Show that the Dirichlet integral

$$\iint x^p y^q \, dA = \frac{p!q!}{(p + q + 2)!} = \frac{B(p + 1, q + 1)}{p + q + 2},$$

where the range of integration is the triangle bounded by the positive x- and y-axes and the line $x + y = 1$.

10.4.8 Show that

$$\int_0^\infty \int_0^\infty e^{-(x^2+y^2+2xy\cos\theta)} \, dx \, dy = \frac{\theta}{2 \sin \theta}.$$

What are the limits on θ?
Hint. Consider oblique xy coordinates.

ANS. $-\pi < \theta < \pi$.

10.4.9 Evaluate (using the beta function)

(a) $\displaystyle\int_0^{\pi/2} \cos^{1/2}\theta \, d\theta = \frac{(2\pi)^{3/2}}{16[(\frac14)!]^2}$,

(b) $\displaystyle\int_0^{\pi/2} \cos^n\theta \, d\theta = \int_0^{\pi/2} \sin^n\theta \, d\theta = \frac{\sqrt{\pi}\,[(n-1)/2]!}{2(n/2)!}$

$$= \begin{cases} \dfrac{(n-1)!!}{n!!} & \text{for } n \text{ odd,} \\[2mm] \dfrac{\pi}{2} \cdot \dfrac{(n-1)!!}{n!!} & \text{for } n \text{ even.} \end{cases}$$

10.4.10 Evaluate $\int_0^1 (1 - x^4)^{-1/2} \, dx$ as a beta function.

ANS. $\dfrac{[(\frac14)!]^2 \cdot 4}{(2\pi)^{1/2}} = 1.311028777$.

10.4.11 Given

$$J_\nu(z) = \frac{2}{\pi^{1/2}(\nu - \frac12)!}\left(\frac{z}{2}\right)^\nu \int_0^{\pi/2} \sin^{2\nu}\theta \cos(z\cos\theta) \, d\theta, \qquad \Re(\nu) > -\tfrac12,$$

show, with the aid of beta functions, that this reduces to the Bessel series

$$J_\nu(z) = \sum_{s=0}^\infty (-1)^s \frac{1}{s!(s + \nu)!}\left(\frac{z}{2}\right)^{2s+\nu},$$

identifying the initial J_ν as an integral representation of the Bessel function, J_ν (Section 11.1).

10.4.12   Given that the associated Legendre polynomial
$$P_m^m(x) = (2m - 1)!!(1 - x^2)^{m/2},$$
Section 12.5, show that

(a) $\displaystyle\int_{-1}^{1} [P_m^m(x)]^2 \, dx = \frac{2}{(2m + 1)} (2m)!, \qquad m = 0, 1, 2, \dots$.

(b) $\displaystyle\int_{-1}^{1} [P_m^m(x)]^2 \frac{dx}{1 - x^2} = 2 \cdot (2m - 1)!, \qquad m = 1, 2, 3, \dots$.

10.4.13   Show that

(a) $\displaystyle\int_0^1 (x^2)^{s+1/2}(1 - x^2)^{-1/2} \, dx = \frac{(2s)!!}{(2s + 1)!!},$

(b) $\displaystyle\int_0^1 (x^2)^p(1 - x^2)^q \, dx = \frac{1}{2} \frac{(p - \frac{1}{2})! \, q!}{(p + q + \frac{1}{2})!}.$

10.4.14   A particle of mass m moving in a symmetric potential that is well described by $V(x) = A|x|^n$ has a total energy $\frac{1}{2}m(dx/dt)^2 + V(x) = E$. Solving for dx/dt and integrating we find that the period of motion is
$$\tau = 2\sqrt{2m} \int_0^{x_{max}} \frac{dx}{(E - Ax^n)^{1/2}},$$
where x_{max} is a classical turning point given by $Ax_{max}^n = E$. Show that
$$\tau = \frac{2}{n} \sqrt{\frac{2\pi m}{E}} \left(\frac{E}{A}\right)^{1/n} \frac{\Gamma(1/n)}{\Gamma(1/n + \frac{1}{2})}.$$

10.4.15   Referring to Exercise 10.4.14,

(a) Determine the limit as $n \to \infty$ of
$$\frac{2}{n} \sqrt{\frac{2\pi m}{E}} \left(\frac{E}{A}\right)^{1/n} \frac{\Gamma(1/n)}{\Gamma(1/n + \frac{1}{2})}.$$

(b) Find $\lim_{n \to \infty} \tau$ from the behavior of the integrand, $(E - Ax^n)^{-1/2}$.

(c) Investigate the behavior of the physical system (potential well) as $n \to \infty$. Obtain the period from inspection of this limiting physical system.

10.4.16   Show that
$$\int_0^{\infty} \frac{\sinh^\alpha x}{\cosh^\beta x} \, dx = \frac{1}{2} B\left(\frac{\alpha + 1}{2}, \frac{\beta - \alpha}{2}\right), \qquad -1 < \alpha < \beta.$$
Hint. Let $\sinh^2 x = u.$

10.4.17   The beta distribution of probability theory has a probability density
$$f(x) = \frac{\Gamma(\alpha + \beta)}{\Gamma(\alpha)\Gamma(\beta)} x^{\alpha-1}(1 - x)^{\beta-1},$$
with x restricted to the interval $(0, 1)$. Show that

(a) $\langle x \rangle (\text{mean}) = \dfrac{\alpha}{\alpha + \beta}.$

(b) $\sigma^2 (\text{variance}) \equiv \langle x^2 \rangle - \langle x \rangle^2 = \dfrac{\alpha\beta}{(\alpha + \beta)^2(\alpha + \beta + 1)}.$

10.4.18   From

$$\lim_{n \to \infty} \frac{\int_0^{\pi/2} \sin^{2n} \theta \, d\theta}{\int_0^{\pi/2} \sin^{2n+1} \theta \, d\theta} = 1$$

derive the Wallis formula for π:

$$\frac{\pi}{2} = \frac{2 \cdot 2}{1 \cdot 3} \cdot \frac{4 \cdot 4}{3 \cdot 5} \cdot \frac{6 \cdot 6}{5 \cdot 7} \cdot \ldots .$$

10.4.19   Tabulate the beta function $B(p, q)$ for p and $q = 1.0(0.1)2.0$, independently. *Check value.* $B(1.3, 1.7) = 0.40774$.

10.4.20   (a) Write a subroutine that will calculate the incomplete beta function $B_x(p, q)$. For $0.5 < x \leq 1$ you will find it convenient to use the relation

$$B_x(p, q) = B(p, q) - B_{1-x}(q, p).$$

(b) Tabulate $B_x(\frac{3}{2}, \frac{3}{2})$. Spot check your results by using the Gauss–Legendre quadrature.

10.5 THE INCOMPLETE GAMMA FUNCTIONS AND RELATED FUNCTIONS

Generalizing the Euler definition of the gamma function (Eq. (10.5)), we define the incomplete gamma functions by the *variable* limit integrals

$$\gamma(a, x) = \int_0^x e^{-t} t^{a-1} \, dt, \qquad \Re(a) > 0$$

and

$$\Gamma(a, x) = \int_x^\infty e^{-t} t^{a-1} \, dt. \tag{10.68}$$

Clearly, the two functions are related, for

$$\gamma(a, x) + \Gamma(a, x) = \Gamma(a). \tag{10.69}$$

The choice of employing $\gamma(a, x)$ or $\Gamma(a, x)$ is purely a matter of convenience. If the parameter a is a positive integer, Eq. (10.68) may be integrated completely to yield

$$\gamma(n, x) = (n - 1)! \left(1 - e^{-x} \sum_{s=0}^{n-1} \frac{x^s}{s!} \right)$$

$$\Gamma(n, x) = (n - 1)! \, e^{-x} \sum_{s=0}^{n-1} \frac{x^s}{s!}, \qquad n = 1, 2, \ldots . \tag{10.70}$$

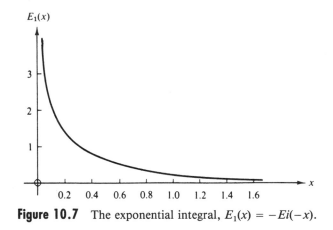

Figure 10.7   The exponential integral, $E_1(x) = -Ei(-x)$.

For nonintegral a a power-series expansion of $\gamma(a, x)$ for small x and an asymptotic expansion of $\Gamma(a, x)$ are developed in Sections 5.7 and 5.10.

$$\gamma(a, x) = x^a \sum_{n=0}^{\infty} (-1)^n \frac{x^n}{n!(a + n)},$$

$$\Gamma(a, x) = x^{a-1}e^{-x} \sum_{n=0}^{\infty} \frac{(a - 1)!}{(a - 1 - n)!} \cdot \frac{1}{x^n} \qquad (10.71)$$

$$= x^{a-1}e^{-x} \sum_{n=0}^{\infty} (-1)^n \frac{(n - a)!}{(-a)!} \cdot \frac{1}{x^n}.$$

These incomplete gamma functions may also be expressed quite elegantly in terms of confluent hypergeometric functions (compare Section 13.6).

EXPONENTIAL INTEGRAL

Although the incomplete gamma function $\Gamma(a, x)$ in its general form (Eq. (10.68)) is only infrequently encountered in physical problems, a special case is quite common and very useful. We define the exponential integral by[1]

$$-Ei(-x) \equiv \int_x^{\infty} \frac{e^{-t}}{t} dt = E_1(x). \qquad (10.72)$$

(See Fig. 10.7.) To obtain a series expansion for small x, we proceed as follows. Then

$$E_1(x) = \Gamma(0, x)$$

$$= \lim_{a \to 0} [\Gamma(a) - \gamma(a, x)]. \qquad (10.73)$$

---

[1] The appearance of the two minus signs in $-Ei(-x)$ is an historical monstrosity. AMS-55 denotes this integral as $E_1(x)$.

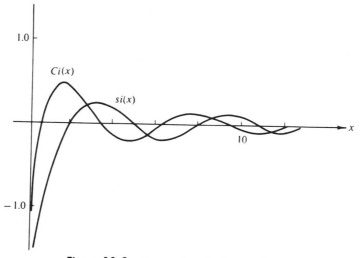

Figure 10.8   Sine and cosine integrals.

Caution is needed here, for the integral in Eq. (10.72) diverges logarithmically as $x \to 0$. We may split the divergent term in the series expansion for $\gamma(a, x)$,

$$E_1(x) = \lim_{a \to 0}\left[\frac{a\Gamma(a) - x^a}{a}\right] - \sum_{n=1}^{\infty}\frac{(-1)^n x^n}{n \cdot n!}. \tag{10.74}$$

Using l'Hospital's rule (Exercise 5.6.9) and

$$\frac{d}{da}\{a\Gamma(a)\} = \frac{d}{da}a! = \frac{d}{da}e^{\ln(a!)} = a!\,\mathsf{F}(a), \tag{10.74a}$$

and then Eq. (10.40),[2] we obtain

$$E_1(x) = -\gamma - \ln x - \sum_{n=1}^{\infty}\frac{(-1)^n x^n}{n \cdot n!}, \tag{10.75}$$

useful for small x. An asymptotic expansion is given in Section 5.10.

Further special forms related to the exponential integral are the sine integral, cosine integral (Fig. 10.8), and logarithmic integral defined by[3]

$$si(x) = -\int_x^{\infty}\frac{\sin t}{t}\,dt$$

$$Ci(x) = -\int_x^{\infty}\frac{\cos t}{t}\,dt \tag{10.76}$$

$$li(x) = \int_0^x \frac{du}{\ln u} = Ei(\ln x).$$

[2] $dx^a/da = x^a \ln x$.
[3] Another sine integral is given by $Si(x) = si(x) + \pi/2$.

By transforming from real to imaginary argument, we can show that

$$si(x) = \frac{1}{2i}[Ei(ix) - Ei(-ix)] = \frac{1}{2i}[E_1(ix) - E_1(-ix)], \qquad (10.77)$$

whereas

$$Ci(x) = \frac{1}{2}[Ei(ix) + Ei(-ix)] = -\frac{1}{2}[E_1(ix) + E_1(-ix)], \qquad |\arg x| < \frac{\pi}{2}. \qquad (10.78)$$

Adding these two relations, we obtain

$$Ei(ix) = Ci(x) + i\,si(x), \qquad (10.79)$$

to show that the relation among these integrals is exactly analogous to that among e^{ix}, $\cos x$, and $\sin x$. In terms of E_1

$$E_1(ix) = -Ci(x) + i\,si(x).$$

Asymptotic expansions of $Ci(x)$ and $si(x)$ are developed in Section 5.10. Power-series expansions about the origin for $Ci(x)$, $si(x)$, and $li(x)$ may be obtained from those for the exponential integral, $E_1(x)$, or by direct integration, Exercise 10.5.10. The exponential, sine, and cosine integrals are tabulated in AMS-55, Chapter 5.

ERROR INTEGRALS

The error integrals

$$erf\ z = \frac{2}{\sqrt{\pi}}\int_0^z e^{-t^2}\,dt$$

$$(10.80a)$$

$$erfc\ z = 1 - erf\ z = \frac{2}{\sqrt{\pi}}\int_z^\infty e^{-t^2}\,dt$$

(normalized so that $erf\ \infty = 1$) are introduced in Section 5.10 (Fig. 10.9). Asymptotic forms are developed there. From the general form of the integrands and Eq. (10.6) we expect that $erf\ z$ and $erfc\ z$ may be written as

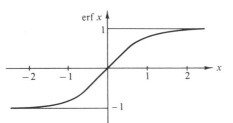

Figure 10.9 Error function, $erf\ x$.

incomplete gamma functions with $a = \frac{1}{2}$. The relations are

$$\text{erf } z = \pi^{-1/2}\gamma(\tfrac{1}{2}, z^2)$$

$$\text{erfc } z = \pi^{-1/2}\Gamma(\tfrac{1}{2}, z^2).$$

(10.80b)

The power-series expansion of erf z follows directly from Eq. (10.71).

EXERCISES

10.5.1   Show that

$$\gamma(a, x) = e^{-x} \sum_{n=0}^{\infty} \frac{(a-1)!}{(a+n)!} x^{a+n}.$$

(a) By repeatedly integrating by parts.
(b) Demonstrate this relation by transforming it into Eq. (10.71).

10.5.2   Show that

(a) $\dfrac{d^m}{dx^m}[x^{-a}\gamma(a, x)] = (-1)^m x^{-a-m}\gamma(a+m, x),$

(b) $\dfrac{d^m}{dx^m}[e^x\gamma(a, x)] = e^x \dfrac{\Gamma(a)}{\Gamma(a-m)}\, \gamma(a-m, x).$

10.5.3   Show that $\gamma(a, x)$ and $\Gamma(a, x)$ satisfy the recurrence relations
(a) $\gamma(a+1, x) = a\gamma(a, x) - x^a e^{-x},$
(b) $\Gamma(a-1, x) = a\Gamma(a, x) + x^a e^{-x}.$

10.5.4   The potential produced by a $1s$ hydrogen electron is (Exercise 12.8.6) given by

$$V(\mathbf{r}) = \frac{q}{4\pi\varepsilon_0 a_0}\left\{\frac{1}{2r}\gamma(3, 2r) + \Gamma(2, 2r)\right\}.$$

(a) For $r \ll 1$ show that

$$V(\mathbf{r}) = \frac{q}{4\pi\varepsilon_0 a_0}\left\{1 - \frac{2}{3}r^2 + \cdots\right\}.$$

(b) For $r \gg 1$ show that

$$V(\mathbf{r}) = \frac{q}{4\pi\varepsilon_0 a_0}\cdot\frac{1}{r}.$$

Here r is a pure number, the number of Bohr radii, a_0.
Note. For computation at intermediate values of r, Eqs. (10.70) are convenient.

10.5.5   The potential of a $2p$ hydrogen electron is found to be (Exercise 12.8.7)

$$V(\mathbf{r}) = \frac{1}{4\pi\varepsilon_0}\cdot\frac{q}{24a_0}\left\{\frac{1}{r}\gamma(5, r) + \Gamma(4, r)\right\}$$

$$-\frac{1}{4\pi\varepsilon_0}\cdot\frac{q}{120a_0}\left\{\frac{1}{r^3}\gamma(7, r) + r^2\Gamma(2, r)\right\}P_2(\cos\theta).$$

Here r is expressed in units of a_0, the Bohr radius. $P_2(\cos\theta)$ is a Legendre polynomial (Section 12.1).

(a) For $r \ll 1$, show that

$$V(\mathbf{r}) = \frac{1}{4\pi\varepsilon_0} \cdot \frac{q}{a_0}\left\{\frac{1}{4} - \frac{1}{120}r^2 P_2(\cos\theta) + \cdots\right\}.$$

(b) For $r \gg 1$, show that

$$V(\mathbf{r}) = \frac{1}{4\pi\varepsilon_0} \cdot \frac{q}{a_0 r}\left\{1 - \frac{6}{r^2}P_2(\cos\theta) + \cdots\right\}.$$

10.5.6   Prove that the exponential integral

$$\int_x^\infty \frac{e^{-t}}{t}\,dt = -\gamma - \ln x - \sum_{n=1}^\infty \frac{(-1)^n x^n}{n\cdot n!}.$$

γ is the Euler–Mascheroni constant.

10.5.7   Show that $E_1(z)$ may be written as

$$E_1(z) = e^{-z}\int_0^\infty \frac{e^{-zt}}{1+t}\,dt.$$

Show also that we must impose the condition $|\arg z| \le \pi/2$.

10.5.8   Related to the exponential integral (Eq. (10.72)) by a simple change of variable is the function

$$E_n(x) = \int_1^\infty \frac{e^{-xt}}{t^n}\,dt.$$

Show that $E_n(x)$ satisfies the recurrence relation

$$E_{n+1}(x) = \frac{1}{n}e^{-x} - \frac{x}{n}E_n(x), \qquad n = 1, 2, 3, \dots.$$

10.5.9   With $E_n(x)$ defined in Exercise 10.5.8, show that $E_n(0) = 1/(n-1)$, $n > 1$.

10.5.10   Develop the following power-series expansions

(a) $si(x) = -\dfrac{\pi}{2} + \displaystyle\sum_{n=0}^\infty \frac{(-1)^n x^{2n+1}}{(2n+1)(2n+1)!}.$

(b) $Ci(x) = \gamma + \ln x + \displaystyle\sum_{n=1}^\infty \frac{(-1)^n x^{2n}}{2n(2n)!}.$

10.5.11   An analysis of a center-fed linear antenna leads to the expression

$$\int_0^x \frac{1-\cos t}{t}\,dt.$$

Show that this is equal to $\gamma + \ln x - Ci(x)$.

10.5.12   Using the relation

$$\Gamma(a) = \gamma(a, x) + \Gamma(a, x),$$

show that if $\gamma(a, x)$ satisfies the relations of Exercise 10.5.2, then $\Gamma(a, x)$ must satisfy the same relations.

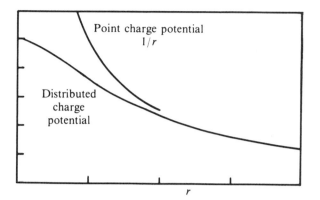

Figure 10.10 Distributed charge potential produced by a 1S hydrogen electron, Exercise 10.15.14.

10.5.13 (a) Write a subroutine that will calculate the incomplete gamma functions: $\gamma(n, x)$ and $\Gamma(n, x)$ for n a positive integer. Spot check $\Gamma(n, x)$ by Gauss–Laguerre quadratures—Appendix 2.

(b) Tabulate $\gamma(n, x)$ and $\Gamma(n, x)$ for $x = 0.0(0.1)1.0$ and $n = 1, 2,$ and 3.

10.5.14 Calculate the potential produced by a 1s hydrogen electron (Exercise 10.5.4) (Fig. 10.10). Tabulate $V(r)/(q/4\pi\varepsilon_0 a_0)$ for $x = 0.0(0.1)4.0$. Check your calculations for $r \ll 1$ and for $r \gg 1$ by calculating the limiting forms given in Exercise 10.5.4.

10.5.15 Using Eqs. (5.204) and (10.75), calculate the exponential integral $E_1(x)$ for

(a) $x = 0.2(0.2)1.0$,

(b) $x = 6.0(2.0)10.0$.

Program your own calculation but check each value, using a library subroutine if available. Also check your calculations at each point by a Gauss–Laguerre quadrature.

You should find that the power-series converges rapidly and yields high precision for small x. The asymptotic series, even for $x = 10$, yields relatively poor accuracy.

Check values. $E_1(1.0) = 0.219384$

$E_1(10.0) = 4.15697 \times 10^{-6}$.

10.5.16 The two expressions for $E_1(x)$, (1) Eq. (5.204), an asymptotic series and (2) Eq. (10.75), a convergent power series, provide a means of calculating the Euler–Mascheroni constant γ to high accuracy. Using double precision, calculate γ from Eq. (10.75) with $E_1(x)$ evaluated by Eq. (5.204).

Hint. As a convenient choice take x in the range 10 to 20. (Your choice of x will set a limit on the accuracy of your result.) To minimize errors in the alternating series of Eq. (10.75), accumulate the positive and negative terms separately.

ANS. For $x = 10$ and "double precision" $\gamma = 0.5772\,1566$.

ADDITIONAL READINGS

ABRAMOWITZ, M., and I. A. STEGUN, eds., *Handbook of Mathematical Functions with Formulas, Graphs, and Mathematical Tables* (AMS-55). Washington, DC: National Bureau of Standards (1972). Contains a wealth of information about gamma functions, incomplete gamma functions, exponential integrals, error functions, and related functions—Chapters 4 to 6.

ARTIN, E., *The Gamma Function* (translated by M. Butler). New York: Holt, Rinehart and Winston (1964). Demonstrates that if a function $f(x)$ is smooth (log convex) and equal to $(n - 1)!$ when $x = n$, it is the gamma function.

DAVIS, H. T., *Tables of the Higher Mathematical Functions*. Bloomington, IN: Principia Press (1933). Volume I contains extensive information on the gamma function and the polygamma functions.

GRADSHTEYN, I. S., and I. M. RYZHIK, *Table of Integrals, Series, and Products*. New York and London: Academic Press (1980).

LUKE, Y. L., *The Special Functions and Their Approximations*, Vol. 1. New York and London: Academic Press (1969).

LUKE, Y. L., *Mathematical Functions and Their Approximations*. New York: Academic Press (1975). This is an updated supplement to *Handbook of Mathematical Functions with Formulas, Graphs, and Mathematical Tables* (AMS-55). Chapter 1 deals with the gamma function. Chapter 4 treats the incomplete gamma function and a host of related functions.

11 BESSEL FUNCTIONS

11.1 BESSEL FUNCTIONS OF THE FIRST KIND, $J_\nu(x)$

Bessel functions appear in a wide variety of physical problems. In Section 8.3 separation of the Helmholtz or wave equation in circular cylindrical coordinates led to Bessel's equation. In Section 11.7 we will see that the Helmholtz equation in spherical polar coordinates also leads to a form of Bessel's equation. Bessel functions may also appear in integral form—integral representations. This may result from integral transforms (Chapter 15) or from the mathematical elegance of starting the study of Bessel functions with Hankel functions, Section 11.4.

Bessel functions and closely related functions form a rich area of mathematical analysis with many representations, many interesting and useful properties, and many interrelations. Some of the major interrelations are developed in Section 11.1 and in succeeding sections. Note that Bessel functions are not restricted to Chapter 11. The asymptotic forms are developed in Section 7.4 as well as in Section 11.6. The confluent hypergeometric representations appear in Section 13.6.

GENERATING FUNCTION, INTEGRAL ORDER, $J_n(x)$

Although Bessel functions are of interest primarily as solutions of differential equations, it is instructive and convenient to develop them from a completely different approach, that of the generating function.[1] This approach also has the advantage of focusing on the functions themselves rather than

---

[1] Generating functions have already been used in Chapter 5. In Section 5.6 the generating function $(1 + x)^n$ generated the binomial coefficients. In Section 5.9 the generating function $x(e^x - 1)^{-1}$ generated the Bernoulli numbers.

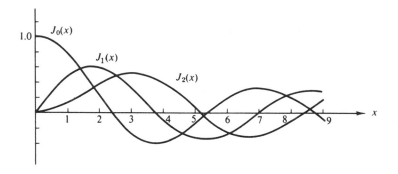

Figure 11.1 Bessel functions, $J_0(x)$, $J_1(x)$, and $J_2(x)$.

on the differential equations they satisfy. Let us introduce a function of two variables,

$$g(x, t) = e^{(x/2)(t-1/t)}. \tag{11.1}$$

Expanding this function in a Laurent series (Section 6.5), we obtain

$$e^{(x/2)(t-1/t)} = \sum_{n=-\infty}^{\infty} J_n(x)t^n. \tag{11.2}$$

It is instructive to compare Eq. (11.2) with the equivalent Eqs. (11.23) and (11.25).

The coefficient of t^n, $J_n(x)$, is defined to be a Bessel function of the first kind of integral order n. Expanding the exponentials, we have a product of Maclaurin series in $xt/2$ and $-x/2t$, respectively,

$$e^{xt/2} \cdot e^{-x/2t} = \sum_{r=0}^{\infty} \left(\frac{x}{2}\right)^r \frac{t^r}{r!} \sum_{s=0}^{\infty} (-1)^s \left(\frac{x}{2}\right)^s \frac{t^{-s}}{s!}. \tag{11.3}$$

For a given s we get t^n $(n \geq 0)$ from $r = n + s$

$$\left(\frac{x}{2}\right)^{n+s} \frac{t^{n+s}}{(n+s)!} (-1)^s \left(\frac{x}{2}\right)^s \frac{t^{-s}}{s!}. \tag{11.4}$$

The coefficient of t^n is then[2]

$$J_n(x) = \sum_{s=0}^{\infty} \frac{(-1)^s}{s!(n+s)!} \left(\frac{x}{2}\right)^{n+2s} = \frac{x^n}{2^n n!} - \frac{x^{n+2}}{2^{n+2}(n+1)!} + \cdots. \tag{11.5}$$

This series form exhibits the behavior of the Bessel function $J_n(x)$ for small x and permits numerical evaluation of $J_n(x)$. The results for J_0, J_1, and J_2 are shown in Fig. 11.1. From Section 5.3 the error in using only a finite number

---

[2] From the steps leading to this series and from its convergence characteristics it should be clear that this series may be used with x replaced by z and with z any point in the finite complex plane.

of terms in numerical evaluation is less than the first term omitted. For instance, if we want $J_n(x)$ to $\pm 1\%$ accuracy, the first term alone of Eq. (11.5) will suffice, provided the ratio of the second term to the first is less than 1% (in magnitude) or $x < 0.2(n + 1)^{1/2}$. The Bessel functions oscillate but are *not* periodic—except in the limit as $x \to \infty$ (Section 11.6). The amplitude of $J_n(x)$ is not constant but decreases asymptotically as $x^{-1/2}$.

Equation (11.5) actually holds for $n < 0$, also giving

$$J_{-n}(x) = \sum_{s=0}^{\infty} \frac{(-1)^s}{s!(s - n)!} \left(\frac{x}{2}\right)^{2s-n}, \tag{11.6}$$

which amounts to replacing n by $-n$ in Eq. (11.5). Since n is an integer (here), $(s - n)! \to \infty$ for $s = 0, \ldots, (n - 1)$. Hence the series may be considered to start with $s = n$. Replacing s by $s + n$, we obtain

$$J_{-n}(x) = \sum_{s=0}^{\infty} \frac{(-1)^{s+n}}{s!(n + s)!} \left(\frac{x}{2}\right)^{n+2s}, \tag{11.7}$$

showing immediately that $J_n(x)$ and $J_{-n}(x)$ are not independent but are related by

$$J_{-n}(x) = (-1)^n J_n(x) \qquad \text{(integral } n\text{)}. \tag{11.8}$$

These series expressions (Eqs. (11.5) and (11.6)) may be used with n replaced by ν to *define* $J_\nu(x)$ and $J_{-\nu}(x)$ for nonintegral ν (compare Exercise 11.1.7).

RECURRENCE RELATIONS

The recurrence relations for $J_n(x)$ and its derivatives may all be obtained by operating on the series, Eq. (11.5), although this requires a bit of clairvoyance (or a lot of trial and error). Verification of the known recurrence relations is straightforward, Exercise 11.1.7. Here it is convenient to obtain them from the generating function, $g(x, t)$. Differentiating Eq. (11.1) partially with respect to t, we find that

$$\frac{\partial}{\partial t} g(x, t) = \frac{1}{2}x\left(1 + \frac{1}{t^2}\right) e^{(x/2)(t-1/t)}$$

$$= \sum_{n=-\infty}^{\infty} n J_n(x) t^{n-1}, \tag{11.9}$$

and substituting Eq. (11.2) for the exponential and equating the coefficients of like powers of t,[3] we obtain

$$J_{n-1}(x) + J_{n+1}(x) = \frac{2n}{x} J_n(x). \tag{11.10}$$

This is a three-term recurrence relation. Given J_0 and J_1, for example, J_2 (and any other integral order J_n) may be computed.

---

[3] This depends on the fact that the power-series representation is unique (Sections 5.7, 6.5).

With the opportunities offered by modern digital computers (and the demands they levy), Eq. (11.10) has acquired an interesting new application. In computing a numerical value of $J_N(x_0)$ for a given x_0, one could use Eq. (11.5) for small x, or the asymptotic form, Eq. (11.144) of Section 11.6 for large x. A better way, in terms of accuracy and machine utilization, is to use the recurrence relation, Eq. (11.10), and work *down*.[4] With $n \gg N$ and $n \gg x_0$, assume

$$J_{n+1}(x_0) = 0 \quad \text{and} \quad J_n(x_0) = \alpha,$$

where α is some *small* number. Then Eq. (11.10) leads to $J_{n-1}(x_0)$, $J_{n-2}(x_0)$, and so on, and finally, to $J_0(x_0)$. Since α is arbitrary, the J_n's are all off by a common factor. This factor is determined by the condition

$$J_0(x_0) + 2 \sum_{m=1}^{\infty} J_{2m}(x_0) = 1. \tag{11.10a}$$

(Set $t = 1$ in Eq. (11.2).) The accuracy of this calculation is checked by trying again at $n' = n + 3$. This technique yields the desired $J_N(x_0)$ and all the lower integral index J's down to J_0. This is the technique employed by the FORTRAN subroutine BESJ; it avoids the fatal accumulation of rounding errors in a recursion relation that works *up*.

High-speed, high-precision numerical computation is more or less an art. Modifications and refinements of this and other numerical techniques are being proposed year by year. For information on the current "state of the art" the student will have to go to the literature, and this means primarily to the journal *Mathematics of Computation*.

Differentiating Eq. (11.1) partially with respect to x, we have

$$\frac{\partial}{\partial x} g(x, t) = \frac{1}{2} \left(1 - \frac{1}{t} \right) e^{(x/2)(t-1/t)}$$

$$= \sum_{n=-\infty}^{\infty} J_n'(x) t^n. \tag{11.11}$$

Again, substituting in Eq. (11.2) and equating the coefficients of like powers of t, we obtain the result

$$J_{n-1}(x) - J_{n+1}(x) = 2J_n'(x). \tag{11.12}$$

As a special case of this general recurrence relation,

$$J_0'(x) = -J_1(x). \tag{11.13}$$

Adding Eqs. (11.10) and (11.12) and dividing by 2, we have

$$J_{n-1}(x) = \frac{n}{x} J_n(x) + J_n'(x). \tag{11.14}$$

---

[4] I. A. Stegun and M. Abramowitz, Generation of Bessel functions on high speed computers. *Math. Tables Aids Comput.* **11**, 255–257 (1957).

Multiplying by x^n and rearranging terms produces

$$\frac{d}{dx}[x^n J_n(x)] = x^n J_{n-1}(x). \tag{11.15}$$

Subtracting Eq. (11.12) from (11.10) and dividing by 2 yields

$$J_{n+1}(x) = \frac{n}{x} J_n(x) - J_n'(x). \tag{11.16}$$

Multiplying by x^{-n} and rearranging terms, we obtain

$$\frac{d}{dx}[x^{-n} J_n(x)] = -x^{-n} J_{n+1}(x). \tag{11.17}$$

BESSEL'S DIFFERENTIAL EQUATION

Suppose we consider a set of functions $Z_\nu(x)$ which satisfies the basic recurrence relations (Eqs. (11.10) and (11.12)), but with ν not necessarily an integer and Z_ν not necessarily given by the series (Eq. (11.5)). Equation (11.14) may be rewritten ($n \to \nu$) as

$$xZ_\nu'(x) = xZ_{\nu-1}(x) - \nu Z_\nu(x). \tag{11.18}$$

On differentiating with respect to x, we have

$$xZ_\nu''(x) + (\nu + 1)Z_\nu' - xZ_{\nu-1}' - Z_{\nu-1} = 0. \tag{11.19}$$

Multiplying by x and then subtracting Eq. (11.18) multiplied by ν gives us

$$x^2 Z_\nu'' + xZ_\nu' - \nu^2 Z_\nu + (\nu - 1)xZ_{\nu-1} - x^2 Z_{\nu-1}' = 0. \tag{11.20}$$

Now we rewrite Eq. (11.16) and replace n by $\nu - 1$.

$$xZ_{\nu-1}' = (\nu - 1)Z_{\nu-1} - xZ_\nu. \tag{11.21}$$

Using this to eliminate $Z_{\nu-1}$ and $Z_{\nu-1}'$ from Eq. (11.20), we finally get

$$x^2 Z_\nu'' + xZ_\nu' + (x^2 - \nu^2)Z_\nu = 0. \tag{11.22}$$

This is just Bessel's equation. Hence *any* functions, $Z_\nu(x)$, that satisfy the recurrence relations (Eqs. (11.10) and (11.12), (11.14) and (11.16), or (11.15) and (11.17)) satisfy Bessel's equation; that is, the unknown Z_ν are Bessel functions. In particular, we have shown that the functions $J_n(x)$, defined by our generating function, satisfy Bessel's equation. If the argument is $k\rho$ rather than x, Eq. (11.22) becomes

$$\rho^2 \frac{d^2}{d\rho^2} Z_\nu(k\rho) + \rho \frac{d}{d\rho} Z_\nu(k\rho) + (k^2\rho^2 - \nu^2)Z_\nu(k\rho) = 0. \tag{11.22a}$$

INTEGRAL REPRESENTATION

A particularly useful and powerful way of treating Bessel functions employs integral representations. If we return to the generating function (Eq. (11.2)), and substitute $t = e^{i\theta}$,

$$e^{ix\sin\theta} = J_0(x) + 2(J_2(x)\cos 2\theta + J_4(x)\cos 4\theta + \cdots)$$

$$+ 2i(J_1(x)\sin\theta + J_3(x)\sin 3\theta + \cdots), \tag{11.23}$$

in which we have used the relations

$$J_1(x)e^{i\theta} + J_{-1}(x)e^{-i\theta} = J_1(x)(e^{i\theta} - e^{-i\theta})$$

$$= 2iJ_1(x)\sin\theta, \tag{11.24}$$

$$J_2(x)e^{2i\theta} + J_{-2}(x)e^{-2i\theta} = 2J_2(x)\cos 2\theta,$$

and so on.

In summation notation

$$\cos(x\sin\theta) = J_0(x) + 2\sum_{n=1}^{\infty} J_{2n}(x)\cos(2n\theta),$$

$$\tag{11.25}$$

$$\sin(x\sin\theta) = 2\sum_{n=1}^{\infty} J_{2n-1}(x)\sin[(2n-1)\theta],$$

equating real and imaginary parts, respectively. It might be noted that angle θ (in radians) has no dimensions. Likewise $\sin\theta$ has no dimensions and the function $\cos(x\sin\theta)$ is perfectly proper from a dimensional point of view.

By employing the orthogonality properties of cosine and sine,[5]

$$\int_0^{\pi} \cos n\theta \cos m\theta \, d\theta = \frac{\pi}{2}\delta_{nm} \tag{11.26a}$$

$$\int_0^{\pi} \sin n\theta \sin m\theta \, d\theta = \frac{\pi}{2}\delta_{nm}, \tag{11.26b}$$

in which n and m are *positive* integers (zero is excluded),[6] we obtain

$$\frac{1}{\pi}\int_0^{\pi} \cos(x\sin\theta)\cos n\theta \, d\theta = \begin{cases} J_n(x), & n \text{ even}, \\ 0, & n \text{ odd}, \end{cases} \tag{11.27}$$

$$\frac{1}{\pi}\int_0^{\pi} \sin(x\sin\theta)\sin n\theta \, d\theta = \begin{cases} 0, & n \text{ even}, \\ J_n(x), & n \text{ odd}. \end{cases} \tag{11.28}$$

---

[5] They are eigenfunctions of a self-adjoint equation (linear oscillator equation) and satisfy appropriate boundary conditions (compare Sections 9.2 and 14.1).

[6] Equations (11.26a) and (11.26b) hold for either m or $n = 0$. If *both* m and $n = 0$, the constant in (11.26a) becomes π; the constant in Eq. (11.26b) becomes 0.

If these two equations are added together,

$$J_n(x) = \frac{1}{\pi} \int_0^\pi [\cos(x \sin \theta) \cos n\theta + \sin(x \sin \theta) \sin n\theta] \, d\theta$$

$$= \frac{1}{\pi} \int_0^\pi \cos(n\theta - x \sin \theta) \, d\theta, \qquad n = 0, 1, 2, 3, \ldots. \tag{11.29}$$

As a special case,

$$J_0(x) = \frac{1}{\pi} \int_0^\pi \cos(x \sin \theta) \, d\theta. \tag{11.30}$$

Noting that $\cos(x \sin \theta)$ repeats itself in all four quadrants ($\theta_1 = \theta$, $\theta_2 = \pi - \theta$, $\theta_3 = \pi + \theta$, $\theta_4 = -\theta$), we may write Eq. (11.30) as

$$J_0(x) = \frac{1}{2\pi} \int_0^{2\pi} \cos(x \sin \theta) \, d\theta. \tag{11.30a}$$

On the other hand, $\sin(x \sin \theta)$ reverses its sign in the third and fourth quadrants so that

$$\frac{1}{2\pi} \int_0^{2\pi} \sin(x \sin \theta) \, d\theta = 0. \tag{11.30b}$$

Adding Eq. (11.30a) and i times Eq. (11.30b), we obtain the complex exponential representation

$$J_0(x) = \frac{1}{2\pi} \int_0^{2\pi} e^{ix \sin \theta} \, d\theta = \frac{1}{2\pi} \int_0^{2\pi} e^{ix \cos \theta} \, d\theta. \tag{11.30c}$$

This integral representation (Eq. (11.29)) may be obtained somewhat more directly by employing contour integration (compare Exercise 11.1.16).[7] Many other integral representations exist (compare Exercise 11.1.18).

Example 11.1.1 Fraunhofer Diffraction, Circular Aperture

In the theory of diffraction through a circular aperture we encounter the integral

$$\Phi \sim \int_0^a \int_0^{2\pi} e^{ibr \cos \theta} \, d\theta r \, dr \tag{11.31}$$

for Φ, the amplitude of the diffracted wave.[8] Here θ is an azimuth angle in the plane of the circular aperture of radius a, and α is the angle defined by a

---

[7] For $n = 0$ a simple integration over θ from 0 to 2π will convert Eq. (11.23) into Eq. (11.30c).

[8] The exponent $ibr \cos \theta$ gives the phase of the wave on the distant screen at angle α relative to the phase of the wave incident on the aperture at the point (r, θ). The imaginary exponential form of this integrand means that the integral is technically a Fourier transform, Chapter 15. In general, the Fraunhofer diffraction pattern is given by the Fourier transform of the aperture.

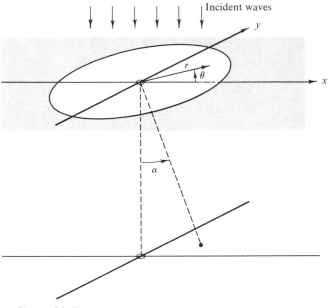

Figure 11.2   Fraunhofer diffraction—circular aperture.

point on a screen below the circular aperture relative to the normal through the center point. The parameter b is given by

$$b = \frac{2\pi}{\lambda} \sin \alpha, \tag{11.32}$$

with λ the wavelength of the incident wave. The other symbols are defined by Fig. 11.2. From Eq. (11.30c) we get[9]

$$\Phi \sim 2\pi \int_0^a J_0(br) r \, dr. \tag{11.33}$$

Equation (11.15) enables us to integrate Eq. (11.33) immediately to obtain

$$\Phi \sim \frac{2\pi ab}{b^2} J_1(ab) \sim \frac{\lambda a}{\sin \alpha} J_1 \left(\frac{2\pi a}{\lambda} \sin \alpha \right). \tag{11.34}$$

The intensity of the light in the diffraction pattern is proportional to Φ^2 and

$$\Phi^2 \sim \left\{ \frac{J_1[(2\pi a/\lambda) \sin \alpha]}{\sin \alpha} \right\}^2. \tag{11.35}$$

[9] We could also refer to Exercise 11.1.16(b).

Table 11.1   **Zeros of the Bessel Functions and Their First Derivatives**

Number of zero	$J_0(x)$	$J_1(x)$	$J_2(x)$	$J_3(x)$	$J_4(x)$	$J_5(x)$
1	2.4048	3.8317	5.1356	6.3802	7.5883	8.7715
2	5.5201	7.0156	8.4172	9.7610	11.0647	12.3386
3	8.6537	10.1735	11.6198	13.0152	14.3725	15.7002
4	11.7915	13.3237	14.7960	16.2235	17.6160	18.9801
5	14.9309	16.4706	17.9598	19.4094	20.8269	22.2178
	$J_0'(x)$	$J_1'(x)$	$J_2'(x)$	$J_3'(x)$		
1	3.8317	1.8412	3.0542	4.2012		
2	7.0156	5.3314	6.7061	8.0152		
3	10.1735	8.5363	9.9695	11.3459		

Note. $J_0'(x) = -J_1(x)$.

From Table 11.1, which lists the zeros of the Bessel functions and their first derivatives,[10] the expression 11.35 will have a zero at

$$\frac{2\pi a}{\lambda} \sin \alpha = 3.8317 \ldots \tag{11.36}$$

or

$$\sin \alpha = \frac{3.8317\lambda}{2\pi a}. \tag{11.37}$$

For green light $\lambda = 5.5 \times 10^{-5}$ cm. Hence, if $a = 0.5$ cm,

$$\alpha \approx \sin \alpha = 6.7 \times 10^{-5} \text{ (radian)} \approx 14 \text{ seconds of arc}, \tag{11.38}$$

which shows that the bending or spreading of the light ray is extremely small. If this analysis had been known in the seventeenth century, the arguments against the wave theory of light would have collapsed.

In mid-twentieth century this same diffraction pattern appears in the scattering of nuclear particles by atomic nuclei—a striking demonstration of the wave properties of the nuclear particles.

A further example of the use of Bessel functions and their roots is provided by the electromagnetic resonant cavity, Example 11.1.2 that follows, and the example and exercises of Section 11.2.

Example 11.1.2 Cylindrical Resonant Cavity

In the interior of a resonant cavity electromagnetic waves oscillate with a time dependence $e^{-i\omega t}$. Maxwell's equations lead to

$$\nabla \times \nabla \times \mathbf{E} = \alpha^2 \mathbf{E}$$

[10] Additional roots of the Bessel functions and their first derivatives may be found in C. L. Beattie, Table of first 700 zeros of Bessel functions. *Bell Syst. Tech. J.* **37**, 689 (1958), and *Bell Monogr.* **3055**.

for the space part of the electric field with $\alpha^2 = \omega^2 \varepsilon_0 \mu_0$ (Example 1.9.2). With $\nabla \cdot \mathbf{E} = 0$ (vacuum, no charges),

$$\nabla^2 \mathbf{E} + \alpha^2 \mathbf{E} = 0.$$

Separating variables in circular cylindrical coordinates (Section 2.4), we find that the z-component (E_z, space part only) satisfies the scalar Helmholtz equation

$$\nabla^2 E_z + \alpha^2 E_z = 0, \tag{11.39}$$

where $\alpha^2 = \omega^2 \varepsilon_0 \mu_0 = \omega^2/c^2$. Further,

$$(E_z)_{mnk} = \sum_{m,n} J_m(\gamma_{mn}\rho) \, e^{\pm im\varphi} [a_{mn} \sin kz + b_{mn} \cos kz]. \tag{11.40}$$

The parameter k is a separation constant introduced in splitting off the z dependence of $E_z(\rho, \varphi, z)$. Similarly, m entered in splitting off the φ dependence. γ enters as $\alpha^2 - k^2$ and is quantized by the requirement that γa be a root of the Bessel function J_m (Eq. (11.43) which follows). Then the n in γ_{mn} designates the nth root of J_m.

For the end surfaces at $z = 0$ and $z = l$ (as in Fig. 11.3), let us set $a_{mn} = 0$, and

$$k = \frac{p\pi}{l}, \qquad p = 0, 1, 2, \ldots . \tag{11.41}$$

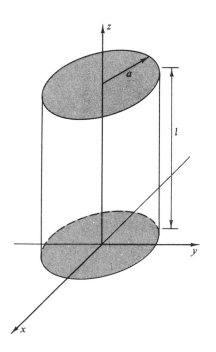

Figure 11.3   Cylindrical resonant cavity.

Maxwell's equations then guarantee that the tangential electric fields E_ρ and E_φ will vanish at $z = 0$ and l. This is the transverse magnetic or TM mode of oscillation. We have

$$\gamma^2 = \frac{\omega^2}{c^2} - k^2$$

$$= \frac{\omega^2}{c^2} - \frac{p^2\pi^2}{l^2}. \tag{11.42}$$

But there is the usual boundary condition that $E_z(\rho = a) = 0$. Hence we must set

$$\gamma_{mn} = \frac{\alpha_{mn}}{a}, \tag{11.43}$$

where α_{mn} is the nth zero of J_m.

The result of the two boundary conditions and the separation constant m^2 is that the angular frequency of our oscillation depends on three discrete parameters

$$\omega_{mnp} = c\sqrt{\frac{\alpha_{mn}^2}{a^2} + \frac{p^2\pi^2}{l^2}}, \qquad \begin{cases} m = 0, 1, 2 \cdots \\ n = 1, 2, 3 \cdots \\ p = 0, 1, 2 \cdots . \end{cases} \tag{11.44}$$

These are the allowable resonant frequencies for our TM mode. The TE mode of oscillation is the topic of Exercise 11.1.26.

ALTERNATE APPROACHES

Bessel functions are introduced here by means of a generating function, Eq. (11.2) Other approaches are possible. Listing the various possibilities, we have

1. Generating function (magic), Eq. (11.2).
2. Series solution of Bessel's differential equation, Section 8.5.
3. Contour integrals: Some writers prefer to start with contour integral definitions of the Hankel functions, Section 7.4 and 11.4, and develop the Bessel function $J_\nu(x)$ from the Hankel functions.
4. Direct solution of physical problems: Example 11.1.1, Fraunhofer diffraction with a circular aperture, illustrates this. Incidentally, Eq. (11.31) can be treated by series expansion, if desired. Feynman[11] develops Bessel functions from a consideration of cavity resonators.

In case the generating function seems too arbitrary, it can be derived from a contour integral, Exercise 11.1.16, or from the Bessel function recurrence relations, Exercise 11.1.6. Note that the contour integral is not limited to integer ν, thus providing a starting point for developing Bessel functions.

[11] R. P. Feynman, R. B. Leighton, and M. Sands, *The Feynman Lectures on Physics*, Vol. II. Reading, MA: Addison–Wesley (1964), Chapter 23.

BESSEL FUNCTIONS OF NONINTEGRAL ORDER

These different approaches are not exactly equivalent. The generating function approach is very convenient for deriving two recurrence relations, Bessel's differential equation, integral representations, addition theorems (Exercise 11.1.2), and upper and lower bounds (Exercise 11.1.1). However, the reader will probably have noticed that the generating function defined only Bessel functions of integral order, J_0, J_1, J_2, and so on. This is a limitation of the generating function approach which can be avoided by using the contour integral in Exercise 11.1.16 instead, thus leading to the approach (3) above. But the Bessel function of the first kind $J_\nu(x)$, may easily be defined for nonintegral ν by using the series (Eq. (11.5)) as a new definition.

The recurrence relations may be verified by substituting in the series form of $J_\nu(x)$ (Exercise 11.1.7). From these relations Bessel's equation follows. In fact, if ν is not an integer, there is actually an important simplification. It is found that J_ν and $J_{-\nu}$ are independent, for no relation of the form of Eq. (11.8) exists. On the other hand, for $\nu = n$, an integer, we need another solution. The development of this second solution and an investigation of its properties form the subject of Section 11.3.

EXERCISES

11.1.1   From the product of the generating functions $g(x, t) \cdot g(x, -t)$ show that
$$1 = [J_0(x)]^2 + 2[J_1(x)]^2 + 2[J_2(x)]^2 + \cdots$$
and therefore that $|J_0(x)| \leq 1$ and $|J_n(x)| \leq 1/\sqrt{2}$, $n = 1, 2, 3, \ldots$.
Hint. Use uniqueness of power series, Section 5.7.

11.1.2   Using a generating function $g(x, t) = g(u + v, t) = g(u, t) \cdot g(v, t)$, show that

(a) $J_n(u + v) = \sum\limits_{s = -\infty}^{\infty} J_s(u) \cdot J_{n-s}(v)$,

(b) $J_0(u + v) = J_0(u)J_0(v) + 2 \sum\limits_{s = 1}^{\infty} J_s(u)J_{-s}(v)$.

These are addition theorems for the Bessel functions.

11.1.3   Using only the generating function
$$e^{(x/2)(t - 1/t)} = \sum_{n = -\infty}^{\infty} J_n(x)t^n$$
and not the explicit series form of $J_n(x)$, show that $J_n(x)$ has odd or even parity according to whether n is odd or even, that is,[12]
$$J_n(x) = (-1)^n J_n(-x).$$

[12] This is easily seen from the series form (Eq. (11.5)).

11.1.4 Derive the Jacobi–Anger expansion

$$e^{iz\cos\theta} = \sum_{m=-\infty}^{\infty} i^m J_m(z) e^{im\theta}.$$

This is an expansion of a plane wave in a series of cylindrical waves.

11.1.5 Show that

(a) $\cos x = J_0(x) + 2 \sum_{n=1}^{\infty} (-1)^n J_{2n}(x),$

(b) $\sin x = 2 \sum_{n=1}^{\infty} (-1)^{n+1} J_{2n+1}(x).$

11.1.6 To help remove the generating function from the realm of magic, show that it can be *derived* from the recurrence relation, Eq. (11.10).

Hint. 1. Assume a generating function of the form

$$g(x, t) = \sum_{m=-\infty}^{\infty} J_m(x) t^m.$$

2. Multiply Eq. (11.10) by t^n and sum over n.
3. Rewrite the preceding result as

$$\left(t + \frac{1}{t}\right) g(x, t) = \frac{2t}{x} \frac{\partial g(x, t)}{\partial t}.$$

4. Integrate and adjust the function of integration (a function of x) so that the coefficient of t^0 is $J_0(x)$ as given by Eq. (11.5).

11.1.7 Show, by direct differentiation, that

$$J_\nu(x) = \sum_{s=0}^{\infty} \frac{(-1)^s}{s!(s+\nu)!} \left(\frac{x}{2}\right)^{\nu+2s}$$

satisfies the two recurrence relations

$$J_{\nu-1}(x) + J_{\nu+1}(x) = \frac{2\nu}{x} J_\nu(x),$$

$$J_{\nu-1}(x) - J_{\nu+1}(x) = 2J_\nu'(x),$$

and Bessel's differential equation

$$x^2 J_\nu''(x) + x J_\nu'(x) + (x^2 - \nu^2) J_\nu(x) = 0.$$

11.1.8 Prove that

(a) $\dfrac{\sin x}{x} = \displaystyle\int_0^{\pi/2} J_0(x \cos \theta) \cos \theta \, d\theta,$

(b) $\dfrac{1 - \cos x}{x} = \displaystyle\int_0^{\pi/2} J_1(x \cos \theta) \, d\theta.$

Hint. The definite integral

$$\int_0^{\pi/2} \cos^{2s+1}\theta \, d\theta = \frac{2 \cdot 4 \cdot 6 \cdots (2s)}{1 \cdot 3 \cdot 5 \cdots (2s + 1)}$$

may be useful.

11.1.9   Show that

$$J_0(x) = \frac{2}{\pi} \int_0^1 \frac{\cos xt}{\sqrt{1 - t^2}} dt.$$

This integral is a Fourier cosine transform (compare Section 15.3). The corresponding Fourier sine transform,

$$J_0(x) = \frac{2}{\pi} \int_1^\infty \frac{\sin xt}{\sqrt{t^2 - 1}} dt,$$

is established in Section 11.4, using a Hankel function integral representation.

11.1.10   Derive

$$J_n(x) = (-1)^n x^n \left(\frac{d}{x\, dx}\right)^n J_0(x).$$

Hint. Try mathematical induction.

11.1.11   Show that between any two consecutive zeros of $J_n(x)$ there is one and only one zero of $J_{n+1}(x)$.
Hint. Equations (11.15) and (11.17) may be useful.

11.1.12   An analysis of antenna radiation patterns for a system with a circular aperture involves the equation

$$g(u) = \int_0^1 f(r) J_0(ur) r\, dr.$$

If $f(r) = 1 - r^2$, show that

$$g(u) = \frac{2}{u^2} J_2(u).$$

11.1.13   The differential cross section in a nuclear scattering experiment is given by $d\sigma/d\Omega = |f(\theta)|^2$. An approximate treatment leads to

$$f(\theta) = \frac{-ik}{2\pi} \int_0^{2\pi} \int_0^R \exp[ik\rho \sin \theta \sin \varphi] \rho\, d\rho\, d\varphi.$$

Here θ is an angle through which the scattered particle is scattered. R is the nuclear radius. Show that

$$\frac{d\sigma}{d\Omega} = (\pi R^2) \frac{1}{\pi} \left[\frac{J_1(kR \sin \theta)}{\sin \theta} \right]^2$$

11.1.14   A set of functions $C_n(x)$ satisfies the recurrence relations

$$C_{n-1}(x) - C_{n+1}(x) = \frac{2n}{x} C_n(x),$$

$$C_{n-1}(x) + C_{n+1}(x) = 2C_n'(x).$$

(a) What linear second-order differential equation does the $C_n(x)$ satisfy?
(b) By a change of variable transform your differential equation into Bessel's equation. This suggests that $C_n(x)$ may be expressed in terms of Bessel functions of transformed argument.

11.1.15 A particle (mass m) is contained in a right circular cylinder (pillbox) of radius R and height H. The particle is described by a wave function satisfying the Schrödinger wave equation

$$-\frac{\hbar^2}{2m}\nabla^2\psi(\rho, \varphi, z) = E\psi(\rho, \varphi, z)$$

and the condition that the wave function go to zero over the surface of the pillbox. Find the lowest (zero point) permitted energy.

$$ANS. \quad E = \frac{\hbar^2}{2m}\left[\left(\frac{z_{pq}}{R}\right)^2 + \left(\frac{n\pi}{H}\right)^2\right],$$

where z_{pq} is the qth zero of J_p, the index p fixed by the azimuthal dependence.

$$E_{\min} = \frac{\hbar^2}{2m}\left[\left(\frac{2.405}{R}\right)^2 + \left(\frac{\pi}{H}\right)^2\right].$$

11.1.16 (a) Show by direct differentiation and substitution that

$$J_\nu(x) = \frac{1}{2\pi_i}\int_C e^{(x/2)(t-1/t)}t^{-\nu-1}\,dt$$

or that the equivalent equation

$$J_\nu(x) = \frac{1}{2\pi_i}\left(\frac{x}{2}\right)^\nu\int e^{s-x^2/4s}s^{-\nu-1}\,ds$$

satisfies Bessel's equation. C is the contour shown in Fig. 11.4. The negative real axis is cut line.

Hint. Show that the total integrand (after substituting in Bessel's differential equation) may be written as a total derivative:

$$\frac{d}{dt}\left\{\exp\left[\frac{x}{2}\left(t - \frac{1}{t}\right)\right]t^{-\nu}\left[\nu + \frac{x}{2}\left(t + \frac{1}{t}\right)\right]\right\}.$$

(b) Show that the first integral (with n an integer) may be transformed into

$$J_n(x) = \frac{1}{2\pi}\int_0^{2\pi} e^{i(x\sin\theta - n\theta)}\,d\theta = \frac{i^{-n}}{2\pi}\int_0^{2\pi} e^{i(x\cos\theta + n\theta)}\,d\theta.$$

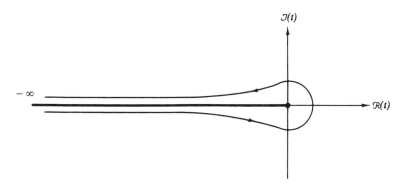

Figure 11.4 Bessel function contour.

11.1.17   The contour C in Exercise 11.1.16 is deformed to the path $-\infty$ to -1, unit circle $e^{-i\pi}$ to $e^{i\pi}$, and finally -1 to $-\infty$. Show that

$$J_\nu(x) = \frac{1}{\pi} \int_0^\pi \cos(\nu\theta - x\sin\theta)\, d\theta$$

$$- \frac{\sin\nu\pi}{\pi} \int_0^\infty e^{(-\nu\theta - x\sinh\theta)}\, d\theta.$$

This is Bessel's integral.

Hint. The negative values of the variable of integration u may be handled by using

$$u = te^{\pm i\pi}.$$

11.1.18   (a) Show that

$$J_\nu(x) = \frac{2}{\pi^{1/2}(\nu - \frac{1}{2})!} \left(\frac{x}{2}\right)^\nu \int_0^{\pi/2} \cos(x\sin\theta)\cos^{2\nu}\theta\, d\theta,$$

where $\nu > -\frac{1}{2}$.

Hint. Here is a chance to use series expansion and term-by-term integration. The formulas of Section 10.4 will prove useful.

(b) Transform the integral in part (a) into

$$J_\nu(x) = \frac{1}{\pi^{1/2}(\nu - \frac{1}{2})!} \left(\frac{x}{2}\right)^\nu \int_0^\pi \cos(x\cos\theta)\sin^{2\nu}\theta\, d\theta$$

$$= \frac{1}{\pi^{1/2}(\nu - \frac{1}{2})!} \left(\frac{x}{2}\right)^\nu \int_0^\pi e^{\pm ix\cos\theta}\sin^{2\nu}\theta\, d\theta$$

$$= \frac{1}{\pi^{1/2}(\nu - \frac{1}{2})!} \left(\frac{x}{2}\right)^\nu \int_{-1}^1 e^{\pm ipx}(1 - p^2)^{\nu-1/2}\, dp.$$

These are alternate integral representation of $J_\nu(x)$.

11.1.19   (a) From

$$J_\nu(x) = \frac{1}{2\pi i} \left(\frac{x}{2}\right)^\nu \int t^{-\nu-1} e^{t - x^2/4t}\, dt$$

derive the recurrence relation

$$J_\nu'(x) = \frac{\nu}{x} J_\nu(x) - J_{\nu+1}(x).$$

(b) From

$$J_\nu(x) = \frac{1}{2\pi i} \int t^{-\nu-1} e^{(x/2)(t - 1/t)}\, dt$$

derive the recurrence relation

$$J_\nu'(x) = \tfrac{1}{2}[J_{\nu-1}(x) - J_{\nu+1}(x)].$$

11.1.20   Show that the recurrence relation

$$J_n'(x) = \tfrac{1}{2}[J_{n-1}(x) - J_{n+1}(x)].$$

follows directly from differentiation of

$$J_n(x) = \frac{1}{\pi} \int_0^\pi \cos(n\theta - x\sin\theta)\, d\theta.$$

11.1.21   Evaluate

$$\int_0^\infty e^{-ax} J_0(bx)\, dx, \qquad a, b > 0.$$

Actually the results hold for $a \geq 0$, $-\infty < b < \infty$. This is a Laplace transform of J_0.

Hint. Either an integral representation of J_0 or a series expansion will be helpful.

11.1.22   Using trigonometric forms, verify that

$$J_0(br) = \frac{1}{2\pi} \int_0^{2\pi} e^{ibr\sin\theta}\, d\theta.$$

11.1.23   (a) Plot the intensity (Φ^2 of Eq. (11.35)) as a function of $(\sin\alpha/\lambda)$ along a diameter of the circular diffraction pattern. Locate the first two minima.

   (b) What fraction of the total light *intensity* falls within the central maximum?

Hint. $[J_1(x)]^2/x$ may be written as a derivative and the area integral of the intensity integrated by inspection.

11.1.24   The fraction of light incident on a circular aperture (normal incidence) that is transmitted is given by

$$T = 2 \int_0^{2ka} J_2(x)\, \frac{dx}{x} - \frac{1}{2ka} \int_0^{2ka} J_2(x)\, dx.$$

Here a is the radius of the aperture, and k is the wave number, $2\pi/\lambda$. Show that

(a) $T = 1 - \dfrac{1}{ka} \sum_{n=0}^{\infty} J_{2n+1}(2ka),$

(b) $T = 1 - \dfrac{1}{2ka} \displaystyle\int_0^{2ka} J_0(x)\, dx.$

11.1.25   The amplitude $U(\rho, \varphi, t)$ of a vibrating circular membrane of radius a satisfies the wave equation

$$\nabla^2 U - \frac{1}{v^2} \frac{\partial^2 U}{\partial t^2} = 0.$$

Here v is the phase velocity of the wave fixed by the elastic constants and whatever damping is imposed.

(a) Show that a solution is

$$U(\rho, \varphi, t) = J_m(k\rho)(a_1 e^{im\varphi} + a_2 e^{-im\varphi})(b_1 e^{i\omega t} + b_2 e^{-i\omega t}).$$

(b) From the Dirichlet boundary condition, $J_m(ka) = 0$, find the allowable values of the wavelength λ ($k = 2\pi/\lambda$).

Note. There are other Bessel functions besides J_m but they all diverge at $\rho = 0$. This is shown explicitly in Section 11.3. The divergent behavior is actually implicit in Eq. (11.6).

11.1.26   Example 11.1.2 describes the TM modes of electromagnetic cavity oscillation. The transverse electric (TE) modes differ in that we work from the z component of the magnetic induction $\mathbf{B}$:

$$\nabla^2 B_z + \alpha^2 B_z = 0$$

with boundary conditions

$$B_z(0) = B_z(l) = 0 \quad \text{and} \quad \left.\frac{\partial B_z}{\partial \rho}\right|_{\rho = a} = 0.$$

Show that the TE resonant frequencies are given by

$$\omega_{mnp} = c\sqrt{\frac{\beta_{mn}^2}{a^2} + \frac{p^2 \pi^2}{l^2}}, \qquad p = 1, 2, 3, \dots.$$

11.1.27 Plot the three lowest TM and the three lowest TE angular resonant frequencies, ω_{mnp}, as a function of the radius/length (a/l) ratio for $0 \le a/l \le 1.5$.
Hint. Try plotting ω^2 (in units of c^2/a^2) versus $(a/l)^2$. Why this choice?

11.1.28 A thin conducting disk of radius a carries a charge q. Show that the potential is described by

$$\varphi(r, z) = \frac{q}{4\pi\varepsilon_0 a} \int_0^\infty e^{-k|z|} J_0(kr) \frac{\sin ka}{k} \, dk,$$

where J_0 is the usual Bessel function and r and z are the familiar cylindrical coordinates.
Note. This is a difficult problem. One approach is through Fourier transforms such as Exercise 15.3.11. For a discussion of the physical problem see Jackson (*Classical Electrodynamics*).

11.1.29 Show that

$$\int_0^a x^m J_n(x) \, dx, \qquad m \ge n \ge 0.$$

(a) is integrable in terms of Bessel functions and powers of x [such as $a^p J_q(a)$] for $m + n$ odd;
(b) may be reduced to integrated terms plus $\int_0^a J_0(x) \, dx$ for $m + n$ even.

11.1.30 Show that

$$\int_0^{\alpha_{0n}} \left(1 - \frac{y}{\alpha_{0n}}\right) J_0(y) y \, dy = \frac{1}{\alpha_{0n}} \int_0^{\alpha_{0n}} J_0(y) \, dy.$$

Here α_{0n} is the nth root of $J_0(y)$. This relation is useful in computation (Exercise 11.2.11). The expression on the right is easier and quicker to evaluate—and much more accurate. Taking the difference of two terms in the expression on the left leads to a large relative error.

11.1.31 Write a program that will compute successive roots of the Bessel function $J_n(x)$, that is, α_{ns}, where $J_n(\alpha_{ns}) = 0$. Tabulate the first five roots of J_0, J_1, and of J_2.
Hint. See Appendix 1 for root-finding techniques and recommendations.
Check value. $\alpha_{12} = 7.01559.$

11.1.32 The circular aperature diffraction amplitude Φ of Eq. (17.35) is proportional to $f(z) = J_1(z)/z$. The corresponding single slit diffraction amplitude is proportional to $g(z) = \sin z/z$.

(a) Calculate and plot $f(z)$ and $g(z)$ for $z = 0.0(0.2)12.0$.

(b) Locate the two lowest values of z ($z > 0$) for which $f(z)$ takes on an extreme value. Calculate the corresponding values of $f(z)$.

(c) Locate the two lowest values of z ($z > 0$) for which $g(z)$ takes on an extreme value. Calculate the corresponding values of $g(z)$.

11.1.33 Calculate the electrostatic potential of a charged disk $\varphi(r, z)/(q/4\pi\varepsilon_0 a)$ from the integral form of Exercise 11.1.28. Calculate the potential for $r/a = 0.0(0.5)2.0$ and $z/a = 0.25(0.25)1.25$. Why is $z/a = 0$ omitted? Exercise 12.3.17 is a spherical harmonic version of this same problem.

Hint. Try a Gauss–Laguerre quadrature, Appendix 2.

11.2 ORTHOGONALITY

If Bessel's equation, Eq. (11.22a), is divided by ρ, we see that it becomes self-adjoint, and therefore by the Sturm–Liouville theory, Section 9.2, the solutions are expected to be orthogonal—if we can arrange to have appropriate boundary conditions satisfied. To take care of the boundary conditions, for a finite interval $[0, a]$, we introduce parameters a and $\alpha_{\nu m}$ into the argument of J_ν to get $J_\nu(\alpha_{\nu m}\rho/a)$. Here a is the upper limit of the cylindrical radial coordinate ρ. From Eq. (11.22a)

$$\rho\frac{d^2}{d\rho^2}J_\nu\left(\alpha_{\nu m}\frac{\rho}{a}\right) + \frac{d}{d\rho}J_\nu\left(\alpha_{\nu m}\frac{\rho}{a}\right) + \left(\frac{\alpha_{\nu m}^2\rho}{a^2} - \frac{\nu^2}{\rho}\right)J_\nu\left(\alpha_{\nu m}\frac{\rho}{a}\right) = 0. \quad (11.45)$$

Changing the parameter $\alpha_{\nu m}$ to $\alpha_{\nu n}$, we find that $J_\nu(\alpha_{\nu n}\rho/a)$ satisfies

$$\rho\frac{d^2}{d\rho^2}J_\nu\left(\alpha_{\nu n}\frac{\rho}{a}\right) + \frac{d}{d\rho}J_\nu\left(\alpha_{\nu n}\frac{\rho}{a}\right) + \left(\frac{\alpha_{\nu n}^2\rho}{a^2} - \frac{\nu^2}{\rho}\right)J_\nu\left(\alpha_{\nu n}\frac{\rho}{a}\right) = 0. \quad (11.45a)$$

Proceeding as in Section 9.2, we multiply Eq. (11.45) by $J_\nu(\alpha_{\nu n}\rho/a)$ and Eq. (11.45a) by $J_\nu(\alpha_{\nu m}\rho/a)$ and subtract, obtaining

$$J_\nu\left(\alpha_{\nu n}\frac{\rho}{a}\right)\frac{d}{d\rho}\left[\rho\frac{d}{d\rho}J_\nu\left(\alpha_{\nu m}\frac{\rho}{a}\right)\right] - J_\nu\left(\alpha_{\nu m}\frac{\rho}{a}\right)\frac{d}{d\rho}\left[\rho\frac{d}{d\rho}J_\nu\left(\alpha_{\nu n}\frac{\rho}{a}\right)\right]$$

$$= \frac{\alpha_{\nu n}^2 - \alpha_{\nu m}^2}{a^2}\rho J_\nu\left(\alpha_{\nu m}\frac{\rho}{a}\right)J_\nu\left(\alpha_{\nu n}\frac{\rho}{a}\right). \quad (11.46)$$

Integrating from $\rho = 0$ to $\rho = a$, we obtain

$$\int_0^a J_\nu\left(\alpha_{\nu n}\frac{\rho}{a}\right)\frac{d}{d\rho}\left[\rho\frac{d}{d\rho}J_\nu\left(\alpha_{\nu m}\frac{\rho}{a}\right)\right]d\rho$$

$$- \int_0^a J_\nu\left(\alpha_{\nu m}\frac{\rho}{a}\right)\frac{d}{d\rho}\left[\rho\frac{d}{d\rho}J_\nu\left(\alpha_{\nu n}\frac{\rho}{a}\right)\right]d\rho$$

$$= \frac{\alpha_{\nu n}^2 - \alpha_{\nu m}^2}{a^2}\int_0^a J_\nu\left(\alpha_{\nu m}\frac{\rho}{a}\right)J_\nu\left(\alpha_{\nu n}\frac{\rho}{a}\right)\rho\, d\rho. \quad (11.47)$$

Upon integrating by parts, we see that the left-hand side of Eq. (11.47) becomes

$$\left| \rho J_\nu \left(\alpha_{\nu n} \frac{\rho}{a} \right) \frac{d}{d\rho} J_\nu \left(\alpha_{\nu m} \frac{\rho}{a} \right) \right|_0^a - \left| \rho J_\nu \left(\alpha_{\nu m} \frac{\rho}{a} \right) \frac{d}{d\rho} J_\nu \left(\alpha_{\nu n} \frac{\rho}{a} \right) \right|_0^a . \qquad (11.48)$$

For $\nu \geq 0$ the factor ρ guarantees a zero at the lower limit, $\rho = 0$. Actually the lower limit on the index ν may be extended down to $\nu > -1$, Exercise 11.2.4.[1] At $\rho = a$, each expression vanishes if we choose the parameters $\alpha_{\nu n}$ and $\alpha_{\nu m}$ to be zeros or roots of J_ν; that is, $J_\nu(\alpha_{\nu m}) = 0$. The subscripts now become meaningful: $\alpha_{\nu m}$ is the mth zero of J_ν.

With this choice of parameters, the left-hand side vanishes (the Sturm–Liouville boundary conditions are satisfied) and for $m \neq n$

$$\int_0^a J_\nu \left(\alpha_{\nu m} \frac{\rho}{a} \right) J_\nu \left(\alpha_{\nu n} \frac{\rho}{a} \right) \rho \, d\rho = 0. \qquad (11.49)$$

This gives us orthogonality over the interval $[0, a]$.

NORMALIZATION

The normalization integral may be developed by returning to Eq. (11.48), setting $\alpha_{\nu n} = \alpha_{\nu m} + \varepsilon$, and taking the limit $\varepsilon \to 0$ (compare Exercise 11.2.2). With the aid of the recurrence relation, Eq. (11.16), the result may be written as

$$\int_0^a \left[J_\nu \left(\alpha_{\nu m} \frac{\rho}{a} \right) \right]^2 \rho \, d\rho = \frac{a^2}{2} [J_{\nu+1}(\alpha_{\nu m})]^2. \qquad (11.50)$$

BESSEL SERIES

If we assume that the set of Bessel functions $J_\nu(\alpha_{\nu m} \rho/a)$ (ν fixed, $m = 1, 2, 3, \ldots$) is complete, then any well-behaved but otherwise arbitrary function $f(\rho)$ may be expanded in a Bessel series (Bessel–Fourier or Fourier–Bessel)

$$f(\rho) = \sum_{m=1}^{\infty} c_{\nu m} J_\nu \left(\alpha_{\nu m} \frac{\rho}{a} \right), \qquad 0 \leq \rho \leq a, \qquad \nu > -1. \qquad (11.51)$$

The coefficients $c_{\nu m}$ are determined by using Eq. (11.50),

$$c_{\nu m} = \frac{2}{a^2 [J_{\nu+1}(\alpha_{\nu m})]^2} \int_0^a f(\rho) J_\nu \left(\alpha_{\nu m} \frac{\rho}{a} \right) \rho \, d\rho. \qquad (11.52)$$

[1] The case $\nu = -1$ reverts to $\nu = +1$, Eq. (11.8).

A similar series expansion involving $J_\nu(\beta_{\nu m}\rho/a)$ with $(d/d\rho) J_\nu(\beta_{\nu m}\rho/a)|_{\rho=a}$ $= 0$ is included in Exercises 11.2.3 and 11.2.6(b).

Example 11.2.1 Electrostatic Potential in a Hollow Cylinder

From Table 8.3 of Section 8.3 (with α replaced by k) our solution of Laplace's equation in circular cylindrical coordinates is a linear combination of

$$\psi_{km}(\rho, \varphi, z) = P_{km}(\rho)\Phi_m(\varphi)Z_k(z)$$
$$= J_m(k\rho) \cdot [a_m \sin m\varphi + b_m \cos m\varphi] \cdot [c_1 e^{kz} + c_2 e^{-kz}]. \quad (11.53)$$

The particular linear combination is determined by the boundary conditions to be satisfied.

Our cylinder here has a radius a and a height l. The top end section has a potential distribution $\psi(\rho, \varphi)$. Elsewhere on the surface the potential is zero.[2] The problem is to find the electrostatic potential

$$\psi(\rho, \varphi, z) = \sum_{k,m} \psi_{km}(\rho, \varphi, z) \quad (11.54)$$

everywhere in the interior.

For convenience, the circular cylindrical coordinates are placed as shown in Fig. 11.3. Since $\psi(\rho, \varphi, 0) = 0$, we take $c_1 = -c_2 = \frac{1}{2}$. The z dependence becomes $\sinh kz$, vanishing at $z = 0$. The requirement that $\psi = 0$ on the cylindrical sides is met by requiring the separation constant k to be

$$k = k_{mn} = \alpha_{mn}/a, \quad (11.55)$$

where the first subscript m gives the index of the Bessel function, whereas the second subscript identifies the particular zero of J_m.

The electrostatic potential becomes

$$\psi(\rho, \varphi, z) = \sum_{m=0}^{\infty} \sum_{n=1}^{\infty} J_m\left(\alpha_{mn}\frac{\rho}{a}\right)$$
$$\cdot [a_{mn} \sin m\varphi + b_{mn} \cos m\varphi] \cdot \sinh\left(\alpha_{mn}\frac{z}{a}\right). \quad (11.56)$$

Equation (11.56) is a double series: a Bessel series in ρ and a Fourier series in φ.

At $z = l$, $\psi = \psi(\rho, \varphi)$, a known function of ρ and φ. Therefore

$$\psi(\rho, \varphi) = \sum_{m=0}^{\infty} \sum_{n=1}^{\infty} J_m\left(\alpha_{mn}\frac{\rho}{a}\right)$$
$$\cdot [a_{mn} \sin m\varphi + b_{mn} \cos m\varphi] \cdot \sinh\left(\alpha_{mn}\frac{l}{a}\right). \quad (11.57)$$

---

[2] If $\psi = 0$ at $z = 0, l$, but $\psi \neq 0$ for $\rho = a$, the modified Bessel functions, Section 11.5, are involved.

The constants a_{mn} and b_{mn} are evaluated by using Eqs. (11.49) and (11.50) and the corresponding equations for $\sin \varphi$ and $\cos \varphi$ (Example 9.2.1 and Eqs. (14.7) to (14.9)). We find[3]

$$\left.\begin{array}{c} a_{mn} \\ b_{mn} \end{array}\right\} = 2\left[\pi a^2 \sinh\left(\alpha_{mn}\frac{l}{a} \right) J_{m+1}^2(\alpha_{mn}) \right]^{-1}$$

$$\cdot \int_0^{2\pi} \int_0^a \psi(\rho, \varphi) J_m\left(\alpha_{mn}\frac{\rho}{a} \right) \left\{ \begin{array}{c} \sin m\varphi \\ \cos m\varphi \end{array} \right\} \rho \, d\rho \, d\varphi. \qquad (11.58)$$

These are definite integrals, that is, numbers. Substituting back into Eq. (11.56) the series is specified and the potential $\psi(\rho, \varphi, z)$ is determined. The problem is solved.

CONTINUUM FORM

The Bessel series, Eq. (11.51), and Exercise 11.2.6 apply to expansions over the finite interval $[0, a]$. If $a \to \infty$, then the series forms may be expected to go over into integrals. The discrete roots α_{vm} become a continuous variable α. A similar situation is encountered in the Fourier series, Section 14.2. The development of the Bessel integral from the Bessel series is left as Exercise 11.2.8.

For operations with a continuum of Bessel functions, $J_v(\alpha\rho)$, a key relation is the Bessel function closure equation

$$\int_0^\infty J_v(\alpha\rho) J_v(\alpha'\rho)\rho \, d\rho = \frac{1}{\alpha}\delta(\alpha - \alpha'), \qquad v > -\tfrac{1}{2}. \qquad (11.59)$$

This may be proved by the use of Hankel transforms, Section 15.1. An alternate approach, starting from a relation similar to Eq. (9.82), is given by Morse and Feshbach, Section 6.3.

A second kind of orthogonality (varying the index) is developed for spherical Bessel functions in Section 11.7.

EXERCISES

11.2.1   (a) Show that

$$(a^2 - b^2)\int_0^P J_v(ax) J_v(bx)x \, dx = P[bJ_v(aP) J_v'(bP) - aJ_v'(aP) J_v(bP)],$$

with

$$J_v'(aP) = \frac{d}{d(ax)} J_v(ax)\Big|_{x = P}.$$

(b) $\int_0^P [J_v(ax)]^2 x \, dx = \frac{P^2}{2}\left\{ [J_v'(aP)]^2 + \left(1 - \frac{v^2}{a^2 P^2} \right)[J_v(aP)]^2 \right\}, \qquad v > -1.$

[3] If $m = 0$, the factor 2 is omitted (compare Eq. (14.8)).

These two integrals are usually called the first and second Lommel integrals.

Hint. We have the development of the orthogonality of the Bessel functions as an analogy.

11.2.2 Show that

$$\int_0^a \left[J_\nu \left(\alpha_{\nu m} \frac{\rho}{a} \right) \right]^2 \rho \, d\rho = \frac{a^2}{2} [J_{\nu+1}(\alpha_{\nu m})]^2, \qquad \nu > -1.$$

Here $\alpha_{\nu m}$ is the mth zero of J_ν.

Hint. With $\alpha_{\nu n} = \alpha_{\nu m} + \varepsilon$, expand $J_\nu[(\alpha_{\nu m} + \varepsilon)\rho/a]$ about $\alpha_{\nu m}\rho/a$ by a Taylor expansion.

11.2.3 (a) If $\beta_{\nu m}$ is the mth zero of $(d/d\rho)J_\nu(\beta_{\nu m}\rho/a)$, show that the Bessel functions are orthogonal over the interval $[0, a]$ with an orthogonality integral

$$\int_0^a J_\nu \left(\beta_{\nu m} \frac{\rho}{a} \right) J_\nu \left(\beta_{\nu n} \frac{\rho}{a} \right) \rho \, d\rho = 0, \qquad m \neq n, \nu > -1.$$

(b) Derive the corresponding normalization integral ($m = n$).

$$\text{ANS.} \quad \frac{a^2}{2} \left(1 - \frac{\nu^2}{\beta_{\nu m}^2} \right) [J_\nu(\beta_{\nu m})]^2, \qquad \nu > -1.$$

11.2.4 Verify that the orthogonality equation, Eq. (11.49), and the normalization equation, Eq. (11.50), hold for $\nu > -1$.

Hint. Using power-series expansions, examine the behavior of Eq. (11.48) as $\rho \to 0$.

11.2.5 From Eq. (11.49) develop a proof that $J_\nu(z)$, $\nu > -1$, has no complex roots.

Hint.

(a) Use the series form of $J_\nu(z)$ to exclude pure imaginary roots.

(b) Assume $\alpha_{\nu m}$ to be complex and take $\alpha_{\nu n}$ to be $\alpha_{\nu m}^*$.

11.2.6 (a) In the series expansion

$$f(\rho) = \sum_{m=1}^\infty c_{\nu m} J_\nu \left(\alpha_{\nu m} \frac{\rho}{a} \right), \qquad 0 \leq \rho \leq a, \qquad \nu > -1,$$

with $J_\nu(\alpha_{\nu m}) = 0$, show that the coefficients are given by

$$c_{\nu m} = \frac{2}{a^2 [J_{\nu+1}(\alpha_{\nu m})]^2} \int_0^a f(\rho) J_\nu \left(\alpha_{\nu m} \frac{\rho}{a} \right) \rho \, d\rho.$$

(b) In the series expansion

$$f(\rho) = \sum_{m=1}^\infty d_{\nu m} J_\nu \left(\beta_{\nu m} \frac{\rho}{a} \right), \qquad 0 \leq \rho \leq a, \qquad \nu > -1,$$

with $(d/d\rho)J_\nu(\beta_{\nu m}\rho/a)|_{\rho=a} = 0$, show that the coefficients are given by

$$d_{\nu m} = \frac{2}{a^2 (1 - \nu^2/\beta_{\nu m}^2)[J_\nu(\beta_{\nu m})]^2} \int_0^a f(\rho) J_\nu \left(\beta_{\nu m} \frac{\rho}{a} \right) \rho \, d\rho.$$

11.2.7   A right circular cylinder has an electrostatic potential of $\psi(\rho, \varphi)$ on both ends. The potential on the curved cylindrical surface is zero. Find the potential at all interior points.
Hint. Choose your coordinate system and adjust your z dependence to exploit the symmetry of your potential.

11.2.8   For the continuum case, show that Eqs. (11.51) and (11.52) are replaced by

$$f(\rho) = \int_0^\infty a(\alpha) J_\nu(\alpha\rho)\, d\alpha,$$

$$a(\alpha) = \alpha \int_0^\infty f(\rho) J_\nu(\alpha\rho)\rho\, d\rho.$$

Hint. The corresponding case for sines and cosines is worked out in Section 15.2. These are Hankel transforms. A derivation for the special case $\nu = 0$ is the topic of Exercise 15.1.1.

11.2.9   A function $f(x)$ is expressed as a Bessel series:

$$f(x) = \sum_{n=1}^\infty a_n J_m(\alpha_{mn} x),$$

with α_{mn} the nth root of J_m. Prove the Parseval relation

$$\int_0^1 [f(x)]^2 x\, dx = \tfrac{1}{2} \sum_{n=1}^\infty a_n^2 [J_{m+1}(\alpha_{mn})]^2.$$

11.2.10   Prove that

$$\sum_{n=1}^\infty (\alpha_{mn})^{-2} = \frac{1}{4(m+1)}.$$

Hint. Expand x^m in a Bessel series and apply the Parseval relation.

11.2.11   A right circular cylinder of length l has a potential
$$\psi(z = \pm l/2) = 100(1 - \rho/a),$$
where a is the radius. The potential over the curved surface (side) is zero. Using the Bessel series from Exercise 11.2.7, calculate the electrostatic potential for $\rho/a = 0.0(0.2)1.0$ and $z/l = 0.0(0.1)0.5$. Take $a/l = 0.5$.
Hint. From Exercise 11.1.30 you have

$$\int_0^{\alpha_{0n}} \left(1 - \frac{y}{\alpha_{0n}}\right) J_0(y) y\, dy.$$

Show that this equals

$$\frac{1}{\alpha_{0n}} \int_0^{\alpha_{0n}} J_0(y)\, dy.$$

Numerical evaluation of this latter form rather than the former is both faster and more accurate.
Note. For $\rho/a = 0.0$ and $z/l = 0.5$ the convergence is slow, 20 terms giving only 98.4 rather than 100.

$\qquad\qquad\qquad$ *Check value.*   For $\rho/a = 0.4$ and $z/l = 0.3$,
$\qquad\qquad\qquad\qquad\qquad\qquad$ $\psi = 24.558$.

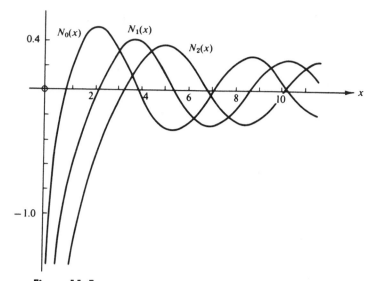

Figure 11.5   Neumann functions, $N_0(x)$, $N_1(x)$, and $N_2(x)$.

11.3 NEUMANN FUNCTIONS, BESSEL FUNCTIONS OF THE SECOND KIND, $N_\nu(x)$

From the theory of differential equations it is known that Bessel's equation has two independent solutions. Indeed, for nonintegral order ν we have already found two solutions and labeled them $J_\nu(x)$ and $J_{-\nu}(x)$, using the infinite series (Eq. (11.5)). The trouble is that when ν is integral Eq. (11.8) holds and we have but one independent solution. A second solution may be developed by the methods of Section 8.6. This yields a perfectly good second solution of Bessel's equation but is not the usual standard form.

DEFINITION AND SERIES FORM

As an alternate approach, we take the particular linear combination of $J_\nu(x)$ and $J_{-\nu}(x)$

$$N_\nu(x) = \frac{\cos \nu\pi J_\nu(x) - J_{-\nu}(x)}{\sin \nu\pi}. \tag{11.60}$$

This is the Neumann function (Fig. 11.5).[1] For nonintegral ν, $N_\nu(x)$ clearly satisfies Bessel's equation, for it is a linear combination of known solutions, $J_\nu(x)$ and $J_{-\nu}(x)$. Substituting the power series Eq. (11.6) for $n \to \nu$ (see Exercise 11.1.7) yields

$$N_\nu(x) = -\frac{(\nu - 1)!}{\pi}\left(\frac{2}{x}\right)^\nu + \cdots,^{2} \tag{11.61}$$

---

[1] In AMS-55 and in most mathematics tables, this is labeled $Y_\nu(x)$.
[2] Note that this limiting form applies to *both* integral and nonintegral values of the index ν.

for $v > 0$. However, for integral v, $v = n$, Eq. (11.8) applies and Eq. (11.60) becomes indeterminate. The definition of $N_v(x)$ was chosen deliberately for this indeterminate property. Again substituting the power series and evaluating $N_v(x)$ for $v \to 0$ by l'Hospital's rule for indeterminate forms, we obtain the limiting value

$$N_0(x) = \frac{2}{\pi}(\ln x + \gamma - \ln 2) + O(x^2) \tag{11.62}$$

for $n = 0$ and $x \to 0$, using

$$v!(-v)! = \frac{\pi v}{\sin \pi v} \tag{11.63}$$

from Eq. (10.32). The first and third terms in Eq. (11.62) come from using $(d/dv)(x/2)^v = (x/2)^v \ln(x/2)$, while γ comes from $(d/dv)v!$ for $v \to 0$ using Eqs. (10.38) and (10.40). For $n > 0$ we obtain similarly

$$N_n(x) = -(n-1)!(2/x)^n/\pi + \cdots + \frac{2}{\pi}\left(\frac{x}{2}\right)^n \frac{1}{n!} \ln\left(\frac{x}{2}\right) + \cdots. \tag{11.64}$$

Equations (11.62) and (11.64) exhibit the logarithmic dependence that was to be expected. This, of course, verifies the independence of J_n and N_n.

OTHER FORMS

As with all the other Bessel functions, $N_v(x)$ has integral representations. For $N_0(x)$ we have

$$N_0(x) = -\frac{2}{\pi}\int_0^\infty \cos(x \cosh t)\, dt$$

$$= -\frac{2}{\pi}\int_1^\infty \frac{\cos(xt)}{(t^2 - 1)^{1/2}}\, dt, \qquad x > 0.$$

These forms can be derived as the imaginary part of the Hankel representations of Exercise 11.4.7. The latter form is a Fourier cosine transform.

To verify that $N_v(x)$, our Neumann function (Fig. 11.5) or Bessel function of the second kind, actually does satisfy Bessel's equation for integral n, we may proceed as follows. L'Hospital's rule applied to Eq. (11.60) yields

$$N_n(x) = \frac{(d/dv)[\cos v\pi J_v(x) - J_{-v}(x)]}{(d/dv)\sin v\pi}\bigg|_{v=n}$$

$$= \frac{-\pi \sin n\pi J_n(x) + [\cos n\pi \partial J_v/\partial v - \partial J_{-v}/\partial v]}{\pi \cos n\pi}\bigg|_{v=n}$$

$$= \frac{1}{\pi}\left[\frac{\partial J_v(x)}{\partial v} - (-1)^n \frac{\partial J_{-v}(x)}{\partial v}\right]\bigg|_{v=n}. \tag{11.65}$$

Differentiating Bessel's equation for $J_{\pm\nu}(x)$ with respect to ν, we have

$$x^2 \frac{d^2}{dx^2}\left(\frac{\partial J_{\pm\nu}}{\partial\nu}\right) + x\frac{d}{dx}\left(\frac{\partial J_{\pm\nu}}{\partial\nu}\right) + (x^2 - \nu^2)\frac{\partial J_{\pm\nu}}{\partial\nu} = 2\nu J_{\pm\nu}. \qquad (11.66)$$

Multiplying the equation for $J_{-\nu}$ by $(-1)^\nu$, subtracting from the equation for J_ν (as suggested by Eq. (11.65)), and taking the limit $\nu \to n$, we obtain

$$x^2 \frac{d^2}{dx^2}N_n + x\frac{d}{dx}N_n + (x^2 - n^2)N_n = \frac{2n}{\pi}[J_n - (-1)^n J_{-n}]. \qquad (11.67)$$

For $\nu = n$, an integer, the right-hand side vanishes by Eq. (11.8) and $N_n(x)$ is seen to be a solution of Bessel's equation. The most general solution for any ν can therefore be written as

$$y(x) = AJ_\nu(x) + BN_\nu(x). \qquad (11.68)$$

It is seen from Eqs. (11.62) and (11.64) that N_n diverges at least logarithmically. Any boundary condition that requires the solution to be finite at the origin [as in our vibrating circular membrane (Section 11.1)] automatically excludes $N_n(x)$. Conversely, in the absence of such a requirement $N_n(x)$ must be considered.

To a certain extent the definition of the Neumann function $N_n(x)$ is arbitrary. Equations (11.62) and (11.64) contain terms of the form $a_n J_n(x)$. Clearly, any finite value of the constant a_n would still give us a second solution of Bessel's equation. Why should a_n have the particular value implicit in Eqs. (11.62) and (11.64)? The answer involves the asymptotic dependence developed in Section 11.6. If J_n corresponds to a cosine wave, then N_n corresponds to a sine wave. This simple and convenient asymptotic phase relationship is a consequence of the particular admixture of J_n in N_n.

RECURRENCE RELATIONS

Substituting Eq. (11.60) for $N_\nu(x)$ (nonintegral ν) or Eq. (11.61) (integral ν) into the recurrence relations (Eqs. (11.10) and (11.12) for $J_n(x)$, we see immediately that $N_\nu(x)$ satisfies these same recurrence relations. This actually constitutes another proof that N_ν is a solution. Note carefully that the converse is not necessarily true. All solutions need not satisfy the same recurrence relations. An example of this sort of trouble appears in Section 11.5.

WRONSKIAN FORMULAS

From Section 8.6 and Exercise 9.1.4 we have the Wronskian formula[3] for solutions of the Bessel equation

$$u_\nu(x)v_\nu'(x) - u_\nu'(x)v_\nu(x) = \frac{A_\nu}{x}, \qquad (11.69)$$

---

[3] This result depends on $P(x)$ of Section 8.5 being equal to $p'(x)/p(x)$, the corresponding coefficient of the self-adjoint form of Section 9.1.

in which A_ν is a parameter that depends on the particular Bessel functions $u_\nu(x)$ and $v_\nu(x)$ being considered. It is a constant in the sense that it is independent of x. Consider the special case

$$u_\nu(x) = J_\nu(x), \qquad v_\nu(x) = J_{-\nu}(x), \tag{11.70}$$

$$J_\nu J'_{-\nu} - J'_\nu J_{-\nu} = \frac{A_\nu}{x}. \tag{11.71}$$

Since A_ν is a constant, it may be identified at any convenient point such as $x = 0$. Using the first terms in the series expansions (Eqs. (11.5) and (11.6)), we obtain

$$
\begin{aligned}
J_\nu &\to \frac{x^\nu}{2^\nu \nu!}, & J_{-\nu} &\to \frac{2^\nu x^{-\nu}}{(-\nu)!}, \\
J'_\nu &\to \frac{\nu x^{\nu-1}}{2^\nu \nu!}, & J'_{-\nu} &\to \frac{-\nu 2^\nu x^{-\nu-1}}{(-\nu)!}.
\end{aligned}
\tag{11.72}
$$

Substitution into Eq. (11.69) yields

$$
\begin{aligned}
J_\nu(x)J'_{-\nu}(x) - J'_\nu(x)J_{-\nu}(x) &= \frac{-2\nu}{x\nu!(-\nu)!} \\
&= -\frac{2 \sin \nu\pi}{\pi x},
\end{aligned}
\tag{11.73}
$$

using Eq. (10.63). Note that A_ν vanishes for integral ν, as it must, since the nonvanishing of the Wronskian is a test of the independence of the two solutions. By Eq. (11.73), J_n and J_{-n} are clearly linearly dependent.

Using our recurrence relations, we may readily develop a large number of alternate forms, among which are

$$J_\nu J_{-\nu+1} + J_{-\nu}J_{\nu-1} = \frac{2 \sin \nu\pi}{\pi x}, \tag{11.74}$$

$$J_\nu J_{-\nu-1} + J_{-\nu}J_{\nu+1} = -\frac{2 \sin \nu\pi}{\pi x}, \tag{11.75}$$

$$J_\nu N'_\nu - J'_\nu N_\nu = \frac{2}{\pi x}, \tag{11.76}$$

$$J_\nu N_{\nu+1} - J_{\nu+1}N_\nu = -\frac{2}{\pi x}. \tag{11.77}$$

Many more will be found in the references given.

The reader will recall that in Chapter 8 Wronskians were of great value in two respects: (1) in establishing the linear independence or linear dependence of solutions of differential equations and (2) in developing an integral form of a second solution. Here the specific forms of the Wronskians and Wronskian-derived combinations of Bessel functions are useful primarily to illustrate the general behavior of the various Bessel functions. Wronskians are of great use in checking tables of Bessel functions. In Chapter 16 Wronskians reappear in connection with Green's functions.

Example 11.3.1 Coaxial Wave Guides

We are interested in an electromagnetic wave confined between the concentric, conducting cylindrical surfaces $\rho = a$ and $\rho = b$. Most of the mathematics is worked out in Section 8.3 and Example 11.1.2. To go from the standing wave of these examples to the traveling wave here, we let $a_{mn} = ib_{mn}$ in Eq. (11.40) and obtain

$$E_z = \sum_{m,n} b_{mn} J_m(\gamma\rho) \, e^{\pm im\varphi} e^{i(kz-\omega t)}. \tag{11.78}$$

Additional properties of the components of the electromagnetic wave in the simple cylindrical wave guide are explored in Exercises 11.3.8 and 11.3.9. For the coaxial wave guide one generalization is needed. The origin $\rho = 0$ is now excluded ($0 < a \le \rho \le b$). Hence the Neumann function $N_m(\gamma\rho)$ may *not* be excluded. $E_z(\rho, \varphi, z, t)$ becomes

$$E_z = \sum_{m,n} [b_{mn} J_m(\gamma\rho) + c_{mn} N_m(\gamma\rho)] \, e^{\pm im\varphi} \, e^{i(kz-\omega t)}. \tag{11.79}$$

With the condition

$$H_z = 0, \tag{11.80}$$

we have the basic equations for a TM (transverse magnetic) wave.

The (tangential) electric field must vanish at the conducting surfaces (Dirichlet boundary condition) or

$$b_{mn} J_m(\gamma a) + c_{mn} N_m(\gamma a) = 0. \tag{11.81}$$

$$b_{mn} J_m(\gamma b) + c_{mn} N_m(\gamma b) = 0. \tag{11.82}$$

These transcendental equations may be solved for $\gamma(\gamma_{mn})$ and the ratio c_{mn}/b_{mn}. From Example 11.1.2,

$$k^2 = \omega^2 \mu_0 \varepsilon_0 - \gamma^2 = \frac{\omega^2}{c^2} - \gamma^2. \tag{11.83}$$

Since k^2 must be positive for a real wave, the minimum frequency that will be propagated (in this TM mode) is

$$\omega = \gamma c, \tag{11.84}$$

with γ fixed by the boundary conditions, Eqs. (11.81) and (11.82). This is the cutoff frequency of the wave guide.

There is also a TE (transverse electric) mode with $E_z = 0$, and H_z given by Eq. (11.79). Then we have Neumann boundary conditions in place of Eqs. (11.81) and (11.82). Finally, for the coaxial guide (*not* for the plain cylindrical guide, $a = 0$), a TEM (transverse electromagnetic) mode, $E_z = H_z = 0$, is possible. This corresponds to a plane wave as in free space.

The simpler cases (no Neumann functions, simpler boundary conditions) of a circular wave guide are included as Exercises 11.3.8 and 11.3.9.

To conclude this discussion of Neumann functions, we introduce the Neumann function, $N_\nu(x)$, for the following reasons:

1. It is a second, independent solution of Bessel's equation, which completes the general solution.
2. It is required for specific physical problems such as electromagnetic waves in coaxial cables and quantum mechanical scattering theory.
3. It leads to a Green's function for the Bessel equation (Sections 9.5 and 8.7).
4. It leads directly to the two Hankel functions (Section 11.4).

EXERCISES

11.3.1   Prove that the Neumann functions N_n (with n an integer) satisfy the recurrence relations

$$N_{n-1}(x) + N_{n+1}(x) = \frac{2n}{x} N_n(x)$$

$$N_{n-1}(x) - N_{n+1}(x) = 2N_n'(x).$$

Hint. These relations may be proved by differentiating the recurrence relations for J_ν or by using the limit form of N_ν but *not* dividing everything by zero.

11.3.2   Show that

$$N_{-n}(x) = (-1)^n N_n(x).$$

11.3.3   Show that

$$N_0'(x) = -N_1(x).$$

11.3.4   If Y and Z are any two solutions of Bessel's equation, show that

$$Y_\nu(x)Z_\nu'(x) - Y_\nu'(x)Z_\nu(x) = \frac{A_\nu}{x},$$

in which A_ν may depend on ν but is independent of x. This is really a special case of Exercise 9.1.4.

11.3.5   Verify the Wronskian formulas

$$J_\nu(x)J_{-\nu+1}(x) + J_{-\nu}(x)J_{\nu-1}(x) = \frac{2 \sin \nu\pi}{\pi x},$$

$$J_\nu(x)N_\nu'(x) - J_\nu'(x)N_\nu(x) = \frac{2}{\pi x}.$$

11.3.6   As an alternative to letting x approach zero in the evaluation of the Wronskian constant, we may invoke uniqueness of power series (Section 5.7). The coefficient of x^{-1} in the series expansion of $u_\nu(x)v_\nu'(x) - u_\nu'(x)v_\nu(x)$ is then A_ν. Show by series expansion that the coefficients of x^0 and x^1 of $J_\nu(x)J_{-\nu}'(x) - J_\nu'(x)J_{-\nu}(x)$ are each zero.

11.3.7   (a) By differentiating and substituting into Bessel's differential equation, show that

$$\int_0^\infty \cos(x \cosh t)\, dt$$

is a solution.

Hint. You can rearrange the final integral as

$$\int_0^\infty \frac{d}{dt} \{x \sin(x \cosh t) \sinh t\}\, dt.$$

(b) Show that

$$N_0(x) = -\frac{2}{\pi} \int_0^\infty \cos(x \cosh t)\, dt$$

is linearly independent of $J_0(x)$.

11.3.8   A cylindrical wave guide has radius r_0. Find the nonvanishing components of the electric and magnetic fields for

(a) TM_{01}, transverse magnetic wave ($H_z = H_\rho = E_\varphi = 0$),
(b) TE_{01}, transverse electric wave ($E_z = E_\rho = H_\varphi = 0$).

The subscripts 01 indicate that the longitudinal component (E_z or H_z) involves J_0 and the boundary condition is satisfied by the *first* zero of J_0 or J_0'.

Hint. All components of the wave have the same factor: $\exp i(kz - \omega t)$.

11.3.9   For a given mode of oscillation the *minimum* frequency that will be passed by a circular cylindrical wave guide (radius r_0) is

$$\nu_{min} = \frac{c}{\lambda_c},$$

in which λ_c is fixed by the boundary condition

$$J_n\left(\frac{2\pi r_0}{\lambda_c}\right) = 0 \qquad \text{for } TM_{nm} \text{ mode,}$$

$$J_n'\left(\frac{2\pi r_0}{\lambda_c}\right) = 0 \qquad \text{for } TE_{nm} \text{ mode.}$$

The subscript n denotes the order of the Bessel function and m indicates the zero used. Find this cut-off wavelength, λ_c for the three TM and three TE modes with the longest cut-off wavelengths. Explain your results in terms of the graph of J_0, J_1, and J_2 (Fig. 11.1).

11.3.10   Write a program that will compute successive roots of the Neumann function $N_n(x)$, that is α_{ns}, where $N_n(\alpha_{ns}) = 0$. Tabulate the first five roots of N_0, N_1, and N_2. Check your values for the roots against those listed in AMS-55 (Chapter 9).

Hint. See Appendix 1 for root-finding techniques and recommendations.

Check value. $\alpha_{12} = 5.42968$.

11.3.11   For the case $m = 0$, $a = 1$, and $b = 2$ the coaxial wave guide boundary conditions lead to

$$f(x) = \frac{J_0(2x)}{N_0(2x)} - \frac{J_0(x)}{N_0(x)}$$

(Fig. 11.6).

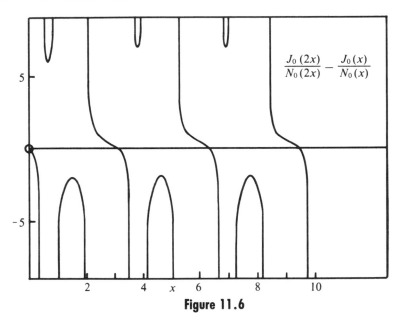

$$\frac{J_0(2x)}{N_0(2x)} - \frac{J_0(x)}{N_0(x)}$$

Figure 11.6

(a) Calculate $f(x)$ for $x = 0.0(0.1)10.0$ and plot $f(x)$ versus x to find the approximate location of the roots.

(b) Call a root-finding subroutine to determine the first three roots to higher precision.

ANS.   3.1230, 6.2734, 9.4182.

Note. The higher roots can be expected to appear at intervals whose length approaches π. Why? AMS-55, Section 9.5, gives an approximate formula for the roots. The function $g(x) = J_0(x)N_0(2x) - J_0(2x)N_0(x)$ is much better behaved than $f(x)$ previously discussed.

11.4 HANKEL FUNCTIONS

Many authors prefer to introduce the Hankel functions by means of integral representations and then use them to define the Neumann function, $N_\nu(z)$. An outline of this approach is given at the end of this section.

DEFINITIONS

As we have already obtained the Neumann function by more elementary (and less powerful) techniques, we may use it to define the Hankel functions, $H_\nu^{(1)}(x)$ and $H_\nu^{(2)}(x)$:

$$H_\nu^{(1)}(x) = J_\nu(x) + iN_\nu(x) \tag{11.85}$$

and

$$H_\nu^{(2)}(x) = J_\nu(x) - iN_\nu(x) \tag{11.86}$$

This is exactly analogous to taking

$$e^{\pm i\theta} = \cos\theta \pm i\sin\theta. \tag{11.87}$$

For real arguments $H_\nu^{(1)}$ and $H_\nu^{(2)}$ are complex conjugates. The extent of the analogy will be seen even better when the asymptotic forms are considered (Section 11.6). Indeed, it is their asymptotic behavior that makes the Hankel functions useful.

Series expansion of $H_\nu^{(1)}(x)$ and $H_\nu^{(2)}(x)$ may be obtained by combining Eqs. (11.5) and (11.63). Often only the first term is of interest; it is given by

$$H_0^{(1)}(x) \approx i\frac{2}{\pi}\ln x + 1 + i\frac{2}{\pi}(\gamma - \ln 2) + \cdots, \tag{11.88}$$

$$H_\nu^{(1)}(x) \approx -i\frac{(\nu - 1)!}{\pi}\left(\frac{2}{x}\right)^\nu + \cdots, \qquad \nu > 0, \tag{11.89}$$

$$H_0^{(2)}(x) \approx -i\frac{2}{\pi}\ln x + 1 - i\frac{2}{\pi}(\gamma - \ln 2) + \cdots, \tag{11.90}$$

$$H_\nu^{(2)}(x) \approx i\frac{(\nu - 1)!}{\pi}\left(\frac{2}{x}\right)^\nu + \cdots, \qquad \nu > 0. \tag{11.91}$$

Since the Hankel functions are linear combinations (with constant coefficients) of J_ν and N_ν, they satisfy the same recurrence relations (Eqs. (11.10) and (11.12)).

$$H_{\nu-1}(x) + H_{\nu+1}(x) = \frac{2\nu}{x}H_\nu(x), \tag{11.92}$$

$$H_{\nu-1}(x) - H_{\nu+1}(x) = 2H_\nu'(x), \tag{11.93}$$

for both $H_\nu^{(1)}(x)$ and $H_\nu^{(2)}(x)$.

A variety of Wronskian formulas can be developed:

$$H_\nu^{(2)}H_{\nu+1}^{(1)} - H_\nu^{(1)}H_{\nu+1}^{(2)} = \frac{4}{i\pi x}, \tag{11.94}$$

$$J_{\nu-1}H_\nu^{(1)} - J_\nu H_{\nu-1}^{(1)} = \frac{2}{i\pi x}, \tag{11.95}$$

$$J_\nu H_{\nu-1}^{(2)} - J_{\nu-1}H_\nu^{(2)} = \frac{2}{i\pi x}. \tag{11.96}$$

Example 11.4.1 Cylindrical Traveling Waves

As an illustration of the use of Hankel functions, consider a two-dimensional wave problem similar to the vibrating circular membrane of Exercise 11.1.25. Now imagine that the waves are generated at $r = 0$ and move outward to infinity. We replace our standing waves by traveling ones. The differential equation remains the same, but the boundary conditions change. We now demand that for large r the solution behave like

$$U \to e^{i(kr-\omega t)} \tag{11.97}$$

to describe an outgoing wave. As before, k is the wave number. This assumes, for simplicity, that there is no azimuthal dependence, that is, no angular momentum, or $m = 0$. In Sections 7.4 and 11.6, $H_0^{(1)}(kr)$ is shown to have the asymptotic behavior (for $r \to \infty$)

$$H_0^{(1)}(kr) \to e^{ikr}. \tag{11.98}$$

This boundary condition at infinity then determines our wave solution as

$$U(r, t) = H_0^{(1)}(kr) \, e^{-i\omega t}. \tag{11.99}$$

This solution diverges as $r \to 0$, which is just the behavior to be expected with a source at the origin.

The choice of a *two*-dimensional wave problem to illustrate the Hankel function $H_0^{(1)}(z)$ is not accidental. Bessel functions may appear in a variety of ways, such as in the separation of conical coordinates. However, they enter most commonly from the radial equations from the separation of variables in the Helmholtz equation in cylindrical and in spherical polar coordinates. We have taken a degenerate form of cylindrical coordinates for this illustration. Had we used spherical polar coordinates (spherical waves), we should have encountered index $v = n + \frac{1}{2}$, n an integer. These special values yield the spherical Bessel functions to be discussed in Section 11.7.

CONTOUR INTEGRAL REPRESENTATION OF THE HANKEL FUNCTIONS

The integral representation (Schlaefli integral)

$$J_v(x) = \frac{1}{2\pi i} \int e^{(x/2)(t-1/t)} \frac{dt}{t^{v+1}} \tag{11.100}$$

may easily be established as a Cauchy integral for $v = n$, an integer [recognizing that the numerator is the generating function (Eq. (11.1)) and integrating around the origin]. If v is not an integer, the integrand is not single-valued and a cut line is needed in our complex plane. Choosing the negative real axis as the cut line and using the contour shown in Fig. 11.7, we can extend Eq. (11.100) to nonintegral v. Substituting Eq. (11.100) into Bessel's differential equation, we can represent the combined integrand by an exact differential that vanishes as $t \to \infty \, e^{\pm i\pi}$ (compare Exercise 11.1.16).

We now deform the contour so that it approaches the origin along the positive real axis, as shown in Fig. 11.8. This particular approach guarantees that the exact differential mentioned will vanish as $t \to 0$ because of the $e^{-x/2t}$

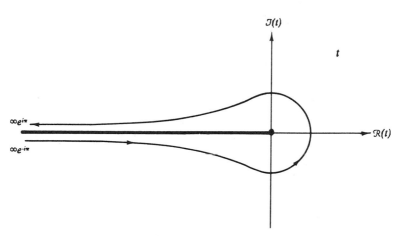

Figure 11.7 Bessel function contour.

factor. Hence each of the separate portions $\infty\, e^{-i\pi}$ to 0 and 0 to $\infty\, e^{i\pi}$ is a solution of Bessel's equation. We define

$$H_\nu^{(1)}(x) = \frac{1}{\pi i} \int_0^{\infty e^{i\pi}} e^{(x/2)(t-1/t)} \frac{dt}{t^{\nu+1}} \tag{11.101}$$

$$H_\nu^{(2)}(x) = \frac{1}{\pi i} \int_{\infty e^{-i\pi}}^0 e^{(x/2)(t-1/t)} \frac{dt}{t^{\nu+1}}. \tag{11.102}$$

These expressions are particularly convenient because they may be handled by the method of steepest descents (Section 7.4). $H_\nu^{(1)}(x)$ has a saddle point at $t = +i$, whereas $H_\nu^{(2)}(x)$ has a saddle point at $t = -i$.

The problem of relating Eqs. (11.101) and (11.102) to our earlier definition

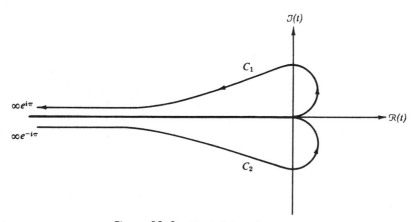

Figure 11.8 Hankel function contours.

of the Hankel function (Eqs. (11.85) and (11.86)) remains. Since Eqs. (11.100) to (11.102) combined yield

$$J_\nu(x) = \tfrac{1}{2}[H_\nu^{(1)}(x) + H_\nu^{(2)}(x)] \tag{11.103}$$

by inspection, we need only show that

$$N_\nu(x) = \frac{1}{2i}[H_\nu^{(1)}(x) - H_\nu^{(2)}(x)]. \tag{11.104}$$

This may be accomplished by the following steps:

1.   With the substitutions $t = e^{i\pi}/s$ for $H_\nu^{(1)}$ and $t = e^{-i\pi}/s$ for $H_\nu^{(2)}$, we obtain

$$H_\nu^{(1)}(x) = e^{-i\nu\pi}H_{-\nu}^{(1)}(x), \tag{11.105}$$

$$H_\nu^{(2)}(x) = e^{i\nu\pi}H_{-\nu}^{(2)}(x). \tag{11.106}$$

2.   From Eqs. (11.103) ($\nu \to -\nu$), (11.105), and (11.106),

$$J_{-\nu}(x) = \tfrac{1}{2}[e^{i\nu\pi}H_\nu^{(1)}(x) + e^{-i\nu\pi}H_\nu^{(2)}(x)]. \tag{11.107}$$

3.   Finally, substitute J_ν (Eq. (11.103)) and $J_{-\nu}$ (Eq. (11.107)) into the defining equation for N_ν, Eq. (11.60). This leads to Eq. (11.104) and establishes the contour integrals Eqs. (11.101) and (11.102) as the Hankel functions.

Integral representations have appeared before: Eq. (10.35) for $\Gamma(z)$ and various representations of $J_\nu(z)$ in Section 11.1. With these integral representations of the Hankel functions, it is perhaps appropriate to ask why we are interested in integral representations. There are at least four reasons. The first is simply aesthetic appeal—some people find them attractive. Second, the integral representations help to distinguish between two linearly independent solutions. In Fig. 11.6, the contours C_1 and C_2 cross *different* saddle points (Section 7.4). For the Legendre functions the contour for $P_n(z)$ (Fig. 12.11) and that for $Q_n(z)$ encircle *different* singular points.

Third, the integral representations facilitate manipulations, analysis, and the development of relations among the various special functions. Fourth, and probably most important of all, the integral representations are extremely useful in developing asymptotic expansions. One approach, the method of steepest descents, appears in Section 7.4. A second approach, the direct expansion of an integral representation is given in Section 11.6 for the modified Bessel function $K_\nu(z)$. This same technique may be used to obtain asymptotic expansions of the confluent hypergeometric functions, M and U—Exercise 13.6.13.

In conclusion, the Hankel functions are introduced here for the following reasons:

1.   As analogs of $e^{\pm ix}$ they are useful for describing traveling waves.
2.   They offer an alternate (contour integral) and a rather elegant definition of Bessel functions.
3. $H_\nu^{(1)}$ is used to define the modified Bessel function K_ν of Section 11.5.

EXERCISES

11.4.1   Verify the Wronskian formulas

(a) $J_\nu(x)H_\nu^{(1)'}(x) - J_\nu'(x)H_\nu^{(1)}(x) = 2i/\pi x$,

(b) $J_\nu(x)H_\nu^{(2)'}(x) - J_\nu'(x)H_\nu^{(2)}(x) = -2i/\pi x$,

(c) $N_\nu(x)H_\nu^{(1)'}(x) - N_\nu'(x)H_\nu^{(1)}(x) = -2/\pi x$,

(d) $N_\nu(x)H_\nu^{(2)'}(x) - N_\nu'(x)H_\nu^{(2)}(x) = -2/\pi x$,

(e) $H_\nu^{(1)}(x)H_\nu^{(2)'}(x) - H_\nu^{(1)'}(x)H_\nu^{(2)}(x) = -4i/\pi x$,

(f) $H_\nu^{(2)}(x)H_{\nu+1}^{(1)}(x) - H_\nu^{(1)}(x)H_{\nu+1}^{(2)}(x) = 4/i\pi x$,

(g) $J_{\nu-1}(x)H_\nu^{(1)}(x) - J_\nu(x)H_{\nu-1}^{(1)}(x) = 2/i\pi x$.

11.4.2   Show that the integral forms

(a) $\dfrac{1}{i\pi}\displaystyle\int_{0C_1}^{\infty e^{i\pi}} e^{(x/2)(t-1/t)}\,\dfrac{dt}{t^{\nu+1}} = H_\nu^{(1)}(x)$,

(b) $\dfrac{1}{i\pi}\displaystyle\int_{\infty e^{-i\pi}C_2}^{0} e^{(x/2)(t-1/t)}\,\dfrac{dt}{t^{\nu+1}} = H_\nu^{(2)}(x)$,

satisfy Bessel's differential equation. The contours C_1 and C_2 are shown in Fig. 11.8.

11.4.3   Using the integrals and contours given in problem 11.4.2, show that

$$\frac{1}{2i}[H_\nu^{(1)}(x) - H_\nu^{(2)}(x)] = N_\nu(x).$$

11.4.4   Show that the integrals in Exercise 11.4.2 may be transformed to yield

(a) $H_\nu^{(1)}(x) = \dfrac{1}{\pi i}\displaystyle\int_{C_3} e^{x\sinh\gamma - \nu\gamma}\,d\gamma$,

(b) $H_\nu^{(2)}(x) = \dfrac{1}{\pi i}\displaystyle\int_{C_4} e^{x\sinh\gamma - \nu\gamma}\,d\gamma$

(see Fig. 11.9).

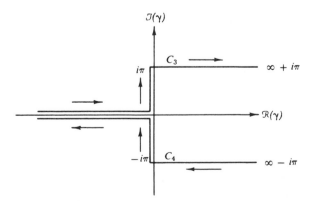

Figure 11.9   Hankel function contours.

11.4.5   (a) Transform $H_0^{(1)}(x)$, Eq. (11.101), into

$$H_0^{(1)}(x) = \frac{1}{i\pi} \int_C e^{ix \cosh s} \, ds,$$

where the contour C runs from $-\infty - i\pi/2$ through the origin of the s-plane to $\infty + i\pi/2$.

(b) Justify rewriting $H_0^{(1)}(x)$ as

$$H_0^{(1)}(x) = \frac{2}{i\pi} \int_0^{\infty + i\pi/2} e^{ix \cosh s} \, ds.$$

(c) Verify that this integral representation actually satisfies Bessel's differential equation. (The $i\pi/2$ in the upper limit is not essential. It serves as a convergence factor. We can replace it by $ia\pi/2$ and take the limit.)

11.4.6   From

$$H_0^{(1)}(x) = \frac{2}{i\pi} \int_0^{\infty} e^{ix \cosh s} \, ds$$

show that

(a) $J_0(x) = \dfrac{2}{\pi} \displaystyle\int_0^{\infty} \sin(x \cosh s) \, ds.$

(b) $J_0(x) = \dfrac{2}{\pi} \displaystyle\int_1^{\infty} \dfrac{\sin(xt)}{\sqrt{t^2 - 1}} \, dt.$

This last result is a Fourier sine transform.

11.4.7   From (see Exercises 11.4.4 and 11.4.5)

$$H_0^{(1)}(x) = \frac{2}{i\pi} \int_0^{\infty} e^{ix \cosh s} \, ds$$

show that

(a) $N_0(x) = -\dfrac{2}{\pi} \displaystyle\int_0^{\infty} \cos(x \cosh s) \, ds.$

(b) $N_0(x) = -\dfrac{2}{\pi} \displaystyle\int_1^{\infty} \dfrac{\cos(xt)}{\sqrt{t^2 - 1}} \, dt.$

These are the integral representations in Section 11.3 (Other Forms). This last result is a Fourier cosine transform.

11.5 MODIFIED BESSEL FUNCTIONS, $I_\nu(x)$ and $K_\nu(x)$

The Helmholtz equation,

$$\nabla^2 \psi + k^2 \psi = 0,$$

separated in circular cylindrical coordinates, leads to Eq. (11.22a), the Bessel equation. Equation (11.22a) is satisfied by the Bessel and Neumann functions $J_\nu(k\rho)$ and $N_\nu(k\rho)$ and any linear combination such as the Hankel functions $H_\nu^{(1)}(k\rho)$ and $H_\nu^{(2)}(k\rho)$. Now the Helmholtz equation describes the space part of wave phenomena. If instead we have a diffusion problem, then the Helmholtz equation is replaced by

$$\nabla^2 \psi - k^2 \psi = 0. \tag{11.108}$$

The analog to Eq. (11.22a) is

$$\rho^2 \frac{d^2}{d\rho^2} Y_\nu(k\rho) + \rho \frac{d}{d\rho} Y_\nu(k\rho) - (k^2\rho^2 + \nu^2) Y_\nu(k\rho) = 0. \tag{11.109}$$

The Helmholtz equation may be transformed into the diffusion equation by the transformation $k \to ik$. Similarly, $k \to ik$ changes Eq. (11.22a) into Eq. (11.109) and shows that

$$Y_\nu(k\rho) = Z_\nu(ik\rho).$$

The solutions of Eq. (11.109) are Bessel functions of imaginary argument. To obtain a solution that is regular at the origin, we take Z_ν as the regular Bessel function J_ν. It is customary (and convenient) to choose the normalization so that

$$Y_\nu(k\rho) = I_\nu(x) \equiv i^{-\nu} J_\nu(ix). \tag{11.110}$$

(Here the variable $k\rho$ is being replaced by x for simplicity.) Often this is written as

$$I_\nu(x) = e^{-\nu\pi i/2} J_\nu(xe^{i\pi/2}). \tag{11.111}$$

I_0 and I_1 are shown in Fig. 11.10.

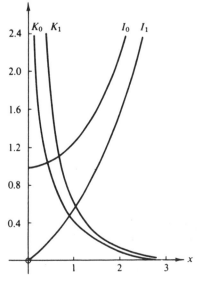

Figure 11.10 Modified Bessel functions.

SERIES FORM

In terms of infinite series this is equivalent to removing the $(-1)^s$ sign in Eq. (11.5) and writing

$$
I_\nu(x) = \sum_{s=0}^{\infty} \frac{1}{s!(s+\nu)!} \left(\frac{x}{2}\right)^{2s+\nu},
$$

$$
I_{-\nu}(x) = \sum_{s=0}^{\infty} \frac{1}{s!(s-\nu)!} \left(\frac{x}{2}\right)^{2s-\nu},
$$

(11.112)

The extra $i^{-\nu}$ normalization cancels the i^ν from each term and leaves $I_\nu(x)$ real. For integral ν this yields

$$
I_n(x) = I_{-n}(x).
$$

(11.113)

RECURRENCE RELATIONS

The recurrence relations satisfied by $I_\nu(x)$ may be developed from the series expansions, but it is perhaps easier to work from the existing recurrence relations for $J_\nu(x)$. Let us replace x by $-ix$ and rewrite Eq. (11.110) as

$$
J_\nu(x) = i^\nu I_\nu(-ix).
$$

(11.114)

Then Eq. (11.10) becomes

$$
i^{\nu-1} I_{\nu-1}(-ix) + i^{\nu+1} I_{\nu+1}(-ix) = \frac{2\nu}{x} i^\nu I_\nu(-ix).
$$

Replacing x by ix, we have a recurrence relation for $I_\nu(x)$,

$$
I_{\nu-1}(x) - I_{\nu+1}(x) = \frac{2\nu}{x} I_\nu(x).
$$

(11.115)

Equation (11.12) transforms to

$$
I_{\nu-1}(x) + I_{\nu+1}(x) = 2 I_\nu'(x).
$$

(11.116)

These are the recurrence relations used in Exercise 11.1.14.

It is worth emphasizing that although two recurrence relations, Eqs. (11.115) and (11.116) or Exercise 11.5.7, specify the second-order differential equation, the converse is not true. The differential equation does not uniquely fix the recurrence relations. Equations (11.115) and (11.116) and Exercise 11.5.7 provide an example.

From Eq. (11.113) it is seen that we have but one independent solution when ν is an integer, exactly as in the Bessel functions J_ν. The choice of a second, independent solution of Eq. (11.108) is essentially a matter of convenience. The second solution given here is selected on the basis of its asymptotic behavior—as shown in the next section. The confusion of choice and notation for this solution is perhaps greater than anywhere else in this field.[1] Many

---

[1] A discussion and comparison of notations will be found in *Math. Tables Aids Comput.* **1**, 207–308 (1944).

authors[2] choose to define a second solution in terms of the Hankel function $H_\nu^{(1)}(x)$ by

$$K_\nu(x) \equiv \frac{\pi}{2} i^{\nu+1} H_\nu^{(1)}(ix)$$

$$= \frac{\pi}{2} i^{\nu+1}[J_\nu(ix) + iN_\nu(ix)]. \tag{11.17}$$

The factor $i^{\nu+1}$ makes $K_\nu(x)$ real when x is real. Using Eqs. (11.60) and (11.110), we may transform Eq. (11.117) to[3]

$$K_\nu(x) = \frac{\pi}{2} \frac{I_{-\nu}(x) - I_\nu(x)}{\sin \nu\pi}, \tag{11.118}$$

analogous to Eq. (11.60) for $N_\nu(x)$. The choice of Eq. (11.117) as a definition is somewhat unfortunate in that the function $K_\nu(x)$ does *not* satisfy the same recurrence relations as $I_\nu(x)$ (compare Exercises 11.5.7 and 11.5.8). To avoid this annoyance other authors[4] have included an additional factor of cosine $n\pi$. This permits K_ν to satisfy the same recurrence relations as I_ν, but it has the disadvantage of making $K_\nu = 0$ for $\nu = \frac{1}{2}, \frac{3}{2}, \frac{5}{2}, \dots$.

The series expansion of $K_\nu(x)$ follows directly from the series form of $H_\nu^{(1)}(ix)$. The lowest order terms are (cf. Eqs. (11.61) and (11.62))

$$K_0(x) = -\ln x - \gamma + \ln 2 + \cdots,$$
$$K_\nu(x) = 2^{\nu-1}(\nu - 1)! x^{-\nu} + \cdots. \tag{11.119}$$

Because the modified Bessel function I_ν is related to the Bessel function J_ν, much as sinh is related to sine, I_ν and the second solution K_ν are sometimes referred to as hyperbolic Bessel functions. K_0 and K_1 are shown in Fig. 11.10. $I_0(x)$ and $K_0(x)$ have the integral representations

$$I_0(x) = \frac{1}{\pi} \int_0^\pi \cosh(x \cos\theta) \, d\theta \tag{11.120}$$

$$K_0(x) = \int_0^\infty \cos(x \sinh t) \, dt = \int_0^\infty \frac{\cos(xt) \, dt}{(t^2 + 1)^{1/2}}, \qquad x > 0. \tag{11.121}$$

Equation (11.120) may be derived from Eq. (11.30) for $J_0(x)$ or may be taken as a special case of Exercise 11.5.4, $\nu = 0$. The integral representation of K_0, Eq. (11.121), is a Fourier transform and may best be derived with Fourier transforms, Chapter 15, or with Green's functions Section 8.7. A variety of other forms of integral representations (including $\nu \neq 0$) appear in the exercises. These integral representations are useful in developing asymptotic forms (Section 11.6) and in connection with Fourier transforms, Chapter 15.

---

[2] Watson, Morse and Feshbach, Jeffreys and Jeffreys (without the $\pi/2$).
[3] For integral index n we take the limit as $\nu \to n$.
[4] Whittaker and Watson.

To put the modified Bessel functions $I_\nu(x)$ and $K_\nu(x)$ in proper perspective, we introduce them here because:

1. These functions are solutions of the frequently encountered modified Bessel equation.
2. They are needed for specific physical problems such as diffusion problems.
3. $K_\nu(x)$ provides a Green's function, Section 8.7.
4. $K_\nu(x)$ leads to a convenient determination of asymptotic behavior (Section 11.6).

EXERCISES

11.5.1   Show that

$$e^{(x/2)(t+1/t)} = \sum_{n=-\infty}^{\infty} I_n(x)t^n,$$

thus generating modified Bessel functions, $I_n(x)$.

11.5.2   Verify the following identities

(a) $1 = I_0(x) + 2 \sum_{n=1}^{\infty} (-1)^n I_{2n}(x),$

(b) $e^x = I_0(x) + 2 \sum_{n=1}^{\infty} I_n(x),$

(c) $e^{-x} = I_0(x) + 2 \sum_{n=1}^{\infty} (-1)^n I_n(x),$

(d) $\cosh x = I_0(x) + 2 \sum_{n=1}^{\infty} I_{2n}(x),$

(e) $\sinh x = 2 \sum_{n=1}^{\infty} I_{2n-1}(x).$

11.5.3   (a) From the generating function of Exercise 11.5.1 show that

$$I_n(x) = \frac{1}{2\pi i} \oint \exp[(x/2)(t + 1/t)] \frac{dt}{t^{n+1}}.$$

(b) For $n = \nu$, not an integer, show that the preceding integral representation may be generalized to

$$I_\nu(x) = \frac{1}{2\pi i} \int_C \exp[(x/2)(t + 1/t)] \frac{dt}{t^{\nu+1}}.$$

The contour C is the same as that for $J_\nu(x)$, Fig. 11.7.

11.5.4 For $v > -\frac{1}{2}$ show that $I_v(z)$ may be represented by

$$I_v(z) = \frac{1}{\pi^{1/2}(v-\frac{1}{2})!} \left(\frac{z}{2}\right)^v \int_0^\pi e^{\pm z\cos\theta} \sin^{2v}\theta \, d\theta$$

$$= \frac{1}{\pi^{1/2}(v-\frac{1}{2})!} \left(\frac{z}{2}\right)^v \int_{-1}^1 e^{\pm zp}(1-p^2)^{v-1/2} \, dp$$

$$= \frac{2}{\pi^{1/2}(v-\frac{1}{2})!} \left(\frac{z}{2}\right)^v \int_0^{\pi/2} \cosh(z\cos\theta)\sin^{2v}\theta \, d\theta.$$

11.5.5 A cylindrical cavity has a radius a and a height l, Fig. 11.3. The ends, $z = 0$ and l, are at zero potential. The cylindrical walls, $\rho = a$, have a potential $V = V(\varphi, z)$.

(a) Show that the electrostatic potential $\Phi(\rho, \varphi, z)$ has the functional form

$$\Phi(\rho, \varphi, z) = \sum_{m=0}^\infty \sum_{n=1}^\infty I_m(k_n\rho) \sin k_n z \cdot (a_{mn}\sin m\varphi + b_{mn}\cos m\varphi),$$

where $k_n = n\pi/l$.

(b) Show that the coefficients a_{mn} and b_{mn} are given by[5]

$$\left.\begin{array}{c} a_{mn} \\ b_{mn} \end{array}\right\} = \frac{2}{\pi l I_m(k_n a)} \int_0^{2\pi}\int_0^l V(\varphi, z)\sin k_n z \cdot \left\{\begin{array}{c}\sin m\varphi \\ \cos m\varphi\end{array}\right\} dz\, d\varphi.$$

Hint. Expand $V(\varphi, z)$ as a double series and use the orthogonality of the trigonometric functions.

11.5.6 Verify that $K_v(x)$ is given by

$$K_v(x) = \frac{\pi}{2}\frac{I_{-v}(x) - I_v(x)}{\sin v\pi}$$

and from this show that

$$K_v(x) = K_{-v}(x).$$

11.5.7 Show that $K_v(x)$ satisfies the recurrence relations

$$K_{v-1}(x) - K_{v+1}(x) = -\frac{2v}{x}K_v(x),$$

$$K_{v-1}(x) + K_{v+1}(x) = -2K_v'(x).$$

11.5.8 If $\mathcal{K}_v = e^{v\pi i}K_v$, show that $\mathcal{K}_v$ satisfies the *same* recurrence relations as I_v.

11.5.9 For $v > -\frac{1}{2}$ show that $K_v(z)$ may be represented by

$$K_v(z) = \frac{\pi^{1/2}}{(v-\frac{1}{2})!}\left(\frac{z}{2}\right)^v \int_0^\infty e^{-z\cosh t}\sinh^{2v} t\, dt, \qquad -\frac{\pi}{2} < \arg z < \frac{\pi}{2}$$

$$= \frac{\pi^{1/2}}{(v-\frac{1}{2})!}\left(\frac{z}{2}\right)^v \int_1^\infty e^{-zp}(p^2-1)^{v-1/2}\, dp.$$

11.5.10 Show that $I_v(x)$ and $K_v(x)$ satisfy the Wronskian relation

$$I_v(x)K_v'(x) - I_v'(x)K_v(x) = -\frac{1}{x}.$$

This result is quoted in Section 8.7 in the development of a Green's function.

[5] When $m = 0$, the 2 in the coefficient is replaced by 1.

11.5.11   If $r = (x^2 + y^2)^{1/2}$, prove that

$$\frac{1}{r} = \frac{2}{\pi} \int_0^\infty \cos(xt)K_0(yt)\, dt.$$

This is a Fourier cosine transform of K_0.

11.5.12   (a) Verify that

$$I_0(x) = \frac{1}{\pi} \int_0^\pi \cosh(x \cos \theta)\, d\theta$$

satisfies the *modified* Bessel equation, $\nu = 0$.
(b) Show that this integral contains no admixture of $K_0(x)$, the irregular second solution.
(c) Verify the normalization factor $1/\pi$.

11.5.13   Verify that the integral representations

$$I_n(z) = \frac{1}{\pi} \int_0^\pi e^{z\cos t} \cos(nt)\, dt,$$

$$K_\nu(z) = \int_0^\infty e^{-z\cosh t} \cosh(\nu t)\, dt, \qquad \mathfrak{R}(z) > 0,$$

satisfy the modified Bessel equation by direct substitution into that equation. How can you show that the first form does not contain an admixture of K_n, that the second form does not contain an admixture of I_ν? How can you check the normalization?

11.5.14   Derive the integral representation

$$I_n(x) = \frac{1}{\pi} \int_0^\pi e^{x \cos \theta} \cos(n\theta)\, d\theta.$$

Hint. Start with the corresponding integral representation of $J_n(x)$. Equation (11.20) is a special case of this representation.

11.5.15   Show that

$$K_0(z) = \int_0^\infty e^{-z\cosh t}\, dt$$

satisfies the modified Bessel equation. How can you establish that this form is linearly independent of $I_0(z)$?

11.5.16   Show that

$$e^{ax} = I_0(a)T_0(x) + 2 \sum_{n=1}^\infty I_n(a)T_n(x), \qquad -1 \le x \le 1.$$

$T_n(x)$ is the nth-order Chebyshev polynomial, Section 13.3 and 13.4.
Hint. Assume a Chebyshev series expansion. Using the orthogonality and normalization of the $T_n(x)$, solve for the coefficients of the Chebyshev series.

11.5.17   (a) Write a double precision subroutine to calculate $I_n(x)$ to 12-decimal place accuracy for $n = 0, 1, 2, 3, \ldots$ and $0 \le x \le 1$. Check your results against the 10 place values given in AMS-55, Table 9.11.

(b) Referring to Exercise 11.5.16, calculate the coefficients in the Chebyshev expansions of $\cosh x$ and of $\sinh x$.

Note. An alternate calculation of these coefficients is one of the topics of Section 13.4.

11.5.18 The cylindrical cavity of Exercise 11.5.5 has a potential along the cylinder walls

$$V(z) = \begin{cases} 100z/l, & 0 \le z/l \le 1/2 \\ 100(1 - z/l), & 1/2 \le z/l \le 1. \end{cases}$$

With the radius–height ratio $a/l = 0.5$, calculate the potential for $z/l = 0.1(0.1)0.5$ and $p/a = 0.0(0.2)1.0$.

Check value.   For $z/l = 0.3$ and $p/a = 0.8$, $V = 26.396$.

11.6 ASYMPTOTIC EXPANSIONS

Frequently in physical problems there is a need to know how a given Bessel or modified Bessel function behaves for large values of the argument, that is, the asymptotic behavior. This is one occasion when computers are not very helpful. One possible approach is to develop a power-series solution of the differential equation, as in Section 8.5, but now using negative powers. This is the Stokes's method, Exercise 11.6.5. The limitation is that starting from some positive value of the argument (for convergence of the series), we do not know what mixture of solutions or multiple of a given solution we have. The problem is to relate the asymptotic series (useful for large values of the variable) to the power series or related definition (useful for small values of the variable). This relationship can be established by introducing a suitable *integral representation* and then using either the method of steepest descent, Section 7.4, or the direct expansion as developed in this section.

EXPANSION OF AN INTEGRAL REPRESENTATION, $K_\nu(z)$

As a direct approach, consider the integral representation (Exercise 11.5.9)

$$K_\nu(z) = \frac{\pi^{1/2}}{(\nu - \frac{1}{2})!} \left(\frac{z}{2}\right)^\nu \int_1^\infty e^{-zx}(x^2 - 1)^{\nu - 1/2}\, dx, \qquad \nu > -\frac{1}{2}. \quad (11.122)$$

For the present let us take z to be real, although Eq. (11.136) may be established for $-\pi/2 < \arg z < \pi/2$ ($\Re(z) > 0$). We have three problems: (1) to show that K_ν as given in Eq. (11.122) actually satisfies the modified Bessel equation (11.108); (2) to show that the regular solution I_ν is absent; and (3) to show that Eq. (11.122) has the proper normalization.

1. The fact that Eq. (11.122) is a solution of the modified Bessel equation may be verified by direct substitution. We obtain

$$z^{\nu+1} \int_1^\infty \frac{d}{dx} [e^{-zx}(x^2 - 1)^{\nu + 1/2}]\, dx = 0,$$

which transforms the combined integrand into the derivative of a function that vanishes at both end points. Hence the integral is some linear combination of I_ν and K_ν.

2. The rejection of the possibility that this solution contains I_ν constitutes Exercise 11.6.1.

3. The normalization may be verified by substituting $x = 1 + t/z$.

$$\frac{\pi^{1/2}}{(\nu - \frac{1}{2})!} \left(\frac{z}{2}\right)^\nu \int_1^\infty e^{-zx}(x^2 - 1)^{\nu - 1/2}\, dx$$

$$= \frac{\pi^{1/2}}{(\nu - \frac{1}{2})!} \left(\frac{z}{2}\right)^\nu e^{-z} \int_0^\infty e^{-t}\left(\frac{t^2}{z^2} + \frac{2t}{z}\right)^{\nu - 1/2} \frac{dt}{z} \qquad (11.123a)$$

$$= \frac{\pi^{1/2}}{(\nu - \frac{1}{2})!} \frac{e^{-z}}{2^\nu z^\nu} \int_0^\infty e^{-t}t^{2\nu - 1}\left(1 + \frac{2z}{t}\right)^{\nu - 1/2} dt, \qquad (11.123b)$$

taking out t^2/z^2 as a factor. This substitution has changed the limits of integration to a more convenient range and has isolated the negative exponential dependence, e^{-z}. The integral in Eq. (11.123b) may be evaluated for $z = 0$ to yield $(2\nu - 1)!$. Then, using the duplication formula (Section 10.4), we have

$$\lim_{z \to 0} K_\nu(z) = \frac{(\nu - 1)!2^{\nu - 1}}{z^\nu}, \qquad \nu > 0, \qquad (11.124)$$

in agreement with Eq. (11.119), which thus checks the normalization.[1]

Now to develop an asymptotic series for $K_\nu(z)$, we may rewrite (11.123a) as

$$K_\nu(z) = \sqrt{\frac{\pi}{2z}} \frac{e^{-z}}{(\nu - \frac{1}{2})!} \int_0^\infty e^{-t}t^{\nu - 1/2}\left(1 + \frac{t}{2z}\right)^{\nu - 1/2} dt \qquad (11.125)$$

(taking out $2t/z$ as a factor).

We expand $(1 + t/2z)^{\nu - 1/2}$ by the binomial theorem to obtain

$$K_\nu(z) = \sqrt{\frac{\pi}{2z}} \frac{e^{-z}}{(\nu - \frac{1}{2})!} \sum_{r=0}^\infty \frac{(\nu - \frac{1}{2})!}{r!(\nu - r - \frac{1}{2})!} (2z)^{-r} \int_0^\infty e^{-t}t^{\nu + r - 1/2}\, dt.$$

$$(11.126)$$

Term-by-term integration (valid for asymptotic series) yields the desired asymptotic expansion of $K_\nu(z)$.

$$K_\nu(z) \sim \sqrt{\frac{\pi}{2z}} e^{-z}\left[1 + \frac{(4\nu^2 - 1^2)}{1!8z} + \frac{(4\nu^2 - 1^2)(4\nu^2 - 3^2)}{2!(8z)^2} + \cdots\right].$$

$$(11.127)$$

Although the integral of Eq. (11.122), integrating along the real axis, was convergent only for $-\pi/2 < \arg z < \pi/2$, Eq. (11.127) may be extended to

---

[1] For $\nu = 0$ the integral diverges logarithmically in agreement with the logarithmic divergence of $K_0(z)$ (Section 11.5).

$-3\pi/2 < \arg z < 3\pi/2$. Considered as an infinite series, Eq. (11.127) is actually divergent.[2] However, this series is asymptotic in the sense that for large enough z, $K_\nu(z)$ may be approximated to any fixed degree of accuracy. (Compare Section 5.10 for a definition and discussion of asymptotic series.)

It is convenient to rewrite Eq. (11.127) as

$$K_\nu(z) = \sqrt{\frac{\pi}{2z}} e^{-z}[P_\nu(iz) + iQ_\nu(iz)], \tag{11.128}$$

where

$$P_\nu(z) \sim 1 - \frac{(\mu - 1)(\mu - 9)}{2!(8z)^2} + \frac{(\mu - 1)(\mu - 9)(\mu - 25)(\mu - 49)}{4!(8z)^4} - \cdots, \tag{11.129a}$$

$$Q_\nu(z) \sim \frac{\mu - 1}{1!(8z)} - \frac{(\mu - 1)(\mu - 9)(\mu - 25)}{3!(8z)^3} + \cdots, \tag{11.129b}$$

and

$$\mu = 4\nu^2.$$

It should be noted that although $P_\nu(z)$ of Eq. (11.129a) and $Q_\nu(z)$ of Eq. (11.129b) have alternating signs, the series for $P_\nu(iz)$ and $Q_\nu(iz)$ of Eq. (11.128) have all signs positive. Finally, for z large, P_ν dominates.

Then with the asymptotic form of $K_\nu(z)$, Eq. (11.128), we can obtain expansions for all other Bessel and hyperbolic Bessel functions by defining relations:

1. From

$$\frac{\pi}{2} i^{\nu+1} H_\nu^{(1)}(iz) = K_\nu(z) \tag{11.130}$$

we have

$$H_\nu^{(1)}(z) = \sqrt{\frac{2}{\pi z}} \exp i\left[z - \left(\nu + \frac{1}{2}\right)\frac{\pi}{2}\right]$$
$$\cdot [P_\nu(z) + iQ_\nu(z)], \qquad -\pi < \arg z < 2\pi. \tag{11.131}$$

2. The second Hankel function is just the complex conjugate of the first (for real argument),

$$H_\nu^{(2)}(z) = \sqrt{\frac{2}{\pi z}} \exp -i\left[z - \left(\nu + \frac{1}{2}\right)\frac{\pi}{2}\right]$$
$$\cdot [P_\nu(z) - iQ_\nu(z)], \qquad -2\pi < \arg z < \pi. \tag{11.132}$$

An alternate derivation of the asymptotic behavior of the Hankel functions appears in Section 7.4 as an application of the method of steepest descents.

---

[2] Our binomial expansion is valid only for $t < 2z$ and we have integrated t out to infinity. The exponential decrease of the integrand prevents a disaster but the resultant series is still only asymptotic, not convergent. By Table 8.3, $z = \infty$ is an essential singularity of the Bessel (and modified Bessel) equations. Fuchs's theorem does not guarantee a convergent series and we do not get a convergent series.

3. Since $J_\nu(z)$ is the real part of $H_\nu^{(1)}(z)$,

$$J_\nu(z) = \sqrt{\frac{2}{\pi z}} \left\{ P_\nu(z) \cos\left[z - \left(\nu + \frac{1}{2} \right) \frac{\pi}{2} \right] \right.$$

$$\left. - Q_\nu(z) \sin\left[z - \left(\nu + \frac{1}{2} \right) \frac{\pi}{2} \right] \right\}, \qquad -\pi < \arg z < \pi. \qquad (11.133)$$

4. The Neumann function is the imaginary part of $H_\nu^{(1)}(z)$, or

$$N_\nu(z) = \sqrt{\frac{2}{\pi z}} \left\{ P_\nu(z) \sin\left[z - \left(\nu + \frac{1}{2} \right) \frac{\pi}{2} \right] \right.$$

$$\left. + Q_\nu(z) \cos\left[z - \left(\nu + \frac{1}{2} \right) \frac{\pi}{2} \right] \right\}, \qquad -\pi < \arg z < \pi. \qquad (11.134)$$

5. Finally, the regular hyperbolic or modified Bessel function $I_\nu(z)$ is given by

$$I_\nu(z) = i^{-\nu} J_\nu(iz) \qquad (11.135)$$

or

$$I_\nu(z) = \frac{e^z}{\sqrt{2\pi z}} [P_\nu(iz) - iQ_\nu(iz)], \qquad -\frac{\pi}{2} < \arg z < \frac{\pi}{2}. \qquad (11.136)$$

This completes our determination of the asymptotic expansions. However, it is perhaps worth noting the primary characteristics. Apart from the ubiquitous $z^{-1/2}$, J_ν and N_ν behave as cosine and sine, respectively. The zeros are *almost* evenly spaced at intervals of π; the spacing becomes exactly π in the limit as $z \to \infty$. The Hankel functions have been defined to behave like the imaginary exponentials, and the modified Bessel functions, I_ν and K_ν, go into the positive and negative exponentials. This asymptotic behavior may be sufficient to eliminate immediately one of these functions as a solution for a physical problem. We should also note that the asymptotic series $P_\nu(z)$ and $Q_\nu(z)$, Eqs. (11.129a) and (11.129b), terminate for $\nu = \pm 1/2, \pm 3/2, \ldots$ and become polynomials (in negative powers of z). For these special values of ν the asymptotic approximations become exact solutions.

It is of some interest to consider the accuracy of the asymptotic forms, taking just the first term, for example (Fig. 11.11),

$$J_n(x) \approx \sqrt{\frac{2}{\pi x}} \cos\left[x - \left(n + \frac{1}{2} \right) \left(\frac{\pi}{2} \right) \right]. \qquad (11.137)$$

Clearly, the condition for the validity of Eq. (11.137) is that the sine term be negligible; that is,

$$8x \gg 4n^2 - 1. \qquad (11.138)$$

For n or $\nu > 1$ the asymptotic region may be far out.

As pointed out in Section 11.3, the asymptotic forms may be used to evaluate the various Wronskian formulas (compare Exercise 11.6.3).

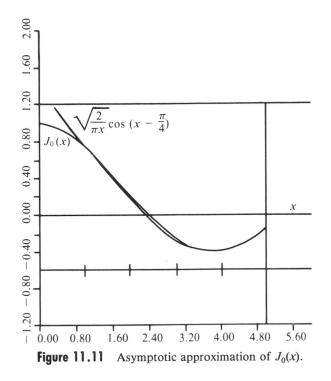

Figure 11.11   Asymptotic approximation of $J_0(x)$.

NUMERICAL EVALUATION

When a program in a high-speed computing machine calls for one of the Bessel or modified Bessel functions, the programmer has two alternatives: to store all the Bessel functions and tell the computer how to locate the required value or to instruct the computer to simply calculate the needed value. The first alternative would be fairly slow and would place unreasonable demands on the storage capacity. Thus our programmer adopts the "compute it yourself" alternative.

The computation of $J_n(x)$ using the recurrence relation, Eq. (11.10), is discussed in Section 11.1. For N_n, I_n, and K_n the preferred methods are the series if x is small and the asymptotic forms (with many terms in the series of negative powers) if x is large. The criteria of large and small may vary as shown in Table 11.2.

EXERCISES

11.6.1   In checking the normalization of the integral representation of $K_\nu(z)$ (Eq. (11.122)), we assumed that $I_\nu(z)$ was not present. How do we know that the integral representation (Eq. (11.122)) does not yield $K_\nu(z) + \varepsilon I_\nu(z)$ with $\varepsilon \neq 0$?

Table 11.2 Equations for the Computation of Neumann and the Modified Bessel Functions

	Power Series		Asymptotic Series	
$N_n(x)$	Eq. (11.63),	$x \le 4$	Eq. (11.134),	$x > 4$
$I_n(x)$	Eq. (11.112),	$x \le 12$ or $\le n$	Eq. (11.136),	$x > 12$ and $> n$
$K_n(x)$	Eq. (11.119),	$x \le 1$	Eq. (11.127),	$x > 1$

In actual practice, it is found convenient to limit the series (power or asymptotic) computation of $N_n(x)$ and $K_n(x)$ to $n = 0, 1$. Then $N_n(x)$, $n \ge 2$ is computed using the recurrence relation, Eq. (11.10). $K_n(x)$, $n \ge 2$ is computed using the recurrence relations of Exercise 11.5.7. $I_n(x)$ could be handled this way, if desired, but direct application of the power series or asymptotic series is feasible for all values of n and x.

11.6.2   (a) Show that

$$y(z) = z^\nu \int e^{-zt}(t^2 - 1)^{\nu - 1/2}\, dt$$

satisfies the modified Bessel equation, provided the contour is chosen so that

$$e^{-zt}(t^2 - 1)^{\nu + 1/2}$$

has the same value at the initial and final points of the contour.

(b) Verify that the contours shown in Fig. 11.12 are suitable for this problem.

11.6.3   Use the asymptotic expansions to verify the following Wronskian formulas:

(a) $J_\nu(x)J_{-\nu-1}(x) + J_{-\nu}(x)J_{\nu+1}(x) = -2 \sin \nu\pi/\pi x,$

(b) $J_\nu(x)N_{\nu+1}(x) - J_{\nu+1}(x)N_\nu(x) = -2/\pi x,$

(c) $J_\nu(x)H_{\nu-1}^{(2)}(x) - J_{\nu-1}(x)H_\nu^{(2)}(x) = 2/i\pi x,$

(d) $I_\nu(x)K_\nu'(x) - I_\nu'(x)K_\nu(x) = -1/x,$

(e) $I_\nu(x)K_{\nu+1}(x) + I_{\nu+1}(x)K_\nu(x) = 1/x.$

11.6.4   From the asymptotic form of $K_\nu(z)$, Eq. (11.127), derive the asymptotic form of $H_\nu^{(1)}(z)$, Eq. (11.131). Note particularly the phase, $(\nu + \tfrac{1}{2})\pi/2$.

11.6.5   Stokes's method.

(a) Replace the Bessel function in Bessel's equation by $x^{-1/2}y(x)$ and show that $y(x)$ satisfies

$$y''(x) + \left(1 - \frac{\nu^2 - \frac{1}{4}}{x^2}\right)y(x) = 0.$$

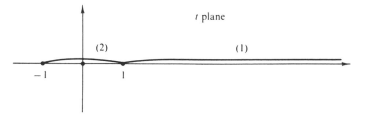

Figure 11.12   Modified Bessel function contours.

(b) Develop a power-series solution with *negative* powers of x starting with the assumed form

$$y(x) = e^{ix} \sum_{n=0}^{\infty} a_n x^{-n}.$$

Determine the recurrence relation giving a_{n+1} in terms of a_n. Check your result against the asymptotic series, Eq. (11.131).
(c) From the results of Section 7.4 determine the initial coefficient, a_0.

11.6.6   Calculate the first 15 partial sums of $P_0(x)$ and $Q_0(x)$, Eqs. (11.129a) and (11.129b). Let x vary from 4 to 10 in unit steps. Determine the number of terms to be retained for maximum accuracy and the accuracy achieved as a function of x. Specifically, how small may x be without raising the error above 3×10^{-6}?

ANS. $x_{min} = 6.$

11.6.7   (a) Using the asymptotic series (partial sums) $P_0(x)$ and $Q_0(x)$ determined in Exercise 11.6.6, write a function subprogram FCT(X) that will calculate $J_0(x)$, x real, for $x \geq x_{min}$.
(b) Test your function by comparing it with the $J_0(x)$ (tables or computer library subroutine) for $x = x_{min}(10)x_{min} + 10$.

Note. A more accurate and perhaps simpler asymptotic form for $J_0(x)$ is given in AMS-55, Eq. (9.4.3).

11.7 SPHERICAL BESSEL FUNCTIONS

When the Helmholtz equation is separated in spherical coordinates the radial equation has the form

$$r^2 \frac{d^2R}{dr^2} + 2r \frac{dR}{dr} + [k^2r^2 - n(n+1)]R = 0. \tag{11.139}$$

This is Eq. (8.65) of Section 8.3. The parameter k enters from the original Helmholtz equation while $n(n+1)$ is a separation constant. From the behavior of the polar angle function (Legendre's equation, Sections 8.5 and 12.7), the separation constant must have this form, with n a non-negative integer. Equation (11.139) has the virtue of being self-adjoint but clearly it is *not* Bessel's equation. However, if we substitute

$$R(kr) = \frac{Z(kr)}{(kr)^{1/2}},$$

Equation (11.139) becomes

$$r^2 \frac{d^2Z}{dr^2} + r \frac{dZ}{dr} + \left[k^2r^2 - \left(n + \frac{1}{2} \right)^2 \right] Z = 0, \tag{11.140}$$

which *is* Bessel's equation. Z is a Bessel function of order $n + \frac{1}{2}$ (n an integer). Because of the importance of spherical coordinates, this combination,

$$\frac{Z_{n+1/2}(kr)}{(kr)^{1/2}},$$

occurs quite often.

DEFINITIONS

It is convenient to label these functions spherical Bessel functions with the following defining equations

$$j_n(x) = \sqrt{\frac{\pi}{2x}} J_{n+1/2}(x),$$

$$n_n(x) = \sqrt{\frac{\pi}{2x}} N_{n+1/2}(x) = (-1)^{n+1} \sqrt{\frac{\pi}{2x}} J_{-n-1/2}(x),^1$$

$$h_n^{(1)}(x) = \sqrt{\frac{\pi}{2x}} H_{n+1/2}^{(1)}(x) = j_n(x) + i n_n(x),$$ (11.141)

$$h_n^{(2)}(x) = \sqrt{\frac{\pi}{2x}} H_{n+1/2}^{(2)}(x) = j_n(x) - i n_n(x).$$

These spherical Bessel functions (Figs. 11.13 and 11.14) can be expressed in series form by using the series (Eq. (11.5)) for J_n, replacing n with $n + \frac{1}{2}$:

$$J_{n+1/2}(x) = \sum_{s=0}^{\infty} \frac{(-1)^s}{s!(s + n + \frac{1}{2})!} \left(\frac{x}{2}\right)^{2s+n+1/2}.$$ (11.142)

Using the Legendre duplication formula,

$$z!(z + \tfrac{1}{2})! = 2^{-2z-1}\pi^{1/2}(2z + 1)!,$$ (11.143)

we have

$$j_n(x) = \sqrt{\frac{\pi}{2x}} \sum_{s=0}^{\infty} \frac{(-1)^s 2^{2s+2n+1}(s + n)!}{\pi^{1/2}(2s + 2n + 1)!s!} \left(\frac{x}{2}\right)^{2s+n+1/2}$$

$$= 2^n x^n \sum_{s=0}^{\infty} \frac{(-1)^s(s + n)!}{s!(2s + 2n + 1)!} x^{2s}.$$ (11.144)

Now $N_{n+1/2}(x) = (-1)^{n+1} J_{-n-1/2}(x)$ and from Eq. (11.5) we find that

$$J_{-n-1/2}(x) = \sum_{s=0}^{\infty} \frac{(-1)^s}{s!(s - n - \frac{1}{2})!} \left(\frac{x}{2}\right)^{2s-n-1/2}.$$ (11.145)

---

[1] This is possible because $\cos(n + \frac{1}{2})\pi = 0$.

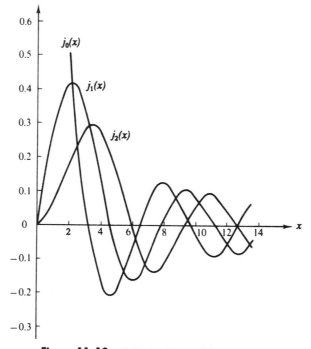

Figure 11.13   Spherical Bessel functions.

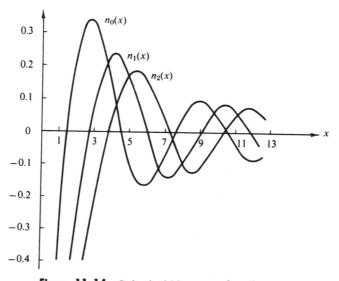

Figure 11.14   Spherical Neumann functions.

This yields

$$n_n(x) = (-1)^{n+1} \frac{2^n \pi^{1/2}}{x^{n+1}} \sum_{s=0}^{\infty} \frac{(-1)^s}{s!(s-n-\frac{1}{2})!} \left(\frac{x}{2}\right)^{2s}.$$ (11.146)

The Legendre duplication formula can be used again to give

$$n_n(x) = \frac{(-1)^{n+1}}{2^n x^{n+1}} \sum_{s=0}^{\infty} \frac{(-1)^s(s-n)!}{s!(2s-2n)!} x^{2s}.$$ (11.147)

These series forms, Eqs. (11.144) and (11.147), are useful in three ways: (1) limiting values as $x \to 0$, (2) closed form representations for $n = 0$, and, as an extension of this, (3) an indication that the spherical Bessel functions are closely related to sine and cosine.

For the special case $n = 0$ we find from Eq. (11.144)

$$j_0(x) = \sum_{s=0}^{\infty} \frac{(-1)^s}{(2s+1)!} x^{2s}$$

$$= \frac{\sin x}{x},$$ (11.148)

whereas for n_0, Eq. (11.147) yields

$$n_0(x) = -\frac{\cos x}{x}.$$ (11.149)

From the definition of the spherical Hankel functions (Eq. (11.141)),

$$h_0^{(1)}(x) = \frac{1}{x}(\sin x - i \cos x) = -\frac{i}{x} e^{ix}$$

$$h_0^{(2)}(x) = \frac{1}{x}(\sin x + i \cos x) = \frac{i}{x} e^{-ix}.$$ (11.150)

Equations (11.148) and (11.149) suggest expressing the spherical Bessel functions as combinations of sine and cosine. The appropriate combinations can be developed from the power-series solutions, Eqs. (11.144) and (11.147), but this approach is awkward. Actually the trigonometric forms are already available as the asymptotic expansion of Section 11.6. From Eqs. (11.131) and (11.129a)

$$h_n^{(1)}(z) = \sqrt{\frac{\pi}{2z}} H_{n+1/2}^{(1)}(z)$$

$$= (-i)^{n+1} \frac{e^{iz}}{z} \{P_{n+1/2}(z) + iQ_{n+1/2}(z)\}.$$ (11.151)

Now $P_{n+1/2}$ and $Q_{n+1/2}$ are *polynomials*. This means that Eq. (11.151) is mathematically exact, not simply an asymptotic approximation. We obtain

$$h_n^{(1)}(z) = (-i)^{n+1} \frac{e^{iz}}{z} \sum_{s=0}^{n} \frac{i^s}{s!(8z)^s} \frac{(2n+2s)!!}{(2n-2s)!!}$$

$$= (-i)^{n+1} \frac{e^{iz}}{z} \sum_{s=0}^{n} \frac{i^s}{s!(2z)^s} \frac{(n+s)!}{(n-s)!}. \tag{11.152}$$

Often a factor $(-i)^n = (e^{-i\pi/2})^n$ will be combined with the e^{iz} to give $e^{i(z-n\pi/2)}$. For z real $j_n(z)$ is the real part of this, $n_n(z)$ the imaginary part, and $h_n^{(2)}(z)$ the complex conjugate.

Specifically,

$$h_1^{(1)}(x) = e^{ix}\left(-\frac{1}{x} - \frac{i}{x^2}\right) \tag{11.153a}$$

$$h_2^{(1)}(x) = e^{ix}\left(\frac{i}{x} - \frac{3}{x^2} - \frac{3i}{x^3}\right) \tag{11.153b}$$

$$j_1(x) = \frac{\sin x}{x^2} - \frac{\cos x}{x},$$

$$j_2(x) = \left(\frac{3}{x^3} - \frac{1}{x}\right)\sin x - \frac{3}{x^2}\cos x, \tag{11.154}$$

$$n_1(x) = -\frac{\cos x}{x^2} - \frac{\sin x}{x},$$

$$n_2(x) = -\left(\frac{3}{x^3} - \frac{1}{x}\right)\cos x - \frac{3}{x^2}\sin x, \tag{11.155}$$

and so on.

LIMITING VALUES

For $x \ll 1,$[2] Eqs. (11.144) and (11.147) yield

$$j_n(x) \approx \frac{2^n n!}{(2n+1)!}x^n = \frac{x^n}{(2n+1)!!} \tag{11.156}$$

$$n_n(x) \approx \frac{(-1)^{n+1}}{2^n} \cdot \frac{(-n)!}{(-2n)!}x^{-n-1}$$

$$= -\frac{(2n)!}{2^n n!}x^{-n-1} = -(2n-1)!!x^{-n-1}. \tag{11.157}$$

The transformation of factorials in the expressions for $n_n(x)$ employs Exercise 10.1.3. The limiting values of the spherical Hankel functions go as $\pm in_n(x)$.

[2] The condition that the second term in the series be negligible compared to the first is actually $x \ll 2[(2n+2)(2n+3)/(n+1)]^{1/2}$ for $j_n(x)$.

The asymptotic values of j_n, n_n, $h_n^{(2)}$, and $h_n^{(1)}$ may be obtained from the Bessel asymptotic forms, Section 11.6. We find

$$j_n(x) \sim \frac{1}{x} \sin\left(x - \frac{n\pi}{2}\right), \tag{11.158}$$

$$n_n(x) \sim -\frac{1}{x} \cos\left(x - \frac{n\pi}{2}\right), \tag{11.159}$$

$$h_n^{(1)}(x) \sim (-i)^{n+1} \frac{e^{ix}}{x} = (-i)\frac{e^{i(x-n\pi/2)}}{x}, \tag{11.160a}$$

$$h_n^{(2)}(x) \sim i^{n+1} \frac{e^{-ix}}{x} = (i)\frac{e^{-i(x-n\pi/2)}}{x}. \tag{11.160b}$$

The condition for these spherical Bessel forms is that $x \gg n(n + 1)/2$. From these asymptotic values we see that $j_n(x)$ and $n_n(x)$ are appropriate for a description of *standing spherical waves*; $h_n^{(1)}(x)$ and $h_n^{(2)}(x)$ corresponding to *traveling spherical waves*. If the time dependence for the traveling waves is taken to be $e^{-i\omega t}$, then $h_n^{(1)}(x)$ yields an outgoing traveling spherical wave, $h_n^{(2)}(x)$ an incoming wave. Radiation theory in electromagnetism and scattering theory in quantum mechanics provide many applications.

RECURRENCE RELATIONS

The recurrence relations to which we now turn provide a convenient way of developing the higher-order spherical Bessel functions. These recurrence relations may be derived from the series, but as with the modified Bessel functions, it is easier to substitute into the known recurrence relations (Eqs. (11.10) and (11.12)). This gives

$$f_{n-1}(x) + f_{n+1}(x) = \frac{2n + 1}{x} f_n(x), \tag{11.161}$$

$$nf_{n-1}(x) - (n + 1)f_{n+1}(x) = (2n + 1)f_n'(x). \tag{11.162}$$

Rearranging these relations (or substituting into Eqs. (11.15) and (11.17)), we obtain

$$\frac{d}{dx}[x^{n+1}f_n(x)] = x^{n+1}f_{n-1}(x) \tag{11.163}$$

$$\frac{d}{dx}[x^{-n}f_n(x)] = -x^{-n}f_{n+1}(x). \tag{11.164}$$

Here f_n may represent j_n, n_n, $h_n^{(1)}$, or $h_n^{(2)}$.

The specific forms, Eqs. (11.154) and (11.155), may also be readily obtained from Eq. (11.164).

By mathematical induction we may establish the Rayleigh formulas

$$j_n(x) = (-1)^n x^n \left(\frac{d}{x\,dx}\right)^n \left(\frac{\sin x}{x}\right), \tag{11.165}$$

$$n_n(x) = -(-1)^n x^n \left(\frac{d}{x\,dx}\right)^n \left(\frac{\cos x}{x}\right). \tag{11.166}$$

$$h_n^{(1)}(x) = -i(-1)^n x^n \left(\frac{d}{x\,dx}\right)^n \left(\frac{e^{ix}}{x}\right)$$

$$\tag{11.167}$$

$$h_n^{(2)}(x) = i(-1)^n x^n \left(\frac{d}{x\,dx}\right)^n \left(\frac{e^{-ix}}{x}\right).$$

NUMERICAL COMPUTATION

The spherical Bessel and modified Bessel functions may be computed using the same techniques described in Sections 11.1 and 11.6 or evaluating the Bessel functions. For $j_n(x)$ and $i_n(x)^3$ it is convenient to use Eq. (11.161) and Exercise 11.7.18 and work *downward*, as is done for $J_n(x)$. Normalization is accomplished by comparing with the known forms of $j_0(x)$ and $i_0(x)$, Eq. (11.15) and Exercise 11.7.15. For $n_n(x)$ and $k_n(x)$, Eq. (11.161) and Exercise 11.7.19 are used again, but this time working *upward*, starting with the known forms of $n_0(x)$, $n_1(x)$, $k_0(x)$, and $k_1(x)$, Eq. (11.155) and Exercise 11.7.17.

ORTHOGONALITY

We may take the orthogonality integral for the ordinary Bessel functions (Eq. (11.50)),

$$\int_0^a J_\nu\left(\alpha_{\nu p}\frac{\rho}{a}\right) J_\nu\left(\alpha_{\nu q}\frac{\rho}{a}\right)\rho\,d\rho = \frac{a^2}{2}[J_{\nu+1}(\alpha_{\nu p})]^2\delta_{pq} \tag{11.168}$$

and substitute in the expression for j_n to obtain

$$\int_0^a j_n\left(\alpha_{np}\frac{\rho}{a}\right) j_n\left(\alpha_{nq}\frac{\rho}{a}\right)\rho^2\,d\rho = \frac{a^3}{2}[j_{n+1}(\alpha_{np})]^2\delta_{pq}. \tag{11.169}$$

Here α_{np} and α_{nq} are roots of j_n.

This represents orthogonality with respect to the roots of the Bessel functions. An illustration of this sort of orthogonality is provided later in this section by the problem of a particle in a sphere. Equation (11.170) guarantees orthogonality of the wave functions $j_n(r)$ for fixed n. (If n varies, the spherical harmonic will provide orthogonality.)

---

[3] The spherical modified Bessel functions, $i_n(x)$ and $k_n(x)$, are defined in Exercise 11.7.15.

Example 11.7.1   Particle in a Sphere

An illustration of the use of the spherical Bessel functions is provided by the problem of a quantum mechanical particle in a sphere of radius a. Quantum theory requires that the wave function ψ, describing our particle, satisfy

$$-\frac{\hbar^2}{2m}\nabla^2\psi = E\psi, \qquad (11.170)$$

and the boundary conditions (1) $\psi(r \le a)$ remains finite, (2) $\psi(a)=0$. This corresponds to a potential $V = 0, r \le a$, and $V = \infty, r > a$. Here $\hbar$ is Planck's constant divided by 2π, m, the mass of our particle, and E, its energy. Let us determine the *minimum* value of the energy for which our wave equation has an acceptable solution. Equation (11.170) is just Helmholtz's equation with a radial part (compare Section 8.3 for separation of variables):

$$\frac{d^2R}{dr^2} + \frac{2}{r}\frac{dR}{dr} + \left[k^2 - \frac{n(n+1)}{r^2} \right]R = 0, \qquad (11.171)$$

with $k^2 = 2mE/\hbar^2$. Hence by Eq. (11.139), with $n = 0$,

$$R = Aj_0(kr) + Bn_0(kr).$$

We choose the orbital angular momentum index $n = 0$, for any angular dependence would raise the energy. The spherical Neumann function is rejected because of its divergent behavior at the origin. To satisfy the second boundary condition (for all angles), we require

$$ka = \frac{\sqrt{2mE}}{\hbar}a = \alpha, \qquad (11.172)$$

where α is a root of j_0, that is, $j_0(\alpha) = 0$. This has the effect of limiting the allowable energies to a certain discrete set or, in other words, application of boundary condition (2) quantizes the energy E. The smallest α is the first zero of j_0,

$$\alpha = \pi$$

and

$$E_{min} = \frac{\pi^2\hbar^2}{2ma^2} = \frac{h^2}{8ma^2}, \qquad (11.173)$$

which means that for any finite sphere the particle energy will have a positive minimum or zero-point energy. This is an illustration of the Heisenberg uncertainty principle for Δp with $\Delta r \le a$.

In solid state physics, astrophysics, and other areas of physics we may wish to know how many different solutions (energy states) correspond to energies less than or equal to some fixed energy E_0. For a cubic volume (Exercise 2.6.5) the problem is fairly simple. The considerably more difficult spherical case is worked out by R. H. Lambert, *Am. J. Phys.* **36**, 417, 1169 (1968).

The relevant orthogonality relation for the $j_n(kr)$ can be derived from the integral given in Exercise 11.7.23.

Another form, orthogonality with respect to the indices, may be written as

$$\int_{-\infty}^{\infty} j_m(x)j_n(x)\, dx = 0, \qquad m \neq n,\ m, n \geq 0. \qquad (11.174)$$

The proof is left as Exercise 11.7.10. If $m = n$ (compare Exercise 11.7.11), we have

$$\int_{-\infty}^{\infty} [j_n(x)]^2\, dx = \frac{\pi}{2n + 1}. \qquad (11.175)$$

Most physical applications of orthogonal Bessel and spherical Bessel functions involve orthogonality with varying roots and an interval $[0, a]$, Eqs. (11.168) and (11.169) and Exercise 11.7.23 for continuous energy eigenvalues.

The spherical Bessel functions will enter again in connection with spherical waves, but further consideration is postponed until the corresponding angular functions, the Legendre functions, have been introduced.

EXERCISES

11.7.1   Show that if

$$n_n(x) = \sqrt{\frac{\pi}{2x}}\, N_{n+1/2}(x),$$

it automatically equals

$$(-1)^{n+1}\sqrt{\frac{\pi}{2x}}\, J_{-n-1/2}(x).$$

11.7.2   Derive the trigonometric-polynomial forms of $j_n(z)$ and $n_n(z)$.[4]

(a) $\displaystyle j_n(z) = \frac{1}{z}\sin\!\left(z - \frac{n\pi}{2}\right) \sum_{s=0}^{[n/2]} \frac{(-1)^s(n + 2s)!}{(2s)!(2z)^{2s}(n - 2s)!}$

$\displaystyle + \frac{1}{z}\cos\!\left(z - \frac{n\pi}{2}\right) \sum_{s=0}^{[(n-1)/2]} \frac{(-1)^s(n + 2s + 1)!}{(2s + 1)!(2z)^{2s+1}(n - 2s - 1)!}.$

(b) $\displaystyle n_n(z) = \frac{(-1)^{n+1}}{z}\cos\!\left(z + \frac{n\pi}{2}\right) \sum_{s=0}^{[n/2]} \frac{(-1)^s(n + 2s)!}{(2s)!(2z)^{2s}(n - 2s)!}$

$\displaystyle + \frac{(-1)^{n+1}}{z}\sin\!\left(z + \frac{n\pi}{2}\right) \sum_{s=0}^{[(n-1)/2]} \frac{(-1)^s(n + 2s + 1)!}{(2s + 1)!(2z)^{2s+1}(n - 2s - 1)!}.$

[4] The upper limit on the summation $[n/2]$ means the largest *integer* that does not exceed $n/2$.

11.7.3   Use the integral representation of $J_\nu(x)$,

$$J_\nu(x) = \frac{1}{\pi^{1/2}(\nu - \frac{1}{2})!} \left(\frac{x}{2}\right)^\nu \int_{-1}^{1} e^{\pm ixp}(1 - p^2)^{\nu - 1/2}\, dp,$$

to show that the spherical Bessel functions $j_n(x)$ are expressible in terms of trigonometric functions; that is, for example,

$$j_0(x) = \frac{\sin x}{x},$$

$$j_1(x) = \frac{\sin x}{x^2} - \frac{\cos x}{x}.$$

11.7.4   (a) Derive the recurrence relations

$$f_{n-1}(x) + f_{n+1}(x) = \frac{2n + 1}{x} f_n(x),$$

$$n f_{n-1}(x) - (n + 1)f_{n+1}(x) = (2n + 1)f_n'(x),$$

satisfied by the spherical Bessel functions, $j_n(x)$, $n_n(x)$, $h_n^{(1)}(x)$, and $h_n^{(2)}(x)$.

(b) Show, from these two recurrence relations, that the spherical Bessel function $f_n(x)$ satisfies the differential equation

$$x^2 f_n''(x) + 2x f_n'(x) + [x^2 - n(n + 1)]f_n(x) = 0.$$

11.7.5   Prove by mathematical induction that

$$j_n(x) = (-1)^n x^n \left(\frac{d}{x\, dx}\right)^n \left(\frac{\sin x}{x}\right)$$

for n an arbitrary nonnegative integer.

11.7.6   From the discussion of orthogonality of the spherical Bessel functions, show that a Wronskian relation for $j_n(x)$ and $n_n(x)$ is

$$j_n(x)n_n'(x) - j_n'(x)n_n(x) = \frac{1}{x^2}.$$

11.7.7   Verify

$$h_n^{(1)}(x)h_n^{(2)\prime}(x) - h_n^{(1)\prime}(x)h_n^{(2)}(x) = -\frac{2i}{x^2}.$$

11.7.8   Verify Poisson's integral representation of the spherical Bessel function,

$$j_n(z) = \frac{z^n}{2^{n+1}n!} \int_0^\pi \cos(z \cos \theta) \sin^{2n+1} \theta\, d\theta.$$

11.7.9   Show that

$$\int_0^\infty J_\mu(x)J_\nu(x)\frac{dx}{x} = \frac{2}{\pi} \frac{\sin[(\mu - \nu)\pi/2]}{\mu^2 - \nu^2}, \qquad \mu + \nu > -1.$$

11.7.10   Derive Eq. (11.174):

$$\int_{-\infty}^\infty j_m(x)j_n(x)\, dx = 0, \qquad \begin{matrix} m \neq n \\ m, n \geq 0. \end{matrix}$$

11.7.11   Derive Eq. (11.175):

$$\int_{-\infty}^\infty [j_n(x)]^2\, dx = \frac{\pi}{2n + 1}.$$

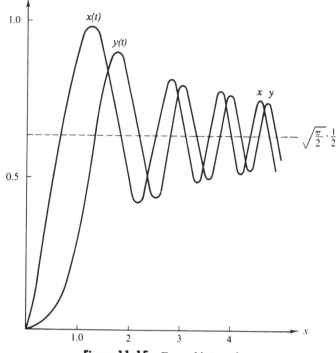

Figure 11.15 Fresnel integrals.

11.7.12 Set up the orthogonality integral for $j_L(kr)$ in a sphere of radius R with the boundary condition

$$j_L(kR) = 0.$$

The result is used in classifying electromagnetic radiation according to its angular momentum.

11.7.13 The Fresnel integrals (Fig. 11.15) occurring in diffraction theory are given by

$$x(t) = \int_0^t \cos(v^2)\, dv,$$

$$y(t) = \int_0^t \sin(v^2)\, dv.$$

Show that these integrals may be expanded in series of spherical Bessel functions,

$$x(s) = \tfrac{1}{2}\int_0^s j_{-1}(u)u^{1/2}\, du = s^{1/2}\sum_{n=0}^{\infty} j_{2n}(s),$$

$$y(s) = \tfrac{1}{2}\int_0^s j_0(u)u^{1/2}\, du = s^{1/2}\sum_{n=0}^{\infty} j_{2n+1}(s).$$

Hint. To establish the equality of the integral and the sum, you may wish to work with their derivatives. The spherical Bessel analogs of Eqs. (11.12) and (11.14) are helpful.

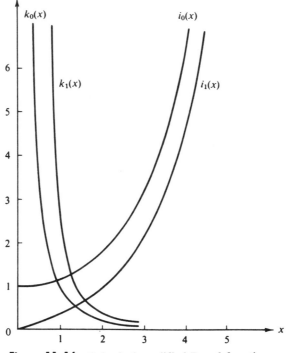

Figure 11.16 Spherical modified Bessel functions.

11.7.14 A hollow sphere of radius a (Helmholtz resonator) contains standing sound waves. Find the minimum frequency of oscillation in terms of the radius a and the velocity of sound v. The sound waves satisfy the wave equation

$$\nabla^2 \psi = \frac{1}{v^2} \frac{\partial^2 \psi}{\partial t^2}$$

and the boundary condition

$$\frac{\partial \psi}{\partial r} = 0, \qquad r = a.$$

This is a Neumann boundary condition. Example 11.7 has the same differential equation but with a Dirichlet boundary condition.

ANS. $v_{\min} = 0.3313 v/a,$
$\lambda_{\max} = 3.018 a.$

11.7.15 Defining the spherical modified Bessel functions (Fig. 11.16) by

$$i_n(x) = \sqrt{\frac{\pi}{2x}} I_{n+1/2}(x),$$

$$k_n(x) = \sqrt{\frac{2}{\pi x}} K_{n+1/2}(x),$$

show that

$$i_0(x) = \frac{\sinh x}{x}$$

$$k_0(x) = \frac{e^{-x}}{x}.$$

Note that the numerical factors in the definitions of i_n and k_n are *not* identical.

11.7.16   (a) Show that the parity of $i_n(x)$ is $(-1)^n$.
(b) Show that $k_n(x)$ has no definite parity.

11.7.17   Show that the spherical modified Bessel functions satisfy the following relations:

(a) $i_n(x) = i^{-n}j_n(ix),$

$\quad k_n(x) = -(i)^n h_n^{(1)}(ix),$

(b) $i_{n+1}(x) = x^n \dfrac{d}{dx}(x^{-n}i_n),$

$\quad k_{n+1}(x) = -x^n \dfrac{d}{dx}(x^{-n}k_n),$

(c) $i_n(x) = x^n\left(\dfrac{d}{x\,dx}\right)^n \dfrac{\sinh x}{x},$

$\quad k_n(x) = (-1)^n x^n\left(\dfrac{d}{x\,dx}\right)^n \dfrac{e^{-x}}{x}.$

11.7.18   Show that the recurrence relations for $i_n(x)$ and $k_n(x)$ are

(a) $i_{n-1}(x) - i_{n+1}(x) = \dfrac{2n+1}{x} i_n(x),$

$\quad ni_{n-1}(x) + (n+1)i_{n+1}(x) = (2n+1)i_n'(x),$

(b) $k_{n-1}(x) - k_{n+1}(x) = -\dfrac{2n+1}{x} k_n(x),$

$\quad nk_{n-1}(x) + (n+1)k_{n+1}(x) = -(2n+1)k_n'(x).$

11.7.19   Derive the limiting values for the spherical modified Bessel functions

(a) $i_n(x) \approx \dfrac{x^n}{(2n+1)!!}$

$\quad k_n(x) \approx \dfrac{(2n-1)!!}{x^{n+1}}, \quad x \ll 1.$

(b) $i_n(x) \sim \dfrac{e^x}{2x}$

$\quad k_n(x) \sim \dfrac{e^{-x}}{x}, \quad x \gg n(n+1)/2.$

11.7.20   Show that the Wronskian of the spherical modified Bessel functions is given by

$$i_n(x)k_n'(x) - i_n'(x)k_n(x) = -\frac{1}{x^2}.$$

11.7.21   A quantum particle is trapped in a "square" well of radius a. The Schrödinger equation potential is

$$V(r) = \begin{cases} -V_0, & 0 \le r < a \\ 0, & r > a. \end{cases}$$

The particle's energy E is negative (an eigenvalue).

(a) Show that the radial part of the wave function is given by $j_l(k_1 r)$ for $0 \le r < a$ and $k_l(k_2 r)$ for $r > a$. (We require that $\psi(0)$ and $\psi(\infty)$ be finite.) Here $k_1^2 = 2M(E + V_0)/\hbar^2$, $k_2^2 = -2ME/\hbar^2$, and l is the angular momentum (n in Eq. (11.139)).

(b) The boundary condition at $r = a$ is that the wave function $\psi(r)$ and its first derivative be continuous. Show that this means

$$\left.\frac{(d/dr)j_l(k_1 r)}{j_l(k_1 r)}\right|_{r=a} = \left.\frac{(d/dr)k_l(k_2 r)}{k_l(k_2 r)}\right|_{r=a}.$$

This equation determines the energy eigenvalues.

Note. This is a generalization of Example 9.1.2.

11.7.22   The quantum mechanical radial wave function for a scattered wave is given by

$$\psi_k = \frac{\sin(kr + \delta_0)}{kr},$$

where k is the wave number, $k = \sqrt{2mE/\hbar}$, and δ_0 is the scattering phase shift. Show that the normalization integral is

$$\int_0^\infty \psi_k(r)\psi_{k'}(r)r^2\,dr = \frac{\pi}{2k}\delta(k - k').$$

Hint. You can use a sine representation of the Dirac delta function. See Exercise 15.3.8.

11.7.23   Derive the spherical Bessel function closure relation

$$\frac{2a^2}{\pi}\int_0^\infty j_n(ar)j_n(br)r^2\,dr = \delta(a - b).$$

Note. An interesting derivation involving Fourier transforms, the Rayleigh plane wave expansion, and spherical harmonics has been given by P. Ugincius, *Am. J. Phys.* **40**, 1690 (1972).

11.7.24   (a) Write a subroutine that will generate the spherical Bessel functions, $j_n(x)$, that is, will generate the numerical value of $j_n(x)$ given x and n.

Note. One possibility is to use the explicit known forms of j_0 and j_1 and to develop the higher index j_n by repeated application of the recurrence relation.

(b) Check your subroutine by an independent calculation such as Eq. (11.153). If possible, compare the machine time needed for this check with the time required for your subroutine.

11.7.25 The wave function of a particle in a sphere (Example 11.7.1) with angular momentum l is $\psi(r, \theta, \varphi) = A j_l((\sqrt{2ME}/\hbar)r) Y_l^m(\theta, \varphi)$. The $Y_l^m(\theta, \varphi)$ is a spherical harmonic, described in Section 12.6. From the boundary condition $\psi(a, \theta, \varphi) = 0$ or $j_l((\sqrt{2ME}/\hbar)a) = 0$ calculate the 10 lowest energy states. Disregard the m degeneracy ($2l + 1$ values of m for each choice of l). Check your results against AMS-55, Table 10.6.

Hint. You can use your spherical Bessel subroutine and a root-finding subroutine.

$$\text{Check values.} \quad j_l(\alpha_{ls}) = 0,$$
$$\alpha_{01} = 3.1416$$
$$\alpha_{11} = 4.4934$$
$$\alpha_{21} = 5.7635$$
$$\alpha_{02} = 6.2832.$$

11.7.26 Let Example 11.7.1 be modified so that the potential is a finite V_0 outside ($r > a$).

(a) For $E < V_0$ show that
$$\psi_{\text{out}}(r, \theta, \varphi) \sim k_l\left(\frac{\sqrt{2M(V_0 - E)}}{\hbar} r\right).$$

(b) The new boundary conditions to be satisfied at $r = a$ are
$$\psi_{\text{in}}(a, \theta, \varphi) = \psi_{\text{out}}(a, \theta, \varphi)$$
$$\frac{\partial}{\partial r} \psi_{\text{in}}(a, \theta, \varphi) = \frac{\partial}{\partial r} \psi_{\text{out}}(a, \theta, \varphi)$$

or
$$\frac{1}{\psi_{\text{in}}} \frac{\partial \psi_{\text{in}}}{\partial r}\bigg|_{r=a} = \frac{1}{\psi_{\text{out}}} \frac{\partial \psi_{\text{out}}}{\partial r}\bigg|_{r=a}.$$

For $l = 0$ show that the boundary condition at $r = a$ leads to
$$f(E) = k\left\{\cot ka - \frac{1}{ka}\right\} + k'\left\{1 + \frac{1}{k'a}\right\}$$
$$= 0,$$
where $k = \sqrt{2ME}/\hbar$ and $k' = \sqrt{2M(V_0 - E)}/\hbar$.

(c) With $a = 1\hbar^2/Me^2$ (Bohr radius) and $V_0 = 4Me^4/2\hbar^2$, compute the possible bound states, $(0 < E < V_0)$.

Hint. Call a root-finding subroutine after you know the approximate location of the roots of
$$f(E), \quad (0, V_0).$$

(d) Show that when $a = 1\hbar^2/Me^2$ the minimum value of V_0 for which a bound state exists is $V_0 = 2.4674Me^4/2\hbar^2$.

11.7.27 In some nuclear stripping reactions the differential cross section is proportional to $j_l(x)^2$, where l is the angular momentum. The location of the maximum on the curve of experimental data permits a determination of l, if the location of the (first) maximum of $j_l(x)$ is known. Compute the location of the first maximum of $j_1(x)$, $j_2(x)$, and $j_3(x)$.

Note. For better accuracy look for the first zero of $j_l'(x)$. Why is this more accurate than direct location of the maximum?

ADDITIONAL READINGS

McBRIDE, E. B., *Obtaining Generating Functions*. New York: Springer-Verlag (1971). An introduction to methods of obtaining generating functions.

WATSON, G. N., *A Treatise on the Theory of Bessel Functions*, 2nd ed. Cambridge: Cambridge University Press (1952). This is the definitive text on Bessel functions and their properties. Although difficult reading, it is invaluable as the ultimate reference.

WATSON, G. N., *A Treatise on the Theory of Bessel Functions*, 1st ed. Cambridge: Cambridge University Press (1922). See also the references listed at the end of Chapter 13.

12 LEGENDRE FUNCTIONS

12.1 GENERATING FUNCTION

Legendre polynomials may appear in many different mathematical and physical situations: (1) They may originate as solutions of the Legendre differential equation which we have already encountered in the separation of variables (Section 8.3) for Laplace's equation, Helmholtz's equation, and similar differential equations in spherical polar coordinates. (2) They may enter as a consequence of a Rodrigues' formula (Section 12.4). (3) They may be constructed as a consequence of demanding a complete, orthogonal set of functions over the interval $[-1, 1]$ (Gram–Schmidt orthogonalization, Section 9.3). (4) In quantum mechanics they (really the spherical harmonics, Sections 12.6 and 12.7) represent angular momentum eigenfunctions. (5) They may be generated by a generating function. We introduce Legendre polynomials here by way of a generating function.

PHYSICAL BASIS—ELECTROSTATICS

As with Bessel functions, it is convenient to introduce the Legendre polynomials by means of a generating function. However, a direct physical interpretation is possible. Consider an electric charge q placed on the z-axis at $z = a$. As shown in Fig. 12.1, the electrostatic potential of charge q is

$$\varphi = \frac{1}{4\pi\varepsilon_0} \cdot \frac{q}{r_1} \qquad \text{(SI units)}. \qquad (12.1)$$

Our problem is to express the electrostatic potential in terms of the spherical polar coordinates r and θ (the coordinate φ is absent because of symmetry

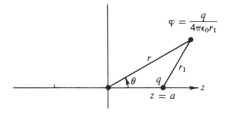

Figure 12.1 Electrostatic potential. Charge q displaced from origin.

about the z-axis). Using the law of cosines, we obtain

$$\varphi = \frac{q}{4\pi\varepsilon_0}(r^2 + a^2 - 2ar\cos\theta)^{-1/2}. \qquad (12.2)$$

LEGENDRE POLYNOMIALS

Consider the case of $r > a$ or, more precisely, $r^2 > |a^2 - 2ar\cos\theta|$. The radical in Eq. (12.2) may be expanded in a binomial series and then rearranged in powers of (a/r). The Legendre polynomial $P_n(\cos\theta)$ (see Fig. 12.2) may be defined as the coefficient of the nth power in

$$\varphi = \frac{q}{4\pi\varepsilon_0 r} \sum_{n=0}^{\infty} P_n(\cos\theta)\left(\frac{a}{r}\right)^n. \qquad (12.3)$$

Dropping the factor $q/4\pi\varepsilon_0 r$ and using x and t instead of $\cos\theta$ and a/r, respectively, we have

$$g(t, x) = (1 - 2xt + t^2)^{-1/2} = \sum_{n=0}^{\infty} P_n(x)t^n, \qquad |t| < 1. \qquad (12.4)$$

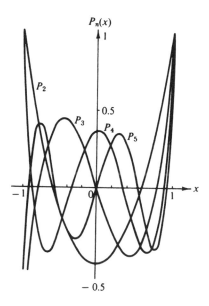

Figure 12.2 Legendre polynomials, $P_2(x)$, $P_3(x)$, $P_4(x)$, and $P_5(x)$.

Equation (12.4) is our generating function. In the next section it is shown that $|P_n(\cos \theta)| \le 1$, which means that the series expansion (Eq. (12.4)) is convergent for $|t| < 1$.[1] Indeed, the series is convergent for $|t| = 1$ except for $|x| = 1$.

In physical applications Eq. (12.4) often appears in the vector form (see Section 8.7)

$$\frac{1}{|\mathbf{r}_1 - \mathbf{r}_2|} = \frac{1}{r_>} \sum_{n=0}^{\infty} \left(\frac{r_<}{r_>}\right)^n P_n(\cos \theta), \tag{12.4a}$$

where

$$\left.\begin{array}{l} r_> = |\mathbf{r}_1| \\ r_< = |\mathbf{r}_2| \end{array}\right\} \qquad \text{for } |\mathbf{r}_1| > |\mathbf{r}_2|, \tag{12.4b}$$

and

$$\left.\begin{array}{l} r_> = |\mathbf{r}_2| \\ r_< = |\mathbf{r}_1| \end{array}\right\} \qquad \text{for } |\mathbf{r}_2| > |\mathbf{r}_1|. \tag{12.4c}$$

Using the binomial theorem (Section 5.6) and Exercise 10.1.15, we expand the generating function as

$$(1 - 2xt + t^2)^{-1/2} = \sum_{n=0}^{\infty} \frac{(2n)!}{2^{2n}(n!)^2} (2xt - t^2)^n$$

$$= 1 + \sum_{n=1}^{\infty} \frac{(2n-1)!!}{(2n)!!} (2xt - t^2)^n. \tag{12.5}$$

For the first few Legendre polynomials, say, P_0, P_1, and P_2, we need the coefficients of t^0, t^1, and t^2. These powers of t appear only in the terms $n = 0$, 1, and 2 and hence we may limit our attention to the first three terms of the infinite series:

$$\frac{0!}{2^0(0!)^2} (2xt - t^2)^0 + \frac{2!}{2^2(1!)^2} (2xt - t^2)^1 + \frac{4!}{2^4(2!)^2} (2xt - t^2)^2$$

$$= 1t^0 + xt^1 + \left(\frac{3}{2}x^2 - \frac{1}{2}\right)t^2 + \mathcal{O}(t^3).$$

Then, from Eq. (12.4) (and uniqueness of power series)

$$P_0(x) = 1, \qquad P_1(x) = x, \qquad P_2(x) = \frac{3}{2}x^2 - \frac{1}{2}.$$

We repeat this limited development in a vector framework later in this section.

---

[1] Note that the series in Eq. (12.3) is convergent for $r > a$ even though the binomial expansion involved is valid only for $r > (a^2 + 2ar)^{1/2}$, $\cos \theta = -1$.

In employing a general treatment, we find that the binomial expansion of the $(2xt - t^2)^n$ factor yields the double series

$$(1 - 2xt + t^2)^{-1/2} = \sum_{n=0}^{\infty} \frac{(2n)!}{2^{2n}(n!)^2} t^n \sum_{k=0}^{n} (-1)^k \frac{n!}{k!(n-k)!} (2x)^{n-k} t^k$$

$$= \sum_{n=0}^{\infty} \sum_{k=0}^{n} (-1)^k \frac{(2n)!}{2^{2n} n! k!(n-k)!} (2x)^{n-k} t^{n+k}. \qquad (12.6)$$

From Eq. (5.64) of Section 5.4 (rearranging the order of summation), Eq. (12.6) becomes

$$(1 - 2xt + t^2)^{-1/2} = \sum_{n=0}^{\infty} \sum_{k=0}^{[n/2]} (-1)^k \frac{(2n-2k)!}{2^{2n-2k} k!(n-k)!(n-2k)!} \cdot (2x)^{n-2k} t^n, \qquad (12.7)$$

with the t^n independent of the index k.[2] Now, equating our two power series (Eqs. (12.4) and (12.7)) term by term, we have[3]

$$P_n(x) = \sum_{k=0}^{[n/2]} (-1)^k \frac{(2n-2k)!}{2^n k!(n-k)!(n-2k)!} x^{n-2k}. \qquad (12.8)$$

Hence, for n even, P_n has only even powers of x and even parity (see Eq. (12.37)), and odd powers and odd parity for odd n.

LINEAR ELECTRIC MULTIPOLES

Returning to the electric charge on the z-axis, we demonstrate the usefulness and power of the generating function by adding a charge $-q$ at $z = -a$, as shown in Fig. 12.3. The potential becomes

$$\varphi = \frac{q}{4\pi\varepsilon_0} \left(\frac{1}{r_1} - \frac{1}{r_2} \right), \qquad (12.9)$$

and by using the law of cosines, we have

$$\varphi = \frac{q}{4\pi\varepsilon_0 r} \left\{ \left[1 - 2\left(\frac{a}{r}\right) \cos\theta + \left(\frac{a}{r}\right)^2 \right]^{-1/2} - \left[1 + 2\left(\frac{a}{r}\right) \cos\theta + \left(\frac{a}{r}\right)^2 \right]^{-1/2} \right\},$$

$$(r > a).$$

Clearly, the second radical is like the first, except that a has been replaced by $-a$. Then, using Eq. (12.4), we obtain

$$\varphi = \frac{q}{4\pi\varepsilon_0 r} \left[\sum_{n=0}^{\infty} P_n(\cos\theta)\left(\frac{a}{r}\right)^n - \sum_{n=0}^{\infty} P_n(\cos\theta)(-1)^n\left(\frac{a}{r}\right)^n \right]$$

$$= \frac{2q}{4\pi\varepsilon_0 r} \left[P_1(\cos\theta)\left(\frac{a}{r}\right) + P_3(\cos\theta)\left(\frac{a}{r}\right)^3 + \cdots \right]. \qquad (12.10)$$

---

[2] $[n/2] = n/2$ for n even, $(n-1)/2$ for n odd.

[3] Equation (12.8) starts with x^n. By changing the index, we can transform it into a series that starts with x^0 for n even and x^1 for n odd. These ascending series are given as hypergeometric functions in Eqs. (13.104) and (13.105), Section 13.5.

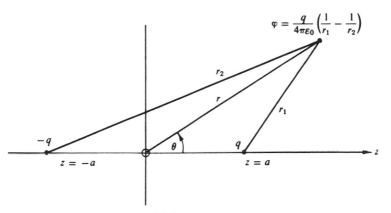

Figure 12.3 Electric dipole.

The first term (and dominant term for $r \gg a$) is

$$\varphi = \frac{2aq}{4\pi\varepsilon_0} \cdot \frac{P_1(\cos \theta)}{r^2}, \qquad (12.11)$$

which is the usual electric dipole potential. Here $2aq$ is the dipole moment (Fig. 12.3).

This analysis may be extended by placing additional charges on the z-axis so that the P_1 term, as well as the P_0 (monopole) term, is canceled. For instance, charges of q at $z = a$ and $z = -a$, $-2q$ at $z = 0$ give rise to a potential whose series expansion starts with $P_2(\cos \theta)$. This is a linear electric quadrupole. Two linear quadrupoles may be placed so that the quadrupole term is canceled, but the P_3, the octupole term, survives.

VECTOR EXPANSION

We consider the electrostatic potential produced by a distributed charge $\rho(\mathbf{r}_2)$:

$$\varphi(\mathbf{r}_1) = \frac{1}{4\pi\varepsilon_0} \int \frac{\rho(\mathbf{r}_2)}{|\mathbf{r}_1 - \mathbf{r}_2|} \, d\tau_2. \qquad (12.12a)$$

This expression has already been encountered in Sections 1.16 and 8.7. Taking the denominator of the integrand, using first the law of cosines and then a binomial expansion, yields (see Fig. 1.37)

$$
\begin{aligned}
\frac{1}{|\mathbf{r}_1 - \mathbf{r}_2|} &= (r_1^2 - 2\mathbf{r}_1 \cdot \mathbf{r}_2 + r_2^2)^{-1/2} \\
&= \frac{1}{r_1}\left[1 + \left(-\frac{2\mathbf{r}_1 \cdot \mathbf{r}_2}{r_1^2} + \frac{r_2^2}{r_1^2}\right)\right]^{-1/2}, \qquad \text{for } r_1 > r_2 \\
&= \frac{1}{r_1}\left[1 + \frac{\mathbf{r}_1 \cdot \mathbf{r}_2}{r_1^2} - \frac{1}{2}\frac{r_2^2}{r_1^2} + \frac{3}{2}\frac{(\mathbf{r}_1 \cdot \mathbf{r}_2)^2}{r_1^4} + \mathcal{O}\left(\frac{r_2}{r_1}\right)^3\right]. \qquad (12.12b)
\end{aligned}
$$

(For $r_1 = 1$, $r_2 = t$, and $\mathbf{r}_1 \cdot \mathbf{r}_2 = xt$, Eq. (12.12b) reduces to the generating function, Eq. (12.4).)

The first term in the square bracket, 1, yields a potential

$$\varphi_0(\mathbf{r}_1) = \frac{1}{4\pi\varepsilon_0} \frac{1}{r_1} \int \rho(\mathbf{r}_2) \, d\tau_2. \tag{12.12c}$$

The integral is just the total charge. This part of the total potential is an electric *monopole*.

The second term yields

$$\varphi_1(\mathbf{r}_1) = \frac{1}{4\pi\varepsilon_0} \frac{\mathbf{r}_1 \cdot}{r_1^3} \int \mathbf{r}_2 \rho(\mathbf{r}_2) \, d\tau_2, \tag{12.12d}$$

where the integral is the dipole moment whose charge density $\rho(\mathbf{r}_2)$ is weighted by a moment arm $\mathbf{r}_2$. We have an electric *dipole* potential. For atomic or nuclear states of definite parity $\rho(\mathbf{r}_2)$ is an even function and the dipole integral is identically zero.

The last two terms, both of order $(r_2/r_1)^2$, may be handled by using Cartesian coordinates

$$(\mathbf{r}_1 \cdot \mathbf{r}_2)^2 = \sum_{i=1}^{3} x_{1i} x_{2i} \sum_{j=1}^{3} x_{1j} x_{2j}.$$

Rearranging variables to keep the x_2's inside the integral yields

$$\varphi_2(\mathbf{r}_1) = \frac{1}{4\pi\varepsilon_0} \frac{1}{2r_1^5} \sum_{i,j=1}^{3} x_{1i} x_{1j} \int [3x_{2i} x_{2j} - \delta_{ij} r_2^2] \rho(\mathbf{r}_2) \, d\tau_2. \tag{12.12e}$$

This is the electric *quadrupole* term. We note that the square bracket in the integrand forms a symmetric, zero trace tensor.

A general electrostatic multipole expansion can also be developed by using Eq. (12.12a) for the potential $\varphi(r_1)$ and replacing $1/(4\pi|\mathbf{r}_1 - \mathbf{r}_2|)$ by Green's function, Eq. (8.187). This yields the potential $\varphi(\mathbf{r}_1)$ as a (double) series of the spherical harmonics $Y_l^m(\theta_1, \varphi_1)$ and $Y_l^m(\theta_2, \varphi_2)$.

Before leaving multipole fields, perhaps we should emphasize three points. First, an electric (or magnetic) multipole has an absolute significance only if all lower-order terms vanish. For instance, the potential of one charge q at $z = a$ was expanded in a series of Legendre polynomials. Although we may refer to the $P_1(\cos\theta)$ term in this expansion as a dipole term, it should be remembered that this term exists only because of our choice of coordinates. We actually have a monopole, $P_0(\cos\theta)$.

Second, in physical systems we do not encounter pure multipoles. As an example, the potential of the finite dipole (q at $z = a$, $-q$ at $z = -a$) contained a $P_3(\cos\theta)$ term. These higher-order terms may be eliminated by shrinking the multipole to a point multipole, in this case keeping the product qa constant ($a \to 0$, $q \to \infty$) to maintain the same dipole moment.

Third, the multipole theory is not restricted to electrical phenomena. Planetary configurations are described in terms of mass multipoles, Sections 12.3 and 12.5. Gravitational radiation depends on the time behavior of mass quadrupoles. (The gravitational radiation field is a *tensor* field. The radiation units, gravitons, carry *two* units of angular momentum.)

It might also be noted that a multipole expansion is actually a decomposition into the irreducible representations of the rotation group (Section 4.2).

EXTENSION TO ULTRASPHERICAL POLYNOMIALS

The generating function, $g(t, x)$, used here is actually a special case of a more general generating function,

$$\frac{1}{(1 - 2xt + t^2)^\alpha} = \sum_{n=0}^{\infty} C_n^{(\alpha)}(x) t^n. \tag{12.13}$$

The coefficients $C_n^{(\alpha)}(x)$ are the ultraspherical polynomials (proportional to the Gegenbauer polynomials). For $\alpha = 1/2$ this equation reduces to Eq. (12.4); that is; $C_n^{(1/2)}(x) = P_n(x)$. The cases $\alpha = 0$ and $\alpha = 1$ are considered in Chapter 13 in connection with the Chebyshev polynomials.

EXERCISES

12.1.1   Develop the electrostatic potential for the array of charges shown. This is a linear electric quadrupole (Fig. 12.4).

12.1.2   Calculate the electrostatic potential of the array of charges shown (Fig. 12.5). Here is an example of two equal but oppositely directed dipoles. The dipole contributions cancel. The octupole terms do not cancel.

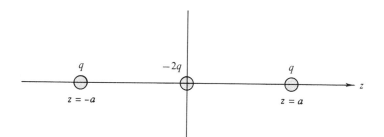

Figure 12.4   Linear electric quadrupole.

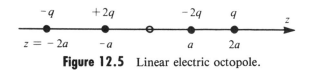

Figure 12.5   Linear electric octopole.

12.1.3   Show that the electrostatic potential produced by a charge q at $z = a$ for $r < a$ is

$$\varphi(\mathbf{r}) = \frac{q}{4\pi\varepsilon_0 a} \sum_{n=0}^{\infty} \left(\frac{r}{a}\right)^n P_n(\cos\theta).$$

12.1.4   Using $\mathbf{E} = -\nabla\varphi$, determine the components of the electric field corresponding to the (pure) electric dipole potential

$$\varphi(\mathbf{r}) = \frac{2aqP_1(\cos\theta)}{4\pi\varepsilon_0 r^2}.$$

Here it is assumed that $r \gg a$.

$$ANS. \quad E_r = +\frac{4aq\cos\theta}{4\pi\varepsilon_0 r^3},$$

$$E_\theta = +\frac{2aq\sin\theta}{4\pi\varepsilon_0 r^3},$$

$$E_\varphi = 0.$$

12.1.5   A point electric dipole of strength $p^{(1)}$ is placed at $z = a$; a second point electric dipole of equal but opposite strength is at the origin. Keeping the product $p^{(1)}a$ constant, let $a \to 0$. Show that this results in a point electric quadrupole.
Hint. Exercise 12.2.5 (when proved) will be helpful.

12.1.6   A point charge q is in the interior of a hollow conducting sphere of radius r_0. The charge q is displaced a distance a from the center of the sphere. If the conducting sphere is grounded, show that the potential in the interior produced by q and the distributed induced charge is the same as that produced by q and its image charge q'. The image charge is at a distance $a' = r_0^2/a$ from the center, collinear with q and the origin (Fig. 12.6).
Hint. Calculate the electrostatic potential for $a < r_0 < a'$. Show that the potential vanishes for $r = r_0$ if we take $q' = -qr_0/a$.

12.1.7   Prove that

$$P_n(\cos\theta) = (-1)^n \frac{r^{n+1}}{n!} \frac{\partial^n}{\partial z^n}\left(\frac{1}{r}\right).$$

Hint. Compare the Legendre polynomial expansion of the generating function $(a \to \Delta z$, Fig. 12.1) with a Taylor series expansion of $1/r$, where z dependence of r changes from z to $z - \Delta z$ (Fig. 12.7).

12.1.8   By differentiation and direct substitution of the series form, Eq. (12.8), show that $P_n(x)$ satisfies the Legendre differential equation. Note that there is no restriction upon x. We may have any x, $-\infty < x < \infty$ and indeed any z in the entire finite complex plane.

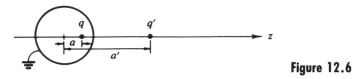

Figure 12.6

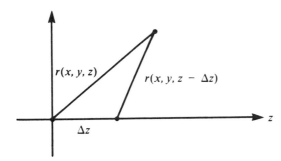

Figure 12.7

12.1.9   The Chebyshev polynomials (type II) are generated by (Eq. (13.62), Section 13.3)

$$\frac{1}{1 - 2xt + t^2} = \sum_{n=0}^{\infty} U_n(x)t^n.$$

Using the techniques of Section 5.4 for transforming series, develop a series representation of $U_n(x)$.

$$ANS. \quad U_n(x) = \sum_{k=0}^{[n/2]} (-1)^k \frac{(n-k)!}{k!(n-2k)!} (2x)^{n-2k}.$$

12.2 RECURRENCE RELATIONS AND SPECIAL PROPERTIES

RECURRENCE RELATIONS

The Legendre polynomial generating function provides a convenient way of deriving the recurrence relations[1] and some special properties. If our generating function (Eq. (12.4)) is differentiated with respect to t, we obtain

$$\frac{\partial g(t, x)}{\partial t} = \frac{x - t}{(1 - 2xt + t^2)^{3/2}} = \sum_{n=0}^{\infty} nP_n(x)t^{n-1}. \qquad (12.14)$$

By substituting Eq. (12.4) into this and rearranging terms, we have

$$(1 - 2xt + t^2) \sum_{n=0}^{\infty} nP_n(x)t^{n-1} + (t - x) \sum_{n=0}^{\infty} P_n(x)t^n = 0. \qquad (12.15)$$

The left-hand side is a power series in t. Since this power series vanishes for all values of t, we may put the coefficient of each power of t equal to zero, that is, our power series is unique (Section 5.7). This may be done easily by separating the individual summations and using distinctive summation indices,

$$\sum_{m=0}^{\infty} mP_m(x)t^{m-1} - \sum_{n=0}^{\infty} 2nxP_n(x)t^n + \sum_{s=0}^{\infty} sP_s(x)t^{s+1}$$

$$+ \sum_{s=0}^{\infty} P_s(x)t^{s+1} - \sum_{n=0}^{\infty} xP_n(x)t^n = 0. \qquad (12.16)$$

[1] We can also apply the explicit series form (Eq. (12.8)) directly.

Table 12.1 Legendre Polynomials

$P_0(x) = 1$

$P_1(x) = x$

$P_2(x) = \frac{1}{2}(3x^2 - 1)$

$P_3(x) = \frac{1}{2}(5x^3 - 3x)$

$P_4(x) = \frac{1}{8}(35x^4 - 30x^2 + 3)$

$P_5(x) = \frac{1}{8}(63x^5 - 70x^3 + 15x)$

$P_6(x) = \frac{1}{16}(231x^6 - 315x^4 + 105x^2 - 5)$

$P_7(x) = \frac{1}{16}(429x^7 - 693x^5 + 315x^3 - 35x)$

$P_8(x) = \frac{1}{128}(6435x^8 - 12012x^6 + 6930x^4 - 1260x^2 + 35)$

Now letting $m = n + 1$, $s = n - 1$, we find

$$(2n + 1)xP_n(x) = (n + 1)P_{n+1}(x) + nP_{n-1}(x), \qquad n = 1, 2, 3, \ldots. \tag{12.17}$$

This is another three-term recurrence relation similar to (but not identical to) the recurrence relation for Bessel functions. With this recurrence relation we may easily construct the higher Legendre polynomials. If we take $n = 1$ and insert the easily found values of $P_0(x)$ and $P_1(x)$ (Exercise 12.1.7 or Eq. (12.8)), we obtain

$$3xP_1(x) = 2P_2(x) + P_0(x) \tag{12.18}$$

or

$$P_2(x) = \frac{1}{2}(3x^2 - 1). \tag{12.19}$$

This process may be continued indefinitely, the first few Legendre polynomials are listed in Table 12.1.

Cumbersome as it may appear at first, this technique is actually more efficient for a large digital computer than is direct evaluation of the series (Eq. (12.8)). For greater stability (to avoid undue accumulation and magnification of round off error), Eq. (12.17) is rewritten as

$$P_{n+1}(x) = 2xP_n(x) - P_{n-1}(x) - [xP_n(x) - P_{n-1}(x)]/(n + 1). \tag{12.17a}$$

One starts with $P_0(x) = 1$, $P_1(x) = x$, and computes the *numerical* values of all the $P_n(x)$ for a given value of x up to the desired $P_N(x)$. The values of $P_n(x)$, $0 \leq n < N$ are available as a fringe benefit.

DIFFERENTIAL EQUATIONS

More information about the behavior of the Legendre polynomials can be obtained if we now differentiate Eq. (12.4) with respect to x. This gives

$$\frac{\partial g(t, x)}{\partial x} = \frac{t}{(1 - 2xt + t^2)^{3/2}} = \sum_{n=0}^{\infty} P_n'(x)t^n \tag{12.20}$$

or

$$(1 - 2xt + t^2) \sum_{n=0}^{\infty} P_n'(x)t^n - t \sum_{n=0}^{\infty} P_n(x)t^n = 0. \qquad (12.21)$$

As before, the coefficient of each power of t is set equal to zero and we obtain

$$P_{n+1}'(x) + P_{n-1}'(x) = 2xP_n'(x) + P_n(x). \qquad (12.22)$$

A more useful relation may be found by differentiating Eq. (12.17) with respect to x and multiplying by 2. To this we add $(2n + 1)$ times Eq. (12.22), canceling the P_n' term. The result is

$$P_{n+1}'(x) - P_{n-1}'(x) = (2n + 1)P_n(x). \qquad (12.23)$$

From Eqs. (12.22) and (12.23) numerous additional equations may be developed,[2] including

$$P_{n+1}'(x) = (n + 1)P_n(x) + xP_n'(x), \qquad (12.24)$$

$$P_{n-1}'(x) = -nP_n(x) + xP_n'(x), \qquad (12.25)$$

$$(1 - x^2)P_n'(x) = nP_{n-1}(x) - nxP_n(x), \qquad (12.26)$$

$$(1 - x^2)P_n'(x) = (n + 1)xP_n(x) - (n + 1)P_{n+1}(x). \qquad (12.27)$$

By differentiating Eq. (12.26) and using Eq. (12.25) to eliminate $P_{n-1}'(x)$, we find that $P_n(x)$ satisfies the linear, second-order differential equation

$$(1 - x^2)P_n''(x) - 2xP_n'(x) + n(n + 1)P_n(x) = 0. \qquad (12.28)$$

The previous equations, Eqs. (12.22) to (12.27), are all first-order differential equations, but with polynomials of two different indices. The price for having all indices alike is a second-order differential equation. Equation (12.28) is Legendre's differential equation. We now see that the polynomials $P_n(x)$ generated by the expansion of $(1 - 2xt + t^2)^{-1/2}$ satisfy Legendre's equation which, of course, is why they are called Legendre polynomials.

In Eq. (12.28) differentiation is with respect to $x(x = \cos \theta)$. Frequently, we encounter Legendre's equation expressed in terms of differentiation with

[2] Using the equation number in parentheses to denote the entire equation, we may write the derivatives as

$$2 \cdot \frac{d}{dx}(12.17) + (2n + 1) \cdot (12.22) \Rightarrow (12.23)$$

$$\frac{1}{2}\{(12.22) + (12.23)\} \Rightarrow (12.24)$$

$$\frac{1}{2}\{(12.22) - (12.23)\} \Rightarrow (12.25)$$

$$(12.24)_{n \to n-1} + x \cdot (12.25) \Rightarrow (12.26)$$

$$\frac{d}{dx}(12.26) + n \cdot (12.25) \Rightarrow (12.28).$$

respect to θ,

$$\frac{1}{\sin \theta} \frac{d}{d\theta}\left(\sin \theta \frac{dP_n(\cos \theta)}{d\theta}\right) + n(n + 1)P_n(\cos \theta) = 0. \qquad (12.29)$$

SPECIAL VALUES

Our generating function provides still more information about the Legendre polynomials. If we set $x = 1$, Eq. (12.4) becomes

$$\frac{1}{(1 - 2t + t^2)^{1/2}} = \frac{1}{1 - t}$$

$$= \sum_{n=0}^{\infty} t^n, \qquad (12.30)$$

using a binomial expansion. But

$$\frac{1}{(1 - 2tx + t^2)^{1/2}_{x=1}} = \sum_{n=0}^{\infty} P_n(1)t^n.$$

Comparing the two series expansions (uniqueness of power series, Section 5.7), we have

$$P_n(1) = 1. \qquad (12.31)$$

If we let $x = -1$, the parity (see Eq. (12.37)) shows that

$$P_n(-1) = (-1)^n. \qquad (12.32)$$

For obtaining these results, we find that the generating function is more convenient than the explicit series form.

If we take $x = 0$, using the binomial expansion

$$(1 + t^2)^{-1/2} = 1 - \tfrac{1}{2}t^2 + \tfrac{3}{8}t^4 + \cdots + (-1)^n \frac{1 \cdot 3 \cdots (2n - 1)}{2^n n!} t^{2n} + \cdots,$$

$$\qquad (12.33)$$

we have[3]

$$P_{2n}(0) = (-1)^n \frac{1 \cdot 3 \cdots (2n - 1)}{2^n n!} = (-1)^n \frac{(2n - 1)!!}{(2n)!!} = \frac{(-1)^n (2n)!}{2^{2n}(n!)^2}$$

$$\qquad (12.34)$$

$$P_{2n+1}(0) = 0, \qquad n = 0, 1, 2, \ldots. \qquad (12.35)$$

These results also follow from Eq. (12.8) by inspection.

---

[3] The double factorial notation is defined in Section 10.1.

$$(2n)!! = 2 \cdot 4 \cdot 6 \cdots (2n). \qquad (2n - 1)!! = 1 \cdot 3 \cdot 5 \cdots (2n - 1).$$

PARITY

Some of these results are special cases of the parity property of the Legendre polynomials. We refer once more to Eqs. (12.4) and (12.8). If we replace x by $-x$ and t by $-t$, the generating function is unchanged. Hence

$$g(t, x) = g(-t, -x)$$

$$= [1 - 2(-t)(-x) + (-t)^2]^{-1/2}$$

$$= \sum_{n=0}^{\infty} P_n(-x)(-t)^n$$

$$= \sum_{n=0}^{\infty} P_n(x)t^n. \tag{12.36}$$

Comparing these two series, we have

$$P_n(-x) = (-1)^n P_n(x); \tag{12.37}$$

that is, the polynomial functions are odd or even (with respect to $x = 0$, $\theta = \pi/2$) according to whether the index n is odd or even. This is the parity[4] or reflection property that plays such an important role in quantum mechanics. For central forces the index n is a measure of the orbital angular momentum, thus linking parity and orbital angular momentum.

The reader will see this parity property confirmed by the series solution and for the special values tabulated in Table 12.1. It might also be noted that Eq. (12.37) may be predicted by inspection of Eq. (12.17), the recurrence relation. Specifically, if $P_{n-1}(x)$ and $xP_n(x)$ are even, then $P_{n+1}(x)$ must be even.

UPPER AND LOWER BOUNDS FOR $P_n(\cos\theta)$

Finally, in addition to these results, our generating function enables us to set an upper limit on $|P_n(\cos\theta)|$. We have

$$(1 - 2t\cos\theta + t^2)^{-1/2} = (1 - te^{i\theta})^{-1/2}(1 - te^{-i\theta})^{-1/2}$$

$$= (1 + \tfrac{1}{2}te^{i\theta} + \tfrac{3}{8}t^2 e^{2i\theta} + \cdots)$$

$$\cdot(1 + \tfrac{1}{2}te^{-i\theta} + \tfrac{3}{8}t^2 e^{-2i\theta} + \cdots), \tag{12.38}$$

with all coefficients *positive*. Our Legendre polynomial, $P_n(\cos\theta)$, still the coefficient of t^n, may now be written as a sum of terms of the form

$$a_m(e^{im\theta} + e^{-im\theta})/2 = a_m\cos m\theta \tag{12.39a}$$

---

[4] In spherical polar coordinates the inversion of the point (r, θ, φ) through the origin is accomplished by the transformation $[r \to r, \theta \to \pi - \theta$, and $\varphi \to \varphi \pm \pi]$. Then, $\cos\theta \to \cos(\pi - \theta) = -\cos\theta$, corresponding to $x \to -x$ (compare Exercise 2.5.8).

with all the a_m *positive* . Then

$$P_n(\cos \theta) = \sum_{m=0 \text{ or } 1}^{n} a_m \cos m\theta. \tag{12.39b}$$

This series (Eq. (12.39b)) is clearly a maximum when $\theta = 0$ and $\cos m\theta = 1$. But for $x = \cos \theta = 1$, Eq. (12.31) shows that $P_n(1) = 1$. Therefore

$$|P_n(\cos \theta)| \le P_n(1) = 1. \tag{12.39c}$$

A fringe benefit of Eq. (12.39b) is that it shows that our Legendre polynomial is a linear combination of $\cos m\theta$. This means that the Legendre polynomials form a complete set for any functions that may be expanded by a Fourier cosine series (Section 14.1) over the interval $(0, \pi)$.

In this section various useful properties of the Legendre polynomials are derived from the generating function, Eq. (12.4). The explicit series representation, Eq. (12.8), offers an alternate and sometimes superior approach.

EXERCISES

12.2.1   Given the series

$$\alpha_0 + \alpha_2 \cos^2 \theta + \alpha_4 \cos^4 \theta + \alpha_6 \cos^6 \theta = a_0 P_0 + a_2 P_2 + a_4 P_4 + a_6 P_6.$$

Express the coefficients α_i as a column vector **α** and the coefficients a_i as a column vector **a** and determine the matrices A and B such that

$$\text{A}\boldsymbol{\alpha} = \text{a} \quad \text{and} \quad \text{B}\text{a} = \boldsymbol{\alpha}.$$

Check your computation by showing that AB = **1** (unit matrix). Repeat for the odd case

$$\alpha_1 \cos \theta + \alpha_3 \cos^3 \theta + \alpha_5 \cos^5 \theta + \alpha_7 \cos^7 \theta = a_1 P_1 + a_3 P_3 + a_5 P_5 + a_7 P_7.$$

Note. $P_n(\cos \theta)$ and $\cos^n \theta$ are tabulated in terms of each other in AMS-55.

12.2.2   By differentiating the generating function, $g(t, x)$, with respect to t, multiplying by $2t$, and then adding $g(t, x)$, show that

$$\frac{1 - t^2}{(1 - 2tx + t^2)^{3/2}} = \sum_{n=0}^{\infty} (2n + 1)P_n(x)t^n.$$

This result is useful in calculating the charge induced on a grounded metal sphere by a point charge q.

12.2.3   (a) Derive Eq. (12.27)

$$(1 - x^2)P_n'(x) = (n + 1)xP_n(x) - (n + 1)P_{n+1}(x).$$

(b) Write out the relation of Eq. (12.27) to preceding equations in symbolic form analogous to the symbolic forms for Eqs. (12.23) to (12.26).

12.2.4   A point electric octupole may be constructed by placing a point electric quadrupole (pole strength $p^{(2)}$ in the z-direction) at $z = a$ and an equal but opposite point electric quadrupole at $z = 0$ and then letting $a \to 0$, subject to $p^{(2)}a$ = constant. Find the electrostatic potential corresponding to a point electric octupole. Show from the construction of the point electric octupole that the corresponding potential may be obtained by differentiating the point quadrupole potential.

12.2.5   Operating in *spherical polar coordinates*, show that

$$\frac{\partial}{\partial z}\left[\frac{P_n(\cos\theta)}{r^{n+1}}\right] = -(n+1)\frac{P_{n+1}(\cos\theta)}{r^{n+2}}.$$

This is the key step in the mathematical argument that the derivative of one multipole leads to the next higher multipole.
Hint. Compare Exercise 2.5.12.

12.2.6   From

$$P_L(\cos\theta) = \frac{1}{L!}\frac{\partial^L}{\partial t^L}(1 - 2t\cos\theta + t^2)^{-1/2}\big|_{t=0}$$

show that

$$P_L(1) = 1, \qquad P_L(-1) = (-1)^L.$$

12.2.7   Prove that

$$P_n'(1) = \frac{d}{dx}P_n(x)\big|_{x=1} = \frac{1}{2}n(n+1).$$

12.2.8   Show that $P_n(\cos\theta) = (-1)^n P_n(-\cos\theta)$ by use of the recurrence relation relating P_n, P_{n+1}, and P_{n-1} and your knowledge of P_0 and P_1.

12.2.9   From Eq. (12.38) write out the coefficient of t^2 in terms of $\cos n\theta$, $n \leq 2$. This coefficient is $P_2(\cos\theta)$.

12.2.10   Write a program that will generate the coefficients a_s in the polynomial form of the Legendre polynomial,

$$P_n(x) = \sum_{s=0}^{n} a_s x^s.$$

12.2.11   (a) Calculate $P_{10}(x)$ over the range $[0, 1]$ and plot your results.
(b) Calculate precise (at least to five decimal places) values of the five positive roots of $P_{10}(x)$. Compare your values with the values listed in AMS-55 (Table 25.4).

Hint. See Appendix 1 for root-finding techniques.

12.2.12   (a) Calculate the *largest* root of $P_n(x)$ for $n = 2(1)50$.
(b) Develop an approximation for the largest root from the hypergeometric representation of $P_n(x)$ (Section 13.4) and compare your values from part (a) with your hypergeometric approximation. Compare also with the values listed in AMS-55 (Table 25.4).

12.2.13   (a) From Exercise 12.2.1 and AMS-55 (Table 22.9) develop the 6×6 matrix B that will transform a series of even order Legendre polynomials through $P_{10}(x)$ into a power series $\sum_{n=0}^{5}\alpha_{2n}x^{2n}$.
(b) Calculate A as B^{-1}. Check the elements of A against the values listed in AMS-55 (Table 22.9).
(c) By using matrix multiplication, transform some even power series $\sum_{n=0}^{5}\alpha_{2n}x^{2n}$ into a Legendre series.

12.2.14   Write a subroutine that will transform a finite power series $\sum_{n=0}^{N} a_n x^n$ into a Legendre series $\sum_{n=0}^{N} b_n P_n(x)$. Use the recurrence relation Eq. (12.17) and follow the technique outlined in Section 13.3 for a Chebyshev series.

12.3 ORTHOGONALITY

Legendre's differential equation (12.28) may be written in the form

$$\frac{d}{dx}[(1 - x^2)P_n'(x)] + n(n + 1)P_n(x) = 0, \tag{12.40}$$

showing clearly that it is self-adjoint. Subject to satisfying certain boundary conditions, then, it is known that the solutions $P_n(x)$ will be orthogonal. Upon comparing Eq. (12.40) with (9.6) and (9.8) we see that the weight function $w(x) = 1$, $\mathcal{L} = (d/dx)(1 - x^2)(d/dx)$, $p(x) = 1 - x^2$ and the eigenvalue $\lambda = n(n + 1)$. The integration limits on x are ± 1, where $p(\pm 1) = 0$. Then for $m \neq n$, Eq. (9.34) becomes

$$\int_{-1}^{1} P_n(x)P_m(x)\, dx = 0,^1 \tag{12.41}$$

$$\int_{0}^{\pi} P_n(\cos \theta)P_m(\cos \theta) \sin \theta\, d\theta = 0, \tag{12.42}$$

showing that $P_n(x)$ and $P_m(x)$ are orthogonal for the interval $[-1, 1]$. This orthogonality may also be demonstrated quite readily by using Rodrigues' definition of $P_n(x)$ (compare Section 12.4, Exercise 12.4.2).

We shall need to evaluate the integral (Eq. (12.41)) when $n = m$. Certainly it is no longer zero. From our generating function

$$(1 - 2tx + t^2)^{-1} = \left[\sum_{n=0}^{\infty} P_n(x)t^n \right]^2. \tag{12.43}$$

Integrating from $x = -1$ to $x = +1$, we have

$$\int_{-1}^{1} \frac{dx}{1 - 2tx + t^2} = \sum_{n=0}^{\infty} t^{2n} \int_{-1}^{1} [P_n(x)]^2\, dx; \tag{12.44}$$

the cross terms in the series vanish by means of Eq. (12.42). Using $y = 1 - 2tx + t^2$, we obtain

$$\int_{-1}^{1} \frac{dx}{1 - 2tx + t^2} = \frac{1}{2t} \int_{(1-t)^2}^{(1+t)^2} \frac{dy}{y} = \frac{1}{t} \ln\left(\frac{1 + t}{1 - t}\right). \tag{12.45}$$

---

[1] In Section 9.4 such integrals are intepreted as inner products in a linear vector (function) space. Alternate notations are

$$\int_{-1}^{1} P_n(x)P_m(x)\, dx \equiv \langle P_n(x) | P_m(x) \rangle$$

$$\equiv (P_n(x), P_m(x)).$$

The $\langle \ \rangle$ form, popularized by Dirac, is common in physics literature. The ( ) form is more common in the mathematics literature.

Expanding this in a power series (Exercise 5.4.1) gives us

$$\frac{1}{t} \ln\left(\frac{1 + t}{1 - t}\right) = 2 \sum_{n=0}^{\infty} \frac{t^{2n}}{2n + 1}. \tag{12.46}$$

Since our power-series representation is known to be unique, we must have

$$\int_{-1}^{1} [P_n(x)]^2 \, dx = \frac{2}{2n + 1}. \tag{12.47}$$

Combining Eq. (12.42) with (12.47) we have the orthonormality condition

$$\int_{-1}^{1} P_m(x)P_n(x) \, dx = \frac{2\delta_{m,n}}{2n + 1}. \tag{12.48}$$

We shall return to this result in Section 12.6 when we construct the orthonormal spherical harmonics.

EXPANSION OF FUNCTIONS, LEGENDRE SERIES

In addition to orthogonality, the Sturm–Liouville theory shows that the Legendre polynomials form a complete set. Let us assume, then, that the series

$$\sum_{n=0}^{\infty} a_n P_n(x) = f(x), \tag{12.49}$$

in the sense of convergence in the mean (Section 9.4) in the interval $[-1, 1]$. This demands that $f(x)$ and $f'(x)$ be at least sectionally continuous in this interval. The coefficients a_n are found by multiplying the series by $P_m(x)$ and integrating term by term. Using the orthogonality property expressed in Eqs. (12.42) and (12.48), we obtain

$$\frac{2}{2m + 1} a_m = \int_{-1}^{1} f(x)P_m(x) \, dx. \tag{12.50}$$

We replace the variable of integration x by t and the index m by n. Then, substituting into Eq. (12.49), we have

$$f(x) = \sum_{n=0}^{\infty} \frac{2n + 1}{2} \left(\int_{-1}^{1} f(t)P_n(t) \, dt\right) P_n(x). \tag{12.51}$$

This expansion in a series of Legendre polynomials is usually referred to as a Legendre series.[2] Its properties are quite similar to the more familiar Fourier series (Chapter 14). In particular, we can use the orthogonality property, (Eq. (12.48)), to show that the series is unique.

On a more abstract (and more powerful) level, Eq. (12.51) gives the representation of $f(x)$ in the linear vector space of Legendre polynomials (a Hilbert space, Section 9.4).

---

[2] Note that Eq. (12.50) gives a_m as a *definite* integral, that is, a number for a given $f(x)$.

From the viewpoint of integral transforms (Chapter 15) Eq. (12.50) may be considered a finite Legendre transform of $f(x)$. Equation (12.51) is then the inverse transform. It may also be interpreted in terms of the *projection operators* of quantum theory. We may take

$$\mathcal{P}_m \equiv P_m(x) \frac{2m+1}{2} \int_{-1}^{1} P_m(t)[\ ]\, dt$$

as an (integral) operator, ready to operate on $f(t)$. [The $f(t)$ would go in the square bracket as a factor in the integrand.] Then, from Eq. (12.50)

$$\mathcal{P}_m f(t) = a_m P_m(x).^3$$

The operator $\mathcal{P}_m$ projects out the mth component of the function f.

Equation (12.3), which leads directly to the generating function definition of Legendre polynomials, is a Legendre expansion of $1/r_1$. This Legendre expansion of $1/r_1$ or $1/r_{12}$ appears in several exercises of Section 12.8. Going beyond a simple Coulomb field, the $1/r_{12}$ is often replaced by a potential $V(|\mathbf{r}_1 - \mathbf{r}_2|)$ and the solution of the problem is again effected by a Legendre expansion.

The Legendre series, Eq. (12.49), has been treated as a *known* function $f(x)$ that we arbitrarily chose to expand in a series of Legendre polynomials. Sometimes the origin and nature of the Legendre series is different. In the next examples we consider *unknown* functions we know can be represented by a Legendre series because of the differential equation the unknown functions satisfy. As before, the problem is to determine the unknown coefficients in the series expansion. Here, however, the coefficients are not found by Eq. (12.50). Rather, they are determined by demanding that the Legendre series match a known solution at a boundary. These are boundary value problems.

Example 12.3.1 Earth's Gravitational Field

An example of a Legendre series is provided by the description of the earth's gravitational potential U (for exterior points), neglecting azimuthal effects. With

$$R = \text{equatorial radius}$$

$$= 6378.1 \pm 0.1 \text{ km}$$

$$\frac{GM}{R} = 62.494 \pm 0.001 \text{ km}^2/\text{sec}^2,$$

we write

$$U(r, \theta) = \frac{GM}{R}\left[\frac{R}{r} - \sum_{n=2}^{\infty} a_n \left(\frac{R}{r}\right)^{n+1} P_n(\cos\theta)\right], \qquad (12.52)$$

---

[3] The dependent variables are arbitrary. Here x came from the x in $\mathcal{P}_m$ while t is a dummy variable of integration.

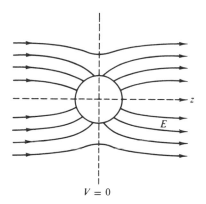

$V = 0$

Figure 12.8 Conducting sphere in a uniform field.

a Legendre series. Artificial satellite motions have shown that

$$a_2 = (1,082,635 \pm 11) \times 10^{-9},$$

$$a_3 = (-2,531 \pm 7) \times 10^{-9},$$

$$a_4 = (-1,600 \pm 12) \times 10^{-9}.$$

This is the famous pear-shaped deformation of the earth. Other coefficients have been computed through $n = 20$. The reader might note that P_1 is omitted, since it would represent a displacement and not a deformation.

More recent satellite data permit a determination of the longitudinal dependence of the earth's gravitational field. Such dependence may be described by a Laplace series (Section 12.6).

Example 12.3.2 Sphere in a Uniform Field

Another illustration of the use of Legendre polynomials is provided by the problem of a neutral conducting sphere (radius r_0) placed in a (previously) uniform electric field (Fig. 12.8). The problem is to find the new, perturbed, electrostatic potential. Calling the electrostatic potential[4] V,

$$\nabla^2 V = 0, \tag{12.53}$$

Laplace's equation. We select spherical polar coordinates because of the spherical shape of the conductor. (This will simplify the application of the boundary condition at the surface of the conductor.) Separating variables and glancing at Table 8.2 if necessary, we can write the unknown potential $V(r, \theta)$ in the region outside the sphere as a linear combination of solutions,

$$V(r, \theta) = \sum_{n=0}^{\infty} a_n r^n P_n(\cos \theta) + \sum_{n=0}^{\infty} b_n \frac{P_n(\cos \theta)}{r^{n+1}}. \tag{12.54}$$

---

[4] It should be emphasized that this is not a presentation of a Legendre series expansion of a known $V(\cos \theta)$. Here we are back to *boundary value* problems.

No φ-dependence appears because of the axial symmetry of our problem. (The center of the conducting sphere is taken as the origin and the z-axis is oriented parallel to the original uniform field.)

It might be noted here that n is an integer, *because* only for integral n is the θ dependence well behaved at $\cos \theta = \pm 1$. For nonintegral n the solutions of Legendre's equation diverge at the ends of the interval $[-1, 1]$, the poles $\theta = 0$, π of the sphere (compare Example 5.2.4 and Exercises 5.2.15 and 8.5.5). It is for this same reason that the *second* solution of Legendre's equation, Q_n, is also excluded.

Now we turn to our (Dirichlet) boundary conditions to determine the unknown a_n's and b_n's of our series solution, Eq. (12.54). If the original unperturbed electrostatic field is E_0, we require, as one boundary condition,

$$V(r \to \infty) = -E_0 z = -E_0 r \cos \theta$$

$$= -E_0 r P_1(\cos \theta). \tag{12.55}$$

Since our Legendre series is unique, we may equate coefficients of $P_n(\cos \theta)$ in Eq. (12.54) $(r \to \infty)$ and Eq. (12.55) to obtain

$$a_n = 0, \qquad n > 1 \text{ and } n = 0,$$

$$a_1 = -E_0. \tag{12.56}$$

If $a_n \neq 0$ for $n > 1$, these terms would dominate at large r and the boundary condition (Eq. (12.55)) could not be satisfied.

As a second boundary condition, we may choose the conducting sphere and the plane $\theta = \pi/2$ to be at zero potential, which means that Eq. (12.54) now becomes

$$V(r = r_0) = \frac{b_0}{r_0} + \left(\frac{b_1}{r_0^2} - E_0 r_0 \right) P_1(\cos \theta) + \sum_{n=2}^{\infty} b_n \frac{P_n(\cos \theta)}{r_0^{n+1}} = 0 \tag{12.57}$$

In order that this may hold for all values of θ, each coefficient of $P_n(\cos \theta)$ must vanish.[5] Hence

$$b_0 = 0,[6]$$

$$b_n = 0, \qquad n \geq 2, \tag{12.58}$$

whereas

$$b_1 = E_0 r_0^3. \tag{12.59}$$

---

[5] Again, this is equivalent to saying that a series expansion in Legendre polynomials (or any complete orthogonal set) is unique.

[6] The coefficient of P_0 is b_0/r_0. We set $b_0 = 0$ since there is no net charge on the sphere. If there is a net charge q, then $b_0 \neq 0$.

The electrostatic potential (outside the sphere) is then

$$V = -E_0 r P_1(\cos \theta) + \frac{E_0 r_0^3}{r^2} P_1(\cos \theta)$$

$$= -E_0 r P_1(\cos \theta)\left(1 - \frac{r_0^3}{r^3}\right). \tag{12.60}$$

In Section 1.16 it was shown that a solution of Laplace's equation that satisfied the boundary conditions over the entire boundary was unique. The electrostatic potential V, as given by Eq. (12.60), is a solution of Laplace's equation. It satisfies our boundary conditions and therefore is the solution of Laplace's equation for this problem.

It may further be shown (Exercise 12.3.13) that there is an induced surface charge density

$$\sigma = -\varepsilon_0 \left.\frac{\partial V}{\partial r}\right|_{r=r_0} = 3\varepsilon_0 E_0 \cos \theta \tag{12.61}$$

on the surface of the sphere and an induced electric dipole moment (Exercise 12.3.13)

$$P = 4\pi r_0^3 \varepsilon_0 E_0. \tag{12.62}$$

Example 12.3.3 Electrostatic Potential of a Ring of Charge

As a further example, consider the electrostatic potential produced by a conducting ring carrying a total electric charge q (Fig. 12.9). From electrostatics (and Section 1.14) the potential ψ satisfies Laplace's equation. Separating variables in spherical polar coordinates (compare Table 8.1), we obtain

$$\psi(r, \theta) = \sum_{n=0}^{\infty} a_n \frac{a^n}{r^{n+1}} P_n(\cos \theta), \qquad r > a. \tag{12.63a}$$

Here a is the radius of the ring that is assumed to be in the $\theta = \pi/2$ plane. There is no φ (azimuthal) dependence because of the cylindrical symmetry of the system.

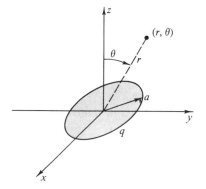

Figure 12.9 Charged, conducting ring.

The terms with positive exponent radial dependence have been rejected since the potential must have an asymptotic behavior

$$\psi \sim \frac{q}{4\pi\varepsilon_0} \cdot \frac{1}{r}, \qquad r \gg a. \tag{12.63b}$$

The problem is to determine the coefficients a_n in Eq. (12.63a). This may be done by evaluating $\psi(r, \theta)$ at $\theta = 0$, $r = z$, and comparing with an independent calculation of the potential from Coulomb's law. In effect, we are using a boundary condition along the z-axis. From Coulomb's law (with all charge equidistant),

$$\psi(r, \theta) = \frac{q}{4\pi\varepsilon_0} \cdot \frac{1}{(z^2 + a^2)^{1/2}}, \qquad \begin{cases} \theta = 0 \\ r = z, \end{cases}$$

$$= \frac{q}{4\pi\varepsilon_0 z} \sum_{s=0}^{\infty} (-1)^s \frac{(2s)!}{2^{2s}(s!)^2} \left(\frac{a}{z}\right)^{2s}, \qquad z > a. \tag{12.63c}$$

The last step uses the result of Exercise 10.1.15. Now, Eq. (12.63a) evaluated at $\theta = 0$, $r = z$ (with $P_n(1) = 1$), yields

$$\psi(r, \theta) = \sum_{n=0}^{\infty} a_n \frac{a^n}{z^{n+1}}, \qquad r = z. \tag{12.63d}$$

Comparing Eqs. (12.63c) and (12.63d), we get $a_n = 0$ for n odd. Setting $n = 2s$, we have

$$a_{2s} = \frac{q}{4\pi\varepsilon_0} (-1)^s \frac{(2s)!}{2^{2s}(s!)^2}, \tag{12.63e}$$

and our electrostatic potential $\psi(r, \theta)$ is given by

$$\psi(r, \theta) = \frac{q}{4\pi\varepsilon_0 r} \sum_{s=0}^{\infty} (-1)^s \frac{(2s)!}{2^{2s}(s!)^2} \left(\frac{a}{r}\right)^{2s} P_{2s}(\cos\theta), \qquad r > a. \tag{12.63f}$$

The magnetic analog of this problem appears in Section 12.5—Example 12.5.1.

EXERCISES

12.3.1   You have constructed a set of orthogonal functions by the Gram–Schmidt process (Section 9.3), taking $u_n(x) = x^n$, $n = 0, 1, 2, \ldots$, in increasing order with $w(x) = 1$ and an interval $-1 \le x \le 1$. Prove that the nth such function constructed is proportional to $P_n(x)$.
Hint. Use mathematical induction.

12.3.2   Expand the Dirac delta function in a series of Legendre polynomials, using the interval $-1 \le x \le 1$.

12.3.3   Verify the Dirac delta function expansions

$$\delta(1 - x) = \sum_{n=0}^{\infty} \frac{2n + 1}{2} P_n(x)$$

$$\delta(1 + x) = \sum_{n=0}^{\infty} (-1)^n \frac{2n + 1}{2} P_n(x).$$

These expressions appear in a resolution of the Rayleigh plane wave expansion (Exercise 12.4.7) into incoming and outgoing spherical waves. *Note.* Assume that the *entire* Dirac delta function is covered when integrating over $[-1, 1]$.

12.3.4   Neutrons (mass 1) are being scattered by a nucleus of mass $A(A > 1)$. In the center of the mass system the scattering is isotropic. Then, in the lab system the average of the cosine of the angle of deflection of the neutron is

$$\langle \cos \psi \rangle = \frac{1}{2} \int_0^\pi \frac{A \cos \theta + 1}{(A^2 + 2A \cos \theta + 1)^{1/2}} \sin \theta \, d\theta.$$

Show, by expansion of the denominator, that $\langle \cos \psi \rangle = 2/3A$.

12.3.5   A particular function $f(x)$ defined over the interval $[-1, 1]$ is expanded in a Legendre series over this same interval. Show that the expansion is unique.

12.3.6   A function $f(x)$ is expanded in a Legendre series $f(x) = \sum_{n=0}^{\infty} a_n P_n(x)$. Show that

$$\int_{-1}^{1} [f(x)]^2 \, dx = \sum_{n=0}^{\infty} 2 \frac{a_n^2}{2n + 1}.$$

This is the Legendre form of the Fourier series Parseval identity, Exercise 14.4.2. It also illustrates Bessel's inequality, Eq. (9.72), becoming an equality for a complete set.

12.3.7   Derive the recurrence relation

$$(1 - x^2)P_n'(x) = nP_{n-1}(x) - nxP_n(x)$$

from the Legendre polynomial generating function.

12.3.8   Evaluate $\int_0^1 P_n(x) \, dx$.

   ANS. $n = 2s$;   1 for $s = 0$, 0 for $s > 0$,
        $n = 2s + 1$; $P_{2s}(0)/(2s + 2) = (-1)^s(2s - 1)!!/(2s + 2)!!$

   Hint. Use a recurrence relation to replace $P_n(x)$ by derivatives and then integrate by inspection! Alternatively, you can integrate the generating function.

12.3.9   (a) For

$$f(x) = \begin{cases} +1, & 0 < x < 1 \\ -1, & -1 < x < 0, \end{cases}$$

   show that

$$\int_{-1}^{1} [f(x)]^2 \, dx = 2 \sum_{n=0}^{\infty} (4n + 3) \left[\frac{(2n - 1)!!}{(2n + 2)!!} \right]^2.$$

   (b) By testing the series, prove that the series is convergent.

12.3.10   Prove that

$$\int_{-1}^{1} x(1 - x^2) P_n' P_m' \, dx = 0, \qquad\qquad \text{unless } m = n \pm 1,$$

$$= \frac{2n(n^2 - 1)}{4n^2 - 1} \delta_{m,n-1}, \qquad \text{if } m < n.$$

$$= \frac{2n(n + 2)(n + 1)}{(2n + 1)(2n + 3)} \delta_{m,n+1}, \qquad \text{if } m > n.$$

12.3.11   The amplitude of a scattered wave is given by

$$f(\theta) = \lambdabar \sum_{l=0}^{\infty} (2l + 1) \exp[i\delta_l] \sin \delta_l P_l(\cos \theta).$$

Here θ is the angle of scattering, l the angular momentum, and δ_l the phase shift produced by the central potential that is doing the scattering. The total cross section is $\sigma_{\text{tot}} = \int f^*(\theta) f(\theta) \, d\Omega$. Show that

$$\sigma_{\text{tot}} = 4\pi\lambdabar^2 \sum_{l=0}^{\infty} (2l + 1) \sin^2 \delta_l.$$

12.3.12   The coincidence counting rate, $W(\theta)$, in a gamma-gamma angular correlation experiment has the form

$$W(\theta) = \sum_{n=0}^{\infty} a_{2n} P_{2n}(\cos \theta).$$

Show that data in the range $\pi/2 \leq \theta \leq \pi$ can, in principle, define the function, $W(\theta)$ (and permit a determination of the coefficients a_{2n}). This means that although data in the range $0 \leq \theta < \pi/2$ may be useful as a check, they are not essential.

12.3.13   A conducting sphere of radius r_0 is placed in an initially uniform electric field, $\mathbf{E}_0$. Show the following:

(a) The induced surface charge density is

$$\sigma = 3\varepsilon_0 E_0 \cos \theta.$$

(b) The induced electric dipole moment is

$$P = 4\pi r_0^3 \varepsilon_0 E_0.$$

The induced electric dipole moment can be calculated either from the surface charge [part (a)], or by noting that the final electric field $\mathbf{E}$ is the result of superimposing a dipole field on the original uniform field.

12.3.14   A charge q is displaced a distance a along the z-axis from the center of a spherical cavity of radius R.

(a) Show that the electric field averaged over the volume $a \leq r \leq R$ is zero.

(b) Show that the electric field averaged over the volume $0 \leq r \leq a$ is

$$\mathbf{E} = \hat{z} E_z = -\hat{z} \frac{q}{4\pi\varepsilon_0 a^2} \qquad \text{(SI units)}$$

$$= -\hat{z} \frac{nqa}{3\varepsilon_0},$$

where n is the number of such displaced charges per unit volume. This is a basic calculation in the polarization of a dielectric.

Hint. $\mathbf{E} = -\nabla\varphi$.

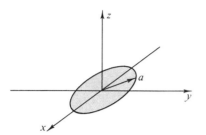

Figure 12.10 Charged, conducting disk.

12.3.15   Determine the electrostatic potential (Legendre expansion) of a circular ring of electric charge for $r < a$.

12.3.16   Calculate the electric field produced by the charged conducting ring of Example 12.3.3 for

(a) $r > a$,
(b) $r < a$.

12.3.17   As an extension of Example 12.3.3, find the potential $\psi(r, \theta)$ produced by a charged conducting disk, Fig. 12.10, for $r > a$, the radius of the disk.
   The charge density σ (on each side of the disk) is

$$\sigma(\rho) = \frac{q}{4\pi a(a^2 - \rho^2)^{1/2}}, \qquad \rho^2 = x^2 + y^2.$$

Hint. The definite integral you get can be evaluated as a beta function, Section 10.4. For more details see Section 5.03 of Smythe in Additional Readings.

$$\text{ANS.} \quad \psi(r, \theta) = \frac{q}{4\pi\varepsilon_0 r} \sum_{l=0}^{\infty} (-1)^l \frac{1}{2l+1} \left(\frac{a}{r}\right)^{2l} P_{2l}(\cos \theta).$$

12.3.18   From the result of Exercise 12.3.17 calculate the potential of the disk. Since you are violating the condition $r > a$, justify your calculation carefully.
   Hint. You may run into the series given in Exercise 5.2.9.

12.3.19   The hemisphere defined by $r = a$, $0 \leq \theta < \pi/2$ has an electrostatic potential $+V_0$. The hemisphere $r = a$, $\pi/2 < \theta \leq \pi$ has an electrostatic potential $-V_0$. Show that the potential at *interior* points is

$$V = V_0 \sum_{n=0}^{\infty} \frac{4n+3}{2n+2} \left(\frac{r}{a}\right)^{2n+1} P_{2n}(0) P_{2n+1}(\cos \theta)$$

$$= V_0 \sum_{n=0}^{\infty} (-1)^n \frac{(4n+3)(2n-1)!!}{(2n+2)!!} \left(\frac{r}{a}\right)^{2n+1} P_{2n+1}(\cos \theta).$$

Hint. You need Exercise 12.3.8.

12.3.20   A conducting sphere of radius a is divided into two electrically separate hemispheres by a thin insulating barrier at its equator. The top hemisphere is maintained at a potential V_0, the bottom hemisphere at $-V_0$.

(a) Show that the electrostatic potential *exterior* to the two hemispheres is

$$V(r, \theta) = V_0 \sum_{s=0}^{\infty} (-1)^s (4s + 3) \frac{(2s-1)!!}{(2s+2)!!} \left(\frac{a}{r}\right)^{2s+2} P_{2s+1}(\cos \theta).$$

(b) Calculate the electric charge density σ on the outside surface. Note that your series diverges at $\cos \theta = \pm 1$ as you expected from the inifite capacitance of this system (zero thickness for the insulating barrier).

$$ANS. \quad \sigma = \varepsilon_0 E_n = -\varepsilon_0 \frac{\partial V}{\partial r}\bigg|_{r=a}$$

$$= \varepsilon_0 V_0 \sum_{s=0}^{\infty} (-1)^s (4s + 3) \frac{(2s-1)!!}{(2s)!!} P_{2s+1}(\cos \theta).$$

12.3.21   In the notation of Section 9.4 $\langle x | \varphi_s \rangle = \sqrt{(2s + 1)/2} P_s(x)$, a Legendre polynomial is renormalized to unity. Explain how $|\varphi_s\rangle\langle\varphi_s|$ acts as a projection operator. In particular, show that if $|f\rangle = \sum_n a'_n |\varphi_n\rangle$, then

$$|\varphi_s\rangle\langle\varphi_s | f\rangle = a'_s |\varphi_s\rangle.$$

12.3.22   Expand x^8 as a Legendre series. Determine the Legendre coefficients from Eq. (12.50),

$$a_m = \frac{2m + 1}{2} \int_{-1}^{1} x^8 P_m(x)\, dx.$$

Check your values against AMS-55, Table 22.9. This illustrates the expansion of a simple function. Actually if $f(x)$ is expressed as a power series, the technique of Exercise 12.2.14 is both faster and more accurate.
Hint. Gaussian quadrature can be used to evaluate the integral.

12.3.23   Calculate and tabulate the electrostatic potential created by a ring of charge, Example 12.3.3, for $r/a = 1.5(0.5)5.0$ and $\theta = 0°(15°)90°$. Carry terms through $P_{22}(\cos \theta)$.
Note. The convergence of your series will be slow for $r/a = 1.5$. Truncating the series at P_{22} limits you to about a four significant figure accuracy.
            Check value.   For $r/a = 2.5$ and $\theta = 60°$, $\psi = 0.40272(q/4\pi\varepsilon_0 r)$.

12.3.24   Calculate and tabulate the electrostatic potential created by a charged disk, Exercise 12.3.17, for $r/a = 1.5(0.5)5.0$ and $\theta = 0°(15°)90°$. Carry terms through $P_{22}(\cos \theta)$.
            Check value.   For $r/a = 2.0$ and $\theta = 15°$, $\psi = 0.46638(q/4\pi\varepsilon_0 r)$.

12.3.25   Calculate the first five (nonvanishing) coefficients in the Legendre series expansion of $f(x) = 1 - |x|$ using Eq. (12.51)—numerical integration. Actually these coefficients can be obtained in closed form. Compare your coefficients with those obtained from Exercise 13.4.4.
ANS. $a_0 = 0.5000$, $a_2 = -0.6250$, $a_4 = 0.1875$, $a_6 = -0.1016$, $a_8 = 0.0664$.

12.3.26   Calculate and tabulate the exterior electrostatic potential created by the two charged hemispheres of Exercise 12.3.20, for $r/a = 1.5(0.5)5.0$ and $\theta = 0°(15°)90°$. Carry terms through $P_{23}(\cos \theta)$.
            Check value.   For $r/a = 2.0$ and $\theta = 45°$, $V = 0.27066V_0$.

12.3.27   (a) Given $f(x) = 2.0$, $|x| < 0.5$; 0, $0.5 < |x| < 1.0$. Expand $f(x)$ in a Legendre series and calculate the coefficients a_n through a_{80} (analytically).
(b) Evaluate $\sum_{n=0}^{80} a_n P_n(x)$ for $x = 0.400(0.005)0.600$. Plot your results.

Note. This illustrates the Gibbs phenomenon of Section 14.5 and the danger of trying to calculate with a series expansion in the vicinity of a discontinuity.

12.4 ALTERNATE DEFINITIONS OF LEGENDRE POLYNOMIALS

RODRIGUES' FORMULA

The series form of the Legendre polynomials (Eq. (12.8)) of Section 12.1 may be transformed as follows. From Eq. (12.8)

$$P_n(x) = \sum_{r=0}^{[n/2]} (-1)^r \frac{(2n - 2r)!}{2^n r!(n - r)!(n - 2r)!} x^{n-2r}. \tag{12.64}$$

For n an integer

$$P_n(x) = \sum_{r=0}^{[n/2]} (-1)^r \frac{1}{2^n r!(n - r)!} \left(\frac{d}{dx}\right)^n x^{2n-2r}$$

$$= \frac{1}{2^n n!} \left(\frac{d}{dx}\right)^n \sum_{r=0}^{n} \frac{(-1)^r n!}{r!(n - r)!} x^{2n-2r}. \tag{12.64a}$$

Note the extension of the upper limit. The reader is asked to show in Exercise 12.4.1 that the additional terms $[n/2] + 1$ to n in the summation contribute nothing. However, the effect of these extra terms is to permit the replacement of the new summation by $(x^2 - 1)^n$ (binomial theorem once again) to obtain

$$P_n(x) = \frac{1}{2^n n!} \left(\frac{d}{dx}\right)^n (x^2 - 1)^n. \tag{12.65}$$

This is Rodrigues' formula. It is useful in proving many of the properties of the Legendre polynomials such as orthogonality. A related application is seen in Exercise 12.4.3. The Rodrigues definition is extended in Section 12.5 to define the associated Legendre functions. In Section 12.7 it is used to identify the orbital angular momentum eigenfunctions.

SCHLAEFLI INTEGRAL

Rodrigues' formula provides a means of developing an integral representation of $P_n(z)$. Using Cauchy's integral formula (Section 6.4)

$$f(z) = \frac{1}{2\pi i} \oint \frac{f(t)}{t - z} dt \tag{12.66}$$

with

$$f(z) = (z^2 - 1)^n, \tag{12.67}$$

we have

$$(z^2 - 1)^n = \frac{1}{2\pi i} \oint \frac{(t^2 - 1)^n}{t - z} dt. \tag{12.68}$$

Differentiating n times with respect to z and multiplying by $1/2^n n!$ gives

$$P_n(z) = \frac{1}{2^n n!} \frac{d^n}{dz^n} (z^2 - 1)^n = \frac{2^{-n}}{2\pi i} \oint \frac{(t^2 - 1)^n}{(t - z)^{n+1}} dt, \tag{12.69}$$

with the contour enclosing the point $t = z$.

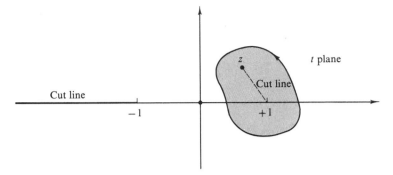

Figure 12.11   Schlaefli integral contour.

This is the Schlaefli integral. Margenau and Murphy[1] use this to derive the recurrence relations we obtained from the generating function.

The Schlaefli integral may readily be shown to satisfy Legendre's equation by differentiation and direct substitution (Fig. 12.11). We obtain

$$(1 - z^2)\frac{d^2 P_n}{dz^2} - 2z\frac{dP_n}{dz} + n(n + 1)P_n = \frac{n + 1}{2^n 2\pi i}\oint \frac{d}{dt}\left[\frac{(t^2 - 1)^{n+1}}{(t - z)^{n+2}}\right] dt.$$

(12.70)

For integral n our function $(t^2 - 1)^{n+1}/(t - z)^{n+2}$ is single-valued, and the integral around the closed path vanishes. The Schlaefli integral may also be used to define $P_\nu(z)$ for nonintegral ν integrating around the points $t = z$, $t = 1$, but not crossing the cut line -1 to $-\infty$. We could equally well encircle the points $t = z$ and $t = -1$, but this would lead to nothing new. A contour about $t = +1$ and $t = -1$ will lead to a second solution $Q_\nu(z)$, Section 12.10.

EXERCISES

12.4.1   Show that *each* term in the summation

$$\sum_{r = [n/2]+1}^{n}\left(\frac{d}{dx}\right)^n \frac{(-1)^r n!}{r!(n - r)!}x^{2n-2r}$$

vanishes (r and n integral).

12.4.2   Using Rodrigues's formula, show that the $P_n(x)$ are orthogonal and that

$$\int_{-1}^{1} [P_n(x)]^2\, dx = \frac{2}{2n + 1}.$$

Hint. Use Rodrigues' formula and integrate by parts.

[1] H. Margenau and G. M. Murphy, *The Mathematics of Physics and Chemistry*, 2nd ed., Princeton, NJ: Van Nostrand (1956), Section 3.5.

12.4.3 Show that $\int_{-1}^{1} x^m P_n(x)\,dx = 0$ when $m < n$.
Hint. Use Rodrigues' formula or expand x^m in Legendre polynomials.

12.4.4 Show that

$$\int_{-1}^{1} x^n P_n(x)\,dx = \frac{2^{n+1} n!\, n!}{(2n+1)!}.$$

Note. You are expected to use Rodrigues' formula and integrate by parts but also see if you can get the result from Eq. (12.8) by inspection.

12.4.5 Show that

$$\int_{-1}^{1} x^{2r} P_{2n}(x)\,dx = \frac{2^{2n+1}(2r)!\,(r+n!)}{(2r+2n+1)!\,(r-n)!}, \qquad r \geq n.$$

12.4.6 As a generalization of Exercises 12.4.4 and 12.4.5, show that the Legendre expansions of x^s are

(a) $$x^{2r} = \sum_{n=0}^{r} \frac{2^{2n}(4n+1)(2r)!\,(r+n)!}{(2r+2n+1)!\,(r-n)!} P_{2n}(x), \qquad s = 2r,$$

(b) $$x^{2r+1} = \sum_{n=0}^{r} \frac{2^{2n+1}(4n+3)(2r+1)!\,(r+n+1)!}{(2r+2n+3)!\,(r-n)!} P_{2n+1}(x), \qquad s = 2r+1.$$

12.4.7 A plane wave may be expanded in a series of spherical waves by the Rayleigh equation

$$e^{ikr\cos\gamma} = \sum_{n=0}^{\infty} a_n j_n(kr) P_n(\cos\gamma).$$

Show that $a_n = i^n(2n+1)$.

Hint. 1. Use the orthogonality of the P_n to solve for $a_n j_n(kr)$.
2. Differentiate n times with respect to (kr) and set $r = 0$ to eliminate the r-dependence.
3. Evaluate the remaining integral by Exercise 12.4.4.

Note. This problem may also be treated by noting that both sides of the equation satisfy the Helmholtz equation. The equality can be established by showing that the solutions have the same behavior at the origin and also behave alike at large distances. A "by inspection" type solution is developed in Section 8.7 using Green's functions.

12.4.8 Verify the Rayleigh equation of Exercise 12.4.7 by starting with the following steps:

1. Differentiate with respect to (kr) to establish

$$\sum_n a_n j_n'(kr) P_n(\cos\gamma) = i \sum_n a_n j_n(kr) \cos\gamma\, P_n(\cos\gamma).$$

2. Use a recurrence relation to replace $\cos\gamma\, P_n(\cos\gamma)$ by a linear combination of P_{n-1} and P_{n+1}.
3. Use a recurrence relation to replace j_n' by a linear combination of j_{n-1} and j_{n+1}.

12.4.9 From Exercise 12.4.7 show that

$$j_n(kr) = \frac{1}{2i^n} \int_{-1}^{1} e^{ikr\mu} P_n(\mu) \, d\mu.$$

This means that (apart from constant factors) the spherical Bessel function $j_n(kr)$ is the Fourier transform of the Legendre polynomial $P_n(\mu)$.

12.4.10 The Legendre polynomials and the spherical Bessel functions are related by

$$j_n(z) = \tfrac{1}{2}(-i)^n \int_0^{\pi} e^{iz\cos\theta} P_n(\cos\theta) \sin\theta \, d\theta, \qquad n = 0, 1, 2, \ldots.$$

Verify this relation by transforming the right-hand side into

$$\frac{z^n}{2^{n+1} n!} \int_0^{\pi} \cos(z\cos\theta) \sin^{2n+1}\theta \, d\theta$$

and using Exercise 11.7.9.

12.4.11 By direct evaluation of the Schlaefli integral show that $P_n(1) = 1$.

12.4.12 Explain why the contour of the Schlaefli integral, Eq. (12.69), is chosen to enclose the points $t = z$ and $t = 1$ when $n \to \nu$, not an integer.

12.4.13 In numerical work (such as the Gauss–Legendre quadrature of Appendix 2) it is useful to establish that $P_n(x)$ has n real zeros in the interior of $[-1, 1]$. Show that this is so.
Hint. Rolle's theorem shows that the first derivative of $(x^2 - 1)^{2n}$ has one zero in the interior of $[-1, 1]$. Extend this argument to the second, third, and ultimately to the nth derivative.

12.5 ASSOCIATED LEGENDRE FUNCTIONS

When Helmholtz's equation is separated in spherical polar coordinates (Section 8.3), one of the separated ordinary differential equations is the associated Legendre equation

$$\frac{1}{\sin\theta} \frac{d}{d\theta}\left(\sin\theta \frac{dP_n^m(\cos\theta)}{d\theta}\right) + \left[n(n+1) - \frac{m^2}{\sin^2\theta}\right] P_n^m(\cos\theta) = 0.$$
$$(12.71)$$

With $x = \cos\theta$, this becomes

$$(1 - x^2)\frac{d^2}{dx^2} P_n^m(x) - 2x\frac{d}{dx} P_n^m(x) + \left[n(n+1) - \frac{m^2}{1-x^2}\right] P_n^m(x) = 0.$$
$$(12.72)$$

Only if the azimuthal separation constant $m^2 = 0$ do we have Legendre's equation, Eq. (12.28).

One way of developing the solution of the associated Legendre equation is to start with the regular Legendre equation and convert it into the associated Legendre equation by using multiple differentiation. These multiple differentiations are suggested by the generation of associated Legendre polynomials, and spherical harmonics of the next Section 12.6 more generally, in Section 4.3 using raising or lowering operators of Eq. (4.69) repeatedly. For their derivative form see Exercise 12.6.8. We take Legendre's equation

$$(1 - x^2)P_n'' - 2xP_n' + n(n + 1)P_n = 0,$$ (12.73)

and with the help of Leibniz's formula[1] differentiate m times. The result is

$$(1 - x^2)u'' - 2x(m + 1)u' + (n - m)(n + m + 1)u = 0,$$ (12.74)

where

$$u \equiv \frac{d^m}{dx^m} P_n(x).$$ (12.75)

Equation (12.74) is not self-adjoint. To put it into self-adjoint form and convert the weighting function to 1, we replace $u(x)$ by

$$v(x) = (1 - x^2)^{m/2}u(x) = (1 - x^2)^{m/2}\frac{d^m P_n(x)}{dx^m}.$$ (12.76)

Solving for u and differentiating, we obtain

$$u' = \left(v' + \frac{mxv}{1 - x^2}\right)(1 - x^2)^{-m/2},$$ (12.77)

$$u'' = \left[v'' + \frac{2mxv'}{1 - x^2} + \frac{mv}{1 - x^2} + \frac{m(m + 2)x^2v}{(1 - x^2)^2}\right] \cdot (1 - x^2)^{-m/2}.$$ (12.78)

Substituting into Eq. (12.74), we find that the new function v satisfies the differential equation

$$(1 - x^2)v'' - 2xv' + \left[n(n + 1) - \frac{m^2}{1 - x^2}\right]v = 0,$$ (12.79)

which is the associated Legendre equation reducing to Legendre's equation, as it must when m is set equal to zero. Expressed in spherical polar coordinates, the associated Legendre equation is

$$\frac{1}{\sin \theta} \frac{d}{d\theta}\left(\sin \theta \frac{dv}{d\theta}\right) + \left[n(n + 1) - \frac{m^2}{\sin^2 \theta}\right]v = 0.$$ (12.80)

ASSOCIATED LEGENDRE FUNCTIONS

The regular solutions, relabeled $P_n^m(x)$, are

$$v \equiv P_n^m(x) = (1 - x^2)^{m/2}\frac{d^m}{dx^m} P_n(x).$$ (12.81)

[1] Leibniz's formula for the nth derivative of a product is

$$\frac{d^n}{dx^n}[A(x)B(x)] = \sum_{s=0}^{n} \binom{n}{s} \frac{d^{n-s}}{dx^{n-s}}A(x)\frac{d^s}{dx^s}B(x), \qquad \binom{n}{s} = \frac{n!}{(n - s)!s!},$$

a binomial coefficient.

Table 12.2 Associated Legendre Functions

$$P_1^1(x) = (1 - x^2)^{1/2} = \sin \theta$$

$$P_2^1(x) = 3x(1 - x^2)^{1/2} = 3 \cos \theta \sin \theta$$

$$P_2^2(x) = 3(1 - x^2) = 3 \sin^2 \theta$$

$$P_3^1(x) = \tfrac{3}{2}(5x^2 - 1)(1 - x^2)^{1/2} = \tfrac{3}{2}(5 \cos^2 \theta - 1) \sin \theta$$

$$P_3^2(x) = 15x(1 - x^2) = 15 \cos \theta \sin^2 \theta$$

$$P_3^3(x) = 15(1 - x^2)^{3/2} = 15 \sin^3 \theta$$

$$P_4^1(x) = \tfrac{5}{2}(7x^3 - 3x)(1 - x^2)^{1/2} = \tfrac{5}{2}(7 \cos^3 \theta - 3 \cos \theta) \sin \theta$$

$$P_4^2(x) = \tfrac{15}{2}(7x^2 - 1)(1 - x^2) = \tfrac{15}{2}(7 \cos^2 \theta - 1) \sin^2 \theta$$

$$P_4^3(x) = 105x(1 - x^2)^{3/2} = 105 \cos \theta \sin^3 \theta$$

$$P_4^4(x) = 105(1 - x^2)^2 = 105 \sin^4 \theta$$

These are the associated Legendre functions.[2] Since the highest power of x in $P_n(x)$ is x^n, we must have $m \le n$ (or the m-fold differentiation will drive our function to zero). In quantum mechanics the requirement that $m \le n$ has the physical interpretation that the expectation value of the square of the z-component of the angular momentum is less than or equal to the expectation value of the square of the angular momentum vector $\mathbf{L}$,

$$\langle L_z^2 \rangle \le \langle L^2 \rangle \equiv \int \psi_{lm}^* \mathbf{L}^2 \psi_{lm} \, d^3 r.$$

From the form of Eq. (12.81) we might expect m to be nonnegative, differentiating a negative number of times not having been defined. However, if $P_n(x)$ is expressed by Rodrigues' formula, this limitation on m is relaxed and we may have $-n \le m \le n$, negative as well as positive values of m being permitted. Using Leibniz's differentiation formula once again, the reader may show (Exercise 12.5.1) that $P_n^m(x)$ and $P_n^{-m}(x)$ are related by

$$P_n^{-m}(x) = (-1)^m \frac{(n - m)!}{(n + m)!} P_n^m(x). \tag{12.81a}$$

From our definition of the associated Legendre functions, $P_n^m(x)$,

$$P_n^0(x) = P_n(x). \tag{12.82}$$

In addition, we may develop Table 12.2.

As with the Legendre polynomials, a generating function for the associated Legendre functions does exist:

$$\frac{(2m)!(1 - x^2)^{m/2}}{2^m m!(1 - 2tx + t^2)^{m+1/2}} = \sum_{s=0}^{\infty} P_{s+m}^m(x)t^s. \tag{12.83}$$

However, because of its more cumbersome form and lack of any direct physical application, it is seldom used.

---

[2] Occasionally (as in AMS-55), the reader will find the associated Legendre functions defined with an additional factor of $(-1)^m$. This $(-1)^m$ seems an unnecessary complication at this point. It will be included in the definition of the spherical harmonics $Y_n^m(\theta, \varphi)$ in Section 12.6.

RECURRENCE RELATIONS

As expected, the associated Legendre functions satisfy recurrence relations. Because of the existence of two indices instead of just one, we have a wide variety of recurrence relations:

$$P_n^{m+1} - \frac{2mx}{(1 - x^2)^{1/2}} P_n^m + [n(n + 1) - m(m - 1)]P_n^{m-1} = 0, \quad (12.84)$$

$$(2n + 1)xP_n^m = (n + m)P_{n-1}^m + (n - m + 1)P_{n+1}^m, \quad (12.85)$$

$$(2n + 1)(1 - x^2)^{1/2}P_n^m = P_{n+1}^{m+1} - P_{n-1}^{m+1}$$
$$= (n + m)(n + m - 1)P_{n-1}^{m-1}$$
$$- (n - m + 1)(n - m + 2)P_{n+1}^{m-1}, \quad (12.86)$$

$$(1 - x^2)^{1/2}P_n^{m\prime} = \tfrac{1}{2}P_n^{m+1} - \tfrac{1}{2}(n + m)(n - m + 1)P_n^{m-1}. \quad (12.87)$$

These relations, and many other similar ones, may be verified by use of the generating function (Eq. (12.4)), by substitution of the series solution of the associated Legendre equation (12.79) or reduction to the Legendre polynomial recurrence relations, using Eq. (12.81), As an example of the last method, consider the third equation in the preceding set. It is similar to Eq. (12.23):

$$(2n + 1) P_n(x) = P'_{n+1}(x) - P'_{n-1}(x). \quad (12.88)$$

Let us differentiate this Legendre polynomial recurrence relation m times to obtain

$$(2n + 1)\frac{d^m}{dx^m} P_n(x) = \frac{d^m}{dx^m} P'_{n+1}(x) - \frac{d^m}{dx^m} P'_{n-1}(x)$$

$$= \frac{d^{m+1}}{dx^{m+1}} P_{n+1}(x) - \frac{d^{m+1}}{dx^{m+1}} P_{n-1}(x). \quad (12.89)$$

Now multiplying by $(1 - x^2)^{(m+1)/2}$ and using the definition of $P_n^m(x)$, we obtain Eq. (12.86).

PARITY

The parity relation satisfied by the associated Legendre functions may be determined by examination of the defining equation (12.81). As $x \to -x$, we already know that $P_n(x)$ contributes a $(-1)^n$. The m-fold differentiation yields a factor of $(-1)^m$. Hence we have

$$P_n^m(-x) = (-1)^{n+m}P_n^m(x). \quad (12.90)$$

A glance at Table 12.2 verifies this for $1 \le m \le n \le 4$.

Also, from the definition in Eq. (12.81)

$$P_n^m(\pm 1) = 0, \quad \text{for } m \neq 0. \quad (12.91)$$

ORTHOGONALITY

The orthogonality of the $P_n^m(x)$ follows from the differential equation just as in $P_n(x)$ (Section 12.3); the term $-m^2/(1 - x^2)$ cancels out, assuming m is the same in both cases. However, it is instructive to demonstrate the orthogonality by another method, a method that will also provide the normalization constant.

Using the definition in Eq. (12.81) and Rodrigues' formula (Eq. (12.65)) for $P_n(x)$, we find

$$\int_{-1}^{1} P_p^m(x)P_q^m(x)\, dx = \frac{(-1)^m}{2^{p+q}p!\,q!} \int_{-1}^{1} X^m \frac{d^{p+m}}{dx^{p+m}} X^p \frac{d^{q+m}}{dx^{q+m}} X^q\, dx. \quad (12.92)$$

The function X is given by $X \equiv (x^2 - 1)$. If $p \neq q$, let us assume that $p < q$. Notice that the superscript m is the same for both functions. This is an essential condition. The technique is to integrate repeatedly by parts; all the integrated parts will vanish as long as there is a factor $X = x^2 - 1$. Let us integrate $q + m$ times to obtain

$$\int_{-1}^{1} P_p^m(x)P_q^m(x)\, dx = \frac{(-1)^m(-1)^{q+m}}{2^{p+q}p!\,q!} \int_{-1}^{1} \frac{d^{q+m}}{dx^{q+m}} \left(X^m \frac{d^{p+m}}{dx^{p+m}} X^p \right) X^q\, dx. \quad (12.93)$$

The integrand on the right-hand side is now expanded by Leibniz's formula to give

$$X^q \frac{d^{q+m}}{dx^{q+m}} \left(X^m \frac{d^{p+m}}{dx^{p+m}} X^p \right)$$

$$= X^q \sum_{i=0}^{i=q+m} \frac{(q+m)!}{i!(q+m-i)!} \frac{d^{q+m-i}}{dx^{q+m-i}} X^m \frac{d^{p+m+i}}{dx^{p+m+i}} X^p. \quad (12.94)$$

Since the term X^m contains no power of x greater than x^{2m}, we must have

$$q + m - i \leq 2m \quad (12.95)$$

or the derivative will vanish. Similarly,

$$p + m + i \leq 2p. \quad (12.96)$$

In the solution of these equations for the index i the conditions for a nonzero result are

$$i \geq q - m, \qquad i \leq p - m. \quad (12.97)$$

If $p < q$, as assumed, there is no solution and the integral vanishes. The same result obviously must follow if $p > q$.

For the remaining case, $p = q$, we may still have the single term corresponding to $i = q - m$. Putting Eq. (12.94) into Eq. (12.93), we have

$$\int_{-1}^{1} [P_q^m(x)]^2\, dx = \frac{(-1)^{q+2m}(q+m)!}{2^{2q}q!\,q!(2m)!(q-m)!} \int_{-1}^{1} X^q \left(\frac{d^{2m}}{dx^{2m}} X^m \right) \left(\frac{d^{2q}}{dx^{2q}} X^q \right) dx. \quad (12.98)$$

Since

$$X^m = (x^2 - 1)^m = x^{2m} - mx^{2m-2} + \cdots, \qquad (12.99)$$

$$\frac{d^{2m}}{dx^{2m}} X^m = (2m)!, \qquad (12.100)$$

Eq. (12.98) reduces to

$$\int_{-1}^{1} [P_q^m(x)]^2 \, dx = \frac{(-1)^{q+2m}(2q)!(q+m)!}{2^{2q}q!\,q!(q-m)!} \int_{-1}^{1} X^q \, dx. \qquad (12.101)$$

The integral on the right is just

$$(-1)^q \int_0^{\pi} \sin^{2q+1}\theta \, d\theta = \frac{(-1)^q 2^{2q+1} q!\,q!}{(2q+1)!} \qquad (12.102)$$

(compare Exercise 10.4.9). Combining Eqs. (12.101) and (12.102), we have the orthogonality integral

$$\int_{-1}^{1} P_p^m(x) P_q^m(x) \, dx = \frac{2}{2q+1} \cdot \frac{(q+m)!}{(q-m)!} \delta_{p,q} \qquad (12.103)$$

or, in spherical polar coordinates,

$$\int_0^{\pi} P_p^m(\cos\theta) P_q^m(\cos\theta) \sin\theta \, d\theta = \frac{2}{2q+1} \cdot \frac{(q+m)!}{(q-m)!} \delta_{p,q}. \qquad (12.104)$$

The orthogonality of the Legendre polynomials is actually a special case of this result, obtained by setting m equal to zero; that is, for $m = 0$, Eq. (12.103) reduces to Eqs. (12.43) and (12.48). In both Eqs. (12.103) and (12.104) our Sturm-Liouville theory of Chapter 9 could provide the Kronecker delta. A special calculation, such as the analysis here, is required for the normalization constant.

The orthogonality of the associated Legendre functions over the same interval and with the same weighting factor as the Legendre polynomials does not contradict the uniqueness of the Gram-Schmidt construction of the Legendre polynomials, Example 9.3.1. Table 12.2 suggests (and Section 12.4 verifies) that $\int_{-1}^{1} P_p^m(x) P_q^m(x) \, dx$ may be written as

$$\int_{-1}^{1} p_p^m(x) \, p_q^m(x)(1 - x^2)^m \, dx.$$

Here

$$p_p^m(x)(1 - x^2)^{m/2} = P_p^m(x).$$

The functions $p_p^m(x)$ may be constructed by the Gram-Schmidt procedure with the weighting function $w(x) = (1 - x^2)^m$.

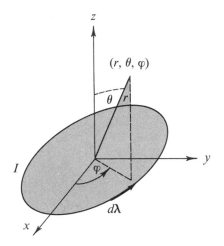

Figure 12.12 Circular current loop.

It is possible to develop an orthogonality relation for associated Legendre functions of the same lower index but different upper index. We find

$$\int_{-1}^{1} P_n^m(x) P_n^k(x) (1 - x^2)^{-1} \, dx = \frac{(n + m)!}{m(n - m)!} \delta_{m, k}.$$ (12.105)

Note that a new weighting factor, $(1 - x^2)^{-1}$, has been introduced. This form is essentially a mathematical curiosity. In physical problems orthogonality of the φ dependence ties the two upper indices together and leads to Eq. (12.104).

Example 12.5.1 Magnetic Induction Field of a Current Loop

Like the other differential equations of mathematical physics, the associated Legendre equation is likely to pop up quite unexpectedly. As an illustration, consider the magnetic induction field **B** and magnetic vector potential **A** created by a single circular current loop in the equatorial plane (Fig. 12.12).

We know from electromagnetic theory that the contribution of current element $I \, d\lambda$ to the magnetic vector potential is

$$d\mathbf{A} = \frac{\mu_0}{4\pi} \frac{I \, d\lambda}{r}.$$ (12.106)

(This follows from Exercise 1.14.4). Equation (12.106), plus the symmetry of our system, shows that **A** has only a $\hat{\varphi}$-component and that the component is independent of φ,[3]

$$\mathbf{A} = \hat{\varphi} A_\varphi(r, \theta).$$ (12.107)

By Maxwell's equations

$$\nabla \times \mathbf{H} = \mathbf{J} \qquad (\partial \mathbf{D}/\partial t = 0, \text{ SI units}).$$ (12.108)

---

[3] Pair off corresponding current elements $I \, d\lambda(\varphi_1)$ and $I \, d\lambda(\varphi_2)$, where $\varphi - \varphi_1 = \varphi_2 - \varphi$.

Since

$$\mu_0 \mathbf{H} = \mathbf{B} = \boldsymbol{\nabla} \times \mathbf{A}, \tag{12.109}$$

we have

$$\boldsymbol{\nabla} \times \boldsymbol{\nabla} \times \mathbf{A} = \mu_0 \mathbf{J}, \tag{12.110}$$

where $\mathbf{J}$ is the current density. In our problem $\mathbf{J}$ is zero everywhere except in the current loop. Therefore, away from the loop,

$$\boldsymbol{\nabla} \times \boldsymbol{\nabla} \times \hat{\boldsymbol{\varphi}} A_\varphi(r, \theta) = 0, \tag{12.111}$$

using Eq. (12.107).

From the expression for the curl in spherical polar coordinates (Section 2.5), we obtain (Example 2.5.2)

$$\boldsymbol{\nabla} \times \boldsymbol{\nabla} \times \hat{\boldsymbol{\varphi}} A_\varphi(r, \theta) = \hat{\boldsymbol{\varphi}} \left[-\frac{\partial^2 A_\varphi}{\partial r^2} - \frac{2}{r}\frac{\partial A_\varphi}{\partial r} - \frac{1}{r^2}\frac{\partial^2 A_\varphi}{\partial \theta^2} - \frac{1}{r^2}\frac{\partial}{\partial \theta}(\cot\theta A_\varphi) \right]$$

$$= 0. \tag{12.112}$$

Letting $A_\varphi(r, \theta) = R(r)\Theta(\theta)$ and separating variables, we have

$$r^2 \frac{d^2 R}{dr^2} + 2\frac{dR}{dr} - n(n+1)R = 0, \tag{12.113}$$

$$\frac{d^2\Theta}{d\theta^2} + \cot\theta \frac{d\Theta}{d\theta} + n(n+1)\Theta - \frac{\Theta}{\sin^2\theta} = 0. \tag{12.114}$$

The second equation is the associated Legendre equation (12.80) with $m = 1$, and we may immediately write

$$\Theta(\theta) = P_n^1(\cos\theta). \tag{12.115}$$

The separation constant $n(n+1)$ was chosen to keep this solution well behaved.

By trial, letting $R(r) = r^\alpha$, we find that $\alpha = n, -n-1$. The first possibility is discarded, for our solution must vanish as $r \to \infty$. Hence

$$A_{\varphi n} = \frac{b_n}{r^{n+1}} P_n^1(\cos\theta) = c_n\left(\frac{a}{r}\right)^{n+1} P_n^1(\cos\theta) \tag{12.116}$$

and

$$A_\varphi(r, \theta) = \sum_{n=1}^\infty c_n\left(\frac{a}{r}\right)^{n+1} P_n^1(\cos\theta) \qquad (r > a). \tag{12.117}$$

Here a is the radius of the current loop.

Since A_φ must be invariant to reflection in the equatorial plane by the symmetry of our problem,

$$A_\varphi(r, \cos\theta) = A_\varphi(r, -\cos\theta), \tag{12.118}$$

the parity property of $P_n^m(\cos\theta)$ (Eq. (12.90)) shows that $c_n = 0$ for n even.

To complete the evaluation of the constants, we may use Eq. (12.117) to calculate B_z along the z-axis $[B_z = B_r(r, \theta = 0)]$ and compare with the expression obtained from the Biot and Savart law. This is the same technique that is used in Example 12.3.3. We have (compare Eq. (2.47))

$$B_r = \nabla \times \mathbf{A}|_r$$

$$= \frac{1}{r \sin \theta} \left[\frac{\partial}{\partial \theta} (\sin \theta \, A_\varphi) \right]$$

$$= \frac{\cot \theta}{r} A_\varphi + \frac{1}{r} \frac{\partial A_\varphi}{\partial \theta}. \tag{12.119}$$

Using

$$\frac{\partial P_n^1(\cos \theta)}{\partial \theta} = -\sin \theta \, \frac{dP_n^1(\cos \theta)}{d(\cos \theta)}$$

$$= -\frac{1}{2} P_n^2 + \frac{n(n + 1)}{2} P_n^0 \tag{12.120}$$

(Eq. (12.87)) and then Eq. (12.84) with $m = 1$,

$$P_n^2(\cos \theta) - \frac{2 \cos \theta}{\sin \theta} P_n^1(\cos \theta) + n(n + 1)P_n(\cos \theta) = 0, \tag{12.121}$$

we obtain

$$B_r(r, \theta) = \sum_{n=1}^{\infty} c_n n(n + 1) \frac{a^{n+1}}{r^{n+2}} P_n(\cos \theta), \qquad r > a \tag{12.122}$$

(for all θ). In particular, for $\theta = 0$,

$$B_r(r, 0) = \sum_{n=1}^{\infty} c_n n(n + 1) \frac{a^{n+1}}{r^{n+2}}. \tag{12.123}$$

We may also obtain

$$B_\theta(r, \theta) = -\frac{1}{r} \frac{\partial(r A_\varphi)}{\partial r}$$

$$= \sum_{n=1}^{\infty} c_n n \frac{a^{n+1}}{r^{n+2}} P_n^1(\cos \theta), \qquad r > a. \tag{12.124}$$

The Biot and Savart law states that

$$d\mathbf{B} = \frac{\mu_0}{4\pi} I \frac{d\boldsymbol{\lambda} \times \hat{\mathbf{r}}}{r^2} \qquad \text{(SI units)}. \tag{12.125}$$

We now integrate over the perimeter of our loop (radius a). The geometry is shown in Fig. 12.13. The resulting magnetic induction field is $\hat{z} B_z$, along

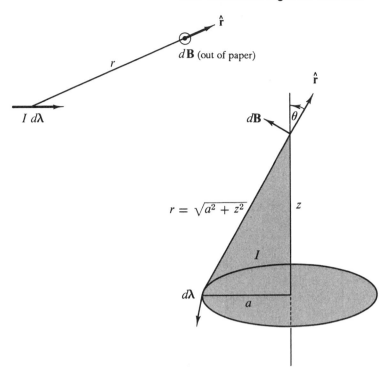

Figure 12.13 Law of Biot and Savart applied to a circular loop.

the z-axis, with

$$B_z = \frac{\mu_0 I}{2} a^2 (a^2 + z^2)^{-3/2}$$

$$= \frac{\mu_0 I}{2} \frac{a^2}{z^3} \left(1 + \frac{a^2}{z^2} \right)^{-3/2}. \qquad (12.126)$$

Expanding by the binomial theorem, we obtain

$$B_z = \frac{\mu_0 I}{2} \frac{a^2}{z^3} \left[1 - \frac{3}{2} \left(\frac{a}{z} \right)^2 + \frac{15}{8} \left(\frac{a}{z} \right)^4 - \cdots \right]$$

$$= \frac{\mu_0 I}{2} \frac{a^2}{z^3} \sum_{s=0}^{\infty} (-1)^s \frac{(2s+1)!!}{(2s)!!} \left(\frac{a}{z} \right)^{2s}, \qquad z > a. \qquad (12.127)$$

Equating Eqs. (12.123) and (12.127) term by term (with $r = z$),[4] we find

$$c_1 = \frac{\mu_0 I}{4}, \qquad c_3 = -\frac{\mu_0 I}{16}, \qquad c_2 = c_4 = \cdots = 0.$$

$$\qquad\qquad\qquad\qquad\qquad\qquad\qquad\qquad\qquad\qquad\qquad (12.128)$$

$$c_n = (-1)^{(n-1)/2} \frac{\mu_0 I}{2n(n+1)} \times \frac{(n/2)!}{[(n-1)/2]!(\frac{1}{2})!}, \qquad n \text{ odd.}$$

[4] The descending power series is also unique.

Equivalently, we may write

$$c_{2n+1} = (-1)^n \frac{\mu_0 I}{2^{2n+2}} \cdot \frac{(2n)!}{n!(n+1)!} = (-1)^n \frac{\mu_0 I}{2} \cdot \frac{(2n-1)!!}{(2n+2)!!} \tag{12.129}$$

and

$$A_\varphi(r, \theta) = \left(\frac{a}{r}\right)^2 \sum_{n=0}^{\infty} c_{2n+1} \left(\frac{a}{r}\right)^{2n} P_{2n+1}^1(\cos \theta), \tag{12.130}$$

$$B_r(r, \theta) = \frac{a^2}{r^3} \sum_{n=0}^{\infty} c_{2n+1}(2n+1)(2n+2)\left(\frac{a}{r}\right)^{2n} P_{2n+1}(\cos \theta), \tag{12.131}$$

$$B_\theta(r, \theta) = \frac{a^2}{r^3} \sum_{n=0}^{\infty} c_{2n+1}(2n+1)\left(\frac{a}{r}\right)^{2n} P_{2n+1}^1(\cos \theta). \tag{12.132}$$

These fields may be described in closed form by the use of elliptic integrals. Exercise 5.8.4 is an illustration of this approach. A third possibility is direct integration of Eq. (12.106) by expanding the factor $1/r$ as a Legendre polynomial generating function. The current is specified by Dirac delta functions. These methods have the advantage of yielding the constants c_n directly.

A comparison of magnetic current loop dipole fields and finite electric dipole fields may be of interest. For the magnetic current loop dipole the preceding analysis gives

$$B_r(r, \theta) = \frac{\mu_0 I}{2} \frac{a^2}{r^3} \left[P_1 - \frac{3}{2}\left(\frac{a}{r}\right)^2 P_3 + \cdots \right], \tag{12.133}$$

$$B_\theta(r, \theta) = \frac{\mu_0 I}{4} \frac{a^2}{r^3} \left[P_1^1 - \frac{3}{4}\left(\frac{a}{r}\right)^2 P_3^1 + \cdots \right]. \tag{12.134}$$

From the finite electric dipole potential of Section 12.1 we have

$$E_r(r, \theta) = \frac{qa}{\pi\varepsilon_0 r^3} \left[P_1 + 2\left(\frac{a}{r}\right)^2 P_3 + \cdots \right], \tag{12.135}$$

$$E_\theta(r, \theta) = \frac{qa}{2\pi\varepsilon_0 r^3} \left[P_1^1 + \left(\frac{a}{r}\right)^2 P_3^1 + \cdots \right]. \tag{12.136}$$

The two fields agree in form as far as the leading term is concerned ($r^{-3}P_1$), and this is the basis for calling them both dipole fields.

As with electric multipoles, it is sometimes convenient to discuss *point* magnetic multipoles (see Fig. 12.14). For the dipole case, Eqs. (12.133) and (12.134), the point dipole is formed by taking the limit $a \to 0$, $I \to \infty$ with Ia^2 held constant. With **n** a unit vector normal to the current loop (positive sense by right-hand rule, Section 1.10) the magnetic moment **m** is given by **m** = $nI\pi a^2$.

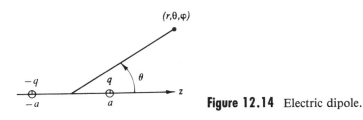

Figure 12.14 Electric dipole.

EXERCISES

12.5.1   Prove that

$$P_n^{-m}(x) = (-1)^m \frac{(n-m)!}{(n+m)!} P_n^m(x),$$

where $P_n^m(x)$ is defined by

$$P_n^m(x) = \frac{1}{2^n n!} (1 - x^2)^{m/2} \frac{d^{n+m}}{dx^{n+m}} (x^2 - 1)^n.$$

Hint. One approach is to apply Leibniz's formula to $(x + 1)^n (x - 1)^n$.

12.5.2   Show that

$$P_{2n}^1(0) = 0$$

$$P_{2n+1}^1(0) = (-1)^n \frac{(2n+1)!}{(2^n n!)^2} = (-1)^n \frac{(2n+1)!!}{(2n)!!},$$

by each of the three methods:

(a) use of recurrence relations,
(b) expansion of the generating function,
(c) Rodrigues' formula.

12.5.3   Evaluate $P_n^m(0)$.

$$\text{ANS.} \quad P_n^m(0) = \begin{cases} (-1)^{(n-m)/2} \dfrac{(n+m)!}{2^n((n-m)/2)!((n+m)/2)!}, & n+m \text{ even,} \\ 0, & n+m \text{ odd.} \end{cases}$$

Also, $P_n^m(0) = (-1)^{(n-m)/2} \dfrac{(n+m-1)!!}{(n-m)!!}$, $n+m$ even.

12.5.4   Show that

$$P_n^n(\cos \theta) = (2n - 1)!! \sin^n \theta, \qquad n = 0, 1, 2, \ldots.$$

12.5.5   Derive the associated Legendre recurrence relation

$$P_n^{m+1}(x) - \frac{2mx}{(1-x^2)^{1/2}} P_n^m(x) + [n(n+1) - m(m-1)]P_n^{m-1}(x) = 0.$$

12.5.6   Develop a recurrence relation that will yield $P_n^1(x)$ as

$$P_n^1(x) = f_1(x, n)P_n(x) + f_2(x, n)P_{n-1}(x).$$

Follow either (a) or (b).

(a) Derive a recurrence relation of the preceding form. Give $f_1(x, n)$ and $f_2(x, n)$ explicitly.

(b) Find the appropriate recurrence relation in print.

   (1) Give the source.
   (2) Verify the recurrence relation.

$$ANS. \quad P_n^1(x) = -\frac{nx}{(1 - x^2)^{1/2}} P_n + \frac{n}{(1 - x^2)^{1/2}} P_{n-1}.$$

12.5.7   Show that

$$\sin \theta P_n'(\cos \theta) = P_n^1(\cos \theta).$$

12.5.8   Show that

(a) $\int_0^\pi \left(\frac{dP_n^m}{d\theta} \frac{dP_{n'}^m}{d\theta} + \frac{m^2 P_n^m P_{n'}^m}{\sin^2 \theta} \right) \sin \theta \, d\theta = \frac{2n(n + 1)}{2n + 1} \frac{(n + m)!}{(n - m)!} \delta_{nn'},$

(b) $\int_0^\pi \left(\frac{P_n^1}{\sin \theta} \frac{dP_{n'}^1}{d\theta} + \frac{P_{n'}^1}{\sin \theta} \frac{dP_n^1}{d\theta} \right) \sin \theta \, d\theta = 0.$

These integrals occur in the theory of scattering of electromagnetic waves by spheres.

12.5.9   As a repeat of Exercise 12.3.10, show, using associated Legendre functions, that

$$\int_{-1}^1 x(1 - x^2)P_n'(x)P_m'(x) \, dx = \frac{n + 1}{2n + 1} \cdot \frac{2}{2n - 1} \cdot \frac{n!}{(n - 2)!} \delta_{m, n-1}$$

$$+ \frac{n}{2n + 1} \cdot \frac{2}{2n + 3} \cdot \frac{(n + 2)!}{n!} \delta_{m, n+1}.$$

12.5.10   Evaluate

$$\int_0^\pi \sin^2 \theta P_n^1(\cos \theta) \, d\theta.$$

12.5.11   The associated Legendre polynomial $P_n^m(x)$ satisfies the self-adjoint differential equation

$$(1 - x^2)P_n^{m''}(x) - 2xP_n^{m'}(x) + \left[n(n + 1) - \frac{m^2}{1 - x^2} \right] P_n^m(x) = 0.$$

From the differential equations for $P_n^m(x)$ and $P_n^k(x)$ show that

$$\int_{-1}^1 P_n^m(x)P_n^k(x) \frac{dx}{1 - x^2} = 0$$

for $k \neq m$.

12.5.12   Determine the vector potential of a magnetic quadrupole by differentiating the magnetic dipole potential.

$$ANS. \quad \mathbf{A}_{MQ} = \frac{\mu_0}{2} (Ia^2)(dz)\hat{\varphi} \frac{P_2^1(\cos \theta)}{r^3} + \text{higher-order terms.}$$

$$\mathbf{B}_{MQ} = \mu_0(Ia^2)(dz)\left[\hat{\mathbf{r}} \frac{3P_2(\cos \theta)}{r^4} + \hat{\theta} \frac{P_2^1(\cos \theta)}{r^4} \right].$$

This corresponds to placing a current loop of radius a at $z = dz$, an oppositely directed current loop at $z = -dz$, and letting $a \to 0$ subject to $(dz)x$ (dipole strength) equal constant.

Another approach to this problem would be to integrate dA (Eq. (12.106), to expand the denominator in a series of Legendre polynomials, and to use the Legendre polynomial addition theorem (Section 12.8).

12.5.13   A single loop of wire of radius a carries a current I.

(a) Find the magnetic induction $\mathbf{B}$ for $r < a$.

(b) Calculate the integral of the magnetic flux $(\mathbf{B} \cdot d\sigma)$ over the area of the current loop, that is

$$\int_0^a \int_0^{2\pi} B_\theta\left(r, \theta = \frac{\pi}{2}\right) d\varphi \, r \, dr.$$

$$ANS. \quad \infty.$$

The earth is within such a ring current in which I approximates millions of amperes arising from the drift of charged particles in the Van Allen belt.

12.5.14   (a) Show that in the point dipole limit the magnetic induction field of the current loop becomes

$$B_r(r, \theta) = \frac{\mu_0}{2\pi} \frac{m}{r^3} P_1(\cos \theta)$$

$$B_\theta(r, \theta) = \frac{\mu_0}{4\pi} \frac{m}{r^3} P_1^1(\cos \theta),$$

with $m = I\pi a^2$.

(b) Compare these results with the magnetic induction of the point magnetic dipole of Exercise 1.8.17. take $\mathbf{m} = \hat{\mathbf{z}}m$.

12.5.15   A uniformly charged spherical shell is rotating with constant angular velocity.

(a) Calculate the magnetic induction $\mathbf{B}$ along the axis of rotation outside the sphere.

(b) Using the vector potential series of Section 12.5, find $\mathbf{A}$ and then $\mathbf{B}$ for all space outside the sphere.

12.5.16   In the liquid drop model of the nucleus the spherical nucleus is subjected to small deformations. Consider a sphere of radius r_0 that is deformed so that its new surface is given by

$$r = r_0[1 + \alpha_2 P_2(\cos \theta)].$$

Find the area of the deformed sphere through terms of order α_2^2.

Hint.

$$dA = \left[r^2 + \left(\frac{dr}{d\theta} \right)^2 \right]^{1/2} r \sin \theta \, d\theta \, d\varphi.$$

ANS. $A = 4\pi r_0^2[1 + \frac{4}{5}\alpha_2^2 + \mathcal{O}(\alpha_2^3)].$

Note. The area element dA follows from noting that the line element ds for fixed φ is given by

$$ds = (r^2 \, d\theta^2 + dr^2)^{1/2} = (r^2 + (dr/d\theta)^2)^{1/2} \, d\theta.$$

12.5.17   A nuclear particle is in a potential $V(r, \theta, \varphi) = 0$ for $0 \le r < a$ and ∞ for $r > a$. The particle is described by a wave function $\psi(r, \theta, \varphi)$ which satisfies the wave equation

$$-\frac{\hbar^2}{2M} \nabla^2 \psi + V_0 \psi = E\psi$$

and the boundary condition

$$\psi(r = a) = 0.$$

Show that for the energy E to be a minimum there must be no angular dependence in the wave function; that is, $\psi = \psi(r)$.
Hint. The problem centers on the boundary condition on the radial function.

12.5.18   (a) Write a subroutine to calculate the numerical value of the associated Legendre function $P_N^1(x)$ for given values of N and x.
Hint. With the known forms of P_1^1 and P_2^1 you can use the recurrence relation Eq. (12.85) to generate P_N^1, $N > 2$.
(b) Check your subroutine by having it calculate $P_N^1(x)$ for $x = 0.0(0.5)1.0$ and $N = 1(1)10$. Check these numerical values against the known values of $P_N^1(0)$ and $P_N^1(1)$ and against the tabulated values of $P_N^1(0.5)$.

12.5.19   Calculate the magnetic vector potential of a current loop, Example 12.5.1. Tabulate your results for $r/a = 1.5(0.5)5.0$ and $\theta = 0°(15°)90°$. Include terms in the series expansion, Eq. (12.130), until the absolute values of the terms drop below the leading term by a factor of 10^5 or more.
Note. This associated Legendre expansion can be checked by comparison with the elliptic integral solution, Exercise 5.8.4.

Check value.   For $r/a = 4.0$ and $\theta = 20°$,
$A_\varphi/\mu_0 I = 4.9398 \times 10^{-3}.$

12.6 SPHERICAL HARMONICS

In the separation of variables of (1) Laplace's equation, (2) Helmholtz's or the space-dependence of the classical wave equation, and (3) the Schrödinger wave equation for central force fields,

$$\nabla^2 \psi + k^2 f(r)\psi = 0, \tag{12.137}$$

the angular dependence, coming entirely from the Laplacian operator, is[1]

$$\frac{\Phi(\varphi)}{\sin\theta}\frac{d}{d\theta}\left(\sin\theta\frac{d\Theta}{d\theta}\right) + \frac{\Theta(\theta)}{\sin^2\theta}\frac{d^2\Phi(\varphi)}{d\varphi^2} + n(n+1)\Theta(\theta)\Phi(\varphi) = 0.$$

$$(12.138)$$

AZIMUTHAL DEPENDENCE—ORTHOGONALITY

The separated azimuthal equation is

$$\frac{1}{\Phi(\varphi)}\frac{d^2\Phi(\varphi)}{d\varphi^2} = -m^2, \qquad (12.139)$$

with solutions

$$\Phi(\varphi) = e^{-im\varphi}, \ e^{im\varphi}, \qquad (12.140)$$

which readily satisfy the orthogonal condition

$$\int_0^{2\pi} e^{-im_1\varphi}e^{im_2\varphi}\,d\varphi = 2\pi\delta_{m_1,m_2}. \qquad (12.141)$$

Notice that it is the product $\Phi_{m_1}^*(\varphi)\Phi_{m_2}(\varphi)$ that is taken and that $*$ is used to indicate the complex conjugate function. This choice is not required, but it is convenient for quantum mechanical calculations. We could have used

$$\Phi = \sin m\varphi, \qquad \cos m\varphi \qquad (12.142)$$

and the conditions of orthogonality that form the basis for Fourier series (Chapter 14). For applications such as describing the earth's gravitational or magnetic field $\sin m\varphi$ and $\cos m\varphi$ would be the preferred choice (see Example 12.6.1).

In electrostatics and most other physical problems we require m to be an integer in order that $\Phi(\varphi)$ may be a single-valued function of the azimuth angle. In quantum mechanics the question is much more involved because the observable quantity that must be single-valued is the square of the magnitude of the wave function, $\Phi^*\Phi$. However, it can be shown that we must still have m integral. Compare footnote in Section 8.3.

By means of Eq. (12.141),

$$\Phi_m = \frac{1}{\sqrt{2\pi}}e^{im\varphi} \qquad (12.143)$$

is orthonormal (orthogonal and normalized) with respect to integration over the azimuth angle φ.

POLAR ANGLE DEPENDENCE

Splitting off the azimuthal dependence, the polar angle dependence (θ) leads to the associated Legendre equation (12.80), which is satisfied by the

---

[1] For a separation constant of the form $n(n+1)$ with n an integer, a Legendre equation series solution becomes a polynomial. Otherwise both series solutions diverge, Exercise 8.5.5.

associated Legendre functions; that is, $\Theta(\theta) = P_n^m(\cos \theta)$. To include negative values of m, we use Rodrigues' formula, Eq. (12.65), in the definition of $P_n^m(\cos \theta)$. This leads to

$$P_n^m(\cos \theta) = \frac{1}{2^n n!} (1 - x^2)^{m/2} \frac{d^{m+n}}{dx^{m+n}} (x^2 - 1)^n, \qquad -n \leq m \leq n.$$

(12.144)

$P_n^m(\cos \theta)$ and $P_n^{-m}(\cos \theta)$ are related as indicated in Exercise 12.5.1. An advantage of this approach over simply defining $P_n^m(\cos \theta)$ for $0 \leq m \leq n$ and requiring that $P_n^{-m} = P_n^m$ is that the recurrence relations valid for $0 \leq m \leq n$ remain valid for $-n \leq m < 0$.

Normalizing the associated Legendre function by Eq. (12.103), we obtain the orthonormal function

$$\mathcal{P}_n^m(\cos \theta) = \sqrt{\frac{2n + 1}{2} \frac{(n - m)!}{(n + m)!}} P_n^m(\cos \theta), \qquad -n \leq m \leq n.$$

(12.145)

SPHERICAL HARMONICS

The function $\Phi_m(\varphi)$ (Eq. (12.143)) is orthonormal with respect to the azimuthal angle φ, whereas the function $\mathcal{P}_n^m(\cos \theta)$ (Eq. (12.145)) is orthonormal with respect to the polar angle θ. We take the product of the two and define

$$Y_n^m(\theta, \varphi) \equiv (-1)^m \sqrt{\frac{2n + 1}{4\pi} \frac{(n - m)!}{(n + m)!}} P_n^m(\cos \theta)e^{im\varphi} \qquad (12.146)$$

to obtain functions of two angles (and two indices) which are orthonormal over the spherical surface. These $Y_n^m(\theta, \varphi)$ are spherical harmonics. The complete orthogonality integral becomes

$$\int_{\varphi=0}^{2\pi} \int_{\theta=0}^{\pi} Y_{n_1}^{m_1*}(\theta, \varphi) Y_{n_2}^{m_2}(\theta, \varphi) \sin \theta \, d\theta \, d\varphi = \delta_{n_1, n_2} \delta_{m_1, m_2}. \qquad (12.47)$$

The extra $(-1)^m$ included in the defining equation of $Y_n^m(\theta, \varphi)$ deserves some comment. It is clearly legitimate, since Eq. (12.137) is linear and homogeneous. It is not necessary, but in moving on to certain quantum mechanical calculations, particularly in the quantum theory of angular momentum (Section 12.7), it is most convenient. The factor $(-1)^m$ is a phase factor, often called the Condon–Shortley phase, after the authors of a classic text on atomic spectroscopy. The effect of this $(-1)^m$ (Eq. (12.146) and the $(-1)^m$ of Eq. (12.81a) for $P_n^{-m}(\cos \theta)$ is to introduce an alternation of sign among the *positive m* spherical harmonics. This is shown in Table 12.3.

The functions $Y_n^m(\theta, \varphi)$ acquired the name "spherical harmonics" first because they are defined over the surface of a sphere with θ the polar angle and φ the azimuth. The "harmonic" was included because solutions of Laplace's equation were called harmonic functions and $Y_n^m(\theta, \varphi)$ is the angular part of such a solution.

**Table 12.3 Spherical Harmonics
(Condon–Shortley Phase)**

$$Y_0^0(\theta, \varphi) = \frac{1}{\sqrt{4\pi}}$$

$$Y_1^1(\theta, \varphi) = -\sqrt{\frac{3}{8\pi}}\,\sin\theta e^{i\varphi}$$

$$Y_1^0(\theta, \varphi) = \sqrt{\frac{3}{4\pi}}\,\cos\theta$$

$$Y_1^{-1}(\theta, \varphi) = +\sqrt{\frac{3}{8\pi}}\,\sin\theta e^{-i\varphi}$$

$$Y_2^2(\theta, \varphi) = \sqrt{\frac{5}{96\pi}}\,3\sin^2\theta e^{2i\varphi}$$

$$Y_2^1(\theta, \varphi) = -\sqrt{\frac{5}{24\pi}}\,3\sin\theta\cos\theta e^{i\varphi}$$

$$Y_2^0(\theta, \varphi) = \sqrt{\frac{5}{4\pi}}\left(\frac{3}{2}\cos^2\theta - \frac{1}{2}\right)$$

$$Y_2^{-1}(\theta, \varphi) = +\sqrt{\frac{5}{24\pi}}\,3\sin\theta\cos\theta e^{-i\varphi}$$

$$Y_2^{-2}(\theta, \varphi) = \sqrt{\frac{5}{96\pi}}\,3\sin^2\theta e^{-2i\varphi}$$

In the framework of quantum mechanics Eq. (12.138) becomes an orbital angular momentum equation and the solution $Y_L^M(\theta, \varphi)$ (n replaced by L, m, by M) is an angular momentum eigenfunction: L being the angular momentum quantum number and M the z-axis projection of L. These relationships are developed in detail in Sections 4.3 and 12.7.

LAPLACE SERIES, FUNDAMENTAL EXPANSION THEOREM

Part of the importance of spherical harmonics lies in the completeness property, a consequence of the Sturm–Liouville form of Laplace's equation. This property, in this case, means that any function $f(\theta, \varphi)$ (with sufficient continuity properties) evaluated over the surface of the sphere can be expanded in a uniformly convergent double series of spherical harmonics[2] (Laplace's series):

$$f(\theta, \varphi) = \sum_{m,n} a_{mn} Y_n^m(\theta, \varphi). \tag{12.148}$$

[2] For a proof of this fundamental theorem see E. W. Hobson, *The Theory of Spherical and Ellipsoidal Harmonics*. New York: Chelsea (1955), Chapter VII. If $f(\theta, \varphi)$ is discontinuous we may still have convergence in the mean, Section 9.4.

Table 12.4 Gravity Field Coefficients, Eq. (12.149)

	Earth	Moon	Mars
C_{20}	1.083×10^{-3}	$(0.200 \pm 0.002) \times 10^{-3}$	$(1.96 \pm 0.01) \times 10^{-3}$
C_{22}	0.16×10^{-5}	$(2.4 \pm 0.5) \times 10^{-5}$	$(-5 \pm 1) \times 10^{-5}$
S_{22}	-0.09×10^{-5}	$(0.5 \pm 0.6) \times 10^{-5}$	$(3 \pm 1) \times 10^{-5}$

C_{20} represents an equatorial bulge, whereas C_{22} and S_{22} represent an azimuthal dependence of the gravitational field.

If $f(\theta, \varphi)$ is known, the coefficients can be immediately found by the use of the orthogonality integral. Within the framework of the theory of linear vector spaces, the completeness of the spherical harmonics follows from Weierstrass's theorem.

Example 12.6.1 Laplace Series—Gravity Fields

The gravity fields of the earth, moon, and Mars have been described by a Laplace series with real eigenfunctions:

$$U(r, \theta, \varphi) = \frac{GM}{R} \left[\frac{R}{r} - \sum_{n=2}^{\infty} \sum_{m=0}^{n} \left(\frac{R}{r} \right)^{n+1} \left\{ C_{nm} Y_{mn}^{e}(\theta, \varphi) + S_{nm} Y_{mn}^{o}(\theta, \varphi) \right\} \right].$$

$$(12.149)$$

Here M is the mass of the body, R the equatorial radius. The real functions Y_{mn}^{e} and Y_{mn}^{o} are defined by

$$Y_{mn}^{e}(\theta, \varphi) = P_{n}^{m}(\cos \theta) \cos m\varphi$$

$$Y_{mn}^{o}(\theta, \varphi) = P_{n}^{m}(\cos \theta) \sin m\varphi.$$

For applications such as this the real trigonometric forms are preferred to the imaginary exponential form of $Y_{L}^{M}(\theta, \varphi)$. Satellite measurements have led to the numerical values shown in Table 12.4.

EXERCISES

12.6.1 Show that the parity of $Y_{L}^{M}(\theta, \varphi)$ is $(-1)^{L}$. Note the disappearance of any M dependence.

Hint. For the parity operation in spherical polar coordinates see Section 2.5 and a footnote in Section 12.2.

12.6.2 Prove that

$$Y_{L}^{M}(0, \varphi) = \left(\frac{2L + 1}{4\pi} \right)^{1/2} \delta_{M0}.$$

12.6.3   In the theory of Coulomb excitation of nuclei we encounter $Y_L^M(\pi/2, 0)$. Show that

$$Y_L^M\left(\frac{\pi}{2}, 0\right) = \left(\frac{2L + 1}{4\pi}\right)^{1/2} \frac{[(L - M)!((L + M)!]^{1/2}}{(L - M)!!(L + M)!!}(-1)^{(L+M)/2}$$

for $L + M$ even

$$= 0 \quad \text{for } L + M \text{ odd.}$$

Here

$$(2n)!! = 2n(2n - 2)\cdots 6 \cdot 4 \cdot 2,$$

$$(2n + 1)!! = (2n + 1)(2n - 1)\cdots 5 \cdot 3 \cdot 1.$$

12.6.4   (a) Express the elements of the quadrupole moment tensor $x_i x_j$ as a linear combination of the spherical harmonics Y_2^m (and Y_0^0).
         Note. The tensor $x_i x_j$ is *reducible*. The Y_0^0 indicates the presence of a scalar component.
   (b) The quadrupole moment tensor is usually defined as

$$Q_{ij} = \int (3x_i x_j - r^2 \delta_{ij})\rho(\mathbf{r}) \, d\tau,$$

with $\rho(\mathbf{r})$ the charge density. Express the components of $(3x_i x_j - r^2\delta_{ij})$ in terms of $r^2 Y_2^M$.
   (c) What is the significance of the $-r^2\delta_{ij}$ term?
         Hint. Compare Section 2.9.

12.6.5   The orthogonal azimuthal functions yield a useful representation of the Dirac delta function. Show that

$$\delta(\varphi_1 - \varphi_2) = \frac{1}{2\pi} \sum_{m=-\infty}^{\infty} \exp[im(\varphi_1 - \varphi_2)].$$

12.6.6   Derive the spherical harmonic closure relation

$$\sum_{l=0}^{\infty} \sum_{m=-l}^{+l} Y_l^m(\theta_1, \varphi_1)Y_l^{m*}(\theta_2, \varphi_2) = \frac{1}{\sin \theta_1}\delta(\theta_1 - \theta_2)\delta(\varphi_1 - \varphi_2)$$

$$= \delta(\cos \theta_1 - \cos \theta_2)\delta(\varphi_1 - \varphi_2).$$

12.6.7   The quantum mechanical angular momentum operators $L_x \pm iL_y$ are given by

$$L_x + iL_y = e^{i\varphi}\left(\frac{\partial}{\partial\theta} + i\cot\theta\frac{\partial}{\partial\varphi}\right),$$

$$L_x - iL_y = -e^{-i\varphi}\left(\frac{\partial}{\partial\theta} - i\cot\theta\frac{\partial}{\partial\varphi}\right).$$

Show that

(a) $(L_x + iL_y)Y_L^M(\theta, \varphi), = +\sqrt{(L - M)(L + M + 1)}\,Y_L^{M+1}(\theta, \varphi)$

(b) $(L_x - iL_y)Y_L^M(\theta, \varphi) = +\sqrt{(L + M)(L - M + 1)}\,Y_L^{M-1}(\theta, \varphi).$

12.6.8   With $L_\pm$ given by

$$L_\pm = L_x \pm iL_y = \pm e^{\pm i\varphi} \left[\frac{\partial}{\partial \theta} \pm i \cot \theta \frac{\partial}{\partial \varphi} \right],$$

show that

(a) $Y_l^m = \sqrt{\dfrac{(l + m)!}{(2l)!(l - m)!}} \, (L_-)^{l-m} Y_l^l,$

(b) $Y_l^m = \sqrt{\dfrac{(l - m)!}{(2l)!(l + m)!}} \, (L_+)^{l+m} Y_l^{-l}.$

12.6.9   In some circumstances it is desirable to replace the imaginary exponential of our spherical harmonic by sine or cosine. Morse and Feschbach define

$$Y_{mn}^e = P_n^m(\cos \theta) \cos m\varphi,$$
$$Y_{mn}^0 = P_n^m(\cos \theta) \sin m\varphi,$$

where

$$\int_0^{2\pi} \int_0^\pi [Y_{mn}^{e \text{ or } 0}(\theta, \varphi)]^2 \sin \theta \, d\theta \, d\varphi = \frac{4\pi}{2(2n + 1)} \frac{(n + m)!}{(n - m)!} \qquad \text{for } n = 1, 2, 3, \ldots$$
$$= 4\pi \qquad \text{for } n = 0 \ (Y_{00}^0 \text{ does not exist}).$$

These spherical harmonics are often named according to the patterns of their positive and negative regions on the surface of a sphere—zonal harmonics for $m = 0$, sectoral harmonics for $m = n$, and tesseral harmonics for $0 < m < n$. For Y_{mn}^e, $n = 4$, $m = 0, 2, 4$, indicate on a diagram of a hemisphere (one diagram for each spherical harmonic) the regions in which the spherical harmonic is positive.

12.6.10   A function $f(r, \theta, \varphi)$ may be expressed as a Laplace series

$$f(r, \theta, \varphi) = \sum_{l, m} a_{lm} r^l Y_l^m(\theta, \varphi).$$

with $\langle \ \rangle_{\text{sphere}}$ used to mean the average over a sphere (centered on the origin), show that

$$\langle f(r, \theta, \varphi) \rangle_{\text{sphere}} = f(0, 0, 0).$$

12.7 ORBITAL ANGULAR MOMENTUM OPERATORS

Now we return to the specific *orbital* angular momentum operators L_x, L_y, and L_z of quantum mechanics introduced in Section 4.3. Equation (4.68) becomes

$$L_z \psi_{LM}(\theta, \varphi) = M \psi_{LM}(\theta, \varphi),$$

and we want to show that

$$\psi_{LM}(\theta, \varphi) = \langle \theta, \varphi \, | \, LM \rangle = Y_L^M(\theta, \varphi)$$

are the eigenfunctions $|LM\rangle$ of L^2 and L_z of Section 4.3 in spherical polar coordinates, the spherical harmonics. The explicit form of $L_z = -i\partial/\partial\varphi$ from

Exercise 2.5.13 indicates that ψ_{LM} has a φ dependence of $\exp(iM\varphi)$—with M an integer to keep ψ_{LM} single-valued. And if M is an integer, then L is an integer also.

To determine the θ dependence of $\psi_{LM}(\theta, \varphi)$, we proceed in two main steps: (1) the determination of $\psi_{LL}(\theta, \varphi)$ and (2) the development of $\psi_{LM}(\theta, \varphi)$ in terms of ψ_{LL} with the phase fixed by ψ_{LO}.

Let

$$\psi_{LM}(\theta, \varphi) = \Theta_{LM}(\theta)e^{iM\varphi}. \tag{12.150}$$

From $L_+\psi_{LL} = 0$, L being the largest M, using the form of L_+ given in Exercises 2.5.14 and 12.6.7, we have

$$e^{i(L+1)\varphi}\left[\frac{d}{d\theta} - L\cot\theta\right]\Theta_{LL}(\theta) = 0, \tag{12.151}$$

and

$$\psi_{LL}(\theta, \varphi) = c_L \sin^L\theta \, e^{iL\varphi}. \tag{12.152}$$

Normalizing, we obtain

$$c_L^* c_L \int_0^{2\pi}\int_0^\pi \sin^{2L+1}\theta \, d\theta \, d\varphi = 1. \tag{12.153}$$

The θ integral may be evaluated as a beta function (Exercise 10.4.9) and

$$|c_L| = \sqrt{\frac{(2L+1)!!}{4\pi(2L)!!}} = \frac{\sqrt{(2L)!}}{2^L L!}\sqrt{\frac{2L+1}{4\pi}}. \tag{12.154}$$

This completes our first step.

To obtain the ψ_{LM}, $M \neq \pm L$, we return to the ladder operators. From Eqs. (4.83) and (4.84) (J_+ replaced by L_+ and J_- replaced by L_-),

$$\psi_{LM}(\theta, \varphi) = \sqrt{\frac{(L+M)!}{(2L)!(L-M)!}}(L_-)^{L-M}\psi_{LL}(\theta, \varphi),$$

$$\psi_{LM}(\theta, \varphi) = \sqrt{\frac{(L-M)!}{(2L)!(L+M)!}}(L_+)^{L+M}\psi_{L,-L}(\theta, \varphi). \tag{12.155}$$

Again, note that the relative phases are set by the ladder operators. L_+ and L_- operating on $\Theta_{LM}(\theta)e^{iM\varphi}$ may be written as

$$L_+\Theta_{LM}(\theta)e^{iM\varphi} = e^{i(M+1)\varphi}\left[\frac{d}{d\theta} - M\cot\theta\right]\Theta_{LM}(\theta)$$

$$= -e^{i(M+1)\varphi}\sin^{1+M}\theta\frac{d}{d(\cos\theta)}\sin^{-M}\theta\,\Theta_{LM}(\theta),$$

$$L_-\Theta_{LM}(\theta)e^{iM\varphi} = -e^{i(M-1)\varphi}\left[\frac{d}{d\theta} + M\cot\theta\right]\Theta_{LM}(\theta) \tag{12.156}$$

$$= e^{i(M-1)\varphi}\sin^{1-M}\theta\frac{d}{d(\cos\theta)}\sin^M\theta\,\Theta_{LM}(\theta).$$

Repeating these operations n times yields

$$(L_+)^n \Theta_{LM}(\theta) e^{iM\varphi} = (-1)^n e^{i(M+n)\varphi} \sin^{n+M} \theta \frac{d^n}{d(\cos \theta)^n} \sin^{-M} \theta \, \Theta_{LM}(\theta),$$

$$(L_-)^n \Theta_{LM}(\theta) e^{iM\varphi} = e^{i(M-n)\varphi} \sin^{n-M} \theta \frac{d^n}{d(\cos \theta)^n} \sin^M \theta \, \Theta_{LM}(\theta).$$

$$(12.157)$$

From Eq. (12.155)

$$\psi_{LM}(\theta, \varphi) = c_L \sqrt{\frac{(L+M)!}{(2L)!(L-M)!}} \, e^{iM\varphi} \sin^{-M} \theta \frac{d^{L-M}}{d(\cos \theta)^{L-M}} \sin^{2L} \theta,$$

$$(12.158)$$

and for $M = -L$

$$\psi_{L,-L}(\theta, \varphi) = \frac{c_L}{(2L)!} e^{-iL\varphi} \sin^L \theta \frac{d^{2L}}{d(\cos \theta)^{2L}} \sin^{2L} \theta$$

$$= (-1)^L c_L \sin^L \theta e^{-iL\varphi}. \qquad (12.159)$$

Note the characteristic $(-1)^L$ phase of $\psi_{L,-L}$ relative to $\psi_{L,L}$. This $(-1)^L$ enters from

$$\sin^{2L} \theta = (1 - x^2)^L = (-1)^L (x^2 - 1)^L. \qquad (12.160)$$

Combining Eqs. (12.155), (12.157), and (12.159), we obtain

$$\psi_{LM}(\theta, \varphi) = (-1)^L c_L \sqrt{\frac{(L-M)!}{(2L)!(L+M)!}} (-1)^{L+M} e^{iM\varphi} \sin^M \theta \frac{d^{L+M}}{d(\cos \theta)^{L+M}} \sin^{2L} \theta.$$

$$(12.161)$$

Equations (12.158) and (12.161) agree that

$$\psi_{L0}(\theta, \varphi) = c_L \frac{1}{\sqrt{(2L)!}} \frac{d^L}{(d \cos \theta)^L} \sin^{2L} \theta. \qquad (12.162)$$

Using Rodrigues's formula, Eq. (12.65), we have

$$\psi_{L0}(\theta, \varphi) = (-1)^L c_L \frac{2^L L!}{\sqrt{(2L)!}} P_L(\cos \theta)$$

$$= (-1)^L \frac{c_L}{|c_L|} \sqrt{\frac{2L+1}{4\pi}} P_L(\cos \theta). \qquad (12.163)$$

The last equality follows from Eq. (12.154). We now demand that $\psi_{L0}(0, 0)$ be real and positive. Therefore

$$c_L = (-1)^L |c_L| = (-1)^L \frac{\sqrt{(2L)!}}{2^L L!} \sqrt{\frac{2L+1}{4\pi}}. \qquad (12.164)$$

With $(-1)^L c_L / |c_L| = 1$, $\psi_{L0}(\theta, \varphi)$ in Eq. (12.163) may be identified with the spherical harmonic $Y_L^O(\theta, \varphi)$ of Section 12.6.

When we substitute $(-1)^L c_L$ into Eq. (12.161),

$$\psi_{LM}(\theta, \varphi) = \frac{\sqrt{(2L)!}}{2^L L!} \sqrt{\frac{2L+1}{4\pi}} \sqrt{\frac{(L-M)!}{(2L)!(L+M)!}} (-1)^{L+M}$$

$$\cdot e^{iM\varphi} \sin^M \theta \frac{d^{L+M}}{d(\cos \theta)^{L+M}} \sin^{2L} \theta$$

$$= \sqrt{\frac{2L+1}{4\pi}} \sqrt{\frac{(L-M)!}{(L+M)!}} e^{iM\varphi} (-1)^M$$

$$\cdot \left\{ \frac{1}{2^L L!} (1-x^2)^{M/2} \frac{d^{L+M}}{dx^{L+M}} (x^2-1)^L \right\}, \qquad x = \cos \theta, \ M \geq 0.$$
$$(12.165)$$

The expression in the curly bracket is identified as the associated Legendre function, (Eq. (12.144), and we have

$$\psi_{LM}(\theta, \varphi) = Y_L^M(\theta, \varphi)$$

$$= (-1)^M \sqrt{\frac{2L+1}{4\pi} \cdot \frac{(L-M)!}{(L+M)!}} \cdot P_L^M(\cos \theta) e^{iM\varphi}, \qquad M \geq 0,$$
$$(12.166)$$

in complete agreement with Section 12.6. Then by Eq. (12.81a), Y_L^M for negative superscript is given by

$$Y_L^{-M}(\theta, \varphi) = (-1)^M Y_L^{M*}(\theta, \varphi). \qquad (12.167)$$

Our angular momentum eigenfunctions $\psi_{LM}(\theta, \varphi)$ are identified with the spherical harmonics. The phase factor $(-1)^M$ is associated with the positive values of M and is seen to be a consequence of the ladder operators.

Our development of spherical harmonics here may be considered a portion of Lie algebra—related to group theory—Section 4.3.

EXERCISES

12.7.1   Using the known forms of L_+ and L_- (Exercises 2.5.14 and 12.6.7), show that

$$\int Y_L^{M*} L_-(L_+ Y_L^M) \, d\Omega = \int (L_+ Y_L^M)^* (L_+ Y_L^M) \, d\Omega.$$

12.7.2   Derive the relations

(a) $\psi_{LM}(\theta, \varphi) = \sqrt{\dfrac{(L+M)!}{(2L)!(L-M)!}} (L_-)^{L-M} \psi_{LL}(\theta, \varphi),$

(b) $\psi_{Lm}(\theta, \varphi) = \sqrt{\dfrac{(L-M)!}{(2L)!(L+M)!}} (L_+)^{L+M} \psi_{L,-L}(\theta, \varphi).$

12.7.3   Derive the multiple operator equations

(a) $(L_+)^n \Theta_{LM}(\theta) e^{iM\varphi} = (-1)^n e^{i(M+n)\varphi} \sin^{n+M} \theta \dfrac{d^n}{d(\cos \theta)^n} \sin^{-M} \theta \, \Theta_{LM}(\theta),$

(b) $(L_-)^n \Theta_{LM}(\theta) e^{iM\varphi} = e^{i(M-n)\varphi} \sin^{n-M} \theta \dfrac{d^n}{d(\cos \theta)^n} \sin^{M} \theta \, \Theta_{LM}(\theta).$

Hint. Try mathematical induction.

12.7.4   Show, using $(L_-)^n$, that

$$Y_L^{-M}(\theta, \varphi) = (-1)^M Y_L^{M*}(\theta, \varphi).$$

12.7.5   Verify by explicit calculation that

(a) $L_+ Y_1^0(\theta, \varphi) = -\sqrt{\dfrac{3}{4\pi}} \sin \theta \, e^{i\varphi} = \sqrt{2} \, Y_1^1(\theta, \varphi),$

(b) $L_- Y_1^0(\theta, \varphi) = +\sqrt{\dfrac{3}{4\pi}} \sin \theta \, e^{-i\varphi} = \sqrt{2} \, Y_1^{-1}(\theta, \varphi).$

The signs (Condon–Shortley phase) are a consequence of the ladddder operators, L_+ and L_-.

12.8 THE ADDITION THEOREM FOR SPHERICAL HARMONICS

TRIGONOMETRIC IDENTITY

In the following discussion (θ_1, φ_1) and (θ_2, φ_2) denote two different directions in our spherical coordinate system, separated by an angle γ (Fig. 12.15). These angles satisfy the trigonometric identity

$$\cos \gamma = \cos \theta_1 \cos \theta_2 + \sin \theta_1 \sin \theta_2 \cos(\varphi_1 - \varphi_2), \qquad (12.168)$$

which is perhaps most easily proved by vector methods (compare Chapter 1).

The addition theorem, then, asserts that

$$P_n(\cos \gamma) = \frac{4\pi}{2n + 1} \sum_{m = -n}^{n} (-1)^m Y_n^m(\theta_1, \varphi_1) Y_n^{-m}(\theta_2, \varphi_2), \qquad (12.169)$$

or equivalently,

$$P_n(\cos \gamma) = \frac{4\pi}{2n + 1} \sum_{m = -n}^{n} Y_n^m(\theta_1, \varphi_1) Y_n^{m*}(\theta_2, \varphi_2).^1 \qquad (12.170)$$

In terms of the associated Legendre functions the addition theorem is

$$P_n(\cos \gamma) = P_n(\cos \theta_1) P_n(\cos \theta_2)$$

$$+ 2 \sum_{m = 1}^{n} \frac{(n - m)!}{(n + m)!} P_n^m(\cos \theta_1) P_n^m(\cos \theta_2) \cos m(\varphi_1 - \varphi_2).$$

$$(12.171)$$

Equation (12.168) is a special case of Eq. (12.171).

---

[1] The asterisk may go on *either* spherical harmonic.

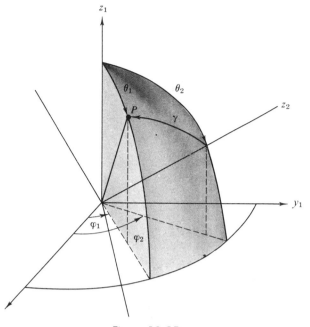

Figure 12.15

DERIVATION OF ADDITION THEOREM

We now derive Eq. (12.170). Let $g(\theta, \varphi)$ be a function that may be expanded in a Laplace series

$$g(\theta_1, \varphi_1) = Y_n^m(\theta_1, \varphi_1) \qquad \text{relative to } x_1, y_1, z_1$$

$$= \sum_{m=-n}^{n} a_{nm} Y_n^m(\gamma, \psi) \qquad \text{relative to } x_2, y_2, z_2. \qquad (12.172)$$

We write no summation over n in Eqs. (12.172) and (12.176) because, as a Legendre polynomial, $P_n(\cos \gamma)$ is an eigenfunction of $\mathbf{L}^2$ with eigenvalue $n(n + 1)$. Actually the choice of the 0 of the azimuth angle ψ is irrelevant. At $\gamma = 0$ we have

$$g(\theta_1, \varphi_1)|_{\gamma=0} = a_{n0} \left(\frac{2n + 1}{4\pi} \right)^{1/2}, \qquad (12.173)$$

since $P_n(1) = 1$, where $P_n^m(1) = 0$ $(m \neq 0)$. Multiplying Eq. (12.172) by $Y_n^{0*}(\gamma, \psi)$ and integrating over the sphere, we obtain

$$\int g(\theta_1, \varphi_1) Y_n^{0*}(\gamma, \psi) \, d\Omega_{\gamma, \psi} = a_{n0}. \qquad (12.174)$$

Now, using Eq. (12.172), we may rewrite Eq. (12.174) as

$$\int Y_n^m(\theta, \varphi_1) Y_n^{0*}(\gamma, \psi) \, d\Omega = a_{n0}. \qquad (12.175)$$

As for Eq. (12.172), we assume that $P_n(\cos \gamma)$ has an expansion of the form

$$P_n(\cos \gamma) = \sum_{m=-n}^{n} b_{nm} Y_n^m(\theta_1, \varphi_1), \qquad (12.176)$$

where the b_{nm} will, of course, depend on θ_2, φ_2, that is, on the orientation of the z_2-axis. Multiplying by $Y_n^{m*}(\theta_1, \varphi_1)$ and integrating with respect to θ_1 and φ_1 over the sphere, we have

$$\int P_n(\cos \gamma) Y_n^{m*}(\theta_1, \varphi_1) \, d\Omega_{\theta_1, \varphi_1} = b_{nm}. \qquad (12.177)$$

In terms of spherical harmonics Eq. (12.177) becomes

$$\left(\frac{4\pi}{2n+1}\right)^{1/2} \int Y_n^0(\gamma, \psi) Y_n^{m*}(\theta_1, \varphi_1) \, d\Omega = b_{nm}. \qquad (12.178)$$

Note that the subscripts have been dropped from the solid angle element $d\Omega$. Since the range of integration is over all solid angles, the choice of polar axis is irrelevant. Then in a comparison of Eqs. (12.175) and (12.178),

$$b_{nm}^* = a_{n0} \left(\frac{4\pi}{2n+1}\right)^{1/2}$$

$$= \frac{4\pi}{2n+1} g(\theta_1, \varphi_1)|_{\gamma=0} \qquad \text{by Eq. (12.173)}$$

$$= \frac{4\pi}{2n+1} Y_n^m(\theta_2, \psi_2) \qquad \text{by Eq. (12.172).} \qquad (12.179)$$

The change in subscripts occurs because

$$\begin{aligned} \theta_1 &\to \theta_2 \\ \varphi_1 &\to \varphi_2 \end{aligned} \qquad \text{for } \gamma \to 0.$$

Substituting back into Eq. (12.176), we obtain Eq. (12.170), thus proving our addition theorem.

The reader familiar with group theory will find a much more elegant proof of Eq. (12.170) by using the rotation group.[2] This is Exercise 4.2.10.

One application of the addition theorem is in the construction of a Green's function for the three-dimensional Laplace equation in spherical polar coordinates. If the source is on the polar axis at the point $(r = a, \theta = 0, \varphi = 0)$, then by Eq. (12.4a)

$$\frac{1}{R} = \frac{1}{|\mathbf{r} - \hat{z}a|} = \sum_{n=0}^{\infty} P_n(\cos \gamma) \frac{a^n}{r^{n+1}}, \qquad r > a$$

$$= \sum_{n=0}^{\infty} P_n(\cos \gamma) \frac{r^n}{a^{n+1}}, \qquad r < a. \qquad (12.180)$$

[2] Compare M. E. Rose, *Elementary Theory of Angular Momentum*. New York: Wiley (1957).

Rotating our coordinate system to put the source at (a, θ_2, φ_2) and the point of observation at (r, θ_1, φ_1), we obtain

$$G(r, \theta_1, \varphi_1, a, \theta_2, \varphi_2)$$

$$= \frac{1}{R}$$

$$= \sum_{n=0}^{\infty} \sum_{m=-n}^{n} \frac{4\pi}{2n+1} Y_n^{m*}(\theta_1, \varphi_1) Y_n^m(\theta_2, \varphi_2) \frac{a^n}{r^{n+1}}, \qquad r > a,$$

$$= \sum_{n=0}^{\infty} \sum_{m=-n}^{n} \frac{4\pi}{2n+1} Y_n^{m*}(\theta_1, \varphi_1) Y_n^m(\theta_2, \varphi_2) \frac{r^n}{a^{n+1}}, \qquad r < a. \qquad (12.181)$$

In Section 16.6 this argument is reversed to provide another derivation of the Legendre polynomial addition theorem.

EXERCISES

12.8.1 In proving the addition theorem, we assumed that $Y_n^k(\theta_1, \varphi_1)$ could be expanded in a series of $Y_n^m(\theta_2, \varphi_2)$ in which m varied from $-n$ to $+n$ but n was held fixed. What arguments can you develop to justify summing only over the upper index m and *not* over the lower index n?
Hints. One possibility is to examine the homogeneity of the Y_n^m, that is, Y_n^m may be expressed entirely in terms of the form $\cos^{n-p} \theta \sin^p \theta$ or $x^{n-p-s} y^p z^s / r^n$. Another possibility is to examine the behavior of the Legendre equation under rotation of the coordinate system.

12.8.2 An atomic electron with angular momentum L and magnetic quantum number M has a wave function

$$\psi(r, \theta, \varphi) = f(r) Y_L^M(\theta, \varphi).$$

Show that the sum of the electron densities in a given complete shell is spherically symmetric; that is, $\sum_{M=-L}^{L} \psi^*(r, \theta, \varphi) \psi(r, \theta, \varphi)$ is independent of θ and φ.

12.8.3 The potential of an electron at point $\mathbf{r}_e$ in the field of Z protons at $\mathbf{r}_p$ is

$$\varphi = -\frac{e^2}{4\pi\varepsilon_0} \sum_{p=1}^{Z} \frac{1}{|\mathbf{r}_e - \mathbf{r}_p|}.$$

Show that this may be written as

$$\varphi = -\frac{e^2}{4\pi\varepsilon_0 r_e} \sum_{p=1}^{Z} \sum_{L, M} \left(\frac{r_p}{r_e}\right)^L \frac{4\pi}{2L+1} Y_L^{M*}(\theta_p, \varphi_p) Y_L^M(\theta_e, \varphi_e),$$

where $r_e > r_p$. How should φ be written for $r_e < r_p$?

12.8.4 Two protons are *uniformly* distributed within the same spherical volume. If the coordinates of one element of charge are $(r_1, \theta_1, \varphi_1)$ and the coordinates of the other are $(r_2, \theta_2, \varphi_2)$ and r_{12} is the distance between them, the element of energy of repulsion will be given by

$$d\psi = \rho^2 \frac{dv_1 \, dv_2}{r_{12}} = \rho^2 \frac{r_1^2 \, dr_1 \sin\theta_1 \, d\theta_1 \, d\varphi_1 \, r_2^2 \, dr_2 \sin\theta_2 \, d\theta_2 \, d\varphi_2}{r_{12}}.$$

Here

$$\rho = \frac{\text{charge}}{\text{volume}} = \frac{3e}{4\pi R^3}, \qquad \text{charge density,}$$

$$r_{12}^2 = r_1^2 + r_2^2 - 2r_1 r_2 \cos\gamma.$$

Calculate the total electrostatic energy (of repulsion) of the two protons. This calculation is used in accounting for the mass difference in "mirror" nuclei, such as O^{15} and N^{15}.

$$ANS. \quad \text{For } r_2 > r_1, \quad \left.\begin{array}{c} \dfrac{3}{5}\dfrac{e^2}{R} \\[2mm] \\ r_2 < r_1, \quad \dfrac{3}{5}\dfrac{e^2}{R} \end{array}\right\rbrace \quad \dfrac{6}{5}\dfrac{e^2}{R} \text{ (total).}$$

This is *double* that required to create a uniformly charged sphere because we have two separate cloud charges interacting *not* one charge interacting with itself (with permutation of pairs *not* considered).

12.8.5 Each of the two 1s electrons in helium may be described by a hydrogenic wave function

$$\psi(\mathbf{r}) = \left(\frac{Z^3}{\pi a_0^3}\right)^{1/2} e^{-Zr/a_0}$$

in the absence of the other electron. Here Z, the atomic number , is 2. The symbol a_0 is the Bohr radius, $\hbar^2/me^2$. Find the mutual potential energy of the two electrons given by

$$\int \psi^*(\mathbf{r}_1)\psi^*(\mathbf{r}_2) \frac{e^2}{r_{12}} \psi(\mathbf{r}_1)\psi(\mathbf{r}_2) d^3 r_1 \, d^3 r_2 .$$

$$ANS. \quad \frac{5e^2 Z}{8a_0}.$$

Note. $d^3 r_1 = r_1^2 \, dr_1 \sin\theta_1 \, d\theta_1 \, d\varphi_1$,

$\qquad r_{12} = |\mathbf{r}_1 - \mathbf{r}_2|$.

12.8.6 The probability of finding a 1s hydrogen electron is a volume element $r^2 \, dr \sin\theta \, d\theta \, d\varphi$ is

$$\frac{1}{\pi a_0^3} \exp[-2r/a_0] r^2 \, dr \sin\theta \, d\theta \, d\varphi.$$

Find the corresponding electrostatic potential. Calculate the potential from

$$V(\mathbf{r}_1) = \frac{q}{4\pi\varepsilon_0} \int \frac{\rho(\mathbf{r}_2)}{r_{12}} d^3 r_2 ,$$

with $\mathbf{r}_1$ *not* on the z-axis. Expand r_{12}. Apply the Legendre polynomial addition theorem and show that the angular dependence of $V(\mathbf{r}_1)$ drops out.

$$ANS. \quad V(\mathbf{r}_1) = \frac{q}{4\pi\varepsilon_0}\left\{\frac{1}{2r_1}\gamma\left(3,\frac{2r_1}{a_0}\right) + \frac{1}{a_0}\Gamma\left(2,\frac{2r_1}{a_0}\right)\right\}.$$

12.8.7   A hydrogen electron in a $2p$ orbit has a charge distribution

$$\rho = \frac{q}{64\pi a_0^5} r^2 e^{-r/a_0} \sin^2 \theta,$$

where a_0 is the Bohr radius, $\hbar^2/me^2$. Find the electrostatic potential corresponding to this charge distribution.

12.8.8   The electric current density produced by a $2p$ electron in a hydrogen atom is

$$\mathbf{J} = \hat{\varphi} \frac{q\hbar}{32 m a_0^5} e^{-r/a_0} r \sin \theta.$$

Using

$$\mathbf{A}(\mathbf{r}_1) = \frac{\mu_0}{4\pi} \int \frac{\mathbf{J}(\mathbf{r}_2)}{|\mathbf{r}_1 - \mathbf{r}_2|} d^3 r_2,$$

find the magnetic vector potential produced by this hydrogen electron. *Hint.* Resolve into Cartesian components. Use the addition theorem to eliminate γ, the angle included between $\mathbf{r}_1$ and $\mathbf{r}_2$.

12.8.9   (a) As a Laplace series and as an example of Eq. (9.80) (now with complex functions), show that

$$\delta(\Omega_1 - \Omega_2) = \sum_{n,m} Y_n^{m*}(\theta_2, \varphi_2) Y_n^m(\theta_1, \varphi_1).$$

(b) Show also that this *same* Dirac delta function may be written as

$$\delta(\Omega_1 - \Omega_2) = \sum_n \frac{2n + 1}{4\pi} P_n(\cos \gamma).$$

Now, if you can justify equating the summations over n *term by term*, you have an alternate derivation of the spherical harmonic addition theorem.

12.9 INTEGRALS OF THE PRODUCT OF THREE SPHERICAL HARMONICS

Frequently in quantum mechanics we encounter integrals of the general form

$$\langle Y_{L_1}^{M_1} | Y_{L_2}^{M_2} | Y_{L_3}^{M_3} \rangle = \int Y_{L_1}^{M_1*} Y_{L_2}^{M_2} Y_{L_3}^{M_3} \, d\Omega \tag{12.182}$$

in which the integration is over all solid angles. The first factor in the integrand may come from the wave function of a final state and the third factor from an initial state, whereas the middle factor may represent an operator that is being evaluated or whose "matrix element" is being determined.

By using group theoretical methods, as in the quantum theory of angular momentum, we may give a general expression for the forms listed. The analysis involves the vector-addition of Clebsch–Gordan coefficients from Section 4.4, which have been tabulated. Three general restrictions appear.[1] (1) The integral vanishes unless the *vector* sum of the L's (angular momentum) is zero, $|L_1 - L_3| \le L_2 \le L_1 + L_3$. (2) The integral vanishes unless $M_2 + M_3 = M_1$. Here we have the theoretical foundation of the vector model of atomic spectroscopy. (3) Finally, the integral vanishes unless the product $Y_{L_1}^{M_1*} Y_{L_2}^{M_2} Y_{L_3}^{M_3}$ is even, that is, unless $L_1 + L_2 + L_3$ is an even integer. This is a parity conservation law.

Let us show some details of this general and powerful approach using Section 4.4. The Wigner–Eckart theorem applied to the matrix element in Eq. (12.182) yields

$$\langle Y_{L_1}^{M_1} | Y_{L_2}^{M_2} | Y_{L_3}^{M_3} \rangle = (-1)^{L_2-L_3+L_1} C(L_2 L_3 L_1 \,|\, M_2 M_3 M_1)$$
$$\cdot \langle Y_{L_1} | |Y_{L_2}| |Y_{L_3} \rangle / \sqrt{(2L_1 + 1)}, \qquad (12.183)$$

where the double bars denote the reduced matrix element that no longer depends on the M_i. The selection rules (1) and (2) mentioned above follow directly from the Clebsch–Gordan coefficient in Eq. (12. 183). Next we use Eq. (12.183) for $M_1 = M_2 = M_3 = 0$ in conjunction with Eq. (12.146) for $m = 0$, which yields

$$\langle Y_{L_1}^0 | Y_{L_2}^0 | Y_{L_3}^0 \rangle = (-1)^{L_2-L_3+L_1} C(L_2 L_3 L_1 \,|\, 000)$$
$$\cdot \langle Y_{L_1} | |Y_{L_2}| |Y_{L_3} \rangle / \sqrt{(2L_1 + 1)}$$
$$= \sqrt{(2L_1 + 1)(2L_2 + 1)(2L_3 + 1)} / (4\pi)^{3/2}$$
$$\cdot \int_{-1}^{1} P_{L_1}(x) P_{L_2}(x) P_{L_3}(x)\, dx, \qquad (12.184)$$

where $x = \cos\theta$. By elementary methods it can be shown that

$$\int_{-1}^{1} P_{L_1}(x) P_{L_2}(x) P_{L_3}(x)\, dx \sim C(L_2 L_3 L_1 \,|\, 000)^2. \qquad (12.185)$$

Substituting Eq. (12.185) into (12.184) we obtain

$$\langle Y_{L_1} | |Y_{L_2}| |Y_{L_3} \rangle \sim C(L_2 L_3 L_1 \,|\, 000). \qquad (12.186)$$

The parity selection rule (3) above follows from Eq. (12.186) in conjunction with the phase relation

$$C(L_2 L_3 L_1 \,|\, -M_2, -M_3, -M_1) = (-1)^{L_2+L_3-L_1} C(L_2 L_3 L_1 \,|\, M_2 M_3 M_1), \qquad (12.187)$$

for which we refer the reader to the references.[1] The reader will note that the vector-addition coefficients are developed in terms of the Condon–Shortley

[1] E.U. Condon and G. H. Shortley, *The Theory of Atomic Spectra.* Cambridge: Cambridge University Press (1951); M. E. Rose, *Elementary Theory of Angular Momentum.* New York: Wiley (1957); A. Edmonds, *Angular Momentum in Quantum Mechanics.* Princeton, NJ: Princeton University Press (1957); E. P. Wigner, *Group Theory and Its Applications to Quantum Mechanics* (translated by J. J. Griffin). New York: Academic Presss (1959).

phase convention in which the $(-1)^m$ of Eq. (12.146) is associated with the positive m.

It is possible to evaluate many of the commonly encountered integrals of this form with the techniques already developed. The integration over azimuth may be carried out by inspection.

$$\int_0^{2\pi} e^{-iM_1\varphi} e^{iM_2\varphi} e^{iM_3\varphi} \, d\varphi = 2\pi\delta_{M_2+M_3-M_1,0}. \tag{12.188}$$

Physically this corresponds to the conservation of the z-component of angular momentum.

APPLICATION OF RECURRENCE RELATIONS

A glance at Table 12.3 will show that the θ-dependence of $Y_{L_2}^{M_2}$, that is, $P_{L_2}^{M_2}(\theta)$ can be expressed in terms of $\cos\theta$ and $\sin\theta$. However, a factor of $\cos\theta$ or $\sin\theta$ may be combined with the $Y_{L_3}^{M_3}$ factor by using the associated Legendre polynomial recurrence relations. For instance, from Eqs. (12.85) and (12.86) we get

$$\cos\theta \, Y_L^M = +\left[\frac{(L-M+1)(L+M+1)}{(2L+1)(2L+3)}\right]^{1/2} Y_{L+1}^M$$

$$+\left[\frac{(L-M)(L+M)}{(2L-1)(2L+1)}\right]^{1/2} Y_{L-1}^M \tag{12.189}$$

$$e^{i\varphi}\sin\theta \, Y_L^M = -\left[\frac{(L+M+1)(L+M+2)}{(2L+1)(2L+3)}\right]^{1/2} Y_{L+1}^{M+1}$$

$$+\left[\frac{(L-M)(L-M-1)}{(2L-1)(2L+1)}\right]^{1/2} Y_{L-1}^{M+1} \tag{12.190}$$

$$e^{-i\varphi}\sin\theta \, Y_L^M = +\left[\frac{(L-M+1)(L-M+2)}{(2L+1)(2L+3)}\right]^{1/2} Y_{L+1}^{M-1}$$

$$-\left[\frac{(L+M)(L+M-1)}{(2L-1)(2L+1)}\right]^{1/2} Y_{L-1}^{M-1}. \tag{12.191}$$

Using these equations, we obtain

$$\int Y_{L_1}^{M_1*} \cos\theta \, Y_L^M \, d\Omega = \left[\frac{(L-M+1)(L+M+1)}{(2L+1)(2L+3)}\right]^{1/2} \delta_{M_1,M}\delta_{L_1,L+1}$$

$$+\left[\frac{(L-M)(L+M)}{(2L-1)(2L+1)}\right]^{1/2} \delta_{M_1,M}\delta_{L_1,L-1}. \tag{12.192}$$

The occurrence of the Kronecker delta $(L_1, L\pm 1)$ is an aspect of the conservation of angular momentum. Physically, this integral arises in a consideration of ordinary atomic electromagnetic radiation (electric dipole). It leads to the

754   12 LEGENDRE FUNCTIONS

familiar selection rule that transitions to an atomic level with orbital angular momentum quantum number L_1 can originate only from atomic levels with quantum numbers $L_1 - 1$ or $L_1 + 1$. The application to expressions such as

$$\text{quadrupole moment} \sim \int Y_L^{M*} P_2(\cos\theta) Y_L^M \, d\Omega$$

is more involved but perfectly straightforward.

EXERCISES

12.9.1   Verify

(a) $\int Y_L^M Y_0^0 Y_L^{M*} \, d\Omega = \dfrac{1}{\sqrt{4\pi}}$,

(b) $\int Y_L^M Y_1^0 Y_{L+1}^{M*} \, d\Omega = \sqrt{\dfrac{3}{4\pi}} \sqrt{\dfrac{(L+M+1)(L-M+1)}{(2L+1)(2L+3)}}$,

(c) $\int Y_L^M Y_1^1 Y_{L+1}^{M+1*} \, d\Omega = \sqrt{\dfrac{3}{8\pi}} \sqrt{\dfrac{(L+M+1)(L+M+2)}{(2L+1)(2L+3)}}$,

(d) $\int Y_L^M Y_1^1 Y_{L-1}^{M+1*} \, d\Omega = -\sqrt{\dfrac{3}{8\pi}} \sqrt{\dfrac{(L-M)(L-M-1)}{(2L-1)(2L+1)}}$.

These integrals were used in an investigation of the angular correlation of internal conversion electrons.

12.9.2   Show that

(a) $\displaystyle\int_{-1}^{1} x P_L(x) P_N(x) \, dx = \begin{cases} \dfrac{2(L+1)}{(2L+1)(2L+3)}, & N = L+1, \\[4mm] \dfrac{2L}{(2L-1)(2L+1)}, & N = L-1, \end{cases}$

(b) $\displaystyle\int_{-1}^{1} x^2 P_L(x) P_N(x) \, dx = \begin{cases} \dfrac{2(L+1)(L+2)}{(2L+1)(2L+3)(2L+5)}, & N = L+2, \\[4mm] \dfrac{2(2L^2+2L-1)}{(2L-1)(2L+1)(2L+3)}, & N = L, \\[4mm] \dfrac{2L(L-1)}{(2L-3)(2L-1)(2L+1)}, & N = L-2. \end{cases}$

12.9.3   Since $xP_n(x)$ is a polynomial (degree $n+1$), it may be represented by the Legendre series

$$xP_n(x) = \sum_{s=0}^{\infty} a_s P_s(x).$$

(a) Show that $a_s = 0$ for $s < n-1$ and $s > n+1$.

(b) Calculate a_{n-1}, a_n, and a_{n+1} and show that you have reproduced the recurrence relation, Eq. 12.17.

Note. This argument may be put in a general form to demonstrate the existence of a three-term recurrence relation for any of our complete sets of orthogonal polynomials:

$$x\varphi_n = a_{n+1}\varphi_{n+1} + a_n\varphi_n + a_{n-1}\varphi_{n-1}.$$

12.9.4   Show that Eq. (12.192) is a special case of Eq. (12.183) and derive the reduced matrix element $\langle Y_{L_1}||Y_1||Y_L\rangle$.

$$ANS. \quad \langle Y_{L_1}||Y_1||Y_L\rangle = (-1)^{L_1+1-L}C(1LL_1|000)\frac{\sqrt{3(2L+1)}}{4\pi}.$$

12.10 LEGENDRE FUNCTIONS OF THE SECOND KIND, $Q_n(x)$

In all the analysis so far in this chapter we have been dealing with one solution of Legendre's equation, the solution $P_n(\cos\theta)$, which is regular (finite) at the two singular points of the differential equation, $\cos\theta = \pm 1$. From the general theory of differential equations it is known that a second solution exists. We develop this second solution, Q_n, by a series solution of Legendre's equation. Later a closed form will be obtained.

SERIES SOLUTIONS OF LEGENDRE'S EQUATION
To solve

$$\frac{d}{dx}\left[(1-x^2)\frac{dy}{dx}\right] + n(n+1)y = 0 \tag{12.193}$$

we proceed as in Chapter 8, letting[1]

$$y = \sum_{\lambda=0}^{\infty} a_\lambda x^{k+\lambda}, \tag{12.194}$$

with

$$y' = \sum_{\lambda=0}^{\infty} (k+\lambda)a_\lambda x^{k+\lambda-1}, \tag{12.195}$$

$$y'' = \sum_{\lambda=0}^{\infty} (k+\lambda)(k+\lambda-1)a_\lambda x^{k+\lambda-2}. \tag{12.196}$$

Substitution into the original differential equation gives

$$\sum_{\lambda=0}^{\infty} (k+\lambda)(k+\lambda-1)a_\lambda x^{k+\lambda-2}$$

$$+ \sum_{\lambda=0}^{\infty} [n(n+1) - 2(k+\lambda) - (k+\lambda)(k+\lambda-1)]a_\lambda x^{k+\lambda} = 0. \tag{12.197}$$

The *indicial equation* is

$$k(k-1) = 0, \tag{12.198}$$

---
[1] Note that x may be replaced by the complex variable z.

with solutions $k = 0, 1$. We try first $k = 0$ with $a_0 = 1, a_1 = 0$. Then our series is described by the recurrence relation

$$(\lambda + 2)(\lambda + 1)a_{\lambda+2} + [n(n + 1) - 2\lambda - \lambda(\lambda - 1)]a_\lambda = 0, \qquad (12.199)$$

which becomes

$$a_{\lambda+2} = -\frac{(n + \lambda + 1)(n - \lambda)}{(\lambda + 1)(\lambda + 2)}a_\lambda. \qquad (12.200)$$

Labeling this series p_n, we have

$$p_n(x) = 1 - \frac{n(n + 1)}{2!}x^2 + \frac{(n - 2)n(n + 1)(n + 3)}{4!}x^4 + \cdots. \qquad (12.201)$$

The second solution of the indicial equation, $k = 1$, with $a_0 = 1$, $a_1 = 0$, leads to the recurrence relation

$$a_{\lambda+2} = -\frac{(n + \lambda + 2)(n - \lambda - 1)}{(\lambda + 2)(\lambda + 3)}a_\lambda. \qquad (12.202)$$

Labeling this series q_n, we obtain

$$q_n(x) = x - \frac{(n - 1)(n + 2)}{3!}x^3 + \frac{(n - 3)(n - 1)(n + 2)(n + 4)}{5!}x^5 - \cdots.$$
$$(12.203)$$

Our general solution of Eq. (12.193), then, is

$$y_n(x) = A_n p_n(x) + B_n q_n(x), \qquad (12.204)$$

provided we have convergence. From Gauss's test, Section 5.2 (see Example 5.2.4), we do not have convergence at $x = \pm 1$. To get out of this difficulty, we set the separation constant n equal to an integer (Exercise 8.5.5) and convert the infinite series into a polynomial.

For n, a positive even integer (or zero), series p_n terminates, and with a proper choice of a normalizing factor (selected to obtain agreement with the definition of $P_n(x)$ in Section 12.1)

$$P_n(x) = (-1)^{n/2}\frac{n!}{2^n[(n/2)!]^2}p_n(x)$$

$$= (-1)^s\frac{(2s)!}{2^{2s}(s!)^2}p_{2s}(x) = (-1)^s\frac{(2s - 1)!!}{(2s)!!}p_{2s}(x), \qquad \text{for } n = 2s.$$
$$(12.205)$$

If n is a positive odd integer, series q_n terminates after a finite number of terms, and we write

$$P_n(x) = (-1)^{(n-1)/2}\frac{n!}{2^{n-1}\{[n - 1)/2]!\}^2}q_n(x)$$

$$= (-1)^s\frac{(2s + 1)!}{2^{2s}(s!)^2}q_{2s+1}(x) = (-1)^s\frac{(2s + 1)!!}{(2s)!!}q_{2s+1}(x), \qquad \text{for } n = 2s + 1.$$
$$(12.206)$$

Note that these expressions hold for all real values of x, $-\infty < x < \infty$, and for complex values in the finite complex plane. The constants that multiply p_n and q_n are chosen to make P_n agree with Legendre polynomials given by the generating function.

Equations (12.201) and (12.203) may still be used with $n = \nu$, not an integer, but now the series no longer terminates, and the range of convergence becomes $-1 < x < 1$. The end points, $x = \pm 1$ are *not* included.

It is sometimes convenient to reverse the order of the terms in the series. This may be done by putting

$$s = \frac{n}{2} - \lambda \qquad \text{in the first form of } P_n(x), \qquad n \text{ even,}$$

$$s = \frac{n-1}{2} - \lambda \qquad \text{in the second form of } P_n(x), \quad n \text{ odd,}$$

so that Eqs. (12.205) and (12.206) become

$$P_n(x) = \sum_{s=0}^{[n/2]} (-1)^s \frac{(2n - 2s)!}{2^n s!(n - s)!(n - 2s)!} x^{n-2s}, \qquad (12.207)$$

where the upper limit $s = n/2$ (for n even) or $(n - 1)/2$ (for n odd). This reproduces Eq. (12.8) of Section 12.1, which is obtained directly from the generating function. This agreement with Eq. (12.8) is the reason for the particular choice of normalization in Eqs. (12.205) and (12.206).

$Q_n(x)$ FUNCTIONS OF THE SECOND KIND

It will be noticed that we have used only p_n for n even and q_n for n odd (because they terminated for this choice of n). We may now define a second solution of Legendre's equation (Fig. 12.16) by

$$Q_n(x) = (-1)^{n/2} \frac{[(n/2)!]^2 2^n}{n!} q_n(x) = (-1)^s \frac{(2s)!!}{(2s - 1)!!} q_{2s}(x), \qquad \text{for } n \text{ even, } n = 2s,$$
$$(12.208)$$

$$Q_n(x) = (-1)^{(n+1)/2} \frac{\{[(n - 1)/2]!\}^2 2^{n-1}}{n!} p_n(x)$$

$$= (-1)^{s+1} \frac{(2s)!!}{(2s + 1)!!} p_{2s+1}(x), \qquad \text{for } n \text{ odd, } n = 2s + 1. \qquad (12.209)$$

This choice of normalizing factors forces Q_n to satisfy the same recurrence relations as P_n. This may be verified by substituting Eqs. (12.208) and (12.209) into Eqs. (12.17) and (12.26). Inspection of the (series) recurrence relations (Eqs. (12.200) and (12.202)), that is, by the Cauchy ratio test, shows that $Q_n(x)$ will converge for $-1 < x < 1$. If $|x| \geq 1$, these series forms of our second solution *diverge*. A solution in a series of negative powers of x can be developed for the region $|x| > 1$ (Fig. 12.17) but we proceed to a closed form solution that can be used over the entire complex plane (apart from the singular points $x = \pm 1$ and with care on cut lines).

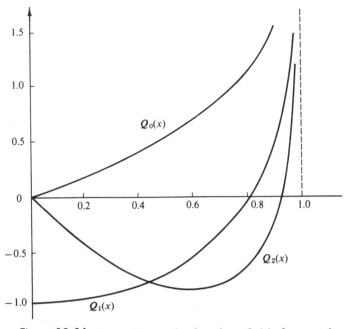

Figure 12.16   Second Legendre function, $Q_n(x)$, $0 \le x < 1$.

CLOSED FORM SOLUTIONS

Frequently, a closed form of the second solution, $Q_n(z)$, is desirable. This may be obtained by the method discussed in Section 8.6. We write

$$Q_n(z) = P_n(z) \left\{ A_n + B_n \int^z \frac{dx}{(1 - x^2)[P_n(x)]^2} \right\}, \qquad (12.210)$$

in which the constant A_n replaces the evaluation of the integral at the arbitrary lower limit. Both constants, A_n and B_n, may be determined for special cases.

For $n = 0$, Eq. (12.210) yields

$$\begin{aligned}
Q_0(z) &= P_0(z) \left\{ A_0 + B_0 \int^z \frac{dx}{(1 - x^2)[P_0(x)]^2} \right\} \\
&= A_0 + B_0 \frac{1}{2} \ln \frac{1 + z}{1 - z} \\
&= A_0 + B_0 \left(z + \frac{z^3}{3} + \frac{z^5}{5} + \cdots + \frac{z^{2s+1}}{2s + 1} + \cdots \right), \qquad (12.211)
\end{aligned}$$

the last expression following from a Maclaurin expansion of the logarithm. Comparing this with the series solution (Eq. (12.203)), we obtain

$$Q_0(z) = q_0(z) = z + \frac{z^3}{3} + \frac{z^5}{5} + \cdots + \frac{z^{2s+1}}{2s + 1} + \cdots, \qquad (12.212)$$

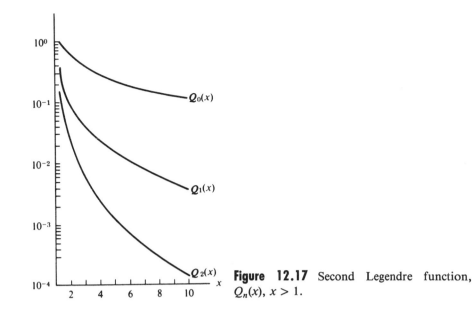

Figure 12.17 Second Legendre function, $Q_n(x)$, $x > 1$.

we have $A_0 = 0$, $B_0 = 1$. Similar results follow for $n = 1$. We obtain

$$Q_1(z) = z\left[A_1 + B_1 \int^z \frac{dx}{(1 - x^2)x^2} \right]$$

$$= A_1 z + B_1 z\left(\frac{1}{2}\ln\frac{1 + z}{1 - z} - \frac{1}{z}\right). \qquad (12.213)$$

Expanding in a power series and comparing with $Q_1(z) = -p_1(z)$, we have $A_1 = 0$, $B_1 = 1$. Therefore we may write

$$Q_0(z) = \frac{1}{2}\ln\frac{1 + z}{1 - z}$$

$$\qquad (12.214)$$

$$Q_1(z) = \frac{1}{2}z\ln\frac{1 + z}{1 - z} - 1, \qquad |z| < 1.$$

Perhaps the best way of determining the higher-order $Q_n(z)$ is to use the recurrence relation (Eq. (12.17)), which may be verified for both $x^2 < 1$ and for $x^2 > 1$ by substituting in the series forms. This recurrence relation technique yields

$$Q_2(z) = \frac{1}{2}P_2(z)\ln\frac{1 + z}{1 - z} - \frac{3}{2}P_1(z). \qquad (12.215)$$

Repeated application of the recurrence formula leads to

$$Q_n(z) = \frac{1}{2} P_n(z) \ln \frac{1+z}{1-z} - \frac{2n-1}{1 \cdot n} P_{n-1}(z) - \frac{2n-5}{3(n-1)} P_{n-3}(z) - \cdots .$$
(12.216)

From the form $\ln [(1+z)/(1-z)]$ it will be seen that for real z these expressions hold in the range $-1 < x < 1$. If we wish to have closed forms valid outside this range, we need only replace

$$\ln \frac{1+x}{1-x} \quad \text{by} \quad \ln \frac{z+1}{z-1}.$$

When using the latter form, valid for large z, we take the line interval $-1 \le x \le 1$ as a cut line. Values of $Q_n(x)$, on the cut line, are customarily assigned by the relation

$$Q_n(x) = \frac{1}{2} [Q_n(x + i0) + Q_n(x - i0)],$$
(12.217)

the arithmetic average of approaches from the positive imaginary side and from the negative imaginary side. The reader will note that for $z \to x > 1$, $z - 1 \to (1 - x)e^{\pm i\pi}$. The result is that for *all* z, except on the real axis $-1 \le x \le 1$, we have

$$Q_0(z) = \frac{1}{2} \ln \frac{z+1}{z-1},$$
(12.218)

$$Q_1(z) = \frac{1}{2} z \ln \frac{z+1}{z-1} - 1,$$
(12.219)

and so on.

For convenient reference some special values of $Q_n(z)$ are given.

1. $Q_n(1) = \infty$, from the logarithmic term (Eq. (12.216)).
2. $Q_n(\infty) = 0$. This is best obtained from a representation of $Q_n(x)$ as a series of negative powers of x, Exercise 12.10.4.
3. $Q_n(-z) = (-1)^{n+1} Q_n(z)$. This follows from the series form. It may also be derived by using $Q_0(z)$, $Q_1(z)$ and the recurrence relation (Eq. (12.17)).
4. $Q_n(0) = 0$, for n even, by (3).

5.
$$Q_n(0) = (-1)^{(n+1)/2} \frac{\{[(n-1)/2]!\}^2}{n!} 2^{n-1}$$

$$= (-1)^{s+1} \frac{(2s)!!}{(2s+1)!!}, \qquad \text{for } n \text{ odd}, n = 2s + 1.$$

This last result comes from the series form (Eq. (12.209)) with $p_n(0) = 1$.

EXERCISES

12.10.1   Derive the parity relation for $Q_n(x)$.

12.10.2   From Eqs. (12.205) and (12.206) show that

(a) $P_{2n}(x) = \dfrac{(-1)^n}{2^{2n-1}} \sum\limits_{s=0}^{n} (-1)^s \dfrac{(2n + 2s - 1)!}{(2s)!(n + s - 1)!(n - s)!} x^{2s}.$

(b) $P_{2n+1}(x) = \dfrac{(-1)^n}{2^{2n}} \sum\limits_{s=0}^{n} (-1)^s \dfrac{(2n + 2s + 1)!}{(2s + 1)!(n + s)!(n - s)!} x^{2s+1}.$

Check the normalization by showing that one term of each series agrees with the corresponding term of Eq. (12.8).

12.10.3   Show that

(a) $Q_{2n}(x) = (-1)^n 2^{2n} \sum\limits_{s=0}^{n} (-1)^s \dfrac{(n + s)!(n - s)!}{(2s + 1)!(2n - 2s)!} x^{2s+1}$

$\qquad + 2^{2n} \sum\limits_{s=n+1}^{\infty} \dfrac{(n + s)!(2s - 2n)!}{(2s + 1)!(s - n)!} x^{2s+1}, \qquad |x| < 1.$

(b) $Q_{2n+1}(x) = (-1)^{n+1} 2^{2n} \sum\limits_{s=0}^{n} (-1)^s \dfrac{(n + s)!(n - s)!}{(2s)!(2n - 2s + 1)!} x^{2s}$

$\qquad + 2^{2n+1} \sum\limits_{s=n+1}^{\infty} \dfrac{(n + s)!(2s - 2n - 2)!}{(2s)!(s - n - 1)!} x^{2s}, \qquad |x| < 1.$

12.10.4   (a) Starting with the assumed form

$$Q_n(x) = \sum_{\lambda=0}^{\infty} b_{-\lambda} x^{k-\lambda},$$

show that

$$Q_n(x) = b_0 x^{-n-1} \sum_{s=0}^{\infty} \frac{(n + s)!(n + 2s)!(2n + 1)!}{s!(n!)^2(2n + 2s + 1)!} x^{-2s}.$$

(b) The standard choice of b_0 is

$$b_0 = \frac{2^n(n!)^2}{(2n + 1)!}.$$

Show that this choice of b_0 brings this negative power series form of $Q_n(x)$ into agreement with the closed form solutions.

12.10.5   Verify that the Legendre functions of the second kind, $Q_n(x)$, satisfy the same recurrence relations as $P_n(x)$, both for $|x| < 1$ and for $|x| > 1$.

$$(2n + 1)xQ_n(x) = (n + 1)Q_{n+1}(x) + nQ_{n-1}(x),$$
$$(2n + 1)Q_n(x) = Q'_{n+1}(x) - Q'_{n-1}(x).$$

12.10.6   (a) Using the recurrence relations, prove (independently of the Wronskian relation) that

$$n[P_n(x)Q_{n-1}(x) - P_{n-1}(x)Q_n(x)] = P_1(x)Q_0(x) - P_0(x)Q_1(x).$$

(b) By direct substitution show that the right-hand side of this equation equals 1.

12.10.7   (a) Write a subroutine that will generate $Q_n(x)$ and lower index Q's based on the recurrence relation for these Legendre functions of the second kind. Take x to be within $(-1, 1)$—excluding the end-points.
Hint. Take $Q_0(x)$ and $Q_1(x)$ to be known.

(b) Test your subroutine for accuracy by computing $Q_{10}(x)$ and comparing with the values tabulated in AMS-55 (Chapter 8).

12.11 VECTOR SPHERICAL HARMONICS

Most of our attention in this chapter has been directed toward solving the equations of scalar fields such as the electrostatic field. This was done primarily because the scalar fields are easier to handle than vector fields! However, with scalar field problems under firm control, more and more attention is being paid to vector field problems.

Maxwell's equations for the vacuum, where the external current and charge densities vanish, lead to the wave (or vector Helmholtz) equation for the vector potential **A**. In a partial wave expansion of **A** in spherical polar coordinates we want to use angular eigenfunctions that are vectors. To this end we write the coordinate unit vectors $\hat{\mathbf{x}}$, $\hat{\mathbf{y}}$, $\hat{\mathbf{z}}$ in spherical notation (see Section 4.4)

$$\hat{\mathbf{e}}_{+1} = -\frac{\hat{\mathbf{x}} + i\hat{\mathbf{y}}}{\sqrt{2}}, \qquad \hat{\mathbf{e}}_0 = \hat{\mathbf{z}}, \qquad \hat{\mathbf{e}}_{-1} = \frac{\hat{\mathbf{x}} - i\hat{\mathbf{y}}}{\sqrt{2}}, \qquad (12.220)$$

so that $\hat{\mathbf{e}}_m$ form a spherical tensor of rank 1. If we couple the spherical harmonics with the $\hat{\mathbf{e}}_m$ to total angular momentum **J** using the relevant Clebsch–Gordan coefficients we are led to the vector spherical harmonics

$$\mathbf{Y}_{JLM_J}(\theta, \varphi) = \sum_{m, M} C(L1J \mid MmM_J)Y_L^M(\theta, \varphi)\hat{\mathbf{e}}_m. \qquad (12.221)$$

It is obvious that they obey the orthogonality relations

$$\int \mathbf{Y}_{JLM_J}^*(\theta, \varphi) \cdot \mathbf{Y}_{J'L'M_J'}(\theta, \varphi) \, d\Omega = \delta_{JJ'} \delta_{LL'} \delta_{M_JM_J'}. \qquad (12.222)$$

Given J, the selection rules of angular momentum coupling tell us that L can only take on the values $J + 1$, J, and $J - 1$. If we look up the Clebsch–Gordan coefficients and invert Eq. (12.221) we get

$$\hat{\mathbf{r}}Y_L^M(\theta, \varphi) = -\left[\frac{L+1}{2L+1}\right]^{1/2} \mathbf{Y}_{LL+1M} + \left[\frac{L}{2L+1}\right]^{1/2} \mathbf{Y}_{LL-1M} \qquad (12.223)$$

displaying the vector character of the **Y**'s and the orbital angular momentum contents, $L + 1$ and $L - 1$, or $\hat{\mathbf{r}}Y_L^M$.

Under the parity operations (coordinate inversion) the vector spherical harmonics transform as

$$\mathbf{Y}_{LL+1M}(\theta', \varphi') = (-1)^{L+1}\mathbf{Y}_{LL+1M}(\theta, \varphi),$$

$$\mathbf{Y}_{LL-1M}(\theta', \varphi') = (-1)^{L+1}\mathbf{Y}_{LL-1M}(\theta, \varphi), \qquad (12.224)$$

$$\mathbf{Y}_{LLM}(\theta', \varphi') = (-1)^{L}\mathbf{Y}_{LLM}(\theta, \varphi),$$

where

$$\theta' = \pi - \theta$$
$$\varphi' = \pi + \varphi. \qquad (12.225)$$

The vector spherical harmonics are useful in a further development of the gradient (Eq. (2.44)), divergence (Eq. (2.45)) and curl (Eq. (2.47)) operators in spherical polar coordinates:

$$\nabla[F(r)Y_L^M(\theta, \varphi)] = -\left[\frac{L+1}{2L+1}\right]^{1/2}\left[\frac{d}{dr} - \frac{L}{r}\right]F\mathbf{Y}_{LL+1M}$$

$$+ \left[\frac{L}{2L+1}\right]^{1/2}\left[\frac{d}{dr} + \frac{L+1}{r}\right]F\mathbf{Y}_{LL-1M}, \quad (12.226)$$

$$\nabla \cdot [F(r)\mathbf{Y}_{LL+1M}(\theta, \varphi)] = -\left(\frac{L+1}{2L+1}\right)^{1/2}\left[\frac{dF}{dr} + \frac{L+2}{r}F\right]Y_L^M(\theta, \varphi), \quad (12.227)$$

$$\nabla \cdot [F(r)\mathbf{Y}_{LL-1M}(\theta, \varphi)] = \left(\frac{L}{2L+1}\right)^{1/2}\left[\frac{dF}{dr} - \frac{L-1}{r}F\right]Y_L^M(\theta, \varphi), \quad (12.228)$$

$$\nabla \cdot [F(r)\mathbf{Y}_{LLM}(\theta, \varphi)] = 0. \qquad (12.229)$$

$$\nabla \times [F(r)\mathbf{Y}_{LL+1M}] = i\left[\frac{L}{2L+1}\right]^{1/2}\left[\frac{dF}{dr} + \frac{L+2}{r}F\right]\mathbf{Y}_{LLM}, \qquad (12.230)$$

$$\nabla \times [F(r)\mathbf{Y}_{LLM}] = i\left(\frac{L}{2L+1}\right)^{1/2}\left[\frac{dF}{dr} - \frac{L}{r}F\right]\mathbf{Y}_{LL+1M}$$

$$+ i\left(\frac{L+1}{2L+1}\right)^{1/2}\left[\frac{dF}{dr} + \frac{(L+1)}{r}F\right]\mathbf{Y}_{LL-1M}, \qquad (12.231)$$

$$\nabla \times [F(r)\mathbf{Y}_{LL-1M}] = i\left[\frac{L+1}{2L+1}\right]^{1/2}\left[\frac{dF}{dr} - \frac{L-1}{r}F\right]\mathbf{Y}_{LLM}. \qquad (12.232)$$

If we substitute Eq. (12.223) into the radial component $\hat{r}\partial/\partial r$ of the gradient operator, for example, we obtain both dF/dr terms in Eq. (12.226). For a complete derivation of Eqs. (12.231) to (12.232) we refer to the literature.[1] These relations play an important role in the partial wave expansion of classical and quantum electrodynamics.

The definitions of the vector spherical harmonics given here are dictated by convenience, primarily in quantum mechanical calculations, in which the angular momentum is a significant parameter.

Further examples of the usefulness and power of the vector spherical harmonics will be found in Blatt and Weisskopf, in Morse and Feshbach, and in Jackson's *Classical Electrodynamics*, 2nd ed. J. Wiley & Sons (1975), which uses vector spherical harmonics in a description of multipole radiation and related electromagnetic problems.

Vector spherical harmonics may be developed as the result of coupling L units of orbital angular momentum and 1 unit of spin angular momentum. An extension, coupling L units of orbital angular momentum and 2 units of spin angular momentum to form *tensor* spherical harmonics, is presented by Mathews.[2] The major application of tensor spherical harmonics is in the investigation of gravitational radiation.

EXERCISES

12.11.1 Construct the $l = 0$, $m = 0$ and $l = 1$, $m = 0$ vector spherical harmonics.

$$ANS. \quad \mathbf{Y}_{010} = -\hat{\mathbf{r}}(4\pi)^{-1/2}$$
$$\mathbf{Y}_{000} = 0$$
$$\mathbf{Y}_{120} = -\hat{\mathbf{r}}(2\pi)^{-1/2}\cos\theta - \hat{\boldsymbol{\theta}}(8\pi)^{-1/2}\sin\theta$$
$$\mathbf{Y}_{110} = \hat{\boldsymbol{\varphi}}i(3/8\pi)^{1/2}\sin\theta$$
$$\mathbf{Y}_{100} = \hat{\mathbf{r}}(4\pi)^{-1/2}\cos\theta - \hat{\boldsymbol{\theta}}(4\pi)^{-1/2}\sin\theta.$$

12.11.2 Verify that the parity of $\mathbf{V}_{LM}$ is $(-1)^{L+1}$, the parity of $\mathbf{X}_{LM}$ is $(-1)^L$, and that of $\mathbf{W}_{LM}$ is $(-1)^{L+1}$. What happened to the M-dependence of the parity?
Hint. $\hat{\mathbf{r}}$ and $\hat{\boldsymbol{\varphi}}$ have odd parity; $\hat{\boldsymbol{\theta}}$ has even parity (compare Exercise 2.5.8).

12.11.3 Verify the orthonormality of the vector spherical harmonics $\mathbf{Y}_{JLM_J}$.

12.11.4 In *Classical Electrodynamics*, 2nd ed., Jackson defines $\mathbf{Y}_{LLM}$ by the equation

$$\mathbf{Y}_{LLM}(\theta, \varphi) = \frac{1}{\sqrt{L(L+1)}} \mathbf{L} Y_L^M(\theta, \varphi),$$

in which the angular momentum operator $\mathbf{L}$ is given by

$$\mathbf{L} = -i(\mathbf{r} \times \nabla).$$

Show that this definition agrees with Eq. (12.221).

[1] E. H. Hill, Theory of vector spherical harmonics. *Am. J. Phys.* **22**, 211 (1954); also J. M. Blatt and V. Weisskopf, *Theoretical Nuclear Physics*. New York: Wiley (1952). Note that Hill assigns phases in accordance with the Condon–Shortley phase convention (Section 4.4). In Hill's notation $\mathbf{X}_{LM} = \mathbf{Y}_{LLM}$, $\mathbf{V}_{LM} = \mathbf{Y}_{LL+1M}$, $\mathbf{W}_{LM} = \mathbf{Y}_{LL-1M}$.
[2] J. Mathews, Gravitational multipole radiation. In *In Memoriam* (H.P. Robertson, ed.). Philadelphia: Society for Industrial and Applied Mathematics (1963).

12.11.5   Show that

$$\sum_{M=-L}^{L} \mathbf{Y}^*_{LLM}(\theta, \varphi) \cdot \mathbf{Y}_{LLM}(\theta, \varphi) = \frac{2L + 1}{4\pi}.$$

Hint. One way is to use Exercise 12.11.4 with **L** expanded in Cartesian coordinates using the raising and lowering operators of Section 4.3.

12.11.6   Show that

$$\int \mathbf{Y}_{LLM} \cdot (\hat{\mathbf{r}} \times \mathbf{Y}_{LLM}) \, d\Omega = 0.$$

The integrand represents an interference term in electromagnetic radiation that contributes to angular distributions but not to total intensity.

ADDITIONAL READINGS

HOBSON, E. W., *The Theory of Spherical and Ellipsoidal Harmonics*. New York: Chelsea (1955). This is a very complete reference, which is the classic text on Legendre polynomials and all related functions.

SMYTHE, W. R., *Static and Dynamic Electricity*, 3rd ed. New York: McGraw-Hill (1969).

See also the references listed in Sections 4.4, 12.9, and at the end of Chapter 13.

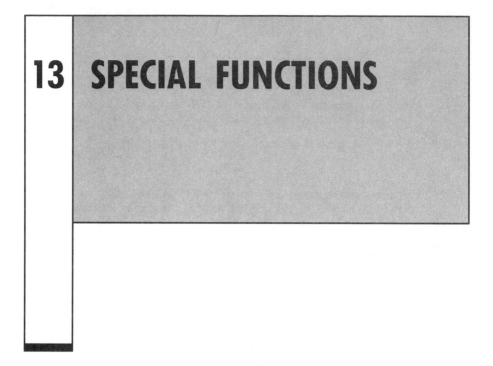

13 | SPECIAL FUNCTIONS

In this chapter we shall study four sets of orthogonal polynomials, Hermite, Laguerre, and Chebyshev[1] of first and second kinds. Although these four sets are of less importance in mathematical physics than the Bessel and Legendre functions of Chapters 11 and 12, they are used occasionally and therefore deserve at least a little attention. For example, Hermite polynomials occur in solutions of the simple harmonic oscillator of quantum mechanics and Laguerre polynomials in wave functions of the hydrogen atom. Because the general mathematical techniques duplicate those of the preceding two chapters, the development of these functions is only outlined. Detailed proofs, along the lines of Chapters 11 and 12, are left to the reader. To conclude the chapter, we express these polynomials and other functions in terms of hypergeometric and confluent hypergeometric functions.

13.1 HERMITE FUNCTIONS

GENERATING FUNCTIONS—HERMITE POLYNOMIALS

The Hermite polynomials (Fig. 13.1), $H_n(x)$, may be defined by the generating function[2]

$$g(x, t) = e^{-t^2 + 2tx} = \sum_{n=0}^{\infty} H_n(x) \frac{t^n}{n!}. \qquad (13.1)$$

---

[1] This is the spelling choice of AMS-55. However, a variety of forms such as Tschebyscheff is encountered.

[2] A derivation of this Hermite generating function is outlined in Exercise 13.1.3.

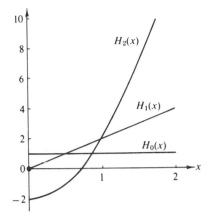

Figure 13.1 Hermite polynomials.

RECURRENCE RELATIONS

Note the absence of a superscript, which distinguishes it from the unrelated Hankel functions. From the generating function we find that the Hermite polynomials satisfy the recurrence relations

$$H_{n+1}(x) = 2xH_n(x) - 2nH_{n-1}(x) \tag{13.2}$$

and

$$H_n'(x) = 2nH_{n-1}(x). \tag{13.3}$$

Equation (13.2) may be obtained by differentiating the generating function with respect to t; differentiation with respect to x leads to Eq. (13.3).

Direct expansion of the generating function easily gives $H_0(x) = 1$ and $H_1(x) = 2x$. Then Eq. (13.2) permits the construction of any $H_n(x)$ desired (integral n). For convenient reference the first several Hermite polynomials are listed in Table 13.1.

Special values of the Hermite polynomials follow from the generating function; that is,

$$H_{2n}(0) = (-1)^n \frac{(2n)!}{n!} \tag{13.4}$$

$$H_{2n+1}(0) = 0. \tag{13.5}$$

Table 13.1 Hermite Polynomials

$H_0(x) = 1$
$H_1(x) = 2x$
$H_2(x) = 4x^2 - 2$
$H_3(x) = 8x^3 - 12x$
$H_4(x) = 16x^4 - 48x^2 + 12$
$H_5(x) = 32x^5 - 160x^3 + 120x$
$H_6(x) = 64x^6 - 480x^4 + 720x^2 - 120$

We also obtain from the generating function the important parity relation

$$H_n(x) = (-1)^n H_n(-x). \tag{13.6}$$

ALTERNATE REPRESENTATIONS

Differentiation of the generating function[3] n times with respect to t and then setting t equal to zero yields

$$H_n(x) = (-1)^n e^{x^2} \frac{d^n}{dx^n} (e^{-x^2}). \tag{13.7}$$

This gives us a Rodrigues representation of $H_n(x)$. A second representation may be obtained by using the calculus of residues (Chapter 7). If we multiply Eq. (13.1) by t^{-m-1} and integrate around the origin, only the term with $H_m(x)$ will survive:

$$H_m(x) = \frac{m!}{2\pi i} \oint t^{-m-1} e^{-t^2+2tx} \, dt. \tag{13.8}$$

Also, from Eq. (13.1) we may write our Hermite polynomial $H_n(x)$ in series form:

$$H_n(x) = (2x)^n - \frac{2\,n!}{(n-2)!2!} (2x)^{n-2} + \frac{4\,n!}{(n-4)!4!} (2x)^{n-4} 1 \cdot 3 \cdots$$

$$= \sum_{s=0}^{[n/2]} (-2)^s (2x)^{n-2s} \binom{n}{2s} 1 \cdot 3 \cdot 5 \cdots (2s-1)$$

$$= \sum_{s=0}^{[n/2]} (-1)^s (2x)^{n-2s} \frac{n!}{(n-2s)!s!}. \tag{13.9}$$

This terminates for integral n and yields our Hermite polynomial.

ORTHOGONALITY

The recurrence relations (Eqs. (13.2) and (13.3) lead to the second-order linear differential equation

$$H_n''(x) - 2xH_n'(x) + 2nH_n(x) = 0, \tag{13.10}$$

which is clearly *not* self-adjoint.

To put Eq. (13.10) in self-adjoint form, we multiply by $\exp(-x^2)$, Exercise 9.1.2. This leads to the orthogonality integral

$$\int_{-\infty}^{\infty} H_m(x)H_n(x) e^{-x^2} \, dx = 0, \qquad m \neq n, \tag{13.10a}$$

[3] Rewrite the generating function as $g(x,t) = e^{x^2} e^{-(t-x)^2}$. Note that

$$\frac{\partial}{\partial t} e^{-(t-x)^2} = -\frac{\partial}{\partial x} e^{-(t-x)^2}.$$

with the weighting function $\exp(-x^2)$ a consequence of putting the differential equation into self-adjoint form. The interval $(-\infty, \infty)$ is selected to satisfy the Hermitian operator boundary conditions, Section 9.1. It is sometimes convenient to absorb the weighting function into the Hermite polynomials. We may define

$$\varphi_n(x) = e^{-x^2/2}H_n(x) \tag{13.11}$$

with $\varphi_n(x)$ no longer a polynomial.

Substitution into Eq. (13.10) yields the differential equation for $\varphi_n(x)$,

$$\varphi_n''(x) + (2n + 1 - x^2)\varphi_n(x) = 0. \tag{13.12}$$

This is the differential equation for a quantum mechanical, simple harmonic oscillator which is perhaps the most important single application of the Hermite polynomials. Equation (13.12) is self-adjoint and the solutions $\varphi_n(x)$ are orthogonal for the interval $(-\infty < x < \infty)$ with a unit weighting function.

The problem of normalizing these functions remains. Proceeding as in Section 12.3, we multiply Eq. (13.1) by itself and then by e^{-x^2}. This yields

$$e^{-x^2} e^{-s^2+2sx} e^{-t^2+2tx} = \sum_{m, n = 0}^{\infty} e^{-x^2}H_m(x)H_n(x) \frac{s^m t^n}{m!n!}. \tag{13.13}$$

When we integrate over x from $-\infty$ to $+\infty$ the cross terms of the double sum drop out because of the orthogonality property[4]

$$\sum_{n=0}^{\infty} \frac{(st)^n}{n!n!} \int_{-\infty}^{\infty} e^{-x^2}[H_n(x)]^2 dx = \int_{-\infty}^{\infty} e^{-x^2-s^2+2sx-t^2+2tx} dx$$

$$= \int_{-\infty}^{\infty} e^{-(x-s-t)^2} e^{2st} dx$$

$$= \pi^{1/2} e^{2st} = \pi^{1/2} \sum_{n=0}^{\infty} \frac{2^n(st)^n}{n!}. \tag{13.14}$$

By equating coefficients of like powers of st, we obtain

$$\int_{-\infty}^{\infty} e^{-x^2}[H_n(x)]^2 dx = 2^n\pi^{1/2}n!. \tag{13.15}$$

QUANTUM MECHANICAL SIMPLE HARMONIC OSCILLATOR

As already indicated, the Hermite polynomials are used in analyzing the quantum mechanical simple harmonic oscillator. For a potential energy $V = \frac{1}{2}Kz^2 = \frac{1}{2}m\omega^2 z^2$ (force $\mathbf{F} = -\nabla V = -Kz$), the Schrödinger wave equation is

$$-\frac{\hbar^2}{2m} \nabla^2\Psi(z) + \frac{1}{2}Kz^2\Psi(z) = E\Psi(z). \tag{13.16}$$

[4] The cross terms $(m \neq n)$ may be left in, if desired. Then, when the coefficients of $s^\alpha t^\beta$ are equated, the orthogonality will be apparent.

Our oscillating particle has mass m and total energy E. By use of the abbreviations

$$x = \alpha z \qquad \text{with} \qquad \alpha^4 = \frac{mK}{\hbar^2} = \frac{m^2\omega^2}{\hbar^2},$$

$$\lambda = \frac{2E}{\hbar}\left(\frac{m}{K}\right)^{1/2} = \frac{2E}{\hbar\omega},$$

(13.17)

in which ω is the angular frequency of the corresponding classical oscillator, Eq. (13.16) becomes [with $\Psi(z) = \Psi(x/\alpha) = \psi(x)$]

$$\frac{d^2\psi(x)}{dx^2} + (\lambda - x^2)\psi(x) = 0.$$

(13.18)

This is Eq. (13.12) with $\lambda = 2n + 1$. Hence (Fig. 13.2),

$$\psi_n(x) = 2^{-n/2}\pi^{-1/4}(n!)^{-1/2}\,e^{-x^2/2}H_n(x) \qquad \text{(normalized)}.$$

(13.19)

The requirement that n be an integer is dictated by the boundary conditions of the quantum mechanical system,

$$\lim_{z \to \pm\infty} \Psi(z) = 0.$$

Specifically, if $n \to \nu$, not an integer, a power-series solution of Eq. (13.10) (Exercise 8.5.6) shows that $H_\nu(x)$ will behave as $x^\nu e^{x^2}$ for large x. The

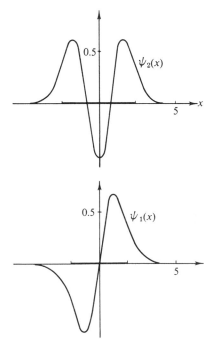

Figure 13.2 Quantum mechanical oscillator wave functions: the heavy bar on the x-axis indicates the allowed range of the classical oscillator with the same total energy.

functions $\psi_\nu(x)$ and $\Psi_\nu(z)$ will therefore blow up at infinity, and it will be impossible to normalize the wave function $\Psi(z)$. With this requirement, energy E becomes

$$E = (n + \tfrac{1}{2})\hbar\omega. \tag{13.20}$$

As n ranges over integral values $(n \geq 0)$, we see that the energy is quantized and that there is a minimum or zero point energy

$$E_{\min} = \tfrac{1}{2}\hbar\omega. \tag{13.21}$$

This zero point energy is an aspect of the uncertainty principle, a purely quantum phenomenon.

RAISING AND LOWERING OPERATORS

An alternate treatment of the quantum mechanical oscillator found in many quantum mechanics texts employs raising and lowering operators:

$$\frac{1}{\sqrt{2}}\left(x - \frac{d}{dx}\right)\psi_n(x) = (n + 1)^{1/2}\psi_{n+1}(x), \tag{13.22a}$$

$$\frac{1}{\sqrt{2}}\left(x + \frac{d}{dx}\right)\psi_n(x) = n^{1/2}\psi_{n-1}(x) \tag{13.22b}$$

Often in quantum mechanics the raising operator is labeled a *creation* operator, $\hat{a}^\dagger$, and the lowering operator an *annihilation* operator, $\hat{a}$. The wave function ψ_n (actually given by Eq. (13.19)) is unknown. The development is similar to the use of the raising and lowering operators presented in Section 4.3. The minimum energy or ground state wave function, ψ_0, satisfies the equation

$$\left(x + \frac{d}{dx}\right)\psi_0(x) = 0. \tag{13.23a}$$

Normalized to unity,

$$\psi_0(x) = \pi^{-1/4}e^{-x^2/2}, \tag{13.23b}$$

in agreement with Eq. (13.19). The excited state wave functions, ψ_1, ψ_2, and so on, are then generated by the raising operator—Eq. (13.22a). The verification of these raising and lowering operators, Eqs. (13.22a) and (13.22b), is left as Exercise 13.1.16.

In quantum mechanical problems, particularly in molecular spectrosopy, a number of integrals of the form

$$\int_{-\infty}^{\infty} x^r e^{-x^2} H_n(x)H_m(x)\,dx$$

are needed. Examples for $r = 1$ and $r = 2$ (with $n = m$) are included in the exercises at the end of this section. A large number of other examples are contained in Wilson, Decius, and Cross.[5]

The oscillator potential has also been employed extensively in calculations of nuclear structure (nuclear shell model) quark models of hadrons, and the nuclear force.

There is a second independent solution of Eq. (13.10). This Hermite function of the second kind is an infinite series (Sections 8.5, 8.6) and of no physical interest, at least not yet.

EXERCISES

13.1.1   Assume the Hermite polynomials are known as solutions of the differential equation (13.10) and from this the recurrence relation, Eq. (13.3), and the values of $H_n(0)$ are also known.

(a) Assume the existence of a generating function

$$g(x, t) = \sum_{n=0}^{\infty} H_n(x)t^n/n!.$$

(b) Differentiate $g(x, t)$ with respect to x and using the recurrence relation develop a first-order differential equation for $g(x, t)$.

(c) Integrate with respect to x, holding t fixed.

(d) Evaluate $g(0, t)$ using Eqs. (13.4) and (13.5). Finally, show that

$$g(x, t) = \exp(-t^2 + 2tx).$$

13.1.2   In developing the properties of the Hermite polynomials, you could start at a number of different points such as:

1.   Hermite differential equations, Eq. (13.10),
2.   Rodrigues' formula, Eq. (13.7),
3.   Integral representation, Eq. (13.8),
4.   Generating function, Eq. (13.1),
5.   Gram–Schmidt construction of a complete set of orthogonal polynomials over $(-\infty, \infty)$ with a weighting factor of $\exp(-x^2)$, Section 9.3.

Outline how you can go from any one of these starting points to all the other points.

13.1.3   From the generating function show that

$$H_n(x) = \sum_{s=0}^{[n/2]} (-1)^s \frac{n!}{(n-2s)!s!} (2x)^{n-2s}.$$

13.1.4   From the generating function derive the recurrence relations

$$H_{n+1}(x) = 2xH_n(x) - 2nH_{n-1}(x),$$
$$H_n'(x) = 2nH_{n-1}(x).$$

---

[5] E. B. Wilson, Jr., J. C. Decius, and P. C. Cross, *Molecular Vibrations.* New York: McGraw–Hill (1955).

13.1.5 Prove that

$$\left(2x - \frac{d}{dx}\right)^n 1 = H_n(x).$$

Hint. Check out the first couple of examples and then use mathematical induction.

13.1.6 Prove that

$$|H_n(x)| \le |H_n(ix)|.$$

13.1.7 Rewrite the series form of $H_n(x)$, Eq. (13.9), as an *ascending* power series.

$$ANS. \quad H_{2n}(x) = (-1)^n \sum_{s=0}^{n} (-1)^s (2x)^{2s} \frac{(2n)!}{(2s)!(n-s)!},$$

$$H_{2n+1}(x) = (-1)^n \sum_{s=0}^{n} (-1)^s (2x)^{2s+1} \frac{(2n+1)!}{(2s+1)!(n-s)!}.$$

13.1.8 (a) Expand x^{2r} in a series of even order Hermite polynomials.
(b) Expand x^{2r+1} in a series of odd order Hermite polynomials.

$$ANS. \quad (a) \quad x^{2r} = \frac{(2r)!}{2^{2r}} \sum_{n=0}^{r} \frac{H_{2n}(x)}{(2n)!(r-n)!}$$

$$(b) \quad x^{2r+1} = \frac{(2r+1)!}{2^{2r+1}} \sum_{n=0}^{r} \frac{H_{2n+1}(x)}{(2n+1)!(r-n)!}, \quad r = 0, 1, 2, \ldots.$$

Hint. Use a Rodrigues representation of $H_{2n}(x)$ and integrate by parts.

13.1.9 Show that

$$(a) \quad \int_{-\infty}^{\infty} H_n(x) \exp[-x^2/2] \, dx = \begin{cases} 2\pi n!/(n/2)!, & n \text{ even} \\ 0, & n \text{ odd.} \end{cases}$$

$$(b) \quad \int_{-\infty}^{\infty} xH_n(x) \exp[-x^2/2] \, dx = \begin{cases} 0, & n \text{ even} \\ 2\pi \frac{(n+1)!}{((n+1)/2)!}, & n \text{ odd.} \end{cases}$$

13.1.10 Show that

$$\int_{-\infty}^{\infty} x^m e^{-x^2} H_n(x) \, dx = 0 \quad \text{for } m \text{ an integer,} \quad 0 \le m \le n - 1.$$

13.1.11 The transition probability between two oscillator states, m and n, depends on

$$\int_{-\infty}^{\infty} xe^{-x^2} H_n(x)H_m(x) \, dx.$$

Show that this integral equals $\pi^{1/2} 2^{n-1} n! \, \delta_{m, n-1} + \pi^{1/2} 2^n (n+1)! \, \delta_{m, n+1}$. This result shows that such transitions can occur only between states of adjacent energy levels, $m = n \pm 1$.
Hint. Multiply the generating function (Eq. (13.1)) by itself using two different sets of variables (x, s) and (x, t). Alternatively, the factor x may be eliminated by the recurrence relation Eq. (13.2).

13.1.12 Show that

$$\int_{-\infty}^{\infty} x^2 e^{-x^2} H_n(x) H_n(x)\, dx = \pi^{1/2} 2^n n! \left(n + \frac{1}{2}\right).$$

This integral occurs in the calculation of the mean-square displacement of our quantum oscillator.

Hint. Use the recurrence relation Eq. (13.2) and the orthogonality integral.

13.1.13 Evaluate

$$\int_{-\infty}^{\infty} x^2 \exp[-x^2] H_n(x) H_m(x)\, dx$$

in terms of n and m and appropriate Kronecker delta functions.

ANS. $2^{n-1}\pi^{1/2}(2n + 1)n!\delta_{n,m} + 2^n\pi^{1/2}(n + 2)!\delta_{n+2,m} + 2^{n-2}\pi^{1/2}n!\delta_{n-2,m}$.

13.1.14 Show that

$$\int_{-\infty}^{\infty} x^r \exp[-x^2] H_n(x) H_{n+p}(x)\, dx = \begin{cases} 0, & p > r \\ 2^n\pi^{1/2}(n + r)!, & p = r. \end{cases}$$

n, p, and r are nonnegative integers.

Hint. Use the recurrence relation, Eq. (13.2), p times.

13.1.15 (a) Using the Cauchy integral formula, develop an integral representation of $H_n(x)$ based on Eq. (13.1) with the contour enclosing the point $z = -x$.

$$ANS. \quad H_n(x) = \frac{n!}{2\pi i} e^{x^2} \oint \frac{e^{-z^2}}{(z + x)^{n+1}}\, dz.$$

(b) Show by direct substitution that this result satisfies the Hermite equation.

13.1.16 With

$$\psi_n(x) = e^{-x^2/2} H_n(x)/(2^n n! \pi^{1/2})^{1/2},$$

verify that

$$\hat{a}_n \psi_n(x) = \frac{1}{\sqrt{2}}\left(x + \frac{d}{dx}\right)\psi_n(x) = n^{1/2}\psi_{n-1}(x),$$

$$\hat{a}_n^\dagger \psi_n(x) = \frac{1}{\sqrt{2}}\left(x - \frac{d}{dx}\right)\psi_n(x) = (n + 1)^{1/2}\psi_{n+1}(x).$$

Note. The usual quantum mechanical operator approach establishes these raising and lowering properties *before* the form of $\psi_n(x)$ is known.

13.1.17 (a) Verify the operator identity

$$x - \frac{d}{dx} = -\exp[x^2/2]\frac{d}{dx}\exp[-x^2/2].$$

(b) The normalized simple harmonic oscillator wave function is

$$\psi_n(x) = (\pi^{1/2}2^n n!)^{-1/2} \exp[-x^2/2]H_n(x).$$

Show that this may be written as

$$\psi_n(x) = (\pi^{1/2}2^n n!)^{-1/2}\left(x - \frac{d}{dx}\right)^n \exp[-x^2/2].$$

Note. This corresponds to an *n*-fold application of the raising operator of Exercise 13.1.16.

13.1.18 (a) Show that the simple oscillator Hamiltonian (from Eq. (13.18)) may be written as

$$H = -\frac{1}{2}\frac{d^2}{dx^2} + \frac{1}{2}x^2 = \frac{1}{2}(\hat{a}\hat{a}^\dagger + \hat{a}^\dagger\hat{a}).$$

Hint. Express *E* in units of $\hbar\omega$.

(b) Using the creation—annihilation operator formulation of part (a)— show that

$$H\psi(x) = (n + \tfrac{1}{2})\psi(x).$$

This means the energy eigenvalues are $E = (n + \tfrac{1}{2})(\hbar\omega)$, in agreement with Eq. (13.20).

13.1.19 Write a program that will generate the coefficients a_s in the polynomial form of the Hermite polynomial, $H_n(x) = \sum_{s=0}^{n} a_s x^s$.

13.1.20 A function $f(x)$ is expanded in an Hermite series:

$$f(x) = \sum_{n=0}^{\infty} a_n H_n(x).$$

From the orthogonality and normalization of the Hermite polynomials the coefficient a_n is given by

$$a_n = \frac{1}{2^n \pi^{1/2} n!} \int_{-\infty}^{\infty} f(x) H_n(x) e^{-x^2} dx.$$

For $f(x) = x^8$ determine the Hermite coefficients a_n by the Gauss–Hermite quadrature (Appendix 2). Check your coefficients against AMS-55, Table 22.12.

13.1.21 (a) In analogy with Exercise 12.2.13 set up the matrix of even Hermite polynomial coefficients that will transform an even Hermite series into an even power series:

$$B = \begin{pmatrix} 1 & -2 & 12 & \cdots \\ 0 & 4 & -48 & \cdots \\ 0 & 0 & 16 & \cdots \\ \vdots & \vdots & \vdots & \vdots \end{pmatrix}.$$

Extend B to handle an even polynomial series through $H_8(x)$.

(b) Invert your matrix to obtain matrix A which will transform an even power series (through x^8) into a series of even Hermite polynomials. Check the elements of A against those listed in AMS-55 (Table 22.12).

(c) Finally, using matrix multiplication, determine the Hermite series equivalent to $f(x) = x^8$.

13.1.22 Write a subroutine that will transform a finite power series $\sum_{n=0}^{N} a_n x^n$ into an Hermite series $\sum_{n=0}^{N} b_n H_n(x)$. Use the recurrence relation Eq. (13.2) and follow the technique outlined in Section 13.4 for a Chebyshev series.

Note. Both Exercises 13.1.21 and 13.1.22 are faster and more accurate than the Gaussian quadrature, Exercise 13.1.20, if $f(x)$ is available as a power series.

13.1.23   Write a subroutine for evaluating Hermite polynomial matrix elements of the form

$$M_{pqr} = \int_{-\infty}^{\infty} H_p(x)H_q(x)x^r e^{-x^2}\, dx,$$

using the 10-point Gauss–Hermite quadrature (for $p + q + r \le 19$). Include a parity check and set equal to zero the integrals with odd parity integrand. Also, check to see if r is in the range $|p - q| \le r \le p + q$. Otherwise $M_{pqr} = 0$. Check your results against the specific cases listed in Exercises 13.1.11, 13.1.12, 13.1.13, and 13.1.14.

13.1.24   Calculate and tabulate the normalized linear oscillator wave functions

$$\psi_n(x) = 2^{-n/2}\pi^{-1/4}(n!)^{-1/2}H_n(x)\exp(-x^2/2) \qquad \text{for } x = 0.0(0.1)5.0$$

and $n = 0(1)5$. If a plotting routine is available, plot your results.

13.2 LAGUERRE FUNCTIONS

DIFFERENTIAL EQUATION—LAGUERRE POLYNOMIALS

If we start with the appropriate generating function, it is possible to develop the Laguerre polynomials in exact analogy with the Hermite polynomials. Alternatively, a series solution may be developed by the methods of Section 8.5. Instead, to illustrate a different technique, let us start with Laguerre's differential equation and obtain a solution in the form of a contour integral, as we did with the modified Bessel function $K_\nu(x)$ (Section 11.6). From this integral representation a generating function will be derived.

Laguerre's differential equation is

$$xy''(x) + (1 - x)y'(x) + ny(x) = 0. \tag{13.24}$$

We shall attempt to represent y, or rather y_n, since y will depend on n, by the contour integral

$$y_n(x) = \frac{1}{2\pi i}\oint \frac{e^{-xz/(1-z)}}{(1-z)z^{n+1}}\, dz. \tag{13.25a}$$

The contour includes the origin but does not enclose the point $z = 1$. By differentiating the exponential in Eq. (13.25a) we obtain

$$y_n'(x) = -\frac{1}{2\pi i}\oint \frac{e^{-xz/(1-z)}}{(1-z)^2 z^n}\, dz, \tag{13.25b}$$

$$y_n''(x) = \frac{1}{2\pi i}\oint \frac{e^{-xz/(1-z)}}{(1-z)^3 z^{n-1}}\, dz. \tag{13.25c}$$

Substituting into the left-hand side of Eq. (13.24), we obtain

$$\frac{1}{2\pi i}\oint \left[\frac{x}{(1-z)^3 z^{n-1}} - \frac{1-x}{(1-z)^2 z^n} + \frac{n}{(1-z)z^{n+1}}\right] e^{-xz/(1-z)}\, dz,$$

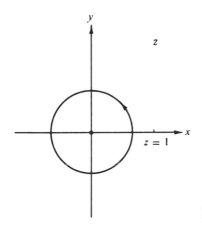

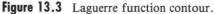

Figure 13.3 Laguerre function contour.

which is equal to

$$\frac{1}{2\pi i} \oint \frac{d}{dz} \left[\frac{e^{-xz/(1-z)}}{(1-z)z^n} \right] dz. \tag{13.26}$$

If we integrate our perfect differential around a contour chosen so that the final value equals the initial value (Fig. 13.3), the integral will vanish, thus verifying that $y_n(x)$ (Eq. (13.25a)) is a solution of Laguerre's equation.

It has become customary to define $L_n(x)$, the Laguerre polynomial (Fig. 13.4), by[1]

$$L_n(x) = \frac{1}{2\pi i} \oint \frac{e^{-xz/(1-z)}}{(1-z)z^{n+1}} dz. \tag{13.27}$$

This is exactly what we would obtain from the series

$$g(x, z) = \frac{e^{-xz/(1-z)}}{1-z} = \sum_{n=0}^{\infty} L_n(x)z^n, \qquad |z| < 1 \tag{13.28}$$

if we multiplied by z^{-n-1} and integrated around the origin. As in the development of the calculus of residues (Section 7.2), only the z^{-1} term in the series survives. On this basis we identify $g(x, z)$ as the generating function for the Laguerre polynomials.

With the transformation

$$\frac{xz}{1-z} = s - x \qquad \text{or} \qquad z = \frac{s-x}{s}, \tag{13.29}$$

$$L_n(x) = \frac{e^x}{2\pi i} \oint \frac{s^n e^{-s}}{(s-x)^{n+1}} ds, \tag{13.30}$$

---

[1] Other definitions of $L_n(x)$ are in use. The definitions here of the Laguerre polynomial $L_n(x)$ and the associated Laguerre polynomial $L_n^k(x)$ agree with AMS-55 (Chapter 22).

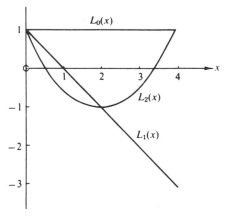

Figure 13.4 Laguerre polynomials.

the new contour enclosing the point $s = x$ in the s-plane. By Cauchy's integral formula (for derivatives)

$$L_n(x) = \frac{e^x}{n!} \frac{d^n}{dx^n} (x^n e^{-x}) \qquad \text{(integral } n), \qquad (13.31)$$

giving Rodrigues' formula for Laguerre polynomials. From these representations of $L_n(x)$ we find the series form (for integral n),

$$L_n(x) = \frac{(-1)^n}{n!}\left[x^n - \frac{n^2}{1!}x^{n-1} + \frac{n^2(n-1)^2}{2!}x^{n-2} - \cdots + (-1)^n n! \right],$$

$$= \sum_{m=0}^{n} (-1)^m \frac{n!}{(n-m)!m!m!} x^m = \sum_{s=0}^{n} (-1)^{n-s} \frac{n! x^{n-s}}{(n-s)!(n-s)!s!}$$

$$(13.32)$$

and the specific polynomials listed in Table 13.2 (Exercise 13.2.1).

By differentiating the generating function in Eq. (13.28), with respect to x and z, we obtain the recurrence relations

$$(n + 1)L_{n+1}(x) = (2n + 1 - x)L_n(x) - nL_{n-1}(x), \qquad (13.33)$$

$$xL'_n(x) = nL_n(x) - nL_{n-1}(x). \qquad (13.34)$$

Table 13.2 Laguerre Polynomials

$L_0(x) = 1$
$L_1(x) = -x + 1$
$2!L_2(x) = x^2 - 4x + 2$
$3!L_3(x) = -x^3 + 9x^2 - 18x + 6$
$4!L_4(x) = x^4 - 16x^3 + 72x^2 - 96x + 24$
$5!L_5(x) = -x^5 + 25x^4 - 200x^3 + 600x^2 - 600x + 120$
$6!L_6(x) = x^6 - 36x^5 + 450x^4 - 2400x^3 + 5400x^2 - 4320x + 720$

Equation (13.33), modified to read

$$L_{n+1}(x) = 2L_n(x) - L_{n-1}(x)$$
$$- [(1 + x)L_n(x) - L_{n-1}(x)]/(n + 1), \tag{13.33a}$$

for reasons of economy and numerical stability, is used for machine computation of numerical values of $L_n(x)$. The computer starts with known numerical values of $L_0(x)$ and $L_1(x)$, Table 13.2, and works up step by step—in milliseconds. This is the same technique discussed for computing Legendre polynomials, Section 12.2.

Also, from Eq. (13.28) we find the special value

$$L_n(0) = 1. \tag{13.35}$$

As may be seen from the form of the generating function, the form of Laguerre's differential equation, or from Table 13.2, the Laguerre polynomials have neither odd nor even symmetry (parity).

The Laguerre differential equation is not self-adjoint and the Laguerre polynomials, $L_n(x)$, do not by themselves form an orthogonal set. However, following the method of Section 9.1, we may multiply Eq. (13.24) by e^{-x} (Exercise 9.1.1) and obtain

$$\int_0^\infty e^{-x} L_m(x) L_n(x)\, dx = \delta_{m,n}. \tag{13.36}$$

This orthogonality is a consequence of the Sturm–Liouville theory, Section 9.1. The normalization follows from the generating function. It is sometimes convenient to define orthogonalized Laguerre functions (with unit weighting function) by

$$\varphi_n(x) = e^{-x/2} L_n(x). \tag{13.37}$$

Our new orthonormal function $\varphi_n(x)$ satisfies the differential equation

$$x\varphi_n''(x) + \varphi_n'(x) + \left(n + \frac{1}{2} - \frac{x}{4}\right)\varphi_n(x) = 0, \tag{13.38}$$

which is seen to have the Sturm–Liouville form (self-adjoint). Note that it is the boundary conditions in the Sturm–Liouville theory that fix our interval as $(0 \le x < \infty)$.

ASSOCIATED LAGUERRE POLYNOMIALS

In many applications, particularly in quantum theory, we need the associated Laguerre polynomials defined by[2]

$$L_n^k(x) = (-1)^k \frac{d^k}{dx^k} [L_{n+k}(x)]. \tag{13.39}$$

[2] Some authors use $\mathcal{L}_{n+k}^k(x) = (d^k/dx^k)[L_{n+k}(x)]$. Hence our $L_n^k(x) = (-1)^k \mathcal{L}_{n+k}^k(x)$.

From the series form of $L_n(x)$

$$L_0^k(x) = 1$$

$$L_1^k(x) = -x + k + 1$$

$$L_2^k(x) = \frac{x^2}{2} - (k + 2)x + \frac{(k + 2)(k + 1)}{2}.$$

(13.40)

In general,

$$L_n^k(x) = \sum_{m=0}^{n} (-1)^m \frac{(n + k)!}{(n - m)!(k + m)!m!} x^m, \qquad k > -1. \quad (13.41)$$

A generating function may be developed by differentiating the Laguerre generating function k times. Adjusting the index to L_{n+k}, we obtain

$$\frac{e^{-xz/(1-z)}}{(1 - z)^{k+1}} = \sum_{n=0}^{\infty} L_n^k(x)z^n, \qquad |z| < 1. \quad (13.42)$$

From this

$$L_n^k(0) = \frac{(n + k)!}{n!k!}. \quad (13.43)$$

Recurrence relations can easily be derived from the generating function or by differentiating the Laguerre polynomial recurrence relations. Among the numerous possibilities are

$$(n + 1)L_{n+1}^k(x) = (2n + k + 1 - x)L_n^k(x) - (n + k)L_{n-1}^k(x) \quad (13.44)$$

$$xL_n^{k\prime}(x) = nL_n^k(x) - (n + k)L_{n-1}^k(x). \quad (13.45)$$

From these or from differentiating Laguerre's differential equation k times we have the associated Laguerre equation

$$xL_n^{k\prime\prime}(x) + (k + 1 - x)L_n^{k\prime}(x) + nL_n^k(x) = 0. \quad (13.46)$$

When associated Laguerre polynomials appear in a physical problem it is usually because that physical problem involves Eq. (13.46).

A Rodrigues representation of the associated Laguerre polynomial is

$$L_n^k(x) = \frac{e^x x^{-k}}{n!} \frac{d^n}{dx^n} (e^{-x} x^{n+k}). \quad (13.47)$$

The reader will note that all these formulas for $L_n^k(x)$ reduce to the corresponding expressions for $L_n(x)$ when $k = 0$.

The associated Laguerre equation (13.46) is not self-adjoint but it can be put in self-adjoint form by multiplying by $e^{-x}x^k$, which becomes the weighting function (Section 9.1). We obtain

$$\int_0^{\infty} e^{-x}x^k L_n^k(x)L_m^k(x)\, dx = \frac{(n + k)!}{n!} \delta_{m,n}. \quad (13.48)$$

Equation (13.48) shows the same orthogonality interval $(0, \infty)$ as that for the Laguerre polynomials, but with a new weighting function we have a new set of orthogonal polynomials, the associated Laguerre polynomials.

By letting $\psi_n^k(x) = e^{-x/2} x^{k/2} L_n^k(x)$, $\psi_n^k(x)$ satisfies the self-adjoint equation

$$x \psi_n^{k\,\prime\prime}(x) + \psi_n^{k\,\prime}(x) + \left(-\frac{x}{4} + \frac{2n + k + 1}{2} - \frac{k^2}{4x} \right) \psi_n^k(x) = 0. \quad (13.49)$$

The $\psi_n^k(x)$ are sometimes called Laguerre *functions*. Equation (13.36) is the special case $k = 0$.

A further useful form is given by defining[3]

$$\Phi_n^k(x) = e^{-x/2} x^{(k+1)/2} L_n^k(x). \quad (13.50)$$

Substitution into the associated Laguerre equation yields

$$\Phi_n^{k\,\prime\prime}(x) + \left(-\frac{1}{4} + \frac{2n + k + 1}{2x} - \frac{k^2 - 1}{4x^2} \right) \Phi_n^k(x) = 0. \quad (13.51)$$

The corresponding normalization integral is

$$\int_0^\infty e^{-x} x^{k+1} L_n^k(x)\, L_n^k(x)\, dx = \frac{(n + k)!}{n!} (2n + k + 1). \quad (13.52)$$

The reader may show that the $\Phi_n^k(x)$ do *not* form an orthogonal set (except with x^{-1} as a weighting function) because of the x^{-1} in the term $(2n + k + 1)/2x$.

The Laguerre functions $L_\nu^\mu(x)$ in which the indices ν and μ are *not* integers may be defined using the confluent hypergeometric functions of Section 13.6.

Example 13.2.1 The Hydrogen Atom

Perhaps the most important single application of the Laguerre polynomials is in the solution of the Schrödinger wave equation for the hydrogen atom. This equation is

$$-\frac{\hbar^2}{2m} \nabla^2 \psi - \frac{Ze^2}{r} \psi = E\psi, \quad (13.53)$$

in which $Z = 1$ for hydrogen, 2 for singly ionized helium, and so on. Separating variables, we find that the angular dependence of ψ is $Y_L^M(\theta, \varphi)$. The radial part, $R(r)$, satisfies the equation

$$-\frac{\hbar^2}{2m} \frac{1}{r^2} \frac{d}{dr} \left(r^2 \frac{dR}{dr} \right) - \frac{Ze^2}{r} R + \frac{\hbar^2}{2m} \frac{L(L + 1)}{r^2} R = ER. \quad (13.54)$$

---

[3] This corresponds to modifying the function ψ in Eq. (13.49) to eliminate the first derivative (compare Exercise 8.6.11).

By use of the abbreviations

$$\rho = \alpha r \quad \text{with } \alpha^2 = -\frac{8mE}{\hbar^2}, \quad E < 0,$$

$$\lambda = \frac{2mZe^2}{\alpha\hbar^2},$$

(13.55)

Eq. (13.54) becomes

$$\frac{1}{\rho^2}\frac{d}{d\rho}\left(\rho^2\frac{d\chi(\rho)}{d\rho}\right) + \left(\frac{\lambda}{\rho} - \frac{1}{4} - \frac{L(L+1)}{\rho^2}\right)\chi(\rho) = 0,$$

(13.56)

where $\chi(\rho) = R(\rho/\alpha)$. A comparison with Eq. (13.51) for $\Phi_n^k(x)$ shows that Eq. (13.56) is satisfied by

$$\rho\chi(\rho) = e^{-\rho/2}\rho^{L+1}L_{\lambda-L-1}^{2L+1}(\rho),$$

(13.57)

In which k is replaced by $2L + 1$ and n by $\lambda - L - 1$.

We must restrict the parameter λ by requiring it to be an integer n, $n = 1, 2, 3, \ldots$[4] This is necessary because the Laguerre function of nonintegral n would diverge as $\rho^n e^\rho$, which is unacceptable for our physical problem in which

$$\lim_{r \to \infty} R(r) = 0.$$

This restriction on λ, imposed by our boundary condition, has the effect of quantizing the energy

$$E_n = -\frac{Z^2me^4}{2n^2\hbar^2}.$$

(13.58)

The negative sign enters because we are dealing here with bound states, $E = 0$, corresponding to an electron that is just able to escape to infinity. Using this result for E_n, we have

$$\alpha = 2\frac{me^2}{\hbar^2}\cdot\frac{Z}{n} = \frac{2Z}{na_0}, \quad \rho = \frac{2Z}{na_0}r$$

(13.59)

with

$$a_0 = \frac{\hbar^2}{me^2} \quad \text{the Bohr radius.}$$

The final normalized hydrogen wave function may be written as

$$\psi_{nLM}(r, \theta, \varphi) = \left[\left(\frac{2Z}{na_0}\right)^3\frac{(n-L-1)!}{2n(n+L)!}\right]^{1/2} e^{-\alpha r/2}(\alpha r)^L L_{n-L-1}^{2L+1}(\alpha r) Y_L^M(\theta, \varphi).$$

(13.60)

[4] This is the conventional notation for λ. It is *not* the same n as the index n in $\Phi_n^k(x)$.

EXERCISES

13.2.1   Show with the aid of the Leibniz formula, that the series expansion of $L_n(x)$ (Eq. (13.32)) follows from the Rodrigues representation (Eq. (13.31)).

13.2.2   (a) Using the explicit series form (Eq. (13.32)) show that

$$L'_n(0) = -n$$
$$L''_n(0) = \tfrac{1}{2}n(n-1).$$

(b) Repeat without using the explicit series form of $L_n(x)$.

13.2.3   From the generating function derive the Rodrigues representation

$$L_n^k(x) = \frac{e^x x^{-k}}{n!} \frac{d^n}{dx^n}(e^{-x}x^{n+k}).$$

13.2.4   Derive the normalization relation (Eq. (13.48)) for the associated Laguerre polynomials.

13.2.5   Expand x^r in a series of associated Laguerre polynomials $L_n^k(x)$, k fixed and n ranging from 0 to r (or to ∞ if r is not an integer).
Hint. The Rodrigues form of $L_n^k(x)$ will be useful.

$$ANS. \quad x^r = (r+k)!r! \sum_{n=0}^{r} \frac{(-1)^n L_n^k(x)}{(n+k)!(r-n)!}, \qquad 0 \le x \le \infty.$$

13.2.6   Expand e^{-ax} in a series of associated Laguerre polynomials $L_n^k(x)$, k fixed and n ranging from 0 to ∞.

(a) Evaluate directly the coefficients in your assumed expansion.
(b) Develop the desired expansion from the generating function.

$$ANS. \quad e^{-ax} = \frac{1}{(1+a)^{1+k}} \sum_{n=0}^{\infty} \left(\frac{a}{1+a}\right)^n L_n^k(x), \qquad 0 \le x < \infty.$$

13.2.7   Show that

$$\int_0^{\infty} e^{-x}x^{k+1}L_n^k(x)L_n^k(x)\, dx = \frac{(n+k)!}{n!}(2n+k+1).$$

Hint. Note that

$$xL_n^k = (2n+k+1)L_n^k - (n+k)L_{n-1}^k - (n+1)L_{n+1}^k.$$

13.2.8   Assume that a particular problem in quantum mechanics has led to the differential equation

$$\frac{d^2y}{dx^2} - \left[\frac{k^2-1}{4x^2} - \frac{2n+k+1}{2x} + \frac{1}{4}\right]y = 0.$$

Write $y(x)$ as

$$y(x) = A(x)B(x)C(x)$$

with the requirement that

(a) $A(x)$ be a *negative* exponential giving the required asymptotic behavior of $y(x)$ and

(b) $B(x)$ be a *positive* power of x giving the behavior of $y(x)$ for $x \ll 1$. Determine $A(x)$ and $B(x)$. Find the relation between $C(x)$ and the associated Laguerre polynomial.

$$\text{ANS.} \quad A(x) = e^{-x/2}$$
$$B(x) = x^{(k+1)/2},$$
$$C(x) = L_n^k(x).$$

13.2.9   From Eq. (13.60) the normalized radial part of the hydrogenic wave function is

$$R_{nL}(r) = \left[\alpha^3 \frac{(n - L - 1)!}{2n(n + L)!} \right]^{1/2} e^{-\alpha r/2} (\alpha r)^L L_{n-L-1}^{2L+1}(\alpha r),$$

in which $\alpha = 2Z/na_0 = 2Zme^2/n\hbar^2$. Evaluate

(a) $\langle r \rangle = \displaystyle\int_0^\infty r R_{nL}(\alpha r) R_{nL}(\alpha r) r^2 \, dr$, (b) $\langle r^{-1} \rangle = \displaystyle\int_0^\infty r^{-1} R_{nL}(\alpha r) R_{nL}(\alpha r) r^2 \, dr$.

The quantity $\langle r \rangle$ is the average displacement of the electron from the nucleus, whereas $\langle r^{-1} \rangle$ is the average of the reciprocal displacement.

$$\text{ANS.} \qquad \langle r \rangle = \frac{a_0}{2} [3n^2 - L(L + 1)]$$

$$\langle r^{-1} \rangle = \frac{1}{n^2 a_0}.$$

13.2.10   Derive the recurrence relation for the hydrogen wave function expectation values.

$$\frac{s + 2}{n^2} \langle r^{s+1} \rangle - (2s + 3)a_0 \langle r^s \rangle + \frac{s + 1}{4} [(2L + 1)^2 - (s + 1)^2] a_0^2 \langle r^{s-1} \rangle = 0$$

with $s \geq -2L - 1$. $\langle r^s \rangle \equiv \overline{r^s}$.
Hint. Transform Eq. (13.56) into a form analogous to Eq. (13.51). Multiply by $\rho^{s+2} u' - c\rho^{s+1} u$. Here $u = \rho\Phi$. Adjust c to cancel terms that do not yield expectation values.

13.2.11   The hydrogen wave functions, Eq. (13.60), are mutually orthogonal as they should be, since they are eigenfunctions of the self-adjoint Schrödinger equation.

$$\int \psi_{n_1 L_1 M_1}^* \psi_{n_2 L_2 M_2} r^2 \, dr \, d\Omega = \delta_{n_1 n_2} \delta_{L_1 L_2} \delta_{m_1 m_2}.$$

Yet the radial integral has the (misleading) form

$$\int_0^\infty e^{-\alpha r/2} (\alpha r)^L L_{n_1-L-1}^{2L+1}(\alpha r) \, e^{-\alpha r/2} (\alpha r)^L L_{n_2-L-1}^{2L+1}(\alpha r)^2 \, dr,$$

which *appears* to match Eq. (13.52) and not the associated Laguerre orthogonality relation, Eq. (13.48). How do you resolve this paradox?

ANS. The parameter α is dependent on n. The first three α's previously shown are $2Z/n_1 a_0$. The last three are $2Z/n_2 a_0$. For $n_1 = n_2$ Eq. (13.52) applies. For $n_1 \neq n_2$ neither Eq. (13.48) nor Eq. (13.52) is applicable.

13.2.12   A quantum mechanical analysis of the Stark effect (parabolic coordinate) leads to the differential equation

$$\frac{d}{d\xi}\left(\xi\frac{du}{d\xi}\right) + \left(\frac{1}{2}E\xi + L - \frac{m^2}{4\xi} - \frac{1}{4}F\xi^2\right)u = 0.$$

Here F is a measure of the perturbation energy introduced by an external electric field. Find the unperturbed wave functions ($F = 0$) in terms of associated Laguerre polynomials.

ANS. $u(\xi) = e^{-\varepsilon\xi/2}\xi^{m/2}L_p^m(\varepsilon\xi)$, with $\varepsilon = \sqrt{-2E} > 0$,
$p = \alpha/\varepsilon - (m + 1)/2$, a nonnegative integer.

13.2.13   The wave equation for the three-dimensional harmonic oscillator is

$$-\frac{\hbar^2}{2M}\nabla^2\psi + \tfrac{1}{2}M\omega^2r^2\psi = E\psi.$$

Here ω is the angular frequency of the corresponding classical oscillator. Show that the radial part of ψ (in spherical polar coordinates) may be written in terms of associated Laguerre functions of argument (βr^2), where $\beta = M\omega/\hbar$.

Hint. As in Exercise 13.2.8, split off radial factors of r^l and $e^{-\beta r^2/2}$. The associated Laguerre function will have the form $L_{1/2(n-l-1)}^{l+1/2}(\beta r^2)$.

13.2.14   Write a program that will generate the coefficients a_s in the polynomial form of the Laguerre polynomial, $L_n(x) = \sum_{s=0}^n a_s x^s$.

13.2.15   Write a subroutine that will transform a finite power series $\sum_{n=0}^N a_n x^n$ into a Laguerre series $\sum_{n=0}^N b_n L_n(x)$. Use the recurrence relation Eq. (13.33).

13.2.16   Tabulate $L_{10}(x)$ for $x = 0.0(0.1)30.0$. This will include the 10 roots of L_{10}. Beyond $x = 30.0$, $L_{10}(x)$ is monotonic increasing. If a plotting subroutine is available, plot your results.

Check value.   Eighth root = 16.279.

13.2.17   Determine the 10 roots of $L_{10}(x)$ using a root-finding subroutine (compare Appendix 1). You may use your knowledge of the approximate location of the roots or develop a search routine to look for the roots. The 10 roots of $L_{10}(x)$ are the evaluation points for the 10-point Gauss–Laguerre quadrature (compare Appendix 2). Check your values by comparing with AMS-55 (Table 25.9).

13.2.18   Calculate the coefficients of a Laguerre series expansion ($L_n(x)$, $k = 0$) of the exponential e^{-x}. Evaluate the coefficients by the Gauss–Laguerre quadrature (compare Eq. (9.65)). Check your results against the values given in Exercise 13.2.6.

Note. Direct application of the Gauss–Laguerre quadrature with $f(x) = L_n(x)e^{-x}$ gives poor accuracy because of the extra e^{-x}. Try a change of variable, $y = 2x$, so that the function appearing in the integrand will be simply $L_n(y/2)$.

13.2.19   (a) Write a subroutine to calculate the Laguerre matrix elements

$$M_{mnp} = \int_0^\infty L_m(x)L_n(x)x^p\,e^{-x}\,dx.$$

Include a check that the condition $|m - n| \leq p \leq m + n$. (If p is outside this range, $M_{mnp} = 0$. Why?)

Note. A 10-point Gauss–Laguerre quadrature will give accurate results for $m + n + p \leq 19$.

(b) Call your subroutine to calculate a variety of Laguerre matrix elements. Check M_{mn1} against Exercise 13.2.7.

13.2.20   Write a subroutine to calculate the numerical value of $L_n^k(x)$ for specified values of n, k, and x. Require that n and k be nonnegative integers and x be ≥ 0.

Hint. Starting with known values of L_0^k and $L_1^k(x)$, we may use the recurrence relation, Eq. (13.44), to generate $L_n^k(x)$, $n = 2, 3, 4, \ldots$.

13.2.21   Write a program to calculate the normalized hydrogen radial wave function $\psi_{nL}(r)$. This is ψ_{nLM} of Eq. (13.60), omitting the spherical harmonic $Y_L^M(\theta, \varphi)$. Take $Z = 1$ and $a_0 = 1$ (which means that r is being expressed in units of Bohr radii). Accept n and L as input data. Tabulate $\psi_{nL}(r)$ for $r = 0.0(0.2)R$ with R taken large enough to exhibit the significant features of ψ. This means roughly $R = 5$ for $n = 1$, 10 for $n = 2$, and 30 for $n = 3$.

13.3 CHEBYSHEV (TSCHEBYSCHEFF) POLYNOMIALS

In this section two types of Chebyshev polynomials are developed as special cases of ultraspherical polynomials. Their properties follow from the ultraspherical polynomial generating function. The primary importance of the Chebyshev polynomials is in numerical analysis.

GENERATING FUNCTIONS

In Section 12.1 the generating function for the ultraspherical or Gegenbauer polynomials

$$\frac{1}{(1 - 2xt + t^2)^\alpha} = \sum_{n=0}^{\infty} C_n^{(\alpha)}(x)t^n, \qquad |x| < 1, \quad |t| < 1 \qquad (13.61)$$

was mentioned, with $\alpha = \frac{1}{2}$ giving rise to the Legendre polynomials. In this section we first take $\alpha = 1$ and then $\alpha = 0$ to generate two sets of polynomials known as the Chebyshev polynomials.

TYPE II

With $\alpha = 1$ and $C_n^{(1)}(x) = U_n(x)$, Eq. (13.61) gives

$$\frac{1}{(1 - 2xt + t^2)} = \sum_{n=0}^{\infty} U_n(x)t^n, \qquad |x| < 1, \quad |t| < 1. \qquad (13.62)$$

These functions, $U_n(x)$, generated by $(1 - 2xt + t^2)^{-1}$ are labeled Chebyshev polynomials type II. Although these polynomials have few applications in mathematical physics, one unusual application is in the development of four-dimensional spherical harmonics used in angular momentum theory.

TYPE I

With $\alpha = 0$ there is difficulty. Indeed, our generating function reduces to the constant 1. We may avoid this problem by first differentiating Eq. (13.61) with respect to t. This yields

$$\frac{-\alpha(-2x + 2t)}{(1 - 2xt + t^2)^{\alpha+1}} = \sum_{n=1}^{\infty} nC_n^{(\alpha)}(x)t^{n-1} \tag{13.63}$$

or

$$\frac{x - t}{(1 - 2xt + t^2)^{\alpha+1}} = \sum_{n=1}^{\infty} \frac{n}{2}\left[\frac{C_n^{(\alpha)}(x)}{\alpha}\right] t^{n-1}. \tag{13.64}$$

We *define* $C_n^{(0)}(x)$ by

$$C_n^{(0)}(x) = \lim_{\alpha \to 0} \frac{C_n^{(0)}(x)}{\alpha}. \tag{13.65}$$

The purpose of differentiating with respect to t was to get α in the denominator and to create an indeterminate form. Now multiplying Eq. (13.64) by $2t$ and adding $1 = (1 - 2xt + t^2)/(1 - 2xt + t^2)$, we obtain

$$\frac{1 - t^2}{1 - 2xt + t^2} = 1 + 2 \sum_{n=1}^{\infty} \frac{n}{2} C_n^{(0)}(x)t^n. \tag{13.66}$$

We define $T_n(x)$ by

$$T_n(x) = \begin{cases} 1, & n = 0 \\ \dfrac{n}{2} C_n^{(0)}(x), & n > 0. \end{cases} \tag{13.67}$$

Notice the special treatment for $n = 0$. This is similar to the treatment of $n = 0$ term in the Fourier series. Also, note carefully that $C_n^{(0)}$ is the limit indicated in Eq. (13.65) and not a literal substitution of $\alpha = 0$ into the generating function series. With these new labels,

$$\frac{1 - t^2}{1 - 2xt + t^2} = T_0(x) + 2 \sum_{n=1}^{\infty} T_n(x)t^n, \qquad |x| \leq 1, \quad |t| < 1. \tag{13.68}$$

We call $T_n(x)$ the Chebyshev polynomials, type I. The reader should be warned that the notation for these functions differs from reference to reference. There is almost no general agreement. Here we follow the usage of AMS-55.

These Chebyshev polynomials (type I), which combine useful features of (1) the Fourier series and (2) orthogonal polynomials, are of great interest in numerical computation. For example, a least-squares approximation minimizes the average squared error. An approximation using Chebyshev polynomials allows a larger average squared error but may keep extreme errors down.

Differentiating the generating function (Eqs. (13.62) and (13.68)) with respect to t, we obtain recurrence relations

$$T_{n+1}(x) - 2xT_n(x) + T_{n-1}(x) = 0, \tag{13.69}$$

$$U_{n+1}(x) - 2xU_n(x) + U_{n-1}(x) = 0 \tag{13.70}$$

(see Table 13.3).

Then, using the generating functions for the first few values of n and these recurrence relations for the higher-order polynomials, we get Tables 13.4 and 13.5 (see also Figs. 13.5 and 13.6).

As with the Hermite polynomials, Section 13.1, the recurrence relations, Eqs. (13.69) and (13.70), together with the known values of $T_0(x)$, $T_1(x)$, $U_0(x)$, and $U_1(x)$, provide a convenient—that is, for a high-speed electronic computer—means of getting the numerical value of any $T_n(x_0)$ or $U_n(x_0)$, with x_0 a given number.

Table 13.3 Orthogonal Polynomial Recurrence Relation[a]
$$P_{n+1}(x) = (A_n x + B_n)P_n(x) - C_n P_{n-1}(x)$$

	$P_n(x)$	A_n	B_n	C_n
Legendre	$P_n(x)$	$\dfrac{2n + 1}{n + 1}$	0	$\dfrac{1}{n + 1}$
Chebyshev I	$T_n(x)$	2	0	1
Shifted Chebyshev I	$T_n^*(x)$	4	-2	1
Chebyshev II	$U_n(x)$	2	0	1
Shifted Chebyshev II	$U_n^*(x)$	4	-2	1
Laguerre Associated Laguerre	$L_n^{(k)}(x)$	$-\dfrac{1}{n + 1}$	$\dfrac{2n + k + 1}{n + 1}$	$\dfrac{n + k}{n + 1}$
Hermite	$H_n(x)$	2	0	$2n$

[a] P_n is any orthogonal polynomial.

Table 13.4 Chebyshev Polynomials, Type I

$T_0 = 1$
$T_1 = x$
$T_2 = 2x^2 - 1$
$T_3 = 4x^3 - 3x$
$T_4 = 8x^4 - 8x^2 + 1$
$T_5 = 16x^5 - 20x^3 + 5x$
$T_6 = 32x^6 - 48x^4 + 18x^2 - 1$

Table 13.5 Chebyshev Polynomials, Type II

$U_0 = 1$
$U_1 = 2x$
$U_2 = 4x^2 - 1$
$U_3 = 8x^3 - 4x$
$U_4 = 16x^4 - 12x^2 + 1$
$U_5 = 32x^5 - 32x^3 + 6x$
$U_6 = 64x^6 - 80x^4 + 24x^2 - 1$

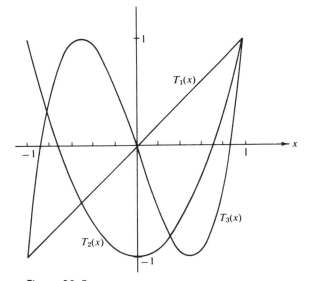

Figure 13.5 Chebyshev polynomials, $T_n(x)$.

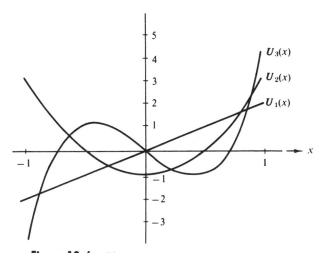

Figure 13.6 Chebyshev polynomials, $U_n(x)$.

Again, from the generating functions, we have the special values

$$T_n(1) = 1$$
$$T_n(-1) = (-1)^n$$
$$T_{2n}(0) = (-1)^n$$
$$T_{2n+1}(0) = 0$$

(13.71)

$$U_n(1) = n + 1$$
$$U_n(-1) = (-1)^n(n + 1)$$
$$U_{2n}(0) = (-1)^n$$
$$U_{2n+1}(0) = 0.$$

(13.72)

The parity relations for T_n and U_n are

$$T_n(x) = (-1)^n T_n(-x)$$

(13.73)

$$U_n(x) = (-1)^n U_n(-x).$$

(13.74)

Rodrigues' representations of $T_n(x)$ and $U_n(x)$ are

$$T_n(x) = \frac{(-1)^n \pi^{1/2}(1 - x^2)^{1/2}}{2^n(n - \frac{1}{2})!} \frac{d^n}{dx^n}[(1 - x^2)^{n-1/2}],$$

(13.75)

and

$$U_n(x) = \frac{(-1)^n(n + 1)\pi^{1/2}}{2^{n+1}(n + \frac{1}{2})!(1 - x^2)^{1/2}} \frac{d^n}{dx^n}[(1 - x^2)^{n+1/2}].$$

(13.76)

RECURRENCE RELATIONS—DERIVATIVES

From the generating functions for $T_n(x)$ and $U_n(x)$ differentiation with respect to x leads to a variety of recurrence relations involving derivatives. Among the more useful equations are

$$(1 - x^2)T_n'(x) = -nxT_n(x) + nT_{n-1}(x),$$

(13.77)

and

$$(1 - x^2)U_n'(x) = -nxU_n(x) + (n + 1)U_{n-1}(x).$$

(13.78)

From Eqs. (13.69) and (13.77), $T_n(x)$ the Chebyshev polynomial type I satisfies

$$(1 - x^2)T_n''(x) - xT_n'(x) + n^2 T_n(x) = 0.$$

(13.79)

$U_n(x)$ the Chebyshev polynomial of type II satisfies

$$(1 - x^2)U_n''(x) - 3xU_n'(x) + n(n + 2)U_n(x) = 0.$$

(13.80)

The ultraspherical equation

$$(1 - x^2)\frac{d^2}{dx^2}C_n^{(\alpha)}(x) - (2\alpha + 1)x\frac{d}{dx}C_n^{(\alpha)}(x) + n(n + 2\alpha)C_n^{(\alpha)}(x) = 0$$

(13.81)

is a generalization of these differential equations, reducing to Eq. (13.79) for $\alpha = 0$ and Eq. (13.80) for $\alpha = 1$ (and to Legendre's equation for $\alpha = \frac{1}{2}$).

TRIGONOMETRIC FORM

At this point in the development of the properties of the Chebyshev solutions it is beneficial to change variables replacing x by $\cos\theta$. With $x = \cos\theta$ and $d/dx = (-1/\sin\theta)(d/d\theta)$. Equation (13.79) becomes

$$\frac{d^2 T_n}{d\theta^2} + n^2 T_n = 0, \tag{13.82}$$

the simple harmonic oscillator equation with solutions $\cos n\theta$ and $\sin n\theta$. The special values (boundary conditions) identify

$$T_n = \cos n\theta = \cos n(\text{arc}\cos x). \tag{13.83a}$$

A second linearly independent solution of Eqs. (13.79) and (13.82) is labeled

$$V_n = \sin n\theta = \sin n(\text{arc}\cos x). \tag{13.83b}$$

The solutions of the type II Chebyshev equation, Eq. (13.80), become

$$U_n = \frac{\sin(n+1)\theta}{\sin\theta} \tag{13.84a}$$

$$W_n = \frac{\cos(n+1)\theta}{\sin\theta}. \tag{13.84b}$$

The two sets of solutions, type I and type II, are related by

$$V_n(x) = (1 - x^2)^{1/2} U_{n-1}(x) \tag{13.85a}$$

$$W_n(x) = (1 - x^2)^{-1/2} T_{n+1}(x). \tag{13.85b}$$

As already seen from generating functions, $T_n(x)$ and $U_n(x)$ are polynomials. Clearly, $V_n(x)$ and $W_n(x)$ are *not* polynomials.
From

$$T_n(x) + iV_n(x) = \cos n\theta + i\sin n\theta$$

$$= (\cos\theta + i\sin\theta)^n$$

$$= [x + i(1 - x^2)^{1/2}]^n, \qquad |x| \le 1 \tag{13.86}$$

we obtain expansions

$$T_n(x) = x^n - \binom{n}{2} x^{n-2}(1 - x^2) + \binom{n}{4} x^{n-4}(1 - x^2)^2 - \cdots, \tag{13.87a}$$

and

$$V_n(x) = \sqrt{1 - x^2}\left[\binom{n}{1} x^{n-1} - \binom{n}{3} x^{n-3}(1 - x^2) + \binom{n}{5} x^{n-5}(1 - x^2)^2 - \cdots\right]. \tag{13.87b}$$

Here the binomial coefficient $\binom{n}{m}$ is given by

$$\binom{n}{m} \equiv \frac{n!}{m!(n-m)!}.$$

From the generating functions, or from the differential equations, power-series representations are

$$T_n(x) = \frac{n}{2} \sum_{m=0}^{[n/2]} (-1)^m \frac{(n-m-1)!}{m!(n-2m)!} (2x)^{n-2m} \qquad (13.88a)$$

and

$$U_n(x) = \sum_{m=0}^{[n/2]} (-1)^m \frac{(n-m)!}{m!(n-2m)!} (2x)^{n-2m}. \qquad (13.88b)$$

ORTHOGONALITY

If Eq. (13.79) is put into self-adjoint form (Section 9.1), we obtain $w(x) = (1 - x^2)^{-1/2}$ as a weighting factor. For Eq. (13.81) the corresponding weighting factor is $(1 - x^2)^{+1/2}$. The resulting orthogonality integrals are

$$\int_{-1}^{1} T_m(x)T_n(x)(1-x^2)^{-1/2} \, dx = \begin{cases} 0, & m \neq n, \\ \dfrac{\pi}{2}, & m = n \neq 0, \\ \pi, & m = n = 0, \end{cases} \qquad (13.89)$$

$$\int_{-1}^{1} V_m(x)V_n(x)(1-x^2)^{-1/2} \, dx = \begin{cases} 0, & m \neq n, \\ \dfrac{\pi}{2}, & m = n \neq 0, \\ 0, & m = n = 0, \end{cases} \qquad (13.90)$$

$$\int_{-1}^{1} U_m(x)U_n(x)(1-x^2)^{1/2} \, dx = \frac{\pi}{2}\delta_{m,n}, \qquad (13.91)$$

and

$$\int_{-1}^{1} W_m(x)W_n(x)(1-x^2)^{1/2} \, dx = \frac{\pi}{2}\delta_{m,n}. \qquad (13.92)$$

This orthogonality is a direct consequence of the Sturm–Liouville theory, Chapter 9. The normalization values may best be obtained by using $x = \cos\theta$ and converting these four integrals into Fourier normalization integrals (for the half integral $[0, \pi]$).

EXERCISES

13.3.1   Another Chebyshev generating function is

$$\frac{1-xt}{1-2xt+t^2} = \sum_{n=0}^{\infty} X_n(x)t^n, \qquad |t| < 1.$$

How is $X_n(x)$ related to $T_n(x)$ and $U_n(x)$?

13.3.2 Given

$$(1 - x^2)U_n''(x) - 3xU_n'(x) + n(n + 2)U_n(x) = 0,$$

show that $V_n(x)$ satisfies

$$(1 - x^2)V_n''(x) - xV_n'(x) + n^2V_n(x) = 0,$$

which is Chebyshev's equation.

13.3.3 Show that the Wronskian of $T_n(x)$ and $V_n(x)$ is given by

$$T_n(x)V_n'(x) - T_n'(x)V_n(x) = -\frac{n}{(1 - x^2)^{1/2}}.$$

This verifies that T_n and $V_n(n \neq 0)$ are independent solutions of Eq. (13.79). Conversely, for $n = 0$, we do not have linear independence. What happens at $n = 0$? Where is the "second" solution?

13.3.4 Show that $W_n(x) = (1 - x^2)^{-1/2}T_{n+1}(x)$ is a solution of

$$(1 - x^2)W_n''(x) - 3xW_n'(x) + n(n + 2)W_n(x) = 0.$$

13.3.5 Evaluate the Wronskian of $U_n(x)$ and $W_n(x) = (1 - x^2)^{-1/2}T_{n+1}(x)$.

13.3.6 $V_n(x) = (1 - x^2)^{1/2}U_{n-1}(x)$ is not defined for $n = 0$. Show that a second and independent solution of the Chebyshev differential equation for $T_n(x)$, $(n = 0)$ is $V_0(x) = $ arc cos x (or arc sin x).

13.3.7 Show that $V_n(x)$ satisfies the same three-term recurrence relation as $T_n(x)$ (Eq. (13.69)).

13.3.8 Verify the series solutions for $T_n(x)$ and $U_n(x)$ (Eqs. (13.88a) and (13.88b)).

13.3.9 Transform the series form of $T_n(x)$, Eq. (13.88a), into an *ascending* power series.

$$ANS. \quad T_{2n}(x) = (-1)^n n \sum_{m=0}^{n} (-1)^m \frac{(n + m - 1)!}{(n - m)!(2m)!}(2x)^{2m},$$

$$T_{2n+1}(x) = (-1)^n \frac{2n + 1}{2} \sum_{m=0}^{n} (-1)^m \frac{(n + m)!}{(n - m)!(2m + 1)!}(2x)^{2m+1}.$$

13.3.10 Rewrite the series form of $U_n(x)$, Eq. (13.88b), as an ascending power series.

$$ANS. \quad U_{2n}(x) = (-1)^n \sum_{m=0}^{n} (-1)^m \frac{(n + m)!}{(n - m)!(2m)!}(2x)^{2m},$$

$$U_{2n+1}(x) = (-1)^n \sum_{m=0}^{n} (-1)^m \frac{(n + m + 1)!}{(n - m)!(2m + 1)!}(2x)^{2m+1}.$$

13.3.11 Derive the Rodrigues representation of $T_n(x)$.

$$T_n(x) = \frac{(-1)^n \pi^{1/2}(1 - x^2)^{1/2}}{2^n(n - \frac{1}{2})!} \frac{d^n}{dx^n}[(1 - x^2)^{n-1/2}].$$

Hint. One possibility is to use the hypergeometric function relation

$$_2F_1(a, b, c; z) = (1 - z)^{-a} {}_2F_1\left(a, c - b, c; \frac{-z}{1 - z}\right),$$

with $z = (1 - x)/2$. An alternate approach is to develop a first-order differential equation for $y = (1 - x^2)^{n-1/2}$. Repeated differentiation of this equation leads to the Chebyshev equation.

13.3.12   (a) From the differential equation for T_n (in self-adjoint form) show that

$$\int_{-1}^{1} \frac{dT_m(x)}{dx} \frac{dT_n(x)}{dx} (1 - x^2)^{1/2} \, dx = 0, \qquad m \neq n.$$

(b) Confirm the preceding result by showing that

$$\frac{dT_n(x)}{dx} = nU_{n-1}(x).$$

13.3.13   The expansion of a power of x in a Chebyshev series leads to the integral

$$I_{mn} = \int_{-1}^{1} x^m T_n(x) \frac{dx}{\sqrt{1 - x^2}}.$$

(a) Show that this integral vanishes for $m < n$.
(b) Show that this integral vanishes for $m + n$ odd.

13.3.14   Evaluate the integral

$$I_{mn} = \int_{-1}^{1} x^m T_n(x) \frac{dx}{\sqrt{1 - x^2}}$$

for $m \geq n$ and $m + n$ even by each of two methods:

(a) Operate with x as the variable replacing T_n by its Rodrigues representation.
(b) Using $x = \cos \theta$, transform the integral to a form with θ as the variable.

$$ANS. \quad I_{mn} = \pi \frac{m!}{(m - n)!} \frac{(m - n - 1)!!}{(m + n)!!}, \qquad m \geq n, \quad m + n \text{ even.}$$

13.3.15   Establish the following bounds, $-1 \leq x \leq 1$:

(a) $|U_n(x)| \leq n + 1$,

(b) $\left| \dfrac{d}{dx} T_n(x) \right| \leq n^2$.

13.3.16   (a) Establish the following bound, $-1 \leq x \leq 1$:

$$V_n(x) = 1.$$

(b) Show that $W_n(x)$ is unbounded in $-1 \leq x \leq 1$.

13.3.17   Verify the orthogonality-normalization integrals for

(a) $T_m(x), \quad T_n(x)$
(b) $V_m(x), \quad V_n(x)$
(c) $U_m(x), \quad U_n(x)$
(d) $W_m(x), \quad W_n(x)$.

Hint. All these can be converted to Fourier orthogonality-normalization integrals.

13.3.18   Show whether

(a) $T_m(x)$ and $V_n(x)$ are or are not orthogonal over the interval $[-1, 1]$ with respect to the weighting factor $(1 - x^2)^{-1/2}$.
(b) $U_m(x)$ and $W_n(x)$ are or are not orthogonal over the interval $[-1, 1]$ with respect to the weighting factor $(1 - x^2)^{1/2}$.

13.3.19   Derive

(a) $T_{n+1}(x) + T_{n-1}(x) = 2xT_n(x)$,
(b) $T_{m+n}(x) + T_{m-n}(x) = 2T_m(x)T_n(x)$,

from the "corresponding" cosine identities.

13.3.20   A number of equations relate the two types of Chebyshev polynomials. As examples show that

$$T_n(x) = U_n(x) - xU_{n-1}(x)$$

and

$$(1 - x^2)U_n(x) = xT_{n+1}(x) - T_{n+2}(x).$$

13.3.21   Show that

$$\frac{dV_n(x)}{dx} = -n\frac{T_n(x)}{\sqrt{1 - x^2}}$$

(a) using the trigonometric forms of V_n and T_n,
(b) using the Rodrigues representation.

13.3.22   Starting with $x = \cos\theta$ and $T_n(\cos\theta) = \cos n\theta$, expand

$$x^k = \left(\frac{e^{i\theta} + e^{-i\theta}}{2}\right)^k$$

and show that

$$x^k = \frac{1}{2^{k-1}}\left[T_k(x) + \binom{k}{1}T_{k-2}(x) + \binom{k}{2}T_{k-4}(x) + \cdots\right],$$

the series in brackets terminating with $\binom{k}{m}T_1(x)$ for $k = 2m + 1$ or $\frac{1}{2}\binom{k}{m}T_0$ for $k = 2m$.

13.3.23   (a) Calculate and tabulate the Chebyshev functions $V_1(x)$, $V_2(x)$, and $V_3(x)$ for $x = -1.0(0.1)1.0$.
   (b) A second solution of the Chebyshev differential equation, Eq. (13.79), for $n = 0$ is $y(x) = \sin^{-1}x$. Tabulate and plot this function over the same range: $-1.0(0.1)1.0$.

13.3.24   Write a program that will generate the coefficients a_s in the polynomial form of the Chebyshev polynomial, $T_n(x) = \sum_{s=0}^n a_s x^s$.

13.3.25   Tabulate $T_{10}(x)$ for $0.00(0.01)1.00$. This will include the five positive roots of T_{10}. If a plotting subroutine is available, plot your results.

13.3.26   Determine the five positive roots of $T_{10}(x)$ by calling a root-finding subroutine (compare Appendix 1). Use your knowledge of the approximate location of these roots from Exercise 13.3.25 or write a search routine to look for the roots. These five positive roots (and their negatives) are the evaluation points of the 10-point Gauss–Chebyshev quadrature method (Appendix 2).

$$x_k = \cos[(2k - 1)\pi/20], \qquad k = 1, 2, 3, 4, 5.$$

13.3.27   Develop the following Chebyshev expansions (for $[-1, 1]$):

(a) $(1 - x^2)^{1/2} = \dfrac{2}{\pi}\left[1 - 2\sum_{s=1}^{\infty}(4s^2 - 1)^{-1}T_{2s}(x)\right].$

(b) $\left.\begin{array}{ll} +1, & 0 < x \le 1 \\ -1, & -1 \le x < 0 \end{array}\right\} = \dfrac{4}{\pi}\sum_{s=0}^{\infty}(-1)^s(2s + 1)^{-1}T_{2s+1}(x).$

13.3.28   (a) For the interval $[-1, 1]$ show that

$$|x| = \frac{1}{2} + \sum_{s=1}^{\infty}(-1)^{s+1}\frac{(2s - 3)!!}{(2s + 2)!!}(4s + 1)P_{2s}(x)$$

$$= \frac{2}{\pi} + \frac{4}{\pi}\sum_{s=1}^{\infty}(-1)^{s+1}\frac{1}{4s^2 - 1}T_{2s}(x).$$

(b) Show that the ratio of the coefficient of $T_{2s}(x)$ to that of $P_{2s}(x)$ approaches $(\pi s)^{-1}$ as $s \to \infty$. This illustrates the relatively rapid convergence of the Chebyshev series.

Hint. Legendre—with the Legendre recurrence relations, rewrite $xP_n(x)$ as a linear combination of derivatives. Chebyshev—the trigonometric substitution $x = \cos\theta$, $T_n(x) = \cos n\theta$ is most helpful.

13.3.29   Show that

$$\frac{\pi^2}{8} = 1 + 2\sum_{s=1}^{\infty}(4s^2 - 1)^{-2}.$$

Hint. Apply Parseval's identity (or the completeness relation) to the results of Exercise 13.3.28.

13.3.30   Show that

(a) $\cos^{-1} x = \dfrac{\pi}{2} - \dfrac{4}{\pi}\sum_{n=0}^{\infty}\dfrac{1}{(2n + 1)^2}T_{2n+1}(x).$

(b) $\sin^{-1} x = \dfrac{4}{\pi}\sum_{n=0}^{\infty}\dfrac{1}{(2n + 1)^2}T_{2n+1}(x).$

13.4 HYPERGEOMETRIC FUNCTIONS

In Chapter 8 the hypergeometric equation[1]

$$x(1 - x)y''(x) + [c - (a + b + 1)x]y'(x) - aby(x) = 0 \qquad (13.93)$$

was introduced as a canonical form of a linear second-order differential equation with regular singularities at $x = 0, 1$, and ∞. One solution is

$y(x) = {}_2F_1(a, b, c; x)$

$$= 1 + \frac{a \cdot b}{c}\frac{x}{1!} + \frac{a(a + 1)b(b + 1)}{c(c + 1)}\frac{x^2}{2!} + \cdots, \qquad c \ne 0, -1, -2, -3, \ldots,$$

$$(13.94)$$

---

[1] This is sometimes called Gauss's differential equation. The solutions then become Gauss functions.

which is known as the hypergeometric function or hypergeometric series. The range of convergence $|x| < 1$ and $x = 1$, for $c > a + b$, and $x = -1$, for $c > a + b - 1$. In terms of the often used Pochhammer symbol

$$(a)_n = a(a+1)(a+2) \cdots (a+n-1) = \frac{(a+n-1)!}{(a-1)!},$$

$$(a)_0 = 1,$$

(13.95)

the hypergeometric function becomes

$$_2F_1(a, b, c; x) = \sum_{n=0}^{\infty} \frac{(a)_n (b)_n}{(c)_n} \frac{x^n}{n!}.$$

(13.96)

In this form the subscripts 2 and 1 become clear. The leading subscript 2 indicates that two Pochhammer symbols appear in the numerator and the final subscript 1 indicates one Pochhammer symbol in the denominator.[2] The confluent hypergeometric function $_1F_1$ with one Pochhammer symbol in the numerator and one in the denominator appears in Section 13.5.

From the form of Eq. (13.94) we see that the parameter c may not be zero or a negative integer. On the other hand, if a or b equals 0 or a negative integer, the series terminates and the hypergeometric function becomes a polynomial.

Many more or less elementary functions can be represented by the hypergeometric function.[3] We find

$$\ln(1 + x) = x \,_2F_1(1, 1, 2; -x).$$

(13.97)

For the complete elliptic integrals K and E

$$K(k^2) = \int_0^{\pi/2} (1 - k^2 \sin^2 \theta)^{-1/2} \, d\theta$$

$$= \frac{\pi}{2} \,_2F_1\left(\frac{1}{2}, \frac{1}{2}, 1; k^2\right),$$

(13.98)

$$E(k^2) = \int_0^{\pi/2} (1 - k^2 \sin^2 \theta)^{1/2} \, d\theta$$

$$= \frac{\pi}{2} \,_2F_1\left(\frac{1}{2}, -\frac{1}{2}, 1; k^2\right).$$

(13.99)

The explicit series forms and other properties of the elliptic integrals are developed in Section 5.8.

---

[2] The Pochhammer symbol is often useful in other expressions involving factorials, for instance,

$$(1 - z)^{-a} = \sum_{n=0}^{\infty} (a)_n z^n / n!, \qquad |z| < 1.$$

[3] With three parameters, a, b, and c, we can represent almost anything.

The hypergeometric equation as a second-order linear differential equation has a second independent solution. The usual form is

$$y(x) = x^{1-c} \, _2F_1(a + 1 - c, b + 1 - c, 2 - c; x), \qquad c \neq 2, 3, 4, \ldots . \qquad (13.100)$$

The reader may show (Exercise 13.4.1) that if c is an integer either the two solutions coincide or (barring a rescue by integral a or integral b) one of the solutions will blow up. In such a case the second solution is expected to include a logarithmic term.

Alternate forms of the hypergeometric equation include

$$(1 - z^2)\frac{d^2}{dz^2} y\left(\frac{1 - z}{2}\right) - [(a + b + 1)z$$

$$-(a + b + 1 - 2c)]\frac{d}{dz} y\left(\frac{1 - z}{2}\right) - aby\left(\frac{1 - z}{2}\right) = 0, \qquad (13.101)$$

$$(1 - z^2)\frac{d^2}{dz^2} y(z^2) - \left[(2a + 2b + 1)z + \frac{1 - 2c}{z}\right]\frac{d}{dz} y(z^2) - 4aby(z^2) = 0. \qquad (13.102)$$

CONTIGUOUS FUNCTION RELATIONS

The parmeters a, b, and c enter in the same way as the parameter n of Bessel, Legendre, and other special functions. As we found with these functions, we expect recurrence relations involving unit changes in the parameters a, b, and c. The usual nomenclature for the hypergeometric functions in which *one* parameter changes by $+$ or -1 is a "contiguous function." Generalizing this term to include simultaneous unit changes in more than one parameter, we find 26 functions contiguous to $_2F_1(a, b, c; x)$. Taking them two at a time, we can develop the formidable total of 325 equations among the contiguous functions. One typical example is

$$(a - b)\{c(a + b - 1) + 1 - a^2 - b^2 + [(a - b)^2 - 1](1 - x)\} \, _2F_1(a, b, c; x)$$

$$= (c - a)(a - b + 1)b \, _2F_1(a - 1, b + 1, c; x)$$

$$+ (c - b)(a - b - 1)a \, _2F_1(a + 1, b - 1, c; x). \qquad (13.103)$$

Another contiguous function relation appears in Exercise 13.4.10.

HYPERGEOMETRIC REPRESENTATIONS

Since the ultraspherical equation (13.81) in Section 13.3 is a special case of Eq. (13.93), we see that ultraspherical functions (and Legendre and Chebyshev functions) may be expressed as hypergeometric functions. For the ultraspherical function we obtain

$$T_n^\beta(x) = \frac{(n + 2\beta)!}{2^\beta n! \beta!} \, _2F_1\left(-n, n + 2\beta + 1, 1 + \beta; \frac{1 - x}{2}\right). \qquad (13.104)$$

For Legendre and associated Legendre functions

$$P_n(x) = {}_2F_1\left(-n, n + 1, 1; \frac{1 - x}{2}\right).$$ (13.105)

$$P_n^m(x) = \frac{(n + m)!}{(n - m)!} \frac{(1 - x^2)^{m/2}}{2^m m!} {}_2F_1\left(m - n, m + n + 1, m + 1; \frac{1 - x}{2}\right).$$ (13.106)

Alternate forms are

$$P_{2n}(x) = (-1)^n \frac{(2n)!}{2^{2n} n! n!} {}_2F_1\left(-n, n + \frac{1}{2}, \frac{1}{2}; x^2\right)$$

$$= (-1)^n \frac{(2n - 1)!!}{(2n)!!} {}_2F_1\left(-n, n + \frac{1}{2}, \frac{1}{2}; x^2\right),$$ (13.107)

$$P_{2n+1}(x) = (-1)^n \frac{(2n + 1)!}{2^{2n} n! n!} x\, {}_2F_1\left(-n, n + \frac{3}{2}, \frac{3}{2}; x^2\right)$$

$$= (-1)^n \frac{(2n + 1)!!}{(2n)!!} x\, {}_2F_1\left(-n, n + \frac{3}{2}, \frac{3}{2}; x^2\right).$$ (13.108)

In terms of hypergeometric functions the Chebyshev functions become

$$T_n(x) = {}_2F_1\left(-n, n, \frac{1}{2}; \frac{1 - x}{2}\right),$$ (13.109)

$$U_n(x) = (n + 1)\, {}_2F_1\left(-n, n + 2, \frac{3}{2}; \frac{1 - x}{2}\right),$$ (13.110)

$$V_n(x) = \sqrt{1 - x^2}\, n\, {}_2F_1\left(-n + 1, n + 1, \frac{3}{2}; \frac{1 - x}{2}\right).$$ (13.111)

The leading factors are determined by direct comparison of complete power series, comparison of coefficients of particular powers of the variable, or evaluation at $x = 0$ or 1, and so on.

The hypergeometric series may be used to define functions with nonintegral indices. The physical applications are minimal.

EXERCISES

13.4.1   (a) For c, and integer, and a and b nonintegral, show that
$$_2F_1(a, b, c; x) \qquad \text{and} \qquad x^{1-c}\, {}_2F_1(a + 1 - c, b + 1 - c, 2 - c; x)$$
yield only one solution to the hypergeometric equation.

(b) What happens if a is an integer, say, $a = -1$, and $c = -2$?

13.4.2   Find the Legendre, Chebyshev I, and Chebyshev II recurrence relations corresponding to the contiguous hypergeometric function equation (13.103).

13.4.3   Transform the following polynomials into hypergeometric functions of argument x^2. (a) $T_{2n}(x)$; (b) $x^{-1}T_{2n+1}(x)$; (c) $U_{2n}(x)$; (d) $x^{-1}U_{2n+2}(x)$.

$\quad\quad$ ANS.   (a) $T_{2n}(x) = (-1)^n {}_2F_1(-n, n, \frac{1}{2}; x^2)$.

$\quad\quad\quad\quad$ (b) $x^{-1}T_{2n+1}(x) = (-1)^n(2n + 1)\,{}_2F_1(-n, n + 1, \frac{3}{2}; x^2)$.

$\quad\quad\quad\quad$ (c) $U_{2n}(x) = (-1)^n {}_2F_1(-n, n + 1, \frac{1}{2}; x^2)$.

$\quad\quad\quad\quad$ (d) $x^{-1}U_{2n+1}(x) = (-1)^n(2n + 2)\,{}_2F_1(-n, n + 2, \frac{3}{2}; x^2)$.

13.4.4   Derive or verify the leading factor in the hypergeometric representations of the Chebyshev functions.

13.4.5   Verify that the Legendre function of the second kind, $Q_\nu(z)$, is given by

$$Q_\nu(z) = \frac{\pi^{1/2}\nu!}{(\nu + \frac{1}{2})!(2z)^{\nu+1}}\,{}_2F_1\left(\frac{\nu}{2} + \frac{1}{2}, \frac{\nu}{2} + 1, \frac{\nu}{2} + \frac{3}{2}; z^{-2}\right),$$

$$|z| > 1, \quad\quad |\arg z| < \pi, \quad\quad \nu \neq -1, -2, -3, \ldots .$$

13.4.6   Analogous to the incomplete gamma function, we may define an incomplete beta function by

$$B_x(a, b) = \int_0^x t^{a-1}(1 - t)^{b-1}\, dt.$$

Show that

$$B_x(a, b) = a^{-1}x^a {}_2F_1(a, 1 - b, a + 1; x).$$

13.4.7   Verify the integral representation

$$_2F_1(a, b, c; z) = \frac{\Gamma(c)}{\Gamma(b)\Gamma(c - b)}\int_0^1 t^{b-1}(1 - t)^{c-b-1}(1 - tz)^{-a}\, dt.$$

What restrictions must you place on the parameters b and c, on the variable z?

Note. The restriction on $|z|$ can be dropped—analytic continuation. For nonintegral a the real axis in the z-plane 1 to ∞ is a cut line.

Hint. The integral is suspiciously like a beta function and can be expanded into a series of beta functions.

$\quad\quad$ ANS. $\Re(c) > \Re(b) > 0$, and $|z| < 1$.

13.4.8   Prove that

$$_2F_1(a, b, c; 1) = \frac{\Gamma(c)\Gamma(c - a - b)}{\Gamma(c - a)\Gamma(c - b)}, \quad\quad c \neq 0, -1, -2, \ldots, \quad\quad c > a + b.$$

Hint. Here is a chance to use the integral representation, Exercise 13.4.7.

13.4.9   Prove that

$$_2F_1(a, b, c; x) = (1 - x)^{-a}\,{}_2F_1\left(a, c - b, c; \frac{-x}{1 - x}\right).$$

Hint. Try an integral representation.

Note. This relation is useful in developing a Rodrigues representation of $T_n(x)$ (compare Exercise 13.3.11).

13.4.10   Verify

$$_2F_1(-n, b, c; 1) = \frac{(c - b)_n}{(c)_n}.$$

Hint. Here is a chance to use the contiguous function relation $[2a - c + (b - a)x]F(a, b, c; x) = a(1 - x)F(a + 1, b, c; x) - (c - a)F(a - 1, b, c; x)$ and mathematical induction. Alternatively, you can use the integral representation and the beta function.

13.5 CONFLUENT HYPERGEOMETRIC FUNCTIONS

The confluent hypergeometric equation[1]

$$xy''(x) + (c - x)y'(x) - ay(x) = 0 \tag{13.112}$$

may be obtained from the hypergeometric equation of Section 13.4 by merging two of its singularities. The resulting equation has a regular singularity at $x = 0$ and an irregular one at $x = \infty$. One solution of the confluent hypergeometric equation is

$$y(x) = {}_1F_1(a, c; x) = M(a, c; x)$$

$$= 1 + \frac{a}{c}\frac{x}{1!} + \frac{a(a + 1)}{c(c + 1)}\frac{x^2}{2!} + \cdots, \quad c \neq 0, -1, -2, \ldots. \tag{13.113}$$

This solution is convergent for all finite x (or complex z). In terms of the Pochhammer symbols, we have

$$M(a, c; x) = \sum_{n=0}^{\infty} \frac{(a)_n}{(c)_n} \frac{x^n}{n!}. \tag{13.114}$$

Clearly, $M(a, c; x)$ becomes a polynomial if the parameter a is 0 or a negative integer. Numerous more or less elementary functions may be represented by the confluent hypergeometric function. Examples are the error function and the incomplete gamma function.

$$\text{erf}(x) = \frac{2}{\pi^{1/2}} \int_0^x e^{-t^2}\, dt = \frac{2}{\pi^{1/2}} xM\left(\frac{1}{2}, \frac{3}{2}; -x^2\right), \tag{13.115}$$

$$\gamma(a, x) = \int_0^x e^{-t}t^{a-1}\, dt = a^{-1}x^a M(a, a + 1; -x),$$

$$\text{from Eq. (10.71), } \Re(a) > 0. \tag{13.116}$$

Clearly, this coincides with the first solution for $c = 1$. The error function and the incomplete gamma function are discussed further in Section 10.5.

---

[1] This is often called Kummer's equation. The solutions, then, are Kummer functions.

A second solution of Eq. (13.112) is given by

$$y(x) = x^{1-c}M(a + 1 - c, 2 - c; x), \qquad c \neq 2, 3, 4, \dots . \quad (13.117)$$

The standard form of the second solution of Eq. (13.132) is a linear combination of Eqs. (13.113) and (13.117):

$$U(a, c; x) = \frac{\pi}{\sin \pi c} \left[\frac{M(a, c; x)}{(a - c)!(c - 1)!} - \frac{x^{1-c}M(a + 1 - c, 2 - c; x)}{(a - 1)!(1 - c)!} \right].$$
$$(13.118)$$

Note the resemblance to our definition of the Neumann function, Eq. (11.60). As with our Neumann function, Eq. (11.60), this definition of $U(a, c; x)$ becomes indeterminate in this case for c an integer.

An alternate form of the confluent hypergeometric equation that will be useful later is obtained by changing the independent variable from x to x^2.

$$\frac{d^2}{dx^2} y(x^2) + \left[\frac{2c - 1}{x} - 2x \right] \frac{d}{dx} y(x^2) - 4ay(x^2) = 0. \quad (13.119)$$

As with the hypergeometric functions, contiguous functions exist in which the parameters a and c are changed by ± 1. Including the cases of simultaneous changes in the two parameters,[2] we have eight possibilities. Taking the original function and pairs of the contiguous functions, we can develop a total of 28 equations.[3]

INTEGRAL REPRESENTATIONS

It is frequently convenient to have the confluent hypergeometric functions in integral form. We find (Exercise 13.5.10)

$$M(a, c; x) = \frac{\Gamma(c)}{\Gamma(a)\Gamma(c - a)} \int_0^1 e^{xt} t^{a-1}(1 - t)^{c-a-1} \, dt, \qquad \Re(c) > \Re(a) > 0,$$
$$(13.120)$$

$$U(a, c; x) = \frac{1}{\Gamma(a)} \int_0^\infty e^{-xt} t^{a-1}(1 + t)^{c-a-1} \, dt, \qquad \Re(x) > 0, \Re(a) > 0.$$
$$(13.121)$$

Three important techniques for deriving or verifying integral representations are as follows:

1. Transformation of generating function expansions and Rodrigues representations: The Bessel and Legendre functions provide examples of this approach.
2. Direct integration to yield a series: This direct technique is useful for a Bessel function representation (Exercise 11.1.18) and a hypergeometric integral (Exercise 13.4.7).

[2] Slater refers to these as associated functions.
[3] The recurrence relations for Bessel, Hermite, and Laguerre functions are special cases of these equations.

3. (a) Verification that the integral representation satisfies the differential equation. (b) Exclusion of the other solution. (c) Verification of normalization. This is the method used in Section 11.5 to establish an integral representation of the modified Bessel function, $K_\nu(z)$. It will work here to establish Eqs. (13.120) and (13.121).

BESSEL AND MODIFIED BESSEL FUNCTIONS

Kummer's first formula,

$$M(a, c; x) = e^x M(c - a, c; -x), \qquad (13.122)$$

is useful in representing the Bessel and modified Bessel functions. The formula may be verified by series expansion or use of an integral representation (compare Exercise 13.5.10).

As expected from the form of the confluent hypergeometric equation and the character of its singularities, the confluent hypergeometric functions are useful in representing a number of the special functions of mathematical physics. For the Bessel functions

$$J_\nu(x) = \frac{e^{-ix}}{\nu!} \left(\frac{x}{2}\right)^\nu M\left(\nu + \frac{1}{2}, 2\nu + 1; 2ix\right), \qquad (13.123)$$

whereas for the modified Bessel functions of the first kind,

$$I_\nu(x) = \frac{e^{-x}}{\nu!} \left(\frac{x}{2}\right)^\nu M\left(\nu + \frac{1}{2}, 2\nu + 1; 2x\right). \qquad (13.124)$$

HERMITE FUNCTIONS

The Hermite functions are given by

$$H_{2n}(x) = (-1)^n \frac{(2n)!}{n!} M\left(-n, \frac{1}{2}; x^2\right), \qquad (13.125)$$

$$H_{2n+1}(x) = (-1)^n \frac{2(2n + 1)!}{n!} x M\left(-n, \frac{3}{2}; x^2\right), \qquad (13.126)$$

using Eq. (13.119).

Comparing the Laguerre differential equation with the confluent hypergeometric equation, we have

$$L_n(x) = M(-n, 1; x). \qquad (13.127)$$

The constant is fixed as unity by noting Eq. (13.35) for $x = 0$. For the associated Laguerre functions

$$L_n^m(x) = (-1)^m \frac{d^m}{dx^m} L_{n+m}(x)$$

$$= \frac{(n + m)!}{n!m!} M(-n, m + 1; x). \qquad (13.128)$$

Alternate verification is obtained by comparing Eq. (13.128) with the power-series solution (Eq. (13.41) of Section 13.2). Note that in the hypergeometric form, as distinct from a Rodrigues representation, the indices n and m need not be integers and, if they are not integers, $L_n^m(x)$ will not be a polynomial.

MISCELLANEOUS CASES

There are certain advantages in expressing our special functions in terms of hypergeometric and confluent hypergeometric functions. If the general behavior of the latter functions is known, the behavior of the special functions we have investigated follows as a series of special cases. This may be useful in determining asymptotic behavior or evaluating normalization integrals. The asymptotic behavior of $M(a, c; x)$ and $U(a, c; x)$ may be conveniently obtained from integral representations of these functions, Eqs. (13.120) and (13.121). The further advantage is that the relations between the special functions are clarified. For instance, an examination of Eqs. (13.125), (13.126), and (13.128) suggests that the Laguerre and Hermite functions are related.

The confluent hypergeometric equation (13.112) is clearly not self-adjoint. For this and other reasons it is convenient to define

$$M_{k\mu}(x) = e^{-x/2}x^{\mu+1/2}M(\mu - k + \tfrac{1}{2}, 2\mu + 1; x). \tag{13.129}$$

This new function $M_{k\mu}(x)$ is a Whittaker function which satisfies the self-adjoint equation

$$M''_{k\mu}(x) + \left(-\frac{1}{4} + \frac{k}{x} + \frac{\tfrac{1}{4} - \mu^2}{x^2}\right)M_{k\mu}(x) = 0. \tag{13.130}$$

The corresponding second solution is

$$W_{k\mu}(x) = e^{-x/2}x^{\mu+1/2}U(\mu - k + \tfrac{1}{2}, 2\mu + 1; x). \tag{13.131}$$

EXERCISES

13.5.1 Verify the confluent hypergeometric representation of the error function

$$\text{erf}(x) = \frac{2x}{\pi^{1/2}}M\left(\frac{1}{2}, \frac{3}{2}; -x^2\right).$$

13.5.2 Show that the Fresnel integrals $C(x)$ and $S(x)$ of Exercise 5.10.2 may be expressed in terms of the confluent hypergeometric function as

$$C(x) + iS(x) = xM\left(\frac{1}{2}, \frac{3}{2}; \frac{i\pi x^2}{2}\right).$$

13.5.3 By direct differentiation and substitution verify that

$$y = ax^{-a}\int_0^x e^{-t}t^{a-1}\,dt = ax^{-a}\gamma(a, x)$$

actually does satisfy

$$xy'' + (a + 1 + x)y' + ay = 0.$$

13.5.4   Show that the modified Bessel function of the second kind $K_\nu(x)$ is given by
$$K_\nu(x) = \pi^{1/2}e^{-x}(2x)^\nu U(\nu + \tfrac{1}{2}, 2\nu + 1; 2x).$$

13.5.5   Show that the cosine and sine integrals of Section 10.5 may be expressed in terms of confluent hypergeometric functions as
$$Ci(x) + i\,si(x) = -e^{ix}U(1, 1; -ix).$$

This relation is useful in numerical computation of $Ci(x)$ and $si(x)$ for large values of x.

13.5.6   Verify the confluent hypergeometric form of the Hermite polynomial $H_{2n+1}(x)$ (Eq. (13.126)) by showing that

(a) $H_{2n+1}(x)/x$ satisfies the confluent hypergeometric equation with $a = -n$, $c = \tfrac{3}{2}$ and argument x^2,

(b) $$\lim_{x \to 0} \frac{H_{2n+1}(x)}{x} = (-1)^n \frac{2(2n + 1)!}{n!}.$$

13.5.7   Show that the contiguous confluent hypergeometric function equation,
$$(c - a)M(a - 1, c; x) + (2a - c + x)M(a, c; x) - aM(a + 1, c; x) = 0,$$

leads to the associated Laguerre function recurrence relation (Eq. (13.44)).

13.5.8   Verify the Kummer transformations:

(a) $M(a, c; x) = e^x M(c - a, c; -x)$

(b) $U(a, c; x) = x^{1-c}U(a - c + 1, 2 - c; x).$

13.5.9   Prove that

(a) $$\frac{d^n}{dx^n} M(a, c; x) = \frac{(a)_n}{(b)_n} M(a + n, b + n; x),$$

(b) $$\frac{d^n}{dx^n} U(a, c; x) = (-1)^n (a)_n\, U(a + n, c + n; x).$$

13.5.10   Verify the following integral representations:

(a) $$M(a, c; x) = \frac{\Gamma(c)}{\Gamma(a)\Gamma(c - a)} \int_0^1 e^{xt}t^{a-1}(1 - t)^{c-a-1}\, dt, \quad \mathfrak{R}(c) > \mathfrak{R}(a) > 0.$$

(b) $$U(a, c; x) = \frac{1}{\Gamma(a)} \int_0^\infty e^{-xt}t^{a-1}(1 + t)^{c-a-1}\, dt, \quad \mathfrak{R}(x) > 0, \quad \mathfrak{R}(a) > 0.$$

Under what conditions can you accept $\mathfrak{R}(x) = 0$ in part (b)?

13.5.11   From the integral representation of $M(a, c; x)$, Exercise 13.5.10(a) show that
$$M(a, c; x) = e^x M(c - a, c; -x).$$

Hint. Replace the variable of integration t by $1 - s$ to release a factor e^x from the integral.

13.5.12   From the integral representation of $U(a, c; x)$, Exercise 13.5.10(b), show that the exponential integral is given by
$$E_1(x) = e^{-x}U(1, 1; x).$$

Hint. Replace the variable of integration t in $E_1(x)$ by $x(1 + s)$.

13.5.13   From the integral representations of $M(a, c; x)$ and $U(a, c; x)$ in Exercise 13.5.10 develop asymptotic expansions of

(a) $M(a, c; x)$,
(b) $U(a, c; x)$.

Hint. You can use the technique that was employed with $K_\nu(z)$, Section 11.6.

ANS.

(a) $\dfrac{\Gamma(c)}{\Gamma(a)} \dfrac{e^x}{x^{c-a}} \left\{ 1 + \dfrac{(1 - a)(c - a)}{1!x} + \dfrac{(1 - a)(2 - a)(c - a)(c - a + 1)}{2!x^2} + \cdots \right\}.$

(b) $\dfrac{1}{x^a} \left\{ 1 + \dfrac{a(1 + a - c)}{1!(-x)} + \dfrac{a(a + 1)(1 + a - c)(2 + a - c)}{2!(-x)^2} + \cdots \right\}.$

13.5.14   Show that the Wronskian of the two confluent hypergeometric functions, $M(a, c; x)$ and $U(a, c; x)$ is given by

$$MU' - M'U = -\frac{(c - 1)!}{(a - 1)!} \frac{e^x}{x^c}.$$

What happens if a is 0 or a negative integer?

13.5.15   The Coulomb wave equation (radial part of the Schrödinger wave equation with Coulomb potential) is

$$\frac{d^2y}{d\rho^2} + \left[1 - \frac{2\eta}{\rho} - \frac{L(L + 1)}{\rho^2} \right] y = 0.$$

Show that a regular solution, $y = F_L(\eta, \rho)$, is given by

$$F_L(\eta, \rho) = C_L(\eta)\rho^{L+1}e^{-i\rho}M(L + 1 - i\eta, 2L + 2; 2i\rho).$$

13.5.16   (a) Show that the radial part of the hydrogen wave function, Eq. (13.60), may be written as

$e^{-\alpha r/2}(\alpha r)^L L_{n-L-1}^{2L+1}(\alpha r)$

$$= \frac{(n + L)!}{(n - L - 1)!(2L + 1)!} e^{-\alpha r/2}(\alpha r)^L M(L + 1 - n, 2L + 2; \alpha r).$$

(b) It was assumed previously that the total (kinetic + potential) energy E of the electron was negative. Rewrite the (unnormalized) radial wave function for the free electron $E > 0$.

ANS. $e^{+i\alpha r/2}(\alpha r)^L M(L + 1 - in, 2L + 2, -i\alpha r)$, outgoing wave. This representation provides a powerful alternative technique for the calculation of photoionization and recombination coefficients.

13.5.17   Show that the Laplace transform of $M(a, c; x)$ is

$$\mathcal{L}\{M(a, c; x)\} = \frac{1}{s}{}_2F_1\left(a, 1, c; \frac{1}{s}\right).$$

13.5.18   Evaluate

(a) $\displaystyle\int_0^\infty [M_{k\mu}(x)]^2 \, dx,$ (b) $\displaystyle\int_0^\infty [M_{k\mu}(x)]^2 \frac{dx}{x},$ (c) $\displaystyle\int_0^\infty [M_{k\mu}(x)]^2 \frac{dx}{x^{1-a}},$

where $2\mu = 0, 1, 2, \ldots, k - \mu - \frac{1}{2} = 0, 1, 2, \ldots, a > -2\mu - 1.$

ANS. (a) $(2\mu)!2k$.   (b) $(2\mu)!$.   (c) $(2\mu)!(2k)^a$.

ADDITIONAL READINGS

ABRAMOWITZ, M., and I. A. STEGUN, eds., *Handbook of Mathematical Functions*, Applied Mathematics Series-55 (AMS-55). Washington, DC: National Bureau of Standards (1964). Paperback edition, New York: Dover (1964). Chapter 22 is a detailed summary of the properties and representations of orthogonal polynomials. Other chapters summarize properties of Bessel, Legendre, hypergeometric, and confluent hypergeometric functions and much more.

BUCHHOLZ, H., *The Confluent Hypergeometric Function*. New York: Springer-Verlag (1953). Translated (1969). Buchholz strongly emphasizes the Whittaker rather than the Kummer forms. Applications to a variety of other transcendental functions.

ERDELYI, A., W. MAGNUS, F. OBERHETTINGER, and F. G. TRICOMI, *Higher Transcendental Functions*, 3 vols. New York: McGraw-Hill (1953). Reprinted (1981). A detailed, almost exhaustive listing of the properties of the special functions of mathematical physics.

FOX, L. and I. B. PARKER, *Chebyshev Polynomials in Numerical Analysis*. Oxford: Oxford University Press (1968). A detailed, thorough but very readable account of Chebyshev polynomials and their applications in numerical analysis.

LEBEDEV, N. N., *Special Functions and their Applications* (translated by R. A. Silverman). Englewood Cliffs, NJ: Prentice-Hall (1965). Paperback, New York: Dover (1972).

LUKE, Y. L., *The Special Functions and Their Approximations*. New York: Academic Press (1969). Two volumes: Volume 1 is a thorough theoretical treatment of gamma functions, hypergeometric functions, confluent hypergeometric functions, and related functions. Volume 2 develops approximations and other techniques for numerical work.

LUKE, Y. L., *Mathematical Functions and Their Approximations*. New York: Academic Press (1975). This is an updated supplement to *Handbook of Mathematical Functions with Formulas, Graphs and Mathematical Tables* (AMS-55).

MAGNUS, W., F. OBERHETTINGER, and R. P. SONI, *Formulas and Theorems for the Special Functions of Mathematical Physics*. New York: Springer (1966). This is a new and enlarged edition. An excellent summary of just what the title says, including the topics of Chapters 10 to 13.

RAINVILLE, E. D., *Special Functions*. New York: Macmillan (1960). This book is a coherent, comprehensive account of almost all the special functions of mathematical physics that the reader is likely to encounter.

SANSONE, G., *Orthogonal Functions* (translated by A. H. Diamond). New York: Interscience (1959). Reprinted (1977).

SLATER, L. J., *Confluent Hypergeometric Functions*. Cambridge: Cambridge University Press (1960). This is a clear and detailed development of the properties of the confluent hypergeometric functions and of relations of the confluent hypergeometric equation to other differential equations of mathematical physics.

SNEDDON, I. N., *Special Functions of Mathematical Physics and Chemistry*, 3rd ed. New York: Longman (1980).

WHITTAKER, E. T., and G. N. WATSON, *A Course of Modern Analysis*. Cambridge: Cambridge University Press (1973). The classic text on special functions and real and complex analysis.

14 FOURIER SERIES

14.1 GENERAL PROPERTIES

FOURIER SERIES

A Fourier series may be defined as an expansion of a function or representation of a function in a series of sines and cosines such as

$$f(x) = \frac{a_0}{2} + \sum_{n=1}^{\infty} a_n \cos nx + \sum_{n=1}^{\infty} b_n \sin nx. \qquad (14.1)$$

The coefficients a_0, a_n, and b_n are related to the periodic function $f(x)$ by definite integrals: Eqs. (14.11) and (14.12). You will notice that a_0 is singled out for special treatment by the inclusion of the factor $\frac{1}{2}$. This is done so that Eq. (14.11) will apply to *all* a_n, $n = 0$ as well as $n > 0$.

The conditions imposed on $f(x)$ to make Eq. (14.1) valid are that $f(x)$ has only a finite number of finite discontinuities and only a finite number of extreme values, maxima, and minima in the interval $[0, 2\pi]$.[1] Functions satisfying these conditions may be called piecewise regular. The conditions themselves are known as the Dirichlet conditions. Although there are some functions that do not obey these Dirichlet conditions, they may well be labeled pathological for purposes of Fourier expansions. In the vast majority of physical problems involving a Fourier series these conditions will be satisfied. In most physical problems we shall be interested in functions that are square integrable (in the Hilbert space L^2 of Section 9.4). In this space the sines and cosines form a complete orthogonal set. And this in turn means that Eq. (14.1) is valid in the sense of convergence in the mean.

---

[1] These conditions are *sufficient* but not *necessary*.

Expressing cos nx and sin nx in exponential form, we may rewrite Eq. (14.1) as

$$f(x) = \sum_{n=-\infty}^{\infty} c_n e^{inx} \tag{14.2}$$

in which

$$c_n = \tfrac{1}{2}(a_n - ib_n),$$
$$c_{-n} = \tfrac{1}{2}(a_n + ib_n), \qquad n > 0, \tag{14.3}$$

and

$$c_0 = \tfrac{1}{2}a_0.$$

COMPLETENESS

The problem of establishing completeness may be approached in a number of different ways. One way is to transform the trigonometric Fourier series into exponential form and compare it with a Laurent series. If we expand $f(z)$ in a Laurent series[2] (assuming $f(z)$ is analytic),

$$f(z) = \sum_{n=-\infty}^{\infty} d_n z^n. \tag{14.4}$$

On the unit circle $z = e^{i\theta}$ and

$$f(z) = f(e^{i\theta}) = \sum_{n=-\infty}^{\infty} d_n e^{in\theta}. \tag{14.5}$$

The Laurent expansion on the unit circle (Eq. (14.5)) has the same form as the complex Fourier series (Eq. (14.2)), which shows the equivalence between the two expansions. Since the Laurent series as a power series has the property of completeness, we see that the Fourier functions, e^{inx}, form a complete set. There is a significant limitation here. Laurent series and power series cannot handle discontinuities such as a square wave or the sawtooth wave of Fig. 14.1.

The theory of vector spaces provides a second approach to the completeness of the sines and cosines. Here completeness is established by the Weierstrass theorem for two variables.

The Fourier expansion and the completeness property may be expected, for the functions sin nx, cos nx, e^{inx} are all eigenfunctions of a self-adjoint linear differential equation,

$$y'' + n^2 y = 0. \tag{14.6}$$

We obtain orthogonal eigenfunctions for different values of the eigenvalue n by choosing the interval $[0, p\pi]$, p an integer, to satisfy the boundary conditions in the Sturm–Liouville theory (Chapter 9). If we further choose $p = 2$, the different eigenfunctions for the same eigenvalue n may be

---

[2] Section 6.5.

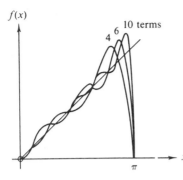

$f(x)$

10 terms

π

x

Figure 14.1 Fourier representation of sawtooth wave.

orthogonal. We have

$$\int_0^{2\pi} \sin mx \sin nx \, dx = \begin{cases} \pi\delta_{m,n}, & m \neq 0, \\ 0, & m = 0, \end{cases} \tag{14.7}$$

$$\int_0^{2\pi} \cos mx \cos nx \, dx = \begin{cases} \pi\delta_{m,n}, & m \neq 0, \\ 2\pi, & m = n = 0, \end{cases} \tag{14.8}$$

$$\int_0^{2\pi} \sin mx \cos nx \, dx = 0 \qquad \text{for all integral } m \text{ and } n. \tag{14.9}$$

Note carefully that any interval $x_0 \leq x \leq x_0 + 2\pi$ will be equally satisfactory. Frequently, we shall use $x_0 = -\pi$ to obtain the interval $-\pi \leq x \leq \pi$. For the complex eigenfunctions $e^{\pm inx}$ orthogonality is usually *defined* in terms of the complex conjugate of one of the two factors,

$$\int_0^{2\pi} (e^{imx})^* e^{inx} \, dx = 2\pi\delta_{m,n}. \tag{14.10}$$

This agrees with the treatment of the spherical harmonics (Section 12.6).

STURM–LIOUVILLE THEORY

The Sturm–Liouville theory guarantees the validity of Eq. (14.1) (for functions satisfying the Dirichlet conditions) and, by use of the orthogonality relations, Eqs. (14.7), (14.8), and (14.9), allows us to compute the expansion coefficients

$$a_n = \frac{1}{\pi} \int_0^{2\pi} f(t) \cos nt \, dt, \tag{14.11}$$

$$b_n = \frac{1}{\pi} \int_0^{2\pi} f(t) \sin nt \, dt, \qquad n = 0, 1, 2, \ldots. \tag{14.12}$$

This, of course, is subject to the requirement that the integrals exist. They do if $f(t)$ is piecewise continuous (or square integrable). Substituting Eqs. (14.11) and (14.12) into Eq. (14.1), we write our Fourier expansion as

$$f(x) = \frac{1}{2\pi} \int_0^{2\pi} f(t) \, dt$$

$$+ \frac{1}{\pi} \sum_{n=1}^{\infty} \left(\cos nx \int_0^{2\pi} f(t) \cos nt \, dt + \sin nx \int_0^{2\pi} f(t) \sin nt \, dt \right)$$

$$= \frac{1}{2\pi} \int_0^{2\pi} f(t) \, dt + \frac{1}{\pi} \sum_{n=1}^{\infty} \int_0^{2\pi} f(t) \cos n(t - x) \, dt, \qquad (14.13)$$

the first (constant) term being the average value of $f(x)$ over the interval $[0, 2\pi]$. Equation (14.13) offers one approach to the development of the Fourier integral and Fourier transforms, Section 15.1.

Another way of describing what we are doing here is to say that $f(x)$ is part of an infinite-dimensional Hilbert space, with the orthogonal $\cos nx$ and $\sin nx$ as the basis. (They can always be renormalized to unity if desired.) The statement that $\cos nx$ and $\sin nx$ ($n = 0, 1, 2, \ldots$) span this Hilbert space is equivalent to saying that they form a complete set. Finally, the expansion coefficients a_n and b_n correspond to the projections of $f(x)$ with the integral inner products (Eqs. (14.11) and (14.12)) playing the role of the dot product of Section 1.3. These points are outlined in Section 9.4.

SAWTOOTH WAVE

An idea of the convergence of a Fourier series and the error in using only a finite number of terms in the series may be obtained by considering the expansion of

$$f(x) = \begin{cases} x, & 0 \le x < \pi, \\ x - 2\pi, & \pi < x \le 2\pi. \end{cases} \qquad (14.14)$$

This is a sawtooth wave, and for convenience we shall shift our interval from $[0, 2\pi]$ to $[-\pi, \pi]$. In this interval we have simply $f(x) = x$. Using Eqs. (14.11) and (14.12), we may show the expansion to be

$$f(x) = x = 2 \left[\sin x - \frac{\sin 2x}{2} + \frac{\sin 3x}{3} - \cdots + (-1)^{n+1} \frac{\sin nx}{n} + \cdots \right].$$

$$(14.15)$$

Figure 14.1 shows $f(x)$ for $0 \le x < \pi$ for the sum of 4, 6, and 10 terms of the series. Three features deserve comment.

1. There is a steady increase in the accuracy of the representation as the number of terms included is increased.
2. All the curves pass through the midpoint $f(x) = 0$ at $x = \pi$.

3. In the vicinity of $x = \pi$ there is an overshoot that persists and shows no sign of diminishing.

As a matter of incidental interest, setting $x = \pi/2$ in Eq. (14.15) provides an alternate derivation of Leibniz's formula, Exercise 5.7.6.

BEHAVIOR OF DISCONTINUITIES

The behavior at $x = \pi$ is an example of a general rule that at a finite discontinuity the series converges to the arithmetic mean. For a discontinuity at $x = x_0$ the series yields

$$f(x_0) = \tfrac{1}{2}[f(x_0+) + f(x_0-)], \qquad (14.16)$$

the arithmetic mean of the right and left approaches to $x = x_0$. A general proof using partial sums, as in Section 14.5, is given by Jeffreys and by Carslaw. The proof may be simplified by the use of Dirac delta functions— Exercise 14.5.1.

The overshoot just before $x = \pi$ is an example of the Gibbs phenomenon, discussed in Section 14.5.

SUMMATION OF A FOURIER SERIES

Usually in this chapter we shall be concerned with finding the coefficients of the Fourier expansion of a known function. Occasionally, we may wish to reverse this process and determine the function represented by a given Fourier series.

Consider the series $\sum_{n=1}^{\infty} (1/n) \cos nx$, $x \in (0, 2\pi)$. Since this series is only conditionally convergent (and diverges at $x = 0$), we take

$$\sum_{n=1}^{\infty} \frac{\cos nx}{n} = \lim_{r \to 1} \sum_{n=1}^{\infty} \frac{r^n \cos nx}{n}, \qquad (14.17)$$

absolutely convergent for $|r| < 1$. Our procedure is to try forming power series by transforming the trigonometric functions into exponential form:

$$\sum_{n=1}^{\infty} \frac{r^n \cos nx}{n} = \frac{1}{2} \sum_{n=1}^{\infty} \frac{r^n e^{inx}}{n} + \frac{1}{2} \sum_{n=1}^{\infty} \frac{r^n e^{-inx}}{n}. \qquad (14.18)$$

Now these power series may be identified as Maclaurin expansions of $-\ln(1 - z)$, $z = re^{ix}$, re^{-ix} (Eq. (5.95)), and

$$\sum_{n=1}^{\infty} \frac{r^n \cos nx}{n} = -\frac{1}{2} [\ln(1 - re^{ix}) + \ln(1 - re^{-ix})]$$

$$= -\ln[(1 + r^2) - 2r \cos x]^{1/2}. \qquad (14.19)$$

Letting $r = 1$,

$$\sum_{n=1}^{\infty} \frac{\cos nx}{n} = -\ln(2 - 2\cos x)^{1/2}$$

$$= -\ln\left(2\sin\frac{x}{2}\right), \qquad x \in (0, 2\pi).^3 \qquad (14.20)$$

Both sides of this expression diverge as $x \to 0$ and 2π.

EXERCISES

14.1.1 A function $f(x)$ (quadratically integrable) is to be represented by a *finite* Fourier series. A convenient measure of the accuracy of the series is given by the integrated square of the deviation

$$\Delta_p = \int_0^{2\pi} \left[f(x) - \frac{a_0}{2} - \sum_{n=1}^{p} (a_n \cos nx + b_n \sin nx) \right]^2 dx.$$

Show that the requirement that Δ_p be minimized, that is,

$$\frac{\partial \Delta_p}{\partial a_n} = 0, \qquad \frac{\partial \Delta_p}{\partial b_n} = 0,$$

for all n, leads to choosing a_n and b_n, as given in Eqs. (14.11) and (14.12). *Note.* Your coefficients a_n and b_n are independent of p. This independence is a consequence of orthogonality and would not hold for powers of x, fitting a curve with polynomials.

14.1.2 In the analysis of a complex waveform (ocean tides, earthquakes, musical tones, etc.) it might be more convenient to have the Fourier series written as

$$f(x) = \frac{a_0}{2} + \sum_{n=1}^{\infty} \alpha_n \cos(nx - \theta_n).$$

Show that this is equivalent to Eq. (14.1) with

$$a_n = \alpha_n \cos \theta_n, \qquad \alpha_n^2 = a_n^2 + b_n^2,$$
$$b_n = \alpha_n \sin \theta_n, \qquad \tan \theta_n = b_n/a_n.$$

Note. The coefficients α_n^2 as a function of n define what is called the *power spectrum*. The importance of α_n^2 lies in its invariance under a shift in the phase θ_n.

14.1.3 A function $f(x)$ is expanded in an exponential Fourier series.

$$f(x) = \sum_{n=-\infty}^{\infty} c_n e^{inx}.$$

If $f(x)$ is real, $f(x) = f^*(x)$, what restriction is imposed on the coefficients c_n?

14.1.4 Assuming that $\int_{-\pi}^{\pi} f(x)\, dx$ and $\int_{-\pi}^{\pi} [f(x)]^2\, dx$ are finite, show that

$$\lim_{m\to\infty} a_m = 0, \qquad \lim_{m\to\infty} b_m = 0.$$

^{3} The limits may be shifted to $[-\pi, \pi]$ (and $x \neq 0$) using $|x|$ on the right-hand side.

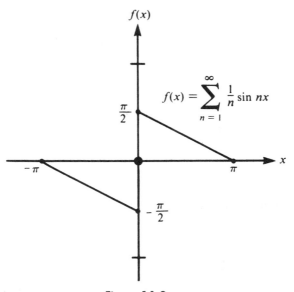

Figure 14.2

Hint. Integrate $[f(x) - s_n(x)]^2$, where $s_n(x)$ is the *n*th partial sum and use Bessel's inequality, Section 9.4. For our *finite* interval the assumption that $f(x)$ is square integrable ($\int_{-\pi}^{\pi} |f(x)|^2 \, dx$ is finite) implies that $\int_{-\pi}^{\pi} |f(x)| \, dx$ is also finite. The converse does not hold.

14.1.5 Apply the summation technique of this section to show that

$$\sum_{n=1}^{\infty} \frac{\sin nx}{n} = \begin{cases} \frac{1}{2}(\pi - x), & 0 < x \leq \pi \\ -\frac{1}{2}(\pi + x), & -\pi \leq x < 0 \end{cases}$$

(Fig. 14.2).

14.1.6 Sum the trigonometric series

$$\sum_{n=1}^{\infty} (-1)^{n+1} \frac{\sin nx}{n}$$

and show that it equals $x/2$.

14.1.7 Sum the trigonometric series

$$\sum_{n=0}^{\infty} \frac{\sin(2n + 1)x}{2n + 1},$$

and show that it equals

$$\begin{cases} \pi/4, & 0 < x < \pi \\ -\pi/4, & -\pi < x < 0. \end{cases}$$

14.1.8 Calculate the sum of the finite Fourier sine series for the sawtooth wave, $f(x) = x$, $(-\pi, \pi)$, Eq. (14.15). Use 4-, 6-, 8-, and 10-term series and $x/\pi = 0.00(0.02)1.00$. If a plotting routine is available, plot your results and compare with Fig. (14.1).

14.2 ADVANTAGES, USES OF FOURIER SERIES

DISCONTINUOUS FUNCTION

One of the advantages of a Fourier representation over some other representation, such as a Taylor series, is that it may represent a discontinuous function. An example is the sawtooth wave in the preceding section. Other examples are considered in Section 14.3 and in the exercises.

PERIODIC FUNCTIONS

Related to this advantage is the usefulness of a Fourier series in representing a periodic function. If $f(x)$ has a period of 2π, perhaps it is only natural that we expand it in a series of functions with period 2π, $2\pi/2$, $2\pi/3$, ... This guarantees that if our periodic $f(x)$ is represented over one interval $[0, 2\pi]$ or $[-\pi, \pi]$ the representation holds for all finite x.

At this point we may conveniently consider the properties of symmetry. Using the interval $[-\pi, \pi]$, $\sin x$ is odd and $\cos x$ is an even function of x. Hence, by Eqs. (14.11) and (14.12),[1] if $f(x)$ is odd, all $a_n = 0$ and if $f(x)$ is even all $b_n = 0$. In other words,

$$f(x) = \frac{a_0}{2} + \sum_{n=1}^{\infty} a_n \cos nx, \qquad f(x) \text{ even}, \qquad (14.21)$$

$$f(x) = \sum_{n=1}^{\infty} b_n \sin nx, \qquad f(x) \text{ odd}. \qquad (14.22)$$

Frequently these properties are helpful in expanding a given function.

We have noted that the Fourier series is periodic. This is important in considering whether Eq. (14.1) holds outside the initial interval. Suppose we are given only that

$$f(x) = x, \qquad 0 \le x < \pi \qquad (14.23)$$

and are asked to represent $f(x)$ by a series expansion. Let us take three of the infinite number of possible expansions.

1. If we assume a Taylor expansion, we have

$$f(x) = x, \qquad (14.24)$$

a one-term series. This (one-term) series is defined for all finite x.

2. Using the Fourier cosine series (Eq. (14.21)), we predict that

$$f(x) = -x, \qquad -\pi < x \le 0,$$
$$f(x) = 2\pi - x, \qquad \pi < x < 2\pi. \qquad (14.25)$$

3. Finally, from the Fourier sine series (Eq. (14.22)), we have

$$f(x) = x, \qquad -\pi < x \le 0,$$
$$f(x) = x - 2\pi, \qquad \pi < x < 2\pi. \qquad (14.26)$$

---

[1] With the range of integration $-\pi \le x \le \pi$.

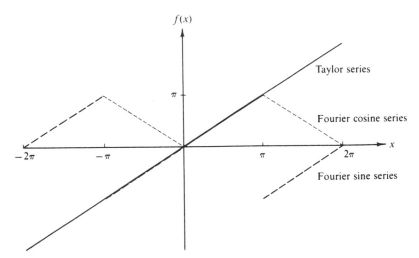

Figure 14.3 Comparison of Fourier cosine series, Fourier sine series, and Taylor series.

These three possibilities, Taylor series, Fourier cosine series, and Fourier sine series, are each perfectly valid in the original interval $[0, \pi]$. Outside, however, their behavior is strikingly different (compare Fig. 14.3). Which of the three, then, is correct? This question has no answer, unless we are given more information about $f(x)$. It may be any of the three or none of them. Our Fourier expansions are valid over the basic interval. Unless the function $f(x)$ is known to be periodic with a period equal to our basic interval, or $(1/n)$th of our basic interval, there is no assurance whatever that the representation (Eq. (14.1)) will have any meaning outside the basic interval.

It should be noted that the set of functions $\cos nx$, $n = 0, 1, 2, \ldots$, forms a complete orthogonal set over $[0, \pi]$. Similarly, the set of functions $\sin nx$, $n = 1, 2, 3, \ldots$, forms a complete orthogonal set over this same interval. Unless forced by boundary conditions or a symmetry restriction, the choice of which set to use is arbitrary.

In addition to the advantages of representing discontinuous and periodic functions, there is a third very real advantage in using a Fourier series. Suppose that we are solving the equation of motion of an oscillating particle subject to a periodic driving force. The Fourier expansion of the driving force then gives us the fundamental term and a series of harmonics. The (linear) differential equation may be solved for each of these harmonics individually, a process that may be much easier than dealing with the original driving force. Then, as long as the differential equation is linear, all the solutions may be added together to obtain the final solution.[2] This is more than just a clever mathematical trick.

---

[2] One of the nastier features of nonlinear differential equations is that this principle of superposition is not valid.

It corresponds to finding the response of the system to the fundamental frequency and to each of the harmonic frequencies.

One question that is sometimes raised is: "Were the harmonics there all along or were they created by our Fourier analysis?" One answer compares the functional resolution into harmonics with the resolution of a vector into rectangular components. The components may have been present in the sense that they may be isolated and observed, but the resolution is certainly not unique. Hence many authorities prefer to say that the harmonics were created by our choice of expansion. Other expansions in other sets of orthogonal functions would give different results. For further discussion the reader should consult a series of notes and letters in the *American Journal of Physics*.[3]

CHANGE OF INTERVAL

So far attention has been restricted to an interval of length 2π. This restriction may easily be relaxed. If $f(x)$ is periodic with a period $2L$, we may write

$$f(x) = \frac{a_0}{2} + \sum_{n=1}^{\infty} \left[a_n \cos \frac{n\pi x}{L} + b_n \sin \frac{n\pi x}{L} \right], \tag{14.27}$$

with

$$a_n = \frac{1}{L} \int_{-L}^{L} f(t) \cos \frac{n\pi t}{L} \, dt, \qquad n = 0, 1, 2, 3, \ldots, \tag{14.28}$$

$$b_n = \frac{1}{L} \int_{-L}^{L} f(t) \sin \frac{n\pi t}{L} \, dt, \qquad n = 1, 2, 3, \ldots, \tag{14.29}$$

replacing x in Eq. (14.1) with $\pi x/L$ and t in Eqs. (14.11) and (14.12) with $\pi t/L$. (For convenience the interval in Eqs. (14.11) and (14.12) is shifted to $-\pi \le t \le \pi$.) The choice of the symmetric interval $(-L, L)$ is *not* essential. For $f(x)$ periodic with a period of $2L$, any interval $(x_0, x_0 + 2L)$ will do. The choice is a matter of convenience or literally personal preference.

EXERCISES

14.2.1 The boundary conditions (such as $\psi(0) = \psi(l) = 0$) may suggest solutions of the form $\sin(n\pi x/l)$ and eliminate the corresponding cosines.

(a) Verify that the boundary conditions used in the Sturm–Liouville theory are satisfied for the interval $(0, l)$. Note that this is only half the usual Fourier interval.

(b) Show that the set of functions $\varphi_n(x) = \sin(n\pi x/l)$, $n = 1, 2, 3, \ldots$ satisfies an orthogonality relation

$$\int_0^l \varphi_m(x)\varphi_n(x) \, dx = \frac{l}{2}\delta_{mn}, \qquad n > 0.$$

[3] B. L. Robinson, Concerning frequencies resulting from distortion. *Am. J. Phys.* **21**, 391 (1953); F. W. Van Name, Jr., Concerning frequencies resulting from distortion. *ibid.* **22**, 94 (1954).

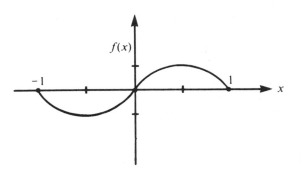

Figure 14.4

14.2.2   (a) Expand $f(x) = x$ in the interval $(0, 2L)$. Sketch the series you have found (right-hand side of Ans.) over $(-2L, 2L)$.

$$ANS. \quad x = L - \frac{2L}{\pi} \sum_{n=1}^{\infty} \frac{1}{n} \sin\left(\frac{n\pi x}{L}\right).$$

(b) Expand $f(x) = x$ as a sine series in the *half* interval $(0, L)$. Sketch the series you have found (right-hand side of Ans.) over $(-2L, 2L)$.

$$ANS. \quad x = \frac{2L}{\pi} \sum_{n=1}^{\infty} (-1)^{n+1} \sin\left(\frac{n\pi x}{L}\right).$$

14.2.3   In some problems it is convenient to approximate $\sin \pi x$ over the interval $[0, 1]$ by a parabola $ax(1 - x)$, where a is a constant. To get a feeling for the accuracy of this approximation, expand $4x(1 - x)$ in a Fourier sine series:

$$f(x) = \begin{cases} 4x(1 - x), & 0 \le x \le 1 \\ 4x(1 + x), & -1 \le x \le 0 \end{cases} = \sum_{n=1}^{\infty} b_n \sin n\pi x$$

$$ANS. \quad b_n = \frac{32}{\pi^3} \cdot \frac{1}{n^3}, \qquad n \text{ odd}$$

$$b_n = 0, \qquad n \text{ even}$$

(Fig. 14.4).

14.3 APPLICATIONS OF FOURIER SERIES

Example 14.3.1 Square Wave—High Frequencies

One simple application of Fourier series, the analysis of a "square" wave (Fig. 14.5) in terms of its Fourier components, may occur in electronic circuits designed to handle sharply rising pulses. Suppose that our wave is defined by

$$f(x) = 0, \quad -\pi < x < 0,$$

$$f(x) = h, \quad 0 < x < \pi.$$

(14.30)

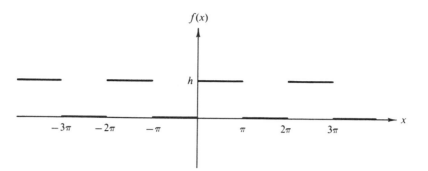

Figure 14.5 Square wave.

From Eqs. (14.11) and (14.12) we find

$$a_0 = \frac{1}{\pi} \int_0^\pi h \, dt = h, \tag{14.31}$$

$$a_n = \frac{1}{\pi} \int_0^\pi h \cos nt \, dt = 0, \qquad n = 1, 2, 3, \ldots, \tag{14.32}$$

$$b_n = \frac{1}{\pi} \int_0^\pi h \sin nt \, dt = \frac{h}{n\pi}(1 - \cos n\pi); \tag{14.33}$$

$$b_n = \frac{2h}{n\pi}, \qquad n \text{ odd}, \tag{14.34}$$

$$b_n = 0, \qquad n \text{ even}. \tag{14.35}$$

The resulting series is

$$f(x) = \frac{h}{2} + \frac{2h}{\pi}\left(\frac{\sin x}{1} + \frac{\sin 3x}{3} + \frac{\sin 5x}{5} + \cdots\right). \tag{14.36}$$

Except for the first term which represents an average of $f(x)$ over the interval $[-\pi, \pi]$, all the cosine terms have vanished. Since $f(x) - h/2$ is odd, we have a Fourier sine series. Although only the odd terms in the sine series occur, they fall only as n^{-1}. This is similar to the convergence (or lack of convergence) of the harmonic series. Physically this means that our square wave contains a lot of high-frequency components. If the electronic apparatus will not pass these components, our square wave input will emerge more or less rounded off, perhaps as an amorphous blob.

Example 14.3.2 Full Wave Rectifier

As a second example, let us ask how well the output of a full wave rectifier approaches pure direct current (Fig. 14.6). Our rectifier may be thought of as having passed the positive peaks of an incoming sine wave and inverting the

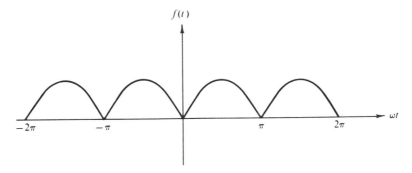

Figure 14.6   Full wave rectifier.

negative peaks. This yields

$$f(t) = \sin \omega t, \qquad 0 < \omega t < \pi,$$
$$f(t) = -\sin \omega t, \qquad -\pi < \omega t < 0.$$

(14.37)

Since $f(t)$ defined here is even, no terms of the form $\sin n\omega t$ will appear. Again, from Eqs. (14.11) and (14.12), we have

$$a_0 = \frac{1}{\pi} \int_{-\pi}^{0} -\sin \omega t \, d(\omega t) + \frac{1}{\pi} \int_{0}^{\pi} \sin \omega t \, d(\omega t)$$

$$= \frac{2}{\pi} \int_{0}^{\pi} \sin \omega t \, d(\omega t) = \frac{4}{\pi},$$

(14.38)

$$a_n = \frac{2}{\pi} \int_{0}^{\pi} \sin \omega t \cos n\omega t \, d(\omega t)$$

$$= -\frac{2}{\pi} \frac{2}{n^2 - 1}, \qquad n \text{ even}$$

$$= 0, \qquad\qquad n \text{ odd}.$$

(14.39)

Note carefully that $[0, \pi]$ is not an orthogonality interval for both sines and cosines together and we do not get zero for even n. The resulting series is

$$f(t) = \frac{2}{\pi} - \frac{4}{\pi} \sum_{n = 2,4,6,\ldots}^{\infty} \frac{\cos n\omega t}{n^2 - 1}.$$

(14.40)

The original frequency ω has been eliminated. The lowest frequency oscillation is 2ω. The high-frequency components fall off as n^{-2}, showing that the full wave rectifier does a fairly good job of approximating direct current. Whether this good approximation is adequate depends on the particular application. If the remaining ac components are objectionable, they may be further suppressed by appropriate filter circuits.

These two examples bring out two features characteristic of Fourier expansions.[1]

1. If $f(x)$ has discontinuities (as in the square wave in Example 14.3.1), we can expect the nth coefficient to be decreasing as $1/n$. Convergence is relatively slow.[2]
2. If $f(x)$ is continuous (although possibly with discontinuous derivatives as in the full wave rectifier of example 14.3.2), we can expect the nth coefficient to be decreasing as $1/n^2$.

Example 14.3.3 Infinite Series, Riemann Zeta Function

As a final example, we consider the purely mathematical problem of expanding x^2. Let

$$f(x) = x^2, \quad -\pi < x < \pi. \tag{14.41}$$

By symmetry all $b_n = 0$. For the a_n's we have

$$a_0 = \frac{1}{\pi} \int_{-\pi}^{\pi} x^2 \, dx = \frac{2\pi^2}{3}, \tag{14.42}$$

$$a_n = \frac{2}{\pi} \int_0^{\pi} x^2 \cos nx \, dx$$

$$= \frac{2}{\pi} \cdot (-1)^n \frac{2\pi}{n^2}$$

$$= (-1)^n \frac{4}{n^2}. \tag{14.43}$$

From this we obtain

$$x^2 = \frac{\pi^2}{3} + 4 \sum_{n=1}^{\infty} (-1)^n \frac{\cos nx}{n^2}. \tag{14.44}$$

As it stands, Eq. (14.44) is of no particular importance, but if we set $x = \pi$,

$$\cos n\pi = (-1)^n \tag{14.45}$$

and Eq. (14.44) becomes[3]

$$\pi^2 = \frac{\pi^2}{3} + 4 \sum_{n=1}^{\infty} \frac{1}{n^2} \tag{14.46}$$

or

$$\frac{\pi^2}{6} = \sum_{n=1}^{\infty} \frac{1}{n^2} \equiv \zeta(2), \tag{14.47}$$

[1] G. Raisbeck, Order of magnitude of Fourier coefficients. *Am. Math. Mon.* **62**, 149–155 (1955).
[2] A technique for improving the rate of convergence is developed in the exercises of Section 14.4.
[3] Note that the point $x = \pi$ is *not* a point of discontinuity.

thus yielding the Riemann zeta function, $\zeta(2)$, in closed form (in agreement with the Bernoulli number result of Section 5.9). From our expansion of x^2 and expansions of other powers of x numerous other infinite series can be evaluated. A few are included in the subsequent list of exercises.

	Fourier Series	Reference
1.	$\sum\limits_{n=1}^{\infty} \dfrac{1}{n} \sin nx = \begin{cases} -\frac{1}{2}(\pi + x), & -\pi \le x < 0 \\ \frac{1}{2}(\pi - x), & 0 \le x < \pi \end{cases}$	Exercise 14.1.5 Exercise 14.3.3
2.	$\sum\limits_{n=1}^{\infty} (-1)^{n+1} \dfrac{1}{n} \sin nx = \frac{1}{2}x, \quad -\pi < x < \pi$	Exercise 14.1.6 Exercise 14.3.2
3.	$\sum\limits_{n=0}^{\infty} \dfrac{1}{2n+1} \sin(2n+1)x = \begin{cases} -\pi/4, & -\pi < x < 0 \\ +\pi/4, & 0 < x < \pi \end{cases}$	Exercise 14.1.7 Eq. (14.36)
4.	$\sum\limits_{n=1}^{\infty} \dfrac{1}{n} \cos nx = -\ln\left[2\sin\left(\dfrac{|x|}{2}\right) \right], \quad -\pi < x < \pi$	Eq. (14.20) Exercise 14.3.15
5.	$\sum\limits_{n=1}^{\infty} (-1)^n \dfrac{1}{n} \cos nx = -\ln\left[2\cos\left(\dfrac{x}{2}\right) \right], \quad -\pi < x < \pi$	Exercise 14.3.15
6.	$\sum\limits_{n=0}^{\infty} \dfrac{1}{2n+1} \cos(2n+1)x = \dfrac{1}{2}\ln\left[\cot\dfrac{|x|}{2} \right], \quad -\pi < x < \pi$	

The square wave Fourier series from Eq. (14.36) and item (3) above,

$$g(x) = \sum_{n=0}^{\infty} \frac{\sin(2n+1)x}{2n+1} = (-1)^m \frac{\pi}{4}, \qquad m\pi < x < (m+1)\pi, \quad (14.48)$$

can be used to derive Riemann's functional equation for the zeta function. Its defining Dirichlet series can be written in various forms:

$$\zeta(s) = \sum_{n=1}^{\infty} n^{-s} = 1 + \sum_{n=1}^{\infty} (2n)^{-s} + \sum_{n=1}^{\infty} (2n+1)^{-s}$$

$$= 2^{-s}\zeta(s) + \sum_{n=0}^{\infty} (2n+1)^{-s}$$

implying

$$\lambda(s) = \sum_{n=0}^{\infty} (2n+1)^{-s} = (1 - 2^{-s})\zeta(s). \qquad (14.49)$$

Here s is a complex variable. Both Dirichlet series converge for $\sigma = \mathrm{Re}\, s > 1$. Alternatively, using Eq. (14.49), we have

$$\eta(s) = \sum_{n=1}^{\infty} (-1)^{n-1} n^{-s} = \sum_{n=0}^{\infty} (2n+1)^{-s} - \sum_{n=1}^{\infty} (2n)^{-s} = (1 - 2^{1-s})\zeta(s),$$
$$(14.50)$$

which converges already for $\mathrm{Re}\, s > 0$ using the Leibniz convergence criterion (see Section 5.3).

Another approach to Dirichlet series starts from Euler's integral for the Γ function

$$\int_0^\infty y^{s-1} e^{-ny}\, dy = n^{-s} \int_0^\infty y^{s-1} e^{-y}\, dy = n^{-s}\Gamma(s), \tag{14.51}$$

which may be summed to yield the integral representation for the zeta function:

$$\int_0^\infty \frac{y^{s-1}}{e^y - 1}\, dy = \zeta(s)\Gamma(s). \tag{14.52}$$

If we combine the alternative forms of Eq. (14.51),

$$\int_0^\infty y^{s-1} e^{-iny}\, dy = n^{-s}\Gamma(s) e^{-i\pi s/2},$$

$$\int_0^\infty y^{s-1} e^{iny}\, dy = n^{-s}\Gamma(s) e^{i\pi s/2},$$

we obtain

$$\int_0^\infty y^{s-1} \sin(ny)\, dy = n^{-s}\Gamma(s) \sin\frac{\pi s}{2}. \tag{14.53}$$

Summing Eq. (14.53) over odd n's yields

$$\int_0^\infty g(y) y^{s-1}\, dy = (1 - 2^{-s-1})\zeta(s + 1)\Gamma(s) \sin\frac{\pi s}{2}, \tag{14.54}$$

using Eqs. (14.48) and (14.49). This relation is at the heart of the functional equation. If we divide the integration range into intervals $m\pi < y < (m + 1)\pi$ and substitute Eq. (14.48) into (14.54) we find

$$
\begin{aligned}
\int_0^\infty g(y) y^{s-1}\, dy &= \frac{\pi}{4} \sum_{m=0}^\infty (-1)^m \int_{m\pi}^{(m+1)\pi} y^{s-1}\, dy \\
&= \frac{\pi^{s+1}}{4s} \left\{ \sum_{m=1}^\infty (-1)^m [(m + 1)^s - m^s] + 1 \right\} \\
&= \frac{\pi^{s+1}}{2s} (1 - 2^{s+1})\zeta(-s),
\end{aligned}
\tag{14.55}
$$

using Eq. (14.50). The series in Eq. (14.55) converges for $\text{Re}\, s < 1$. Comparing Eqs. (14.54) and (14.55) for $0 < \sigma = \text{Re}\, s < 1$, we get

$$\frac{\pi^{s+1}}{2s} (1 - 2^{s+1})\zeta(-s) = (1 - 2^{-s-1})\zeta(s + 1)\Gamma(s) \sin\frac{\pi s}{2},$$

which can be rewritten as

$$\zeta(1 - s) = 2(2\pi)^{-s}\zeta(s)\Gamma(s) \cos\frac{\pi s}{2}. \tag{14.56}$$

This functional equation provides an analytic continuation of $\zeta(s)$ into the negative half plane in s. For $s \to 1$ the pole of $\zeta(s)$ and the zero of $\cos(\pi s/2)$ cancel in Eq. (14.56), so that $\zeta(0) = -1/2$ results. Since $\cos(\pi s/2) = 0$ for $s = 2m + 1 =$ odd integer, Eq. (14.56) gives $\zeta(-2m) = 0$, the trivial zeros of the zeta function for $m = 1, 2, \ldots$. All other zeros must lie in the "critical strip" $0 < \sigma = \operatorname{Re} s < 1$. They are closely related to the distribution of prime numbers because the prime number product for $\zeta(s)$ (see Section 5.9) can be converted into a Dirichlet series over prime powers for $\zeta'/\zeta = d \ln \zeta(s)/ds$. Using the inverse Mellin transform (see Section 16.2) yields the relation

$$\sum_{\substack{p^m < x, p = \text{prime} \\ m = 1,2,\ldots}} \ln p = -(2\pi i)^{-1} \int_{\sigma - i\infty}^{\sigma + i\infty} \frac{\zeta'(s)}{\zeta(s)s} x^s \, ds \qquad (14.57)$$

for $\sigma > 1$, which is a cornerstone of analytic number theory. Since zeros of $\zeta(s)$ become simple poles of ζ'/ζ, the asymptotic distribution of prime numbers is directly related by Eq. (14.57) to the zeros of the Riemann zeta function. Riemann conjectured that all zeros lie on the line $\sigma = 1/2$, i.e., have the form $1/2 + it$ with real t. If so, one could shift the line of integration to the left to $\sigma = 1/2 + \varepsilon$, the simple pole of $\zeta(s)$ at $s = 1$ giving rise to the residue x, while the integral along the line $\sigma = 1/2 + \varepsilon$ is of order $O(x^{1/2+\varepsilon})$. Hence the remarkably small remainder in the asymptotic estimate

$$\sum_{p < x} \ln p \sim x + O(x^{1/2+\varepsilon}), \qquad x \to \infty$$

would result for arbitrarily small ε. This is equivalent to the estimate for the number of primes below x

$$\pi(x) = \sum_{p < x} 1 = \int_2^x (\ln t)^{-1} \, dt + O(x^{1/2+\varepsilon}), \qquad x \to \infty.$$

In fact, numerical studies have shown that the first 1.5×10^9 zeros are simple and lie all on the critical line $\sigma = 1/2$. For more details the reader is referred to the classic text by E. C. Titchmarsh and D. R. Heath-Brown, *The Theory of the Riemann Zeta-Function*. Oxford: Clarendon Press (1986); H. M. Edwards, *Riemann's Zeta Function*. New York: Academic Press (1974); J. Van de Lune, H. J. J. Te Riele, and D. T. Winter, On the zeros of the Riemann zeta function in the critical strip. IV. *Math. Comput.* **47**, 667 (1986).

More recently the statistics of the zeros ρ of the Riemann zeta function on the critical line played a prominent role in the development of theories of chaos (see Chapter 18 for an introduction). Assuming that there is a quantum mechanical system whose energies are the imaginary parts of the ρ's, then primes label the primitive periodic orbits of the associated classically chaotic system. For this case Gutzwiller's trace formula, which relates quantum energy levels and classical periodic orbits, plays a central role and can be better understood using properties of the zeta function and primes. For more details see Sections 12.6 and 12.7 by J. Keating, in *The Nature of Chaos* (T. Mullin, ed.). Oxford: Clarendon Press (1993).

COMPLEX VARIABLES—ABEL'S THEOREM

Consider a function $f(z)$ represented by a convergent power series

$$f(z) = \sum_{n=0}^{\infty} c_n z^n = \sum_{n=0}^{\infty} c_n r^n e^{in\theta}. \tag{14.58}$$

This is our Fourier exponential series, Eq. (14.2). Separating real and imaginary parts

$$u(r, \theta) = \sum_{n=0}^{\infty} c_n r^n \cos n\theta$$

$$v(r, \theta) = \sum_{n=1}^{\infty} c_n r^n \sin n\theta, \tag{14.59a}$$

the Fourier cosine and sine series. Abel's theorem asserts that if $u(1, \theta)$ and $v(1, \theta)$ are convergent for a given θ, then

$$u(1, \theta) + iv(1, \theta) = \lim_{r \to 1} f(re^{i\theta}). \tag{14.59b}$$

An application of this appears as Exercise 14.3.15.

EXERCISES

14.3.1   Develop the Fourier series representation of

$$f(t) = \begin{cases} 0, & -\pi \le \omega t \le 0, \\ \sin \omega t, & 0 \le \omega t \le \pi. \end{cases}$$

This is the output of a simple half-wave rectifier. It is also an approximation of the solar thermal effect that produces "tides" in the atmosphere.

$$ANS. \quad f(t) = \frac{1}{\pi} + \frac{1}{2} \sin \omega t - \frac{2}{\pi} \sum_{\substack{n=2,4,6,\ldots \\ \text{even}}}^{\infty} \frac{\cos n\omega t}{n^2 - 1}.$$

14.3.2   A sawtooth wave is given by

$$f(x) = x, \quad -\pi < x < \pi.$$

Show that

$$f(x) = 2 \sum_{n=1}^{\infty} \frac{(-1)^{n+1}}{n} \sin nx.$$

14.3.3   A different sawtooth wave is described by

$$f(x) = \begin{cases} -\frac{1}{2}(\pi + x), & -\pi \le x < 0 \\ +\frac{1}{2}(\pi - x), & 0 < x \le \pi. \end{cases}$$

Show that $f(x) = \sum_{n=1}^{\infty} (\sin nx/n)$.

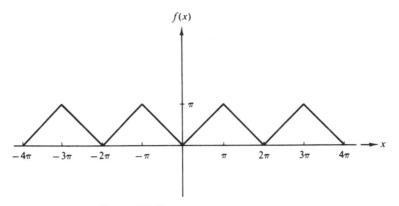

Figure 14.7   Triangular wave.

14.3.4   A triangular wave (Fig. 14.7) is represented by

$$f(x) = \begin{cases} x, & 0 < x < \pi \\ -x, & -\pi < x < 0. \end{cases}$$

Represent $f(x)$ by a Fourier series.

$$ANS. \quad f(x) = \frac{\pi}{2} - \frac{4}{\pi} \sum_{\substack{n = 1,3,5,\dots \\ odd}} \frac{\cos nx}{n^2}.$$

14.3.5   Expand

$$f(x) = \begin{cases} 1, & x^2 < x_0^2 \\ 0, & x^2 > x_0^2 \end{cases}$$

in the interval $[-\pi, \pi]$.
Note. This variable width square wave is of some importance in electronic music.

14.3.6   A metal cylindrical tube of radius a is split lengthwise into two nontouching halves. The top half is maintained at a potential $+V$, the bottom half at a potential $-V$ (Fig. 14.8). Separate the variables in Laplace's equation and solve for the electrostatic potential for $r \le a$. Observe the resemblance between your solution for $r = a$ and the Fourier series for a square wave.

14.3.7   A metal cylinder is placed in a (previously) uniform electric field, E_0, the axis of the cylinder perpendicular to that of the original field.

(a) Find the perturbed electrostatic potential.
(b) Find the induced surface charge on the cylinder as a function of angular position.

14.3.8   Transform the Fourier expansion of a square wave, Eq. (14.36), into a power series. Show that the coefficients of x^1 form a *divergent* series. Repeat for the coefficients of x^3.

A power series cannot handle a discontinuity. These infinite coefficients are the result of attempting to beat this basic limitation on power series.

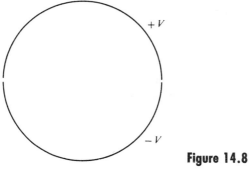

Figure 14.8

14.3.9   (a) Show that the Fourier expansion of $\cos ax$ is

$$\cos ax = \frac{2a \sin a\pi}{\pi} \left\{ \frac{1}{2a^2} - \frac{\cos x}{a^2 - 1^2} + \frac{\cos 2x}{a^2 - 2^2} - \cdots \right\},$$

$$a_n = (-1)^n \frac{2a \sin a\pi}{\pi(a^2 - n^2)}.$$

(b) From the preceding result show that

$$a\pi \cot a\pi = 1 - 2 \sum_{p=1}^{\infty} \zeta(2p)a^{2p}.$$

This provides an alternate derivation of the relation between the Riemann zeta function and the Bernoulli numbers, Eq. (5.151).

14.3.10   Derive the Fourier series expansion of the Dirac delta function $\delta(x)$ in the interval $-\pi < x < \pi$.

(a) What significance can be attached to the constant term?

(b) In what region is this representation valid?

(c) With the identity

$$\sum_{n=1}^{N} \cos nx = \frac{\sin(Nx/2)}{\sin(x/2)} \cos\left[\left(N + \frac{1}{2}\right)x/2 \right],$$

show that your Fourier representation of $\delta(x)$ is consistent with Eq. (1.189).

14.3.11   Expand $\delta(x - t)$ in a Fourier series. Compare your result with the bilinear form of Eq. (1.189).

$$ANS. \quad \delta(x - t) = \frac{1}{2\pi} + \frac{1}{\pi} \sum_{n=1}^{\infty} (\cos nx \cos nt + \sin nx \sin nt)$$

$$= \frac{1}{2\pi} + \frac{1}{\pi} \sum_{n=1}^{\infty} \cos n(x - t).$$

14.3.12   Verify that

$$\delta(\varphi_1 - \varphi_2) = \frac{1}{2\pi} \sum_{m=-\infty}^{\infty} e^{im(\varphi_1 - \varphi_2)}$$

is a Dirac delta function by showing that it satisfies the definition of a Dirac delta function:

$$\int_{-\pi}^{\pi} f(\varphi_1) \frac{1}{2\pi} \sum_{m=-\infty}^{\infty} e^{im(\varphi_1-\varphi_2)} \, d\varphi_1 = f(\varphi_2).$$

Hint. Represent $f(\varphi_1)$ by an exponential Fourier series.
Note. The continuum analog of this expression is developed in Section 15.2. The most important application of this expression is in the determination of Green's functions, Section 8.7.

14.3.13   (a) Using

$$f(x) = x^2, \quad -\pi < x < \pi,$$

show that

$$\sum_{n=1}^{\infty} \frac{(-1)^{n+1}}{n^2} = \frac{\pi^2}{12} = \eta(2).$$

(b) Using the Fourier series for a triangular wave developed in Exercise 14.3.4, show that

$$\sum_{n=1}^{\infty} \frac{1}{(2n-1)^2} = \frac{\pi^2}{8} = \lambda(2).$$

(c) Using

$$f(x) = x^4, \quad -\pi < x < \pi.$$

show that

$$\sum_{n=1}^{\infty} \frac{1}{n^4} = \frac{\pi^4}{90} = \zeta(4),$$

$$\sum_{n=1}^{\infty} \frac{(-1)^{n+1}}{n^4} = \frac{7\pi^4}{720} = \eta(4).$$

(d) Using

$$f(x) = \begin{cases} x(\pi - x), & 0 < x < \pi, \\ x(\pi + x), & -\pi < x < 0, \end{cases}$$

derive

$$f(x) = \frac{8}{\pi} \sum_{n=1,3,5,\ldots \atop \text{odd}}^{\infty} \frac{\sin nx}{n^3}$$

and show that

$$\sum_{n=1,3,5,\ldots \atop \text{odd}}^{\infty} (-1)^{(n-1)/2} n^{-3} = 1 - \frac{1}{3^3} + \frac{1}{5^3} - \frac{1}{7^3} + \cdots = \frac{\pi^3}{32} = \beta(3).$$

(e) Using the Fourier series for a square wave, show that

$$\sum_{n=1,3,5,\ldots \atop \text{odd}}^{\infty} (-1)^{(n-1)/2} n^{-1} = 1 - \frac{1}{3} + \frac{1}{5} - \frac{1}{7} + \cdots = \frac{\pi}{4} = \beta(1).$$

This is Leibniz's formula for π, obtained by a different technique in Exercise 5.7.6.
Note. The $\eta(2)$, $\eta(4)$, $\lambda(2)$, $\beta(1)$, and $\beta(3)$ functions are defined by the indicated series. General definitions appear in Sections 5.9 and 14.3.

14.3.14   (a) Find the Fourier series representation of
$$f(x) = \begin{cases} 0, & -\pi < x \le 0 \\ x, & 0 \le x < \pi. \end{cases}$$

(b) From the Fourier expansion show that
$$\frac{\pi^2}{8} = 1 + \frac{1}{3^2} + \frac{1}{5^2} + \cdots.$$

14.3.15   Let $f(z) = \ln(1 + z) = \sum_{n=1}^{\infty} (-1)^{n+1} z^n/n$. (This series converges to $\ln(1 + z)$ for $|z| \le 1$, except at the point $z = -1$.)

(a) From the imaginary parts show that
$$\ln\left(2\cos\frac{\theta}{2}\right) = \sum_{n=1}^{\infty} (-1)^{n+1}\frac{\cos n\theta}{n}, \qquad -\pi < \theta < \pi.$$

(b) Using a change of variable, transform part (a) into
$$-\ln\left(2\sin\frac{\varphi}{2}\right) = \sum_{n=1}^{\infty}\frac{\cos n\varphi}{n}, \qquad 0 < \varphi < 2\pi.$$

14.3.16   A symmetric triangular pulse of adjustable height and width is described by
$$f(x) = \begin{cases} a(1 - x/b), & 0 \le |x| \le b \\ 0, & b \le |x| \le \pi. \end{cases}$$

(a) Show that the Fourier coefficients are
$$a_0 = \frac{ab}{\pi}, \qquad a_n = \frac{2ab}{\pi}(1 - \cos nb)/(nb)^2.$$

Sum the finite Fourier series through $n = 10$ and through $n = 100$ for $x/\pi = 0(1/9)1$. Take $a = 1$ and $b = \pi/2$.

(b) Call a Fourier analysis subroutine (if available) to calculate the Fourier coefficients of $f(x)$, a_0 through a_{10}.

14.3.17   (a) Using a Fourier analysis subroutine, calculate the Fourier cosine coefficients a_0 through a_{10} of
$$f(x) = [1 - (x/\pi)^2]^{1/2}, \qquad x \in [-\pi, \pi].$$

(b) Spot check by calculating some of the preceding coefficients by direct numerical quadrature.
Check values. $a_0 = 0.785$, $a_2 = 0.284$.

14.3.18   Using a Fourier analysis subroutine, calculate the Fourier coefficients through a_{10} and b_{10} for

(a) a full-wave rectifier, Example 14.3.2,

(b) a half-wave rectifier, Exercise 14.3.1. Check your results against the analytic forms given (Eq. (14.39) and Exercise 14.3.1).

14.4 PROPERTIES OF FOURIER SERIES

CONVERGENCE

It might be noted, first, that our Fourier series should not be expected to be uniformly convergent if it represents a discontinuous function. A uniformly convergent series of continuous functions ($\sin nx$, $\cos nx$) always yields a continuous function (compare Section 5.5). If, however, (a) $f(x)$ is continuous, $-\pi \le x \le \pi$, (b) $f(-\pi) = f(+\pi)$, and (c) $f'(x)$ is sectionally continuous, the Fourier series for $f(x)$ will converge uniformly. These restrictions do not demand that $f(x)$ be periodic, but they will be satisfied by continuous, differentiable, periodic functions (period of 2π). For a proof of uniform convergence the reader is referred to the literature.[1] With or without a discontinuity in $f(x)$, the Fourier series will yield convergence in the mean, Section 9.4.

INTEGRATION

Term-by-term integration of the series

$$f(x) = \frac{a_0}{2} + \sum_{n=1}^{\infty} a_n \cos nx + \sum_{n=1}^{\infty} b_n \sin nx \tag{14.60}$$

yields

$$\int_{x_0}^{x} f(x)\, dx = \frac{a_0 x}{2}\bigg|_{x_0}^{x} + \sum_{n=1}^{\infty} \frac{a_n}{n} \sin nx \bigg|_{x_0}^{x} - \sum_{n=1}^{\infty} \frac{b_n}{n} \cos nx \bigg|_{x_0}^{x}. \tag{14.61}$$

Clearly, the effect of integration is to place an additional power of n in the denominator of each coefficient. This results in more rapid convergence than before. Consequently, a convergent Fourier series may always be integrated term by term, the resulting series converging uniformly to the integral of the original function. Indeed, term-by-term integration may be valid even if the original series (Eq. (14.60)) is not itself convergent! The function $f(x)$ need only be integrable. A discussion will be found in Jeffreys and Jeffreys, Section 14.06.

Strictly speaking, Eq. (14.61) may not be a Fourier series; that is, if $a_0 \ne 0$, there will be a term $\frac{1}{2}a_0 x$. However,

$$\int_{x_0}^{x} f(x)\, dx - \tfrac{1}{2}a_0 x \tag{14.62}$$

will still be a Fourier series.

DIFFERENTIATION

The situation regarding differentiation is quite different from that of integration. Here the word is *caution*. Consider the series for

$$f(x) = x, \qquad -\pi < x < \pi. \tag{14.63}$$

We readily find (compare Exercise 14.3.2) that the Fourier series is

$$x = 2 \sum_{n=1}^{\infty} (-1)^{n+1} \frac{\sin nx}{n}, \qquad -\pi < x < \pi. \tag{4.64}$$

[1] See, for instance, R. V. Churchill, *Fourier Series and Boundary Value Problems*. New York: McGraw–Hill (1941), Section 38.

Differentiating term by term, we obtain

$$1 = 2 \sum_{n=1}^{\infty} (-1)^{n+1} \cos nx, \tag{14.65}$$

which is not convergent! *Warning.* Check your derivative.

For a triangular wave (Exercise 14.3.4), in which the convergence is more rapid (and uniform),

$$f(x) = \frac{\pi}{2} - \frac{4}{\pi} \sum_{n=1,\,\text{odd}}^{\infty} \frac{\cos nx}{n^2}. \tag{14.66}$$

Differentiating term by term

$$f'(x) = \frac{4}{\pi} \sum_{n=1,\,\text{odd}}^{\infty} \frac{\sin nx}{n}, \tag{14.67}$$

which is the Fourier expansion of a square wave

$$f'(x) = \begin{cases} 1, & 0 < x < \pi, \\ -1, & -\pi < x < 0. \end{cases} \tag{14.68}$$

Inspection of Fig. 14.7 verifies that this is indeed the derivative of our triangular wave.

As the inverse of integration, the operation of differentiation has placed an additional factor n in the numerator of each term. This reduces the rate of convergence and may, as in the first case mentioned, render the differentiated series divergent.

In general, term-by-term differentiation is permissible under the same conditions listed for uniform convergence.

EXERCISES

14.4.1   Show that integration of the Fourier expansion of $f(x) = x$, $-\pi < x < \pi$, leads to

$$\frac{\pi^2}{12} = \sum_{n=1}^{\infty} (-1)^{n+1} n^{-2}$$

$$= 1 - \tfrac{1}{4} + \tfrac{1}{9} - \tfrac{1}{16} + \cdots.$$

14.4.2   Parseval's identity.

(a) Assuming that the Fourier expansion of $f(x)$ is uniformly convergent, show that

$$\frac{1}{\pi} \int_{-\pi}^{\pi} [f(x)]^2 \, dx = \frac{a_0^2}{2} + \sum_{n=1}^{\infty} (a_n^2 + b_n^2).$$

This is Parseval's identity. It is actually a special case of the completeness relation, Eq. (9.73).

(b) Given

$$x^2 = \frac{\pi^2}{3} + 4 \sum_{n=1}^{\infty} \frac{(-1)^n \cos nx}{n^2}, \qquad -\pi \le x \le \pi,$$

apply Parseval's identity to obtain $\zeta(4)$ in closed form.

(c) The condition of uniform convergence is not necessary. Show this by applying the Parseval identity to the square wave

$$f(x) = \begin{cases} -1, & -\pi < x < 0 \\ 1, & 0 < x < \pi \end{cases}$$

$$= \frac{4}{\pi} \sum_{n=1}^{\infty} \frac{\sin(2n-1)x}{2n-1}.$$

14.4.3   Show that integrating the Fourier expansion of the Dirac delta function (Exercise 14.3.10) leads to the Fourier representation of the square wave, Eq. (14.36), with $h = 1$.

Note. Integrating the constant term $(1/2\pi)$ leads to a term $x/2\pi$. What are you going to do with this?

14.4.3A   Integrate the Fourier expansion of the unit step function

$$f(x) = \begin{cases} 0, & -\pi < x < 0 \\ x, & 0 < x < \pi. \end{cases}$$

Show that your integrated series agrees with Exercise 14.3.14.

14.4.4   In the interval $(-\pi, \pi)$,

$$\delta_n(x) = n, \qquad \text{for } |x| < \frac{1}{2n},$$

$$0, \qquad \text{for } |x| > \frac{1}{2n}$$

(Fig. 14.9).

(a) Expand $\delta_n(x)$ as a Fourier cosine series.

(b) Show that your Fourier series agrees with a Fourier expansion of $\delta(x)$ in the limit as $n \to \infty$.

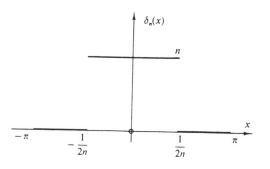

Figure 14.9 Rectangular pulse.

14.4.5   Confirm the delta function nature of your Fourier series of Exercise 14.4.4 by showing that for any $f(x)$ that is finite in the interval $[-\pi, \pi]$ and continuous at $x = 0$,

$$\int_{-\pi}^{\pi} f(x)[\text{Fourier expansion of } \delta_\infty(x)] \, dx = f(0).$$

14.4.6   (a) Show that the Dirac delta function $\delta(x - a)$, expanded in a Fourier sine series in the half interval $(0, L)$, $(0 < a < L)$, is given by

$$\delta(x - a) = \frac{2}{L} \sum_{n=1}^{\infty} \sin\left(\frac{n\pi a}{L}\right) \sin\left(\frac{n\pi x}{L}\right).$$

Note that this series actually describes

$$-\delta(x + a) + \delta(x - a) \text{ in the interval } (-L, L).$$

(b) By integrating both sides of the preceding equation from 0 to x, show that the cosine expansion of the square wave

$$f(x) = \begin{cases} 0, & 0 \leq x < a \\ 1, & a < x < L, \end{cases}$$

is

$$f(x) = \frac{2}{\pi} \sum_{n=1}^{\infty} \frac{1}{n} \sin\left(\frac{n\pi a}{L}\right) - \frac{2}{\pi} \sum_{n=1}^{\infty} \frac{1}{n} \sin\left(\frac{n\pi a}{L}\right) \cos\left(\frac{n\pi x}{L}\right), \quad 0 \leq x < L.$$

(c) Verify that the term

$$\frac{2}{\pi} \sum_{n=1}^{\infty} \frac{1}{n} \sin\left(\frac{n\pi a}{L}\right) \text{ is } \langle f(x) \rangle.$$

14.4.7   Verify the Fourier cosine expansion of the square wave, Exercise 14.4.6(b), by direct calculation of the Fourier coefficients.

14.4.8   (a) A string is clamped at both ends $x = 0$ and $x = L$. Assuming small amplitude vibrations, we find that the amplitude $y(x, t)$ satisfies the wave equation

$$\frac{\partial^2 y}{\partial x^2} = \frac{1}{v^2} \frac{\partial^2 y}{\partial t^2}.$$

Here v is the wave velocity. The string is set in vibration by a sharp blow at $x = a$. Hence we have

$$y(x, 0) = 0$$

$$\frac{\partial y(x, t)}{\partial t} = Lv_0 \delta(x - a) \qquad \text{at } t = 0.$$

The constant L is included to compensate for the dimensions (inverse length) of $\delta(x - a)$. With $\delta(x - a)$ given by Exercise 14.4.6(a), solve the wave equation subject to these initial conditions.

$$ANS. \quad y(x, t) = \frac{2v_0 L}{\pi v} \sum_{n=1}^{\infty} \frac{1}{n} \sin\frac{n\pi a}{L} \sin\frac{n\pi x}{L} \sin\frac{n\pi v t}{L}.$$

(b) Show that the transverse velocity of the string $\partial y(x, t)/\partial t$ is given by

$$\frac{\partial y(x\,t)}{\partial t} = 2v_0 \sum_{n=1}^{\infty} \sin\frac{n\pi a}{L} \sin\frac{n\pi x}{L} \cos\frac{n\pi v t}{L}.$$

14.4.9   A string, clamped at $x = 0$ and at $x = l$, is vibrating freely. Its motion is described by the wave equation

$$\frac{\partial^2 u(x, t)}{\partial t^2} = v^2 \frac{\partial^2 u(x, t)}{\partial x^2}.$$

Assume a Fourier expansion of the form

$$u(x, t) = \sum_{n=1}^{\infty} b_n(t) \sin \frac{n\pi x}{l}$$

and determine the coefficients $b_n(t)$. The initial conditions are

$$u(x, 0) = f(x) \quad \text{and} \quad \frac{\partial}{\partial t} u(x, 0) = g(x).$$

Note. This is only half the conventional Fourier orthogonality integral interval. However, as long as only the sines are included here, the Sturm–Liouville boundary conditions are still satisfied and the functions are orthogonal.

ANS. $b_n(t) = A_n \cos \dfrac{n\pi v t}{l} + B_n \sin \dfrac{n\pi v t}{l}$,

$$A_n = \frac{2}{l} \int_0^l f(x) \sin \frac{n\pi x}{l} \, dx, \quad B_n = \frac{2}{n\pi v} \int_0^l g(x) \sin \frac{n\pi x}{l} \, dx.$$

14.4.10   (a) Continuing the vibrating string problem, Exercise 14.4.9, the presence of a resisting medium will damp the vibrations according to the equation

$$\frac{\partial^2 u(x, t)}{\partial t^2} = v^2 \frac{\partial^2 u(x, t)}{\partial x^2} - k \frac{\partial u(x, t)}{\partial t}.$$

Assume a Fourier expansion

$$u(x, t) = \sum_{n=1}^{\infty} b_n(t) \sin \frac{n\pi x}{l}$$

and again determine the coefficients $b_n(t)$. Take the initial and boundary conditions to be the same as in Exercise 14.4.9. Assume the damping to be small.

(a) Repeat but assume the damping to be large.

ANS.

(a) $b_n(t) = e^{-kt/2}\{A_n \cos \omega_n t + B_n \sin \omega_n t\}$,

$$A_n = \frac{2}{l} \int_0^l f(x) \sin \frac{n\pi x}{l} \, dx,$$

$$B_n = \frac{2}{\omega_n l} \int_0^l g(x) \sin \frac{n\pi x}{l} \, dx + \frac{k}{2\omega_n} A_n, \quad \omega_n^2 = \left(\frac{n\pi v}{l}\right)^2 - \left(\frac{k}{2}\right)^2 > 0.$$

(b) $b_n(t) = e^{-kt/2}\{A_n \cosh \sigma_n t + B_n \sinh \sigma_n t\}$,

$$A_n = \frac{2}{l} \int_0^l f(x) \sin \frac{n\pi x}{l} \, dx,$$

$$B_n = \frac{2}{\sigma_n l} \int_0^l g(x) \sin \frac{n\pi x}{l} \, dx + \frac{k}{2\sigma_n} A_n, \quad \sigma_n^2 = \left(\frac{k}{2}\right)^2 - \left(\frac{n\pi v}{l}\right)^2 > 0.$$

14.4.11 Find the charge distribution over the interior surfaces of the semicircles of Exercise 14.3.6.

Note. You obtain a divergent series and this Fourier approach fails. Using conformal mapping techniques, we may show the charge density to be proportional to csc θ. Does csc θ have a Fourier expansion?

14.4.12 Given

$$\varphi_1(x) = \sum_{n=1}^{\infty} \frac{\sin nx}{n} = \begin{cases} -\dfrac{1}{2}(\pi + x), & -\pi \le x < 0 \\[2ex] \dfrac{1}{2}(\pi - x) & 0 < x \le \pi, \end{cases}$$

show by integrating that

$$\varphi_2(x) \equiv \sum_{n=1}^{\infty} \frac{\cos nx}{n^2} = \begin{cases} \dfrac{(\pi + x)^2}{4} - \dfrac{\pi^2}{12}, & -\pi \le x \le 0 \\[2ex] \dfrac{(\pi - x)^2}{4} - \dfrac{\pi^2}{12} & 0 \le x \le \pi, \end{cases}$$

14.4.13 Given

$$\psi_{2s}(x) = \sum_{n=1}^{\infty} \frac{\sin nx}{n^{2s}}$$

$$\psi_{2s+1}(x) = \sum_{n=1}^{\infty} \frac{\cos nx}{n^{2s+1}}.$$

Develop the following recurrence relations:

(a) $\psi_{2s}(x) = \displaystyle\int_0^x \psi_{2s-1}(x)\, dx$

(b) $\psi_{2s+1}(x) = \zeta(2s + 1) - \displaystyle\int_0^x \psi_{2s}(x)\, dx.$

Note. These functions $\psi_n(x)$ and the $\varphi_n(x)$ of the preceding exercise are known as Clausen functions. In theory they may be used to improve the rate of convergence of a Fourier series. As with the series of Chapter 5, there is always the question of how much analytical work we do and how much arithmetic work we demand that the computer do. As machines become steadily more powerful, the balance progressively shifts so that we are doing less and demanding that the machines do more.

14.4.14 Show that

$$f(x) = \sum_{n=1}^{\infty} \frac{\cos nx}{n + 1}$$

may be written as

$$f(x) = \psi_1(x) - \varphi_2(x) + \sum_{n=1}^{\infty} \frac{\cos nx}{n^2(n + 1)}.$$

Note. $\psi_1(x)$ and $\varphi_2(x)$ are defined in the preceding exercises.

14.5 GIBBS PHENOMENON

The Gibbs phenomenon is an overshoot, a peculiarity of the Fourier series and other eigenfunction series at a simple discontinuity. An example is seen in Fig. 14.1.

SUMMATION OF SERIES

In Section 14.1 the sum of the first several terms of the Fourier series for a sawtooth wave was plotted (Fig. 14.1). Now we develop analytic method of summing the first r terms of our Fourier series.

From Eq. (14.13)

$$a_n \cos nx + b_n \sin nx = \frac{1}{\pi} \int_{-\pi}^{\pi} f(t) \cos n(t - x)\, dt. \qquad (14.69)$$

Then the rth partial sum becomes[1]

$$s_r(x) = \frac{a_0}{2} + \sum_{n=1}^{r} (a_n \cos nx + b_n \sin nx)$$

$$= \Re \frac{1}{\pi} \int_{-\pi}^{\pi} f(t) \left[\frac{1}{2} + \sum_{n=1}^{r} e^{-i(t-x)n} \right] dt. \qquad (14.70)$$

Summing the finite series of exponentials (geometric progression),[2] we obtain

$$s_r(x) = \frac{1}{2\pi} \int_{-\pi}^{\pi} f(t) \frac{\sin(r + \frac{1}{2})(t - x)}{\sin \frac{1}{2}(t - x)}\, dt. \qquad (14.71)$$

This is convergent at all points, including $t = x$. The factor

$$\frac{(2\pi)^{-1} \sin[(r + \frac{1}{2})(t - x)]}{\sin \frac{1}{2}(t - x)}$$

is the Dirichlet kernel mentioned in Section 1.15 as a Dirac delta distribution.

SQUARE WAVE

For convenience of numerical calculation we consider the behavior of the Fourier series that represents the periodic square wave

$$f(x) = \begin{cases} \dfrac{h}{2}, & 0 < x < \pi, \\[2mm] -\dfrac{h}{2}, & -\pi < x < 0. \end{cases} \qquad (14.72)$$

This is essentially the square wave used in Section 14.3, and we immediately

---

[1] It is of some interest to note that this series also occurs in the analysis of the diffraction grating (r slits).

[2] Compare Exercise 6.1.7 with initial value $n = 1$.

see that the solution is

$$f(x) = \frac{2h}{\pi} \left(\frac{\sin x}{1} + \frac{\sin 3x}{3} + \frac{\sin 5x}{5} + \cdots \right). \tag{14.73}$$

Applying Eq. (14.71) to our square wave (Eq. (14.72)), we have the sum of the first r terms (plus $\frac{1}{2}a_0$, which is zero here).

$$s_r(x) = \frac{h}{4\pi} \int_0^\pi \frac{\sin(r + \frac{1}{2})(t - x)}{\sin\frac{1}{2}(t - x)} dt - \frac{h}{4\pi} \int_{-\pi}^0 \frac{\sin(r + \frac{1}{2})(t - x)}{\sin\frac{1}{2}(t - x)} dt$$

$$= \frac{h}{4\pi} \int_0^\pi \frac{\sin(r + \frac{1}{2})(t - x)}{\sin\frac{1}{2}(t - x)} dt - \frac{h}{4\pi} \int_0^\pi \frac{\sin(r + \frac{1}{2})(t + x)}{\sin\frac{1}{2}(t + x)} dt. \tag{14.74}$$

This last result follows from the transformation

$$t = -t \qquad \text{in the second integral.}$$

Replacing $t - x$ in the first term with s and $t + x$ in the second term with s, we obtain

$$s_r(x) = \frac{h}{4\pi} \int_{-x}^{\pi-x} \frac{\sin(r + \frac{1}{2})s}{\sin\frac{1}{2}s} ds - \frac{h}{4\pi} \int_x^{\pi+x} \frac{\sin(r + \frac{1}{2})s}{\sin\frac{1}{2}s} ds. \tag{14.75}$$

The intervals of integration are shown in Fig. 14.10 (top). Because the integrands have the same mathematical form, the integrals for x to $\pi - x$ cancel leaving the integral ranges shown in the bottom portion of Fig. 14.10:

$$s_r(x) = \frac{h}{4\pi} \int_{-x}^x \frac{\sin(r + \frac{1}{2})s}{\sin\frac{1}{2}s} ds - \frac{h}{4\pi} \int_{\pi-x}^{\pi+x} \frac{\sin(r + \frac{1}{2})s}{\sin\frac{1}{2}s} ds. \tag{14.76}$$

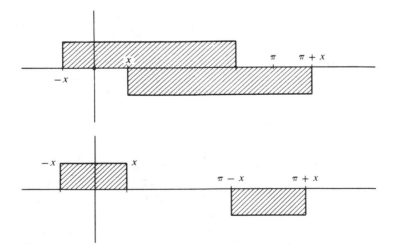

Figure 14.10   Intervals of integration—Eq. (14.75).

Consider the partial sum in the vicinity of the discontinuity at $x = 0$. As $x \to 0$, the second integral becomes negligible, and we associate the first integral with the discontinuity at $x = 0$. Using $(r + \frac{1}{2}) = p$ and $ps = \xi$, we obtain

$$s_r(x) = \frac{h}{2\pi} \int_0^{px} \frac{\sin \xi}{\sin(\xi/2p)} \cdot \frac{d\xi}{p}. \tag{14.77}$$

CALCULATION OF OVERSHOOT

Our partial sum, $s_r(x)$, starts at zero when $x = 0$ (in agreement with Eq. (14.16)) and increases until $\xi = ps = \pi$, at which point the numerator, $\sin \xi$, goes negative. For large r, and therefore for large p, our denominator remains positive. We get the maximum value of the partial sum by taking the upper limit $px = \pi$. Right here we see that x, the location of the overshoot maximum, is inversely proportional to the number of terms taken

$$x = \frac{\pi}{p} \approx \frac{\pi}{r}.$$

The maximum value of the partial sum is then

$$s_r(x)_{\text{max}} = \frac{h}{2} \cdot \frac{1}{\pi} \int_0^\pi \frac{\sin \xi \, d\xi}{\sin(\xi/2p)p} \approx \frac{h}{2} \cdot \frac{2}{\pi} \int_0^\pi \frac{\sin \xi}{\xi} d\xi. \tag{14.78}$$

In terms of the sine integral, $si(x)$ of Section 10.5,

$$\int_0^\pi \frac{\sin \xi}{\xi} d\xi = \frac{\pi}{2} + si(\pi). \tag{14.79}$$

The integral is clearly greater than $\pi/2$, since it can be written as

$$\left(\int_0^\infty - \int_\pi^{3\pi} - \int_{3\pi}^{5\pi} - \cdots \right) \frac{\sin \xi}{\xi} d\xi = \int_0^\pi \frac{\sin \xi}{\xi} d\xi. \tag{14.80}$$

We saw in Section 7.2 that the integral from 0 to ∞ is $\pi/2$. From this integral we are subtracting a series of *negative* terms. A Gaussian quadrature (Appendix 2) or a power-series expansion and term-by-term integration yields

$$\frac{2}{\pi} \int_0^\pi \frac{\sin \xi}{\xi} d\xi = 1.1789797 \ldots, \tag{14.81}$$

which means that the Fourier series tends to overshoot the positive corner by some 18 percent and to undershoot the negative corner by the same amount, as suggested in Fig. 14.11. The inclusion of more terms (increasing r) does nothing to remove this overshoot but merely moves it closer to the point of discontinuity. The overshoot is the Gibbs phenomenon, and because of it the Fourier series representation may be highly unreliable for precise numerical work, especially in the vicinity of a discontinuity.

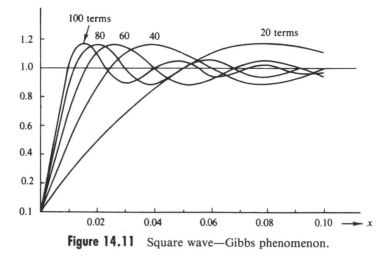

Figure 14.11   Square wave—Gibbs phenomenon.

The Gibbs phenomenon is not limited to the Fourier series. It occurs with other eigenfunction expansions. Exercise 12.3.27 is an example of the Gibbs phenomenon for a Legendre series. For more details, see W. J. Thompson, Fourier series and the Gibbs phenomenon, *Am. J. Phys.* **60**, 425 (1992).

EXERCISES

14.5.1   With the partial sum summation techniques of this section, show that at a discontinuity in $f(x)$ the Fourier series for $f(x)$ takes on the arithmetic mean of the right- and left-hand limits:

$$f(x_0) = \tfrac{1}{2}[f(x_0+) + f(x_0-)].$$

In evaluating $\lim_{r \to \infty} s_r(x_0)$ you may find it convenient to identify part of the integrand as a Dirac delta function.

14.5.2   Determine the partial sum, s_n, of the series in Eq. (14.73) by using

(a) $\dfrac{\sin mx}{m} = \displaystyle\int_0^x \cos my \, dy$

and

(b) $\displaystyle\sum_{p=1}^{n} \cos(2p - 1)y = \dfrac{\sin 2ny}{2 \sin y}$.

Do you agree with the result given in Eq. (14.79).

14.5.3   Evaluate the finite step function series, Eq. (14.73), $h = 2$, using 100, 200, 300, 400, and 500 terms for $x = 0.0000(0.0005)0.0200$. Sketch your results (five curves) or if a plotting routine is available, plot your results.

14.5.4   (a) Calculate the value of the Gibbs's phenomenon integral

$$I = \frac{2}{\pi} \int_0^\pi \frac{\sin t}{t} \, dt$$

by numerical quadrature accurate to 12 significant figures.

(b) Check your result by (1) expanding the integrand as a series, (2) integrating term by term, and (3) evaluating the integrated series. This calls for double precision calculation.

ANS. $I = 1.1789\,7974\,4472$.

14.6 DISCRETE ORTHOGONALITY—DISCRETE FOURIER TRANSFORM

For many physicists the Fourier transform is automatically the continuous Fourier transform of Chapter 15. The use of the electronic digital computer, however, necessarily replaces a continuum of values by a discrete set; an integration is replaced by a summation. The continuous Fourier transform becomes the discrete Fourier transform and an appropriate topic for this chapter.

ORTHOGONALITY OVER DISCRETE POINTS

The orthogonality of the trignometric functions and the imaginary exponentials is expressed in Eqs.(14.7) to (14.10). This is the usual orthogonality for functions: *integration* of a product of functions over the orthogonality interval. The sines, cosines, and imaginary exponentials have the remarkable property that they are also orthogonal over a series of discrete, equally spaced points over the period (the orthogonality interval).

Consider a set of $2N$ time values

$$t_k = 0, \frac{T}{2N}, \frac{2T}{2N}, \ldots, \frac{(2N-1)T}{2N} \tag{14.82}$$

for the time interval $(0, T)$. Then

$$t_k = \frac{kT}{2N}, \qquad k = 0, 1, 2, \ldots, 2N - 1. \tag{14.83}$$

We shall prove that the exponential functions $\exp(2\pi i p t_k / T)$ and $\exp(2\pi i q t_k / T)$ satisfy an orthogonality relation over the discrete points t_k:

$$\sum_{k=0}^{2N-1} [\exp(2\pi i p t_k / T)]^* \exp(2\pi i q t_k / T) = 2N \delta_{p, q \pm 2nN}. \tag{14.84}$$

Here n, p, and q are all integers.

Replacing $q - p$ by s, we find that the left-hand side of Eq. (14.84) becomes

$$\sum_{k=0}^{2N-1} \exp(2\pi i s t_k / T) = \sum_{k=0}^{2N-1} \exp(2\pi i s k / 2N).$$

This right-hand side is obtained by using Eq. (14.83) to replace T. This is a finite geometric series with an initial term 1 and a ratio

$$r = \exp(\pi i s/N).$$

From Eq. (5.7)

$$\sum_{k=0}^{2N-1} \exp(2\pi i s t_k/T) = \begin{cases} \dfrac{1 - r^{2N}}{1 - r} = 0, & r \neq 1 \\[2mm] 2N, & r = 1, \end{cases} \tag{14.85}$$

establishing Eq. (14.84), our basic orthogonality relation. The upper value zero, is a consequence of

$$r^{2N} = \exp(2\pi i s) = 1$$

for s an integer. The lower value, $2N$, for $r = 1$ corresponds to $p = q$.

The orthogonality of the corresponding trigonometric functions is left as Exercise 14.6.1.

DISCRETE FOURIER TRANSFORM

To simplify the notation slightly and to make more direct contact with physics, we introduce the (reciprocal) ω-space, angular frequency, with

$$\omega_p = 2\pi p/T, \qquad p = 0, 1, 2, \ldots, 2N - 1. \tag{14.86}$$

We make p range over the same integers as k. The exponential $\exp(\pm 2\pi i p t_k/T)$ of Eq. (14.84) becomes $\exp(\pm i\omega_p t_k)$. The choice of whether to use the $+$ or the $-$ sign is a matter of convenience or convention. In quantum mechanics the negative sign is selected when expressing the time dependence.

Consider a function of time defined (measured) at the discrete time values t_k. We may construct

$$F(\omega_p) = \frac{1}{2N} \sum_{k=0}^{2N-1} f(t_k) e^{i\omega_p t_k}. \tag{14.87}$$

Employing the orthogonality relation, we obtain

$$\frac{1}{2N} \sum_{p=0}^{2N-1} (e^{i\omega_p t_m})^* e^{i\omega_p t_k} = \delta_{mk}, \tag{14.88}$$

and then replacing subscript m by k, we find that the amplitudes, $f(t_k)$, become

$$f(t_k) = \sum_{p=0}^{2N-1} F(\omega_p) e^{-i\omega_p t_k}. \tag{14.89}$$

The time function $f(t_k)$, $k = 0, 1, 2, \ldots, 2N - 1$, and the frequency function $F(\omega_p)$, $p = 0, 1, 2, \ldots, 2N - 1$, are *discrete* Fourier transforms of each other.[1] Compare Eqs. (14.87) and (14.89) with the corresponding continuous Fourier transforms Eqs. (15.22) and (15.23) of Chapter 15.

---

[1] The two transform equations may be symmetrized with a resulting $(2N)^{-1/2}$ in each equation if desired.

LIMITATIONS

Taken as a pair of mathematical relations, the discrete Fourier transforms are exact. We can say that the $2N$ $2N$-component vectors $\exp(-i\omega_p t_k)$, $k = 0, 1, 2, \ldots, 2N - 1$, form a complete set[2] spanning the t_k-space. Then $f(t_k)$ in Eq. (14.89) is simply a particular linear combination of these vectors. Alternatively, we may take the $2N$ measured components $f(t_k)$ as defining a $2N$ component vector in t_k-space. Then, Eq. (14.87) yields the $2N$ component vector $F(\omega_p)$ in the *reciprocal* ω_p-space. Equations (14.87) and (14.89) become matrix equations with $\exp(i\omega_p t_k)/(2N)^{1/2}$ the elements of a unitary matrix.

The limitations of the discrete Fourier transform arise when we apply Eqs. (14.87) and (14.89) to physical systems and attempt physical interpretation and the generalization $F(\omega_p) \to F(\omega)$. Example 14.6.1 illustrates the problems that can occur. The most important precaution to be taken to avoid trouble is to take N sufficiently large so that there is no angular frequency component of a higher angular frequency than $\omega_N = 2\pi N/T$. For details on errors and limitations in the use of the discrete Fourier transform the reader is referred to Bergland and Hamming.

Example 14.6.1 Discrete Fourier Transform—Aliasing

Consider the relatively simple case of $T = 2\pi$, $N = 2$, and $f(t_k) = \cos t_k$. From

$$t_k = kT/4 = k\pi/2, \qquad k = 0, 1, 2, 3 \tag{14.90}$$

$f(t_k) = \cos(t_k)$ is represented by the four-component vector

$$f(t_k) = (1, 0, -1, 0). \tag{14.91}$$

The frequencies, ω_p are given by Eq. (14.86):

$$\omega_p = 2\pi p/T = p. \tag{14.92}$$

Clearly, $\cos t_k$ implies a $p = 1$ component and no other frequency components. The transformation matrix

$$(2N)^{-1} \exp(i\omega_p t_k) = (2N)^{-1} \exp(ipk\pi/2)$$

becomes

$$\frac{1}{4} \begin{pmatrix} 1 & 1 & 1 & 1 \\ 1 & i & -1 & -i \\ 1 & -1 & 1 & -1 \\ 1 & -i & -1 & i \end{pmatrix}. \tag{14.93}$$

Note that the $2N \times 2N$ matrix has only $2N$ independent components. It is the repetition of values that makes the fast Fourier transform technique possible.

Operating on column vector $f(t_k)$, we find that this matrix yields a column vector

$$F(\omega_p) = (0, \tfrac{1}{2}, 0, \tfrac{1}{2}). \tag{14.94}$$

---

[2] By Eq. (14.85) these vectors are orthogonal and are therfore linearly independent.

Apparently, there is a $p = 3$ frequency component present. We reconstruct $f(t_k)$ by Eq. (14.89), obtaining

$$f(t_k) = \tfrac{1}{2}e^{-it_k} + \tfrac{1}{2}e^{-3it_k}. \tag{14.95}$$

Taking real parts, we can rewrite the equation as

$$f(t_k) = \tfrac{1}{2}\cos t_k + \tfrac{1}{2}\cos 3t_k. \tag{14.96}$$

Obviously, this result, Eq. (14.96), is not identical with our original $f(t_k) = \cos t_k$. But $\cos t_k = \tfrac{1}{2}\cos t_k + \tfrac{1}{2}\cos 3t_k$ at $t_k = 0$, $\pi/2$, π; and $3\pi/2$. The $\cos t_k$ and $\cos 3t_k$ mimic each other because of the limited number of data points (and the particular choice of data points). This error of one frequency mimicking another is known as *aliasing*. The problem can be minimized by taking more data points.

FAST FOURIER TRANSFORM

The fast Fourier transform is a particular way of factoring and rearranging the terms in the sums of the discrete Fourier transform. Brought to the attention of the scientific community by Cooley and Tukey,[3] its importance lies in the drastic reduction in the number of numerical operations required. Because of the tremendous increase in speed achieved (and reduction in cost), the fast Fourier transform has been hailed as one of the few really significant advances in numerical analysis in the past few decades.

For N time values (measurements) a direct calculation of a discrete Fourier transform would mean about N^2 multiplications. For N a power of 2 the fast Fourier transform technique of Cooley and Tukey cuts the number of multiplications required to $(N/2)\log_2 N$. If $N = 1024$ ($=2^{10}$), the fast Fourier transform achieves a computational reduction by a factor of over 200. This is why the fast Fourier transform is called fast and why it has literally revolutionized the digital processing of waveforms.

The fast Fourier transform should be available at every computation center. Details on the internal operation will be found in the paper by Cooley and Tukey and in the paper by Bergland.[4]

EXERCISES

14.6.1   Derive the trigonometric forms of discrete orthogonality corresponding to Eq. (14.84):

$$\sum_{k=0}^{2N-1} \cos(2\pi pt_k/T)\sin(2\pi qt_k/T) = 0$$

[3] J. W. Cooley and J. W. Tukey, *Math. Comput.* **19**, 297 (1965).

[4] G. D. Bergland, A guided tour of the fast Fourier transform, *IEEE Spectrum*, July, pp. 41–52 (1969); see also, W. H. Press, B. P. Flannery, S. A. Teukolsky, and W. T. Vetterling, *Numerical Recipes*, 2nd ed. Cambridge: Cambridge University Press (1992), Section 12.3.

$$\sum_{k=0}^{2N-1} \cos(2\pi p t_k/T)\cos(2\pi q t_k/T) = \begin{cases} 0, & p \neq q \\ N & p = q \neq 0, N \\ 2N, & p = q = 0, N \end{cases}$$

$$\sum_{k=0}^{2N-1} \sin(2\pi p t_k/T)\sin(2\pi q t_k/T) = \begin{cases} 0, & p \neq q \\ N, & p = q \neq 0, N \\ 0, & p = q = 0, N \end{cases}$$

Hint. Trigonometric identities such as

$$\sin A \cos B = \tfrac{1}{2}[\sin(A + B) + \sin(A - B)]$$

are useful.

14.6.2   Equation (14.84) exhibits orthogonality summing over time points. Show that we have the same orthogonality summing over frequency points.

$$\frac{1}{2N}\sum_{p=0}^{2N-1}(e^{i\omega_p t_m})^* e^{i\omega_p t_k} = \delta_{mk}.$$

14.6.3   Show, in detail, how to go from

$$F(\omega_p) = \frac{1}{2N}\sum_{k=0}^{2N-1} f(t_k)e^{i\omega_p t_k}$$

to

$$f(t_k) = \sum_{p=0}^{2N-1} F(\omega_p)e^{-i\omega_p t_k}.$$

14.6.4   The functions $f(t_k)$ and $F(\omega_p)$ are discrete Fourier transforms of each other. Derive the following symmetry relations:

(a) If $f(t_k)$ is real, $F(\omega_p)$ is Hermitian symmetric; that is,

$$F(\omega_p) = F^*\left(\frac{4\pi N}{T} - \omega_p\right).$$

(b) If $f(t_k)$ is pure imaginary,

$$F(\omega_p) = -F^*\left(\frac{4\pi N}{T} - \omega_p\right).$$

Note. The symmetry of part (a) is an illustration of *aliasing.* The frequency $4\pi N/T - \omega_p$ masquerades as the frequency ω_p.

14.6.5   Given $N = 2$, $T = 2\pi$, and $f(t_k) = \sin t_k$.
(a) Find $F(\omega_p)$, $p = 0, 1, 2, 3$.
(b) Reconstruct $f(t_k)$ from $F(\omega_p)$ and exhibit the aliasing of $\omega_1 = 1$ and $\omega_3 = 3$.

$$ANS. \quad (a) \ F(\omega_p) = (0, i/2, 0, -i/2)$$
$$(b) \ f(t_k) = \tfrac{1}{2}\sin t_k - \tfrac{1}{2}\sin 3t_k.$$

14.6.6 Show that the Chebyshev polynomials $T_m(x)$ satisfy a discrete orthogonality relation

$$\tfrac{1}{2}T_m(-1)T_n(-1) + \sum_{s=1}^{N-1} T_m(x_s)T_n(x_s) + \tfrac{1}{2}T_m(1)T_n(1) = \begin{cases} 0, & m \neq n \\ N/2, & m = n \neq 0 \\ N, & m = n = 0. \end{cases}$$

Here, $x_s = \cos \theta_s$, where the $(N + 1)\theta_s$'s are equally spaced along the θ axis:

$$\theta_s = \frac{s\pi}{N}, \qquad s = 0, 1, 2, \dots, N.$$

ADDITIONAL READINGS

CARSLAW, H. S., *Introduction to the Theory of Fourier's Series and Integrals*, 2nd ed. London: Macmillan (1921); 3rd ed., paperback, New York: Dover (1952). This is a detailed and classic work, which includes a considerable discussion of Gibbs phenomenon in Chapter IX.

HAMMING, R. W., *Numerical Methods for Scientists and Engineers*, 2nd ed. New York: McGraw-Hill (1973). Chapter 33 provides an excellent description of the fast Fourier transform.

JEFFREYS, H., and B. S. JEFFREYS, *Methods of Mathematical Physics*, 3rd ed. Cambridge: Cambridge University Press (1966).

KUFNER, A., and J. KADLEC, *Fourier Series*. London: Iliffe (1971). This book is a clear account of Fourier series in the context of Hilbert space.

LANCZOS, C. *Applied Analysis*. Englewood Cliffs, NJ: Prentice-Hall (1956). The book gives a well-written presentation of the Lanczos convergence technique (which suppresses the Gibbs phenomenon oscillations). This and several other topics are presented from the point of view of a mathematician who wants useful numerical results and not just abstract existence theorems.

OBERHETTINGER, F., *Fourier Expansions, A Collection of Formulas*. New York and London: Academic Press (1973).

ZYGMUND, A., *Trigonometric Series*. Cambridge: Cambridge University Press (1977). The volume contains an extremely complete exposition, including relatively recent results in the realm of pure mathematics.

$$e^{i\theta} = \cos\theta + i\sin\theta$$

15 INTEGRAL TRANSFORMS

15.1 INTEGRAL TRANSFORMS

Frequently in mathematical physics we encounter pairs of functions related by an expression of the form

$$g(\alpha) = \int_a^b f(t)K(\alpha, t)\, dt. \tag{15.1}$$

The function $g(\alpha)$ is called the (integral) transform of $f(t)$ by the kernel $K(\alpha, t)$. The operation may also be described as mapping a function $f(t)$ in t-space into another function $g(\alpha)$ in α-space. This interpretation takes on physical significance in the time–frequency relation of Example 15.3.1 and in the real space–momentum space relations of Section 15.6.

FOURIER TRANSFORM

One of the most useful of the infinite number of possible transforms is the Fourier transform given by

$$g(\alpha) = \frac{1}{\sqrt{2\pi}} \int_{-\infty}^{\infty} f(t)e^{i\alpha t}\, dt. \tag{15.2}$$

Two modifications of this form, developed in Section 15.3, are the Fourier cosine and Fourier sine transforms:

$$g_c(\alpha) = \sqrt{\frac{2}{\pi}} \int_0^{\infty} f(t)\cos\alpha t\, dt, \tag{15.3}$$

$$g_s(\alpha) = \sqrt{\frac{2}{\pi}} \int_0^{\infty} f(t)\sin\alpha t\, dt. \tag{15.4}$$

The Fourier transform is based on the kernel $e^{i\alpha t}$ and its real and imaginary parts taken separately, $\cos \alpha t$ and $\sin \alpha t$. Because these kernels are the functions used to describe waves, Fourier transforms appear frequently in studies of waves and the extraction of information from waves, particularly when phase information is involved. The output of a stellar interferometer, for instance, involves a Fourier transform of the brightness across a stellar disk. The electron distribution in an atom may be obtained from a Fourier transform of the amplitude of scattered X-rays. In quantum mechanics the physical origin of the Fourier relations of Section 15.6 is the wave nature of matter and our description of matter in terms of waves.

LAPLACE, MELLIN, AND HANKEL TRANSFORMS
Three other useful kernels are

$$e^{-\alpha t}, \qquad tJ_n(\alpha t), \qquad t^{\alpha - 1}.$$

These give rise to the following transforms

$$g(\alpha) = \int_0^\infty f(t)e^{-\alpha t}\,dt, \qquad \text{Laplace transform} \tag{15.5}$$

$$g(\alpha) = \int_0^\infty f(t)tJ_n(\alpha t)\,dt, \qquad \text{Hankel transform (Fourier–Bessel)} \tag{15.6}$$

$$g(\alpha) = \int_0^\infty f(t)t^{\alpha - 1}\,dt, \qquad \text{Mellin transform.} \tag{15.7}$$

Clearly, the possible types are unlimited. These transforms have been useful in mathematical analysis and in physical applications. We have actually used the Mellin transform without calling it by name; that is, $g(\alpha) = (\alpha - 1)!$ is the Mellin transform of $f(t) = e^{-t}$. See E. C. Titchmarsh, *Introduction to the Theory of Fourier Integrals*, 2nd ed. New York: Oxford University Press (1937), for more Mellin transforms. Of course, we could just as well say $g(\alpha) = n!/\alpha^{n+1}$ is the Laplace transform of $f(t) = t^n$. Of the three, the Laplace transform is by far the most used. It is discussed at length in Sections 15.8 to 15.12. The Hankel transform, a Fourier transform for a Bessel function explansion, represents a limiting case of a Fourier–Bessel series. It occurs in potential problems in cylindrical coordinates and has been applied extensively in acoustics.

LINEARITY
All these integral transforms are linear; that is,

$$\int_a^b [c_1f_1(t) + c_2f_2(t)]K(\alpha, t)\,dt$$

$$= \int_a^b c_1f_1(t)K(\alpha, t)\,dt + \int_a^b c_2f_2(t)K(\alpha, t)\,dt, \tag{15.8}$$

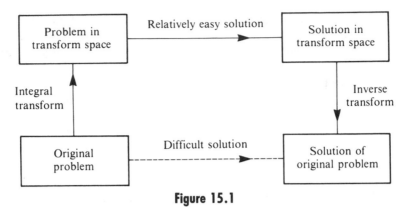

Figure 15.1

$$\int_a^b cf(t)K(\alpha, t)\, dt = c \int_a^b f(t)K(\alpha, t)\, dt, \qquad (15.9)$$

where c_1 and c_2 are constants and $f_1(t)$ and $f_2(t)$ are functions for which the transform operation is defined.

Representing our linear integral transform by the operator $\mathcal{L}$, we obtain

$$g(\alpha) = \mathcal{L}f(t). \qquad (15.10)$$

We expect an inverse operator $\mathcal{L}^{-1}$ exists such that[1]

$$f(t) = \mathcal{L}^{-1}g(\alpha). \qquad (15.11)$$

For our three Fourier transforms $\mathcal{L}^{-1}$ is given in Section 15.3. In general, the determination of the inverse transform is the main problem in using integral transforms. The inverse Laplace transform is discussed in Section 15.12. For details of the inverse Hankel and inverse Mellin transforms the reader is referred to the references at the end of the chapter.

Integral transforms have many special physical applications and interpretations that are noted in the remainder of this chapter. The most common application is outlined in Fig. 15.1. Perhaps an original problem can be solved only with difficulty, if at all, in the original coordinates (space). It often happens that the transform of the problem can be solved relatively easily. Then, the inverse transform returns the solution from the transform coordinates to the original system. Example 15.4.1 and Exercise 15.4.1 illustrate this technique.

---

[1] Expectation is not proof, and here proof of existence is complicated because we are actually in an *infinite*-dimensional Hilbert space. We shall prove existence in the special cases of interest by actual construction.

EXERCISES

15.1.1 The Fourier transforms for a function of two variables are

$$F(u, v) = \frac{1}{2\pi} \int_{-\infty}^{\infty} \int f(x, y)e^{i(ux+vy)} \, dx \, dy,$$

$$f(x, y) = \frac{1}{2\pi} \int_{-\infty}^{\infty} \int F(u, v)e^{-i(ux+vy)} \, du \, dv.$$

Using $f(x, y) = f([x^2 + y^2]^{1/2})$, show that the *zero*-order Hankel transforms

$$F(\rho) = \int_0^{\infty} rf(r)J_0(\rho r) \, dr,$$

$$f(r) = \int_0^{\infty} \rho F(\rho)J_0(\rho r) \, d\rho,$$

are a special case of the Fourier transforms.

This technique may be generalized to derive the Hankel transforms of order ν, $\nu = 0, \frac{1}{2}, 1, \frac{3}{2}, \ldots$ [compare I. N. Sneddon, *Fourier Transforms*. New York: McGraw–Hill (1951)]. A more general approach, valid for $\nu > -\frac{1}{2}$, is presented in Sneddon's *The Use of Integral Transforms* [New York: McGraw–Hill (1972)]. It might also be noted that the Hankel transforms of nonintegral order $\nu = \pm\frac{1}{2}$ reduce to Fourier sine and cosine transforms.

15.1.2 Assuming the validity of the Hankel transform–inverse transform pair of equations

$$g(\alpha) = \int_0^{\infty} f(t)J_n(\alpha t)t \, dt$$

$$f(t) = \int_0^{\infty} g(\alpha)J_n(\alpha t)\alpha \, d\alpha,$$

show that the Dirac delta function has a Bessel integral representation

$$\delta(t - t') = t \int_0^{\infty} J_n(\alpha t)J_n(\alpha t')\alpha \, d\alpha.$$

This expression is useful in developing Green's functions in cylindrical coordinates, where the eigenfunctions are Bessel functions.

15.1.3 From the Fourier transforms, Eqs. (15.22) and 15.23), show that the transformation

$$t \to \ln x$$

$$i\omega \to \alpha - \gamma$$

leads to

$$G(\alpha) = \int_0^{\infty} F(x)x^{\alpha-1} \, dx$$

and

$$F(x) = \frac{1}{2\pi i} \int_{\gamma-i\infty}^{\gamma+i\infty} G(\alpha)x^{-\alpha} \, d\alpha.$$

These are the Mellin transforms. A similiar change of variables is employed in Section 15.12 to derive the inverse Laplace transform.

15.1.4   Verify the following Mellin transforms:

(a) $\displaystyle\int_0^\infty x^{\alpha-1}\sin(kx)\,dx = k^{-\alpha}(\alpha-1)!\sin\frac{\pi\alpha}{2},$ $-1 < \alpha < 1.$

(b) $\displaystyle\int_0^\infty x^{\alpha-1}\cos(kx)\,dx = k^{-\alpha}(\alpha-1)!\cos\frac{\pi\alpha}{2},$ $0 < \alpha < 1.$

Hint. You can force the integrals into a tractable form by inserting a convergence factor e^{-bx} and (after integrating) letting $b \to 0$. Also, $\cos kx + i\sin kx = \exp ikx.$

15.2 DEVELOPMENT OF THE FOURIER INTEGRAL

In Chapter 14 it was shown that Fourier series are useful in representing certain functions (1) over a limited range $[0, 2\pi]$, $[-L, L]$, and so on, or (2) for the infinite interval $(-\infty, \infty)$, *if the function is periodic.* We now turn our attention to the problem of representing a nonperiodic function over the infinite range. Physically this means resolving a single pulse or wave packet into sinusoidal waves.

We have seen (Section 14.2) that for the interval $[-L, L]$ the coefficients a_n and b_n could be written as

$$a_n = \frac{1}{L}\int_{-L}^{L} f(t)\cos\frac{n\pi t}{L}\,dt \tag{15.12}$$

$$b_n = \frac{1}{L}\int_{-L}^{L} f(t)\sin\frac{n\pi t}{L}\,dt. \tag{15.13}$$

The resulting Fourier series is

$$f(x) = \frac{1}{2L}\int_{-L}^{L} f(t)\,dt + \frac{1}{L}\sum_{n=1}^{\infty}\cos\frac{n\pi x}{L}\int_{-L}^{L} f(t)\cos\frac{n\pi t}{L}\,dt$$

$$+ \frac{1}{L}\sum_{n=1}^{\infty}\sin\frac{n\pi x}{L}\int_{-L}^{L} f(t)\sin\frac{n\pi t}{L}\,dt \tag{15.14}$$

or

$$f(x) = \frac{1}{2L}\int_{-L}^{L} f(t)\,dt + \frac{1}{L}\sum_{n=1}^{\infty}\int_{-L}^{L} f(t)\cos\frac{n\pi}{L}(t-x)\,dt. \tag{15.15}$$

We now let the parameter L approach infinity, transforming the finite interval $[-L, L]$ into the infinite interval $(-\infty, \infty)$. We set

$$\frac{n\pi}{L} = \omega, \qquad \frac{\pi}{L} = \Delta\omega, \qquad \text{with } L \to \infty.$$

Then we have

$$f(x) \to \frac{1}{\pi}\sum_{n=1}^{\infty}\Delta\omega\int_{-\infty}^{\infty} f(t)\cos\omega(t-x)\,dt \tag{15.16}$$

or

$$f(x) = \frac{1}{\pi} \int_0^\infty d\omega \int_{-\infty}^\infty f(t) \cos \omega(t - x) \, dt, \qquad (15.17)$$

replacing the infinite sum by the integral over ω. The first term (corresponding to a_0) has vanished, assuming that $\int_{-\infty}^\infty f(t) \, dt$ exists.

It must be emphasized that this result (Eq. (15.17)) is purely formal. It is not intended as a rigorous derivation, but it can be made rigorous (compare I. N. Sneddon, *Fourier Transforms*, Section 3.2). We take Eq. (15.17) as the Fourier integral. It is subject to the conditions that $f(x)$ is (1) piecewise continuous, (2) differentiable, and (3) absolutely integrable—that is, $\int_{-\infty}^\infty |f(x)| \, dx$ is finite.

FOURIER INTEGRAL—EXPONENTIAL FORM

Our Fourier integral (Eq. (15.17)) may be put into exponential form by noting that

$$f(x) = \frac{1}{2\pi} \int_{-\infty}^\infty d\omega \int_{-\infty}^\infty f(t) \cos \omega(t - x) \, dt, \qquad (15.18)$$

whereas

$$\frac{1}{2\pi} \int_{-\infty}^\infty d\omega \int_{-\infty}^\infty f(t) \sin \omega(t - x) \, dt = 0; \qquad (15.19)$$

$\cos \omega(t - x)$ is an even function of ω and $\sin \omega(t - x)$ is an odd function of ω. Adding Eqs. (15.18) and (15.19) (with a factor i), we obtain

$$f(x) = \frac{1}{2\pi} \int_{-\infty}^\infty e^{-i\omega x} \, d\omega \int_{-\infty}^\infty f(t) e^{i\omega t} \, dt. \qquad (15.20)$$

The variable ω introduced here is an arbitrary mathematical variable. In many physical problems, however, it corresponds to the angular frequency ω. We may then interpret Eq. (15.18) or (15.20) as a representation of $f(x)$ in terms of a distribution of infinitely long sinusoidal wave trains of angular frequency ω in which this frequency is a *continuous* variable.

DIRAC DELTA FUNCTION DERIVATION

If the order of integration of Eq. (15.20) is reversed, we may rewrite it as

$$f(x) = \int_{-\infty}^\infty f(t) \left\{ \frac{1}{2\pi} \int_{-\infty}^\infty e^{i\omega(t-x)} \, d\omega \right\} dt. \qquad (15.20a)$$

Apparently the quantity in curly brackets behaves as a delta function— $\delta(t - x)$. We might take Eq. (15.20a) as presenting us with a representation of the Dirac delta function. Alternatively, we take it as a clue to a new derivation of the Fourier integral theorem.

From Eq. (1.77) (shifting the singularity from $t = 0$ to $t = x$)

$$f(x) = \lim_{n \to \infty} \int_{-\infty}^{\infty} f(t)\delta_n(t - x)\, dt, \qquad (15.21a)$$

where $\delta_n(t - x)$ is a sequence defining the distribution $\delta(t - x)$. Note that Eq. (15.21a) assumes that $f(t)$ is continuous at $t = x$.

We take $\delta_n(t - x)$ to be

$$\delta_n(t - x) = \frac{\sin n(t - x)}{\pi(t - x)} = \frac{1}{2\pi} \int_{-n}^{n} e^{i\omega(t-x)}\, d\omega, \qquad (15.21b)$$

using Eq. (1.174). Substituting into Eq. (15.21a), we have

$$f(x) = \lim_{n \to \infty} \frac{1}{2\pi} \int_{-\infty}^{\infty} f(t) \int_{-n}^{n} e^{i\omega(t-x)}\, d\omega\, dt. \qquad (15.21c)$$

Interchanging the order of integration and then taking the limit as $n \to \infty$, we have Eq. (15.20), the Fourier integral theorem.

With the understanding that it belongs under an integral sign as in Eq. (15.21a), the identification

$$\delta(t - x) = \frac{1}{2\pi} \int_{-\infty}^{\infty} e^{i\omega(t-x)}\, d\omega, \qquad (15.21d)$$

provides a very useful representation of the delta function. It is used to great advantage in Sections 15.5 and 15.6.

15.3 FOURIER TRANSFORMS—INVERSION THEOREM

Let us *define* $g(\omega)$, the Fourier transform of the function $f(t)$, by

$$g(\omega) \equiv \frac{1}{\sqrt{2\pi}} \int_{-\infty}^{\infty} f(t)e^{i\omega t}\, dt. \qquad (15.22)$$

EXPONENTIAL TRANSFORM

Then from Eq. (15.20) we have the inverse relation

$$f(x) = \frac{1}{\sqrt{2\pi}} \int_{-\infty}^{\infty} g(\omega)e^{-i\omega x}\, d\omega. \qquad (15.23)$$

It will be noted that Eqs. (15.22) and (15.23) are almost but not quite symmetrical, differing in the sign of i.

Here two points deserve comment. First, the $1/\sqrt{2\pi}$ symmetry is a matter of choice, not of necessity. Many authors will attach the entire $1/2\pi$ factor of Eq. (15.20) to one of the two equations: to Eq. (15.22) or Eq. (15.23). Second, although the Fourier integral Eq. (15.20) has received much attention in the mathematics literature, we shall be primarily interested in the Fourier transform and its inverse. They are the equations with physical significance.

When we move the Fourier transform pair to three-dimensional space, it becomes

$$g(\mathbf{k}) = \frac{1}{(2\pi)^{3/2}} \int f(\mathbf{r}) e^{i\mathbf{k}\cdot\mathbf{r}} d^3r \qquad (15.23a)$$

$$f(\mathbf{r}) = \frac{1}{(2\pi)^{3/2}} \int g(\mathbf{k}) e^{-i\mathbf{k}\cdot\mathbf{r}} d^3k. \qquad (15.23b)$$

The integrals are over all space. Verification, if desired, follows immediately by substituting the left-hand side of one equation into the integrand of the other equation and using the three-dimensional delta function.[1] Equation (15.23b) may be interpreted as an expansion of a function $f(\mathbf{r})$ in a continuum of plane wave eigenfunctions. $g(\mathbf{k})$ then becomes the amplitude of the wave $\exp(-i\mathbf{k}\cdot\mathbf{r})$.

COSINE TRANSFORM

If $f(x)$ is odd or even, these transforms may be expressed in a somewhat different form. Consider, first, $f(x) = f(-x)$, even. Writing the exponential of Eq. (15.22) in trigonometric form, we have

$$g_c(\omega) = \frac{1}{\sqrt{2\pi}} \int_{-\infty}^{\infty} f_c(t)(\cos \omega t + i \sin \omega t)\, dt$$

$$= \sqrt{\frac{2}{\pi}} \int_0^{\infty} f_c(t) \cos \omega t\, dt, \qquad (15.24)$$

the $\sin \omega t$ dependence vanishing on integration over the symmetric interval $(-\infty, \infty)$. Similarly, since $\cos \omega t$ is even, Eq. (15.23) transforms to

$$f_c(x) = \sqrt{\frac{2}{\pi}} \int_0^{\infty} g_c(\omega) \cos \omega x\, d\omega. \qquad (15.25)$$

Equations (15.24) and (15.25) are known as Fourier cosine transforms.

SINE TRANSFORM

The corresponding pair of Fourier sine transforms is obtained by assuming that $f(x) = -f(-x)$, odd, and applying the same symmetry arguments. The

[1] $\delta(\mathbf{r}_1 - \mathbf{r}_2) = \delta(x_1 - x_2)\delta(y_1 - y_2)\delta(z_1 - z_2)$

$= \dfrac{1}{2\pi} \displaystyle\int_{-\infty}^{\infty} \exp[ik_1(x_1 - x_2)]dk_1 \cdot \dfrac{1}{2\pi} \displaystyle\int_{-\infty}^{\infty} \exp[ik_2(y_1 - y_2)]\, dk_2$

$\cdot \dfrac{1}{2\pi} \displaystyle\int_{-\infty}^{\infty} \exp[ik_3(z_1 - z_2)]\, dk_3$

$= \dfrac{1}{(2\pi)^3} \displaystyle\iiint_{-\infty}^{\infty} \exp[i\mathbf{k}\cdot(\mathbf{r}_1 - \mathbf{r}_2)\, d^3k.$

equations are

$$g_s(\omega) = \sqrt{\frac{2}{\pi}} \int_0^\infty f_s(t) \sin \omega t \, dt,^2 \qquad (15.26)$$

$$f_s(x) = \sqrt{\frac{2}{\pi}} \int_0^\infty g_s(\omega) \sin \omega x \, d\omega. \qquad (15.27)$$

From the last equation we may develop the physical interpretation that $f(x)$ is being described by a continuum of sine waves. The amplitude of $\sin \omega x$ is given by $\sqrt{2/\pi}\, g_s(\omega)$, in which $g_s(\omega)$ is the Fourier sine transform of $f_s(x)$. It will be seen that Eq. (15.27) is the integral analog of the summation (Eq. (14.18)). Similar interpretations hold for the cosine and exponential cases.

If we take Eqs. (15.22), (15.24), and (15.26) as the direct integral transforms, described by $\mathcal{L}$ in Eq. (15.10) (Section 15.1), the corresponding inverse transforms, $\mathcal{L}^{-1}$ of Eq. (15.11), are given by Eqs. (15.23), (15.25), and (15.27).

The reader will note that the Fourier cosine transforms and the Fourier sine transforms each involve only positive values (and zero) of the arguments. We use the parity of $f(x)$ to establish the transforms, but once the transforms are established, the behavior of the functions f and g for negative argument is irrelevant. In effect, the transform equations themselves impose a definite parity; even for the Fourier cosine transform and odd for the Fourier sine transform.

Example 15.3.1 Finite Wave Train

An important application of the Fourier transform is the resolution of a finite pulse into sinusoidal waves. Imagine that an infinite wave train $\sin \omega_0 t$ is clipped by Kerr cell or saturable dye cell shutters so that we have

$$f(t) = \begin{cases} \sin \omega_0 t, & |t| < \dfrac{N\pi}{\omega_0}, \\[2mm] 0, & |t| > \dfrac{N\pi}{\omega_0}. \end{cases} \qquad (15.28)$$

This corresponds to N cycles of our original wave train (Fig. 15.2). Since $f(t)$ is odd, we may use the Fourier sine transform (Eq. (15.26)) to obtain

$$g_s(\omega) = \sqrt{\frac{2}{\pi}} \int_0^{N\pi/\omega_0} \sin \omega_0 t \sin \omega t \, dt. \qquad (15.29)$$

Integrating, we find our amplitude function

$$g_s(\omega) = \sqrt{\frac{2}{\pi}} \left[\frac{\sin[(\omega_0 - \omega)(N\pi/\omega_0)]}{2(\omega_0 - \omega)} - \frac{\sin[(\omega_0 + \omega)(N\pi/\omega_0)]}{2(\omega_0 + \omega)} \right]. \qquad (15.30)$$

---

[2] Note that a factor $-i$ has been absorbed into this $g(\omega)$.

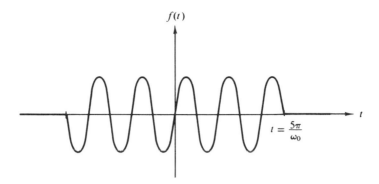

Figure 15.2   Finite wave train.

It is of some considerable interest to see how $g_s(\omega)$ depends on frequency. For large ω_0 and $\omega \approx \omega_0$ only the first term will be of any importance. It is plotted in Fig. 15.3. This is the amplitude curve for the single slit diffraction pattern. There are zeroes at

$$\frac{\omega_0 - \omega}{\omega_0} = \frac{\Delta\omega}{\omega_0} = \pm\frac{1}{N}, \ \pm\frac{2}{N}, \text{ and so on.} \tag{15.31}$$

$g_s(\omega)$ may also be interpreted as a Dirac delta distribution as in Section 1.15. Since the contributions outside the central maximum are small, we may take

$$\Delta\omega = \frac{\omega_0}{N} \tag{15.32}$$

as a good measure of the spread in frequency of our wave pulse. Clearly, if N is large (a long pulse), the frequency spread will be small. On the other hand, if our pulse is clipped short, N small, the frequency distribution will be wider.

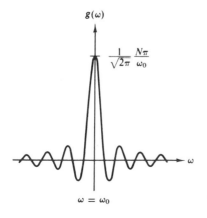

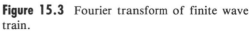

Figure 15.3 Fourier transform of finite wave train.

UNCERTAINTY PRINCIPLE

Here is a classical analog of the famous uncertainty principle of quantum mechanics. If we are dealing with electromagnetic waves,

$$\frac{h\omega}{2\pi} = E, \qquad \text{energy (of our wave pulse or photon)}$$

$$\frac{h\,\Delta\omega}{2\pi} = \Delta E,$$

(15.33)

h being Planck's constant, which represents an uncertainty in the energy of our pulse. There is also an uncertainty in the time, for our wave of N cycles requires $2N\pi/\omega_0$ seconds to pass. Taking

$$\Delta t = \frac{2N\pi}{\omega_0},$$

(15.34)

we have the product of these two uncertainties:

$$\Delta E \cdot \Delta t = \frac{h\,\Delta\omega}{2\pi} \cdot \frac{2\pi N}{\omega_0}$$

$$= h\frac{\omega_0}{2\pi N} \cdot \frac{2\pi N}{\omega_0} = h.$$

(15.35)

The Heisenberg uncertainty principle actually states

$$\Delta E \cdot \Delta t \geq \frac{h}{4\pi},$$

(15.36)

and this is clearly satisfied in our example.

EXERCISES

15.3.1   (a) Show that $g(-\omega) = g^*(\omega)$ is a necessary and sufficient condition for $f(x)$ to be real.
(b) Show that $g(-\omega) = -g^*(\omega)$ is a necessary and sufficient condition for $f(x)$ to be pure imaginary.

Note. The condition of part (a) is used in the development of the dispersion relations of Section 7.3.

15.3.2   Let $F(\omega)$ be the Fourier (exponential) transform of $f(x)$ and $G(\omega)$ the Fourier transform of $g(x) = f(x + a)$. Show that
$$G(\omega) = e^{-ia\omega}F(\omega).$$

15.3.3   The function
$$f(x) = \begin{cases} 1, & |x| < 1 \\ 0, & |x| > 1 \end{cases}$$
is a symmetrical finite step function.

(a) Find the $g_c(\omega)$, Fourier cosine transform of $f(x)$.

(b) Taking the inverse cosine transform, show that

$$f(x) = \frac{2}{\pi} \int_0^\infty \frac{\sin \omega \cos \omega x}{\omega} \, d\omega.$$

(c) From part (b) show that

$$\int_0^\infty \frac{\sin \omega \cos \omega x}{\omega} \, d\omega = \begin{cases} 0, & |x| > 1. \\[6pt] \dfrac{\pi}{4}, & |x| = 1. \\[6pt] \dfrac{\pi}{2}, & |x| < 1. \end{cases}$$

15.3.4 (a) Show that the Fourier sine and cosine transforms of e^{-at} are

$$g_s(\omega) = \sqrt{\frac{2}{\pi}} \frac{\omega}{\omega^2 + a^2}$$

$$g_c(\omega) = \sqrt{\frac{2}{\pi}} \frac{a}{\omega^2 + a^2}.$$

Hint. Each of the transforms can be related to the other by integration by parts.

(b) Show that

$$\int_0^\infty \frac{\omega \sin \omega x}{\omega^2 + a^2} \, d\omega = \frac{\pi}{2} e^{-ax}, \qquad x > 0,$$

$$\int_0^\infty \frac{\cos \omega x}{\omega^2 + a^2} \, d\omega = \frac{\pi}{2a} e^{-ax}, \qquad x > 0.$$

These results may also be obtained by contour integration (Exercise 7.2.14).

15.3.5 Find the Fourier transform of the triangular pulse

$$f(x) = \begin{cases} h(1 - a|x|), & |x| < 1/a, \\ 0, & |x| > 1/a. \end{cases}$$

Note. This function provides another delta sequence with $h = a$ and $a \to \infty$.

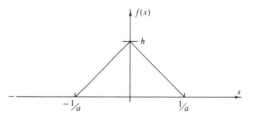

15.3.6 We may define a sequence

$$\delta_n(x) = \begin{cases} n, & |x| < 1/2n, \\ 0, & |x| > 1/2n. \end{cases}$$

(This is Eq. (1.171).) Express $\delta_n(x)$ as a Fourier integral (via the Fourier integral theorem, inverse transform, etc.). Finally, show that we may write

$$\delta(x) = \lim_{n \to \infty} \delta_n(x) = \frac{1}{2\pi} \int_{-\infty}^{\infty} e^{-ikx} \, dk.$$

15.3.7   Using the sequence

$$\delta_n(x) = \frac{n}{\sqrt{\pi}} \exp(-n^2 x^2),$$

show that

$$\delta(x) = \frac{1}{2\pi} \int_{-\infty}^{\infty} e^{-ikx} \, dk.$$

Note. Remember that $\delta(x)$ is defined in terms of its behavior as part of an integrand—Section 1.15, especially Eqs. (1.177) and (1.178).

15.3.8   Derive sin and cosine representations of $\delta(t - x)$ that are comparable to the exponential representation, Eq. (15.21d).

$$ANS. \quad \frac{2}{\pi} \int_0^{\infty} \sin \omega t \sin \omega x \, d\omega$$

$$\frac{2}{\pi} \int_0^{\infty} \cos \omega t \cos \omega x \, d\omega.$$

15.3.9   In a resonant cavity an electromagnetic oscillation of frequency ω_0 dies out as

$$A(t) = A_0 e^{-\omega_0 t/2Q} e^{-i\omega_0 t}, \qquad t > 0.$$

(Take $A(t) = 0$ for $t < 0$.)

The parameter Q is a measure of the ratio of stored energy to energy loss per cycle. Calculate the frequency distribution of the oscillation, $a^*(\omega)a(\omega)$, where $a(\omega)$ is the Fourier transform of $A(t)$.

Note. The larger Q is, the sharper your resonance line will be.

$$ANS. \quad a^*(\omega)a(\omega) = \frac{A_0^2}{2\pi} \frac{1}{(\omega - \omega_0)^2 + (\omega_0/2Q)^2}.$$

15.3.10   Prove that

$$\frac{\hbar}{2\pi i} \int_{-\infty}^{\infty} \frac{e^{-i\omega t} \, d\omega}{E_0 - i\Gamma/2 - \hbar\omega} = \begin{cases} \exp(-\Gamma t/2\hbar) \exp(-iE_0 t/\hbar), & t > 0, \\ 0, & t < 0. \end{cases}$$

This Fourier integral appears in a variety of problems in quantum mechanics: WKB barrier penetration, scattering, time-dependent perturbation theory, and so on.

Hint. Try contour integration.

15.3.11   Verify that the following are Fourier integral transforms of one another:

(a) $\sqrt{\dfrac{2}{\pi}} \cdot \dfrac{1}{\sqrt{a^2 - x^2}}, \qquad |x| < a,$

and $J_0(ay)$,

$\qquad 0, \qquad\qquad\qquad |x| > a,$

(b) $0, \qquad\qquad\qquad\qquad |x| < a,$

and $N_0(a|y|)$,

$\qquad -\sqrt{\dfrac{2}{\pi}} \dfrac{1}{\sqrt{x^2 + a^2}}, \qquad |x| > a,$

(c) $\sqrt{\dfrac{\pi}{2}} \cdot \dfrac{1}{\sqrt{x^2 + a^2}}$ and $K_0(a|y|)$.

(d) Can you suggest why $I_0(ay)$ is not included in this list?

Hint. J_0, N_0, and K_0 may be transformed most easily by using an exponential representation, reversing the order of integration, and employing the Dirac delta function exponential representation (Section 15.2). These cases can be treated equally well as Fourier cosine transforms.

Note. The K_0 relation appears as a consequence of a Green's function equation in Exercise 8.7.14.

15.3.12 A calculation of the magnetic field of a circular current loop in circular cylindrical coordinates leads to the integral

$$\int_0^\infty \cos kz \, k \, K_1(ka) \, dk.$$

Show that this integral is equal to

$$\frac{\pi a}{2(z^2 + a^2)^{3/2}},$$

Hint. Try differentiating Exercise 15.3.11(c).

15.3.13 As an extension of Exercise 15.3.11, show that

(a) $\displaystyle\int_0^\infty J_0(y) \, dy = 1$,

(b) $\displaystyle\int_0^\infty N_0(y) \, dy = 0$,

(c) $\displaystyle\int_0^\infty K_0(y) \, dy = \frac{\pi}{2}$.

15.3.14 The Fourier integral, Eq. (15.18), has been held meaningless for $f(t) = \cos \alpha t$. Show that the Fourier integral can be extended to cover $f(t) = \cos \alpha t$ by use of the Dirac delta function.

15.3.15 Show that

$$\int_0^\infty \sin ka \, J_0(k\rho) \, dk = \begin{cases} (a^2 - \rho^2)^{-1/2}, & \rho < a, \\ 0, & \rho > a. \end{cases}$$

Here a and ρ are positive. The equation comes from the determination of the distribution of charge on an isolated conducting disk, radius a.

Note that the function on the right has an *infinite* discontinuity at $\rho = a$.

Note. A Laplace transform approach appears in Exercise 15.10.8.

15.3.16 The function $f(\mathbf{r})$ has a Fourier exponential transform

$$g(\mathbf{k}) = \frac{1}{(2\pi)^{3/2}} \int f(\mathbf{r}) e^{i\mathbf{k}\cdot\mathbf{r}} \, d^3r = \frac{1}{(2\pi)^{3/2} k^2}.$$

Determine $f(\mathbf{r})$.

Hint. Use spherical polar coordinates in k-space.

ANS. $f(\mathbf{r}) = \dfrac{1}{4\pi r}$.

15.3.17   (a) Calculate the Fourier exponential transform of $f(x) = e^{-a|x|}$.
(b) Calculate the inverse transform by employing the calculus of residues (Section 7.2).

15.3.18   Show that the following are Fourier transforms of each other

$$i^n J_n(t) \quad \text{and} \quad \begin{cases} \sqrt{\dfrac{2}{\pi}} T_n(x)(1 - x^2)^{-1/2}, & |x| < 1 \\ 0, & |x| > 1 \end{cases}$$

$T_n(x)$ is the nth-order Chebyshev polynomial.
Hint. With $T_n(\cos \theta) = \cos n\theta$, the transform of $T_n(x)(1 - x^2)^{-1/2}$ leads to an integral representation of $J_n(t)$.

15.3.19   Show that the Fourier exponential transform of

$$f(\mu) = \begin{cases} P_n(\mu), & |\mu| \le 1 \\ 0, & |\mu| > 1 \end{cases}$$

is $(2i^n/2\pi)j_n(kr)$. Here $P_n(\mu)$ is a Legendre polynomial and $j_n(kr)$ is a spherical Bessel function.

15.3.20   Show that the three-dimensional Fourier exponential transform of a radially symmetric function may be rewritten as a Fourier sine transform:

$$\frac{1}{(2\pi)^{3/2}} \int_{-\infty}^{\infty} f(r)e^{i\mathbf{k}\cdot\mathbf{r}}\, d^3x = \frac{1}{k}\sqrt{\frac{2}{\pi}} \int_0^{\infty} [rf(r)] \sin kr\, dr.$$

15.3.21   (a) Show that $f(x) = x^{-1/2}$ is a *self-reciprocal* under both Fourier cosine and sine transforms; that is,

$$\sqrt{\frac{2}{\pi}} \int_0^{\infty} x^{-1/2} \cos xt\, dx = t^{-1/2}$$

$$\sqrt{\frac{2}{\pi}} \int_0^{\infty} x^{-1/2} \sin xt\, dx = t^{-1/2}.$$

(b) Use the preceding results to evaluate the Fresnel integrals $\int_0^{\infty} \sin(y^2)\, dy$ and $\int_0^{\infty} \sin(y^2)\, dy$.

15.4 FOURIER TRANSFORM OF DERIVATIVES

In Section 15.1, Fig. 15.1 outlines the overall technique of using Fourier transforms and inverse transforms to solve a problem. Here we take an initial step in solving a differential equation—obtaining the Fourier transform of a derivative.

Using the exponential form, we determine that the Fourier transform of $f(x)$ is

$$g(\omega) = \frac{1}{\sqrt{2\pi}} \int_{-\infty}^{\infty} f(x)e^{i\omega x}\, dx \tag{15.37}$$

and for $df(x)/dx$

$$g_1(\omega) = \frac{1}{\sqrt{2\pi}} \int_{-\infty}^{\infty} \frac{df(x)}{dx} e^{i\omega x}\, dx. \tag{15.38}$$

Integrating Eq. (15.38) by parts, we obtain

$$g_1(\omega) = \frac{e^{i\omega x}}{\sqrt{2\pi}} f(x)\Big|_{-\infty}^{\infty} - \frac{i\omega}{\sqrt{2\pi}} \int_{-\infty}^{\infty} f(x)e^{i\omega x}\, dx. \tag{15.39}$$

If $f(x)$ vanishes[1] as $x \to \pm\infty$, we have

$$g_1(\omega) = -i\omega\, g(\omega); \tag{15.40}$$

that is, the transform of the derivative is $(-i\omega)$ times the transform of the original function. This may readily be generalized to the nth derivative to yield

$$g_n(\omega) = (-i\omega)^n g(\omega), \tag{15.41}$$

provided all the integrated parts vanish as $x \to \pm\infty$. This is the power of the Fourier transform, the reason it is so useful in solving (partial) differential equations. The operation of differentiation has been replaced by a multiplication.

Example 15.4.1 Wave Equation

This technique may be used to advantage in handling partial differential equations. To illustrate the technique let us derive a familiar expression of elementary physics. An infinitely long string is vibrating freely. The amplitude y of the (small) vibrations satisfies the wave equation

$$\frac{\partial^2 y}{\partial x^2} = \frac{1}{v^2}\frac{\partial^2 y}{\partial t^2}. \tag{15.42}$$

We shall assume an initial condition

$$y(x, 0) = f(x). \tag{15.43}$$

Applying our Fourier transform, which means multiplying by $e^{i\alpha x}$ and integrating over x, we obtain

$$\int_{-\infty}^{\infty} \frac{\partial^2 y(x, t)}{\partial x^2} e^{i\alpha x}\, dx = \frac{1}{v^2}\int_{-\infty}^{\infty} \frac{\partial^2 y(x, t)}{\partial t^2} e^{i\alpha x}\, dx \tag{15.44}$$

or

$$(-i\alpha)^2 Y(\alpha, t) = \frac{1}{v^2}\frac{\partial^2 Y(\alpha, t)}{\partial t^2}. \tag{15.45}$$

Here we have used

$$Y(\alpha, t) = \frac{1}{\sqrt{2\pi}}\int_{-\infty}^{\infty} y(x, t)e^{i\alpha x}\, dx \tag{15.46}$$

and Eq. (15.41) for the second derivative. Note that the integrated part of Eq. (15.39) vanishes. The wave has not yet gone to ∞. Since no derivatives with

[1] Apart from cases such as Exercise 15.3.6, $f(x)$ must vanish as $x \to \pm\infty$ in order for the Fourier transform of $f(x)$ to exist.

respect to α appear, Eq. (15.45) is actually an ordinary differential equation—in fact, the linear oscillator equation. This transformation, from a partial to an ordinary differential equation, is a significant achievement. We solve Eq. (15.45) subject to the appropriate initial conditions. At $t = 0$, applying Eq. (15.43), Eq. (15.46) reduces to

$$Y(\alpha, 0) = \frac{1}{\sqrt{2\pi}} \int_{-\infty}^{\infty} f(x)e^{i\alpha x} \, dx$$

$$= F(\alpha). \tag{15.47}$$

The general solution of Eq. (15.45) in exponential form is

$$Y(\alpha, t) = F(\alpha)e^{\pm iv\alpha t}. \tag{15.48}$$

Using the inversion formula (Eq. (15.23)), we have

$$y(x, t) = \frac{1}{\sqrt{2\pi}} \int_{-\infty}^{\infty} Y(\alpha, t)e^{-i\alpha x} \, d\alpha \tag{15.49}$$

and, by Eq. (15.48),

$$y(x, t) = \frac{1}{\sqrt{2\pi}} \int_{-\infty}^{\infty} F(\alpha)e^{-i\alpha(x \mp vt)} \, d\alpha. \tag{15.50}$$

Since $f(x)$ is the Fourier inverse transform of $F(\alpha)$,

$$y(x, t) = f(x \mp vt), \tag{15.51}$$

corresponding to waves advancing in the $+x$- and $-x$-directions, respectively.

The particular linear combinations of waves is given by the boundary condition of Eq. (15.43) and some other boundary condition such as a restriction on $\partial y/\partial t$.

The accomplishment of the Fourier transform here deserves special emphasis. Our Fourier transform converted a partial differential equation into an ordinary differential equation, where the "degree of transcendence" of the problem was reduced. In Section 15.9 Laplace transforms are used to convert ordinary differential equations (with constant coefficients) into algebraic equations. Again, the degree of transcendence is reduced. The problem is simplified—as outlined in Fig. 15.1.

EXERCISES

15.4.1   The one-dimensional Fermi age equation for the diffusion of neutrons slowing down in some medium (such as graphite) is

$$\frac{\partial^2 q(x, \tau)}{\partial x^2} = \frac{\partial q(x, \tau)}{\partial \tau}.$$

Here q is the number of neutrons that slow down, falling below some given energy per second per unit volume. The Fermi age, τ, is a measure of the energy loss.

If $q(x, 0) = S\delta(x)$, corresponding to a plane source of neutrons at $x = 0$, emitting S neutrons per unit area per second, derive the solution

$$q = S\frac{e^{-x^2/4\tau}}{\sqrt{4\pi\tau}}.$$

Hint. Replace $q(x, \tau)$ with

$$p(k, \tau) = \frac{1}{\sqrt{2\pi}} \int_{-\infty}^{\infty} q(x, \tau) e^{ikx}\, dx.$$

This is analogous to the diffusion of heat in an infinite medium.

15.4.2   Equation (15.41) yields

$$g_2(\omega) = -\omega^2 g(\omega)$$

for the Fourier transform of the second derivative of $f(x)$. The condition $f(x) \to 0$ for $x \to \pm\infty$ may be relaxed slightly. Find the *least* restrictive condition for the preceding equation for $g_2(\omega)$ to hold.

$$ANS. \quad \left[\frac{df(x)}{dx} - i\omega f(x)\right]e^{i\omega x}\bigg|_{-\infty}^{\infty} = 0.$$

15.4.3   The one-dimensional neutron diffusion equation with a (plane) source is

$$-D\frac{d^2\varphi(x)}{dx^2} + K^2 D\varphi(x) = Q\delta(x),$$

where $\varphi(x)$ is the neutron flux, $Q\delta(x)$ is the (plane) source at $x = 0$, and D and K^2 are constants. Apply a Fourier transform. Solve the equation in transform space. Transform your solution back into x-space.

$$ANS. \quad \varphi(x) = \frac{Q}{2KD}e^{-|Kx|}.$$

15.4.4   For a point source at the origin the three-dimensional neutron diffusion equation becomes

$$-D\nabla^2\varphi(\mathbf{r}) + K^2 D\varphi(\mathbf{r}) = Q\delta(\mathbf{r}).$$

Apply a three-dimensional Fourier transform. Solve the transformed equation. Transform the solution back into $\mathbf{r}$-space.

15.4.5   (a) Given that $F(\mathbf{k})$ is the three-dimensional Fourier transform of $f(\mathbf{r})$ and $F_1(\mathbf{k})$ is the three-dimensional Fourier transform of $\nabla f(\mathbf{r})$, show that

$$F_1(\mathbf{k}) = (-i\mathbf{k})F(\mathbf{k}).$$

This is a three-dimensional generalization of Eq. (15.40).

(b) Show that the three-dimensional Fourier transform of $\nabla \cdot \nabla f(\mathbf{r})$ is

$$F_2(\mathbf{k}) = (-i\mathbf{k})^2 F(\mathbf{k}).$$

Note. Vector $\mathbf{k}$ is a vector in the transform space. In Section 15.6 we shall have $\hbar\mathbf{k} = \mathbf{p}$, linear momentum.

15.5 CONVOLUTION THEOREM

We shall employ convolutions to solve differential equations, to normalize momentum wave functions (Section 15.6), and to investigate transfer functions (Section 15.7).

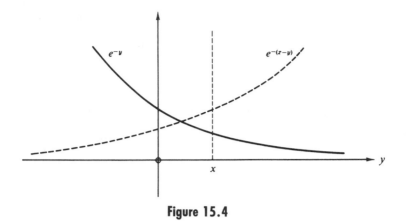

Figure 15.4

Let us consider two functions $f(x)$ and $g(x)$ with Fourier transforms $F(t)$ and $G(t)$, respectively. We define the operation

$$f * g \equiv \frac{1}{\sqrt{2\pi}} \int_{-\infty}^{\infty} g(y)f(x - y)\, dy \qquad (15.52)$$

as the *convolution* of the two functions f and g over the interval $(-\infty, \infty)$. This form of an integral appears in probability theory in the determination of the probability density of two random, independent variables. Our solution of Poisson's equation, Eq. (8.99), may be interpreted as a convolution of a charge distribution, $\rho(\mathbf{r}_2)$, and a weighting function, $(4\pi\varepsilon_0 |\mathbf{r}_1 - \mathbf{r}_2|)^{-1}$. In other works this is sometimes referred to as the *Faltung*, to use the German term for "folding."[1] We now transform the integral in Eq. (15.52) by introducing the Fourier transforms:

$$\int_{-\infty}^{\infty} g(y)f(x - y)\, dy = \frac{1}{\sqrt{2\pi}} \int_{-\infty}^{\infty} g(y) \int_{-\infty}^{\infty} F(t)e^{-it(x-y)}\, dt\, dy$$

$$= \frac{1}{\sqrt{2\pi}} \int_{-\infty}^{\infty} F(t) \left[\int_{-\infty}^{\infty} g(y)e^{ity}\, dy \right] e^{-itx}\, dt \qquad (15.53)$$

$$= \int_{-\infty}^{\infty} F(t)G(t)e^{-itx}\, dt,$$

interchanging the order of integration and transforming $g(y)$. This result may be interpreted as follows: The Fourier inverse transform of a *product* of Fourier transforms is the convolution of the original functions, $f * g$.

---

[1] For $f(y) = e^{-y}$, $f(y)$ and $f(x - y)$ are plotted in Fig. 15.4. Clearly, $f(y)$ and $f(x - y)$ are mirror images of each other in relation to the vertical line $y = x/2$, that is, we could generate $f(x - y)$ by folding over $f(y)$ on the line $y = x/2$.

For the special case $x = 0$ we have

$$\int_{-\infty}^{\infty} F(t)G(t)\, dt = \int_{-\infty}^{\infty} f(-y)g(y)\, dy. \tag{15.54}$$

The minus sign in $-y$ suggests that modifications be tried. We now do this with g^* instead of g using a different technique.

PARSEVAL'S RELATION

Results analogous to Eqs. (15.53) and (15.54) may be derived for the Fourier sine and cosine transforms (Exercises 15.5.1 and 15.5.2). Equation (15.54) and the corresponding sine and cosine convolutions are often labeled "Parseval's relations" by analogy with Parseval's theorem for Fourier series (Chapter 14, Exercise 14.4.42).

The Parseval relation[2,3]

$$\int_{-\infty}^{\infty} F(\omega)G^*(\omega)\, d\omega = \int_{-\infty}^{\infty} f(t)g^*(t)\, dt, \tag{15.55}$$

may be derived very beautifully using the Dirac delta function representation, Eq. (15.21d). We have

$$\int_{-\infty}^{\infty} f(t)g^*(t)\, dt = \int_{-\infty}^{\infty} \frac{1}{\sqrt{2\pi}} \int_{-\infty}^{\infty} F(\omega)e^{-i\omega t}\, d\omega \cdot \frac{1}{\sqrt{2\pi}} \int_{-\infty}^{\infty} G^*(x)e^{ixt}\, dx\, dt, \tag{15.56}$$

with attention to the complex conjugation in the $G^*(x)$ to $g^*(t)$ transform. Integrating over t first, and using Eq. (15.21d), we obtain

$$\int_{-\infty}^{\infty} f(t)g^*(t)\, dt = \int_{-\infty}^{\infty} F(\omega) \int_{-\infty}^{\infty} G^*(x)\delta(x - \omega)\, dx\, d\omega$$

$$= \int_{-\infty}^{\infty} F(\omega)G^*(\omega)\, d\omega, \tag{15.57}$$

our desired Parseval relation. If $f(t) = g(t)$, then the integrals in the Parseval relation are normalization integrals (Section 9.4). Equation (15.57) guarantees that if a function $f(t)$ is normalized to unity, its transform $F(\omega)$ is likewise normalized to unity. This is extremely important in quantum mechanics as developed in the next section.

It may be shown that the Fourier transform is a unitary operation (in the Hilbert space L^2, square integrable functions). The Parseval relation is a reflection of this unitary property—analogous to Exercise 3.4.26 for matrices.

In Fraunhofer diffraction optics the diffraction pattern (amplitude) appears as the transform of the function describing the aperture (compare Exercise

---

[2] Note that all arguments are positive in contrast to Eq. (15.54).

[3] Some authors prefer to restrict Parseval's name to series and refer to Eq. (15.55) as Rayleigh's theorem.

15.5.5). With intensity proportional to the square of the amplitude the Parseval relation implies that the energy passing through the aperture seems to be somewhere in the diffraction pattern—a statement of the conservation of energy.

Parseval's relations may be developed independently of the inverse Fourier transform and then used rigorously to derive the inverse transform. Details are given by Morse and Feshbach,[4] Section 4.8 (see also Exercise 15.5.4).

EXERCISES

15.5.1   Work out the convolution equation corresponding to Eq. (15.53) for
(a) Fourier sine transforms
$$\frac{1}{2} \int_{-\infty}^{\infty} g(y)f(x-y)\, dy = -\int_{0}^{\infty} F_s(s)G_s(s) \cos sx\, ds,$$
where f and g are odd functions.
(b) Fourier cosine transforms
$$\frac{1}{2} \int_{-\infty}^{\infty} g(y)f(x-y)\, dy = \int_{0}^{\infty} F_c(s)G_c(s) \cos sx\, ds,$$
where f and g are even functions.

15.5.2 $F(\rho)$ and $G(\rho)$ are the Hankel transforms of $f(r)$ and $g(r)$, respectively (Exercise 15.1.1). Derive the Hankel transform Parseval relation:
$$\int_{0}^{\infty} F^*(\rho)G(\rho)\rho\, d\rho = \int_{0}^{\infty} f^*(r)g(r)r\, dr.$$

15.5.3   Show that for both Fourier sine and Fourier cosine transforms Parseval's relation has the form
$$\int_{0}^{\infty} F(t)G(t)\, dt = \int_{0}^{\infty} f(y)g(y)\, dy.$$

15.5.4   Starting from Parseval's relation (Eq. (15.54)), let $g(y) = 1, 0 \le y \le \alpha$, and zero elsewhere. From this derive the Fourier inverse transform (Eq. (15.23)). *Hint.* Differentiate with respect to α.

15.5.5   (a) A rectangular pulse is described by
$$f(x) = \begin{cases} 1, & |x| < a \\ 0, & |x| > a. \end{cases}$$
Show that the Fourier exponential transform is
$$F(t) = \sqrt{\frac{2}{\pi}} \frac{\sin at}{t}.$$
Here is the single slit diffraction problem of physical optics. The slit is described by $f(x)$. The diffraction pattern *amplitude* is given by the Fourier transform $F(t)$.

[4] P. M. Morse and H. Feshbach, *Methods of Theoretical Physics*. New York: McGraw–Hill (1953).

(b) Use the Parseval relation to evaluate

$$\int_{-\infty}^{\infty} \frac{\sin^2 t}{t^2}\, dt.$$

This integral may also be evaluated by using the calculus of residues, Exercise 7.2.12.

ANS. (b) π.

15.5.6 Solve Poisson's equation $\nabla^2 \psi(\mathbf{r}) = -\rho(\mathbf{r})/\varepsilon_0$ by the following sequence of operations:

(a) Take the Fourier transform of both sides of this equation. Solve for the Fourier transform of $\psi(\mathbf{r})$.

(b) Carry out the Fourier inverse transform by using a three-dimensional analog of the convolution theorem, Eq. (15.53).

15.5.7 (a) Given $f(x) = 1 - |x/2|$, $-2 \le x \le 2$ and zero elsewhere, show that the Fourier transform of $f(x)$ is

$$F(t) = \sqrt{\frac{2}{\pi}} \left(\frac{\sin t}{t} \right)^2.$$

(b) Using the Parseval relation, evaluate

$$\int_{-\infty}^{\infty} \left(\frac{\sin t}{t} \right)^4 dt.$$

ANS. (b) $\dfrac{2\pi}{3}$.

15.5.8 With $F(t)$ and $G(t)$ the Fourier transforms of $f(x)$ and $g(x)$, respectively, show that

$$\int_{-\infty}^{\infty} |f(x) - g(x)|^2\, dx = \int_{-\infty}^{\infty} |F(t) - G(t)|^2\, dt.$$

If $g(x)$ is an approximation to $f(x)$, the preceding relation indicates that the mean square deviation in t-space is equal to the mean square deviation in x-space.

15.5.9 Use the Parseval relation to evaluate

(a) $\displaystyle\int_0^{\infty} \frac{d\omega}{(\omega^2 + a^2)^2}$,

(b) $\displaystyle\int_0^{\infty} \frac{\omega^2\, d\omega}{(\omega^2 + a^2)^2}$.

Hint. Compare Exercise 15.3.4.

ANS. (a) $\dfrac{\pi}{4a^3}$,

(b) $\dfrac{\pi}{4a}$.

15.6 MOMENTUM REPRESENTATION

In advanced dynamics and in quantum mechanics linear momentum and spacial position occur on an equal footing. In this section we shall start with the usual space distribution and derive the corresponding momentum distribution. For the one-dimensional case our wave function $\psi(x)$, a solution of the Schrödinger wave equation, has the following properties:

1. $\psi^*(x)\psi(x)\,dx$ is the probability of finding the quantum particle between x and $x + dx$ and

2.
$$\int_{-\infty}^{\infty} \psi^*(x)\psi(x)\,dx = 1,$$
(15.58)

corresponding to *one* particle (along the x-axis).

In addition, we have

3.
$$\langle x \rangle = \int_{-\infty}^{\infty} \psi^*(x)\,x\,\psi(x)\,dx$$
(15.59)

for the *average* position of the particle along the x-axis. This is often called an expectation value.

We want a function $g(p)$ that will give the same information about the momentum.

1. $g^*(p)g(p)\,dp$ is the probability that our quantum particle has a momentum between p and $p + dp$.

2.
$$\int_{-\infty}^{\infty} g^*(p)\,g(p)\,dp = 1.$$
(15.60)

3.
$$\langle p \rangle = \int_{-\infty}^{\infty} g^*(p)\,p\,g(p)\,dp.$$
(15.61)

As subsequently shown, such a function is given by the Fourier transform of our space function $\psi(x)$. Specifically,[1]

$$g(p) = \frac{1}{\sqrt{2\pi\hbar}} \int_{-\infty}^{\infty} \psi(x)e^{-ipx/\hbar}\,dx$$
(15.62)

$$g^*(p) = \frac{1}{\sqrt{2\pi\hbar}} \int_{-\infty}^{\infty} \psi^*(x)e^{ipx/\hbar}\,dx.$$
(15.63)

[1] The $\hbar$ may be avoided by using the wave number k, $p = k\hbar$ (and $\mathbf{p} = \mathbf{k}\hbar$), so that

$$\varphi(k) = \frac{1}{(2\pi)^{1/2}} \int \psi(x)e^{-ikx}\,dx.$$

An example of this notation appears in Section 16.1.

The corresponding three-dimensional momentum function is

$$g(\mathbf{p}) = \frac{1}{(2\pi\hbar)^{3/2}} \int\int\int_{-\infty}^{\infty} \psi(\mathbf{r}) e^{-i\mathbf{r}\cdot\mathbf{p}/\hbar} \, d^3r.$$

To verify Eqs. (15.62) and (15.63), let us check on properties 2 and 3.

Property 2, the normalization, is automatically satisfied as a Parseval relation, Eq. (15.55). If the space function $\psi(x)$ is normalized to unity, the momentum function $g(p)$ is also normalized to unity.

To check on property 3, we must show that

$$\langle p \rangle = \int_{-\infty}^{\infty} g^*(p) p \, g(p) \, dp = \int_{-\infty}^{\infty} \psi^*(x) \frac{\hbar}{i} \frac{d}{dx} \psi(x) \, dx, \qquad (15.64)$$

where $(\hbar/i)(d/dx)$ is the momentum operator in the space representation. We replace the momentum functions by Fourier transformed space functions, and the first integral becomes

$$\frac{1}{2\pi\hbar} \int\int\int_{-\infty}^{\infty} p e^{-ip(x-x')/\hbar} \psi^*(x')\psi(x) \, dp \, dx' \, dx. \qquad (15.65)$$

Now

$$p e^{-ip(x-x')/\hbar} = \frac{d}{dx}\left[-\frac{\hbar}{i} e^{-ip(x-x')/\hbar} \right]. \qquad (15.66)$$

Substituting into Eq. (15.65) and integrating by parts, holding x' and p constant, we obtain

$$\langle p \rangle = \int\int_{-\infty}^{\infty} \left[\frac{1}{2\pi\hbar} \int_{-\infty}^{\infty} e^{-ip(x-x')/\hbar} \, dp \right] \cdot \psi^*(x') \frac{\hbar}{i} \frac{d}{dx} \psi(x) \, dx' \, dx. \qquad (15.67)$$

Here we assume $\psi(x)$ vanishes as $x \to \pm\infty$, eliminating the integrated part. Again using the Dirac delta function, Eq. (15.21c), Eq. (15.67) reduces to Eq. (15.64) to verify our momentum representation. The reader will note that technically we have employed the inverse Fourier transform in Eq. (15.62). This was chosen deliberately to yield the proper sign in Eq. (15.67).

Example 15.6.1 Hydrogen Atom

The hydrogen atom ground state[2] may be described by the spacial wave function

$$\psi(\mathbf{r}) = \left(\frac{1}{\pi a_0^3} \right)^{1/2} e^{-r/a_0}, \qquad (15.68)$$

a_0 being the Bohr radius, $\hbar^2/me^2$. We now have a three-dimensional wave

---

[2] See E. V. Ivash, A momentum representation treatment of the hydrogen atom problem. *Am. J. Phys.* **40**, 1095 (1972) for a momentum representation treatment of the hydrogen atom, $l = 0$ states.

function. The transform corresponding to Eq. (15.62) is

$$g(\mathbf{p}) = \frac{1}{(2\pi\hbar)^{3/2}} \int \psi(\mathbf{r}) e^{-i\mathbf{p}\cdot\mathbf{r}/\hbar} \, d^3r. \qquad (15.69)$$

Substituting Eq. (15.68) into Eq. (15.69) and using

$$\int e^{-ar+i\mathbf{b}\cdot\mathbf{r}} \, d^3r = \frac{8\pi a}{(a^2 + b^2)^2}, \qquad (15.70)$$

we obtain the hydrogenic momentum wave function

$$g(\mathbf{p}) = \frac{2^{3/2}}{\pi} \frac{a_0^{3/2}\hbar^{5/2}}{(a_0^2 p^2 + \hbar^2)^2}. \qquad (15.71)$$

Such momentum functions have been found useful in problems like Compton scattering from atomic electrons, the wavelength distribution of the scattered radiation, depending on the momentum distribution of the target electrons.

The relation between the ordinary space representation and the momentum representation may be clarified by considering the basic commutation relations of quantum mechanics. We can go from a classical Hamiltonian to the Schrödinger wave equation by requiring that momentum p and position x *not* commute. Instead, we require that

$$[p, x] \equiv (px - xp) = -i\hbar. \qquad (15.72)$$

For the multidimensional case Eq. (15.72) is replaced by

$$[p_i, x_j] = -i\hbar\delta_{ij}. \qquad (15.73)$$

The Schrödinger (space) representation is obtained by using

$$x_j \rightarrow x_j,$$
$$p_i \rightarrow -i\hbar \frac{\partial}{\partial x_i}, \qquad (x)$$

replacing the momentum by a partial space derivative. The reader will easily see that

$$[p, x]\psi(x) = -i\hbar\psi(x). \qquad (15.74)$$

However, Eq. (15.72) can equally well be satisfied by using

$$x_j \rightarrow i\hbar \frac{\partial}{\partial p_j},$$
$$p_i \rightarrow p_i. \qquad (p)$$

This is the momentum representation. Then

$$[p, x]g(p) = -i\hbar g(p). \qquad (15.75)$$

Hence the representation (x) is not unique; (p) is an alternate possibility.

In general, the Schrödinger representation (x) leading to the Schrödinger wave equation is more convenient because the potential energy V is generally given as a function of position $V(x, y, z)$. The momentum representation (p) usually leads to an integral equation (compare Chapter 16 for the pros and cons of the integral equations). For an exception, consider the harmonic oscillator.

Example 15.6.2 Harmonic Oscillator

The classical Hamiltonian (kinetic energy + potential energy = total energy) is

$$H(p, x) = \frac{p^2}{2m} + \frac{1}{2} kx^2 = E. \tag{15.76}$$

where k is the Hooke's law constant.

In the Schrödinger representation we obtain

$$-\frac{\hbar^2}{2m} \frac{d^2\psi(x)}{dx^2} + \frac{1}{2} kx^2 \psi(x) = E\psi(x). \tag{15.77}$$

For total energy E equal to $\sqrt{(k/m)}\, \hbar/2$ there is a solution (Section 13.1)

$$\psi(x) = e^{-(\sqrt{mk}/2\hbar)x^2}. \tag{15.78}$$

The momentum representation leads to

$$\frac{p^2}{2m} g(p) - \frac{\hbar^2 k}{2} \frac{d^2 g(p)}{dp^2} = Eg(p). \tag{15.79}$$

Again, for

$$E = \sqrt{\frac{k}{m}} \frac{\hbar}{2} \tag{15.80}$$

the momentum wave equation (15.79) is satisfied by

$$g(p) = e^{-p^2/(2\hbar\sqrt{mk})}. \tag{15.81}$$

Either representation, space or momentum (and an infinite number of other possibilities), may be used, depending on which is more convenient for the particular problem under attack.

The demonstration that $g(p)$ is the momentum wave function corresponding to Eq. 15.78—that it is the Fourier inverse transform of Eq. (15.78)—is left as Exercise 15.6.3.

EXERCISES

15.6.1   The function $e^{i\mathbf{k}\cdot\mathbf{r}}$ describes a plane wave of momentum $\mathbf{p} = \hbar\mathbf{k}$ normalized to unit density. (Time dependence of $e^{-i\omega t}$ is assumed.) Show that these plane wave functions satisfy an orthogonality relation

$$\int (e^{i\mathbf{k}\cdot\mathbf{r}})^* e^{i\mathbf{k}'\cdot\mathbf{r}}\, dx\, dy\, dz = (2\pi)^3 \delta(\mathbf{k} - \mathbf{k}').$$

15.6.2   An infinite plane wave in quantum mechanics may be represented by the function

$$\psi(x) = e^{ip'x/\hbar}.$$

Find the corresponding momentum distribution function. Note that it has an infinity and that $\psi(x)$ is not normalized.

15.6.3   A linear quantum oscillator in its ground state has a wave function

$$\psi(x) = a^{-1/2}\pi^{-1/4}e^{-x^2/2a^2}.$$

Show that the corresponding momentum function is

$$g(p) = a^{1/2}\pi^{-1/4}\hbar^{-1/2}e^{-a^2p^2/2\hbar^2}.$$

15.6.4   The nth excited state of the linear quantum oscillator is described by

$$\psi_n(x) = a^{-1/2}2^{-n/2}\pi^{-1/4}(n!)^{-1/2}e^{-x^2/2a^2}H_n(x/a);$$

where $H_n(x/a)$ is the nth Hermite polynomial, Section 13.1. As an extension of Exercise 15.6.3, find the momentum function corresponding to $\psi_n(x)$.
Hint. $\psi_n(x)$ may be represented by $\mathcal{L}_+^n\psi_0(x)$, where $\mathcal{L}_+$ is the raising operator, Exercise 13.1.16.

15.6.5   A free particle in quantum mechanics is described by a plane wave

$$\psi_k(x, t) = e^{i[kx-(\hbar k^2/2m)t]}.$$

Combining waves of adjacent momentum with an amplitude weighting factor $\varphi(k)$, we form a wave packet

$$\Psi(x, t) = \int_{-\infty}^{\infty} \varphi(k)e^{i[kx-(\hbar k^2/2m)t]}\, dk.$$

(a) Solve for $\varphi(k)$ given that

$$\Psi(x, 0) = e^{-x^2/2a^2}.$$

(b) Using the known value of $\varphi(k)$, integrate to get the explicit form of $\Psi(x, t)$. Note that this wave packet diffuses or spreads out with time.

$$ANS. \quad \Psi(x, t) = \frac{e^{-\{x^2/2[(a^2+(i\hbar/m)t]\}}}{[1 + (i\hbar t/ma^2)]^{1/2}}.$$

Note. An interesting discussion of this problem from the evolution operator point of view is given by S. M. Blinder, Evolution of a Gaussian wave-packet. *Am. J. Phys.* **36**, 525 (1968).

15.6.6   Find the time-dependent momentum wave function $g(k, t)$ corresponding to $\Psi(x, t)$ of Exercise 15.6.5. Show that the momentum wave packet $g^*(k, t)g(k, t)$ is *independent* of time.

15.6.7   The deuteron, Example 9.1.2, may be described reasonably well with a Hulthen wave function

$$\psi(\mathbf{r}) = A[e^{-\alpha r} - e^{-\beta r}]/r,$$

with A, α, and β constants. Find $g(\mathbf{p})$ the corresponding momentum function.
Note. The Fourier transform may be rewritten as Fourier sine and cosine transforms or as a Laplace transform, Section 15.8.

15.6.8 The nuclear form factor $F(k)$ and the charge distribution $\rho(r)$ are three-dimensional Fourier transforms of each other:

$$F(k) = \frac{1}{(2\pi)^{3/2}} \int \rho(r) e^{i\mathbf{k}\cdot\mathbf{r}} \, d^3r.$$

If the measured form factor is

$$F(k) = (2\pi)^{-3/2}\left(1 + \frac{k^2}{a^2}\right)^{-1},$$

find the corresponding charge distribution.

ANS. $\rho(r) = \dfrac{a^2}{4\pi}\dfrac{e^{-ar}}{r}.$

15.6.9 Check the normalization of the hydrogen momentum wave function

$$g(\mathbf{p}) = \frac{2^{3/2}}{\pi}\frac{a_0^{3/2}\hbar^{5/2}}{(a_0^2 p^2 + \hbar^2)^2}$$

by direct evaluation of the integral

$$\int g^*(\mathbf{p})g(\mathbf{p}) \, d^3p.$$

15.6.10 With $\psi(\mathbf{r})$ a wave function in ordinary space and $\varphi(\mathbf{p})$ the corresponding momentum function, show that

(a) $\dfrac{1}{(2\pi\hbar)^{3/2}} \displaystyle\int \mathbf{r}\psi(\mathbf{r}) e^{-i\mathbf{r}\cdot\mathbf{p}/\hbar} \, d^3r = i\hbar\nabla_p\varphi(\mathbf{p}),$

(b) $\dfrac{1}{(2\pi\hbar)^{3/2}} \displaystyle\int \mathbf{r}^2\psi(\mathbf{r}) e^{-i\mathbf{r}\cdot\mathbf{p}/\hbar} \, d^3r = (i\hbar\nabla_p)^2\varphi(\mathbf{p}).$

Note. ∇_p is the gradient in momentum space:

$$\hat{\mathbf{x}}\frac{\partial}{\partial p_x} + \hat{\mathbf{y}}\frac{\partial}{\partial p_y} + \hat{\mathbf{z}}\frac{\partial}{\partial p_z}.$$

These results may be extended to any positive integer power of r and therefore to any (analytic) function that may be expanded as a Maclaurin series in r.

15.6.11 The ordinary space wave function $\psi(\mathbf{r}, t)$ satisfies the time-dependent Schrödinger equation

$$i\hbar\frac{\partial\psi(\mathbf{r}, t)}{\partial t} = -\frac{\hbar^2}{2m}\nabla^2\psi + V(\mathbf{r})\psi.$$

Show that the corresponding time-dependent momentum function satisfies the analogous equation

$$i\hbar\frac{\partial\varphi(\mathbf{p}, t)}{\partial t} = \frac{p^2}{2m}\varphi + V(i\hbar\nabla_p)\varphi.$$

Note. Assume that $V(\mathbf{r})$ may be expressed by a Maclaurin series and use Exercise 15.6.10.
$V(i\hbar\nabla_p)$ is the same function of the variable $i\hbar\nabla_p$ that $V(\mathbf{r})$ is of the variable $\mathbf{r}$.

15.6.12   The one-dimensional time-independent Schrödinger wave equation is

$$-\frac{\hbar^2}{2m}\frac{d^2\psi(x)}{dx^2} + V(x)\psi(x) = E\psi(x).$$

For the special case of $V(x)$ an analytic function of x, show that the corresponding momentum wave equation is

$$V\left(i\hbar\frac{d}{dp}\right)g(p) + \frac{p^2}{2m}g(p) = Eg(p).$$

Derive this momentum wave equation from the Fourier transform, Eq. (15.62), and its inverse. Do not use the substitution $x \rightarrow i\hbar(d/dp)$ directly.

15.7 TRANSFER FUNCTIONS

A time-dependent electrical pulse may be regarded as built-up as a superposition of plane waves of many frequencies. For angular frequency ω we have a contribution

$$F(\omega)e^{i\omega t}.$$

Then the complete pulse may be written as

$$f(t) = \frac{1}{2\pi}\int_{-\infty}^{\infty} F(\omega)e^{i\omega t}\,d\omega. \tag{15.82}$$

Because the angular frequency ω is related to the linear frequency ν by

$$\nu = \frac{\omega}{2\pi},$$

it is customary to associate the entire $1/2\pi$ factor with this integral.

But if ω is a frequency, what about the negative frequencies? The negative ω's may be looked on as a mathematical device to avoid dealing with two functions ($\cos \omega t$ and $\sin \omega t$) separately (compare Section 14.1).

Because Eq. (15.82) has the form of a Fourier transform, we may solve for $F(\omega)$ by writing the inverse transform

$$F(\omega) = \int_{-\infty}^{\infty} f(t)e^{-i\omega t}\,dt. \tag{15.83}$$

Equation (15.83) represents a *resolution* of the pulse $f(t)$ into its angular frequency components. Equation (15.82) is a *synthesis* of the pulse from its components.

Consider some device such as a servomechanism or a stereo amplifier Fig. 15.5) with an input $f(t)$ and an output $g(t)$. For an input of a single frequency ω, $f_\omega(t) = e^{i\omega t}$, the amplifier will alter the amplitude and may also change the phase. The changes will probably depend on the frequency. Hence

$$g_\omega(t) = \varphi(\omega)f_\omega(t). \tag{15.84}$$

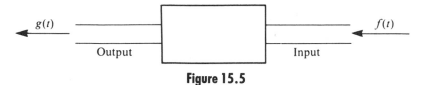

Figure 15.5

This amplitude and phase modifying function, $\varphi(\omega)$, is called a *transfer* function. It usually will be complex:

$$\varphi(\omega) = u(\omega) + iv(\omega), \tag{15.85}$$

where the functions $u(\omega)$ and $v(\omega)$ are real.

In Eq. (15.84) we assume that the transfer function $\varphi(\omega)$ is independent of input amplitude and of the presence or absence of any other frequency components. That is, we are assuming a linear mapping of $f(t)$ onto $g(t)$. Then the total output may be obtained by integrating over the entire input, as modified by the amplifier

$$g(t) = \frac{1}{2\pi} \int_{-\infty}^{\infty} \varphi(\omega)F(\omega)e^{i\omega t} \, d\omega. \tag{15.86}$$

The transfer function is characteristic of the amplifier. Once the transfer function is known (measured or calculated), the output $g(t)$ can be calculated for any input $f(t)$. Let us consider $\varphi(\omega)$ as the Fourier (inverse) transform of some function $\Phi(t)$

$$\varphi(\omega) = \int_{-\infty}^{\infty} \Phi(t)e^{-i\omega t} \, dt. \tag{15.87}$$

Then Eq. (15.86) is the Fourier transform of two inverse transforms. From Section 15.5 we obtain the convolution

$$g(t) = \int_{-\infty}^{\infty} f(\tau)\Phi(t - \tau) \, d\tau. \tag{15.88}$$

Interpreting Eq. (15.88), we have an input—a "cause"—$f(\tau)$, modified by $\Phi(t - \tau)$, producing an output—an "effect"—$g(t)$. Adopting the concept of *causality*—that the cause precedes the effect—we must require $\tau < t$. We do this by requiring

$$\Phi(t - \tau) = 0, \qquad \tau > t. \tag{15.89}$$

Then Eq. (15.88) becomes

$$g(t) = \int_{-\infty}^{t} f(\tau)\Phi(t - \tau) \, d\tau. \tag{15.90}$$

The adoption of Eq. (15.89) has profound consequences here and equivalently in the dispersion theory, Section 7.3.

SIGNIFICANCE OF $\Phi(t)$

To see the significance of Φ, let $f(\tau)$ be a sudden impulse starting at $\tau = 0$,

$$f(\tau) = \delta(\tau),$$

where $\delta(\tau)$ is a Dirac delta distribution on the positive side of the origin. Then Eq. (15.90) becomes

$$g(t) = \int_{-\infty}^{t} \delta(\tau)\Phi(t - \tau) \, d\tau$$

$$g(t) = \begin{cases} \Phi(t), & t > 0 \\ 0, & t < 0. \end{cases} \tag{15.91}$$

This identifies $\Phi(t)$ as the output function corresponding to a unit impulse at $t = 0$. Equation (15.91) also serves to establish that $\Phi(t)$ is real. Our original transfer function gives the steady-state output corresponding to a unit amplitude single frequency input. $\Phi(t)$ and $\varphi(\omega)$ are Fourier transforms of each other.

From Eq. (15.87) we now have

$$\varphi(\omega) = \int_{0}^{\infty} \Phi(t)e^{-i\omega t} \, dt, \tag{15.92}$$

with the lower limit set equal to zero by causality (Eq. (15.89)). With $\Phi(t)$ real from Eq. (15.91) we separate real and imaginary parts and write

$$u(\omega) = \int_{0}^{\infty} \Phi(t) \cos \omega t \, dt$$

$$v(\omega) = -\int_{0}^{\infty} \Phi(t) \sin \omega t \, dt, \qquad t > 0. \tag{15.93}$$

From this we see that the real part of $\varphi(\omega)$, $u(\omega)$, is even, whereas the imaginary part of $\varphi(\omega)$, $v(\omega)$, is odd:

$$u(-\omega) = u(\omega)$$

$$v(-\omega) = -v(\omega).$$

Compare this result with Exercise 15.3.1.

Interpreting Eq. (15.93) as Fourier cosine and sine transforms, we have

$$\Phi(t) = \frac{2}{\pi} \int_{0}^{\infty} u(\omega) \cos \omega t \, d\omega$$

$$= -\frac{2}{\pi} \int_{0}^{\infty} v(\omega) \sin \omega t \, d\omega. \qquad t > 0. \tag{15.94}$$

Combining Eqs. (15.93) and (15.94), we obtain

$$v(\omega) = -\int_{0}^{\infty} \sin \omega t \left\{ \frac{2}{\pi} \int_{0}^{\infty} u(\omega') \cos \omega' t \, d\omega' \right\} dt, \tag{15.95}$$

showing that if our transfer function has a real part, it will also have an imaginary part (and vice versa). Of course, this assumes that the Fourier transforms exist, thus excluding cases such as $\Phi(t) = 1$.

The imposition of causality has led to a mutual interdependence of the real and imaginary parts of the transfer function. The reader should compare this with the results of the dispersion theory of Section 7.3, also involving causality.

It may be helpful to show that the parity properties of $u(\omega)$ and $v(\omega)$ require $\Phi(t)$ to vanish for negative t. Inverting Eq. (15.87), we have

$$\Phi(t) = \frac{1}{2\pi} \int_{-\infty}^{\infty} [u(\omega) + i\,v(\omega)][\cos \omega t + i \sin \omega t]\, d\omega. \qquad (15.96)$$

With $u(\omega)$ even and $v(\omega)$ odd, Eq. (15.96) becomes

$$\Phi(t) = \frac{1}{\pi} \int_{0}^{\infty} u(\omega) \cos \omega t\, d\omega - \frac{1}{\pi} \int_{0}^{\infty} v(\omega) \sin \omega t\, d\omega, \qquad t > 0. \qquad (15.97)$$

From Eq. (15.94)

$$\int_{0}^{\infty} u(\omega) \cos \omega t\, d\omega = -\int_{0}^{\infty} v(\omega) \sin \omega t\, d\omega, \qquad t > 0. \qquad (15.98)$$

If we reverse the sign of t, $\sin \omega t$ reverses sign and from Eq. (15.97)

$$\Phi(t) = 0, \qquad t < 0$$

(demonstrating the internal consistency of our analysis).

EXERCISE

15.7.1   Derive the convolution

$$g(t) = \int_{-\infty}^{\infty} f(\tau)\Phi(t - \tau)\, d\tau.$$

15.8 ELEMENTARY LAPLACE TRANSFORMS

DEFINITION

The Laplace transform $f(s)$ or $\mathcal{L}$ of a function $F(t)$ is defined by[1]

$$f(s) = \mathcal{L}\{F(t)\} = \lim_{a \to \infty} \int_{0}^{a} e^{-st} F(t)\, dt = \int_{0}^{\infty} e^{-st} F(t)\, dt. \qquad (15.99)$$

---

[1] This is sometimes called a one-sided Laplace transform; the integral from $-\infty$ to $+\infty$ is referred to as a two-sided Laplace transform. Some authors introduce an additional factor of s. This extra s appears to have little advantage and continually gets in the way (compare Jeffreys and Jeffreys, Section 14.13 for additional comments). Generally, we take s to be real and positive. It is possible to have s complex provided $\mathfrak{R}(s) > 0$.

A few comments on the existence of the integral might be in order. The infinite integral of $F(t)$,

$$\int_0^\infty F(t)\, dt,$$

need not exist. For instance, $F(t)$ may diverge exponentially for large t. However, if there is some *constant* s_0 such that

$$|e^{-s_0 t} F(t)| \le M, \tag{15.100}$$

a positive constant for sufficiently large t, $t > t_0$, the Laplace transform (Eq. (15.99)), will exist for $s > s_0$; $F(t)$ is said to be of exponential order. As a counterexample, $F(t) = e^{t^2}$ does not satisfy the condition given by Eq. (15.100) and is *not* of exponential order. $\mathcal{L}\{e^{t^2}\}$ does *not* exist.

The Laplace transform may also fail to exist because of a sufficiently strong singularity in the function $F(t)$ as $t \to 0$; that is,

$$\int_0^\infty e^{-st} t^n\, dt$$

diverges at the origin for $n \le -1$. The Laplace transform $\mathcal{L}\{t^n\}$ does not exist for $n \le -1$.

Since, for two functions $F(t)$ and $G(t)$, for which the integrals exist

$$\mathcal{L}\{aF(t) + bG(t)\} = a\mathcal{L}\{F(t)\} + b\mathcal{L}\{G(t)\}, \tag{15.101}$$

the operation denoted by $\mathcal{L}$ is *linear*.

ELEMENTARY FUNCTIONS

To introduce the Laplace transform, let us apply the operation to some of the elementary functions. In all cases we assume that $F(t) = 0$ for $t < 0$.

$$F(t) = 1, \quad t > 0.$$

Then

$$\mathcal{L}\{1\} = \int_0^\infty e^{-st}\, dt = \frac{1}{s}, \quad \text{for } s > 0. \tag{15.102}$$

Again, let

$$F(t) = e^{kt}, \quad t > 0.$$

The Laplace transform becomes

$$\mathcal{L}\{e^{kt}\} = \int_0^\infty e^{-st} e^{kt}\, dt = \frac{1}{s - k}, \quad \text{for } s > k. \tag{15.103}$$

Using this relation, we may easily obtain the Laplace transform of certain other functions. Since

$$\cosh kt = \tfrac{1}{2}(e^{kt} + e^{-kt}),$$
$$\sinh kt = \tfrac{1}{2}(e^{kt} - e^{-kt}), \tag{15.104}$$

we have

$$\mathcal{L}\{\cosh kt\} = \frac{1}{2}\left(\frac{1}{s-k} + \frac{1}{s+k}\right) = \frac{s}{s^2 - k^2},$$

$$\mathcal{L}\{\sinh kt\} = \frac{1}{2}\left(\frac{1}{s-k} - \frac{1}{s+k}\right) = \frac{k}{s^2 - k^2},$$

(15.105)

both valid for $s > k$. We have the relations

$$\cos kt = \cosh ikt,$$

$$\sin kt = -i \sinh ikt.$$

(15.106)

Using Eqs. (15.105) with k replaced by ik, we find that the Laplace transforms are

$$\mathcal{L}\{\cos kt\} = \frac{s}{s^2 + k^2},$$

$$\mathcal{L}\{\sin kt\} = \frac{k}{s^2 + k^2},$$

(15.107)

both valid for $s > 0$. Another derivation of this last transform is given in the next section. Note that $\lim_{s \to 0} \mathcal{L}\{\sin kt\} = 1/k$. The Laplace transform assigns a value of $1/k$ to $\int_0^\infty \sin kt \, dt$.

Finally, for $F(t) = t^n$, we have

$$\mathcal{L}\{t^n\} = \int_0^\infty e^{-st} t^n \, dt,$$

which is just the factorial function. Hence

$$\mathcal{L}\{t^n\} = \frac{n!}{s^{n+1}}, \qquad s > 0, n > -1.$$

(15.108)

The reader will note that in all these transforms we have the variable s in the denominator—negative powers of s. In particular, $\lim_{s \to \infty} f(s) = 0$. The significance of this point is that if $f(s)$ involves positive powers of $s(\lim_{s \to \infty} f(s) \to \infty)$, then no inverse transform exists.

INVERSE TRANSFORM

There is little importance to these operations unless we can carry out the inverse transform, as in Fourier transforms. That is, with

$$\mathcal{L}\{F(t)\} = f(s),$$

then

$$\mathcal{L}^{-1}\{f(s)\} = F(t).$$

(15.109)

Taken literally, this inverse transform is *not* unique. Two functions $F_1(t)$ and $F_2(t)$ may have the same transform, $f(s)$. However, in this case

$$F_1(t) - F_2(t) = N(t),$$

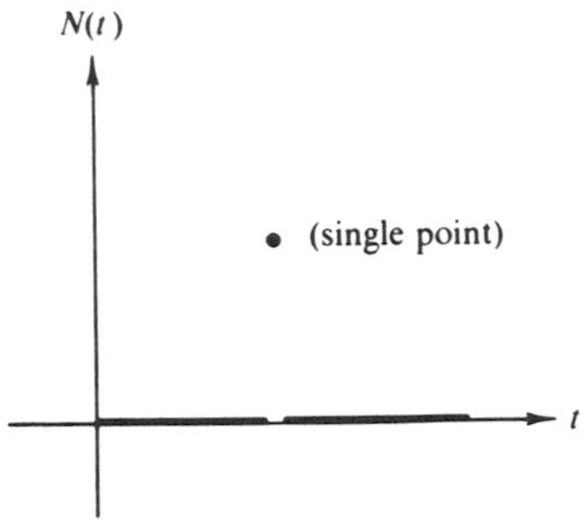

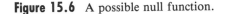

$N(t)$

• (single point)

t

Figure 15.6 A possible null function.

where $N(t)$ is a null function (Fig. 15.6) indicating that

$$\int_0^{t_0} N(t)\, dt = 0.$$

for all positive t_0. This result is known as Lerch's theorem. Therefore to the physicist and engineer $N(t)$ may almost always be taken as zero and the inverse operation becomes unique.

The inverse transform can be determined in various ways. (1) A table of transforms can be built-up and used to carry out the inverse transformation exactly as a table of logarithms can be used to look up antilogarithms. The preceding transforms constitute the embryonic beginnings of such a table. For a more complete set of Laplace transforms see Table 15.2 or AMS-55, Chapter 29. Employing partial fraction expansions and various operational theorems, which are considered in succeeding sections, may facilitate use of the tables. There is some justification for suspecting that these tables are probably of more value in solving textbook exercises than in solving real-world problems. (2) A general technique for $\mathcal{L}^{-1}$ will be developed in Section 15.12 by using the calculus of residues. (3) The difficulties and the possibilities of a numerical approach—numerical inversion—are considered at the end of this section.

PARTIAL FRACTION EXPANSION

Utilization of a table of transforms (or inverse transforms) is facilitated by expanding $f(s)$ in *partial fractions*.

Frequently $f(s)$, our transform, occurs in the form $g(s)/h(s)$, where $g(s)$ and $h(s)$ are polynomials with no common factors, $g(s)$ being of lower degree than $h(s)$. If the factors of $h(s)$ are all linear and distinct, then by the theory of partial fractions we may write

$$f(s) = \frac{c_1}{s - a_1} + \frac{c_2}{s - a_2} + \cdots + \frac{c_n}{s - a_n}, \tag{15.110}$$

where the c_i's are independent of s. The a_i's are the roots of $h(s)$. If any one of the roots, say a_1, is multiple (occurring m times), then $f(s)$ has the form

$$f(s) = \frac{c_{1,m}}{(s-a_1)^m} + \frac{c_{1,m-1}}{(s-a_1)^{m-1}} + \cdots + \frac{c_{1,1}}{s-a_1} + \sum_{i=2}^n \frac{c_i}{s-a_i}. \quad (15.111)$$

Finally, if one of the factors is quadratic, $(s^2 + ps + q)$, the numerator, instead of being a simple constant, will have the form

$$\frac{as + b}{s^2 + ps + q}.$$

There are various ways of determining the constants introduced. For instance, in Eq. (15.110) we may multiply through by $(s - a_i)$ and obtain

$$c_i = \lim_{s \to a_i} (s - a_i)f(s). \quad (15.112)$$

In elementary cases a direct solution is often the easiest.

Example 15.8.1 Partial Fraction Expansion
Let

$$f(s) = \frac{k^2}{s(s^2 + k^2)} = \frac{c}{s} + \frac{as + b}{s^2 + k^2}. \quad (15.113)$$

Putting the right side of the equation over a common denominator and equating like powers of s in the numerator, we obtain

$$\frac{k^2}{s(s^2 + k^2)} = \frac{c(s^2 + k^2) + s(as + b)}{s(s^2 + k^2)}, \quad (15.114)$$

$$c + a = 0, \qquad s^2,$$
$$b = 0, \qquad s^1,$$

and

$$ck^2 = k^2, \qquad s^0.$$

Solving these ($s \neq 0$), we have

$$c = 1,$$
$$b = 0,$$
$$a = -1,$$

giving

$$f(s) = \frac{1}{s} - \frac{s}{s^2 + k^2}, \quad (15.115)$$

and

$$\mathcal{L}^{-1}\{f(s)\} = 1 - \cos kt \quad (15.116)$$

by Eqs. (15.102) and (15.107).

Example 15.8.2 A Step Function

As one application of Laplace transforms, consider the evaluation of

$$F(t) = \int_0^\infty \frac{\sin tx}{x} dx. \tag{15.117}$$

Suppose we take the Laplace transform of this definite (and improper) integral:

$$\mathcal{L}\left\{ \int_0^\infty \frac{\sin tx}{x} dx \right\} = \int_0^\infty e^{-st} \int_0^\infty \frac{\sin tx}{x} dx \, dt. \tag{15.118}$$

Now interchanging the order of integration (which must be justified!),[2] we get

$$\int_0^\infty \frac{1}{x} \left[\int_0^\infty e^{-st} \sin tx \, dt \right] dx = \int_0^\infty \frac{dx}{s^2 + x^2}, \tag{15.119}$$

since the factor in square brackets is just the Laplace transform of $\sin tx$. From the integral tables

$$\int_0^\infty \frac{dx}{s^2 + x^2} = \frac{1}{s} \tan^{-1}\left(\frac{x}{s}\right)\Big|_0^\infty = \frac{\pi}{2s} = f(s). \tag{15.120}$$

By Eq. (15.102) we carry out the inverse transformation to obtain

$$F(t) = \frac{\pi}{2}, \qquad t > 0, \tag{15.121}$$

in agreement with an evaluation by the calculus of residues (Section 7.2). It has been assumed that $t > 0$ in $F(t)$. For $F(-t)$ we need note only that $\sin(-tx) = -\sin tx$, giving $F(-t) = -F(t)$. Finally, if $t = 0$, $F(0)$ is clearly zero. Therefore

$$\int_0^\infty \frac{\sin tx}{x} dx = \begin{cases} \dfrac{\pi}{2}, & t > 0 \\[2mm] 0, & t = 0 \\[2mm] -\dfrac{\pi}{2}, & t < 0. \end{cases} \tag{15.122}$$

$$= \frac{\pi}{2} [2u(t) - 1].$$

Note that $\int_0^\infty (\sin tx/x) \, dx$, taken as a function of t, describes a step function (Fig. 15.7), a step of height π at $t = 0$. This is consistent with Eq. (1.174).

The technique in the preceding example was to (1) introduce a second integration—the Laplace transform, (2) reverse the order of integration and integrate, and (3) take the inverse Laplace transform. There are many opportunities where this technique of reversing the order of integration can be applied and proved very useful. Exercise 15.8.6 is a variation of this.

---

[2] See Jeffreys and Jeffreys (1966), Chapter 1 (uniform convergence of integrals).

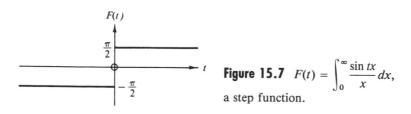

Figure 15.7 $F(t) = \int_0^\infty \dfrac{\sin tx}{x}\,dx$, a step function.

NUMERICAL INVERSION

As an integration the Laplace transform is a highly stable operation—stable in the sense that small fluctuations (or errors) in $F(t)$ are averaged out in the determination of the area under a curve. Also, the weighting factor, e^{-st}, means that the behavior of $F(t)$ at large t is effectively ignored—unless s is small. As a result of these two effects, a large change in $F(t)$ at large t indicates a very small, perhaps insignificant change, in $f(s)$. In contrast to the Laplace transform operation, going from $f(s)$ to $F(t)$, is highly unstable. A tiny change in $f(s)$ may result in a wild variation of $F(t)$. All significant figures may disappear. In a matrix formulation the matrix is ill-conditioned with respect to inversion.

There is no general, completely satisfactory numerical method for inverting Laplace transforms. However, if we are willing to restrict attention to relatively smooth functions, various possibilities open up. Bellman, Kalaba, and Lockett[3] convert the Laplace transform to a Mellin transform ($x = e^{-t}$) and use numerical quadrature based on shifted Legendre polynomials, $P_n^*(x) = P_n(1 - 2x)$. The key step is analytic inversion of the resulting matrix. Kyrlov and Skoblya[4] focus on evaluation of the Bromwich integral (Section 15.12). As one technique, they replace the integrand with an interpolating polynomial of negative powers and integrate analytically.

EXERCISES

15.8.1   Prove that
$$\lim_{s \to \infty} sf(s) = \lim_{t \to +0} F(t).$$
Hint. Assume that $F(t)$ can be expressed as $F(t) = \sum_{n=0}^\infty a_n t^n$.

15.8.2   Show that
$$\frac{1}{\pi} \lim_{s \to 0} \mathcal{L}\{\cos xt\} = \delta(x).$$

---

[3] R. Bellman, R. E. Kalaba, and J. A. Lockett, *Numerical Inversion of the Laplace Transforms*. New York: American Elsevier (1966).

[4] V. I. Krylov and N. S. Skoblya, *Handbook of Numerical Inversion of Laplace Transforms* (translated by D. Louvish). Jerusalem: Israel Program for Scientific Translations (1969).

15.8.3   Verify that

$$\mathcal{L}\left\{\frac{\cos at - \cos bt}{b^2 - a^2}\right\} = \frac{s}{(s^2 + a^2)(s^2 + b^2)}, \qquad a^2 \neq b^2.$$

15.8.4   Using partial fraction expansions, show that

(a) $\mathcal{L}^{-1}\left\{\dfrac{1}{(s + a)(s + b)}\right\} = \dfrac{e^{-at} - e^{-bt}}{b - a}, \qquad a \neq b.$

(b) $\mathcal{L}^{-1}\left\{\dfrac{s}{(s + a)(s + b)}\right\} = \dfrac{ae^{-at} - be^{-bt}}{a - b}, \qquad a \neq b.$

15.8.5   Using partial fraction expansions, show that

(a) $\mathcal{L}^{-1}\left\{\dfrac{1}{(s^2 + a^2)(s^2 + b^2)}\right\} = -\dfrac{1}{a^2 - b^2}\left(\dfrac{\sin at}{a} - \dfrac{\sin bt}{b}\right), \qquad a^2 \neq b^2,$

(b) $\mathcal{L}^{-1}\left\{\dfrac{s^2}{(s^2 + a^2)(s^2 + b^2)}\right\} = \dfrac{1}{a^2 - b^2}\{a \sin at - b \sin bt\}, \qquad a^2 \neq b^2.$

15.8.6   The electrostatic potential of a charged conducting disk is known to have the general form (circular cylindrical coordinates)

$$\Phi(\rho, z) = \int_0^\infty e^{-k|z|} J_0(k\rho) f(k)\, dk,$$

with $f(k)$ unknown. At large distances ($z \to \infty$) the potential must approach the Coulomb potential $Q/4\pi\varepsilon_0 z$. Show that

$$\lim_{k \to 0} f(k) = \frac{q}{4\pi\varepsilon_0}.$$

Hint. You may set $\rho = 0$ and assume a Maclaurin expansion of $f(k)$ or, using e^{-kz}, construct a delta sequence.

15.8.7   Show that

(a) $\displaystyle\int_0^\infty \frac{\cos s}{s^\nu}\, ds = \frac{\pi}{2(\nu - 1)! \cos(\nu\pi/2)}, \qquad 0 < \nu < 1,$

(b) $\displaystyle\int_0^\infty \frac{\sin s}{s^\nu}\, ds = \frac{\pi}{2(\nu - 1)! \sin(\nu\pi/2)}, \qquad 0 < \nu < 2.$

Why is ν restricted to (0, 1) for (a), to (0, 2) for (b)? These integrals may be interpreted as Fourier transforms of $s^{-\nu}$ and as Mellin transforms of $\sin s$ and $\cos s$.

Hint. Replace $s^{-\nu}$ by a Laplace transform integral: $\mathcal{L}\{t^{\nu-1}\}/(\nu - 1)!$. Then integrate with respect to s. The resulting integral can be treated as a beta function (Section 10.4).

15.8.8   A function $F(t)$ can be expanded in a power series (Maclaurin); that is,

$$F(t) = \sum_{n=0}^\infty a_n t^n.$$

Then

$$\mathcal{L}\{F(t)\} = \int_0^\infty e^{-st} \sum_n a_n t^n\, dt$$

$$= \sum_n a_n \int_0^\infty e^{-st} t^n\, dt.$$

Show that $f(s)$, the Laplace transform of $F(t)$, contains no powers of s greater than s^{-1}. Check your result by calculating $\mathcal{L}\{\delta(t)\}$ and comment intelligently on this fiasco.

15.9 LAPLACE TRANSFORM OF DERIVATIVES

Perhaps the main application of Laplace transforms is in converting differential equations into simpler forms that may be solved more easily. It will be seen, for instance, that coupled differential equations with constant coefficients transform to simultaneous linear algebraic equations.

Let us transform the first derivative of $F(t)$:

$$\mathcal{L}\{F'(t)\} = \int_0^\infty e^{-st}\frac{dF(t)}{dt}\,dt.$$

Integrating by parts, we obtain

$$\mathcal{L}\{F'(t)\} = e^{-st}F(t)\Big|_0^\infty + s\int_0^\infty e^{-st}F(t)\,dt$$

$$= s\mathcal{L}\{F(t)\} - F(0). \tag{15.123}$$

Strictly speaking, $F(0) = F(+0)$[1] and dF/dt is required to be at least piecewise continuous for $0 \le t < \infty$. Naturally, both $F(t)$ and its derivative must be such that the integrals do not diverge. Incidentally, Eq. (15.123) provides another proof of Exercise 15.8.1.

An extension gives

$$\mathcal{L}\{F^{(2)}(t)\} = s^2\mathcal{L}\{F(t)\} - sF(+0) - F'(+0), \tag{15.124}$$

$$\mathcal{L}\{F^{(n)}(t)\} = s^n\mathcal{L}\{F(t)\} - s^{n-1}F(+0) - s^{n-2}F'(+0) - \cdots - F^{(n-1)}(+0) \tag{15.125}$$

The Laplace transform like the Fourier transform replaces differentiation with multiplication. In the following examples differential equations become algebraic equations. Here is the power and the utility of the Laplace transform. But see Example 15.10.3 for what may happen if the coefficients are not constant.

Note carefully how the initial conditions, $F(+0)$, $F'(+0)$, and so on, are incorporated into the transform. Equation (15.124) may be used to derive $\mathcal{L}\{\sin kt\}$. We use the identity

$$-k^2\sin kt = \frac{d^2}{dt^2}\sin kt. \tag{15.126}$$

[1] Zero is approached from the *positive* side.

Then applying the Laplace transform operation, we have

$$-k^2 \mathcal{L}\{\sin kt\} = \mathcal{L}\left\{\frac{d^2}{dt^2} \sin kt\right\}$$

$$= s^2 \mathcal{L}\{\sin kt\} - s \sin(0) - \frac{d}{dt} \sin kt\big|_{t=0}. \qquad (15.127)$$

Since $\sin(0) = 0$ and $d/dt \sin kt\big|_{t=0} = k$,

$$\mathcal{L}\{\sin kt\} = \frac{k}{s^2 + k^2}, \qquad (15.128)$$

verifying Eq. (15.107).

Example 15.9.1 Simple Harmonic Oscillator

As a simple but reasonably physical example, consider a mass m oscillating under the influence of an ideal spring, spring constant k. As usual, friction is neglected. Then Newton's second law becomes

$$m\frac{d^2 X(t)}{dt^2} + kX(t) = 0; \qquad (15.129)$$

also

$$X(0) = X_0,$$

$$X'(0) = 0.$$

Applying the Laplace transform, we obtain

$$m\mathcal{L}\left\{\frac{d^2 X}{dt^2}\right\} + k\mathcal{L}\{X(t)\} = 0, \qquad (15.130)$$

and by use of Eq. (15.124) this becomes

$$ms^2 x(s) - msX_0 + k\,x(s) = 0, \qquad (15.131)$$

$$x(s) = X_0 \frac{s}{s^2 + \omega_0^2}, \qquad \text{with } \omega_0^2 \equiv \frac{k}{m}. \qquad (15.132)$$

From Eq. (15.107) this is seen to be the transform of $\cos \omega_0 t$, which gives

$$X(t) = X_0 \cos \omega_0 t, \qquad (15.133)$$

as expected.

Example 15.9.2 Earth's Nutation

A somewhat more involved example is proved by the nutation of the earth's poles (force-free precession). Treating the earth as a rigid (oblate) spheroid, the Euler equations of motion reduce to

$$\frac{dX}{dt} = -aY$$

$$\frac{dY}{dt} = +aX,$$

(15.134)

where $a \equiv [(I_z - I_x)/I_z]\omega_z$,

$X = \omega_x$,

$Y = \omega_y$ with angular velocity vector $\omega = (\omega_x, \omega_y, \omega_z)$ (Fig. 15.8),

I_z = moment of inertial about the z-axis and $I_y = I_x$ moment of inertia about x- (or y-) axis.

The z-axis coincides with the axis of symmetry of the earth. It differs from the axis for the earth's daily rotation, ω, by some 15 meters, measured at the poles. Transformation of these coupled differential equations yields

$$s\,x(s) - X(0) = -a\,y(s),$$

$$s\,y(s) - Y(0) = a\,x(s).$$

(15.135)

Combining to eliminate $y(s)$, we have

$$s^2 x(s) - sX(0) + aY(0) = -a^2 x(s)$$

or

$$x(s) = X(0)\frac{s}{s^2 + a^2} - Y(0)\frac{a}{s^2 + a^2}.$$

(15.136)

Hence

$$X(t) = X(0)\cos at - Y(0)\sin at.$$

(15.137)

Similarly,

$$Y(t) = X(0)\sin at + Y(0)\cos at.$$

(15.138)

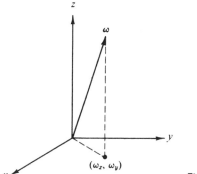

Figure 15.8

This is seen to be a rotation of the vector (X, Y) counterclockwise (for $a > 0$) about the z-axis with angle $\theta = at$ and angular velocity a.

A direct interpretation may be found by choosing the time axis so that $Y(0) = 0$. Then

$$X(t) = X(0) \cos at,$$
$$Y(t) = X(0) \sin at,$$
(15.139)

which are the parametric equations for rotation of (X, Y) in a circular orbit of radius $X(0)$, with angular velocity a in the counterclockwise sense.

In the case of the earth's angular velocity vector $X(0)$ is about 15 meters, whereas a, as defined here, corresponds to a period $(2\pi/a)$ of some 300 days. Actually because of departures from the idealized rigid body assumed in setting up Euler's equations, the period is about 427 days.[2]

If in Eq. (15.134) we set

$$X(t) = L_x,$$
$$Y(t) = L_y,$$

where L_x and L_y = the x- and y-components of the angular momentum $\mathbf{L}$,
$\quad a = -g_L B_z,$
$\quad g_L$ = gyromagnetic ratio,
$\quad B_z$ = magnetic field (along the z-axis),

Eq. (15.134) describes the Larmor precession of charged bodies in a uniform magnetic field, B_z.

DIRAC DELTA FUNCTION

For use with differential equations one further transform is helpful—the Dirac delta function:[3]

$$\mathcal{L}\{\delta(t - t_0)\} = \int_0^\infty e^{-st}\delta(t - t_0)\, dt = e^{-st_0}, \qquad \text{for } t_0 \geq 0, \qquad (15.140)$$

and for $t_0 = 0$

$$\mathcal{L}\{\delta(t)\} = 1, \qquad (15.141)$$

where it is assumed that we are using a representation of the delta function such that

$$\int_0^\infty \delta(t)\, dt = 1, \qquad \delta(t) = 0, \qquad \text{for } t > 0. \qquad (15.142)$$

[2] D. Menzel, ed., *Fundamental Formulas of Physics.* Englewood Cliffs, NJ: Prentice–Hall (1955), p. 695.
[3] Strictly speaking, the Dirac delta function is undefined. However, the integral over it is well defined. This approach is developed in Section 1.16 using delta sequences.

As an alternate method, $\delta(t)$ may be considered the limit as $\varepsilon \to 0$ of $F(t)$, where

$$F(t) = \begin{cases} 0, & t < 0, \\ \varepsilon^{-1}, & 0 < t < \varepsilon, \\ 0, & t > \varepsilon. \end{cases} \tag{15.143}$$

By direct calculation

$$\mathcal{L}\{F(t)\} = \frac{1 - e^{-\varepsilon s}}{\varepsilon s}. \tag{15.144}$$

Taking the limit of the integral (instead of the integral of the limit), we have

$$\lim_{\varepsilon \to 0} \mathcal{L}\{F(t)\} = 1$$

or Eq. (15.141)

$$\mathcal{L}\{\delta(t)\} = 1.$$

This delta function is frequently called the impulse function because it is so useful in describing impulsive forces, that is, forces lasting only a short time.

Example 15.9.3 Impulsive Force

Newton's second law for impulsive force acting on a particle of mass m becomes

$$m \frac{d^2 X}{dt^2} = P \delta(t), \tag{15.145}$$

where P is a constant.

Transforming, we obtain

$$ms^2 x(s) - msX(0) - mX'(0) = P. \tag{15.146}$$

For a particle starting from rest $X'(0) = 0.$[4] We shall also take $X(0) = 0$. Then

$$x(s) = \frac{P}{ms^2}, \tag{15.147}$$

and

$$X(t) = \frac{P}{m} t, \tag{15.148}$$

$$\frac{dX(t)}{dt} = \frac{P}{m}, \quad \text{a constant.} \tag{15.149}$$

The effect of the impulse $P \delta(t)$ is to transfer (instantaneously) P units of linear momentum to the particle.

---

[4] This really should be $X'(+0)$. To include the effect of the impulse, consider that the impulse will occur at $t = \varepsilon$ and let $\varepsilon \to 0$.

A similar analysis applies to the ballistic galvanometer. The torque on the galvanometer is given initially by ki, in which i is a pulse of current and k is a proportionality constant. Since i is of short duration, we set

$$ki = kq\,\delta(t), \qquad\qquad (15.150)$$

where q is the total charge carried by the current i. Then, with I the moment of inertia,

$$I\frac{d^2\theta}{dt^2} = kq\,\delta(t), \qquad\qquad (15.151)$$

and transforming as before, we find that the effect of the current pulse is a transfer of kq units of *angular* momentum to the galvanometer.

EXERCISES

15.9.1   Use the expression for the transform of a second derivative to obtain the transform of $\cos kt$.

15.9.2   A mass m is attached to one end of an unstretched spring, spring constant k. At time $t = 0$ the free end of the spring experiences a constant acceleration a, away from the mass. Using Laplace transforms,

(a) Find the position x of m as a function of time.
(b) Determine the limiting form of $x(t)$ for small t.

$$ANS. \quad (a)\ x = \frac{1}{2}at^2 - \frac{a}{\omega^2}(1 - \cos\omega t)$$

$$\omega^2 = \frac{k}{m},$$

$$(b)\ x = \frac{a\omega^2}{4!}t^4, \qquad \omega t \ll 1.$$

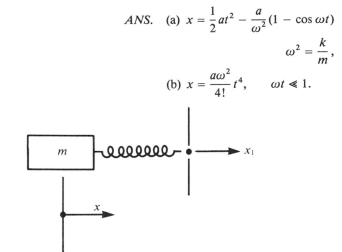

15.9.3   Radioactive nuclei decay according to the law

$$\frac{dN}{dt} = -\lambda N,$$

N being the concentration of a given nuclide and λ, the particular decay constant. This equation may be interpreted as stating that the rate of decay is proportional to the number of these radioactive nuclei present. They all decay independently.

In a radioactive series of n different nuclides, starting with N_1,

$$\frac{dN_1}{dt} = -\lambda_1 N_1,$$

$$\frac{dN_2}{dt} = \lambda_1 N_1 - \lambda_2 N_2, \qquad \text{and so on}$$

$$\frac{dN_n}{dt} = \lambda_{n-1} N_{n-1}, \qquad \text{stable.}$$

Find $N_1(t)$, $N_2(t)$, and $N_3(t)$, $n = 3$, with $N_1(0) = N_0$, $N_2(0) = N_3(0) = 0$.

ANS. $N_1(t) = N_0 e^{-\lambda_1 t}$,

$$N_2(t) = N_0 \frac{\lambda_1}{\lambda_2 - \lambda_1} (e^{-\lambda_1 t} - e^{-\lambda_2 t}),$$

$$N_3(t) = N_0 \left(1 - \frac{\lambda_2}{\lambda_2 - \lambda_1} e^{-\lambda_1 t} + \frac{\lambda_1}{\lambda_2 - \lambda_1} e^{-\lambda_2 t} \right).$$

Find an approximate expression for N_2 and N_3, valid for small t when $\lambda_1 \approx \lambda_2$.

ANS. $N_2 \approx N_0 \lambda_1 t$

$$N_3 \approx \frac{N_0}{2} \lambda_1 \lambda_2 t^2.$$

Find approximate expressions for N_2 and N_3, valid for large t, when

(a) $\lambda_1 \gg \lambda_2$,   ANS.   (a) $N_2 \approx N_0 e^{-\lambda_2 t}$

(b) $\lambda_1 \ll \lambda_2$. $N_3 \approx N_0 (1 - e^{-\lambda_2 t})$, $\lambda_1 t \gg 1$.

$$\text{(b)} \ N_2 \approx N_0 \frac{\lambda_1}{\lambda_2} e^{-\lambda_1 t},$$

$$N_3 \approx N_0 (1 - e^{-\lambda_1 t}), \qquad \lambda_2 t \gg 1.$$

15.9.4   The formation of an isotope in a nuclear reactor is given by

$$\frac{dN_2}{dt} = nv\sigma_1 N_{10} - \lambda_2 N_2(t) - nv\sigma_2 N_2(t).$$

Here the product nv is the neutron flux, neutrons per cubic centimeter, times centimeters per second mean velocity; σ_1 and σ_2 (cm²) are measures of the probability of neutron absorption by the original isotope, concentration N_{10}, which is assumed constant and the newly formed isotope, concentration N_2, respectively. The radioactive decay constant for the isotope is λ_2.

(a) Find the concentration N_2 of the new isotope as a function of time.
(b) If the original element is Eu^{153}, $\sigma_1 = 400$ barns $= 400 \times 10^{-24} \, \mathrm{cm}^2$, $\sigma_2 = 1000$ barns $= 1000 \times 10^{-24} \, \mathrm{cm}^2$, and $\lambda_2 = 1.4 \times 10^{-9} \, \mathrm{sec}^{-1}$. If $N_{10} = 10^{20}$ and $(nv) = 10^9 \, \mathrm{cm}^{-2} \, \mathrm{sec}^{-1}$, find N_2, the concentration of Eu^{154} after one year of continuous irradiation. Is the assumption that N_1 is constant justified?

15.9.5   In a nuclear reactor Xe^{135} is formed as both a direct fission product and a decay product of I^{135}, half-life, 6.7 hours. The half-life of Xe^{135} is 9.2 hours.

As Xe^{135} strongly absorbs thermal neutrons thereby "poisoning" the nuclear reactor, its concentration is a matter of great interest. The relevant equations are

$$\frac{dN_I}{dt} = \gamma_I \varphi \sigma_f N_U - \lambda_I N_I,$$

$$\frac{dN_X}{dt} = \lambda_I N_I + \gamma_X \varphi \sigma_f N_U - \lambda_X N_X - \varphi \sigma_X N_X.$$

Here N_I = concentration of I^{135} (Xe^{135}, U^{235}). Assume N_U = constant.
γ_I = yield of I^{135} per fission = 0.060,
γ_X = yield of Xe^{135} direct from fission = 0.003,
$\lambda_I = I^{135}$ (Xe^{135}) decay constant = $\dfrac{\ln 2}{t_{1/2}} = \dfrac{0.693}{t_{1/2}}$,
σ_f = thermal neutron fission cross section for U^{235},
σ_X = thermal neutron absorption cross section for Xe^{135} = 3.5×10^6 barns,
$$= 3.5 \times 10^{-18} \text{ cm}^2.$$
(σ_I, the absorption cross section of I^{135} is negligible.)
φ = neutron flux = neutrons/cm^3 × mean velocity (cm/sec).

(a) Find $N_X(t)$ in terms of neutron flux φ and the product $\sigma_f N_U$.
(b) Find $N_X(t \to \infty)$.
(c) After N_X has reached equilibrium, the reactor is shut down, $\varphi = 0$. Find $N_X(t)$ following shut down. Notice the increase in N_X, which may for a few hours interfere with starting the reactor up again.

15.10 OTHER PROPERTIES

SUBSTITUTION

If we replace the parameter s by $s - a$ in the definition of the Laplace transform (Eq. (15.99)), we have

$$f(s - a) = \int_0^\infty e^{-(s-a)t} F(t) \, dt = \int_0^\infty e^{-st} e^{at} F(t) \, dt$$

$$= \mathcal{L}\{e^{at} F(t)\}. \tag{15.152}$$

Hence the replacement of s with $s - a$ corresponds to multiplying $F(t)$ by e^{at} and conversely. This result can be used to good advantage in extending our table of transforms. From Eq. (15.107) we find immediately that

$$\mathcal{L}\{e^{at} \sin kt\} = \frac{k}{(s - a)^2 + k^2}; \tag{15.153}$$

also

$$\mathcal{L}\{e^{at} \cos kt\} = \frac{s - a}{(s - a)^2 + k^2}, \qquad s > a.$$

Example 15.10.1 Damped Oscillator

These expressions are useful when we consider an oscillating mass with damping proportional to the velocity. Equation (15.129), with such damping added, becomes

$$mX''(t) + bX'(t) + kX(t) = 0, \tag{15.154}$$

in which b is a proportionality constant. Let us assume that the particle starts from rest at $X(0) = X_0$, $X'(0) = 0$. The transformed equation is

$$m[s^2 x(s) - sX_0] + b[s\, x(s) - X_0] + k\, x(s) = 0 \tag{15.155}$$

and

$$x(s) = X_0 \frac{ms + b}{ms^2 + bs + k}. \tag{15.156}$$

This may be handled by completing the square of the denominator,

$$s^2 + \frac{b}{m} s + \frac{k}{m} = \left(s + \frac{b}{2m}\right)^2 + \left(\frac{k}{m} - \frac{b^2}{4m^2}\right). \tag{15.157}$$

If the damping is small, $b^2 < 4\,\mathrm{km}$, the last term is positive and will be denoted by ω_1^2.

$$x(s) = X_0 \frac{s + b/m}{(s + b/2m)^2 + \omega_1^2}$$

$$= X_0 \frac{s + b/2m}{(s + b/2m)^2 + \omega_1^2} + X_0 \frac{(b/2m\omega_1)\omega_1}{(s + b/2m)^2 + \omega_1^2}. \tag{15.158}$$

By Eq. (15.153)

$$X(t) = X_0 e^{-(b/2m)t}\left(\cos \omega_1 t + \frac{b}{2m\omega_1} \sin \omega_1 t\right)$$

$$= X_0 \frac{\omega_0}{\omega_1} e^{-(b/2m)t} \cos(\omega_1 t - \varphi), \tag{15.159}$$

where

$$\tan \varphi = \frac{b}{2m\omega_1},$$

$$\omega_0^2 = \frac{k}{m}.$$

Of course, as $b \to 0$, this solution goes over to the undamped solution, (Section 15.9).

RLC ANALOG

It is worth noting the similarity between this damped simple harmonic oscillation of a mass on a spring and an *RLC* circuit (resistance, inductance,

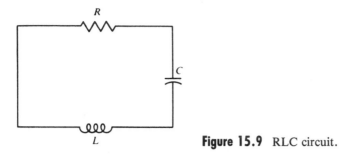

Figure 15.9 RLC circuit.

and capacitance) (Fig. 15.9). At any instant the sum of the potential differences around the loop must be zero (Kirchhoff's law, conservation of energy). This gives

$$L\frac{dI}{dt} + RI + \frac{1}{C}\int^t I\,dt = 0. \tag{15.160}$$

Differentiating the current I with respect to time (to eliminate the integral), we have

$$L\frac{d^2I}{dt^2} + R\frac{dI}{dt} + \frac{1}{C}I = 0. \tag{15.161}$$

If we replace $I(t)$ with $X(t)$, L with m, R with b, C^{-1} with k, Eq. (15.161) is identical with the mechanical problem. It is but one example of the unification of diverse branches of physics by mathematics. A more complete discussion will be found in Olson's book.[1]

TRANSLATION

This time let $f(s)$ be multiplied by e^{-bs}, $b > 0$.

$$e^{-bs}f(s) = e^{-bs}\int_0^\infty e^{-st}F(t)\,dt$$

$$= \int_0^\infty e^{-s(t+b)}F(t)\,dt. \tag{15.162}$$

Now let $t + b = \tau$. Equation (15.162) becomes

$$e^{-bs}f(s) = \int_0^\infty e^{-s\tau}F(\tau - b)\,d\tau$$

$$= \int_0^\infty e^{-s\tau}F(\tau - b)u(\tau - b)\,d\tau, \tag{15.163}$$

where $u(\tau - b)$ is the unit step function. This relation is often called the "Heaviside shifting theorem" (Fig. 15.10).

[1] H. F. Olson, *Dynamical Analogies*. New York: Van Nostrand (1943).

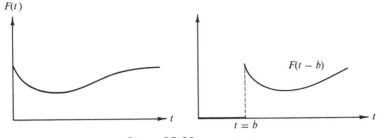

Figure 15.10 Translation.

Since $F(t)$ is assumed to be equal to zero for $t < 0$, $F(\tau - b) = 0$ for $0 \le \tau < b$. Therefore we can extend the lower limit to zero without changing the value of the integral. Then, noting that τ is only a variable of integration, we obtain

$$e^{-bs}f(s) = \mathcal{L}\{F(t - b)\}. \tag{15.164}$$

Example 15.10.2 Electromagnetic Waves

The electromagnetic wave equation with $E = E_y$ or E_z, a transverse wave propagating along the x-axis, is

$$\frac{\partial^2 E(x, t)}{\partial x^2} - \frac{1}{v^2} \frac{\partial^2 E(x, t)}{\partial t^2} = 0, \tag{15.165}$$

Transforming this equation with respect to t, we get

$$\frac{\partial^2}{\partial x^2} \mathcal{L}\{E(x, t)\} - \frac{s^2}{v^2} \mathcal{L}\{E(x, t)\} + \frac{s}{v^2} E(x, 0) + \frac{1}{v^2} \left. \frac{\partial E(x, t)}{\partial t} \right|_{t=0} = 0. \tag{15.166}$$

If we have the initial condition $E(x, 0) = 0$ and

$$\left. \frac{\partial E(x, t)}{\partial t} \right|_{t=0} = 0,$$

then

$$\frac{\partial^2}{\partial x^2} \mathcal{L}\{E(x, t)\} = \frac{s^2}{v^2} \mathcal{L}\{E(x, t)\}. \tag{15.167}$$

The solution (of this *ordinary* differential equation) is

$$\mathcal{L}\{E(x, t)\} = c_1 e^{-(s/v)x} + c_2 e^{+(s/v)x}. \tag{15.168}$$

The "constants" c_1 and c_2 are obtained by additional boundary conditions. They are constant with respect to x but may depend on s. If our wave remains finite as $x \to \infty$, $\mathcal{L}\{E(x, t)\}$ will also remain finite. Hence $c_2 = 0$.

If $E(0, t)$ is denoted by $F(t)$, then $c_1 = f(s)$ and

$$\mathcal{L}\{E(x, t)\} = e^{-(s/v)x}f(s). \tag{15.169}$$

From the translation property (Eq. (15.164)) we find immediately that

$$
E(x, t) = \begin{cases} F\left(t - \dfrac{x}{v}\right), & t \geq \dfrac{x}{v}, \\ 0, & t < \dfrac{x}{v}. \end{cases} \tag{15.170}
$$

Differentiation and substitution into Eq. (15.165) verifies Eq. (15.170). Our solution represents a wave (or pulse) moving in the positive x-direction with velocity v. Note that for $x > vt$ the region remains undisturbed; the pulse has not had time to get there. If we had wanted a signal propagated along the negative x-axis, c_1 would have been set equal to 0 and we would have obtained

$$
E(x, t) = \begin{cases} F\left(t + \dfrac{x}{v}\right), & t \geq -\dfrac{x}{v}, \\ 0, & t < -\dfrac{x}{v}, \end{cases} \tag{15.171}
$$

a wave along the negative x-axis.

DERIVATIVE OF A TRANSFORM

When $F(t)$, which is at least piecewise continuous, and s are chosen so that $e^{-st}F(t)$ converges exponentially for large s, the integral

$$
\int_0^\infty e^{-st}F(t)\, dt
$$

is uniformly convergent and may be differentiated (under the integral sign) with respect to s. Then

$$
f'(s) = \int_0^\infty (-t)e^{-st}F(t)\, dt = \mathcal{L}\{-t\,F(t)\}. \tag{15.172}
$$

Continuing this process, we obtain

$$
f^{(n)}(s) = \mathcal{L}\{(-t)^n F(t)\}. \tag{15.173}
$$

All the integrals so obtained will be uniformly convergent because of the decreasing exponential behavior of $e^{-st}F(t)$.

This same technique may be applied to generate more transforms. For example,

$$
\mathcal{L}\{e^{kt}\} = \int_0^\infty e^{-st}e^{kt}\, dt
$$

$$
= \frac{1}{s - k}, \qquad s > k. \tag{15.174}
$$

Differentiating with respect to s (or with respect to k), we obtain

$$\mathcal{L}\{te^{kt}\} = \frac{1}{(s-k)^2}, \qquad s > k. \tag{15.175}$$

Example 15.10.3 Bessel's Equation

An interesting application of a differentiated Laplace transform appears in the solution of Bessel's equation with $n = 0$. From Chapter 11 we have

$$x^2 y''(x) + xy'(x) + x^2 y(x) = 0. \tag{15.176}$$

Dividing by x and substituting $t = x$ and $F(t) = y(x)$ to agree with the present notation, we see that the Bessel equation becomes

$$tF''(t) + F'(t) + tF(t) = 0. \tag{15.177}$$

We need a regular solution, in particular, $F(0) = 1$. From Eq. (15.177) with $t = 0$, $F'(+0) = 0$. Also, we assume that our unknown $F(t)$ has a transform. Transforming and using Eqs. (15.123), (15.124) and (15.172), we have

$$-\frac{d}{ds}[s^2 f(s) - s] + s f(s) - 1 - \frac{d}{ds} f(s) = 0. \tag{15.178}$$

Rearranging Eq. (15.178), we obtain

$$(s^2 + 1)f'(s) + s f(s) = 0 \tag{15.179}$$

or

$$\frac{df}{f} = -\frac{s\,ds}{s^2 + 1}, \tag{15.180}$$

a first-order differential equation. By integration,

$$\ln f(s) = -\tfrac{1}{2}\ln(s^2 + 1) + \ln C, \tag{15.181}$$

which may be rewritten as

$$f(s) = \frac{C}{\sqrt{s^2 + 1}}. \tag{15.182}$$

To make use of Eq. (15.108), we expand $f(s)$ in a series of negative powers of s, convergent for $s > 1$:

$$\begin{aligned}
f(s) &= \frac{C}{s}\left(1 + \frac{1}{s^2}\right)^{-1/2} \\
&= \frac{C}{s}\left[1 - \frac{1}{2s^2} + \frac{1 \cdot 3}{2^2 \cdot 2! s^4} - \cdots + \frac{(-1)^n (2n)!}{(2^n n!)^2 s^{2n}} + \cdots\right].
\end{aligned} \tag{15.183}$$

Inverting, term by term, we obtain

$$F(t) = C \sum_{n=0}^{\infty} \frac{(-1)^n t^{2n}}{(2^n n!)^2}. \tag{15.184}$$

When C is set equal to 1, as required by the initial condition $F(0) = 1$, $F(t)$ is just $J_0(t)$, our familiar Bessel function of order zero. Hence

$$\mathcal{L}\{J_0(t)\} = \frac{1}{\sqrt{s^2 + 1}}. \tag{15.185}$$

Note that we assumed $s > 1$. The proof for $s > 0$ is left as a problem.

It is perhaps worth noting that this appliation was successful and relatively easy because we took $n = 0$ in Bessel's equation. This made it possible to divide out a factor of x (or t). If this had not been done, the terms of the form $t^2 F(t)$ would have introduced a second derivative of $f(s)$. The resulting equation would have been no easier to solve than the original one.

When we go beyond linear differential equations with constant coefficients, the Laplace transform may still be applied, but there is no guarantee that it will be helpful.

The application to Bessel's equation, $n \neq 0$, will be found in the references. Alternatively, we can show that

$$\mathcal{L}\{J_n(at)\} = \frac{a^{-n}(\sqrt{s^2 + a^2} - s)^n}{\sqrt{s^2 + a^2}} \tag{15.186}$$

by expressing $J_n(t)$ as an infinite series and transforming term by term.

INTEGRATION OF TRANSFORMS

Again, with $F(t)$ at least piecewise continuous and x large enough so that $e^{-xt}F(t)$ decreases exponentially (as $x \to \infty$), the integral

$$f(x) = \int_0^\infty e^{-xt}F(t)\,dt \tag{15.187}$$

is uniformly convergent with respect to x. This justifies reversing the order of integration in the following equation:

$$\int_s^b f(x)\,dx = \int_s^b \int_0^\infty e^{-xt}F(t)\,dt\,dx$$

$$= \int_0^\infty \frac{F(t)}{t}(e^{-st} - e^{-bt})\,dt, \tag{15.188}$$

on integrating with respect to x. The lower limit s is chosen large enough so that $f(s)$ is within the region of uniform convergence. Now letting $b \to \infty$, we have

$$\int_s^\infty f(x)\,dx = \int_0^\infty \frac{F(t)}{t}e^{-st}\,dt$$

$$= \mathcal{L}\left\{\frac{F(t)}{t}\right\}, \tag{15.189}$$

provided that $F(t)/t$ is finite at $t = 0$ or diverges less strongly than t^{-1} (so that $\mathcal{L}\{F(t)/t\}$ will exist).

LIMITS OF INTEGRATION—UNIT STEP FUNCTION

The actual limits of integration for the Laplace transform may be specified with the (Heaviside) unit step function

$$u(t - k) = \begin{cases} 0, & t < k \\ 1, & t > k. \end{cases}$$

For instance,

$$\mathcal{L}\{u(t - k)\} = \int_k^\infty e^{-st}\, dt$$

$$= \frac{1}{s} e^{-ks}.$$

A rectangular pulse of width k and unit height is described by $F(t) = u(t) - u(t - k)$. Taking the Laplace transform, we obtain

$$\mathcal{L}\{u(t) - u(t - k)\} = \int_0^k e^{-st}\, dt$$

$$= \frac{1}{s}(1 - e^{-ks}).$$

The unit step function is also used in Eq. (15.163) and could be invoked in Exercise 15.10.13.

EXERCISES

15.10.1   Solve Eq. 15.154, which describes a damped simple harmonic oscillator for $X(0) = X_0$, $X'(0) = 0$, and

(a) $b^2 = 4\,km$ (critically damped),
(b) $b^2 > 4\,km$ (overdamped).

$$ANS. \quad (a)\ X(t) = X_0 e^{-(b/2m)t}\left(1 + \frac{b}{2m}t\right).$$

15.10.2   Solve Eq. (15.154), which describes a damped simple harmonic oscillator for $X(0) = 0$, $X'(0) = v_0$, and

(a) $b^2 < 4\,km$ (underdamped),
(b) $b^2 = 4\,km$ (critically damped),

$$ANS. \quad (a)\ X(t) = \frac{v_0}{\omega_1} e^{-(b/2m)t} \sin \omega_1 t,$$

$$(b)\ X(t) = v_0 t e^{-(b/2m)t}.$$

(c) $b^2 > 4\,km$ (overdamped).

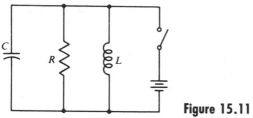

Figure 15.11 Ringing circuit.

15.10.3   The motion of a body falling in a resisting medium may be described by

$$m \frac{d^2X(t)}{dt^2} = mg - b \frac{dX(t)}{dt}$$

when the retarding force is proportional to the velocity. Find $X(t)$ and $dX(t)/dt$ for the initial conditions

$$X(0) = \frac{dX}{dt}\bigg|_{t=0} = 0.$$

15.10.4   Ringing circuit. In certain electronic circuits resistance, inductance, and capacitance are placed in the plate circuit in parallel (Fig. 15.11). A constant voltage is maintained across the parallel elements, keeping the capacitor charged. At time $t = 0$ the circuit is disconnected from the voltage source. Find the voltages across the parallel elements R, L, and C as a function of time. Assume R to be *large*.

Hint. By Kirchhoff's laws

$$I_R + I_C + I_L = 0 \qquad \text{and} \qquad E_R = E_C = E_L,$$

where

$$E_R = I_R R,$$

$$E_C = \frac{q_0}{C} + \frac{1}{C} \int_0^t I_C \, dt,$$

$$E_L = L \frac{dI_L}{dt},$$

$$q_0 = \text{initial charge of capacitor.}$$

With the DC impedance of $L = 0$, let $I_L(0) = I_0$, $E_L(0) = 0$. This means $q_0 = 0$.

15.10.5   With $J_0(t)$ expressed as a contour integral, apply the Laplace transform operation, reverse the order of integration, and thus show that

$$\mathcal{L}\{J_0(t)\} = (s^2 + 1)^{-1/2}, \qquad \text{for } s > 0.$$

15.10.6   Develop the Laplace transform of $J_n(t)$ from $\mathcal{L}\{J_0(t)\}$ by using the Bessel function recurrence relations.

Hint. Here is a chance to use mathematical induction.

15.10.7 A calculation of the magnetic field of a circular current loop in circular cylindrical coordinates leads to the integral

$$\int_0^\infty e^{-kz} k J_1(ka)\, dk, \qquad \Re(z) \geq 0.$$

Show that this integral is equal to $a/(z^2 + a^2)^{3/2}$.

15.10.8 The electrostatic potential of a point charge q at the origin in circular cylindrical coordinates is

$$\frac{q}{4\pi\varepsilon_0} \int_0^\infty e^{-kz} J_0(k\rho)\, dk = \frac{q}{4\pi\varepsilon_0} \cdot \frac{1}{(\rho^2 + z^2)^{1/2}}, \qquad \Re(z) \geq 0.$$

From this relation show that the Fourier cosine and sine transforms of $J_0(k\rho)$ are

(a) $\sqrt{\dfrac{\pi}{2}} F_c\{J_0(k\rho)\} = \displaystyle\int_0^\infty J_0(k\rho) \cos k\zeta\, dk = \begin{cases} (\rho^2 - \zeta^2)^{-1/2}, & \rho > \zeta, \\ 0, & \rho < \zeta. \end{cases}$

(b) $\sqrt{\dfrac{\pi}{2}} F_s\{J_0(k\rho)\} = \displaystyle\int_0^\infty J_0(k\rho) \sin k\zeta\, dk = \begin{cases} 0, & \rho > \zeta, \\ (\zeta^2 - \rho^2)^{-1/2}, & \rho < \zeta. \end{cases}$

Hint. Replace z by $z + i\zeta$ and take the limit as $z \to 0$.

15.10.9 Show that

$$\mathcal{L}\{I_0(at)\} = (s^2 - a^2)^{-1/2}, \qquad s > a.$$

15.10.10 Verify the following Laplace transforms:

(a) $\mathcal{L}\{j_0(at)\} = \mathcal{L}\left\{\dfrac{\sin at}{at}\right\} = \dfrac{1}{a}\cot^{-1}\left(\dfrac{s}{a}\right)$,

(b) $\mathcal{L}\{n_0(at)\}$ does not exist.

(c) $\mathcal{L}\{i_0(at)\} = \mathcal{L}\left\{\dfrac{\sinh at}{at}\right\} = \dfrac{1}{2a}\ln\dfrac{s+a}{s-a} = \dfrac{1}{a}\coth^{-1}\left(\dfrac{s}{a}\right)$,

(d) $\mathcal{L}\{k_0(at)\}$ does not exist.

15.10.11 Develop a Laplace transform solution of Laguerre's equation

$$tF''(t) + (1 - t)F'(t) + nF(t) = 0.$$

Note that you need a derivative of a transform and a transform of derivatives. Go as far as you can with $n = n$; then (and only then) set $n = 0$.

15.10.12 Show that the Laplace transform of the Laguerre polynomial $L_n(at)$ is given by

$$\mathcal{L}\{L_n(at)\} = \frac{(s-a)^n}{s^{n+1}}, \qquad s > 0.$$

15.10.13 Show that

$$\mathcal{L}\{E_1(t)\} = \frac{1}{s}\ln(s + 1), \qquad s > 0,$$

where

$$E_1(t) = \int_t^\infty \frac{e^{-\tau}\, d\tau}{\tau} = \int_1^\infty \frac{e^{-xt}}{x}\, dx.$$

$E_1(t)$ is the exponential-integral function.

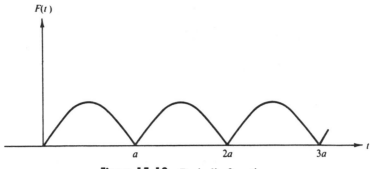

Figure 15.12 Periodic function.

15.10.14 (a) From Eq. (15.189) show that

$$\int_0^\infty f(x)\,dx = \int_0^\infty \frac{F(t)}{t}\,dt,$$

provided the integrals exist.

(b) From the preceding result show that

$$\int_0^\infty \frac{\sin t}{t}\,dt = \frac{\pi}{2}$$

in agreement with Eqs. (15.122) and (7.51).

15.10.15 (a) Show that

$$\mathcal{L}\left\{\frac{\sin kt}{t}\right\} = \cot^{-1}\left(\frac{s}{k}\right).$$

(b) Using this result (with $k = 1$), prove that

$$\mathcal{L}\{si(t)\} = -\frac{1}{s}\tan^{-1}s,$$

where

$$si(t) = -\int_t^\infty \frac{\sin x}{x}\,dx, \qquad \text{the sine integral.}$$

15.10.16 If $F(t)$ is periodic (Fig. 15.12) with a period a so that $F(t + a) = F(t)$ for all $t \ge 0$, show that

$$\mathcal{L}\{F(t)\} = \frac{\int_0^a e^{-st}F(t)\,dt}{1 - e^{-as}},$$

with the integration now over only the *first period* of $F(t)$.

15.10.17 Find the Laplace transform of the square wave (period a) defined by

$$F(t) = \begin{cases} 1, & 0 < t < a/2 \\ 0, & a/2 < t < a \end{cases}$$

$$ANS. \quad f(s) = \frac{1}{s}\cdot\frac{1 - e^{-as/2}}{1 - e^{-as}}.$$

15.10.18 Show that

(a) $\mathcal{L}\{\cosh at \cos at\} = \dfrac{s^3}{s^4 + 4a^4}$,

(b) $\mathcal{L}\{\cosh at \sin at\} = \dfrac{as^2 + 2a^3}{s^4 + 4a^4}$,

(c) $\mathcal{L}\{\sinh at \cos at\} = \dfrac{as^2 - 2a^3}{s^4 + 4a^4}$,

(d) $\mathcal{L}\{\sinh at \sin at\} = \dfrac{2a^2 s}{s^4 + 4a^4}$.

15.10.19 Show that

(a) $\mathcal{L}^{-1}\{(s^2 + a^2)^{-2}\} = \dfrac{1}{2a^3} \sin at - \dfrac{1}{2a^2} t \cos at$,

(b) $\mathcal{L}^{-1}\{s(s^2 + a^2)^{-2}\} = \dfrac{1}{2a} t \sin at$,

(c) $\mathcal{L}^{-1}\{s^2(s^2 + a^2)^{-2}\} = \dfrac{1}{2a} \sin at + \dfrac{1}{2} t \cos at$,

(d) $\mathcal{L}^{-1}\{s^3(s^2 + a^2)^{-2}\} = \cos at - \dfrac{a}{2} t \sin at$.

15.10.20 Show that
$$\mathcal{L}\{(t^2 - k^2)^{-1/2} u(t - k)\} = K_0(ks).$$
Hint. Try transforming an integral representation of $K_0(ks)$ into the Laplace transform integral.

15.10.21 The Laplace transform
$$\int_0^\infty e^{-xs} x J_0(x)\, dx = \dfrac{s}{(s^2 + 1)^{3/2}}$$
may be rewritten as
$$\dfrac{1}{s^2} \int_0^\infty e^{-y} y J_0(y/s)\, dy = \dfrac{s}{(s^2 + 1)^{3/2}},$$
which is in Gauss–Laguerre quadrature form. Evaluate this integral for $s = 1.0, 0.9, 0.8, \ldots$ decreasing s in steps of 0.1 until the relative error rises to 10 percent. (The effect of decreasing s is to make the integrand oscillate more rapidly per unit length of y, thus decreasing the accuracy of the numerical quadrature.)

15.10.22 (a) Evaluate
$$\int_0^\infty e^{-kz} k J_1(ka)\, dk$$
by the Gauss–Laguerre quadrature. Take $a = 1$ and $z = 0.1(0.1)1.0$.
(b) From the analytic form, Exercise 15.10.7, calculate the absolute error and the relative error.

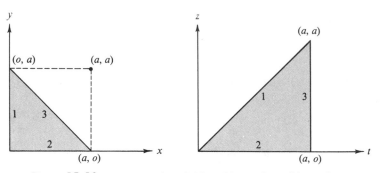

Figure 15.13   Change of variables, (a) xy-plane (b) zt-plane.

15.11 CONVOLUTION OR FALTUNG THEOREM

One of the most important properties of the Laplace transform is that given by the convolution or faltung theorem.[1] We take two transforms

$$f_1(s) = \mathcal{L}\{F_1(t)\} \quad \text{and} \quad f_2(s) = \mathcal{L}\{F_2(t)\} \tag{15.190}$$

and multiply them together. To avoid complications when changing variables, we hold the upper limits finite:

$$f_1(s) \cdot f_2(s) = \lim_{a \to \infty} \int_0^a e^{-sx} F_1(x)\, dx \int_0^{a-x} e^{-sy} F_2(y)\, dy. \tag{15.191}$$

The upper limits are chosen so that the area of integration, shown in Fig. 15.13a, is the shaded triangle, not the square. If we integrate over a square in xy-plane, we have a parallelogram in the tz-plane, which simply adds complications. This modification is permissible because the two integrands are assumed to decrease exponentially. In the limit $a \to \infty$ the integral over the unshaded triangle will give zero contribution. Substituting $x = t - z$, $y = z$ the region of integration is mapped into the triangle shown in Fig. 15.13b. To verify the mapping, map the vertices: $t = x + y$, $z = y$. Using Jacobians to transform the element of area, we have

$$dx\, dy = \begin{vmatrix} \dfrac{\partial x}{\partial t} & \dfrac{\partial y}{\partial t} \\[2mm] \dfrac{\partial x}{\partial z} & \dfrac{\partial y}{\partial z} \end{vmatrix} dt\, dz = \begin{vmatrix} 1 & 0 \\ -1 & 1 \end{vmatrix} dt\, dz \tag{15.192}$$

[1] An alternate derivation employs the Bromwich integral (Section 15.12). This is Exercise 15.12.3.

or $dx\,dy = dt\,dz$. With this substitution Eq. (15.191) becomes

$$f_1(s) \cdot f_2(s) = \lim_{a \to \infty} \int_0^a e^{-st} \int_0^t F_1(t - z)F_2(z)\,dz\,dt$$

$$= \mathscr{L}\left\{\int_0^t F_1(t - z)F_2(z)\,dz\right\}. \tag{15.193}$$

For convenience this integral is represented by the symbol

$$\int_0^t F_1(t - z)F_2(z)\,dz \equiv F_1 * F_2 \tag{15.194}$$

and referred to as the convolution, closely analogous to the Fourier convolution (Section 15.5). If we substitute $w = t - z$, we find

$$F_1 * F_2 = F_2 * F_1, \tag{15.195}$$

showing that the relation is symmetric.

Carrying out the inverse transform, we also find

$$\mathscr{L}^{-1}\{f_1(s) \cdot f_2(s)\} = \int_0^t F_1(t - z)F_2(z)\,dz. \tag{15.196}$$

This can be useful in the development of new transforms or as an alternative to a partial fraction expansion. One immediate application is in the solution of integral equations (Section 16.2). Since the upper limit t is variable, this Laplace convolution is useful in treating Volterra integral equations. The Fourier convolution with fixed (infinite) limits would apply to Fredholm integral equations.

Example 15.11.1 Driven Oscillator with Damping

As one illustration of the use of the convolution theorem, let us return to the mass m on a spring, with damping and a driving force $F(t)$. The equation of motion ((15.129) or (15.154)) now becomes

$$mX''(t) + bX'(t) + kX(t) = F(t). \tag{15.197}$$

Initial conditions $X(0) = 0$, $X'(0) = 0$ are used to simplify this illustration, and the transformed equation is

$$ms^2x(s) + bs\,x(s) + k\,x(s) = f(s) \tag{15.198}$$

or

$$x(s) = \frac{f(s)}{m} \times \frac{1}{(s + b/2m)^2 + \omega_1^2}, \tag{15.199}$$

where $\omega_1^2 \equiv k/m - b^2/4m^2$, as before.

By the convolution theorem (Eq. (15.193) or (15.196),

$$X(t) = \frac{1}{m\omega_1} \int_0^t F(t - z)e^{-(b/2m)z} \sin \omega_1 z \, dz. \qquad (15.200)$$

If the force is impulsive, $F(t) = P\,\delta(t),$[2]

$$X(t) = \frac{P}{m\omega_1} e^{-(b/2m)t} \sin \omega_1 t. \qquad (15.201)$$

P represents the momentum transferred by the impulse and the constant P/m takes the place of an initial velocity $X'(0)$.

If $F(t) = F_0 \sin \omega t$, Eq. (15.200) may be used, but a partial fraction expansion is perhaps more convenient. With

$$f(s) = \frac{F_0 \omega}{s^2 + \omega^2}$$

Eq. (15.199) becomes

$$x(s) = \frac{F_0 \omega}{m} \times \frac{1}{s^2 + \omega^2} \times \frac{1}{(s + b/2m)^2 + \omega_1^2}$$

$$= \frac{F_0 \omega}{m} \left[\frac{a's + b'}{s^2 + \omega^2} + \frac{c's + d'}{(s + b/2m)^2 + \omega_1^2} \right]. \qquad (15.202)$$

The coefficients a', b', c', and d' are independent of s. Direct calculation shows

$$-(a')^{-1} = \frac{b}{m}\omega^2 + \frac{m}{b}(\omega_0^2 - \omega^2)^2,$$

$$-(b')^{-1} = -\frac{m}{b}(\omega_0^2 - \omega^2)\left[\frac{b}{m}\omega^2 + \frac{m}{b}(\omega_0^2 - \omega^2)^2 \right].$$

Since c' and d' will lead to exponentially decreasing terms (transients), they will be discarded here. Carrying out the inverse operation, we find for the steady-state solution

$$X(t) = \frac{F_0}{[b^2\omega^2 + m^2(\omega_0^2 - \omega^2)^2]^{1/2}} \sin(\omega t - \varphi), \qquad (15.203)$$

where

$$\tan \varphi = \frac{b\omega}{m(\omega_0^2 - \omega^2)}.$$

Differentiating the denominator, we find that the amplitude has a maximum when

$$\omega^2 = \omega_0^2 - \frac{b^2}{2m^2} = \omega_1^2 - \frac{b^2}{4m^2}. \qquad (15.204)$$

[2]Note that $\delta(t)$ lies *inside* the interval $[0, t]$.

This is the resonance condition.[3] At resonance the amplitude becomes $F_0/b\omega_1$, showing that the mass m goes into infinite oscillation at resonance if damping is neglected ($b = 0$). It is worth noting that we have had three different characteristic frequencies:

$$\omega_2^2 = \omega_0{}^2 - \frac{b^2}{2m^2},$$

resonance for forced oscillations, with damping,

$$\omega_1^2 = \omega_0^2 - \frac{b^2}{4m^2},$$

free oscillation frequency, with damping,

$$\omega_0^2 = \frac{k}{m},$$

free oscillation frequency, no damping. They coincide *only* if the damping is zero.

Returning to Eqs. (15.197) and (15.199), Eq. (15.197) is our differential equation for the response of a dynamical system to an arbitrary driving force. The final response clearly depends on both the driving force and the characteristics of our system. This dual dependence is separated in the transform space. In Eq. (15.199) the transform of the response (output) appears as the product of two factors, one describing the driving force (input) and the other describing the dynamical system. This latter part, which modfies the input and yields the output, is often called a *transfer function*. Specifically, $[(s + b/2m)^2 + \omega_1^2]^{-1}$ is the transfer function corresponding to this damped oscillator. The concept of a transfer function is of great use in the field of servomechanisms. Often the characteristics of a particular servomechanism are described by giving its transfer function. The convolution theorem then yields the output signal for a particular input signal.

EXERCISES

15.11.1   From the convolution theorem show that

$$\frac{1}{s} f(s) = \mathcal{L}\left\{\int_0^t F(x)\, dx\right\},$$

where $f(s) = \mathcal{L}\{F(t)\}$.

15.11.2   If $F(t) = t^a$ and $G(t) = t^b$, $a > -1$, $b > -1$

(a) Show that the convolution

$$F * G = t^{a+b+1}\int_0^1 y^a(1 - y)^b\, dy.$$

---

[3] The amplitude (squared) has the typical resonance denominator, the Lorentz line shape, Exercise 15.3.9.

(b) By using the convolution theorem, show that

$$\int_0^1 y^a (1-y)^b \, dy = \frac{a!\,b!}{(a+b+1)!}.$$

When replacing a by $a-1$ and b by $b-1$, we have the Euler formula for the beta function (Eq. (10.60)).

15.11.3 Using the convolution integral, calculate

$$\mathcal{L}^{-1}\left\{\frac{s}{(s^2+a^2)(s^2+b^2)}\right\}, \qquad a^2 \neq b^2.$$

15.11.4 An undamped oscillator is driven by a force $F_0 \sin \omega t$. Find the displacement as a function of time. Notice that it is a linear combination of two simple harmonic motions, one with the frequency of the driving force and one with the frequency ω_0 of the free oscillator. (Assume $X(0) = X'(0) = 0$.)

$$\text{ANS.} \quad X(t) = \frac{F_0/m}{\omega^2 - \omega_0^2}\left(\frac{\omega}{\omega_0}\sin \omega_0 t - \sin \omega t\right).$$

Other exercises involving the Laplace convolution appear in Section 16.2.

15.12 INVERSE LAPLACE TRANSFORMATION

BROMWICH INTEGRAL

We now develop an expression for the inverse Laplace transform, $\mathcal{L}^{-1}$, appearing in the equation

$$F(t) = \mathcal{L}^{-1}\{f(s)\}. \tag{15.205}$$

One approach lies in the Fourier transform for which we know the inverse relation. There is a difficulty, however. Our Fourier transformable function had to satisfy the Dirichlet conditions. In particular, we required that

$$\lim_{\omega \to \infty} G(\omega) = 0 \tag{15.206}$$

so that the infinite integral would be well defined.[1] Now we wish to treat functions, $F(t)$, that may diverge exponentially. To surmount this difficulty, we extract an exponential factor, $e^{\gamma t}$, from our (possibly) divergent Laplace function and write

$$F(t) = e^{\gamma t} G(t). \tag{15.207}$$

If $F(t)$ diverges as $e^{\alpha t}$, we require γ to be greater than α so *that $G(t)$ will be convergent*. Now, with $G(t) = 0$ for $t < 0$ and otherwise suitably restricted so that it may be represented by a Fourier integral (Eq. (15.20)),

$$G(t) = \frac{1}{2\pi}\int_{-\infty}^{\infty} e^{iut}\,du \int_0^{\infty} G(v)e^{-iuv}\,dv. \tag{15.208}$$

---

[1] If delta functions are included, $G(\omega)$ may be a cosine. Although this does not satisfy Eq. (15.206), $G(\omega)$ is still bounded.

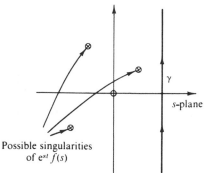

Possible singularities
of $e^{st} f(s)$

Figure 15.14   Singularities of $e^{st}f(s)$.

Using Eq. (15.207), we may rewrite (15.208) as

$$F(t) = \frac{e^{\gamma t}}{2\pi} \int_{-\infty}^{\infty} e^{iut} \, du \int_{0}^{\infty} F(v) e^{-\gamma v} e^{-iuv} \, dv. \qquad (15.209)$$

Now with the change of variable,

$$s = \gamma + iu, \qquad (15.210)$$

the integral over v is thrown into the form of a Laplace transform

$$\int_{0}^{\infty} f(v) e^{-sv} \, dv = f(s); \qquad (15.211)$$

s is now a complex variable and $\Re(s) \geq \gamma$ to guarantee convergence. Notice that the Laplace transform has mapped a function specified on the positive real axis onto the complex plane, $\Re(s) \geq \gamma.$[2]

With γ as a constant, $ds = i \, du$. Substituting Eq. (15.211) into Eq. (15.209), we obtain

$$F(t) = \frac{1}{2\pi i} \int_{\gamma - i\infty}^{\gamma + i\infty} e^{st} f(s) \, ds. \qquad (15.212)$$

Here is our inverse transform. We have rotated the line of integration through $90°$ (by using $ds = i \, du$). The path has become an infinite *vertical* line in the complex plane, the constant γ having been chosen so that all the singularities of $f(s)$ are on the left-hand side (Fig. 15.14).

Equation (15.212), our inverse transformation, is usually known as the Bromwich integral, although sometimes it is referred to as the Fourier–Mellin theorem or Fourier–Mellin integral. This integral may now be evaluated by the regular methods of contour integration (Chapter 7). If $t > 0$, the contour may be closed by an infinite semicircle in the left half-plane. Then by the residue

[2] For a derivation of the inverse Laplace transform using only real variables see C. L. Bohn and R. W. Flynn, Real variable inversion of Laplace transforms: An application in plasma physics. *Am. J. Phys.* **46**, 1250 (1978).

theorem (Section 7.2)

$$F(t) = \sum (\text{residues included for } \Re(s) < \gamma). \qquad (15.213)$$

Possibly this means of evaluation with $\Re(s)$ ranging through negative values seems paradoxical in view of our previous requirement that $\Re(s) \geq \gamma$. The paradox disappears when we recall that the requirement $\Re(s) \geq \gamma$ was imposed to guarantee convergence of the Laplace transform integral that defined $f(s)$. Once $f(s)$ is obtained, we may then proceed to exploit its properties as an analytical function in the complex plane wherever we choose.[3] In effect we are employing analytic continuation to get $\mathcal{L}\{F(t)\}$ in the left half-plane exactly as the recurrence relation for the factorial function was used to extend the Euler integral definition (Eq. (10.5)) to the left half-plane.

Perhaps a pair of examples may clarify the evaluation of Eq. (15.212).

Example 15.12.1 Inversion via Calculus of Residues

If $f(s) = a/(s^2 - a^2)$, then

$$e^{st}f(s) = \frac{ae^{st}}{s^2 - a^2} = \frac{ae^{st}}{(s+a)(s-a)}. \qquad (15.214)$$

The residues may be found by using Exercise 7.1.1 or various other means. The first step is to identify the singularities, the poles. Here we have one simple pole at $s = a$ and another simple pole at $s = -a$. By Exercise 7.1.1 the residue at $s = a$ is $(\frac{1}{2})e^{at}$ and the residue at $s = -a$ is $(-\frac{1}{2})e^{-at}$. Then

$$\text{Residues} = (\tfrac{1}{2})(e^{at} - e^{-at}) = \sinh at = F(t) \qquad (15.215)$$

in agreement with Eq. (15.105).

Example 15.12.2

If

$$f(s) = \frac{1 - e^{-as}}{s},$$

then we have

$$e^{st}f(s) = \frac{e^{st}}{s} - e^{-as}\left(\frac{e^{st}}{s}\right). \qquad (15.216)$$

The first term on the right has a simple pole at $s = 0$, residue $= 1$. Then by Eq. (15.213)

$$F_1(t) = \begin{cases} 1, & t > 0, \\ 0, & t < 0, \end{cases}$$

$$= u(t), \qquad (15.217)$$

[3] In numerical work $f(s)$ may well be available only for discrete real, positive values of s. Then numerical procedures are indicated. See Section 15.8 and the reference to Krylov and Skoblya.

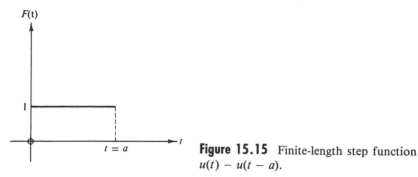

Figure 15.15 Finite-length step function $u(t) - u(t - a)$.

where $u(t)$ is the unit step function. Neglecting the minus sign and the e^{-as}, we find that the second term on the right also has a simple pole at $s = 0$, residue $= 1$. Noting the translation property (Eq. (15.164)), we have

$$F_2(t) = \begin{cases} 1, & t - a > 0, \\ 0, & t - a < 0, \end{cases}$$

$$= u(t - a). \tag{15.218}$$

Therefore

$$F(t) = F_1(t) - F_2(t) = \begin{cases} 0, & t < 0, \\ 1, & 0 < t < a, \\ 0, & t > a, \end{cases}$$

$$= u(t) - u(t - a) \tag{15.219}$$

a step function of unit height and length a (Fig. 15.15).

Two general comments may be in order. First, these two examples hardly begin to show the usefulness and power of the Bromwich integral. It is always available for inverting a complicated transform when the tables prove inadequate.

Second, this derivation is not presented as a rigorous one. Rather, it is given more as a plausibility argument, although it can be made rigorous. The determination of the inverse transform is somewhat similar to the solution of a differential equation. It makes little difference how you get the solution. Guess at it if you want. The solution can always be checked by substitution back into the original differential equation. Similarly, $F(t)$ can (and, to check on careless errors, should) be checked by determining whether by Eq. (15.99)

$$\mathcal{L}\{F(t)\} = f(s).$$

Two alternate derivations of the Bromwich integral are the subjects of Exercises 15.12.1 and 15.12.2.

As a final illustration of the use of the Laplace inverse transform, we have some results from the work of Brillouin and Sommerfeld (1914) in electromagnetic theory.

Example 15.12.3 Velocity of Electromagnetic Waves in a
Dispersive Medium

The group velocity u of traveling waves is related to the phase velocity v by
the equation

$$u = v - \lambda \frac{dv}{d\lambda}. \qquad (15.220)$$

Here λ is the wavelength. In the vicinity of an absorption line (resonance)
$dv/d\lambda$ may be sufficiently negative so that $u > c$ (Fig. 15.16). The question
immediately arises whether a signal can be transmitted faster than c, the
velocity of light in vacuum. This question, which assumes that such a group
velocity is meaningful, is of fundamental importance to the theory of special
relativity.

We need a solution to the wave equation

$$\frac{\partial^2 \psi}{\partial x^2} = \frac{1}{v^2} \frac{\partial^2 \psi}{\partial t^2}, \qquad (15.221)$$

corresponding to a harmonic vibration starting at the origin at time zero. Since
our medium is dispersive, v is a function of the angular frequency. Imagine,
for instance, a plane wave, angular frequency ω, incident on a shutter at the
origin. At $t = 0$ the shutter is (instantaneously) opened, and the wave is permitted to advance along the positive x-axis.

Let us then build up a solution starting at $x = 0$. It is convenient to use the
Cauchy integral formula, Eq. (6.43),

$$\psi(0, t) = \frac{1}{2\pi i} \oint \frac{e^{-izt}}{z - z_0} dz = e^{-iz_0 t}$$

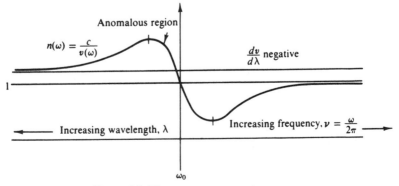

Figure 15.16 Optical dispersion.

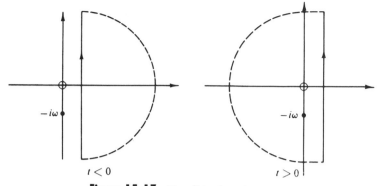

Figure 15.17   Possible closed contours.

(for contour encircling $z = z_0$ in the positive sense). Using $s = -iz$ and $z_0 = \omega$, we obtain

$$\psi(0, t) = \frac{1}{2\pi i} \int_{\gamma-i\infty}^{\gamma+i\infty} \frac{e^{st}}{s + i\omega} \, ds = \begin{cases} 0, & t < 0 \\ e^{-i\omega t}, & t > 0. \end{cases} \tag{15.222}$$

To be complete, the loop integral is over the vertical line $\Re(s) = \gamma$ *and* an infinite semicircle as shown in Fig. 15.17. The location of the infinite semicircle is chosen so that the integral over it vanishes. This means a semicircle in the left half-plane for $t > 0$ and the residue is enclosed. For $t < 0$ we pick the right half-plane and no singularity is enclosed. The fact that this is just the Bromwich integral may be verified by noting that

$$F(t) = \begin{cases} 0, & t < 0, \\ e^{-i\omega t}, & t > 0, \end{cases} \tag{15.223}$$

and applying the Laplace transform. The transformed function $f(s)$ becomes

$$f(s) = \frac{1}{s + i\omega}. \tag{15.224}$$

Our Cauchy–Bromwich integral provides us with the time dependence of a signal leaving the origin at $t = 0$. To include the space dependence, we note that

$$e^{s(t-x/v)}$$

satisfies the wave equation. With this as a clue, we replace t by $t - x/v$ and write a solution

$$\psi(x, t) = \frac{1}{2\pi i} \int_{\gamma-i\infty}^{\gamma+i\infty} \frac{e^{s(t-x/v)}}{s + i\omega} \, ds. \tag{15.225}$$

It was seen in the derivation of the Bromwich integral that our variable s replaces the ω of the Fourier transformation. Hence the wave velocity v becomes a function of s, that is, $v(s)$. Its particular form need not concern us

Table 15.1 Laplace Transform Operations

	Operations	Equation
1. Laplace transform	$f(s) = \mathcal{L}\{F(t)\} = \displaystyle\int_0^\infty e^{-st}F(t)\,dt$	(15.99)
2. Transform of derivative	$sf(s) - F(+0) = \mathcal{L}\{F'(t)\}$ $s^2 f(s) - sF(+0) - F'(+0) = \mathcal{L}\{F''(t)\}$	(15.123)
3. Transform of integral	$\dfrac{1}{s}f(s) = \mathcal{L}\left\{\displaystyle\int_0^t F(x)\,dx\right\}$	(Exercise 15.11.1)
4. Substitution	$f(s-a) = \mathcal{L}\{e^{at}F(t)\}$	(15.152)
5. Translation	$e^{-bs}f(s) = \mathcal{L}\{F(t-b)\}$	(15.164)
6. Derivative of transform	$f^{(n)}(s) = \mathcal{L}\{(-t)^n F(t)\}$	(15.173)
7. Integral of transform	$\displaystyle\int_s^\infty f(x)\,dx = \mathcal{L}\left\{\dfrac{F(t)}{t}\right\}$	(15.189)
8. Convolution	$f_1(s)\cdot f_2(s) = \mathcal{L}\left\{\displaystyle\int_0^t F_1(t-z)F_2(z)\,dz\right\}$	(15.193)
9. Inverse transform, Bromwich integral	$\dfrac{1}{2\pi i}\displaystyle\int_{\gamma - i\infty}^{\gamma + i\infty} e^{st}f(s)\,ds = F(t)$	(15.212)

here. We need only the property

$$\lim_{|s| \to \infty} v(s) = \text{constant}, c. \qquad (15.226)$$

This is suggested by the asymptotic behavior of the curve on the right side of Fig. 15.16.[4]

Evaluating Eq. (15.225) by the calculus of residues, we may close the path of integration by a semicircle in the *right* half-plane, provided

$$t - \frac{x}{c} < 0.$$

Hence

$$\psi(x, t) = 0, \qquad t - \frac{x}{c} < 0, \qquad (15.227)$$

which means that the velocity of our signal cannot exceed the velocity of light in vacuum c. This simple but very significant result was extended by Sommerfeld and Brillouin to show just how the wave advanced in the dispersive medium.

SUMMARY—INVERSION OF LAPLACE TRANSFORM

1. Direct use of tables, Table 15.2, and references; use of partial fractions (Section 15.8) and the operational theorems of Table 15.1.

---

[4] Equation (15.226) follows rigorously from the theory of anomalous dispersion. See also the Kronig–Kramers optical dispersion relations of Section 7.3.

Table 15.2 Laplace Transforms

$f(s)$	$F(t)$	Limitation	Equation
1. 1	$\delta(t)$	Singularity at $+0$	(15.141)
2. $\dfrac{1}{s}$	1	$s > 0$	(15.102)
3. $\dfrac{n!}{s^{n+1}}$	t^n	$s > 0$ $n > -1$	(15.108)
4. $\dfrac{1}{s - k}$	e^{kt}	$s > k$	(15.103)
5. $\dfrac{1}{(s - k)^2}$	te^{kt}	$s > k$	(15.175)
6. $\dfrac{s}{s^2 - k^2}$	$\cosh kt$	$s > k$	(15.105)
7. $\dfrac{k}{s^2 - k^2}$	$\sinh kt$	$s > k$	(15.105)
8. $\dfrac{s}{s^2 + k^2}$	$\cos kt$	$s > 0$	(15.107)
9. $\dfrac{k}{s^2 + k^2}$	$\sin kt$	$s > 0$	(15.107)
10. $\dfrac{s - a}{(s - a)^2 + k^2}$	$e^{at}\cos kt$	$s > a$	(15.153)
11. $\dfrac{k}{(s - a)^2 + k^2}$	$e^{at}\sin kt$	$s > a$	(15.153)
12. $\dfrac{s^2 - k^2}{(s^2 + k^2)^2}$	$t \cos kt$	$s > 0$	(15.172)
13. $\dfrac{2ks}{(s^2 + k^2)^2}$	$t \sin kt$	$s > 0$	(15.172)
14. $(s^2 + a^2)^{-1/2}$	$J_0(at)$	$s > 0$	(15.185)
15. $(s^2 - a^2)^{-1/2}$	$I_0(at)$	$s > a$	(Exercise 15.10.10)
16. $\dfrac{1}{a}\cot^{-1}\left(\dfrac{s}{a}\right)$	$j_0(at)$	$s > 0$	(Exercise 15.10.11)
17. $\left\{\begin{array}{l}\dfrac{1}{2a}\ln\dfrac{s + a}{s - a}\\[2mm]\dfrac{1}{a}\coth^{-1}\left(\dfrac{s}{a}\right)\end{array}\right\}$	$i_0(at)$	$s > a$	(Exercise 15.10.11)
18. $\dfrac{(s - a)^n}{s^{n+1}}$	$L_n(at)$	$s > 0$	(Exercise 15.10.13)
19. $\dfrac{1}{s}\ln(s + 1)$	$E_1(x) = -Ei(-x)$	$s > 0$	(Exercise 15.10.14)
20. $\dfrac{\ln s}{s}$	$-\ln t - C$	$s > 0$	(Exercise 15.12.9)

A more extensive table of Laplace transforms appears in Chapter 29 of AMS-55.

2. Bromwich integral, Eq. (15.212), and the calculus of residues.
3. Numerical inversion, Section 15.8, and references.

EXERCISES

15.12.1 Derive the Bromwich integral from Cauchy's integral formula.
Hint. Apply the inverse transform $\mathcal{L}^{-1}$ to

$$f(s) = \frac{1}{2\pi i} \lim_{\alpha \to \infty} \int_{\gamma - i\alpha}^{\gamma + i\alpha} \frac{f(z)}{s - z} dz,$$

where $f(z)$ is analytic for $\Re(z) \geq \gamma$.

15.12.2 Starting with

$$\frac{1}{2\pi i} \int_{\gamma - i\infty}^{\gamma + i\infty} e^{st} f(s) \, ds,$$

show that by introducing

$$f(s) = \int_0^\infty e^{-sz} F(z) \, dz,$$

we can convert one integral into the Fourier representation of a Dirac delta function. From this derive the inverse Laplace transform.

15.12.3 Derive the Laplace transformation convolution theorem by use of the Bromwich integral.

15.12.4 Find

$$\mathcal{L}^{-1}\left\{\frac{s}{s^2 - k^2}\right\}$$

(a) by a partial fraction expansion,
(b) repeat, using the Bromwich integral.

15.12.5 Find

$$\mathcal{L}^{-1}\left\{\frac{k^2}{s(s^2 + k^2)}\right\}$$

(a) by using a partial fraction expansion,
(b) repeat using the convolution theorem,
(c) repeat using the Bromwich integral.

ANS. $F(t) = 1 - \cos kt$.

15.12.6 Use the Bromwich integral to find the function whose transform is $f(s) = s^{-1/2}$. Note that $f(s)$ has a branch point at $s = 0$. The negative x-axis may be taken as a cut line.

ANS. $F(t) = (\pi t)^{-1/2}$.

15.12.7 Show that

$$\mathcal{L}^{-1}(s^2 + 1)^{-1/2}\} = J_0(t)$$

by evaluation of the Bromwich integral.
Hint. Convert your Bromwich integral into an integral representation of $J_0(t)$. Figure 15.18 shows a possible contour.

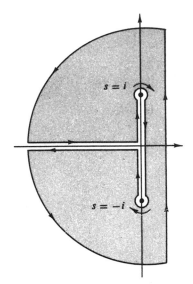

Figure 15.18 A possible contour for the inversion of $J_0(t)$.

15.12.8 Evaluate the inverse Laplace transform

$$\mathcal{L}^{-1}\{(s^2 - a^2)^{-1/2}\}$$

by each of the following methods:

(a) Expansion in a series and term-by-term inversion.

(b) Direct evaluation of the Bromwich integral.

(c) Change of variable in the Bromwich integral: $s = (a/2)(z + z^{-1})$.

15.12.9 Show that

$$\mathcal{L}^{-1}\left\{\frac{\ln s)}{s}\right\} = -\ln t - \gamma,$$

where $\gamma = 0.5772\ldots$, the Euler–Mascheroni constant.

15.12.10 Evaluate the Bromwich integral for

$$f(s) = \frac{s}{(s^2 + a^2)^2}.$$

15.12.11 Heaviside expansion theorem. If the transform $f(s)$ may be written as a ratio

$$f(s) = \frac{g(s)}{h(s)},$$

where $g(s)$ and $h(s)$ are analytic functions, $h(s)$ having simple, isolated zeros at $s = s_i$, show that

$$F(t) = \mathcal{L}^{-1}\left\{\frac{g(s)}{h(s)}\right\} = \sum_i \frac{g(s_i)}{h'(s_i)} e^{s_i t}.$$

Hint. See Exercise 7.1.2.

15.12.12 Using the Bromwich integral, invert $f(s) = s^{-2}e^{-ks}$. Express $F(t) = \mathcal{L}^{-1}\{f(s)\}$ in terms of the (shifted) unit step function $u(t - k)$.

$$ANS. \quad F(t) = (t - k)u(t - k).$$

15.12.13   You have a Laplace transform:

$$f(s) = \frac{1}{(s + a)(s + b)}, \qquad a \neq b.$$

Invert this transform by each of three methods:

(a) Partial fractions and use of tables.
(b) Convolution theorem.
(c) Bromwich integral.

$$ANS. \quad F(t) = \frac{e^{-bt} - e^{-at}}{a - b}.$$

ADDITIONAL READINGS

CHAMPENEY, D. C., *Fourier Transforms and Their Physical Applications*. New York: Academic Press (1973). Fourier transforms are developed in a careful, easy to follow manner. Approximately 60 percent of the book is devoted to applications of interest in physics and engineering.

ERDELYI, A., W. MAGNUS, F. OBERHETTINGER, and F. G. TRICOMI, *Tables of Integral Transforms*, 2 vols. New York: McGraw-Hill (1954). This text contains extensive tables of Fourier sine, cosine, and exponential transforms, Laplace and inverse Laplace transforms, Mellin and inverse Mellin transforms, Hankel transforms, and other more specialized integral transforms.

HANNA, J. R., *Fourier Series and Integrals of Boundary Value Problems*. Somerset, NJ: Wiley (1982). This book is a broad treatment of the Fourier solution of boundary value problems. The concepts of convergence and completeness are given careful attention.

JEFFREYS, H., and B. S. JEFFREYS, *Methods of Mathematical Physics*, 3rd ed. Cambridge: Cambridge University Press (1966).

KRYLOV, V. I., and N. S. SKOBLYA, *Handbook of Numerical Inversion of Laplace Transform*. Jerusalem: Israel Program for Scientific Translations (1969).

LEPAGE, W. R., *Complex Variables and the Laplace Transform for Engineers*. New York: McGraw-Hill (1961); New York: Dover (1980). A complex variable analysis which is carefully developed and then applied to Fourier and Laplace transforms. It is written to be read by students, but intended for the serious student.

MCCOLLUM, P. A., and B. F. BROWN, *Laplace Transform Tables and Theorems*. New York: Holt, Rinehart and Winston (1965).

MILES, J. W., *Integral Transforms in Applied Mathematics*. Cambridge: Cambridge University Press (1971). This is a brief but interesting and useful treatment for the advanced undergraduate. It emphasizes applications rather than abstract mathematical theory.

PAPOULIS, A., *The Fourier Integral and Its Applications*. New York: McGraw-Hill (1962). This is a rigorous development of Fourier and Laplace transforms and has extensive applications in science and engineering.

ROBERTS, G. E., and H. KAUFMAN, *Table of Laplace Transforms*. Philadelphia: Saunders (1966).

SNEDDON, I. N., *Fourier Transforms*. New York: McGraw-Hill (1951). A detailed comprehensive treatment, this book is loaded with applications to a wide variety of fields of modern and classical physics.

SNEDDON, I. H., *The Use of Integral Transforms*. New York: McGraw-Hill (1972). Written for students in science and engineering in terms they can understand, this book covers all the integral transforms mentioned in this chapter as well as in several others. Many applications are included.

VAN DER POL, B., and H. BREMMER, *Operational Calculus Based on the Two-sided Laplace Integral*, 2nd ed. Cambridge: Cambridge University Press (1955). Here is a development based on the integral range $-\infty$ to $+\infty$, rather than the useful 0 to $+\infty$. Chapter V contains a detailed study of the Dirac delta function (impulse function).

WOLF, K. B., *Integral Transforms in Science and Engineering*. New York: Plenum Press (1979). This book is a very comprehensive treatment of integral transforms and their applications.

16 INTEGRAL EQUATIONS

16.1 INTRODUCTION

With the exeption of the integral transforms of the last chapter, we have been considering equations with relations between the unknown function $\varphi(x)$ and one or more of its derivatives. We now proceed to investigate equations containing the unknown function within an integral. As with differential equations, we shall largely confine our attention to linear relations, linear integral equations. Integral equations are classified in two ways:

1. If the limits of integration are fixed, we call the equation a Fredholm equation; if one limit is variable, it is a Volterra equation.
2. If the unknown function appears *only* under the integral sign, we label it "first kind." If it appears both inside and outside the integral, it is labeled, "second kind."

DEFINITIONS

Symbolically, we have Fredholm equation of the first kind,

$$f(x) = \int_a^b K(x, t)\varphi(t)\,dt; \tag{16.1}$$

the Fredholm equation of the second kind,

$$\varphi(x) = f(x) + \lambda \int_a^b K(x, t)\varphi(t)\,dt; \tag{16.2}$$

the Volterra equation of the first kind,

$$f(x) = \int_a^x K(x, t)\varphi(t)\, dt; \tag{16.3}$$

and the Volterra equation of the second kind,

$$\varphi(x) = f(x) + \int_a^x K(x, t)\varphi(t)\, dt. \tag{16.4}$$

In all four cases $\varphi(t)$ is the unknown function. $K(x, t)$, which we call the kernel, and $f(x)$ are assumed to be known. When $f(x) = 0$, the equation is said to be homogeneous.

The reader may wonder, with some justification, why we bother about integral equations. After all, the differential equations have done a rather good job of describing our physical world so far. There are several reasons for introducing integral equations here.

We have placed considerable emphasis on the solution of differential equations *subject to particular boundary conditions*. For instance, the boundary condition at $r = 0$ determines whether the Neumann function $N_n(r)$ is present when Bessel's equation is solved. The boundary condition for $r \to \infty$ determines whether the $I_n(r)$ is present in our solution of the modified Bessel equation. The integral equation relates the unknown function not only to its values at neighboring points (derivatives) but also to its values throughout a region, including the boundary. In a very real sense the boundary conditions are built into the integral equation rather than imposed at the final stage of the solution. It will be seen later, when we construct kernels (Section 16.5), that the form of the kernel depends on the values on the boundary. The integral equation, then, is compact and may turn out to be a more convenient or powerful form than the differential equation. Mathematical problems such as existence, uniqueness, and completeness may often be handled more easily and elegantly in integral form. Finally, whether or not we like it, there are some problems, such as some diffusion and transport phenomena, that cannot be represented by differential equations. If we wish to solve such problems, we are forced to handle integral equations.

An integral equation may also appear as a matter of deliberate choice based on convenience or the need for the mathematical power of an integral equation formulation.

Example 16.1.1 Momentum Representation in Quantum Mechanics

The Schrödinger equation (in ordinary space representation) is

$$-\frac{\hbar^2}{2m}\nabla^2\psi(\mathbf{r}) + V(\mathbf{r})\psi(\mathbf{r}) = E\psi(\mathbf{r}) \tag{16.5}$$

or

$$(\nabla^2 + a^2)\psi(\mathbf{r}) = v(\mathbf{r})\psi(\mathbf{r}), \tag{16.6}$$

where

$$a^2 = \frac{2m}{\hbar^2} E, \qquad v(\mathbf{r}) = \frac{2m}{\hbar^2} V(\mathbf{r}). \tag{16.7}$$

We may generalize Eq. (16.6) to

$$(\nabla^2 + a^2)\psi(\mathbf{r}) = \int v(\mathbf{r}, \mathbf{r}')\psi(\mathbf{r}') \, d^3r'. \tag{16.8}$$

For the special case of

$$v(\mathbf{r}, \mathbf{r}') = v(\mathbf{r}') \, \delta(\mathbf{r} - \mathbf{r}'), \tag{16.9}$$

which represents local interaction Eq. (16.8) reduces to Eq. (16.6). Equation (16.8) is now subject to the Fourier transform (compare Section 15.6).

$$\Phi(\mathbf{k}) = \frac{1}{(2\pi)^{3/2}} \int \psi(\mathbf{r})e^{-i\mathbf{k}\cdot\mathbf{r}} \, d^3r$$

$$\psi(\mathbf{r}) = \frac{1}{(2\pi)^{3/2}} \int \Phi(\mathbf{k})e^{i\mathbf{k}\cdot\mathbf{r}} \, d^3k. \tag{16.10}$$

Here the abbreviation

$$\frac{\mathbf{p}}{\hbar} = \mathbf{k} \qquad \text{(wave number)} \tag{16.11}$$

has been introduced. Developing Eq. (16.10), we obtain

$$\int (\nabla^2 + a^2)\psi(\mathbf{r})e^{-i\mathbf{k}\cdot\mathbf{r}} \, d^3r = \iint v(\mathbf{r}, \mathbf{r}')\psi(\mathbf{r}')e^{-i\mathbf{k}\cdot\mathbf{r}} \, d^3r' \, d^3r. \tag{16.12}$$

Note that the ∇^2 on the left operates only on the $\psi(\mathbf{r})$. Integrating the left-hand side by parts and substituting Eq. (16.10) for $\psi(\mathbf{r}')$ on the right, we get

$$\int (-k^2 + a^2)\psi(\mathbf{r})e^{-i\mathbf{k}\cdot\mathbf{r}}d^3r = (2\pi)^{3/2}(-k^2 + a^2)\Phi(\mathbf{k})$$

$$= \frac{1}{(2\pi)^{3/2}} \iiint v(\mathbf{r},\mathbf{r}')\Phi(\mathbf{k}')e^{-i(\mathbf{k}\cdot\mathbf{r}-\mathbf{k}'\cdot\mathbf{r}')}d^3r' \, d^3rd^3k'. \tag{16.13}$$

If we use

$$f(\mathbf{k}, \mathbf{k}') = \frac{1}{(2\pi)^{3/2}} \iint v(\mathbf{r}, \mathbf{r}')e^{-i(\mathbf{k}\cdot\mathbf{r}-\mathbf{k}'\cdot\mathbf{r}')} \, d^3r' \, d^3r, \tag{16.14}$$

Eq. (16.13) becomes

$$(-k^2 + a^2)\Phi(\mathbf{k}) = \int f(\mathbf{k}, \mathbf{k}')\Phi(\mathbf{k}') \, d^3k', \tag{16.15}$$

a homogeneous Fredholm equation of the second kind in which the parameter a^2 corresponds to the eigenvalue.

For our special but important case of local interaction, application of Eq. (16.9) leads to

$$f(\mathbf{k}, \mathbf{k}') = f(\mathbf{k} - \mathbf{k}').\tag{16.16}$$

This is our momentum representation equivalent to an ordinary static interaction potential in coordinate space. Our momentum function $\Phi(\mathbf{k})$ satisfies the integral equation (Eq. (16.15)). It must be emphasized that all through here we have assumed that the required Fourier integrals exist. For a linear oscillator potential, $V(\mathbf{r}) = r^2$, the required integrals would not exist. Equation (16.10) would lead to divergent oscillations and we would have no Eq. (16.15).

TRANSFORMATION OF A DIFFERENTIAL EQUATION INTO AN INTEGRAL EQUATION

Often we find that we have a choice. The physical problem may be represented by a differential or an integral equation. Let us assume that we have the differential equation and wish to transform it into an integral equation. Starting with a *linear* second-order differential equation

$$y'' + A(x)y' + B(x)y = g(x)\tag{16.17}$$

with initial conditions

$$y(a) = y_0,$$

$$y'(a) = y_0',$$

we integrate to obtain

$$y' = -\int_a^x Ay'\, dx - \int_a^x By\, dx + \int_a^x g\, dx + y_0'.\tag{16.18}$$

Integrating the first integral on the right by parts yields

$$y' = -Ay - \int_a^x (B - A')y\, dx + \int_a^x g\, dx + A(a)y_0 + y_0'.\tag{16.19}$$

Notice how the initial conditions are being absorbed into our new version. Integrating a second time, we obtain

$$y = -\int_a^x Ay\, dx - \int_a^x \int_a^x [B(t) - A'(t)]y(t)\, dt\, dx$$

$$+ \int_a^x \int_a^x g(t)\, dt\, dx + [A(a)y_0 + y_0'](x - a) + y_0.\tag{16.20}$$

To transform this equation into a neater form, we use the relation

$$\int_a^x \int_a^x f(t)\, dt\, dx = \int_a^x (x - t) f(t)\, dt. \qquad (16.21)$$

This may be verified by differentiating both sides. Since the derivatives are equal, the original expressions can differ only by a constant. Letting $x \to a$, the constant vanishes and Eq. (16.21) is established. Applying it to Eq. (16.20), we obtain

$$y(x) = -\int_a^x \{A(t) + (x - t)[B(t) - A'(t)]\} y(t)\, dt$$

$$+ \int_a^x (x - t) g(t)\, dt + [A(a) y_0 + y_0'](x - a) + y_0. \qquad (16.22)$$

If we now introduce the abbreviations

$$K(x, t) = (t - x)[B(t) - A'(t)] - A(t),$$

$$f(x) = \int_a^x (x - t) g(t)\, dt + [A(a) y_0 + y_0'](x - a) + y_0, \qquad (16.23)$$

Eq. (16.22) becomes

$$y(x) = f(x) + \int_a^x K(x, t) y(t)\, dt, \qquad (16.24)$$

which is a Volterra equation of the second kind. This reformulation as a Volterra integral equation offers certain advantages in investigating questions of existence and uniqueness.

Example 16.1.3 Linear Oscillator Equation

As a simple illustration, consider the linear oscillator equation

$$y'' + \omega^2 y = 0 \qquad (16.25)$$

with

$$y(0) = 0$$

$$y'(0) = 1.$$

This yields

$$A(x) = 0,$$

$$B(x) = \omega^2,$$

$$g(x) = 0.$$

Substituting into Eq. (16.22) (or Eqs. (16.23) and (16.24)), we find that the integral equation becomes

$$y(x) = x + \omega^2 \int_0^x (t - x) y(t)\, dt. \qquad (16.26)$$

This integral equation, Eq. (16.26), is equivalent to the original differential equation *plus* the initial conditions. The reader may show that each form is indeed satisfied by $y(x) = (1/\omega) \sin \omega x$.

Let us reconsider the linear oscillator equation (16.25) but now with the boundary conditions

$$y(0) = 0,$$

$$y(b) = 0.$$

Since $y'(0)$ is not given, we must modify the procedure. The first integration gives

$$y' = -\omega^2 \int_0^x y \, dx + y'(0). \tag{16.27}$$

Integrating a second time and again using Eq. (16.21), we have

$$y = -\omega^2 \int_0^x (x - t)y(t) \, dt + y'(0)x. \tag{16.28}$$

To eliminate the unknown $y'(0)$, we now impose the condition $y(b) = 0$. This gives

$$\omega^2 \int_0^b (b - t)y(t) \, dt = by'(0). \tag{16.29}$$

Substituting this back into Eq. (16.28), we obtain

$$y(x) = -\omega^2 \int_0^x (x - t)y(t) \, dt + \omega^2 \frac{x}{b} \int_0^b (b - t)y(t) \, dt. \tag{16.30}$$

Now let us break the interval $[0, b]$ into two intervals $[0, x]$ and $[x, b]$. Since

$$\frac{x}{b}(b - t) - (x - t) = \frac{t}{b}(b - x), \tag{16.31}$$

we find

$$y(x) = \omega^2 \int_0^x \frac{t}{b}(b - x)y(t) \, dt + \omega^2 \int_x^b \frac{x}{b}(b - t)y(t) \, dt. \tag{16.32}$$

Finally, if we define a kernel (Fig. 16.1)

$$K(x, t) = \begin{cases} \dfrac{t}{b}(b - x), & t < x, \\[2mm] \dfrac{x}{b}(b - t), & t > x, \end{cases} \tag{16.33}$$

we have

$$y(x) = \omega^2 \int_0^b K(x, t)y(t) \, dt, \tag{16.34}$$

a homogeneous Fredholm equation of the second kind.

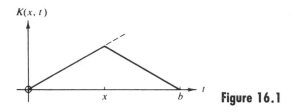

Figure 16.1

Our new kernel, $K(x, t)$, has some interesting properties.

1. It is symmetric, $K(x, t) = K(t, x)$.
2. It is continuous in the sense that

$$\left.\frac{t}{b}(b - x)\right|_{t=x} = \left.\frac{x}{b}(b - t)\right|_{t=x}.$$

3. Its derivative with respect to t is *discontinuous*. As t increases through the point $t = x$, there is a discontintinuity of -1 in $\partial K(x, t)/\partial t$.

According to these properties in Section 8.7 we identify $K(x, t)$ as a Green's function.

In the transformation of a linear, second-order differential equation into an integral equation, the initial or boundary conditions play a decisive role. If we have *initial* conditions (only one end of our interval), the differential equation transforms into a Volterra integral equation. For the case of the linear oscillator equation with *boundary* conditions (both ends of our interval), the differential equation leads to a Fredholm integral equation with a kernel that will be a Green's function.

It might be noted that the reverse transformation (integral equation to differential equation) is not always possible. There exist integral equations for which no corresponding differential equation is known.

EXERCISES

16.1.1   Starting with the differential equation, integrate twice and derive the Volterra integral equation corresponding to

(a) $y''(x) - y(x) = 0$; $y(0) = 0$, $y'(0) = 1$.

$$ANS. \quad y = \int_0^x (x - t)y(t)\, dt + x.$$

(b) $y''(x) - y(x) = 0$; $y(0) = 1$, $y'(0) = -1$.

$$ANS. \quad y = \int_0^x (x - t)y(t)\, dt - x + 1.$$

Check your results with Eq. (16.23).

16.1.2   Derive a Fredholm integral equation corresponding to

$$y''(x) - y(x) = 0; \qquad y(1) = 1,$$
$$y(-1) = 1,$$

(a) by integrating twice,
(b) by forming the Green's function.

$$ANS. \qquad y(x) = 1 - \int_{-1}^{1} K(x, t)y(t)\, dt,$$

$$K(x, t) = \begin{cases} \frac{1}{2}(1 - x)(t + 1), & x > t, \\ \frac{1}{2}(1 - t)(x + 1), & x < t. \end{cases}$$

16.1.3   (a) Starting with the given answers of Exercise 16.1.1, differentiate and recover the original differential equations *and the boundary conditions*.
(b) Repeat for Exercise 16.1.2.

16.1.4   The general second-order linear differential equation with constant coefficients is

$$y''(x) + a_1 y'(x) + a_2 y(x) = 0.$$

Given the boundary conditions

$$y(0) = y(1) = 0,$$

integrate twice and develop the integral equation

$$y(x) = \int_0^1 K(x, t)y(t)\, dt,$$

with

$$K(x, t) = \begin{cases} a_2 t(1 - x) + a_1(x - 1), & t < x, \\ a_2 x(1 - t) + a_1 x, & x < t. \end{cases}$$

Note that $K(x, t)$ is symmetric and continuous if $a_1 = 0$. How is this related to self-adjointness of the differential equation?

16.1.5   Verify that $\int_a^x \int_a^x f(t)\, dt\, dx = \int_a^x (x - t)f(t)\, dt$ for all $f(t)$ (for which the integrals exist).

16.1.6   Given $\varphi(x) = x - \int_0^x (t - x)\varphi(t)\, dt$. Solve this integral equation by converting it to a differential equation (plus boundary conditions) and solving the differential equation (by inspection).

16.1.7   Show that the homogeneous Volterra equation of the second kind

$$\psi(x) = \lambda \int_0^x K(x, t)\psi(t)\, dt$$

has no solution (apart from the trivial $\psi = 0$).
Hint. Develop a Maclaurin expansion of $\psi(x)$. Assume $\psi(x)$ and $K(x, t)$ are differentiable with respect to x as needed.

16.2 INTEGRAL TRANSFORMS, GENERATING FUNCTIONS

Analogous to differentiation, linear differential equations are solved completely in Chapter 8. Analogous to integration, there is no general method available for inverting integral equations. However, certain special cases may be treated with our *integral transforms* (Chapter 15). For convenience these are listed here. If

$$\psi(x) = \frac{1}{\sqrt{2\pi}} \int_{-\infty}^{\infty} e^{ixt}\varphi(t)\, dt,$$

then

$$\varphi(x) = \frac{1}{\sqrt{2\pi}} \int_{-\infty}^{\infty} e^{-ixt}\psi(t)\, dt \qquad \text{(Fourier).} \qquad (16.35)$$

If

$$\psi(x) = \int_{0}^{\infty} e^{-xt}\varphi(t)\, dt,$$

then

$$\varphi(x) = \frac{1}{2\pi i} \int_{\gamma-i\infty}^{\gamma+i\infty} e^{xt}\psi(t)\, dt \qquad \text{(Laplace).} \qquad (16.36)$$

If

$$\psi(x) = \int_{0}^{\infty} t^{x-1}\varphi(t)\, dt,$$

then

$$\varphi(x) = \frac{1}{2\pi i} \int_{\gamma-i\infty}^{\gamma+i\infty} x^{-t}\psi(t)\, dt \qquad \text{(Mellin).} \qquad (16.37)$$

If

$$\psi(x) = \int_{0}^{\infty} t\varphi(t)J_\nu(xt)\, dt,$$

then

$$\varphi(x) = \int_{0}^{\infty} t\psi(t)J_\nu(xt)\, dt \qquad \text{(Hankel).} \qquad (16.38)$$

Actually the usefulness of the integral transform technique extends a bit beyond these four rather specialized forms.

Example 16.2.1 Fourier Transform Solution

Let us consider a Fredholm equation of the first kind with a kernel of the general type $k(x - t)$,

$$f(x) = \int_{-\infty}^{\infty} k(x - t)\varphi(t)\, dt, \qquad (16.39)$$

in which $\varphi(t)$ is our unknown function. *Assuming that the needed transforms exist*, we apply the Fourier convolution theorem (Section 15.5) to obtain

$$f(x) = \int_{-\infty}^{\infty} K(\omega)\Phi(\omega)e^{-i\omega x}\, d\omega. \qquad (16.40)$$

The functions $K(\omega)$ and $\Phi(\omega)$ are the Fourier transforms of $k(x)$ and $\varphi(x)$, respectively. Inverting, by Eq. (16.35), we have

$$K(\omega)\Phi(\omega) = \frac{1}{2\pi}\int_{-\infty}^{\infty} f(x)e^{i\omega x}\,dx = \frac{F(\omega)}{\sqrt{2\pi}}. \tag{16.41}$$

Then

$$\Phi(\omega) = \frac{1}{\sqrt{2\pi}}\cdot\frac{F(\omega)}{K(\omega)}, \tag{16.42}$$

and again inverting we have

$$\varphi(x) = \frac{1}{2\pi}\int_{-\infty}^{\infty}\frac{F(\omega)}{K(\omega)}e^{-i\omega x}\,d\omega. \tag{16.43}$$

For a rigorous justification of this result the reader is invited to follow Morse and Feshbach across complex planes. An extension of this transformation solution appears as Exercise 16.2.1.

Example 16.2.2 Generalized Abel Equation, Convolution Theorem

The generalized Abel equation is

$$f(x) = \int_0^x \frac{\varphi(t)}{(x-t)^\alpha}\,dt, \qquad 0 < \alpha < 1, \qquad \text{with} \begin{cases} f(x) & \text{known,} \\ \varphi(t) & \text{unknown.} \end{cases} \tag{16.44}$$

Taking the Laplace transform of both sides of this equation, we obtain

$$\mathcal{L}\{f(x)\} = \mathcal{L}\left\{\int_0^x \frac{\varphi(t)}{(x-t)^\alpha}\,dt\right\}$$

$$= \mathcal{L}\{x^{-\alpha}\}\mathcal{L}\{\varphi(x)\}, \tag{16.45}$$

the last step following by the Laplace convolution theorem (Section 15.11). Then

$$\mathcal{L}\{\varphi(x)\} = \frac{s^{1-\alpha}\mathcal{L}\{f(x)\}}{(-\alpha)!}. \tag{16.46}$$

Dividing by s,[1] we obtain

$$\frac{1}{s}\mathcal{L}\{\varphi(x)\} = \frac{s^{-\alpha}\mathcal{L}\{f(x)\}}{(-\alpha)!}$$

$$= \frac{\mathcal{L}\{x^{\alpha-1}\}\mathcal{L}\{f(x)\}}{(\alpha-1)!(-\alpha)!}. \tag{16.47}$$

Combining the factorials Eq. (10.32)), and applying the Laplace convolution theorem again, we discover that

$$\frac{1}{s}\mathcal{L}\{\varphi(x)\} = \frac{\sin \pi\alpha}{\pi}\mathcal{L}\left\{\int_0^x \frac{f(t)}{(x-t)^{1-\alpha}}\,dt\right\}. \tag{16.48}$$

---

[1] $s^{1-\alpha}$ does not have an inverse for $0 < \alpha < 1$.

Inverting with the aid of Exercise 15.11.1, we get

$$\int_0^x \varphi(t)\, dt = \frac{\sin \pi\alpha}{\pi} \int_0^x \frac{f(t)}{(x-t)^{1-\alpha}}\, dt, \tag{16.49}$$

and finally, by differentiating,

$$\varphi(x) = \frac{\sin \pi\alpha}{\pi} \frac{d}{dx} \int_0^x \frac{f(t)}{(x-t)^{1-\alpha}}\, dt. \tag{16.50}$$

GENERATING FUNCTIONS

Occasionally, the reader may encounter integral equations that involve generating functions. Suppose we have the admittedly special case

$$f(x) = \int_{-1}^1 \frac{\varphi(t)}{(1 - 2xt + x^2)^{1/2}}\, dt, \qquad -1 \le x \le 1. \tag{16.51}$$

We notice two important features:

1. $(1 - 2xt + x^2)^{-1/2}$ generates the Legendre polynomials.
2. $[-1, 1]$ is the orthogonality interval for the Legendre polynomials.

If we now expand the denominator (property 1) and assume that our unknown $\varphi(t)$ may be written as a series of these same Legendre polynomials,

$$f(x) = \int_{-1}^1 \sum_{n=0}^\infty a_n P_n(t) \sum_{r=0}^\infty P_r(t) x^r\, dt. \tag{16.52}$$

Utilizing the orthogonality of the Legendre polynomials (property 2), we obtain

$$f(x) = \sum_{r=0}^\infty \frac{2a_r}{2r+1} x^r. \tag{16.53}$$

We may identify the a_n's by differentiating n times and then setting $x = 0$.

$$f^{(n)}(0) = n! \frac{2}{2n+1} a_n. \tag{16.54}$$

Hence

$$\varphi(t) = \sum_{n=0}^\infty \frac{2n+1}{2} \frac{f^{(n)}(0)}{n!} P_n(t). \tag{16.55}$$

Similar results may be obtained with the other generating functions (compare Exercise 15.2.9). Actually the technique of expanding in a series of special functions is always available. It is worth a try whenever the expansion is possible (and convenient) and the interval is appropriate.

EXERCISES

16.2.1   The kernel of a Fredholm equation of the second kind,

$$\varphi(x) = f(x) + \lambda \int_{-\infty}^{\infty} K(x, t)\varphi(t)\, dt,$$

is of the form $k(x - t)$.[2] Assuming that the required transforms exist, show that

$$\varphi(x) = \frac{1}{\sqrt{2\pi}} \int_{-\infty}^{\infty} \frac{F(t)e^{-ixt}\, dt}{1 - \sqrt{2\pi}\,\lambda K(t)}.$$

$F(t)$ and $K(t)$ are the Fourier transforms of $f(x)$ and $k(x)$, respectively.

16.2.2   The kernel of a Volterra equation of the first kind,

$$f(x) = \int_{0}^{x} K(x, t)\varphi(t)\, dt,$$

has the form $k(x - t)$. Assuming that the required transforms exist, show that

$$\varphi(x) = \frac{1}{2\pi i} \int_{\gamma - i\infty}^{\gamma + i\infty} \frac{F(s)}{K(s)} e^{xs}\, ds.$$

$F(s)$ and $K(s)$ are the Laplace transforms of $f(x)$ and $k(x)$, respectively.

16.2.3   The kernel of a Volterra equation of the second kind,

$$\varphi(x) = f(x) + \lambda \int_{0}^{x} K(x, t)\varphi(t)\, dt$$

has the form $k(x - t)$. Assuming that the required transforms exist, show that

$$\varphi(x) = \frac{1}{2\pi i} \int_{\gamma - i\infty}^{\gamma + i\infty} \frac{F(s)}{1 - \lambda K(s)} e^{xs}\, ds.$$

16.2.4   Using the Laplace transform solution (Exercise 16.2.3), solve

(a) $\varphi(x) = x + \displaystyle\int_{0}^{x} (t - x)\varphi(t)\, dt.$

$$ANS. \quad \varphi(x) = \sin x,$$

(b) $\varphi(x) = x - \displaystyle\int_{0}^{x} (t - x)\varphi(t)\, dt.$

$$ANS. \quad \varphi(x) = \sinh x.$$

Check your results by substituting back into the original integral equations.

16.2.5   Reformulate the equations of Example 16.2.1 (Eqs. (16.39) to (16.43)), using Fourier cosine transforms.

---

[2] This kernel and a range $0 \le x < \infty$ are the characteristics of integral equations of the Wiener–Hopf type. Details will be found in Chapter 8 of Morse and Feshbach (1953).

16.2.6   Given the Fredholm integral equation,

$$e^{-x^2} = \int_{-\infty}^{\infty} e^{-(x-t)^2} \varphi(t)\, dt,$$

apply the Fourier convolution technique of Example 16.2.1 to solve for $\varphi(t)$.

16.2.7   Solve Abel's equation,

$$f(x) = \int_0^x \frac{\varphi(t)}{(x-t)^\alpha}\, dt, \qquad 0 < \alpha < 1$$

by the following method:

(a) Multiply both sides by $(z - x)^{\alpha-1}$ and integrate with respect to x over the range $0 \le x \le z$.

(b) Reverse the order of integration and evaluate the integral on the right-hand side (with respect to x) by the beta function.

Note.

$$\int_t^z \frac{dx}{(z-x)^{1-\alpha}(x-t)^\alpha} = B(1-\alpha, \alpha)$$

$$= (-\alpha)!(\alpha - 1)!$$

$$= \frac{\pi}{\sin \pi\alpha}.$$

16.2.8   Given the generalized Abel equation with $f(x) = 1$,

$$1 = \int_0^x \frac{\varphi(t)}{(x-t)^\alpha}\, dt, \qquad 0 < \alpha < 1.$$

Solve for $\varphi(t)$ and verify that $\varphi(t)$ is a solution of the preceding equation.

$$ANS. \quad \varphi(t) = \frac{\sin \pi\alpha}{\pi} t^{\alpha-1}.$$

16.2.9   A Fredholm equation of the first kind has a kernel $e^{-(x-t)^2}$:

$$f(x) = \int_{-\infty}^{\infty} e^{-(x-t)^2} \varphi(t)\, dt.$$

Show that the solution is

$$\varphi(x) = \frac{1}{\sqrt{\pi}} \sum_{n=0}^{\infty} \frac{f^{(n)}(0)}{2^n n!} H_n(x),$$

in which $H_n(x)$ is an nth-order Hermite polynomial.

16.2.10   Solve the integral equation

$$f(x) = \int_{-1}^{1} \frac{\varphi(t)}{(1 - 2xt + x^2)^{1/2}}\, dt, \qquad -1 \le x \le 1,$$

for the unknown function $\varphi(t)$ if

(a) $f(x) = x^{2s}$,   (b) $f(x) = x^{2s+1}$.

$$ANS. \quad (a) \ \varphi(t) = \frac{4s+1}{2} P_{2s}(t),$$

$$(b) \ \varphi(t) = \frac{4s+3}{2} P_{2s+1}(t).$$

16.2.11   A Kirchhoff diffraction theory analysis of a laser leads to the integral equation

$$v(\mathbf{r}_2) = \gamma \iint K(\mathbf{r}_1, \mathbf{r}_2) v(\mathbf{r}_1) \, dA.$$

The unknown $v(\mathbf{r}_1)$ gives the geometric distribution of the radiation field over one mirror surface; the range of integration is over the surface of that mirror. For square confocal spherical mirrors the integral equation becomes

$$v(x_2, y_2) = \frac{-i\gamma e^{ikb}}{\lambda b} \int_{-a}^{a} \int_{-a}^{a} e^{-(ik/b)(x_1 x_2 + y_1 y_2)} v(x_1, y_1) \, dx_1 \, dy_1,$$

in which b is the centerline distance between the laser mirrors. This can be put in a somewhat simpler form by the substitutions

$$\frac{kx_i^2}{b} = \xi_i^2, \qquad \frac{ky_i^2}{b} = \eta_i^2, \qquad \text{and} \qquad \frac{ka^2}{b} = \frac{2\pi a^2}{\lambda b} = \alpha^2.$$

(a) Show that the variables separate and we get two integral equations.
(b) Show that the new limits $\pm \alpha$ may be approximated by $\pm \infty$ for a mirror dimension $a \gg \lambda$.
(c) Solve the resulting integral equations.

16.3 NEUMANN SERIES, SEPARABLE (DEGENERATE) KERNELS

Many and probably most integral equations cannot be solved by the specialized integral transform techniques of the preceding section. Here we develop three rather general techniques for solving integral equations. The first, due largely to Neumann, Liouville, and Volterra, develops the unknown function $\varphi(x)$ as a power series in λ, where λ is a given constant. The method is applicable whenever the series converges.

The second method is somewhat restricted because it requires that the two variables appearing in the kernel $K(x, t)$ be separable. However, there are two major rewards: (1) the relation between an integral equation and a set of simultaneous linear algebraic equations is shown explicitly, and (2) the method leads to eigenvalues and eigenfunctions—in close analogy to Section 3.5.

Third, a technique for numerical solution of Fredholm equations of both the first and second kind is outlined. The problem posed by ill-conditioned matrices is emphasized.

NEUMANN SERIES

We solve a linear integral equation of the second kind by successive approximations; our integral equation is the Fredholm equation

$$\varphi(x) = f(x) + \lambda \int_a^b K(x, t) \varphi(t) \, dt \tag{16.56}$$

in which $f(x) \neq 0$. If the upper limit of the integral is a variable (Volterra equation), the following development will still hold, but with minor modifications. Let us try (there is no *guarantee* that it will work) to approximate our unknown function by

$$\varphi(x) \approx \varphi_0(x) = f(x). \tag{16.57}$$

This choice is not mandatory. If you can make a better guess, go ahead and guess. The choice here is equivalent to saying that the integral or the constant λ is small. To improve this first crude approximation, we feed $\varphi_0(x)$ back into the integral, Eq. (16.56), and get

$$\varphi_1(x) = f(x) + \lambda \int_a^b K(x, t)f(t) \, dt. \tag{16.58}$$

Repeating this process of substituting the new $\varphi_n(x)$ back into Eq. (15.56), we develop the sequence

$$\varphi_2(x) = f(x) + \lambda \int_a^b K(x, t_1)f(t_1) \, dt_1$$

$$+ \lambda^2 \int_a^b \int_a^b K(x, t_1)K(t_1, t_2)f(t_2) \, dt_2 \, dt_1 \tag{16.59}$$

and

$$\varphi_n(x) = \sum_{i=0}^n \lambda^i u_i(x), \tag{16.60}$$

where

$$u_0(x) = f(x)$$

$$u_1(x) = \int_a^b K(x, t_1)f(t_1) \, dt_1$$

$$u_2(x) = \int_a^b \int_a^b K(x, t_1)K(t_1, t_2)f(t_2) \, dt_2 \, dt_1 \tag{16.61}$$

$$u_n(x) = \int \int \cdots \int K(x, t_1)K(t_1, t_2) \cdots K(t_{n-1}, t_n) \cdot f(t_n) \, dt_n \cdots dt_1.$$

We expect that our solution $\varphi(x)$ will be

$$\varphi(x) = \lim_{n \to \infty} \varphi_n(x) = \lim_{n \to \infty} \sum_{i=0}^n \lambda^i u_i(x), \tag{16.62}$$

provided that our infinite series converges.

We may conveniently check the convergence by the Cauchy ratio test, Section 5.2, noting that

$$|\lambda^n u_n(x)| \le |\lambda^n| \cdot |f|_{\max} \cdot |K|_{\max}^n \cdot |b - a|^n, \tag{16.63}$$

using $|f|_{\max}$ to represent the *maximum* value of $|f(x)|$ in the interval $[a, b]$ and $|K|_{\max}$ to represent the maximum value of $|K(x, t)|$ in its domain in the x,

t-plane. We have convergence if

$$|\lambda| \cdot |K|_{\max} \cdot |b - a| < 1. \qquad (16.64)$$

Note that $\lambda|u_n(\max)|$ is being used as a *comparison* series. If it converges, our actual series must converge. If this condition is not satisfied, we may or may not have convergence. A more sensitive test is required. Of course, even if the Neumann series diverges, there still may be a solution obtainable by another method.

To see what has been done with this iterative manipulation, we may find it helpful to rewrite the Neumann series solution, Eq. (16.59), in operator form. We start by rewriting Eq. (16.56) as

$$\varphi = \lambda K\varphi + f,$$

where K represents the integration operator $\int_a^b K(x, t)[\ ]\, dt$. Solving for φ, we obtain

$$\varphi = (1 - \lambda K)^{-1}f.$$

Binomial expansion leads to Eq. (16.59). The *convergence* of the Neumann series is a demonstration that the inverse operator $(1 - \lambda K)^{-1}$ exists.

Example 16.3.1 Neumann Series Solution

To illustrate the Neumann method, we consider the integral equation

$$\varphi(x) = x + \tfrac{1}{2}\int_{-1}^{1}(t - x)\varphi(t)\, dt. \qquad (16.65)$$

To start the Neumann series, we take

$$\varphi_0(x) = x. \qquad (16.66)$$

Then

$$\varphi_1(x) = x + \tfrac{1}{2}\int_{-1}^{1}(t - x)t\, dt$$

$$= x + \tfrac{1}{2}|\tfrac{1}{3}t^3 - \tfrac{1}{2}t^2 x|_{-1}^{1}$$

$$= x + \tfrac{1}{3}.$$

Substituting $\varphi_1(x)$ back into Eq. (16.65), we get

$$\varphi_2(x) = x + \frac{1}{2}\int_{-1}^{1}(t - x)t\, dt + \frac{1}{2}\int_{-1}^{1}(t - x)\frac{1}{3}\, dt$$

$$= x + \frac{1}{3} - \frac{x}{3}.$$

Continuing this process of substituting back into Eq. (16.65), we obtain

$$\varphi_3(x) = x + \frac{1}{3} - \frac{x}{3} - \frac{1}{3^2},$$

and by induction

$$\varphi_{2n}(x) = x + \sum_{s=1}^{n} (-1)^{s-1} 3^{-s} - x \sum_{s=1}^{n} (-1)^{s-1} 3^{-s}. \qquad (16.67)$$

Letting $n \to \infty$, we get

$$\varphi(x) = \tfrac{3}{4}x + \tfrac{1}{4}. \qquad (16.68)$$

This solution can (and should) be checked by substituting back into the original equation, Eq. (16.65).

It is interesting to note that our series converged easily even though Eq. (16.64) is *not* satisfied in this particular case. Actually Eq. (16.64) is a rather crude upper bound on λ. It can be shown that a necessary and sufficient condition for the convergence of our series solution is that $|\lambda| < |\lambda_e|$, where λ_e is the eigenvalue of smallest magnitude of the corresponding homogeneous equation $[f(x) = 0)]$. For this particular example $\lambda_e = \sqrt{3}/2$. Clearly, $\lambda = \tfrac{1}{2} < \lambda_e = \sqrt{3}/2$.

One approach to the calculation of time-dependent perturbations in quantum mechanics starts with the integral equation for the evolution operator

$$U(t, t_0) = 1 - \frac{i}{\hbar} \int_{t_0}^{t} V(t_1) U(t_1, t_0) \, dt_1. \qquad (16.69a)$$

Iteration leads to

$$U(t, t_0) = 1 - \frac{i}{\hbar} \int_{t_0}^{t} V(t_1) \, dt_1 + \left(\frac{i}{\hbar}\right)^2 \int_{t_0}^{t} \int_{t_0}^{t_1} V(t_1) V(t_2) \, dt_2 \, dt_1 + \cdots. \qquad (16.69b)$$

The evolution operator is obtained as a series of multiple integrals of the perturbing potential $V(t)$, closely analogous to the Neumann series, Eq. (16.60). For $V = V_0$, independent of t, the evolution operator becomes

$$U(t_1, t_0) = \exp[-i(t - t_0)V_0/\hbar].$$

A second and similar relationship between the Neumann series and quantum mechanics appears when the Schrödinger wave equation for scattering is reformulated as an integral equation. The first term in a Neumann series solution is the incident (unperturbed) wave. The second term is the first-order Born approximation, Eq. (8.212) of Section 8.7.

The Neumann method may also be applied to Volterra integral equations of the second kind, Eq. (16.4) or Eq. (16.56) with the fixed upper limit b replaced by a variable x. In the Volterra case the Neumann series converges for all λ as long as the kernel is square integrable.

SEPARABLE KERNEL

The technique of replacing our integral equation by simultaneous algebraic equations may also be used whenever our kernel $K(x, t)$ is separable in the sense that

$$K(x, t) = \sum_{j=1}^{n} M_j(x)N_j(t), \tag{16.70}$$

where n, the upper limit of the sum, is *finite*. Such kernels are sometimes called degenerate. Our class of separable kernels includes all polynomials and many of the elementary transcendental functions; that is,

$$\cos(t - x) = \cos t \cos x + \sin t \sin x. \tag{16.70a}$$

If Eq. (16.70) is satisfied, substitution into the Fredholm equation of the second kind, Eq. (16.2), yields

$$\varphi(x) = f(x) + \lambda \sum_{j=1}^{n} M_j(x) \int_a^b N_j(t)\varphi(t)\, dt, \tag{16.71}$$

interchanging integration and summation. Now the integral with respect to t is a constant,

$$\int_a^b N_j(t)\varphi(t)\, dt = c_j. \tag{16.72}$$

Hence Eq. (16.71) becomes

$$\varphi(x) = f(x) + \lambda \sum_{j=1}^{n} c_j M_j(x). \tag{16.73}$$

This gives us $\varphi(x)$, our solution, once the constants c_i have been determined. Equation (16.73) further tells us the form of $\varphi(x)$: $f(x)$, plus a linear combination of the x-dependent factors of the separable kernel.

We may find c_i by multiplying Eq. (16.73) by $N_i(x)$ and integrating to eliminate the x-dependence. Use of Eq. (16.72) yields

$$c_i = b_i + \lambda \sum_{j=1}^{n} a_{ij}c_j, \tag{16.74}$$

where

$$b_i = \int_a^b N_i(x)f(x)\, dx.$$
$$\tag{16.75}$$
$$a_{ij} = \int_a^b N_i(x)M_j(x)\, dx.$$

It is perhaps helpful to write Eq. (16.74) in matrix form, with $A = (a_{ij})$:

$$\mathbf{b} = \mathbf{c} - \lambda A\mathbf{c} = (1 - \lambda A)\mathbf{c}, \tag{16.76a}$$

or[1]

$$\mathbf{c} = (1 - \lambda A)^{-1}\mathbf{b}. \tag{16.76b}$$

[1] Notice the similarity to the operator form of the Neumann series.

Equation (16.76a) is equivalent to a set of simultaneous linear algebraic equations

$$(1 - \lambda a_{11})c_1 - \lambda a_{12}c_2 - \lambda a_{13}c_3 - \cdots = b_1,$$

$$-\lambda a_{21}c_1 + (1 - \lambda a_{22})c_2 - \lambda a_{23}c_3 - \cdots = b_2, \qquad (16.77)$$

$$-\lambda a_{31}c_1 - \lambda a_{32}c_2 + (1 - \lambda a_{33})c_3 - \cdots = b_3, \qquad \text{and so on.}$$

If our integral equation is homogeneous, $[f(x) = 0]$, then $\mathbf{b} = 0$. To get a solution, we set the determinant of the coefficients of c_i equal to zero,

$$|1 - \lambda \mathsf{A}| = 0, \qquad (16.78)$$

exactly as in Section 3.5. The roots of Eq. (16.78) yield our eigenvalues. Substituting into $(1 - \lambda \mathsf{A})\mathbf{c} = 0$, we find the c_i's and then Eq. (16.73) gives our solution.

Example 16.3.2

To illustrate this technique for determining eigenvalues and eigenfunctions of the homogeneous Fredholm equation, we consider the simple case

$$\varphi(x) = \lambda \int_{-1}^{1} (t + x)\varphi(t) \, dt. \qquad (16.79)$$

Here

$$M_1 = 1, \qquad M_2(x) = x,$$

$$N_1(t) = t, \qquad N_2 = 1.$$

Equation (16.75) yields

$$a_{11} = a_{22} = 0,$$

$$a_{12} = \tfrac{2}{3},$$

$$a_{21} = 2.$$

Equation (16.78), our secular equation, becomes

$$\begin{vmatrix} 1 & -\dfrac{2\lambda}{3} \\ -2\lambda & 1 \end{vmatrix} = 0. \qquad (16.80)$$

Expanding, we obtain

$$1 - \frac{4\lambda^2}{3} = 0, \qquad \lambda = \pm\frac{\sqrt{3}}{2}. \qquad (16.81)$$

Substituting the eigenvalues $\lambda = \pm\sqrt{3}/2$ into Eq. (16.76), we have

$$c_1 \mp \frac{c_2}{\sqrt{3}} = 0. \qquad (16.82)$$

Finally, with a choice of $c_1 = 1$, Eq. (16.73) gives

$$\varphi_1(x) = \frac{\sqrt{3}}{2}(1 + \sqrt{3}x), \qquad \lambda = \frac{\sqrt{3}}{2}, \tag{16.83}$$

$$\varphi_2(x) = -\frac{\sqrt{3}}{2}(1 - \sqrt{3}x), \qquad \lambda = -\frac{\sqrt{3}}{2}. \tag{16.84}$$

Since our equation is homogeneous, the normalization of $\varphi(x)$ is arbitrary.

If the kernel is not separable in the sense of Eq. (16.70), there is still the possibility that it may be approximated by a kernel that is separable. Then we can get the exact solution of an approximate equation, an equation that approximates the original equation. The solution of the separable approximate kernel problem can then be checked by substituting back into the original, unseparable kernel problem.

NUMERICAL SOLUTION

There is extensive literature on the numerical solution of integral equations, much of it concerns special techniques for certain situations. One method of fair generality is the replacement of the single integral equation by a set of simultaneous algebraic equations. And again matrix techniques are invoked. This simultaneous algebraic equation-matrix approach—is applied here to two different cases. For the homogeneous Fredholm equation of the second kind this method works well. For the Fredholm equation of the first kind the method is a disaster. First we deal with the disaster.

We consider the Fredholm integral equation of the first kind

$$f(x) = \int_a^b K(x, t)\varphi(t)\, dt, \tag{16.84a}$$

with $f(x)$ and $K(x, t)$ known and $\varphi(t)$ unknown. The integral can be evaluated (in principle) by quadrature techniques. For maximum accuracy the Gaussian method (Appendix 2) is recommended (if the kernel is continuous and has continuous derivatives). The numerical quadrature replaces the integral by a summation,

$$f(x_i) = \sum_{k=1}^{n} A_k K(x_i, t_k)\varphi(t_k), \tag{16.84b}$$

with A_k the quadrature coefficients. We abbreviate $f(x_i)$ as f_i, $\varphi(t_k)$ as φ_k, and $A_k K(x_i, t_k)$ as B_{ik}. In effect we are changing from a function description to a vector–matrix description with the n components of the vector (f_i) defined as the values of the function at the n discrete points $[f(x_i)]$. Equation (16.84b) becomes

$$f_i = \sum_{k=1}^{n} B_{ik}\varphi_k,$$

a matrix equation. Inverting (B_{ik}), we obtain

$$\varphi(x_k) = \varphi_k = \sum_{k=1}^{n} B_{ki}^{-1} f_i, \tag{16.84c}$$

and Eq. (16.84a) is solved—in principle. In practice, the quadrature coefficient-kernel matrix is often "ill-conditioned" (with respect to inversion). This means that in the inversion process small (numerical) errors are multiplied by large factors. In the inversion process all significant figures may be lost and Eq. (16.84c) becomes numerical nonsense.

This disaster should not be entirely unexpected. Integration is essentially a smoothing operation. $f(x)$ is relatively insensitive to local variation of $\varphi(t)$. Conversely, $\varphi(t)$ may be exceedingly sensitive to small changes in $f(x)$. Small errors in $f(x)$ or in B^{-1} are magnified and accuracy disappears. This same behavior shows up in attempts to invert Laplace transforms numerically— Section 15.8.

When the quadrature—matrix technique is applied to the integral equation eigenvalue problem, the symmetric kernel, homogeneous Fredholm equation of the second kind,[2]

$$\lambda \varphi(x) = \int_a^b K(x, t)\varphi(t)\, dt, \tag{16.84d}$$

the technique is far more successful. Replacing the integral by a set of simultaneous algebraic equations (numerical quadrature, Appendix 2), we have

$$\lambda \varphi_i = \sum_{k=1}^{n} A_k K_{ik} \varphi_k, \tag{16.84e}$$

with $\varphi_i = \varphi(x_i)$ as before. The points x_i, $i = 1, 2, \ldots, n$ are taken to be the same (numerically) as t_k, $k = 1, 2, \ldots, n$, so that K_{ik} will be symmetric. The system is symmetrized by multiplying by $A_i^{1/2}$ so that

$$\lambda(A_i^{1/2}\varphi_i) = \sum_{k=1}^{n} (A_i^{1/2} K_{ik} A_k^{1/2})(A_k^{1/2}\varphi_k). \tag{16.84f}$$

Replacing $A_i^{1/2}\varphi_i$ by ψ_i and $A_i^{1/2} K_{ik} A_k^{1/2}$ by S_{ik}, we obtain

$$\lambda \psi = S\psi, \tag{16.84g}$$

with S symmetric (since the kernel $K(x, t)$ was assumed symmetric. ψ, of course, has components $\psi_i = \psi(x_i)$. Equation (16.84g) is our matrix eigenvalue equation, Eq. (3.136). The eigenvalues are readily obtained by calling a canned eigenroutine.[3] For kernels such as those of Exercise 16.3.15 and using

---

[2] The eigenvalue λ has been written on the left side, multiplying the eigenfunction, as is customary in matrix analysis (Section 3.5). In this form λ will take on a *maximum value*.

[3] See W. H. Press, B. P. Flannery, S. A. Teukolsky, and W. T. Vetterling, *Numerical Recipes*, 2nd ed. Cambridge: Cambridge University Press (1992), Chapter 11, for details, references and computer codes. The symbolic software *Mathematica* and *Maple* also include matrix functions for computing eigenvalues and eigenvectors.

a 10-point Gauss–Legendre quadrature, the eigenroutine determines the largest eigenvalue to within about 0.5 percent for the cases where the kernel has discontinuities in its derivatives. If the derivatives are continuous, the accuracy is much better.

Linz[4] has described an interesting variational refinement in the determination of λ_{max} to high accuracy. The key to his method is Exercise 17.8.7. The components of the eigenfunction vector are obtained from Eq. (16.84d) with $\varphi(t_k)$ now known and $\varphi_1 = \varphi(x_i)$ generated as required. (The x_i are no longer tied to the t_k.)

EXERCISES

16.3.1   Using the Neumann series, solve

(a) $\varphi(x) = 1 - 2 \int_0^x t\varphi(t)\,dt,$

$ANS.$ (a) $\varphi(x) = e^{-x^2}.$

(b) $\varphi(x) = x + \int_0^x (t - x)\varphi(t)\,dt,$

(c) $\varphi(x) = x - \int_0^x (t - x)\varphi(t)\,dt.$

16.3.2   Solve the equation

$$\varphi(x) = x + \tfrac{1}{2}\int_{-1}^1 (t + x)\varphi(t)\,dt$$

by the separable kernel method. Compare with the Neumann method solution of Section 16.3.

$ANS.$ $\varphi(x) = \tfrac{1}{2}(3x - 1).$

16.3.3   Find the eigenvalues and eigenfunctions of

$$\varphi(x) = \lambda \int_{-1}^1 (t - x)\varphi(t)\,dt.$$

16.3.4   Find the eigenvalues and eigenfunctions of

$$\varphi(x) = \lambda \int_0^{2\pi} \cos(x - t)\varphi(t)\,dt.$$

$ANS.$ $\lambda_1 = \lambda_2 = \dfrac{1}{\pi},$ $\varphi(x) = A\cos x + B\sin x.$

16.3.5   Find the eigenvalues and eigenfunctions of

$$y(x) = \lambda \int_{-1}^1 (x - t)^2 y(t)\,dt.$$

Hint. This problem may be treated by the separable kernel method or by a Legendre expansion.

[4] P. Linz, On the numerical computation of eigenvalues and eigenvectors of symmetric integral equations. *Math. Comput.* **24**, 905 (1970).

16.3.6 If the separable kernel technique of this section is applied to a Fredholm equation of the first kind, (Eq. (16.1)), show that Eq. (16.76) is replaced by

$$\mathbf{c} = \mathbf{A}^{-1}\mathbf{b}.$$

In general the solution for the unknown $\varphi(t)$ is *not* unique.

16.3.7 Solve

$$\psi(x) = x + \int_0^1 (1 + xt)\psi(t)\, dt$$

by each of the following methods:

(a) the Neumann series technique,
(b) the separable kernel technique,
(c) educated guessing.

16.3.8 Use the separable kernel technique to show that

$$\psi(x) = \lambda \int_0^\pi \cos x \sin t\, \psi(t)\, dt$$

has *no* solution (apart from the trivial $\psi = 0$). Explain this result in terms of separability and symmetry.

16.3.9 Solve

$$\varphi(x) = 1 + \lambda^2 \int_0^x (x - t)\varphi(t)\, dt$$

by each of the following methods:

(a) reduction to a differential equation (including establishment of boundary conditions),
(b) the Neumann series,
(c) the use of Laplace transforms.

$$ANS. \quad \varphi(x) = \cosh \lambda x.$$

16.3.10 (a) In Eq. (16.69a) take $V = V_0$, independent of t. Without using Eq. (16.69b), show that Eq. (16.69a) leads directly to

$$U(t - t_0) = \exp[-i(t - t_0)V_0/\hbar].$$

(b) Repeat for Eq. (16.69b) without using Eq. (16.69a).

16.3.11 Given $\varphi(x) = \lambda \int_0^1 (1 + xt)\varphi(t)\, dt$, solve for the eigenvalues and the eigenfunctions by the separable kernel technique.

16.3.12 Knowing the form of the solutions can be a great advantage, for the integral equation

$$\varphi(x) = \lambda \int_0^1 (1 + xt)\varphi(t)\, dt,$$

assume $\varphi(x)$ to have the form $1 + bx$. Substitute into the integral equation. Integrate and solve for b and λ.

16.3.13 The integral equation

$$\varphi(x) = \lambda \int_0^1 J_0(\alpha xt)\varphi(t)\, dt, \qquad J_0(\alpha) = 0$$

is approximated by

$$\varphi(x) = \lambda \int_0^1 [1 - x^2 t^2]\varphi(t)\, dt.$$

Find the minimum eigenvalue λ and the corresponding eigenfunction $\varphi(t)$ of the approximate equation.

ANS. $\lambda_{\min} = 1.112486$, $\varphi(x) = 1.-0.303337x^2$.

16.3.14   You are given the integral equation

$$\varphi(x) = \lambda \int_0^1 \sin \pi x t \, \varphi(t)\, dt.$$

Approximate the kernel by

$$K(x, t) = 4xt(1 - xt) \approx \sin \pi x t.$$

Find the positive eigenvalue and the corresponding eigenfunction for the approximate integral equation.

Note. For $K(x, t) = \sin \pi x t$, $\lambda = 1.6334$.

ANS. $\lambda = 1.5678$

$$\varphi(x) = x - 0.6955x^2$$
$$(\lambda_+ = \sqrt{31} - 4, \lambda_- = -\sqrt{31} - 4)$$

16.3.15   The equation

$$f(x) = \int_a^b K(x, t)\varphi(t)\, dt$$

has a degenerate kernel $K(x, t) = \sum_{i=1}^n M_i(x)N_i(t)$.

(a) Show that this integral equation has no solution unless $f(x)$ can be written as

$$f(x) = \sum_{i=1}^n f_i M_i(x),$$

with the f_i constants.

(b) Show that to any solution $\varphi(x)$ we may add $\psi(x)$, provided $\psi(x)$ is orthogonal to all $N_i(x)$:

$$\int_a^b N_i(x)\psi(x)\, dx = 0 \qquad \text{for all } i.$$

16.3.16   Using numerical quadrature, convert

$$\varphi(x) = \lambda \int_0^1 J_0(\alpha x t)\varphi(t)\, dt, \qquad J_0(\alpha) = 0$$

to a set of simultaneous linear equations.

(a) Find the minimum eigenvalue λ.

(b) Determine $\varphi(x)$ at discrete values of x and plot $\varphi(x)$ versus x. Compare with the approximate eigenfunction of Exercise 16.3.13.

ANS.   (a) $\lambda_{\min} = 1.14502$.

16.3.17   Using numerical quadrature, convert

$$\varphi(x) = \lambda \int_0^1 \sin \pi x t \, \varphi(t)\, dt$$

to a set of simultaneous linear equations.

(a) Find the minimum eigenvalue λ.
(b) Determine $\varphi(x)$ at discrete values of x and plot $\varphi(x)$ versus x. Compare with the approximate eigenfunction of Exercise 16.3.14.

$$ANS. \quad (a) \ \lambda_{\min} = 1.6334.$$

16.3.18   Given a homogeneous Fredholm equation of the second kind

$$\lambda \varphi(x) = \int_0^1 K(x, t) \varphi(t) \, dt.$$

(a) Calculate the largest eigenvalue λ_0. Use the 10-point Gauss–Legendre quadrature technique. For comparison the eigenvalues listed by Linz are given as λ_{exact}.
(b) Tabulate $\varphi(x_k)$, where the x_k are the 10 evaluation points in $[0, 1]$.
(c) Tabulate the ratio

$$\int_0^1 K(x, t) \varphi(t) \, dt / \lambda_0 \varphi(x) \qquad \text{for } x = x_k.$$

This is the test of whether or not you really have a solution.

(a) $K(x, t) = e^{xt}$.

$$ANS. \quad \lambda_{\text{exact}} = 1.35303.$$

(b) $K(x, t) = \begin{cases} \frac{1}{2}x(2 - t), & x < t \\ \frac{1}{2}t(2 - x), & x > t. \end{cases}$

$$ANS. \quad \lambda_{\text{exact}} = 0.24296.$$

(c) $K(x, t) = |x - t|$.

$$ANS. \quad \lambda_{\text{exact}} = 0.34741.$$

(d) $K(x, t) = \begin{cases} x, & x < t \\ t, & x > t. \end{cases}$

$$ANS. \quad \lambda_{\text{exact}} = 0.40528.$$

Note. (1) The evaluation points x_i of Gauss–Legendre quadrature for $[-1, 1]$ may be *linearly* transformed into $[0, 1]$,

$$x_i[0, 1] = \tfrac{1}{2}(x_i[-1, 1] + 1).$$

Then the weighting factors A_i are reduced in proportion to the length of the interval

$$A_i[0, 1] = \tfrac{1}{2}A_i[-1, 1].$$

16.3.19   Using the matrix variational technique of Exercise 17.8.7, refine your calculation of the eigenvalue of Exercise 16.3.18(c) $[K(x, t) = |x - t|]$. Try a 40×40 matrix.
Note. Your matrix should be symmetric so that the (unknown) eigenvectors will be orthogonal.

$$ANS. \quad \text{(40 point Gauss–Legendre quadrature) } 0.34727.$$

16.4 HILBERT–SCHMIDT THEORY

SYMMETRIZATION OF KERNELS

This is the development of the properties of linear integral equations (Fredholm type) with symmetric kernels.

$$K(x, t) = K(t, x). \tag{16.85}$$

Before plunging into the theory, we note that some important nonsymmetric kernels can be symmetrized. If we have the equation

$$\varphi(x) = f(x) + \lambda \int_a^b K(x, t)\rho(t)\varphi(t)\, dt. \tag{16.86}$$

the total kernel is actually $K(x, t)\rho(t)$, clearly not symmetric if $K(x, t)$ alone is symmetric. However, if we multiply Eq. (16.86) by $\sqrt{\rho(x)}$ and substitute

$$\sqrt{\rho(x)}\varphi(x) = \psi(x), \tag{16.87}$$

we obtain

$$\psi(x) = \sqrt{\rho(x)}f(x) + \lambda \int_a^b [K(x, t)\sqrt{\rho(x)\rho(t)}]\psi(t)\, dt, \tag{16.88}$$

with a symmetric total kernel, $K(x, t)\sqrt{\rho(x)\rho(t)}$. We shall meet $\rho(x)$ later as a positive weighting factor in this integral equation Sturm–Liouville theory.

ORTHOGONAL EIGENFUNCTIONS

We now focus on the homogeneous Fredholm equation of the second kind:

$$\varphi(x) = \lambda \int_a^b K(x, t)\varphi(t)\, dt. \tag{16.89}$$

We assume that the kernel $K(x, t)$ is symmetric and real. Perhaps one of the first questions the mathematician might ask about the equation is: "Does it make sense?" or more precisely, "Does an eigenvalue λ satisfying this equation exist?" With the aid of the Schwarz and Bessel inequalities, Courant and Hilbert (Chapter III, Section 4) show that if $K(x, t)$ is continuous, there is at least one such eigenvalue and possibly an infinite number of them.

We show that the eigenvalues, λ, are real and that the corresponding eigenfunctions, $\varphi_i(x)$, are orthogonal. Let λ_i, λ_j be two *different* eigenvalues and $\varphi_i(x)$, $\varphi_j(x)$, the corresponding eigenfunctions. Equation (16.89) then becomes

$$\varphi_i(x) = \lambda_i \int_a^b K(x, t)\varphi_i(t)\, dt, \tag{16.90a}$$

$$\varphi_j(x) = \lambda_j \int_a^b K(x, t)\varphi_j(t)\, dt. \tag{16.90b}$$

If we multiply Eq. (16.90a) by $\lambda_j\varphi_j(x)$, Eq. (16.90b) by $\lambda_i\varphi_i(x)$, and then integrate with respect to x, the two equations become[1]

$$\lambda_j \int_a^b \varphi_i(x)\varphi_j(x)\, dx = \lambda_i\lambda_j \int_a^b \int_a^b K(x, t)\varphi_i(t)\varphi_j(x)\, dt\, dx, \tag{16.91a}$$

$$\lambda_i \int_a^b \varphi_i(x)\varphi_j(x)\, dx = \lambda_i\lambda_j \int_a^b \int_a^b K(x, t)\varphi_j(t)\varphi_i(x)\, dt\, dx. \tag{16.91b}$$

[1] We assume that the necessary integrals exist. For an example of a simple pathological case, see Exercise 16.4.3.

Since we have demanded that $K(x, t)$ by symmetric, Eq. (16.91b) may be rewritten as

$$\lambda_i \int_a^b \varphi_i(x)\varphi_j(x)\,dx = \lambda_i \lambda_j \int_a^b \int_a^b K(x, t)\varphi_i(t)\varphi_j(x)\,dt\,dx. \qquad (16.92)$$

Subtracting Eq. (16.92) from Eq. (16.91a), we obtain

$$(\lambda_j - \lambda_i) \int_a^b \varphi_i(x)\varphi_j(x)\,dx = 0. \qquad (16.93)$$

This has the same form as Eq. (9.34) in the Sturm–Liouville theory. Since $\lambda_i \neq \lambda_j$,

$$\int_a^b \varphi_i(x)\varphi_j(x)\,dx = 0, \qquad i \neq j, \qquad (16.94)$$

proving orthogonality. Note that with a real symmetric kernel no complex conjugates are involved in Eq. (16.94). For the self-adjoint or Hermitian kernel see Exercise 16.4.1.

If the eigenvalue λ_i is degenerate,[2] the eigenfunctions for that particular eigenvalue may be orthogonalized by the Gram–Schmidt method (Section 9.3). Our orthogonal eigenfunctions may, of course, be normalized, and we assume that this has been done. The result is

$$\int_a^b \varphi_i(x)\varphi_j(x)\,dx = \delta_{ij}. \qquad (16.95)$$

To demonstrate that the λ_i are real, we need to get into complex conjugates. Taking the complex conjugate of Eq. (16.90a), we have

$$\varphi_i^*(x) = \lambda_i^* \int_a^b K(x, t)\varphi_i^*(t)\,dt, \qquad (16.96)$$

provided the kernel $K(x, t)$ is real. Now, using Eq. (16.96) instead of Eq. (16.90b), we see that the analysis leads to

$$(\lambda_i^* - \lambda_i) \int_a^b \varphi_i^*(x)\varphi_i(x)\,dx = 0. \qquad (16.97)$$

This time the integral cannot vanish (unless we have the trivial solution, $\varphi_i(x) = 0$) and

$$\lambda_i^* = \lambda_i \qquad (16.98)$$

or λ_i, our eigenvalue, is real.

If readers feel that somehow this state of affairs is vaguely familiar, they are right. This is the *third* time we have passed this way, first with Hermitian matrices, then with Sturm–Liouville (self-adjoint) equations, and now with

---

[2] If more than one distinct eigenfunction corresponds to the same eigenvalue (satisfying Eq. (16.89)), that eigenvalue is said to be degenerate (see Chapter 4).

Hilbert–Schmidt integral equations. The correspondence between the Hermitian matrices and the self-adjoint differential equations shows up in modern physics as the two outstanding formulations of quantum mechanics—the Heisenberg matrix approach and the Schrödinger differential operator approach. In Section 16.5 we shall explore further the correspondence between the Hilbert–Schmidt symmetric kernel integral equations and the Sturm–Liouville self-adjoint differential equations.

The eigenfunctions of our integral equations form a complete set[3] in the sense that any function $g(x)$ that can be generated by the integral

$$g(x) = \int K(x, t)h(t)\, dt,\tag{16.99}$$

in which $h(t)$ is any piecewise continuous function, can be represented by a series of eigenfunctions,

$$g(x) = \sum_{n=1}^{\infty} a_n \varphi_n(x).\tag{16.100}$$

The series converges uniformly and absolutely.

Let us extend this to the kernel, $K(x, t)$, by asserting that

$$K(x, t) = \sum_{n=1}^{\infty} a_n \varphi_n(t),\tag{16.101}$$

and $a_n = a_n(x)$. Substituting into the original integral equation (Eq. (16.89)) and using the orthogonality integral, we obtain

$$\varphi_i(x) = \lambda_i a_i(x).\tag{16.102}$$

Therefore for our homogeneous Fredholm equation of the second kind the kernel may be expressed in terms of the eigenfunctions and eigenvalues by

$$K(x, t) = \sum_{n=1}^{\infty} \frac{\varphi_n(x)\varphi_n(t)}{\lambda_n}\qquad \text{(zero not an eigenvalue).}\tag{16.103}$$

Here we have a bilinear expansion, a linear expansion in $\varphi_n(x)$ and linear in $\varphi_n(t)$. Similar bilinear expansions appear in Section 8.7. It is possible that the expansion given by Eq. (16.101) may not exist. As an illustration of the sort of pathological behavior that may occur, the reader is invited to apply this analysis to

$$\varphi(x) = \lambda \int_0^\infty e^{-xt}\varphi(t)\, dt$$

(compare Exercise 16.4.3).

It should be emphasized that this Hilbert–Schmidt theory is concerned with the establishment of properties of the eigenvalues (real) and eigenfunctions (orthogonality, completeness), properties that may be of great interest and

---

[3] For a proof of this statement see Courant and Hilbert (1953), Chapter III, Section 5.

value. The Hilbert–Schmidt theory does *not* solve the homogeneous integral equation for us any more than the Sturm–Liouville theory of Chapter 9 solved the differential equations. The solutions of the integral equation come from Sections 16.2 and 16.3 (including numerical analysis).

NONHOMOGENEOUS INTEGRAL EQUATION

We need a solution of the nonhomogeneous equation

$$\varphi(x) = f(x) + \lambda \int_a^b K(x, t)\varphi(t)\, dt. \tag{16.104}$$

Let us assume that the solutions of the corresponding homogeneous integral equation are known:

$$\varphi_n(x) = \lambda_n \int_a^b K(x, t)\varphi_n(t)\, dt, \tag{16.105}$$

the solution $\varphi_n(x)$ corresponding to the eigenvalue λ_n. We expand both $\varphi(x)$ and $f(x)$ in terms of this set of eigenfunctions:

$$\varphi(x) = \sum_{n=1}^{\infty} a_n \varphi_n(x) \qquad (a_n \text{ unknown}) \tag{16.106}$$

$$f(x) = \sum_{n=1}^{\infty} b_n \varphi_n(x) \qquad (b_n \text{ known}). \tag{16.107}$$

Substituting into Eq. (16.104), we obtain

$$\sum_{n=1}^{\infty} a_n \varphi_n(x) = \sum_{n=1}^{\infty} b_n \varphi_n(x) + \lambda \int_a^b K(x, t) \sum_{n=1}^{\infty} a_n \varphi_n(t)\, dt. \tag{16.108}$$

By interchanging the order of integration and summation, we may evaluate the integral by Eq. (16.105), and we get

$$\sum_{n=1}^{\infty} a_n \varphi_n(x) = \sum_{n=1}^{\infty} b_n \varphi_n(x) + \lambda \sum_{n=1}^{\infty} \frac{a_n \varphi_n(x)}{\lambda_n}. \tag{16.109}$$

If we multiply by $\varphi_i(x)$ and integrate from $x = a$ to $x = b$, the orthogonality of our eigenfunctions leads to

$$a_i = b_i + \lambda \frac{a_i}{\lambda_i}. \tag{16.110}$$

This can be rewritten as

$$a_i = b_i + \frac{\lambda}{\lambda_i - \lambda} b_i \tag{16.111}$$

which brings us to our solution

$$\varphi(x) = f(x) + \lambda \sum_{i=1}^{\infty} \frac{\int_a^b f(t)\varphi_i(t)\, dt}{\lambda_i - \lambda} \varphi_i(x). \tag{16.112}$$

Here it is assumed that the eigenfunctions, $\varphi_i(x)$, are normalized to unity. *Note that if $f(x) = 0$ there is no solution unless $\lambda = \lambda_i$.* This means that our homogeneous equation has no solution (except the trivial $\varphi(x) = 0$) unless λ is an eigenvalue, λ_i.

In the event that λ for the nonhomogeneous equation (16.104) is equal to one of the eigenvalues, λ_p, of the homogeneous equation, our solution (Eq. (16.112)) blows up. To repair the damage we return to Eq. (16.110) and give the value

$$a_p = b_p + \lambda_p \frac{a_p}{\lambda_p} = b_p + a_p \tag{16.113}$$

special attention. Clearly, a_p drops out and is no longer determined by b_p, whereas $b_p = 0$. This implies that $\int f(x)\varphi_p(x)\, dx = 0$, that is, $f(x)$ is orthogonal to the eigenfunction $\varphi_p(x)$. If this is *not* the case, we have no solution.

Equation (16.111) still holds for $i \neq p$, so we multiply by $\varphi_i(x)$ and sum over $i(i \neq p)$ to obtain

$$\varphi(x) = f(x) + a_p \varphi_p + \lambda_p \sum_{\substack{i=1 \\ i \neq p}}^{\infty}{}' \frac{\int_a^b f(t)\varphi_i(t)\, dt}{\lambda_i - \lambda_p} \varphi_i(x); \tag{16.114}$$

the prime emphasizes that the value $i = p$ is omitted. In this solution the a_p remains as an undetermined constant.[4]

EXERCISES

16.4.1 In the Fredholm equation

$$\varphi(x) = \lambda \int_a^b K(x, t)\varphi(t)\, dt$$

the kernel $K(x, t)$ is self-adjoint or Hermitian.

$$K(x, t) = K^*(t, x).$$

Show that

(a) the eigenfunctions are orthogonal in the sense

$$\int_a^b \varphi_m^*(x)\varphi_n(x)\, dx = 0, \qquad m \neq n \ (\lambda_m \neq \lambda_n),$$

(b) the eigenvalues are real.

16.4.2 Solve the integral equation

$$\varphi(x) = x + \tfrac{1}{2} \int_{-1}^{1} (t + x)\varphi(t)\, dt$$

(compare Exercise 16.3.2) by the Hilbert–Schmidt method.

---

[4] This is like the inhomogeneous linear differential equation. We may add to its solution any constant times a solution of the corresponding homogeneous differential equation.

The application of the Hilbert–Schmidt technique here is somewhat like using a shotgun to kill a mosquito, especially when the equation can be solved in about 15 seconds by expanding in Legendre polynomials.

16.4.3   Solve the Fredholm integral equation

$$\varphi(x) = \lambda \int_0^\infty e^{-xt}\varphi(t)\,dt.$$

Note. A series expansion of the kernel e^{-xt} would permit a separable kernel-type solution (Section 16.3), except that the series is infinite. This suggests an infinite number of eigenvalues and eigenfunctions. If you stop with

$$\varphi(x) = x^{-1/2},$$
$$\lambda = \pi^{-1/2},$$

you will have missed most of the solutions! Show that the normalization integrals of the eigenfunctions do *not* exist. A basic reason for this anomalous behavior is that the range of integration is infinite, making this a "singular" integral equation.

16.4.4   Given

$$y(x) = x + \lambda \int_0^1 xt\,y(t)\,dt.$$

(a) Determine $y(x)$ as a Neumann series.
(b) Find the range of λ for which your Neumann series solution is convergent. Compare with the value obtained from

$$|\lambda| \cdot |K|_{max} < 1.$$

(c) Find the eigenvalue and the eigenfunction of the corresponding homogeneous integral equation.
(d) By the separable kernel method show that the solution is

$$y(x) = \frac{3x}{3 - \lambda}.$$

(e) Find $y(x)$ by the Hilbert–Schmidt method.

16.4.5   In Exercise 16.3.4

$$K(x, t) = \cos(x - t).$$

The (unnormalized) eigenfunctions are $\cos x$ and $\sin x$.

(a) Show that there is a function $h(t)$ such that $K(x, s)$, considered as a function of s alone, may be written as

$$K(x, s) = \int_0^{2\pi} K(s, t)h(t)\,dt.$$

(b) Show that $K(x, t)$ may be expanded as

$$K(x, t) = \sum_{n=1}^{2} \frac{\varphi_n(x)\varphi_n(t)}{\lambda_n}.$$

16.4.6   The integral equation $\varphi(x) = \lambda \int_0^1 (1 + xt)\varphi(t)\,dt$ has eigenvalues $\lambda_1 = 0.7889$ and $\lambda_2 = 15.211$ and eigenfunctions $\varphi_1 = 1 + 0.5352x$ and $\varphi_2 = 1 - 1.8685x$.

(a) Show that these eigenfunctions are orthogonal over the interval [0, 1].

(b) Normalize the eigenfunctions to unity.

(c) Show that

$$K(x, t) = \frac{\varphi_1(x)\varphi_1(t)}{\lambda_1} + \frac{\varphi_2(x)\varphi_2(t)}{\lambda_2}.$$

$$ANS. \quad (b) \; \varphi_1(x) = 0.7831 + 0.4191x$$
$$\varphi_2(x) = 1.8403 -$$
$$3.4386x.$$

16.4.7   An alternate form of the solution to the nonhomogeneous integral equation, Eq. (16.104), is

$$\varphi(x) = \sum_{i=1}^{\infty} \frac{b_i \lambda_i}{\lambda_i - \lambda} \varphi_i(x).$$

(a) Derive this form without using Eq. (16.112).

(b) Show that this form and Eq. (16.112) are equivalent.

16.4.8   (a) Show that the eigenfunctions of Exercise 16.3.5 are orthogonal.

(b) Show that the eigenfunctions of Exercise 16.3.11 are orthogonal.

ADDITIONAL READINGS

BOCHER, M., *An Introduction to the Study of Integral Equations*, Cambridge Tracts in Mathematics and Mathematical Physics, No. 10. New York: Hafner (1960). This is a very helpful introduction to integral equations.

COCHRAN, J. A., *The Analysis of Linear Integral Equations*. New York: McGraw-Hill (1972). This is a comprehensive treatment of linear integral equations which is intended for applied mathematicians and mathematical physicists. It assumes a moderate to high level of mathematical competence on the part of the reader.

COURANT, R., and D. HILBERT, *Methods of Mathematical Physics*, Vol. 1 (English edition). New York: Interscience (1953). This is one of the classic works of mathematical physics. Originally published in German in 1924, the revised English edition is an excellent reference for a rigorous treatment of integral equations, Green's functions, and a wide variety of other topics on mathematical physics.

GOLBERG, M. A., ed., *Solution Methods of Integral Equations*. New York: Plenum Press (1979). This is a set of papers from a conference on integral equations. The initial chapter is excellent for up-to-date orientation and a wealth of current references.

KANWAL, R. P., *Linear Integral Equations*. New York: Academic Press (1971). This book is a detailed but readable treatment of a variety of techniques for solving linear integral equations.

MORSE, P. M., and H. FESHBACH, *Methods of Theoretical Physics*. New York: McGraw-Hill (1953). Chapter 7 is a particularly detailed, complete discussion of Green's functions from the point of view of mathematical physics. Note, however, that Morse and Feshbach frequently choose a source of $4\pi\delta(\mathbf{r} - \mathbf{r}')$ in place of our $\delta(\mathbf{r} - \mathbf{r}')$. Considerable attention is devoted to bounded regions.

STAKGOLD, I., *Green's Functions and Boundary Value Problems*. New York: Wiley (1979).

17 CALCULUS OF VARIATIONS

USES OF THE CALCULUS OF VARIATIONS

Before plunging into this rather different branch of mathematical physics, let us summarize some of its uses in both physics and mathematics.

1. Existing physical theories:

   a. Unification of diverse areas of physics—using energy as a key concept.
   b. Convenience in analysis—Lagrange equations, Section 17.3.
   c. Convenient introduction of constraints, Section 17.7.

2. Starting point for new, complex areas of physics and engineering. In general relativity the geodesic is taken as the minimum path of a light pulse or the free fall path of a particle in curved Riemannian space. Variational principles appear in modern quantum field theory. Variational principles have been applied extensively in modern control theory.

3. Mathematical unification. Variational analysis provides a proof of the completeness of the Sturm–Liouville eigenfunctions, Chapter 9, and establishes a lower bound for the eigenvalues. Similar results follow for the eigenvalues and eigenfunctions of the Hilbert–Schmidt integral equation, Section 16.4.

4. Calculation techniques, Section 17.8. Calculation of the eigenfunctions and eigenvalues of the Sturm–Liouville equation. Integral equation eigenfunctions and eigenvalues may be calculated using numerical quadrature and matrix techniques, Section 16.3.

17.1 ONE-DEPENDENT AND ONE-INDEPENDENT VARIABLE

CONCEPT OF VARIATION

The calculus of variations involves problems in which the quantity to be minimized (or maximized) appears as a stationary integral. As the simplest case, let

$$J = \int_{x_1}^{x_2} f(y, y_x, x)\, dx. \tag{17.1}$$

Here J is the quantity that takes on a stationary value. Under the integral sign, f is a known function of the indicated variables $y(x)$, $y_x(x) \equiv dy(x)/dx$ and x, but the dependence of y on x is not fixed; that is, $y(x)$ is *unknown*. This means that although the integral is from x_1 to x_2, the exact path of integration is not known (Fig. 17.1).

We are to choose the path of integration through points (x_1, y_1) and (x_2, y_2) to minimize J. Strictly speaking, we determine stationary values of J: minima, maxima, or saddle points. In most cases of physical interest the stationary value will be a minimum.

This problem is considerably more difficult than the corresponding problem of a function $y(x)$ in differential calculus. Indeed, there may be no solution. In differential calculus the minimum is determined by comparing $y(x_0)$ with $y(x)$, where x ranges over neighboring points. Here we assume the existence of an optimum path, that is, an acceptable path for which J is stationary, and then compare J for our (unknown) optimum path with that obtained from neighboring paths. In Fig. 17.1 two possible paths are shown. (There are an infinite number of possibilities, of course.) The difference between these two for a given x is called the variation of y, δy, and is conveniently described by introducing a new function $\eta(x)$ to define the arbitrary deformation of the path and a scale factor α to give the magnitude of the variation. The function

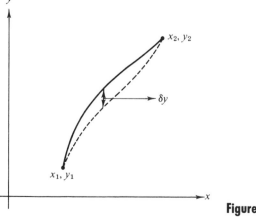

Figure 17.1 A varied path.

$\eta(x)$ is arbitrary except for two restrictions. First,

$$\eta(x_1) = \eta(x_2) = 0, \tag{17.2}$$

which means that all varied paths must pass through the fixed end points. Second, as will be seen shortly, $\eta(x)$ must be differentiable; that is, we may not use

$$\eta(x) = 1, \qquad x = x_0,$$

$$= 0, \qquad x \neq x_0, \tag{17.3}$$

but we can choose $\eta(x)$ to have a form similar to the functions used to represent the Dirac delta function (Chapter 1) so that $\eta(x)$ differs from zero only over an infinitesimal region.[1] Then, with the path described with α and $\eta(x)$,

$$y(x, \alpha) = y(x, 0) + \alpha\eta(x), \tag{17.4}$$

and

$$\delta y = y(x, \alpha) - y(x, 0) = \alpha\eta(x). \tag{17.5}$$

Let us choose $y(x, \alpha = 0)$ as the unknown path that will minimize J. Then $y(x, \alpha)$ describes a neighboring path. In Eq. (17.1), J is now a function[2] of our new parameter α:

$$J(\alpha) = \int_{x_1}^{x_2} f[y(x, \alpha), y_x(x, \alpha), x] \, dx, \tag{17.6}$$

and our condition for an extreme value is that

$$\left[\frac{\partial J(\alpha)}{\partial \alpha}\right]_{\alpha=0} = 0, \tag{17.7}$$

analogous to the vanishing of the derivative dy/dx in differential calculus.

Now the α-dependence of the integral is contained in $y(x, \alpha)$ and $y_x(x, \alpha) = (\partial/\partial x)y(x, \alpha)$. Therefore[3]

$$\frac{\partial J(\alpha)}{\partial \alpha} = \int_{x_1}^{x_2} \left[\frac{\partial f}{\partial y}\frac{\partial y}{\partial \alpha} + \frac{\partial f}{\partial y_x}\frac{\partial y_x}{\partial \alpha}\right] dx. \tag{17.8}$$

From Eq. (17.4)

$$\frac{\partial y(x, \alpha)}{\partial \alpha} = \eta(x) \tag{17.9}$$

$$\frac{\partial y_x(x, \alpha)}{\partial \alpha} = \frac{d\eta(x)}{dx}. \tag{17.10}$$

_____

[1] Compare H. Jeffreys and B. S. Jeffreys, *Methods of Mathematical Physics*, 3rd ed. Cambridge: Cambridge University Press (1966), Chapter 10, for a more complete discussion of this point.

[2] Technically, J is a *functional*, depending on the functions $y(x, \alpha)$ and $y_x(x, \alpha)$: $J[y(x, \alpha)]$, $y_x(x, \alpha)]$.

[3] Note that y and y_x are being treated as *independent* variables.

Equation (17.8) becomes

$$\frac{\partial J(\alpha)}{\partial \alpha} = \int_{x_1}^{x_2} \left(\frac{\partial f}{\partial y} \eta(x) + \frac{\partial f}{\partial y_x} \frac{d\eta(x)}{dx} \right) dx. \tag{17.11}$$

Integrating the second term by parts to get $\eta(x)$ as a common factor, we obtain

$$\int_{x_1}^{x_2} \frac{d\eta(x)}{dx} \frac{\partial f}{\partial y_x} dx = \eta(x) \frac{\partial f}{\partial y_x} \Big|_{x_1}^{x_2} - \int_{x_1}^{x_2} \eta(x) \frac{d}{dx} \frac{\partial f}{\partial y_x} dx. \tag{17.12}$$

The integrated part vanishes by Eq. (17.2) and Eq. (17.11) becomes

$$\int_{x_1}^{x_2} \left[\frac{\partial f}{\partial y} - \frac{d}{dx} \frac{\partial f}{\partial y_x} \right] \eta(x) \, dx = 0. \tag{17.13}$$

In this form α has been set equal to zero corresponding to the solution path and, in effect, is no longer part of the problem.

Occasionally we will see Eq. (17.13) multiplied by α, which gives

$$\int_{x_1}^{x_2} \left(\frac{\partial f}{\partial y} - \frac{d}{dx} \frac{\partial f}{\partial y_x} \right) \delta y \, dx = \alpha \left[\frac{\partial J}{\partial \alpha} \right]_{\alpha = 0}$$

$$= \delta J = 0. \tag{17.14}$$

Since $\eta(x)$ is arbitrary (as already discussed), we may choose it to have the same sign as the bracketed expression whenever the latter differs from zero. Hence the integrand is always nonnegative. Equation (17.13), our condition for the existence of a stationary value, can then be satisfied only if the bracketed term itself is identically zero. The condition for our stationary value is thus a partial differential equation,[4]

$$\frac{\partial f}{\partial y} - \frac{d}{dx} \frac{\partial f}{\partial y_x} = 0, \tag{17.15}$$

known as the Euler equation, which can be expressed in various other forms. Sometimes solutions are missed when they are not twice differentiable as required by Eq. (17.15). An example is Goldschmidt's discontinuous solution of Section 17.2.7.

ALTERNATE FORMS OF EULER EQUATIONS

One other form (Exercise 17.1.1), which is often useful is

$$\frac{\partial f}{\partial x} - \frac{d}{dx} \left(f - y_x \frac{\partial f}{\partial y_x} \right) = 0, \tag{17.16}$$

---

[4] It is important to watch the meaning of $\partial/\partial x$ and d/dx closely. For example, if $f = f[y(x), x]$,

$$\frac{df}{dx} = \frac{\partial f}{\partial x} + \frac{\partial f}{\partial y} \frac{dy}{dx}$$

The first term on the right gives the *explicit* x-dependence. The second term gives the *implicit* x-dependence.

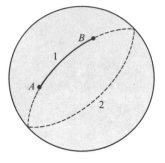

Figure 17.2 Stationary paths over a sphere.

In problems in which $f = f(y, y_x)$ and x does not appear explicitly, Eq. (17.16) reduces to

$$\frac{d}{dx}\left(f - y_x \frac{\partial f}{\partial y_x}\right) = 0 \qquad (17.17)$$

or

$$f - y_x \frac{\partial f}{\partial y_x} = \text{constant.} \qquad (17.18)$$

It is clear that Eq. (17.15) or (17.16) must be satisfied for J to take on a stationary value, that is, for Eq. (17.14) to be satisfied. Equation (17.15) is necessary, but it is by no means sufficient.[5] Courant and Robbins illustrate this very nicely by considering the distance over a sphere between points on the sphere, A and B Fig. 17.2. Path (1), a great circle route, is found from Eq. (17.15). But path (2), the remainder of the great circle through points A and B, also satisfies the Euler equation. Path (2) is a maximum but only if we demand that it be a great circle and then only if we make less than one circuit; that is, path (2) + n complete revolutions is also a solution. If the path is not required to be a great circle, any deviation from (2) will increase the length. This is hardly the property of a local maximum, and that is why it is important to check the properties of solutions of Eq. (17.15) to see if they satisfy the physical conditions of the given problem.

EXERCISES

17.1.1   For $dy/dx \equiv y_x \neq 0$, show the equivalence of the two forms of Euler's equation:

$$\frac{\partial f}{\partial y} - \frac{d}{dx}\frac{\partial f}{\partial y_x} = 0$$

[5] For a discussion of sufficiency conditions and the development of the calculus of variations as a part of modern mathematics see G. M. Ewing, *Calculus of Variations with Applications* [Norton, New York (1969)]. Sufficiency conditions are also covered by Sagan (reference listed at the end of this chapter).

and

$$\frac{\partial f}{\partial x} - \frac{d}{dx}\left(f - y_x \frac{\partial f}{\partial y_x}\right) = 0.$$

17.1.2 Derive Euler's equation by expanding the integrand of

$$J(\alpha) = \int_{x_1}^{x_2} f[y(x, \alpha), y_x(x, \alpha), x]\, dx$$

in powers of α, using a Taylor (Maclaurin) expansion with y and y_x as the two variables (Section 5.6).

Note. The stationary condition is $\partial J(\alpha)/\partial \alpha - 0$, evaluated at $\alpha - 0$. The terms quadratic in α may be useful in establishing the nature of the stationary solution (maximum, minimum, or saddle point).

17.1.3 Find the Euler equation corresponding to Eq. (17.15) if $f = f(y_{xx}, y_x, y, x)$.

$$ANS. \quad \frac{d^2}{dx^2}\left(\frac{\partial f}{\partial y_{xx}}\right) - \frac{d}{dx}\left(\frac{\partial f}{\partial y_x}\right) + \frac{\partial f}{\partial y} = 0,$$

$$\text{with} \qquad \eta(x_1) = \eta(x_2) = 0,$$
$$\eta_x(x_1) = \eta_x(x_2) = 0.$$

17.1.4 The integrand $f(y, y_x, x)$ of Eq. (17.1) has the form

$$f(y, y_x, x) = f_1(x, y) + f_2(x, y)y_x.$$

(a) Show that the Euler equation leads to

$$\frac{\partial f_1}{\partial y} - \frac{\partial f_2}{\partial x} = 0.$$

(b) What does this imply for the dependence of the integral J upon the choice of path?

17.1.5 Show that the condition that

$$J = \int f(x, y)\, dx$$

has a stationary value

(a) leads to $f(x, y)$ independent of y and
(b) yields no information about any x-dependence.

We get no (continuous, differentiable) solution. To be a meaningful variational problem dependence on y_x or higher derivatives is essential.

Note. The situation will change when constraints are introduced (compare Exercise 17.7.7).

17.2 APPLICATIONS OF THE EULER EQUATION

Example 17.2.1 Straight Line

Perhaps the simplest application of the Euler equation is in the determination of the shortest distance between two points in the xy-plane. Since the element of distance is

$$ds = [(dx)^2 + (dy)^2]^{1/2} = [1 + y_x^2]^{1/2}\, dx, \tag{17.19}$$

the distance J, may be written as

$$J = \int_{x_1,y_1}^{x_2,y_2} ds = \int_{x_1}^{x_2} [1 + y_x^2]^{1/2}\, dx. \qquad (17.20)$$

Comparison with Eq. (17.1) shows that

$$f(y, y_x, x) = (1 + y_x^2)^{1/2}. \qquad (17.21)$$

Substituting into Eq. (17.16), we obtain

$$-\frac{d}{dx}\left[\frac{1}{(1 + y_x^2)^{1/2}}\right] = 0 \qquad (17.22)$$

or

$$\frac{1}{(1 + y_x^2)^{1/2}} = C, \qquad \text{a constant.} \qquad (17.23)$$

This is satisfied by

$$y_x = a, \qquad \text{a second constant} \qquad (17.24)$$

and

$$y = ax + b, \qquad (17.25)$$

which is the familiar equation for a straight line. The constants a and b, of course, are chosen so that the line passes through the two points (x_1, y_1) and (x_2, y_2). Hence the Euler equation predicts that the shortest[6] distance between two fixed points is a straight line.

The generalization of this in curved four-dimensional space–time leads to the important relativity concept, the geodesic.

Example 17.2.2 Soap Film

As a second illustration (Fig. 17.3), consider two parallel coaxial wire circles to be connected by a surface of minimum area that is generated by revolving a curve $y(x)$ about the x-axis. The curve is required to pass through fixed end points (x_1, y_1) and (x_2, y_2). The variational problem is to choose the curve $y(x)$ so that the area of the resulting surface will be a minimum.

For the element of area shown in Fig. 17.3

$$dA = 2\pi y\, ds = 2\pi y(1 + y_x^2)^{1/2}\, dx. \qquad (17.26)$$

The variational equation is then

$$J = \int_{x_1}^{x_2} 2\pi y(1 + y_x^2)^{1/2}\, dx. \qquad (17.27)$$

Neglecting the 2π, we obtain

$$f(y, y_x, x) = y(1 + y_x^2)^{1/2}. \qquad (17.28)$$

---

[6] Technically, we have a stationary value. From the α^2 terms it can be identified as a minimum (Exercise 17.2.1).

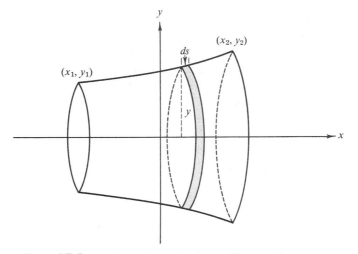

Figure 17.3   Surface of rotation–soap film problem.

Since $\partial f/\partial x = 0$, we may apply Eq. (17.18) directly and get

$$y(1 + y_x^2)^{1/2} - yy_x^2 \frac{1}{(1 + y_x^2)^{1/2}} = c_1, \qquad (17.29)$$

or

$$\frac{y}{(1 + y_x^2)^{1/2}} = c_1. \qquad (17.30)$$

Squaring, we get

$$\frac{y^2}{1 + y_x^2} = c_1^2 \qquad \text{with } c_1^2 \leq y_{\min}^2, \qquad (17.31)$$

and

$$(y_x)^{-1} = \frac{dx}{dy} = \frac{c_1}{\sqrt{y^2 - c_1^2}}. \qquad (17.32)$$

This may be integrated to give

$$x = c_1 \cosh^{-1} \frac{y}{c_1} + c_2. \qquad (17.33)$$

Solving for y, we have

$$y = c_1 \cosh\left(\frac{x - c_2}{c_1}\right) \qquad (17.34)$$

and again c_1 and c_2 are determined by requiring the hyperbolic cosine to pass through the points (x_1, y_1) and (x_2, y_2). Our "minimum" area surface is a special case of a catenary of revolution or a catenoid.

SOAP FILM—MINIMUM AREA

This calculus of variations contains many pitfalls for the unwary. (Remember the Euler equation is a *necessary* condition assuming a *differentiable* solution. The sufficiency conditions are quite involved. See the references for details.) Perhaps respect for some of these hazards may be developed by considering a specific physical problem, for example, the minimum area problem with $(x_1, y_1) = (-x_0, 1)$, $(x_2, y_2) = (+x_0, 1)$. The minimum surface is a soap film stretched over the two rings of unit radius at $x = \pm x_0$. The problem is to predict the curve $y(x)$ assumed by the soap film.

By referring to Eq. (17.34), we find that $c_2 = 0$ by the symmetry of the problem. Then

$$y = c_1 \cosh\left(\frac{x}{c_1}\right), \qquad c_1 \cosh\left(\frac{x_0}{c_1}\right) = 1. \qquad (17.34a)$$

If we take $x_0 = \frac{1}{2}$, we obtain the transcendental equation for c_1,

$$1 = c_1 \cosh\left(\frac{1}{2c_1}\right). \qquad (17.35)$$

We find that this equation has two solutions; $c_1 = 0.2350$, leading to a "deep" curve, and $c_1 = 0.8483$, leading to a "flat" curve. Which is our minimum? Which curve is assumed by the soap film? Before answering these questions, consider the physical situation with the rings moved apart so that $x_0 = 1$. Then Eq. (17.34a) becomes

$$1 = c_1 \cosh\left(\frac{1}{c_1}\right). \qquad (17.36)$$

which has *no real solutions*! The physical significance is that as the unit radius rings were moved out from the origin a point was reached at which the soap film could no longer maintain the same horizontal force over each vertical section. Stable equilibrium was no longer possible. The soap film broke (irreversible process) and formed a circular film over each ring (with a total area of $2\pi = 6.2832 \ldots$). This is the Goldschmidt discontinuous solution.

The next question is: How large may x_0 be and still give a real solution for Eq. (17.34a)?[7] Letting $c_1^{-1} = p$, Eq. (17.34a) becomes,

$$p = \cosh p x_0. \qquad (17.37)$$

To find $x_{0\,max}$ we could solve for x_0 (as in Eq. (17.33)) and then differentiate with respect to p. Finally, with an eye on Fig. 17.4, dx_0/dp would be set equal to zero. Alternatively, direct differentiation of Eq. (17.37) with respect to p yields

$$1 = \sinh p x_0 [x_0 + p\, dx_0/dp].$$

---

[7] From a numerical point of view it is easier to invert the problem. Pick a value of c_1 and solve for x_0. Equation (17.34a) becomes $x_0 = c_1 \cosh^{-1}(1/c_1)$. This has numerical solutions in the range $0 < c_1 \leq 1$.

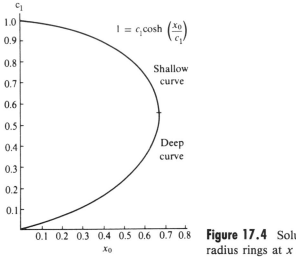

Figure 17.4 Solutions of Eq. (17.34a) for unit radius rings at $x = \pm x_0$.

The requirement that dx_0/dp vanish, leads to

$$1 = x_0 \sinh px_0. \tag{17.38}$$

Equations (17.37) and (17.38) may be combined to form

$$px_0 = \coth px_0 \tag{17.39}$$

with the root

$$px_0 = 1.1997. \tag{17.40}$$

Substituting into Eqs. (17.37) or (17.38), we obtain

$$p = 1.810 \qquad c_1 = 0.5524 \tag{17.41}$$

and

$$x_{0\,\text{max}} = 0.6627. \tag{17.42}$$

Returning to the question of the solution of Eq. (17.35) that describes the soap film, let us calculate the area corresponding to each solution. We have

$$A = 4\pi \int_0^{x_0} y(1 + y_x^2)^{1/2} \, dx = \frac{4\pi}{c_1} \int_0^{x_0} y^2 \, dx \qquad \text{(by Eq. (17.30))}$$

$$= 4\pi c_1 \int_0^{x_0} \left(\cosh \frac{x}{c_1}\right)^2 dx$$

$$= \pi c_1^2 \left[\sinh\left(\frac{2x_0}{c_1}\right) + \frac{2x_0}{c_1}\right]. \tag{17.43}$$

For $x_0 = \frac{1}{2}$, Eq. (17.35) leads to

$$c_1 = 0.2350 \to A = 6.8456,$$

$$c_1 = 0.8483 \to A = 5.9917,$$

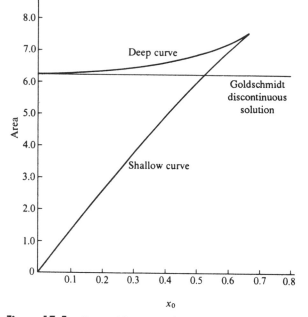

Figure 17.5   Catenoid area (unit radius rings at $x = \pm x_0$).

showing that the former can at most be only a local minimum. A more detailed investigation (compare Bliss, *Calculus of Variations*, Chapter IV) shows that this surface is not even a local minimum. For $x_0 = \frac{1}{2}$ the soap film will be described by the flat curve

$$y = 0.8483 \cosh\left(\frac{x}{0.8483}\right). \tag{17.44}$$

This flat or shallow catenoid (catenary of revolution) will be an absolute minimum for $0 \leq x_0 < 0.528$. However, for $0.528 < x < 0.6627$ its area is greater than that of the Goldschmidt discontinuous solution (6.2832) and it is only a relative minimum (Fig. 17.5).

For an excellent discussion of both the mathematical problems and experiments with soap films, the reader is referred to Courant and Robbins.

EXERCISES

17.2.1   A soap film is stretched across the space between two rings of unit radius centered at $\pm x_0$ on the x-axis and perpendicular to the x-axis. Using the solution developed in Section 17.2, set up the transcendental equations for the condition that x_0 is such that the area of the curved surface of rotation equals the area of the two rings (Goldschmidt discontinuous solution). Solve for x_0 (Fig. 17.6).

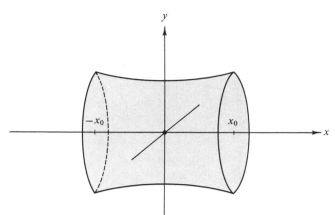

Figure 17.6   Surface of rotation.

17.2.2   In Example 17.2.1 expand $J[y(x, \alpha)] - J[y(x, 0)]$ in powers of α. The term linear in α leads to the Euler equation and to the straight-line solution Eq. (17.25). Investigate the α^2 term and show that the stationary value of J, the straight-line distance, is a *minimum*.

17.2.3   (a) Show that the integral

$$J = \int_{x_1}^{x_2} f(y, y_x, x)\, dx, \qquad \text{with } f = y(x),$$

has *no* extreme values.

(b) If $f(y, y_x, x) = y^2(x)$ find a discontinuous solution similar to the Goldschmidt solution for the soap film problem.

17.2.4   Fermat's principle of optics states that a light ray will follow the path, $y(x)$, for which

$$\int_{x_1, y_1}^{x_2, y_2} n(y, x)\, ds$$

is a minimum when n is the index of refraction. For $y_2 = y_1 = 1$, $-x_1 = x_2 = 1$ find the ray path if

(a) $n = e^y$,

(b) $n = a(y - y_0)$, $y > y_0$.

17.2.5   A frictionless particle moves from point A on the surface of the earth to point B by sliding through a tunnel. Find the differential equation to be satisfied if the transit time is to be a minimum.

Note. Assume the earth to be nonrotating sphere of uniform density.

$$ANS. \quad (\text{Eq. } (17.15))\ r_{\varphi\varphi}(r^3 - ra^2) + r_\varphi^2(2a^2 - r^2) + a^2 r^2 = 0,$$
$$r(\varphi = 0) = r_0,$$
$$r_\varphi(\varphi = 0) = 0,$$
$$r(\varphi = \varphi_A) = a,$$
$$r(\varphi = \varphi_B) = a.$$

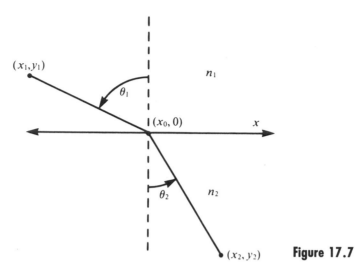

Figure 17.7

$$\text{(Eq. (17.18))} \quad r_\varphi^2 = \frac{a^2 r^2}{r_0^2} \cdot \frac{r^2 - r_0^2}{a^2 - r^2}.$$

The solution of these equations is a hypocycloid, generated by a circle of radius $\frac{1}{2}(a - r_0)$ rolling inside the circle of radius a. The student might like to show that the transit time is

$$t = \pi \frac{(a^2 - r_0^2)^{1/2}}{(ag)^{1/2}}.$$

For details see P. W. Cooper, *Am. J. Phys.* **34**, 68 (1966); G. Venezian *et al.*, *ibid.*, pp. 701–704.

17.2.6   A ray of light follows a straight-line path in a first homogeneous medium, is refracted at an interface, and then follows a new straight-line path in the second medium. Use Fermat's principle of optics to derive Snell's law of refraction:

$$n_1 \sin \theta_1 = n_2 \sin \theta_2.$$

Hint. Keep the points (x_1, y_1) and (x_2, y_2) fixed and vary x_0 to satisfy Fermat (Fig. 17.7). This is *not* an Euler equation problem. (The light path is not differentiable at x_0.)

17.2.7   A second soap film configuration for the unit radius rings at $x = \pm x_0$ consists of a circular disk, radius a, in the $x = 0$ plane and two catenoids of revolution, one joining the disk and each ring. One catenoid may be described by

$$y = c_1 \cosh\left(\frac{x}{c_1} + c_3\right).$$

(a) Impose boundary conditions at $x = 0$ and $x = x_0$.
(b) Although not necessary, it is convenient to require that the catenoids form an angle of 120° where they join the central disk. Express this third boundary condition in mathematical terms.

(c) Show that the total area of catenoids plus central disk is

$$A = c_1^2 \left[\sinh\left(\frac{2x_0}{c_1} + 2c_3\right) + \frac{2x_0}{c_1} \right].$$

Note. Although this soap film configuration is physically realizable and stable, the area is larger than that of the simple catenoid for all ring separations for which both films exist.

$$\text{ANS. (a)} \quad \begin{cases} 1 = c_1 \cosh\left(\dfrac{x_0}{c_1} + c_3\right) \\ a = c_1 \cosh c_3, \end{cases}$$

$$\text{(b)} \quad \frac{dy}{dx} = \tan 30° = \sinh c_3.$$

17.2.8 For the soap film described in Exercise 17.2.7 find (numerically) the maximum value of x_0.

Note. This calls for a pocket calculator with hyperbolic functions or a table of hyperbolic cotangents.

$$\text{ANS.} \quad x_{0\,max} = 0.4078.$$

17.2.9 Find the root of $px_0 = \coth px_0$ (Eq. (17.39)) and determine the corresponding values of p and x_0 (Eqs. (17.41) and (17.42)). Calculate your values to five significant figures.

Hint. Try one of the root-determining subroutines listed in Appendix 1.

17.2.10 For the two-ring soap film problem of this section calculate and tabulate x_0, p, p^{-1}, and A, the soap film area for $px_0 = 0.00(0.02)1.30$.

17.2.11 Find the value of x_0 (to five significant figures) that leads to a soap film area, Eq. (17.43) equal to 2π, the Goldschmidt discontinuous solution.

$$\text{ANS.} \quad x_0 = 0.52770.$$

17.2.12 Find the curve of quickest descent from $(0, 0)$ to (x_0, y_0) for a particle sliding under gravity and without friction. Show that the ratio of times taken by the particle along a straight line joining the two points compared to along the curve of quickest descent is $(1 + 4/\pi^2)^{1/2}$.

Hint. Take y to increase downwards. Apply Eq. (17.18) to obtain $y_x^2 = (1 - c^2y)/c^2y$, where c is an integration constant. Then make the substitution $y = (\sin^2\varphi/2)/c^2$ to parametrize the cycloid and take $(x_0, y_0) = (\pi/2c^2, 1/c^2)$.

17.3 GENERALIZATIONS, SEVERAL DEPENDENT VARIABLES

Our original variational problem, Eq. (17.1), may be generalized in several respects. In this section we consider the integrand, f, to be a function of several *dependent* variables, $y_1(x), y_2(x), y_3(x), \ldots$, all of which depend on x, the independent variable. In Section 17.4 f again will contain only one unknown function y, but y will be a function of several independent variables (over which we integrate). In Section 17.5 these two generalizations are combined. Finally, in Section 17.7 the stationary value is restricted by one or more constraints.

For more than one dependent variable Eq. (17.1) becomes

$$J = \int_{x_1}^{x_2} f[y_1(x), y_2(x), \ldots, y_n(x), y_{1x}(x), y_{2x}(x), \ldots, y_{nx}(x), x] \, dx. \qquad (17.45)$$

As in Section 17.1, we determine the extreme value of J by comparing neighboring paths. Let

$$y_i(x, \alpha) = y_i(x, 0) + \alpha \eta_i(x), \qquad i = 1, 2, \ldots, n, \qquad (17.46)$$

with the η_i independent of one another, but subject to the restrictions discussed in Section 17.1. By differentiating Eq. (17.45) with respect to α and setting $\alpha = 0$, since Eq. (17.7) still applies, we obtain

$$\int_{x_1}^{x_2} \sum_i \left(\frac{\partial f}{\partial y_i} \eta_i + \frac{\partial f}{\partial y_{ix}} \eta_{ix} \right) dx = 0, \qquad (17.47)$$

the subscript x denoting differentiations with respect to x; that is, $y_{ix} = dy_i/dx$, and so on. Again, each of the terms $(\partial f/\partial y_{ix})\eta_{ix}$ is integrated by parts. The integrated part vanishes and Eq. (17.47) becomes

$$\int_{x_1}^{x_2} \sum_i \left(\frac{\partial f}{\partial y_i} - \frac{d}{dx} \frac{\partial f}{\partial y_{ix}} \right) \eta_i \, dx = 0. \qquad (17.48)$$

Since the η_i are arbitrary and *independent* of one another,[1] each of the terms in the sum must vanish *independently*. We have

$$\frac{\partial f}{\partial y_i} - \frac{d}{dx} \frac{\partial f}{\partial (dy_i/dx)} = 0, \qquad i = 1, 2, \ldots, n, \qquad (17.49)$$

a whole set of Euler equations, each of which must be satisfied for an extreme value.

HAMILTON'S PRINCIPLE

The most important application of Eq. (17.45) occurs when the integrand f is taken to be the Lagrangian L. The Langrangian (for nonrelativistic systems; see Exercise 17.3.5 for a relativisitic particle) is defined as the *difference* of kinetic and potential energies of a system.

$$L \equiv T - V. \qquad (17.50)$$

Using time as an independent variable instead of x and $x_i(t)$ as the dependent variables,

$$x \rightarrow t,$$

$$y_i \rightarrow x_i(t),$$

$$y_{ix} \rightarrow \dot{x}_i(t);$$

[1] For example, we could set $\eta_2 = \eta_3 = \eta_4 \cdots = 0$, eliminating all but one term of the sum, and then treat η_1 exactly as in Section 17.1.

$x_i(t)$ is the location and $\dot{x}_i = dx_i/dt$, the velocity of particle i as a function of time. The equation $\delta J = 0$ is then a a mathematical statement of Hamilton's principle of classical mechanics,

$$\delta \int_{t_1}^{t_2} L(x_1, x_2, \ldots, x_n, \dot{x}_1, \dot{x}_2, \ldots, \dot{x}_n; t)\, dt = 0. \tag{17.51}$$

In words, Hamilton's principle asserts that the motion of the system from time t_1 to t_2 is such that the time integral of the Lagrangian L has a stationary value. The resulting Euler equations are usually called the Lagrangian equations of motion,

$$\frac{d}{dt}\frac{\partial L}{\partial \dot{x}_i} - \frac{\partial L}{\partial x_i} = 0. \tag{17.52}$$

These Lagrangian equations can be derived from Newton's equations of motion and Newton's equations can be derived from Lagrange's. The two sets of equations are equally "fundamental."

The Lagrangian formulation has certain valuable advantages over the conventional Newtonian laws. Whereas Newton's equations are vector equations, we see that Lagrange's equations involve only scalar quantities. The coordinates $x_1, x_2, \ldots$ need not be any standard set of coordinates or lengths. They can be selected to match the conditions of the physical problem. The Lagrange equations are invariant with respect to the choice of coordinate system. Newton's equations (in component form) are not invariant. Exercise 2.5.10 shows what happens to $\mathbf{F} = m\mathbf{a}$ resolved in spherical polar coordinates.

Exploiting the concept of energy, we may easily extend the Lagrangian formulation from mechanics to diverse fields such as electrical networks and acoustical systems. Extensions to electromagnetism appear in the exercises. The result is a unity of otherwise separate areas of physics. In the development of new areas the quantization of Lagrangian particle mechanics provided a model for the quantization of electromagnetic fields and led to the modern theory of quantum electrodynamics.

One of the most valuable advantages of the Hamilton principle—Lagrange equation formulation—is the ease in seeing a relation between a symmetry and a conservation law. As an example, let $x_i = \varphi$, an azimuthal angle. If our Lagrangian is independent of φ (i.e., φ is an ignorable coordinate), there are two consequences: (1) an axial (rotational) symmetry and (2) from Eq. (17.52) $\partial L/\partial \dot{\varphi} = $ constant. Physically, this corresponds to the conservation or invariance of a component of angular momentum. Similarly, invariance under translation leads to conservation of linear momentum. Noether's theorem is a generalization of this invariance (symmetry)—the conservation law relation.

Example 17.3.1 Moving Particle—Cartesian Coordinates

Consider Eq. (17.50) which describes one particle with kinetic energy

$$T = \tfrac{1}{2}m\dot{x}^2 \tag{17.53}$$

and potential energy $V(x)$, in which, as usual, the force is given by the negative gradient of the potential,

$$F(x) = -\frac{dV(x)}{dx}.$$ (17.54)

From Eq. (17.52)

$$\frac{d}{dt}(m\dot{x}) - \frac{\partial(-V)}{\partial x} = m\ddot{x} - F(x) = 0,$$ (17.55)

which is simply Newton's second law of motion.

Example 17.3.2 Moving Particle—Circular Cylindrical Coordinates

Now let us describe a moving particle in cylindrical coordinates ($z = 0$)-plane. The kinetic energy is

$$T = \tfrac{1}{2}m(\dot{x}^2 + \dot{y}^2) = \tfrac{1}{2}m(\dot{\rho}^2 + \rho^2\dot{\varphi}^2),$$ (17.56)

and we take $V = 0$.

The transformation of $\dot{x}^2 + \dot{y}^2$ into circular cylindrical coordinates could be carried out by taking $x(\rho, \varphi)$ and $y(\rho, \varphi)$, Eq. (2.28), and differentiating with respect to time and squaring. It is much easier to interpret $\dot{x}^2 + \dot{y}^2$ as v^2 and just write down the components of $\mathbf{v}$ as $\hat{\boldsymbol{\rho}}(ds_\rho/dt) = \hat{\boldsymbol{\rho}}\dot{\rho}$, and so on. (The ds_ρ is an increment of *length*, ρ changing by $d\rho$, φ remaining constant. See Sections 2.1 and 2.4.)

The Lagrangian equations yield

$$\frac{d}{dt}(m\dot{\rho}) - m\rho\dot{\varphi}^2 = 0,$$

$$\frac{d}{dt}(m\rho^2\dot{\varphi}) = 0.$$ (17.57)

The second equation is a simple statement of conservation of angular momentum. The first may be interpreted as radial acceleration[2] equated to centrifugal force. In this sense the centrifugal force is a real force. It is of some interest that this interpretation of centrifugal force as a real force is supported by the general theory of relativity.

EXERCISES

17.3.1 (a) Develop the equations of motion corresponding to $L = \tfrac{1}{2}m(\dot{x}^2 + \dot{y}^2)$.

(b) In what sense do your solutions *minimize* the integral $\int_{t_1}^{t_2} L \, dt$?

Compare the result for your solution with $x = $ const., $y = $ const.

---

[2] Here is a second method of attacking Exercise 2.4.8.

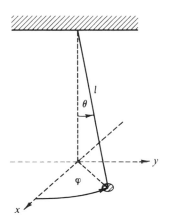

Figure 17.8 Spherical pendulum.

17.3.2 From the Lagrangian equations of motion, Eq. (17.52), show that a system in stable equilibrium has a minimum potential energy.

17.3.3 Write out the Lagrangian equations of motion of a particle in spherical coordinates for potential V equal to a constant. Identify the terms corresponding to (a) centifugal force and (b) Coriolis force.

17.3.4 The spherical pendulum consists of a mass on a wire of length l, free to move in polar angle θ and azimuth angle φ (Fig. 17.8).

(a) Set up the Lagrangian for this physical system.
(b) Develop the Lagrangian equations of motion.

17.3.5 Show that the Lagrangian

$$L = m_0 c^2 \left(1 - \sqrt{1 - \frac{v^2}{c^2}}\right) - V(\mathbf{r})$$

leads to a relativistic form of Newton's second law of motion,

$$\frac{d}{dt}\left(\frac{m_0 v_i}{\sqrt{1 - v^2/c^2}}\right) = F_i,$$

in which $F_i = -\partial V/\partial x_i$.

17.3.6 The Lagrangian for a particle with charge q in an electromagnetic field described by scalar potential φ and vector potential $\mathbf{A}$ is

$$L = \tfrac{1}{2}mv^2 - q\varphi + q\mathbf{A} \cdot \mathbf{v}.$$

Find the equation of motion of the charged particle.
Hint. $(d/dt)A_j = \partial A_j/\partial t + \sum_i(\partial A_j/\partial x_i)\dot{x}_i$. The dependence of the force fields $\mathbf{E}$ and $\mathbf{B}$ upon the potentials φ and $\mathbf{A}$ is developed in Section 1.13 (compare Exercise 1.13.10).

$$ANS. \quad m\ddot{x}_i = q[\mathbf{E} + \mathbf{v} \times \mathbf{B}]_i.$$

17.3.7 Consider a system in which the Lagrangian is given by

$$L(q_i, \dot{q}_i) = T(q_i, \dot{q}_i) - V(q_i),$$

where q_i and $\dot{q}_i$ represent sets of variables. The potential energy V is independent of velocity and neither T nor V have any explicit time dependence.

(a) Show that

$$\frac{d}{dt}\left(\sum_j \dot{q}_j \frac{\partial L}{\partial \dot{q}_j} - L\right) = 0.$$

(b) The constant quantity

$$\sum_j \dot{q}_j \frac{\partial L}{\partial \dot{q}_j} - L$$

defines the Hamiltonian H. Show that under the preceding assumed conditions, $H = T + V$, the total energy.

Note. The kinetic energy T is a quadratic function of the $\dot{q}_i$'s.

17.4 SEVERAL INDEPENDENT VARIABLES

Sometimes the integrand f of Eq. (17.1) will contain one unknown function u which is a function of several independent variables, $u = u(x, y, z)$, for the three-dimensional case. Equation (17.1) becomes

$$J = \iiint f[u, u_x, u_y, u_z, x, y, z]\, dx\, dy\, dz, \tag{17.58}$$

u_x indicating $\partial u/\partial x$, and so on. The variational problem is to find the function $u(x, y, z)$ for which J is stationary,

$$\delta J = \alpha \left.\frac{\partial J}{\partial \alpha}\right|_{\alpha = 0} = 0. \tag{17.59}$$

Generalizing Section 17.1, we let

$$u(x, y, z, \alpha) = u(x, y, z, 0) + \alpha\eta(x, y, z). \tag{17.60}$$

$u(x, y, z, \alpha = 0)$ represents the (unknown) function for which Eq. (17.59) is satisfied, whereas again $\eta(x, y, z)$ is the arbitrary deviation that describes the varied function $u(x, y, z, \alpha)$. This deviation, $\eta(x, y, z)$ is required to be differentiable and to vanish at the end points. Then from Eq. (17.60),

$$u_x(x, y, z, \alpha) = u_x(x, y, z, 0) + \alpha\eta_x, \tag{17.61}$$

and similarly for u_y and u_z.

Differentiating the integral Eq. (17.58)) with respect to the parameter α and then setting $\alpha = 0$, we obtain

$$\left.\frac{\partial J}{\partial \alpha}\right|_{\alpha = 0} = \iiint \left(\frac{\partial f}{\partial u}\eta + \frac{\partial f}{\partial u_x}\eta_x + \frac{\partial f}{\partial u_y}\eta_y + \frac{\partial f}{\partial u_z}\eta_z\right) dx\, dy\, dz = 0. \tag{17.62}$$

Again, we integrate each of the terms $(\partial f/\partial u_i)\eta_i$ by parts. The integrated part vanishes at the end points (because the deviation η is required to go to zero at the end points) and

$$\int\int\int \left(\frac{\partial f}{\partial u} - \frac{\partial}{\partial x}\frac{\partial f}{\partial u_x} - \frac{\partial}{\partial y}\frac{\partial f}{\partial u_y} - \frac{\partial}{\partial z}\frac{\partial f}{\partial u_z}\right)\eta(x, y, z)\, dx\, dy\, dz = 0.^1 \qquad (17.63)$$

Since the variation $\eta(x, y, z)$ is arbitrary, the term in large parentheses may be set equal to zero. This yields the Euler equation for (three) independent variables,

$$\frac{\partial f}{\partial u} - \frac{\partial}{\partial x}\frac{\partial f}{\partial u_x} - \frac{\partial}{\partial y}\frac{\partial f}{\partial u_y} - \frac{\partial}{\partial z}\frac{\partial f}{\partial u_z} = 0. \qquad (17.64)$$

Example 17.4.1 Laplace's Equation

An example of this sort of variational problem is provided by electrostatics. The energy of an electrostatic field is

$$\text{energy density} = \tfrac{1}{2}\varepsilon E^2, \qquad (17.65)$$

in which $\mathbf{E}$ is the usual electrostatic force field. In terms of the static potential φ,

$$\text{energy density} = \tfrac{1}{2}\varepsilon(\nabla\varphi)^2. \qquad (17.66)$$

Now let us impose the requirement that the electrostatic energy (associated with the field) in a given volume be a minimum. (Boundary conditions on $\mathbf{E}$ and φ must still be satisfied.) We have the volume integral[2]

$$J = \int\int\int (\nabla\varphi)^2\, dx\, dy\, dz$$

$$= \int\int\int (\varphi_x^2 + \varphi_y^2 + \varphi_z^2)\, dx\, dy\, dz. \qquad (17.67)$$

With

$$f(\varphi, \varphi_x, \varphi_y, \varphi_z, x, y, z) = \varphi_x^2 + \varphi_y^2 + \varphi_z^2, \qquad (17.68)$$

the function φ replacing the u of Eq. (17.64), Euler's equation (Eq. (17.64)) yields

$$-2(\varphi_{xx} + \varphi_{yy} + \varphi_{zz}) = 0 \qquad (17.69)$$

or

$$\nabla^2\varphi(x, y, z) = 0, \qquad (17.70)$$

which is just Laplace's equation of electrostatics.

---

[1] Again, it is imperative that the precise meaning of partial derivatives be understood fully. Specifically, in Eq. (17.63), $\partial/\partial x$ is a partial derivative, in that y and z are constant. But $\partial/\partial x$ is also a total derivative in that it acts on *implicit* x-dependence as well as on *explicit* x-dependence. In this sense

$$\frac{\partial}{\partial x}\left(\frac{\partial f}{\partial u_x}\right) = \frac{\partial^2 f}{\partial x\,\partial u_x} + \frac{\partial^2 f}{\partial u\,\partial u_x}u_x + \frac{\partial^2 f}{\partial u_x^2}u_{xx} + \frac{\partial^2 f}{\partial u_y\,\partial u_x}u_{xy} + \frac{\partial^2 f}{\partial u_z\,\partial u_x}u_{xz}.$$

[2] Remember that the subscript x indicates the x-partial derivative, *not* an x-component.

Closer investigation shows that this stationary value is indeed a minimum. Thus the demand that the field energy be minimized leads to Laplace's equation.

EXERCISES

17.4.1 The Lagrangian for a vibrating string (small amplitude vibrations) is

$$L = \int (\tfrac{1}{2}\rho u_t^2 - \tfrac{1}{2}\tau u_x^2)\, dx,$$

where ρ is the (constant) linear mass density and τ is the (constant) tension. The x-integration is over the length of the string. Show that application of Hamilton's principle to the Lagrangian density (the integrand), now with *two* independent variables, leads to the classical wave equation

$$\frac{\partial^2 u}{\partial x^2} = \frac{\rho}{\tau}\frac{\partial^2 u}{\partial t^2}.$$

17.4.2 Show that the stationary value of the total energy of the electrostatic field of Example 17.4.1 is a *minimum*.
Hint. Use Eq. (17.61) and investigate the α^2 terms.

17.5 MORE THAN ONE DEPENDENT, MORE THAN ONE INDEPENDENT VARIABLE

In some cases our integrand f contains more than one dependent variable *and* more than one independent variable. Consider

$$f = f[p(x, y, z), p_x, p_y, p_z, q(x, y, z), q_x, q_y, q_z, r(x, y, z), r_x, r_y, r_z, x, y, z].$$
(17.71)

We proceed as before with

$$p(x, y, z, \alpha) = p(x, y, z, 0) + \alpha\xi(x, y, z),$$

$$q(x, y, z, \alpha) = q(x, y, z, 0) + \alpha\eta(x, y, z),$$
(17.72)

$$r(x, y, z, \alpha) = r(x, y, z, 0) + \alpha\zeta(x, y, z), \qquad \text{and so on.}$$

Keeping in mind that ξ, η, and ζ are independent of one another, as were the η_i in Section 17.3, the same differentiation and then integration by parts leads to

$$\frac{\partial f}{\partial p} - \frac{\partial}{\partial x}\frac{\partial f}{\partial p_x} - \frac{\partial}{\partial y}\frac{\partial f}{\partial p_y} - \frac{\partial}{\partial z}\frac{\partial f}{\partial p_z} = 0,$$
(17.73)

with similar equations for functions q and r. Replacing $p\, q, r, \ldots$ with y_i and $x, y, z, \ldots$ with x_y, we can put Eq. (17.73) in a more compact form:

$$\frac{\partial f}{\partial y_i} - \sum_j \frac{\partial}{\partial x_j}\left(\frac{\partial f}{\partial y_{ij}}\right) = 0, \qquad i = 1, 2, \ldots,$$
(17.73a)

in which

$$y_{ij} \equiv \frac{\partial y_i}{\partial x_j}.$$

An application of Eq. (17.73) appears in Section 17.7.

RELATION TO PHYSICS

The calculus of variations as developed so far provides a convenient and perhaps elegant description of wide variety of physical phenomena. The physics includes ordinary mechanics. Section 17.3; relativistic mechanics, Exercise 17.3.5; electrostatics, Example 17.4.1; and electromagnetic theory in Exercise 17.5.1. The convenience and elegance should not be minimized, but at the same time the student should be aware that in these cases the calculus of variations has only provided an alternate description of what was already known. It has *not* provided any new physics.

The situation does change with incomplete theories. If the basic physics is not yet known a postulated variational principle can be a useful starting point.

EXERCISE

17.5.1   The Lagrangian (per unit volume) of an electromagnetic field with a charge density ρ is given by

$$L = \frac{1}{2}\left(\varepsilon_0 E^2 - \frac{B^2}{\mu_0}\right) - \rho\varphi + \rho\mathbf{v}\cdot\mathbf{A}.$$

Show that Lagrange's equations lead to two of Maxwell's equations. (The remaining two are a consequence of the definition of $\mathbf{E}$ and $\mathbf{B}$ in terms of $\mathbf{A}$ and φ.) This Lagrangian density comes from a scalar expression in Section 4.6.

Hint. Take A_1, A_2, A_3, and φ as *dependent* variables, x, y, z, and t as *independent* variables. $\mathbf{E}$ and $\mathbf{B}$ are given in terms of $\mathbf{A}$ and φ by Eq. (4.142).

17.6 LAGRANGIAN MULTIPLIERS

In this section the concept of a constraint is introduced. To simplify the treatment, the constraint appears as a simple function rather than as an integral. In this section we are not concerned with the calculus of variations, but in Section 17.7 the constraints, with our newly developed Lagrangian multipliers, are incorporated into the calculus of variations.

Consider a function of three independent variables, $f(x, y, z)$. For the function f to be a maximum (or extreme)[1]

$$df = 0. \tag{17.74}$$

[1] Including a four-dimensional saddle point.

The necessary and sufficient condition for this is

$$\frac{\partial f}{\partial x} = \frac{\partial f}{\partial y} = \frac{\partial f}{\partial z} = 0, \tag{17.75}$$

in which

$$df = \frac{\partial f}{\partial x} dx + \frac{\partial f}{\partial y} dy + \frac{\partial f}{\partial z} dz. \tag{17.76}$$

Often in physical problems the variables x, y, z are subjected to constraints so that they are no longer all independent. It is possible, at least in principle, to use each constraint to eliminate one variable and to proceed with a new and smaller set of independent variables.

The use of Lagrangian multipliers is an alternate technique that may be applied when this elimination of variables is inconvenient or undesirable. Let our equation of constraint be

$$\varphi(x, y, z) = 0. \tag{17.77}$$

from which

$$d\varphi = \frac{\partial \varphi}{\partial x} dx + \frac{\partial \varphi}{\partial y} dy + \frac{\partial \varphi}{\partial z} dz = 0. \tag{17.78}$$

Returing to Eq. (17.74), we see that Eq. (17.75) no longer follows because there are now only two independent variables. If we take x and y as these independent variables, dz is no longer arbitrary. However, we may add Eq. (17.76) and a multiple of Eq. (17.78) to obtain

$$df + \lambda d\varphi = \left(\frac{\partial f}{\partial x} + \lambda \frac{\partial \varphi}{\partial x}\right) dx + \left(\frac{\partial f}{\partial y} + \lambda \frac{\partial \varphi}{\partial y}\right) dy + \left(\frac{\partial f}{\partial z} + \lambda \frac{\partial \varphi}{\partial z}\right) dz = 0. \tag{17.79}$$

Our Lagrangian multiplier λ is chosen so that

$$\frac{\partial f}{\partial z} + \lambda \frac{\partial \varphi}{\partial z} = 0, \tag{17.80}$$

assuming that $\partial \varphi / \partial z \neq 0$. Equation (17.79) now becomes

$$\left(\frac{\partial f}{\partial x} + \lambda \frac{\partial \varphi}{\partial x}\right) dx + \left(\frac{\partial f}{\partial y} + \lambda \frac{\partial \varphi}{\partial y}\right) dy = 0. \tag{17.81}$$

However, we took dx and dy to be arbitrary and the quantities in parentheses must vanish,

$$\frac{\partial f}{\partial x} + \lambda \frac{\partial \varphi}{\partial x} = 0,$$

$$\frac{\partial f}{\partial y} + \lambda \frac{\partial \varphi}{\partial y} = 0. \tag{17.82}$$

When Eqs. (17.80) and (17.82) are satisfied, $df = 0$ and f is an extremum. Notice that there are now four unknowns: x, y, z, and λ. The fourth equation is, of course, the constraint (17.77). Actually we want only x, y, and z; λ need not be determined. For this reason λ is sometimes called Lagrange's undetermined multiplier. This method will fail if all the coefficients of λ vanish at the extremum, $\partial\varphi/\partial x$, $\partial\varphi/\partial y$, $\partial\varphi/\partial z = 0$. It is then impossible to solve for λ.

The reader might note that from the form of Eqs. (17.80) and (17.82), we could identify f as the function taking an extreme value subject to φ, the constraint or identify f as the constraint and φ as the function.

If we have a set of constraints φ_k, then Eqs. (17.80) and (17.82) become

$$\frac{\partial f}{\partial x_i} + \sum_k \lambda_k \frac{\partial \varphi_k}{\partial x_i} = 0, \qquad i = 1, 2, \ldots, n,$$

with a separate Lagrange multiplier λ_k for each φ_k.

Example 17.6.1 Particle in a Box

As an example of the use of Lagrangian multipliers, consider the quantum mechanical problem of a particle (mass m) in a box. The box is a rectangular parallelepiped with sides a, b, and c. The ground state energy of the particle is given by

$$E = \frac{h^2}{8m}\left(\frac{1}{a^2} + \frac{1}{b^2} + \frac{1}{c^2}\right). \tag{17.83}$$

We seek the shape of the box that will minimize the energy E, subject to constraint that the volume is constant,

$$V(a, b, c) = abc = k. \tag{17.84}$$

With $f(a, b, c) = E(a, b, c)$ and $\varphi(a, b, c) = abc - k = 0$, we obtain

$$\frac{\partial E}{\partial a} + \lambda \frac{\partial V}{\partial a} = -\frac{h^2}{4ma^3} + \lambda bc = 0. \tag{17.85}$$

Also,

$$-\frac{h^2}{4mb^3} + \lambda ac = 0,$$

$$-\frac{h^2}{4mc^3} + \lambda ab = 0.$$

Multiplying the first of these expressions by a, the second by b, and the third by c, we have

$$\lambda abc = \frac{h^2}{4ma^2} = \frac{h^2}{4mb^2} = \frac{h^2}{4mc^2}. \tag{17.86}$$

Therefore our solution is

$$a = b = c, \qquad \text{a cube.} \tag{17.87}$$

Notice that λ has not been determined but follows from Eq. (17.86).

Example 17.6.2 Cylindrical Nuclear Reactor

A further example is provided by the nuclear reactor theory. Suppose a (thermal) nuclear reactor is to have the shape of a right circular cylinder of radius R and height H. Neutron diffusion theory supplies a constraint:

$$\varphi(R, H) = \left(\frac{2.4048}{R}\right)^2 + \left(\frac{\pi}{H}\right)^2 = \text{constant.}^2 \qquad (17.88)$$

We wish to minimize the volume of the reactor

$$f(R, H) = \pi R^2 H. \qquad (17.89)$$

Application of Eq. (17.82) leads to

$$\frac{\partial f}{\partial R} + \lambda \frac{\partial \varphi}{\partial R} = 2\pi RH - 2\lambda \frac{(2.4048)^2}{R^3} = 0,$$

$$\frac{\partial f}{\partial H} + \lambda \frac{\partial \varphi}{\partial H} = \pi R^2 - 2\lambda \frac{\pi^2}{H^3} = 0. \qquad (17.90)$$

By multiplying the first of these equations by $R/2$ and the second by H, we obtain

$$\pi R^2 H = \lambda \frac{(2.4048)^2}{R^2} = \lambda \frac{2\pi^2}{H^2} \qquad (17.91)$$

or

$$H = \frac{\sqrt{2}\pi R}{2.4048} = 1.847R, \qquad (17.92)$$

for the minimum volume right-circular cylindrical reactor.

Strictly speaking, we have found only an extremum. Its identification as a minimum follows from a consideration of the original equations.

EXERCISES

The following problems are to be solved by using Lagrangian multipliers.

17.6.1   The ground state energy of a particle in a pillbox (right-circular cylinder) is given by

$$E = \frac{\hbar^2}{2m} \left(\frac{(2.4048)^2}{R^2} + \frac{\pi^2}{H^2}\right),$$

in which R is the radius and H, the height of the pillbox. Find the ratio of R to H that will minimize the energy for a fixed volume.

17.6.2   Find the ratio of R(radius) to H(height) that will minimize the total surface area of a right-circular cylinder of fixed volume.

---

[2] 2.4048 . . . is the lowest root of Bessel function $J_0(R)$ (compare Section 11.1).

17.6.3   The U.S. Post Office limits first class mail to Canada to a total of 36 inches, length plus girth. Using a Lagrange multiplier, find the maximum volume and the dimensions of a (rectangular parallelepiped) package subject to this constraint.

17.6.4   A thermal nuclear reactor is subject to the constraint

$$\varphi(a, b, c) = \left(\frac{\pi}{a}\right)^2 + \left(\frac{\pi}{b}\right)^2 + \left(\frac{\pi}{c}\right)^2 = B^2, \quad \text{a constant.}$$

Find the ratios of the sides of the rectangular parallelepiped reactor of minimum volume.

ANS. $a = b = c$, cube.

17.6.5   For a simple lens of focal length f the object distance p and the image distance q are related by $1/p + 1/q = 1/f$.
Find the minimum object-image distance $(p + q)$ for fixed f. Assume real object and image (p and q both positive).

17.6.6   You have an ellipse $(x/a)^2 + (y/b)^2 = 1$. Find the inscribed rectangle of maximum area. Show that the ratio of the area of the maximum area rectangle to the area of the ellipse is $(2/\pi) = 0.6366$.

17.6.7   A rectangular parallelepiped is incribed in an ellipsoid of semiaxes a, b, and c. Maximize the volume of the inscribed rectangular parallelepiped. Show that the ratio of the maximum volume to the volume of the ellipsoid is $2/\pi\sqrt{3} \approx 0.367$.

17.6.8   A *deformed* sphere has a radius given by $r = r_0\{\alpha_0 + \alpha_2 P_2(\cos\theta)\}$, where $\alpha_0 \approx 1$ and $\alpha_2 \approx 0$. From Exercise 12.5.14 the area and volume are

$$A = 4\pi r_0^2 \alpha_0^2 \left\{1 + \frac{4}{5}\left(\frac{\alpha_2}{\alpha_0}\right)^2\right\}$$

$$V = \frac{4\pi r_0^3}{3} \alpha_0^3 \left\{1 + \frac{3}{5}\left(\frac{\alpha_2}{\alpha_0}\right)^2\right\}.$$

Terms of order α_2^3 have been neglected.

(a) With the constraint that the enclosed volume be held constant, that is, $V = 4\pi r_0^3/3$, show that bounding surface of minimum area is a sphere, ($\alpha_0 = 1$, $\alpha_2 = 0$).
(b) With the constraint that the area of the bounding surface be held constant; that is, $A = 4\pi r_0^2$. Show that the enclosed volume is a maximum when the surface is a sphere.

17.6.9   Find the maximum value of the directional derivative of $\varphi(x, y, z)$,

$$\frac{d\varphi}{ds} = \frac{\partial\varphi}{\partial x}\cos\alpha + \frac{\partial\varphi}{\partial y}\cos\beta + \frac{\partial\varphi}{\partial z}\cos\gamma.$$

subject to the constraint

$$\cos^2\alpha + \cos^2\beta + \cos^2\gamma = 1.$$

ANS. $\left(\dfrac{d\varphi}{ds}\right) = |\nabla\varphi|.$

Note concerning the following exercises:

In a quantum-mechanical system there are g_i distinct quantum states between energy E_i and $E_i + dE_i$. The problem is to describe how n_i particles are distributed among these states subject to two constraints:

(a) fixed number of particles,

$$\sum_i n_i = n.$$

(b) fixed total energy,

$$\sum_i n_i E_i = E.$$

17.6.10 For identical particles obeying the Pauli exclusion principle the probability of a given arrangement is

$$W_{FD} = \prod_i \frac{g_i!}{n_i!(g_i - n_i)!}.$$

Show that maximizing W_{FD} subject to a fixed number of particles and fixed total energy leads to

$$n_i = \frac{g_i}{e^{\lambda_1 + \lambda_2 E_i} + 1}.$$

With $\lambda_1 = -E_0/kT$ and $\lambda_2 = 1/kT$, this yields Fermi–Dirac statistics.

Hint. Try working with $\ln W$ and using Stirling's formula, Section 10.3. The justification for *differentiation* with respect to n_i is that we are dealing here with a large number of particles, $\Delta n_i/n_i \ll 1$.

17.6.11 For identical particles but no restriction on the number in a given state the probability of a given arrangement is

$$W_{BE} = \prod_i \frac{(n_i + g_i - 1)!}{n_i!(g_i - 1)!}.$$

Show that maximizing W_{BE}, subject to a fixed number of particles and fixed total energy, leads to

$$n_i = \frac{g_i}{e^{\lambda_1 + \lambda_2 E_i} - 1}.$$

With $\lambda_1 = -E_0/kT$ and $\lambda_2 = 1/kT$, this yields Bose–Einstein statistics.

Note. Assume that $g_i \gg 1$.

17.6.12 Photons satisfy W_{BE} and the constraint that total energy is constant. They clearly do *not* satisfy the fixed number constraint. Show that eliminating the fixed number constraint leads to the foregoing result but with $\lambda_1 = 0$.

17.7 VARIATION SUBJECT TO CONSTRAINTS

As in the preceding sections, we seek the path that will make the integral

$$J = \int f\left(y_i, \frac{\partial y_i}{\partial x_j}, x_j\right) dx_j \tag{17.93}$$

stationary. This is the general case in which x_j represents a set of independent variables and y_i, a set of dependent variables. Again,

$$\delta J = 0. \tag{17.94}$$

Now, however, we introduce one or more constraints. This mens that the y_i's are no longer independent of each other. Not all the η_i's may be varied arbitrarily and Eqs. (17.62) or (17.73a) would not apply. The constraint may have the form

$$\varphi_k(y_i, x_j) = 0, \tag{17.95}$$

as in Section 17.6. In this case we may multiply by a function of x_j, say, $\lambda_k(x_j)$ and integrate over the same range as in Eq. (17.93) to obtain

$$\int \lambda_k(x_j)\varphi_k(y_i, x_j)\, dx_j = 0. \tag{17.96}$$

Then clearly

$$\delta \int \lambda_k(x_j)\varphi_k(y_i, x_j)\, dx_j = 0. \tag{17.97}$$

Alternatively, the constraint may appear in the form of an integral

$$\int \varphi_k(y_i, \partial y_i/\partial x_j, x_j)\, dx_j = \text{constant}. \tag{17.98}$$

We may introduce any *constant* Lagrangian multiplier and again Eq. (17.97) follows—now with λ a constant.

In either case, by adding Eqs. (17.94) and (17.97), possibly with more than one constraint, we obtain

$$\delta \int \left[f\left(y_i, \frac{\partial y_i}{\partial x_j}, x_j\right) + \sum_k \lambda_k \varphi_k(y_i, x_j) \right] dx_j = 0. \tag{17.99}$$

The Lagrangian multiplier λ_k may depend on x_j when $\varphi(y_i, x_j)$ is given in the form of Eq. (17.95).

Treating the entire integrand as a new function

$$g\left(y_i, \frac{\partial y_i}{\partial x_j}, x_j\right),$$

we obtain

$$g\left(y_i, \frac{\partial y_i}{\partial x_j}, x_j\right) = f + \sum_k \lambda_k \varphi_k. \tag{17.100}$$

If we have Ny_i's $(i = 1, 2, \ldots, N)$ and m constraints $(k = 1, 2, \ldots, m)$, $N - m$ of the η_i's may be taken as arbitrary. For the remaining $m\eta_i$'s, the λ's may, in principle, be chosen so that the remaining Euler–Lagrange equations are satisfied, completely analogous to Eq. (17.80). The result is that our composite function g must satisfy the usual Euler–Lagrange equations

$$\frac{\partial g}{\partial y_i} - \sum_j \frac{\partial}{\partial x_j} \frac{\partial g}{\partial(\partial y_i/\partial x_j)} = 0, \qquad (17.101)$$

with one such equation for each dependent variable y_i (compare Eqs. (17.64) and (17.73)). These Euler equations and the equations of constraint are then solved simultaneously to find the function yielding a stationary value.

LAGRANGIAN EQUATIONS

In the absence of constraints Lagrange's equations of motion (Eq. (17.52)) were found to be[1]

$$\frac{d}{dt}\frac{\partial L}{\partial \dot{q}_i} - \frac{\partial L}{\partial q_i} = 0,$$

with t (time) the one independent variable and $q_i(t)$ (particle position) a set of dependent variables. Usually the generalized coordinates q_i are chosen to eliminate the forces of constraint, but this is not necessary and not always desirable. In the presence of constraints φ_k Hamilton's principle is

$$\delta \int \left[L(q_i, \dot{q}_i, t) + \sum_k \lambda_k(t)\varphi_k(q_i, t) \right] dt = 0, \qquad (17.102)$$

and the constrained Lagrangian equations of motion are

$$\frac{d}{dt}\frac{\partial L}{\partial \dot{q}_i} - \frac{\partial L}{\partial q_i} = \sum_k a_{ik}\lambda_k. \qquad (17.103)$$

Usually $\varphi_k = \varphi_k(q_i, t)$, independent of the generalized velocities $\dot{q}_i$. In this case the coefficient a_{ik} is given by

$$a_{ik} = \frac{\partial \varphi_k}{\partial q_i}. \qquad (17.104)$$

If q_i is a length, then $a_{ik}\lambda_k$ (no summation) represents the force of the kth constraint in the q_i-direction, appearing in Eq. (17.103) in exactly the same way as $-\partial V/\partial q_i$.

Example 17.7.1 Simple Pendulum

To illustrate, consider the simple pendulum, a mass m, constrained by a wire of length l to swing in an arc (Fig. 17.9). In the absence of the one constraint

$$\varphi_1 = r - l = 0 \qquad (17.105)$$

there are two generalized coordinates r and θ (motion in vertical plane). The Lagrangian is

$$L = T - V = \tfrac{1}{2}m(\dot{r}^2 + r^2\dot{\theta}^2) + mgr\cos\theta. \qquad (17.106)$$

---

[1] The symbol q is customary in advanced mechanics. It serves to emphasize that the variable is not necessarily a Cartesian variable (and not necessarily a length).

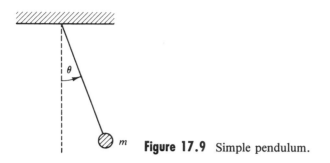

Figure 17.9 Simple pendulum.

taking the potential V to be zero when the pendulum is horizontal, $\theta = \pi/2$. By Eq. (17.103) the equations of motion are

$$\frac{d}{dt}\frac{\partial L}{\partial \dot{r}} - \frac{\partial L}{\partial r} = \lambda_1 \qquad (a_{r_1} = 1, \, a_{\theta_1} = 0),$$

$$\frac{d}{dt}\frac{\partial L}{\partial \dot{\theta}} - \frac{\partial L}{\partial \theta} = 0,$$

(17.107)

or

$$\frac{d}{dt}(m\dot{r}) - mr\dot{\theta}^2 - mg\cos\theta = \lambda_1,$$

$$\frac{d}{dt}(mr^2\dot{\theta}) + mgr\sin\theta = 0.$$

(17.108)

Substituting in the equation of constraint ($r = l$, $\dot{r} = 0$), we have

$$ml\dot{\theta}^2 + mg\cos\theta = -\lambda_1,$$

$$ml^2\ddot{\theta} + mgl\sin\theta = 0.$$

(17.109)

The second equation may be solved for $\theta(t)$ to yield simple harmonic motion if the amplitude is small ($\sin\theta = \theta$), whereas the first equation expresses $-\lambda_1$, the tension in the wire in terms of θ and $\dot{\theta}$.

Note that since the equation of constraint, Eq. (17.105), is in the form of Eq. (17.95), the Lagrange multiplier λ may be (and here is) a function of t (or of θ).

Example 17.7.2 Sliding off a Log

Closely related to this is the problem of a particle sliding on a cylindrical surface. The object is to find the critical angle θ_c at which the particle flies off from the surface. This critical angle is the angle at which the radial force of constraint goes to zero (Fig. 17.10).

We have

$$L = T - V = \tfrac{1}{2}m(\dot{r}^2 + r^2\dot{\theta}^2) - mgr\cos\theta \qquad (17.110)$$

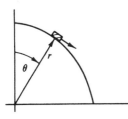

Figure 17.10 A particle sliding on a cylindrical surface.

and the one equation of constraint

$$\varphi_1 = r - l = 0. \tag{17.111}$$

Proceeding as in Example 17.7.1 with $a_{r_1} = 1$,

$$m\ddot{r} - mr\dot{\theta}^2 + mg \cos \theta = \lambda_1(\theta),$$
$$mr^2\ddot{\theta} + 2mr\dot{r}\dot{\theta} - mgr \sin \theta = 0, \tag{17.112}$$

in which the constraining force $\lambda_1(\theta)$ is a function of the angle θ.[2] Since $r = l$, $\ddot{r} = \dot{r} = 0$, Eq. (17.112) reduces to

$$-ml\dot{\theta}^2 + mg \cos \theta = \lambda_1(\theta), \tag{17.113a}$$

$$ml^2\ddot{\theta} - mgl \sin \theta = 0. \tag{17.113b}$$

Differentiating Eq. (17.113a) with respect to time and remembering that

$$\frac{df(\theta)}{dt} = \frac{df(\theta)}{d\theta} \dot{\theta}, \tag{17.114}$$

we obtain

$$-2ml\ddot{\theta} - mg \sin \theta = \frac{d\lambda_1(\theta)}{d\theta}. \tag{17.115}$$

Using Eq. (17.113b) to eliminate the $\ddot{\theta}$ term and then integrating, we have

$$\lambda_1(\theta) = 3mg \cos \theta + C. \tag{17.116}$$

Since

$$\lambda_1(0) = mg, \tag{17.117}$$

$$C = -2mg. \tag{17.118}$$

The particle m will stay on the surface as long as the force of constraint is non-negative, that is, as long as the surface has to push outward on the particle

$$\lambda_1(\theta) = 3mg \cos \theta - 2mg \geq 0. \tag{17.119}$$

---

[2] Note carefully that λ_1 is the *radial* force exerted by the cylinder on the particle. Consideration of the physical problem should show that λ_1 must depend on the angle θ. We permitted $\lambda = \lambda(t)$. Now we are replacing the time-dependence by an (unknown) angular dependence.

The critical angle lies where $\lambda_1(\theta_c) = 0$, the force of constraint going to zero. From Eq. (17.119))

$$\cos \theta_c = \tfrac{2}{3}, \qquad \text{or} \qquad \theta_c = 48°11' \tag{17.120}$$

from the vertical. At this angle (neglecting all friction) our particle takes off.

It must be admitted that this result can be obtained more easily by considering a varying centripetal force furnished by the radial component of the gravitational force. The example was chosen to illustrate the use of Lagrange's undetermined multiplier without confusing the reader with a complicated physical system.

Example 17.7.3 The Schrödinger Wave Equation

As a final illustration of a constrained minimum, let us find the Euler equations for the quantum mechanical problem

$$\delta \iiint \psi^*(x, y, z)\, H\psi(x, y, z)\, dx\, dy\, dz = 0, \tag{17.121}$$

with the constraint

$$\iiint \psi^*\psi\, dx\, dy\, dz = 1. \tag{17.122}$$

Equation (17.121) is a statement that the energy of the system is stationary, H being the quantum mechanical Hamiltonian for a particle of mass m, a differential operator,

$$H = -\frac{\hbar^2}{2m}\nabla^2 + V(x, y, z). \tag{17.123}$$

Equation (17.122), the constraint, is the condition that there will be exactly one particle present; ψ is the usual wave function, a dependent variable, and ψ^*, its complex conjugate, is treated as a second[3] dependent variable.

The integrand in Eq. (17.121) involves second derivatives, which can be converted to first derivatives by integrating by parts:

$$\int \psi^* \frac{\partial^2 \psi}{\partial x^2}\, dx = \psi^* \frac{\partial \psi}{\partial x}\bigg| - \int \frac{\partial \psi^*}{\partial x}\frac{\partial \psi}{\partial x}\, dx. \tag{17.124}$$

We assume either periodic boundary conditions (as in the Sturm–Liouville theory, Chapter 9) or that the volume of integration is so large that ψ and ψ^* vanish strongly[4] at the boundary. Then the integrated part vanishes and Eq. (17.121) may be rewritten as

$$\delta \iiint \left[\frac{\hbar^2}{2m}\nabla\psi^* \cdot \nabla\psi + V\psi^*\psi\right] dx\, dy\, dz = 0. \tag{17.125}$$

[3] Compare Section 6.1.
[4] $\lim_{r\to\infty} r^{1/2}\psi(r) = 0.$

The function g of Eq. (17.100) is

$$g = \frac{\hbar^2}{2m} \nabla \psi^* \cdot \nabla \psi + V\psi^*\psi - \lambda\psi^*\psi$$

$$= \frac{\hbar^2}{2m} (\psi_x^*\psi_x + \psi_y^*\psi_y + \psi_z^*\psi_z) + V\psi^*\psi - \lambda\psi^*\psi, \qquad (17.126)$$

again using the subscript x to denote $\partial/\partial x$. For $y_i = \psi^*$, Eq. (17.101) becomes

$$\frac{\partial g}{\partial \psi^*} - \frac{\partial}{\partial x}\frac{\partial g}{\partial \psi_x^*} - \frac{\partial}{\partial y}\frac{\partial g}{\partial \psi_y^*} - \frac{\partial}{\partial z}\frac{\partial g}{\partial \psi_z^*} = 0.$$

This yields

$$V\psi - \lambda\psi - \frac{\hbar^2}{2m}(\psi_{xx} + \psi_{yy} + \psi_{zz}) = 0$$

or

$$-\frac{\hbar^2}{2m}\nabla^2\psi + V\psi = \lambda\psi. \qquad (17.127)$$

Reference to Eq. (17.123) enables us to identify λ physically as the energy of the quantum mechanical system. With this interpretation, Eq. (17.127) is the celebrated Schrödinger wave equation. This variational approach is more than just a matter of academic curiosity. It provides a very powerful method of obtaining approximate solutions of the wave equation (Rayleigh–Ritz variational method, Section 17.8).

EXERCISES

17.7.1   A particle, mass m, is on a frictionless horizontal surface. It is constained to move so that $\theta = \omega t$ (rotating radial arm, no friction). With the initial conditions

$$t = 0, \qquad r = r_0, \qquad \dot{r} = 0,$$

(a) find the radial positions as a function of time.

                                      ANS. $r(t) = r_0 \cosh \omega t.$

(b) find the force exerted on the particle by the constraint.

                ANS. $F^{(c)} = 2m\dot{r}\omega = 2mr_0\omega^2 \sinh \omega t.$

17.7.2   A point mass m is moving over a flat, horizontal, frictionless plane. The mass is constrained by a string to move radially inward at a constant rate. Using plane polar coordinates (ρ, φ), $\rho = \rho_0 - kt$

(a) Set up the Lagrangian.

(b) Obtain the constrained Lagrange equations.

(c) Solve the φ-dependent Lagrange equation to obtain $\omega(t)$, the angular velocity. What is the physical significance of the constant of integration that you get from your "free" integration?

(d) Using the $\omega(t)$ from part (b), solve the ρ-dependent (constrained) Lagrange equation to obtain $\lambda(t)$. In other words, explain what is happening to the *force* of constraint as $\rho \to 0$.

17.7.3 A flexible cable is suspended from two fixed points. The length of the cable is fixed. Find the curve that will minimize the total gravitational potential energy of the cable.

ANS. Hyperbolic cosine.

17.7.4 A fixed volume of water is rotating in a cylinder with constant angular velocity ω. Find the curve of the water surface that will minimize the total potential energy of the water in the combined gravitational-centrifugal force field.

ANS. Parabola.

17.7.5 (a) Show that for a fixed-length perimeter the figure with maximum area is a circle.

(b) Show that for a fixed area the curve with minimum perimeter is a circle.
Hint. The radius of curvature R is given by

$$R = (r^2 + r_\theta^2)^{3/2}/(rr_{\theta\theta} - 2r_\theta^2 - r^2).$$

Note. The problems of this section, variation subject to constraints, are often called *isoperimetric*. The term arose from problems of maximizing area subject to a fixed perimeter—as in Exercise 17.7.5(a).

17.7.6 Show that requiring J, given by

$$J = \int_a^b (p(x)y_x^2 - q(x)y^2)\, dx,$$

to have a stationary value subject to the normalizing condition

$$\int_a^b y^2 w(x)\, dx = 1$$

leads to the Sturm–Liouville equation of Chapter 9:

$$\frac{d}{dx}\left(p\,\frac{dy}{dx}\right) + qy + \lambda wy = 0.$$

Note. The boundary condition

$$py_x y\big|_a^b = 0$$

is used in Section 9.1 in establishing the Hermitian property of the operator.

17.7.7 Show that requiring J, given by

$$J = \int_a^b \int_a^b K(x, t)\varphi(x)\varphi(t)\, dx\, dt,$$

to have a stationary value subject to the normalizing condition

$$\int_a^b \varphi^2(x)\, dx = 1$$

leads to the Hilbert–Schmidt integral equation, Eq. (16.89).
Note. The kernel $K(x, t)$ is symmetric.

17.8 RAYLEIGH–RITZ VARIATIONAL TECHNIQUE

Exercise 17.7.6 opens up a relation between the calculus of variations and eigenfunction–eigenvalue problems. We may rewrite the expression of Exercise 17.7.6 as

$$F[y(x)] = \frac{\int_a^b (py_x^2 - qy^2)\, dx}{\int_a^b y^2 w\, dx}, \tag{17.128}$$

in which the constraint appears in the denominator as a usual normalizing condition. The quantity F, a function of the function $y(x)$, is sometimes called a *functional*. It is homogeneous in y and independent of the normalization of y. Since the denominator is constant (for normalized functions), the stationary values of J correspond to the stationary values of F. Then from Exercise 17.7.6 when $y(x)$ is such that J and F take on a stationary value, the optimum function $y(x)$ satisfies the Sturm–Liouville equation

$$\frac{d}{dx}\left(p\frac{dy}{dx}\right) + qy + \lambda wy = 0, \tag{17.129}$$

with λ the eigenvalue (*not* a Lagrangian multiplier). Integrating the first term in the numerator of Eq. (17.128) by parts and using the *boundary condition*,

$$py_x y|_a^b = 0, \tag{17.130}$$

we obtain

$$F[y(x)] = -\int_a^b y\left\{\frac{d}{dx}\left(p\frac{dy}{dx}\right) + qy\right\} dx \Big/ \int_a^b y^2 w\, dx. \tag{17.131}$$

Then substituting in Eq. (17.129), the stationary values of $F[y(x)]$ are given by

$$F[y_n(x)] = \lambda_n, \tag{17.132}$$

with λ_n the eigenvalue corresponding to the eigenfunction y_n. Equation (17.132) with F given by either Eq. (17.128) or (17.131) forms the basis of the Rayleigh–Ritz method for the computation of eigenfunctions and eigenvalues.

GROUND STATE EIGENFUNCTION

Suppose that we seek to compute the ground state eigenfunction y_0 and eigenvalue[1] λ_0 of some complicated atomic or nuclear system. The classical example for which no exact solution exists is the helium atom problem. The eigenfunction y_0 is *unknown*, but we shall assume we can make a pretty good guess at an approximate function y, so that mathematically we may write[2]

---

[1] This means that λ_0 is the lowest eigenvalue. It is clear from Eq. (17.128) that if $p(x) \geq 0$ and $q(x) \leq 0$ (compare Table 9.1), then $F[y(x)]$ has a lower bound and this lower bound is non-negative. Recall from Section 9.1 that $w(x) \geq 0$.

[2] We are guessing at the *form* of the function. The normalization is irrelevant.

$$y = y_0 + \sum_{i=1}^{\infty} c_i y_i. \tag{17.133}$$

The c_i's are small quantities. (How small depends on how good our guess was.) The y_i's are orthonormalized eigenfunctions (also unknown), and therefore our trial function y is not normalized.

Substituting the approximate function y into Eq. (17.131) and noting that

$$\int_a^b y_i \left\{ \frac{d}{dx} \left(p \frac{dy_j}{dx} \right) + q y_j \right\} dx = -\lambda_i \delta_{ij}, \tag{17.134}$$

$$F[y(x)] = \frac{\lambda_0 + \sum\limits_{i=1}^{\infty} c_i^2 \lambda_i}{1 + \sum\limits_{i=1}^{\infty} c_i^2}. \tag{17.135}$$

Here we have taken the eigenfunctions to be orthogonal—since they are solutions of the Sturm–Liouville equation, Eq. (17.129). We also assume that y_0 is nondegenerate. Now, if we expand the denominator of Eq. (17.135) by the binomial theorem and discard terms of order c_i^4,

$$F[y(x)] = \lambda_0 + \sum_{i=1}^{\infty} c_i^2 (\lambda_i - \lambda_0). \tag{17.136}$$

Equation (17.136) contains two important results.

1. Whereas the error in the eigenfunction y was $O(c_i)$, the error in λ is $O(c_i^2)$. Even a poor approximation of the eigenfunctions may yield an accurate calculation of the eigenvalue.
2. If λ_0 is the lowest eigenvalue (ground state), then since $\lambda_i - \lambda_0 > 0$,

$$F[y(x)] = \lambda \geq \lambda_0, \tag{17.137}$$

or our approximation is always on the high side becoming lower, converging on λ_0 as our approximate eigenfunction y improves ($c_i \to 0$). Note that Eq. (17.137) is a direct consequence of Eq. (17.135) independent of our binomial approximation. More directly, without the expansion in Eq. (17.136), $F[y(x)]$ in Eq. (17.135) is the positively weighted average of the λ_i's and, therefore, must be no smaller than the smallest λ_i, to wit, λ_0. Often parameters in y may be varied to minimize F and thereby improve the estimate of the ground state energy λ_0.

Example 17.8.1 Vibrating String

A vibrating string, clamped at $x = 0$ and 1, satisfies the eigenvalue equation

$$\frac{d^2 y}{dx^2} + \lambda y = 0, \tag{17.138}$$

and the boundary condition $y(0) = y(1) = 0$. For this simple example the student will recognize immediately that $y_0(x) = \sin \pi x$ (unnormalized) and $\lambda_0 = \pi^2$. But let us try out the Rayleigh–Ritz technique.

With one eye on the boundary conditions, we try

$$y(x) = x(1 - x).$$ (17.139)

Then with $p = 1$ and $w = 1$, Eq. (17.128) yields

$$F[y(x)] = -\frac{\int_0^1 (1 - 2x)^2\, dx}{\int_0^1 x^2(1 - x)^2\, dx} = \frac{1/3}{1/30} = 10.$$ (17.140)

This result, $\lambda = 10$, is a fairly good approximation (1.3% error)[3] of $\lambda_0 = \pi^2 = 9.8696$. The reader may have noted that $y(x)$, Eq. (17.139), is not normalized to unity. The denominator in $F[y(x)]$ compensates for the lack of unit normalization. F may also be calculated from Eq. (17.131) since Eq. (17.130) is satisfied by y from Eq. (17.139).

In the usual scientific calculation the eigenfunction would be improved by introducing more terms and adjustable parameters such as

$$y = x(1 - x) + a_2 x^2(1 - x)^2.$$ (17.141)

It is convenient to have the additional terms orthogonal, but it is not necessary. The parameter a_2 is adjusted to *minimize* $F[y(x)]$. In this case, choosing $a_2 = 1.1353$ drives $F[y(x)]$ down to 9.8697, very close to the exact eigenvalue value.

EXERCISES

17.8.1   From Eq. (17.128) develop in detail the argument that $\lambda \geq 0$. Explain the circumstances under which $\lambda = 0$ and illustrate with several examples.

17.8.2   An unknown function satisfies the differential equation

$$y'' + \left(\frac{\pi}{2}\right)^2 y = 0$$

and the boundary conditions

$$y(0) = 1, \qquad y(1) = 0.$$

(a) Calculate the approximation

$$\lambda = F[y_{\text{trial}}]$$

for

$$y_{\text{trial}} = 1 - x^2.$$

(b) Compare with the exact eigenvalue.

ANS.   (a) $\lambda = 2.5$
(b) $\lambda/\lambda_{\text{exact}} = 1.013$.

---

[3] The closeness of the fit may be checked by a Fourier sine expansion (compare Exercise 14.2.3 over the half interval [0, 1] or, equivalently, over the interval $[-1, 1]$, with $y(x)$ taken to be odd). Because of the even symmetry relative to $x = 1/2$, only odd n terms appear:

$$y(x) = x(1 - x) = \left(\frac{8}{\pi^3}\right)\left[\sin \pi x + \frac{\sin 3\pi x}{3^3} + \frac{\sin 5\pi x}{5^3} + \cdots\right].$$

17.8.3   In Exercise 17.8.2 use a trial function

$$y = 1 - x^n.$$

(a) Find the value of n that will minimize $F[y_{\text{trial}}]$.
(b) Show that the optimum value of n drives the ratio $\lambda/\lambda_{\text{exact}}$ down to 1.003.

ANS. (a) $n = 1.7247$.

17.8.4   A quantum mechanical particle in a sphere (Example 11.7.1) satisfies

$$\nabla^2 \psi + k^2 \psi = 0,$$

with $\kappa^2 = 2mE/\hbar^2$. The boundary condition is that $\psi(r = a) = 0$, where a is the radius of the sphere. For the ground state [where $\psi = \psi(r)$] try an approximate wave function

$$\psi_a(r) = 1 - \left(\frac{r}{a}\right)^2$$

and calculate an approximate eigenvalue k_a^2.
Hint. To determine $p(r)$ and $w(r)$, put your equation in self-adjoint form (in spherical polar coordinates).

$$ANS. \quad k_a^2 = \frac{10.5}{a^2}$$

$$k_{\text{exact}}^2 = \frac{\pi^2}{a^2}.$$

17.8.5   The wave equation for the quantum mechanical oscillator may be written as

$$\frac{d^2\psi(x)}{dx^2} + (\lambda - x^2)\psi(x) = 0,$$

with $\lambda = 1$ for the ground state (Eq. (13.18)). Take

$$\psi_{\text{trial}} = \begin{cases} 1 - (x^2/a^2), & x^2 \le a^2 \\ 0, & x^2 > a^2 \end{cases}$$

for the ground-state wave function (with a^2 an adjustable parameter) and calculate the corresponding ground-state energy. How much error do you have?
Note. Your parabola is really not a very good approximation to a Gaussian exponential. What improvements can you suggest?

17.8.6   The Schrödinger equation for a central potential may be written as

$$\mathcal{L}u(r) + \frac{\hbar^2 l(l + 1)}{2Mr^2} u(r) = E u(r).$$

The $l(l + 1)$ term comes from splitting off the angular dependence (Section 2.5). Treating this term as a perturbation, use your variational technique to show that $E > E_0$, where E_0 is the energy eigenvalue of $\mathcal{L}u_0 = E_0 u_0$ corresponding to $l = 0$. This means that the minimum energy state will have $l = 0$, zero angular momentum.
Hint. You can expand $u(r)$ as $u_0(r) + \sum_{i=1}^{\infty} c_i u_i$, where $\mathcal{L}u_i = E_i u_i$, $E_i > E_0$.

17.8.7   In the *matrix* eigenvector, eigenvalue equation

$$\mathbf{A}\mathbf{r}_i = \lambda_i \mathbf{r}_i,$$

where A is an $n \times n$ Hermitian matrix. For simplicity, assume that its n real eigenvalues (Section 3.5) are distinct, λ_1 being the largest. If $\mathbf{r}$ is an approximation to $\mathbf{r}_1$,

$$\mathbf{r} = \mathbf{r}_1 + \sum_{i=2}^{n} \delta_i \mathbf{r}_i,$$

show that

$$\frac{\mathbf{r}^\dagger \mathbf{A}\mathbf{r}}{\mathbf{r}^\dagger \mathbf{r}} \leq \lambda_1$$

and that the error in λ_1 is of the order $|\delta_i|^2$. Take $|\delta_i| \ll 1$.
Hint. The n $\mathbf{r}_i$ form a *complete* orthogonal set spanning the n-dimensional (complex) space.

17.8.8   The variational solution of Example 17.8.1 may be refined by taking $y = x(1 - x) + a_2 x^2 (1 - x)^2$. Using the numerical quadrature, calculate $\lambda_{\text{approx}} = F[y(x)]$, Eq. (17.128), for a fixed value of a_2. Vary a_2 to minimize λ. Calculate the value of a_2 that minimizes λ and λ itself to five significant figures. Compare your eigenvalue λ with π^2.

ADDITIONAL READINGS

BLISS, G. A., *Calculus of Variations*. The Mathematical Association of America, Open Court Publishing Co. IL: LaSalle (1925). As one of the older texts, this is still a valuable reference for details of problems such as minimum area problems.

COURANT, R., and H. ROBBINS, *What Is Mathematics?* 2nd ed. New York: Oxford University Press (1979). Chapter VII contains a fine discussion of the calculus of variations, including soap film solutions to minimum area problems.

LANCZOS, C., *The Variational Principles of Mechanics*, 4th ed. Toronto: University of Toronto Press (1970). This book is a very complete treatment of variational principles and their applications to the development of classical mechanics.

SAGAN, H., *Boundary and Eigenvalue Problems in Mathematical Physics*. New York: Wiley (1961). This delightful text could also be listed as a reference for Sturm–Liouville theory, Legendre and Bessel functions, and Fourier Series. Chapter 1 is an introduction to the calculus of variations with applications to mechanics. Chapter 7 picks up the calculus of variations again and applies it to eigenvalue problems.

SAGAN, H., *Introduction to the Calculus of Variations*. New York: McGraw-Hill (1969). This is an excellent introduction to the modern theory of the calculus of variations which is more sophisticated and complete than his 1961 text. Sagan covers sufficiency conditions and relates the calculus of variations to problems of space technology.

WEINSTOCK, R., *Calculus of Variations*. New York: McGraw-Hill (1952). Paperback, New York: Dover. A detailed, systematic development of the calculus of variations and applications to Sturm–Liouville theory and physical problems in elasticity, electrostatics, and quantum mechanics.

YOURGRAU, W., and S. MANDELSTAM, *Variational Principles in Dynamics and Quantum Theory*, 3rd ed. Philadelphia: Saunders (1968). New York: Dover (1979). This is a comprehensive, authoritative treatment of variational principles. The discussions of the historical development and the many metaphysical pitfalls are of particular interest.

18 NONLINEAR METHODS AND CHAOS

18.1 INTRODUCTION

The origin of nonlinear dynamics goes back to the work of the renowned French mathematician Henri Poincaré on celestial mechanics at the turn of the century. Classical mechanics is, in general, nonlinear in its dependence on the coordinates of the particles and the velocities. An example are vibrations with a nonlinear restoring force. The Navier–Stokes equations are nonlinear which makes hydrodynamics difficult to handle. For almost four centuries however, following the lead of Galilei, Newton, and others, physicists have focused on predictable, effectively linear responses of classical systems which usually have linear and nonlinear properties.

Poincaré was the first to understand the possibility of completely irregular or "chaotic" behavior of solutions of nonlinear differential equations that are characterized by an extreme sensitivity to initial conditions: Given slightly different initial conditions, from errors in measurements for example, solutions can grow exponentially apart with time, so that the system soon becomes effectively unpredictable or "chaotic." This property of chaos is often called the "butterfly" effect and will be discussed in Section 18.3. Since the rediscovery of this effect by Lorenz in meteorology in the early 1960s, the field of nonlinear dynamics has grown tremendously. Thus, nonlinear dynamics and chaos theory now have entered the mainstream of physics.

Numerous examples of nonlinear systems have been found to display irregular behavior. Surprisingly, order in the sense of quantitative similarities as universal properties or other regularities may arise spontaneously in chaos; a first example, Feigenbaum's universal numbers α and δ will occur in Section 18.2. Dynamical chaos is not a rare phenomenon but ubiquitous in nature. It includes irregular shapes of clouds, coast lines, and other landscapes, which are examples of fractals to be discussed in Section 18.3, and turbulent flow of fluids, water dripping from a faucet, and the weather, of course. The damped, driven pendulum is among the simplest systems displaying chaotic motion. We take the time as the independent dynamic variable.

Necessary conditions for chaotic motion in dynamical systems described by *first order* differential equations are

(i) at least three dynamical variables, and
(ii) one or more nonlinear terms coupling two or several of them.

As in classical mechanics, the space of the three or more time-dependent dynamical variables of a system of coupled differential equations is called its *phase space*. In such deterministic systems, trajectories in phase space are not allowed to cross. If they did, the system would have a choice at each intersection and would not be deterministic. In two dimensions such nonlinear systems allow only for fixed points. An example is a damped pendulum whose second derivative $\ddot{\theta} = f(\dot{\theta}, \theta)$ can be written as two first-order derivatives $\omega = \dot{\theta}, \dot{\omega} = f(\omega, \theta)$ involving just two dynamic variables $\omega(t)$ and $\theta(t)$. In the undamped case, there will only be periodic motion and equilibrium points. With three or more dynamic variables (e.g., *damped*, *driven* pendulum written as first-order ODEs again), more complicated nonintersecting trajectories are possible. These include chaotic motion and are called *deterministic chaos*.

A central theme in chaos is the evolution of *complex* forms from the repetition of *simple*, but *nonlinear* operations; this is being recognized as a fundamental organizing principle of nature. While nonlinear differential equations are a natural place in physics for chaos to occur, the mathematically simpler iteration of nonlinear functions provides a quicker entry to chaos theory, which we will pursue first in Section 18.2.

18.2 THE LOGISTIC MAP

The nonlinear, one-dimensional iteration or difference equation

$$x_{n+1} = \mu x_n(1 - x_n), \qquad x_n \in [0, 1]; \quad 1 < \mu < 4 \tag{18.1}$$

is patterned after the nonlinear differential equation $dx/dt = \mu x(1 - x)$, used by P. F. Verhulst in 1845 to model the development of a breeding population whose generations do not overlap. The density of the population at time n is x_n. The linear term simulates the birth rate and the nonlinear term the death rate of the species in a constant environment controlled by the parameter μ.

The quadratic function $f_\mu(x) = \mu x(1 - x)$ is chosen because it has one maximum in the interval $[0, 1]$ and is zero at the endpoints, $f_\mu(0) = 0 = f_\mu(1)$. The maximum at $x_m = 1/2$ is determined from

$$f_\mu'(x_m) = \mu(1 - 2x_m) = 0, \tag{18.2}$$

where $f_\mu(1/2) = \mu/4$. Varying the single parameter μ controls a rich and complex behavior including one-dimensional chaos as we shall see. More parameters or additional variables are hardly necessary at this point to increase the complexity. In a rather qualitative sense the simple *logistic* map of Eq. (18.1) is representative of many dynamical systems in biology, chemistry, and physics.

Starting with some $x_0 \in [0, 1]$ in Eq. (18.1) we get x_1, then x_2, ... Plotting $f_\mu(x) = \mu x(1 - x)$ along with the diagonal, straight vertical lines to show the intersections with the curve f_μ and horizontal lines to convert $f_\mu(x_i) = x_{i+1}$ to the next x-coordinate, we can construct Fig. (18.1). Choosing $\mu (= 2$ in Fig. 18.1) and $x_0 (= 0.1)$, the vertical line $x = x_0$ meets the curve $f_\mu(x)$ in $x_1 (= 0.18)$, a horizontal line from (x_0, x_1) intersects the diagonal in (x_1, x_1). The vertical line through (x_1, x_1) meets the curve in $x_2 (= 0.2952$ in Fig. 18.1), etc.

The x_i converge toward the fixed point $(0.5, 0.5)$. At the fixed point x^*, or *attractor*, the iteration stops so that

$$f_\mu(x^*) = \mu x^*(1 - x^*) = x^*, \qquad \text{i.e., } x^* = 1 - 1/\mu. \qquad (18.3)$$

For any initial x_0 satisfying $f_\mu(x_0) < x^*$, or $0 < x_0 < 1/\mu$, the x_i converge to the attractor x^*; the interval $(0, 1/\mu)$ defines a *basin of attraction* for the fixed point x^*. (For $\mu > 1$ and $x_0 < 0$ or $x_0 > 1$, it is easy to verify graphically or analytically that the $x_i \rightarrow -\infty$. The origin $x = 0$ is a *repellent* fixed point since $f_\mu'(0) = \mu > 1$.) The attractor x^* is *stable* provided the slope $|f_\mu'(x^*)| < 1$, or $1 < \mu < 3$.

When

$$f_\mu'(x^*) = \mu(1 - 2x^*) = 2 - \mu = -1$$

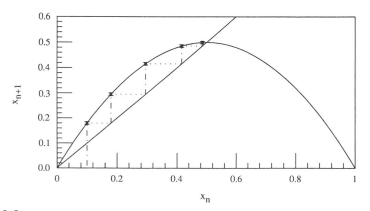

Figure 18.1   Cycle $(x_0, x_1, \ldots)$ for the logistic map for $\mu = 2$, starting value $x_0 = 0.1$ and attractor $x^* = 1/2$.

is reached for $\mu = 3$, two fixed points occur. Thus, for $\mu > 3$ the stable attractor *bifurcates* into two fixed points x_2^*. They can be located by solving

$$x_2^* = f_\mu(f_\mu(x_2^*)) = \mu^2 x_2^*(1 - x_2^*)[1 - \mu x_2^*(1 - x_2^*)],$$

where we abbreviate $f^{(1)}(x) = f_\mu(x)$, $f^{(2)}(x) = f_\mu(f_\mu(x))$ for the second iterate, etc. Each x_2^* is a point of period 2 and invariant under two iterations of the map f_μ. A point x_n is defined as a *periodic point of period n* for f_μ if $f^{(n)}(x_0) = x_0$ but $f^{(i)}(x_0) \neq x_0$ for $0 < i < n$. The bifurcation for $\mu = 3$, where the doubling occurs, is called a *pitchfork* bifurcation (see Fig. 18.2) because of its shape.

For $\mu = 1 + \sqrt{6} \sim 3.45$, which derives from $df_\mu(f_\mu)/dx = -1$, each branch of fixed points bifurcates again so that $x_4^* = f^{(4)}(x_4^*)$, i.e., has period 4. With increasing period doublings it becomes impossible to obtain analytic solutions. The iterations are better done numerically on a programmable pocket calculator or a personal computer whose rapid improvements (computer-driven graphics, in particular) and wide distribution in the 1970s and 1980s has speeded up the development of chaos theory. The sequence of bifurcations continues with ever longer periods until we reach $\mu_\infty = 3.5699456\ldots$, where an infinite number of bifurcations occur. Near bifurcation points fluctuations, rounding errors in initial conditions, etc., play an increasing role because the system has to choose between two possible branches and becomes much more sensitive to small perturbations. In the present case the x_n never repeat. The bands of fixed points x^* begin forming a continuum (shown dark in Fig. 18.2): This is where chaos starts.

This increasing period doubling is the route to chaos for the logistic map that is characterized by a universal constant δ called a Feigenbaum number. If the first bifurcation occurs at $\mu_1 = 3$, the second at $\mu_2 = 3.45$, ... then the

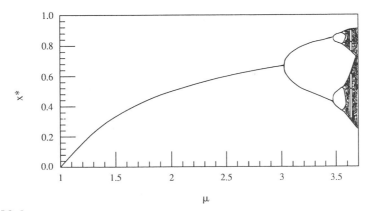

Figure 18.2   Part of the bifurcation plot for the logistic map: Fixed points x^* versus μ.

ratio of spacings between the μ_n converges to δ:

$$\lim_{n \to \infty} \frac{\mu_n - \mu_{n-1}}{\mu_{n+1} - \mu_n} = \delta = 4.66920161 \ldots \tag{18.4}$$

From the bifurcation plot Fig. 18.2 we obtain

$$(\mu_2 - \mu_1)/(\mu_3 - \mu_2) = (3.45 - 3.00)/(3.54 - 3.45) = 5.0$$

as a first approximation for the dimensionless δ. The corresponding critical period-$2n$ points x_n^* lead to another universal and dimensionless quantity

$$\lim_{n \to \infty} \frac{x_n^* - x_{n-1}^*}{x_{n+1}^* - x_n^*} = \alpha = 2.5029 \ldots \tag{18.5}$$

Again reading off Fig. 18.2 approximate values x_n^* we obtain

$$(0.44 - 0.67)/(0.37 - 0.44) = 3.3$$

as a first approximation for α.

The Feigenbaum number δ is universal for the route to chaos via period doublings for all maps with a *quadratic* maximum like the logistic map. It is an example of order in chaos. Experience shows that its validity is even wider including two-dimensional (dissipative) systems and twice continuously differentiable functions with subharmonic bifurcations.[1] When the maps behave like $|x - x_m|^{1+\varepsilon}$ near their maximum x_m for some ε between 0 and 1, the Feigenbaum number will depend on the exponent ε; thus $\delta(\varepsilon)$ varies between $\delta(1)$ given in Eq. (18.4) for quadratic maps to $\delta(0) = 2$ for $\varepsilon = 0$.[2]

EXERCISES

18.2.1   Show that $x^* = 1$ is a nontrivial fixed point of the map $x_{n+1} = x_n \exp[r(1 - x_n)]$ with a slope $1 - r$, so that the equilibrium is stable if $0 < r < 2$.

18.2.2   Draw a bifurcation diagram for the exponential map of Exercise 18.2.1 for $r > 1.9$.

18.2.3   Determine fixed points of the cubic map $x_{n+1} = ax_n^3 + (1 - a)x_n$ for $0 < a < 4$ and $0 < x_n < 1$.

18.2.4   Write the time delayed logistical map $x_{n+1} = \mu x_n(1 - x_{n-1})$ as a two-dimensional map: $x_{n+1} = \mu x_n(1 - y_n)$, $y_{n+1} = x_n$, and determine some of its fixed points.

18.2.5   Show that the second bifurcation for the logistical map that leads to cycles of period 4 is located at $\mu = 1 + \sqrt{6}$.

---

[1] More details and computer codes for the logistic map are given by G. L. Baker and J. P. Gollub, *Chaotic Dynamics: An Introduction.* Cambridge: Cambridge University Press (1990).

[2] For other maps and a discussion of the fascinating history how chaos became again a hot research topic, see D. Holton and R. M. May, in *The Nature of Chaos* (T. Mullin, ed.). Oxford: Clarendon Press (1993), Section 5, p. 95; and Gleick's *Chaos* (1987).

18.3 SENSITIVITY TO INITIAL CONDITIONS AND PARAMETERS

LYAPUNOV EXPONENTS

In Section 18.2 we described how, as we approach the period doubling accumulation parameter value $\mu_\infty = 3.5699 \ldots$ from below, the period $n + 1$ of cycles $(x_0, x_1, \ldots, x_n)$ with $x_{n+1} = x_0$ gets longer. It is also easy to check that the distances

$$d_n = |f^{(n)}(x_0 + \varepsilon) - f^{(n)}(x_0)| \tag{18.6}$$

grow as well for small $\varepsilon > 0$. From experience with chaotic behavior we expect this distance to increase exponentially with $n \to \infty$, i.e., $d_n/\varepsilon = e^{\lambda n}$, or

$$\lambda = \frac{1}{n} \ln\left(\frac{|f^{(n)}(x_0 + \varepsilon) - f^{(n)}(x_0)|}{\varepsilon}\right), \tag{18.7}$$

where λ is a *Lyapunov exponent* for the cycle. For $\varepsilon \to 0$ we may rewrite Eq. (18.7) in terms of derivatives as

$$\lambda = \frac{1}{n} \ln\left|\frac{df^{(n)}(x_0)}{dx}\right| = \frac{1}{n}\sum_{i=0}^{n} \ln|f'(x_i)|, \tag{18.8}$$

using the chain rule of differentiation for $df^{(n)}(x)/dx$, where

$$\frac{df^{(2)}(x_0)}{dx} = \frac{df_\mu}{dx}\bigg|_{x=f_\mu(x_0)} \frac{df_\mu}{dx}\bigg|_{x=x_0} = f_\mu'(x_1)f_\mu'(x_0) \tag{18.9}$$

and $f_\mu' = df_\mu/dx$, etc. Our Lyapunov exponent has been calculated at the point x_0. Let us repeat the procedure for several initial points that are randomly chosen. If we average those Lyapunov exponents, we obtain the *average Lyapunov exponent* for the sample. This average value is often called and taken as the Lyapunov exponent.

The Lyapunov exponent λ is a quantitative measure of chaos: A one-dimensional iterated function like the logistic map has *chaotic* cycles $(x_0, x_1, \ldots)$ for the parameter μ if *the average Lyapunov exponent is positive* for that value of μ. Any such initial point x_0 is called a *strange* or *chaotic attractor* (the shaded region in Fig. 18.2). For cycles of finite period λ is negative. This is the case for $\mu < 3$, for $\mu < \mu_\infty$ and even in the periodic window at $\mu \sim 3.627$ inside the chaotic region of Fig. 18.2. At bifurcation points, $\lambda = 0$. For $\mu > \mu_\infty$ the Lyapunov exponent is positive except in the periodic windows, where $\lambda < 0$, and λ grows with μ. In other words, the system becomes more chaotic as the control parameter μ increases.

In the chaos region of the logistic map there is a scaling law for the average Lyapunov exponent (we do not derive it),

$$\lambda(\mu) = \lambda_0(\mu - \mu_\infty)^{\ln 2/\ln \delta}, \tag{18.10}$$

where $\ln 2/\ln \delta \sim 0.445$, δ is the universal Feigenbaum number of Section 18.2 and λ_0 is a constant. This relation (18.10) is reminiscent of a physical observable at a (second-order) phase transition. The exponent in Eq. (18.10) is a universal number; the Lyapunov exponent plays the role of an *order parameter*, while $\mu - \mu_\infty$ is the analog of $T - T_c$, where T_c is the *critical* temperature at which the phase transition occurs.

FRACTALS

In chaotic systems often new geometric objects with intricate shapes appear that are called fractals based on their noninteger dimension. Fractals are irregular geometric objects that exist at many scales so that their smaller parts resemble their larger parts. Intuitively a fractal is a set which is (approximately) *self-similar* under magnification. The dimension is typically not integral. A set of attracting points with noninteger dimension is called a *strange* attractor.

We need a quantitative measure of dimensionality in order to describe fractals. Unfortunately, there are several definitions with usually different numerical values, none of which has become yet a standard. For strictly self-similar sets one measure suffices, of course. More complicated (e.g., only approximately self-similar) sets require more measures for their complete description. The simplest is the *box-counting dimension* due to Kolmogorov and Hausdorff. For a one-dimensional set, cover the curve by line segments of length R. In two dimensions the boxes are squares of area R^2, in three dimensions cubes of volume R^3, etc. Count the number $N(R)$ of boxes needed to cover the set. Letting R go to zero we expect N to scale as $N(R) \sim R^{-d}$. Taking the logarithm the *box-counting dimension* is defined as

$$d = \lim_{R \to 0} [-\ln N(R)/\ln R]. \qquad (18.11)$$

For example, in a two-dimensional space a single point is covered by one square so that $\ln N(R) = 0$ and $d = 0$. A finite set of isolated points also has dimension $d = 0$. For a differentiable curve of length L, $N(R) \sim L/R$ as $R \to 0$ so that $d = 1$ from Eq. (18.11) as expected.

Let us now construct a more irregular set, the *Koch* curve. We start with a line segment of unit length in Fig. 18.3 and remove the middle 1/3. Then we replace it with two segments of length 1/3 which form a triangle in Fig. 18.3. We iterate this procedure with each segment ad infinitum. The resulting Koch curve is infinitely long and is nowhere differentiable because of the infinitely many discontinuous changes of slope. At the nth step each line segment has length $R_n = 3^{-n}$ and there are $N(R_n) = 4^n$ segments. Hence its dimension is $d = \ln 4/\ln 3 = 1.26 \ldots$, which is more than a curve but less than a surface. As the Koch curve results from iteration of the first step, it is strictly self-similar.

For the logistic map the box-counting dimension at a period-doubling accumulation point μ_∞ is 0.5388 ... which is a universal number for iterations of functions in one variable with a quadratic maximum. To see roughly how this comes about, consider the pairs of line segments originating from successive bifurcation points for a given parameter μ in the chaos regime (see Fig. 18.2). Imagine removing the interior space from the chaotic bands. When

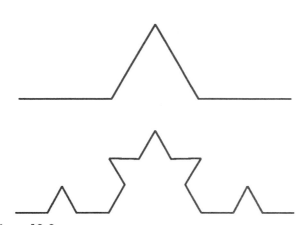

Figure 18.3   Construction of the Koch curve by iterations.

we go to the next bifurcation the relevant scale parameter is $\alpha = 2.5029 \ldots$ from Eq. (18.5). Suppose we need 2^n line segments of length R to cover 2^n bands. In the next stage we then need 2^{n+1} segments of length R/α to cover the bands. This yields a dimension $d = -\ln(2^n/2^{n+1})/\ln \alpha = 0.4498 \ldots$. This crude estimate can be improved by taking into account that the width between neighboring pairs of line segments differs by $1/\alpha$ (see Fig. 18.2). The improved estimate 0.543 is closer to 0.5388 .... This example suggests that when the fractal set does not have a simple self-similar structure, then the box-counting dimension depends on the box construction method.

EXERCISE

18.3.1   Use a programmable pocket calculator (or a personal computer with BASIC or FORTRAN or symbolic software such as Mathematica or Maple) to obtain the iterates x_i of an initial $0 < x_0 < 1$ and $f'_\mu(x_i)$ for the logistic map. Then calculate the Lyapunov exponent. Show that for $\mu < \mu_\infty$ the Lyapunov exponent $\lambda = 0$ at bifurcation points and negative elsewhere, while for $\mu > \mu_\infty$ it is positive except in periodic windows.
Hint. See Fig. 9.3 of Hilborn in Additional Readings.

18.4 NONLINEAR DIFFERENTIAL EQUATIONS

In the introductory Section 18.1 we mentioned nonlinear differential equations (abbreviated as *NDEs*) as the natural place in physics for chaos to occur, but continued with the simpler iteration of nonlinear functions of one variable

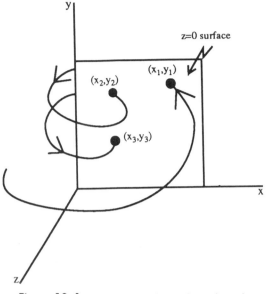

Figure 18.4   Schematic of a Poincaré section.

(maps). Here we briefly address the much broader area of NDEs and the far greater complexity in the behavior of their solutions. However, maps and systems of solutions of NDEs are closely related. The latter can often be analyzed in terms of discrete maps. One prescription is the so-called *Poincaré section* of a system of NDE solutions. Placing a surface transverse into a trajectory (of a solution of a NDE) intersects the surface in a series of points with increasing time, e.g., in Fig. 18.4 $(x(t_1), y(t_1)) = (x_1, y_1), (x_2, y_2)$, etc.

Let us start with a few classical examples of NDEs. In Chapter 8 we have already discussed the soliton solution of the nonlinear Korteweg–de Vries PDE.

BERNOULLI AND RICCATI EQUATIONS

Bernoulli equations are also nonlinear having the form

$$y'(x) = p(x)y(x) + q(x)[y(x)]^n, \tag{18.12}$$

where p and q are real functions and $n \neq 0$, 1 to exclude first-order linear ODEs. If we substitute

$$u(x) = [y(x)]^{1-n}, \tag{18.13}$$

then Eq. (18.12) becomes a first-order linear ODE

$$u' = (1 - n)y^{-n}y' = (1 - n)[p(x)u(x) + q(x)], \tag{18.14}$$

which we can solve as described in Section 8.2.

Riccati equations are quadratic in $y(x)$:

$$y' = p(x)y^2 + q(x)y + r(x),\qquad(18.15)$$

where $p \neq 0$ to exclude linear ODEs and $r \neq 0$ to exclude Bernoulli equations. There is no general method for solving Riccati equations. However, when a special solution $y_0(x)$ of Eq. (18.15) is known by guess or inspection, then the general solution $y = y_0 + u$ leads to the Bernoulli equation

$$u' = pu^2 + (2py_0 + q)u,\qquad(18.16)$$

because substitution of Eq. (18.15) for y_0 removes $r(x)$ from Eq. (18.16).

Just as for Riccati equations there are no general methods for obtaining exact solutions of nonlinear ordinary differential equations. It is more important to develop methods for finding the approximate behavior of solutions. In Chapter 8 we described that power series solutions of ODEs exist except (possibly) at regular or essential singularities, which are directly given by local analysis of the coefficient functions of the ODE. Such local analysis provides us with the asymptotic behavior of solutions as well.

FIXED AND MOVABLE SINGULARITIES, SPECIAL SOLUTIONS

Solutions of NDEs also have such singular points which are independent of the initial or boundary conditions and are called *fixed singularities*. In addition they may have *spontaneous* or *movable* singularities that vary with the initial or boundary conditions. They complicate the (asymptotic) analysis of NDEs. This point is illustrated by a comparison of the linear ODE

$$y' + y/(x - 1) = 0,$$

which has the obvious regular singularity at $x = 1$, with the NDE $y' = y^2$. Both have the same solution with initial condition $y(0) = 1$, namely $y(x) = 1/(1 - x)$. For $y(0) = 2$, though, the pole in the solution $y(x) = 2/(1 - 2x)$ of the NDE has moved to $x = 1/2$.

For a second-order ODE we have a complete description of (the asymptotic behavior of) its solutions when (that of) two linearly independent solutions are known. For NDEs there may still be *special solutions* whose asymptotic behavior is not obtainable from two independent solutions. This is another *characteristic* property of NDEs which we illustrate again by a simple example. The general solution of the NDE $y'' = yy'/x$ is given by

$$y(x) = 2c_1 \tan(c_1 \ln x + c_2) - 1,\qquad(18.17)$$

where c_i are integration constants. An obvious special solution is $y = c_3 =$ constant which cannot be obtained from Eq. (18.17) for any choice of the parameters c_1, c_2. Note that using the substitution $x = e^t$, $Y(t) = y(e^t)$ so that $x\,dy/dx = dY/dt$, we obtain the ODE $Y'' = Y'(Y + 1)$. This ODE can be integrated once to give $Y' = \frac{1}{2}Y^2 + Y + c$ with $c = 2(c_1^2 + 1/4)$ an integration constant, and again according to Section 8.2 to lead to the solution of Eq. (18.17).

AUTONOMOUS DIFFERENTIAL EQUATIONS

Differential equations which do not explicitly contain the independent variable, taken to be the time t here, are called *autonomous*. Verhulst's NDE $\dot{y} = dy/dt = \mu y(1 - y)$, which we encountered briefly in Section 18.2 as motivation for the logistic map, is a special case of this wide and important class of ODEs. For one dependent variable $y(t)$ they can be written as

$$\dot{y} = f(y), \tag{18.18a}$$

and for several dependent variables as a system

$$\dot{y}_i = f_i(y_1, y_2, \ldots, y_n), \qquad i = 1, 2, \ldots, n \tag{18.18b}$$

with sufficiently differentiable functions f, f_i. A solution of Eq. (18.18) is a curve or trajectory $y(t)$ for $n = 1$ and in general a trajectory $(y_1(t), y_2(t), \ldots, y_n(t))$ in an n-dimensional (so-called) *phase space*. As we discussed already in the introductory Section 18.1 two trajectories cannot cross because of the uniqueness of the solutions of ODEs. Clearly, solutions of the algebraic system

$$f_i(y_1, y_2, \ldots, y_n) = 0 \tag{18.19}$$

are special points in phase space, where the position vector $(y_1, y_2, \ldots, y_n)$ does not move on the trajectory; they are called *critical* (or *fixed*) *points*. Again because of the uniqueness of the solutions a critical point can only be approached asymptotically (for $t \to \infty$, say) but cannot be actually reached. It turns out that a local analysis of solutions near critical points leads to an understanding of the global behavior of the solutions. Let us look at a simple example.

For Verhulst's ODE $y(1 - y) = 0$ gives $y = 0$ and $y = 1$ as the critical points; for the logistic map they are repellent fixed points as $|f'_\mu| > 1$ for both. A local analysis near $y = 0$ suggests neglecting the y^2 term and solving $\dot{y} = \mu y$ instead. Integrating $\int dy/y = \mu t + \ln c$ gives the solution $y(t) = ce^{\mu t}$ which diverges as $t \to \infty$, so that $y = 0$ is an *unstable* critical point. [Note that for $\mu < 0$ the critical point $y = 0$ would be stable leading to a converging $y \sim e^{\mu t}$ solution.] Similarly at $y = 1$, $\int dy/(1 - y) = \mu t - \ln c$ leads to $y(t) = 1 - ce^{-\mu t} \to 1$ for $t \to \infty$. Hence $y = 1$ is a *stable* critical point. As the ODE is separable its general solution is given by

$$\int \frac{dy}{y(1 - y)} = \int dy[y^{-1} + (1 - y)^{-1}] = \ln[y/(1 - y)] = \mu t + \ln c.$$

Hence $y(t) = ce^{\mu t}/[1 + ce^{\mu t}]$ for $t \to \infty$ converges to 1, thus confirming the local analysis. Let us summarize various cases.

GLOBAL AND LOCAL BEHAVIOR—SUMMARY

In one-dimensional phase space (y-line) a solution $y(t)$ can exhibit the following types of global behavior:

1. As $t \to \infty$ a trajectory may approach a critical point;

2. it may diverge to $+\infty$ or $-\infty$,
3. it may stay at a critical point for all t.

Near a critical point there are the following types of local behavior:

1. All trajectories converge toward the critical point, which then is stable $(\cdots \rightarrow \cdot \leftarrow \cdots)$.
2. All trajectories may move away from the critical point, which then is unstable $(\cdots \leftarrow \cdot \rightarrow \cdots)$.
3. Trajectories on one side of the critical point may converge toward it and diverge from it on the other side. Such a critical point is a *saddle* point $(\cdots \rightarrow \cdot \rightarrow \cdots)$. An example is $y = 0$ for $\dot{y} = y^2$ and $-\infty < t < \infty$.

In two dimensions there are additional global possibilities:

4. The trajectory may describe a closed orbit.
5. The trajectory may approach a closed orbit (spiraling inward or outward toward the orbit) as $t \rightarrow \infty$.

In two or more dimensions the local behavior of a trajectory near a critical point is also more varied in general: At a stable critical point all trajectories may approach the critical point along straight lines or spiral inward (toward the *stable spiral node*), or may follow a more complicated path, or none of these. If all time-reversed trajectories move toward the critical point in spirals as $t \rightarrow -\infty$, then the critical point is called an unstable spiral point or *spiral repellor*. When some trajectories approach the critical point while others move away from it, then it is called a saddle point. When all trajectories form closed orbits about the critical point, it is called a *center*.

In three or more dimensional NDEs chaotic motion may occur, often when a constant of the motion (an energy integral for NDEs defined by a Hamiltonian, for example) restricts the trajectories to a finite volume in phase space and when there are no critical points. One characteristic signal for chaos is when for each trajectory there are nearby ones, some of which move away from it, while others approach it with increasing time. The notion of exponential divergence of nearby trajectories is made quantitative by the Lyapunov exponent λ. If two nearby trajectories are at a distance d_0 at time $t = 0$, diverge with a distance $d(t)$ at a later time, then $d(t) = d_0 e^{\lambda t}$ holds. A positive λ value is characteristic of chaotic behavior (see Section 18.3 for more details).

There are many routes to chaos. Some include sequences of bifurcations and period doublings, others sudden changes in trajectories associated with several critical points such as *global bifurcations*. Often they involve changes in basins of attraction and/or other global structures. Bifurcations that are linked to sudden changes in the qualitative behavior of dynamical systems at a single fixed point are called *local bifurcations*. Rather sudden changes from regular to random behavior are characteristic of bifurcations, as is sensitive dependence on initial conditions: Nearby initial conditions can lead to very different long term behavior.

ADDITIONAL READINGS

AMANN, H., *Ordinary Differential Equations: An Introduction To Nonlinear Analysis.* Berlin and New York: de Gruyter (1990).

BAKER, G. L., and J. P. GOLLUB, *Chaotic Dynamics: An Introduction.* Cambridge: Cambridge University Press (1990).

BENDER, C. M., and S. A. ORSZAG, *Advanced Mathematical Methods For Scientists and Engineers.* New York: McGraw-Hill (1978), Chapter 4 in particular.

BERGÉ, P., Y. POMEAU, and C. VIDAL, *Order within Chaos.* New York: Wiley (1984).

CVITANOVIC, P., *Universality in Chaos.* Bristol: Adam Hilger (1984).

DEVANEY, R. L., *An Introduction to Chaotic Dynamical Systems.* Menlo Park, CA: Benjamin/Cummings (1986).

EARNSHAW, J. C. and D. HAUGHEY, Lyapunov exponents for pedestrians. *Am. J. Phys.* **61**, 401 (1993).

GLEICK, J., *Chaos.* New York: Penguin Books (1987).

HILBORN, R. C., *Chaos and Nonlinear Dynamics.* New York: Oxford University Press (1994).

INFELD, E., and G. ROWLANDS, *Nonlinear Waves, Solitons and Chaos.* Cambridge: Cambridge University Press (1990).

JORDAN, D. W., and P. SMITH, *Nonlinear Ordinary Differential Equations*, 2nd ed. Oxford: Oxford University Press (1987).

LYAPUNOV, A. M., *The General Problem of the Stability of Motion.* Bristol, PA: Taylor & Francis (1992).

MANDELBROT, B. B., *The Fractal Geometry of Nature.* San Francisco: W. H. Freeman (1977).

MOON, F. C., *Chaotic and Fractal Dynamics.* New York: Wiley (1992).

SACHDEV, P. L., *Nonlinear Differential Equations and their Applications.* New York: Dekker (1991).

TUFILLARO, N. B., T. ABBOTT, and J. REILLY, *An Experimental Approach to Nonlinear Dynamics and Chaos.* Redwood City, CA: Addison-Wesley (1992).

APPENDIX 1:
Real Zeros of a Function

The demand for the values of the real zeros of a function occurs frequently in mathematical physics. Examples include the boundary conditions on the solution of a coaxial wave guide problem, Example 11.3.1, eigenvalue problems in quantum mechanics such as the deuteron with a square well potential, Example 9.1.2, and the location of the evaluation points in Gaussian quadrature (Appendix 2).

Most methods require close initial guesses of the zero or root. How close depends on how wildly your function is varying and what accuracy you demand. All are methods for refining a good initial value. To obtain the good initial value and to locate pathological features that must be avoided (such as discontinuities or singularities), you should make a reasonably detailed graph of the function. There is no real substitute for a graph. Exercise 11.3.12 emphasizes this point.

Newton's method is often presented in differential calculus because it assumes the function $f(x)$ to have a continuous first derivative and requires its computation. It is a method to *avoid* because it is treacherous: It may fail to converge or may converge to the wrong root. When iterated it may lead to chaos for suitable initial guesses. We therefore do not discuss it in detail and refer to Chapter 9 of Press *et al.* in the references.

BISECTION METHOD

This method assumes only that $f(x)$ is *continuous*. It requires that initial values x_l and x_r straddle the zero being sought. Thus $f(x_l)$ and $f(x_r)$ will have opposite signs, making the product $f(x_l) \cdot f(x_r)$ negative. In the simplest form of the bisection method, take the midpoint $x_m = \frac{1}{2}(x_l + x_r)$ and test to see which interval $[x_l, x_m]$ or $[x_m, x_r]$ contains the zero. The easiest test is to see if one product, say, $f(x_m) \cdot f(x_r) < 0$. If this product is negative, then the root is in the upper half interval $[x_m, x_r]$, if positive, then the root must be in the lower half interval $[x_l, x_m]$. Remember, we are assuming $f(x)$ to be continuous.

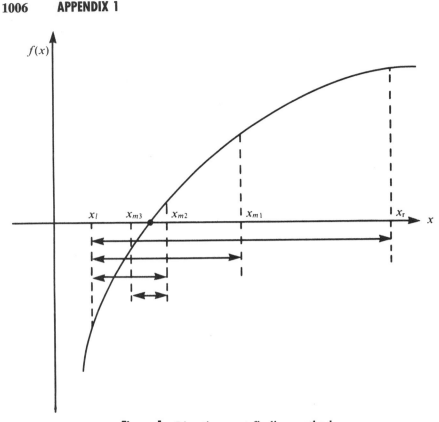

Figure 1   Bisection root-finding method.

The interval containing the zero is relabeled $[x_l, x_r]$ and the bisecting continues (as in Fig. 1) until the root is located to the desired degree of accuracy. Of course, the better the initial choice of x_l and x_r is, the fewer will be the bisections required. However, as explained subsequently, it is important to specify the maximum number of bisections that will be permitted.

This bisection technique may not have the elegance of Newton's method, but it is reasonably fast and much more reliable—almost foolproof if you avoid discontinuous functions, such as $f(x) = 1/(x - a)$, shown in Fig. 2. Again, there is no substitute for knowing the detailed local behavior of your function in the vicinity of your supposed root.

In general, the bisection method is *recommended*.

TWO WARNINGS
1. Since the computer carries only a finite number of significant figures we cannot expect to calculate a zero with infinite precision. It is necessary to specify some tolerance. When the root is located to within this tolerance the subroutine returns control to the main calling program.

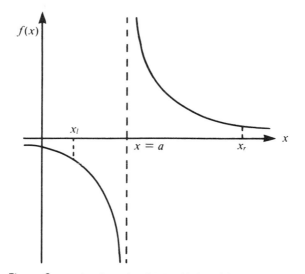

Figure 2   A simple pole, $f(x_l) \cdot f(x_r) < 0$ but no root.

2. All the approaches mentioned here are iteration techniques. How many times do you iterate? How do you decide to stop? It is possible to program the iteration so that it continues until the desired accuracy is obtained. The danger is that some factor may prevent reasonable convergence. Then your tolerance is never achieved and you have an infinite loop. It is far safer to specify in advance a maximum number of iterations. Thus these subroutines will stop when either a zero is determined to within your specified tolerance or the number of iterations reaches your specified maximum—whichever occurs first. With a simple bisection technique the selection of a number of iterations depends on the initial spread $x_r - x_l$ and on the precision you demand. Each iteration will cut the range by a factor of 2. Since $2^{10} = 1024 \approx 10^3$, 10 iterations should add 3 significant figures, 20 should add 6 significant figures to the location of the root.

EXERCISES

1.1   Write a *simple* bisection root determination subroutine that will determine a simple real root once you have straddled it. Test your subroutine by determining the roots of one or more polynomials or elementary transcendental functions.

1.2   Try the bisection method to locate a root of the following functions

(a) $f(x) = x^2 + 1$, and $x_0 = 0.9, 1.0$
(b) $f(x) = (x^2 + 1)^{1/2}$, $x_0 = 0.9, 1.0$

(c) $f(x) = \sin x$, $x_0 = 1.0, 1.1, 1.2$
(d) $f(x) = \tanh x$, $x_0 = 0.9, 1.0, 1.1$.

1.3 The theory of free radial oscillations of a homogeneous earth leads to an equation

$$\tan x = \frac{x}{1 - a^2 x^2}.$$

The parameter a depends on the velocities of the primary and secondary waves. For $a = 1.0$, find the first three positive roots of this equation.

ANS. $x_1 = 2.7437$ $x_2 = 6.1168$ $x_3 = 9.3166$.

1.4 (a) Using the Bessel function $J_0(x)$ generated by function BESSJ0(x) of Press *et al.* in the references, locate consecutive roots of $J_0(x)$: α_n and α_{n+1} for $n = 5, 10, 15, \ldots, 30$. Tabulate α_n, α_{n+1}, $(\alpha_{n+1} - \alpha_n)$ and $(\alpha_{n+1} - \alpha_n)/\pi$. Note how this last ratio is approaching unity.
(b) Compare your values of α_n with values calculated from McMahon's expansion, AMS-55, Eq. (9.5.12).

ADDITIONAL READINGS

HAMMING, R. W., *Introduction to Applied Numerical Analysis.* New York: McGraw-Hill (1971), especially Chapter 2. In terms of the author's insight into numerical computation and his ability to communicate to the average reader, this book is unexcelled.

PRESS, W. H., B. P. FLANNERY, S. A. TEUKOLSKY, and W. T. VETTERLING, *Numerical Recipes*, 2nd ed. Cambridge: Cambridge University Press (1992).

APPENDIX 2:
Gaussian Quadrature

INTERPOLATORY FORMULAS

The problem is to find the numerical value of a definite integral

$$I = \int_a^b f(x)w(x)\,dx.$$

We *approximate* our integral by a finite sum

$$\int_a^b f(x)w(x)\,dx \approx \sum_{k=1}^{n} A_k f(x_k). \tag{A2.1}$$

The sum in Eq. (A2.1) contains $2n + 1$ parameters:

$$n\ x_k\text{'s, points for evaluating } f(x)$$

$$n\ A_k\text{'s, coefficients}$$

and

$$1 \text{ the choice of } n \text{ itself.}$$

We proceed by replacing $f(x)$ by an *interpolating* polynomial $P(x)$ of degree $n - 1$ and a remainder term:

$$f(x) = P(x) + r(x). \tag{A2.2}$$

$P(x)$ is fitted to $f(x)$ at the n x_k $[P(x_k) = f(x_k)]$ by the choice

$$P(x) = \sum_{k=1}^{n} \frac{\alpha(x)}{(x - x_k)\alpha'(x_k)} f(x_k), \tag{A2.3}$$

where $\alpha(x)$ is a completely factored nth-degree polynomial,

$$\alpha(x) = (x - x_1)(x - x_2) \cdots (x - x_n). \tag{A2.4}$$

Note that

$$\lim_{x \to x_k} \frac{\alpha(x)}{(x - x_k)\alpha'(x_k)} = 1. \tag{A2.5}$$

For $f(x)$ a polynomial of degree $n - 1$ the remainder term $r(x)$ is zero and Eq. (A2.3) becomes an identity. Specifically (using Eq. (A2.5)), $P(x_k) = f(x_k)$, the $(n - 1)$-degree polynomial is fitted to $f(x)$ at $n x_k$.

When the integral of the remainder term is small

$$\int_a^b f(x)w(x)\,dx \approx \int_a^b P(x)w(x)\,dx$$

$$= \int_a^b \sum_{k=1}^n f(x_k) \frac{\alpha(x)}{(x - x_k)\alpha'(x_k)} w(x)\,dx \qquad (A2.6)$$

using Eq. (A2.3). Interchanging summation and integration, we obtain

$$\int_a^b f(x)w(x)\,dx \approx \sum_{k=1}^n f(x_k) \int_a^b \frac{\alpha(x)}{(x - x_k)\alpha'(x_k)} w(x)\,dx$$

$$= \sum_{k=1}^n A_k f(x_k). \qquad (A2.7)$$

Quadrature formulas of this type are labeled *interpolatory*. Since every polynomial $f(x)$ of degree $n - 1$ may be represented exactly [$r(x) = 0$] by our n-point-fit interpolating polynomial $P(x)$, Eq. (A2.7) is exact for such polynomial functions, $f(x)$.

The locations of the x_k, the zeros of $\alpha(x)$ in Eq. (A2.7) have not been specified. Taking them to be equally spaced leads to the various Newton–Cotes formulas. Of these Simpson's rule (Eq. (A2.8)) is probably the best known and, among the simpler formulas, it is the most accurate.

$$\int_a^b f(x)\,dx \approx \frac{h}{3}\{f(a) + 4f(a + h) + 2f(a + 2h) + 4f(a + 3h)$$

$$+ 2f(a + 4h) + \cdots + 4f(b - h) + f(b)\}. \qquad (A2.8)$$

Here h is the distance between the equally spaced points, $h = x_2 - x_1 = x_3 - x_2$, and so on. Equation (A2.8) may be considered a sum of three-point fits

$$\int_c^{c+2h} f(x)\,dx \approx \frac{h}{3}\{f(c) + 4f(c + h) + f(c + 2h)\}, \qquad (A2.9)$$

which is expected to be exact if $f(x)$ is of degree ≤ 2 over the interval $[c, c + 2h]$.

Actually Simpson's rule is better than this. An analysis of the error shows that the error in Simpson's rule is given by $-h^5 f^{(4)}(\xi)/90$ where ξ is a point in $[c, c + 2h]$. For $f(x) = x^3$, $f^{(4)}(x) = 0$ and Simpson's rule is *exact for cubic equations*. The reader may verify this by showing that $\int_0^b x^3\,dx$ is given exactly by Eq. (A2.8).

This result may be interpreted as a consequence of symmetry principles: (1) the coefficients in Simpson's rule are symmetric with respect to the middle x_k; 1, 4, 1 for Eq. (A2.9). (2) For Simpson's rule $n = 3$, odd and x^3 is an odd function. If we set $c = -h$, $c + h = 0$, then both sides of Eq. (A2.9) vanish—by (anti) symmetry. This additional degree of precision appears for each of the Newton–Coles formulas where n is odd.

GAUSSIAN QUADRATURE

It was pointed out by Gauss that the locations of x_k represent unused parameters that may be used to improve the accuracy of Eq. (A2.7), that greater precision can be obtained if the zeros of $\alpha(x)$ are *not* equally spaced but are chosen as follows.

Take the x_k so that our completely factored nth-degree polynomial $\alpha(x)$ is the nth-degree polynomial which is orthogonal to all lower degree polynomials over $[a, b]$ with respect to the weighting factor $w(x)$. The most frequently encountered combinations of interval and weighting factor are those in Table 9.3.[1] The x_k's therefore are the n zeros of the nth-degree polynomials—Legendre, Hermite, Laguerre, Chebyshev, and so on. Both the x_k's and the corresponding coefficients A_k are tabulated in AMS-55, Chapter 25. Computing subroutines exist in both single and double precision for the Legendre, Laguerre, and Hermite cases.

We shall prove that this choice of x_k (zeros of the appropriate nth-degree orthogonal polynomial) makes the quadrature formula (A2.7) *exact* for $f(x)$ a polynomial of degree $\leq 2n - 1$. Here is the power of this Gaussian choice. (Taking the x_k equally spaced (Newton–Cotes) is exact only for $f(x)$ a polynomial of degree $\leq n - 1$, n even or $\leq n$, n odd.)

Proofs of the necessity and sufficiency of this choice of orthogonal polynomial roots follow.

THEOREM: A necessary and sufficient condition that an interpolatory formula of the form of Eq. (A2.7) be exact for all polynomials of degree $\leq 2n - 1$ is that $\alpha(x)$ be orthogonal with respect to $w(x)$ over the interval $[a, b]$ to all polynomials of degree $\leq n - 1$.

Necessity. Assume Eq. (A2.7) is exact for $f(x)$ any polynomial of degree $\leq 2n - 1$. Let $Q_1(x)$ be any polynomial of degree $\leq n - 1$. Then $f(x) = \alpha(x)Q_1(x)$ is a polynomial of degree $\leq 2n - 1$. By simple substitution, we have

$$\int_a^b f(x)w(x)\, dx = \int_a^b \alpha(x)Q_1(x)w(x)\, dx, \qquad (A2.10a)$$

and since Eq. (A2.7) is assumed exact for this degree polynomial integrand,

$$\int_a^b \alpha(x)Q_1(x)w(x)\, dx = \sum_{k=1}^n A_k \alpha(x_k)Q_1(x_k)$$

$$= 0. \qquad (A2.10b)$$

The final $= 0$ follows because $\alpha(x_k) = 0$, Eq. (2.4). But this is a statement that

---

[1] If a and b are finite, the interval $[a, b]$ can always be transformed to $[-1, 1]$ by the *linear* transformation $t = [2x - (a + b)]/(a - b)$, $x = [(b - a)t + (b + a)]/2$. Then $\int_a^b f(x)\, dx = \int_{-1}^1 f(t)\, dt$.

our nth-degree polynomial $w(x)$ is orthogonal to all polynomials $Q_1(x)$ of degree $< n - 1$.

Sufficiency. Assume the orthogonality of $\alpha(x)$ to all polynomials of degree $\leq n - 1$. Let $f(x)$ be a polynomial of degree $\leq 2n - 1$. Dividing $f(x)$ by $\alpha(x)$ we obtain

$$\frac{f(x)}{\alpha(x)} = Q_2(x) + \frac{\rho(x)}{\alpha(x)} \tag{A2.11}$$

or

$$f(x) = \alpha(x)Q_2(x) + \rho(x), \tag{A2.12}$$

with $Q_2(x)$ and $\rho(x)$ polynomials of degree $\leq n - 1$. Integrating yields

$$\int_a^b f(x)w(x)\, dx = \int_a^b \alpha(x)Q_2(x)w(x)\, dx + \int_a^b \rho(x)w(x)\, dx. \tag{A2.13}$$

The first integral on the right vanishes because of our postulated orthogonality. Then, because the degree of $\rho(x)$ is $\leq n - 1$, Eq. (A2.7) (which is interpolatory) is exact and we have

$$\int_a^b f(x)w(x)\, dx = \sum_{k=1}^n A_k \rho(x_k). \tag{2.14}$$

Since $\alpha(x_k) = 0$, Eq. (A2.12) yields

$$\rho(x_k) = f(x_k).$$

Therefore

$$\int_a^b f(x)w(x)\, dx = \sum_{k=1}^n A_k f(x_k), \tag{A2.15}$$

exact. This is Eq. (A2.7), exact for $f(x)$, any polynomial of degree $\leq 2n - 1$.

As a specific example of Eq. (A2.15), consider the case where $[a, b] = [-1, 1]$ with $w(x) = 1$. The polynomials orthogonal over this interval with respect to this weighting function are the Legendre polynomials of Chapter 12. For the choice $n = 10$ the x_k are the 10 roots of $P_{10}(x)$. The values of A_k are given in principle by Eq. (A2.7). A more convenient expression is derived by Krylov[2].

Finally, with the numerical values of A_k and x_k, Eq. (A2.15) becomes

$$\int_{-1}^1 f(x)\, dx = +0.06667134\, f(+0.97390652)$$
$$+0.14945134\, f(+0.86506336)$$
$$+0.21908636\, f(+0.67940956)$$
$$+0.26926671\, f(+0.43339539)$$
$$+0.29552422\, f(+0.14887433)$$

[2] Tabulations of the A_k and x_k are found in the references that follow and in AMS-55 (Chapter 25).

$$+0.2955\,2422\,f(-0.1488\,7433)$$

$$+0.2692\,6671\,f(-0.4333\,9539)$$

$$+0.2190\,8636\,f(-0.6794\,0956)$$

$$+0.1494\,5134\,f(-0.8650\,6336)$$

$$+0.0666\,7134\,f(-0.9739\,0652),\qquad (A2.16)$$

exact (to the number of digits listed) for $f(x)$ a polynomial of degree ≤ 19.

The actual usefulness of Gaussian integration is contingent upon two factors (1) the availability of computers and (2) the availability of the values of $f(x)$ at $x = x_k$. This generally means that $f(x)$ be expressed in closed form or approximated in some convenient form so that $f(x_k)$ may readily be calculated. If $f(x)$ is given only as equally spaced tabulated values, Simpson's rule is probably the best choice for the numerical integration.

Warning. Our fundamental assumption is that $f(x)$ can be accurately represented by a $(2n - 1)$-degree polynomial with n reasonably small. If $f(x)$ has a singularity in the integration interval, this assumption of a polynomial representation is obviously not valid. Even if $f(x)$ remains finite, the presence of an infinite slope means our assumption is poor and that numerical accuracy will be relatively low. Exercise A2.7 illustrates these points.

EXERCISES

2.1   (a) Verify Eq. (A2.5).
       (b) With $P(x)$ a polynomial of degree $\leq n - 1$ and $\alpha(x)$ given by Eq. (A2.4), verify that

$$P(x) = \sum_{k=1}^{n} \frac{\alpha(x)}{(x - x_k)\alpha'(x_k)} P(x_k).$$

2.2   Using a 10-point Gauss–Legendre subroutine, evaluate

$$\int_0^1 x^n \, dx \qquad \text{for } n = 0(1)40.$$

Tabulate the computed value of the integral, the exact value, and the relative error. Plot log (relative error) versus n.

2.3   Using a 10-point Gauss–Laguerre subroutine, evaluate

$$\int_0^\infty x^n e^{-x} \, dx \qquad \text{for } n = 0(1)25.$$

Tabulate the computed value of the integral, the exact value, and the relative error. Plot log (relative error) versus n.

2.4   Using a 10-point Gauss–Hermite subroutine, evaluate

$$\int_{-\infty}^{\infty} x^n e^{-x^2} dx \qquad \text{for } n = 0(2)22.$$

Tabulate the computed value of the integral, the exact value, and the relative error. Plot log (relative error) versus n.

2.5   (a) Write a double precision Gauss–Chebyshev subroutine that will evaluate integrals of the form

$$\int_{-1}^{1} \frac{f(x)}{(1 - x^2)^{1/2}} dx$$

using 20 points, the 20 roots of the Chebyshev polynomial, $T_{20}(x)$. These roots and the coefficients A_k are tabulated by Stroud and Secrest (1966).
   (b) Check your subroutine by using it to compute

$$\int_{-1}^{1} x^{2n}(1 - x^2)^{-1/2} dx$$

for $n = 0(2)30$. Tabulate the computed value of the integral, the exact value, and the relative error. Plot log (relative error) versus n.

2.6   Evaluate

$$4 \int_{0}^{1} \frac{dx}{1 + x^2},$$

using Gauss–Legendre quadrature. How many evaluation points are needed to obtain a result accurate to 5 significant figures? to 12 significant figures?
   ANS.   4 point Gauss–Laguerre quadrature $\Rightarrow 5$ significant figures
   12 points $\Rightarrow 12$ significant figures.

2.7   From Exercise 10.2.11 the Euler–Mascheroni constant γ may be written as

1. $\gamma = -\int_{0}^{\infty} \ln r e^{-r} dr$

2. $\gamma = 1.0 - \int_{0}^{\infty} r \ln r e^{-r} dr$

   [32 points $\Rightarrow 3$ significant figures]

3. $\gamma = 1.5 - 0.5 \int_{0}^{\infty} r^2 \ln r e^{-r} dr.$

   (a) Explain why Gauss–Laguerre quadrature should *not* be attempted on the first integral.
   (b) Evaluate (2) and (3) using a 32-point Gauss–Laguerre quadrature and explain the very limited accuracy of your results.

2.8   (a) Evaluate the integral

$$I = \int_{-\infty}^{\infty} \frac{e^{-x^2} dx}{1 + |x|}$$

using Gauss–Hermite quadrature formulas for several values of n (number of evaluation points).

(b) Rewrite the integral as

$$I = 2 \int_0^\infty \frac{e^{-x^2+x}}{1+x} e^{-x} dx,$$

and evaluate by Gauss–Laguerre quadrature for several values of n.

ANS.   (b) 1.2103.

ADDITIONAL READINGS

DAVIS, P. J., and P. RABINOWITZ, *Methods of Numerical Integration.* Orlando, FL: Academic Press (1975).

GARCIA, A. L., *Numerical Methods for Physics.* Englewood Cliffs, NJ: Prentice-Hall (1994).

KRYLOV, V. I. (translated by A. H. Stroud), *Approximate Calculation of Integrals.* New York: Macmillan (1962). This is a very clearly written book, which covers virtually all aspects of the approximate calculation of integrals, and is an excellent discussion of Gaussian and other methods of numerical quadrature. Tables of evaluation points and weighting factors are also included.

STROUD, A. H. *Numerical Quadrature and Solution of Ordinary Differential Equations*, Applied Mathematics Series, Vol. 10. New York: Springer-Verlag (1974). As an excellent discussion of Gaussian and other methods of numerical quadrature, this volume also includes tables of evaluation points and weighting factors.

STROUD, A. H., and D. SECREST, *Gaussian Quadrature Formulas.* Englewood, NJ: Prentice-Hall (1966). This is a valuable book primarily because it contains extensive tables of x_k and A_k for a wide variety of intervals and weighting factors.

GENERAL REFERENCES

1. E. T. WHITTAKER and G. N. WATSON, *A Course of Modern Analysis*, 4th ed. Cambridge: Cambridge University Press (1962), paperback. Although this is the oldest (original edition 1902) of the references, it still is the classic reference. It leans strongly to pure mathematics, as of 1902, with full mathematical rigor.

2. P. M. MORSE and H. FESHBACH, *Methods of Theoretical Physics*, 2 vols. New York: McGraw-Hill (1953). This work presents the mathematics of much of theoretical physics in detail but at a rather advanced level. It is recommended as the outstanding source of information for supplementary reading and advanced study.

3. H. S. JEFFREYS and B. S. JEFFREYS, *Methods of Mathematical Physics*, 3rd ed. Cambridge: Cambridge University Press (1956). This is a scholarly treatment of a wide range of mathematical analysis, in which considerable attention is paid to mathematical rigor. Applications are to classical physics and to geophysics.

4. R. COURANT and D. HILBERT, *Methods of Mathematical Physics*, Vol. I (1st English edition). New York: Wiley (Interscience) (1953). As a reference book for mathematical physics, it is particularly valuable for existence theorems and discussions of areas such as eigenvalue problems, integral equations, and calculus of variations.

5. F. W. BYRON, JR and R. W. FULLER, *Mathematics of Classical and Quantum Physics*, Reading, MA: Addison-Wesley (1969). This is an advanced text that presupposes a moderate knowledge of mathematical physics.

6. C. M. BENDER and S. A. ORSZAG, *Advanced Mathematical Methods for Scientists and Engineers*. New York: McGraw-Hill (1978).

7. *Handbook of Mathematical Functions with Formulas, Graphs, and Mathematical Tables*, Applied Mathematics Series-55 (AMS-55). Washington, DC: National Bureau of Standards, U.S. Department of Commerce (1964). As a tremendous compilation of just what the title says, this is an extremely useful reference.

Additional more specialized references are listed at the end of each chapter.

INDEX